COLLEGE MATHEMATICS
THROUGH APPLICATIONS

COLLEGE MATHEMATICS THROUGH APPLICATIONS

JOHN C. PETERSON
Chattanooga State Technical Community College

WILLIAM J. WAGNER

STEPHEN S. WILLOUGHBY
University of Arizona
Consulting Author

Delmar Publishers Inc.

I(T)P˙ An International Thomson Publishing Company

Albany • Bonn • Boston • Cincinnati • Detroit • London • Madrid • Melbourne
Mexico City • New York • Pacific Grove • Paris • San Francisco • Singapore • Tokyo
Toronto • Washington

NOTICE TO THE READER

Delmar Staff
Publisher: Alar E. Elken
Acquisitions Editor: Paul Shepardson
Developmental Editor: Ohlinger Publishing Services
Production Manager: Larry Main
Art and Design Coordinator: Mary Beth Vought

COPYRIGHT © 1998
By Delmar Publishers Inc.
an International Thomson Publishing Company

The ITP logo is a trademark under license.

Printed in the United States of America

For more information, contact:

Delmar Publishers Inc.
3 Columbia Circle, Box 15015
Albany, NY 12212-5015

International Thomson Publishing Europe
Berkshire House 168-173
High Holborn
London, WC1V 7AA
England

Thomas Nelson Australia
102 Dodds Street
South Melbourne, 3205
Victoria, Australia

Nelson Canada
1120 Birchmont Road
Scarborough, Ontario
Canada, M1K 5G4

International Thomson Editores
Campos Eliseos 385, Piso 7
Col Polanco
11560 Mexico D F Mexico

International Thomson Publishing GmbH
Konigswinterer Strasse 418
53227 Bonn
Germany

International Thomson Publishing Asia
221 Henderson Road
#05-10 Henderson Building
Singapore 0315

International Thomson Publishing—Japan
Hirakawacho Kyowa Building, 3F
2-2-1 Hirakawacho
Chiyoda-ku, Tokyo 102
Japan

ISBN 0-7668-0207-8

Contents

PREFACE   **xv**

1 LINEAR EQUATIONS   **0**

Chapter Project—The Mathematical Crystal Ball   **2**

1.1   Take a Walk with the CBL   4
        Looking at the Data, 6

1.2   Distance, Time, and Two Kinds of Rate   7
        Velocity and Speed, 7

1.3   Equation of a Line: Slope and y-Intercept   15
        Slope of a Straight Line, 15
        Applications of Slope, 17
        Intercepts of a Graph, 21
        Fitting a Line to the Data—The Slope-Intercept Form
        of the Line, 23
        The Constant Velocity Equation, 26

1.4   Three More Forms of the Linear Equation   31
        Using the Slope and a Point, 31
        Using Two Points, 33
        Using the Angle of the Line, 33
        Decimal Precision and Decimal Accuracy, 39

1.5   A Linear Model That Uses All the Points   45
        Fitting a Regression Line to Data, 49

        Chapter 1 Summary and Review   55

        Chapter 1 Test   58

2 QUADRATIC FUNCTIONS   **60**

Chapter Project—The Bouncing Ball   **62**

2.1   Velocity That Varies   63
        Graphs That Tell a Story, 63
        Average Speed and Average Velocity, 64

2.2   Follow the Bouncing Ball   69
        CBL and the Bouncing Ball, 69
        Looking at One Bounce, 72
        Quadratic Regression, 73
        Thinking about Quadratic Regression, 74

2.3   Introduction to Functions   75
        Introduction to Functions, 75
        Naming a Function, 77
        How Functions Are Defined, 79
        Graphing Functions, 80
        Looking More Closely at a Graph, 82

2.4   Working with Parabolas   83
        Roots of a Quadratic Function, 84
        Roots and Factors, 85
        Graphs of Quadratic Equations, 90

2.5   Projectile Motion   93
        A Falling Object, 94
        Transforming Data, 95
        Giving it a Toss, 96
        Comparing Transformed Graphs, 99
        Another Look at Coefficients a and b, 99
        Thinking about the Quadratic Model, 101

2.6   Velocity at an Instant   105
        Return to the Bouncing Ball, 111

        Chapter 2 Summary and Review   113

        Chapter 2 Test   115

3   MODELS OF PERIODIC DATA: INTRODUCING
    TRIGONOMETRY   116

    Chapter Project—Analyzing the Touchtone Phone   118

3.1   Introduction to the Trigonometric Functions   119
        Definitions of the Trigonometric Functions, 120
        Calculations with Triangles, 121

3.2   Graphs and Roots of Trigonometric Functions   131
        Angles Larger than 90°, 132
        Drawing the Graphs, 135
        The Peculiar Tangent Function, 137
        Negative Angles and Other Strange Things, 139
        Modeling Alternating Currents, 141
        Solving Trigonometric Equations, 142
        Roots of Trigonometric Functions, 147

3.3   Period, Amplitude, Frequency, and Roots   149
        Period and Cycle, 149
        Frequency, 151
        Amplitude, 152
        Solving More Advanced Trigonometric Equations, 157
        Roots of Transformed Trigonometric Functions, 159

3.4   Vertical and Horizontal Translations   164
           Vertical Shift or Vertical Translation, 164
           Phase Shift, 165
           Phase Angle, 166
           Horizontal Shifts or Translations of Functions, 167
           Fitting Data to a Sinusoidal Curve, 171

3.5   Modeling Sound Waves   177
           Radians and Degrees, 177
           Modeling Vibrations, 180
           Modeling Sounds, 182

3.6   Modeling Wave Combinations   188
           More about Alternating Current, 188
           Alternating Current and the Addition of Trigonometric
           Functions, 190
           Combining Waves of Different Frequency, 191
           Modeling Musical Notes and Chords, 194

       Chapter 3 Summary and Review   204

       Chapter 3 Test   209

**4   MATHEMATICAL MODELS IN GEOMETRY: IMAGES IN TWO
       AND THREE DIMENSIONS   210**

       Chapter Project—What's the Best Shape?   212

4.1   Volume and Surface Area   214
           Volume, 214
           Surface Area, 217
           Exploring Max/Min Problems, 219
           First Attempt at the Minimum Area of a Cylinder, 221

4.2   Triangles Are Everywhere   226
           Why Is the Area of a Triangle Half the Base × Height?, 226
           Trigonometry and the Areas of Triangles, 227
           The Law of Sines, 229
           The Pythagorean Theorem, 233
           Distance on a Graph, 235
           The Law of Cosines, 237

4.3   Pythagorean Theorem and Circles   242
           Definition of a Circle and Some of Its Parts, 242
           Circles Not Centered at the Origin, 245
           Pythagorean Trigonometry Identities, 246
           Circumference and Area of a Circle, 248

4.4   Definition and Use of Radians   253
           Radians, 254
           Summary of Differences: Radians vs. Degrees, 258
           Moving in Circles, 259

4.5   Prisms, Pyramids, and Other 3-D Figures   262
           Prisms, 262
           Pyramids, 263
           Cylinders, 265
           Cones, 266
           Solution of the Chapter Project's Cylinder Problem, 267

Chapter 4 Summary and Review   271

Chapter 4 Test   274

5   MODELING MOTION IN TWO DIMENSIONS   276

Chapter Project—What's the Best Angle?   278

Preliminary Analysis, 278

5.1   Flight Trajectories   280
Looking at Trajectories, 281
Vectors and the Components of Velocity, 281
The Influence of Gravity, 283

5.2   Parametric Equations   289
Graphing Parametric Equations, 290
Trigonometry, Circles, and Parametric Equations, 292
Parametric Equations of Ellipses, 294
Plotting Planetary Trajectories, 295
Modeling Planetary Velocity, 297
Modeling Other Gravity-Influenced Trajectories, 298

5.3   Vectors   304
Vectors in Navigation, 304
Modeling Vector Sums with Triangles, 306
Modeling Vector Sums with Parallelograms, 308
Vectors and Gravity: Motion on a Ramp, 313
Modeling Impedance in RC Circuits, 316

5.4   Vectors and Complex Numbers   323
Types of Numbers, 323
Imaginary Numbers, 328
Arithmetic of Complex Numbers, 332
Geometry of Complex Numbers, 334
Polar Form for a Complex Number, 337
Phasors and Complex Numbers, 340

5.5   Solving the Best-Angle Problem   342
Solution That Uses Algebra and Trigonometry, 344
What If the Beginning and Ending Heights Are Different, 345

Chapter 5 Summary and Review   348

Chapter 5 Test   352

6   POLAR GRAPHING AND ELEMENTARY PROGRAMMING   354

Chapter Project—Programming a Robot Arm   356

6.1   Review of Two-Dimensional Graphing   358
Connecting Cartesian and Parametric Graphing, 358
Parametric Equations and the Oscilloscope, 365
Lissajous Curves, 367

6.2   Polar Coordinates   369
Plotting Points in Polar Coordinates, 369
Plotting Polar Equations, 370
Patterns in Polar Graphs, 377
Patterns in n-Leafed Roses and Their Relatives, 378
Connecting Polar and Rectangular Equations, 380

6.3   Applications of Polar Graphing: Ellipses   385
Ellipses and Eccentricity, 385
Plotting Orbits of Comets, 387
Polar Equations of Ellipses, 388
Other Applications of Ellipses, 390

6.4   Understanding *TI-83* Programs   396
Introduction to *TI-83* Programming, 397
How the Program XYGRABBR Works, 399

6.5   Writing *TI-83* Programs   401
Entering a Program, 401
Editing an Existing Program, 402

6.6   Solving the Chapter Project   406
Plotting Curves One Point at a Time, 407
Plotting Connected Curves, 409

Chapter 6 Summary and Review   413

Chapter 6 Test   417

7   MODELING WITH SEQUENCES AND SERIES   418

Chapter Project—Estimating Cross-Sectional Areas   420

Preliminary Analysis, 420

7.1   Sequences   424
Defining Sequences Directly, 425
Defining Sequences Recursively, 432

7.2   Limits of Sequences   438
Sequences and Chaos Theory, 442
A Sequence of Areas, 443

7.3   Looking More Closely at Limits   449
Three Important Sequences, 449
The Algebra of Limits, 453

7.4   Applications of Geometric Sequences   459
Interest on a Savings Account, 459
Bouncing Ball, 466
Charging and Discharging a Capacitor, 469
Radioactivity and Half-Life, 474
Exponents and Notes in a Musical Scale, 477

7.5   Arithmetic and Geometric Series   484
A Legendary Bonus, 484
Limits of Geometric Series, 489
Charging a Capacitor, 491
Annuities, 493
Arithmetic Series, 494

7.6   Estimating Areas under Curves   497
Riemann Sums, 497
Estimation with Middle Rectangles, 500
A Program to Estimate Areas, 502
How the Program Works, 504

Chapter 7 Summary and Review   508

Chapter 7 Test   512

8   MODELING WITH ALGEBRAIC FUNCTIONS   **514**

Chapter Project—Seismographs and Pendulums   **516**

Preliminary Analysis, 516

8.1   Power Functions: $f(x) = kx^n$ 522
Algebraic Functions, 523
Exploring the Graphs of $f(x) = kx^n$, 524
Summary: Shapes of Power Function Graphs, 531

8.2   Variation and Power Functions   535
A Legend about Gifts and Gold, 535
Direct Variation, 535
Variation of Light Intensity with Distance, 540
Inverse Variation, 542

8.3   Polynomial Functions   548
Roots of Polynomials, 549
Factored Form of a Polynomial, 551
Limits and Graphs of Polynomial Functions, 553

8.4   Application of Polynomials   557
How do Calculators Compute Trigonometric Values?, 557
The nth Term of a Sequence, 559

8.5   Patterns in Slopes of Power Functions   569
Plotting a Tangent to a Curve, 570
A Program that Draws Tangents, 574
Modeling the Slope Function, 576

8.6   Analyzing Rational Functions   579
Definition and Examples, 579
Shifted Forms of Power Functions, 579
Variations on Familiar Functions, 586
Solving a Light Intensity Problem with Graph, Table, and Algebra, 590

8.7   Solving the Chapter Project   595

Chapter 8 Summary and Review   599

Chapter 8 Test   604

9   EXPONENTIAL AND LOGARITHMIC FUNCTIONS   **606**

Chapter Project—Modeling Temperature Changes   **608**

Preliminary Analysis, 608

9.1   Exponential Functions   612
Graphs of Exponential Functions, 614
Intercepts and Asymptotes in Exponential Functions, 616

9.2   Logarithms and Inverse Functions   617
What Are Logarithms?, 617
Properties of Logarithms, 620
Why Study Logarithms?, 623
Inverse Functions, 625

9.3   Natural Logarithms and the Number e 629
            Continuous Growth and the Number e, 631
            Solving Exponential Equations, 635
            Charging a Capacitor and the Number e, 637
            Converting any Exponential Function to an Exponential Function
            with Base e, 639
            Exponential Damping, 640
            The Slope Function for Exponential and Logarithmic
            Functions, 643

9.4   Using Logarithmic Scales   648
            The Mathematics of the Logarithmic Scale, 649
            Semilog Graphing on the Graphing Calculator, 657

9.5   Analysis of the Cooling Data   658
            Exponential Regression, 658
            Slope Function of Data, 659
            Vertical Transformation of the Data, 660
            The Accuracy of the Model, 661
            Using a Semilog Plot of the Data, 664

        Chapter 9 Summary and Review   666

        Chapter 9 Test   670

10 SYSTEMS OF EQUATIONS AND INEQUALITIES   **672**

Chapter Project—Maximizing the Profit   **674**

            Preliminary Analysis, 674

10.1   Review: Equations in One Variable   676
            Solving Equations Without a Calculator, 677
            When to Use the Calculator?, 678
            Solving Equations With a Calculator, 679

10.2   Solving Two Equations in Two Variables   686
            Solving Two Equations: The Addition Method, 687
            Solving Two Equations: The Graphical Method, 688
            Solving Two Equations: The Substitution Method, 690
            Questions About the Addition Method, 692
            Application: Kirchhoff's Laws of Circuit Analysis, 695
            How Many Solutions are Possible?, 698

10.3   Inequalities in One Variable   703
            Representing the Solution Of an Inequality With a Graph, 704
            Solving an Inequality with Algebra, 705
            Solving an Inequality With a Calculator, 709
            Solving Inequalities: Binary Logic and the Step Function, 711
            Binary Logic and the Manufacturing Problem, 713

10.4   Inequalities in Two Variables   716
            The Graph of a System of Inequalities, 716
            Graphing Systems of Inequalities on the Calculator, 720
            Graphing the Systems for the Chapter Project, 721

10.5   Linear Programming   727
            What is Linear Programming?, 728
            When the Maximum Occurs at an Unacceptable Point, 729
            Adding a New Constraint, 731

Chapter 10 Summary and Review   734

Chapter 10 Test   736

11 MATRICES AND 3-D GRAPHING   738

Chapter Project—Scaling and Rotating Points   740

Preliminary Analysis, 740

11.1   Introduction to the Transformation of Points   742
Scaling, 743
Rotation, 745
Rotation and Scaling Together, 747
The Trigonometry of Rotation, 749

11.2   Introduction to Matrix Algebra   755
Matrix Multiplication, 756
Matrix Algebra on the *TI-83*, 759
Which Matrices Can Be Multiplied?, 761
Choosing the Dimensions of a Matrix, 762
Matrices and Transformation of Points in Two Dimensions, 762

11.3   Matrices and Equation Solving   771
Representing Systems with Matrices, 772
Solving a Matrix Equation, 773

11.4   Graphing in Three Dimensions   786
Cartesian Coordinates in Three Dimensions, 786
Computing Distances in Space, 788
Motion in 3-D, 791

11.5   Plotting Equations in 3-D   794
Solving Systems in Three or More Dimensions, 799

11.6   Curves and Surfaces in 3-D   805
Parametric Equations in 3-D, 805
Spherical Coordinates, 807
Cylindrical Coordinates, 810

11.7   Rotations and Translations in Three Dimensions   813
Scaling in Three Dimensions, 814
Rotation in Three Dimensions, 815

Chapter 11 Summary and Review   818

Chapter 11 Test   825

12 MODELING WITH PROBABILITY AND STATISTICS   826

Chapter Project—Assessing Quality on the Assembly Line   828

Preliminary Analysis, 828

12.1   Looking at Data   832
Numerical Analysis—Failure Rate, 832
Numerical Analysis—Throwing Two Dice, 835
Analyzing Data with the *TI-83*, 836
Representing Data Graphically—The Histogram, 837
Representing Data Graphically—The Box Plot, 840

12.2   Modeling Uncertainty with Probability   845
        Translating English to Mathematics, 847
        How Many Ways Can It Happen—Adding Probabilities, 848
        Expected Number: How Well Can We Predict?, 850

12.3   Probability of Binary Events   858
        Independent Events—Multiplying Probabilities, 858
        Counting Events, 864

12.4   Analyzing Games of Chance   869
        Card Games, 870
        Dice Games, Odds, and Expected Return, 870
        Roulette, 874
        State Lotteries, 876

12.5   Solving the Chapter Project   880

        Chapter 12 Summary and Review   884

        Chapter 12 Test   889

INDEX OF APPLICATIONS   891

INDEX   893

PREFACE

Introduction

Students in technical and engineering technology programs require a mathematics curriculum that focuses on the real environments in which they will apply their knowledge and the tools they will employ there, without degenerating into a set of rules and algorithms. Their mathematics education should be intellectually challenging and should lay a foundation for further learning and development.

College Mathematics Through Applications has been developed to be the first text in a series that will focus on mathematics for these students. It covers advanced algebra, trigonometry, geometry, and intuitive calculus and explores these topics through applications. The series will use workplace-based applications as the cornerstone of the instruction and will involve students in developing solutions and methods. Inspired equally by the world of work and the current reform movement in mathematics education, we have looked hard at the traditional content of these courses and have chosen topics that are used in a wide variety of technical fields and that are intellectually rich. The presentation and classroom activities are designed to be interesting yet challenging to all students.

We expect that each student will have a graphing calculator, but just giving a calculator to students or letting calculators creep into the classroom doesn't bring the benefits. Therefore, we have fully integrated the calculator into this text. The use of technology cannot replace thinking, but it should reduce mathematical error.

Approach

The goals of our presentation are for students to

- ▶ Understand how mathematics is used in the workplace
- ▶ Understand the limitations of tools, simulations, and mathematical methods
- ▶ Develop intuition about whether results do or do not make sense
- ▶ Not be held back by traditional prerequisites or barriers
- ▶ Learn to apply mathematics in real settings
- ▶ Review prerequisite concepts in context
- ▶ Use available technology to develop deep understanding of concepts

To reach these goals we have laid the following philosophical and pedagogical foundations:

- Learning in the context of real applications promotes retention and understanding.
- Mathematical content should reflect actual workplace needs.
- Students learn better by doing, writing, and discussing.
- Mathematical instruction should use the technology to perform traditional computations.
- Calculators should take over much of the machinery of calculations, allowing students to concentrate on a problem and focus on concepts.
- Content should be presented using the "rule of four": ideas are presented and students work in symbolic, graphic, and numeric methods and are then asked to express their ideas and answers in writing.
- Students who communicate their mathematical understanding through written and oral responses to well-designed, thought-provoking questions and problems will gain valuable workplace skills.

In addition, the text has these common threads:

- Applications and real data are used whenever possible.
- Equations and functions are used as models of data.
- Students assess the accuracy and reasonableness of results.
- Technology provides alternative methods for approximating solutions.
- Technology is used in the classroom every day.
- Intuitive calculus is woven throughout.

Features

Chapters begin with a Project (usually workplace-based), and the goal of the chapter is to learn the mathematics necessary to solve the problems posed by this Project. The world of work doesn't present problems in a neat, organized way, so to better prepare students for the workplace, these projects are designed to force students to organize their thoughts and decide what the problem is asking them to do. Good problem solvers get information and skills as they need them, so at several points in the chapter students are asked to relate the mathematics they have learned to the solution of the Project. By the end of the chapter they will have learned how to complete the Project.

The chapters also include these additional features:

- Frequent Activities and Calculator Labs are designed to get students involved in performing experiments, collecting and analyzing data, and forming conclusions— as they will have to do in their future careers.
- Technology—graphing calculators and an interactive CD—is integrated throughout the text to allow students to explore more advanced mathematics concepts.
- Numerous examples and exercises show how the mathematics relates to different technical fields.
- A chapter summary, review exercises, and a chapter test conclude every chapter.
- Intuitive ideas underlying many mathematics concepts are used, including the intuitive foundations of calculus.
- Applications and actual data are used whenever possible to emphasize the usefulness of mathematics.
- The text uses a spiral approach to anticipate and reinforce instruction.

Acknowledgments

We would like to thank the following reviewers for their assistance with the project:

Stephen Backman, Director of Education, ITT Technical Institute, Murray, UT
Curt Baragar, ITT Technical Institute, Grand Rapids
Jeff Birkholz, ITT Technical Institute, Knoxville
Jacqueline Bollenbacher, ITT Technical Institute, Fort Wayne
Francis Booth, ITT Technical Institute, Tampa
Russ Duty, ITT Technical Institute, Hayward, CA
Julia Hassett, DeVry Institute of Technology
Peter Hovanec, ITT Technical Institute, Tampa
Kevin Jensen, ITT Technical Institute, Murray, UT
Patrick Kiely, ITT Technical Institute, Tampa
Robert Kimball, Wake Technical Community College
James McDonald, Springfield Technical Community College
Edward Nichols, Chattanooga State Technical Community College
Ellena Reda, Dutchess Community College
Khaled Sakalla, ITT Technical Institute, Knoxville
James Sarli, ITT Technical Institute, Phoenix
Brian Taylor, ITT Technical Institute, Murray, UT

In addition we want to recognize the support of David Parker, Director of Education at ITT Technical Institute in Hayward, CA, for allowing us to field test parts of the text at his school. We'd also like to thank the students who participated in this field test for their valuable feedback.

We also appreciate the work of the following professors who checked the accuracy of the text:

Sherri Barnes, Chattanooga State Technical Community College
Mark Lancaster, Chattanooga State Technical Community College
Edward Nichols, Chattanooga State Technical Community College

We would like to thank Marsha Schoonover for her excellent work on the Solutions Manual and Kathy McKenzie for her hard work on the Test Bank. We also wish to thank Texas Instruments for providing input for some of the illustrations contained in this book.

Our wives, Marla and Linda, have been patient and understanding and have greatly contributed to making this book possible. Their editorial comments and their willingness to listen and offer suggestions made this a better book.

We would also like to especially thank our developmental editor, Monica Ohlinger. Her suggestions, prodding, understanding, and sense of humor have all contributed immeasurably to this book. We also need to recognize Bob Lynch who envisioned this project and saw it through its first steps. Paul Shepardson took over from Bob and has quietly and effectively guided us to this stage of development.

Finally, we wish to thank the people who have been in charge of the text's production: Larry Main of Delmar and Mimi Jett of ETP Harrison.

COLLEGE MATHEMATICS
THROUGH APPLICATIONS

Linear Equations
Making Mathematical Models of Data

Topics You'll Learn or Review

- Describing how the slope of a line is related to the graph of the line and the equation of the line
- Describing the trigonometric connection between the slope of a line and the angle the line makes with the *x*-axis
- Using the connection between a constant velocity experiment and the slope and *y*-intercept of the distance vs. time graph
- Writing the equation of a straight line given any of the following:
  - the slope and the *y*-intercept
  - one point and the slope
  - two points
  - one point and the inclination angle of the line
- Using a mathematical model to study real events and relationships; that is,
  - using a linear model of data to estimate missing values in the data
  - analyzing how well a linear model fits the data

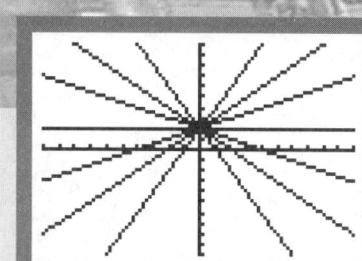

Calculator Skills You'll Need

- Doing calculations on the home screen
- Using the ⬛2nd⬛ key to reach calculator functions named in yellow
- Working with graphs to
  - enter a function in the Y= screen and graph it
  - turn graphs of functions on and off
  - use Trace to determine the coordinates of points
  - change screen dimensions with the WINDOW menu
- Using the MODE menu to switch between radians and degrees

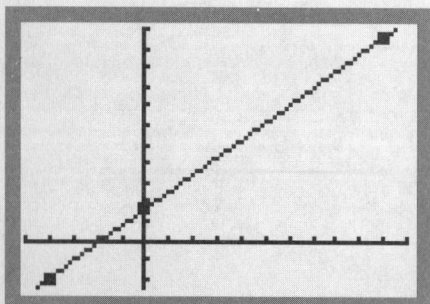

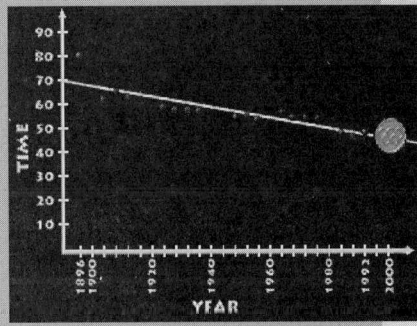

Mathematics You'll Use

▶ Doing calculations involving formulas, including decimals and negative numbers
▶ Calculating average speed, given distance and time traveled
▶ Graphing on a Cartesian coordinate system
▶ Plotting points, reading graphs, and understanding the following terminology:
 • x-axis, y-axis
 • x-coordinate, y-coordinate
 • origin
 • y-intercept and x-intercept of a graph
▶ Making a graph from a table of values for x and y
▶ Calculating the slope of a line
▶ Calculating with positive and negative numbers

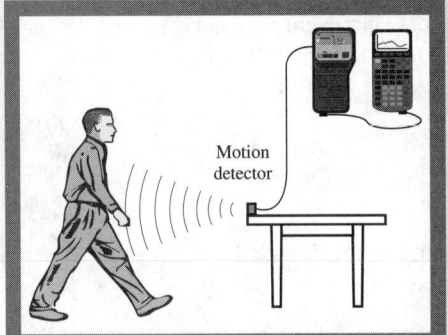

Calculator Skills You'll Learn

▶ Entering data into lists on the calculator
▶ Saving lists of data by giving them names
▶ Defining and producing a scatterplot from two lists
▶ Turning scatterplots on and off and changing their characteristics
▶ Calculating and graphing the linear regression equation line to fit data

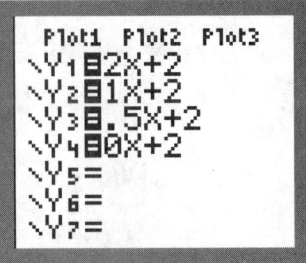

What You'll Do in This Chapter

Throughout this book you will study different mathematical models and will see how to apply them to real world situations. In this chapter the mathematical models will be those whose graphs are straight lines. We will look at historical data, data that come from motion experiments, and other data related to science and technology. In later chapters we will go more deeply into these and other models. As you move through this chapter you are going to learn how to answer questions like the following:

▶ What is a mathematical model and how is it useful?
▶ What kind of motion can be modeled with a linear function?
▶ What is the best way to build a linear model of a set of data with and without a calculator?

Chapter Project—The Mathematical Crystal Ball

Can you predict the future by looking at the past? The mathematics you will learn in this chapter can help you estimate the winning time for the men's 100-meter freestyle swimming event in the 1996 Olympics by looking at past performance. Here is the data for previous Olympic competitions:

Year	1896	1904	1908	1912	1920	1924	1928	1932	1936	1948	1952
Time (sec)	82.2	62.8	65.6	63.4		59.0	58.6	58.2	57.6	57.3	57.4

Year	1956	1960	1964	1968	1972	1976	1980	1984	1988	1992	1996
Time (sec)	55.4		53.4	52.2	51.2	50.0		49.8	48.6	49.02	

We have omitted the results for 1920, 1960, 1980, and 1996. Your job is to see how close you can come to estimating these missing times. (This event was not part of the 1900 Olympics and there were no Olympics in 1916, 1940, and 1944.)

Activity 1.1
Graphing Olympic Swimming Data

1. Use the blank graph in Figure 1.1 to plot the winning times provided.

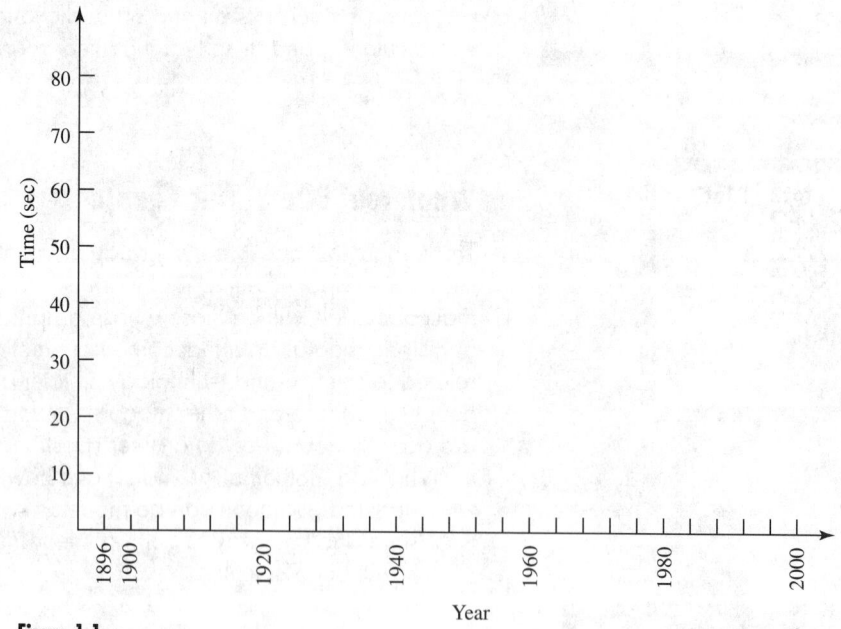

Figure 1.1

2. Estimate the missing times.
3. Explain in writing how you made these estimates and how confident you are of their accuracy.

Although you may not have used the word *model* in your answer to question 3, you were actually using a **mathematical model** when you estimated the missing numbers based on the numbers (or, the data) you were given. The assumptions you make about the data form your mathematical model of the data. In this chapter you'll learn how to build mathematical models of data in several ways. This book will introduce you to many different mathematical models that are used to analyze situations which arise commonly on the job and in other parts of your life.

We will return to the problem of estimating the Olympic records throughout this chapter as you learn new methods for making estimates.

Example 1.1

Enter the Olympic swimming data into your calculator and plot the data.

Solution   Access the `Stat Edit` screen; then enter the years into list L1 and the times into list L2. Do not type the years where no time is given. The first few years are shown in the left column of Figure 1.2. If you skip one of the entries, you can put it anywhere in the list. The first part of the list of times is shown in column L2 of Figure 1.2.

Figure 1.2

Note: If Your Lists Are Not Blank . . .

Press the right arrow to move to the right until you see two blank lists.

If there still are not two blank lists available, you will need to clear two of the existing lists. For instructions on clearing these lists, see page 12–12 of the *TI-83 Graphing Calculator Guidebook*.

Before we can graph these points, we must tell the calculator how to plot them. Access the STAT PLOTS menu shown in Figure 1.3 and press **ENTER** to select `Plot1`.

In `Plot1` we want the `Xlist` to be L1 and the `Ylist` to be L2. Change the mark to a square in order to make your data points larger and easier to see. These settings are shown in Figure 1.4.

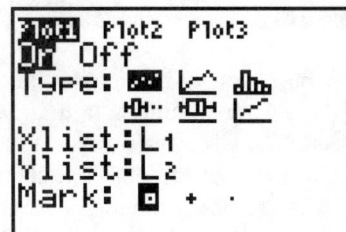

Figure 1.3

Before you press **GRAPH**, turn off or erase any entries in the Y= menu.

Before graphing this data we need to change the graph window settings and make sure that only these points will be plotted. To know what values to use for Xmin, Xmax, Ymin, and Ymax, you need to look at your data. Since the years on the x-axis go from 1896 to 1992, we will use window settings that include these extremes and provide a little extra viewing space: $1890 \leq x \leq 2000$. Using similar thinking for the times, we let $40 \leq y \leq 90$. We use the settings shown in Figure 1.5.

Press **GRAPH** to see a graph of your data. It will look something like Figure 1.6.

To check that you have entered the data correctly, press **TRACE**. You will see a screen like Figure 1.7, in which the cursor is blinking on the point that represents

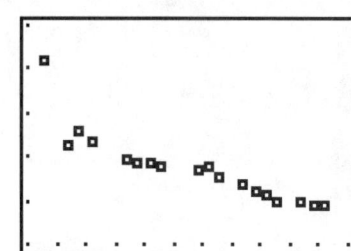

Figure 1.4   **Figure 1.5**   **Figure 1.6**

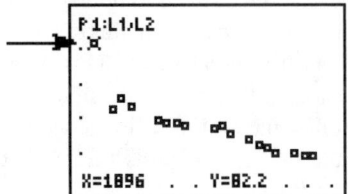

Figure 1.7

the first year in list L1, in this case 1896. Move the cursor from point to point by pressing the ▶ key to check the numbers on the calculator screen against the numbers in the textbook.

We will want to use these lists later in the chapter, so we need to save the data in the lists. In order to save these data for future use, it is useful to give the two lists new names so that lists L1 and L2 can be reused. To do this we will use the *TI-83*'s store command to store the years in a list called YEAR and store the winning times in a list called TIME. ■

1.1   Take a Walk with the CBL

The graph of the Olympic records for the men's 100-meter freestyle swimming event (Figure 1.6 and the graph you drew in Activity 1.1) appears to fall roughly along a straight line. Therefore, the first mathematical models you will use will help you find a straight line that describes this data. Before we learn how to find this line, we need to review some material about linear functions: mathematical expressions whose graphs are straight lines. You may already have studied linear functions, but this chapter approaches these concepts in a different way—one we think will help you increase your understanding of these fundamental concepts.

In order to better understand linear functions, you are going to use a laboratory device, shown in Figure 1.8, to obtain some data that can be modeled with a straight line. You will use an instrument that can collect distance and time data, and then your calculator can graph these data to give a picture of the motion. This instrument, called Calculator-Based Laboratory™ (CBL™) System, can attach to a calculator such as a *TI-82*, *TI-83*, or *TI-85*. In your experiments the motion detector and calculator should be set on a firm surface like a table, as shown in Figure 1.9.

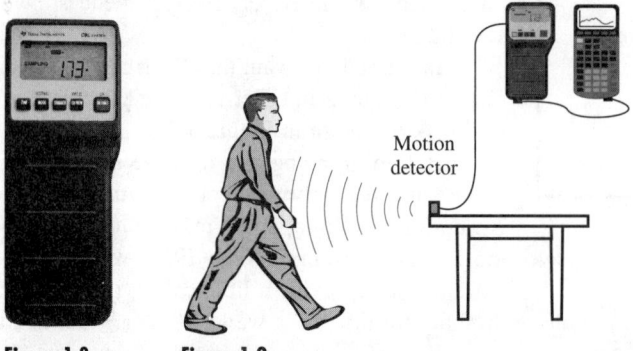

Figure 1.8   **Figure 1.9**

Note: The Motion Detector

The motion detector sends out short bursts of 40-kilohertz (kHz) ultrasonic waves and "listens" for the echo of these waves returning to it after reflecting off a wall, a person, or another object. The CBL system converts the return time for the reflected wave into the distance of the object from the motion detector. The motion detector you will use has a minimum range of about 0.5 meters (1.5 feet) and a maximum range of about 6 meters (30 feet).

Activity 1.2
Take a Hike

This activity can be carried out in groups of two or more, or by the whole class.   In this activity someone will walk toward or away from the motion detector. The CBL will translate the motion into a graph of distance vs. time such as the one shown in Figure 1.10.

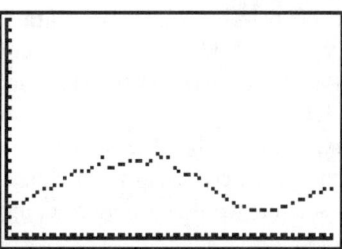

Figure 1.10

In Figure 1.10 the vertical axis measures distance from the motion detector from 0 to 20 feet and the horizontal axis measures time from 0 to 6 seconds. These limits are set by the program HIKER.

Procedures

1. Position a person, the hiker, in front of the motion detector. The hiker should be between 1.5 ft and 20 ft away from the motion detector. Make sure that both the CBL and the calculator are turned on; then select the program HIKER on the calculator.

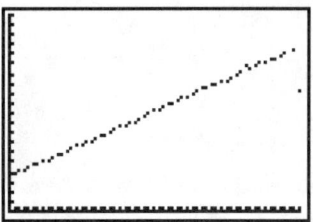

Figure 1.11

2. Instruct the hiker to walk toward or away from the motion detector when given the signal. The hiker should continue walking as long as he or she hears the motion detector clicking (about 6 seconds) or until he or she is told to stop.

    Look at the calculator screen. A few seconds after the clicking stops, a graph similar to the one in Figure 1.10 will appear.

3. Try several different experiments in order to produce different patterns in the graph. For example, try to make a straight line that goes up, a straight line that goes down, a horizontal line, a vertical line, a ∨-shape, a ∧, or a ∨∨.

    As you experiment with the CBL and the motion detector, try to answer the following questions:

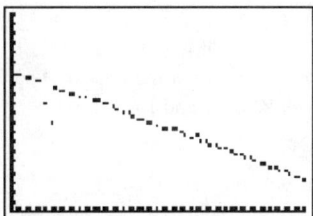

Figure 1.12

(a) What kind of motion produces a straight line on the distance vs. time graph?

(b) When a straight line graph is produced, how does the steepness of the line change as the speed of the walker changes?

(c) Describe the differences in walking that would produce the graphs in Figure 1.11 and Figure 1.12. Ignore the data points lying apart from the main pattern of the points; they occurred when reflection from a closer object was picked up and can be common in CBL motion experiments.

(d) How could you tell from the graph where the walker was when the experiment began?

(e) Describe the differences in walking that would produce the two graphs in Figure 1.13.

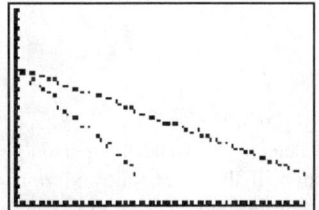

Figure 1.13

(f) What kind of walking can produce a horizontal line?

(g) What kind of walking can produce a vertical line?

(h) Write a description of the motion that produced Figure 1.10.

The activities in the next section will help you to improve your answers to these questions and will increase your understanding of these concepts.

Looking at the Data

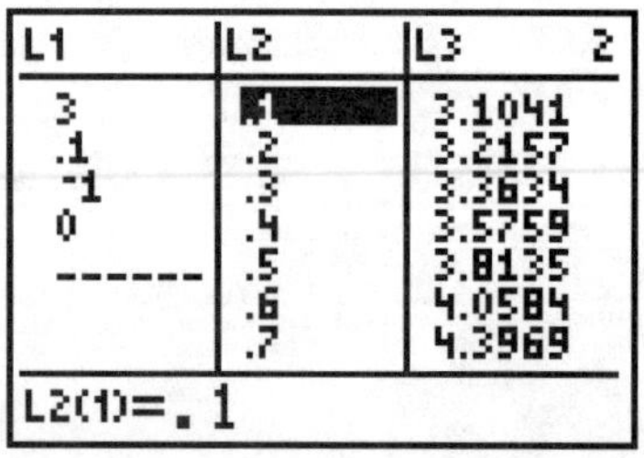

Figure 1.10 was produced using the HIKER program and the CBL. The program directed the CBL to measure distance every tenth of a second for six seconds and to write the resulting 60 times and 60 distances into two lists (called L2 and L3) in the calculator. These 60 pairs of values are **data** from the experiment. You can look at the data after you run a CBL experiment controlled by the HIKER program by pressing ⬤ STAT 1 and then moving the right arrow until L2 and L3 are visible, as shown in Figure 1.14.

After the experiment, the HIKER program automatically displays the graph. This kind of graph, which consists of individual data points, is called a **scatterplot**.

In the next section we'll see how to compute the speed of the walker from a CBL experiment.

Figure 1.14

Section 1.1 Exercises

1. **(a)** Make a list of the following data for the Olympic results in the men's pole vault. **(b)** Graph the data on your calculator. Does the data look linear? **(c)** Save the data as lists in your calculator with names YEAR2 and PV.

Year	1896	1900	1904	1908	1912	1920	1924	1928
Height (cm)	330.2	330.2	349.89	370.84	394.97		394.97	419.73

Year	1932	1936	1948	1952	1956	1960	1964	1968
Height (cm)	431.16	434.97	431.16	454.66	455.93		509.91	539.75

Year	1972	1976	1980	1984	1988	1992	1996
Height (cm)	549.91	549.91		574.68	589.92	579.75	

2. **(a)** Make a list of the following data for the Olympic results in the women's 100-meter dash. **(b)** Graph the data on your calculator. Does the data look linear? **(c)** Save the data as lists in your calculator with names YEAR3 and DASH.

Year	1928	1932	1936	1948	1952	1956	1960	1964
Time (sec)	12.2	11.9	11.5	11.9	11.5	11.5		11.4

Year	1968	1972	1976	1980	1984	1988	1992	1996
Time (sec)	11.08	11.07	11.08		10.97	10.54	10.82	

3. **(a)** Write a sentence telling how someone should walk in the HIKER experiment in order to produce a graph in the shape of a ∪. **(b)** Have someone follow your description in (a) and walk in front of the motion detector. **(c)** Did this person's hike produce the desired graph? If not, try to determine if your directions were clear enough, if the hiker followed your directions, or if your directions were not correct. **(d)** Based on your answers in (c), make any needed changes in (a) and repeat part (b).

4. (a) Write a sentence telling how someone should walk in the HIKER experiment in order to produce a graph in the shape of a ∩. **(b)** Have someone follow your description in (a) and walk in front of the motion detector. **(c)** Did this person's hike produce the desired graph? If not, try to determine if your directions were clear enough, if the hiker followed your directions, or if your directions were not correct. **(d)** Based on your answers in (c), make any needed changes in (a) and repeat part (b).

5. Write a sentence telling how someone should walk in the HIKER experiment in order to produce a graph in the shape of a ⌐‾\.

6. Write a sentence telling how someone should walk in the HIKER experiment in order to produce a graph in the shape of a __/.

7. Write a sentence telling how someone should walk in the HIKER experiment in order to produce a graph in the shape of a ⊃.

8. Write a sentence telling how someone should walk in the HIKER experiment in order to produce a graph in the shape of a ⊂.

9. Look at the graph in Figure 1.15. This graph was produced by a HIKER experiment. Describe how the person who produced this graph walked.

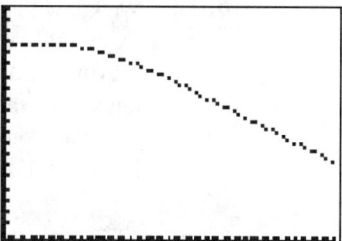

Figure 1.15

10. Look at the graph in Figure 1.16. This graph was produced by a HIKER experiment. Describe how the person who produced this graph walked.

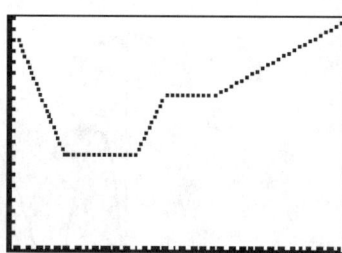

Figure 1.16

1.2 Distance, Time, and Two Kinds of Rate

In this section we will begin a more extensive study of lines and motion. First you will learn the technical difference between velocity and speed. Then you will see how to estimate the average speed of a hiker during a walking experiment by reading coordinates of points from the scatterplot.

Velocity and Speed

Let's begin with speed. Assume that an object is moving in a straight line. If it moves the same distance in each successive unit of time, it is said to move with **constant speed**. For example, if a car in cruise control on a level stretch of interstate highway covers 88 feet of highway each second that it moves, then it has a constant speed of 88 ft/sec (60 miles per hour).

However, you are usually unable to maintain a constant speed in a car, as we see in the following example.

Example 1.2   Application

You leave your house at 7:00 and drive to work. You park your car in the company parking lot at 7:24. If the total distance that you drive from home to work is 8.3 miles, what was your average speed?

Solution   We know that you did not drive at a constant speed during this 24-minute period. We know that your speed was 0 mph when you began, that it increased as you started moving, and that it slowed down to 0 when you parked the car. It would not be unusual if there were several other times when you stopped and started or slowed down as you drove from home to work. One possible graph of your trip is shown in Figure 1.17. So, what was your average speed?

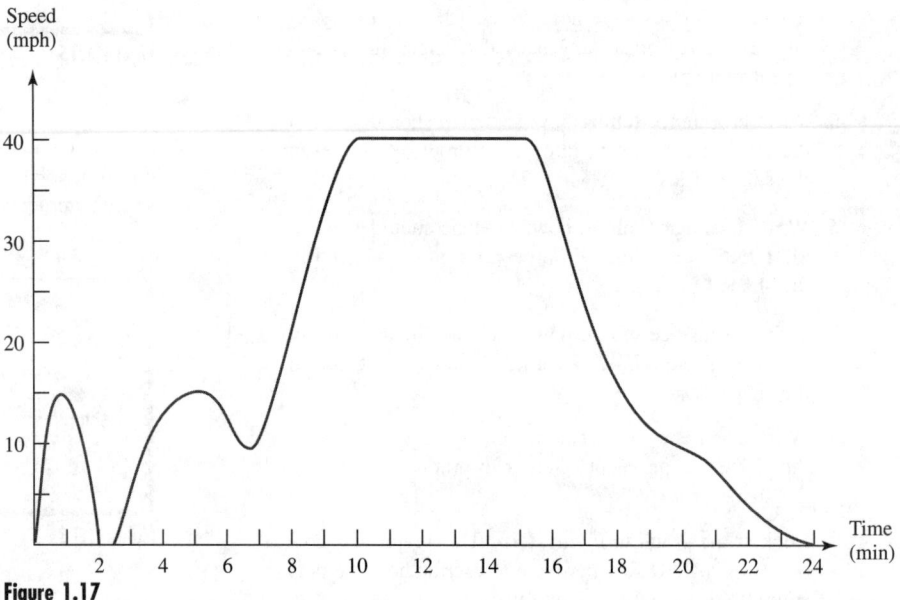

Figure 1.17

One way to get the average speed is to divide the total distance you drove, 8.3 miles, by the time it took you to travel this distance, 24 minutes. This calculation would give the following answer in miles per minute (mi/min)

$$\frac{8.3 \text{ mi}}{24 \text{ min}} = \frac{8.3}{24} \text{ mi/min}$$

$$\approx 0.34583 \text{ mi/min}$$

However we do not normally use the unit mi/min. Instead we use miles per hour (mi/hr or mph). To convert the answer to the more common unit, miles per hour, we would need to convert the minutes part of the answer to hours. In the original problem the 24 min was in the denominator. Since 60 min = 1 hr, we multiply

$$\frac{8.3 \text{ mi}}{24 \text{ min}} = \frac{8.3 \text{ mi}}{24 \text{ min}} \times \frac{60 \text{ min}}{1 \text{ hr}}$$

$$= 20.75 \text{ mi/hr}$$

Thus your average speed from home to work on this day was 20.75 mi/hr.   ■

Note: Units

Writing out the units, as in the above solution, can help you set up the calculation correctly. We multiplied by $\frac{60 \text{ min}}{1 \text{ hr}}$ so that the minute (min) units would "cancel."

Whether the speed is constant or not, the average speed of a moving object is defined as

average speed = distance traveled divided by elapsed time

This formula can also be thought of as $d = rt$, which translates into words as "**d**istance traveled on a trip equals the average **r**ate times the **t**ime taken for the trip." Using algebraic manipulation, this equation becomes $r = \dfrac{d}{t}$, rate equals distance divided by time.

Notice that the direction the object is moving does not matter. Common units for speed are miles per hour (mi/hr or mph), meters per second (m/s or mps), kilometers per hour (km/h or kph), and feet per second (ft/sec or fps).

Velocity, on the other hand, takes both the speed *and* the direction of the movement into account. In this chapter we are studying motion in a straight line, so we can think of a positive velocity as movement in one direction and a negative velocity as movement in the other direction. For example, on an east–west highway, you might choose to think of east as a positive direction and west as negative. Thus a car that is moving toward the east at a speed of 65 mph has a velocity of 65 mph. However a car that is moving toward the west at a speed of 72 mph has a velocity of −72 mph.

Note: West Is Not Always Negative

It is important to remember that choosing to call motion toward the east positive and motion toward the west negative is a decision made for the problem you are studying at the moment. There is not a permanent or universal connection between east/west and positive/negative. The point is that negative velocity is the direction opposite to what is considered positive.

Example 1.3

Compute the average speed and velocity from the scatterplot in Figure 1.18.

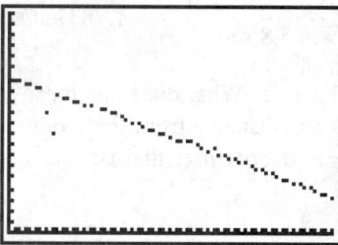

Figure 1.18

Solution We'll begin with finding the average speed.

In the graph shown in Figure 1.18 distance is measured in the *y*-direction (vertically) and time is measured in the *x*-direction (horizontally). We must choose two points on the graph and figure out the distance traveled between them and the time elapsed between them.

By using the Trace feature of the calculator, you can move the cursor to the leftmost x- and y-coordinates, as shown in Figure 1.19. In the point shown, the time is $x = 0.1$ second and the distance is $y = 13.8263$ feet.

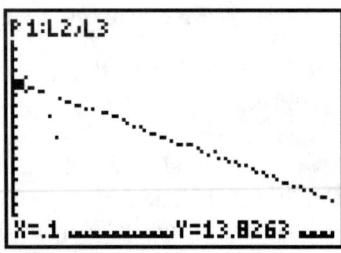

Figure 1.19   **Figure 1.20**

Now we need another point. Where should it be? We will use a point near the end of the walking time. In Figure 1.20 you can see that the time is 5.9 seconds and the distance is 3.48409 feet.

To compute the average speed we divide the change in distance between the points by the change in time:

A bar over a word such as speed *can mean average. Thus* $\overline{\text{speed}}$ *means* average speed.

$$\overline{\text{speed}} = \frac{13.8263 \text{ ft} - 3.48409 \text{ ft}}{5.9 \text{ sec} - 0.1 \text{ sec}}$$

$$= \frac{10.34221 \text{ ft}}{5.8 \text{ sec}} \approx 1.7831 \text{ ft/sec}$$

Next we'll calculate the average velocity represented by this graph. In finding the speed we subtracted the smaller numbers from the larger numbers to get the change in distance and the change in time. With velocity we must be careful to subtract the numbers in order: for example, the second number minus the first, because the sign of the answer makes a difference in its meaning.

$$\overline{\text{velocity}} = \frac{3.48409 \text{ ft} - 13.8263 \text{ ft}}{5.9 \text{ sec} - 0.1 \text{ sec}}$$

$$= \frac{-10.34221 \text{ ft}}{5.8 \text{ sec}} \approx -1.7831 \text{ ft/sec}$$

The velocity is negative. What does this mean? Since the distance along the y-axis is defined here as the distance from the motion detector, a negative velocity means that this distance is *decreasing*; that is, the motion is *toward* the motion detector. ∎

Definition: Average Velocity

If an object is at position d_1 at time t_1, and later it is at position d_2 at time t_2, then the average velocity from time t_1 to time t_2 is defined as follows:

$$\overline{v} = \frac{d_2 - d_1}{t_2 - t_1} \quad \text{or} \quad \overline{v} = \frac{d_1 - d_2}{t_1 - t_2}$$

This definition does not depend on the path that the object travels between these two points. That is, the path that connects the two points in Figure 1.21 does not have to be straight.

The average velocity formula can also be written as

$$\overline{v} = \frac{\Delta d}{\Delta t}$$

The symbols Δd and Δt are abbreviations for "change in d" and "change in t." Δ (delta) is the Greek letter D and is used in mathematics to mean "change in." Δd is pronounced "delta dee."

From now on, whenever we want to measure velocity v over a time interval, you should interpret that as finding the average velocity $\overline{v}$.

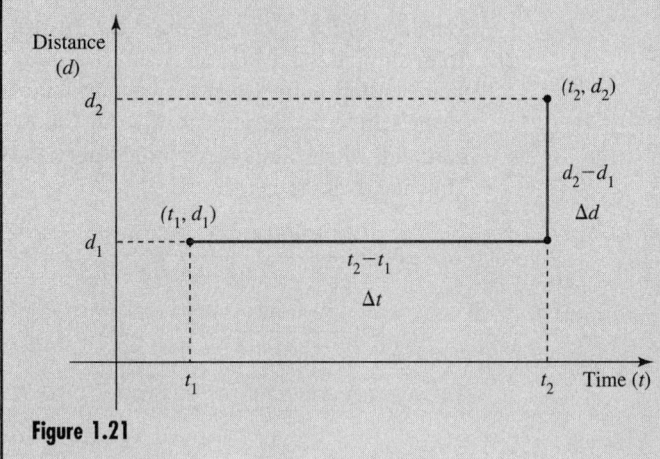

Figure 1.21

When computing the speed of a trip where directions are reversed, you have to remember to divide the total distance of the trip by the total time of the trip. Example 1.4 will look at computing speed and velocity in this type of situation.

Example 1.4

Marsha and Ed get on Interstate 95 at Kittery, Maine and drive northeast at an average speed of 72 mph. After traveling 108 miles they reach Augusta and conduct some business. Two hours later they retrace their trip on I-95 until they reach their office in Portland 56 miles southwest of Augusta. Because of some construction delays this part of the trip has an average speed of 50 mph. Excluding the two hours they spent conducting business, what is (a) their average speed for the trip from Kittery to Portland, and (b) their average velocity for this trip?

Solutions (a) The first segment of the trip is the 108 miles from Kittery to Augusta. Since their average speed for this portion of the trip was 72 mph, this part of the trip must have taken $\frac{108 \text{ miles}}{72 \text{ mph}} = 1.5$ hours. The part of the trip from Augusta to Portland took $\frac{56 \text{ miles}}{50 \text{ mph}} = 1.12$ hours. Thus we have the following data:

Trip segment	Distance (mi)	Time (hr)
Kittery to Augusta	108	1.5
Augusta to Portland	56	1.12
Total trip	164	2.62

The average speed is the total distance divided by the total time. Thus we have

$$\overline{\text{speed}} = \frac{164}{2.62} \approx 62.6$$

Marsha and Ed's average speed for this trip was 62.6 mph.

(b) To determine the average velocity, we need to consider how far they were from their starting point when they got to the office. If we think of the first portion of the trip as being in a positive direction, then the 56 miles from Augusta to Portland, when they reversed the direction, would be represented by −56 miles. Thus the average velocity is

$$\overline{v} = \frac{108 \text{ miles} + (-56) \text{ miles}}{2.62 \text{ hours}}$$

$$= \frac{52 \text{ miles}}{2.62 \text{ hours}} \approx 19.8 \text{ mph}$$

The average velocity for this trip was 19.8 mph.

Section 1.2 Exercises

1. *Transportation* A car on the interstate has its cruise control set at 67 mph. How far does the car travel in 1 hour 45 minutes?

2. *Transportation* A car travels 106 km. If the average speed was 29 m/s, how many hours were required for the trip?

3. *Sports technology* Sound travels through the air at an average speed of 340 m/s. You are standing at the end of a 100-m race when you see the smoke indicating that the starter's gun has been fired. How long will it take for you to hear the sound of the gun? (Assume that no time elapsed between the firing of the gun and your seeing the smoke.)

4. *Transportation* A truck traveled 875 miles. The average speed for the entire trip was 56.5 mph. How long was the trip?

5. *Transportation* An airplane takes off from an airport and lands 3 hours later at another airport 1320 miles away. What was the airplane's average speed in miles per hour?

6. *Transportation* An airplane takes off from New York's Kennedy Airport and lands 7 hours later at London's Heathrow Airport 3455 miles away. What was the airplane's average speed in miles per hour?

7. *Transportation* An airplane takes off from the San Francisco Airport and lands 6 hours 40 minutes later at Atlanta's Hartsfield Airport 2139 miles away. What was the airplane's average speed in miles per hour?

8. *Transportation* An airplane takes off from the Dallas airport and lands 2 hours 5 minutes later at the Denver airport 643 miles away. What was the airplane's average speed in miles per hour?

9. *Sports technology* In 1994 Martin Buser and his dog team won the 1151 mile Iditarod Trail Sled Dog Race in 10 days, 13 hours, 2 minutes, 39 seconds. What was their average speed in miles per hour if you assume that they only rested a total of 40 hours during the race?

10. *Sports technology* In 1994 Sterling Martin won the Daytona 500 NASCAR auto race with an average speed of 156.931 mph. How long did it take him to complete the race?

11. *Transportation* At noon a car enters Interstate 70, heading west from Kansas City, Kansas. At 3:35 P.M. the car exits the interstate in Salina, Kansas 172 miles away. If we assume that east is positive, what was the car's average speed and average velocity?

12. *Transportation* At noon a car enters Interstate 25, heading north from Pueblo, Colorado. At 1:25 P.M. the car exits the interstate in Salina, Kansas 107 miles away. If we assume that north is positive, what was the car's average speed and average velocity?

13. *Transportation* John gets on Interstate 70 at Burlington, Colorado and drives due west at an average speed of 72 mph. After traveling 73 mi he reaches the Limon exit and realizes that he has gone too far. He turns around and travels due east 23 miles back to the Arriba exit at an average speed of 66 mph. For the whole trip from Burlington to Arriba, what is **(a)** his average velocity and **(b)** his average speed? (*Hint:* Assume that it took no time to turn around.)

14. *Transportation* Sandy gets on Interstate 24 at Nashville, Tennessee and drives southeast at an average speed of 75 mph. After traveling 97 mi he reaches the Monteagle exit and realizes that he has gone too far. He turns around and travels northwest 23 miles back to the Manchester exit at an average speed of 68 mph. For the whole trip from Nashville to Manchester, what is **(a)** his average velocity and **(b)** his average speed?

15. A straight stretch of pavement is marked off in 100-m intervals. Students are stationed on a hill that overlooks the pavement and measure the time for a car to pass each of the markers. The following data are obtained:

Distance (m)	0	100	200	300	400	500
Time (sec)	0	1.8	3.7	5.5	7.3	9.2

(a) Plot a graph with distance as the vertical axis and time as the horizontal axis.

(b) What was the average speed of the car during the time it was observed in km/hr and mi/sec?

16. *Police science* A straight stretch of interstate highway is marked off in 1-mile intervals. A highway patrol officer flying in an airplane measured the time for a car to pass each of the markers. The following data were obtained:

Distance (mi)	0	1	2	3	4	5	6
Time (sec)	0	42	85	128	170	215	260

(a) Plot a graph with distance as the vertical axis and time as the horizontal axis.

(b) What was the average speed of the car in mi/sec and mi/hr?

17. Figures 1.22 and 1.23 are two graphs that show the result of a HIKER experiment. What was the average speed and average velocity of this hiker between the two marked times? Assume that the time is in seconds and the distance is in feet.

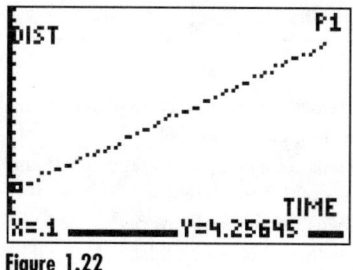

Figure 1.22

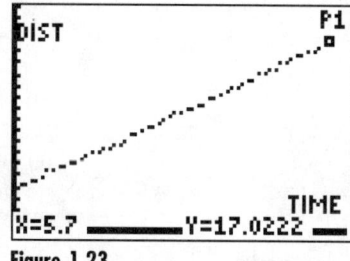

Figure 1.23

18. Figures 1.24 and 1.25 show two graphs that show the result of a HIKER experiment. The line is horizontal to the left of the 1.1 mark in Figure 1.24. What was the average speed and average velocity of this hiker between the two marked times? Assume that the time is in seconds and the distance is in feet.

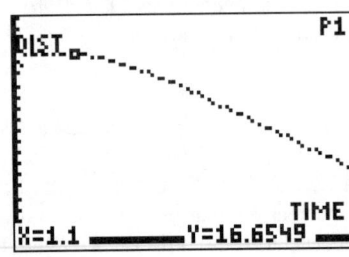

Figure 1.24

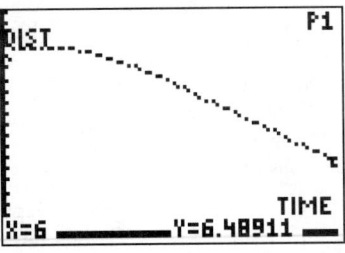

Figure 1.25

19. The graph in Figure 1.26 represents a person's distance from a motion detector during a 13-sec interval. **(a)** Describe what the person was doing during these 13 sec. **(b)** What was the person's average velocity?

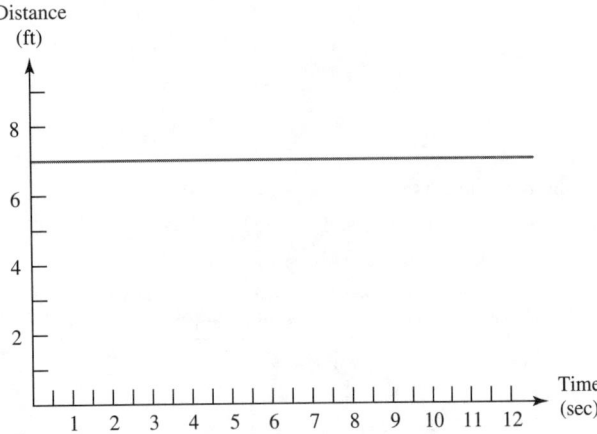

Figure 1.26

20. The graph in Figure 1.27 shows the distances of two different people from a motion detector during a 12-sec interval. **(a)** What was the velocity of the person who followed path l_1? **(b)** What was the velocity of the person who followed path l_2?

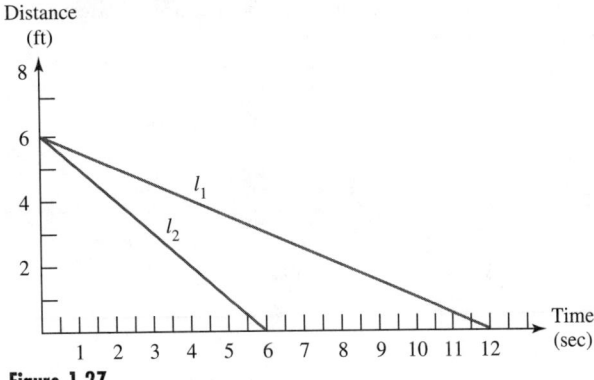

Figure 1.27

21. The graph in Figure 1.28 represents a person's distance from a motion detector during a 10-sec interval.

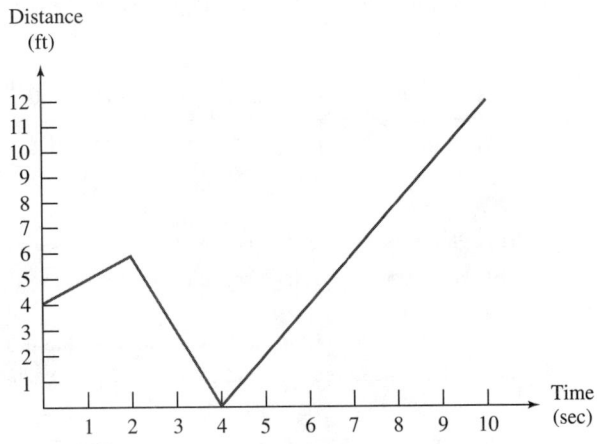

Figure 1.28

(a) Complete the following table:

Time (sec)	0	1	2	3	4	5
Distance (ft)	4				0	

Time (sec)	6	7	8	9	10
Distance (ft)					

(b) What was the person's average speed and velocity for the first two seconds?

(c) What was the person's average speed and velocity from second 2 to second 4?

(d) What was the person's average speed and velocity from second 4 to second 10?

(e) What was the person's average speed and velocity for the entire trip?

22. Monica left her house to walk to a friend's house. When she was partway there it started to rain. Because she did not have an umbrella she decided to return to her house. Figure 1.29 shows a graph of her walk.

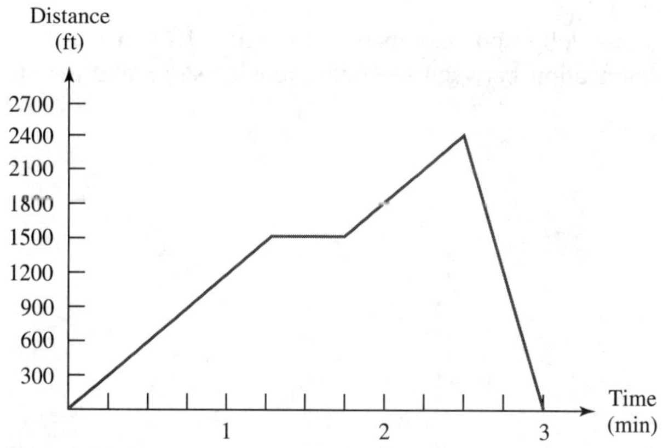

Time (min)	0	0.25	0.5	0.75	1	1.25	1.5
Distance (ft)							

Time (min)	1.75	2	2.25	2.5	2.75	3
Distance (ft)						

Figure 1.29

(a) Complete the following table:

(b) What was her average velocity for the first 1.25 min?

(c) What was her average velocity from 1.25 min to 1.75 min?

(d) What was her average velocity from 2.5 min to 3.0 min?

(e) What was her average velocity for the entire walk?

(f) What was her average speed for the entire walk?

(g) When do you think it started to rain?

(h) What happened from 1.25 min to 1.75 min?

1.3   Equation of a Line: Slope and y-Intercept

There are several ways to write equations of lines that could serve as mathematical models for data that appear to be linear. The first involves two features of a straight line graph: the **slope** and **y-intercept**. You will see that slope is closely related to the definition of velocity discussed in the previous section. When you know how to estimate the slope and the y-intercept of a straight line, you will be able to write its equation. In this section you will write your first mathematical models for data.

Slope of a Straight Line

We begin with the definition of the slope of a line.

Definition: Slope

The **slope** m of a straight line is a measure of the steepness of the line with respect to the x-axis. If (x_1, y_1) and (x_2, y_2) are two different points on a line, then the slope of a line is defined by

$$m = \frac{y_2 - y_1}{x_2 - x_1} = \frac{\Delta y}{\Delta x}$$

provided that x_1 is not equal to x_2.

The subtraction could be done in the opposite order to get the same result:

$$m = \frac{y_1 - y_2}{x_1 - x_2}$$

When y and x are expressed in units, such as miles and hours, the slope should be given in the appropriate units, such as miles per hour. Again, the symbol Δ is used to mean "change in." In addition you see in Figure 1.30 the words *rise* and *run*. They are common synonyms for Δy and Δx. It is sometimes helpful to think of the slope as *rise over run* or $\frac{\text{rise}}{\text{run}}$.

Figure 1.30 illustrates this definition. Compare it to Figure 1.21, the distance vs. time graph; notice the connection between the definitions of slope and average velocity.

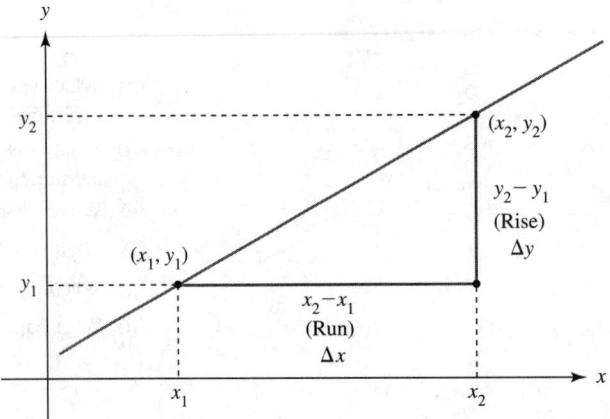

Figure 1.30

Example 1.5

Plot the points $(2, 5)$ and $(4, -8)$, draw the line through these points, and find the slope of the line that passes through them.

Solution   The two points and the line through them are shown in Figure 1.31.

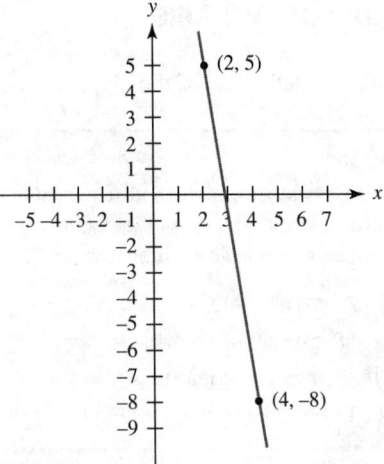

Figure 1.31

To determine the slope we let $x_1 = 2$, $x_2 = 4$, $y_1 = 5$, and $y_2 = -8$. Then, using the above formula for the slope, we find that the slope is

$$m = \frac{y_2 - y_1}{x_2 - x_1}$$

$$= \frac{-8 - 5}{4 - 2} = \frac{-13}{2} = -6.5$$

The slope of the line through the points $(2, 5)$ and $(4, -8)$ is -6.5. ■

Example 1.6

Plot the points $(-4, 4)$ and $(-1, 7)$, draw the line through these points, and find the slope of the line that passes through them.

Solution The two points and the line through them are in Figure 1.32.

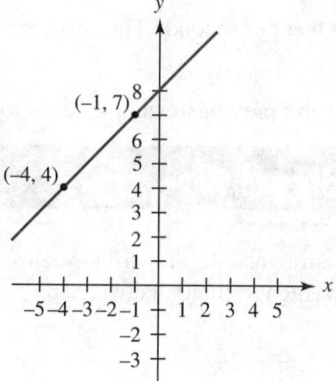

Figure 1.32

To determine the slope we let $x_1 = -4$, $x_2 = -1$, $y_1 = 4$, and $y_2 = 7$. Then, using the above formula for the slope, we find that the slope is

$$m = \frac{y_2 - y_1}{x_2 - x_1}$$

$$= \frac{7 - 4}{-1 - (-4)} = \frac{3}{3} = 1$$

The slope of the line through the points $(-4, 4)$ and $(-1, 7)$ is 1. ■

Applications of Slope

Example 1.7

Compute the slope of the line between the two points marked in Figures 1.33 and 1.34. The units of measurement are: distance in feet and time in seconds. Use the correct units in the answer.

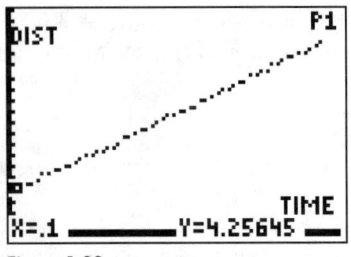

Figure 1.33

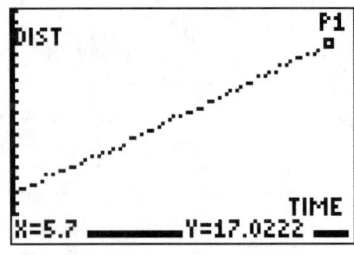

Figure 1.34

Solution From Figure 1.33 we see that $x_1 = 0.1$ and $y_1 = 4.25645$ and from Figure 1.34 we have $x_2 = 5.7$ and $y_2 = 17.0222$. Thus the slope m is

$$
\begin{aligned}
m &= \frac{y_2 - y_1}{x_2 - x_1} \\
&= \frac{17.0222 \text{ ft} - 4.25645 \text{ ft}}{5.7 \text{ s} - 0.1 \text{ s}} \\
&= \frac{12.76575 \text{ ft}}{5.6 \text{ s}} \approx 2.280 \text{ ft/s}
\end{aligned}
$$

The slope is 2.280 feet per second. This answer is also the average velocity between $t = 0.1$ and $t = 5.7$.   ∎

The result of the previous example leads to the following generalization.

Average Velocity and Slope

The slope of a straight line drawn between two points in a distance vs. time graph is the average velocity between the times associated with those two points.

Depending on the units used in the example, slope will have different meanings and different names. People in construction refer to the steepness of a roof as the *roof pitch*. Truck drivers refer to the *grade* of a hill, and you might see a sign as you start down a steep hill that says something like, "WARNING! 4% GRADE NEXT 3 MILES." In some fields of study, **declination** refers to a line with negative slope while **inclination** refers to a line with positive slope.

Example 1.8   Application

An automobile engine thermostat is being tested. The thermostat is marked 92°C but the thermometer used in the test is marked in degrees F. If the thermostat water is heated, the thermostat should open at 92°C. Water freezes at 32°F, which is 0°C, and water boils at 212°F, or 100°C. The relationship between the Celsius and Fahrenheit scales for measuring temperature can be drawn as a straight line. The freezing and boiling points of water can be represented as the points (0°C, 32°F) and (100°C, 212°F).

(a) Graph these two points on the Cartesian coordinate system.
(b) Draw the line through the points.
(c) Determine the slope of this line.
(d) Determine the thermometer reading if the thermostat opens at 92°C.

Solutions (a) The two points are plotted in Figure 1.35. Notice that for this problem the horizontal axis represents the Celsius temperatures and the vertical axis represents the Fahrenheit temperatures.

(b) The line through these two points is shown in Figure 1.36.

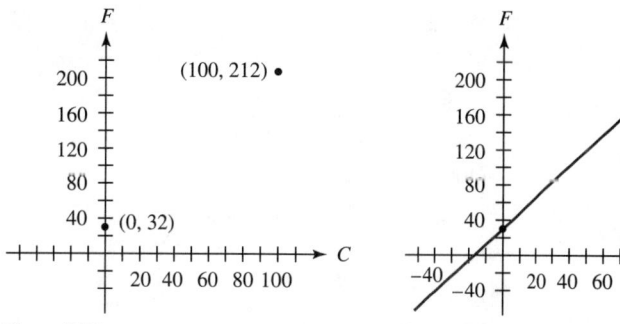

Figure 1.35   **Figure 1.36**

(c) To determine the slope of this line we need two points. If we let $x_1 = 0$, $y_1 = 32$, $x_2 = 100$, and $y_2 = 212$, then we calculate the slope of this line as

$$m = \frac{y_2 - y_1}{x_2 - x_1} = \frac{212 - 32}{100 - 0}$$

$$= \frac{180}{100} = \frac{9}{5} = 1.8$$

Thus this line has a slope of 1.8°F per °C. That is, for every change in one degree Celsius there is a change of 1.8 degrees Fahrenheit.

(d) We are given the Celsius temperature of 92° and want to find its corresponding Fahrenheit temperature. In other words, we want to find the missing coordinate of the point $(92, y)$. Let's use one of the two points we were given and compute the slope. We just found out that the slope is 1.8, so if we use the point $(0, 32)$, then we need to solve the equation $\frac{y - 32}{92 - 0} = 1.8$. We can solve this equation as follows

$$\frac{y - 32}{92 - 0} = 1.8$$

$$\frac{y - 32}{92} = 1.8$$

$$y - 32 = (1.8)(92)$$

$$y - 32 = 165.6$$

$$y = 165.6 + 32 = 197.6$$

The thermometer should read 197.6°F   ∎

Example 1.9   Application

A trucker approaches a hill and sees a sign that says, "6% GRADE NEXT 5 MILES." How much lower can we expect the highway to be at the bottom of the hill than at the top?

Solution

Before we can begin we need to know what is meant by a 6% grade. Remember that 6 percent means "6 per hundred." In this instance, since the truck is going downhill, it means that the road will drop 6 ft for every 100 ft (you could also say that it will drop 6 miles for every 100 miles) of horizontal distance the truck covers. Thus if $\Delta x = 100$ then $\Delta y = -6$. Why is Δy negative? The truck is going downhill, and on the *y*-axis *down* is in a negative direction. Thus we see that the hill has a slope of

$$m = \frac{\Delta y}{\Delta x} = \frac{-6}{100} = -6\% = -0.06$$

Our goal here is to find the value of *y* if we know the value of *x*. We can rewrite the definition of slope, $m = \dfrac{\Delta y}{\Delta x}$, as

$$\Delta y = m \cdot \Delta x$$

Therefore our solution comes from substituting 5 for Δx and -0.06 for *m* in the above equation:

$$\Delta y = -0.06 \times 5 \text{ miles} = -0.3 \text{ mile}$$

Thus, in this 5-mile stretch of highway, the road will drop 0.3 mi. Since there are 5280 feet in one mile, the vertical drop works out to be

$$-0.3 \text{ mi} \cdot \frac{5280 \text{ ft}}{1 \text{ mi}} = -1584 \text{ ft}$$

While the road may curve, we will think of this as a straight stretch of highway. In Example 3.9, we look at what happens if the road is curved.

Example 1.10   *Chapter Project*

What is the slope of a line that would model the Olympic men's 100-meter freestyle swimming records given in the Chapter Project? Estimate by how many seconds the winning time will decrease between 1992 and 1996. Estimate the winning time in 1996.

Solution

We need to choose two points that seem to be representative. We will select points that are far apart in years in order to help reduce error. We'll choose 1896 and 1992. Since we placed the years on the horizontal axis, we have $x_1 = 1896$ and $x_2 = 1992$. The corresponding *y*-values are $y_1 = 82.2$ and $y_2 = 49.0$; then the slope becomes

$$m = \frac{49.0 - 82.2}{1992 - 1896} = \frac{-33.2}{96} \approx -0.346$$

The units of this slope are seconds per year. The value of this slope can be interpreted as follows. Its negative value means that the race times decrease (Δy is negative) as the years increase (Δx is positive). Its numerical value means that the times are falling at a rate of 0.346 seconds every year. Using the formula

$$\Delta y = m \cdot \Delta x$$

and substituting -0.346 for the slope *m* and 4 for Δx, we reach the conclusion that the decrease will be $0.346 \times 4 \approx 1.38$ seconds between 1992 and 1996.

Using this reasoning we could conclude that the 1996 winning time would have been $49.0 - 1.38 = 47.62$ seconds. We shall see at the end of the chapter what the actual time was in that year.

There are other methods for making this estimate. We will explore some of them later in this chapter.

Intercepts of a Graph

A second part of a straight line that is needed to build a linear mathematical model is a point. One of the most convenient points to select is the y-intercept. Like the slope, the y-intercept is important in studying various kinds of graphs and experiments.

The intercepts of any graph (straight or not) are defined as the point or points where the graph crosses the axes. A y intercept is any point where the graph crosses the y-axis and an x-intercept is any point where the graph crosses the x-axis. Since the value of x is 0 all along the y-axis, then the y-intercept is the point on the graph where $x = 0$. Likewise the x-intercept is the point where $y = 0$. Sometimes the y-intercept refers to the y-coordinate of the point.

Look at the four graphs in Figures 1.37–1.40. In Figure 1.37 the curve, in this case a line, has an x-intercept at a and a y-intercept at b. The curve in Figure 1.38 has no x-intercepts and one y-intercept at c. Figure 1.39 has two x-intercepts at e and g and two y-intercepts at d and f. The last curve, shown in Figure 1.40, has no y-intercepts and one x-intercept at h.

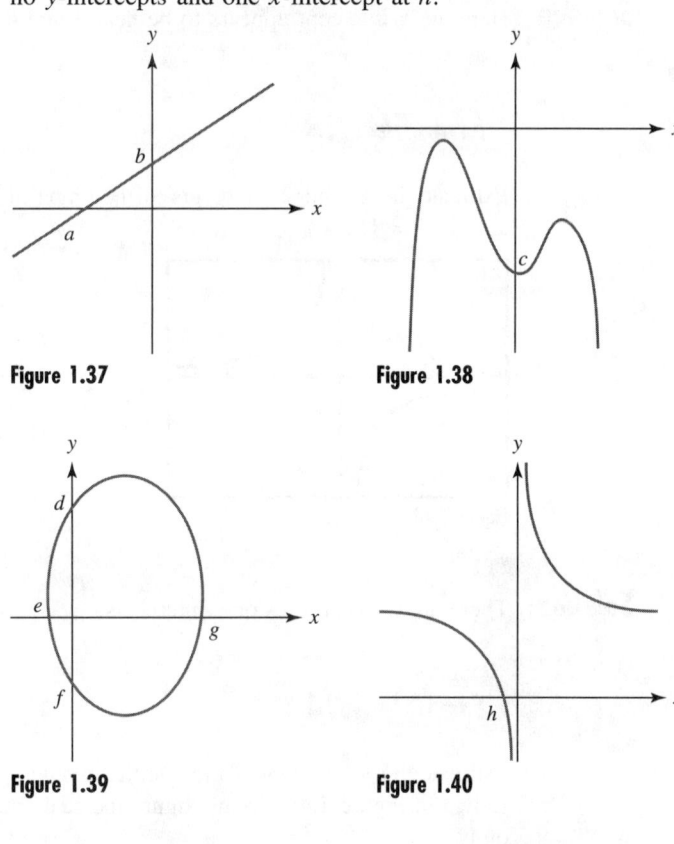

Figure 1.37

Figure 1.38

Figure 1.39

Figure 1.40

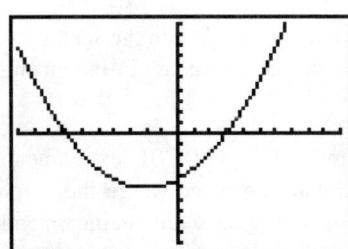

Figure 1.41

Example 1.11

Estimate the x- and y-intercepts of the graph in Figure 1.41. In the figure, Xscl = 1 and Yscl = 1.

Solution   This graph crosses the x-axis twice, so it has two x-intercepts. The x-intercepts are −7 and 3. There is only one y-intercept. It appears to be at −4.   ■

Example 1.12

Estimate the x- and y-intercepts of the graph in Figure 1.42. In the figure, Xscl = 1 and Yscl = 1.

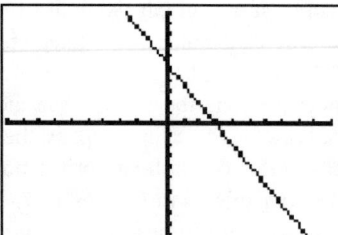

Figure 1.42

Solution   Here the x-intercept appears to be near 3 and the y-intercept is near 5.   ■

Example 1.13

Estimate the x- and y-intercepts of the graph in Figure 1.43. In the figure, Xscl = 1 and Yscl = 1.

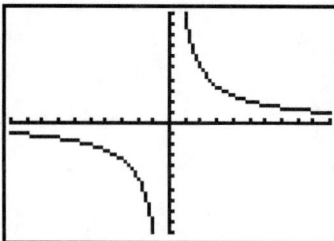

Figure 1.43

Solution   There are no visible x- or y-intercepts.   ■

Example 1.14

Estimate the y-intercept of the line that would approximate the data in the scatterplot shown in Figure 1.44. In this figure the scale marks on the y-axis are in increments of 1.

Solution   These data are from a CBL experiment (see Example 1.3) and CBL experiments do not include $x = 0$. Therefore we can only estimate the place where the graph would cross the y-axis. Since the scale marks on the y-axis are in increments of 1, a reasonable estimate is 14 feet. What is the physical meaning of this number?

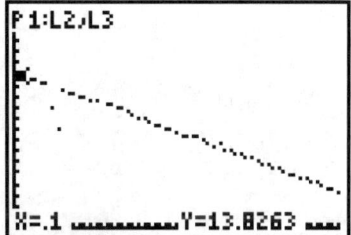

Figure 1.44

Definition: The y-Intercept In Distance-Time Experiments

The y-intercept on a distance vs. time plot of an experiment is the distance of the moving object from the motion detector at the beginning of the experiment because the experiment begins at $t = 0$. On a calculator graph, $x = 0$ takes the place of $t = 0$.

Fitting a Line to the Data—The Slope-Intercept Form of the Line

Now that you have seen how to find the slope and y-intercept of a line in a variety of circumstances, we can put them together to form the equation of the line that could be used as a mathematical model for a set of data from an experiment, a survey, or, as in the case of Olympic records, from history. Scientists and engineers often seek a single equation that can approximate the data they have collected. The process is called **modeling data** or **fitting curves to data**. In this book you will study several ways to fit curves to data and you will learn to judge how well the curve fits the data.

In the next activity you will fit a line to the distance vs. time data we have been working with. You will use the slope and y-intercept that you have estimated in previous examples.

Definition: Slope-Intercept Equation of a Line

If you know the slope and y-intercept of a line, then the equation of the line can be written as

$$y = mx + b$$

where m is the slope of the line and b is the y-intercept of the line.

Figure 1.45 can help make sense of the above statement. There we have drawn a line with y-intercept b and slope m. We can say that every time x changes by 1, y changes by m. What can be said about the y-coordinate of the point where $x = 3$? The answer is $y = 3m + b$. You can extend this reasoning to say that for any value of x the equation of the line is $y = xm + b$ or $y = mx + b$.

Example 1.15

If the slope of a line is -3 and the y-intercept is 4, what is the equation of the line?

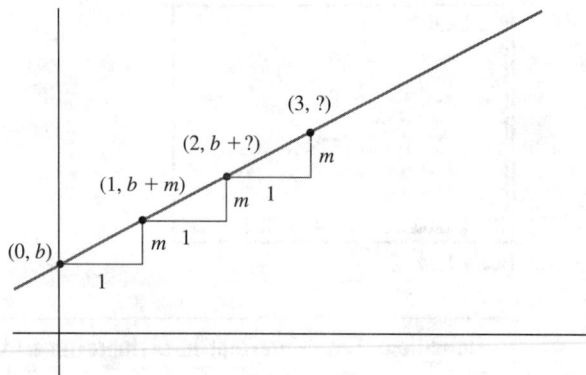

Figure 1.45

Solution   Since the slope is -3, we have $m = -3$. A y-intercept of 4 means that $b = 4$. Substituting these values in the equation $y = mx + b$, we get the equation of the line as $y = -3x + 4$.   ∎

Example 1.16

If a line passes through the point $(2, -5)$ and has a y-intercept of -8.6, what is the equation of the line?

Solution   We have the y-intercept, so we know that $b = -8.6$. We have to determine the slope. Two points are needed for the slope. We can use the y-intercept, -8.6, as the second point $(0, -8.6)$. From the two points we find that the slope is

$$m = \frac{y_2 - y_1}{x_2 - x_1} = \frac{-8.6 - (-5)}{0 - 2}$$

$$= \frac{-3.6}{-2} = 1.8$$

Substituting $m = 1.8$ and $b = -8.6$ into the equation $y = mx + b$, we get the equation of the line as $y = 1.8x - 8.6$.   ∎

Example 1.17

If a line passes through the points $(0, -3)$ and $(4, 7)$, what is the equation of the line?

Solution   The first point $(0, -3)$ is the y-intercept itself, so we know that the y-intercept is -3. The slope is $\frac{7-(-3)}{4-0} = \frac{10}{4} = 2.5$. Therefore the equation of the line is $y = 2.5x - 3$.   ∎

Example 1.18   Application

What is the equation of the line connecting the Celsius and Fahrenheit values in Example 1.8 on page 18?

Solution We computed the slope as 1.8°F per °C. What is the y-intercept? Actually we are looking for the "F-intercept," since the variable F was plotted on the vertical axis. Look at the values in the earlier example. One of the pairs is $(0, 32)$. Therefore the F-intercept is 32 and the equation is

$$F = 1.8C + 32$$

Calculator Lab 1.1

What Does Slope Tell about the Line?

In this Calculator Lab you will graph several lines on your calculator and study the relationship between the slope of the lines and their steepness.

Procedures

1. Draw the graph of the lines with slopes of 2, 1, 0.5, and 0, and y-intercept 2 by entering the equations shown in Figure 1.46. Be sure to turn off all data plots on your calculator before graphing.

2. Set the dimensions of the graph window to ZSquare (Zoom Square), which means that the scales are the same in both directions. Press GRAPH and you will see a screen matching Figure 1.47. To learn which graph is connected to which equation, press TRACE; then press the right arrow a few times so the blinking cursor ☒ is visible. In Figure 1.48 you can see the blinking cursor near the top middle of the window. The equation is shown in the upper left corner. To move to another line, press the up or down arrow.

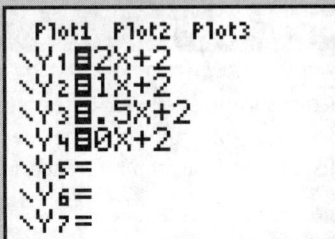

Figure 1.46

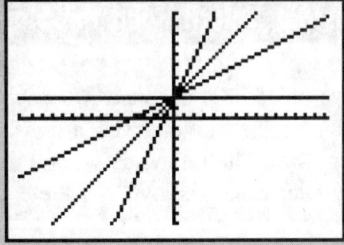

Figure 1.47

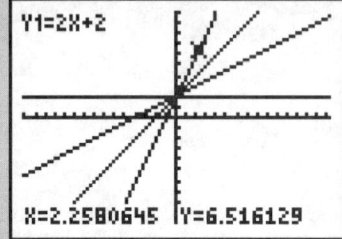

Figure 1.48

3. Now graph the lines with slopes of -2, -1, and -0.5 and y-intercept 2. Be sure to use the (-) (negative) key and not the - (subtract) key when entering the negative slopes. The result should look like Figure 1.49.

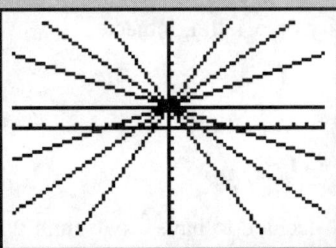

Figure 1.49

4. Answer the following questions about the graphs on your calculator screen.

   (a) In what direction does a line go if the slope is positive?
   (b) In what direction does a line go if the slope is negative?

(c) Describe a line whose slope is 0.

(d) Two lines have slopes equal to 0.5 and 2. Which line is steeper?

(e) Two lines have slopes equal to −0.5 and −2. Which line is steeper?

(f) True or false? If the slope of line l_1 is greater than the slope of line l_2, then l_1 is steeper than l_2.

(g) There are three pairs of perpendicular lines produced in Figure 1.49. What can you say about the slopes of perpendicular lines? Test your answer by multiplying the slopes of two perpendicular lines. Test your answer by graphing other pairs to see if you have found a guaranteed way to produce perpendicular lines.

The Constant Velocity Equation

We have seen that if the velocity is constant between two times, then the slope of the line joining those two points is equal to the constant velocity. We have also seen that the y-intercept in a distance vs. time experiment equals the distance between the moving object and the motion detector at the start of the experiment. These concepts together with the slope-intercept form of the equation of a line lead to the following summary.

Constant Velocity and the Slope-Intercept Equation

If an object at a distance d_0 away from a measuring device or other reference point begins moving with a constant velocity v_0 in a straight line toward or away from the reference point, then its distance d from the reference point at any later time t is given by the equation

$$d = v_0 t + d_0$$

This is a line with slope v_0 and y-intercept d_0. The subscript 0 (zero) in the symbols d_0 and v_0 are commonly used in science and engineering to indicate the value of the variable at $t = 0$. In Chapter 2 we will add to this equation a term that reflects the force of gravity on a falling object.

Example 1.19

Chris and Pat decided to have a two-mile walking race. On the day of the race, Pat showed up on crutches with a sprained ankle. Pat insisted that they still have the race, but added, "Normally I can walk about 4.5 miles per hour, the same as you. With these crutches the best I can do over two miles is 1 mph, so I'll need a head start. I figure that a 1.5-mile head start will be about right." Chris agreed to give Pat that much head start; they synchronized their watches and drove to the 1.5-mile mark. Chris drove back to the start and the two began walking at the same instant.

During the race Chris averaged 4.5 mph and Pat averaged 1.1 mph, a little better than he predicted.

(a) Who will win the race?
(b) If Chris wins, when will he pass Pat and how far is it to the finish line at that point?
(c) If Pat wins, what is the margin of victory (to the nearest ten feet)?

Solutions   (a) To find out who won we will use the formula Time $= \dfrac{\text{Distance}}{\text{Rate}}$ to compute the time it took for each person to reach the finish line (since Pat had a 1.5 mi head start, he only raced 0.5 mi):

$$\text{Pat:}\quad \frac{\text{Distance}}{\text{Rate}} = \frac{0.5 \text{ mi}}{1.1 \text{ mi/hr}} = 0.45 \text{ hr}$$

$$\text{Chris:}\quad \frac{\text{Distance}}{\text{Rate}} = \frac{2 \text{ mi}}{4.5 \text{ mi/hr}} = 0.44 \text{ hr}$$

Since 0.44 hr $<$ 0.45 hr, Chris was the first to get to the finish line and is the winner.

(b) To find out when Chris passed Pat, we will use a graph and $y = mx + b$, the slope-intercept form for the equation of a line. We can think of this problem in either of the two ways shown in Figures 1.50 and 1.51. Figure 1.50 graphs each walker's time vs. his distance from the finish line. Figure 1.51 graphs each walker's time vs. his distance from Chris's starting point.

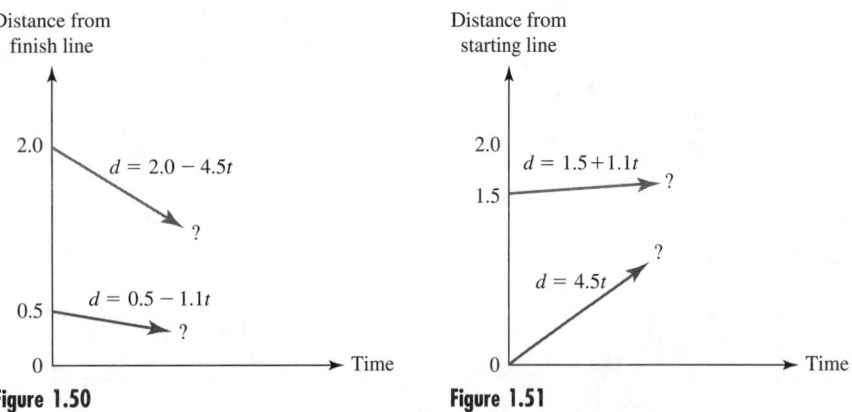

Figure 1.50   **Figure 1.51**

You should be able to say why the lines in Figure 1.50 have a negative slope and the ones in Figure 1.51 have a positive slope. In both cases the problem is to find the point where the lines meet. At this point the two walkers will have the same time coordinate, t, and the same distance coordinate, d. That is, they will be in the same place at the same time. We'll solve the problem represented in Figure 1.50 and let you show in Exercise 35 that the lines in Figure 1.51 give the same answer.

The function for Chris's progress is $d = 2.0 - 4.5t$ and $d = 0.5 - 1.1t$ is the function for Pat's progress. There are two ways to find the point of intersection of these two lines. One is by graphing and the other is by algebra. We will solve the problem using each method.

Graphing   The graphs of these two functions are in Figure 1.52, which doesn't provide the detail we need. Zooming in we obtain Figure 1.53 and see that indeed Chris passed Pat before the finish line. (If Chris had been the loser, then the graph would have shown the two lines meeting *below* the x-axis.) According to the graph in Figure 1.53, the two walkers met 0.441 hours (about 26.5 minutes) after the start, and at that time they were about 0.013 miles (about 69 feet) from the finish line.

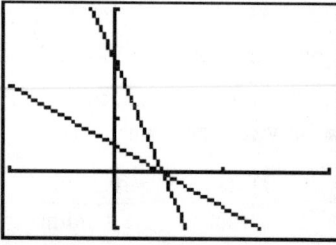

Figure 1.52   **Figure 1.53**

Algebra   The idea is to find the values of d and t that satisfy both equations. You will get more practice with this concept later in this book. The equations are $d = 2.0 - 4.5t$ and $d = 0.5 - 1.1t$.

Since we want to find the point (t, d) that satisfies both equations, we can substitute one equation's expression for d into the other equation and solve for t. This gives the equation

$$2.0 - 4.5t = 0.5 - 1.1t$$
$$-3.4t = -1.5 \qquad \text{Subtract 2.0 from both sides and add } 1.1t \text{ to both sides.}$$
$$t = \frac{-1.5}{-3.4} \approx 0.441 \text{ hr}$$

To find the value of d where the lines cross, we substitute 0.441 for t in one of the original equations. We'll choose the first equation, $d = 2.0 - 4.5t$.

$$d = 2.0 - 4.5 \times 0.441$$
$$= 2.0 - 1.9845 = 0.0155$$

Thus the lines cross when the runners are about 0.0155 miles (82 feet) from the finish line.   ■

In this example, the answer computed by algebra is more accurate since it gave the exact answer $\frac{-1.5}{-3.4}$. With a graph there will often be an error in your estimate of the coordinates of the point of intersection. However the graphing solution can be made more accurate by zooming in a few more times.

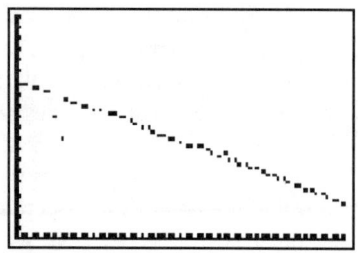

Figure 1.54

Example 1.20

Write a linear model for the CBL experiment shown in Figure 1.54.

Solution   The figure shows the data for which we have found the slope and y-intercept in previous examples. In Example 1.3 (page 9) we estimated the slope to be -1.7831 ft/sec. In Example 1.14 (pages 22–23) we estimated the y-intercept to be 14. Therefore the equation of a line that could be a model for this data is

$$y = -1.7831x + 14$$

The graph of this line is superimposed on the data in Figure 1.55. The line lies over most of the data points. You can see that this is a very good fit to the data, and also that the walker was able to walk at a very constant velocity.   ■

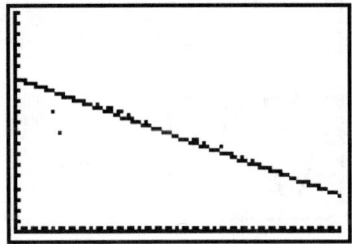

Figure 1.55

Example 1.21

Write a linear model of the Olympic data given in the Chapter Project (page 2) using the slope-intercept equation of the line.

Solution   To answer this problem you need estimates of the slope and y-intercept of the line that might fit the data. You estimated the slope in Example 1.10. However the y-intercept poses an interesting problem. To cross the y-axis we need the point where $x = 0$. But the x-axis is scaled in calendar years, so we would be looking for an estimate of the swimming record for the year 0, almost 2000 years ago. There are a number of problems with making such an estimate. For now let's just say that the slope-intercept form of the line is not useful for modeling data like this.

There are many ways to solve this problem. One of them is to rename year 1896 to be year 0 and subtract 1896 from all the other years. Then the record for 1896 would be the y-intercept. This method is called **transforming the data** and will be discussed in Chapter 2. For now we'll move on.   ■

Section 1.3 Exercises

In Exercises 1–8 plot each of the given points, draw the line through the points, and determine the slope of that line.

1. $(3, -5)$ and $(-4, 2)$

2. $(-4, -1)$ and $(5, -3)$

3. $(-2, 3)$ and $(4, 3)$

4. $(5, 1)$ and $(5, -4)$

5. $(1, 4)$ and $(-5, 2)$

6. $(-1, -5)$ and $(2, 7)$

7. $(-2, -3)$ and $(-2, 5)$

8. $(6, -2)$ and $(-5, -2)$

In Exercises 9–16 determine an equation in slope-intercept form for the line that satisfies the given information.

9. $m = \frac{1}{2}$, $b = 4.2$

10. $m = 3.1$, $b = -2.5$

11. $m = 1.5$, passes through the point $(-2, -5)$

12. $m = -2.7$, passes through the point $(3, -4)$

13. y-intercept of 19 and passes through the point $(-2, -5)$

14. y-intercept of -3.5 and passes through the point $(-5, 1)$

15. passes through the points $(6.5, -0.6)$ and $(-2.5, 3)$

16. passes through the points $(-2.2, 1)$ and $(-0.2, -3.7)$

In Exercises 17–20 estimate the x- and y-intercepts of the graphs. In each figure, Xscl = 1 and Yscl = 1.

17.

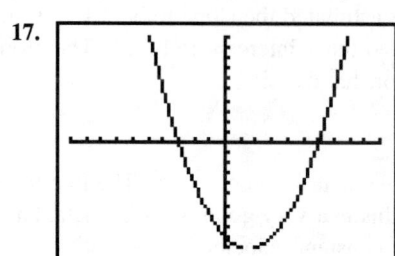

19.

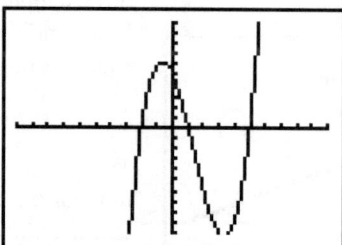

18.

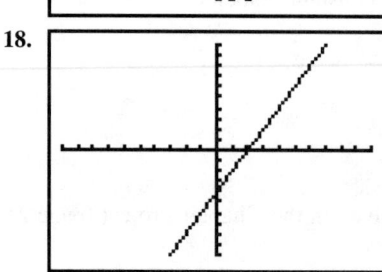

20.

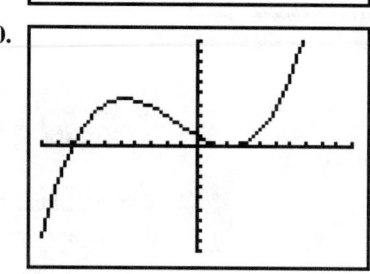

21. In Exercise Set 1.1, Exercise 1, you saved the following data for the Olympic results in the men's pole vault as lists YEAR2 and PV.

Year	1896	1900	1904	1908	1912	1920	1924	1928
Height (cm)	330.2	330.2	349.89	370.84	394.97		394.97	419.73

Year	1932	1936	1948	1952	1956	1960	1964	1968
Height (cm)	431.16	434.97	431.16	454.66	455.93		509.91	539.75

Year	1972	1976	1980	1984	1988	1992	1996
Height (cm)	549.91	549.91		574.68	589.92	579.75	

Estimate the slope and y-intercept for these data.

22. In Exercise Set 1.1, Exercise 2, you saved the following data for the Olympic results in the women's 100-meter dash as YEAR3 and DASH.

Year	1928	1932	1936	1948	1952	1956	1960	1964
Time (sec)	12.2	11.9	11.5	11.9	11.5	11.5		11.4

Year	1968	1972	1976	1980	1984	1988	1992	1996
Time (sec)	11.08	11.07	11.08		10.97	10.54	10.82	

Estimate the slope and y-intercept for these data.

23. *Hydrology* The pressure P at a depth h in a liquid is a linear function of the density of the liquid ρ. In a certain liquid, at 4 ft the pressure is 275 lb/ft^2. At 10 ft the pressure is 486.2 lb/ft^2. Write an equation in slope-intercept form for the pressure in terms of the depth.

24. *Electronics* The resistance of a circuit element is found to increase by 0.006 Ω for every 1°C increase in temperature over a wide range of temperatures. If the resistance is 3.500 Ω at 0°C,

    (a) write the equation relating the resistance R to the temperature T.

    (b) determine R when $T = 17$°C.

25. *Physics* The instantaneous velocity v of an object under constant acceleration a during an elapsed time t is given by $v = v_0 + at$, where v_0 is its initial velocity. If an object has an initial velocity of 12 ft/sec and a velocity of 24 ft/sec after 8 seconds of constant acceleration,

    (a) write the equation relating velocity to time.

    (b) sketch its graph of $0 \le t \le 12$ sec.

    (c) what is the physical meaning for the point where this graph intersects the vertical, v, axis?

26. *Industrial management* The ambient temperature at an assembly line increased throughout the afternoon. The table below shows the temperature at several times during the afternoon.

Time, t	12:30 P.M.	3:30 P.M.	5:00 P.M.
Temperature, T (°F)	74	82	86

    (a) Change the times to the number of hours elapsed since 12:00 P.M.

    (b) Plot the data in (a) on your calculator. Put time on the horizontal axis.

    (c) Write an equation for T in terms of the time t, in hours, elapsed since 12:00 P.M.

    (d) What was the temperature at 2:30 P.M.?

27. In Exercises 3 and 8 the slopes were 0. Look at those lines and describe what it means for a line to have a slope of 0.

28. In Exercises 4 and 7 the slopes were not defined. Look at those lines and describe what it means for a line to have a slope that is not defined.

In Exercises 29–32, solve each system of equations (a) graphically and (b) algebraically.

29. $x = 5 - 2y$
      $3x - 4y = -8$

30. $y = 8x + 12$
      $2x - 0.75y = 1$

31. $y = 0.6 - x$
     $y = 3.5x - 5.7$

32. $a = 2.4b + 12.85$
     $a = -1.6b + 1.85$

33. (a) Determine the time t and distance d when the following two equations intersect.

$$\begin{cases} d = 1.5 + 1.1t \\ d = 4.5t \end{cases}$$

    (b) These two equations were from Example 1.19. Interpret your answers in (a) as if they are answers for the example.

34. Use the calculator program GUESSLIN to review slope and y-intercept of lines. When you run the program it will draw a line on the calculator screen. When you think that you know the slope and y-intercept of that line, press ⟨ENTER⟩. The calculator will ask you for your answers and tell you if they are correct. Press ⟨ENTER⟩ again for another line.

35. Solve the problem represented in Figure 1.51. That is, find the values of d and t where these two lines intersect.

1.4   Three More Forms of the Linear Equation

We ended the previous section with bad news: the slope-intercept form of the line was not useful for modeling data like we have for Olympic records without transforming the data. This gap will be filled by two new versions of the straight line equation: the point-slope form, in which you know the slope and one point, and the two point form, in which you know two points on the line. Finally, in the next section, you will see a linear modeling method that uses *all* the points in the data to build a model.

Using the Slope and a Point

The first method of building a linear equation is the most widely used when you have a small amount of information about the line. It is more useful than the slope-intercept form when the y-intercept is not part of the data.

The Point-Slope Form of the Linear Equation

If you are given m, the slope of a line, and (x_1, y_1), the coordinates of one point on the line, the equation of the line is

$$y = y_1 + m(x - x_1)$$

This equation can be understood by referring back to the formula $\Delta y = m\Delta x$. Looking at Figure 1.56, think of the expression $x - x_1$ as being the *run* from x_1 to x. Then, to reach the point (x, y) from (x_1, y_1), you will see a *rise* of $\Delta y = m(x - x_1)$. Add this *rise* to y_1 and you get the above equation for the point-slope form of a linear equation.

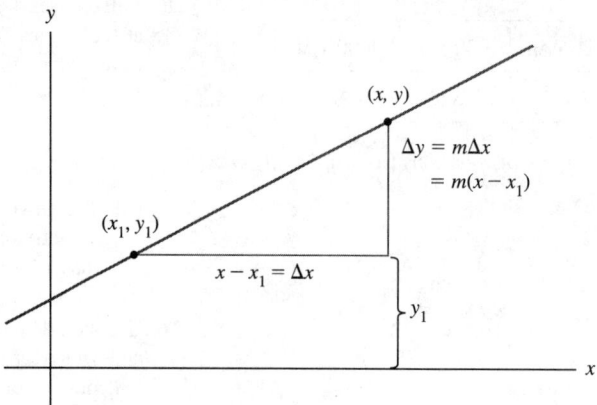

Figure 1.56

Example 1.22

Write the equation of a line that passes through $(-2, 3)$ with a slope of -0.52.

Solution   Taking x_1 and y_1 from the point $(-2, 3)$, we have $x_1 = -2$ and $y_1 = 3$. We are given $m = -0.52$, so

$$y = y_1 + m(x - x_1)$$
$$= 3 + (-0.52)(x - -2)$$
$$= 3 - 0.52(x + 2)$$

The equation $y = 3 - 0.52(x + 2)$ could be entered directly in the calculator.

If further simplification is desired, the equation could be manipulated as follows:

$$y = 3 - 0.52(x + 2)$$
$$= 3 - 0.52x - 0.52(2)$$
$$= -0.52x + 3 - 1.04$$
$$= -0.52x + 1.96$$

Either form of the line is acceptable and both can be used in a calculator to graph the line. The advantage of the simplified form is that it shows the slope and the y-intercept.   ■

Using Two Points

The next method of finding a linear model for a set of data is closely related to the point-slope form. You used this form in the last section in Examples 1.16 and 1.17.

The Two-Point Form of the Linear Equation

If the points (x_1, y_1) and (x_2, y_2) are two points on a line, then the equation of the line through these two points is

$$y = y_1 + m(x - x_1) \quad \text{or} \quad y = y_2 + m(x - x_2)$$

where

$$m = \frac{y_2 - y_1}{x_2 - x_1}$$

This is just a matter of using the two points to compute the slope and then using the point-slope form of the equation. Notice that either one of the points can be used in the equation.

Example 1.23

Find an equation of the line that passes through the points $(-22, -51)$ and $(31, -18)$.

Solution   The first step is to compute the slope, which is

$$m = \frac{-18 - (-51)}{31 - (-22)} = \frac{33}{53} \approx 0.623$$

Then we take either of the given points and use the point-slope form. If we use the second point, then $x_2 = 31$ and $y_2 = -18$. This produces the equation

$$y = y_2 + m(x - x_2)$$
$$= -18 + 0.623(x - 31)$$

which simplifies to

$$y = 0.623x - 37.313$$ ■

Using the Angle of the Line

Although we have not had a situation in which the angle of the line has been measured, we introduce this method here in order to complete the methods for finding the equation of a line.

The symbols α and β are the lowercase Greek letters *alpha* and *beta*. These are the first two letters of the Greek alphabet.

First look at the two lines pictured in Figure 1.57. Line A intersects the x-axis at an angle of α and line B intersects the x-axis at an angle of β.   Together, a point on the line and the angle completely describe a line. Imagine one of the lines swiveling around the y-intercept. As the angle changes, the slope changes. There must be a relationship between the angle and the slope. What is it?

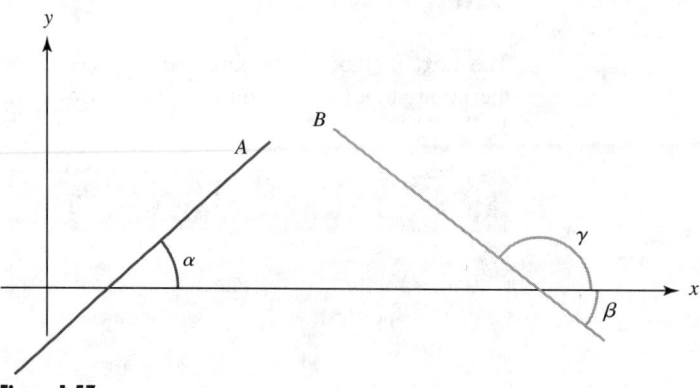

Figure 1.57

Trigonometry has the answer. Later we will spend more time on trigonometry, but for now we'll take a peek at one corner of it—the **tangent ratio**. There are six trigonometric functions and they can be defined in various ways. One way is with a right triangle such as the one shown in Figure 1.58.

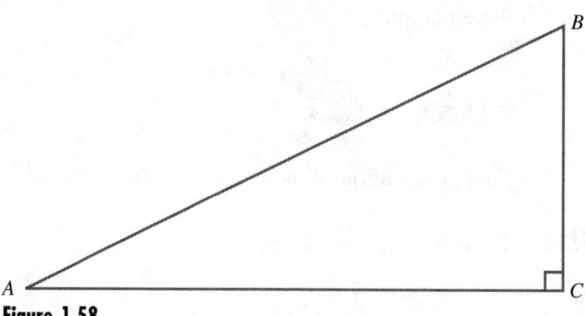

Figure 1.58

The angles of the triangle are named A, B, and C, with C a $90°$ angle. The legs of the triangle (the two shorter sides) are $\overline{AC}$ and $\overline{BC}$ and the hypotenuse (the longest side and the one opposite the $90°$ angle) is $\overline{AB}$. The tangent of angle A is defined as the ratio $\dfrac{BC}{AC}$. This relationship is usually abbreviated as $\tan A = \dfrac{BC}{AC}$.

Look at the tangent ratio. Now look at the legs $\overline{BC}$ and $\overline{AC}$. This should look familiar. There is a triangle like this in the definition of slope. Look at Figure 1.30 on page 16. BC and AC are like Δy and Δx, and so if $\overline{AC}$ is on or parallel to the x-axis, the tangent ratio is the slope of the line that contains the hypotenuse $\overline{AB}$.

The one difficulty comes in using the correct angle, since any line makes four angles when it crosses another line. Look at Figure 1.57. The angles marked α and β are the best ones to use: acute angles formed by a line and the *positive* x-axis. We measure these angles according to the direction the positive x-axis is rotated to reach the other side of the angle.   Tradition says that a counterclockwise rotation is a positive direction and a clockwise turn is a negative direction. Thus, in Figure 1.57, α is a positive angle and β is a negative angle. (There are times when it is more convenient to use the *supplement* of an angle. In the case of β, you would use the angle marked γ. We will discuss this more in a later chapter.)

The symbol γ is the lowercase Greek letter *gamma*. It is the third letter of the Greek alphabet.

In summary, the slopes of the lines in Figure 1.57 can be written as $m = \tan\alpha$ for line A and $m = \tan\beta$ or $m = \tan\gamma$ for line B. Since the slope and the angle are related, we can proceed with the definition of another form of a linear equation.

The Angle-Point Form of the Linear Equation

If a line passes through the point (x_1, y_1) and makes an angle of α with the positive x-axis, then the equation of the line is

$$y = y_1 + (x - x_1)\tan\alpha$$

or $y = x\tan\alpha + b$

where b is the y-intercept.

Example 1.24

A line passes through the point $(2, -3)$ and makes an angle of $50°$ with the positive x-axis. What is the equation of the line?

Solution   The slope of the line is $\tan 50°$. Using a calculator, you must first make sure it is ready to use angles measured in degrees. On the *TI-83*, you change to degrees in the MODE menu. After the calculator is in degree mode, move to the home screen by pressing ⬛2nd⬛ [QUIT]. To approximate $\tan 50°$, press ⬛TAN⬛ 50 ⬛)⬛ ⬛ENTER⬛. The result is shown in Figure 1.59.

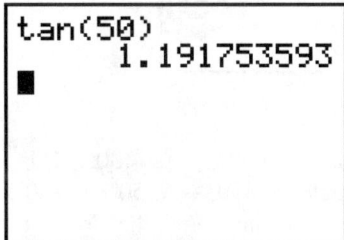

Figure 1.59

Since we now have the slope, it remains only to apply the point-slope form of the linear equation. The equation of this line is $y \approx -3 + 1.19(x - 2)$ or, when written in slope-intercept form, $y \approx 1.19x - 5.38$. ∎

Example 1.25

A line passes through the point $(2, -3)$ and makes an angle of $130°$ with the positive x-axis. What is the equation of the line?

```
tan(50)
        1.191753593
tan(130)
        -1.191753593
█
```

Figure 1.60

Solution   The slope of the line is tan 130°. To approximate tan 130°, press ▄▄TAN▄▄ 130 ▄▄)▄▄ ▄ENTER▄. The result is shown in Figure 1.60.

(There is a reason that the values of tan 50° and tan 130° differ only in their signs. You'll learn the reason for that when you study trigonometry in Chapter 3.)

Again, since we now have the slope, we can apply the point-slope form of the linear equation to obtain $y \approx -3 - 1.19(x - 2)$. This is $y \approx -1.19x - 0.62$ when it is written in slope-intercept form.   ■

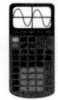

Example 1.26

A line passes through the point $(2, -3)$ and makes an angle of $-50°$ with the positive x-axis. What is the equation of the line?

Solution   The slope of the line is $\tan(-50°)$ and the approximate value of $\tan(-50°)$ is shown in Figure 1.61.

```
tan(50)
        1.191753593
tan(130)
        -1.191753593
tan(-50)
        -1.191753593
```

Figure 1.61

Notice that $\tan(-50°) = \tan 130°$. This is because the difference in the measures of their angles is $130° - (-50°) = 180°$.

Again, since we now have the slope, it remains only to apply the point-slope form of the linear equation. The equation of this line is $y \approx -3 - 1.19(x - 2)$ or, when written in slope-intercept form, $y \approx -1.19x - 0.62$.   ■

Example 1.27   Application

An access ramp to a building makes a 5° angle with the ground. What is the slope of this ramp?

Solution   The slope is tan 5°. Using a calculator, you should obtain $\tan 5° \approx 0.0875$. Thus the slope of this ramp is about 0.0875.   ■

Example 1.28 Application

If the access ramp in Example 1.27 runs 20 ft along the ground, how high is it when it reaches the building?

Solution Figure 1.62 is a drawing of the problem as it was described. In Figure 1.63 we have pictured the problem as a rectangular coordinate system with the bottom of the ramp along the positive x-axis and starting at the origin. The top of the ramp is at the point marked B. We know that the x-coordinate of B is 20. We want to find the y-coordinate of B.

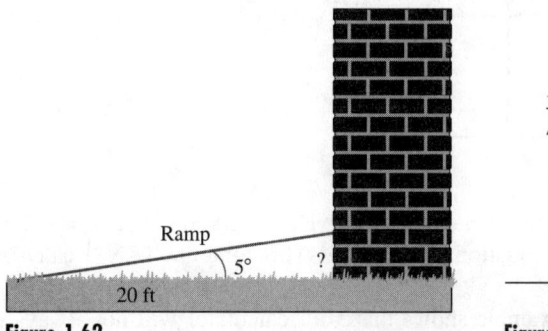

Figure 1.62

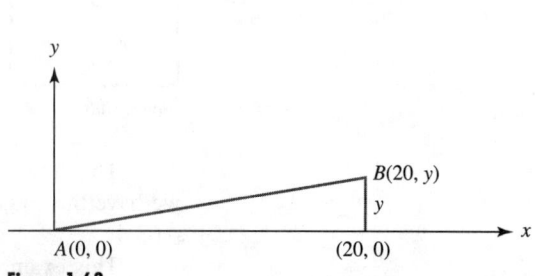

Figure 1.63

The ramp starts at the origin, so we know that the equation of the line that forms the top of the ramp is $y = mx$ or $y = (\tan 5°)x$. Since $x = 20$ we have

$$y = (\tan 5°)(20)$$
$$\approx (0.0875)(20) = 1.75$$

Thus the top of the ramp is about 1.75 ft or 21 in. above the ground. ∎

Example 1.29

Determine the angle that the line $y = 2x + 7$ makes with the x-axis.

Solution The line is graphed in Figure 1.64 and has a slope of 2. We know that the slope of a line is the tangent of the angle α that the line makes with the x-axis, so $m = \tan \alpha$. Here we have $m = 2$. Thus we have the equation $2 = \tan \alpha$.

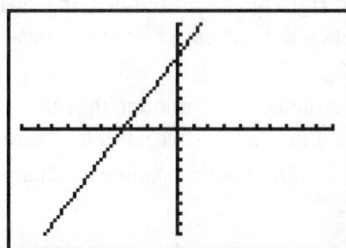

Figure 1.64

We want the angle whose tangent is 2. To determine this angle, use the yellow label above the **TAN** key on your calculator: TAN^{-1}. On the home screen, press **2nd** $[\text{TAN}^{-1}]$ 2 **)** and you will see that the angle is approximately 63.4°. ∎

Example 1.30

Determine the angle that the line $y = -2x + 7$ makes with the x-axis.

Solution   The line $y = -2x + 7$, along with the line from the previous example, $y = 2x + 7$, is graphed in Figure 1.65.

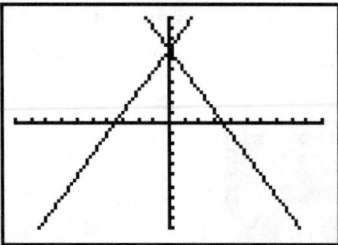

Figure 1.65

This line has a slope of -2. Here we know that $m = -2$, and since $m = \tan \alpha$, we have the equation $-2 = \tan \alpha$. This time the TAN^{-1} calculation gives the result $-63.4°$.

This example shows that your calculator will not always give you the answer you expect. ∎

Note: A Caution about Measuring Angles

If you measure an angle α that a line makes with the x-axis, the slope of the line will equal $\tan \alpha$ only if the scales on the x- and y-axes are the same. That is, the same distance on each axis would equal the same number. One way to produce axes on the *TI-83* that fit this description is to press ZOOM, then 5 (ZSquare).

Example 1.31   Application

θ (*theta*) is another Greek letter.

A small boat leaves a dock and heads due east at 8.3 kn (knots*). A current of 2.4 kn (north) is at right angles to the heading of the boat, as shown in Figure 1.66. At what angle θ does the current change the heading of the boat?

Solution   The angle θ that indicates the amount that this boat is deflected from its heading is the slope of the line that forms its actual direction. Thus $\tan \theta = \frac{2.4}{8.3}$, so $\theta = \tan^{-1}\left(\frac{2.4}{8.3}\right) \approx 16.13°$. The boat's heading is changed by about $16.13°$.

---

*1 kn ≈ 1.15 mi/hr

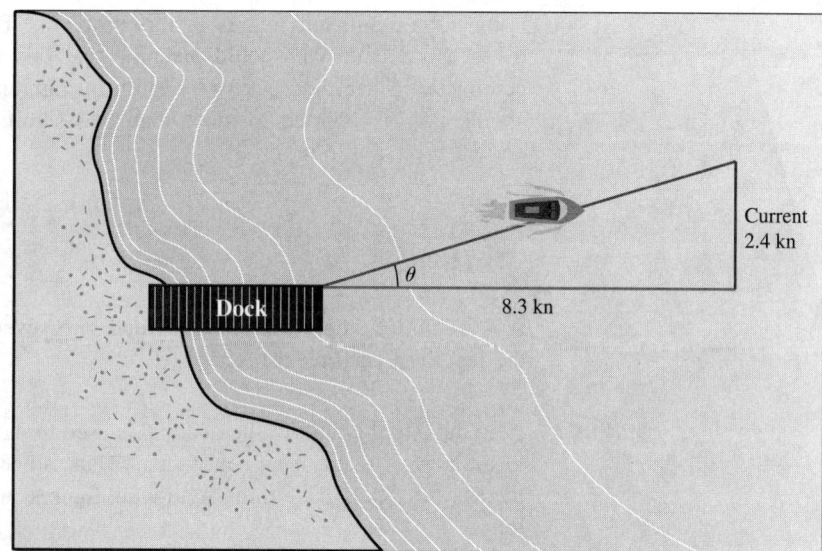

Figure 1.66

Decimal Precision and Decimal Accuracy

When a calculator does a problem like $\frac{48}{7}$, the answer produced (6.857142857 in this case on a *TI-83*) might contain more decimal places than you should use. The calculator can give an answer whose accuracy is nine places as in this example. However the precision of this answer, which means how many places make sense in the context of the problem from which it came, is another matter. You should use and report no more precision than makes sense. The next two examples illustrate this.

Example 1.32

$48 is divided equally among seven people. How much does each person get?

Solution   The answer $6.857142857 does not make sense. Since money in the United States is measured to the nearest cent (one hundredth of a dollar, or two decimal places), it makes sense to report the answer as $6.86. Note however that $7 \cdot \$6.86 = \$48.02$, so in reality each person could not get exactly $6.86. Maybe one should answer $6.85 so that everyone could get the same amount with some left over. The context of the problem would tell you what answer to give.

Example 1.33

A 48-cm-long board is to be divided into seven pieces of equal length. What is the length of each piece?

Solution   Again, it would not make sense to answer 6.857142857 cm, since that number is accurate to a billionth of a centimeter, which is about the radius of an atom. We should consider the proper precision of the answer. To do so we must consider the number of significant figures in this problem. The length of the original board is given as 48 cm. We should presume that this number is accurate to the nearest centimeter. Since there are two significant figures in the original length, the answer should not be reported to more than two significant figures. Thus 6.9 cm is the correct precision to report.   ∎

Example 1.34

A 48.0-cm-long board is to be divided into seven pieces of equal length. What is the length of each piece?

Solution   Here the 0 to the right of the decimal is used to indicate that this board was measured to the nearest tenth of an centimeter. Thus, since 48.0 has three significant figures, the answer should have three significant figures, and so each piece would be 6.86 cm long.   ∎

Handling precision, accuracy, and significant figures can be quite complicated. We will return to these ideas throughout the book so that you will gradually become familiar with them. It doesn't hurt to keep your calculator's decimal accuracy set at its highest, or to Float; this gives the highest accuracy possible, but reports $\frac{1}{2}$ as 0.5, not 0.500000000. What hurts is to believe that all those decimal places are worth reporting. In the examples in this book we will usually tell you what accuracy to use. If we don't say differently, use three places.

Example 1.35

Using the two-point form for the equation of a line, write a linear model for the distance vs. time data that is shown in Figures 1.67 and 1.68 below. (These were shown earlier as Figures 1.19 and 1.20.)

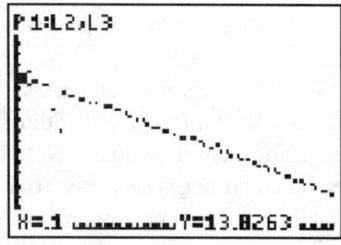

Figure 1.67

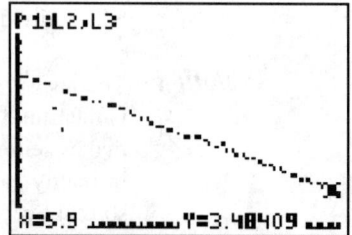

Figure 1.68

Solution   Two points, (0.1, 13.8263) and (5.9, 3.48409), are shown in the figures, but other points could be found by looking at the lists for the data, as shown in Figure 1.14 on page 6. In general the model will be more accurate when you use points that are far apart.

The slope, as computed in Example 1.3, is $m = -1.7831$ ft/sec. Choosing the second point to write the equation, we have

$$y = 3.48409 - 1.7831(x - 5.9)$$

or when simplified,

$$y \approx -1.7831x + 14.004$$

In Example 1.20 on page 29, we obtained

$$y = -1.7831x + 14$$

As you can see, there is very little difference. This equation would match the data just as well as the one shown in Figure 1.55.

According to the rules for significant figures, the answer to this example should be $y = -1.8x + 14$. ∎

Example 1.36   *Chapter Project*

Write a linear model of the Olympic data in the Chapter Project using two points.

Solution   We attempted this problem at the end of the previous section. We were unsuccessful because of difficulty in estimating the y-intercept. Since we have many points, the two-point form of the line will lead to an answer. The question becomes, "Which points to use?"

We will solve this one following some earlier advice to "use points that are far apart." The points then are $(1896, 82.2)$ and $(1992, 49.02)$. The slope is

$$\frac{49.02 - 82.2}{1992 - 1896} = \frac{-33.18}{96} \approx -0.346$$

While 49.02, 1992, and 1896 have four significant figures, 82.2 has only three, so the answer is rounded off to three significant figures.

and using the first point, the equation is

$$y = 82.2 - 0.346(x - 1896)$$
$$= 82.2 - 0.346x - 0.346(-1896)$$
$$= -0.346x + 738$$

∎

In the next example we'll see how well this model fits the data.

Example 1.37

On the graph you made in Activity 1.1, plot the equation we found in Example 1.35.

Solution   You will need to calculate at least two points in order to plot this line. However, to be sure you are correct, always plot a third point. If the three points fall on a line, you can be fairly sure you did not make a mistake.

We must choose three values of x (the year), which will be on your graph. That is, they should be between 1896 and 1996. Let's pick the three for which we do not have times: 1920, 1960, and 1980.

If $x = 1920$, what is y? You can substitute 1920 for x in either form of the equation that was found in the previous example. Let's pick the slope-intercept form.

$$y = -0.346 \cdot 1920 + 738$$
$$= -664.32 + 738 = 73.68 \approx 73.7 \text{ sec}$$

That is, this line estimates that the winning time in 1920 would have been 73.7 seconds. The actual time was 60.4 seconds. Your estimate was probably closer to 60.4 seconds than 73.7. That is because you could look at all the times, but this equation was developed as if you only knew the times for 1896 and 1992.

The other two points that are estimated from this equation are (1960, 59.8) and (1980, 52.9).

Now locate these three points on your graph, and with a ruler draw a straight line that includes them.   ■

In the next section we will see a method of fitting a line to data that will almost certainly be more accurate than the ones we have used so far because it uses *all* the points. The bad news is that the method requires long calculations. Fortunately the *TI-83* does the calculations with just a few keystrokes.

Section 1.4 Exercises

In Exercises 1–8 write an equation of the line that satisfies the given information.

1. The line passes through the points $(-1, 5)$ and $(4, -7)$.

2. The line passes through the point $(3, -2)$ and has a slope of $-\frac{5}{3}$.

3. The line passes through the point $(4, 2)$ and makes an angle of $72°$ with the positive x-axis.

4. The line has a slope of $\frac{2}{5}$ and a y-intercept of -3.

5. The line passes through the point $(-2, 6)$ and has a slope of 1.3.

6. The line passes through the point $(-3, -1)$ and makes an angle of $125.7°$ with the positive x-axis.

7. The line has a slope of -0.25 and a y-intercept of 3.15.

8. The line passes through the points $(-2.1, -3.7)$ and $(5.6, -3.2)$.

9. Determine the angle that the line $x - y = 2$ makes with the positive x-axis.

10. Determine the angle that the line $y = 0.75x - 7$ makes with the positive x-axis.

11. Figures 1.69 and 1.70 are two graphs that show the results of a HIKER experiment.

    (a) What is an equation of the line that passes through these two points?

    (b) What was the approximate velocity of this hiker?

    (c) How far from the motion detector was the hiker when the motion detector was turned on?

    (d) Was the hiker moving toward or away from the motion detector?

    (e) Write one sentence that uses all your answers to parts (b)–(d).

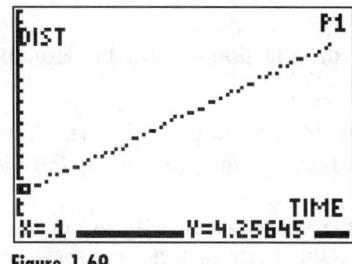

Figure 1.69

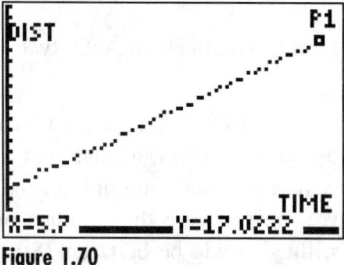

Figure 1.70

12. Figures 1.71 and 1.72 are two graphs that show the result of a HIKER experiment.

    (a) What is an equation of the line that passes through these two points?

    (b) What was the approximate velocity of this hiker?

    (c) How far from the motion detector was the hiker when the motion detector was turned on?

(d) Was the hiker moving toward or away from the motion detector?

(e) Write one sentence that uses all your answers to parts (b)–(d).

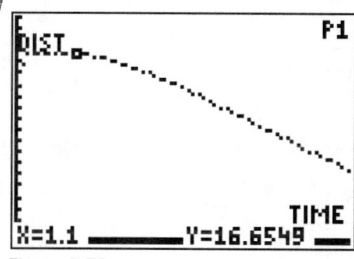
Figure 1.71

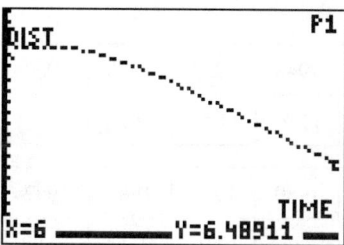
Figure 1.72

13. The graph in Figure 1.73 represents the distance a person was from a motion detector over a 13-sec interval.

  (a) What is an equation of this line?

  (b) What was the approximate velocity of this hiker?

  (c) How far from the motion detector was the hiker when the motion detector was turned on?

  (d) Was the hiker moving toward or away from the motion detector?

  (e) Write one sentence that uses all your answers to parts (b)–(d).

14. The graph in Figure 1.74 shows the distances of two different people from a motion detector over a 12-sec interval.

  (a) What are the equations of these lines?

  (b) What was the approximate velocity of these hikers?

  (c) How far from the motion detector was each hiker when the motion detector was turned on?

  (d) Were the hikers moving toward or away from the motion detector?

  (e) Write one sentence for each hiker that uses all your answers to parts (b)–(d).

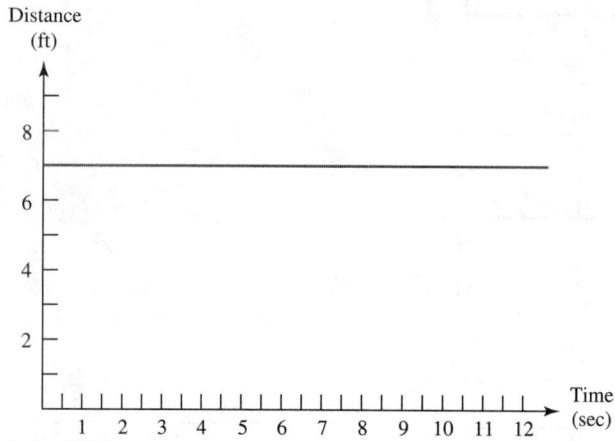
Figure 1.73

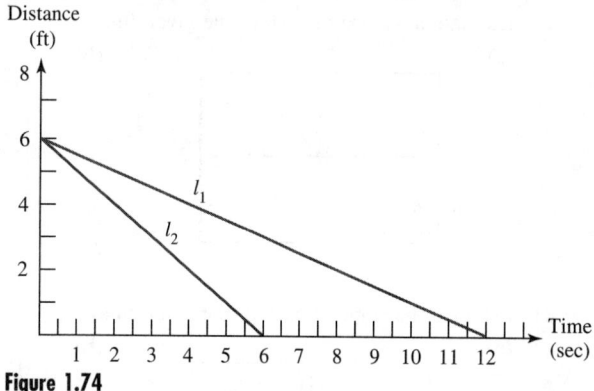
Figure 1.74

15. In Exercise Set 1.1, Exercise 1, you saved the following data for the Olympic results in the men's pole vault as lists as YEAR2 and PV. Pick two of these points and write a linear model of these data.

Year	1896	1900	1904	1908	1912	1920	1924	1928
Height (cm)	330.2	330.2	349.89	370.84	394.97		394.97	419.73

Year	1932	1936	1948	1952	1956	1960	1964	1968
Height (cm)	431.16	434.97	431.16	454.66	455.93		509.91	539.75

Year	1972	1976	1980	1984	1988	1992	1996
Height (cm)	549.91	549.91		574.68	589.92	579.75	

16. In Exercise Set 1.1, Exercise 1, you saved the following data for the Olympic results in the women's 100-meter dash as YEAR3 and DASH. Pick two of these points and write a linear model of these data.

Year	1928	1932	1936	1948	1952	1956	1960	1964
Time (sec)	12.2	11.9	11.5	11.9	11.5	11.5		11.4

Year	1968	1972	1976	1980	1984	1988	1992	1996
Time (sec)	11.08	11.07	11.08		10.97	10.54	10.82	

17. Consider the linear equation $y = 3x - 5$. Give the coordinates of three points on this line.

18. Consider the linear equation $2y + x = -2$. Give the coordinates of three points on this line.

19. Estimate the slope of each of the given lines.

(a)
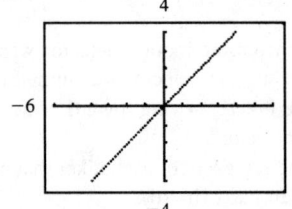

(b)
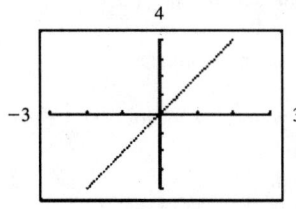

20. Estimate the slope of each of the given lines.

(a)
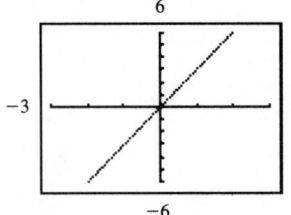

(b)
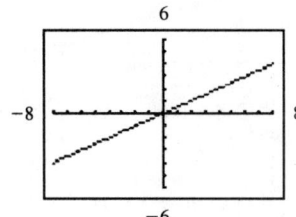

21. Estimate the slope of each of the given lines.

(a)
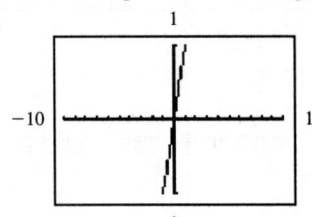

(b)
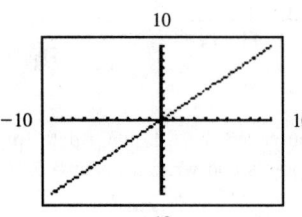

(c)
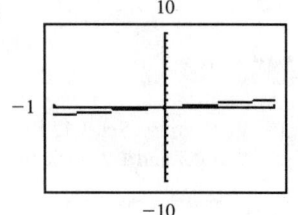

22. Estimate the slope of each of the given lines.

(a)
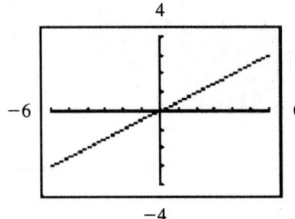

(b)
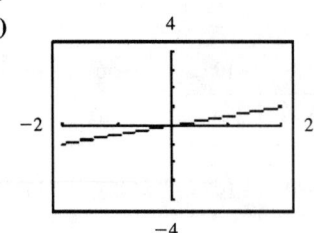

(c)
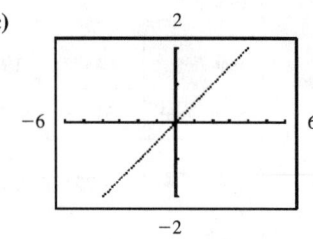

23. *Engineering technology* Within certain limits, the force
    F needed to extend a shock spring is a linear function
    of the amount a the spring is extended.
    (a) Use the following test data to write a linear model
        that relates F and a.
    (b) Predict the force needed to extend the spring to
        3.8 in.
    (c) How far will the spring extend if a force of 100 lb
        is applied?

a (in.)	1 4	3 0
F (lb)	58	162

24. *Engineering technology* The force F needed to extend
    a shock spring is a linear function of the amount a the
    spring is extended.
    (a) Use the following test data to write a linear model
        that relates F and a.
    (b) Predict the force needed to extend the spring to
        0.11 m.
    (c) How far will the spring extend if a force of 100 N
        is applied?

a (m)	0.06	0.14
F (N)	32	128

25. *Engineering technology* As the temperature of a solid
    increases, its length increases according to the equation
    $\Delta l = \alpha l_0 \Delta T$, where: Δl is the change in length of the
    solid due to the change in temperature ΔT; l_0 is the
    initial length of the solid; and α is a constant called
    the coefficient of linear expansion. (The coefficient
    of linear expansion is different for different materials.)
    For steel $\alpha = 0.000012/°C$. Suppose the initial length
    of a steel girder is 200 m. Complete the following
    table of the increase in temperature vs. the change in
    length.

ΔT (°C)	10	20	30	40	50
Δl (m)		0.048			

26. *Engineering technology*   A team of engineering
    technology students is trying to determine the coeffi-
    cient of linear expansion for concrete. The team pours
    a slab of concrete when the temperature is 85°F. Sev-
    eral months later they measure the slab when the tem-
    perature is about 10°F and find that the slab is 0.002 cm
    shorter. Write a linear model that satisfies these data.

27. *Electricity* The electrical resistance R, in ohms (Ω),
    of a material is a linear function of the temperature T
    in °C. The following experimental data was obtained
    for an aluminum wire. Write a linear model of these
    data.

T (°C)	0	20
R (Ω)	0.300	0.323

28. *Business* A CAD designer buys a new software pro-
    gram for $735. For tax purposes the designer uses
    linear depreciation over three years. After three years
    the program will have to be replaced with an upgrade
    and hence will have no value.
    (a) Write a linear equation that gives the value V of
        this program during its 3 years t of use.
    (b) What is the value of the program after 15 months?

29. *Navigation* An airplane is heading due east, accord-
    ing to the compass, at 250 mph. A crosswind of
    40 mph is blowing directly out of the south. At what
    angle is the wind pushing the plane off its compass
    heading?

30. *Construction* The roof in Figure 1.75 has a rise of 4 ft
    and a run of 14 ft. Find the angle that the rafter makes
    with the horizontal.

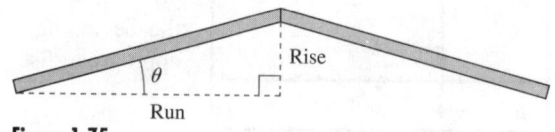

Figure 1.75

1.5

A Linear Model That Uses All the Points

In this section we will continue the study of lines, use lines to predict results, use
results to create lines, and see if we can realistically use lines to predict anything.
You have seen several methods that can be used to produce the equation of a line

from a little information: two points on the line, a point and the slope, the slope and y-intercept, and one point and the angle the line makes with the x-axis.

Now you will see a method that looks at *all* the data points available and calculates the line that best fits the points. The method is called **linear regression**. Its drawback is that the calculations are lengthy if you try to do them by hand. In this section you will see how to use your calculator for linear regression. In other chapters of this book you will see other kinds of regression that are used to fit data that does appear to fall in a straight line. Also later in the book you will learn more about how linear regression works.

In Example 1.18 you saw how to write an equation that related temperatures in degrees Fahrenheit and degrees Celsius. That equation was determined from the slope and the y-intercept. The result was

$$F = 1.8C + 32$$

In the next three examples you will see how to enter data into the calculator so the data can be used to produce a linear regression line to fit the data. In the first example you will enter the Celsius-Fahrenheit data and graph the points; in the second example you will check to see if the graph is correct; and in the third example you will apply linear regression to the data.

Example 1.38   *Application*

Plot the three Celsius-Fahrenheit points from the table below on the *TI-83*.

Celsius	−40	0	100
Fahrenheit	−40	32	212

Solution   Apply the calculator procedure given in Example 1.1. Your graph will look something like Figure 1.76. The window dimensions we used are

$$-50 \le x \le 110 \quad \text{and} \quad -50 \le y \le 220$$

By using the Trace function and moving the cursor from one point to another you can compare the coordinates you see on the screen with those in the table at the beginning of this example. If there are differences, make the changes in the appropriate list. ∎

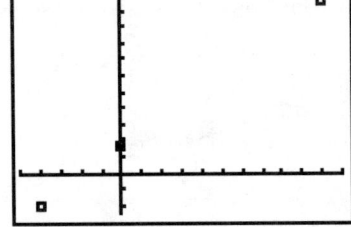

Figure 1.76

The next step will use linear regression to produce an equation for a line through these points. You have already seen one way to find this equation, the result is written at the beginning of this section. We will compare that equation with the one that results from linear regression.

Example 1.39

Use linear regression to find the equation of the line through the points graphed in the previous example.

Solution   Since we entered the points in the preceding example, we can go directly to the regression calculation. Begin by pressing ▄STAT▄ to access the statistics main menu.

Celsius	−40	0	100
Fahrenheit	−40	32	212

Press to move to the CALC (or calculation) menu, shown in Figure 1.77. There you see a list of seven of the 13 types of statistical calculations your *TI-83* can perform. The fourth item in the list, 4:LinReg(ax+b), is an abbreviation for linear regression.

Press [4] to select LinReg(ax+b) and to write that command on the home screen. Press [2nd] [L1] [,] [2nd] [L2] to tell the calculator which lists to do the regression on (Figure 1.78 shows the home screen before you press [ENTER]), and press [ENTER] to do the calculation.

```
EDIT CALC TESTS
1█1-Var Stats
2:2-Var Stats
3:Med-Med
4:LinReg(ax+b)
5:QuadReg
6:CubicReg
7↓QuartReg
```

Figure 1.77

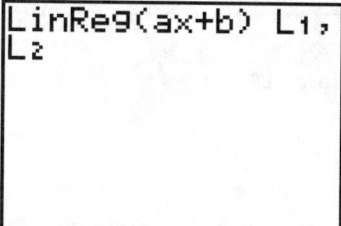

Figure 1.78

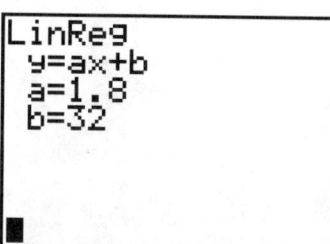

Figure 1.79

The results of the regression are shown in Figure 1.79. The values of a and b in the regression equation $y = ax + b$ are reported as $a = 1.8$ and $b = 32$, so the regression equation is $y = 1.8x + 32$. Since the y-values are the Fahrenheit temperatures and the x-values are the Celsius temperatures, we can write this regression equation as $F = 1.8C + 32$. ■

The regression equation is exactly the same as the equation produced in the earlier example using the slope and the y-intercept. The next step is to plot the regression line on the same graph and see how well the line fits the data.

Introduction to Linear Regression

Entering the Lists. The command LinReg(ax+b) L1, L2 tells the calculator that L1 contains the *x* values and L2 has the *y* values.

What Does Regression Do? A regression calculation finds the equation of the line that comes closest to all the points in the data. If it can't pass through all the points, it makes the best compromise. Think of the points that do not touch the line. As shown in Figure 1.80, the vertical distance from each of these points to the line measures the amount by which the line missed that point. Points which are closer have a smaller distance and therefore a smaller error. The idea is to add up these errors and make the sum of all the errors as small as possible. The regression line is the line that makes the total error

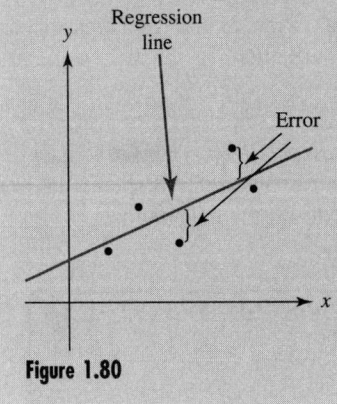

the smallest. Because error can be positive or negative depending on whether the point is above or below the line—and for other mathematical reasons—the process actually tries to minimize the sum of the *squares* of the errors so that only positive numbers are added.

Figure 1.80

Example 1.40

Plot the regression line $y = 1.8x + 32$ on the same graph with the data that was plotted in Example 1.38.

Solution Since we have the equation we use the Y= screen of the *TI-83*. Press [Y=]. The cursor is blinking at the right of the equal sign in \Y1. If there is an equation there, press [CLEAR] to remove it. Now type in 1.8x+32 to match the result shown in Figure 1.81. Use the [X,T,θ,n] key to enter the X variable in the *TI-83*. Press [GRAPH] and you will see Figure 1.82.

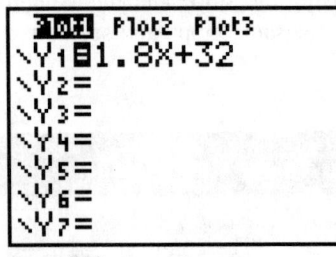

Figure 1.81

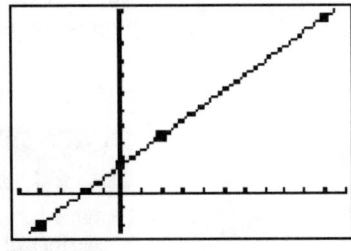

Figure 1.82

Why does the line match the points exactly? The original points were not produced from an experiment but from knowledge about the relationship between the two temperature scales. With experimental data such as distance vs. time points gathered from a CBL or historic data like Olympic records over the years, the points will not lie on a straight line, and the line produced by linear regression will be an approximation of the points. Before we move on to using regression with experimental or historic data, we'll look at a common use of any fitted line: estimating additional values.

Example 1.41

In the equation $y = 1.8x + 32$, what y-value corresponds to $x = 20$? Check your work by entering the two values into the lists you used and show that the point lies on the graph.

Solution   To answer this question, substitute 20 for x in the equation $y = 1.8x + 32$. The result is

$$y = 1.8(20) + 32$$
$$= 36 + 32 = 68$$

That is, the temperature 20°C corresponds to 68°F. Enter these two values in the lists L1 and L2 (or whatever lists you used in Example 1.38) and you should see that the new point lies on the line as shown in Figure 1.83.

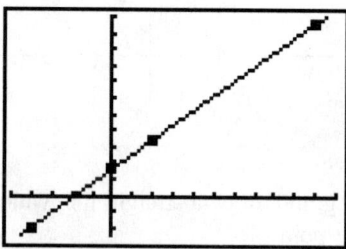

Figure 1.83

Example 1.42

Complete the table below for Celsius and Fahrenheit temperatures and graph the points on your *TI-83*. Do these calculations without the calculator. Check your answers by showing that the points lie on the same line.

Celsius	−40	−20	0	20	25	30	35	40	100
Fahrenheit	−40		32	68					212

Solution   Substitute each of the Celsius temperatures into the equation $y = 1.8x + 32$ and then put the resulting pairs of numbers into the calculator lists. The complete table is shown below and the graph looks like Figure 1.84.

Celsius	−40	−20	0	20	25	30	35	40	100
Fahrenheit	−40	−4	32	68	77	86	95	104	212

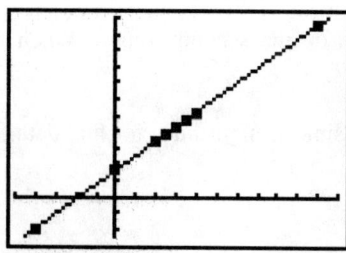

Figure 1.84

Fitting a Regression Line to Data

In the last example you fit a regression line to points that came from a linear formula. The regression line naturally fit the points perfectly. The real use of regression is for finding an equation that fits a set of data points that come from life, not from an equation. Once we have that regression equation we can use it to predict future results. In the next activity you will get an idea of the meaning of the regression line.

Activity 1.3

Figure 1.85 contains two identical graphs with five data points. You will make two attempts to fit a line to these points and then compare the total errors.

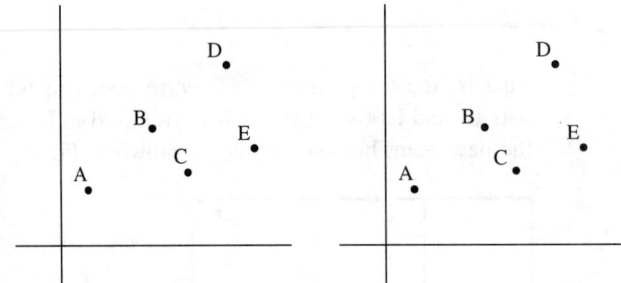

Figure 1.85

1. Using the first graph, draw a straight line which seems to you a good approximation of the five points.
2. With a ruler, measure the errors and enter your results in Table 1.1. For each point the error is the vertical distance from the point to the line (as shown in Figure 1.80). It does not matter if the error is positive (point is above the line) or negative (point is below the line); just enter the absolute value of the error.

Table 1.1

Line 1	Error	Line 2	Error
Point A		Point A	
Point B		Point B	
Point C		Point C	
Point D		Point D	
Point E		Point E	
Total Error		Total Error	

3. Add the five errors to give the total error for this line.
4. Repeat the above steps for a different line drawn on the second graph. Which line is better?

   The regression line is the best choice of all possible straight lines for that data because it is the one with the lowest amount of total error.*

---

*Actually, the formulas used by the calculator are based on finding the lowest total of the *squares* of the errors, not the *absolute values* of the errors, because using squares makes the formulas simpler.

Example 1.43

Make a linear regression line from the CBL data that is graphed in Figures 1.19 and 1.20. Graph the regression line along with the data points.

Solution   Before you begin, turn off the data plots and function graphs that were in use for the previous examples.

When you do a CBL experiment under the control of the HIKER program, the time and distance data go into lists L2 and L3 respectively. Follow the steps in Example 1.39 and enter the linear regression command as

$$\text{LinReg(ax+b) L2, L3}$$

Before plotting the graph, change the graph window to look like Figure 1.86.

The regression equation is shown in Figure 1.87. Compare this with the result from Example 1.35.

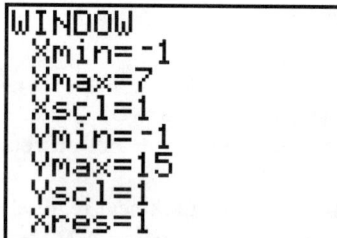

Figure 1.86   **Figure 1.87**

To plot this equation you could copy the expression $y = -1.81948055x + 14.33225234$ into the Y= screen of the *TI-83*. However, since the distance data is only reported to 3 significant figures, you need only to enter the equation as $y = -1.82x + 14.3$. The resulting graph is shown in Figure 1.88. As you saw in Figure 1.55, this fit is also extremely good even though there is a little difference in the equation developed there. In Figure 1.89, which shows a portion of the upper left part of the graph, you can see that most of the points are clustered around the line, while two of them are quite far away.

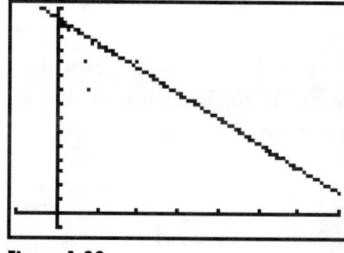
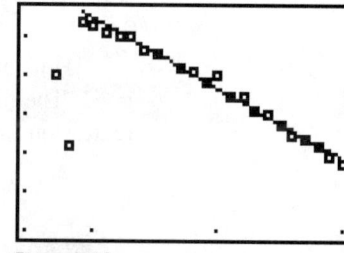

Figure 1.88   **Figure 1.89**   ■

Next we will return to the Olympic swimming records and see how regression helps us predict the times that were not provided in the table.

Example 1.44   *Chapter Project*

Produce and plot a linear regression line for the men's 100-meter freestyle swimming event in the Olympics.

Solution   Early in this chapter you entered the Olympic men's freestyle 100-meter swimming records into your *TI-83* using the names YEAR and TIME for the two lists. Now we will use them.

First, deselect or clear any equations that are on the Y= screen; then turn off all plots except Plot1. Change Plot1 to match Figure 1.90. Finally, change the graphing window to look like Figure 1.91. Your graph should look like Figure 1.92.

Figure 1.90

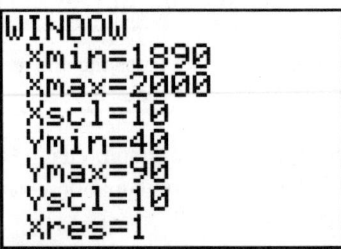

Figure 1.91

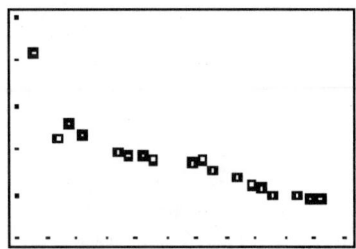

Figure 1.92

To make the linear regression line, press **STAT**, select the CALC menu, and press **4** to select LinReg(ax+b). Enter the lists named YEAR and TIME by selecting them from the LIST menu. Before you press **ENTER** the home screen looks like Figure 1.93.

Press **ENTER** to produce the linear regression line. The result is approximately equal to the following equation: $y = -0.2332x + 511.2$. Press **Y=** and enter this equation; then press **GRAPH**. The result is shown in Figure 1.94.

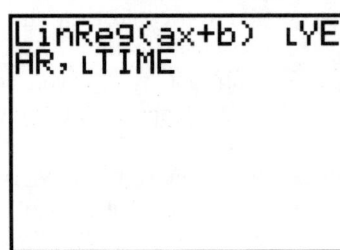

Figure 1.93

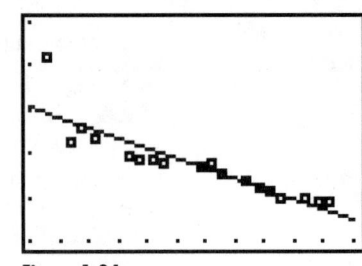

Figure 1.94

How good is this fit? For some years it looks good and for others it is not so good. The first point $(1896, 82.2)$ has the largest error. Now let's see how well the regression line makes predictions.

Example 1.45

Use the regression line for the Olympic swimming records to predict the times for 1920, 1960, 1980, and 1996.

Solution   Substitute each of the years 1920, 1960, 1980, and 1996 for x in the regression equation, $y = -0.2332x + 511.2$.

The results are $(1920, 63.5)$, $(1960, 54.1)$, $(1980, 49.5)$, and $(1996, 45.7)$.

The actual times were $(1920, 60.4)$, $(1960, 55.2)$, $(1980, 50.4)$, and $(1996, 48.74)$. Compare these estimates with your own.

Example 1.46

Discuss the accuracy of the linear regression line for Olympic swimming records and its usefulness for predicting the future.

Solution   One of the four estimates are within one second of the actual result. An error of one second out of approximately 50 seconds is an error of about 2 in 100 or about 2%. For some purposes 2% error is quite satisfactory.

However there are serious flaws in using a linear regression line with this kind of data. Two of these flaws relate to the y-intercept (in this instance, the TIME-intercept) and the x-intercept (here, the YEAR-intercept). Figure 1.95 shows the Olympic men's swimming results and the regression line so that you can see the x- and y-intercepts.

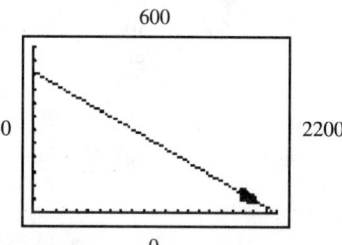

600

0   2200

0

Figure 1.95

Think about it. The y-intercept of the regression equation is 511.2. This means that in the year zero (1896 years before the first modern Olympics) the estimated time for the 100-meter swim is 511.2 seconds, or about 8.5 minutes. While there certainly are people today who cannot swim 100 meters in 8.5 minutes, it is likely that the best swimmers 2000 years ago could beat that time by a large margin.

But the x-intercept (remember, the YEAR-intercept is where TIME = 0) presents an even more ridiculous estimate. Without calculating this value, you probably can see that this doesn't make sense: TIME = 0 means that the 100-meter swim is finished in 0 seconds. That is impossible. But the linear regression equation does produce a prediction for when TIME = 0. You can find that by solving the regression equation for the value of x when $y = 0$:

$$0 = -0.2332x + 511.2$$
$$0.2332x = 511.2$$
$$\text{or } x = \frac{511.2}{0.2332} \approx 2192$$

There may be a lot of miracles by the year 2192, but swimming 100 meters in 0 seconds is not one of them!

To carry this to an even more extreme possibility, according to this formula anyone who wins the 100-meter freestyle swimming event after the year 2192 will finish with a negative time—which means that they would finish before the race began.   ■

There are two important lessons about fitting straight lines to data:

1. Think carefully before you use any equation that is designed to fit data if you are trying to estimate results that are too far beyond the limits of the data.

2. Think carefully about whether a straight line is the best model for your data. Think, for example, about how useful a linear regression line would be for the data displayed in Figure 1.14.

In the next chapter we will discuss data that does not look like a straight line, and we will consider a mathematical model that is not linear.

Section 1.5 Exercises

 1. Some people who are new to the Celsius temperature scale use a conversion formula that is simpler to remember and to calculate with than the actual formula; it is "double Celsius and add 30" to get Fahrenheit.

    (a) Write this sentence as an equation that begins $F =$.

    (b) Plot this equation with the Celsius-Fahrenheit data in Example 1.42 and discuss how well you can estimate temperatures with it. In particular, discuss where this simpler formula gives good estimates to the correct answer and where it does not.

2. In Example 1.39 you found that the linear regression equation for Fahrenheit vs. Celsius was $F = 1.8C + 32$. In that example you had the calculator compute LinReg(ax+b) L1, L2.

    (a) Use the calculator to determine the linear regression for Celsius vs. Fahrenheit. That is, compute LinReg(ax+b) L2, L1.

    (b) How does the answer to (a) compare to the previous answer of $F = 1.8C + 32$?

 3. For the men's 100-meter swimming results, store the list YEAR in L1 and TIME in L2. Delete the data for 1896 by pressing `STAT` `ENTER` and moving the cursor to 1896. Pressing `DEL` erases the number 1896. Next, delete the time for 1896, which is 82.2, by moving the cursor to 82.2 and pressing `DEL`.

    (a) Compute the linear regression model on these shorter lists.

    (b) Use this regression line to compute the expected results for the years 1920, 1960, 1980, and 1996.

    (c) Does the resulting regression line do a better job of matching the remainder of the data? Explain why you think it does or does not do a better job of matching the remainder of the data.

    (d) Explain why this was a sensible point to remove?

 4. In Exercise Set 1.1, Exercise 1, you made a list of the following data for the Olympic results in the men's pole vault and saved the data as lists YEAR2 and PV. The data are repeated below.

Year	1896	1900	1904	1908	1912	1920	1924	1928
Height (cm)	330.2	330.2	349.89	370.84	394.97		394.97	419.73

Year	1932	1936	1948	1952	1956	1960	1964	1968
Height (cm)	431.16	434.97	431.16	454.66	455.93		509.91	539.75

Year	1972	1976	1980	1984	1988	1992	1996
Height (cm)	549.91	549.91		574.68	589.92	579.75	

    (a) Compute the linear regression model on these lists.

    (b) Use this regression line to compute the expected results for the years 1920, 1960, 1980, and 1996.

5. In Exercise Set 1.1, Exercise 2, you made a list of the following data for the Olympic results in the women's 100-meter dash and saved the data as lists YEAR3 and DASH. The data are repeated below.

Year	1928	1932	1936	1948	1952	1956	1960	1964
Time (sec)	12.2	11.9	11.5	11.9	11.5	11.5		11.4

Year	1968	1972	1976	1980	1984	1988	1992	1996
Time (sec)	11.08	11.07	11.08		10.97	10.54	10.82	

(a) Compute the linear regression model on these lists.

(b) Use this regression line to compute the expected results for the years 1960, 1980, and 1996.

 6. Ask each person in class to remove his or her right shoe. Measure the length of each person's right foot to the nearest 0.5 cm. Record that person's shoe size (the length, not the width). For example, a man's right foot that measures 26.5 cm should wear a size $9\frac{1}{2}$ shoe. Make one list for the men and one for the women. Take the longer of the lists and

(a) plot the data on a graphing calculator.

(b) use linear regression to find the equation of the line for the points graphed in (a).

(c) find the slope of the line in (b).

(d) find the y-intercept of this line.

(e) graph the line from (b) on the calculator.

(f) explain why some points are not on the line.

●●●●●●●●● Chapter 1 Summary and Review

Topics You Learned or Reviewed

▶ The slope of a line is the same as the tangent of the angle the line makes with the positive x-axis.

▶ In a distance vs. time graph from a constant velocity experiment, the slope of the line, m, is the same as the constant velocity, and the y-intercept gives an approximation of the starting distance.

▶ You can write the equation of a straight line given any of the following, where m is the slope of the line, b is the y-intercept, (x_1, y_1) and (x_2, y_2) are two points on the line, and α is the angle the line makes with the positive x-axis:

- the slope and the y-intercept: $y = mx + b$
- one point and the slope: $y = y_1 + m(x - x_1)$
- two points: $y = y_1 + m(x - x_1)$ or $y = y_2 + m(x - x_2)$ where $m = \dfrac{y_2 - y_1}{x_2 - x_1}$
- one point and the angle the line makes with the positive x-axis: $y = y_1 + (\tan \alpha)(x - x_1)$

▶ You can use a mathematical model to study real events and relationships; that is,

- estimating missing values in the data by use of a linear regression model data.
- analyzing how well a linear model fits the data.

Review Exercises

In Exercises 1–8 write a linear equation that satisfies the given data.

1. $m = -2.7$, $b = 12.3$.

2. $m = \frac{15}{11}$ and the line passes through $(-1, \frac{2}{3})$.

3. The line passes through $(2.6, 3.1)$ and $(-2.8, -7.7)$.

4. The line passes through the point $(5, 7)$ and makes an angle of $57°$ with the positive x-axis.

5. $m = -\frac{2}{5}$ and the line passes through $(-\frac{1}{3}, \frac{1}{4})$.

6. $m = 0.28$ and $b = -3.1$.

7. The line passes through the point $(-2.7, 3.5)$ and makes an angle of $32°$ with the positive x-axis.

8. The line passes through the points $(-1.25, 7.38)$ and $(4.75, -10.62)$.

9. The data in the table below is for the Olympic results in the women's 200-meter dash.

Year	1948	1952	1956	1960	1964	1968	1972
Time (sec)	24.4	23.7	23.4		23	22.5	22.4

Year	1976	1980	1984	1988	1992	1996
Time (sec)	22.37		21.81	21.34	21.81	

(a) Compute the slope using the first and last points.

(b) Write the equation of the line through the first and last points.

(c) Compute the linear regression model on these lists.

(d) What is the slope of the regression line?

(e) What is the y-intercept of the regression line?

(f) Use this regression line to compute the expected results for the years 1960, 1980, and 1996.

10. The data in the table below is for the average speed of some of the winners of the Indianapolis 500-mile automobile race.

Year	1915	1920	1925	1930	1935	1940	1950	1955
Speed (mph)	89.840	88.618	101.127	100.448	106.240	114.277	124.002	128.213

Year	1960	1965	1970	1975	1980	1985	1990	1995
Speed (mph)	138.767	150.686	155.749	149.213	142.862	152.982	185.981	153.616

(a) Compute the slope using the first and last points.

(b) Write the equation of the line through the first and last points.

(c) Compute the linear regression model on this list.

(d) What is the slope of the regression line?

(e) What is the y-intercept of the regression line?

(f) Use this regression line to compute the expected results for the year 1996.

(g) Explain why your answer in (f) differs from the actual answer for 1996 of 147.956 mph?

In Exercises 11–12 estimate the x- and y-intercepts of the graphs. In each figure, Xscl = 1 and Yscl = 1.

11.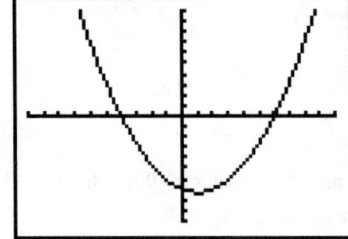

12.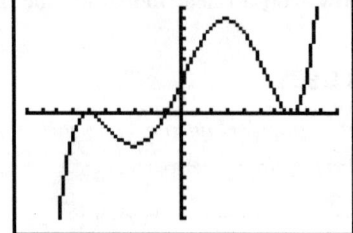

13. *Sports technology* Zalkin Halmay of Hungary won the 1904 Olympic men's 50-meter freestyle swimming event in 28.0 seconds. What was his average velocity in **(a)** meters per second and **(b)** kilometers per hour?

14. *Sports technology* Amy Van Dyken of the United States won the 1996 Olympic women's 50-meter freestyle swimming event in 24.87 seconds. What was her average velocity in **(a)** meters per second and **(b)** kilometers per hour?

15. Look at the graph in Figure 1.96. This graph was produced by a HIKER experiment. Describe how the person who produced this graph walked.

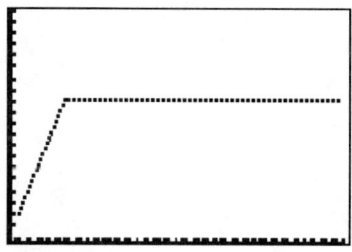

Figure 1.96

16. *Transportation* Paul gets on state highway 330 just outside of Des Moines, Iowa and drives northeast at an average speed of 67 mph. After traveling 28 mi he reaches U.S. Highway 30 and realizes that he has gone too far. He turns around and travels southwest 6 miles back to Melbourne at an average speed of 72 mph. For the whole trip from Des Moines to Melbourne, what is **(a)** his average velocity and **(b)** his average speed?

17. Figures 1.97–1.99 are three graphs that show the result of a HIKER experiment.

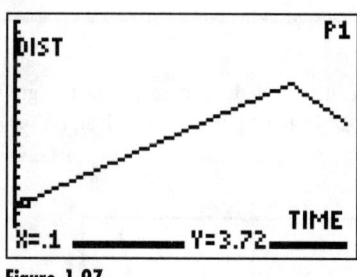

Figure 1.97

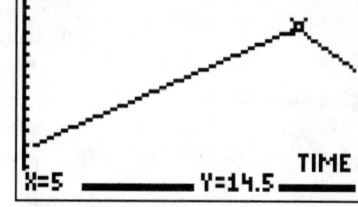

Figure 1.98

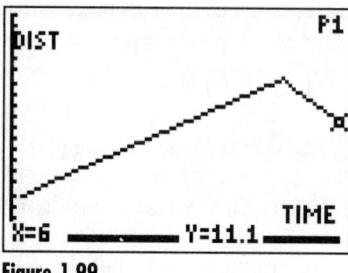

Figure 1.99

(a) What is an equation of the line that passes through the two points when $x = 0.1$ and when $x = 5$?

(b) What were the approximate velocity and speed of this hiker from $x = 0.1$ to $x = 5$?

(c) How far from the motion detector was the hiker when the motion detector was turned on?

(d) What is an equation of the line that passes through the two points when $x = 5$ and when $x = 6$?

(e) What were the approximate velocity and speed of this hiker from $x = 5$ to when $x = 6$?

(f) What were the approximate velocity and speed of this hiker from $x = 0.1$ and when $x = 6$?

(g) Write one sentence that uses all your answers to parts (b), (c), (e), and (f).

18. *Construction* An electrician must bend a pipe to make a 125-cm rise in a 175-cm horizontal distance, as shown in Figure 1.100. **(a)** What is the angle at x? **(b)** What is the angle at y?

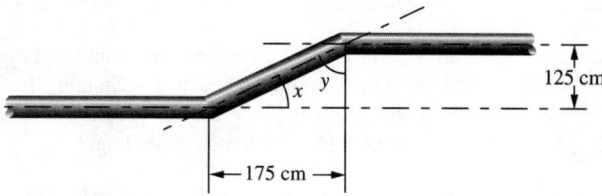

Figure 1.100

19. *Construction* Find angle α in the taper shown in Figure 1.101.

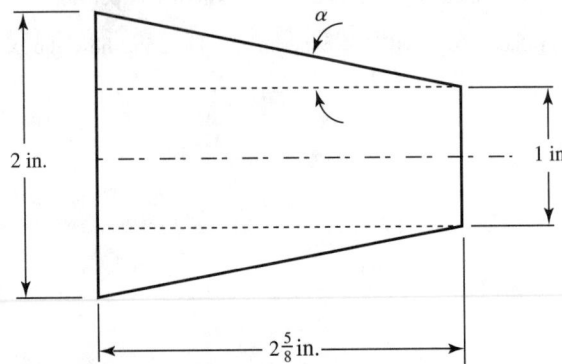

Figure 1.101

20. *Air traffic control* An airplane is being tracked on radar by a radar antenna located on top of the control tower. The airplane was last tracked at 32.50 km on a direct line for the control tower at an angle of elevation of 12.7°. What is the approximate altitude of the airplane?

● ● ● ● ● ● ● ● Chapter 1 Test

In Exercises 1–4 write a linear equation that satisfies the given data.

1. $m = 4.3$, $b = -3.17$.

2. $m = 0.75$ and the line passes through $(-1.2, 3.25)$.

3. The line passes through $(-6.2, 1.5)$ and $(8.4, -7.2)$.

4. The line passes through the point $(-2, 4)$ and makes an angle of 43° with the positive x-axis.

5. The data in the table below is for the Olympic results in the men's 50-meter freestyle swimming event. (This event was not held from 1908–1984.)

Year	1904	1988	1992	1996
Time (sec)	28	22.14	21.91	

(a) Compute the linear regression model on these lists.

(b) What is the slope of the regression line?

(c) What is the y-intercept of the regression line?

(d) Use this regression line to compute the expected results for 1996.

(e) Explain why your answer in (d) differs from the actual answer for 1996 of 22.13 seconds?

6. Estimate the x- and y-intercepts of the graph in Figure 1.102. In the figure, Xscl = 1 and Yscl = 1.

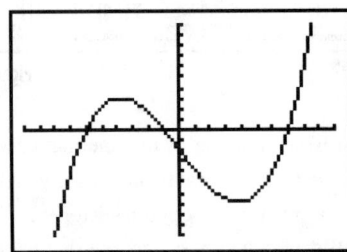

Figure 1.102

7. Aleksandr Popov of Russia won the 1996 Olympic men's 50-meter freestyle swimming event in 22.13 seconds. What was his average velocity in **(a)** meters per second and **(b)** kilometers per hour?

8. Kristin Otto of East Germany won the 1988 Olympic women's 50-meter freestyle swimming in 25.49 seconds. What was her average velocity in **(a)** meters per second and **(b)** kilometers per hour?

9. Look at the graph in Figure 1.103. This graph was produced by a HIKER experiment. Describe how the person who produced this graph walked.

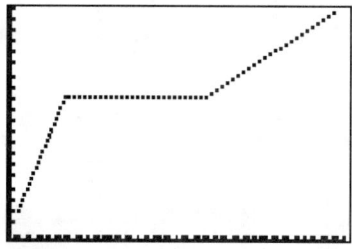

Figure 1.103

10. Figures 1.104 and 1.105 are two graphs that show the result of a HIKER experiment.
    (a) What is an equation of the line that passes through these two points?
    (b) What was the approximate velocity of this hiker?
    (c) How far from the motion detector was the hiker when the motion detector was turned on?
    (d) Was the hiker moving toward or away from the motion detector?
    (e) Write one sentence that uses all your answers to parts (b)–(d).

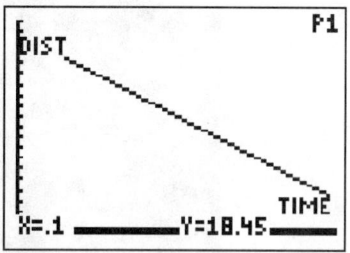

Figure 1.104   **Figure 1.105**

11. The design for a television tower calls for a cable to be attached to the tower so that the cable makes an angle of 35° with the level ground. The construction people want to attach the connecting bolt to the tower before it is raised into place. They know that the cable will be anchored to the ground at a point that is 75 ft from the base of the tower. How high on the tower should the connecting bolt be placed?

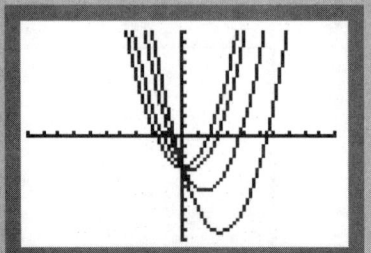

CHAPTER 2

Quadratic Functions
Introduction to Nonlinear Models

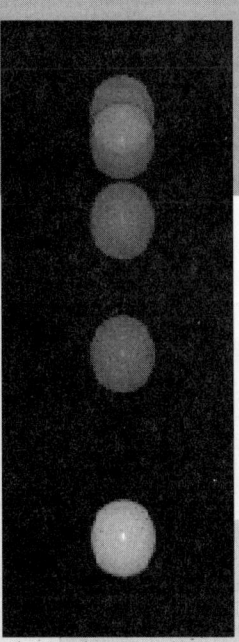

Topics You'll Learn or Review

- A mathematical model for motion influenced by gravity
- Fitting a quadratic function to data
- Solving quadratic equations with different methods:
  - numeral estimates
  - graphical estimates
  - quadratic formula
- Graphing quadratic functions:
  - the relationship between the graph of a quadratic function and its roots
  - the relationship between the factored form of a quadratic function and its roots
  - the relationship between the coefficients of a quadratic function and its graph
  - the relationship between the coefficients of a quadratic function and projectile motion
- Using parametric equations to model gravitational motion
- The meaning of the term *instantaneous velocity*

Calculator Skills You'll Learn

- Working with data:
  - working with a portion of a list ("Select" feature)
  - transforming data by subtracting the same number from every item in the list
  - doing quadratic regression
- More features of the MODE menu

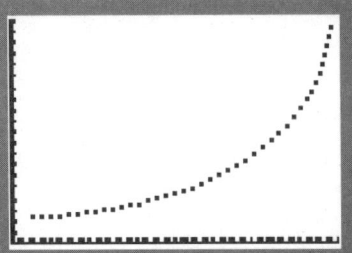

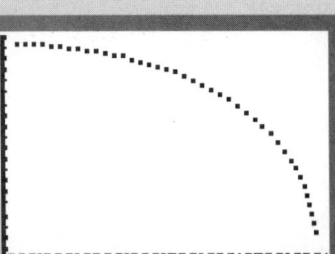

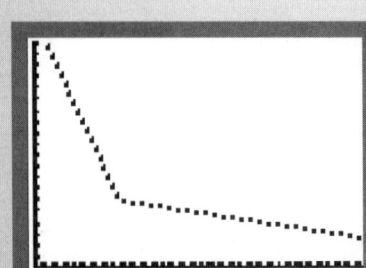

Mathematics You'll Use

▶ Calculations involving quadratic polynomial functions formulas, including decimals and negative numbers
▶ Writing the equation of a line from a point and the slope

Calculator Skills You'll Need

▶ Skills used in Chapter 1, plus use of [2nd] [ENTRY] and [2nd] [INS] to modify and repeat calculations
▶ Using the [ALPHA] key and the [2nd] [ALPHA] key combination to enter the letters printed in green above the keys
▶ Using the MODE menu to change the decimal accuracy
▶ Entering nonlinear functions in the Y= screen, including use of [(-)], [x^2], [^], and [2nd] [$\sqrt{}$] keys
▶ Zooming on graphs

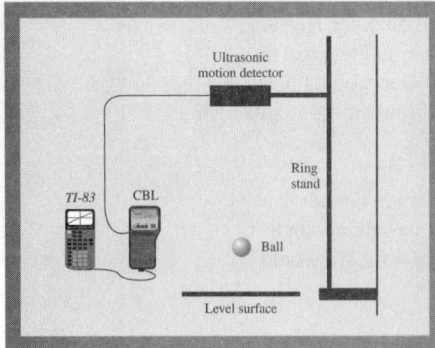

What You'll Do in This Chapter

In Chapter 1 you studied linear models, the simplest kind of mathematical model. In this chapter you will expand your knowledge of mathematical models to include quadratic models. For over 300 years quadratic models have been essential to the study of motion influenced by gravity or any constant force. As you study this chapter you are going to learn how to answer questions like the following:

▶ What kind of motion can be modeled with a quadratic function?
▶ How can I build a quadratic model of data produced in a motion experiment?
▶ What is the difference between the trajectory of an object in motion and the graph of distance vs. time?
▶ What does it mean to ask, "How fast is a moving object traveling at a particular instant?"

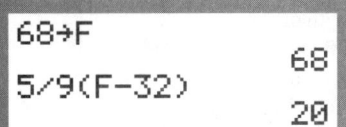

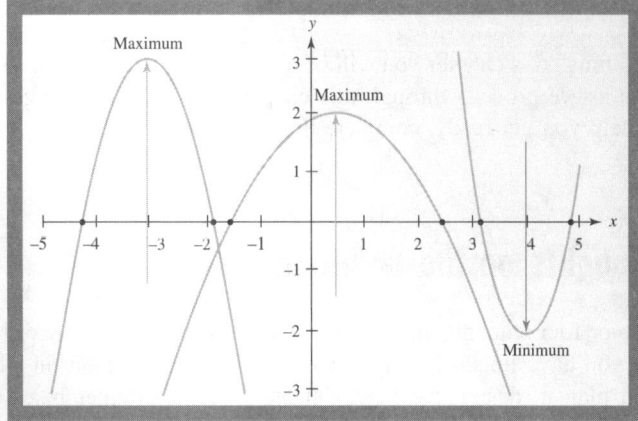

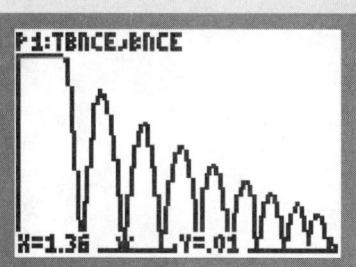

Chapter Project—The Bouncing Ball

Big Bouncers, Inc.

"Let us put a ball in your fun"

To: All employees
From: Management
Subject: New product test

We are getting ready to begin production of our new product, the Bigger Bounce Ball (BBB). The marketing department wants to sell this ball to athletes using the theme, "More bounce than the rest." Our product liability insurance wants us to devise a test that will protect us from a truth-in-advertising lawsuit.

You are to design a test that will ensure that the statement in our ads is true. We know that our main competitor claims that its ball rebounds with a velocity of 20 ft/s when dropped from a height of 8 ft. We would like you to design two tests: a height test and a time test.

- The *height test* would establish a minimum rebound height that any BBB must reach between the first and second bounce. We would know that any ball that reaches this height would have a rebound velocity of at least 20 ft/s when dropped from a height of 8 ft.
- The *time test* would establish a minimum time for the length of the first bounce. Any ball that is dropped from a height of 8 ft and takes longer than this time--from the instant it hits the ground the first time until it hits the ground the second time--would have a rebound velocity of 20 ft/s or more.

Please keep us informed of your progress in designing these tests.

By the time you finish this chapter you will be able to provide an answer to this memo. Several times as we proceed through this chapter we will tell management of our progress. To help you get ready, complete the following activity.

Activity 2.1
Preliminary Thoughts on the Bouncing Ball

With a group of three or four students, discuss what needs to be done to answer the memo above. When you have finished, write a memo to the management in which you outline how you plan to design the height and time tests. Remember, since this is just a proposed plan, management will not be surprised if your plan does not completely work and if you later decide to try different ideas. Try to answer the following questions.

1. How high do you think a ball will bounce (on the first bounce) if you know the ball is dropped from a height of 8 ft and has a rebound velocity of 20 ft/sec? Will it bounce as high as 8 ft? How will the second and third bounces compare? Does it matter to your design of the test described in the memo?

2. If the BBB has a rebound velocity of more than 20 ft/sec, will it bounce higher or lower than the competitor's ball?

3. Does the type of surface on which the ball is dropped make a difference? If so, what do we do about our competitor, since we do not know the surface it uses for testing?

4. How long do you think the ball will be in the air between the first and second bounce if it is dropped from a height of 8 ft and has a rebound velocity of 20 ft/sec? How will the time between the second and third bounces compare?

5. If the BBB has a rebound velocity of more than 20 ft/sec, will the time of its first bounce last longer than the first bounce of the competition's ball?

2.1   Velocity That Varies

Up to this point we have emphasized motion that has had constant velocity during the time of the experiment. The bouncing ball is an example of motion that is not constant. In this chapter we will look at several types of nonconstant motion.

Graphs That Tell a Story

In Chapter 1 you were able to look at a graph produced by the HIKER program and describe how a person would need to walk in order to produce such a graph. In each of those graphs, the hiker walked at a constant velocity. We'll begin our discussion of nonconstant motion with a look at more complex motion experiments. In the examples in this section we look at graphs formed by putting together several parts that contain constant motion.

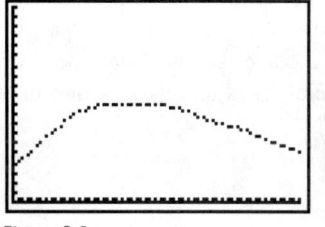

Figure 2.1

Example 2.1

Describe how to use the CBL and the HIKER program to walk in such a way as to produce a graph similar to the graph shown in Figure 2.1.

Solution Walk away from the motion detector, stop for a while, then turn and walk back at a slower speed to your starting point.   ∎

Example 2.2

Sketch a graph that would result from the motion described below. The y-axis of your graph should show the distance from the starting point; that is, the y-intercept is equal to zero. Show approximate units on your graph.

"I left my house at 11 A.M. and drove 8 miles to college, stayed there all day, and then left for home at 5 P.M. There was more traffic when I drove home, so I couldn't travel as fast as I did in the morning."

Solution   The graph in Figure 2.2 shows the simplest graph that describes this motion. It assumes that you did not change speed during either trip. Notice that the graph shows a faster speed in the morning than in the afternoon. The parts of the graph that represent the driving need not be straight.

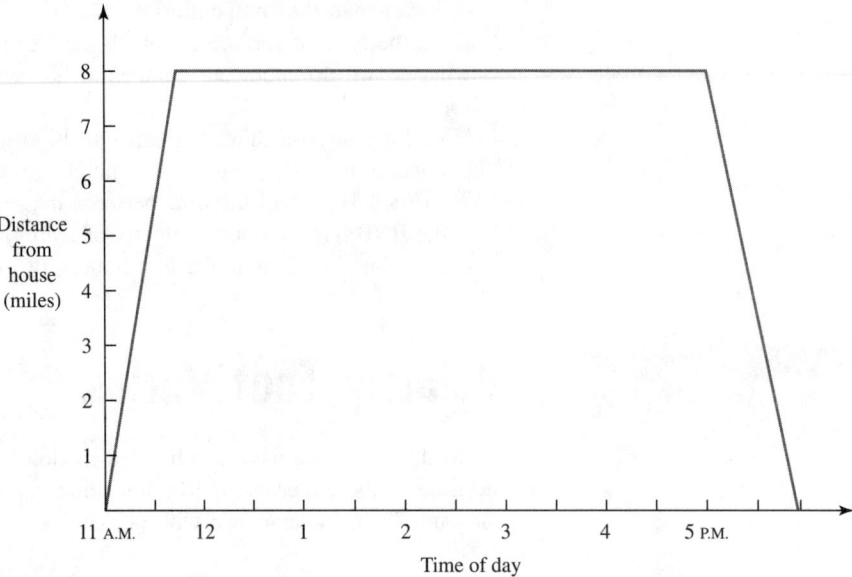

Figure 2.2

Average Speed and Average Velocity

When velocity changes, as it does in the examples in this section, we have to be careful when answering the question "How fast did you travel?" The question has to be clarified. Does the questioner want to know how fast you went on a part of your trip or the average rate for the entire trip? Is the conversation about speed or velocity? The following examples will show you several ways to answer a question such as "How fast did you travel?"

Example 2.3

I left my house in Washington, D.C. at 11:30 A.M. and drove to Baltimore at an average speed of 58 miles per hour. (See Figure 2.3.) I stayed in Baltimore for 1 hour to have lunch and then drove to Philadelphia at an average speed of 62 mph. I stayed in Philadelphia for 3 hours and then drove back to Washington at an average speed of 60 mph. The distance from Washington to Baltimore is 37 miles, the distance from Baltimore to Philadelphia is 96 miles, and the distance from Philadelphia to Washington is 133 miles.

(a) Sketch a distance traveled vs. time graph for this trip. Show mileage on the vertical axis and time on the horizontal axis. Show the exact times for arrival and departure from each city.

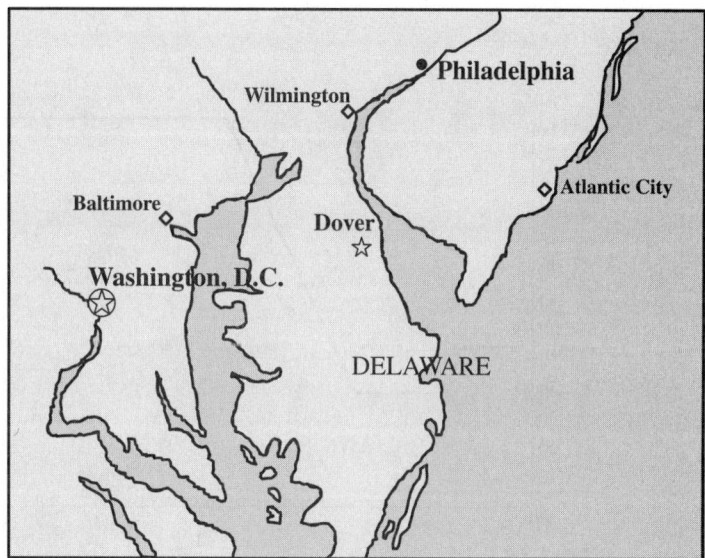

Figure 2.3

(b) Excluding the lunch break and the three hours in Philadelphia, what was the average speed for the trip? Why isn't this number equal to the average of the three speeds, which would be 60 mph?

(c) At the conclusion of the lunch break in Baltimore, what was my average speed up to then (including the time spent at lunch)? What was the average velocity over the same time interval?

(d) Including the lunch break, what was the average speed between home and the arrival in Philadelphia? What was the average velocity over the same period?

(e) What was the average speed between the end of the lunch break and the arrival back home? What was the average velocity for this time interval?

(f) What was the average velocity for the entire trip?

Solutions (a) Driving the 37 miles from Washington to Baltimore at 58 mph it took $\frac{37}{58} \approx$ 0.6 hour. (To make the graph easier to produce, all times have been rounded to the nearest 0.1 hour, or 6 minutes.) Thus I left home at 11.5 and arrived in Baltimore at 12.1. Since I spent one hour in Baltimore, I departed Baltimore at 13.1 (1.1 P.M.). It took $\frac{96}{62} \approx 1.5$ hours to drive from Baltimore to Philadelphia, so I arrived in Philadelphia at 14.6 (2.6 P.M.). Three hours later, at 17.6 (5.6 P.M.), I departed Philadelphia. Since it took $\frac{133}{60} \approx 2.2$ hours to drive home, I arrived home at 19.8 (7.8 P.M.). The resulting graph should look something like Figure 2.4.

(b) The average speed for the driving portion of the trip from Washington to Philadelphia was the distance traveled divided by the time that I spent driving.

$$\text{Total distance traveled:}\quad 37 + 96 + 133 = 266 \text{ miles}$$
$$\text{Total driving time:}\quad 19.8 - 11.5 - 4 = 4.3 \text{ hours}$$

So the average speed was $\frac{266}{4.3} \approx 61.9$ mph. The answer is not equal to the average of the individual speeds because we used the definition of average speed: distance divided by time.

(c) At the end of my lunch break in Baltimore, I had traveled a distance of 37 miles. It had been 1.6 hours since I left home, so the average speed was $\frac{37}{1.6} \approx 23.1$ mph.

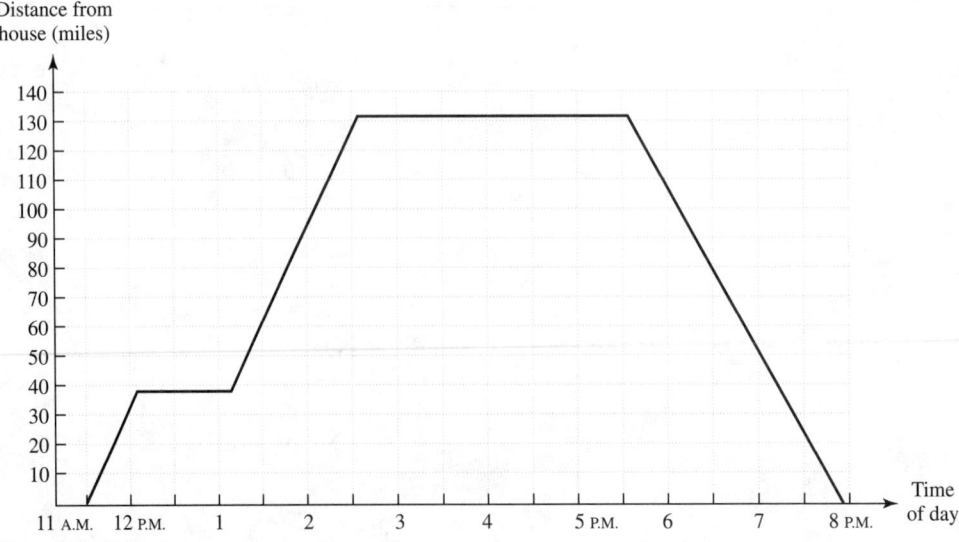

Figure 2.4

The velocity and the speed are the same over this time interval because travel was in only one direction.

(d) The distance from Washington to Philadelphia is 133 miles. It took 3.1 hours for me to get to Philadelphia, so the average speed was about 42.9 mph.

The average velocity for the trip from Washington to Philadelphia is equal to the slope of the line segment drawn between the point $(11.5, 0)$, when the trip began, and the point $(14.6, 133)$, when I got to Philadelphia. We could also think of this as the distance from Washington to Philadelphia divided by the time it took to get to Philadelphia after I started. The slope is $\frac{133-0}{14.6-11.5} = 42.9$, or 42.9 mph. The speed and the velocity are the same over this interval because the travel was always in the same direction.

(e) From the end of the lunch break until I arrived home, I traveled a distance of 229 miles. Since it took 6.7 hours to travel this distance, the average speed was $\frac{229}{6.7} \approx 34.2$ mph.

The velocity is the slope of the line segment between the points $(13.1, 37)$, when I left Baltimore, and $(19.8, 0)$, when I arrived home. So the average velocity was $\frac{0-37}{19.8-13.1} = \frac{-37}{6.7} \approx -5.5$ mph.

A negative average velocity may seem strange. However it means that the greater part of the time was spent traveling in the homeward (negative) direction. In examples that have to do with driving, average speed will usually be of more interest than average velocity. Since average velocity indicates direction, it can give some useful insight about motion.

(f) The average velocity for the entire trip is the slope of the line segment between $(11.5, 0)$ and $(19.8, 0)$. The slope is $\frac{0-0}{19.8-11.5} = \frac{0}{8.3} = 0$. Thus the average velocity for the entire trip was 0 mph.

In general the average velocity of a trip that starts and ends at the same point is 0, no matter how many stops and different speeds there are on the trip.

Now that we have looked at variable motion, the next section will look at the graph of a bouncing ball when its motion is captured by a CBL. As you can imagine, the height and velocity of a bouncing ball are continuously changing. Thus we will have to look beyond linear models.

Section 2.1 Exercises

In Exercises 1–10 (a) write a description of each graph using the words slope and velocity in your description, and (b) use the CBL and the HIKER *program to walk in such a way as to produce a similar graph.*

1.

6.
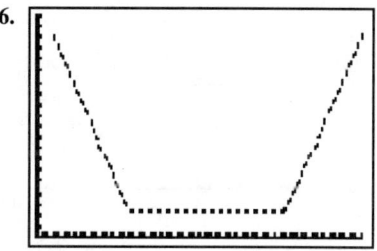

2.
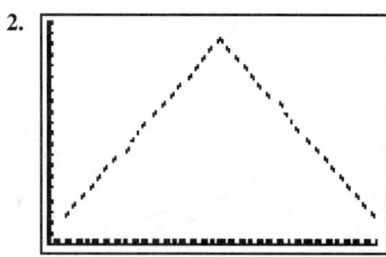

7.
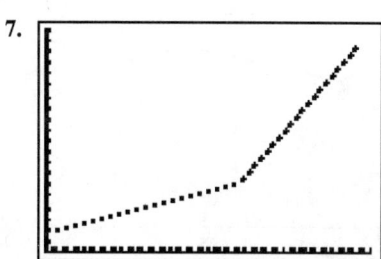

3.

8.
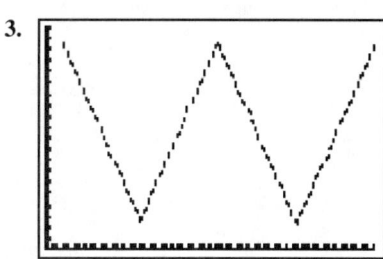

4.
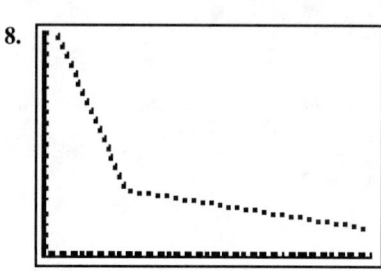

9.
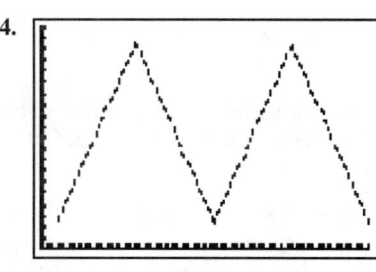

5.
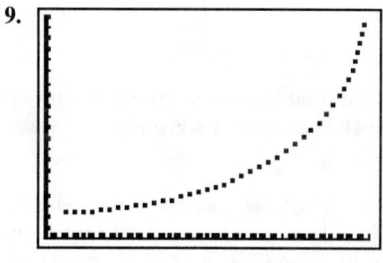

10.
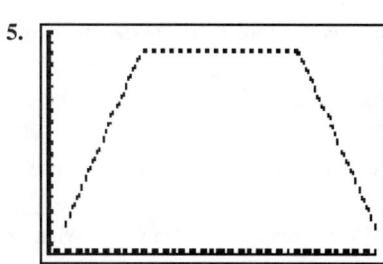

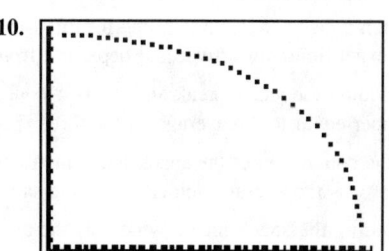

In Exercises 11–16 compute the average velocity during the 8-hour period shown on the distance vs. time graph.

11. Distance (miles)

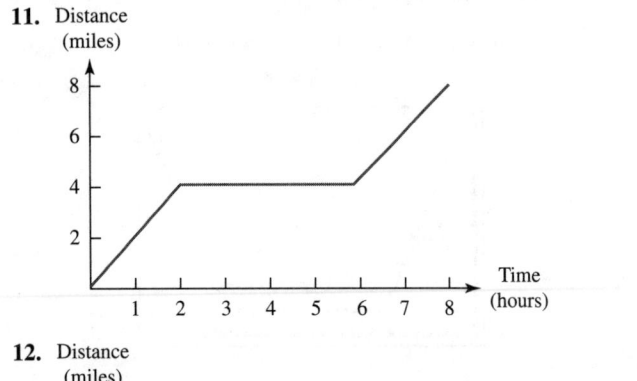

14. Distance (miles)

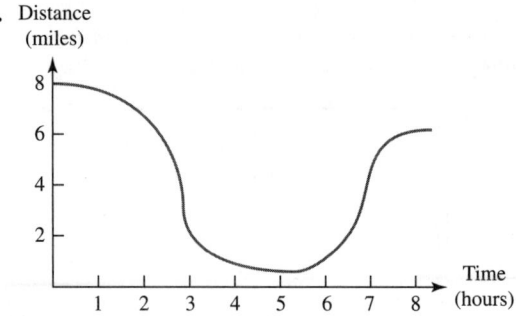

12. Distance (miles)

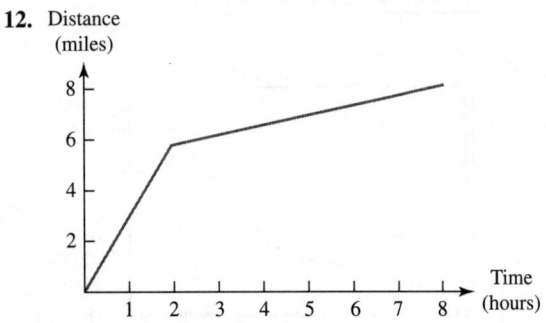

15. Distance (miles)

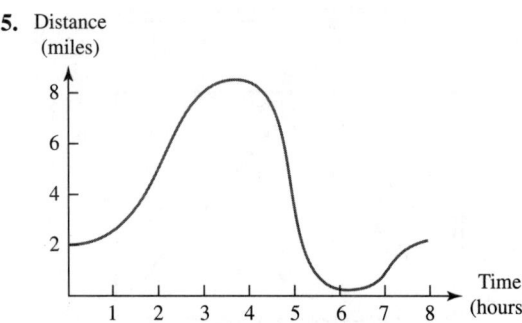

13. Distance (miles)

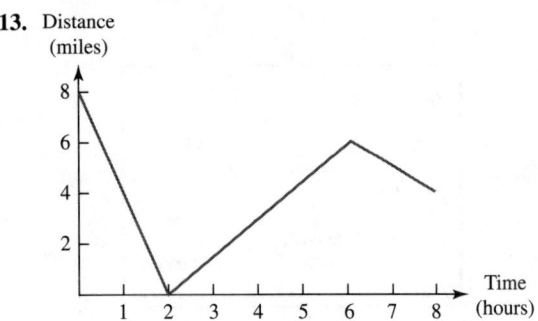

16. Distance (miles)

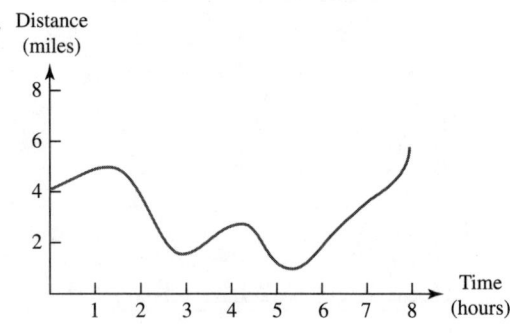

In Exercises 17–18 sketch a graph that would result from the patterns of motion described. The vertical axis of your graph should show the distance from the starting point; that is, the y-intercept is equal to zero. Place approximate units on your axes.

17. Chris walked from home to the store. He stayed in the store for awhile and then walked home.

18. Pat ran from home to the store. She stayed in the store for awhile and then walked slowly home.

19. *Transportation* I left my house in Knoxville, TN at 8:00 P.M. and drove to Athens, TN at an average speed of 56 mph. I stayed in Athens for one-half hour to get gasoline and a snack and then drove to Chattanooga at an average speed of 62 mph. I stayed in Chattanooga for 2 hours and then drove back to Knoxville at an average speed of 65 mph. The distance from Knoxville to Athens is 60 miles and the distance from Athens to Chattanooga is 55 miles.

   (a) Sketch a distance vs. time graph for this trip. Show mileage on the vertical axis and time on the horizontal axis. Show the exact times for arrival and departure from each city.

   (b) Excluding the snack break and the two hours in Chattanooga, what was the average speed for the trip? Why isn't this number equal to the average of the three speeds, which would be 61 mph?

   (c) At the conclusion of the snack break in Athens, what was my average speed up to then? Include the time spent eating. What was the average velocity over the same time interval?

   (d) Including the snack break, what was the average speed between home and the arrival in Chattanooga? What was the average velocity over the same period?

(e) What was the average speed between the end of the snack break and the arrival back home? What was the average velocity for this time interval?

(f) What was the average velocity for the entire trip?

20. *Transportation*   If I travel a total distance of 60 miles and drive the first half of the distance at 30 mph, how fast must I travel in the second half of the distance to average 60 mph for the entire trip?

2.2   Follow the Bouncing Ball

The first example of nonconstant motion will be that of a bouncing ball. The CBL with a motion detector attached is well suited for this situation. When you see distance vs. time graphs that represent a bouncing ball, think about how the velocity is changing:

▶ When is the velocity positive? When is it negative?
▶ Can the velocity ever be zero?
▶ How quickly does the velocity change from one value to another?

CBL and the Bouncing Ball

In Activity 2.2 you will hold a ball under a motion detector and drop it from a height of about 50 centimeters (20 inches) to the floor. As the ball moves, the program BOUNCE will measure the height of the ball above the floor and produce a graph like the one shown in Figure 2.5.* The BOUNCE program measures height in meters and time in seconds.

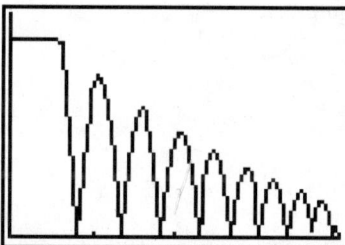

Figure 2.5

As you can see, the motion displayed in Figure 2.5 is not linear. In fact, except for the flat part at the beginning, it is not even made up of straight segments.

Activity 2.2
That's the Way the Ball Bounces

Use a ball such as a racquetball or tennis ball and a setup like the one shown in Figure 2.6. Hold the ball about 50 cm (20 in.) above a hard level surface. Make sure

---

*This experiment is described in detail on pages 42–44 of *CBL System Experiment Workbook*, published by Texas Instruments. The text of the experiment is also available on the World Wide Web at the location: http://www.ti.com/calc/docs/cbl1_p2.htm.

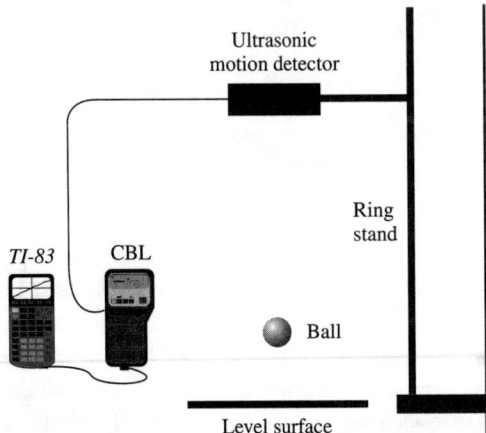

Figure 2.6

that the ball is held directly under the motion detector. After four or five seconds, you should obtain a figure like the one shown in Figure 2.5. Look at your results and answer items 1–5 below. We will think of a bounce as the period from the instant the ball hits the level surface until it hits the surface again.

1. Describe the path of the ball during one bounce.
2. Describe the velocity of the ball during one bounce.
3. In each bounce the ball reaches a highest point, called the maximum height. Describe how the maximum height changes throughout the experiment.
4. Does the time that the ball is in the air change with each bounce?
5. Make a sketch of the velocity of the ball during one bounce. The graph should show velocity on the vertical axis and time on the horizontal axis, like the one shown in Figure 2.7.

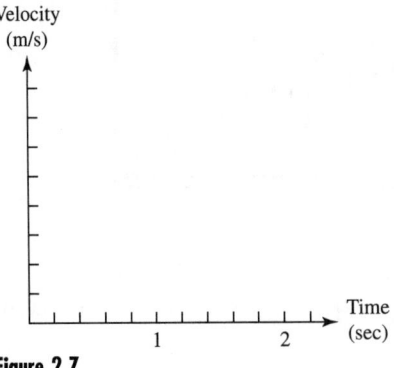

Figure 2.7

You are going to use your results from the BOUNCE experiment, so you need to save them. Save the times under the name TBNCE and the corresponding heights of the ball using the name BNCE.

The next three Calculator Labs will use your results from the BOUNCE experiment. The answers and figures in these Calculator Labs are the result of an experiment by the authors. The results from your BOUNCE experiment should provide similar results, but your actual answers will probably be different.

Before you begin these labs, graph BNCE vs. TBNCE. You should obtain a graph of BNCE vs. TBNCE like the one in Figure 2.5.

| **Calculator Lab 2.1** | **BNCE vs. TBNCE** |

In this Calculator Lab you will use your graph of BNCE vs. TBNCE to answer the following questions. The values for BNCE (height) are in meters and TBNCE (time) are in seconds.

1. At what time does the ball first touch the ground?
2. What is the maximum height it reaches after the first bounce?
3. How long does it take to reach the maximum height after the first bounce?
4. How long is the ball in the air between the first and second bounces?
5. What is the average velocity between the time of the first bounce and the time the ball reaches its maximum height after the first bounce?
6. What is the average velocity between the time the ball reaches its maximum height after the first bounce and the time of the second bounce?

Procedures   Change the calculator's decimal precision to 2 places. This change will make some of the screens easier to read. We do not need full precision when working with this experimental data.

1. Trace along the graph of BNCE vs. TBNCE to move the cursor to the point where the ball hits the ground the first time. The authors' result, shown in Figure 2.8, has $x = 0.8$ and $y = 0.01$. Even though $y \neq 0$, it is the smallest value for y in this part of the graph, so it is the one we select. So it takes about 0.8 sec for the ball to first touch the ground.

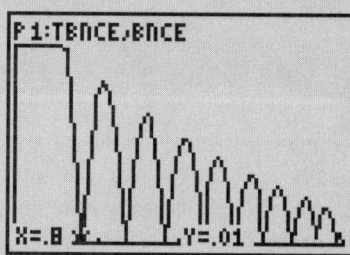

Figure 2.8

2. Move the cursor to the top of the first bounce and read off the y-coordinate (see Figure 2.9). The maximum height after the first bounce is about 0.36 meters.

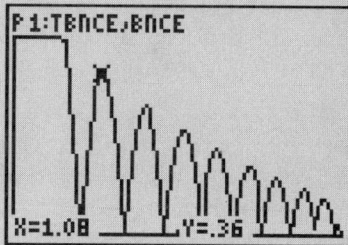

Figure 2.9

3. The time it takes to reach this maximum height after the ball first hits the ground is the difference between the time of the first bounce (0.8 seconds) and the time of the maximum height (1.08 seconds), or $1.08 - 0.8 = 0.28$ seconds.

4. Note the time when the ball bounces the second time, as in Figure 2.10. Subtract the time the ball first hits the ground (0.8) from this number. For the authors' experiment the answer was $1.36 - 0.8 = 0.56$ seconds.

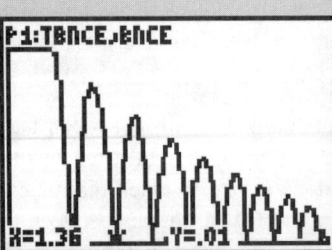

Figure 2.10

5. Average velocity between two points in a distance vs. time graph is the slope of the line joining the two points. The coordinates of the maximum height after the first bounce are $(1.08, 0.36)$. The coordinates of the first bounce are $(0.8, 0.01)$. The average velocity between these two points is $\frac{0.36-0.01}{1.08-0.8} = \frac{0.35}{0.28} = 1.25$ m/s.

6. The average velocity between the time the ball reaches its maximum height between the first and second bounces and the time of the second bounce is $\frac{0.01-0.36}{1.36-1.08} = \frac{-0.35}{0.28} = -1.25$ m/s.

Are you surprised that the answers in items 5 and 6 above are the same except for the sign?

Looking at One Bounce

The graph in Figure 2.10 extends over eight different bounces. We want to isolate one of the bounces and study it independently of the others. The *TI-83* has a special feature called Select which allows you to cut out all the data you don't want so that you can focus on the data you do want.

Calculator Lab 2.2	**Selecting the First Bounce**

Use the `Select(` feature of the *TI-83* to study only the first bounce of the ball. We will think of the first bounce as the period from the instant the ball hits the level surface until it hits the surface again.

Procedures

1. To select only the points of the first bounce, press `2nd` [LIST] and select OPS from the top of the LIST menu. The Select feature is number 8 on this menu. Choose `Select(` and press `2nd` [L1] `,` `2nd` [L2] `)` to make the home screen display the line

$$\text{Select(L1, L2)}$$

Then press `ENTER`.

2. Move the cursor to the leftmost point you want to select—in this case the bottom of the first bounce (about $x = 0.80$ seconds). The y-coordinate will be close to zero; press (ENTER) to mark that as the left bound. Move the cursor to the right until it reaches the next point where the y-coordinate is zero or close to zero; press (ENTER) to select this point as the right bound. The graph will change to look like Figure 2.11.

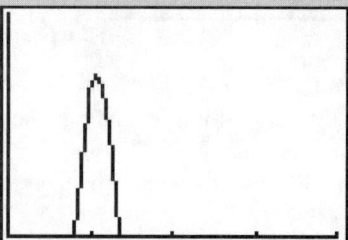

Figure 2.11

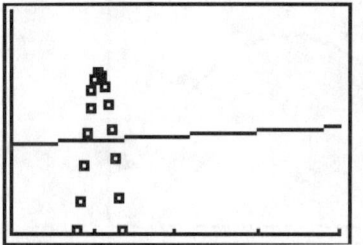

Figure 2.12

Clearly these data points are not linear. If you completed a linear regression on these data, the line would be something like the one shown in Figure 2.12. (We changed the definition of Plot1 from a dot to a square and the plot type from xy-line to scatter to show the data points more clearly in Figure 2.12.)

Quadratic Regression

Because a linear regression did not produce a very good fit, we need to see if a different type of curve will fit these points. In addition to linear regression there are several other types of regression built into most graphing calculators. The next kind of regression we will study is quadratic regression. While linear regression fits a line of the form $y = ax + b$, a quadratic regression fits a curve of the form $y = ax^2 + bx + c$, where $a \neq 0$. That is, it adds one more term, the x^2 term, to the regression equation. We will study quadratic equations more completely starting with Section 2.3.

Calculator Lab 2.3

Quadratic Regression

Fit a quadratic regression equation to the data in Figure 2.11.

Procedures

1. Press (STAT) and select the CALC menu. Choose QuadReg from this menu and indicate the two lists you want to use in the regression to fit a quadratic to the data. The completed QuadReg command should be: QuadReg L1,L2. (We use L1 and L2 because these are the two lists where the Select(command stored the data. Remember that the first list named in a regression command is the one plotted on the x-axis; the second is plotted on the y-axis.) When the command is executed, the result should look like Figure 2.13.

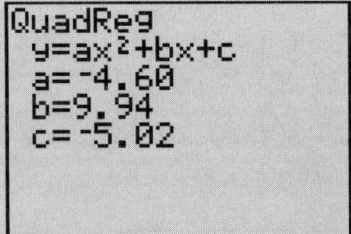

Figure 2.13

2. Substitute the values for a, b, and c in the quadratic equation $y = ax^2 + bx + c$. Then enter this equation, $y = -4.6x^2 + 9.94x - 5.02$, in the Y= screen. The graph will look something like Figure 2.14. The fit appears to be quite close.

Figure 2.15 gives a closer look at a part of the graph, as does the calculator's zoom feature. You can see that there is some error here, but the worst error (distance between the top point and the line) is about 2.8%.

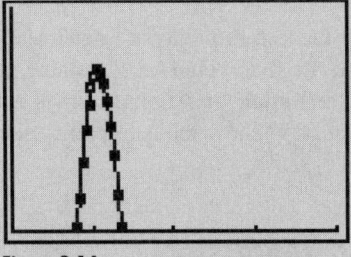

Figure 2.14

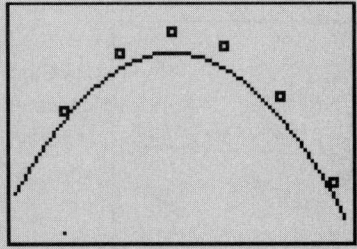

Figure 2.15

3. By zooming in (Figure 2.15) we can see that the fit is not as close as it appears to be in Figure 2.14. The difference between the regression curve and the actual data values is of the same type of error as that which resulted when we studied linear regression.
4. We will do more with this data from the first complete bounce of the ball later in this chapter. For now, the last thing to do with the lists L1 and L2 is to save their contents into two lists with the names TBNC1 and BNCE1.

Thinking about Quadratic Regression

Quadratic regression can prove to be quite useful when data appear curved rather than straight. However there are many curves that are not quadratic, so we must be careful when applying quadratic regression. It turns out that a quadratic equation like $y = ax^2 + bx + c$ is particularly useful to model some kinds of motion. In the 17th century an Englishman named Isaac Newton discovered that quadratic expressions produce good mathematical models for motion that is affected by gravity (or any constant force). That is, the motion of falling objects, thrown objects, planets, and comets all can be studied with quadratic expressions. So when the data come from an experiment in which an object is affected by gravity, think about using quadratic regression to model the results.

Section 2.2 Exercises

In Exercises 1–7 use your lists BNCE *and* TBNCE *and the* Select(*feature to study the second bounce of the ball you dropped in Activity 2.2.*

1. At what time does the ball touch the ground the third time?

2. What is the maximum height reached after the second bounce?

3. How long does it take to reach the maximum height after the second bounce?

4. How long is the ball in the air between the second and third bounces? Was this the same as the time between the first and second bounces?

5. What is the average velocity between the time of the second bounce and the time the ball reaches the maximum height after the second bounce?

6. What is the average velocity between the time the ball reaches the maximum height after the second bounce and the third bounce?

7. Use quadratic regression to determine a quadratic equation that approximates the second bounce and plot the regression curve.

Exercises 8–15 involve data gathered by tossing a large ball such as a basketball straight up over a motion detector and catching it while it is still in the air. For these exercises either (a) collect your own data or (b) use the authors' data. To collect your own data, use a CBL and the program BALLDROP, *and toss a large ball straight up over the motion detector. Make sure to protect the motion detector from being hit by the ball. The authors' time data (in seconds for this experiment) is in the list named* TBALL, *and the height (in feet) is in a list* BALL. *The graph that results in this type of ball toss is shown in Figure 2.16.*

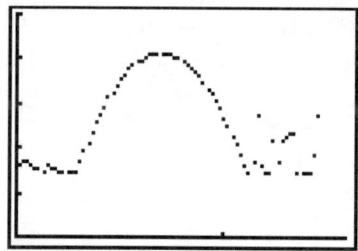

Figure 2.16

8. Use quadratic regression on the data and plot the regression curve with the data. Why does this regression curve fit so badly?

9. Use Select(to study only the part of the experiment in which the ball was moving. (In the data provided, this is between about 0.29 seconds and 1.13 seconds.)

10. At what time does the ball reach its maximum height?

11. What is the maximum height reached by the ball?

12. How long is the ball in the air?

13. What is the average velocity going up to the maximum height?

14. What is the average velocity coming down from the maximum height?

15. Use quadratic regression to determine a quadratic equation on the part of the curve you selected in Exercise 9.

2.3 Introduction to Functions

The study of functions is a large part of mathematics. In Chapter 1 we talked about linear functions and also included a brief glimpse at trigonometric functions. In the previous section you were introduced to quadratic functions. Now you will learn some of the language and symbolism connected to the study of functions.

Introduction to Functions

A **function** may be thought of as a rule that assigns each member of one group to one member of another group. For example, $y = 2x + 5$ is a rule for a function. Here x represents numbers in the first group and y represents numbers in the second group. This rule tells how to get the value of y for a particular value of x. The rule says, "Multiply the value of x by 2 and then add 5." Another rule for a function is $t = 5$. This rule says, "For each member of the first group assign the number 5 from the second group."

You can also think of a function as a machine that computes an output from an input. Look at Figure 2.17. The function has been given a name L and it has the recipe, or rule of computation, $2x + 5$. The result of the computation is represented by the symbol $L(x)$, pronounced "L of x."

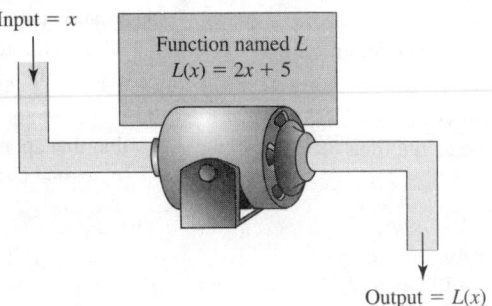

Figure 2.17

In the language of the first description of a function, the set of all allowable inputs is the first group and the set of all outputs is the second group. In mathematical terminology the set of inputs is called the **domain** of the function and the set of outputs is called the **range**.

Example 2.4

If $L(x) = 2x + 5$, compute (a) $L(3)$, (b) $L(0)$, (c) $L(-5)$, and (d) $L(\text{time})$.

Solutions   The idea here is to substitute the number or symbol in parentheses for each x in the recipe or rule of computation.

(a) To compute $L(3)$ we substitute 3 for the x in $2x + 5$. This produces

$$L(3) = 2 \cdot 3 + 5$$
$$= 6 + 5 = 11$$

Thus $L(3) = 11$.

(b) $L(0)$ is evaluated by replacing the x with 0 and then computing the result.

$$L(0) = 2 \cdot 0 + 5$$
$$= 0 + 5 = 5$$

So $L(0) = 5$.

(c) For $L(-5)$ we replace the x with -5. The result is

$$L(-5) = 2(-5) + 5$$
$$= -10 + 5 = -5$$

(d) In this part the word time is used to replace the x. This solution does not give a number as the answer; instead we get $L(\text{time}) = 2(\text{time}) + 5$. ■

Example 2.5

If a quadratic function is defined by $f(x) = 3x^2 - 2x + 5$, compute the values of (a) $f(4)$, (b) $f(-1)$, and (c) $f(b)$.

Solutions   Once again we substitute the number or symbol in parentheses for each x in the recipe or rule of computation.

(a) To compute $f(4)$ we substitute 4 for the x in $3x^2 - 2x + 5$. This produces

$$f(4) = 2x^2 - 2x + 5$$
$$= 3 \cdot 4^2 - 2 \cdot 4 + 5$$
$$= 3 \cdot 16 - 2 \cdot 4 + 5$$
$$= 48 - 8 + 5 = 45$$

Hence we see that $f(4) = 45$.

(b) $f(-1)$ is evaluated by substituting -1 for each x. This produces

$$f(-1) = 3(-1)^2 - 2(-1) + 5$$
$$= 3 \cdot 1 + 2 + 5 = 10$$

Thus $f(-1) = 10$.

(c) Here we replace each x with the letter b and obtain

$$f(b) = 3(b)^2 - 2(b) + 5$$
$$= 3b^2 - 2b + 5$$

So $f(b) = 3b^2 - 2b + 5$.

Naming a Function

If you write $y = 2x + 5$, you are describing a function because you are giving a rule for computing a value of y for any value of x. If you write $L(x) = 2x + 5$, you are giving that same function a name, L. Another way to write this same function is $y = L(x)$.

The name of a function can be a single letter such as L, f, g, or V. A function name can also be a group of letters such as ln, log, tan, or abs. In the next example you will explore a few of the functions built into your calculator.

Many mathematical functions are built into your calculator. Some have their own keys: for example, **LOG**, **SIN**, **COS**, **TAN**, and **2nd** [$\sqrt{}$]. (You saw how to use the **TAN** key in the last chapter.) Other functions can be found in menus. Calculations with these functions are often done on the home screen.

The following Calculator Lab and Example 2.6 will show how to use some of these functions. Don't worry if you do not understand how the function works—you will learn that later in the book. The main purpose of these examples is to show how to use the calculator. Press **2nd** [QUIT] to reach the home screen.

Calculator Lab 2.4	

Logging On

If $y = \log x$, what are (a) $\log(1000)$ and (b) $\log 0.0001$?

Procedures

In order for your answers to match the ones in the book, make sure that the decimal mode is set to `Float`.

1. The log function is accessed by pressing the `LOG` key on the calculator.* You need to enter the number, the right parenthesis, and then press `ENTER` to complete the computation. So to have the calculator evaluate $\log(1000)$, press `LOG` 1000 `)` `ENTER`. The result, as shown in Figure 2.18, should be 3.
2. Press `LOG` 0 `.` 0001 `)` `ENTER`. The result, as shown in Figure 2.19, is -4.

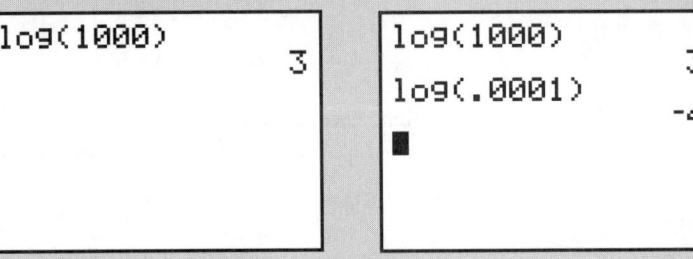

Figure 2.18   Figure 2.19

Example 2.6

(a) If $y = \cos x$, compute $\cos(180°)$.
(b) If $f(x) = \sqrt{x}$, compute $f(5600)$.
(c) For $g(x) = \sqrt{x} - 5$, determine $g(2025)$.
(d) If $h(x) = |x|$, evaluate $h(-500)$.

Solutions

(a) The `COS` key is an abbreviation for the cosine function. Like the tangent function in the last chapter, it is a trigonometric function. The value of $\cos(180°)$ is shown in Figure 2.20 and should be -1. If you got an answer of about $-.598$, then your calculator was in radian mode rather than degree mode.

(b) To access the square root key you need first to press the `2nd` key and then the `x²` key, because $\sqrt{\ }$ is printed in yellow above the `x²` key. The approximate result, 74.83314774, is shown in Figure 2.21.

(c) For $g(x) = \sqrt{x} - 5$, notice that only the x is under the square root symbol. To determine $g(2025)$, press `2nd` $[\sqrt{\ }]$ 2025 `)` `−` 5 `ENTER`. The right parenthesis,), after 2025 is used to tell the calculator to compute the square root of 2025 and then subtract 5. The result should be 40, as shown in Figure 2.22.

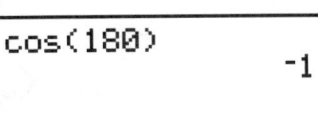

Figure 2.20

---

*Pressing the `LOG` key on a *TI-83* calculator produces `log(` on the screen. On other calculators, pressing the `LOG` key displays `log` without a left parenthesis.

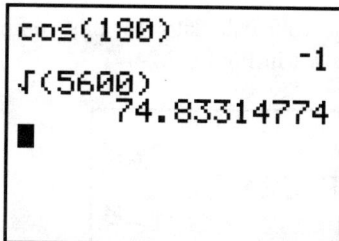

Figure 2.21

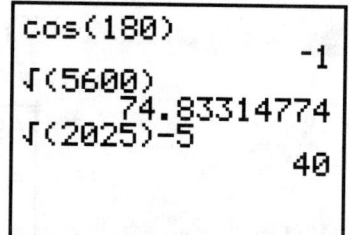

Figure 2.22

(d) Here $h(x) = |x|$ is the absolute value function. This function is on one of the calculator's menus and is found by pressing the ▄MATH▄ key and selecting the NUM menu. The result, 500, is shown in Figure 2.23.

Figure 2.23

How Functions Are Defined

From the previous examples you can see that sometimes a function is defined in terms of a *computation*, as in $L(x) = 2x + 5$ and $f(x) = 3x^2 - 2x + 5$. Sometimes a function can be defined in *words*, such as "Square root of x is the number that when multiplied by itself gives the answer x." Functions can also be described by pairs of numbers and by graphs. Later in this section we will look at these ways of defining functions.

You will study many different functions in this book. As we have already seen with linear and quadratic functions, different functions form suitable mathematical models for different kinds of data.

Example 2.7

Compute the value of the function $C(F) = \frac{5}{9}(F - 32)$ when F has the value 68. Do this computation (a) by hand and (b) with a calculator.

Solutions   (a) Substitute 68 where you see F.

$$C(68) = \frac{5}{9}(68 - 32)$$

$$= \frac{5}{9}(36) \qquad \text{Divide 36 by 9}$$

$$= 5 \cdot 4 = 20$$

(b) If you use your calculator for this computation you should obtain the result of 20, as shown in Figure 2.24.

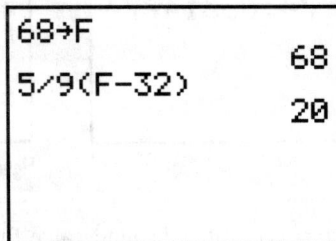

Figure 2.24

Graphing Functions

You have already graphed linear and quadratic functions. To graph functions on a graphing calculator you used the Y= screen. You have also used a table of values to help graph a function. In the next several examples we will review these methods for linear and quadratic functions and then use a graphing calculator to graph some other functions.

Example 2.8

Complete the following table of values for $q(t) = t^2 - 4t - 5$ and then draw the graph by hand.

t	−6	−4	−2	0	2	4	6
$q(t)$							

Solution   To find points on the graph we calculate a value of q for each of the values of t. For example, if $t = -6$, then

$$q(-6) = (-6)^2 - 4(-6) - 5$$

$$= 36 + 24 - 5 = 55$$

Thus the entry in the table for $q(t)$ when $t = -6$ is 55.

t	−6	−4	−2	0	2	4	6
$q(t)$	55						

Complete the table by hand and then check each of your calculations with the calculator.

Graphing this function is similar to other graphing you have done: For each pair of numbers for t and $q(t)$ in the table, plot a point on the graph. Then connect the dots in a curve, as shown in Figure 2.25.

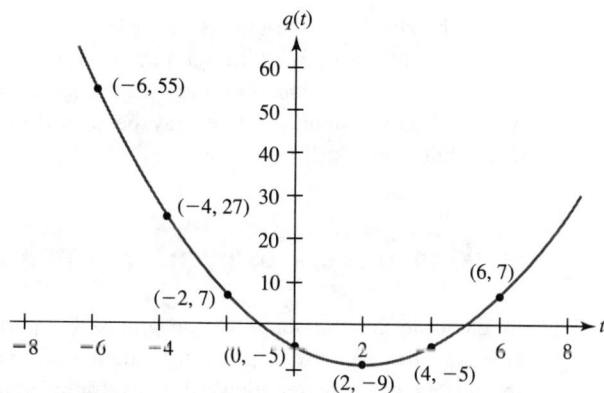

Figure 2.25

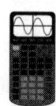

Example 2.9

Use a graphing calculator to draw the graph of the functions (a) $L(w) = 2w + 5$, (b) $f(y) = y^2 - y - 6$, (c) $G(t) = \sqrt{9t}$, (d) $y = \dfrac{10}{x}$, (e) $y = \log x$, and (f) $g(x) = 10^x$.

Solutions   All the results were graphed using the ZStandard window.

(a) To graph $L(w) = 2w + 5$, enter the right-hand side of this equation in the Y= screen. When this is entered into the calculator you must use the variable x rather than w, because the graphing calculator requires an x in order to graph a function. When you press **GRAPH** you should obtain the result shown in Figure 2.26.

(b) The graph of $f(y) = y^2 - y - 6$ is shown in Figure 2.27.

(c) The graph of $G(t) = \sqrt{9t}$ is shown in Figure 2.28.

(d) The graph of $y = \dfrac{10}{x}$ is shown in Figure 2.29.

(e) The graph of $y = \log x$ is shown in Figure 2.30.

(f) The graph of $g(x) = 10^x$ is shown in Figure 2.31.

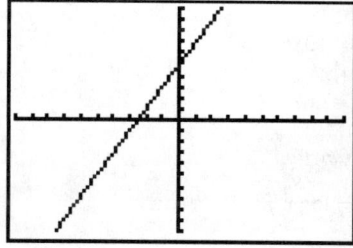

Figure 2.26

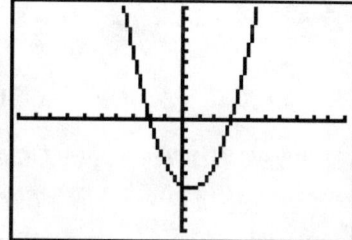

Figure 2.27

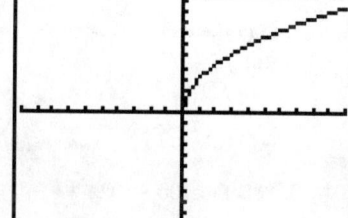

Figure 2.28

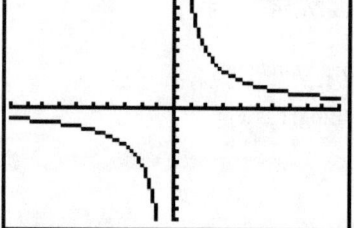

Figure 2.29

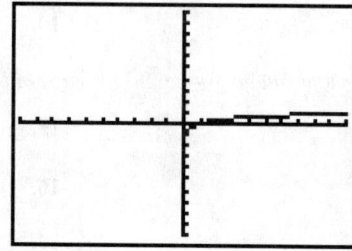

Figure 2.30

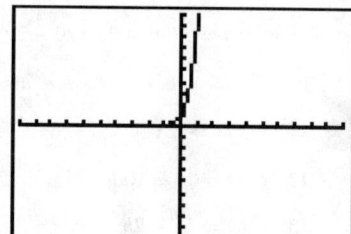

Figure 2.31

The shapes of the graphs of these functions should become familiar to you as you gain experience working with them. Try to remember the shape of the graph of each function. The location and specific shape will change, but the general shape will not. For instance, a line may be in a different location and have a different slope, but it will still be a line.

Looking More Closely at a Graph

In Example 2.9 the shape of the graphs for parts (e) and (f) were difficult to see clearly. You can use a graphing calculator's Zoom features or different window settings to get a better view. For example, you could use the ZoomFit feature on the graph shown in Figure 2.30. Then you would obtain a better picture of the graph of the function $y = \log x$ for values of x between -10 and 10—as shown in Figure 2.32.

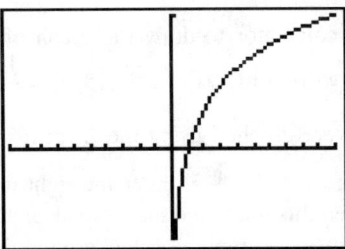

Figure 2.32

Section 2.3 Exercises

In Exercises 1–2 evaluate each of the given functions at the indicated values.

1. $f(x) = 4x^2 - 3x + 2$
(a) $f(-5)$,
(b) $f(1.6)$

2. $g(x) = -3x^2 + 5x - 7$
(a) $g(4)$,
(b) $g(-2.6)$

 In Exercises 3–10 use your calculator to evaluate $f(100)$ for each of the given functions.

3. $f(x) = 100 - 2x$

4. $f(x) = 100 \div 2x$

5. $f(x) = \dfrac{3x - 5}{2}$

6. $f(x) = x(30 - 2x)(20 - 2x)$

7. $f(x) = \log x$

8. $f(x) = \sqrt{x}$

9. $f(x) = |x| - 3$

10. $f(x) = \sqrt{676 - x}$

 In Exercises 11–18 use your calculator to evaluate $g(-0.625)$ for each of the given functions.

11. $g(x) = -2x + 5$

12. $g(x) = |x - 100|$

13. $g(h) = h^2 - 2h + 5$

14. $g(t) = \dfrac{t^2 - 9}{(t - 3)(t + 3)}$

15. $g(x) = \log(1 - x)$

16. $g(x) = \sqrt{-10x}$

17. $g(x) = \sqrt{0.6475 + x}$

18. $g(x) = 10^x$

In Exercises 19–28 use your calculator to (a) graph each of the functions and (b) approximate all the x-intercepts.

19. $f(x) = 9x^2 - 5x - 4$

20. $g(x) = 0.2x^2 - 0.44x - 0.15$

21. $h(t) = 2\sqrt{t} - 4$

22. $j(w) = \log(w^2) - 2$

23. $F(b) = |b^2 - 4| - 3$

24. $G(v) = \dfrac{v^2 - 9}{v^2 - 3v - 4}$

25. $V(t) = 2^t - \sqrt{5t^2} - 1$

26. $y(p) = \sqrt{|p - 3|} - 2$

27. $f(x) = \text{int}(x)$ [*Hint:* int(x) is the calculator's method for writing the greatest integer function. Mathematicians also write int(x) as [x] where [x] is the largest integer that is smaller than or equal to x. Press **MATH** ▶ to select the NUM menu. Then press **5** to select the greatest integer (int) function.]

28. $f(x) = 4\cos\left(\dfrac{\pi x}{2}\right)$ [*Hint:* Before you graph this, press **MODE** to change from degrees to radians. Press **2nd** [π] to enter the number π After you have finished this exercise, change your calculator back to degree mode.]

2.4

Working with Parabolas

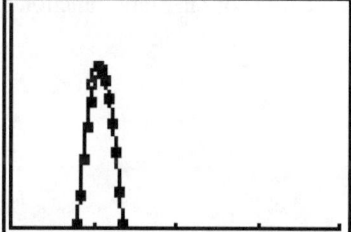

Figure 2.33

Figures 2.33 and 2.34 are examples of graphs of quadratic functions: one opens downward and one opens upward.

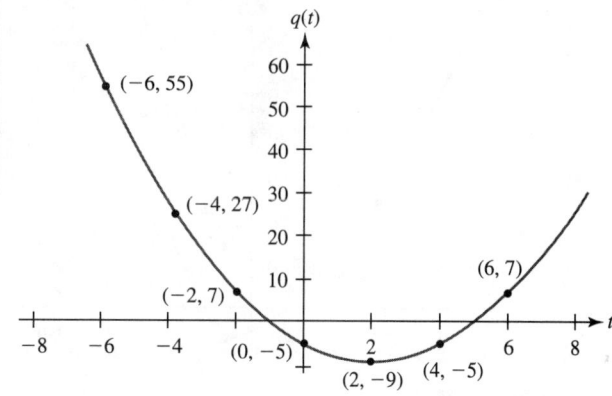

Figure 2.34

> **Definition: Quadratic Function**
> Any function f given by $f(x) = ax^2 + bx + c$ with $a \neq 0$ is called a **quadratic function**. The numbers a, b, and c are called the coefficients of the function.

The graphs in both figures have the characteristic shape, called a **parabola** or **parabolic curve**. In this section you will learn how to predict the shape, direction, and location of a parabola by looking at the quadratic expression that defines it. You will also see how to find the x- and y-intercepts of a quadratic function and how to find its high point (the **maximum**) or low point (the **minimum**) of the curve.

The Bouncing Ball assignment from the Chapter Project requires that you study three aspects of the parabola:

▶ the time between bounces
▶ the height of the bounce
▶ the velocity with which the ball rebounds

We'll study the first two aspects in this section. Let's begin with the time between bounces.

The x-intercepts of the graph in Figure 2.33 are the times when the ball hits the ground. Actually t-intercept would be a more accurate phrase, since in this figure the horizontal axis measures time. However we'll continue to use the phrase x-intercept for the point or points where the graph crosses the horizontal axis.

Roots of a Quadratic Function

You saw in Chapter 1 that the x-intercept of a line is a point where $y = 0$. In the language of functions, if $f(x)$ is a function then the x-intercept is the point where $f(x) = 0$. This point is called the **root** or **zero** of the function.

If $f(x)$ is a linear function and we write $f(x) = ax + b$, then the root can be found by solving the equation $ax + b = 0$ for x. As long as a does not equal zero, the linear function $f(x)$ will have exactly one root at $x = -\dfrac{b}{a}$. But a parabola is different. In the two quadratic graphs shown in Figures 2.33 and 2.34, you can see that each has two x-intercepts or roots. As you'll see later, not all quadratic graphs have two x-intercepts.

In Figure 2.35 the roots of the function $q(t) = t^2 - 4t - 5$ are marked. In the following example we will look at one way to determine roots of quadratic functions.

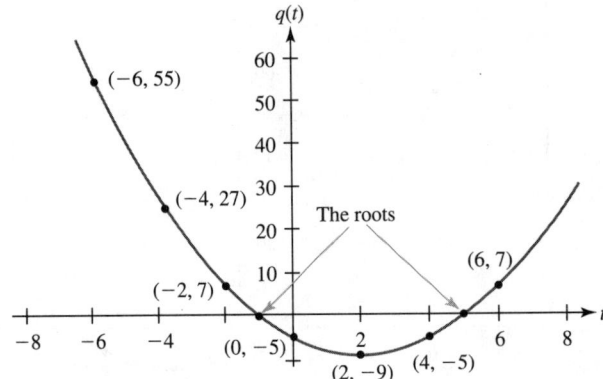

Figure 2.35

Example 2.10

Estimate the roots of the quadratic function $q(t) = t^2 - 4t - 5$ and check your answer.

Solution   Looking at the graph, it appears that good estimates of the roots are $t = -1$ and $t = 5$.

We need to check and see how close our estimates are to the correct roots. Since roots are solutions to the equation $q(t) = 0$, we must substitute each of our estimated values into the function and see if we get a result of zero when the right-hand side is simplified.

We will first check our estimate of $t = -1$. Substituting -1 for t in $q(t) = t^2 - 4t - 5$, we obtain

$$q(-1) = (-1)^2 - 4(-1) - 5$$
$$= 1 + 4 - 5 = 0$$

Since $q(-1) = 0$, we have verified that -1 is a root of q.

Next we check our estimate for the second root by substituting 5 for t in $q(t) = t^2 - 4t - 5$, which produces

$$q(5) = 5^2 - 4 \cdot 5 - 5$$
$$= 25 - 20 - 5 = 0$$

Since $q(5) = 0$, we have verified that 5 is a root of q, and hence that both our estimates of the roots were correct. ■

Roots and Factors

In Chapter 1 you saw that the equation of a linear function could be written in several ways. Here are two of them:

▶ the slope-intercept form of a line, $y = mx + b$
▶ the point-slope form of a line, $y = y_1 + m(x - x_1)$

Each of these could be written to show it is a function of x by replacing the symbol y with the symbol $f(x)$.

There are several ways to write a quadratic function. One way, the **general form of a quadratic function**, is $f(x) = ax^2 + bx + c$ with $a \neq 0$; you saw it earlier in this chapter. Another form is the **factored form of a quadratic function**:

$$f(x) = a(x - r_1)(x - r_2)$$

You have seen factoring in other mathematics courses. Many numbers and expressions can be written in factored form, as shown in the next example.

Example 2.11

Factor (a) 45, (b) $35xy^2$, and (c) $x^2 - 5x + 6$.

Solutions   (a) $45 = 3 \cdot 3 \cdot 5$ The prime factors of 45 are 3, 3, and 5.
(b) $35xy^2 = 5 \cdot 7 \cdot x \cdot y \cdot y$ The prime factors are 5, 7, x, y, and y.
(c) $x^2 - 5x + 6 = (x - 2)(x - 3)$ The factors are $x - 2$ and $x - 3$.   ■

In each of these examples, if you multiply what's on the right-hand side of the equal sign, you get what is on the left-hand side. Thus

$$(x - 2)(x - 3) = (x - 2)x - (x - 2)3$$
$$= x^2 - 2x - 3x + 6$$
$$= x^2 - 5x + 6$$

The roots of a quadratic equation are related to its factored form. Remember that the roots of a function $f(x)$ are the solutions of the equation $f(x) = 0$. Consider a quadratic function written in the factored form as $f(x) = a(x - r_1)(x - r_2) = 0$. The roots will be the solutions to the quadratic equation

$$a(x - r_1)(x - r_2) = 0$$

The only way a multiplication problem can have a result of zero is if one or more of the factors is equal to zero. In this case that means that one or more of the following must be true:

$$a = 0 \quad \text{or} \quad x - r_1 = 0 \quad \text{or} \quad x - r_2 = 0$$

Since this is a quadratic function, we know that $a \neq 0$. If $x - r_1 = 0$ then $x = r_1$, and if $x - r_2 = 0$ then $x = r_2$. Therefore the following is true about the roots of a quadratic function.

Roots and the Factored Form of a Quadratic Function

If a quadratic function is in factored form $f(x) = a(x - r_1)(x - r_2)$, then the roots are $x = r_1$ and $x = r_2$.

Example 2.12

If the factored form of the function $z(t) = 6t^2 + 3t - 45$ is $z(t) = (2t - 5)(3t + 9)$, what are the roots of the function? Check your answers on a graph.

Solution   The roots are the solutions of $z(t) = 0$ or $(2t - 5)(3t + 9) = 0$. There are two parts of the solution. One is when $2t - 5 = 0$ or $t = \frac{5}{2} = 2.5$. The other is when $3t + 9 = 0$ or when $t = \frac{-9}{3} = -3$.

To check this on a graph, plot the function $y = 6t^2 + 3t - 45$. Do not graph $y = (2t - 5)(3t + 9)$ because the function might not have been factored correctly. Using ZStandard the graph looks like the one shown in Figure 2.36. Even though the bottom of the parabola is out of sight below, the roots are shown on the screen. To locate one of the roots, move the trace cursor until it is near one of the x-intercepts, as shown in Figure 2.37.

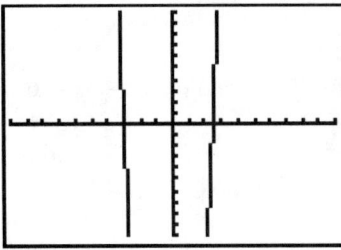

Figure 2.36

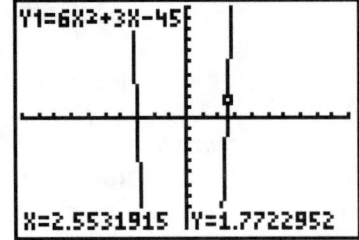

Figure 2.37

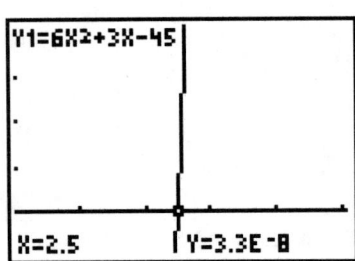

Figure 2.38

If you find that the trace cursor jumps too far each time, use the zoom cursor only.

Let's take a closer look at the graph near the root. Press 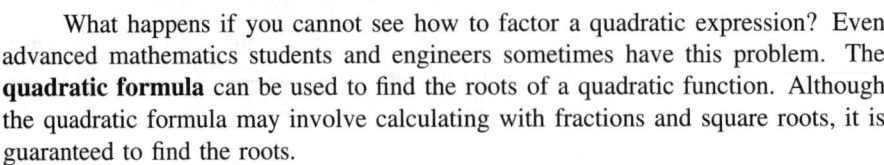 to select Zoom In. Figure 2.38 shows how close you can get. Look at the numbers at the bottom of the screen. The value of x is 2.5, which matches one of the roots. The value of y is written as 3.3E⁻8. This is how the Texas Instruments graphing calculators, like many computer systems and other calculators, write scientific notation. 3.3E⁻8 means 3.3×10^{-8}, which is the same as 0.000000033. This number is close enough to zero to seem to verify that 2.5 is a root of z.

Now *you* try to locate the other root. Before you begin, press ZOOM 6 to return to the standard graph window.   ∎

What happens if you cannot see how to factor a quadratic expression? Even advanced mathematics students and engineers sometimes have this problem. The **quadratic formula** can be used to find the roots of a quadratic function. Although the quadratic formula may involve calculating with fractions and square roots, it is guaranteed to find the roots.

Quadratic Formula for Roots of a Quadratic Function

The two roots of the function $f(x) = ax^2 + bx + c$ with $a \neq 0$ are

$$x = \frac{-b + \sqrt{b^2 - 4ac}}{2a} \qquad x = \frac{-b - \sqrt{b^2 - 4ac}}{2a}$$

The only difference between these roots is the sign in front of the square root symbol, so the formula is sometimes written as

$$x = \frac{-b \pm \sqrt{b^2 - 4ac}}{2a}$$

The symbol $\pm$ is pronounced "plus or minus," and means that there are really two formulas: one with a plus sign and one with a minus sign.

Example 2.13

Use the quadratic formula to compute the roots of the function $z(t) = 6t^2 + 3t - 45$.

Solution   The coefficients are $a = 6$, $b = 3$, and $c = -45$. Substituting these values in the quadratic formula, we see that one root is

$$t = \frac{-b + \sqrt{b^2 - 4ac}}{2a}$$

$$= \frac{-3 + \sqrt{3^2 - 4 \cdot 6(-45)}}{2 \cdot 6}$$

$$= \frac{-3 + \sqrt{9 + 1080}}{12}$$

$$= \frac{-3 + \sqrt{1089}}{12}$$

$$= \frac{-3 + 33}{12} = \frac{30}{12} = 2.5$$

Once you find one root, the other is easier to find; change the plus sign to a minus sign in the last step that has a square root symbol:

$$t = \frac{-3 - \sqrt{1089}}{12}$$

$$= \frac{-3 - 33}{12} = \frac{-36}{12} = -3$$

Now we'll apply this technique to the quadratic function that we fit to the bouncing ball in Figures 2.14 and 2.33. When we computed the regression on these data (see Figure 2.13) we obtained the equation $y = -4.6x^2 + 9.94x - 5.02$. Since this regression equation is used to approximate the height h of the ball at time t, we would write this function as $h(t) = -4.6t^2 + 9.94t - 5.02$.

Example 2.14

Find the roots of the quadratic function $h(t) = -4.6t^2 + 9.94t - 5.02$.

Solution   The coefficients are $a = -4.6$, $b = 9.94$, and $c = -5.02$. Substituting these values in the quadratic formula produces

$$t = \frac{-9.94 + \sqrt{9.94^2 - 4(-4.6)(-5.02)}}{2(-4.6)}$$

$$= \frac{-9.94 + \sqrt{98.8036 - 92.368}}{-9.2}$$

$$= \frac{-9.94 + \sqrt{6.4356}}{-9.2}$$

$$\approx \frac{-9.94 + 2.54}{-9.2} \approx 0.8$$

Thus we see that the regression equation indicates that one bounce occurs when t is about 0.8 sec. The other bounce occurs at

$$t = \frac{-9.94 - \sqrt{6.4356}}{-9.2}$$

$$\approx \frac{-9.94 - 2.54}{-9.2} \approx 1.36 \text{ sec}$$

These are the same answers we got in Calculator Lab 2.1 (see Figures 2.8 and 2.10).
∎

Example 2.15

How long is the ball in the air in the previous example?

Solution   The total time in the air is the difference between the time of the first bounce and the time of the second bounce. Thus the total time in the air is $1.36 - 0.80 = 0.56$ sec.
∎

So far all the quadratic functions we have looked at have had two roots. The next two quadratic functions will be different.

Example 2.16

Find the roots of the quadratic function $r(x) = -2x^2 + 12x - 18$.

Solution   The coefficients are $a = -2$, $b = 12$, and $c = -18$. One of the roots is

$$x = \frac{-12 + \sqrt{12^2 - 4(-2)(-18)}}{2(-2)}$$

$$= \frac{-12 + \sqrt{144 - 144}}{-4}$$

$$= \frac{-12 + \sqrt{0}}{-4} = \frac{-12 + 0}{-4} = 3$$

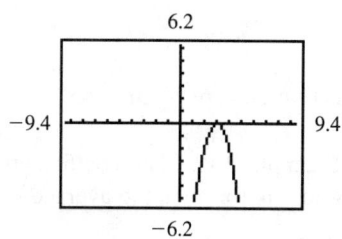

Figure 2.39

The other root is at $x = \frac{-12 - \sqrt{0}}{-4} = \frac{-12 + 0}{-4} = 3$. This answer is the same as the first root. The graph of r in Figure 2.39 shows that the parabola seems to have only one x-intercept when $x = 3$. ■

Example 2.17

Find the roots of the quadratic function $G(x) = x^2 - 5x + 9$.

Solution The coefficients are $a = 1$, $b = -5$, and $c = 9$. One of the roots is

$$x = \frac{5 + \sqrt{25 - 4 \cdot 1 \cdot 9}}{2} = \frac{5 + \sqrt{-11}}{2}$$

But the square root of -11 is not a real number. The calculator gives either an error message or a complex number. (We will discuss complex numbers in Chapter 5.) What's going on? A look at the graph in Figure 2.40 shows that the parabola does not cross the x-axis, so there are no real roots. ■

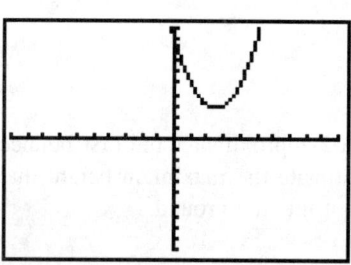

Figure 2.40

Finally, in order to study the height to which the ball in Activity 2.1 bounces, we need to know the value of x at the top (or bottom) of the parabola. If you look at the parabolas we have been studying, and those in Figure 2.41, you can see that the top or bottom of the graph—called the **vertex** (plural **vertices**) of the parabola—occurs at the x-value halfway between the two roots.

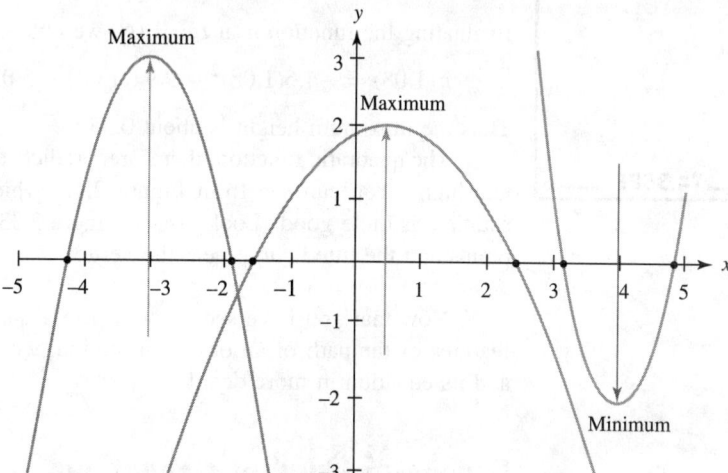

Figure 2.41

The number halfway between two numbers is the **average**, which is computed as the sum of the numbers divided by 2. Let's see what happens when we add the two roots given by the quadratic formula:

$$\frac{-b + \sqrt{b^2 - 4ac}}{2a} + \frac{-b - \sqrt{b^2 - 4ac}}{2a} = \frac{-b + \sqrt{b^2 - 4ac} - b - \sqrt{b^2 - 4ac}}{2a}$$

This looks like an impossible mess until you see that the sum of the two square roots (as shown by the gray shading) is zero, leaving $\frac{-2b}{2a} = -\frac{b}{a}$ for the sum of the two roots. Now we divide this sum by 2 to get the average, $-\frac{b}{2a}$.

Example 2.18

What is the average of the roots of the quadratic function $zt = 6t^2 + 3t - 45$?

Solution   This is the same quadratic function we studied in Example 2.13. The coefficients are $a = 6$, $b = 3$, and $c = -45$. Therefore according to this formula, the average of the two roots is $-\dfrac{b}{2a} = -\dfrac{3}{2 \cdot 6} = -\dfrac{3}{12} = -\dfrac{1}{4} = -0.25$.

In Example 2.13 we found that the roots were 2.5 and -3. The average of these two roots is $\dfrac{2.5 + (-3)}{2} = \dfrac{-0.5}{2} = -0.25$. As you can see, the formula worked!   ∎

Example 2.19

The quadratic function $h(t) = -4.6t^2 + 9.94t - 5.02$ approximates the first bounce of the ball in Activity 2.1. Use this function to estimate the maximum height that the ball reaches between the first and second times it hits the ground.

Solution   The coefficients are $a = -4.6$, $b = 9.94$, and $c = -5.02$. Therefore the vertex is

$$t = -\frac{b}{2a} = -\frac{9.94}{2(-4.6)} = \frac{9.94}{9.2} \approx 1.08$$

Evaluating the function h at $t = 1.08$, we obtain

$$h(1.08) = -4.6(1.08)^2 + 9.94(1.08) - 5.02 \approx 0.35$$

Thus the maximum height is about 0.35 m.

The quadratic function, therefore, predicts that the first bounce will be 0.35 meters high. You can see from Figure 2.42, which shows the actual data, that this estimate is quite good. Look also at Figure 2.33 on page 83, which shows the data points and the fitted curve near the vertex.   ∎

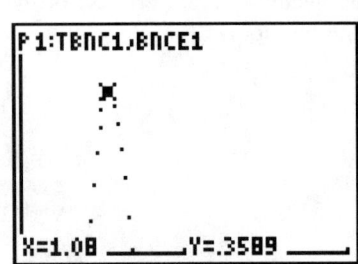

Figure 2.42

Now that you have seen how to use a quadratic model to predict certain key features of the path of an object in motion, we will move on to study the parabola and its equation in more detail.

Graphs of Quadratic Equations

You now know the following about graphs and roots of quadratic equations:

▶ How to make a graph by hand.
▶ How to graph quadratic functions on a graphing calculator.
▶ The shape of a quadratic graph is called a parabola.
▶ The solutions to the equation $f(x) = 0$ are called the roots of the function.
▶ If a quadratic function in factored form is $f(x) = a(x - r_1)(x - r_2)$, then the roots are r_1 and r_2.
▶ If a quadratic function is in the form $f(x) = ax^2 + bx + c$ with $a \neq 0$, then
  • the roots can be calculated using the quadratic formula.
  • the average of the two roots is $-\dfrac{b}{2a}$.

▶ The vertex of a parabola occurs when the value of x is the average of the two roots.

You also know the following about the graph of a linear function $f(x) = ax + b$:

▶ The slope is a and the y-intercept is b.
▶ The root is $-\dfrac{b}{a}$ if $a \neq 0$.

Next you will do a calculator experiment to find out how the coefficients a, b, and c of a quadratic function influence its graph. Calculator Lab 2.5 will show how to do this experiment with the coefficient b. You will then be able to do similar experiments for a and c.

Calculator Lab 2.5

Controlling Variables

In this Calculator Lab you will systematically change the coefficient b in the quadratic function $y = ax^2 + bx + c$ and describe the effect that the value of b has on the shape and location of the parabola.

First let's think about how to run an experiment in which you want to find the effect of one variable among several. For example, if you wanted to learn whether using a gasoline additive or replacing the spark plugs would make your car run better, you would try one and then the other—not both together. That is, you would conduct your experiment by **controlling the variables**.

To figure out the effect of the coefficient b on a parabola's graph, we should keep a and c fixed while we vary b. One way to do that is to plot the graphs of several quadratic functions that have the same coefficients a and c but different values of b. In this example we will graph $y = x^2 + bx - 3$ and pick values of b to be positive, negative, and zero in an orderly fashion. In particular we will let b assume the values of -5, -3, -1, 0, 1, 3, and 5.

Procedures

1. Begin by graphing the quadratic when b is zero. That is, graph $y = x^2 - 0x - 3$ (or $y = x^2 - 3$). Set the viewing window to ZStandard by pressing ⬤ZOOM⬤ ⬤6⬤. The result is shown in Figure 2.43.

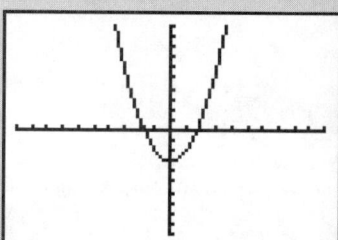

Figure 2.43

2. Keep this graph and graph the three quadratic functions where b is negative. That is, do not clear $y = x^2 - 0x - 3$ from the Y= list, but graph $y = x^2 - 5x - 3$, $y = x^2 - 3x - 3$, and $y = x^2 - 1x - 3$ on the same set of axes. The result is shown in Figure 2.44.
3. Add the graphs of the three quadratic functions where b is positive. All seven parabolas are shown in Figure 2.45.

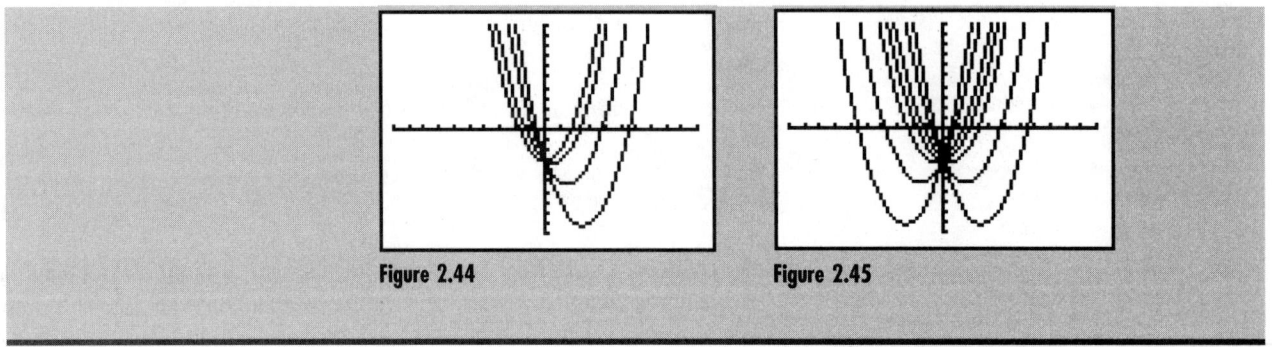

Figure 2.44   **Figure 2.45**

Here are some things to notice from looking closely at these curves. First, the coefficient b does *not* affect the following:

▸ All the parabolas have the same shape.
▸ All the parabolas open upward.
▸ All of the curves have the same y-intercept. (While this is hard to see from the graphs, the point $(0, -3)$ is on all seven graphs.)

Next, the coefficient b *does* seem to affect the location of the vertex.

▸ The vertices of the parabolas with positive values of b are to the left of the y-axis, and the vertices of those parabolas with negative values of b are to the right of the y-axis.

Actually, this last statement is related to what you know about the vertex of a parabola. Since the x-coordinate of the vertex is at $x = -\dfrac{b}{2a}$, the last conclusion is true only if a is a positive number.

▸ As the absolute value of b decreases (from -5 to -3 to -1 to 0, and from 5 to 3 to 1 to 0), the vertex moves upward on the graph.

Do you think this conclusion would be true if a is a negative number?

There is one quite surprising thing to say about this set of parabolas: All the vertices of the parabolas are on a parabola whose equation is $y = -x^2 + 0x - 3$ (as shown in Figure 2.46). Why do you think that is true?

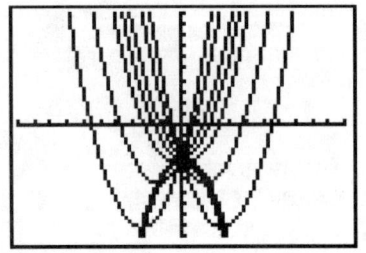

Figure 2.46

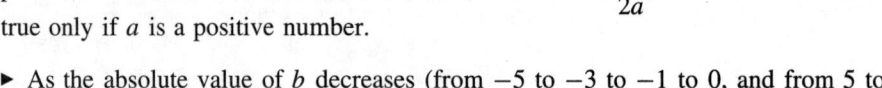

Section 2.4 Exercises

 For each function in Exercises 1–6, (a) graph the function, (b) use your calculator to estimate its roots, and (c) use the quadratic formula to determine its roots.

1. $f(x) = 9x^2 - 3x - 12$

2. $g(x) = 0.3x^2 - 4.65x - 27$

3. $h(t) = 2t^2 - 5t + 3$

4. $j(w) = 4w^2 - 13w - 35$

5. $m(n) = n^2 + 0.8n + 0.16$

6. $k(r) = 9r^2 - 6r + 1$

 In Exercises 7–10 determine the roots of the given functions.

7. $f(x) = 4(x - 5.2)(x + 7.1)$

8. $f(x) = 0.25(x + 0.51)(x - 4.37)$

9. $f(x) = 6.1(3x - 7)(2x - 5)$

10. $f(x) = (4x + 3)(5x + 12)$

For each quadratic function in Exercises 11–16, (a) calculate the x-coordinate of the vertex, (b) determine the y-coordinate of the vertex, and (c) graph the function and use Zoom and Trace to approximate the coordinates of the vertex. (Note: These are the same functions that were in Exercises 1–6.)

11. $f(x) = 9x^2 - 3x - 12$

12. $g(x) = 0.3x^2 - 4.65x - 27$

13. $h(t) = 2t^2 - 5t + 3$

14. $j(w) = 4w^2 - 13w - 35$

15. $m(n) = n^2 + 0.8n + 0.16$

16. $k(r) = 9r^2 - 6r + 1$

 In Exercises 17–19 use the lists BNCE *and* TBNCE *from Section 2.2 to answer the following questions about the second bounce of the ball you dropped in Activity 2.2.*

17. Use quadratic regression to determine a quadratic function that approximates the second bounce.

18. What are the roots of the quadratic function in Exercise 17?

19. What are the x- and y-coordinates of the vertex of the quadratic function in Exercise 17?

 In Exercises 20–22 use the data from Exercises 9–15 of Section 2.2, where you selected a portion of the results obtained from tossing a large ball straight up over the motion detector.

20. Use quadratic regression to determine a quadratic function that approximates the tossed ball.

21. What are the roots of the quadratic function in Exercise 20?

22. What are the x- and y-coordinates of the vertex of the quadratic function in Exercise 20?

23. We saw that the maximum or minimum point of the parabola occurs at the point whose x-coordinate is equal to the average of the roots—that is, when $x = -\dfrac{b}{2a}$. Does this relationship still hold true if the roots are not real numbers, as in Figure 2.40? Why or why not?

 24. Use your calculator to experiment with the family of quadratic functions $y = ax^2 + bx + c$. What effect does the coefficient a have on the shape, location, and direction (opening up or down) of the parabola? Make sure you test negative and positive values of the coefficient as well as different numerical values.

 25. Use your calculator to experiment with the family of quadratic functions $y = ax^2 + bx + c$. What effect does the coefficient c have on the shape, location, and direction (opening up or down) of the parabola? Make sure you test negative and positive values of the coefficient as well as different numerical values.

26. Write a memo to management in which you describe what you have learned about the bouncing ball and how you can use this information to design the height and time tests they requested in the Chapter Project.

2.5

Projectile Motion

In the previous sections you saw that quadratic equations are good models of data produced by motion that is influenced by gravity. Motion like this comes in many forms: bouncing balls, falling apples, thrown baseballs, kicked footballs, launched rockets, and soaring fireworks. Objects that are given an initial velocity, which can be 0, and then move through the air under the influence of gravity and air resistance are called **projectiles**. The material in this chapter can be applied to the study of projectile motion.

In Chapter 1 you learned that the coefficients of a linear function tell you something about velocity and starting point in a constant velocity experiment.

Constant Velocity and Linear Functions

If $f(t) = at + b$ is the model of a motion experiment, then a is the velocity and b is the beginning location of the object. In many of our experiments, b is the beginning distance from the motion detector.

In this section you will see what the coefficients of the quadratic function $f(t) = at^2 + bt + c$ can tell you about projectile motion.

A Falling Object

The simplest kind of projectile motion is illustrated by a falling object such as an apple dropping from a tree. A large flat object that falls fairly straight makes a good target for the motion detector. We used a book. Figure 2.47 shows how to set up the CBL experiment, and Figure 2.48 shows a graph of the data that resulted from the experiment. The distance is measured in feet and time is measured in seconds. The book was dropped from a height that was a little over 5 feet above the motion detector and was caught about 1.5 feet above it. (If you do this experiment, be sure the catcher's hands are kept out of the view of the motion detector.)

Motion
detector

Figure 2.47

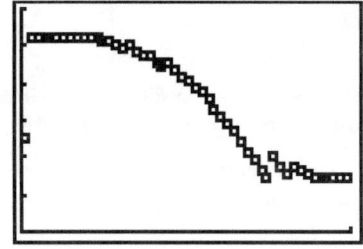

Figure 2.48

Calculator Lab 2.6

Regression for a Falling Book

Compute the quadratic regression model for a falling book experiment. You did something like this in Calculator Lab 2.3 on page 73.

Procedures
1. Conduct a falling book experiment as described above. You might want to save your data using the file names BOOK and TBOOK.
2. Select the data at the beginning and the end of the time during which the book was falling. Figure 2.49 shows the data we selected.
3. Enter the command QuadReg L1,L2 [ENTER]; Figure 2.50 shows the resulting equation.
4. Plot the graph by entering the regression equation in the Y= menu. Figure 2.51 shows the regression equation plotted with the data.

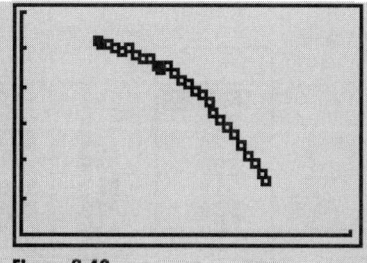

Figure 2.49

Figure 2.50

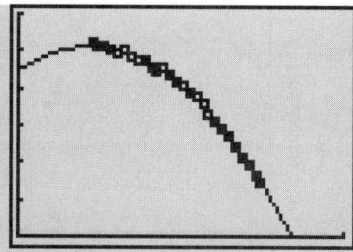

Figure 2.51

The fit for the authors' data is quite good, but you can see that you should be careful when you try to apply this model to the data. For example, if you used the regression equation to predict where the object was before it began to fall or after it had stopped, you would get incorrect answers. Figure 2.52 shows all the original data from the experiment and a quadratic regression equation for the selected portion of the data.

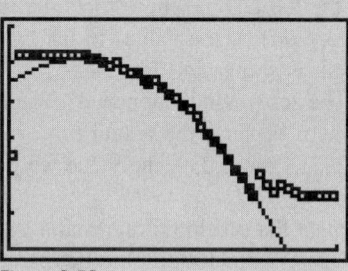

Figure 2.52

Transforming Data

Our experiment would have been more successful if we could have dropped the book at the very beginning of the experiment when $t = 0$. We can make up for this by changing the data so that it does begin at $t = 0$. In certain cases like this one, we can actually change the data from an experiment so the mathematics becomes simpler to analyze.

For example, suppose we want to run an experiment to study the rate at which boiling water cools after it is removed from the heat source. The time of day when the experiment begins is not important; but the total time for the experiment (the *elapsed time*) is important. Therefore, if the cooling begins at 2 P.M. and we measured the temperature of the water every hour thereafter, we would have to subtract 2 hours from each of the time measurements to get the elapsed time. This is demonstrated by the two drawings in Figure 2.53.

The same principle applies to the falling book. We must subtract the starting time from all the other time measurements in order to focus on the actual time that the book was in the air.

When did the book begin to fall? If we knew that, we could subtract that number from all the other time measurements. To find out, press **STAT** **ENTER** to see the lists L1 and L2, as shown in Figure 2.54. L1 is the list of the selected times and L2 is the list of the distances.

Although your selected times may be different, you can see that the first time that we selected, rounded to three decimal places, was $t = 0.230$. We need to

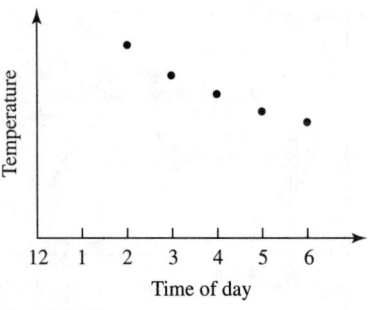

Figure 2.53

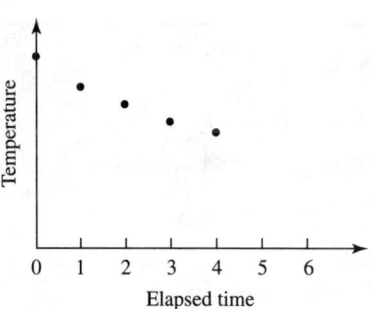

Figure 2.54

Figure 2.55

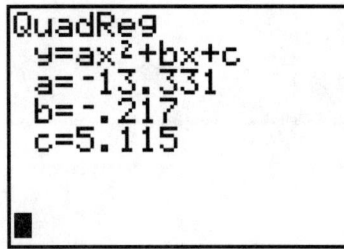

Figure 2.56

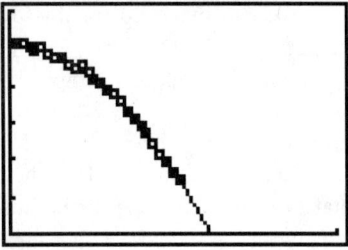

Figure 2.57

subtract that time from each of the measurements. To do that, return to the home screen and enter the command

> `2nd` [L1] `−` 0.230 `STO→` `2nd` [L1]

This command subtracts the time at which the book was dropped, about 0.230, from each item in the list `L1` and then stores the new values in the list `L1`. With this one command, every one of the values in list `L1` is transformed by subtraction.

The result is shown in Figure 2.55. Look at the first value in the new list. We expected it to be zero. What happened? The first value, $-2.870E^-4$ or -0.000287, is not quite zero because the actual time value of the first selected data point was closer to 0.229713 not 0.230, the value we got by changing the decimal accuracy to 3 places.

Now repeat the quadratic regression command you used previously. The result is shown in Figure 2.56. If you plot the regression equation you will see Figure 2.57.

There are many circumstances in which transforming data will produce more useful results. In this case we produced a quadratic model (Figure 2.56) that is more useful because it models data that starts at $t = 0$.

The quadratic regression function, $h(t) = -13.331t^2 - 0.217t + 5.115$, gives the height of a projectile t seconds after it has been dropped from a height of about 5 feet.

What can the coefficients -13.331, -0.217, and 5.115 tell us about the experiment? First think about $h(0)$, the height at $t = 0$. It is calculated as

$$h(0) = -13.331(0^2) - 0.217(0) + 5.115$$
$$= 0 - 0 + 5.115 = 5.115$$

Therefore the constant term c is equal to the distance from the motion detector at the beginning of the experiment. In Chapter 1 we found a similar result for a constant velocity motion: the constant term b in the linear function $d(t) = at + b$ is the distance from the motion detector at the beginning of the experiment.

We'll move on to a study of the meaning of the coefficients a and b later in this chapter.

Giving it a Toss

Because we simply dropped the book and did not throw it up or down, its initial velocity was zero. We turn next to projectile motion that begins with an initial velocity. In this experiment, first conducted for Exercises 9–15 of Section 2.2, we tossed a basketball a few feet off the ground. However the mathematics involved could also be used to study the height of a rocket after launch or the distance traveled by a home run in a baseball game.

Calculator Lab 2.7

Quadratic Regression for a Basketball

Fit a quadratic regression model to the motion portion of data lists TBALL (the time in seconds) and BALL (the height of the basketball, in feet, above the motion detector).

Procedures

1. Determine the values to use in the WINDOW menu. You haven't had to do this before without assistance. One way to determine the smallest and largest values to use in the WINDOW menu is to use a calculator function that reports the minimum and maximum values of a list.

   (a) From the LIST MATH submenu press `ENTER` or `1` to take the first part of the min(command to the home screen then insert the name TBALL from the `2nd` [LIST] menu and press `)` `ENTER`.
   (b) Repeat these steps for the BALL list.
   (c) Then repeat steps (a) and (b) using the max(command.

   Your results will look like Figure 2.58. Notice that the line min(LTBALL) has scrolled off the top of the screen.

Figure 2.58

2. In the WINDOW menu, enter values for Xmin and Ymin that are smaller than the minimum and values for Xmax and Ymax that are larger than the maximums. (We used 0 for Ymin to give a little more space at the bottom of the screen; this allows all the points to be visible when selecting the data.)

3. Use Select(L1,L2) to enter the command that will put your selected data into the lists L1 and L2. The graph of the selected data will look like Figure 2.59.

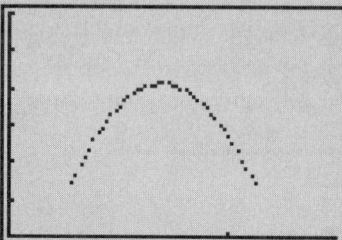

Figure 2.59

4. Transform the list of times (now stored in L1) so that it begins at $t = 0$. This time use min(L1) to transform the time data all in one step with the command

$$L1 - min(L1)\ \boxed{STO►}\ L1$$

The MODE has been changed to 3 decimal points.

As you can see from Figure 2.60, using the min(command has the added benefit of actually making the first value in the time list zero, since the *actual* value of the minimum time is subtracted not approximated.

Figure 2.60

Figure 2.61

5. The selected and transformed data is now stored in the lists L1 and L2. Do a quadratic regression to obtain a result like the one shown in Figure 2.61. Notice that the value of the coefficient c tells us that the ball was tossed from about 1.4 feet above the motion detector.

6. Store the transformed data in the lists NTBAL and NBALL. The letter N is used to indicate that the lists are *New* lists based on TBALL and BALL.

7. Plot the data and the regression function.

   Rather than type the regression equation directly into the Y= screen, we will use a method that can protect you from typing errors. Press [Y=] and place your cursor on a blank line. Press [VARS] to reach the VARIABLES menu (Figure 2.62), and [5] to select the STATISTICS menu (Figure 2.63). Press [▶] [▶] to select the EQ (Equations) menu and [ENTER] to select the first choice, RegEQ. The regression equation will appear in the Y= screen.

Figure 2.62

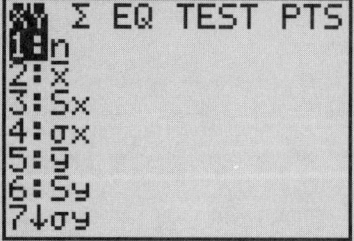

Figure 2.63

8. Press [GRAPH] and you will see the graph in Figure 2.64. (We changed the symbol used for the data to the square so you can see the difference between the data and the regression parabola.)

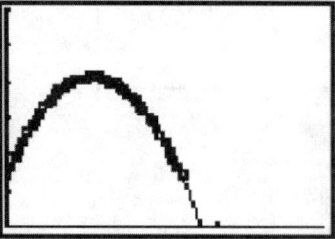

Figure 2.64

Comparing Transformed Graphs

Compare the shapes of the data plots in Figures 2.59 and 2.64. The shapes are the same, but the second graph has been moved horizontally to the left. This shift is caused by subtracting the same number from every data point used as an x-coordinate. You will see the same relationship between Figures 2.51 and 2.57. The regression equations of these plots, shown in Figure 2.50 and Figure 2.56, are

$$y = -13.331x^2 + 5.915x + 4.460$$

and

$$y = -13.331x^2 - 0.217x + 5.115$$

If you compare these equations you'll see that the value of the coefficient a is the same in both cases. Since you have learned that the coefficient a determines the shape, you'd expect that these coefficients would not change when the data is transformed as we have done.

Transforming Data Points Horizontally and Vertically

When the same positive value is subtracted from each of the x-coordinates of the data, the graph is unchanged in shape but shifted horizontally to the left.

When the same positive value is subtracted from each of the y-coordinates of the data, the graph is unchanged in shape but shifted vertically down.

Another Look at Coefficients a and b

The two experiments we have analyzed in this section have the shared attribute of an object moving in air under the influence of gravity. In the first example the book was merely dropped, so the beginning velocity was zero. In the second example we tossed the ball in the air, so the beginning velocity was different from zero.

Do you think the beginning velocity of the basketball was a positive number or a negative number? Think about it: The toss made the ball go up, and up is in the positive direction on a coordinate system. Therefore the beginning velocity must be positive in the tossed-basketball example.

Now compare the regression equations in Figure 2.56 and Figure 2.61:

Dropped book: $h(t) = -13.331t^2 - 0.217t + 5.115$

Tossed basketball: $h(t) = -15.427t^2 + 12.861t + 1.439$

You know that the coefficient c gives the height of the projectile at the beginning of the motion.

The only remaining positive coefficient is $b = 12.861$ for the tossed basketball. For the falling book, the value of b is close to zero (-0.217). It would appear that b is related to the velocity. In the next section we will explore these values in more detail. For now we will tell you that the coefficient b estimates the beginning velocity of the projectile. Therefore the basketball was tossed in the air with a velocity of about 12.9 feet per second. Since the book was dropped, its initial velocity should have been 0. There are several possible reasons why it is not 0. What reasons can you give?

The values of a are different. In a perfect experiment these values would be approximately -16 because a is related to the effect of gravity, which should be the same for any experiment carried out on planet Earth.

Modeling Motion That Is Affected by Gravity

Motion affected by gravity can be described by the function

$$h(t) = a\,t^2 + bt + c$$

where $h(t)$ is the height of the projectile at any time t after the release of the projectile.

a is the effect of gravity. The value of a on Earth is about -16 if the height is measured in feet and time is measured in seconds and -4.9 if height is in meters and time is in seconds. a is negative because height above the ground is taken to be positive and the force of gravity works to *decrease* the height.

b is the velocity of the projectile at $t = 0$ (the beginning velocity) and c is the height of the projectile at $t = 0$ (the beginning height).

Example 2.20

A ball is tossed straight up at 30 ft/sec from a height of 6 feet. (a) How long does the ball take to reach its highest point? (b) How high does the ball go? (c) After how many seconds does the ball hit the ground?

Solutions   From the given information we see that the model of this motion is given by the function $h(t) = -16t^2 + 30t + 6$.

(a) Remember, the vertex of the parabola occurs at $t = -\dfrac{b}{2a} = -\dfrac{30}{-32} = 0.9375$. Therefore the highest point is reached at $t \approx 0.9375$ sec.

(b) The highest point will be at $h(0.9375) = -16(0.9375^2) + 30(0.9375) + 6 = 20.0625$ feet.

(c) To determine when the ball hits the ground we must solve the equation $h(t) = 0$, or $-16t^2 + 30t + 6 = 0$. Using the quadratic formula we have

$$t = \frac{-30 + \sqrt{30^2 - 4(-16)(6)}}{2(-16)}$$

$$= \frac{-30 + \sqrt{1284}}{-32} \approx -0.18$$

The other value of t is $t = \dfrac{-30 - \sqrt{1284}}{-32} \approx 2.06$. Thus the ball hits the ground at either -0.18 sec or at 2.06 sec. Since a negative time is *before* the beginning of the experiment, the ball must hit the ground after 2.06 seconds. ∎

Example 2.21

Answer the questions of the previous example assuming that the ball is thrown in the same way, but this time on the Moon. The Moon's gravity is approximately $\frac{1}{6}$ of Earth's, so the value of a should be $-\frac{16}{6} \approx -2.67$.

Solution   The difference is the value of the coefficient a. The model of this motion is given by the function $h(t) = -2.67t^2 + 30t + 6$.

(a) The highest point is reached at $t = -\dfrac{b}{2a} = -\dfrac{30}{2(-2.67)} \approx 5.62$ sec.

(b) The height at the highest point is $h(5.62) = -2.67(5.62)^2 + 30(5.62) + 6 \approx$ 90.27 feet.

(c) As in Example 2.20, to determine when the ball hits the ground we must solve the equation $-2.67t^2 + 30t + 6 = 0$. Using the quadratic formula produces

$$t = \frac{-30 + \sqrt{30^2 - 4(-2.67)(6)}}{2(-2.67)}$$

$$= \frac{-30 + \sqrt{964.08}}{-5.34} \approx -0.20$$

The other value is $t = \dfrac{-30 - \sqrt{964.08}}{-5.34} \approx 11.43$.

The answer is 11.45 seconds, since a negative time is *before* the beginning of the experiment.   ∎

Thinking about the Quadratic Model

There are two important questions about any mathematical model of real data to consider: "How accurate is it?" and "For what values is it true?"

How Accurate Is This Model?

The accuracy question asks how much we should trust the results of our models. You have seen that quadratic functions fit the projectile motion data quite well. However, like all mathematical models, the quadratic model of motion is not strictly accurate because it ignores some other influences on motion. The most important influence that is ignored by this model is the effect of air friction. Air friction slows down all objects, but it slows down some more than others. Because of the omission of the effect of air friction, the quadratic model would do a better job of describing the motion of a falling apple than that of a falling feather.

For What Values Is the Model True?

In the two previous examples we saw that the quadratic equation produced two values for the time at which the projectile reached the ground: one was negative and one was positive. Since this model assumes that the experiment starts at $t = 0$, we must reject any values of t that are negative. This idea is explored further in the following example.

Example 2.22

A ball is tossed straight up at 30 ft/sec from a height of 6 feet. How high is the ball 5 seconds later?

Solution   The model of this motion is given by the function $h(t) = -16t^2 + 30t + 6$. At $t = 5$ we have

$$h(5) = -16(5^2) + 30 \cdot 5 + 6$$
$$= -400 + 150 + 6 = -244$$

Thus at $t = 5$ the height of the ball is -244 feet.

What could a height of -244 feet mean? To answer this question we must return to the definition of the model itself. $h(t)$ is the height above the ground and $h(t) = -16t^2 + 30t + 6$ describes motion in the air. A height of -244 feet is 244 feet below the ground. This answer doesn't make sense unless the projectile can burrow through earth the same way it moves through the air (or it was thrown off the edge of a cliff). But it is impossible for a projectile to burrow through earth as if it were air, and we are not given any reason to believe that the thrower was at the edge of a cliff. The model must be used only for the time during which the projectile is in the air. That is, $h(t)$ must be a positive number.

Therefore the following restrictions must be added to the model of motion of a projectile:

The model is defined for $t \geq 0$ and $h(t) \geq 0$.

In the mathematical models we will discuss in future chapters, we will always state the restrictions on the models in this way.

Compare the quadratic model of motion in this chapter with the linear model of constant velocity motion in Chapter 1. Notice that once again the coefficient of t relates to the velocity. You already know from general experience that the velocity of an object changes as it moves under the influence of gravity. In the next section we will study the changing velocity of a projectile as it goes up and then down.

Section 2.5 Exercises

Exercises 1–4 require a motion detector and a CBL connected to a calculator with the program BALLDROP. *Remember that the motion detector does not record distances less than 1.5 feet. You may have to repeat some of the experiments until you get satisfactory results. Use the* BALLDROP *program to record your data. Follow the directions of either the book drop or basketball toss for selecting and transforming the data and for calculating the quadratic regression.*

 1. Drop a large ball such as a beach ball or soccer ball from a height of about 6 feet. Use your results to answer the following questions.

    (a) What is the regression equation for this experiment?

    (b) Sketch a graph of the transformed data and the regression function.

    (c) How high was the ball when it was dropped?

    (d) What was the ball's initial velocity?

    (e) What was the gravity constant?

    (f) How high was the ball 1 sec after it was dropped?

    (g) Repeat the experiment with a smaller ball such as a baseball and answer questions (a)–(f) for this ball.

    (h) Which seem to affect the *gravity constant*—the size of the ball, the weight of the ball, the ball's density, or something else? Explain your answer.

 2. Inflate three similar balloons to different sizes. Drop each inflated balloon from a height of about 6 feet. To help the balloon fall straight, attach a small paper clip to the nozzle and hold the nozzle down when you drop the balloon. For each balloon use the BALLDROP program to record your data. Use your results to answer the following questions for *each* balloon.

(a) What is the regression equation for this experiment?

(b) Sketch a graph of the transformed data and the regression function.

(c) How high was the balloon when it was dropped?

(d) What was the balloon's initial velocity?

(e) What was the gravity constant? Why isn't it −16?

(f) How high was the balloon 1 sec after it was dropped?

(g) How does the amount of air in the balloon affect any of the values of a, b, and c in the regression function $f(t) = at^2 + bt + c$?

3. Use the pattern in Figure 2.65 to make three paper helicopters from strips of paper about 2 inches wide and 11 inches long. (Cut on the solid lines and fold on the dotted. Secure the bottom with a piece of tape.) Design each helicopter so it has a different wing span from the other two. Drop each helicopter from a height of about 6 feet. For each helicopter use the BALLDROP program to record your data. Use your results to answer the following questions for *each* helicopter.

(a) What is the regression equation for each experiment?

(b) Sketch a graph of the transformed data and the regression function.

(c) How high was the helicopter when it was dropped?

(d) What was the helicopter's initial velocity?

(e) What was the gravity constant? Why isn't it −16?

(f) How high was the helicopter 1 sec after it was dropped?

(g) Does the wing span of the helicopter affect any of the values of a, b, and c in the regression function $f(t) = at^2 + bt + c$? Explain your answer.

(h) The helicopter drops a short distance before it begins to spin. How does that affect the results?

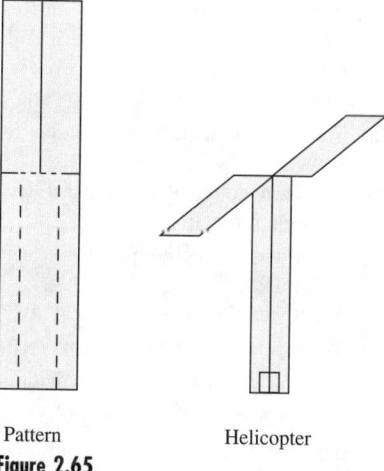

Pattern   Helicopter

Figure 2.65

4. Throw a large ball such as a beach ball or soccer ball into the air as in the basketball toss. This ball, however, should be lighter than a basketball. Use your results to answer the following questions.

(a) What is the regression equation for this experiment?

(b) Sketch a graph of the transformed data and the regression function.

(c) What was the highest point the ball reached after it was tossed?

(d) What was the ball's initial velocity?

(e) What was the gravity constant? Why isn't it −16?

(f) How high was the ball 1 sec after it was thrown?

Exercises 5–6 require a ramp, a motion detector, and a CBL connected to a calculator with the program HIKER. *Remember that the motion detector does not record distances less than 1.5 feet. You may have to repeat some of the experiments until you get satisfactory results. Use the* HIKER *program to record your data. Follow the directions of either the book drop or basketball toss for selecting and transforming the data and for calculating the quadratic regression.*

[To make the ramp you can put some books under the legs at one end of a 6–8-foot-long table as shown in Figure 2.66. The motion detector is placed at one end of the ramp. Two boards have been laid on the ramp to keep the moving object in front of the motion detector.]

Motion detector

Figure 2.66

5. Start the motion detector and roll a small ball such as a racquet ball up the ramp so that it almost reaches the end and then rolls back down the ramp.

(a) What is the regression equation for this experiment?

(b) Sketch a graph of the transformed data and the regression function.

(c) What was the highest point the ball reached after it was rolled?

(d) What was the ball's initial velocity?

(e) What was the gravity constant? Why isn't it -16?

(f) How far was the ball from the motion detector 1 sec after it was rolled?

6. Start the motion detector and roll a toy car up the ramp so that it almost reaches the end and then rolls back down the ramp. Make sure you use a toy car large enough for the motion detector to "see."

(a) What is the regression equation for this experiment?

(b) Sketch a graph of the transformed data and the regression function.

(c) What was the highest point the car reached after it was rolled?

(d) What was the car's initial velocity?

(e) Why isn't the gravity constant -16?

(f) How far was the car from the motion detector 1 sec after it was rolled?

7. *Physics* A missile is fired vertically into the air. The height h, in feet, of the missile above the ground at time t, in seconds, is given by $h(t) = -16t^2 + 600t + 50$.

(a) When does the missile hit the ground?

(b) What is the initial velocity of the missile?

(c) What is the height of the launch pad?

(d) How many seconds after launch does the missile reach the vertex of its flight?

(e) What is the height of the vertex?

8. *Physics* A ball is thrown almost straight up into the air from the top of a building. The height h, in feet, of the ball above the ground at time t, in seconds, is given by $h(t) = -16t^2 + 64t + 192$.

(a) When does the ball hit the ground?

(b) What is the initial velocity of the ball?

(c) What is the height of the building?

(d) How many seconds after it is thrown does the ball reach the vertex of its flight?

(e) What is the height of the vertex?

9. *Physics* A ball is thrown vertically upward from the top of a building that is 234 feet tall.

(a) If the initial velocity is 56 ft/sec, write the quadratic function that describes the height of the ball t seconds after it is thrown.

(b) How long does it take for the ball to return to the top of the building?

(c) When does the ball reach its highest point and how high is that?

10. *Physics* The ball in Exercise 9 missed the building when it came down:

(a) How long from when it is thrown does it take for the ball to hit the ground?

(b) When does the ball reach its highest point and how high is that?

11. The World Trade Center is 411 m high. How long will it take for a ball dropped from the top of the World Trade Center to reach the ground?

12. *Physics* A ball is thrown downward from the top of the 411-m-high World Trade Center with an initial velocity of 1.2 m/s.

(a) Write the quadratic function that describes the height of the ball t seconds after it is thrown.

(b) How long does it take for the ball to hit the ground?

13. *Physics* A ball is thrown upward from the top of the 411-m-high World Trade Center with an initial velocity of 1.2 m/s.

(a) Write the quadratic function that describes the height of the ball t seconds after it is thrown.

(b) How long does it take for the ball to hit the ground?

(c) When does the ball reach its highest point and how high is that?

14. Figure 2.67 is the graph of $f(x) = x^2 - 2x - 3$. Without using your calculator, draw the graph of $f(x - 4)$. Check your answer by graphing $f(x - 4) = (x - 4)^2 - 2(x - 4) - 3$ on your calculator.

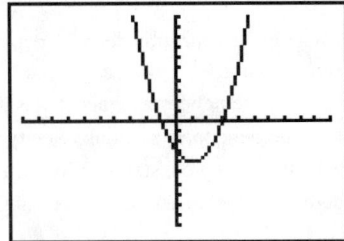

Figure 2.67

15. Figure 2.68 is the graph of $g(x) = x^2 + 2x - 8$. Without using your calculator, draw the graph of $g(x - 3)$. Check your answer by graphing $g(x - 3) = (x - 3)^2 + 2(x - 3) - 8$ on your calculator.

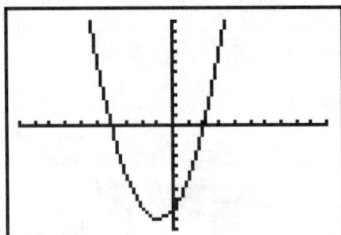

Figure 2.68

16. Figure 2.69 is the graph of $h(x) = x^2 - 3x$. Without using your calculator, draw the graph of $h(x) - 5$. Check your answer by graphing $h(x) - 5 = x^2 - 3x - 5$ on your calculator.

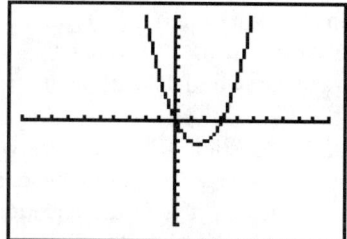

Figure 2.69

17. Figure 2.70 is the graph of $k(x) = 0.1x^4 + 0.1x^3 - 1.2x^2$. Without using your calculator, draw the graph of $k(x)+4$. Check your answer by graphing $k(x)+4 = 0.1x^4 + 0.1x^3 - 1.2x^2 + 4$ on your calculator.

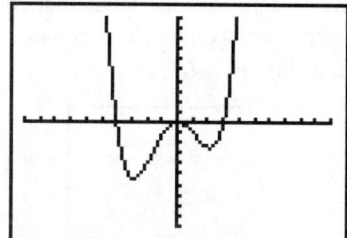

Figure 2.70

18. *Sports technology* The data in the table below shows results in the Olympic men's 50-meter freestyle swimming event. (This event was not held from 1908–1984.)

Year	1904	1988	1992	1996
Time (sec)	28	22.14		21.91

(a) Compute the linear regression equation with decimal place accuracy set to 4 decimal places.

(b) Plot the data and two regression lines: one with 3 decimal places and one with 4 decimal places. Your graph should look something like Figure 2.71.

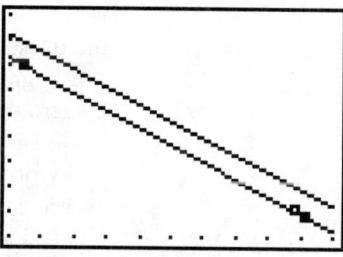

Figure 2.71

The lower line in the figure is the one written with 3 decimal place accuracy and the upper line is written with 4 place accuracy. Normally there should not be such a large difference in the accuracy of two models that differ only in the fourth decimal place. The reason for such a visible difference is that essentially the model is starting at the y-intercept, which in this case is where YEAR = 0, or 2000 years ago. In that great a stretch of time, two lines that are very close together will move apart, as you see in the figure. The solution is to transform the data so that the starting year 1904 is transformed to year 0.

(c) Do that transformation and plot the linear regression lines with 3 and 4 decimal places of accuracy.

(d) Use the four different linear regression equations you have to predict the 1996 winning time in this event. The actual time in 1996 was 22.13 seconds.

(e) Write your opinion about the accuracy of this data transformation.

2.6

Velocity at an Instant

Velocity at a particular instant is a concept that mystified the greatest mathematicians in ancient Greece and continued to pose a problem for another 2000 years. They could compute average velocity during a time interval, just as you can, using the formula

$$v = \frac{\Delta d}{\Delta t}$$

where Δd is the change in distance during the time interval t.

But the Greeks, like us, thought of an instant as a time interval of zero duration, that is, $\Delta t = 0$. But if $\Delta t = 0$, the velocity v cannot be computed because the equation would require dividing by zero. And worse, if $\Delta t = 0$, what can you say about the amount of distance covered in that time interval? It would have to be zero as well, so $\Delta d = 0$. Therefore the equation for v becomes 0 divided by 0. No wonder the smartest people of the ancient world had difficulty with this concept.

When you are driving in a car and ask the question "How fast are we going *right now*?" you are asking a similar question. The speedometer, of course, answers the question. But how can it? If velocity is defined as it is in the equation above and "right now" means *at this instant*, then you too are asking for division of zero by zero.

The bouncing ball memo that began this chapter mentioned the rebound velocity of the ball, which is the velocity it has at the instant it comes off the floor, or velocity at $t = 0$. In this section you'll learn how to give an estimate for the rebound velocity.

We'll begin the investigation by returning to the problem of projectile motion, for which we have a mathematical model. We'll start with the mathematical model of the thrown ball you worked with in the previous section. The equation of the model is shown in Figure 2.72 and the graph in Figure 2.73.

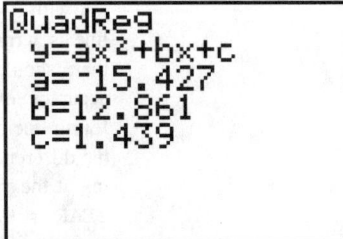

Figure 2.72

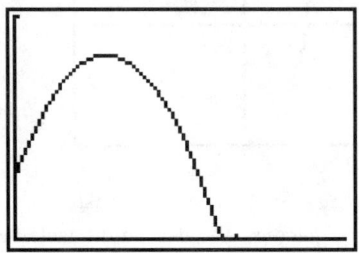

Figure 2.73

Example 2.23

Write a description of the motion of the tossed ball that is graphed in Figure 2.73. Visualize the ball going up and then down. Use phrases or words like *slows down*, *speeds up*, *positive velocity*, *negative velocity*, *gravity*, and *stops*.

Solution   The ball is tossed straight up in the air with a positive velocity. It gradually slows down under the influence of gravity until it reaches its highest point. At the highest point its velocity is zero for an instant; then it begins to fall with a negative velocity. The ball falls faster and faster; that is, the absolute value of the velocity increases until the ball hits the ground—or is caught—and stops moving.

In describing the motion of a tossed ball you thought about velocity as it changes over time. Next we will look at a very small time interval and think about how the velocity changes in that interval. In order to get the most accuracy as we look at tiny intervals of time, we first need to re-do the regression using the highest accuracy a graphing calculator offers, indicated by the Float (for floating point setting) in the MODE menu. ■

Calculator Reminder: Floating Point Setting

The Floating Point setting will display up to 10 digits on the calculator. Using a floating point is usually preferable to setting the number of decimal places to

9 because it handles whole numbers and decimals more efficiently, as shown below.

Calculation	Answer with Floating Point	Answer with 9 Decimal Place Accuracy	Answer with 5 Decimal Place Accuracy
2 ➕ 3	5	5.000000000	5.00000
3 ➗ 4	0.75	0.750000000	0.75000
2000 ➗ 7	285.7142857	285.7142857	285.71429

Calculator Lab 2.8

Quadratic Regression Using the Floating Point Setting

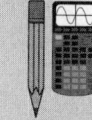

Plot the quadratic regression model for the tossed basketball using the floating point setting on your graphing calculator.

Procedures

1. Change the decimal setting to floating point and produce the regression equation

   ```
   QuadReg NTBALL, NBALL
   ```

2. Press **ENTER** to execute the QuadReg command. The result is shown in Figure 2.74.

```
QuadReg
 y=ax²+bx+c
 a=-15.42704156
 b=12.86050589
 c=1.439379164

■
```

Figure 2.74

3. Plot the graph change using the WINDOW settings shown in Figure 2.75. Then use the procedure shown in Calculator Lab 2.7 (on page 97) to enter the regression equation into the Y= screen, as shown in Figure 2.76. The graph of the regression equation is shown in Figure 2.77.

```
WINDOW
 Xmin=0
 Xmax=1
 Xscl=.1
 Ymin=0
 Ymax=5
 Yscl=1
 Xres=1
```

Figure 2.75

```
Plot1 Plot2 Plot3
\Y1◼-15.42704155
6683X^2+12.86050
5894166X+1.43937
91640594
\Y2=
\Y3=
\Y4=
```

Figure 2.76

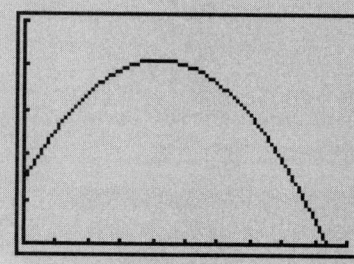

Figure 2.77

The graph and the equation represent the *mathematical model* of the flight of the ball. We are no longer dealing with the data from which this model was developed. However conclusions that we reach from the model can be applied to the motion of the actual ball, as you will see.

4. Take a close look at the graph when $t = 0.25$. Zoom in on the point $t = 0.25$ in the graph in Figure 2.77 until the graph looks straight. Figure 2.78 shows an intermediate step and Figure 2.79 shows the solution.

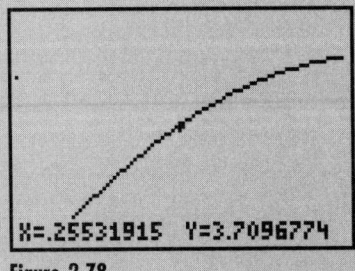

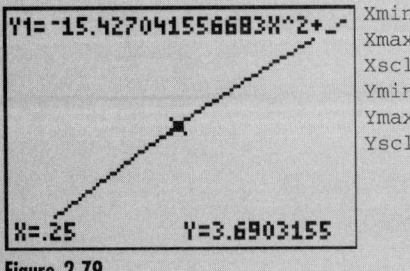

```
Xmin = 0.21875
Xmax = 0.28125
Xscl = 0.1
Ymin = 3.533266129
Ymax = 3.845766129
Yscl = 1
```

Figure 2.78   **Figure 2.79**

5. Think about the "almost straight" line. This line is a distance vs. time graph, so we know that the velocity is "almost constant" during that time interval. What is that velocity? Can you guess whether it is positive or negative? Write your conclusions in a sentence or two.

Example 2.24

Compute the velocity between two points in time on the zoomed "almost straight" graph in Figure 2.79.

Solution   We need to know the coordinates of two nearby points on the graph. The slope of the line between these points will be equal to the average velocity in this time interval.

Press **TRACE** and you will see one point on the graph. Write the x- and y-coordinates of this point. Ours were $(0.25, 3.6903155)$. Now press **▶** or **◀** a few times to reach a new point. Write the new x- and y-coordinates. Ours were $(0.25083112, 3.6945826)$.

The slope of the line between these two points will be an estimate of the velocity between those two times. The slope of our line was

$$\frac{3.6945826 - 3.6903155}{0.25083112 - 0.25} = \frac{0.0042671}{0.00083112} \approx 5.134$$

The answer is positive because the ball is going *up*. The average velocity between the two points $(0.25, 3.6903155)$ and $(0.25083112, 3.6945826)$ is about 5.1342 ft/sec. ■

Look at Δt. It is about 0.0008, or eight ten-thousandths of a second. This is a very small time interval, and in this interval the ball traveled a very small distance, about 0.004 feet (slightly more than $\frac{1}{250}$ of a foot or $\frac{1}{20}$ an inch). But the velocity, about 5-feet per second, was not very small; it was about 3.5-miles per hour, a brisk walking speed.

How is the slope of the "almost straight" portion of the parabola related to the graph of the parabola itself? To answer this question, we will plot the graph of the line that passes through the points you used in the previous example.

We have the slope, so we can use either of the points with the point-slope form of the equation of a line. Now that we have completed the zooming, we can return to 3 decimal place accuracy. Taking the first point we have $y = 5.134(x - 0.25) + 3.690$.

When you graph this line you won't see much difference on the screen because the curve is almost straight. The line almost matches the curve. If you zoom out to get a better look, your graph will look something like Figure 2.80. This figure has the window settings shown in Figure 2.75.

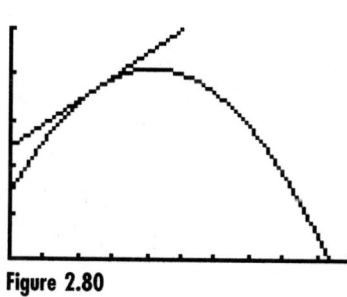

Figure 2.80

This line has an important relationship with the curve. As you see it in Figure 2.80, the line looks like a tangent to the curve. The everyday understanding of tangent is that close to a point on a curve a tangent line touches the curve just once. We know that the line in Figure 2.80 is *not* a tangent because it goes through two points on the curve. A line that goes through two points on a curve is called a **secant**. However it is the line's closeness to being a tangent that is of interest here.

Remember that the slope of the secant we constructed is an approximation of the tangent. Its slope is equal to the average velocity between two points that are very close together. Now imagine that the points we chose were not 0.0008 seconds apart but 800 times closer, or a millionth of a second (a microsecond) apart. A microsecond can be written as either 1 μs, 10^{-6} second, or 0.000001 second.

Example 2.25

Compute the average velocity of the basketball in the microsecond that immediately follows the time $t = 0.25$.

Solution   We know that when $t = 0.25$ then $h = 3.6903155$. Since $\Delta t = 0.000001$, the time for the next point is $t = 0.250001$. What is the value of h at this time? To find out we compute the value of the quadratic regression line, stored in Y1, for $x = 0.250001$. The value of h when $t = 0.250001$, as shown in Figure 2.81, is 3.690320687.

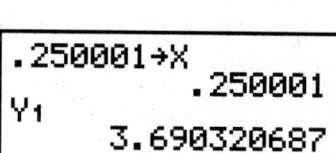

Figure 2.81

Now we have all the numbers necessary to compute the average velocity in the interval between $t = 0.25$ and $t = 0.250001$, one microsecond later. The slope of the secant through these two points is

$$\frac{3.690320687 - 3.6903155}{0.000001} = \frac{0.000005187}{0.000001} = 5.187$$

Thus the average velocity from $t = 0.25$ to $t = 0.250001$ is 5.187 ft/sec.

As in the previous example, this result is quite close to 5 feet per second. We reduced the time interval by a factor of 800, yet the average velocity in the interval changed very little. This small change should not be surprising, since velocity is the slope of a line between two points in a distance vs. time graph, and we say that the parabola representing this motion is "almost straight," which indicates very little change in velocity.

We are getting closer to the meaning of velocity at an instant. We can give an approximate answer to the question by taking a very small time interval and estimating the average velocity in that interval. In the next example you will see more clearly what this means. ∎

Example 2.26

A ball is thrown up in the air from a distance of 8 feet above the ground. If the ball has an initial velocity of 24 ft/sec and $h(t)$ is the height in feet at any time t (in seconds) after it is thrown, then $h(t) = -16t^2 + 24t + 8$. Estimate the velocity at the instant $t = 0.5$ by filling in the table below.

Change in Time $= \Delta t$	Change in Height $= \Delta h$	Average Velocity $= \dfrac{\Delta h}{\Delta t}$
0.1 ($t = 0.5$ to $t = 0.6$)		
0.01 ($t = 0.5$ to $t = 0.51$)		
0.001 ($t = 0.5$ to $t = 0.501$)		
0.0001 ($t = 0.5$ to $t = 0.5001$)		
0.00001 ($t = 0.5$ to $t = 0.50001$)		

Solution   First evaluate $h(0.5)$. You can use your calculator to determine that $h(0.5) = 16$.

For $\Delta t = 0.1$ you must compute $h(0.6)$, since $0.5 + \Delta t = 0.5 + 0.1 = 0.6$. Evaluating we find that $h(0.6) = 16.64$. Thus

$$\Delta h = h(0.6) - h(0.5)$$
$$= 16.64 - 16 = 0.64$$

Enter this value in the empty cell at the top of the middle column. Thus the average velocity in the top line of the table is $\dfrac{\Delta h}{\Delta t} = \dfrac{0.64}{0.1} = 6.4$ ft/sec.

Continuing in this way for the other values of Δt, we obtain the answers shown in the table below.

Change in time $= \Delta t$	Change in height $= \Delta h$	Average velocity $= \dfrac{\Delta h}{\Delta t}$
0.1 ($t = 0.5$ to $t = 0.6$)	0.64	6.4 ft/sec
0.01 ($t = 0.5$ to $t = 0.51$)	0.0784	7.84 ft/sec
0.001 ($t = 0.5$ to $t = 0.501$)	0.007984	7.984 ft/sec
0.0001 ($t = 0.5$ to $t = 0.5001$)	0.00079984	7.9984 ft/sec
0.00001 ($t = 0.5t = 0.5$ to $t = 0.50001$)	0.0000799984	7.99984 ft/sec

You can see that as the time interval shrinks down, the average velocity near $t = 0.5$ seconds seems to get closer and closer to 8 ft/sec. ■

In the last two examples you have seen that for smaller and smaller time intervals the average velocity during these time intervals changed very little. Perhaps you are seeing where this is taking us. Although we cannot compute the velocity *at* an instant, we can compute the average velocity *near* that time. In the ball toss experiment, for example, as we get very close to $t = 0.25$, the value of the average velocity is about 5 feet per second. This value is approximately equal to the slope of the line we drew in Figure 2.80, a line which is approximately tangent to the curve. This discussion leads to the following statement, which will be clarified further throughout this book.

Velocity at an Instant

The velocity at a particular point in time (an instant) is approximately equal to the slope of a secant drawn between the point we are interested in and a point very close to it. This line is very nearly a tangent to the curve at the point in time at which we want to know the velocity.

That is, the slope of the line in Figure 2.80 is approximately equal to the velocity represented by the parabola at that particular point in time.

Return to the Bouncing Ball

The previous discussion about velocity at an instant, which substitutes approximate values for specific values, may not be clear yet. However we are knocking at the door of some very sophisticated concepts. We'll return to this discussion later. For now let's go back to the bouncing ball problem and use these ideas to estimate the velocity of the ball as it bounces from the floor—the rebound velocity.

Example 2.27

Estimate the velocity at the beginning of the first bounce of the experiment represented by the data TBNCE and BNCE.

Solution   You did similar work in the previous section of this chapter. The data is stored in lists named TBNC1 and BNCE1, and the graph looks like Figure 2.82. In Figure 2.83 you can see how the data looks with different settings in the WINDOW menu.

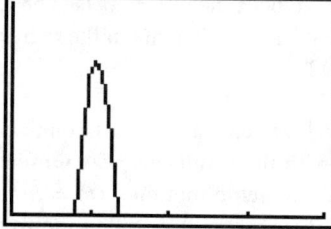

Figure 2.82

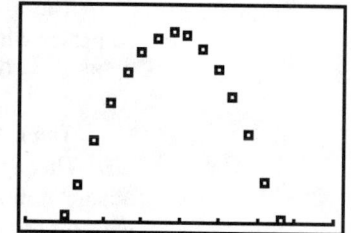

Figure 2.83

Transform the data so that the time begins at $t = 0$ and produce a quadratic regression function $h(t)$ for the transformed data.

Using floating point accuracy we get the regression model written to 5 decimal places:

$$h(t) = -4.59743t^2 + 2.58503t - 0.00992$$

Remember that $h(t)$ gives the height in meters in this experiment.

Since we want the velocity when the ball leaves the ground, we must choose two points on the parabola with one at $t = 0$ and the other at a time very shortly following.

We selected a point 0.0001 away from $t = 0$. That is, you are using the points $(0, h(0))$ and $(0.0001, h(0.0001))$, as shown in Figure 2.84 and Figure 2.85. The rectangle in Figure 2.84 shows the portion that is enlarged in Figure 2.85. Notice that while it may not appear as if the parabola is below the x-axis for positive values, the enlarged graph in Figure 2.85 shows that this is the case.

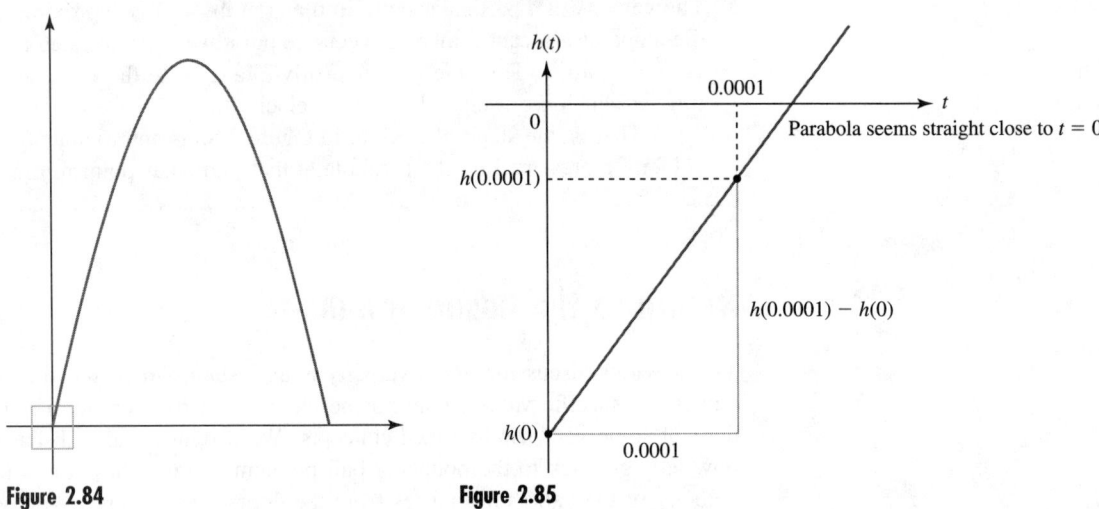

Figure 2.84   **Figure 2.85**

The average velocity near $t = 0$ is an estimate of the actual velocity at $t = 0$. Using the regression model $h(t) = -4.59743t^2 + 2.58503t - 0.00992$ produces

$$\frac{\Delta h}{\Delta t} = \frac{h(0.0001) - h(0)}{0.0001 - 0}$$

$$= \frac{-0.009661543 - (-0.00992)}{0.0001}$$

$$\approx \frac{0.0002584570}{0.0001} = 2.584570$$

The average velocity near $t = 0$ is 2.584570 m/s. Notice how this velocity compares with one of the coefficients in the regression equation $h(t) = -4.59743t^2 + 2.58503t - 0.00992$. ■

The estimated velocity at $t = 0$ is quite close to the value of the coefficient of t. This agrees with the result that was presented earlier: The coefficient b in the quadratic model of projectile motion $f(t) = at^2 + bt + c$ is an approximation of the velocity at $t = 0$. This velocity is called the **initial velocity**. Since we are looking at a single bounce, the initial velocity is equal to the velocity at which it comes up from the floor: the rebound velocity.

Section 2.6 Exercises

1. *Physics*
   (a) Estimate the velocity at $t = 0$ for the basketball toss. Make sure you have transformed the data before you determine the quadratic regression.

   (b) Compare the result with the coefficient b in the quadratic model.

2. Write a report to management telling how to estimate the rebound velocity of a ball.

3. Use the regression formula from Calculator Lab 2.3 to estimate the velocity with which the ball hits the ground at the end of the first bounce. First you must find out at what time the ball hit the ground. This is approximately the last value of t in the transformed list L_1. Then pick a point near, but before, that last point and compute the slope between these two points.

4. Use the lists BNCE and TBNCE from Section 2.2 to answer the following questions for the second bounce of the ball you dropped in Activity 2.2.

(a) What is the quadratic regression equation when the data are transformed so that the bounce begins at $t = 0$?

(b) Estimate the velocity at $t = 0$.

(c) How does the result in (b) compare with the coefficient b in the quadratic model from (a)?

(d) Estimate the velocity when the ball strikes the ground at the end of the second bounce.

5. In the Chapter Project the management asked you to design a *height test* so they would know if a ball dropped from a height of 8 feet had a rebound velocity of 20 ft/sec. Write a report to management describing how you would design such a height test.

6. In the Chapter Project the management asked you to design a *time test* so they would know if a ball dropped from a height of 8 feet had a rebound velocity of 20 ft/sec. Write a report to management describing how you would design such a time test.

●●●●●●●●● Chapter 2 Summary and Review

Topics You Learned or Reviewed

▶ The language and symbolism of functions and how to evaluate linear and quadratic functions.

▶ A quadratic equation is a mathematical model for motion influenced by gravity.

▶ A quadratic equation can be solved by numerical or graphical estimates, or the quadratic formula.

▶ The graph of a quadratic function is called a parabola

▶ The graph of a quadratic function intersects the x-axis at each of its real roots.

▶ A quadratic function has a factor of $x - r$ for each of its roots.

▶ The coefficients of a quadratic function $f(x) = ax^2 + bx + c$ tell the following about its graph:

  • If $a > 0$, then the graph opens upward; if $a < 0$, the graph opens downward. The value of a also affects the width of the graph. If $|a| > 1$,* the graph is narrower than the graph of $y = x^2$; if $|a| < 1$, the graph is wider than the graph of $y = x^2$.

  • If $a > 0$ and $b > 0$, then the vertex of the parabola is to the left of the y-axis; if $a > 0$ and $b < 0$, then the vertex of the parabola is to the right of the y-axis. The opposite is true when $a < 0$.

  • c is the y-intercept of the graph.

▶ If the graph of $f(x) = ax^2 + bx + c$ models projectile motion, then a represents the influence of gravity (or any constant force), b the initial velocity, and c the initial height of the projectile.

▶ Quadratic regression can be used to approximate the function for a set of data points for some motion influenced by gravity.

▶ The instantaneous velocity of an object at some particular time can be estimated by the slope of a secant to a smooth graph of the object's motion through two points on the graph near the particular time.

---

*$|a| > 1$ means $a > 1$ or $a < -1$.

Review Exercises

In Exercises 1–2 (a) write a description of each of the graphs, using the words slope *and* velocity *in your description, and (b) use the* CBL *and the* HIKER *program to walk in such a way as to produce graphs similar to the graphs sketched below.*

1.
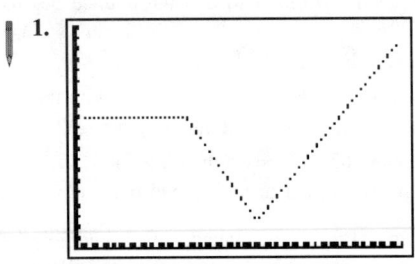

2.
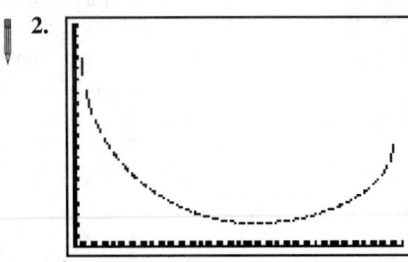

In Exercises 3–4 compute the average velocity from the distance vs. time graph.

3.
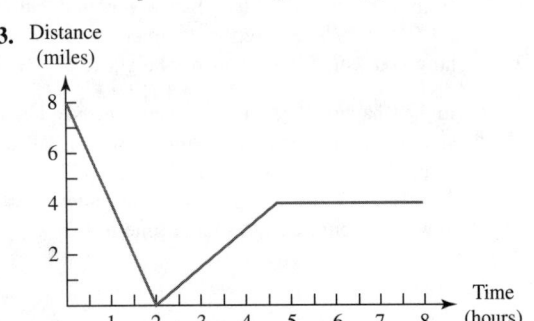

4.
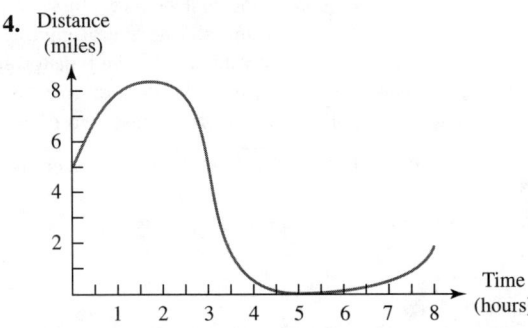

5. *Transportation* I left my house in Knoxville, TN at 8:00 A.M. and drove to Crossville, TN at an average speed of 62 mph. I stayed in Crossville for one-half hour to get gasoline and a snack and then drove to Nashville at an average speed of 72 mph. I stayed in Nashville for $2\frac{1}{2}$ hours and then drove back to Knoxville at an average speed of 70 mph. The distance from Knoxville to Crossville is 70 miles and from Crossville to Nashville it is 108 miles.

   (a) Sketch a distance vs. time graph for this trip. Show mileage on the vertical axis and time on the horizontal axis. Show the exact time for arrival and departure from each city.

   (b) Excluding the snack break and the $2\frac{1}{2}$ hours in Nashville, what was the average speed for the trip? Why isn't this number equal to the average of the three speeds, which would be 68 mph?

   (c) At the conclusion of the snack break in Crossville, what was my average speed up to then? Include the time spent eating. What was the average velocity over the same time interval?

   (d) Including the snack break, what was the average speed between home and the arrival in Nashville? What was the average velocity over the same period?

   (e) What was the average speed between the end of the snack break and the arrival back home? What was the average velocity for this time interval?

   (f) What was the average velocity for the entire trip?

6. Determine the quadratic regression equation for the data in the table below. Write the equation correct to 3 decimal places.

x	0	0.3	0.6	0.9	1.2	1.5	1.8	2.1	2.4	2.7	3.0
y	3	3.55	3.86	4.01	3.98	3.72	3.43	2.71	2.23	0.87	0.05

7. Evaluate $f(x) = 3x^2 - 5x + 7$ at $x = 2$.

8. Use your calculator to evaluate $f(1.732)$ if $f(x) = 3x^2 - 5.48x + 7$.

9. Use your calculator to evaluate $g(-2.7)$ if $g(x) = x + \log(2 - x)$.

10. (a) Graph the function $h(x) = x^2 - 2\sqrt{x^2 - 1}$ and **(b)** approximate all of its roots.

11. (a) Graph the function $j(t) = 0.1t^3 - |t^2 - 1|$ and **(b)** approximate all of its roots.

12. Use the quadratic formula to determine the roots of the function $f(x) = 3.1x^2 - 4x - 7.5$.

13. Determine the coordinates of the vertex of the function $f(x) = 3.1x^2 - 4x - 7.5$.

14. Estimate the slope of the tangent to the graph of $f(x) = 1.3x^2 - 7.4x + 5.2$ when $x = 2.1$.

15. *Sports technology* A football quarterback throws an incomplete pass. If the ball was released when it was 6 ft above the ground and had an initial velocity upward of 66 ft/sec, how long does it take to hit the ground?

16. *Sports technology* What was the highest point above the ground for the football in Exercise 15?

17. *Sports technology* Estimate the velocity of the football in Exercise 15 when it hit the ground.

18. Use the lists BNCE and TBNCE from Section 2.2 to answer the following questions for the third bounce of the ball you dropped in Activity 2.2.

   (a) What is the quadratic regression equation when the data are transformed so the bounce begins at $t = 0$?
   (b) Estimate the velocity at $t = 0$.
   (c) How does the result in (b) compare with the coefficient b in the quadratic model from (a)?
   (d) Estimate the velocity when the ball strikes the ground at the end of the third bounce.

●●●●●●●●● Chapter 2 Test

1. Write a description of the graph sketched below using the words *slope* and *velocity* in your description.

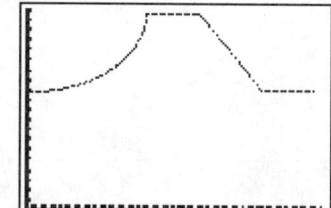

2. Compute the average velocity from the distance vs. time graph shown below.

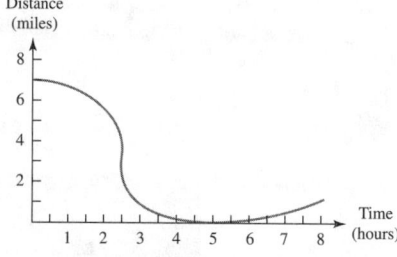

3. I left my house in Cedar Rapids, Iowa at 8:00 A.M. and drove to Newton, Iowa at an average speed of 66 mph. I stayed in Newton for one hour to get gasoline, have my oil changed, and eat lunch, and then drove to Des Moines at an average speed of 72 mph. I stayed in Des Moines for 2 hours and then drove back to Cedar Rapids at an average speed of 78 mph. The distance from Cedar Rapids to Newton is 99 miles and from Newton to Des Moines it is 33 miles.

   (a) Excluding the lunch break and the 2 hours in Des Moines, what was the average speed for the trip?

   (b) Including the lunch break, what was the average speed between home and the arrival in Des Moines? What was the average velocity over the same period?

   (c) What was the average speed between the end of the lunch break and the arrival back home? What was the average velocity for this time interval?

   (d) What was the average velocity for the entire trip?

4. Determine the quadratic regression equation for the data in the table below. Write the equation correct to 3 decimal places.

x	0	0.2	0.6	0.8	1.1	1.4	1.6	2.1	2.5	2.7	3.0
y	4.1	5.1	6.2	6.5	7.3	7.0	6.7	5.7	4.1	2.8	0.7

5. Use your calculator to evaluate $f(7.52)$ if $f(x) = 1.2x^2 - 3.5x + 7$.

6. (a) Graph the function $h(x) = x^2 + x - 2\sqrt{x^2 + 2}$ and (b) approximate all of its roots.

7. (a) Graph the function $j(t) = 0.1t^3 - |t^2 - 8|$ and (b) approximate all of its roots.

8. Use your calculator to evaluate $g(-3.4)$ if $g(x) = x^2 \log(x + 5)$.

9. Use the quadratic formula to determine the roots of the function $f(x) = 2.5x^2 - 3.2x - 6.4$.

10. Determine the coordinates of the vertex of the function $f(x) = 2.7x^2 - 10.8x - 7.1$.

11. An arrow is shot upward into the air with an initial velocity of 196 ft/sec. If the arrow was released when it was 6 ft above the ground, how long does it take to hit the ground?

12. What was the highest point above the ground for the arrow in Exercise 11?

13. Estimate the velocity of the arrow in Exercise 11 when it hit the ground.

Models of Periodic Data
Introducing Trigonometry

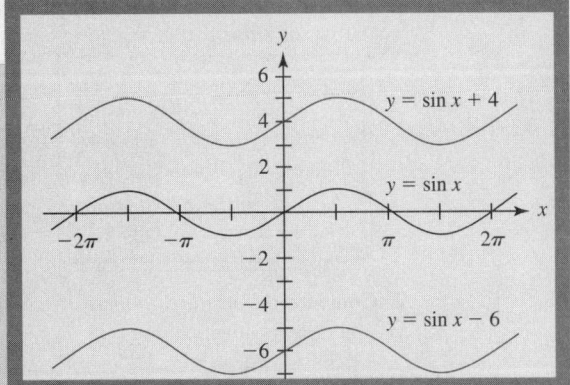

Topics You'll Learn or Review

- Definition of sine, cosine, and tangent in terms of coordinates on a circle
- Definition of sine, cosine, and tangent in terms of ratios of sides of a right triangle
- Using trigonometric functions to determine the sizes of the unknown parts of a right triangle
- Recognizing the graphs of sine, cosine, and tangent functions
- Using symmetry within the circle to compute trigonometric functions of angles larger than 90° and smaller than 0°
- Interpreting the concepts *approaching infinity* and *infinite number of values*
- Finding roots of trigonometric functions and learning how to express all the roots
- Definition of period, amplitude, frequency, phase shift, and vertical translation in trigonometric functions
- Recognizing the effect on the graph of the coefficients A, B, and C in the function $f(x) = A \sin(Bx - C)$
- Writing mathematical models for RC circuits and for musical notes
- Analyzing the sum of two trigonometric functions and applying that information
- Identifying, in symbols and graphs, the envelope function for the function $f(x) = \sin(ax) + \sin(bx)$

Calculator Skills You'll Use

- Graphing and Trace
- Changing the graph window settings to see the relevant portion of a graph
- Graphing styles to highlight a curve

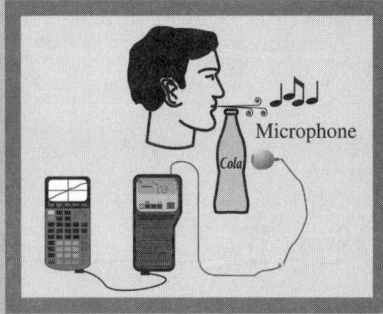

Topics You'll Need to Know

► Geometry
  • angle sum in a triangle
  • complementary and supplementary angles
  • measuring angles in degrees
► Algebra
  • solving proportions like $\frac{3}{x} = 45$ and $\frac{a}{x} = bc$
  • understanding the meaning of the solution of an equation
  • use of symbolism:
    • absolute value $|x|$
    • inequalities $0 \le x \le 30$
    • function notation $f(x)$ and $f(x) + g(x)$
  • computations with slope
  • equations of a straight line
► Graphing
  • using Cartesian coordinates
  • interpreting the meaning of the solution of an equation $f(x) = 0$

Calculator Skills You'll Learn

► Understanding that the calculator can be incorrect or misleading in graphing the tangent function
► Understanding that the calculator can be incorrect or misleading in plotting points of a function like $f(x) = \sin(40x)$
► Determining the least common multiple of two or more positive integers

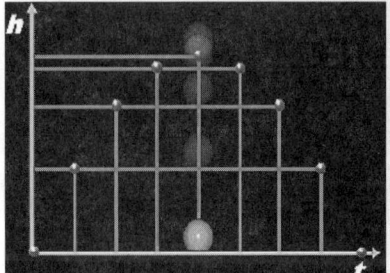

What You'll Do in This Chapter

In Chapters 1 and 2 you studied linear and quadratic functions, which are the essential building blocks for mathematical modeling. You saw that these functions are useful for modeling two types of motion: motion with constant velocity and motion that is influenced by gravity. In this chapter we will extend our study of functions to include trigonometric functions, the most common periodic functions. Trigonometric functions were invented thousands of years ago to help in the measurement of distances and angles on land and in the skies.

You will also see how to use trig functions to model the following phenomena:

► Circular motion and the effects of the motion of the earth and sun on the hours of daylight at different locations on earth
► Vibrations
► Out of phase currents
► Sound waves
► Combinations of two or more sound waves

Chapter Project—Analyzing the Touchtone Phone

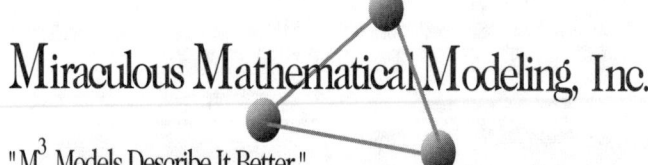

Miraculous Mathematical Modeling, Inc.

"M³ Models Describe It Better"

To: *Research department*
From: *Management*
Subject: *Picturing tones*

Northwest Communications, Inc. has invited us to compete for a research grant to study one aspect of the Touchtone phone codes.

As you know, pressing any button on a Touchtone phone produces a distinctive tone which is recognized by the telephone system hardware. They want us to help them present a graphical version of each tone so that tones could be recognized visually. The model must be one that converts the tone you hear to a distinctive picture.

We know that each tone is made up of two pure notes. Please come up with a way to represent combinations of two notes so that each combination will be visually distinguishable from all of the others.

Activity 3.1
First Look at the Touchtone Problem

The Touchtone system, or the DTMF (dual tone multifrequency) system, uses two tones to represent each key. There is a single-frequency tone for each horizontal row on the keypad and a single-frequency tone for each vertical column of the keypad. When a key is pressed, the note for that row (the numbers labeled *A* in Figure 3.1) and the note for that column (the numbers labeled *B*) are played simultaneously.

Figure 3.1 shows the frequencies that are played for each key. For example, if the number 4 is pressed, notes of frequency 1209 Hz and 770 Hz are played together. For comparison, middle C on a piano has a frequency of 262 Hz.

Complete Table 3.1 for each key on the Touchtone keypad. (In later parts of this chapter you will need to refer back to this table.)

Figure 3.1

Table 3.1 Sum and difference of Touchtone frequencies

Key	A Frequency	B Frequency	A + B	A − B
1				
2				
3				
4				
5				
6				
7				
8				
9				
0				
*				
#				

In this chapter you will learn how to model single-frequency tones with a mathematical function and how to model the result of playing two notes simultaneously. The graphs of these functions will be one way to represent each tone on the telephone keypad with a picture.

We will begin our study of periodic functions by looking at the trigonometric functions named sine, cosine, and tangent—the most widely used periodic functions.

3.1   Introduction to the Trigonometric Functions

There are many periodic functions, but probably the ones used the most are the trigonometric functions. Trigonometric functions are used in the study of subjects as widely varying as the motion of a Slinky, operation of electric motors, and the motion of planets. In Chapter 1 you saw that one of the trigonometric functions, the tangent function, is useful in the study of straight lines. The tangent function was defined in Chapter 1 as the ratio of the two legs of a right triangle. In this chapter we will use a definition that will at first look different from the right triangle definition, but you will soon see that it produces the same result.

The three most commonly used trigonometric functions are sine, cosine, and tangent. These functions are represented on your calculator keypad as **SIN**, **COS**, and **TAN**, since *sin*, *cos*, and *tan* are the abbreviations for these three functions.

Example 3.1

Use your calculator to evaluate (a) sin 30°, (b) cos 30°, and (c) tan 30°.

Solutions Check to see that your calculator is set to degrees and its decimal precision to Float. To evaluate sin 30° press **SIN** 30 **)** **ENTER**.

In a similar manner you can evaluate cos 30° and tan 30°. The answers are shown in Figure 3.2

```
sin(30)
                .5
cos(30)
        .8660254038
tan(30)
        .5773502692
```

Figure 3.2   ■

Note

What is the difference between sin (30°) and sin 30°? The answer is: Only the parentheses—there is no difference in the meaning. You will want to use parentheses when you think someone might not understand the meaning if they were left out.

Definitions of the Trigonometric Functions

The linear and quadratic functions that you used in Chapters 1 and 2 were defined in terms of rules like $y = 2x^2 - 3x + 5$ and $f(x) = \frac{1}{2}x = 7.25$. The definitions of the trigonometric functions use quite different rules.

The functions sine, cosine, and tangent are defined in terms of a circle of radius r drawn so that its center is at the origin of the rectangular coordinate system. In our definition the input to the function is an angle. In Example 3.1 the angle was 30, and because you set the calculator to degrees the angle was 30°. Look at the triangle $\triangle ABC$ shown in Figure 3.3. Think about $\angle A$, or $\angle BAC$, as the input for the functions. The definitions of the three functions are

$$\sin A = \frac{BC}{AB} = \frac{y}{r}$$

$$\cos A = \frac{AC}{AB} = \frac{x}{r}$$

$$\tan A = \frac{BC}{AC} = \frac{y}{x}$$

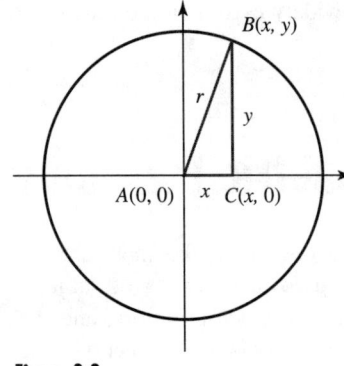

Figure 3.3

Example 3.2

Show that the definition given here for the tangent function is equivalent to the definition given in Chapter 1, page 34. What is the difference between the two definitions?

Solution   The triangle, $\triangle ABC$, is the same in both cases and the definition is the same. There are differences because the definition that uses a triangle will allow $\angle A$ to have values between $0°$ and $90°$, while the definition that uses the circle will allow the $\angle A$ to have values that are any number—positive, negative, or zero.   ∎

Calculations with Triangles

Let's look at $\triangle ABC$ in Figure 3.4. The size of the angle at C is $90°$, a **right angle**, so this triangle is called a **right triangle**. $\overline{AB}$, the side across from the right angle, is the longest side of the triangle and it is called the **hypotenuse**. The two non-right angles, $\angle A$ and $\angle B$, are formed by the hypotenuse and one of the other sides, or **legs**, of the triangle. To help define the trigonometric functions, we will refer to the leg that helps form an angle as the **side adjacent** to the angle. We will call the other leg the **side opposite** the angle. Whether a side is an opposite side or an adjacent side depends on the angle. In Figure 3.4 for example, side $\overline{AC}$ is adjacent to $\angle A$ and is opposite $\angle B$. Similarly, $\overline{BC}$ is adjacent to $\angle B$ and opposite $\angle A$.

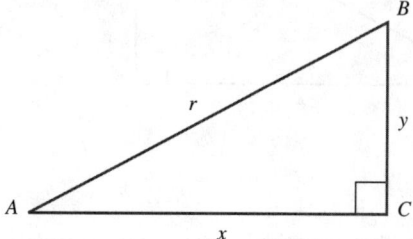

Figure 3.4

Using the phrases *opposite side*, *adjacent side*, and *hypotenuse*, we have the following definitions of sine, cosine, and tangent for $\angle A$ in the right triangle in Figure 3.4.

Right Triangle Definitions of Trigonometric Functions

If $\angle A$ is an acute angle of triangle $\triangle ABC$ with right angle C, then

$$\sin A = \frac{BC}{AB} = \frac{y}{r} = \frac{\text{side opposite } \angle A}{\text{hypotenuse}}$$

$$\cos A = \frac{AC}{AB} = \frac{x}{r} = \frac{\text{side adjacent to } \angle A}{\text{hypotenuse}}$$

$$\tan A = \frac{BC}{AC} = \frac{y}{x} = \frac{\text{side opposite } \angle A}{\text{side adjacent to } \angle A}$$

We can similarly define the sine, cosine, and tangent for $\angle B$ in the same right triangle in Figure 3.4.

$$\sin B = \frac{AC}{AB} = \frac{x}{r} = \frac{\text{side opposite } \angle B}{\text{hypotenuse}}$$

$$\cos B = \frac{BC}{AB} = \frac{y}{r} = \frac{\text{side adjacent to } \angle B}{\text{hypotenuse}}$$

$$\tan B = \frac{AC}{BC} = \frac{x}{y} = \frac{\text{side opposite } \angle B}{\text{side adjacent to } \angle B}$$

Examples 3.3–3.5 will use the phrases *opposite side*, *adjacent side*, and *hypotenuse* to help make these definitions understandable for *any* right triangle.

Example 3.3

For $\triangle ABC$ shown in Figure 3.5, compute the value of each of the three most commonly used trigonometric functions at each of the acute angles.

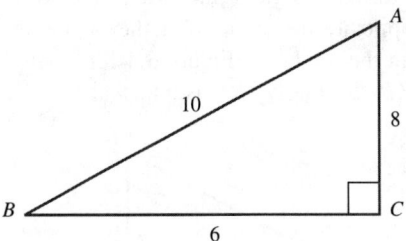

Figure 3.5

Solution   Make sure that you think about the concepts of *opposite side*, *adjacent side*, and *hypotenuse* as you work each of these.

Using $\angle A$:

$$\sin A = \frac{\text{side opposite } \angle A}{\text{hypotenuse}} = \frac{6}{10} = 0.6$$

$$\cos A = \frac{\text{side adjacent to } \angle A}{\text{hypotenuse}} = \frac{8}{10} = 0.8$$

$$\tan A = \frac{\text{side opposite } \angle A}{\text{side adjacent to } \angle A} = \frac{6}{8} = 0.75$$

Using $\angle B$:

$$\sin B = \frac{\text{side opposite } \angle B}{\text{hypotenuse}} = \frac{8}{10} = 0.8$$

$$\cos B = \frac{\text{side adjacent to } \angle B}{\text{hypotenuse}} = \frac{6}{10} = 0.6$$

$$\tan B = \frac{\text{side opposite } \angle B}{\text{side adjacent to } \angle B} = \frac{8}{6} = 1.33$$

Example 3.4

For $\triangle DEF$ shown in Figure 3.6, compute the value of each of the three most commonly used trigonometric functions at each of the acute angles.

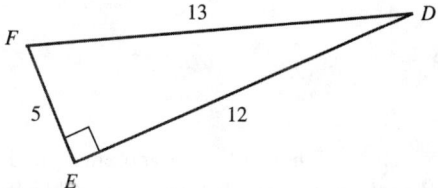

Figure 3.6

Solution Using $\angle D$:

$$\sin D = \frac{\text{side opposite}}{\text{hypotenuse}} = \frac{5}{13} \approx 0.385$$

$$\cos D = \frac{\text{side adjacent}}{\text{hypotenuse}} = \frac{12}{13} \approx 0.923$$

$$\tan D = \frac{\text{side opposite}}{\text{side adjacent}} = \frac{5}{12} \approx 0.417$$

Using $\angle F$:

$$\sin F = \frac{\text{side opposite}}{\text{hypotenuse}} = \frac{12}{13} \approx 0.923$$

$$\cos F = \frac{\text{side adjacent}}{\text{hypotenuse}} = \frac{5}{13} \approx 0.385$$

$$\tan F = \frac{\text{side opposite}}{\text{side adjacent}} = \frac{12}{5} = 2.4$$

It is often easier to refer to the side of a triangle by the lower case version of the angle it is opposite. Thus in Figure 3.7, side a is opposite $\angle A$, side b is opposite $\angle B$, and side c is opposite $\angle C$.

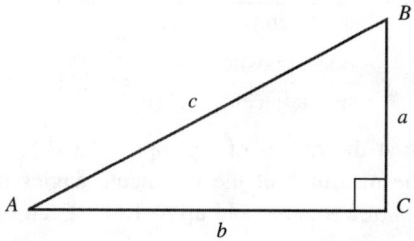

Figure 3.7

Example 3.5

For $\triangle ABC$ shown in Figure 3.8, you are given $c = 25$ and $a = 24$. Compute the value of each of the three most commonly used trigonometric functions at each of the acute angles.

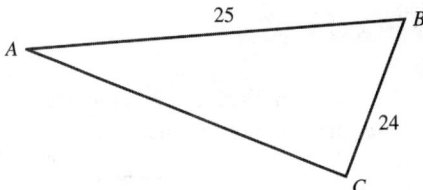

Figure 3.8

Solution   We are given the length of the hypotenuse and the length of one of the legs. We first need to determine the length of the other leg. Since this is a right triangle, we can use the Pythagorean theorem. Since $a^2 + b^2 = c^2$, then

$$b = \sqrt{c^2 - a^2}$$
$$= \sqrt{25^2 - 24^2}$$
$$= \sqrt{625 - 576} = \sqrt{49} = 7$$

The length of c is 7 units.

Since we know the lengths of the three sides, we can evaluate the trigonometric functions of the two acute angles.

Now that we know that $AC = 7$, we can see that Figure 3.8 is not in the correct scale. Figures are often drawn with little regard to the correct scale, since their main purpose is to help us see mathematical relationships.

Using $\angle A$:

$$\sin A = \frac{\text{side opposite}}{\text{hypotenuse}} = \frac{24}{25} = 0.960$$

$$\cos A = \frac{\text{side adjacent}}{\text{hypotenuse}} = \frac{7}{25} = 0.280$$

$$\tan A = \frac{\text{side opposite}}{\text{side adjacent}} = \frac{24}{7} \approx 3.429$$

Using $\angle B$:

$$\sin B = \frac{\text{side opposite}}{\text{hypotenuse}} = \frac{7}{25} = 0.280$$

$$\cos B = \frac{\text{side adjacent}}{\text{hypotenuse}} = \frac{24}{25} = 0.960$$

$$\tan B = \frac{\text{side opposite}}{\text{side adjacent}} = \frac{7}{24} \approx 0.292$$ ■

Look at the results of Examples 3.3–3.5. What is the numerical relationship between the measures of the two acute angles in each triangle? The measures of three angles of a triangle add up to 180°. Each of the triangles in Examples 3.3–3.5 is a right triangle, so the measure of one angle of each triangle is 90°. Therefore the measures of the two acute angles must add up to 90°. Angles that add up to 90° are **complements** of each other and are called **complementary angles**. Thus the two acute angles in the triangles of Examples 3.3–3.5 are complementary angles. Symbolically, if $\angle A$ and $\angle B$ are complementary angles then $A = 90° - B$. Similarly, $B = 90° - A$.

If two angles (such as $\angle A$ and $\angle B$ in Figure 3.5) are complementary angles, then $\sin A = \cos B$ or

$$\sin A = \cos(90° - A)$$

You should be able to explain in words why this is true. It is from the word *complementary* that we have the word *co*sine, since the cosine of an angle is equal to the sine of the angle's complement. Since A and B are complementary angles, we can also write $\sin B = \cos (90° - B)$.

Furthermore, if two angles (such as $\angle A$ and $\angle B$ in Figure 3.5) are complementary angles, then the values of $\tan A$ and $\tan B$ are always **reciprocals** of one another. That is,

$$\frac{1}{\tan A} = \tan B$$

or $\dfrac{1}{\tan A} = \tan(90° - A)$

The next two examples show how trigonometric functions can be used to calculate the length of the sides of a right triangle if one of the sides and the size of one acute angle are known.

Example 3.6

For right triangle $\triangle ABC$ (Figure 3.9), where $AC = 12$ and $\angle B = 24.7°$, determine the lengths of the other two sides and the values of the three most commonly used trigonometric functions with respect to $\angle A$.

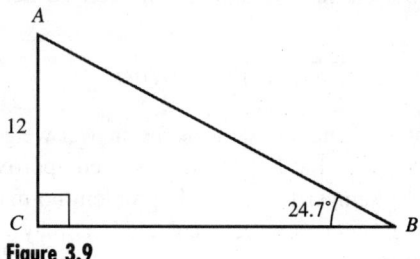

Figure 3.9

Solution   Since angles A and B are complementary angles and $\angle B = 24.7°$, then

$$\angle A + \angle B = 90°$$
$$\angle A + 24.7° = 90°$$
$$\angle A = 90° - 24.7°$$
$$= 65.3°$$

We can now use our calculator to find the values of the three most commonly used trigonometric functions with respect to $\angle A$.

$$\sin A = \sin 65.3° \approx 0.909$$
$$\cos A = \cos 65.3° \approx 0.418$$
$$\tan A = \tan 65.3° \approx 2.174$$

Now to find the lengths of the other two sides we will use $\angle A$ and the length of $\overline{AC}$, the side adjacent to $\angle A$. The two trigonometric functions that use the adjacent side are cosine and tangent. We can use the cosine to determine the length of the hypotenuse.

$$\cos A = \frac{\text{side adjacent } \angle A}{\text{hypotenuse}}$$

$$\cos 65.3° = \frac{AC}{AB} = \frac{12}{AB}$$

$$AB = \frac{12}{\cos 65.3°} \approx \frac{12}{0.418} \approx 28.708$$

Finally we will use the tangent function to find the length of the opposite side.

$$\tan A = \frac{\text{side opposite } \angle A}{\text{side adjacent } \angle A}$$

$$\tan 65.3° = \frac{BC}{AC} = \frac{BC}{12}$$

$$12 \tan 65.3° = BC$$

$$BC \approx 12 \times 2.174 \approx 26.088$$

Thus the lengths of the three sides of the triangle are 12, 26.088, and 28.708. ■

In the previous example, once you found the length of the hypotenuse you could have used the Pythagorean theorem to determine the length of the remaining side:

$$BC = \sqrt{28.708^2 - 12^2} \approx 26.080$$

Notice that this is not the same as the answer we got in Example 3.6. The discrepancy is due to the fact that both solutions used approximate numbers. We had to use an approximate number to solve this problem no matter which method we used. Using approximate numbers in your calculations may magnify discrepancies in your data. For this reason, if at all possible you should use the original numbers given or measured in a problem.

Example 3.7

If A is an acute angle of a right triangle and $\cos A = \frac{15}{17}$, find the values of $\sin A$ and $\tan A$.

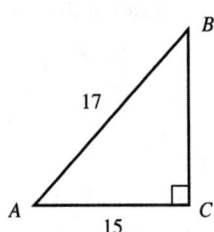

Figure 3.10

Solution   There are two methods you can use to solve this example. One method requires that you find the length of the third side of the triangle and the other method takes advantage of a calculator. We will examine both methods.

Method 1   Since $\cos A = \dfrac{\text{side adjacent } \angle A}{\text{hypotenuse}} = \dfrac{15}{17}$ we can assume that we have a right triangle with a hypotenuse of length 17 and side adjacent to $\angle A$ with length 15, much like the one in Figure 3.10. Using the Pythagorean theorem we can determine the length of the third side, the side opposite $\angle A$, as $a = \sqrt{17^2 - 15^2} = \sqrt{64} = 8$. This solution allows us to determine the values of the two requested trigonometric functions.

$$\sin A = \frac{\text{side opposite } \angle A}{\text{hypotenuse}} = \frac{8}{17}$$

$$\tan A = \frac{\text{side opposite } \angle A}{\text{side adjacent } \angle A} = \frac{8}{15}$$

Method 2   In this method we will find the value of $\angle A$, store it in the calculator, and then find approximate values of $\sin A$ and $\tan A$.

To find the value of $\angle A$ we proceed much as we did in Chapter 1. On your calculator press

Store this result as x by pressing [STO ►] [X,T,θ,n] [ENTER]; then calculate $\sin x \approx 0.4706$ and $\tan x \approx 0.5333$. Check to see if these are the same values you obtained by Method 1. ■

Example 3.8   Application

The distance DE across the lake in Figure 3.11 is not known, but it is possible to place poles in the ground at A, B, and C so that A, B, D, and E are in a straight line and $\angle C = 90°$. If $\angle A$ is measured to be $37°$ and the distance AC is measured to be 870 feet, what is the distance across the lake if $AD = 28'$ and $EB = 14'$?

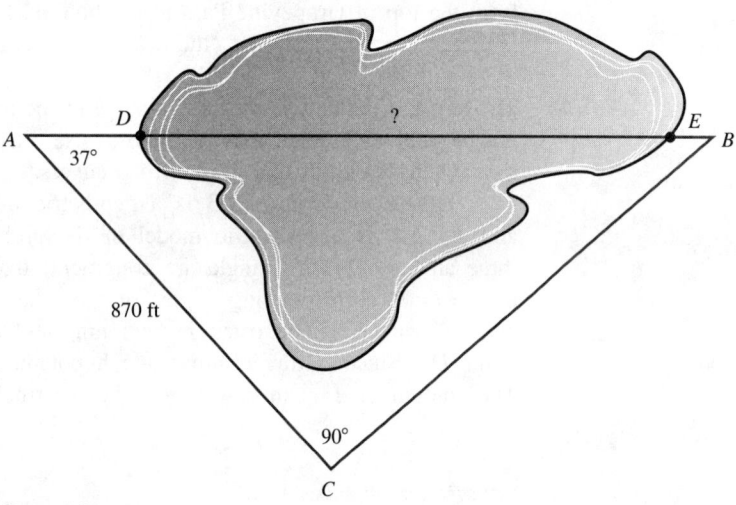

Figure 3.11

Solution   We will first find the length AB and then subtract the lengths of $\overline{AD}$ and $\overline{EB}$. We know that the length of $\overline{AC}$, the side adjacent to $\angle A$, is 870 feet; we want to determine the length of the hypotenuse. The cosine function is the ratio of the adjacent side and the hypotenuse of a right triangle. If we let x represent AB, this leads to the equation $\cos 37° = \dfrac{AC}{AB} = \dfrac{870}{x}$. Solving this equation we obtain

$$\cos 37° = \frac{870}{x}$$

$$0.799 = \frac{870}{x} \qquad \text{Evaluate } \cos 37°$$

$$0.799x = 870 \qquad \text{Multiply by } x$$

$$x = \frac{870}{0.799} \approx 1088.861 \qquad \text{Solve for } x$$

To determine the distance across the lake we subtract the lengths of $\overline{AD}$ and $\overline{EB}$:

$$1088.861 - 28 - 14 = 1046.861$$

If we assume that 870 has three significant figures, we should report the result to the same level of precision. Writing 1046.861 to three significant figures gives 1050 feet. Note that the answer is not 1047 feet, since 1047 has four significant figures. Thus it is about 1050 feet across the lake. ■

The following example is similar to one you encountered in Chapter 1. The solution requires that you do two trigonometric calculations: one to compute the angle and another to compute the hypotenuse.

Example 3.9   Application

Grapevine Pass, north of Los Angeles, is shown in Figure 3.12. The highway through Grapevine Pass averages a 2% grade over a distance of 8.3 miles. If a truck drives from the top of Grapevine Pass to the bottom, how much drop in vertical distance is there? Give the answer to the nearest 100 feet.

Solution   The term 2% grade means that the slope of the road is $\frac{2}{100} = 0.02$. Although it may not be obvious at first, it is possible to use a triangle as a model to represent this road, even though in real life the road curves.*

$\triangle ABC$ in Figure 3.13 is a geometric model of a road with a 2% grade and $\triangle DEF$ is a geometric model of the problem we are asked to solve. All three angles of each triangle are congruent; that is, they have the same measure as the three corresponding angles of the other triangle: $\angle A = \angle D$, $\angle B = \angle E$, and $\angle C = \angle F$. The truck is beginning its descent at point E and ending at point D. Since it travels along the hypotenuse, the distance $ED = 8.3$ miles. The amount of drop in elevation while the truck travels 8.3 miles is equal to the side EF.

---

*No matter how much the road curves, you can think of it as being made up of straight segments (you saw something like this when you zoomed in on a parabola in Chapter 2). Each of these segments can be thought of as the hypotenuse of a triangle, and you can line up all these little triangles to make one big triangle.

Figure 3.12
Photo of Grapevine Pass (Photo courtesy of Phil Borden/PhotoEdit)

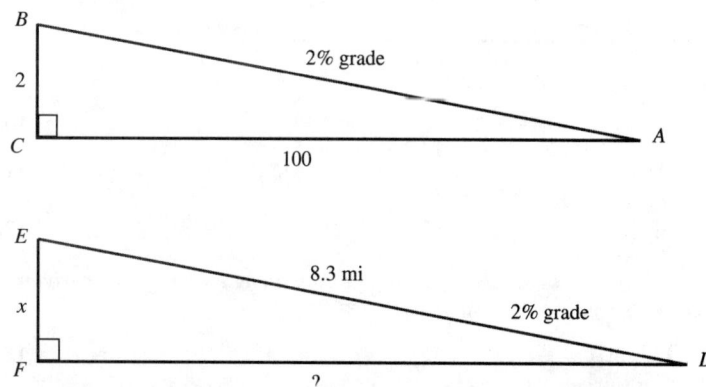

Figure 3.13
Two models for the Grapevine Pass highway

The first step is to use $\triangle ABC$ to calculate $\angle A$. In Chapter 1 you learned to use the (2nd) [TAN^{-1}] key to compute the angle from the value of its tangent. Since $\tan A = 0.02$ we have $\angle A = \tan^{-1}(0.02) \approx 1.146°$, and so $\angle D \approx 1.146°$.

The drop in height EF is the side opposite $\angle D$, and we know the length of the hypotenuse is 8.3 mi. The trigonometric function that uses the hypotenuse and the opposite side is the sine function. Therefore

$$\frac{EF}{8.3} = \sin D$$
$$= \sin 1.146°$$
$$= 0.02$$
$$EF = 0.020 \times 8.3 = 0.166$$

Thus the drop is about 0.166 mi or 876.48 ft. (Multiply 0.166 mi by $\dfrac{5280 \text{ ft}}{1 \text{ mi}}$ to give the number of feet.) The problem asked for vertical distance to the nearest one hundred feet, so the answer is 900 feet. ∎

Section 3.1 Exercises

In Exercises 1–6 use your calculator to evaluate each of the following functions to four decimal places.

1. $\sin 16.3°$

2. $\cos 76.7°$

3. $\tan 36.1°$

4. $\sin 79.4°$

5. $\cos 13.009°$

6. $\tan 85.3°$

In Exercises 7–12 find the sin, cos, *and* tan *of* ∠A *and* ∠B *for the indicated sides of* △ABC. *(See Figure 3.14.)*

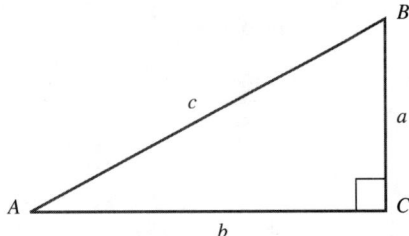

Figure 3.14

7. $a = 15, b = 8, c = 17$

8. $a = 9, b = 40, c = 41$

9. $a = 5, b = 8, c = \sqrt{89}$

10. $a = 12, b = \sqrt{145}, c = 17$

11. $a = 7, b = 12$

12. $b = 9, c = 14$

In Exercises 13–16 let A be an angle of a right triangle with the given trigonometric function. Find the values of the two other functions from the list sin A, cos A, *and* tan A.

13. $\sin A = \frac{12}{37}$

14. $\tan A = \frac{60}{11}$

15. $\tan A = \frac{25}{35}$

16. $\cos A = \frac{12}{23}$

Solve Exercises 17–24.

17. *Transportation* A trucker approaches a hill and sees a sign that says "6% GRADE NEXT 5 MILES." How much lower in miles and in feet can we expect the highway to be at the bottom of the hill than at the top?

18. *Environmental science* A forester measures the length of the shadow of a giant redwood tree when the sun is 63.4° above the horizon. If the tree's shadow is 108.2 ft, how tall is the tree?

19. *Environmental science* An environmentalist wants to know the width of a stream in order to properly set instruments for studying the pollutants in the water. The environmentalist notices a point *A* directly across the stream from a point *C*, as shown in Figure 3.15. The environmentalist walks straight downstream (perpendicular to $\overline{AC}$) from *C* for 47.5 ft to point *B* before the stream turns. From *B* it is determined that $\angle ABC = 52°$. How wide is the river; that is, what is the length of $\overline{AC}$?

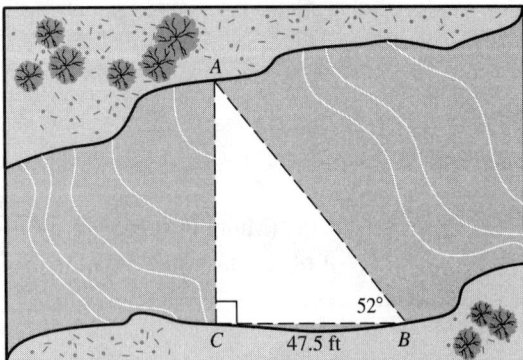

Figure 3.15

20. *Fire science* A ladder company recommends that, for safety purposes, 70° is the maximum angle a ladder should make with the ground as the ladder leans against a building. What is the maximum height that a 36.0-ft ladder can reach without violating the company's recommendation?

21. *Meteorology* The height of a cloud can be measured by shining a searchlight vertically at the cloud. A person stands a known distance away from the light and measures the angle of elevation between the ground and the cloud. If a person stands 750 ft away from the light and the angle of elevation is 83.2°, how high is the cloud?

22. *Electricity*   An electrician must bend a pipe to make a $5\frac{1}{4}$-ft rise in an 8-ft horizontal distance, as shown in Figure 3.16. What is the size of the angles at each bend (angles x and y in Figure 3.16)?

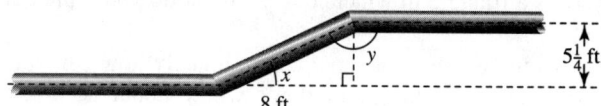

Figure 3.16

23. *Electricity*   A 135-ft-long guy wire reaches from the top of a pole to a point on the ground 80 ft from the foot of the pole? What angle does the wire make with the ground?

24. *Fire science*   A fire truck has a 15.5-m ladder. Because of some rubble from the fire, the fire fighters must place the foot of their ladder 4.5 m from the wall.
   (a)   How high up the wall can the ladder reach?
   (b)   What angle does the ladder make with the level ground under the rubble?

3.2 Graphs and Roots of Trigonometric Functions

In this section you'll see how the sine, cosine, and tangent functions are defined for angles outside the range of 0° to 90°. What you learn here will be the gateway to understanding the many applications of trigonometry that are not connected with right triangles.

Activity 3.2

Figure 3.17 contains four variations of right triangle ABC. In each triangle the length r of the hypotenuse is the same. The following questions are designed to help you determine how the fractions $\dfrac{y}{x}$, $\dfrac{y}{r}$, and $\dfrac{x}{r}$ change as the value of $\angle A$ increases.

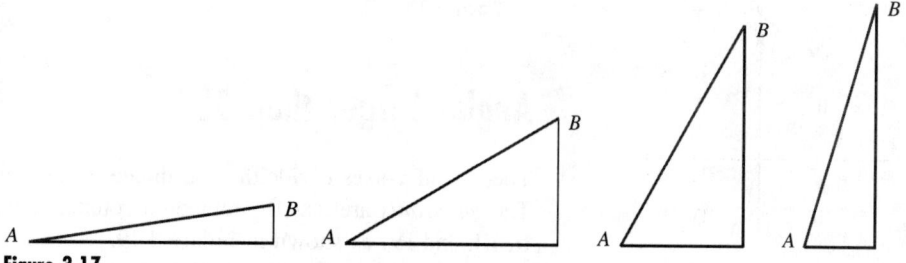

Figure 3.17

   1. As the size of $\angle A$ increases, what happens to the size of $\angle B$?
   2. As the size of $\angle A$ decreases, what happens to the size of $\angle B$?
   3. As the size of $\angle A$ increases, what happens to the value of $\sin A$?
   4. As the size of $\angle A$ increases, what happens to the value of $\cos A$?
   5. As the size of $\angle A$ increases, what happens to the value of $\tan A$?
   6. How big can $\angle A$ get before there is no longer a right triangle?
   7. How small can $\angle A$ get before there is no longer a right triangle?

8. As the size of ∠A approaches very close to zero, what do y and x approach?

9. As the size of ∠A approaches very close to 90°, what do y and x approach?

10. At some point there is a triangle in which $x = y$. What do you think ∠A is equal to in that triangle?

11. Use Figure 3.18 to sketch graphs that show approximately how ∠B, sin A, cos A, and tan A vary with ∠A. At this point don't worry about accuracy of your graph. You will learn enough in this section to make your graphs more accurate later.

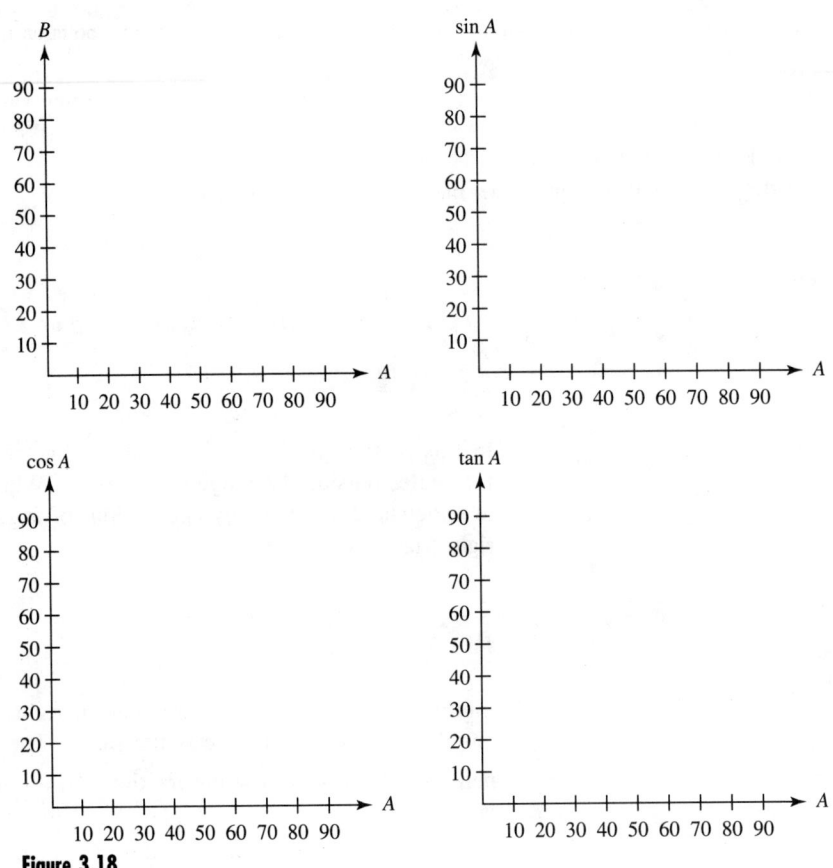

Figure 3.18

Angles Larger than 90°

The x- and y-axes divide the coordinate plane into four sections called **quadrants**.* The *quadrants* are usually numbered counterclockwise with the Roman numerals I, II, III, and IV, as shown in Figure 3.19.

The triangle definitions of the trigonometric functions require the input to the functions to be between 0° and 90°. The circle definition used in Figure 3.3 allows

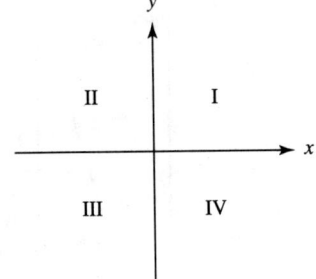

Figure 3.19

---

*The word *quadrant* comes from a Latin word for "four." "Cuatro," the Spanish word for four, comes from the same root.

for other values. Figures 3.20–3.22 show locations of point B in each quadrant. As point B moves counterclockwise around the circle the size of $\angle A$ increases.

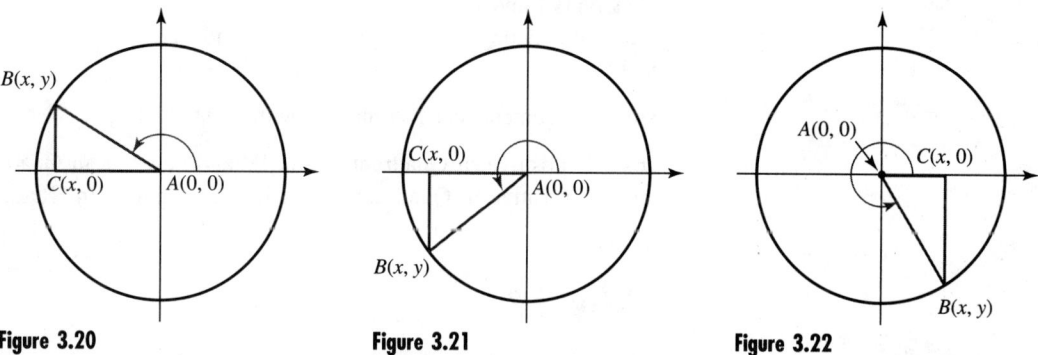

Figure 3.20
Point B is in Quadrant II

Figure 3.21
Point B is in Quadrant III

Figure 3.22
Point B is in Quadrant IV

It is important to remember that the trigonometric functions are defined in terms of x and y (the coordinates of B) and r (the radius of the circle). While r is always a positive number, x and y can be positive, negative, or zero depending on the location of point B. Table 3.2 summarizes the signs of x and y in each quadrant.

Table 3.2   Signs of x, y, and r in Quadrants I–IV

	Quadrant I	Quadrant II	Quadrant III	Quadrant IV
x is	positive	negative	negative	positive
y is	positive	positive	negative	negative
r is	positive	positive	positive	positive

Example 3.10

Use the signs of x, y, and r in Table 3.2 to complete Table 3.3 with a plus ($+$) or a minus sign ($-$) to indicate whether the sign of the function is positive or negative in each quadrant.

Table 3.3   Signs of trigonometric functions in each quadrant

	Point B is in			
	Quadrant I	Quadrant II	Quadrant III	Quadrant IV
$\sin A$ is				
$\cos A$ is				
$\tan A$ is				

Solutions   Remember r is positive in all four quadrants.

▶ Since $\sin A$ is $\dfrac{y}{r}$ and y is positive in Quadrants I and II, $\sin A$ is positive in Quadrants I and II.

▶ Because y is negative in Quadrants III and IV, $\sin A$ is negative in Quadrants III and IV.

A similar argument leads to the following conclusions:

▶ $\cos A$ is positive in Quadrants I and IV and negative in Quadrants II and III.
▶ $\tan A$ is positive in Quadrants I and III and negative in Quadrants II and IV. ■

Activity 3.3

Figure 3.23 shows measures of $\angle A$ marked off on a circle. Measure with a ruler the radius of the circle and each of the y values. Use these numbers to estimate the values of the sine function for each angle. *Don't forget to consider whether the values are positive or negative.* Enter your results in Table 3.4. Enter only the number of decimal places that match the precision of your measurements. As a check after you are done, see what your calculator gives for the value of each function at the various angles. [*Hint:* After you complete the table for the first quadrant, it should be possible to use the symmetry of the circle to fill in all other values without further measurement or calculation.]

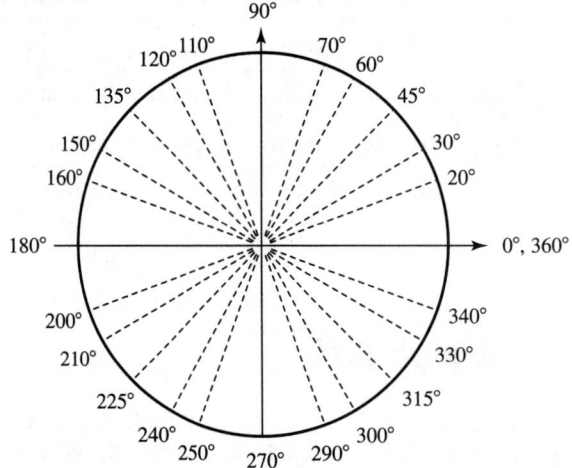

Figure 3.23

Use Table 3.4 to help determine the value of each A where A is the smallest positive angle that satisfies each of the following statements. The first entry has been completed for you.

1. If $\sin 160° = \sin A$ then $A = \underline{\ 20°\ }$ and $160° + A = \underline{\ 180°\ }$.
2. If $\sin 150° = \sin A$ then $A = \underline{\hspace{2em}}$ and $150° + A = \underline{\hspace{2em}}$.
3. If $\sin 135° = \sin A$ then $A = \underline{\hspace{2em}}$ and $135° + A = \underline{\hspace{2em}}$.
4. If $\sin 120° = \sin A$ then $A = \underline{\hspace{2em}}$ and $120° + A = \underline{\hspace{2em}}$.
5. If $\sin 110° = \sin A$ then $A = \underline{\hspace{2em}}$ and $110° + A = \underline{\hspace{2em}}$.

Do you see a general pattern? If so, write the pattern in symbols. Will this pattern still be true for angles larger than $180°$?

Table 3.4

∠A	0°	20°	30°	45°	60°	70°	90°	110°	120°	135°	150°	160°
y												
sin A												

∠A	180°	200°	210°	225°	240°	250°	270°	290°	300°	315°	330°	340°	360°
y													
sin A													

Activity 3.4

Using Figure 3.23, measure with a ruler the radius of the circle and each of the x values. Use these numbers to estimate the values of the cosine function for each angle. Enter your results in Table 3.5.

Use Table 3.5 to help determine the value of each A where A is the smallest positive angle that satisfies each of the following statements.

1. If $\cos 160° = -\cos A$ then $A =$ _____ and $180° - A =$ _____.
2. If $\cos 150° = -\cos A$ then $A =$ _____ and $180° - A =$ _____.
3. If $\cos 135° = -\cos A$ then $A =$ _____ and $180° - A =$ _____.
4. If $\cos 120° = -\cos A$ then $A =$ _____ and $180° - A =$ _____.
5. If $\cos 110° = -\cos A$ then $A =$ _____ and $180° - A =$ _____.

Do you see a general pattern? If so, write the pattern in symbols. Will this pattern still be true for angles larger than 180°?

Table 3.5

∠A	0°	20°	30°	45°	60°	70°	90°	110°	120°	135°	150°	160°
x												
cos A												

∠A	180°	200°	210°	225°	240°	250°	270°	290°	300°	315°	330°	340°	360°
x													
cos A													

Drawing the Graphs

Activity 3.5

Use the values of x and y in Tables 3.4–3.5 to complete Table 3.6.

On graphs like the one shown in Figure 3.24, plot the points you calculated in Tables 3.4–3.6 and connect each set of points with a smooth curve. Label the curves sine, cosine, and tangent.

Table 3.6

∠A	0°	30°	45°	60°	90°	120°	135°	150°
tan A								

∠A	180°	210°	225°	240°	270°	300°	315°	330°	360°
tan A									

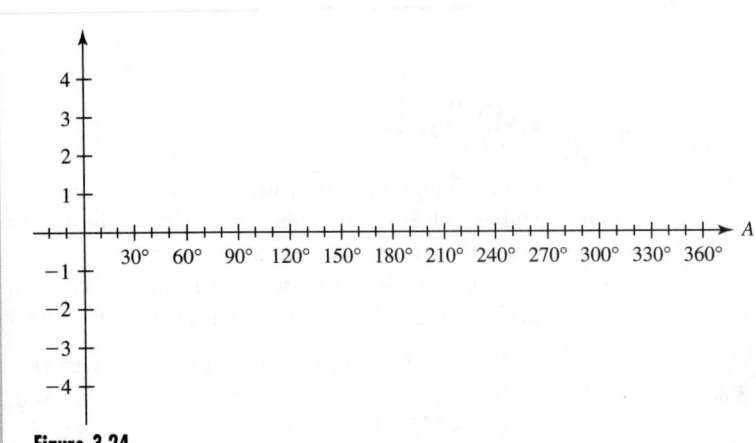

Figure 3.24

Your graphs for Activity 3.5 should look like those drawn by a graphing calculator, as shown in Figures 3.25–3.27. In Figure 3.27 we set Xmax = 352.5° and not 360° to avoid a drawing error by the calculator.

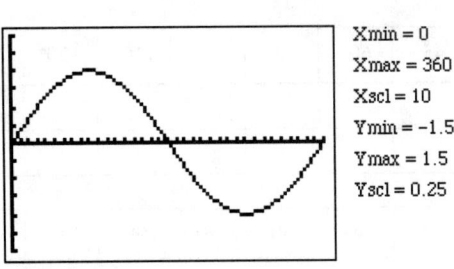

Figure 3.25
Sine graph

Xmin = 0
Xmax = 360
Xscl = 10
Ymin = −1.5
Ymax = 1.5
Yscl = 0.25

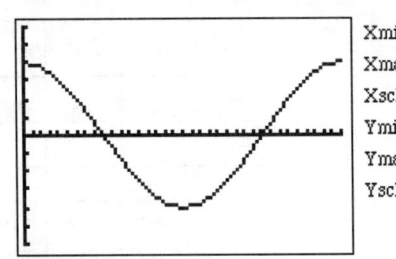

Figure 3.26
Cosine graph

Xmin = 0
Xmax = 360
Xscl = 10
Ymin = −1.5
Ymax = 1.5
Yscl = 0.25

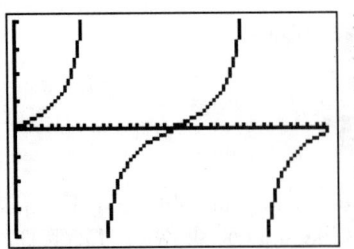

Figure 3.27
Tangent graph

Xmin = 0
Xmax = 352.5
Xscl = 10
Ymin = −4
Ymax = 4
Yscl = 1

The Peculiar Tangent Function

You have already seen that for the angles of 90° and 270° you could not compute the value of the tangent function because the calculation involved dividing by zero. We'll look more closely at what's happening here in Activity 3.1.

Calculator Lab 3.1	**Evaluating the Tangent Function Near 90°**

Use your calculator to determine the values of the tangent function for the values of ∠A given in Table 3.7. Write your answer to the nearest integer (zero decimal places).

Table 3.7

∠A	80°	88°	89°	89.9°	89.99°	89.999°
tan A						

∠A	100°	92°	91°	90.1°	90.01°	90.001°
tan A						

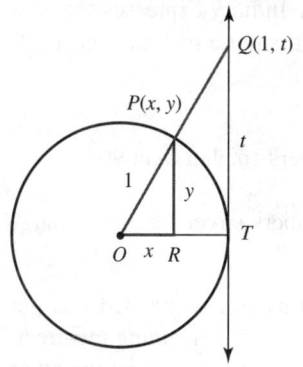

Figure 3.28

Notice that as the size of the angle gets close to 90°, two similar things seem to happen. If the sizes of the angles are less than 90° but getting closer to 90°, the values of the tangent function keep getting larger. If the sizes of the angles are more than 90° but getting closer to 90°, then values of the tangent function are negative and keep getting smaller.

Figures 3.28–3.32 can help you visualize what is going on here. In Figure 3.28 the point O is the origin of the coordinate system; the circle has a radius equal to 1 to simplify our calculations. The line $\overleftrightarrow{TQ}$ is **tangent** to the circle, which means that it touches the circle at only one point, with all other points of the line being outside the circle. It turns out that t, the y-coordinate of point Q, is equal to the value of $\tan(\angle POR)$. Let's proceed to an explanation of why this is the case.

The slope of the line segment $\overline{OQ}$ can be computed two ways: between points O and P and between points O and Q.

$$\text{Slope of segment } OP: \quad \frac{y-0}{x-0} = \frac{y}{x} = \tan(\angle POR)$$

$$\text{Slope of segment } OQ: \quad \frac{t-0}{1-0} = \frac{t}{1} = t$$

Therefore we can conclude that $t = \tan(\angle POR)$. (Make sure you understand why the x-coordinate of point Q is equal to 1).

That is, as point P moves around the circle, the tangent of $\angle POR$ is measured along the line that is tangent to the circle. Now you can see why this function was given the name *tangent*. Figure 3.29 shows that $t = \tan(\angle POR)$ can become very large as point $\angle POR$ gets close to 90°.

What happens when point P moves into the other quadrants? You already know that the tangent function can be positive or negative depending on the location of point P. Figures 3.30–3.32 can help you see the tangent function of angles in each quadrant. Let's look at Figure 3.30 for an example. t is still equal to $\frac{y}{x}$, but in the second quadrant x is negative, so t will be negative. Figures 3.31 and 3.32 show why the tangent function is positive in the third quadrant (remember x and y are both negative, so $\frac{y}{x}$ is positive) and negative again in the fourth quadrant.

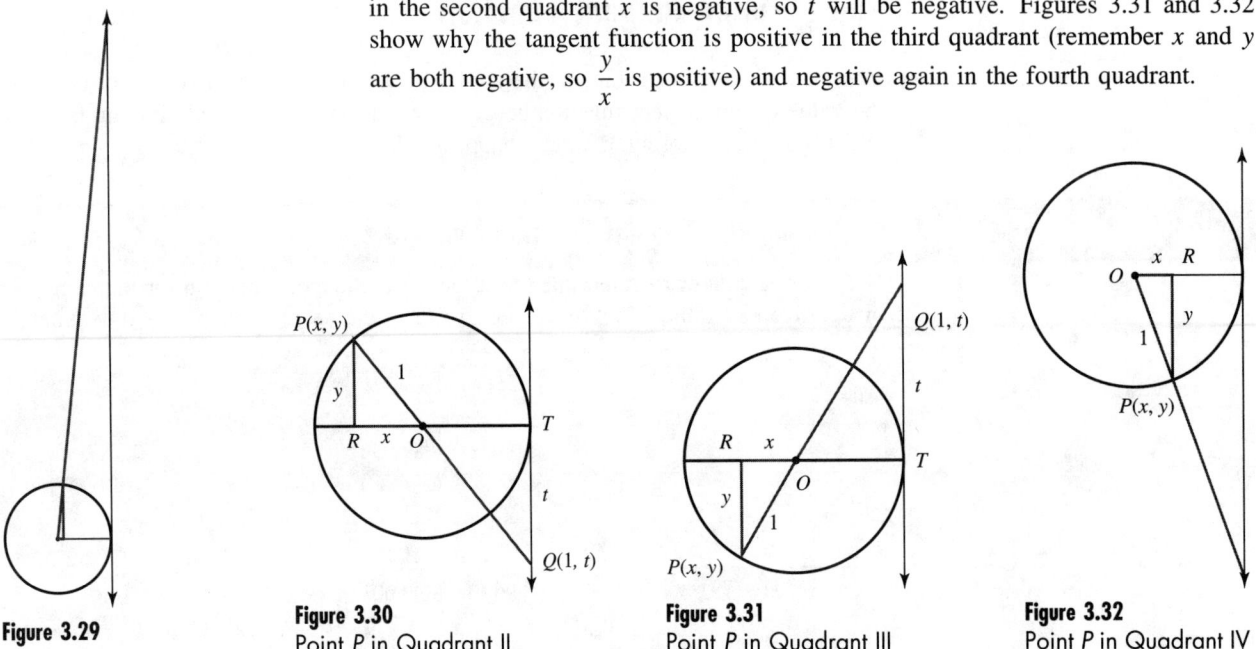

Figure 3.29

Figure 3.30
Point P in Quadrant II

Figure 3.31
Point P in Quadrant III

Figure 3.32
Point P in Quadrant IV

In mathematics we say that when the values of a function like the tangent function get bigger and bigger as the input gets close to a fixed number (in this case the number is 90°), the function is approaching **infinity**. Infinity expresses the idea of something that is larger than all real numbers and so it is not a real number itself. These ideas are summarized in the following statements:

▶ There is no value of the tangent function at 90°.
▶ As $\angle A$ approaches 90° from the left (that is, for numbers smaller than 90°), tan A approaches positive infinity.
▶ As $\angle A$ approaches 90° from the right (that is, for numbers larger than 90°), tan A approaches negative infinity.

On some graphing calculators, when the tangent function is graphed you get something like Figure 3.33. In this case the calculator has done something incorrect. It is not correct to connect the points on one side of 90° with the points on the other side of 90°; that would indicate a value of the tangent function at 90°, and you have already seen that there is no real numerical value there. To make a graphing calculator produce the tangent graph in Figure 3.27, we used **ZOOM** [ZTrig] and changed Xmin to 0. You could also have changed the calculator mode from connected to dot. Such a change would produce the graph in Figure 3.34.

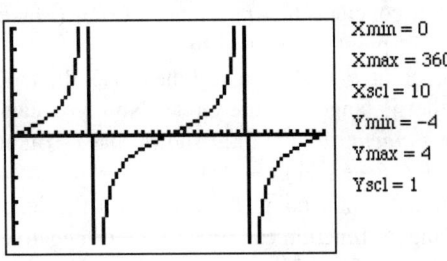

Xmin = 0
Xmax = 360
Xscl = 10
Ymin = –4
Ymax = 4
Yscl = 1

Figure 3.33
Tangent function with drawing errors

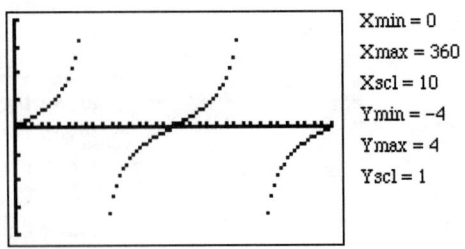

Xmin = 0
Xmax = 360
Xscl = 10
Ymin = –4
Ymax = 4
Yscl = 1

Figure 3.34
Tangent function graphed in dot mode

Negative Angles and Other Strange Things

What does $-50°$ mean? How about $750°$? Although these numbers may seem to have no connection to the applications of trigonometric functions you have seen, we need to define these quantities so that the trigonometric functions can be more widely applicable, especially in electronics.

Definition: Positive and Negative Angles
1. A positive angle is one that opens counterclockwise.
2. A negative angle is one that opens clockwise.
3. An angle that is either larger than $360°$ or smaller than $-360°$ is one that makes more than one trip around the circle.

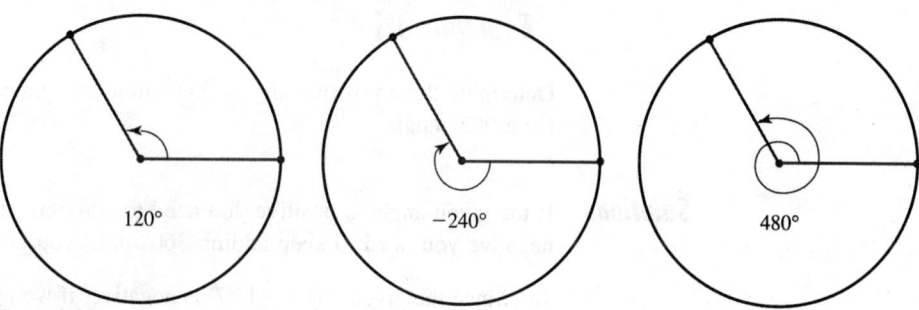

Figure 3.35

Figure 3.35 shows three angles that end at the same point on the circle. The $120°$ angle is the smallest positive angle equivalent to the $-240°$ and $480°$ angles. Since these three angles are equivalent, then

▶ $\sin 120° = \sin(-240°) = \sin 480°$.
▶ $\cos 120° = \cos(-240°) = \cos 480°$.
▶ $\tan 120° = \tan(-240°) = \tan 480°$.
▶ there are an unlimited number of other angles that share the same point.

You can obtain additional angles equivalent to a given angle by adding or subtracting a whole number multiple of $360°$. For example, $480° - 360° = 120°$ and $120° - 360° = -240°$.

Example 3.11

Determine three negative angles that end at the same point as (a) a $-80°$ angle and (b) a $1250°$ angle.

Solutions   If the given angle is negative you need to subtract $360°$ three times. If the given angle is positive you need to keep subtracting $360°$ until you get three negative angles.

(a) Since the given angle $-80°$ is negative, we need to subtract only three values of $360°$. This results in the values of

$$-80° - 360° = -440°$$
$$-440° - 360° = -800°$$
$$-800° - 360° = -1160°$$

(b) Here the given angle of $1250°$ is positive. Since $1250° - 3 \times 360° = 170°$ is positive but less than $360°$, the next time we subtract $360°$ we should get a negative number. The following are three possible values:

$$170° - 360° = -190°$$
$$-190° - 360° = -550°$$
$$-550° - 360° = -910°$$
■

Example 3.12

Determine three positive angles that end at the same point as (a) a $-160°$ angle and (b) a $905°$ angle.

Solutions   If the given angle is positive you need to add $360°$ three times. If the given angle is negative you need to keep adding $360°$ until you get three positive angles.

(a) Since the given angle $-160°$ is negative, if we add $360°$ we get a positive angle. Add $360°$ two more times to produce two other positive angles that end at the same point as a $-160°$ angle.

$$-160° + 360° = 200°$$
$$200° + 360° = 560°$$
$$560° + 360° = 920°$$

(b) Here the given angle of $905°$ is positive. The easiest way to find three other positive angles is to add $360°$ three times. While this does not produce the three smallest positive angles, it does satisfy the question.

$$905° + 360° = 1265°$$
$$1265° + 360° = 1625°$$
$$1625° + 360° = 1985°$$

If, in addition to $1265°$, you wanted to find the three smallest positive angles, you could subtract $360°$ twice to obtain $545°$ and $185°$. ■

Since any negative angle has its smallest positive equivalent angle, you could complete Table 3.8 without any measurement or calculation and use these values to graph the trigonometric functions. Figures 3.36–3.38 show the graphs with the values for negative angles added.

Adding the concept of angles larger than $360°$ and smaller than $-360°$, we can extend the graphs of the trigonometric functions indefinitely to the right and to the left. Functions like these which repeat over and over again are called **periodic**

Table 3.8

∠A	−20°	−70°	−90°	−135°	−150°	−180°	−210°	−270°	−300°	−330°	−360°
sin A			−1								
cos A				−0.707							
tan A								−0.577			

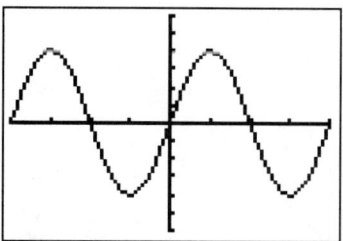

Xmin = −360
Xmax = 360
Xscl = 90
Ymin = −1.5
Ymax = 1.5
Yscl = 0.25

Figure 3.36
Graph of sine function

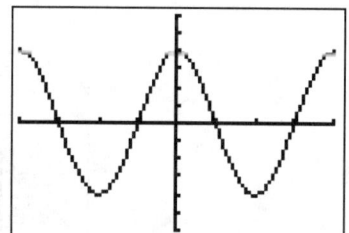

Xmin = −360
Xmax = 360
Xscl = 90
Ymin = −1.5
Ymax = 1.5
Yscl = 0.25

Figure 3.37
Graph of cosine function

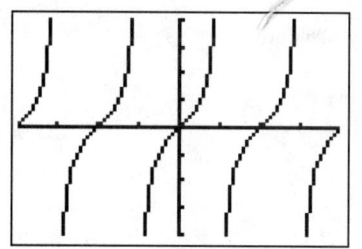

Xmin = −352.5
Xmax = 352.5
Xscl = 90
Ymin = −4
Ymax = 4
Yscl = 1

Figure 3.38
Graph of tangent function

functions. In the next section you will learn more about the applications of periodic functions.

The shape of the sine and cosine functions is called a **sine wave**. Any graph with the general shape of $y = \sin x$ is said to be **sinusoidal**. Any wave that does not have the basic shape of $y = \sin x$ is referred to as **nonsinusoidal**.

There are many periodic functions that are not sinusoidal. Some sinusoidal waves common to electricity are the square, triangular, sawtooth, and rectified waves shown in Figure 3.39.

Modeling Alternating Currents

Electricity can flow through wires in two ways—by **direct current** (dc) and by **alternating current** (ac). An example of direct current is a circuit powered by a battery. In a battery-powered circuit, electrons move through the wire from the negative pole of the battery to the positive pole. The graph of a dc current, as shown in Figure 3.40, is another example of a nonsinusoidal wave.

In an alternating current, electrons flow through the wire but move first in one direction and then back in the opposite direction—they oscillate. The electron movement produces energy. How much the electrons move in each direction and their speed determine the nature of the current.

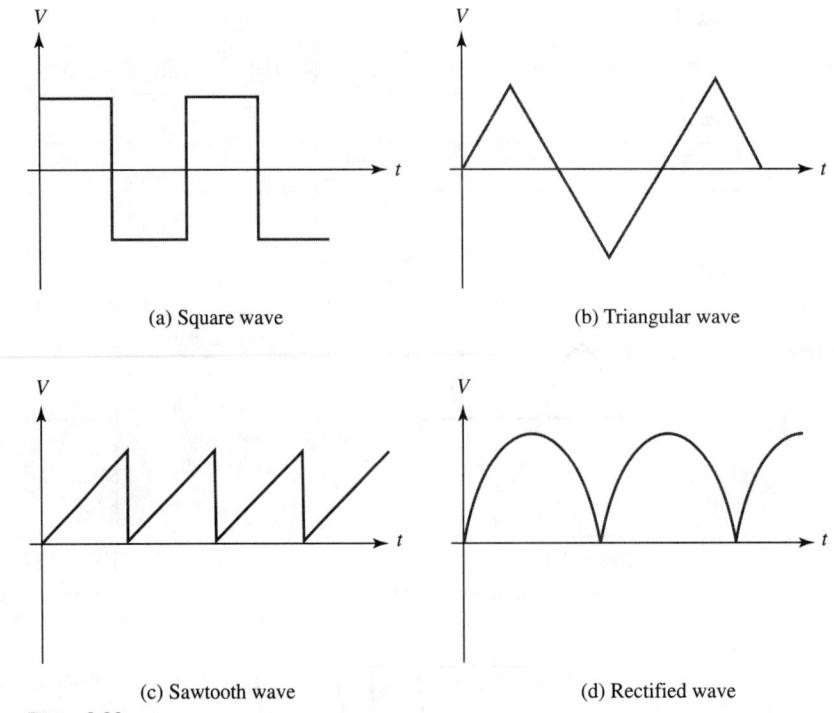

(a) Square wave   (b) Triangular wave

(c) Sawtooth wave   (d) Rectified wave

Figure 3.39
Nonsinusoidal waves

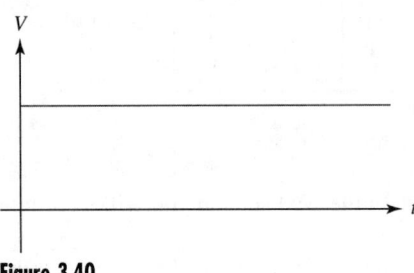

Figure 3.40

Because this back and forth motion of the electrons can be modeled with a sine or cosine function, electronics is one of the most important applications of trigonometry. In the sine wave shown in Figure 3.41, for example, the maximum and minimum points correspond to the farthest distance reached by the electron in each direction and to the measured voltage (V) in the circuit. The time (t) between two consecutive zero points corresponds to the speed of motion of the electrons and to the frequency of the current. In the next section we will look at the specifics of how the voltage and frequency of a circuit correspond to mathematical properties of the sine and cosine functions.

Solving Trigonometric Equations

Thus far in this chapter you have started with an angle and then, by measuring and/or calculating, you have produced the sine, cosine, or tangent of that angle. You made tables like Tables 3.4–3.6. But what if you know the cosine of an angle and need to find the value of the angle that produced that cosine?

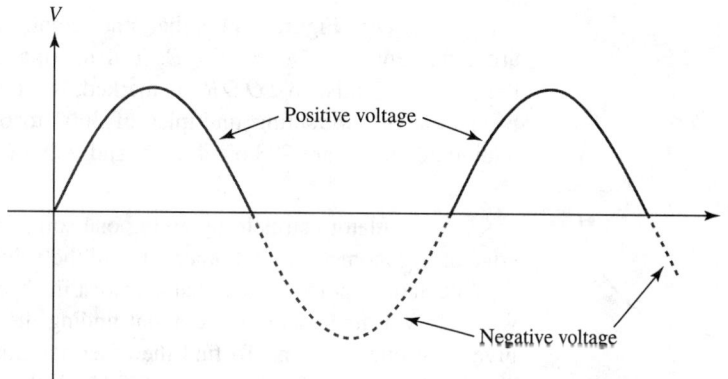

Figure 3.41

Example 3.13

If $\sin A = 0.5$, what is the value of angle A?

Solution   You can see from Tables 3.4 and 3.8 that there is more than one solution. Among positive angles, $A = 30°$ and $A = 150°$ both satisfy the equation $\sin A = 0.5$. Among the negative angles, $A = -210°$ and $A = -330°$ also are solutions to the equation. ∎

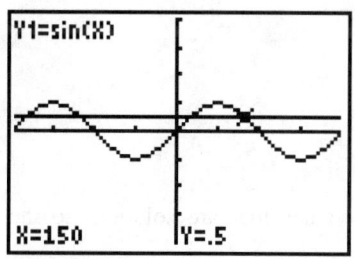

Figure 3.42

If you think about how we defined angles outside the values $0° \leq x \leq 360°$, you can see that there are many solutions. In fact there is no limit to the number of solutions. In mathematical language we say that the number of solutions to the equation $\sin A = 0.5$ is *infinite*.

Another way to look at the solutions of an equation like $\sin A = 0.5$ is with a graph. Figure 3.42 shows the graph of $y = \sin x$ and the graph of $y = 0.5$. The x-coordinates of the points of intersection are all solutions to the equation $\sin A = 0.5$. One of the solutions is shown by the Trace feature of the calculator.

Example 3.14

Give four solutions to the equation $\cos x = 0.4$. Report the solutions in degrees to the nearest tenth degree.

Solutions   The value 0.4 is not in the tables you created earlier in this section. In the years before calculators, people wanting to solve this problem made larger tables or used larger tables made by others. The calculator helps with the solutions to this problem, but you will see that you still have to use your knowledge of trigonometry to find more than one solution.

In Chapter 1 you used the calculator's ⬛2nd⬛ [TAN^{-1}] keys to find the angle a line made with the x-axis when you knew the slope of the line. For this example, the calculator reports the solution $x = \cos^{-1}(0.4) = 66.4°$. Figure 3.43 shows this solution on a circle of radius 10. Other solutions of the equation will correspond to other points on the circle that have the same x-coordinates as point P. There turns out to be only one other point on the circle that has the same x-coordinate as point P. Where is it?

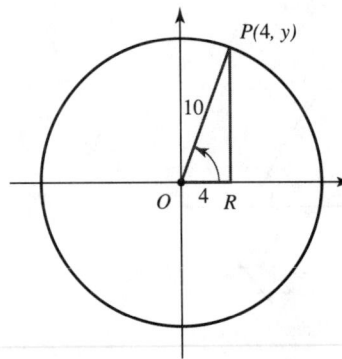

Figure 3.43

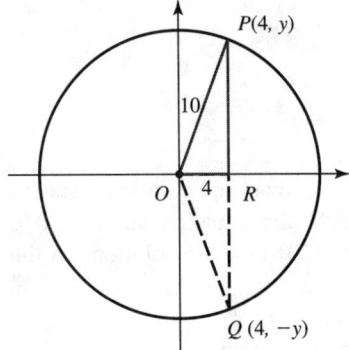

Figure 3.44

Point Q in Figure 3.44 is that other point. Since the triangles OPR and OQR are congruent, $\angle OPR = \angle OQR$. But in trigonometry, angles are measured from the positive x-axis, so $\angle OQR$, as marked, is $-66.4°$. We can find additional angles by adding or subtracting multiples of $360°$ to $66.4°$ and $-66.4°$. Some of these additional angles are $293.6°$, $426.4°$, and $-293.6°$. ∎

A calculator can help by giving one solution, but you have to use your knowledge of trigonometry to get others. Of all the solutions to an equation like $\cos x = a$, the calculator reports the one that is lowest in absolute value. The following example will help you understand more about finding the other solutions after the calculator gives you one of them. To find the other solutions use the calculator's answer and Table 3.9 (the completed version of Table 3.3).

Table 3.9 Signs of trigonometric functions in each quadrant

	Point *B* is in			
	Quadrant I	Quadrant II	Quadrant III	Quadrant IV
sin *A* is	+	+	−	−
cos *A* is	+	−	−	+
tan *A* is	+	−	+	−

 Example 3.15

Write the two angles with the smallest absolute values that are solutions to the equation $\sin x = -0.5$.

Solution The calculator reports that $\sin^{-1}(-0.5) = -30°$; the $-30°$ angle is marked in Figure 3.45. To find the next larger angle we'll use Table 3.9. According to the table, the sine is negative in Quadrants III and IV. Since $-30°$ is in Quadrant IV, the next larger angle must be in the third quadrant. Construct the congruent triangle in the third quadrant as shown in Figure 3.45; then mark the smallest angle that ends in that quadrant. The second solution is $210°$.

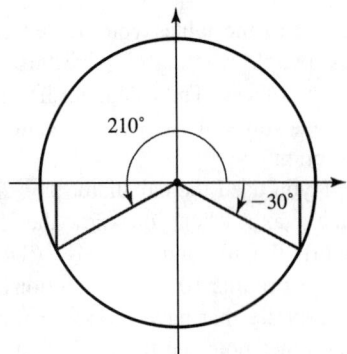

Figure 3.45   ∎

Example 3.16

Write the smallest two positive angles that are solutions to the equation $\cos x = -0.5$.

Solution   The calculator reports that $\cos^{-1}(-0.5) = 120°$, which is in Quadrant II, as shown in Figure 3.46. According to Table 3.9, the cosine is negative in Quadrants II and III, so the next larger angle must end in the third quadrant. Construct the congruent triangle in the third quadrant as shown in Figure 3.46 and mark the smallest angle that ends in that quadrant. The second solution is 240°.

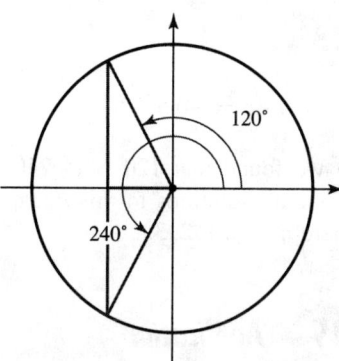

Figure 3.46

Example 3.17

Write the two angles with the smallest absolute values that are solutions to the equation $\tan x = -0.5$.

Solution   The calculator reports $\tan^{-1}(-0.5) = -27°$; that angle is in Quadrant IV, as shown in Figure 3.47. To find the next larger angle, first remember that it must end in the second quadrant, since the tangent is negative in Quadrants II and IV. Construct the congruent triangle in the second quadrant as shown in Figure 3.47 and mark the smallest angle ending in that quadrant. The second solution is 153°.   ∎

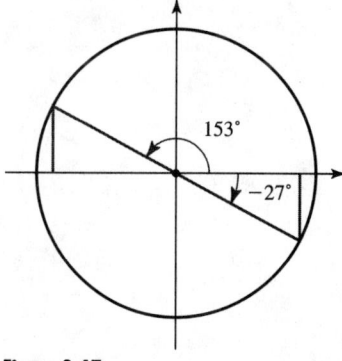

Figure 3.47

In Examples 3.15–3.17 we found two solutions to each of the given equations. How would we find all solutions to an equation such as $\cos x = -0.5$?

If one solution is the angle x, then by adding multiples of 360° to x the angle will terminate at the same point where x terminates. That is, the angles $x + 360°$, $x + 720°$, and $x + 1080°$ are also solutions if x is a solution. Going in the other direction, three solutions are $x - 360°$, $x - 720°$, and $x - 1080°$. That is, if you add or subtract multiples of 360° to a solution, you get more solutions. In mathematical language the solutions are expressed by the formula $x \pm 360n$ where n is one of the numbers $0, 1, 2, 3 \ldots$.

Remember (Examples 3.15–3.17) that there are *two different* terminating segments that represent angles having the same values of sine, cosine, or tangent.

Writing All Roots of Trigonometric Equations

If x_1 and x_2 are two angles that terminate in different segments and x_1 and x_2 produce the same value of sine, cosine, or tangent, then all the angles that have the same value of sine, cosine, or tangent can be written as

$$x_1 \pm 360n \quad \text{and} \quad x_2 \pm 360n$$

where $n = 0, 1, 2, 3, \ldots$.

Example 3.18

Find all solutions to $\cos x = -0.5$.

Solution   In Example 3.16 we found that $120°$ and $240°$ are two solutions of the equation $\cos x = -0.5$. Thus all solutions to this equation are given by $120° \pm 360°n$ and $240° \pm 360°n$ where $n = 0, 1, 2, 3, \ldots$. ∎

Example 3.19   Application

The **second law of refraction** is known as **Snell's law**. It states that as a light ray passes from one medium to another, as shown in Figure 3.48, the ratio of the sine of the angle of incidence θ_i to the sine of the angle of refraction θ_r is a constant μ called the **index of refraction**, with respect to the two mediums. Thus we have

$$\frac{\sin \theta_i}{\sin \theta_r} = \mu$$

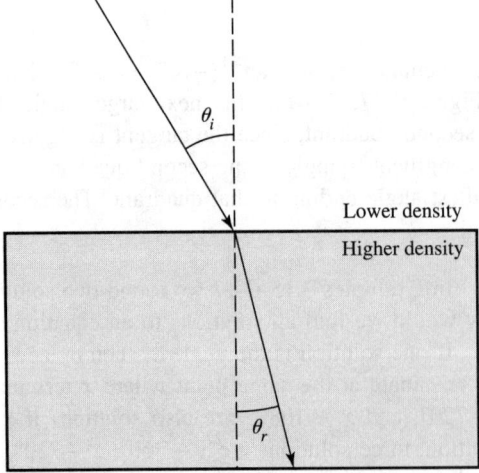

Lower density

Higher density

Figure 3.48

The index of refraction of heat-resistant epoxy relative to air is $\mu = 1.56$. Determine the angle of refraction θ_r of a ray of light that strikes some heat-resistant epoxy with an angle of incidence $\theta_i = 40°$.

Solution Substituting the given information into the equation $\dfrac{\sin \theta_i}{\sin \theta_r} = \mu$ produces $\dfrac{\sin 40°}{\sin \theta_r} =$ 1.56, which can be rewritten as

$$\sin \theta_r = \frac{\sin 40°}{1.56} \approx 0.41204$$

Since $\sin^{-1} 0.41204 \approx 24.3°$, we see that the angle of refraction is about $24.3°$. ∎

Roots of Trigonometric Functions

In Chapter 2 you learned that a root of a function $f(x)$ is a solution of the equation $f(x) = 0$. Roots are quite visible on a graph because they are located at the points where the graph of the function crosses the x-axis. For example, a root of $f(x) = \cos x$ is a solution to the equation $\cos x = 0$.

Example 3.20

Determine the roots of the function $f(x) = \cos x$.

Solution This problem is asking us to solve the equation $\cos x = 0$. Using the techniques of Example 3.16 we can find that two solutions of the equation $\cos x = 0$ are $x_1 = 90°$ and $x_2 = 270°$. Thus all the solutions are $90° \pm 360°n$ and $270° \pm 360°n$ where $n = 0, 1, 2, 3, \ldots$. ∎

Section 3.2 Exercises

In Exercises 1–6 determine the two smallest positive angles that are equivalent to the given angle.

1. $967°$
2. $691°$
3. $1884.2°$
4. $-173°$
5. $-896.4°$
6. $-1996.9°$

7. As a continuation of Activity 3.3, respond to each of the following statements where x is the largest negative angle that satisfies the statement.
   (a) If $\sin 220° = \sin x$ then $x =$ _____ and $220° + x =$ _____.
   (b) If $\sin 261° = \sin x$ then $x =$ _____ and $261° + x =$ _____.

(c) If $\sin 314° = \sin x$ then $x =$ _____ and $314° + x =$ _____.

(d) In general, $\sin(180° - x) =$ _____.

8. As a continuation of Activity 3.4, respond to each of the following statements where x is the largest negative angle that satisfies the statement.
   (a) If $\cos 250° = -\cos x$ then $x =$ _____ and $250° + x =$ _____.
   (b) If $\cos 291° = -\cos x$ then $x =$ _____ and $291° + x =$ _____.
   (c) If $\cos 349° = -\cos x$ then $x =$ _____ and $349° + x =$ _____.
   (d) In general, $\cos(180° - x) =$ _____.

In Exercises 9–14 classify each statement as true, false, *or* can't tell *because you do not have enough information.*

9. $\sin(90° - x) = \cos x$
10. $\cos(90° - x) = \sin x$
11. $\tan(180° - x) = \tan x$
12. $\tan(180° - x) = -\tan x$
13. $\sin(180° + x) = \sin x$
14. $\sin(180° + x) = -\sin x$

The following information is needed for Exercises 15–30. You may have noticed that the value of $\cos 60°$ and $\cos(-60°)$ are both equal to 0.5. For the sine function the values have the opposite sign, since $\sin 60° = 0.8660$ and $\sin(-60°) = -0.8660$. In general the following are true:

▶ $\cos(-x) = \cos x$.

▶ $\sin(-x) = -\sin x$.

▶ A function (like cosine) for which $f(-x) = f(x)$ is called an **even function.**

▶ A function (like sine) for which $f(-x) = -f(x)$ is called an **odd function.**

Given the above, classify each of the functions in Exercises 15–24 as odd, even, or neither.

15. $f(x) = x^2$

16. $g(x) = x^3$

17. $h(x) = x^4 + 10$

18. $i(x) = x^3 + 10$

19. $j(x) = 10 - x^2$

20. $k(x) = 12/x$

21. $m(x) = 2x^3 - 4x$

22. $n(x) = 3x^4 - x^2 - 5$

23. $p(x) = x^2 - x$

24. $q(x) = 2x - 1$

25. Write a linear function that is an odd function.

26. Write a linear function that is an even function.

27. Is the tangent function odd, even, or neither?

28. Is the function given by $y = \dfrac{1}{\sin x}$ odd, even, or neither?

29. Graph each of the following even functions and explain for each how you can tell from the graph that the function is even.
   (a) $f(x) = x^2$
   (b) $g(x) = x^4 - 3x^2$
   (c) $h(x) = x^6 + x^2 - 8$
   (d) $j(x) = \cos x$

30. Graph each of the following odd functions and explain for each how you can tell from the graph that the function is odd.
   (a) $f(x) = x^3$
   (b) $g(x) = x^5 - 3x^3$
   (c) $h(x) = x^5 + x^3 - 3x$
   (d) $j(x) = \tan x$

31. Consider the function $f(x) = \dfrac{12}{x - 3}$.
   (a) For what value(s) of x will f have no value?
   (b) Complete this table for the function f.

x	−3	−2	−1	0	1	2	3	4	5	6
f(x)										

   (c) Use your calculator to investigate the function near $x = 3$. Check values on either side of 3; for example, at $x = 2.99$, $x = 2.999$, $x = 3.01$, and $x = 3.001$.
   (d) Plot the graph by hand on graph paper and check the result with your calculator.

32. Consider the function $g(x) = \dfrac{9}{x^2 - 4}$.
   (a) For what value(s) of x will g have no value?
   (b) Complete the following table for the function f.

x	−3	−2	−1	0	1	2	3	4	5	6
g(x)										

   (c) Use your calculator to investigate the function near $x = -2$. Check values on either side of -2; for example, check at $x = -2.01$, $x = -2.001$, $x = -1.99$, and $x = 1.999$.
   (d) Use your calculator to investigate the function near $x = 2$. Check values on either side of 2; for example, at $x = 1.99$, $x = 1.999$, $x = 2.01$, and $x = 2.001$.
   (e) Plot the graph by hand on graph paper and check the result with your calculator.

In Exercises 33–38 solve each of the trigonometric equations.

33. $\sin x = 0$

34. $\tan x = 0$

35. $\sin x = 0.75$

36. $\cos x = 0.25$

37. $2 \tan x = -5.6$

38. $4 \sin x = -4$

39. *Optics* The index of refraction of styrene polyester is 1.59. If the angle of incidence is 32.0°, what is the angle of refraction?

40. *Optics* The index of refraction of carbon disulfide is 1.669. If the angle of incidence is 47.5°, what is the angle of refraction?

In Exercises 41–46 use the information that the value of the voltage at any instant in an alternating current (ac) is given by the formula $V = V_{max} \sin \theta$ where V_{max} is the maximum voltage and θ is the angle the conductor makes with respect to the parallel lines of force. Similarly, the value of the current at any instant is given by $I = I_{max} \sin \theta$ where I_{max} is the maximum current.

41. *Electronics* If an ac wave has a maximum voltage of 110 V, what is the instantaneous voltage at 70°?

42. *Electronics* If an ac wave has a maximum voltage of 105 mA, what is the instantaneous voltage at 30°?

43. *Electronics* An ac voltage wave has an instantaneous value of 85 V at 48°. What is the maximum value of the wave?

44. *Electronics* What is the maximum current of an ac wave if the instantaneous current is 15 A at 12°?

45. *Electronics* The maximum current of an ac wave is 10 A. At what instant will the instantaneous current be 8.5 A? (Give the smallest positive angle.)

46. *Electronics* The maximum voltage of an ac wave is 110 V. At what instant will the instantaneous voltage be 65 V? (Give the smallest positive angle.)

3.3   Period, Amplitude, Frequency, and Roots

You saw in the previous section that the graphs of the trigonometric functions repeat themselves as you move along the x-axis. Functions like these are called **periodic**. In applications of periodic functions, five terms are often used: period, cycle, amplitude, frequency, and phase shift. We'll address the first four terms in this section.

Period and Cycle

In Section 3.2 we saw that $\sin 120° = \sin(-240°) = \sin 480°$. If we add any integral multiple of 360° to 120° we have an angle that has the same sine value. For example,

$$\sin 120° = \sin(120° + 6 \times 360°) = \sin 2280°$$
$$\text{and } \sin 120° = \sin(120° +{}^-8 \times 360°) = \sin(-2760°)$$

The same reasoning applies to any angle. For instance,

$$\sin 248° = \sin(248° + 7 \times 360°) = \sin 2768°$$
$$\text{and } \sin 37° = \sin(37° +{}^-5 \times 360°) = \sin(-1763°)$$

We can express this concept formally as

$$\sin x = \sin(x + k \cdot 360°)$$

where k is an integer.

Since 360° is the smallest positive angle for which this is true, we say that the **period** of the sine function is 360°. In radians the period of the sine function is 2π. We will talk more about radians in Section 3.5.

> **Definition: Periodic Function**
> If the function f is a periodic function, then
>
> $$f(x) = f(x + p)$$
>
> The smallest positive value of p for which $f(x) = f(x + p)$ is called the **period**. A **cycle** is that portion of the graph of a periodic function contained within one period.

Example 3.21

Sketch one cycle of $y = \sin x$ starting with (a) $-50°$, (b) $30°$, and (c) $0°$.

Solutions   (a) The solution is shown in Figure 3.49.
(b) The solution is shown in Figure 3.50.
(c) The cycle shown in Figure 3.51 is the one that most people associate with one complete cycle of the sine function.

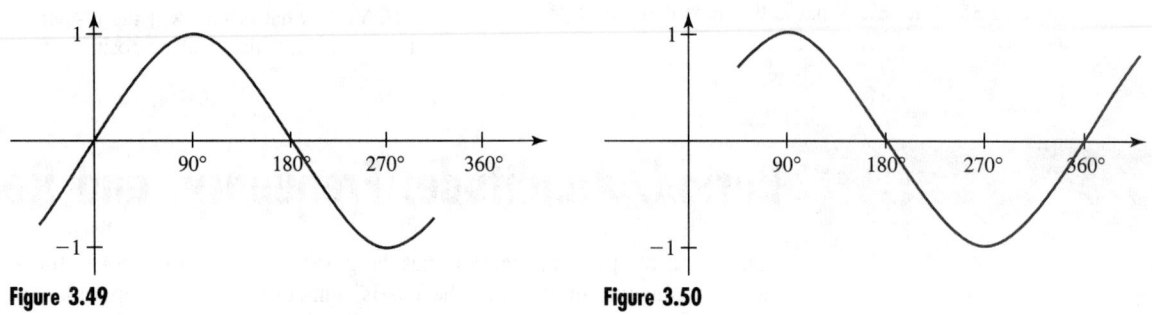

Figure 3.49   **Figure 3.50**

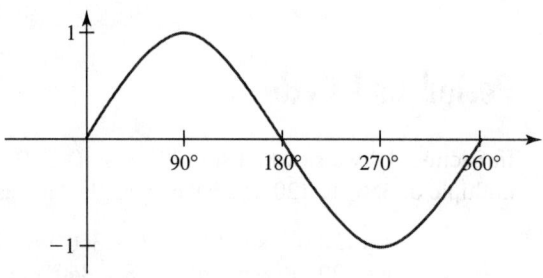

Figure 3.51

Example 3.22

What is the period of the periodic function graphed in Figure 3.52.

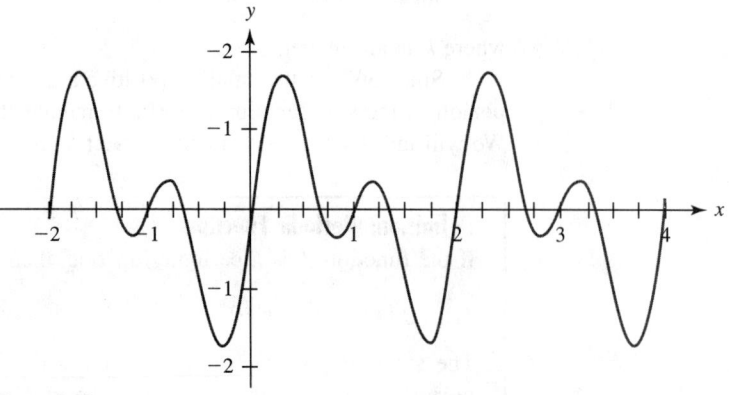

Figure 3.52

Solution   One complete cycle is highlighted in Figure 3.53. Since this cycle begins at 0 and ends at 2, the period is $2 - 0 = 2$ units.

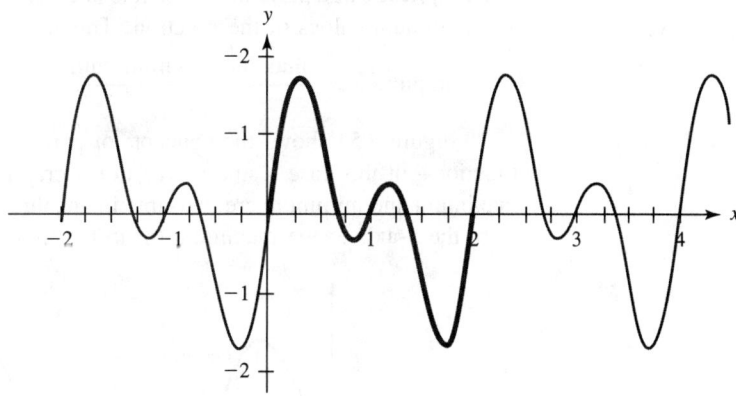

Figure 3.53

Frequency

The **frequency** of a periodic function is the reciprocal of the period. Frequency is measured in cycles per unit. If the x-axis unit is degrees, the frequency is the number of cycles per degree. The metric unit of frequency is **hertz** (Hz) and

$$1 \text{ Hz} = 1 \text{ cycle/sec} = 1 \text{ s}^{-1}$$

Example 3.23

What is the frequency of $y = \sin x$ if x is measured in degrees?

Solution   Since the period of $y = \sin x$ is $360°$, the frequency is

$$\frac{1 \text{ cycle}}{360°} = \frac{1}{360} \text{ cycle/degree}$$

Example 3.24

Find the period of a periodic wave with a frequency of 500 Hz.

Solution   Since frequency $= \dfrac{1}{\text{period}}$, we have period $= \dfrac{1}{\text{frequency}}$. We are given a frequency of 500 Hz, so the period is

$$\frac{1}{500 \text{ Hz}} = 0.002 \text{ sec, or 2 ms}$$

Amplitude

The **amplitude** of a periodic function is one-half the difference between the maximum and minimum values of the function. Thus

$$\text{amplitude} = \frac{\text{maximum} - \text{minimum}}{2}$$

Figure 3.54 shows the concepts of period, cycle, and amplitude for a periodic function—in this case a sine wave. For a graph like the one in Figure 3.54, whose maximum and minimum are the same in absolute value, the amplitude is the distance from the x-axis to the maximum (or to the minimum).

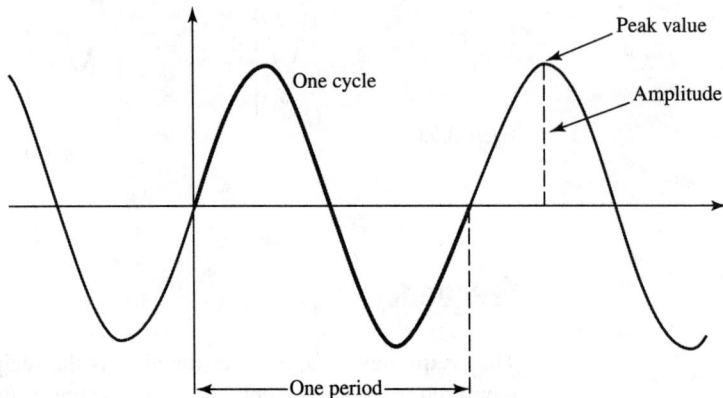

Figure 3.54

Example 3.25

What are the period, amplitude, and frequency of the basic trigonometric functions (a) $f(x) = \sin x$, (b) $g(x) = \cos x$, and (c) $h(x) = \tan x$?

Solutions   Refer back to the graphs you made in Activity 3.5. The period of the sine and cosine functions are both 360° or 2π radians, and the period of the tangent function is 180° or π radians. The frequency of both sine and cosine is $\frac{1}{360°} \approx 0.02778$ cycle/degree. The period of the tangent function is $\frac{1}{180°} \approx 0.05556$ cycle/degree. The amplitude of both sine and cosine is 1. There is no amplitude for the tangent function because it does not have either a maximum or a minimum.   ■

You will see that other forms of these functions can have different values for the amplitude, period, and frequency. For example, the function $f(x) = \sin 2x$ does not have a period of 360°.

Example 3.26   **Application**

What are the period, amplitude, and frequency of the graph in Figure 3.55. This graph is a plot of current (in amps) vs. time (in seconds) of an alternating current circuit.

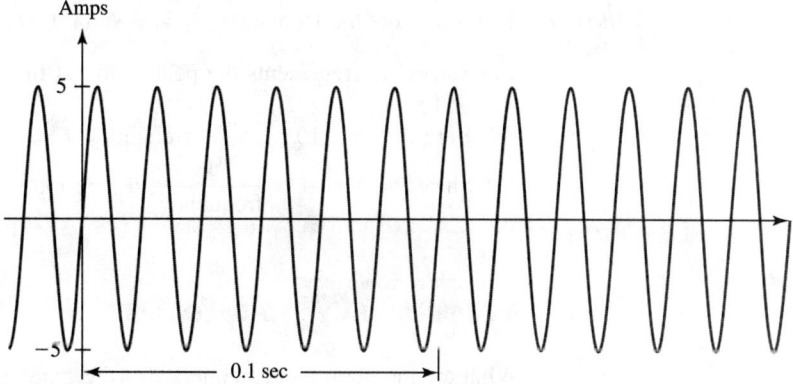

Figure 3.55

Solution

There are six cycles in the time 0.1 seconds. Therefore the period is $\frac{0.1}{6} \approx 0.01667$ seconds per cycle. The frequency is $\dfrac{1}{\text{period}} \approx \dfrac{1}{0.01667} = 60$ cycles per second. Since the hertz is defined as one cycle per second, this graph represents a 60-hertz or 60-Hz circuit.

Looking at the graph it appears that the maximum value is 5 amps and the minimum value is −5 amps. Thus the amplitude is $\frac{5-(-5)}{2} = \frac{10}{2} = 5$. The amplitude is 5 amps.

The equations for the sine wave voltage V and current i can be written in terms of the frequency f as

$$\begin{array}{cc} \text{Degrees} & \text{Radians} \end{array}$$
$$V = V_{\max} \sin(360° ft) = V_{\max} \sin(2\pi ft)$$
$$i = I_{\max} \sin(360° ft) = I_{\max} \sin(2\pi ft)$$

where $V_{\max}$ and $I_{\max}$ are the **peak values** for the voltage and current (see Figure 3.54).

Example 3.27 Application

Given the ac sine wave current $i = 2.5 \sin(21,600°t)$, find (a) the peak value for the current, (b) the frequency, and (c) the period of the wave where t is in seconds.

Solutions

This wave fits the formula $i = I_{\max} \sin(360° ft)$.

(a) Since $I_{\max}$ represents the peak value of the current, this wave has a peak value of 2.5 A.
(b) Here $360° f = 21,600°$, so the frequency $f = \frac{21,600°}{360°} = 60$ Hz.
(c) Since the period $= \dfrac{1}{\text{frequency}}$, the period is $\frac{1}{60}$ sec ≈ 0.0167 s $= 16.7$ ms. ■

Example 3.28 Application

An ac sine wave voltage is given by $V = 150 \sin(110\pi t)$. Find (a) the peak value for the current, (b) the frequency, and (c) the period of the wave where t is in seconds.

Solution This wave fits the formula $V = V_{max} \sin(110\pi t)$.

(a) Since V_{max} represents the peak value of the voltage, this wave has a peak value of 150 V.

(b) Here $2\pi f = 110\pi$, so the frequency $f = \frac{110\pi}{2\pi} = 55$ Hz.

(c) Since the period $= \dfrac{1}{\text{frequency}}$, the period is $\frac{1}{55}$ sec ≈ 0.01818 s ≈ 18.2 ms. ■

Example 3.29   Application

What are the period and frequency of the graph pictured in Figure 3.56? This graph is an electrocardiogram (ECG) of a heartbeat (the large spikes) and a heart flutter (the smaller peaks). The paper for an ECG moves at a constant rate of 25 mm/sec. The lines on the graph paper are 1 mm apart.

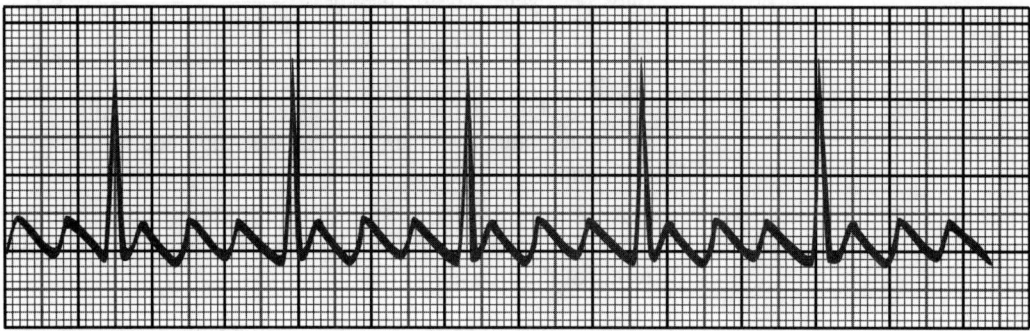

Figure 3.56

Solution Actually there are two periods of interest here—the heartbeat period and the flutter period.

▶ The heartbeat period is the time between large spikes. In any set of actual data like this you must measure a few periods and take the average. We measured the times in the four intervals shown and got the numbers 0.96, 0.92, 0.96, and 0.94 seconds. The average of these is $\frac{0.96+0.92+0.96+0.94}{4} = \frac{3.78}{4} = 0.945$ sec. The heartbeat frequency is $1/0.945 \approx 1.06$ beats per second or $1.06 \times 60 \approx 64$ beats per minute.

▶ The flutter period is estimated in the same way. We got about 0.26 seconds for the period. The flutter frequency is $1/0.26 \approx 3.85$ flutters per second or about 232 flutters per minute. ■

Example 3.30   Application

The graphs in Figure 3.57 show the times of sunrise and sunset and the number of hours of daylight at different times of year in San Francisco. The data was recorded every ten days for two years. What are the (a) period and (b) amplitude for the day-length curve, and (c) why should these curves look like trigonometric functions?

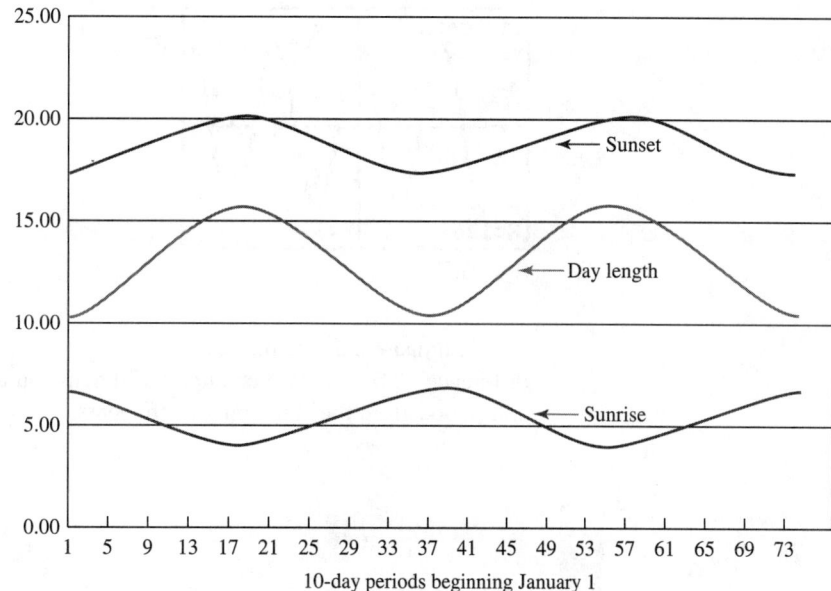

Figure 3.57
Sunrise, sunset, and length of day for San Francisco over two years (Source: http://www.nws.mbay.net/sunset.html#sfo_tide)

Solution   Measurements will be based on estimated readings from the graph.

(a) The period appears to be about 37 units. Since these units are 10 days each, the period is about 370 days. Actually data like this repeats every year, so the period is really about 365.25 days.

(b) Estimating the amplitude requires that we estimate the y-coordinate of the top and bottom points. With this graph the best you can do is to measure them to the nearest hour. We got 16 hours for the maximum and 10 hours for the minimum. The amplitude is therefore about $\frac{16-10}{2} = 3$ hours.

(c) Why should these curves look like trigonometric functions? This is difficult to answer precisely. If you think about how a circle is used in the definition of the trigonometric functions, about how sunset and sunrise are determined by the somewhat circular motion of Earth around the Sun, about the "circular" shape of Earth, and about how Earth's axis of rotation is not perpendicular to a line from the center of the Sun to the center of Earth, then you can see that there might be a connection. What is surprising is how much like a sine or cosine the day-length curve appears. ∎

Example 3.31

What are the period and amplitude of the function $f(x) = 3\cos 2x$?

Solution   The graph of $f(x) = 3\cos 2x$ is shown in Figure 3.58. Use the calculator's Trace function to move to the end of one cycle, as shown in the figure. You can see from the trace coordinates at the bottom of the screen that the period is $180°$ and the amplitude is 3.

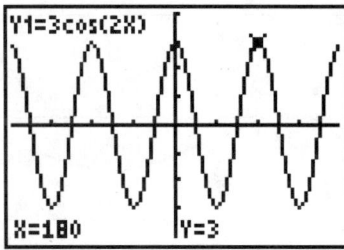

Figure 3.58

Compare these results with what you saw in Example 3.25(b). What causes the difference? The next two examples will help you to see how to predict the amplitude and period from the function's coefficients.

Example 3.32

What are the period and amplitude of the function $f(x) = -\dfrac{\sin(x/3)}{2} = -0.5 \sin\left(\frac{1}{3}x\right)$?

Solution   The graph of $y = -0.5 \sin\left(\frac{1}{3}x\right)$ is shown in Figure 3.59. Notice that the window settings are different from those used in Figure 3.58. You can see from the trace coordinates at the bottom of the screen in Figure 3.59 that the period is 1080°. If you move the trace cursor to one of the high or low points, you will see that the amplitude is 0.5.

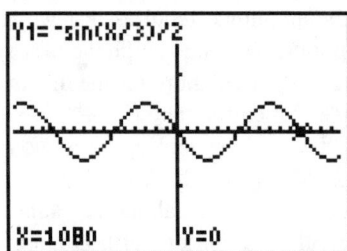

Figure 3.59

Calculator Lab 3.2

Determining Period and Amplitude

Use your calculator to experiment with the function $f(x) = A \sin Bx$ to determine the connection between the coefficients A and B and the period and amplitude of the function.

Procedures   Systematically change the coefficients A and B in the trigonometric function $f(x) = A \sin Bx$; then describe the connection between the coefficients A and B and the period and amplitude of the function.

   As with other experiments you have done, it is better to change one coefficient at a time. When you know what effect it has on the function, change the other coefficient. Record your answers in a table similar to Table 3.10. (You might want to refer to Examples 3.31–3.32.)

1. Write your conclusions for positive whole number values of A and B.
2. Try fractional values of A and B and see if your conclusions fit what you observe. Change your conclusions if necessary.
3. Try negative values of A and B and see if your conclusions fit what you observe. Change your conclusions if necessary.
4. Repeat steps 2 and 3 for the cosine and tangent functions.
5. Write a paragraph that summarizes your results.

Table 3.10   Recording the effects of A and B

A	B	Amplitude	Period
1	1		
3	2		
−0.5	$\frac{1}{3}$		
0.5	−1		

Solving More Advanced Trigonometric Equations

Now that you have studied functions of the form $f(x) = A \sin Bx$, you are ready to learn how to solve equations involving functions of this type. You will see that once again there are many solutions to these equations and that the methods you learned earlier will be useful.

Calculator Lab 3.3

Solving Trigonometric Equations

This calculator lab will show you one way to use your calculator to solve more advanced trigonometric equations by solving the particular equations $3 \sin x = 1.5$ and $3 \sin 2x = 1.5$.

Procedures

1. Use the ZTrig window setting and plot the graphs of (a) $y = 3 \sin x$, (b) $y = 3 \sin 2x$, and (c) $y = 1.5$.
2. Solve the equation $3 \sin x = 1.5$ by using the calculator's Trace feature to determine the x-coordinates of the points of intersection of the graphs of $y = 3 \sin x$ and $y = 1.5$.
3. Solve the equation $3 \sin 2x = 1.5$ by using the calculator's Trace feature to determine the x-coordinates of the points of intersection of the graphs of $y = 3 \sin 2x$ and $y = 1.5$.

4. Enter three solutions in each empty cell of the following table.

Equation	Positive Solutions	Negative Solutions
$3 \sin x = 1.5$		
$3 \sin 2x = 1.5$		

5. Explain how you could have gotten these solutions without using the graphs.
6. Write formulas for all the solutions of each equation.

Equation	Formula for All the Solutions
$3 \sin x = 1.5$	
$3 \sin 2x = 1.5$	

Example 3.33

Determine to the nearest degree all the solutions of the equation $5 \sin 3x = 3$.

Solution   Begin by dividing both sides of the equation by 5 in order to change the equation to $\sin 3x = \frac{3}{5}$ or $\sin 3x = 0.6$.

Next use the calculator's $\sin^{-1}$ function ([2nd] [SIN^{-1}]) to estimate a value for $3x$. You should obtain $3x = \sin^{-1}(0.6) \approx 36.9°$.

Draw a congruent triangle in the second quadrant (where sine is also positive) to find another solution of $\sin 3x = 0.6$. See Figure 3.60. The second angle is approximately $180° - 36.9° = 143.1°$.

Thus the formula for all the answers in terms of $3x$ are

$$3x \approx 36.9° \pm 360°n \text{ and } 3x \approx 143.1° \pm 360°n$$

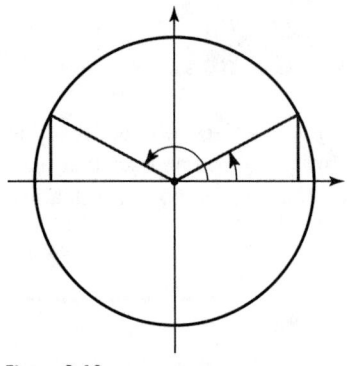

Figure 3.60

where $n = 0, 1, 2, 3, \ldots$. But these are the answers for $3x$ not x. To find the answers for x we need to divide these formulas by 3. Dividing the first formula, $3x \approx 36.9° \pm 360°n$, by 3 produces

$$x \approx \frac{36.9}{3} \pm 120°n \approx 12.3° \pm 120°n$$

which to the nearest degree is $12° \pm 120°n$. Some of the solutions are $x = 12°, 132°, 252°$ and $x = -108°, -228°, -348°$.

Dividing the second formula, $3x \approx 143.1° \pm 360°n$, by three we obtain

$$x \approx \frac{143.1}{3} \pm 120°n$$

$$\approx 47.7° \pm 120°n$$

which to the nearest degree is $48° \pm 120°n$. Some of the solutions are $x \approx 48°, 168°, 288°$ and $x \approx -72°, -192°, -312°$.   ∎

You will use these same ideas in the final part of this section as you study the roots of equations of the form $y = A \sin Bx$.

Roots of Transformed Trigonometric Functions

Changing the value of B in functions such as $y = \sin Bx$, $y = \cos Bx$, and $y = \tan Bx$ has the effect of horizontally stretching or shrinking the graphs of these functions. What does all this stretching and shrinking do to the roots of a trigonometric function? Let's explore the roots of the function $f(x) = \cos Bx$ where $B = 1$, 2, 3, ... Since the roots of a function f are the x-intercepts where $f(x) = 0$, this is the same as asking where the graph of f touches the x-axis.

| Calculator Lab 3.4 | The Effect of Stretching and Shrinking on the Location of Roots |

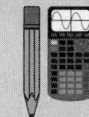

In this lab you will write three consecutive roots, including one negative root, for each of several functions of the form $f(x) = \cos Bx$. You will then estimate the spacing between the roots and look for a relationship between the spacing and the value of B in the function.

Procedures

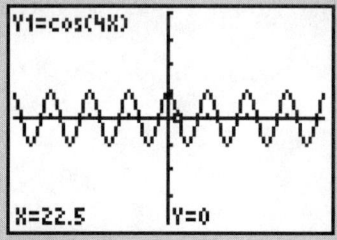

Figure 3.61

1. Write three consecutive roots, including one negative root, for each of the functions (a) $f_1(x) = \cos x$, (b) $f_2(x) = \cos 2x$, (c) $f_3(x) = \cos 3x$, and (d) $f_4(x) = \cos 4x$. If you graph these functions using the ZTrig window setting you can read the roots off the graph. For example, in Figure 3.61 you can see how the Trace function reveals that $x = 22.5$ is one of the roots of $f_4(x) = \cos 4x$. The first two rows of Table 3.11 have been completed for you. Some roots of the other two functions have also been provided.

Table 3.11 Roots of some cosine functions

Function	Some Roots	Spacing between Roots
$\cos x$	$-90°, 90°, 270°$	$180°$
$\cos 2x$	$-45°, 45°, 135°$	$90°$
$\cos 3x$	$-30°, 30°$	
$\cos 4x$	$22.5°$	
$\cos(-4x)$		
$\cos(-6x)$		
$\cos\left(\frac{1}{2}x\right)$		
$\cos 12x$		
$\cos Bx$		

2. Write three consecutive roots, including one negative root, for each of the functions (a) $f_1(x) = \cos(-4x)$, (b) $f_2(x) = \cos(-6x)$, (c) $f_3(x) = \cos\left(\frac{1}{2}x\right)$, and (d) $f_4(x) = \cos 12x$.
3. Determine the spacing between two consecutive roots on each row and enter this number in the last column.

4. Look for some patterns. Use them to complete the last row of the table.
5. Write a short paragraph expressing your patterns in words.

There were two patterns to figure out in Calculator Lab 3.4: the location of the first positive root and the spacing between roots. You should have noticed the following patterns.

1. The first positive root of $\cos Bx$ is $\dfrac{90°}{|B|}$ because that root occurs at the point where $|B|x = 90°$.

2. The spacing appears to be $\dfrac{180°}{|B|}$ because the period of the function $f(x) = \cos Bx$ is $\dfrac{360°}{|B|}$ and there are two roots within each period of the sine or cosine function. Therefore there are two roots within a distance of $\dfrac{360°}{|B|}$; so the spacing between the roots is $\dfrac{360°}{|B|} \div 2 = \dfrac{180°}{|B|}$.

The missing portions of Table 3.11 in Example 3.4 are

Function	First Positive Root	Spacing between Roots
$\cos(-4x)$	$\dfrac{90°}{|-4|} = 22.5°$	$\dfrac{180°}{|-4|} = 45°$
$\cos(-6x)$	$\dfrac{90°}{|-6|} = 15°$	$\dfrac{180°}{|-6|} = 30°$
$\cos\left(\frac{1}{2}x\right)$	$\dfrac{90°}{1/2} = 180°$	$\dfrac{180°}{1/2} = 360°$
$\cos(12x)$	$\dfrac{90°}{12} = 7.5°$	$\dfrac{180°}{12} = 15°$
$\cos Bx$	$\dfrac{90°}{|B|}$	$\dfrac{180°}{|B|}$

These results are important because sometimes you cannot rely on the graphing calculator and its Trace function. For example, Figures 3.62 and 3.63 show that with the ZTrig setting the graph of $\cos(30x)$ is unreadable and $\cos(40x)$ is inaccurate. You can see that the graph of $y = \cos(40x)$ is incorrect because it should have a

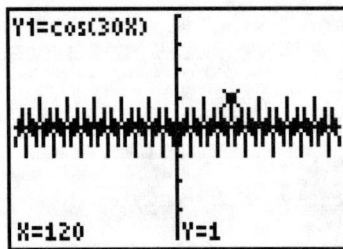

Figure 3.62
The graph of $y = \cos(30x)$ is unreadable in the ZTrig setting

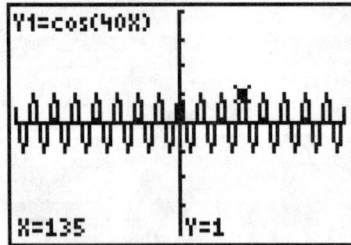

Figure 3.63
The graph of $y = \cos(40x)$ is inaccurate in the ZTrig setting

shorter period than the graph of $y = \cos(30x)$. The graph is inaccurate because each plotted point is $7.5°$ apart, but the first two positive roots of $\cos(40x)$ are $\frac{90°}{40} = 2.25°$ and $2.25° + \frac{180°}{40} = 6.75°$. So the first two roots are missed completely. You will have to be alert for such occurrences whenever you plot a graph with a calculator or a computer. It is often a good idea to use mathematics to see if the calculator is giving you the complete and accurate solution.

Finally let's return to the question of a formula for the roots of the function $f(x) = \cos Bx$. Since the first positive root is $\frac{90°}{|B|}$ and the spacing is $\frac{180°}{|B|}$, we modify the formula produced on page 147 to make the following statement about roots of $y = \cos Bx$.

The roots R of the function $f(x) = \cos Bx$ are given by $R = \dfrac{90°}{|B|} + \dfrac{180°}{|B|}(n-1)$ and $R = -\dfrac{90°}{|B|} - \dfrac{180°}{|B|}(n-1)$ where $n = 1, 2, 3, \ldots$.

Example 3.34

Consider the function $f(x) = \cos(30x)$. (a) Write the formulas for the roots of f, (b) list six positive and six negative roots of f, and (c) write a single formula for *all* the roots of f.

Solutions   (a) The positive roots are given by the formula $R = \dfrac{90°}{|B|} + \dfrac{180°}{|B|}(n-1)$. Here $B = 30$, so the positive roots are

$$R = \frac{90°}{30} + \frac{180°}{30}(n-1) = 3° + 6°(n-1)$$

where $n = 1, 2, 3, \ldots$.

The negative roots are described by the formula $R = -\dfrac{90°}{|B|} - \dfrac{180°}{|B|}(n-1)$. Again, with $B = 30$, we obtain the following formula for the negative roots

$$R = -\frac{90°}{30} - \frac{180°}{30}(n-1) = -3° - 6°(n-1)$$

where $n = 1, 2, 3, \ldots$.
(b) The first six positive roots are $3°$, $9°$, $15°$, $21°$, $27°$, and $33°$ and the six largest negative roots are $-3°$, $-9°$, $-15°$, $-21°$, $-27°$, and $-33°$. You may have found some other roots.
(c) A single formula for all the roots is

$$R = 3° \pm 6°(n-1)$$

where $n = 1, 2, 3, \ldots$. ∎

This study of the roots of variations of the cosine function can be applied to the sine and tangent as well. Also you will find that the roots of the cosine function will play a role in the study of musical chords later in this chapter.

Section 3.3 Exercises

In Exercises 1–8 find the (a) amplitude, (b) period, and (c) frequency of each of the given functions and (d) graph two cycles of each function.

1. $y = 5 \sin 2x$

2. $y = 6 \cos(-3x)$

3. $y = -2 \cos 4x$

4. $y = 2 \tan 3x$

5. $y = -4 \sin\left(\frac{1}{2}x\right)$

6. $y = 0.25 \cos\left(\frac{1}{3}x\right)$

7. $y = \frac{1}{5} \tan\left(-\frac{1}{3}x\right)$

8. $y = -\frac{3}{2} \cos(2x)$

In Exercises 9–10 estimate the (a) amplitude, (b) period, and (c) frequency from each of the given sinusoidal waves.

9.

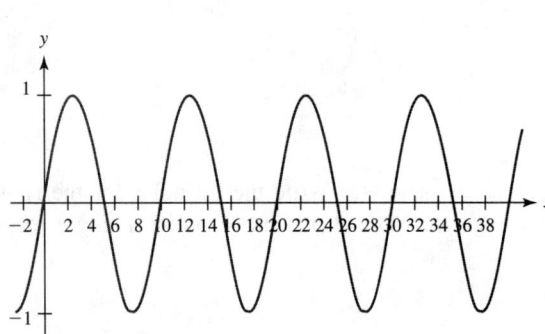

10.

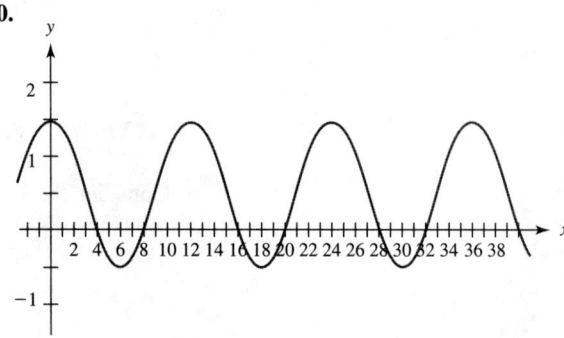

In Exercises 11–16 estimate the (a) amplitude, (b) period, and (c) frequency from each of the given nonsinusoidal waves.

11.

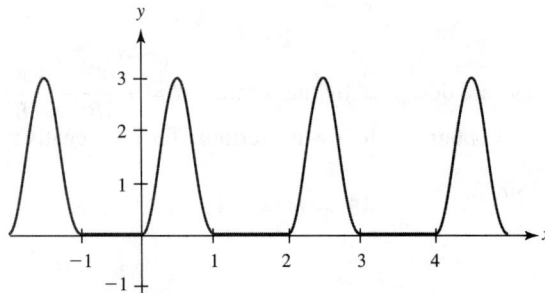

13.

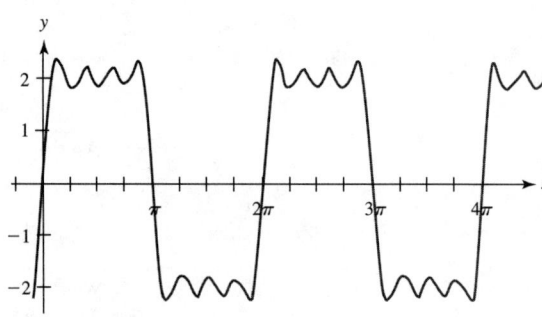

12.

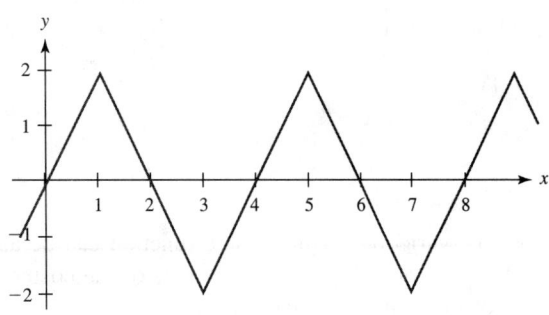

14.

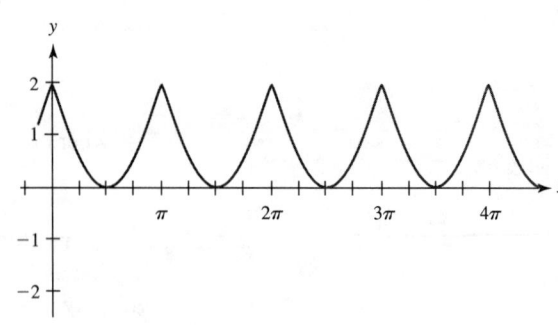

15.

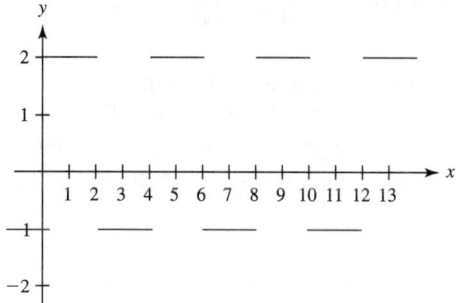

16.

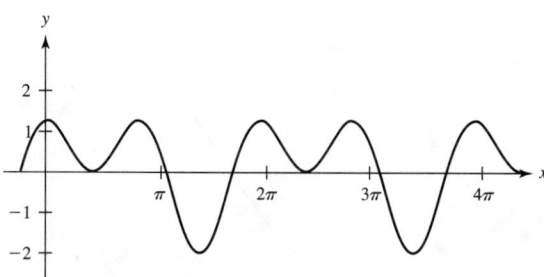

17. *Meteorology* The graphs in Figure 3.64 show the time of sunrise and sunset and the number of hours of daylight at different times of year in Fairbanks, Alaska. The data was recorded every ten days for two years beginning January 1, 1995. What are the **(a)** period and **(b)** amplitude for each of the curves?

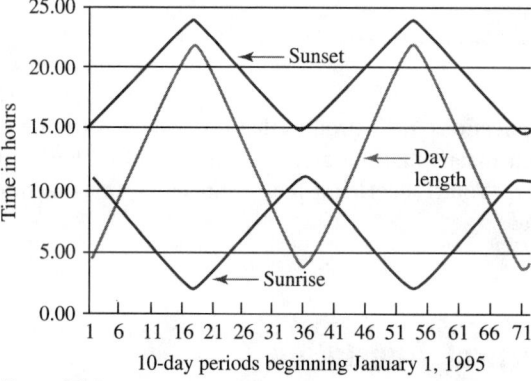

Figure 3.64
Sunrise, sunset, and length of day for Fairbanks, AK, 1995–1996 (Source: http://riemann.usno.navy.mil/AA/data/docs/SunRiseSet.html)

18. *Meteorology* Discuss the relationship between the amplitudes of sunrise and sunset and the amplitude of the day-length curve in Figure 3.64.

19. Give six roots of $\sin(30x)$.

20. Give a formula for all the roots of $\sin(30x)$.

21. *Electronics* Given an ac circuit containing a capacitance C, the voltage across the capacitance is given by $V(t) = 200\sin(21,600°t)$ where V is in volts (V) and time is in seconds. What are **(a)** the amplitude, **(b)** period, and **(c)** frequency of V?

22. *Medical technology* The velocity v in liters per second of airflow during a respiratory cycle for a person at rest is given by the formula $v = 0.85\sin 960°t$ where t is the time in seconds.

 (a) What does it mean when $v > 0$ and when $v < 0$?

 (b) Find the time for one respiratory cycle.

 (c) What is the frequency of a respiratory cycle?

 (d) How many respiratory cycles are there each minute?

23. *Medical technology* An ultrasonic transducer used for medical diagnosis is oscillating at a frequency of 6.7 MHz (or 6.7×10^6 Hz). How much time does each oscillation take?

24. *Meteorology* The graph in Figure 3.65 shows the average monthly temperature for a two-year period in Albany, New York. What is the **(a)** period and **(b)** amplitude of this curve?

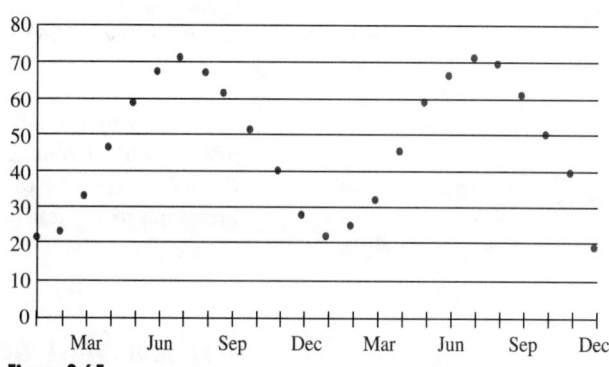

Figure 3.65
Average monthly temperature in Albany, NY

25. Write a memo to the management in which you describe what you have learned about trigonometry and how you can use this information to develop a way to represent combinations of two notes so that each combination is visually distinguished from all the others (as requested in the Chapter Project).

3.4

Vertical and Horizontal Translations

Many real world applications do not produce graphs that begin at the point $(0, 0)$ or travel along the x-axis. For example, the graph of the number of hours of daylight for San Francisco shown in Figure 3.66 begins at the point $(1, 10.6)$ and oscillates between a minimum value of 10.5 hr and a maximum of 15.8 hr. In this section you will see how to model such applications.

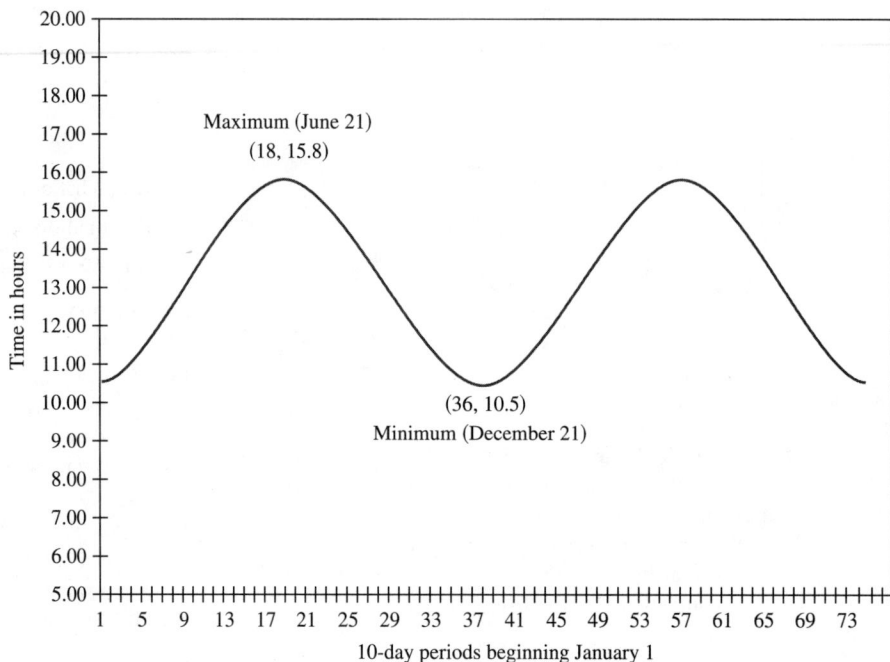

Figure 3.66
Length of day for San Francisco over two years

We will also study applications in electronics that involve two or more voltages that are **out of phase**, which means that the graph of one is shifted horizontally to the left or right of the other. You'll also learn how changes in any function shifts the graph to the left or right.

Vertical Shift or Vertical Translation

Figures 3.67 and 3.68 each show three graphs that are shifted vertically from one another. You studied these in Chapters 1 and 2 and you probably remember that graphs that are shifted this way (vertically) by addition or subtraction are identical in shape. Now look at Figure 3.69. In the middle is the graph of $y = \sin x$. What are the equations of the other two graphs?

The answer is consistent with what you observed for linear and quadratic functions: To move the graph of a function f up or down, add or subtract a constant from the function. For example, the graphs in Figure 3.69 are $y = \sin x$, $y = \sin x + 4$, and $y = \sin x - 6$. The graph of $y = \sin x + 4$ is the result of shifting (or translating) the graph of $y = \sin x$ upward 4 units. In a similar manner, the graph of $y = \sin x - 6$ is the result of shifting (or translating) the graph of $y = \sin x$ downward 6 units.

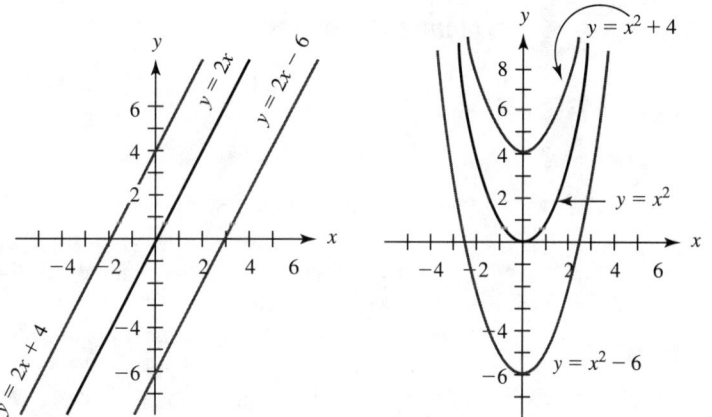

Figure 3.67   **Figure 3.68**

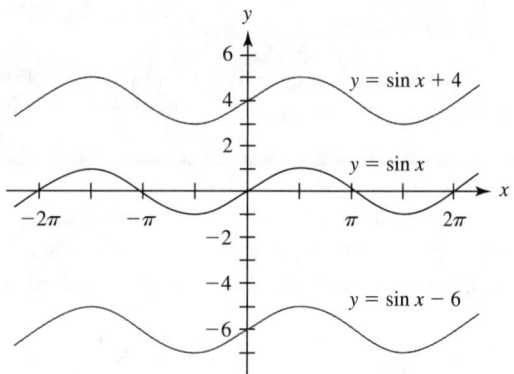

Figure 3.69

Vertical Shift or Translation

If f is a function then the graph of $y = f(x) + D$ translates (or shifts) the graph of $y = f(x)$ up if $D > 0$ and down if $D < 0$. The amount of the shift is $|D|$ units. There is no change in the shape of the graph. We say that replacing $f(x)$ with $f(x) + D$ produces a **vertical translation** or **vertical shift** of the graph of f.

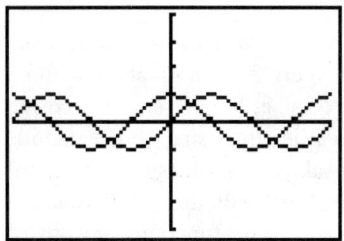

Figure 3.70

Phase Shift

You probably have noticed that the graph of the sine function is the same shape as the cosine function, but one is shifted horizontally from the other. If you look at the graph in Figure 3.70 you might be able to guess how far apart the two functions are. This horizontal distance between two otherwise identical periodic functions is called the **phase shift**. We'll explore the phase shift of the sine and cosine in Calculator Lab 3.5.

Calculator Lab 3.5

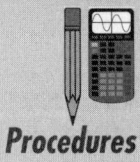

Exploring Phase Shifts

In this Calculator Lab you'll explore the phase shift between the sine and cosine functions.

Procedures

1. Plot $y = \sin x$ and then plot functions defined by expressions like $\cos(x - 20°)$, $\cos(x - 30°)$, and $\cos(x - 40°)$ until you find a graph that is identical to $y = \sin x$. When the curves are identical, the subtracted number is the phase shift between the sine and cosine functions.
2. Repeat this process beginning with $y = \cos x$ and then plotting functions defined by expressions like $\sin(x - 20°)$, $\sin(x - 30°)$, and $\sin(x - 40°)$ until you find a graph identical to $y = \cos x$. When the curves are identical, the subtracted number is the phase shift between the sine and cosine functions.
3. Complete the statements
   (a) $\cos(x - \underline{\hphantom{000}}) = \sin x$
   (b) $\sin(x - \underline{\hphantom{000}}) = \cos x$

   (Actually there are many correct answers. Fill in the ones you found.)
4. Explain why the numbers in the blanks in Question 3 aren't the same.

In Calculator Lab 3.5 you found that the graph of $\cos(x - 90°)$ is the same as the graph of $\sin x$, and we know that $\cos(x - 90°)$ is shifted 90° to the right of $\cos x$. Thus we can say that $\sin x$ is shifted 90° to the right of $\cos x$.

Example 3.35

Graph $y = \sin x$ and $y = \cos(x + 270°)$.

Solution   You should get the same graphs. Thus the graph of $\sin x$ looks the same as the graph of $\cos(x + 270°)$. Since the graph of $y = \cos(x + 270°)$ is the graph of $y = \cos x$ shifted 270° to the left, we can say that the graph of cosine is shifted 270° to the left of sine. ◼

Phase Angle

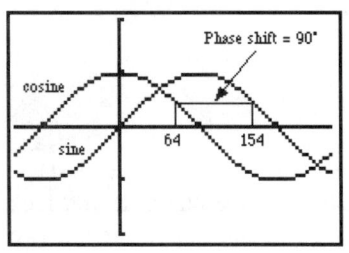

Figure 3.71

In some applications such as electronics, the words *leads* and *lags* are used to describe the relative positions of periodic functions. Look at Figure 3.71, a graph containing sine and cosine curves. Cosine is shifted to the left of sine, but we say that cosine **leads** the sine by 90° because it reaches any point earlier than sine. For example, the peak of cosine occurs at $x = 0°$, but the next peak of sine doesn't occur until $x = 90°$. You could also say that the sine function **lags** or **trails** the cosine function by 90°. While we often hear it said that the sine and cosine functions are **out of phase** by 90°, it is important to know which one is leading.

The formula for a sinusoidal curve can be written in terms of the angular frequency ω as

$$y = A \sin(\omega t + \phi)$$
$$\text{and} \quad y = A \cos(\omega t + \phi)$$

Again A is the amplitude of the motion. The constant ϕ is called the **phase angle**. It tells us at what point in the cycle the motion started at the initial time $t = 0$.

To compare the phase angle between two waves they must have the same frequency; otherwise the relative phase keeps changing. The amplitudes can be different for the two waves. We can compare the phase of two voltages, two currents, or a current with a voltage.

Example 3.36   Application

What is the phase relationship between the sinusoidal waves $v = 15 \sin(\omega t + 30°)$ and $i = 10 \sin(\omega t + 80°)$?

Solution   The phase angle of i is $-80°$ while the phase angle of v is $-30°$. The difference in these phase angles is $|-80° - {}^-30°| = 50°$. The graphs of the two functions, with $\omega = 1$, are shown in Figure 3.72. Notice that i crosses the x-axis at $-80°$ while v crosses at $-30°$. Since i crossed the x-axis $50°$ before v, we say that i leads v by $50°$ or v lags i by $50°$.

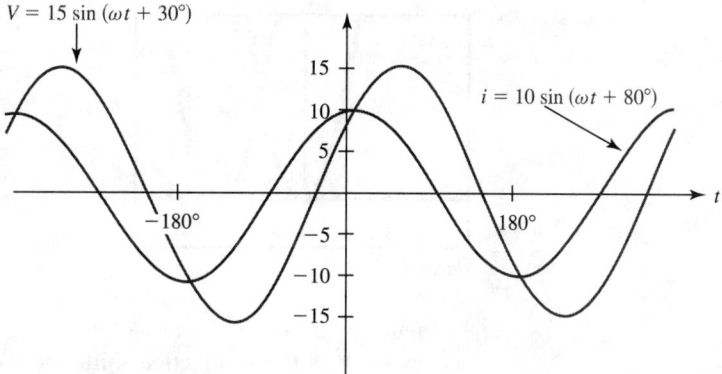

Figure 3.72

Horizontal Shifts or Translations of Functions

In Chapter 2 you saw how to shift a graph to the left by subtracting the same number from every one of the x-value data points. In Calculator Lab 3.5 we saw that the function defined by $\cos(x - 90°)$ shifts the cosine curve $90°$ to the right. We will now see how these two horizontal shifts or **horizontal translations** are related.

In general, if you compare the graph of $y = f(x)$ with the graph of $y = f(x - k)$ you will see that the shapes are identical, but the second one is shifted to the right by k units if $k > 0$. The next example gives practice with this by returning to nonperiodic functions.

Example 3.37

Plot the function whose equation is given by $y = x^2$. Then plot the variations of that function (a) $y = (x-2)^2$, (b) $y = (x-5)^2$, and (c) $y = (x+7)^2$. Explain how you could have predicted the location of these curves.

Solutions   Remember from the discussion of functions in Section 2.3 that if $f(x) = x^2$ then $f(x-2) = (x-2)^2$, $f(x-5) = (x-5)^2$, and $f(x+7) = (x+7)^2$. In Figures 3.73–3.75 the graph of $y = x^2$ is shown in bold.

(a) Figure 3.73 shows the graph of $y = x^2$ and $y = (x-2)^2$. As you can see, $y = (x-2)^2$ is the x^2 function shifted to the **right** 2 units.

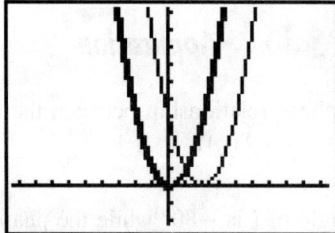

Figure 3.73

(b) Figure 3.74 contains the graphs of $y = x^2$ and $y = (x-5)^2$. Here $y = (x-5)^2$ is the x^2 function shifted to the **right** 5 units.

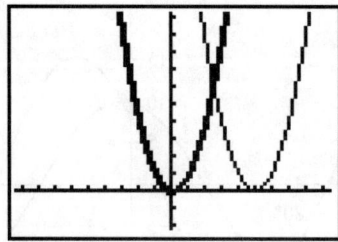

Figure 3.74

(c) In Figure 3.75 are the graphs of $y = x^2$ and $y = (x+7)^2$. In this figure $y = (x+7)^2$ is the x^2 function shifted to the **left** 7 units.

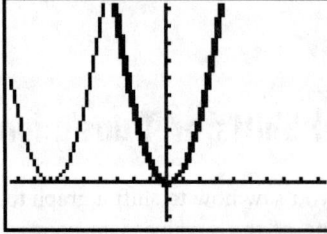

Figure 3.75   ■

In Chapter 2 you saw another way to shift a graph horizontally when you subtracted the first time value in an experiment from all the other measured times so that the experiment would begin at $t = 0$. This process shifted the entire graph to the left without changing its shape. What you see here is just the opposite. We are not subtracting the number k from each x value but graphing the function $f(x - k)$. The difference is illustrated in Figure 3.76. Point $P(3, 9)$ is on $f(x) = x^2$ and point $Q(7, 9)$ is on $g(x) = (x - 4)^2$. Because of the subtraction of 4, point Q is plotted 4 units to the **right** of point P. The same would be true of every point on the graph of function f. Each would have a *twin* on function g, 4 units to the right.

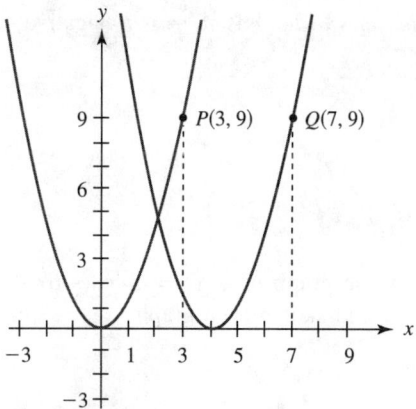

Figure 3.76

You have seen how to interpret the graphs of the function $f(x) = A \sin(Bx)$. Next let's see how to add the horizontal shift. For the function f, how would you interpret $f(x - C)$? Study these examples:

If $f(x) = A \sin(Bx)$ then

$$f(x - 45°) = A \sin[B(x - 45°)]$$
(shift $f(x) = A \sin(Bx)$ to the **right** $45°$)
$$f(x - 180°) = A \sin[B(x - 180°)]$$
(shift $f(x) = A \sin(Bx)$ to the **right** $180°$)
$$f(x + 45°) = A \sin[B(x + 45°)]$$
(shift $f(x) = A \sin(Bx)$ to the **left** $45°$)
$$f(x - P°) = A \sin[B(x - P°)]$$
(shift $f(x) = A \sin(Bx)$ to the **right** $P°$ if $P° > 0$)

We can summarize what you know about the effect of the three letters A, B, and P on the graph of the function $f(x) = A \sin[B(x - P)]$. If we let $BP = C$ we can rewrite this function as

$$f(x) = A \sin(Bx - C)$$

Graphs of $y = A \sin(Bx - C)$ and $y = A \cos(Bx - C)$

Compared with the graph of the basic sine function $y = \sin x$ and the basic cosine function $y = \cos x$, the graphs of $y = A \sin(Bx - C)$ and $y = A \cos(Bx - C)$

▶ have an amplitude of A. You can think of this as stretching or shrinking the graph vertically by a factor of A.

> ▸ have a period of $\dfrac{360°}{|B|}$ and a frequency of $\dfrac{|B|}{360°}$. The graph has $|B|$ cycles in 360°, so if B is larger than 1 or less than -1 the graph is compressed. If B is between -1 and 1 the graph is stretched horizontally.
>
> ▸ are shifted horizontally $\dfrac{C}{B}$ units. The shift is to the **right** if $\dfrac{C}{B}$ is a positive number and to the **left** if $\dfrac{C}{B}$ is a negative number

Example 3.38

Explain how the graph of $g(x) = 2\cos(-6x + 420°)$ compares with the graph of $f(x) = \cos x$ and sketch each graph.

Solution   In function g we have $A = 2$ and $B = -6$. But $420° = -C$ and so $C = -420°$. Since $A = 2$, g has an amplitude of 2. With $B = -6$ the period is $\dfrac{360°}{|B|} = \dfrac{360°}{6} = 60°$. Finally, using $C = -420°$, the graph is shifted $\dfrac{C}{B} = \dfrac{-420°}{-6} = 70°$ or 70° to the right. The graphs of f and g are shown in Figure 3.77.

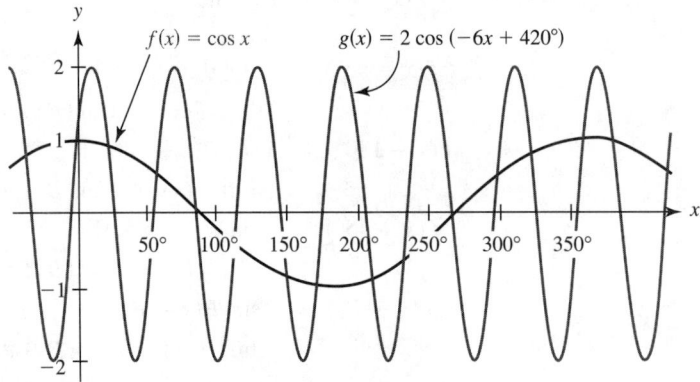

Figure 3.77

Example 3.39

Explain how the graph of $f(x) = 0.7\cos(0.5x - 45°)$ compares with the graph of $k(x) = \cos x$ and sketch each graph.

Solution   In function f we have $A = 0.7$, $B = 0.5$, and $C = 45°$. Thus f has an amplitude of 0.7, a period of $\dfrac{360°}{0.5} = 720°$, and the graph is shifted $\dfrac{45°}{0.5} = 90°$ or 90° to the right. The graphs of f and k are shown in Figure 3.78.

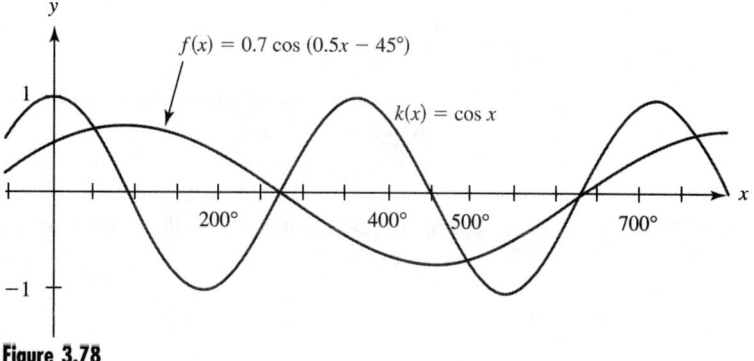

Figure 3.78

Example 3.40

Explain how the graph of $g(x) = 3 \tan(2x + 90°)$ compares with the graph of $h(x) = \tan x$ and sketch each graph.

Solution   Since the tangent function has no amplitude, g has no amplitude. However the value of $A = 3$ does have the effect of stretching the graph of g vertically compared to the graph of h. Since the period of the tangent function is $180°$, the period of g is $\frac{180°}{2} = 90°$. In g we have $C = -90°$, so the graph is shifted $\frac{-90°}{2} = -45°$ or $45°$ to the left. The graphs of g and h are shown in Figure 3.79.

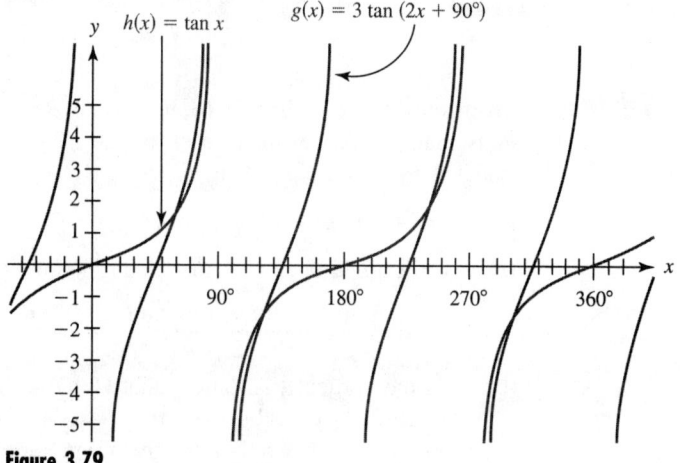

Figure 3.79

Fitting Data to a Sinusoidal Curve

In the next example you will see how to build a mathematical model of data that appears to follow a sine or cosine curve. Such data is referred to as **sinusoidal**. To accomplish this task you will use most of what you know about trigonometry; and you will use your knowledge of coordinate graphing as well. Therefore this example has a double purpose. You will not only see how to model real periodic data, but you will also have a thorough review of trigonometry and coordinate geometry.

Example 3.41

Given the coordinates of the maximum and minimum points of a set of periodic data, write a mathematical model for the data. In Figure 3.80 the coordinates of the maximum and minimum points are given for a set of data that appears to be sinusoidal. Notice that point Q in the figure is the origin of a coordinate system (shown by a dashed line) on which the data appear as a function of the form $y = A \sin(Bx)$. That is, in the dashed coordinate system, the graph of the data is not shifted.

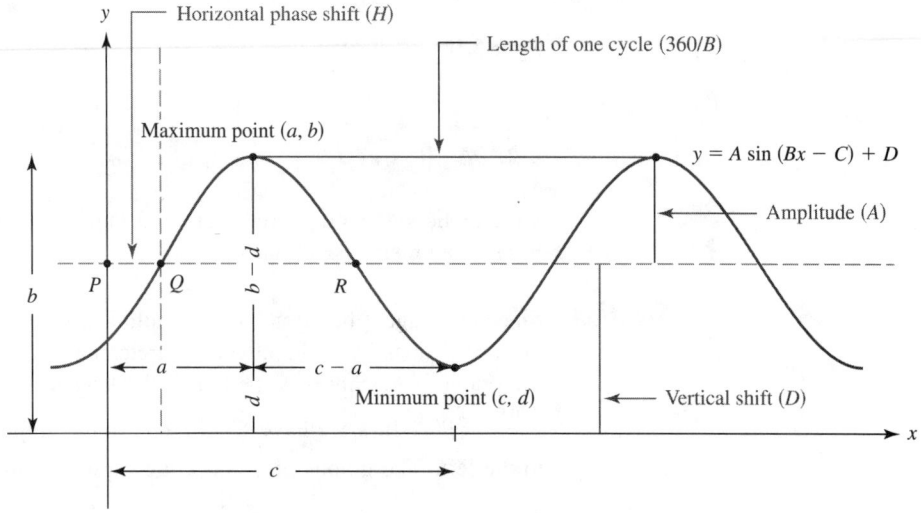

Figure 3.80

Solutions   Given only the coordinates (a, b) and (c, d) and the assumption that the data is sinusoidal, we can estimate each of the needed constants A, B, C, and D in the modeled equation:

$$y = A \sin(Bx - C) + D$$

(a) A is the amplitude of the modeled function. Amplitude is defined on page 152 as

$$A = \frac{\text{maximum} - \text{minimum}}{2} = \frac{b - d}{2}$$

(b) D is the vertical translation; that is, D is the amount by which the graph has been shifted vertically from the x-axis to the axis (shown by the dashed line in the figure) that runs through the middle of the data. Points Q and R have been chosen to be halfway between the maximum point and the minimum point. Therefore the coordinates of point R are $\left(\dfrac{a + c}{2}, \dfrac{b + d}{2}\right)$. You can see from the figure that D is the y-coordinate of point R. Therefore

$$D = \frac{b + d}{2}$$

(c) B is computed from the period or length of the cycle, which is equal to the horizontal distance between maximum points. Since $c - a$ is the horizontal distance between the maximum point and the minimum point and because the modeled curve has the symmetry of a sine wave, the period is twice $(c - a)$. To simplify things we assume that both $c - a$ and B are positive. Thus we can write the period as $\dfrac{360°}{B}$. Therefore $2(c - a) = \dfrac{360°}{B}$, so

$$B = \dfrac{360°}{2(c - a)} = \dfrac{180°}{c - a} \quad \text{in degrees}$$

$$\text{or} \quad B = \dfrac{\pi}{c - a} \quad \text{in radians}$$

(d) C is the product BH because the horizontal phase shift is $H = \dfrac{C}{B}$. In Figure 3.80 H is equal to the distance PQ. You can see from the figure that $PQ = PR - QR$. But PR is the x-coordinate of the point R which was given above as $\frac{a+c}{2}$. Since QR is half of the period, the distance $QR = \frac{1}{2} \cdot 2(c - a) = c - a$. Therefore $H = PQ$ is

$$H = PR - QR = \dfrac{a + c}{2} - (c - a) = \dfrac{a}{2} + \dfrac{c}{2} - c + a$$

$$= \dfrac{3}{2}a - \dfrac{c}{2} = \dfrac{3a - c}{2}$$

Since $B = \dfrac{180°}{c - a}$ and $H = \dfrac{3a - c}{2}$, then

$$C = BH = \dfrac{180°}{c - a} \cdot \dfrac{3a - c}{2}$$

$$= \dfrac{90°(3a - c)}{c - a} \quad \text{in degrees}$$

$$= \dfrac{\pi(3a - c)}{2(c - a)} \quad \text{in radians} \qquad \blacksquare$$

Example 3.42

Write an equation that could model the number of hours of daylight in San Francisco. The graph is reproduced in Figure 3.81.

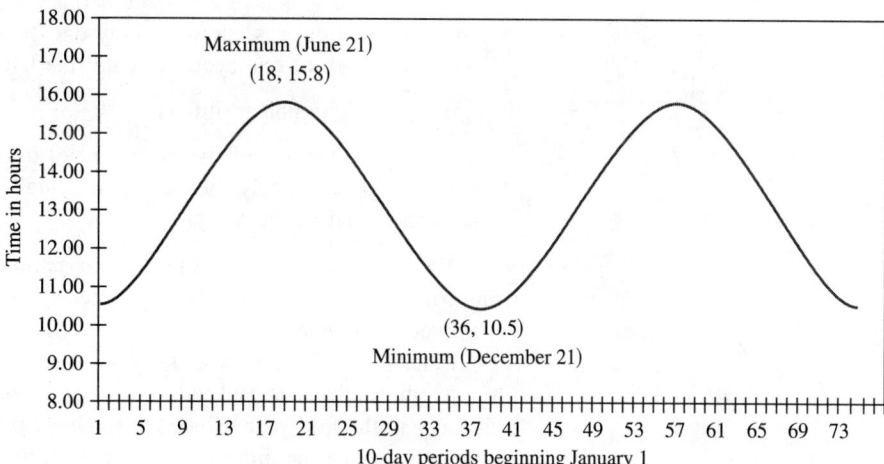

Figure 3.81
Length of day for San Francisco over two years

Solution The equation will be of the form $y = A \sin(Bx - C) + D$. We need to use the coordinates (a, b) and (c, d) to determine A, B, C, and D. First write the values of the coordinates (a, b) and (c, d); they are $a = 18$, $b = 15.8$, $c = 36$, and $d = 10.5$.

Now we follow the steps from Example 3.41.

(a) $A = \dfrac{b - d}{2} = \dfrac{15.8 - 10.5}{2} \approx 2.7$

(b) $D = \dfrac{b + d}{2} = \dfrac{15.8 + 10.5}{2} \approx 13.2$

(c) $B = \dfrac{\pi}{c - a} = \dfrac{\pi}{36 - 18} = \dfrac{\pi}{18}$

(d) $C = \dfrac{\pi(3a - c)}{2(c - a)} = \dfrac{\pi(3 \cdot 18 - 36)}{2(36 - 18)} = \dfrac{\pi}{2}$

Therefore a modeled sine wave for these data is

$$y = A \sin(Bx - C) + D$$
$$= 2.7 \sin\left(\frac{\pi}{18}x - \frac{\pi}{2}\right) + 13.2$$

where y is the number of hours of daylight and x is the number of 10-day intervals. ∎

Let's look at each of the values of A, B, C, and D in Example 3.42 to try to understand them in terms of the actual data.

1. $A = 2.7$.   The amplitude is equal to half the difference between the number of hours in the longest and shortest days. You can understand that this is expected by comparing the result with the amplitude of the graph of $y = \sin x$.
2. $D = 13.2$.   The vertical displacement is equal to the average of the number of hours in the longest and shortest days. This ensures that in the dashed coordinate system the maximum and the minimum points are equally distant from the dashed x-axis.
3. $B = \dfrac{\pi}{18}$.   This value is easier to understand in terms of the frequency, which is the reciprocal of the period. The frequency (cycles per 10 days) is $\dfrac{B}{2\pi} = \dfrac{\pi/18}{2\pi} = \dfrac{1}{36}$. A frequency of $\frac{1}{36}$ means that there is one cycle every 36 10-day intervals, or 360 days. But 360 days is approximately one year, which is what you would expect to be the length of one cycle for data like length of the day throughout the year.
4. $C = \dfrac{\pi}{2}$.   The phase shift is $\dfrac{C}{B} = \dfrac{\pi/2}{\pi/18} = 9$, which is nine 10-day intervals or 90 days. In terms of the calendar, 90 days brings you to the end of March, which is approximately the time (the actual day is March 21) halfway between the longest and shortest days.

You can see then that although we estimated the values of A, B, C, and D using trigonometry and coordinate geometry, the resulting values make sense when considered in context.

Finally let's look at how well the modeled equation fits the actual data. Figure 3.82 shows the data and the model. You can see that the model fits quite well, especially in the first year. Since the modeled period is 360 days and not 365 days, we begin to notice the difference in period in the second year. By the tenth year the horizontal difference between these two curves would be quite large. Some models are good for short periods of time only. Remember our linear models for swimming in Chapter 1?

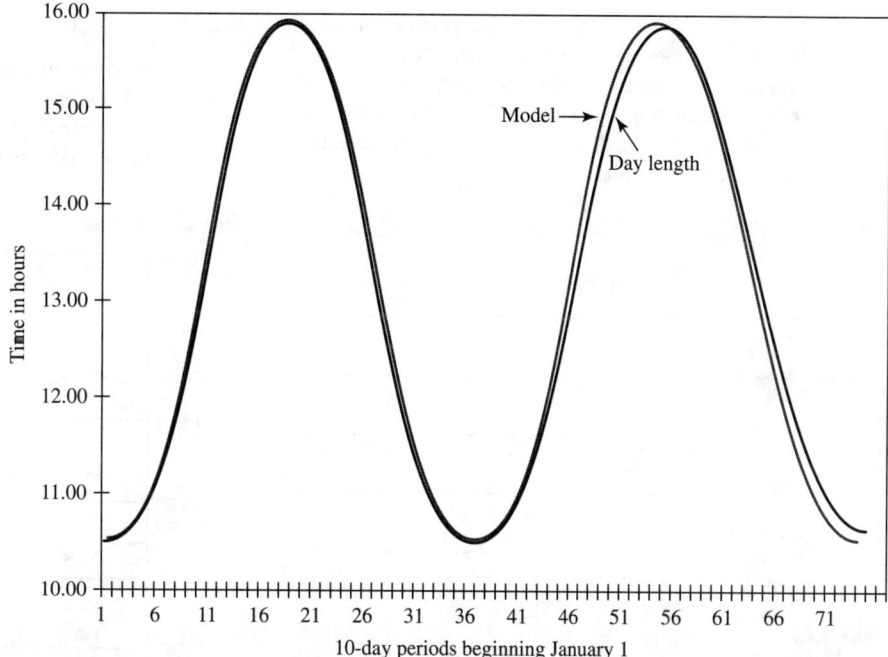

Figure 3.82

Why should the data for number of daylight hours fit so well to a sine wave? The variation in the length of the day is affected by the rotation of Earth around the Sun and by the rotation of any point on Earth (in this case San Francisco) around Earth's axis. Since both of these motions are approximately circular and because the sine and cosine functions are defined in terms of points on a circle, we should not be too surprised to find that trigonometry is very useful in studying this and other planetary variations. The surprise is that the fit is so good—something which may not always happen.

Section 3.4 Exercises

In Exercises 1–6 determine the (a) amplitude, (b) period, (c) frequency, and (d) phase shift and (e) graph two cycles of each of the given functions.

1. $f(x) = 2\sin(3x - 60°)$

2. $g(x) = -\cos(2x + 80°)$

3. $h(x) = \sin(-4x + 40°)$

4. $f(x) = \tan(x + 50°)$

5. $g(x) = -3\cos\left(-\frac{1}{2}x - 20°\right)$

6. $f(x) = \frac{7}{3}\sin(\frac{1}{3}x - 10°)$

In Exercises 7–10 determine the phase angles between the two given sinusoidal waves.

7. *Electronics* $i = 25\sin(\omega t + 50°)$
$v = 10\sin(\omega t + 70°)$

8. *Electronics* $i = -5\sin(\omega t + 30°)$
$v = 7\sin(\omega t - 20°)$

9. *Electronics* $i = -4\cos(\omega t + 30°)$
$v = 8\sin(\omega t - 120°)$

10. *Electronics* $i = 8\cos(\omega t + 10°)$
$v = 3\sin(\omega t - 10°)$

11. *Electronics*   In a certain circuit the maximum values of the voltage and current are 175 V and 115 A, respectively. The frequency is 55 Hz and the current leads the voltage by 30°.

(a) Write the equation for the voltage at any time t.

(b) Write the equation for the current at any time t.

(c) Use your answers from (a) and (b) to plot two cycles of both curves starting with $t = 0$. (Assume $\omega = 1$.)

12. *Meteorology* Figure 3.83 contains a graph of the average daily temperatures for Nashville, TN over a two-year period. What are the **(a)** amplitude, **(b)** period, and **(c)** phase shift from January 1. **(d)** Write a sinusoidal function that fits the data in Figure 3.83. (Assume that the curve is sinusoidal.)

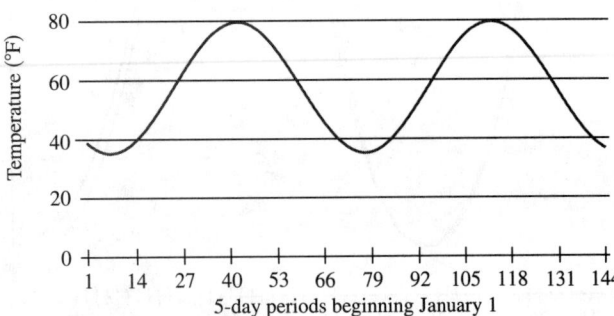

Figure 3.83
Average daily temperature for Nashville, TN over two years (Source:http://205.165.7.67ftproct.chxhtml/normals.txt)

13. *Meteorology* The graph in Figure 3.84 shows the number of hours of daylight at different times of the year in Fairbanks, Alaska. The data was recorded every ten days for two years beginning January 1, 1995. What are the **(a)** amplitude, **(b)** period, and **(c)** phase shift from January 1. **(d)** Write a sinusoidal function that fits the data in Figure 3.83. (Assume that the curve is sinusoidal.)

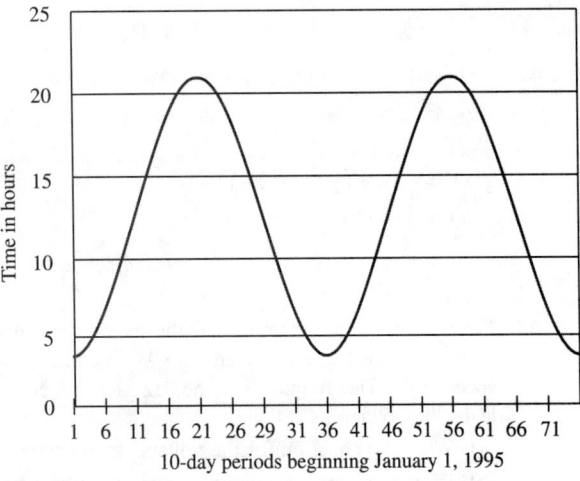

Figure 3.84
Length of day for Fairbanks, AK, 1995–1996 (Source: http://riemann.usno.navy.mil/AA/data/docs/SunRiseSet.html)

In Exercises 14–16 use Figure 3.85.

14. *Meteorology* What are the **(a)** amplitude, **(b)** period, and **(c)** phase shift of the sunrise curve in Figure 3.85? **(d)** Write a sinusoidal function that fits the data in the figure. Assume that the curve is sinusoidal.

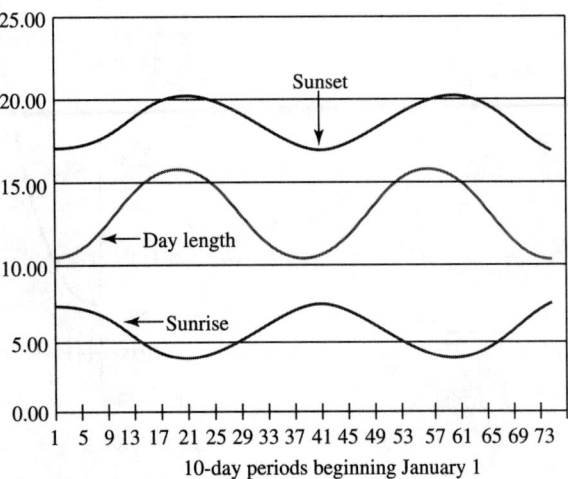

Figure 3.85
Sunrise, sunset, and length of day for San Francisco over two years (Source: http://www.nws.mbay.net/sunset.html#sfo_tide)

15. *Meteorology* What are the **(a)** amplitude, **(b)** period, and **(c)** phase shift of the sunset curve in Figure 3.85? **(d)** Write a sinusoidal function that fits the data in the figure. Assume that the curve is sinusoidal.

16. *Meteorology* Discuss the relationship between the amplitudes of sunrise and sunset and the amplitude of the day length curve in Figure 3.85.

17. *Electronics* Given an ac circuit containing a capacitance C, the voltage across the capacitance is given by $V(t) = 200 \sin(21,600°t)$ where V is in volts (V) and time is in seconds. What are the **(a)** amplitude, **(b)** period, and **(c)** frequency of V?

18. *Electronics* Given an ac circuit containing a capacitance C, the current in the circuit is given by

$$i(t) = 1.0 \sin (21,600°t + 90°)$$

(a) What is the amplitude of the current?
(b) What is the period of the current?
(c) What is the frequency of the current?
(d) By how much does the current lead the voltage if $V(t) = 2.5 \sin(21,600°t)$?

3.5

Modeling Sound Waves

Trigonometry is used in the study of sound and music. Sound is caused by vibrations in the air, which move away from the noise source like ripples made by a stone dropped in a pool of water (see Figure 3.86). Your ear collects these vibrations like a TV satellite dish and channels them to the brain where they are interpreted. These vibrations of air molecules move in what is called a **sound wave**.

Figure 3.86
(Photo courtesy of David Young-Wolff/PhotoEdit)

Radians and Degrees

Before we look at applications that do not involve angles we need to stop and consider units of measure for trigonometric functions. The choice of units makes a big difference in the appearance of the graph. For example, think about an experiment in which a ball is thrown upward beginning at $t = 0$. Figures 3.87 and 3.88 show two graphs made from the same experiment—one in which the time is measured in seconds and one in which it is measured in minutes. Remember, it is one experiment in which one person uses a minute timer and the other a second timer. Both graphs show parabolas beginning at $(0, 0)$, but clearly the choice of units affects the appearance of the graphs. In particular the choice of units changes the roots of the function being studied. In Figure 3.87 the roots are 0 and 30 and in Figure 3.88 the roots are 0 and 0.5. Obviously these are the same amounts of time since 30 seconds equals 0.5 minutes. The point is that the choice of units makes a big difference in the roots and in the appearance of the graph.

Unlike time or distance, in which there are many possible units of measure, for angles there are only two units in common use. You have used one of them

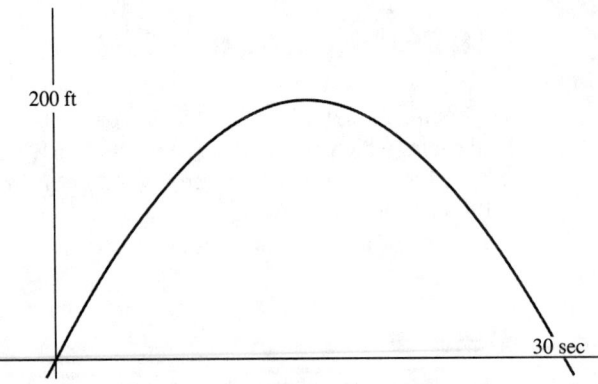

Figure 3.87
Projectile motion experiment measured in feet and seconds

Figure 3.88
Same projectile motion experiment with time measured in minutes

(degrees) and you have seen the other (the **radian**) on your calculator and in this text. We will look at radians more closely in Chapter 4, but for now you need to know that there are π radians in 180°. That is, π radians $= 180°$ and so 1 radian $= \dfrac{180°}{\pi} \approx 57.3°$. To convert between radians and degrees we can use the proportion $\dfrac{180°}{\pi} = \dfrac{D}{R}$ where D is a quantity measured in degrees and R is the same quantity measured in radians.

Example 3.43

Convert $\frac{7}{4}\pi$ radians to degrees.

Solution Using the proportion $\dfrac{180°}{\pi} = \dfrac{D}{R}$ and the fact that we are given $R = \frac{7}{4}\pi$, we have the proportion

$$\frac{180°}{\pi} = \frac{D}{7\pi/4}$$

$$D = \frac{180°}{\pi} \times \frac{7}{4}\pi = 315°$$

Thus $\frac{7}{4}\pi$ radians $= 315°$.   ■

Example 3.44

Convert $120°$ to radians.

Solution Again using the proportion $\dfrac{180°}{\pi} = \dfrac{D}{R}$, we are given $D = 120°$. We substitute this value of D into the proportion and solve for R.

$$\frac{180°}{\pi} = \frac{120°}{R}$$

$$R = 120° \times \frac{\pi}{180°}$$

$$= \frac{2}{3}\pi \approx 2.094$$

Thus $120° = \frac{2}{3}\pi \approx 2.094$ radians. ∎

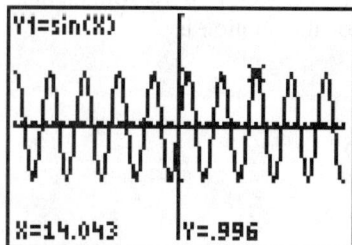

Figure 3.89

Graphs look very different and roots are very different depending on which unit is used. To illustrate this point, Figures 3.89 and 3.90 show $y = \sin x$ plotted on the calculator in radian mode and in degree mode. The window settings are $-30 \le x \le 30$ and $-2 \le y \le 2$. You can see that 14.043 radians (about 800°) is two cycles away from the origin, while 14.043° is quite early in the first cycle. Also there are 19 roots in radians in this interval and only one root in degrees. Different units of measure make a big difference in the roots and in the appearance of the graph!

The radian equations for the sine wave voltage V and current i can be written in terms of the frequency f as

$$v = V_{\text{max}} \sin(2\pi f t)$$

$$i = I_{\text{max}} \sin(2\pi f t)$$

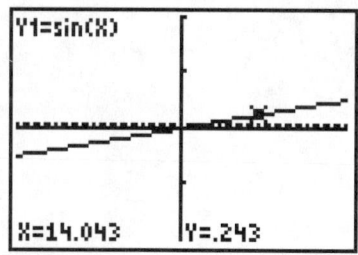

Figure 3.90

where V_{max} and I_{max} are the **peak values** for the voltage and current.

Example 3.45 Application

Given the ac sine wave voltage $V = 1.25 \sin(120\pi t)$, find the (a) peak value for the voltage, (b) frequency, and (c) period of the wave where t is in seconds.

Solutions This wave fits the formula $v = V_{\text{max}} \sin(2\pi f t)$.

(a) Since V_{max} represents the peak value of the voltage, this wave has a peak value of 1.25 A.

(b) Here $2\pi f = 120\pi$, so the frequency $f = \dfrac{120\pi}{2\pi} = 60$ Hz.

(c) Since the period $= \dfrac{1}{\text{frequency}}$, the period is $\frac{1}{60}$ sec ≈ 0.0167 s $= 16.7$ ms. ∎

In the remainder of this chapter we will be dealing with trigonometric functions in which the x-values are measured in seconds. You will see that we can use either degrees or radians to study these experiments, but sometimes, as in the sine regression, you will see that we are forced by the calculator to use radians.

Modeling Vibrations

The motion of a spring is one way to model a simple vibration. To study that motion we set up an experiment using a motion detector and a Slinky™ and recorded the motion for 10 seconds (see Figure 3.91). Figure 3.92 shows the distance vs. time graph that resulted when we used a small (diameter = 1.5 in.) version of the Slinky. Of course you would expect that the graph would go up and down. What you might not expect is how much the graph looks like a sine or cosine curve. To see the connection between the motion of a spring and the sine function, look at the plot of the trigonometric regression equation, which matches quite well to the spring motion (Figure 3.93).* The sinusoidal regression equation for this motion is

$$d = 0.72\sin(5.1t + 0.13) + 3.7$$

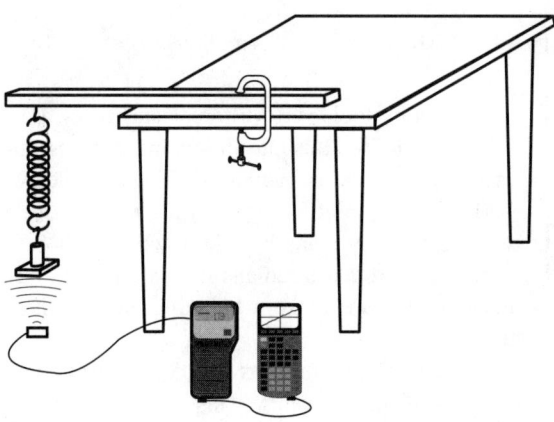

Figure 3.91

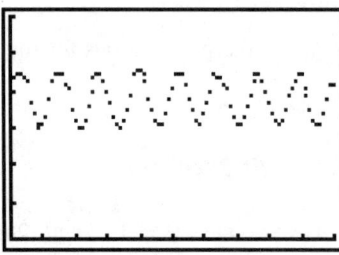

Figure 3.92

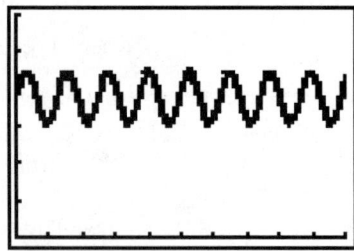

Figure 3.93

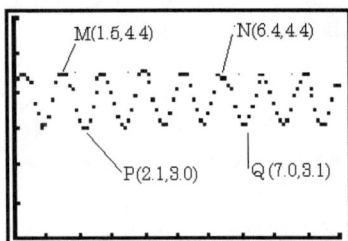

Figure 3.94

M(1.5, 4.4)   N(6.4, 4.4)   P(2.1, 3.0)   Q(7.0, 3.1)

  Example 3.46

Find the (a) period, (b) frequency, (c) amplitude, and (d) vertical displacement of the Slinky data. Figure 3.94 has all the information you need.

*When a TI graphing calculator performs sine regression it uses radians as the unit, not degrees. We'll study radians in Chapter 4.

Solutions   (a) There are four cycles between M and N and four cycles between P and Q. The time (x-coordinate) between M and N is $6.4 - 1.5 = 4.9$ sec; the time between P and Q also is 4.9 sec. Therefore one cycle lasts $\frac{4.9}{4} = 1.225$ sec ≈ 1.2 sec and so the length of one cycle is about 1.2 seconds.

(b) The frequency is $\frac{1}{1.225} \approx 0.82$ cycles per second or about 0.82 Hz.

(c) To get the amplitude you must go back to the original definition of amplitude (see page 152) as one-half the difference between the maximum and minimum values of the function. We need the difference between the y-coordinates of M and P and of those between N and Q. They are 1.4 ft and 1.3 ft, respectively. Therefore we have two estimates of the amplitude: $\frac{1.4}{2} = 0.7$ and $\frac{1.3}{2} = 0.65$. The average of these is $\frac{0.7+0.65}{2} = 0.675$ or 0.68 ft. Thus the amplitude is about 0.68 ft or 8 in. ≈ 20.5 cm.

(d) The vertical displacement is the height of a horizontal line drawn through the "middle" of the graph. Thus we need to calculate the average of the highest and the lowest y-coordinates. The highest y-value is 4.4 and the lowest is 3.0; so the vertical displacement is $\frac{4.4+3.0}{2} = \frac{7.4}{2} = 3.7$. ∎

Example 3.47

Discuss the sinusoidal regression equation of the Slinky data based on what you know about the graph of $y = A \sin(Bx) + D$.

Solution   You have seen that the coefficient A gives the amplitude of the sine function. Our estimate of 0.68 ft is very close to the value of A, which is 0.72. The coefficient A equals the amplitude even if the graph is not near the x-axis.

You have seen that the period is $\dfrac{360°}{|B|}$. But the trigonometric regression equation produced by the calculator uses radians instead of degrees. Since there are 2π radians in $360°$, an estimate of the period from the coefficient B would be $\dfrac{2\pi}{5.1} \approx 1.232$, which is quite close to our estimate from the graph.

Since the vertical displacement is represented by D, we have $D = 3.7$. This is exactly what the calculator's regression equation determined for D.

Combining the above information, we have estimated that a sinusoidal regression line of the Slinky data would be $y = 0.72 \sin(5.1x) + 3.7$. This estimate compares quite favorably with the calculator's equation $d = 0.72 \sin(5.1t + 0.13) + 3.7$. We leave the estimation of the phase shift as an exercise. ∎

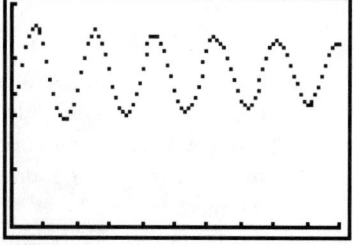

Figure 3.95

The motion of any spring that is stretched and then released will produce a graph that looks like a sine wave. Different kinds of springs will produce graphs with different amplitudes and periods. A tightly coiled spring would vibrate faster and therefore produce a shorter period and higher frequency than the one shown in Figure 3.92. A spring that is stretched out more at the beginning of the experiment will show a higher amplitude. Figure 3.95 shows the graph of an experiment conducted with a large (diameter = 3 in.) plastic Slinky.

The larger Slinky has a longer period because it has about five cycles in 10 seconds compared with about eight cycles for the smaller Slinky. The larger Slinky also has a larger amplitude. There is one other significant difference between these two graphs. You can see in Figure 3.95 that the amplitude decreases between

$t = 0$ and $t = 10$. The vibration of the Slinky is decreasing during that time interval. All vibrations decrease eventually, but the decrease in amplitude of the smaller Slinky is not noticeable in Figure 3.92. You have seen a graph like this before: when the height of the bouncing ball decreased over time. We will study the mathematics of the decreasing height of the bouncing ball and the decreasing amplitude of a vibrating spring in later chapters. It turns out they are both examples of **exponential damping**.

Modeling Sounds

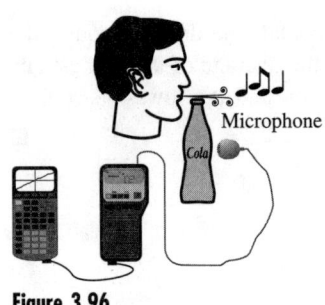

Figure 3.96

Let's move now to experiments with sounds. You will see that sounds behave in a way similar to the vibrations of a spring but with differences in period, frequency, and amplitude. The graphs of loud and soft sounds differ, as do the graph of a high-pitched sound like a whistle and the graph of a low bass note. Let's begin with loud and soft sounds.

We will begin by creating a sound by a familiar technique—blowing across the top of a bottle. A microphone was attached to the CBL system to measure the changes in air pressure caused by the sound waves we produced with the bottle. The experiment was set up as shown in Figure 3.96. In this experiment we used a small perfume bottle. The pressure vs. time graph is shown in Figure 3.97.

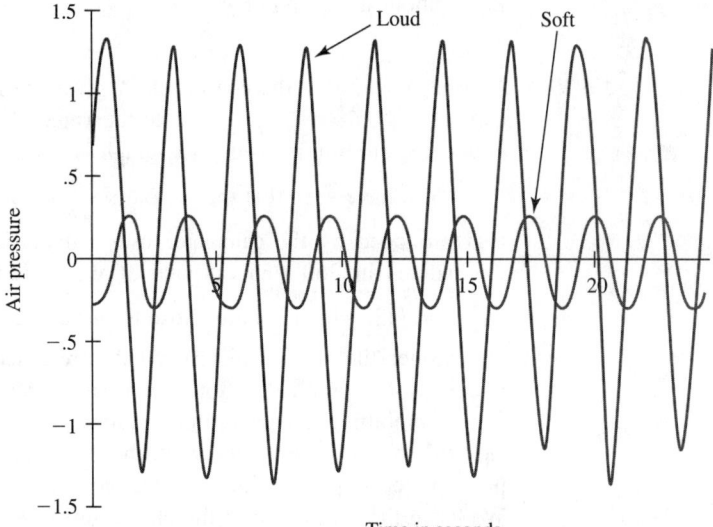

Figure 3.97
Loud and soft sounds from a perfume bottle

If we compare the graphs of the loud and soft sounds in Figure 3.97 we see that the amplitudes are quite different; the softer sound has a smaller amplitude. However the periods are almost identical. You can see this by counting the number of cycles in the time interval or by noticing that the peaks made by the softer sound occur in the same relative locations as those of the louder sounds.

The next experiment shows the result of using five different empty bottles to make sounds: a chemistry test tube, the small perfume bottle in the previous experiment, a 12-ounce beer bottle, a 0.75-liter wine bottle, and a 1.5-liter plastic soft drink bottle. If you have tried to make sounds like this you probably have noticed that the higher notes come from the smaller bottles. The graphs are shown in Figures 3.98–3.102.*

Example 3.48

What are the frequencies of the sounds made by each bottle. Give your answer in hertz (cycles per second).

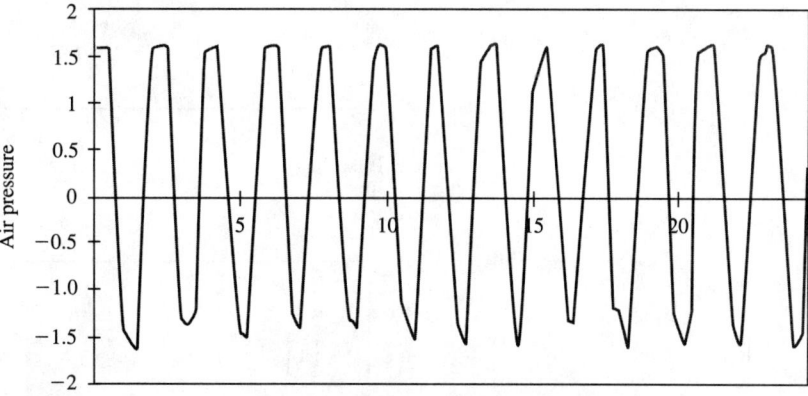

Figure 3.98
Test tube

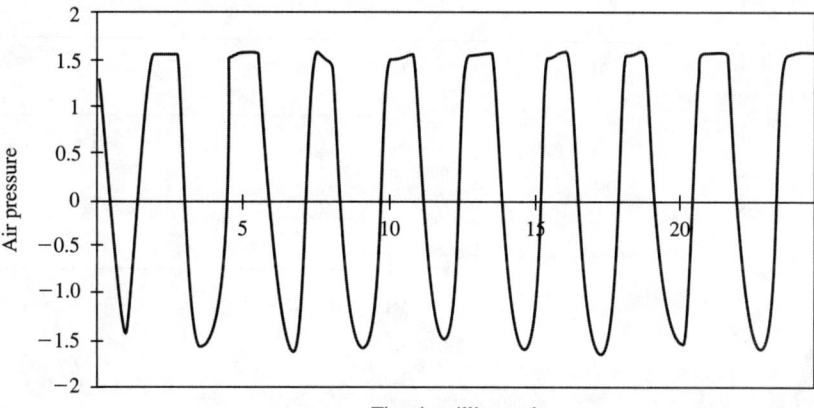

Figure 3.99
Perfume bottle

Solution The strategy for each of the graphs is to count the number of complete cycles shown and then to divide that by the number of seconds covered by these cycles. On each of these graphs, time was measured in milliseconds (ms) where 1 ms = 0.001 second, so 1000 ms = 1 second.

(a) Test Tube: 12 cycles ÷ 23 ms ≈ 0.52 cycles per ms = 520 Hz
(b) Perfume Bottle: 9 cycles ÷ 25 ms = 0.36 cycles per ms = 360 Hz

---

*Although the amplitudes in each of these experiments seem to be the same, if you look at the scale on the *y*-axes you will see that they are all different.

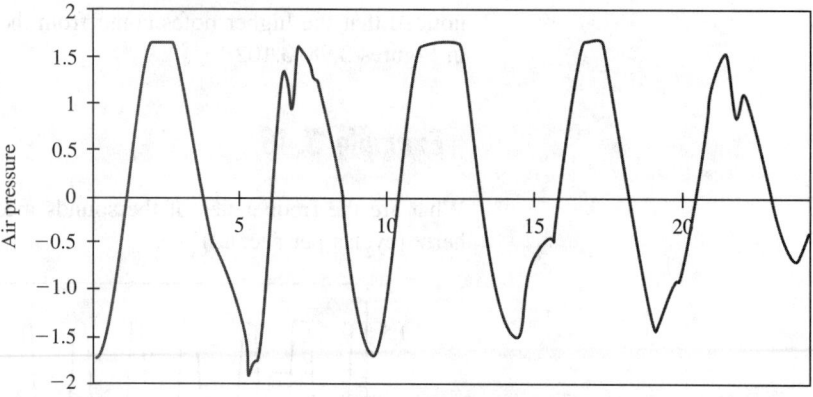

Figure 3.100
Beer bottle

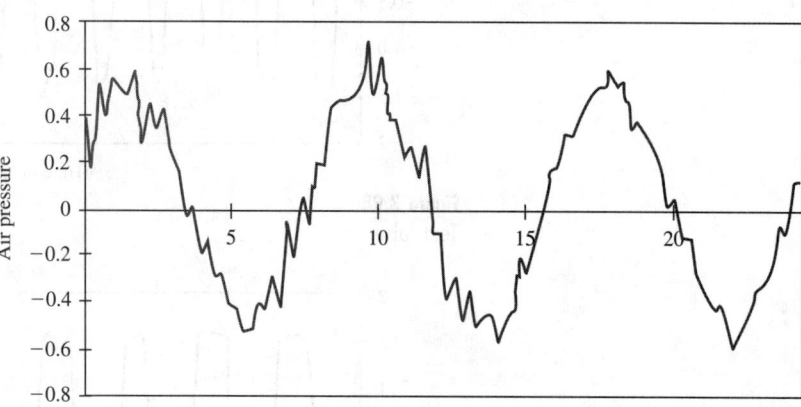

Figure 3.101
Wine bottle

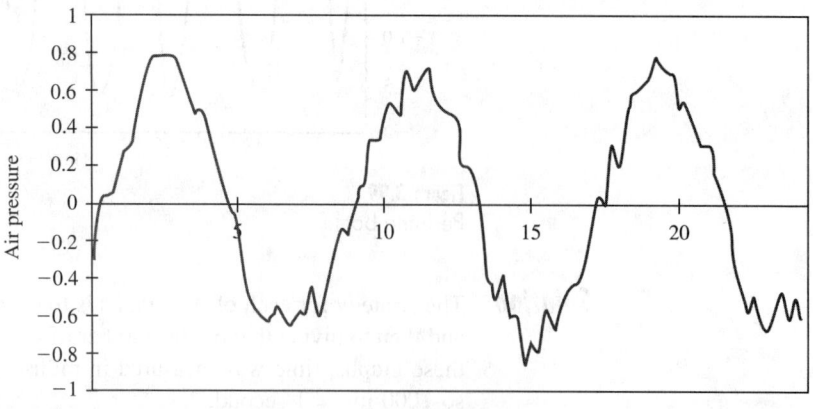

Figure 3.102
Plastic soda bottle

(c) Beer Bottle: 5 cycles ÷ 25 ms = 0.2 cycles per ms = 200 Hz
(d) Wine Bottle: 3 cycles ÷ 25 ms = 0.12 cycles per ms = 120 Hz
(e) Soda Bottle: 2 cycles ÷ 17 ms ≈ 0.12 cycles per ms = 120 Hz

Your answers might be slightly different because there are many ways to estimate where a cycle begins and ends. In general you can see that the higher pitched sounds (from the smaller bottles) have higher frequencies. (To put these frequencies in context, we may note that the human ear can hear sounds in the range of 20 to 20,000 hertz. The piano ranges from 33 to 4224 hertz, with middle C at 262 hertz.

Example 3.49

(a) Write the mathematical model of the two notes that make up the sound produced when the number 1 is pressed on the Touchtone keypad.
(b) Make a graph on the calculator using an interval on the x-axis that shows accurately as much as possible of the graph.

Solutions   We will do the modeling in radians although it could be done in degrees also.

(a) The tones have frequencies of 697 Hz and 1209 Hz. The two sine waves are $y = \sin(B_1 x)$ and $y = \sin(B_2 x)$. The period of the sine function $y = \sin(B_1 x)$ is equal to $\dfrac{2\pi}{|B_1|}$ and the period is $\dfrac{1}{\text{frequency}}$ or $\dfrac{|B_1|}{2\pi}$. Thus we have

$$\frac{|B_1|}{2\pi} = 697 \quad \text{and} \quad \frac{|B_2|}{2\pi} = 1209$$

Therefore $B_1 = 2\pi \times 697$ and $B_2 = 2\pi \times 1209$ and the modeling functions are

$$y = \sin(2\pi \times 697x) \quad \text{and} \quad y = \sin(2\pi \times 1209x)$$

(b) These two functions have much higher frequencies than we have encountered in other graphs, and we have to think a bit in order to get a good picture of the graph. Figure 3.103 shows the graph of $y = \sin(2\pi \times 697x)$ using the `ZTrig` window setting. Figure 3.103 does not present a true picture of the function because not enough points are being graphed. You can see that this is true by looking at Figure 3.104, in which we plotted the same graph for the interval $0 \le x \le 0.131$. In other words, Figure 3.104 shows the graph between $x = 0$ and $x = 0.131$, the first point plotted in Figure 3.103.

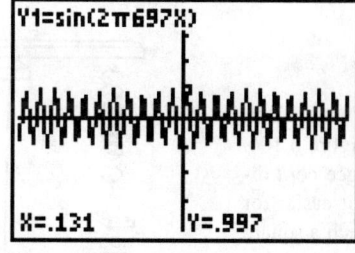

Figure 3.103

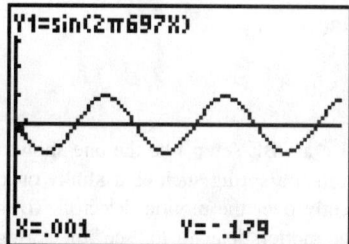

Figure 3.104

The graph in Figure 3.104 looks much better but it doesn't look like a true sine wave. It is negative for positive values of x close to 0 whereas a sine graph should be positive for these values of x. We could try again with a smaller interval, perhaps repeatedly, but that process could go on for some time. In Example 3.50 you'll see how to quickly produce a graph that shows as many cycles as you wish.

Example 3.50

Plot a graph of the functions $y_1 = \sin(2\pi \times 697x)$ and $y_2 = \sin(2\pi \times 1209x)$ showing two cycles of y_1.

Solution   You've learned that the graph of the function $y = \sin(Bx)$ has a period of length $\dfrac{2\pi}{|B|}$.

Therefore the period of the function $y_1 = \sin(2\pi \times 697x)$ is equal to $\dfrac{2\pi}{2\pi \times 697} = \dfrac{1}{697} \approx 0.00143$. If we wanted to show two cycles of this graph we would use values of x in the interval that is twice as long as the period, in other words, $0 \le x \le 0.00286$. Figure 3.105 shows a plot of the two graphs in that interval. ∎

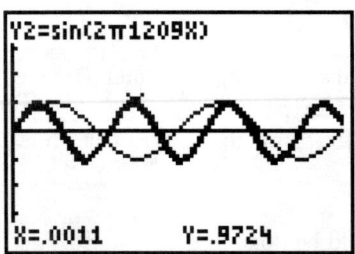

Figure 3.105

You can see that the two tones have frequencies that are almost in a ratio of $1:2$, but not quite. In fact the DTMF (Touchtone) tones are chosen to avoid simple relationships between the frequencies. In the final section of this chapter you will learn how to model the sound produced by two tones played together.

Section 3.5 Exercises

In Exercises 1–6 convert each of the given angle measures from radians to degrees.

1. 2π

2. 1.5π

3. $-3.5\pi = -\frac{7}{2}\pi$

4. $\frac{5}{2}\pi$

5. 2.25

6. 0.35

In Exercises 7–10 convert each of the given angle measures from degrees to radians.

7. $225°$

8. $210°$

9. $-30°$

10. $-540°$

11. Use a CBL setup like the one shown in Figure 3.106. Attach a spring such as a Slinky or a bungee cord directly over the motion detector. To make it easier for the motion detector to "see" the spring, attach a square piece of cardboard about 2–3″ on a side to the bottom of the spring. The object should be light enough so that it does not noticeably stretch the spring. As an alternative attach a smaller mass but tape a piece of cardboard to the bottom of the mass. The cardboard will serve as a target for the motion detector. Pull the spring straight down, turn on the CBL, and start the program SLINKY on the calculator. Release the spring and allow it to oscillate. As soon as the motion of the spring is only up and down (with no side motion), press the TRIGGER button on the CBL. Use the graph that results to answer the following.

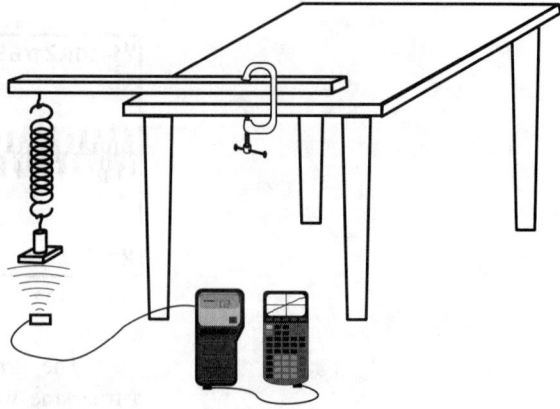

Figure 3.106

(a) What is the amplitude of oscillation?

(b) What is the period of oscillation?

(c) What is the frequency of oscillation?

(d) Perform a sine regression on your data. Remember this formula is in radians not degrees.

(e) What are the amplitude, period, and frequency of the sine regression in (d)?

(f) Compare your results in (a)–(c) with those in (d). How close are these results? If they do not exactly agree, comment on why they do not.

12. Use the same setup as in Exercise 11 but with either a "tighter" spring or a heavier object on the end of the spring. Use the resulting graph to answer the following.

(a) What is the amplitude of oscillation?

(b) What is the period of oscillation?

(c) What is the frequency of oscillation?

(d) Perform a sine regression on your data. Remember this formula is in radians not degrees.

(e) What are the amplitude, period, and frequency of the sine regression in (d)?

(f) Compare your results in (a)–(c) with those in (d). How close are these results? If they do not exactly agree, comment on why they do not.

13. *Sound*   A sound wave is given by $f(t) = 3.25 \sin(75{,}600°t)$ where t is in seconds.

(a) What is the amplitude of f?

(b) What is the period of f?

(c) What is the frequency of f?

(d) Sketch the graph of the sound wave for $0 \le t \le 3.5$.

14. *Sound*   A sound wave is given by $g(t) = 7.35 \sin(63{,}000°t)$ where t is in seconds.

(a) What is the amplitude of g?

(b) What is the period of g?

(c) What is the frequency of g?

(d) Sketch the graph of the sound wave for $0 \le t \le 3.5$.

15. *Sound*   A sound wave is given by $h(t) = 0.055 \sin(400\pi t)$ where t is in seconds. (*Hint*: This function is in radians not degrees.)

(a) What is the amplitude of h?

(b) What is the period of h?

(c) What is the frequency of h?

(d) Sketch the graph of the sound wave for $0 \le t \le 0.01$.

16. *Sound*   A sound wave is given by $k(t) = 1.025 \sin(1200\pi t)$, where t is in seconds. (*Hint*: This function is in radians not degrees.)

(a) What is the amplitude of k?

(b) What is the period of k?

(c) What is the frequency of k?

(d) Sketch the graph of the sound wave for $0 \le t \le 0.01$.

17. *Music* Suppose the following functions are graphs of musical notes.

$$s(t) = 4.2 \sin(20{,}000°t)$$

$$r(t) = 2.7 \sin(40{,}000°t)$$

$$q(t) = 3.4 \sin(35{,}000°t)$$

(a) Which function is related to the loudest sound? Which to the softest?

(b) Which function is related to the highest pitch? The lowest pitch?

18. *Music* Suppose the following functions are graphs of musical notes.

$$s(t) = 4.2 \sin(480\pi t)$$

$$r(t) = 2.7 \sin(960\pi t)$$

$$q(t) = 3.4 \sin(3600\pi t)$$

(a) Which function is related to the loudest sound? To the softest?

(b) Which function is related to the highest pitch? The lowest pitch?

19. *Sound* Write the mathematical model of the two notes that make up the sound produced when the number 3 is pressed on the Touchtone keypad.

20. *Sound* Write the mathematical model of the two notes that make up the sound produced when the number 8 is pressed on the Touchtone keypad.

21. *Sound* Use a CBL setup like the one shown in Figure 3.107. Start the program SOUND on the calculator. Do not press ENTER until told to do so. Ask a student to blow across the open end of a bottle to produce a sound. While the student is making the sound, hold the microphone over the opening of the bottle (about level with the student's nose) and press ENTER. Do not hold the microphone so that the student is blowing directly onto it. You may have to repeat this until you get a satisfactory set of data.

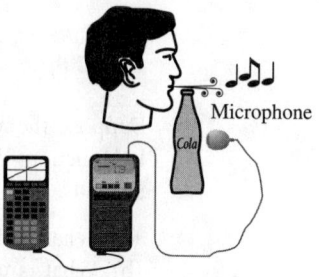

Microphone

Figure 3.107

(a) What is the amplitude of the sound?

(b) What is the period of the sound?

(c) What is the frequency of the sound?

(d) Perform a sine regression on your data. Remember this formula is in radians not degrees.

(e) What are the amplitude, period, and frequency of the sine regression in (d)?

(f) Compare your results in (a)–(c) with those in (d). How close are these results? If they do not exactly agree, comment on why they do not.

22. *Electronics* Electrical power in Europe has a frequency of 50 Hz and the peak value of the voltage at an outlet is 340 V. Write a formula that describes the available voltage at any time t.

23. *Electronics* The hum you hear on a radio when it is not functioning or tuned in properly is a sound wave with a frequency of 60 Hz. Write an equation for this sound. (Assume that the amplitude is 0.5.)

24. **(a)** Describe how you would estimate the phase shift for the Slinky experiment in Examples 3.46–3.47.

(b) Use your answer to (a) and Figure 3.94 to estimate the phase shift for that Slinky experiment.

25. Write a memo to the management in which you describe what you have learned about trigonometry and how you can use this information to develop a way to represent combinations of two notes so that each combination is visually distinguished from all the others, as requested in the Chapter Project.

3.6 Modeling Wave Combinations

In this chapter you have seen that trigonometric functions are useful in modeling electronics, sound waves, vibrations, spring motion, and some aspects of planetary motion. In this section we will study the effect of combining two or more trigonometric functions. We will look at an alternating current example and at the use of trigonometric functions to model the combination of two musical notes to make a chord.

More about Alternating Current

On page 141 we discussed the connection between alternating current, the most common form of electricity, and trigonometry. In Example 3.26 we saw that the amplitude of the function corresponds to the amount of current flowing* and that the frequency of the function gives the frequency of the current. Frequency of a current is measured in hertz (cycles per second).

When two voltage sources are present in a circuit they combine to make one measured voltage. To model such a combination mathematically we add the two functions.

Example 3.51 Application

Suppose the values, in amps, of the currents of two voltage sources are modeled by functions $I_1(t)$ and $I_2(t)$, which give the values of the currents at any time t, in seconds, as $I_1(t) = 20\sin(21600°t)$ and $I_2(t) = 30\sin(21600°t)$.

(a) What is the current represented by each function?
(b) What is the frequency represented by each function?

---

*The amplitude also can correspond to the voltage in the circuit, since voltage and current are proportional quantities. The amplitude of the function, a theoretical result, will always be more than the actual measured voltage.

(c) What are the resulting values of the current and frequency when these two voltages are combined?

Solutions (a) The value of the current is the amplitude of the function, so the currents are 20 and 30 amps.

(b) The frequency is the same for both voltages. The period is $\frac{360°}{21,600°}$, so the frequency is $\frac{21,600°}{360°} = 60$ cycles per second or 60 hertz.

(c) The combined current is the sum of the two functions

$$I_1(t) + I_2(t) = 20\sin(21,600°t) + 30\sin(21,600°t)$$

$$I_1(t) + I_2(t) = 50\sin(21,600°t)$$

Therefore the current is 50 amps and the frequency is the same, 60 hertz.

Figure 3.108 shows a calculator plot of the three functions. The sum is shown in bold. Note that all three functions have the same period and that the amplitude of the sum is equal to the sum of the amplitudes of the original functions.   ∎

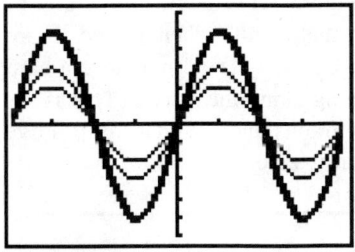

Figure 3.108

Be Careful

Functions like I_1 and I_2 in Example 3.51 add easily because the two frequencies, represented by $\sin(21,600°t)$, are the same. The addition of these functions is, therefore, as simple as $20x + 30x = 50x$. If the frequencies were different the combination would not be so easy, as we will see later.

When you think about what it means to add two or more functions, keep in mind Figure 3.108 and the graph shown in Figure 3.109. They both can remind you that to add two functions you add the y-values at every value of x.

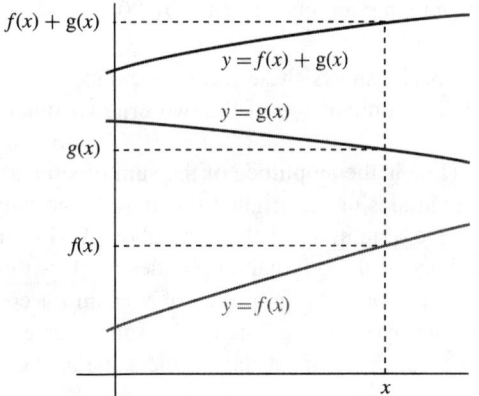

Figure 3.109

All three functions in Example 3.51 are said to be **in phase** because they cross the t-axis at the same points and their maximums and minimums occur at the same values of t. We will now look at adding trigonometric functions that are not in phase.

Example 3.52

What is the result of adding trigonometric functions $\sin x$ and $\sin(x - 180°)$ that are out of phase by 180°?

Solution   Figure 3.110 shows the graph of $y_1 = \sin x$ and Figure 3.111 shows the graph of $y_2 = \sin(x - 180°)$. Figure 3.112 shows the graph that results when y_1 and y_2 are added; that is, Figure 3.112 shows the graph of $y_3 = y_1 + y_2$. You may have a hard time seeing the graph of y_3 because it is a line running along the x-axis. That is, its value is zero throughout. You can look at the two graphs and see that the positive parts are identical to the negative parts, so their sum is zero.

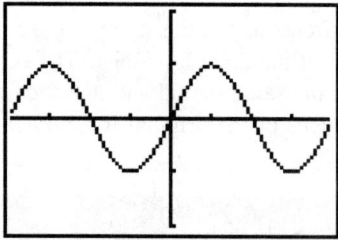

Figure 3.110

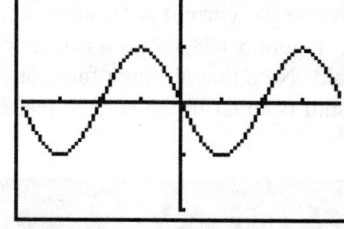

Figure 3.111

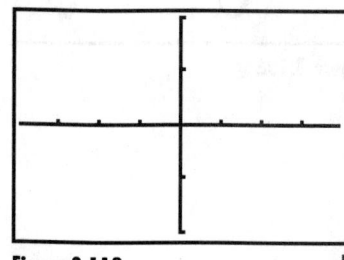

Figure 3.112

The function in Example 3.52, $f(x) = \sin x + \sin(x - 180°)$, is an example of a periodic function that does not have a period. It is periodic because $f(x) = f(x+p)$ for any number p, but it does not have a period because there is no smallest positive value of p where $f(x) = f(x + p)$.

Example 3.53

What is the result of adding the trigonometric functions $\sin x$ and $\cos x$? Remember that $\cos x = \sin(x + 90°)$ and $\sin(x + 90°)$ is a shift of $\sin x$ to the left by 90°, so $\cos x$ and $\sin x$ are out of phase by 90°.

Solution   Figure 3.113 shows these three functions. The sum is in bold print. You can see that it is a combination of the two original functions.

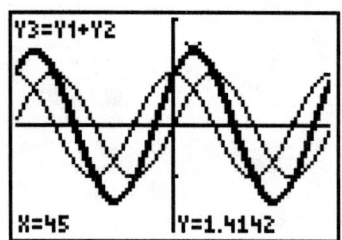

Figure 3.113

How is the amplitude of the sum of $\sin x$ and $\cos x$ in Example 3.53 related to the amplitudes of the original functions? The only thing we can say from this graph is that the amplitude of the sum, about 1.4142 according to Figure 3.113, is more than either of the original amplitudes but less than their sum.

What about the frequency of $y = \sin x + \cos x$? Look carefully at Figure 3.113 where the maximum points of the sum intersect the other curves and you can see that the frequency of the sum is the same as the frequency of the original functions.

Alternating Current and the Addition of Trigonometric Functions

The sum of a sine and a cosine function is important in electronics because it models an alternating current circuit containing a resistor and a capacitor connected in series, as shown in Figure 3.114. Such a circuit is called an **RC circuit**. Because of the way it works, a capacitor reduces the voltage by an amount proportional to its capacitance and produces a voltage that is 90° out of phase from the voltage it receives. A resistor reduces the voltage it receives by an amount proportional to its resistance but does not change the phase. Neither a resistor nor a capacitor changes the frequency of the voltage.

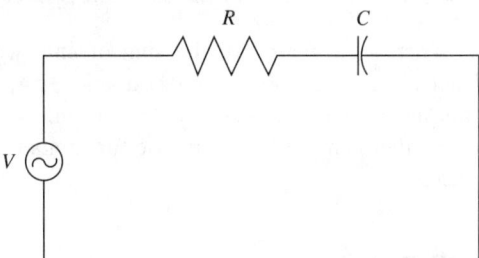

Figure 3.114

The resulting voltage in an RC circuit is found by adding the resistor effect and the capacitor effect. Assuming that the original voltage is expressed in terms of the sine function, the current I in the circuit is found by adding a resistor effect and a capacitor effect, as in the equation

$$I(t) = R\sin(t) + C\cos(t)$$

We will study this equation in more detail in Chapter 4, where you will see how to apply the Pythagorean theorem from geometry to an RC circuit.

Combining Waves of Different Frequency

You have seen in the previous section that different musical notes can be modeled by sine or cosine functions that have different frequencies—for example, $\sin 2x$ and $\sin 3x$. When two notes are played together the resulting sound, called a **chord** in music, can be modeled mathematically by adding together the functions representing the two original notes.

Example 3.54

Compare the amplitude and frequency of the sum of f and g with the amplitudes and frequencies of the functions themselves where $f(x) = \sin x$ and $g(x) = \sin 2x$.

Solution   The original functions are graphed in Figure 3.115. The function $f(x) = \sin x$ is graphed in bold print. You can see that two periods of $\sin 2x$ occur in the space of one period of $\sin x$. Since the period of f is 360°, the period of g is 180°, which is consistent with the fact that the period of $\sin(Bx)$ is $\dfrac{360°}{|B|}$ or, in radians, $\dfrac{2\pi}{|B|}$.

Figure 3.116 shows the sum of the two functions. You can see that the sum appears to be periodic. The calculator's Trace function can be used to estimate the period. For example, we picked two points where the graph of the sum reaches a maximum. One endpoint of a cycle is shown in Figure 3.116. By tracing we see that another end of the same cycle occurs when $x = -300°$. Thus the period of $f(x) = \sin x + \sin 2x$ is $60° - (-300°) = 360°$.

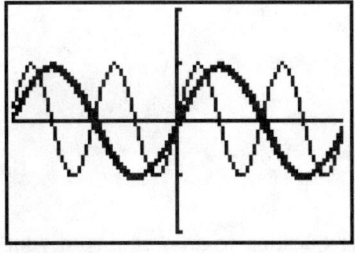

Figure 3.115

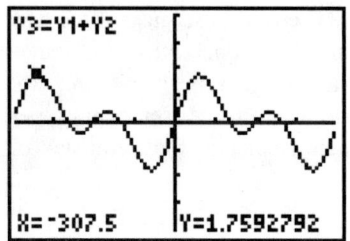

Figure 3.116

To compute the amplitude we need to remember the original definition (from page 152). The amplitude is half the vertical difference between the maximum and the minimum y-values. By tracing we found that a maximum occurs near the point $(52.5°, 1.759)$; a minimum value is near the point $(-52.5°, -1.759)$. Therefore half the difference between these y-values is $\frac{1.759-(-1.759)}{2} = \frac{3.518}{2} = 1.759$. As you have seen before, if the maximum and the minimum values are the same in absolute value then the maximum value of the function is the amplitude.   ■

What is the period of the sum of any two sine functions? We know that it will not always be $360°$ (as in Example 3.54) because the sum in Example 3.52 did not have a period. Activity 3.6 will allow you to explore what happens when you add other pairs of trigonometric functions with the same amplitude and different frequency.

Activity 3.6

Draw graphs of the different combinations of two sine functions shown in Table 3.12 and investigate the properties of the sum of the two functions. Then look at the pattern of the results in your table and answer the following question: If you add two sine functions with the same amplitude but different periods, what can you say about (a) the amplitude and (b) the period of the sum? The first two lines in the table were done in previous examples. Remember that the period of the function $f(x) = \sin(Bx)$ is $\dfrac{360°}{|B|}$ or $\dfrac{2\pi}{|B|}$.

Table 3.12   Experimental results: combining frequencies

$f(x)$	$g(x)$	Period of f	Period of g	Period of $f+g$	Amplitude of $f+g$
$\sin x$	$\sin x$				
$\sin x$	$\sin 2x$				
$\sin x$	$\sin 3x$				
$\sin 3x$	$\sin 3x$				
$\sin 2x$	$\sin 4x$				
$\sin 3x$	$\sin 6x$				
$\sin 4x$	$\sin 8x$				
$\sin 3x$	$\sin 9x$				
$\sin 2x$	$\sin 5x$				
$\sin 3x$	$\sin 8x$				

Solution

We'll take the amplitude first, since that is the easier question to answer. From your table you can see that the amplitude of $f + g$ is always larger than 1.4 but never exceeds 2.0. This is not a very precise answer but it is the best we can do from the information in the table.

The changing values of the period are more difficult to analyze. Try thinking that the period of the sum is equal to the horizontal distance required for the two functions to get back into phase. For example, Figure 3.115 shows that the graphs of $\sin x$ and $\sin 2x$ are in phase at $x = 0°$ and are again in phase at $x = 360°$. From this you could predict that the period of the sum is $360°$.

Let's apply this reasoning to the last pair in the table, $\sin 3x$ and $\sin 8x$. The periods are $120°$ and $45°$, starting out together at $x = 0°$. Where do they next begin a cycle at the same x-value?

In answering this question it will be helpful to think about the two periods as different-sized bricks that you are laying in two rows, as in Figure 3.117. The question then is, "How many of each kind of brick are needed until they line up again?" Since eight 45s and three 120s each make 360, we can say that it requires $360°$ for the pattern to repeat. In mathematical language 360 is the **least common multiple** (lcm) of 45 and 120. That is, 360 is the smallest positive number that can be divided evenly by both 45 and 120.

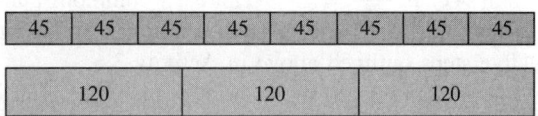

Figure 3.117

If a function has a period it will be the "least common multiple" of the periods of the terms of the function. But the least common multiple is defined for positive integers only. However there is a procedure we can use with many nonnatural numbers:

▶ Change each period to a rational number.
▶ Find the product D of the denominators of all these rational numbers.
▶ Multiply each rational number by D.
▶ Find the lcm of these numbers a and b. On a TI-83 press ⬛MATH NUM 8 a ⬛, b ⬛) ⬛ENTER.
▶ Divide the lcm by D. The answer will be the period, if the function has a period.

Example 3.55

Use your calculator to find the lcm of (a) 9 and 12 and (b) 4, 15, and 18.

Solutions   Remember that to find the lcm of numbers a and b on a TI-83 press ⬛MATH NUM 8 a ⬛, b ⬛) ⬛ENTER.

(a) Press ⬛MATH NUM 8 9 ⬛, 12 ⬛) ⬛ENTER. The result, 36, is shown in Figure 3.118.

(b) The calculator can find the lcm of only two numbers. So find the lcm of two of the numbers and then find the lcm of that answer and the third number. We begin by finding the lcm of 4 and 15 by pressing ⬛MATH NUM 8 4 ⬛, 15 ⬛) ⬛ENTER. The result is 60. Next find the lcm of 60 and 18 by pressing ⬛MATH NUM 8 60 ⬛, 18 ⬛) ⬛ENTER. The result, 180, is shown in Figure 3.119.

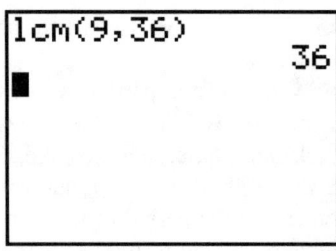

Figure 3.118

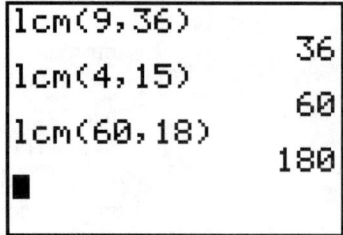

Figure 3.119

Example 3.56

What is the period of $y = \sin(16x) + \sin(24x)$?

Solution   If this function has a period it will be the *least common multiple* of the periods of the terms of the function. The period of $\sin(16x)$ is $\frac{360°}{16} = 22.5°$ and the period of $\sin(24x)$ is $\frac{360°}{24} = 15°$. The least common multiple is defined for natural numbers only and 22.5 is not a natural number, so we will use the brick idea and then the five steps outlined above in Activity 3.6.

Figure 3.120 shows bricks of length 15 in. and 22.5 in. lined up alongside each other. Notice that three 15s are the same length as two 22.5s. Since $3 \times 15 = 45$, the lcm of 22.5 and 15 is 45.

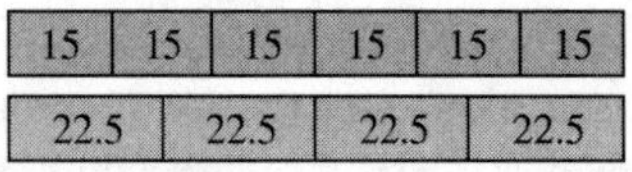

15	15	15	15	15	15

22.5	22.5	22.5	22.5

Figure 3.120

While the brick idea should help you to understand the idea behind least common denominators, it is not always very practical. Here we repeat the five-step process:

▶ Change each period to a rational number: $22.5 = \frac{45}{2}$ and $15 = \frac{15}{1}$.
▶ Find the product D of the denominators of all these rational numbers: $2 \times 1 = 2$ so $D = 2$.
▶ Multiply each rational number by D: $\frac{45}{2} \times 2 = 45$ and $\frac{15}{1} \times 2 = 30$.
▶ Find the lcm of these numbers: $\text{lcm}(45, 30) = 90$.
▶ Divide the lcm by D: $90 \div 2 = 45$. The answer will be the period if the function has a period. If it has a period, the period of $y = \sin(16x) + \sin(24x)$ is $45°$.

Indeed we see that $\sin(16x)$ has 2 cycles in $45°$ and $\sin(24x)$ completes 3 cycles in $45°$. Therefore the period of the sum of $\sin(16x)$ and $\sin(24x)$ is $45°$, as you can see in Figure 3.121. ■

Next you'll see how the addition of sine functions is used in a mathematical model of musical chords.

Modeling Musical Notes and Chords

Musical notes are sounds with specific frequencies that are spaced apart according to a mathematical rule. For example, a sound produced by blowing across the top of a

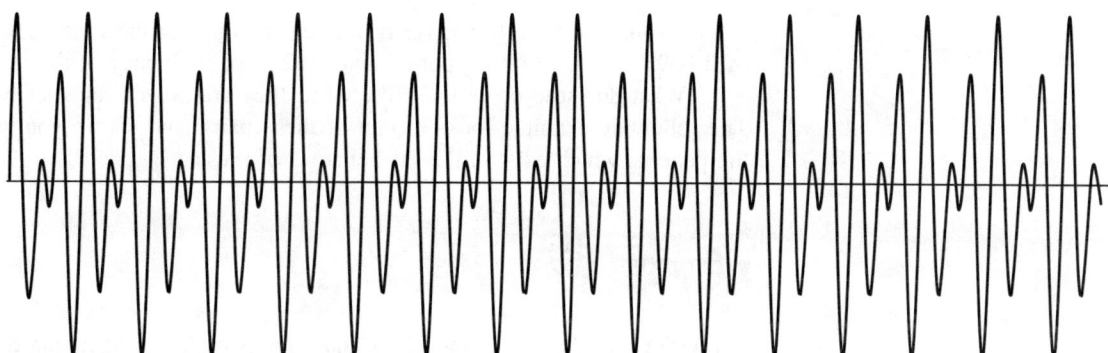

Figure 3.121

bottle may happen to be a musical note if its frequency matches the frequency of a particular note. In most music that is written and performed in Europe and America, the musical notes are organized into a 12-tone scale (shown in Table 3.13). A set of twelve consecutive notes is called an **octave**.

Table 3.13   Frequencies of the notes in one octave

Note	Frequency (Hz)
C (middle C)	262
C sharp	277
D	294
D sharp	311
E	330
F	349
F sharp	370
G	392
G sharp	415
A	440
A sharp	466
B	494
C	523

You can see that the frequency of each note changes by a different amount each time; the difference between the first two frequencies is 15 Hz and the difference between the last two frequencies is 29 Hz. This tells us that the rule for computing the frequency of one note from the previous one is not linear. To get the next frequency higher than any note, multiply its frequency by the 12th root of 2 (about 1.0595). That is, if F_1 and F_2 are the frequencies of consecutive notes then

$$F_2 \approx 1.0595 F_1$$

Since the 12th root of 2 is a number that if multiplied by itself 12 times gives the answer 2, then the 13th note (after 12 multiplications) will have a frequency of twice the first one. You can verify this by multiplying the frequency of Middle C by 1.0595^{12}. The result is approximately 523, the frequency of the next higher C.*

What do these notes look like when they are played together two at a time? The following example looks at one of these pairs, and shows you an example of the patterns which exist in the mathematics of music.

Example 3.57

(a) Write a mathematical model for the notes Middle C and E, and draw the graph that represents these models.
(b) Then write the mathematical model that represents the result of playing these two notes together. (The vertical axis is loudness $L(t)$ and the horizontal axis is time t in seconds.)

Solutions (a) The frequencies are 262 for C and 330 for E, so the two equations are

$$L_C(t) = L_0 \sin(360° \times 262t)$$
$$= L_0 \sin(94,320°t) \quad \text{(in degrees)}$$
$$= L_0 \sin(2\pi \times 262t)$$
$$= L_0 \sin(524\pi t) \quad \text{(in radians)}$$
$$L_E(t) = L_0 \sin(360° \times 330t)$$
$$= L_0 \sin(118,800°t) \quad \text{(in degrees)}$$
$$= L_0 \sin(2\pi \times 330t)$$
$$= L_0 \sin(660\pi t) \quad \text{(in radians)}$$

where L_0 = the loudness of these notes, which we assume here to be the same for both.

(b) The graphs of periodic functions with such large frequencies should be drawn over a very short interval. The graphs in Figure 3.122 are plotted from $t = 0$ to $t = 0.02$ seconds on a *TI-83*. Figure 3.123 shows the same plot done on a computer. You can see that the curves get out of phase before the halfway point in the graph window and are again in phase near the end of the interval. Of course this pattern of being in and out of phase would continue beyond the graph window shown here.

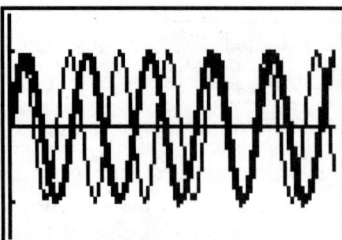

Figure 3.122

---

*The result is not exactly 523 because of rounding error in the numbers in the table. All computations are actually based on the frequency of A (440 Hz). In each octave A is the only note with an exact whole number frequency.

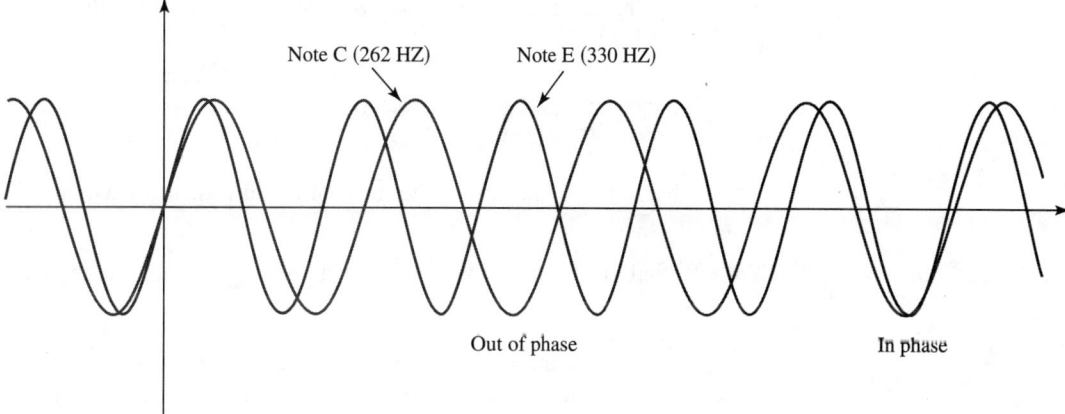

Figure 3.123

Before looking at the graph of the sum of these two notes, think a little about how the sum might look when the curves are out of phase and when they are in phase. Where would the maximum and minimum points be in this interval? Where would the sum be near zero? Figure 3.124 shows the sum as drawn on the *TI-83*, and Figure 3.125 shows the same graphs from a computer graphing program.

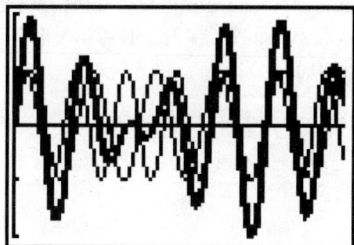

Figure 3.124

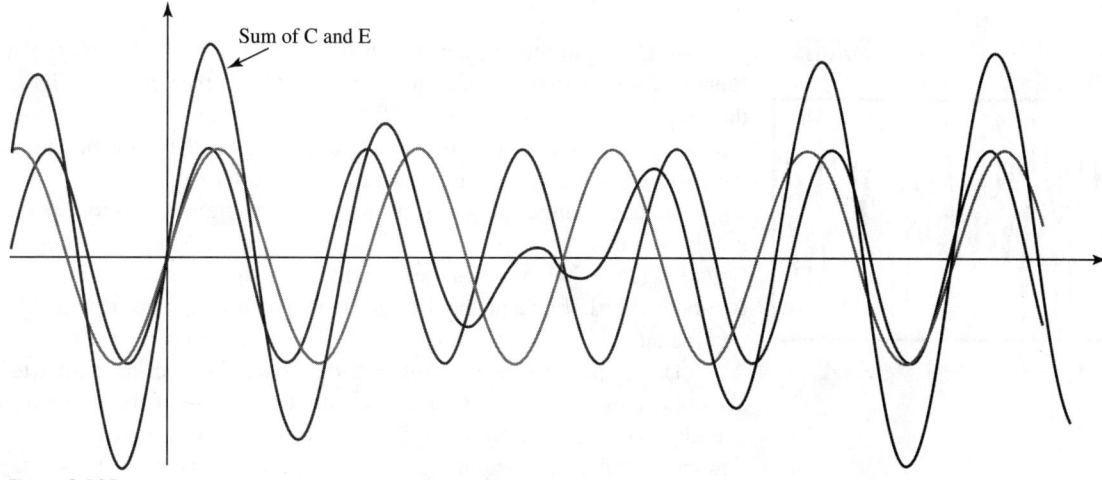

Figure 3.125

In the next examples we will look more closely at the sum of these two notes.

Example 3.58

What is the frequency of the musical chord formed by Middle C and E?

Solution   To find the frequency we first find the period and then use the fact that the frequency is $\dfrac{1}{\text{period}}$.

To find the period we will use the same kind of reasoning we used in Example 3.56. Since Middle C has a frequency of 262 Hz, its period is $\frac{1}{262}$. The frequency of E is 330 Hz, so its period is $\frac{1}{330}$. If this function has a period, it will be the least common multiple of the periods of the terms of the function. The period of $\sin(16x)$ is $\frac{360°}{16} = 22.5°$ and the period of $\sin(24x)$ is $\frac{360°}{24} = 15°$. The least common multiple is defined only for natural numbers and 22.5 is not a natural number, so we will use the steps outlined previously for Activity 3.6.

▶ Change each period to a rational number: The periods are already written as rational numbers, namely $\frac{1}{262}$ and $\frac{1}{330}$.
▶ Find the product D of the denominators of all these rational numbers: $262 \times 330 = 86,460$, so $D = 86,460$.
▶ Multiply each rational number by D: $\frac{1}{262} \times 86,460 = 330$ and $\frac{1}{330} \times 86,460 = 262$.
▶ Find the lcm of these numbers: $\text{lcm}(330, 262) = 43,230$.
▶ Divide the lcm by D: $43,230 \div 86,460 = 0.5$. Thus the period of these two functions is $0.5°$.

Thus the period is $0.5°$ and the frequency is $\frac{1}{0.5} = 2.0$ cycles per degree. ■

Example 3.59

Find a distinctive graphical pattern for the C-E chord.

Solution   First we'll look at the pattern shown within a period by plotting the sum of the two functions across two periods and turning off the other curves. Figure 3.126 shows the graph from $x = -0.5$ to $x = 0.5$. Although you can see that the period on the left of the y-axis appears similar to the period on the right of the y-axis, we will have to look more closely to detect a usable pattern.

We can change the graph by using values from $t = 0$ to $t = 0.06$. The *TI-83* graph is shown in Figure 3.127 and the computer graphing program result is in Figure 3.128. Now you can see a pattern in the figures that is distinctive for this two-note chord. Each of the 12 chords that you could make with Middle C has its own pattern.

The graphs appear to show several cycles, but we know that technically they are not cycles since the period $0.5°$ exceeds the width of the graphing window. You can also see that they are not cycles by looking carefully at Figure 3.128, in which you can see that the beginning of what appears to be a cycle is different for each cycle.

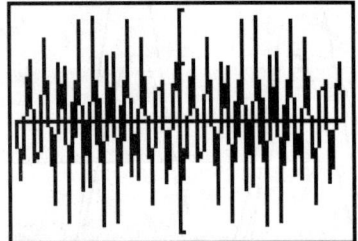

Figure 3.126

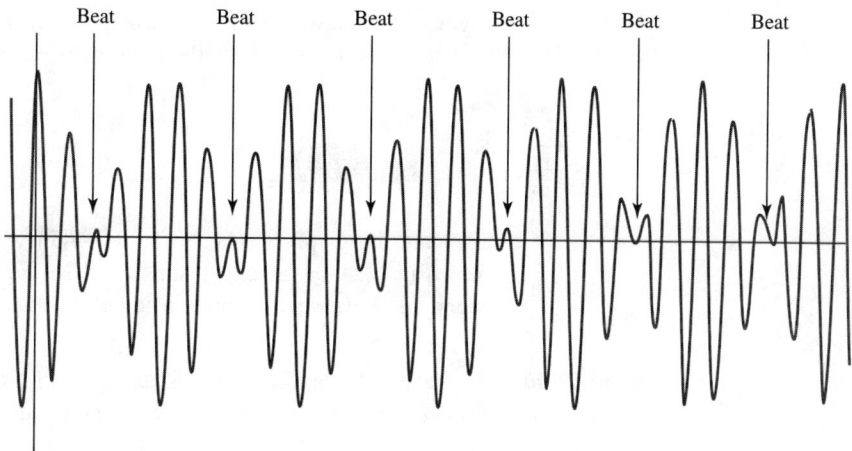

Figure 3.128

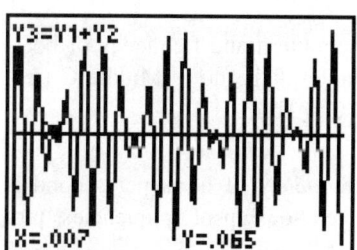

Figure 3.127

But if the regularities in Figure 3.128 are not cycles, what are they? They are called **envelopes**, and each envelope begins and ends with what music theorists call **beats**. Beats are marked off in Figure 3.128. Beats are used by musicians to detect difference in frequency between two notes played together. For example, two guitar players will listen for beats when tuning their guitar strings to the same frequencies. They adjust their strings until they hear no beats when the strings are plucked together.

Example 3.60

Estimate the frequency of the beats (in beats per second) produced by a chord formed by Middle C and E.

Solution To estimate the frequency of beats we need to count the number of beats during a certain time interval. According to the calculator's Trace feature there is a beat at $t = 0.007$, and the third beat after that occurs at about $t = 0.052$. Therefore we have three beats in 0.045 seconds, and the time between beats is $0.045/3 = 0.015$ seconds. The frequency of beats then is estimated to be $1/0.015 \approx 67$ beats per second.

This answer is very close to the difference between the frequencies of the two notes, which is $330 - 262 = 68$. Is this just a coincidence? Let's see.

First we'll see what the mathematics of the two-note chord can tell us about beats. If two functions $f(x) = \sin(ax)$ and $g(x) = \sin(bx)$ are added together, the resulting function

$$h(x) = \sin(ax) + \sin(bx)$$

can be rewritten as an equivalent function, by using a little known property of the sum of two sine functions, which is

$$\sin(ax) + \sin(bx) = 2 \cos\left(\frac{a-b}{2}x\right) \sin\left(\frac{a+b}{2}x\right)$$

This equation says that the sum of two sines can be rewritten as the product of a sine and a cosine. This result is occasionally useful in other parts of mathematics and engineering. However the next example will help you see that this equation reveals something important about the patterns of beats in two-note chords.

Example 3.61

Compare the graph of $j(x) = 2\cos\left(\frac{a-b}{2}x\right)$ with the pattern and frequency of beats shown in $h(x) = \sin(ax) + \sin(bx)$. Use the example of the notes Middle C (frequency 262 Hz) and E above Middle C (frequency 330 Hz).

Solution   We have seen in the previous example that the frequency of beats per second is close to the value of $a - b$. To demonstrate this relationship of frequencies, plot

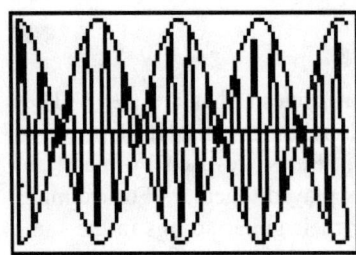

Figure 3.129

$$j(x) = 2\cos\left(\frac{a-b}{2}x\right) \text{ and } -j(x) = -2\cos\left(\frac{a-b}{2}x\right)$$ on the same graph with $h(x) = \sin(ax) + \sin(bx)$. Remember that for these notes $a = 360 \times 262 = 94320$ and $b = 360 \times 330 = 118800$. Figure 3.129 is the result on the calculator and Figure 3.130 shows how we plotted the graphs (we stored values in A and B in the calculator's home screen). Figure 3.131 shows the same two graphs, made with a computer graphing program. A function that sets boundaries for another function is called an **envelope** function. In this case the envelope of $h(x) = \sin(ax) + \sin(bx)$ is formed by the functions $j(x) = 2\cos\left(\frac{a-b}{2}x\right)$ and $-j(x) = -2\cos\left(\frac{a-b}{2}x\right)$.

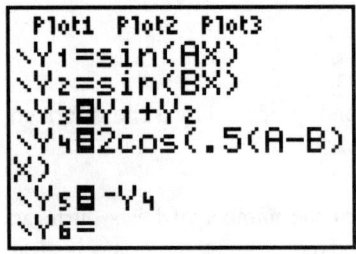

Figure 3.130

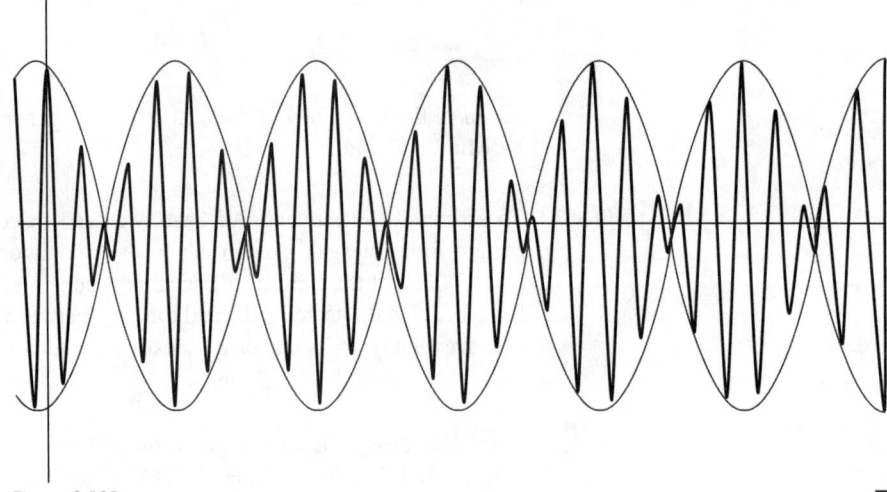

Figure 3.131   ■

You can see in Figure 3.131 that the roots of $j(x) = 2\cos\left(\frac{a-b}{2}x\right)$ correspond to the beats of the chord. Let's look at that function in more detail because we can use it to develop a general way to calculate the beats per second for *any* two-note chord. The discussion requires thinking in steps.

1. For the function $j(x) = 2\cos\left(\frac{a-b}{2}x\right)$, how many roots are there per 360°? In Activity 3.1 you found that $\cos(ax)$ has $2a$ roots per 360°. Therefore $j(x)$ has
$$2 \times \frac{a-b}{2} = a - b \text{ roots per 360°.}$$

2. Since the roots of $j(x)$ fall where the beats are, we can conclude that there are $\dfrac{a-b}{360°}$ beats per second.

3. What is the difference between the frequencies of $\sin(ax)$ and $\sin(bx)$? The frequencies are $\dfrac{a}{360°}$ and $\dfrac{b}{360°}$. The difference in frequency is

$$\frac{a}{360°} - \frac{b}{360°} = \frac{a-b}{360°}$$

4. Therefore the difference between the frequencies of the original notes is the number of beats per second, as the estimation in Example 3.60 made us think.

Scientists who study the ear have discovered that the number of beats per second affects the way we experience a two-note chord. That is, chords are more than just the two notes played together. The frequency of beats, or **beat frequency**, adds an additional dimension and contributes to the degree of enjoyment we get out of a chord. Beats are responsible for some chords sounding better than others to us.

Definition: Beat Frequency

If $y = y_1 + y_2$ where $y_1 = A_1 \sin(2\pi f_1 t) = A_1 \sin(360° f_1 t)$ and $y_2 = A_2 \sin(2\pi f_2 t) = A_2 \sin(360° f_2 t)$, the number of beats per second (the **beat frequency**) is the absolute value of the difference in the frequencies, $|f_2 - f_1|$.

In multi-engine aircraft the engines have to be synchronized so that the sounds don't cause annoying beats, which are heard as throbbing sounds. On some planes this is done electronically and on others the pilot does it by ear, much like a guitar player tuning a string.

Calculator Lab 3.6

Picturing Touchtones

You have now learned enough to plot graphs of the functions that model the combined tones produced by any key on the telephone keypad. The Chapter Project asked for a way to recognize these tones graphically. To test your ability to do this we have made graphs that model each of the twelve combined tones together with the cosine envelope functions of each. Your job is to match our graphs with the numbers on the keypad.

Procedures

1. Develop a way to plot these graphs easily. We suggest using the variables A and B in the calculator's Y= screen, as shown in Figure 3.132. Then change the values of A and B in the home screen with the calculator's Store feature. Notice that Y1 and Y2 are not graphed since we care only about their sum.

2. Choose the keypad number for which the period of the cosine envelope function is the longest. To do that look at the equation of the cosine envelope function:

$$y = 2\cos\left(2\pi \frac{A-B}{2} x\right) = 2\cos(\pi(A-B)x)$$

```
Plot1 Plot2 Plot3
\Y1=sin(2πAX)
\Y2=sin(2πBX)
\Y3=Y1+Y2
\Y4=2cos(π(A-B)X
)
\Y5=-Y4
\Y6=
```

Figure 3.132

Its period is $\dfrac{2\pi}{\pi |A-B|} = \dfrac{2}{|A-B|}$. The period is longest when $|A-B|$ is smallest. Which tone has the smallest value of $|A-B|$? Set the x-interval to the value of the period of that function. Figure 3.133 shows the graph of the function with the longest such period.

3. Set the graph window to graph from $x = 0$ to $x = $ the largest period of the twelve functions. Use the same graph window for each plot. In order to get the largest graphs we have set the values of y to be $-2 \leq y \leq 2$.

4. Plot each of the twelve key tones and match your graphs with the ones in Figures 3.133–3.144.

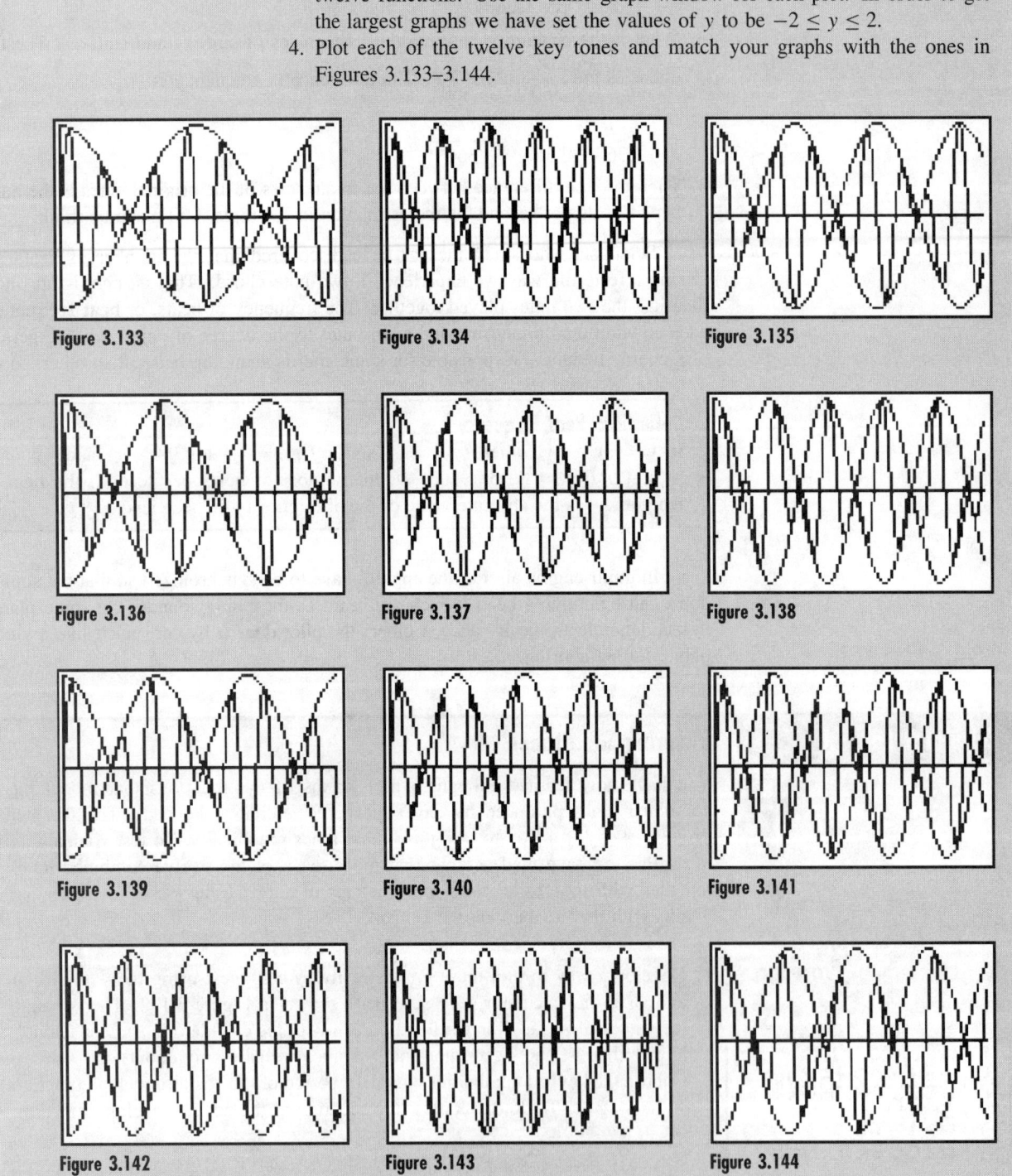

Figure 3.133   Figure 3.134   Figure 3.135

Figure 3.136   Figure 3.137   Figure 3.138

Figure 3.139   Figure 3.140   Figure 3.141

Figure 3.142   Figure 3.143   Figure 3.144

In this section you have seen that trigonometry can be applied to the study of musical notes and chords and that each two-note chord has a graph that can be distinguished from other two-note chords. Throughout the later chapters of this book you will see other applications of trigonometry.

Section 3.6 Exercises

In Exercises 1–4 find the least common multiple of the given numbers.

1. 21 and 35

2. 8 and 18

3. 5, 12, and 26

4. 6, 8, and 21

 For each of the given Exercises 5–11 determine (a) its period if it has one, (b) its amplitude, and (c) graph two cycles of the function making sure to indicate the viewing window.

5. $y = \cos x + \cos 2x$

6. $y = \sin 3x + 3\cos x$

7. $y = 2\sin 3x - 0.5\cos 4x$

8. $y = 1.5\cos 2x - 0.75\sin 5x$

9. $y = 3\sin x + 2\sin 2x + \sin 3x$

10. $y = 3\sin(4x - 180°) + 5\cos(2x + 45°)$

11. $y = 0.5\sin(x - 180°) + 0.25\cos(x + 90°)$

12. *Automotive technology* A function of the form $A_1 \sin(Bt) + A_2 \cos(Bt)$ may be used to define a vibrating system such as an automobile suspension system. One such function is defined by $g(t) = 8\cos(5t) + 6\sin(5t)$.

   (a) Graph two cycles of g.

   (b) What is the amplitude of g?

   (c) What is the period of g?

13. *Automotive technology* The acceleration of a piston is given by the equation $a = 1000(\cos 10t + 0.5\cos 20t)$.

   (a) Graph two cycles of a.

   (b) What is the maximum acceleration and when does it occur?

   (c) Find three values of t where the acceleration reaches the maximum value.

14. *Sound* A sound wave is given by $g(x) = 0.005\sin(400\pi t) + 0.001\sin(1200\pi t)$ where t is in seconds. (This function is in radians.)

   (a) What is the amplitude of g?

   (b) What is the period of g?

   (c) What is the frequency of g?

   (d) Sketch the graph of two cycles of this sound wave.

15. *Electronics* The hum you hear on a radio when it is not functioning or tuned in properly is a sound wave with a frequency of 60 Hz. Write an equation for the loudness L of this hum as a function of t.

16. *Electronics* A **triangular pulse** has applications in timer circuitry. You can approximate a triangular pulse with the following function where x is expressed in radians.

$$y = \sin x + \frac{1}{3^2}\sin 3x + \frac{1}{5^2}\sin 5x + \frac{1}{7^2}\sin 7x + \cdots$$

The more terms you add the closer the graph approaches a triangular pulse.

   (a) Graph two complete cycles of the first three terms of this function. That is, graph $y_1 = \sin x + \frac{1}{3^2}\sin 3x + \frac{1}{5^2}\sin 5x$ with Xmin = 0 and Xmax = 4π.

   (b) Graph two complete cycles of the first six terms of y using the same window settings.

17. *Music* The sound of musical instruments are not simple sine waves but sums of different waves. A clarinet produces a *square* wave that can be approximated by the function

$$y = \sin x + \frac{1}{3}\sin 3x + \frac{1}{5}\sin 5x + \frac{1}{7}\sin 7x + \cdots$$

The more terms you add the closer the graph approaches a square wave.

   (a) Graph two complete cycles of the first three terms of this function. That is, graph $y_1 = \sin x + \frac{1}{3}\sin 3x + \frac{1}{5}\sin 5x$ with Xmin = 0 and Xmax = 4π.

   (b) Graph two complete cycles of the first six terms of y using the same window settings.

18. *Electronics* Electrical power in Europe has a frequency of 50 Hz and the peak value of the voltage at an outlet is 340 V. Write a formula that models the available voltage at any time t.

19. *Sound* Suppose a certain tuning fork produces sounds with frequency 256 Hz and a second produces sounds with frequency 232 Hz. If they are struck simultaneously,

   (a) what is the beat frequency?

   (b) what function describes the sum of these sound waves?

   (c) Write the answer to (b) as a product.

   (d) Graph your answer to (c).

20. *Sound* Two ships have just passed in the night. You are standing on one of the ships when it sounds its horn at a frequency of 100 Hz. The other ship has an identical foghorn, but because of the Doppler effect you hear its horn with an 8% lower pitch. Thus the sound

of the other ship's horn reaches you at a frequency of 92 Hz.
- **(a)** What is the beat frequency?
- **(b)** What function describes the sum of these sound waves?
- **(c)** Write the answer to (b) as a product.
- **(d)** Graph your answer from (c).

21. *Sound* A computer game company wants to design a remote control for a new game system. The remote control would allow a player to move an object on the screen by pressing one of nine buttons. For development purposes the buttons have been named A through I and are arranged as shown below. Each button will produce a sound that will be heard and interpreted by the game unit. However the sounds cannot interfere with someone using a telephone.

A	B	C
D	E	F
G	H	I

- **(a)** Develop a series of tones that will meet the criteria established by the company.

- **(b)** Produce the graph that would result from pressing the B key.
- **(c)** Produce the graph that would result from pressing the G key.
- **(d)** Explain why these sounds will not interfere with someone using a telephone.

22. *Sound* One of the reasons the particular frequencies of the Touchtone telephone were chosen was to avoid **harmonics** (that is, no frequency is a multiple of another, the difference between any two frequencies does not equal any of the frequencies, and the sum of any two frequencies does not equal any of the frequencies).
- **(a)** What are harmonics?
- **(b)** What types of problems would result if harmonics had not been avoided?

23. Write a memo to management in which you answer its request in the Chapter Project. Make sure to describe how you would develop a method to represent combinations of two notes so that each combination is visually distinguished from all the others.

Chapter 3 Summary and Review

Topics You Learned or Reviewed

▶ The trigonometric functions sine, cosine, and tangent are defined in terms of coordinates of the point $P(x, y)$ on a circle of radius r centered at the origin A as

$$\sin A = \frac{y}{r} \qquad \cos A = \frac{x}{r} \qquad \tan A = \frac{y}{x}$$

▶ The trigonometric functions sine, cosine, and tangent are defined in terms of ratios of sides of a right triangle: If $\angle A$ is an acute angle of $\triangle ABC$ with right angle C, then

$$\sin A = \frac{BC}{AB} = \frac{\text{side opposite } \angle A}{\text{hypotenuse}}$$

$$\cos A = \frac{AC}{AB} = \frac{\text{side adjacent to } \angle A}{\text{hypotenuse}}$$

$$\tan A = \frac{BC}{AC} = \frac{\text{side opposite } \angle A}{\text{side adjacent to } \angle A}$$

▶ The trigonometric functions can be used to determine the sizes of the unknown parts of a right triangle.
▶ The graphs of sine, cosine, and tangent functions have characteristic shapes. Any graph that looks like a sine curve is called sinusoidal.
▶ Symmetry within the circle can be used to compute trigonometric functions of angles larger than 90° and smaller than 0°.
▶ How to interpret the concepts *approaching infinity* and *infinite number of values*.
▶ How to find roots of trigonometric functions and how to express all the roots with a formula.

▶ How to solve a trigonometric equation and express all the roots with a formula.

▶ How to define period, amplitude, frequency, phase shift, and vertical translation in trigonometric functions.

▶ Compared with the graph of the basic sine function $y = \sin x$ and the basic cosine function $y = \cos x$, the graphs of $y = A\sin(Bx - C)$ and $y = A\cos(Bx - C)$

- have an amplitude of A.
- have a period of $\dfrac{360°}{|B|} = \dfrac{2\pi}{|B|}$ and a frequency of $\dfrac{|B|}{360°} = \dfrac{|B|}{2\pi}$.
- are shifted horizontally C units. The shift is to the **right** if C is a positive number and to the **left** if C is a negative number.

▶ How to write mathematical models for RC circuits and for musical notes.

▶ The envelope function for the function $f(x) = \sin(ax) + \sin(bx)$ is
$$g(x) = \pm 2\cos\left(\frac{a - b}{2}x\right).$$

Review Exercises

In Exercises 1–4 use your calculator to evaluate each of the following functions to four decimal places.

1. $\sin 43.7°$

2. $\cos 236.25°$

3. $\tan(-167.1°)$

4. $\sin(-330°)$

In Exercises 5–8 find the sin, cos, and tan of $\angle A$ and $\angle B$ for the indicated sides of $\triangle ABC$. (See Figure 3.145.)

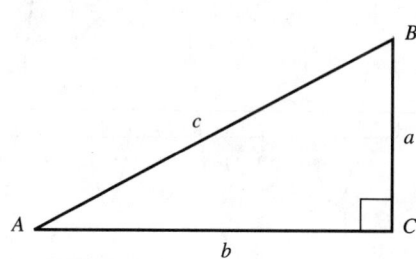

Figure 3.145

5. $a = 55$, $b = 48$, $c = 73$

6. $a = 9$, $b = \sqrt{208}$, $c = 27$

7. $a = 8$, $b = 5$

8. $b = 11$, $c = 15$

In Exercises 9–10 let A be an angle of a right triangle with the given trigonometric function.

9. If $\sin A = \frac{63}{65}$, what are $\cos A$ and $\tan A$?

10. If $\tan A = \frac{15}{8}$, what are $\sin A$ and $\cos A$?

In Exercises 11–12 determine the smallest positive angle that is equivalent to the given angle.

11. $734°$

12. $-981.4°$

Classify each of the functions in Exercises 13–16 as odd, even, or neither.

13. $f(x) = x^2 - 3$

14. $g(x) = x^3 + 2x$

15. $h(x) = x^2 + \cos x$

16. $j(x) = x^3 - \sin x$

17. Consider the function $f(x) = \dfrac{3}{x - 12}$.

   (a) For what value(s) of x will f have no value?

   (b) Complete this table for the function f.

x	−4	−2	−1	0	2	4	6	8	10	12	14
f(x)											

(c) Use your calculator to investigate the function near $x = 12$. Check values on either side of 12; for example, $x = 11.99$, $x = 11.999$, $x = 12.01$ and $x = 12.001$.

(d) Plot the graph by hand on graph paper and check the result with your calculator.

In Exercises 18–23 find formulas for all the solutions of the given equation.

18. $\sin x = 0.4$

19. $\tan x = -1.50$

20. $\cos 4x = -2$

21. $\tan 3x = 1.5$

22. $-2\sin 5x = 0.5$

23. $4\cos 2x = 2.6$

In Exercises 24–27 find the (a) amplitude, (b) period, and (c) frequency of each of the given functions and (d) graph two cycles of each function.

24. $y = 2\sin(5x)$

25. $y = -3\cos(-4x)$

26. $y = 3\tan(2x)$

27. $y = 2.5\sin\left(\frac{1}{3}x\right)$

In Exercises 28–29 estimate the (a) amplitude, (b) period, (c) frequency, and (d) write an equation that models the graph for each of the given sinusoidal waves. Then (e) use your model in (d) to estimate the value of the function at $t = 60°$.

28.

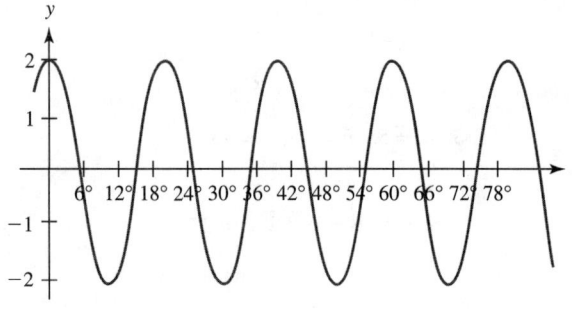

29.

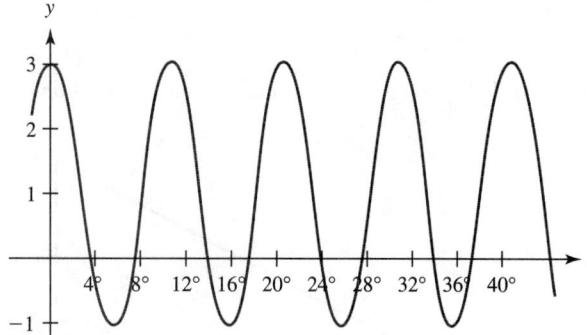

In Exercises 30–35 estimate the (a) amplitude, (b) period, and (c) frequency from each of the given nonsinusoidal waves.

30.

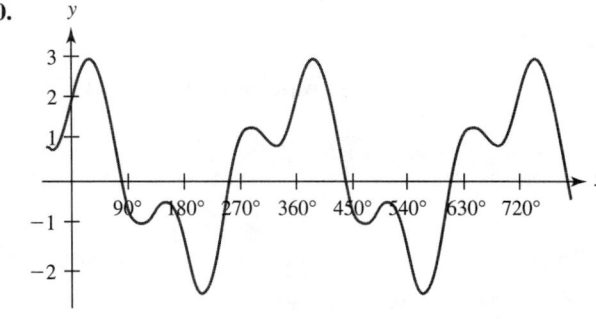

32.

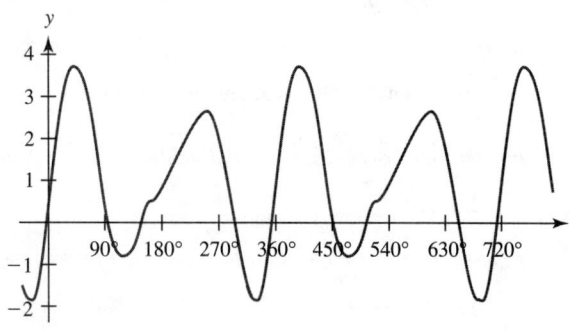

31.

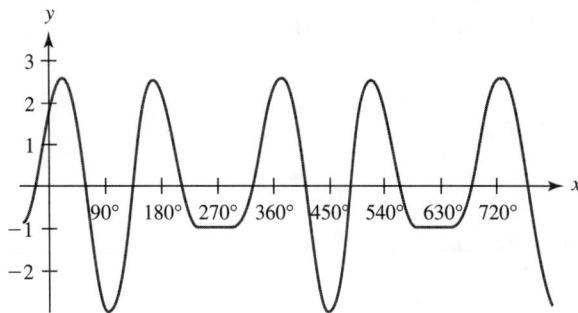

33.

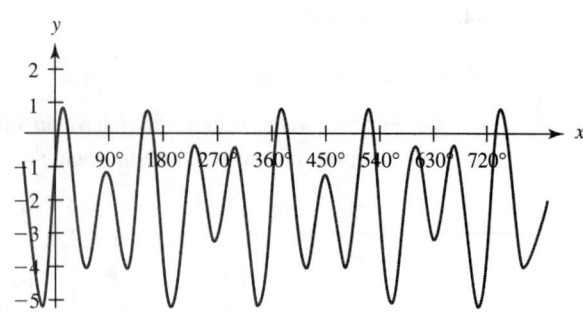

34.

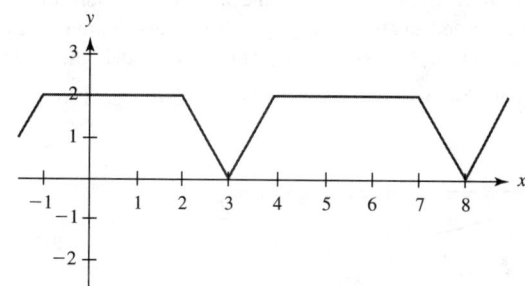

35.

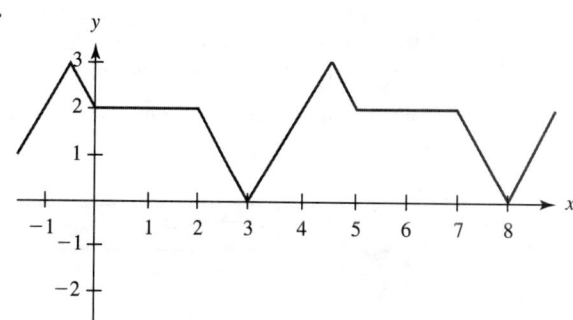

In Exercises 36–39 determine (a) the phase shift and (b) graph two cycles of each of the given functions.

36. $f(x) = 4\sin(2x - 30°)$

37. $g(x) = 0.25\cos(0.5x + 50°)$

38. $h(x) = \sin(-0.25x + 10°)$

39. $f(x) = \tan(2x + 50°)$

40. *Electronics* Determine the phase relationship between

the two sinusoidal waves

$$i = 25\sin(\omega t + 70°)$$

$$v = 10\sin(\omega t + 30°)$$

In Exercises 41–42 convert each of the given angle measures from radians to degrees.

41. $\dfrac{7\pi}{6}$

42. -0.4π

43. -2.45

44. 0.50

In Exercises 45–46 convert each of the given angle measures from degrees to radians.

45. $45°$

46. $-225°$

 For each of the given Exercises 47–50 determine (a) its period if it has one, (b) its amplitude, and (c) graph two cycles of the function making sure to indicate the viewing window.

47. $y = \cos x - 2\cos 3x$

48. $y = \sin(2x) + 0.5\cos x$

49. $y = \sin(2x) - 0.5\cos(4x - 60°)$

50. $y = 0.75\sin(2x + 90°) - 0.5\cos(x - 30°)$

Solve Exercises 51–63.

51. *Transportation* A trucker approaches a hill and sees a sign that says "5% GRADE NEXT 8 MILES." How much lower can we expect the highway to be at the bottom of the hill than at the top?

52. *Police science* A security camera is to be mounted at a height of 8.25 ft on a bank wall that is 18.5 ft from the head teller. The camera is located so that it has a good view of the head teller's customers. Find the angle of depression that the lens makes with the horizontal.

53. *Electronics* In a certain circuit the maximum values of the voltage and current are 185 V and 110 A, respectively. The frequency is 60 Hz and the current leads the voltage by 20°.

   (a) Write the equation for the voltage at any time t. Assume that $C = 0$.

   (b) Write the equation for the current at any time t. Assume that $C = 0$.

   (c) Use your answers from (a) and (b) to plot two cycles of each curve starting with $t = 0$.

54. *Automotive technology* The acceleration of a piston is given by the equation $a = 1000\,(\cos 10t + 0.5\cos 20t)$.

   (a) Graph two cycles of a.

   (b) What is the maximum acceleration? Give three times when it does occur.

55. *Sound*   A sound wave is given by $g(x) = 0.005\sin(400\pi t) + 0.001\sin(1200\pi t)$ where t is in seconds. (This function is in radians.)

   (a) What is the amplitude of g?

   (b) What is the period of g?

   (c) What is the frequency of g?

   (d) Sketch the graph of two cycles of this sound wave.

56. *Meteorology*   The graphs in Figure 3.146 show the time of sunrise and sunset and the number of hours of daylight at different times of year in Honolulu, Hawaii. The data was recorded every ten days for two years beginning January 1, 1995. What is the (a) period and (b) amplitude for each of the curve? (c) Write an equation that models these data.

57. Write an explanation of the similarities and differences between the models for San Francisco (see page 176, Figure 3.85) and for Honolulu.

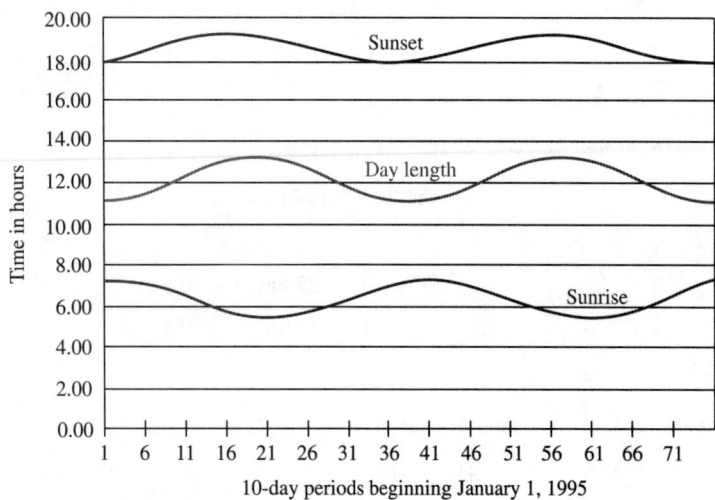

Figure 3.146
Sunrise, sunset, and length of day for Honolulu, HI in 1995–1996 (Source: http://riemann.usno.navy.mil/AA/data/docs/SunRiseSet.html)

58. *Electronics*   Given an ac circuit containing a capacitance C, the voltage across the capacitance is given by $V(t) = 180 \sin(18,000°t)$ where V is in volts (V) and time is in seconds. What are (a) the amplitude, (b) period, and (c) frequency of V?

59. *Electronics*   In Exercise 58 you were given an ac circuit containing a capacitance C. The current in the circuit is given by

$$i(t) = 1.0 \sin(18,000°t + 65°)$$

   (a) What is the amplitude of the current?
   (b) What is the period of the current?
   (c) What is the frequency of the current?
   (d) By how much does the current lead the voltage?

60. *Sound*   A sound wave is given by $f(t) = 2.75 \sin(135°t)$, where t is in seconds.
   (a) What is the amplitude of f?
   (b) What is the period of f?
   (c) What is the frequency of f?
   (d) Sketch the graph of the sound wave for $0 \leq t \leq 3.5$.

61. *Electronics*   A **sawtooth wave** can be approximated with the function

$$y = \sin x + \frac{1}{2}\sin(2x) + \frac{1}{3}\sin(3x) + \frac{1}{4}\sin(4x) + \cdots$$

where x is measured in radians. The more terms you

add, the closer the graph approaches a triangular pulse.

   (a) Graph two complete cycles of the first three terms of this function. That is, graph $y_1 = \sin x + \frac{1}{2}\sin(2x) + \frac{1}{3}\sin(3x)$ with Xmin = 0 and Xmax = 4π.

   (b) Graph two complete cycles of the first six terms of y using the same window settings.

62. Suppose the average monthly temperature in Oklahoma City ranges from a low of 37°F in January to a high of 87°F in July.

   (a) Sketch a rough graph of the average monthly temperatures in Oklahoma City over the course of an entire year.

   (b) Write a possible formula for this graph.

63. *Sound*   Suppose two tuning forks are struck simultaneously and one produces a sound with frequency 60 Hz and the other a sound with frequency 20 Hz.

   (a) What is the beat frequency?

   (b) What function describes the sum of these sound waves?

   (c) Write the answer to (b) as a product.

   (d) Graph your answer from (c).

●●●●●●●● Chapter 3 Test

In Exercises 1–4 use your calculator to evaluate each of the following functions to four decimal places.

1. $\tan 85.3°$

2. $\sin(-125°)$

3. $\cos 0.103$

4. $\tan 5.35$

5. Find the sin, cos, and tan of $\angle A$ and for the indicated sides of right $\triangle ABC$ if $a = \sqrt{152}$, $b = 17$, and $c = 21$.

9. Consider the function $f(x) = \dfrac{3}{x^2 - 9}$.

   (a) For what value(s) of x will f have no value?

   (b) Complete this table for the function f.

x	−5	−4	−3	−2	−1	0	1	2	3	4	5
$f(x)$											

   (c) Use your calculator to investigate the function near $x = -3$. Check values on either side of -3. Explain what seems to be happening.

   (d) Plot the graph with your calculator.

10. Find formulas for all the solutions of the equation $3 \sin(2x) = -1.5$.

11. Consider the function $f(x) = 2 \sin(3x)$. Find the (a) amplitude, (b) period, and (c) frequency of the given function, and (d) graph two cycles of the function.

12. Estimate the (a) amplitude, (b) period, (c) frequency, and (d) vertical displacement for the nonsinusoidal wave shown in Figure 3.147. Then (e) determine the phase shift of the function $f(x) = -2 \cos(3x - 30°)$.

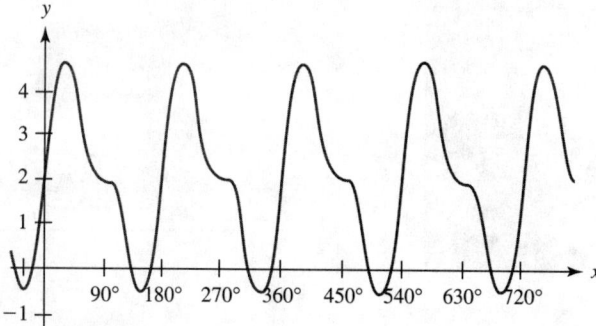

Figure 3.147

6. Let A be an angle of a right triangle with $\sin A = \frac{12}{13}$. Find the values of (a) $\cos A$ and (b) $\tan A$.

7. What is the smallest positive angle that is equivalent to an angle of 1297°?

8. Classify the function $f(x) = x + \tan x$ as odd, even, or neither. Justify your answer.

13. Convert $\dfrac{5\pi}{12}$ from radians to degrees.

14. Convert 115° from degrees to radians.

15. A guy wire needs to be fastened to a power pole at a point that is 25 ft off the ground. Company policy states that the wire should make an angle of 65° with the ground.

   (a) How far out from the base of the pole should the wire be fastened.

   (b) How much wire is needed to reach from the ground to the point where the wire is fastened to the pole?

16. In an ac circuit with only a constant inductance the voltage is given by

$$V = 8 \sin(110\pi t + 0.8\pi)$$

   (a) Determine the amplitude, period, and phase shift of this function.

   (b) Sketch the graph of this function for two periods. Make sure that you label the times on the x-axis where the graph crosses the x-axis.

CHAPTER 4

Mathematical Models in Geometry
Images in Two and Three Dimensions

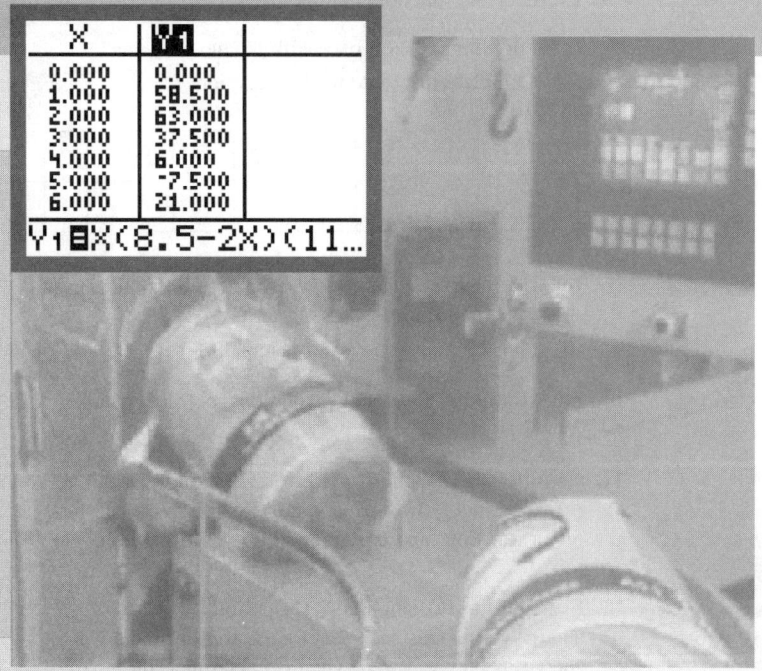

Topics You'll Learn or Review

▶ Use of the formulas for volume and surface area of rectangular solids, cylinders, cones, prisms, and pyramids
▶ Definitions of terms in geometry: circle, radius, tangent, secant, diameter, circumference, prism, pyramid, cylinder, cone
▶ Pythagorean theorem
▶ Law of sines, law of cosines, Pythagorean trigonometric identities
▶ Solve maximum/minimum problems
▶ Use of radians for measuring angles
▶ Angular speed

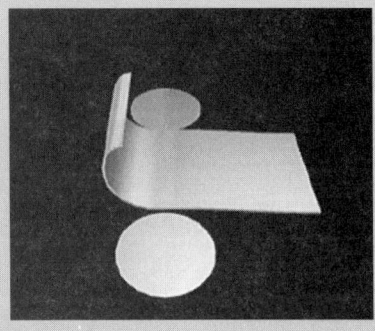

Calculator Skills You'll Learn

▶ When and how to use the zTrig (Zoom Trig) and zSquare (Zoom Square) zoom features
▶ How to approximate the maximum or minimum values of a function using a graph and a table

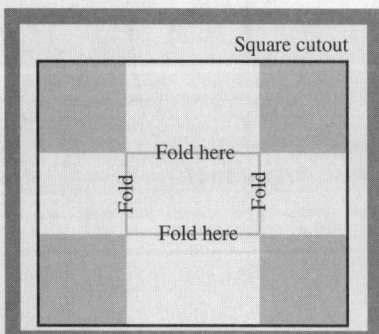

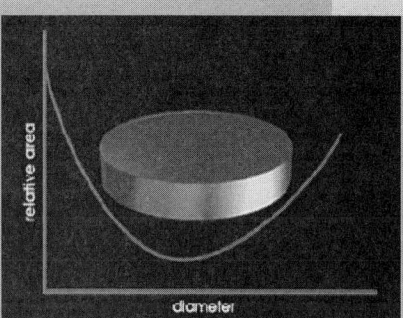

Topics You'll Need to Know

▶ Definitions of terms in geometry: rectangle, square, right angle, right triangle, circle, radius, diameter, circumference
▶ Use of the formulas for the areas of rectangles and squares
▶ Use of unit analysis in calculations when solving problems with units of measure
▶ How to solve the proportion $\dfrac{a}{b} = \dfrac{c}{d}$ for a, b, c, or d

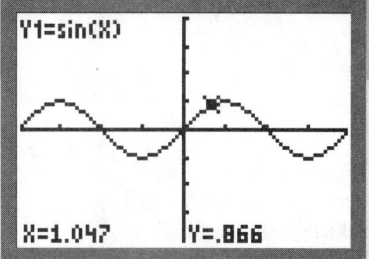

Calculator Skills You'll Use

▶ Use of TABLE and TBLSET to display values of functions
▶ Computing with square roots
▶ Use of π in calculations
▶ Converting decimals to fractions
▶ Graphing with the Path style
▶ Turning functions off and on in the Y= screen
▶ Plotting the graph of, for example, $Y_1 + Y_2$ in the Y= screen

What You'll Do in This Chapter

In this chapter you will learn to apply geometric formulas and concepts to several areas, including finding the most economical way to build packages in the shape of rectangular solids (boxes) and right circular cylinders (cans). You'll learn how to build mathematical models to estimate answers to problems that ask how to produce the largest volume or how to use the smallest amount of material in building packages.

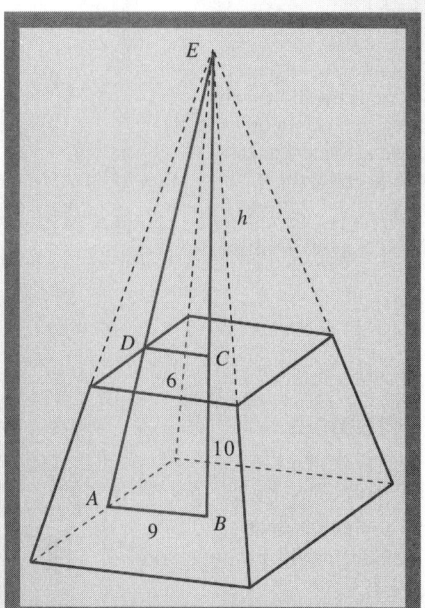

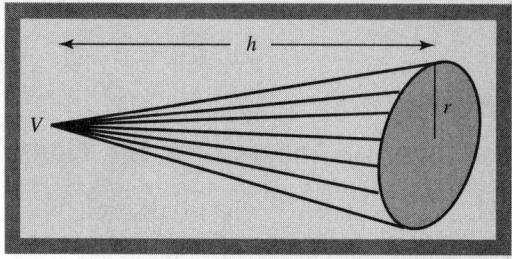

Chapter Project—What's the Best Shape?

Acme
Packaging, Inc.
"If you can make it, we can pack it"

To: Research department
From: Management
Subject: Recycling unused material

We need research in the following two areas:

1. We have carloads of rectangular pieces of sheet metal measuring 20" by 30" left over from the Madison job. The Jefferson job will require open-top rectangular containers. It has been suggested that we make these boxes from the material we have left from the Madison job.

   Boxes would be made by cutting a square out of each corner of the rectangular sheet, folding up the sides, and sealing the joints.

   Please study and report on how the size of the squares cut out from the corners affects the volume and shape of the resulting box. In particular we need answers to the following questions:

   (a) What is the maximum possible volume for an open box made this way?

   (b) What size square should be cut from the corners of the rectangle to produce the box with the largest volume? What are the dimensions of that box?

   (c) What size square should be removed if the box should hold 1000 cm^3? What are the dimensions of this box?

2. One of our customers needs several cylindrical storage tanks with lids, able to each hold 10 000 liters of oil. Each tank will be the same size. In order to save money they want the tanks to use the least possible amount of metal. Please advise us on the dimensions of the tank we should construct.

Activity 4.1
Preliminary Analysis of the Open Box Problem

To begin our analysis of the Chapter Project's open box problems we are going to visualize the situation, make physical models, and graph collected data.

First work on visualizing how Figure 4.1, when folded, could produce something like Figure 4.2. Write answers to the following questions:

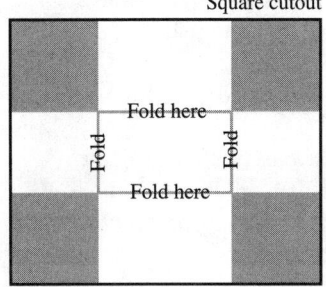

Square cutout

Figure 4.1

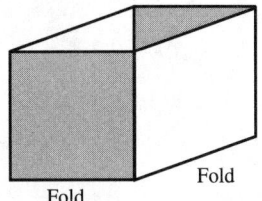

Figure 4.2

1. What would the shape of the container be if the square were very small?
2. What could that container hold?
3. How would you describe the largest cut that could be made?
4. What shape would result if the square were almost as large as it could be?
5. What could that container be used to hold?

The next step after trying to visualize the situation is to make a model. We'll start with a paper model, returning to the mathematical model later in the chapter. A physical model will help you better understand the mathematics.

To make a paper model, begin with a sheet of paper; we'll assume that you will be working with a sheet that is 8.5 by 11 inches, but any size will work just as well. You'll need a ruler, scissors, and tape. Working alone or in a group, follow these steps and record your results on the Acme Packaging Data Sheet below:

1. Decide on a length of square to cut out and call that length x.
2. Cut out the squares, fold the paper, and tape it to make a box.
3. Measure y and z, the other dimensions of the box.
4. Enter the values of x, y, and z into the table on the Acme Packaging Data Sheet.
5. Calculate the volume by multiplying $x \times y \times z$ and enter it into the table.

Acme Packaging, Inc.

"If you can make it, we can pack it"

Data Sheet—Engineering Department

Project: Open Box (Paper Model)
Date: _____

Engineer(s): _____

Dimensions of paper (inches): _____ by _____

Dimensions of paper (cm): _____ by _____

x = measured length of one side of the cutout squares

y = measured length of longer side of the base of the container

z = measured length of shorter side of the base of the container

V = volume of container = $x \times y \times z = xyz$

Trial or Group Number	x	y	z	V

Repeat this process for a different size of cut, or obtain the results of other members or groups in the class. Try to make a variety of shapes and sizes: Make one near the largest possible cut, one near the smallest possible cut, and make others somewhere in the middle.

For the final step in this analysis—the graphing—you will need data from at least five boxes.

On a graph like Figure 4.3 plot, for at least four trials, the graph of volume (V) vs. the side of the square. Estimate the size of the cut that produces the box with the largest volume. Data from one model ($x = 1$ inch and $V = 58.5$ cubic inches) is plotted for you. Follow that example to plot the data you have.

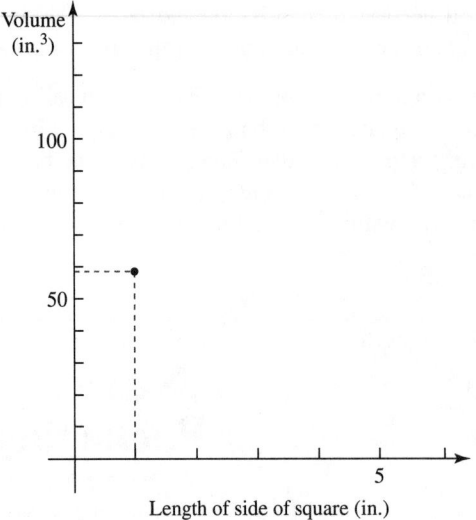

Figure 4.3

In the rest of this chapter we will explore the applications of basic geometric principles to problems like these and others in the areas of trigonometry, estimating distances and areas, and graphing with circles.

Volume and Surface Area

In this section we'll study volume and surface area of rectangular solids, which are more commonly called *boxes*. Then we will build a mathematical model of the open box problem and solve it. You will learn how to find the maximum or minimum value of a function using your graphing calculator.

Volume

Volume is the measure of the amount of space within a three-dimensional figure; it is reported in such units as cubic inches (in.^{3}) or cubic centimeters (cm^3). **Surface area** is the area on the outside of the three-dimensional figure and is measured in such units as square inches (in.^{2}) or square centimeters (cm^2). There are two types of surface area: **lateral surface area** refers to the surface area of all the sides but not any bases, and **total surface area** is the lateral surface area plus the area of the bases.

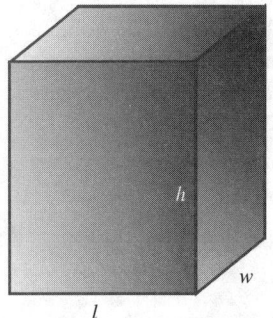

Figure 4.4

When you worked on the problem of cutting corners from paper and folding up the sides to make a box, you were reminded that the volume V of a box is computed with the formula

$$V = l \times w \times h = lwh$$

where l = length, w = width, and h = height (see Figure 4.4). Since it may be unclear which dimensions are the length, width, and height, sometimes we'll use the letters x, y, and z instead of l, w, and h.

An alternative to the volume formula given above comes from thinking of the volume in terms of the area of the base B of the box. Since the base of a box is a rectangle its area is $B = l \times w$, and the volume of the box is

$$V = B \times h = Bh$$

Example 4.1

If one inch of water falls on an acre of land, what is the weight in pounds of all that rain?

Solution Think of the water as filling a box whose height is one inch and whose base area is one acre, as in Figure 4.5. Here are some conversions you will need:

▶ One cubic foot of water weighs about 62.4 pounds.
▶ There are 640 acres in one square mile.
▶ There are 5280 feet in one mile.
▶ There are 12 inches in one foot.

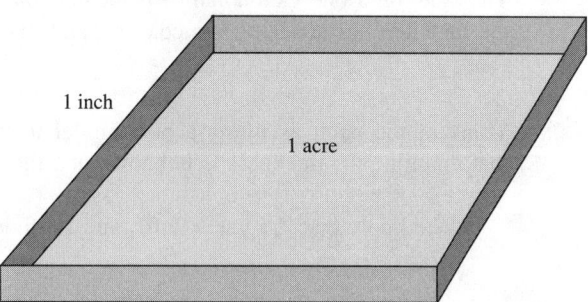

1 inch

1 acre

Figure 4.5

Since we know the weight of one cubic foot of water, we need to find the volume in cubic feet of water that is one inch deep and covers one acre. So the first question is "How many square feet are in one acre?"

$$1 \text{ square mile} = 5280 \text{ ft} \times 5280 \text{ ft}$$
$$= 27{,}878{,}400 \text{ ft}^2$$

$$\text{The number of ft}^2 \text{ in one acre} = \frac{27{,}878{,}400 \text{ ft}^2/\text{mi}^2}{640 \text{ acre/mi}^2}$$

$$= 43{,}560 \frac{\text{ft}^2}{\text{acre}}$$

Thus we can think of this acre of land as a box of water with a base of $43{,}560$ ft^2.

The next question is "How much is the volume of this rain?" The volume is $V = Bh$ where $B = 43,560$ ft^2 and $h = 1$ in. To better understand the volume, the area of the base and the height should both be in the same units. For this example we will use feet. Since 12 in. = 1 ft, then 1 in. = $\frac{1}{12}$ ft. Thus the volume is

$$V = Bh$$

$$= 43,560 \text{ ft}^2 \cdot \frac{1}{12} \text{ ft}$$

$$= 3,630 \text{ ft}^3$$

So the volume of this rain is $3,630$ ft^3.

The last question is the original question of the example: "How much does this water weigh?"

$$\text{The weight} = 3630 \text{ ft}^3 \times 62.4 \frac{\text{lb}}{\text{ft}^3}$$

$$\approx 227,000 \text{ lb}$$

We report only three significant figures because 62.4 has three significant figures.

This weight is over 100 tons on one acre for every inch of rain.   ■

Sometimes it is necessary to find the length of one of the sides of a box given its volume. Then we have to solve either of the volume formulas for one of the other variables, as we'll see in the next example.

Example 4.2   Application

The label on a one-gallon can of paint says that it can cover 400–450 ft^2. Estimate the thickness in inches of this coat of paint. There are about 7.5 gallons in a cubic foot.

Solution

We report two significant figures because 7.5 has two significant figures. However, in the calculations that follow we won't round off 0.133333 until the very last computation in order to ensure the accuracy of our answer.

Think of the paint as filling a box similar to the one shown in Figure 4.5 except that this time the thickness is unknown and the area of the base is between 400 and 450 ft^2.

We know that 7.5 gal = 1 ft^3, and so dividing both sides by 7.5, we get

$$\frac{7.5 \text{ gal}}{7.5} = \frac{1 \text{ ft}^3}{7.5}$$

or $1 \text{ gal} = \frac{1}{7.5} \text{ ft}^3 \approx 0.1333333 \text{ ft}^3 \approx 0.13 \text{ ft}^3$

We know that the area of the base B is between 400 and 450 ft^2. We don't know the height h. But since $V = Bh$ we have

$$V = Bh$$
$$\frac{V}{B} = h \qquad \text{Divide both sides by } B.$$

Thus we have $h = \dfrac{V}{B}$. We already determined that $V = 1$ gal ≈ 0.133333 ft^3. We must calculate the value of h for the two given extreme values of B, namely 400 ft^2 and 450 ft^2. The two results are

$$h = \tfrac{0.133333}{400} \approx 0.0003333 \text{ ft}$$

and $h = \tfrac{0.133333}{450} = 0.0002963 \text{ ft}$

Multiplying these last numbers by 12 (to convert to inches) we get an estimate of between 0.004 in. and 0.0036 in. for the thickness of one coat of paint. How thick is that? Let's compare it with the thickness of ordinary copy paper. A ream (500 sheets) of 20# paper is about $2''$ thick, so a sheet of paper is $\frac{2}{500} = 0.004$ in. or about the same thickness as the coat of paint. ∎

Surface Area

Total surface area is the total area on the outside of a three-dimensional solid. A box like the one shown in Figure 4.4 is made up of six rectangles around its outside, called **faces**. Visualize each of these six faces and think about the area of each. Notice that there are three pairs of equal-sized rectangles (the top and bottom, for example). The two rectangles in each pair have the same area. Thus we have the following formula for the surface area S of a right rectangular box:

$$S = 2lw + 2wh + 2lh$$

If you factor the 2 from each term the formula becomes

$$S = 2(lw + wh + lh)$$

You can decide which of these formulas is easier to remember and use.

The lateral surface area L is the area of the four sides. For the box in Figure 4.4 the lateral surface area is

$$L = 2wh + 2lh$$
$$= 2h(w + l)$$

Next let's look at two special kinds of boxes: one with a square base (Figure 4.6) and one with squares on all six sides (Figure 4.7). From here on we will replace the variables l, w, and h with x, y, and z.

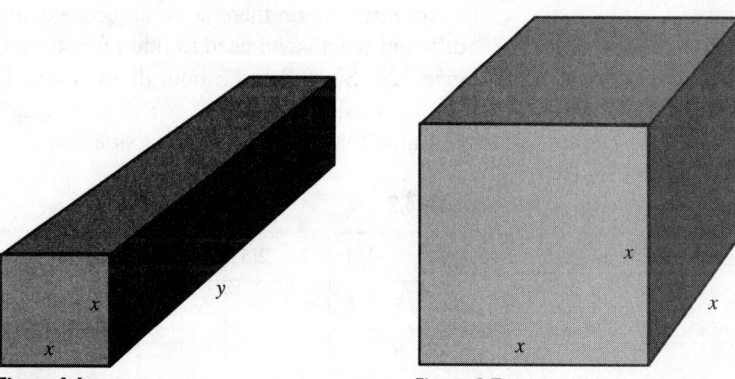

Figure 4.6   **Figure 4.7**

The box with two square faces has a total surface area of

$$S = 2x^2 + 4xy$$
or $$S = 2x(x + 2y)$$

A box with squares on all six sides has all sides equal in length and is called a **cube**; its surface area is

$$S = 6x^2$$

Example 4.3   Application

A rectangular box of cereal is designed to contain 120 in.^{3} of cereal. The box is filled to the top with cereal.

(a) Write the dimensions of three boxes that could contain 120 in.^{3}. Make one of them a cube, one not a cube but with a square base, and one with all three dimensions of different lengths.
(b) Calculate the surface area of each box to the nearest whole tenth of an inch.
(c) Which box requires the least amount of cardboard to build?

Solutions (a) *Cube:* Since the volume is $120 = x^3$, then $x = \sqrt[3]{120}$. Using your calculator you should find that $\sqrt[3]{120} \approx 4.93242$.

Square base but not a cube: There are many correct answers. The value of x could be any positive value smaller than $\sqrt{120} \approx 10.95$. (Why couldn't $x = 11$?) The formula for the volume of a rectangular box with a square base is $V = x^2 y$. We are given $V = 120$, so we have the equation

$$y = \frac{120}{x^2}$$

Use Table 4.1 to determine the value of y for different values of x.

Table 4.1

x	1	2	2.5	3	3.75	4	4.5	5	7.5	10
$y = \dfrac{120}{x^2}$	120	30								

Other box: Again there are many correct answers. For a box with sides of three different lengths we need to find three different positive numbers that multiply to give 120. Since 2 and 3 both divide evenly into 120 we have $120 = 2 \times 3 \times 20$, so one possible answer is $x = 2$, $y = 3$, and $z = 20$. Now completes Table 4.2. (Assume that x is the smallest side and z is the longest.)

Table 4.2

x	2	1	2	2.5	3	3	3.5	4	4	
y	3	6	4	4	5					
z	20									10

(b) *Cube:* The surface area of the cube is $S = 6x^2 \approx 6 \times 4.93242^2 \approx 146.0$ in.^{2}.

Square base: Here the surface area is $S = 2x(x + 2y)$. Use $S = 2x(x + 2y)$ and the completed Table 4.1 to complete Table 4.3.

Other box: The surface area is given by the formula $S = 2(xy + xz + yz)$. Use this formula and the completed Table 4.3 to complete Table 4.4.

(c) The cube has the lowest surface area of these boxes, which seems to indicate that the cube requires the smallest amount of cardboard in its manufacture. Since we did not check all of the possible sizes for boxes with a volume of 120 in.³, we cannot be sure without more mathematics if the cube is the best choice.

Table 4.3

x	1	2	2.5	3	3.75	4	4.5	5	5.25	7.5	10
y	120	30									
S	482	248				152					

Table 4.4

x	2	1	2	2.5	3	3	3.5	4	4	
y	3	6	4	4	5					
z	20									10
S										

Exploring Max/Min Problems

The question posed in the first half of the opening problem is "What size square should be cut out of the corners of a rectangular sheet measuring 20″ by 30″ so that the volume is the largest possible?" In mathematics, business, and engineering one often meets problems of finding the largest (maximum) or smallest (minimum) value of a variable. Such problems are called **max/min problems**. The strategy for solving max/min problems can be broken down into seven steps.

Solving a Max/Min Problem

Step 1: Select a letter to represent what is being changed (in this case the size of the cut). This value is called the **independent variable** because we have control over its possible values.

Step 2: Select a letter to represent what you want to maximize or minimize (in this case volume). This value is called the **dependent variable** because its value depends on the value of the independent variable.

Step 3: Write other variables in the problem in terms of the independent variable.

Step 4: Write a formula that connects the dependent and the independent variables, and only these variables. This is the **mathematical model** of the problem.

Step 5: Decide what values of the independent variable are acceptable. The acceptable values for the independent variable make up what is called the **domain of the model**; so in this step you will determine the **practical domain** of the model.

Step 6: Use a table of values or a graph to locate the largest or smallest value of the dependent variable.

Step 7: Answer the question.

We'll show you how to solve the open box problem for the 8.5″ by 11″ paper you cut and folded (in Activity 4.1) and leave it to you to solve the problem as stated in the opening memo. We'll take the steps one at a time.

Step 1: *Select the independent variable:* Let x = the length in inches of the cut. (See Figure 4.8).

Step 2: *Select the dependent variable:* Let V = the volume of the box.

Step 3: *Write other variables in terms of the independent variable:* Figure 4.9 shows that x is the height of the box. Figure 4.8 shows how the length and width of the box can be written in terms of x.

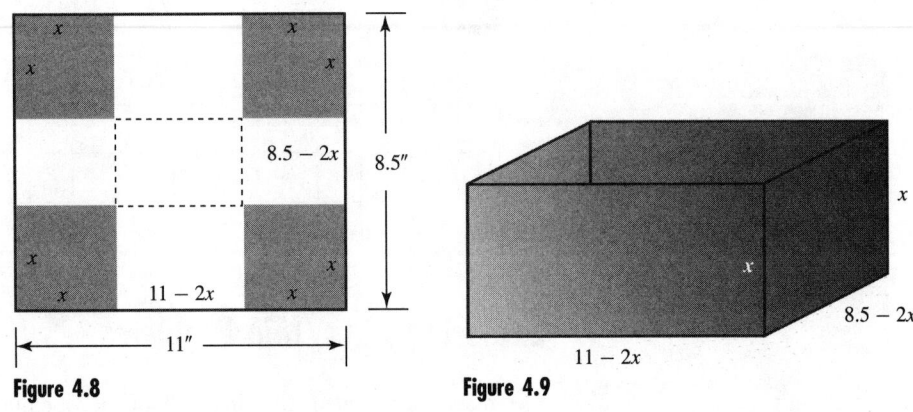

Figure 4.8   **Figure 4.9**

Step 4: *Write a formula connecting the dependent and independent variables:* From the diagrams we can see that

$$V(x) = x(8.5 - 2x)(11 - 2x)$$

Step 5: *Determine the domain of the model (acceptable values of the independent variable):* The length of the cut cannot be a negative number, so $x \geq 0$. Also, for this sheet of paper, the cut cannot be greater than half the shorter length of the paper; that is, $x \leq 4.25$. Putting these together we write

$$0 \leq x \leq 4.25$$

which can be translated as "x is between 0 and 4.25, inclusive." Is this the best domain? What would happen if $x = 0$ or $x = 4.25$? Should we change the domain to $0 < x < 4.25$?

Step 6: *Analyze the table and/or graph:*

(a) *Table:* We'll use a table to estimate the best settings for the graph window. In this case make the table start at $x = 0$ and continue for $x = 1, 2, 3, \ldots$. That is, let ΔTbl $= 1$. Our table, as produced on a *TI-83*, is shown in Figure 4.10. You can see that for $x = 5$ the volume is negative, an impossibility that is consistent with our restriction that x cannot exceed 4.25. Why do you think the volume becomes positive at $x = 6$? From Figure 4.10 we see that the largest volume seems to be between 1 and 3. Change ΔTbl to 0.25 and see if you can get a better approximation for the volume of x that maximizes $V(x)$.

(b) *Graph:* The table helps us to figure out values for the dimensions of the graph window because it shows some of the highest values of V. A graph of V is shown in Figure 4.11.

X	Y1
0.000	0.000
1.000	58.500
2.000	63.000
3.000	37.500
4.000	6.000
5.000	-7.500
6.000	21.000

Y1■X(8.5-2X)(11…

Figure 4.10

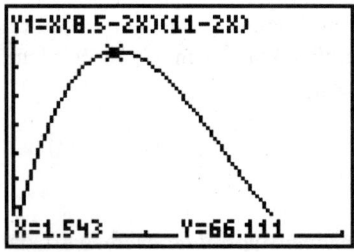

Figure 4.11

Step 7: *Answer the question:* By tracing the graph we see that a cut of approximately 1.6 inches gives the maximum volume. If more accuracy is required you can zoom in near the maximum value of $V(x)$ on the graph or change the table to look more closely at the values near the maximum.

Now use these seven steps to solve a max/min open box problem for a different-sized sheet of paper.

Example 4.4

Estimate the size of squares you should cut out of the corners of a 20″ by 30″ sheet of cardboard to produce the largest volume of the resulting box.

Solution   Follow each of the seven steps described above. To the nearest tenth, the value of x that produces the largest volume is 3.9″.   ■

First Attempt at the Minimum Area of a Cylinder

The second part of the Chapter Project requires you to find the shape of a 10 000-liter cylinder that would use the smallest amount of material. That is, of all the cylinders whose volume is 10 000 liters, which one has the minimum surface area? A **cylinder** is a three-dimensional solid in which both the top and bottom, called the **bases**, have the same size and shape. Figure 4.12 shows three different solids that are all cylinders. The cylinder in Figure 4.12(a) is a **right cylinder** because the side of the cylinder is perpendicular to the base. Figure 4.12(a) is a **right circular cylinder** because the bases are circles and it is a right cylinder. When many people use the term *cylinder* they are talking about a right circular cylinder. How would you measure the total surface area of a cylinder? We'll return to that question in a later section of this chapter. For now we must be content to work a similar problem involving a box.

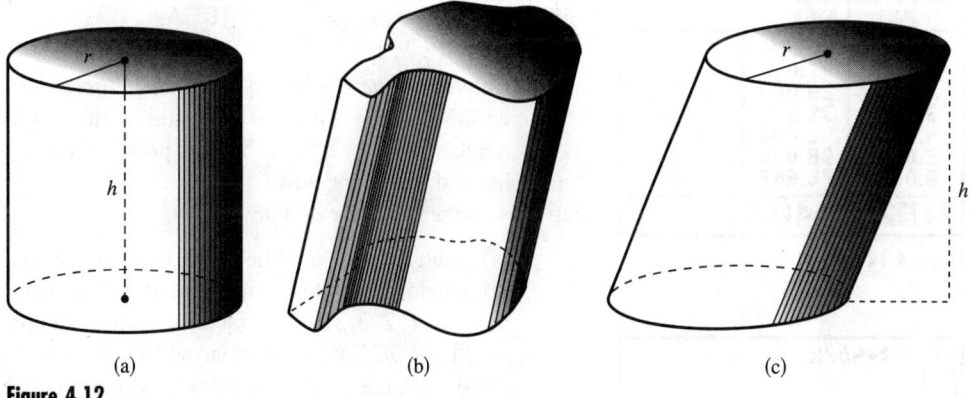

(a)   (b)   (c)

Figure 4.12

Example 4.5

What is the shape of a rectangular solid with a square base and volume of 10 000 liters that has the minimum total surface area?

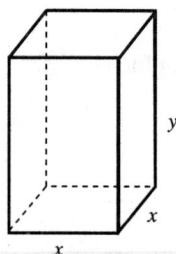

Figure 4.13

Solution There are many differently shaped boxes with the same volume. Our job is to find the one with the minimum surface area. This is another kind of max/min problem, and we'll repeat the seven steps we outlined previously.

Step 1: *Select the independent variable:* Let $x =$ the length of the base (in cm) and y the height of the box (in cm), as shown in Figure 4.13.

Step 2: *Select the dependent variable:* Let $S =$ the surface area of the box.

Step 3: *Write other variables in terms of the independent variable:* Here we have two possible independent variables—x and y. At this time we know how to work with one independent variable only; this means that we need to define one of the variables in terms of the other. We will write y in terms of x. The way to do this is to use the given information that the volume is 10 000 liters and the fact that for this box $V = x^2 y$. Thus $y = \dfrac{V}{x^2}$; but V is in liters and x is in centimeters (cm). One liter is 1000 cm^3, so a volume of 10 000 liters is also 10 000 000 cm^3. Since there are 100 cm in one meter, there are 100^3 cm$^3 = 1\,000\,000$ cm^3 in one cubic meter, so we have a volume of 10 m^3. Since $V = x^2 y$, the equation for y is

$$y = \frac{V}{x^2} = \frac{10}{x^2}$$

where x and y are measured in meters.

Step 4: *Write a formula for the dependent variable:* One formula for the total surface area is

$$S = 2x^2 + 4xy$$

Into this formula we'll substitute $\dfrac{10}{x^2}$ for y:

$$S = 2x^2 + 4x\left(\frac{10}{x^2}\right)$$

$$= 2x^2 + \frac{40}{x}$$

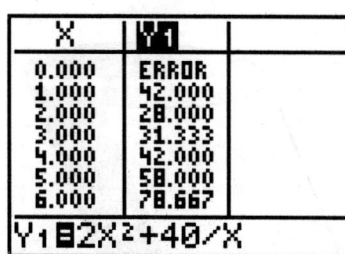

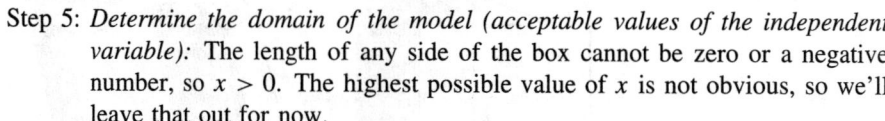

Figure 4.14

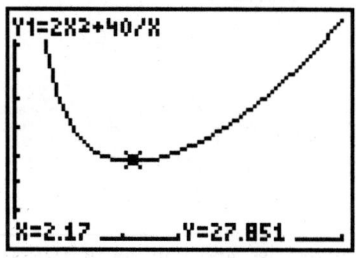

Figure 4.15

Step 5: *Determine the domain of the model (acceptable values of the independent variable):* The length of any side of the box cannot be zero or a negative number, so $x > 0$. The highest possible value of x is not obvious, so we'll leave that out for now.

Step 6: *Analyze the table and/or graph:*

(a) *Table:* We'll use the table to estimate the best settings for the graph window. In this case make the table start at $x = 0$ and continue for $x = 1, 2, 3, \ldots$. Our table is shown in Figure 4.14. The value of $S(0)$ is "ERROR," which is clear when you look at the formula for $S(x)$. If $x = 0$ we have a division by zero which produces an error on the calculator. From the table you can see that a minimum value of surface area occurs around $x = 2$. Changing ΔTbl to 0.1 shows that the minimum surface area is around $x = 2.2$. The graph will reveal more.

(b) *Graph:* The table helps us to figure out values for the dimensions of the graph window because it shows the highest and lowest values of $S(x)$. We used the indicated settings and produced the graph shown in Figure 4.15.

Step 7: *Answer the question:* The graph suggests that the minimum surface area occurs at approximately $x = 2.17$ meters. We are asked for the shape of the minimum area box, so we must find the value of y. Since $y = \dfrac{10}{x^2}$ we have $y = \dfrac{10}{2.17^2} \approx 2.12$. Because x and y are very close in value, the shape of the box is close to a cube. In fact if we zoom in further to get more accuracy, the best value of x is approximately 2.154 meters and the resulting value of y is approximately 2.155 meters. The surface area is about 27.85 m². ∎

We solved the Chapter Project with an unstated assumption that minimum total surface area means minimum cost. Perhaps that is true and perhaps not. In some containers, parts such as the top and bottom are more expensive either to manufacture or to attach. In the next example we take cost of materials into account.

Example 4.6   Application

A container in the shape of a rectangular box with a square base is to be designed to hold 10 m³ and to minimize the cost of materials needed to build the box. If the cost (in dollars per m²) of the square top and square bottom is three times that of each of the other four faces, what dimensions minimize the cost of building the container?

Solution   In this situation you are minimizing the cost rather than minimizing the surface area. However the cost is based on the surface area of the box. We know that the surface area is $S = 2x^2 + 4xy$. Since the cost of the top and bottom is three times that of each of the other faces, to get the total cost we will pretend that we are using triple materials for the top and bottom. Thus the cost C can be expressed as

$$C = 6x^2 + 4xy$$

From here on the solution remains the same as in Example 4.5. The best length of the square face turns out to be $x \approx 1.49$ m, and $y \approx \dfrac{10}{1.49^2} \approx 4.50$. That is, the result is that the container is three times as tall as it is wide. Did you guess that? By the way, the surface area of this box is about 31.26 m². ∎

Section 4.1 Exercises

In Exercises 1–4 write the formulas for the (a) volume and (b) surface area of each rectangular solid.

1. The base is a square. Each side of the square has a length of x, and the height of the solid is double the length of one side of the square.

2. The base is a square. Each side of the square has a length of x, and the height of the solid is 5″ more than x.

3. The height is 2″ longer than the length, and the width is 4″ shorter than the length.

4. The height and length are equal, and the width is 5 cm shorter than the height.

In Exercises 5–8 determine the volume of the rectangular solid with the given dimensions.

5. $l = 3.0$ cm, $w = 4.5$ cm, and $h = 2.4$ cm

7. $l = 3\frac{1}{2}$ in., $w = 2\frac{1}{4}$ in., and $h = 6.0$ in.

6. $l = 6.2$ mm, $w = 2.5$ mm, and $h = 9.1$ mm

8. $l = 6'2''$, $w = 4'6''$, and $h = 3'0''$

9. If the volume of a rectangular solid is 156.06 cm³, and if $w = 4.25$ cm and $h = 5.40$ cm, determine l.

10. If the volume of a rectangular solid is 91 ft³, $l = 4.0$ ft, and $w = 6.5$ ft, determine h.

In Exercises 11–14 determine the (a) lateral surface area L and (b) total surface area S of the rectangular solid with the given dimensions. Assume that the length and width are the dimensions of the base.

11. $l = 4.5$ cm, $w = 7.2$ cm, and $h = 6.1$ cm

12. $l = 2.35$ m, $w = 1.45$ m, and $h = 1.80$ m

13. $l = 6.2$ in., $w = 5.3$ in., and $h = 2.5$ in.

14. $l = 2.25$ ft, $w = 0.75$ ft, and $h = 3.5$ ft

15. *Transportation*   A closed railroad container car in the shape of a rectangular box is 30 ft long, 8 ft wide, and 12 ft high.

    (a) How much can the container car hold?

    (b) How many square feet of aluminum are needed to make the car?

16. *Commercial design*   An open box is to be made by cutting squares of size x cm from the four corners of a rectangle 18 cm long and 12 cm wide. Determine the size of x that will result in a box with the largest volume.

    (a) Write a mathematical model for this problem.

    (b) What is the practical domain for your model?

    (c) What is the size of x that produces a box with the largest volume?

    (d) What are the dimensions of this box?

    (e) What is the volume of this box?

17. *Commercial design*   A closed box is to be made by cutting squares of size x cm from the four corners of a rectangle 25 cm long and 16 cm wide. The box must be at least 19 cm long in order to hold pencils that are 18.8 cm long and 0.8 cm wide. Determine the size of x that will result in a box with the largest volume.

    (a) Write a mathematical model for this problem.

    (b) What is the practical domain for your model?

    (c) What is the size of x that produces a box with the largest volume?

    (d) What are the dimensions of this box?

    (e) How many pencils will this box hold?

18. *Commercial design*   An open box with a capacity of 4,800 in.^{3} is to be made by cutting squares of size x from the four corners of a rectangle. The box is to be twice as long as it is wide.

    (a) What are the dimensions of the box with the smallest surface area?

    (b) What is the smallest surface area?

19. *Industrial design*   A cereal box's inside measurements are 24.6 cm high, 16.6 cm deep, and 5.3 cm wide.

    (a) How much cereal can this box hold?

    (b) If the cereal that is placed in the box has a mass of 0.2 g/cm^3, what is the total mass of the cereal in a full box?

20. *Industrial design*   A box with a square base and an open top is to hold 64 in.^{3}. You want to make this box using the least amount of material.

    (a) Write a mathematical model for this problem.

    (b) What is the practical domain for your model?

    (c) What dimensions, to two decimal places, will require the least amount of material?

    (d) Use your results in (c) to determine the least amount of material that will be needed to make this box.

21. *Industrial design*   A box with a square base and a closed top is to hold 64 in.^{3}. You want to make this box using the least amount of material.

    (a) Write a mathematical model for this problem.

    (b) What is the practical domain for your model?

    (c) What dimensions, to two decimal places, will require the least amount of material?

    (d) Use your results in (c) to determine the least amount of material that will be needed to make this box.

22. *Industrial design*   A metal trash container is to have a capacity of 108 ft^3. To fit with the company's removal method, the length is 1.5 as long as the width. Because of environmental concerns the cost per ft^2 to make the bottom is three times as much as the cost per ft^2 to make the sides and the top. You want to make this trash container for the least cost.

    (a) Write a mathematical model for this problem.

    (b) What is the practical domain for your model?

    (c) What dimensions, to two decimal places, will result in the least cost to make this container?

    (d) Use your results in (c) to determine surface area of the material needed to make this box for the least cost.

23. *Agriculture*   A company has 10 000 m of available fencing. It wants to use the fencing to enclose a rectangular field. If the fence is 1.2 m high, what are the length and width of the largest area that can be enclosed?

24. *Agriculture*   A company has 10 000 m of available fencing. It wants to use the fencing to enclose a rectangular field. One side of the field is bordered by a warehouse and no fencing is needed along this side. If the fence is 1.2 m high, what are the length and width of the largest area that can be enclosed?

25. When we solved the surface area max/min problem in Example 4.5, we substituted $\dfrac{10}{x^2}$ for y in the equation for $S(x)$. If we had chosen y to be the independent variable, we would have had to solve the equation $V = x^2 y$ for x and then made the substitution $x = \sqrt{\dfrac{10}{y}}$.

Show that this approach would yield the same result; that is, that the shape of the best box is a cube of approximately 2.15 meters on a side.

26. Solve Example 4.6 when the cost in dollars per m^2 of the top and bottom are one-third the cost of the sides.

27. *Industrial design* A metal trash container is to have a capacity of 108 ft^3. To fit with the company's removal method, the length is 1.5 as long as the width. Because of environmental concerns the bottom is $\frac{3}{4} = 0.75$ in. thick while the sides and the top are each $\frac{3}{8} = 0.375$ in. thick. You want to make this trash container using the least amount (volume) of material.

(a) Write a mathematical model for this problem.

(b) What is the practical domain for your model?

(c) What dimensions, to two decimal places, will require the least amount of material?

(d) Use your results in (c) to determine the least amount of material (volume) that will be needed to make this trash container.

(e) If this trash container is made from steel and steel weighs 482 lb/ft^3, what is the weight of this container?

28. *Industrial design* A cereal box measures 24.8 cm high, 16.8 cm deep, and 5.5 cm wide. The box is made out of one piece of cardboard. The length of the cardboard is made 1.1 cm longer than necessary. This extra portion is used to glue the two ends of the box together. The top and bottom of the box are each made from two pieces of cardboard that overlap by 1.4 cm in order to seal the top and bottom of the box. What is the area of the piece of cardboard needed to make this box?

29. *Industrial design* A pizza box is to be made from a rectangular piece of cardboard that measures 18″ by 36″. So the cardboard will fold into a box with a lid, six squares are cut from the cardboard (as shown in Figure 4.16). Make a model from a sheet of paper. Follow the procedures in Activity 4.1 (see page 212) to complete Table 4.5 where x, y, and z are the lengths indicated in Figure 4.16.

(a) What is the practical domain for this problem?

(b) Based on the data in Table 4.5, estimate the size of the cut that produces the box with the largest volume.

(c) Write a mathematical model for this problem.

(d) What sizes of pizza would fit in this box?

(e) Is this a good design for a pizza box? If you think so, explain why. If you do not think it is a good design, explain why and suggest some restrictions you might place on the box design to make the answer more practical.

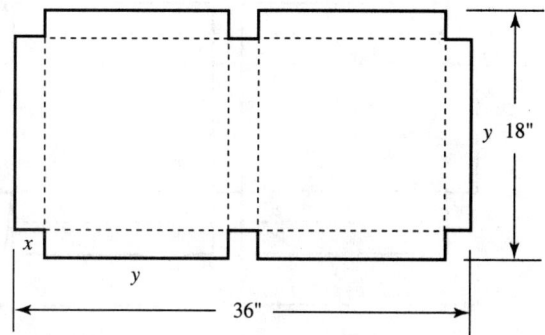

Figure 4.16

Table 4.5

Trial or Group No.	x	y	z	V

30. *Industrial design* A pizza box is to be made from a rectangular piece of cardboard that measures 18″ by 36″. To help keep the box closed, the sides of the box will be folded twice in order to produce a "sleeve" to tuck the flaps on the lid. So the cardboard will fold into a box with a lid, six rectangles are cut from the cardboard (as shown in Figure 4.17). Make a model from a sheet of paper. Follow the procedures in Activity 4.1 (see page 212) to complete Table 4.6 where x, y, and z are the lengths indicated in Figure 4.17.

(a) What is the practical domain for this problem?

(b) Based on the data in Table 4.6, estimate the size of the cut that produces the box with the largest volume.

(c) Write a mathematical model for this problem.

(d) What sizes of pizza would fit in this box?

(e) Is this a good design for a pizza box? If you think so, explain why. If you do not think it is a good design, explain why and suggest some restrictions you might place on the box design to make the answer more practical.

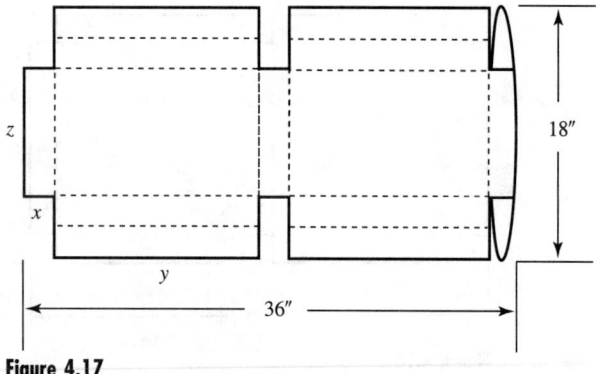

Figure 4.17

Table 4.6

Trial or Group No.	x	y	z	V

4.2   Triangles Are Everywhere

In this section we'll look at triangles in terms of their area and then at the famous Pythagorean theorem about right triangles. These topics lead back to trigonometry and a look at two relationships between the sides and angles of triangles, called the *law of sines* and the *law of cosines*.

Why Is the Area of a Triangle Half the Base × Height?

Look at the triangle in Figure 4.18. The area A of $\triangle ABC$ is computed by the formula

$$A = \frac{1}{2}b \times h = \frac{1}{2}bh = \frac{bh}{2}$$

Why is this true? If you take any triangle and a duplicate of the triangle, you can put them together to make a parallelogram, as shown in Figure 4.19. A **parallelogram** is a four-sided figure with opposite sides parallel. The area of a parallelogram is base × height because you can make a rectangle of the same area out of a parallelogram, as shown in Figure 4.20. Therefore a triangle has an area equal to one-half the base × height.

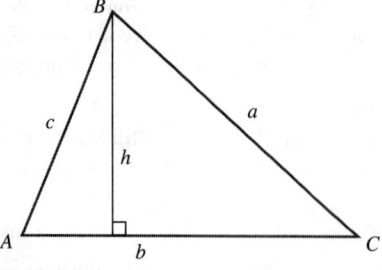

Figure 4.18

The formula that the area of a triangle is one-half the base × height is true for *any* triangle. One result of this formula is illustrated in Figure 4.21, which shows that if you put point B on other locations along a line parallel to $\overline{AC}$, the area will be the same because the height h is the same for any position of point B.

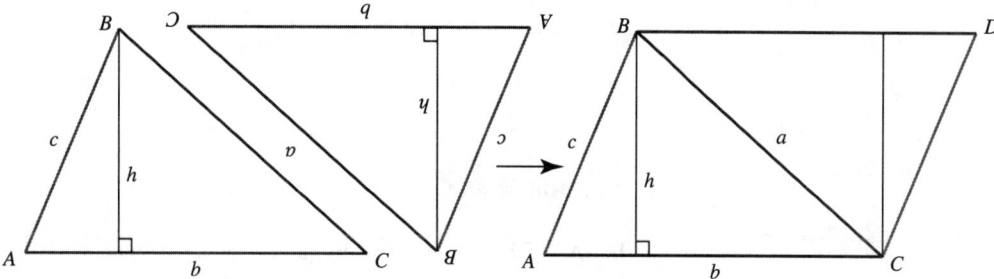

Figure 4.19

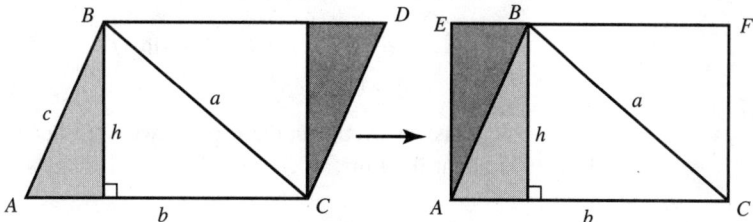

Figure 4.20

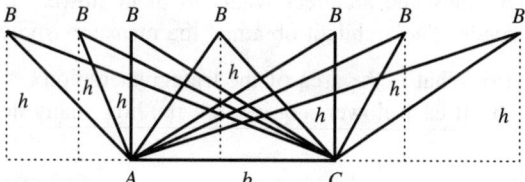

Figure 4.21

Trigonometry and the Areas of Triangles

There are situations in which you will need to compute the area of a triangle when you do not know the values of a base and a height. In Figure 4.22, for example, suppose you don't know h, but you do know the values of a, c, and angle B. You can use trigonometry to compute the height h as $h = a \sin B$. Therefore we have the following formula for the area of a triangle if you know the lengths of two sides

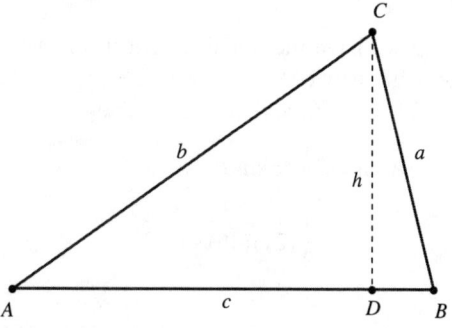

Figure 4.22

and the angle formed by those two sides:

$$\text{Area} = \frac{1}{2}ac\sin B$$

Example 4.7

If $\angle A = 37°$, $b = 1.2$ meters, and $c = 2.5$ meters, what is the area of $\triangle ABC$?

Solution   The area of the triangle is given by the formula

$$\text{Area} = \frac{1}{2}bc\sin A$$
$$= 0.5 \times 1.2 \times 2.5 \times \sin 37°$$
$$\approx 0.90$$

The answer should contain only two significant figures, so the area of this triangle is about 0.90 m². ∎

Example 4.8   Application

A landscape architect wants to plant flowers in a triangular section between two roads. The architect obtained the measures shown in Figure 4.23.

(a) What is the area of the triangular region?
(b) If each flower requires 0.4 ft², how many flowers can be planted in this region?

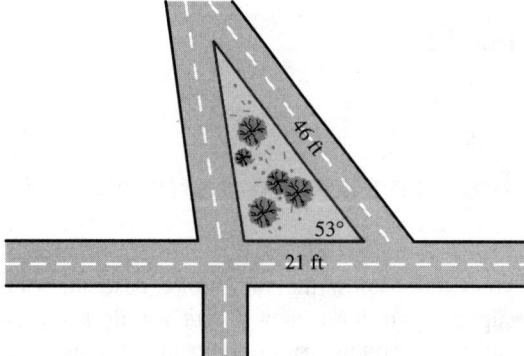

Figure 4.23

Solution   (a) Since we have the lengths, in feet, of two sides of the triangle and the size of the angle between these two sides, we can use the formula $A = \frac{1}{2}ac\sin B$ with $a = 21$, $c = 46$, and $B = 53°$. Thus

$$\text{Area} = \frac{1}{2}ac\sin B$$

$$= \frac{1}{2}(21)(46)\sin 53°$$

$$\approx 385.74$$

The area is about 385.74 ft².

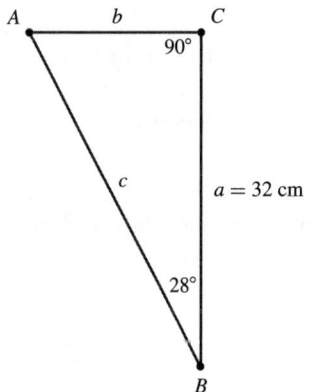

Figure 4.24

(b) If each flower needs 0.4 ft², then the number of flowers is determined by $385.74 \div 0.4 = 964.35$. You can plant 964 flowers in this triangular region. ■

Another situation in which trigonometry comes in handy is in the case of right triangles in which only the length of one side and the size of one of the two acute angles is known. For example, if you know a and $\angle B$ in Figure 4.24, you can find the area by using trigonometry to compute the length of the other leg, b. Since $b = a \tan B$, then the area is

$$\text{Area} = \frac{1}{2}ab$$

$$= \frac{1}{2}a^2 \tan B$$

Remember that this formula is correct only for right triangles.

Example 4.9

In Figure 4.24 compute the area of $\triangle ABC$.

Solution We are given the length of one side as 32 cm, so let $a = 32$. The acute angle is 28°, so $B = 28°$. Then $A = 0.5 \times 32^2 \times \tan 28° \approx 270$. The area is about 270 cm². ■

Two trigonometry laws for triangles also are used to compute the lengths of two sides and sizes of angles of triangles. The first of these laws that we will study is called the law of sines.

The Law of Sines

The law of sines is a relationship among the sides and angles of triangles. The following activity will give you some practical experience with triangles that will help you to understand the law of sines.

Activity 4.2
Law of Sines

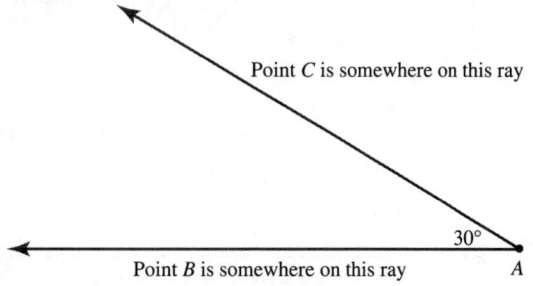

Figure 4.25

1. Using a diagram like the one in Figure 4.25, construct triangle ABC with $AC = 7$ cm and $CB = 5$ cm.

(a) How long is side $\overline{AB}$ in your triangle?

(b) Ask other students what they got for the length of $\overline{AB}$. Did all of them get approximately the same answer?

(c) Is it possible to construct two different triangles that fit this description? If so, draw both of these triangles.

2. Use another copy of the angle in Figure 4.25 to construct the triangles listed below. Begin by choosing a location for the point B on the horizontal ray. Measure the distance AB. *Look for situations in which you can construct more than one triangle to fit the measurements.*

(a) Construct segment $BC = 0.55 \times AB$.

(b) Construct segment $BC = 0.5 \times AB$.

(c) Construct segment $BC = 0.3 \times AB$.

(d) Construct segment $BC = 1.2 \times AB$.

3. If $\dfrac{BC}{AB} = 0.5$, you were able to construct only one triangle. What kind of triangle was it?

4. Write an answer to each of the following questions.

(a) For what values of the ratio $\dfrac{BC}{AB}$ can you construct two triangles? The answer is not just $\dfrac{BC}{AB} = 0.55$. What is the *range* of values for which you can construct two triangles?

(b) For what values can you construct only one triangle?

(c) For what values are you unable to construct any triangles?

5. Use the angle in Figure 4.26 and repeat the constructions that you did in Step 2 above for $\angle A = 30°$. Answer the following questions. (*Hint:* There will be a different value of $\dfrac{BC}{AB}$ that produces just one triangle.)

(a) For what values of the ratio $\dfrac{BC}{AB}$ can you construct two triangles? The answer is not just $\dfrac{BC}{AB} = 0.75$. What is the *range* of values for which you can construct two triangles?

(b) For what values can you construct only one triangle?

(c) For what values are you unable to construct any triangles?

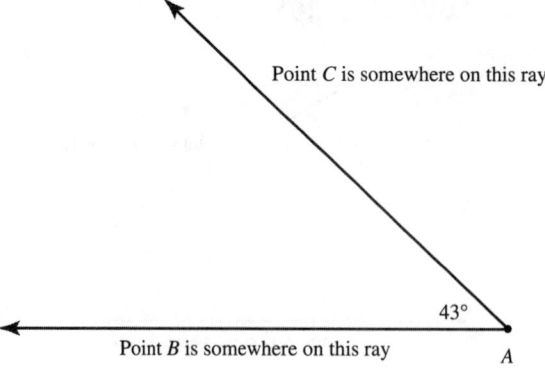

Point C is somewhere on this ray

43°

Point B is somewhere on this ray A

Figure 4.26

6. Compare the answers for $\angle A = 30°$ and for $\angle A = 43°$. For which of the questions are your answers the same and for which are they different?

In Activity 4.2 you have found that for some combinations of sides and angles you can construct two, one, or no triangles when given two sides and an angle opposite one of the sides. The law of sines will explain how to analyze this situation.

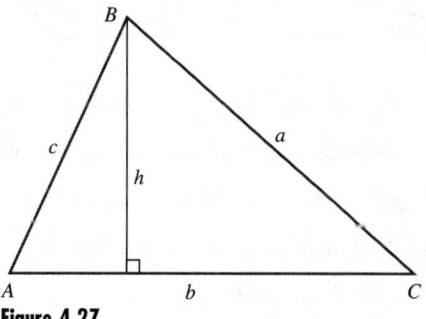

Figure 4.27

Take any triangle with a height drawn to one side (for example, Figure 4.27). The height h divides $\triangle ABC$ into two right triangles. Using the definition of the sine function in a right triangle we have

$$\sin A = \frac{h}{c} \qquad \text{and} \qquad \sin C = \frac{h}{a}$$

and so

$$h = c \sin A \qquad \text{and} \qquad h = a \sin C$$

But that means that

$$c \sin A = a \sin C$$

$$\text{or} \quad \frac{\sin A}{a} = \frac{\sin C}{c}$$

In the same way we could have proved that $\dfrac{\sin A}{a} = \dfrac{\sin B}{b}$. Combining these two proportions we obtain

$$\frac{\sin A}{a} = \frac{\sin B}{b} = \frac{\sin C}{c}$$

This relationship between the sides and angles of a triangle is called the **law of sines.**

Law of Sines

The **law of sines** or **sine law** states that if $\triangle ABC$ is a triangle with sides of lengths a, b, and c and opposite angles A, B, and C, then

$$\frac{\sin A}{a} = \frac{\sin B}{b} = \frac{\sin C}{c}$$

Example 4.10

In Figure 4.28 we are given $\angle A = 14°$, $\angle B = 133°$, and $a = 1.85''$. Determine (a) the measure of angle C and (b) the length of sides $\overline{AC}$ and $\overline{AB}$. Report your answer in appropriate significant figures.

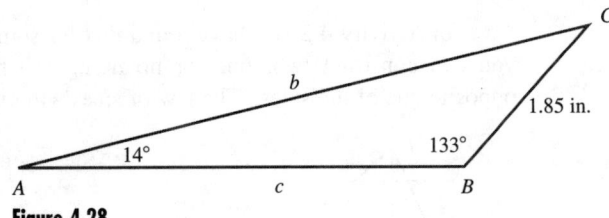

Figure 4.28

Solutions (a) Since the angles of a triangle add up to 180°, then $\angle C + 133° + 14° = 180°$, so

$$\angle C = 180° - (133° + 14°) = 33°$$

(b) From the law of sines we have

$$\frac{\sin A}{a} = \frac{\sin B}{b}$$

$$\frac{\sin 14°}{1.85} = \frac{\sin 133°}{b}$$

This can be rewritten as

$$\frac{b}{\sin 133°} = \frac{1.85}{\sin 14°}$$

Multiplying by $\sin 133°$ we have

$$b = \frac{1.85 \times \sin 133°}{\sin 14°} \approx 5.59273 \approx 5.59$$

Thus $AC = b$ is 5.59 in. In a similar way we get $c \approx 4.16$ in.   ■

Example 4.11

In Activity 4.2 you did some constructions with a triangle like that shown in Figure 4.29. You saw that if the ratio $\dfrac{BC}{AB}$ was equal to 0.55, two triangles were possible. Explain why, using the law of sines.

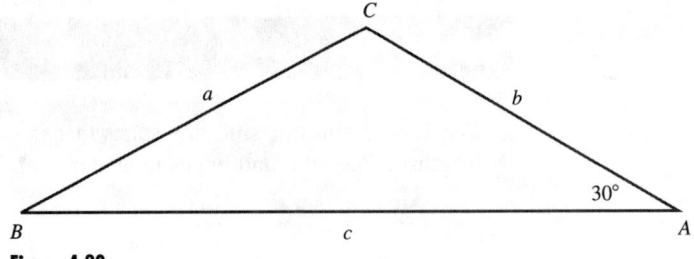

Figure 4.29

Solution For sides a and c the law of sines says

$$\frac{\sin A}{a} = \frac{\sin C}{c}$$

which can be written

$$\sin C = \left(\frac{c}{a}\right)\sin A$$

Since $\angle A = 30°$, then $\sin A = 0.5$. Since $\dfrac{a}{c} = \dfrac{BC}{AB} = 0.55$, then $\dfrac{c}{a} = \dfrac{1}{0.55}$. Therefore we have the result

$$\sin C = \frac{1}{0.55} \times 0.5 = \frac{0.5}{0.55} \approx 0.9091$$

This equation is similar to one you learned to solve in Chapter 3. How many solutions does it have? Of course the equation itself has an infinite number of solutions; but if $\angle C$ must be in a triangle, the largest value it could have is 180°, so the acceptable solutions must be less than 180°. The calculator gives $\angle C \approx 66°$, and the only other acceptable solution is $\angle C \approx 180° - 66° \approx 114°$. One solution is in the first quadrant (given by the calculator) and one in the second quadrant. The two triangles that can result are $\triangle ABC_1$ and $\triangle ABC_2$, as shown in Figure 4.30.

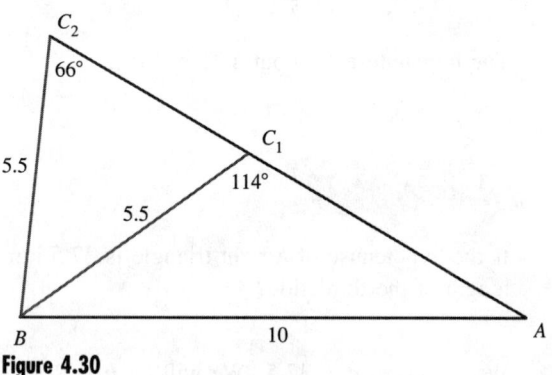

Figure 4.30

The Pythagorean Theorem

Right triangles played a key role in our introduction to trigonometry. The Pythagorean theorem will connect to several other topics we have already met. Although the theorem is identified with Pythagoras, a Greek philosopher who lived on an island in what is now called Italy, it may really have come from the secret society named for him, the Pythagoreans. Over 2500 years ago the Pythagoreans studied the mathematics behind geometry and music.

The theorem states that the square of the longest side of a right triangle (the **hypotenuse**) equals the sum of the squares of the other two sides, the legs. If c is the length of the hypotenuse, then in mathematical terms we have

$$a^2 + b^2 = c^2$$

There are two other forms of the Pythagorean theorem that can be produced from the original:

1. Solving the Pythagorean theorem for c we have

$$c = \sqrt{a^2 + b^2}$$

2. Solving the Pythagorean theorem for either a or b we have

$$a = \sqrt{c^2 - b^2}$$
$$\text{or} \quad b = \sqrt{c^2 - a^2}$$

Example 4.12

If two sides of a right triangle are equal to 2.56 meters and 3.47 meters, what is the length of the hypotenuse?

Solution   For this solution we will keep the full accuracy until the last step, when we report the answer to three significant figures. Since we are to find the length of the hypotenuse, c, the given lengths are for the legs. Let $a = 2.56$ and $b = 3.47$. Thus

$$c = \sqrt{a^2 + b^2}$$
$$= \sqrt{2.56^2 + 3.47^2} = \sqrt{6.5536 + 12.0409}$$
$$= \sqrt{18.5945} \approx 4.31$$

The hypotenuse is about 4.31 m long.   ■

Example 4.13

If the hypotenuse of a right triangle is 37.5 km and one leg is 2.30 km, what is the length of the third side?

Solution   We are given $c = 37.5$. We will let $b = 2.30$ and solve for a. There will be three significant figures in the answer.

$$a = \sqrt{c^2 - b^2}$$
$$= \sqrt{37.5^2 - 2.30^2} = \sqrt{1406.25 - 5.29}$$
$$= \sqrt{1400.96} \approx 37.4$$

The third side of this triangle is about 37.4 km long.   ■

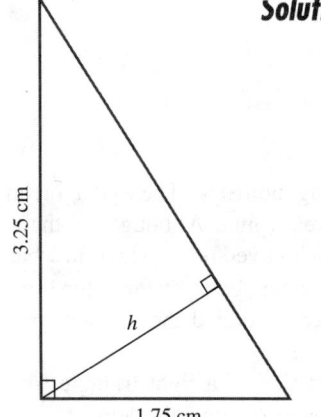

3.25 cm

1.75 cm

Figure 4.31

Example 4.14

In the right triangle pictured in Figure 4.31, what is the length of the distance labeled h?

Solution   The distance h is the height of the triangle if the hypotenuse is considered to be the base. The area of the triangle is $\frac{1}{2}(1.75)(3.25)$, but it is also equal to $\frac{1}{2}h \times$ hypotenuse. The length of the hypotenuse is $\sqrt{1.75^2 + 3.25^2}$. Using these two formulas for the area of the given triangle we have

$$\frac{1}{2}(1.75)(3.25) = \frac{1}{2}h\sqrt{1.75^2 + 3.25^2}$$

Dividing both sides by $\frac{1}{2}$ we have

$$5.6875 = h\sqrt{13.625}$$

Therefore

$$h = \frac{5.6875}{\sqrt{13.625}} \approx 1.54$$

Thus h is about 1.54 cm long.   ■

Distance on a Graph

Our final application of the Pythagorean theorem in this section is its connection to distance on a coordinate graph.

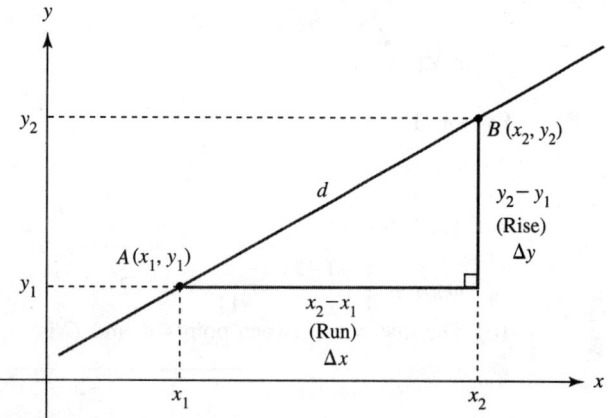

Figure 4.32

Figure 4.32 was used in Chapter 1 to illustrate the computation of the slope of the line that passes through the two points A and B whose coordinates are $A(x_1, y_1)$ and $B(x_2, y_2)$. The slope was computed as $\frac{\Delta y}{\Delta x}$ where $\Delta y = y_2 - y_1$ and $\Delta x = x_2 - x_1$. The distance d between these two points can be computed with the Pythagorean theorem as follows:

$$d(A, B) = \sqrt{(\Delta x)^2 + (\Delta y)^2}$$
$$= \sqrt{(x_2 - x_1)^2 + (y_2 - y_1)^2}$$

Example 4.15

Plot the points $A(1, 5)$, $B(4, 1)$, $C(-4, -5)$, and $D(-7, -1)$ on a graph. Answer all the questions below to three decimal place precision.

(a) Calculate the slope of line segments $\overline{AC}$ and $\overline{BD}$.
(b) Calculate the length of line segments $\overline{AC}$ and $\overline{BD}$.
(c) Write the equations of the lines that contain line segments $\overline{AC}$ and $\overline{BD}$.
(d) Estimate the coordinates of the point of intersection of the line segments $\overline{AC}$ and $\overline{BD}$ in two ways:

  (i) Graph the two lines on your own graph paper.
  (ii) Draw the graphs of the two lines on your calculator and zoom in on the point of intersection until you have the coordinates of the point of intersection to one decimal place.

Solutions   The points are graphed as in Figure 4.33.

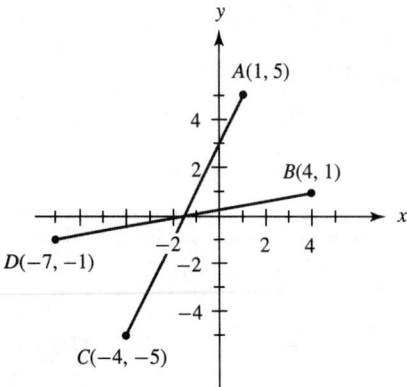

Figure 4.33

(a) The slope of $\overline{AC}$ is $m_{AC} = \dfrac{-5-5}{-4-1} = \dfrac{-10}{-5} = 2$ and the slope of $\overline{BD}$ is

$m_{BD} = \dfrac{1-(-1)}{4-(-7)} = \dfrac{2}{11} \approx 0.182.$

(b) The distance between points A and C is

$$d(A, C) = \sqrt{(-4-1)^2 + (-5-5)^2}$$
$$= \sqrt{(-5)^2 + (-10)^2}$$
$$= \sqrt{25 + 100} = \sqrt{125} \approx 11.18$$

and the distance between B and D is

$$d(B, D) = \sqrt{(4--7)^2 + (1--1)^2}$$
$$= \sqrt{11^2 + 2^2}$$
$$= \sqrt{121 + 4} = \sqrt{125} \approx 11.18$$

(c) In (a) we determined the slope of $\overline{AC}$ as $m_{AC} = 2$. We will use the point-slope form for the equation of a line with the point $(1, 5)$ to obtain the equation $y - 5 = 2(x - 1)$ or $y = 2x + 3$.

Using the slope of $\overline{BD}$, which is $m_{BD} = \frac{2}{11}$, and the point $(4, 1)$ we obtain the equation $y - 1 = \frac{2}{11}(x - 4)$ or $y = \frac{2}{11}x - \frac{8}{11} + 1 = \frac{2}{11}x + \frac{3}{11}$. The decimal approximation of this equation is $y = 0.182x + 0.273$.

(d) The graph of the two lines is shown in Figure 4.34 and, after zooming, the point of intersection is $(-1.5, 0)$.

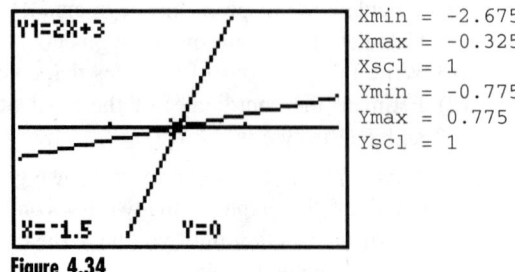

```
Xmin = -2.675
Xmax = -0.325
Xscl = 1
Ymin = -0.775
Ymax = 0.775
Yscl = 1
```

Figure 4.34

Example 4.16

On a graph, plot the points $A(2, 5)$, $B(6, 3)$, and $C(1, -7)$. Draw the segments $\overline{AB}$ and $\overline{BC}$ and compute the slopes of these two segments.

Solution   The three points and the segments $\overline{AB}$ and $\overline{BC}$ are shown in Figure 4.35. The slopes of the two segments are $m_{AB} = \frac{5-3}{2-6} = \frac{2}{-4} = -\frac{1}{2}$ and $m_{BC} = \frac{3--7}{6-1} = \frac{10}{5} = 2.$ ∎

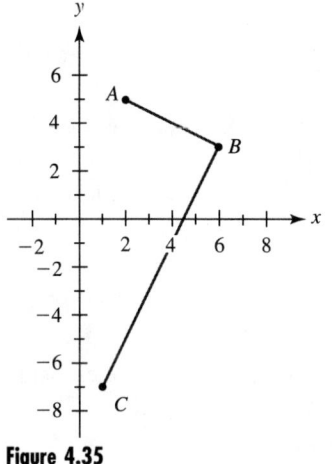

Figure 4.35

Example 4.16 reveals a connection between the slopes of lines and the angle formed by the lines. In Chapter 1 you saw that lines with equal slope are parallel. Now we'll add another piece to this relationship by looking at the slopes of lines that are perpendicular to each other. Look at the slopes of line segments $\overline{AB}$ and $\overline{BC}$. The angle they form is $90°$, so they are perpendicular. Fractions like $\frac{5}{8}$ and $\frac{8}{5}$ are called **reciprocals**, and in Example 4.16 the two slopes $-\frac{1}{2}$ and $\frac{2}{1} = 2$ are **negative reciprocals**. It turns out that this relationship is true for *almost all* pairs of lines that are perpendicular. We can summarize this relationship as follows:

Slopes of Perpendicular Lines

If m_1 and m_2 are slopes of two perpendicular lines, then $m_1 = -\dfrac{1}{m_2}$ or $m_1 m_2 = -1$.

You can see now why we said *almost all* lines. If either m_1 or m_2 is equal to zero (horizontal line), we have a division by zero for the other slope, but dividing by zero is not defined. Since a vertical line does not have a value for the slope, this pair of lines is the exception to the perpendicular line relationship stated above.

The Law of Cosines

There is a version of the Pythagorean theorem that can be applied to *any* triangle with sides a, b, and c. It was not discovered by the Pythagoreans because trigonometry came later. Look at the three triangles in Figure 4.36. In Figure 4.36(a), $\angle C$ is less than $90°$; in Figure 4.36(b), $\angle C$ is equal to $90°$; and in Figure 4.36(c), $\angle C$ is greater

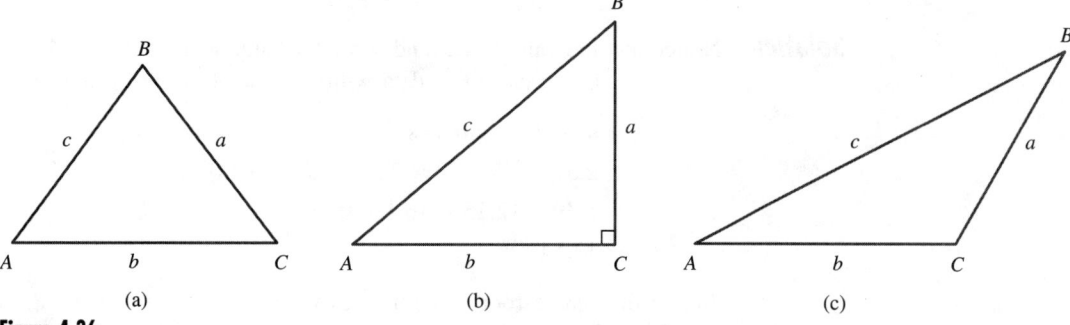

(a)   (b)   (c)

Figure 4.36

than 90°. We have drawn these triangles so that all the sides marked a are the same length and all the sides marked b are the same length. The law of cosines tells how to compute c if the other two sides and the angle C are known.

Law of Cosines

If $\triangle ABC$ is a triangle with sides of lengths a, b, and c and opposite angles A, B, and C, then by the **law of cosines** or **cosine law**

$$a^2 = b^2 + c^2 - 2bc \cos A$$
$$b^2 = a^2 + c^2 - 2ac \cos B$$
$$\text{or} \quad c^2 = a^2 + b^2 - 2ab \cos C$$

(Notice that there are three versions of the cosine law. Each version restates the law so that different parts of the triangle are used.)

Example 4.17

Compute the length of c in Figure 4.36 if $a = 2.3$ miles, $b = 3.5$ miles, and $\angle C = 47°$.

Solution   Using the given data with the law of cosines produces

$$\begin{aligned}
c^2 &= a^2 + b^2 - 2ab \cos C \\
&= 2.3^2 + 3.5^2 - 2 \times 2.3 \times 3.5 \times \cos 47° \\
&\approx 5.29 + 12.25 - 16.1 \times 0.681998 \\
&\approx 6.559826
\end{aligned}$$

Taking the square root of both sides we find that $c \approx 2.561216$. Since the problem requires two significant figures, $c \approx 2.6$ mi.   ∎

Example 4.18

Compute the length of c in Figure 4.36 if $a = 2.3$ miles, $b = 3.5$ miles, and $\angle C = 90°$.

Solution   Notice that the values of a and b are the same as in Example 4.17. Here, however, $\angle C = 90°$. Using the given data with the law of cosines produces

$$\begin{aligned}
c^2 &= a^2 + b^2 - 2ab \cos C \\
&= 2.3^2 + 3.5^2 - 2 \times 2.3 \times 3.5 \times \cos 90° \\
&= 5.29 + 12.25 - 16.1 \times 0 \\
&= 17.54
\end{aligned}$$

Taking the square root of both sides we find that $c \approx 4.188078$. Thus, using two significant figures, we see that $c \approx 4.2$ mi.   ∎

Notice in Example 4.18 that the law of cosines simplified to the Pythagorean theorem because $C = 90°$.

Example 4.19   Application

A 73.25-m tower stands vertically on sloping ground that is inclined 12.7° with the horizontal. From the top of the tower, two cables are fastened to the ground. Each cable is located 27.5 m from the base of the tower. One cable is directly downhill from the tower and the other directly uphill. How long are the cables?

Solution   A drawing of this situation is shown in Figure 4.37. Since the hill makes an angle of 12.7° with the horizontal, the tower makes an angle of $90° - 12.7° = 77.3°$ with the uphill cable and an angle of $90° + 12.7° = 102.7°$ with the downhill cable.

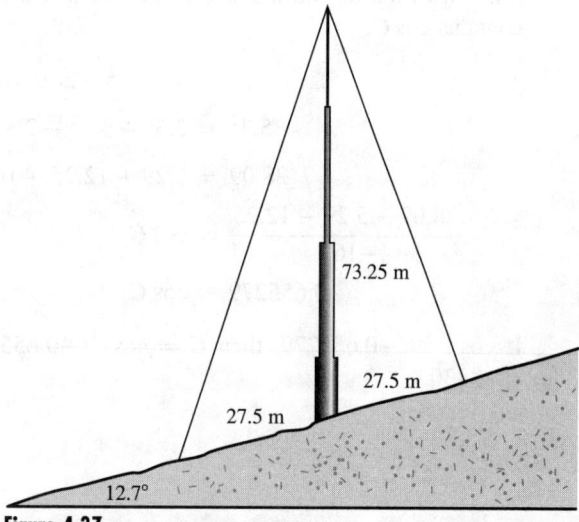

73.25 m

27.5 m

27.5 m

12.7°

Figure 4.37

We will first solve for the length of the uphill cable. We are given the lengths of two sides. One side is 73.25 m (the tower) and the other is 27.5 m (the distance from the tower to where the cable is fastened to the ground). We determined that the angle between these two sides is 77.3°. Since we have the lengths of two sides and the size of the angle between them, we can use the law of cosines. We will let $a = 73.25$, $b = 27.5$, and $\angle C_1 = 77.3°$ and c_1 will be the length of the uphill cable.

$$c_1{}^2 = a^2 + b^2 - 2ab \cos C$$
$$= 73.25^2 + 27.5^2 - 2(73.25)(27.5) \cos 77.3°$$
$$\approx 5236.107104$$

Taking the square root we see that $c_1 \approx 72.36$.

For the downhill cable we again let $a = 73.25$ and $b = 27.5$. The angle is $\angle C_1 = 102.7°$ and c_2 will be the length of the downhill cable.

$$c_2{}^2 = a^2 + b^2 - 2ab \cos C$$
$$= 73.25^2 + 27.5^2 - 2(73.25)(27.5) \cos 102.7°$$
$$\approx 7007.517896$$

Taking the square root we see that $c_2 \approx 83.71$.

The lengths of the two cables are about 72.36 m for the uphill cable and 83.71 m for the downhill cable. ∎

The next example will use the lengths of the three sides of a triangle to determine the size of one of the angles of the triangle.

Example 4.20

If the lengths of the sides of a triangle are $a = 2.3$ km, $b = 3.5$ km, and $c = 5.3$ km, determine the size of $\angle C$.

Solution   Since we want to find the size of $\angle C$ we will use the version of the cosine law that contains $\cos C$.

$$c^2 = a^2 + b^2 - 2ab \cos C$$

$$5.3^2 = 2.3^2 + 3.5^2 - 2 \times 2.3 \times 3.5 \times \cos C$$

$$28.09 = 5.29 + 12.25 - 16.1 \cos C$$

$$\frac{28.09 - 5.29 - 12.25}{-16.1} = \cos C$$

$$-0.655279 \approx \cos C$$

If $\cos C \approx -0.655279$, then $C = \cos^{-1}(-0.655279) \approx 130.9°$. Thus we see that $C \approx 130.9°$. ∎

Section 4.2 Exercises

In Exercises 1–2 find the distance between the given pairs of points.

1. $(-2, 5)$ and $(3, -7)$

2. $(-6, -3)$ and $(4, -2)$

In Exercises 3–6 determine the area of the triangle.

3. $a = 10$ m, $\angle B = 50°$, $c = 8$ m

5. $\angle A = 72°$, $b = 15$ in., $c = 9$ in.,

4. $a = 6$ cm, $b = 12$ cm, $\angle C = 36°$

6. $a = 6$ mm, $b = 8$ mm, $\angle A = 42°$, $\angle B = 57°$

In Exercises 7–14 determine the measures of the three other parts of the triangle (if possible).

7. $\angle A = 86°$, $\angle B = 57°$, $a = 15.2$ cm

11. $a = 1.245$ m, $b = 4.532$ m, $c = 3.882$ m

8. $\angle A = 57°$, $\angle B = 72°$, $c = 25.6$ ft

12. $a = 12.5$ mi, $b = 8.9$ mi, $c = 6.35$ mi

9. $a = 15$ in., $\angle B = 119.5°$, $c = 36$ in.

13. $\angle A = 32.8°$, $\angle C = 101.0°$, $c = 413$ mm

10. $\angle A = 45.3°$, $b = 3.75$ yd, $c = 2.60$ yd

14. $b = 11.37$ cm, $c = 10.10$ cm, $\angle C = 45.55°$

15. *Civil engineering* A high-tension wire is to be strung across a river from tower A to tower C. To determine the distance between these two towers a third point B is located on the same side of the river as point A, 450.0 m from A. If $\angle ABC = 62.8°$ and $\angle BAC = 36.7°$, what is the distance between towers A and C?

16. *Forestry* Forest rangers are located in two observation towers at points A and B. The towers are 0.75 mile apart. A tourist walking from A to B spots a forest fire at point C and reports it via cellular phone to the rangers in the towers. If $\angle CAB = 42.3°$ and $\angle CBA = 59.2°$, how far is the fire from each of the observation towers?

17. *Civil engineering* A hillside has a $14°$ angle of inclination. A vertical flagpole is erected at a point F. A cable is to run from the top of the pole to a point that is 42.5 ft directly downhill from the pole. The cable will make an angle of $32°$ with the hill.

(a) How long is the cable?

(b) How high is the pole?

18. *Recreation* The design for a new playground calls for a children's slide that is 12.5 meters long and inclined $30°$ with the horizontal. To reach the top of the slide the children will climb a ladder that is 6.4 meters long.

(a) Determine the angle of inclination of the ladder.

(b) Draw a sketch that shows the slide, ladder, and ground. (*Hint:* There are two correct answers!)

19. *Civil engineering* It is planned that an electric transmission line will be strung over a heavily wooded area. The transmission line will extend from a tower to be placed at position A to a tower that will be located at position B, as shown in Figure 4.38. When the surveyor is standing at the point where one of the towers will be located, she cannot see the location of the other tower. She is able to find a place, C, from which she can see the locations of the two towers. If $AC = 372.0$ m, $BC = 432.0$ m, and $\angle ACB = 79.6°$, what is the distance between the proposed locations of the two towers?

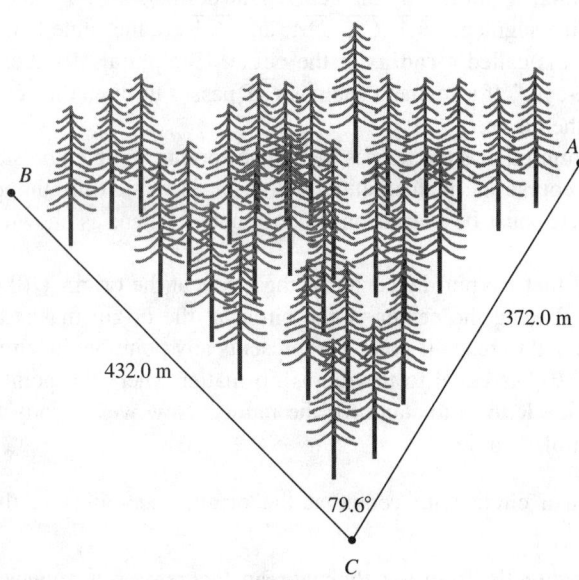

Figure 4.38

20. *Civil engineering* A vertical antenna 96.5 ft high is erected on a hillside that is inclined $15.3°$ with the horizontal. The antenna is held in place by four cables. Each cable is anchored on the hillside 42.25 ft from the base of the antenna, as shown in Figure 4.39. Determine the lengths of the cables that are directly uphill and directly downhill from the antenna.

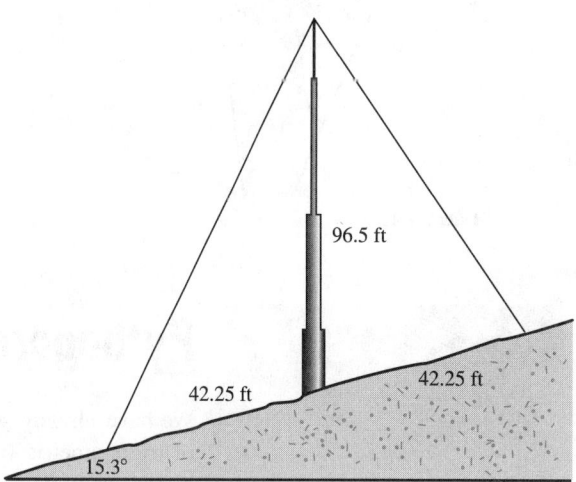

Figure 4.39

21. *Electronics* Two electrical forces F_1 and F_2 act on a particle in an electric field at an angle θ, as indicated in Figure 4.40. If $F_1 = 8.4 \times 10^{-6}$ N, $\theta = 55°$, and $R = 12.7 \times 10^{-6}$ N, determine the magnitude of F_2.

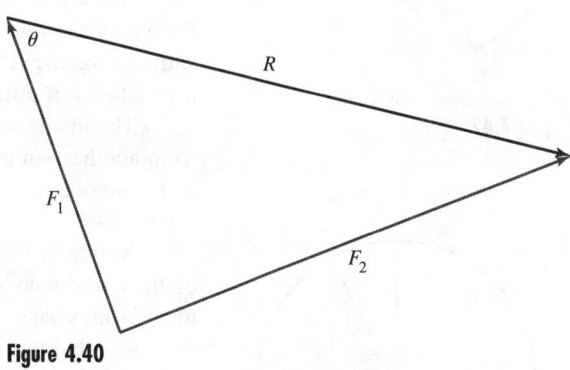

Figure 4.40

22. *Electronics* Two electrical forces F_1 and F_2 act on a particle in an electric field at an angle θ, as indicated in Figure 4.40. If $F_1 = 3.5 \times 10^{-6}$ N, $\theta = 115°$, and $F_2 = 5.3 \times 10^{-6}$ N, determine the magnitude of R.

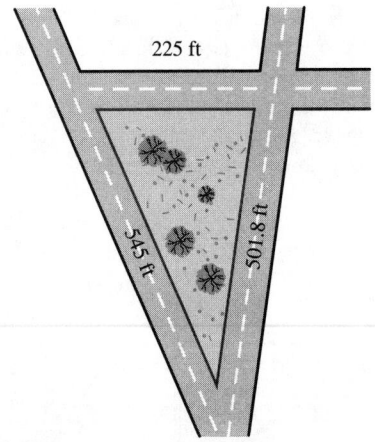

225 ft

545 ft

501.8 ft

Figure 4.41

23. *Land management*   A highway cuts a corner from a parcel of land as shown in Figure 4.41. Find the area in acres of the triangular lot that remains. (1 acre = $43,560$ ft^2.)

24. *Land management*   A plot of land in the shape of a parallelogram has sides of 37 m and 72 m. One of the diagonals of the parallelogram measures 85 m. Determine the area of this plot of land.

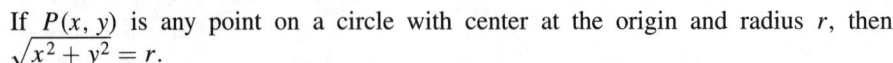

4.3   Pythagorean Theorem and Circles

We have already seen that the circle is essential to the definition and understanding of trigonometric functions. But what is a circle? Most people know what a circle is, but many would have difficulty giving a definition without drawing a picture. In this section you will learn both word and symbol definitions for a circle.

Definition of a Circle and Some of Its Parts

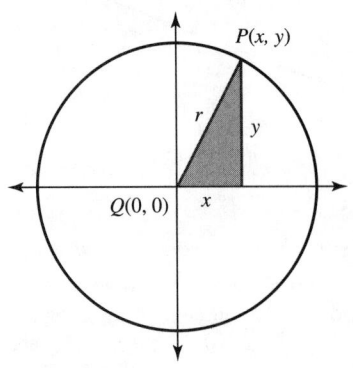

C

Diameter (d)

D

Radius (r)

A

B   E

└─ Sector

Figure 4.42

A **circle** is defined in standard dictionaries as a plane curve all of whose points are the same distance from a point called the **center** of the circle. The points on a circle are said to be equidistant from the center.

Referring to Figure 4.42, if point A is the center and points B, C, D, and E are on the circle, then the line segments $\overline{BA}$, $\overline{CA}$, $\overline{DA}$, and $\overline{EA}$ are the same length. Each of these line segments is called a **radius** of the circle. The plural of radius is **radii**, pronounced "ray-dee-eye." If a segment such as $\overline{CE}$ passes through the center it is called a **diameter** of the circle.

The mathematical definition of a circle uses the idea of points being equidistant from another point, but of course the mathematical definition is written in numbers and symbols. First the circle must be placed on a coordinate system, as shown in Figure 4.43.

Notice in Figure 4.43 that we put the center of the circle at the origin $Q(0, 0)$ of the coordinate system. Placing the center of the circle at the origin makes the mathematics easier to write and to read. If point P represents any point on the circle, we know that the distance PQ is equal to the radius no matter where the point P is on the circle. We'll use the letter r to stand for the radius. Now we're ready for the mathematical definition of a circle:

$P(x, y)$

r

y

$Q(0, 0)$ x

Figure 4.43

If $P(x, y)$ is any point on a circle with center at the origin and radius r, then $\sqrt{x^2 + y^2} = r$.

This equation comes directly from the Pythagorean theorem as it applies to the right triangle in Figure 4.43. You can also think of this result as an application of the distance formula applied to the points Q and P. According to the distance formula, $\sqrt{(x-0)^2 + (y-0)^2} = r$, which can be simplified to $\sqrt{x^2 + y^2} = r$.

The equation $\sqrt{x^2 + y^2} = r$ is often written $x^2 + y^2 = r^2$. In either form this equation is called the **equation of a circle**.

Equation of a Circle Centered at the Origin

A circle of radius r and centered at the origin has the equation

$$x^2 + y^2 = r^2$$

Example 4.21

Suppose that point $A(-3, 6)$ is on a circle whose center is at the point $B(0, 0)$, as shown in Figure 4.44.

(a) What is the equation of the circle?
(b) Write the coordinates of two other points that are on this circle but in different quadrants than A.
(c) Write the coordinates of one point on the x-axis and one point on the y-axis that are both on this circle.
(d) Graph this circle on your calculator.

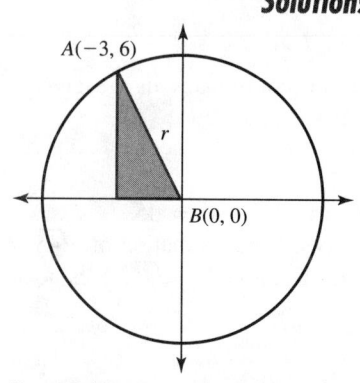

Figure 4.44

Solutions (a) Since the circle is centered at the origin, its equation is $x^2 + y^2 = r^2$. In order to write the equation we need to determine the radius. We are told that A is one point on the circle and B is the center. Thus the radius $r = d(A, B)$. Using the distance formula we find that

$$r = \sqrt{(-3-0)^2 + (6-0)^2}$$
$$= \sqrt{9 + 36} = \sqrt{45} \approx 6.7$$

If $r = \sqrt{45}$ then $r^2 = 45$, so the equation of this circle is $x^2 + y^2 = 45$.

(b) Point $A(-3, 6)$ is in Quadrant II. Since the circle is centered at the origin, we can find three other points just by switching the signs of the coordinates of A. For example, in Quadrant I the point $(3, 6)$ is also on the circle. Similarly, in Quadrant III $(-3, -6)$ and in Quadrant IV $(3, -6)$ are additional points on the circle.

Since $x^2 + y^2$ is the same as $y^2 + x^2$, you can switch x and y and get the following four points that are on the circle: in Quadrant I, $(6, 3)$; Quadrant II, $(-6, 3)$; Quadrant III, $(-6, -3)$; and Quadrant IV, $(6, -3)$.

(c) On the y-axis all the points have $x = 0$. Therefore we have $(0-0)^2 + (y-0)^2 = 45$ or $y = \pm\sqrt{45}$. So the locations of the two points on the y-axis are $\left(0, \sqrt{45}\right)$ and $\left(0, -\sqrt{45}\right)$, which are approximately $(0, 6.7)$ and $(0, -6.7)$. The approximate coordinates of the two points on the circle and on the x-axis are $(6.7, 0)$ and $(-6.7, 0)$.

(d) To enter the equation into the calculator you must solve the equation for y. You have done similar work with the Pythagorean theorem, but there is a slight yet *very important difference* here. When the equation for the circle is solved for y the resulting equation is

$$y = \pm\sqrt{45 - x^2}$$

We have to use the symbol $\pm$ here because for every value of x between $-\sqrt{45}$ and $\sqrt{45}$ there are *two* values of y. That means that you have to enter two equations in the calculator. For example, you might use $Y_1 = \sqrt{45 - x^2}$ and $Y_2 = -\sqrt{45 - x^2}$.

Figure 4.45 shows that the circle does not appear round when the window settings are the same in all directions.* Figure 4.46 shows that the circle appears round with the ZSquare (Zoom Square) setting. The upper and lower portions of the circle do not appear to meet near the x-axis because the point on the x-axis is not a rational number and a pixel on the calculator screen cannot assume its value. Since the *TI-83* uses only 94 points to draw graphs, sometimes incomplete pictures like those in Figures 4.45 and 4.46 will occur.

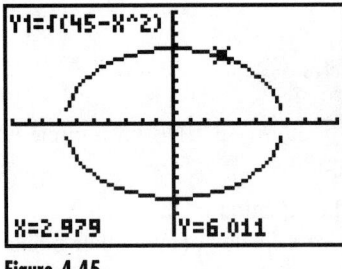

Figure 4.45   **Figure 4.46**

Example 4.22

Is the point $(6.7, 0)$ on the circle $x^2 + y^2 = 45$?

Solution   The best way to check this problem is to substitute the coordinates in the given equation and see if $x^2 + y^2$ is 45. Substituting we obtain

$$6.7^2 + 0^2 = 44.89 + 0 = 44.89$$

Since 44.89 is not 45, the point $(6.7, 0)$ is not on the circle $x^2 + y^2 = 45$.

The point $(\sqrt{45}, 0)$ is on the circle, but 6.7 is only an approximation of $\sqrt{45}$. Even 6.708203932, the best approximation the calculator can give for $\sqrt{45}$, does not produce a point that is actually on the circle.

Example 4.23

Write the equation of a circle whose radius is twice that of the radius of the circle in Example 4.21.

Solution   The radius of the circle in Example 4.21 is $\sqrt{45}$. The radius of a circle whose radius is twice that is $2\sqrt{45}$. So the equation of the second circle is $x^2 + y^2 = \left(2\sqrt{45}\right)^2$ or $x^2 + y^2 = 180$.

---

*We don't see a round circle in the ZStandard graph setting because the shape of the screen of the *TI-83* is not square. Therefore the distance between consecutive numbers is larger in the x-direction than in the y-direction.

Example 4.24 **Application**

Earth's orbit around the Sun is approximately a circle of radius 1.495×10^8 km. If the Sun is placed at the center of a coordinate system, what is the equation of Earth's orbit?

Solution   Since the Sun is at the center of this circle, the circle is centered at $(0, 0)$ and has a radius of 1.495×10^8 km. Thus the equation of Earth's orbit is $x^2 + y^2 = \left(1.495 \times 10^8\right)^2$ or $x^2 + y^2 = 2.235 \times 10^{16}$.   ■

Circles Not Centered at the Origin

What about circles that are not centered at the origin? In Chapter 3 you learned how to write equations of curves that are shifted horizontally or vertically. You learned that the function

$$y = \sin(x - c) + d$$

is the graph of the basic sine function shifted $|c|$ units to the right or left and $|d|$ units up or down. Let's explore that idea in relation to a circle.

Example 4.25

What is the equation of a circle of radius 5 centered at the point $(3, 2)$?

Solution   You know that a circle of radius 5 whose center is at the point $(0, 0)$ has the equations

$$y = \pm\sqrt{25 - x^2}$$

But a circle of radius 5 centered at $(2, 3)$ has been shifted 3 units to the right and 2 units up. Therefore the equations of the new circle are

$$y = \pm\sqrt{25 - (x - 3)^2} + 2$$

(The graph is shown in Figure 4.47).   ■

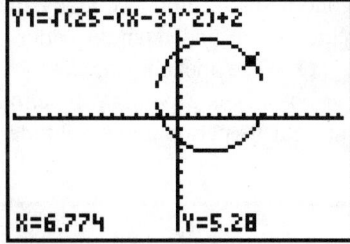

Y1=√(25-(X-3)^2)+2

X=6.774   Y=5.28

Figure 4.47

We can use the distance formula to explain why the equation for the shifted circle is correct.

Think back to the mathematical definition of the circle, which said that if the point $P(x, y)$ represents all the points on the circle, then the distance from P to the center of the circle is equal to the radius. In this case the center is at the point $(3, 2)$ and the radius is 5. Therefore the distance formula tells us that

$$5 = \sqrt{(x - 3)^2 + (y - 2)^2}$$

To solve for y we first solve for $(y - 2)$.

$$25 = (x - 3)^2 + (y - 2)^2$$
$$\text{or} \quad (y - 2)^2 = 25 - (x - 3)^2$$

Taking the square root of both sides produces

$$y - 2 = \pm\sqrt{25 - (x - 3)^2}$$
$$\text{or} \qquad y = \pm\sqrt{25 - (x - 3)^2} + 2$$

In the second step we had the equation $25 = (x - 3)^2 + (y - 2)^2$. If we rewrite this as $(x - 3)^2 + (y - 2)^2 = 5^2$, we have written it in the standard equation of a circle.

Notice that the $(x - 3)^2$ indicates that the x-coordinate of the center is 3 and the $(y - 2)^2$ shows that the y-coordinate is 2. Finally, the 5^2 represents the square of the radius.

Equation of a Circle Centered at (h, k)

A circle of radius r and centered at the point (h, k) has the equation

$$(x - h)^2 + (y - k)^2 = r^2$$

This equation is graphed on the calculator by graphing the equations

$$y = \pm\sqrt{r^2 - (x - h)^2} + k$$

Developing the general equation for a circle used the Pythagorean theorem and the distance formula. In the next section you will see another connection between the Pythagorean theorem and graphing and circles, and you will have an opportunity to review some trigonometry while you learn something new.

Pythagorean Trigonometry Identities

Since the sine and cosine functions were defined in Chapter 3 using a circle and a right triangle, it should be no surprise that the Pythagorean theorem will come up again to show you another relationship in trigonometry. Before we study that relationship we'll make some graphs that will prepare you to understand it better.

Let's look at a new function: $f(x) = \cos^2(x)$. The meaning of $\cos^2 x$ is $(\cos x) \times (\cos x)$ or $(\cos x)^2$. The function $f(x) = \cos^2(x)$ is a periodic function, and once you see its graph you will be able to figure out its period and amplitude.

Calculator Lab 4.1

Pythagorean Trigonometric Identity

This calculator lab explores the expressions $\cos^2 x + \sin^2 x$ and $\cos^2 x - \sin^2 x$.

Procedures   Clear all the functions in the Y= screen. Set the graph window using ZTrig (Zoom Trig); then change Ymin to -2 and Ymax to 2.

1. Plot $y = \cos^2 x$.

   (a) What is the amplitude and period of $\cos^2 x$?
   (b) Describe in words the appearance of its graph.

2. Turn off $\cos^2 x$ in the calculator's Y= screen and repeat 1(a) and (b) with the function $y = \sin^2 x$.

   (a) What is the period and amplitude of $\sin^2 x$?
   (b) Compare its graph with the graph of $\cos^2 x$.
   (c) What is the phase shift between the graphs of $\sin^2 x$ and $\cos^2 x$?

3. Turn off $\sin^2 x$ in the calculator's Y= screen and plot a new function: $y = \cos^2 x + \sin^2 x$.

(a) What appears to be another expression that is equivalent to $\cos^2 x + \sin^2 x$?
(b) Check your guess by computing $\cos^2 x + \sin^2 x$ for two different values of x.
(c) Why do you think this relationship is true?

4. Turn off all the functions in the Y= screen and plot only the function $\cos^2 x - \sin^2 x$.

(a) What appears to be another expression that is equivalent to $\cos^2 x - \sin^2 x$?
(b) Check your guess by computing $\cos^2 x - \sin^2 x$ for two different values of x.

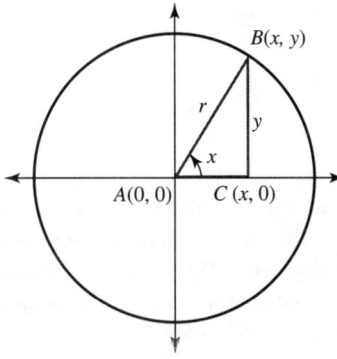

Figure 4.48

To investigate $\cos^2 x + \sin^2 x$ in more detail we'll return to the original definition of sine and cosine in Chapter 3. The original definitions of sine and cosine were based on Figure 3.3, repeated here as Figure 4.48.

If we combine what we know about the Pythagorean theorem,

$$x^2 + y^2 = r^2$$

with the definitions of the trigonometric functions

$$x = r \cos A \quad \text{and} \quad y = r \sin A$$

we have the result

$$(r \cos A)^2 + (r \sin A)^2 = r^2$$
$$r^2 \cos^2 A + r^2 \sin^2 A = r^2$$

Dividing both sides by r^2 we have

$$\frac{r^2 \cos^2 A + r^2 \sin^2 A}{r^2} = \frac{r^2}{r^2}$$

and simplifying we have a **Pythagorean trigonometry identity**:

$$\cos^2 A + \sin^2 A = 1$$

This is one of the basic relationships of trigonometry. You'll see it throughout your future studies of trigonometry. Note that A can be replaced by another expression and the relationship will still be true. For example, in Chapter 3 we studied a function $I_1(t) = 20 \sin(21600t)$. This Pythagorean trigonometry identity also holds for expressions like

$$\cos^2(21600t) + \sin^2(21600t) = 1$$

Figure 4.49 is a figure from Chapter 3 showing that the value of the tangent function is equal to the length of a segment along a tangent line of the circle. Triangle QTO leads to another Pythagorean trigonometric identity.

The Pythagorean theorem says that

$$t^2 + 1^2 = (OQ)^2$$

and we know that $t = \tan(\angle POR)$. But what is OQ? We can find out by writing the cosine of $\angle POR$ from $\triangle QTO$:

$$\cos(\angle POR) = \frac{1}{OQ} \quad \text{or} \quad OQ = \frac{1}{\cos(\angle POR)}$$

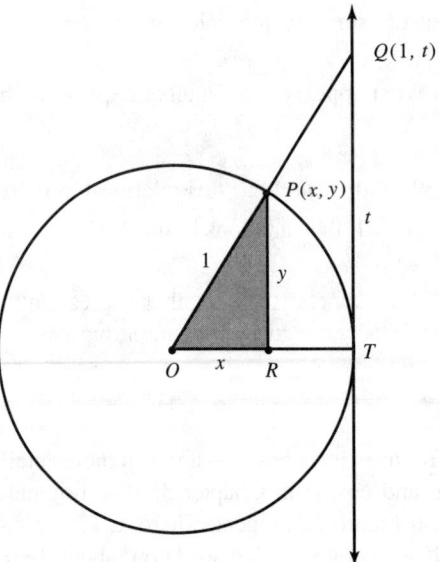

Figure 4.49

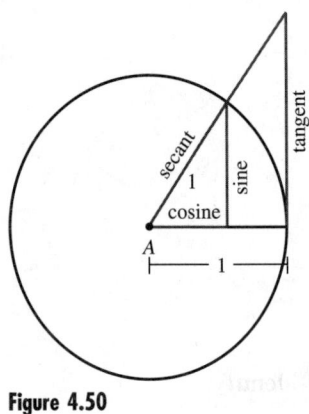

Figure 4.50

Therefore OQ equals the reciprocal of the cosine function. The name for the reciprocal of the cosine function is the **secant function**. The name *secant* was chosen for this function because it is the length of the line segment $\overline{OQ}$, which is part of a secant line of the circle. A secant line is any line that intersects a circle in two points. Using the abbreviation **sec** for the secant function, we have

$$\tan^2(\angle POR) + 1 = \sec^2(\angle POR)$$

Replacing $\angle POR$ with A we have a second Pythagorean trigonometric identity

$$\tan^2 A + 1 = \sec^2 A$$

As with the first Pythagorean trigonometric identity, the angle A can be replaced with any other expression. In Figure 4.50 you can see a way to remember the two Pythagorean trigonometric identities you have learned.

Pythagorean Trigonometric Identities

$$\cos^2 A + \sin^2 A = 1$$

and

$$\tan^2(A) + 1 = \sec^2(A)$$

Circumference and Area of a Circle

Calculating the perimeter and area of a circle presented the same difficulties to the earliest mathematicians as it probably does to you. You can easily imagine how to compute the perimeter of a square or a triangle—just add up the lengths of the sides. But how many sides does a circle have? And how long is each side? We'll show you one way that the ancient Greeks, and probably mathematicians from Asia as well, approached this problem of estimating the perimeter of a circle. The technical term for the perimeter of a circle is **circumference**.

Activity 4.3
Measuring Circles

Find several circular objects to measure. You might choose a wheel, a frying pan, a coffee cup, a garbage can lid, a plate, or a basketball. You can also draw a lot of circles, as we did in Figure 4.51. One way to measure circumference is with a piece of string. Another way is to roll the object through one complete revolution and measure the distance covered.

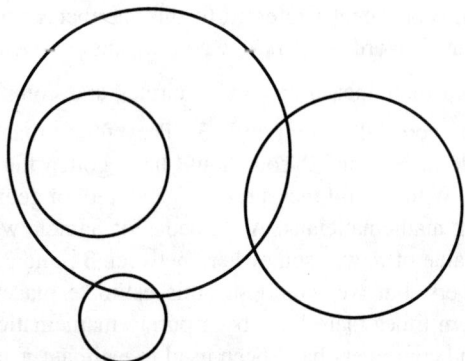

Figure 4.51

1. Complete Table 4.7.

   Table 4.7

Object	Measured Circumference (C)	Measured Diameter (d)	Calculated $\frac{C}{d}$
Circle 1			
Circle 2			
Circle 3			
Circle 4			
Average			

2. After you have completed the table, write about what you can conclude about the entries in the last column.

Ancient mathematicians measured a lot of circles and concluded that circumference was a linear function of the diameter. They had their own ways of writing this idea. We would say that if C is the circumference of a circle of diameter d, then

$$C = md + b \quad \text{where } m \text{ and } b \text{ are constants}$$

But of course if $d = 0$, then $C = 0$ (if diameter is zero the circumference is zero), so we have

$$0 = m \times 0 + b$$

from which we conclude that $b = 0$. Therefore they knew that

$$C = md$$

This equation means that the ratio $\dfrac{C}{d}$ is equal to some constant m.

The next problem was to estimate the value of the constant number represented by m. The Greeks referred to this number with the letter π (pi), the first letter in the Greek word *perifereia*, which means *perimeter*. To estimate the value of π they studied their measurements of circles and computed the ratio $\dfrac{C}{d}$ for many circles, just as you did in Activity 4.3. They found that the average value of this ratio was a little more than 3. You should have gotten the same result.

With careful measurements and a lot of geometric reasoning, one of the greatest Greek mathematicians, Archimedes of Samos, who died in 212 B.C., estimated that the value of π was somewhere between $3\frac{1}{7}$ and $3\frac{10}{71}$. The Greeks did not use decimal numbers, but we would say this estimate places π between 3.141 and 3.143. In modern times there have been purely mathematical ways for estimating the value of π, and computers have been used to estimate π to thousands of decimal places. The value your calculator gives is

$$\pi \approx 3.141592654$$

In the 19th century it was proved that no matter how many decimal places are computed for π there will never be a repeat in the pattern. That is, unlike fractions, whose decimals either terminate (as in $\frac{1}{8} = 0.125$) or repeat (as in $\frac{1}{7} = 0.142857142857\dots$) the decimal part of π will never repeat.

Thus the circumference C of a circle is given by two formulas:

$$C = \pi d$$
$$ = 2\pi r$$

where d = the diameter and r = the radius of the circle.

The Greeks found that the ratio π also plays a role in the area of a circle. The formula for area of a circle is

$$A = \pi r^2$$

Example 4.26

If the circumference of a circle is equal to 20.6 cm, what is the radius?

Solution   The formula is $C = 2\pi r$, so $r = \dfrac{C}{2\pi}$. We know that $C = 20.6$ cm. Therefore

$$r = \frac{20.6}{2\pi} \approx 3.28 \text{ cm.} \qquad \blacksquare$$

Example 4.27

If the area of a circle is 215 acres, what is the radius in miles? (There are 640 acres in one square mile.)

Solution The formula is $A = \pi r^2$, so $r = \sqrt{\dfrac{A}{\pi}}$. We know that $A = \dfrac{215}{640}\dfrac{\text{acres}}{\text{acres/mi}^2} \approx$ 0.33594 mi^2. Therefore $r \approx \sqrt{\dfrac{0.33594}{\pi}} \approx \sqrt{0.106933} \approx 0.327$ mi. ∎

Area and Circumference of a Circle

For a circle of radius r and diameter d we have $d = 2r$. The area and circumference are given by the following formulas.

$$\text{Circumference:}\quad C = 2\pi r$$
$$= \pi d$$
$$\text{Area:}\quad A = \pi r^2$$

where the constant $\pi \approx 3.141592654$.

There is one more feature of a circle to study. In Figure 4.42 you saw the word *sector*. We will do some problems with sectors and then describe their properties.

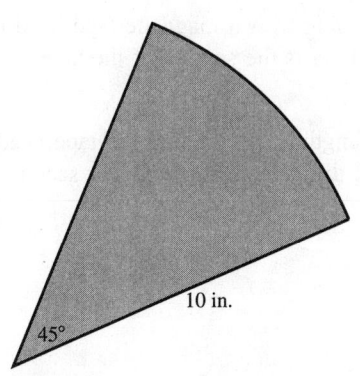

Figure 4.52

Example 4.28 Application

A circular pizza of radius 10 inches is cut into equal wedge-shaped pieces; the angle at the point of each piece is 45° (as shown in Figure 4.52).

(a) Into how many pieces has the pizza been cut?
(b) What is the area of each piece?
(c) What is the length of the crust on each piece?

Solutions (a) $\dfrac{360}{45} = 8$ pieces.

(b) Each of the 8 pieces has equal area, so the area of one piece is $\dfrac{\pi r^2}{8} = \dfrac{\pi \times 100}{8} \approx$ 39 in.^{2}.

(c) Each of the 8 pieces has an equal length of crust, so each piece has $\dfrac{2\pi r}{8} = \dfrac{2\pi \times 10}{8} = 7.9$ in. ∎

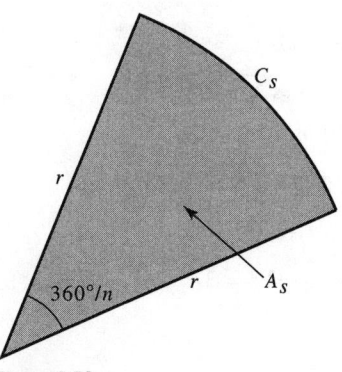

Figure 4.53

When a circle is cut into wedge-shaped pieces (like the pizza in Example 4.28) each piece is called a **sector** and the angle is called a **central angle**. The portion of the circle's circumference that belongs to the sector is called the **arc** of the sector. Figure 4.53 shows one of the sectors that results from dividing a circle of radius r into n equal sectors. This circle forms a fraction $\dfrac{1}{n}$ of the entire circle, so we can easily compute the size of the central angle, the area, and the arc length of the sector.

Area and Arc Length of a Sector

If a sector forms $\dfrac{1}{n}$ of a circle, then the central angle is $\theta = \dfrac{360°}{n} = \dfrac{2\pi}{n}$ and the area A_s and arc length s of a sector are

Degrees	Radians

$$A_s = \frac{\pi r^2}{n} = \frac{\pi}{360°}r^2\theta = \frac{1}{2}r^2\theta$$

and

$$s = \frac{2\pi r}{n} = \frac{\pi}{180°}r\theta = r\theta$$

Example 4.29   Application

The central angle formed by a section of tape passing over a magnetic tape head is 112.5°. If the radius of the head is 3.8 cm, how long is the section of the tape.

Solution   The length of the section of the tape is the arc length of a sector of the tape head. Since the tape head has a radius of 3.8 cm and the central angle of the sector is 112.5°, the arc length is

$$s = \frac{\pi}{180°}r\theta$$

$$= \frac{\pi}{180°}(3.8)(112.5°)$$

$$\approx 7.46$$

The length of the tape section passing over the tape head is about 7.5 cm.   ∎

Section 4.3 Exercises

In Exercises 1–6 find the equation for the circle with the given center C and radius r.

1. $C = (0, 0)$, $r = 2.4$

2. $C = (0, 0)$, $r = 0.45$

3. $C = (5, 1)$, $r = 5.1$

4. $C = (-2, 4)$, $r = \sqrt{17}$

5. $C = (7.9, -3.2)$, $r = \sqrt{51}$

6. $C = (-6.7, -1.4)$, $r = 3$

 In Exercises 7–10, (a) give the center and radius described by each equation and (b) sketch the circle on your calculator.

7. $x^2 + y^2 = 49$

8. $x^2 + y^2 = 6.25$

9. $(x + 2)^2 + (y - 5.3)^2 = 29$

10. $(x - 7.25)^2 + (y + 4.32)^2 = 37.21$

11. What are the circumference and area of a circle with a radius of 7.25 in.?

12. What are the circumference and area of a circle with a diameter of 9.42 cm?

In Exercises 13–18 determine (a) arc length s and (b) area A_s of the sector.

13. The sector is $\frac{1}{10}$ of a circle with radius 9.2 cm.

14. The central angle is 212.5° and the radius is 5.75 in.

15. The central angle is $\frac{\pi}{8} \approx 0.3927$ radians and the radius is 12.95 mm.

16. The sector is $\frac{2}{9}$ of a circle with radius 14.7 mm.

17. The central angle is 57.296° and the radius is 4'3". (Change the radius to all inches or all feet.)

18. The central angle is $\frac{7\pi}{5} \approx 4.3982$ radians and the radius is 5.75 km.

19. *Construction* A rectangular piece of insulation is to be wrapped around a pipe with a diameter of 6.25 in. How wide should you cut the rectangular piece of insulation?

20. *Construction* See Figure 4.54. Find the radius of the smallest circular duct that will rest on the ceiling joists without touching the ceiling.

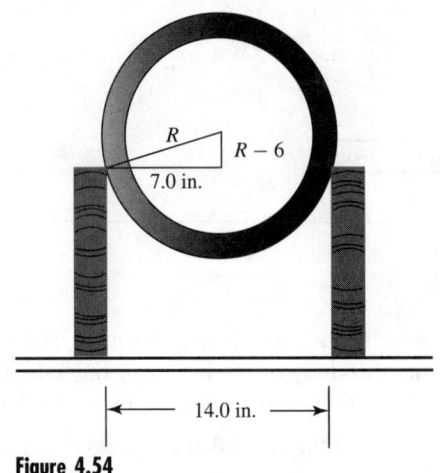

Figure 4.54

21. *Landscape architecture* A landscaper wants to make a trundle wheel that will measure distances in meters. This particular wheel will be designed so that each time the wheel completes one turn it will have moved 2 meters. What should be the radius of the wheel?

22. *Electricity* A coil of bell wire has 65 turns. The diameter of the coil is 0.6 m. How long is the wire on this coil?

23. *Industrial engineering* A Ferris wheel is constructed by placing seats around the outside of a circular frame. If a particular Ferris wheel is designed so that there are 5.75 ft between the rods that support each seat, what radius Ferris wheel would allow for 24 seats.

24. *Civil engineering* A highway makes a 42°18′ = 42.3° turn in a radius of 134 ft. Determine the length of the curve?

25. *Construction* A curved walk is to be constructed 4.0 in. thick and $4\frac{1}{4}$ ft wide. The radius of the outside arc is 18.0 ft and the central angle will be 132°. How many cubic yards of concrete are needed? (27 ft^3 = 1 yd^3.)

26. *Machine design* See Figure 4.55. You need to program a machine to cut the hook-shaped object out of sheets of metal. To do this you need to write equations for the inside and outside arcs with the origin at the point labeled O.

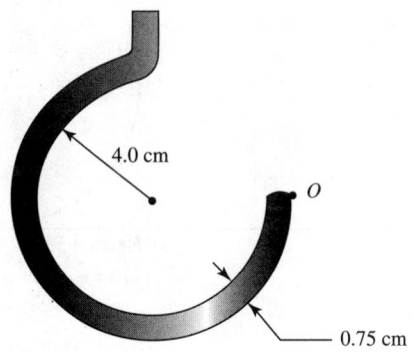

Figure 4.55

27. *Astronomy* Earth's orbit around the Sun is approximately a circle of radius 1.495×10^8 km. The Moon's orbit around Earth is approximately a circle of radius 3.844×10^5 km. If the Sun is placed at the center of a coordinate system, what is the equation of the Moon's orbit around Earth? Assume that Earth is located on the x-axis.

28. Describe how you can use the area of a circle to find the dimensions of a cylindrical storage tank able to hold 10 000 liters of oil (as described in the management memo on page 212).

4.4

Definition and Use of Radians

What are radians? You have seen the word in the MODE menu of your calculator. You learned in Chapter 3 that radians are an alternative to degrees; in Section 4.3 you used radians to determine the arc length and area of a sector; and you have seen that 1 radian $= \dfrac{180°}{\pi} \approx 57.3°$. But certain questions remain. What are radians? And why is there a need for a different unit of measure for angles?

Radians

First let's review the idea of a degree. The measurement itself goes back to astronomers in ancient Syria who divided the circle into 360 equal parts. We call the measure of each of these equal parts one degree. The number 360 proved easy to work with over the years because it has many divisors, thus simplifying arithmetic computation with degrees. Modern mathematicians have found that dividing the circle into $2\pi \approx 6.28$ equal parts simplifies a lot of other calculations. Therefore one radian is the measure of the angle that makes a little less than one-sixth of a complete circle. Figure 4.56 illustrates the relationship between degrees and radians.

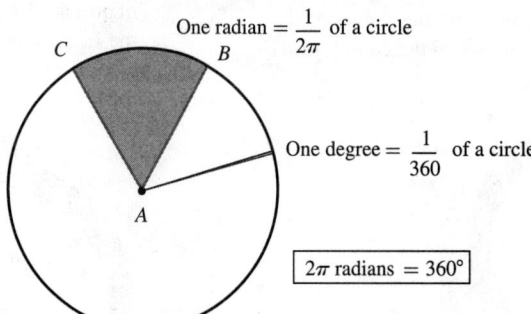

One radian $= \dfrac{1}{2\pi}$ of a circle

One degree $= \dfrac{1}{360}$ of a circle

2π radians $= 360°$

Figure 4.56
One radian and one degree

In a later chapter we'll investigate the calculations that are made easier through use of radians. For now we'll say that the study of the instantaneous rate of change of the function $f(x) = \sin x$ is greatly simplified if angles are measured in radians.

Technically 1 **radian** is the measure of the angle that produces a sector arc $\overset{\frown}{BC}$ (see Figure 4.56) equal in length to the radius AB. Since the circumference $= 2\pi r$, it takes 2π radii to make a complete circumference. Since the length of $\overset{\frown}{BC} = AB$, the radius, it takes 2π arcs equal to $\overset{\frown}{BC}$ to make a complete circumference. Therefore we have the conclusion

The symbol rad is an abbreviation for radian.

2π radians $= 360$ degrees

Example 4.30

Complete Table 4.8 by giving in fractions the radian equivalent of π for each of the given degrees.

Solution   You can convert $30°$ to radians by using the following relationships.

$$R \text{ radians} = 30°$$
$$\text{and} \quad \pi \text{ radians} = 180°$$

Table 4.8   Degrees to radians

Degrees	0	30	45	60	90	120	135	180	360	540	21,600
Radians		$\frac{\pi}{6}$									

If you have trouble simplifying a fraction use the calculator feature that converts to fractions (on the MATH menu).

Therefore $\frac{R}{\pi} = \frac{30°}{180°}$, which simplifies to $R = \pi \times \frac{30°}{180°} = \pi \times \frac{1}{6} = \frac{\pi}{6}$. Thus we see that $30° = \frac{\pi}{6}$ radians.   ∎

Example 4.31

Complete Table 4.9 by giving the degree equivalent for each of the given radian measures.

Table 4.9   Radians to degrees

Degrees		30						
Radians	0	$\frac{\pi}{6}$	$\frac{2\pi}{3}$	$\frac{7\pi}{6}$	$\frac{3\pi}{2}$	$\frac{7\pi}{2}$	8π	660π

Solution   To convert $\frac{\pi}{6}$ to degrees, follow a method similar to the one you used in Example 4.30.

$$\frac{\pi}{6} \text{ radians} = D$$
$$\text{and} \quad \pi \text{ radians} = 180°$$

Therefore $\frac{\pi/6}{\pi} = \frac{D}{180°}$ or $D = \frac{\pi/6}{\pi} \cdot 180°$. This simplifies to $\frac{180°}{6} = 30°$. Again we see that $30° = \frac{\pi}{6}$ radians.   ∎

Although you may sometimes see the symbol (R) used for radians, as in $1^{(R)} \approx 57.3°$, it is usually omitted. This means that $\sin 2.5°$ is asking for the sine of a $2.5°$ angle and that $\sin 2.5$ is the sine of 2.5 radians. Thus $\sin 2.5° \approx 0.0436$ while $\sin 2.5 \approx 0.5985$.

Example 4.32

Plot $f(x) = \sin x$ in radian mode.

Solution   The graph of $\sin x$ will have the same periodic form in degrees or in radians, but you may not recognize it until you set the dimensions of the viewing window correctly. Suppose you change your MODE setting to `Radian` and plot $y = \sin x$. You might get any of the graphs in Figures 4.57–4.59, or something else, depending on the window settings. If you look closely at Figure 4.57 you'll see that it displays an incorrect graph. The x-values are approximately -360 to 360. The calculator computes 94 points in the graph window, so the horizontal distance between points is $\frac{720}{94} = 7.7$. Since 7.7 is larger than 2π, this graph skips over entire cycles.

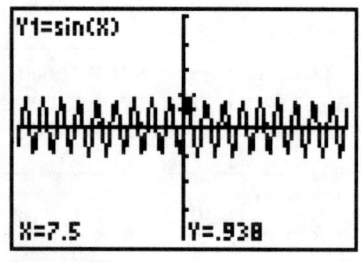

Figure 4.57

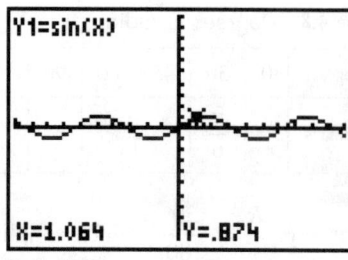

Figure 4.58

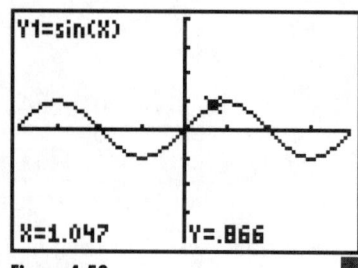

Figure 4.59

Note: ZTrig Window Settings

To plot a function in radian mode, first switch to Radians in the MODE menu, then select ZTrig (ZoomTrig) from the ZOOM menu. The ZTrig settings in radian mode are different from those in degree mode. In radian mode the horizontal dimensions of the ZTrig graph window are $-6.152285613 \leq x \leq 6.152285613$, or approximately $-2\pi \leq x \leq 2\pi$. In degree mode the horizontal dimensions are $-352.5° \leq x \leq 352.5°$.

You might expect the ZTrig setting to use precisely $360°$ or 2π, but these other numbers have been chosen for two reasons. First, they represent the measure of the same angle. Second, because the *TI-83* graphs 94 points within the graph window, the horizontal distance in radians between the points is $\dfrac{6.152285613 - (-6.152285613)}{94} \approx \dfrac{\pi}{24}$, which is $7.5°$. So if you use the ZTrig setting you can be sure that the Trace feature will land on angles like $30°, 45°, 90°, 180°$ since each of these is an even multiple of $7.5°$.

Example 4.33

Write all the roots in radians of $f_1(x) = \sin x$ in the interval $-2\pi \leq x \leq 2\pi$.

Solution It is easy to read off the values given by the Trace function of the calculator. However we want you to practice working with angles written in multiples of π. To give the answers in multiples of π requires remembering that the distance between roots of $f(x) = \sin(Bx)$ is $\dfrac{180°}{|B|}$ in degrees, and therefore in radians the interval is $\dfrac{\pi}{|B|}$. For the function $f_1(x) = \sin x$, $B = 1$ and there are 5 roots: $-2\pi, -\pi, 0, \pi,$ and 2π.

Example 4.34

Write all the roots in radians of $f_2(x) = \sin 2x$ in the interval $-2\pi \leq x \leq 2\pi$.

Solution Here $B = 2$, so the roots are spaced $\dfrac{\pi}{|B|} = \dfrac{\pi}{2}$ units apart. For the function $f_2(x) = \sin 2x$, one root is $x = 0$ and the others are obtained by repeatedly adding and subtracting $\frac{\pi}{2}$. The results are the nine roots $-2\pi, -\frac{3}{2}\pi, -\pi, -\frac{1}{2}\pi, 0, \frac{1}{2}\pi, \pi, \frac{3}{2}\pi,$ and 2π.

In Chapter 3 we looked at ways to solve trigonometric equations. The next example repeats Example 3.33 except that it presents the solutions in radians rather than in degrees.

Example 4.35

Determine to the nearest 0.001 radian all the solutions of the equation $5 \sin 3x = 3$.

Solution   Begin by dividing both sides of the equation by 5 to change the equation to $\sin(3x) = \frac{3}{5}$ or $\sin(3x) = 0.6$.

Make sure the calculator is in radian mode and then use the calculator's $\sin^{-1}$ function (⬛ 2nd ⬛ [SIN^{-1}]) to estimate a value for $3x$. You should obtain $3x = \sin^{-1}(0.6) \approx 0.6435$.

Draw a congruent triangle in the second quadrant, as in Figure 4.60 (where sine is also positive), to find another solution of $\sin(3x) = 0.6$. The second angle is approximately $\pi - 0.6435 = 2.4981$.

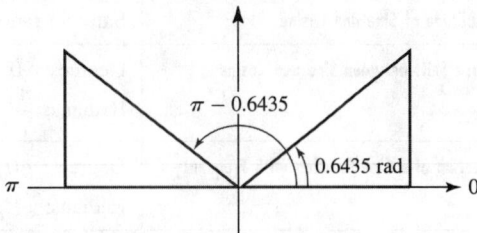

Figure 4.60

Thus the formula for all the answers in terms of $3x$ are

$$3x \approx 0.6435 \pm 2\pi n$$
$$\text{and}\quad 3x \approx 2.4981 \pm 2\pi n$$

where $n = 0, 1, 2, 3 \ldots$.

But these are the answers for $3x$ not x. To find the answers for x we need to divide these formulas by 3. Dividing the first formula, $3x \approx 0.6435 \pm 2\pi n$, by 3 produces

$$x \approx \frac{0.6435}{3} \pm \frac{2\pi}{3} n \approx 0.2145 \pm \frac{2\pi}{3} n$$

which to the nearest 0.001 radian is $0.215 \pm \frac{2\pi}{3} n$. Some of the solutions are $x \approx 0.215$, 2.309, 4.404 and $x \approx -1.879, -3.974, -6.068$.

Dividing the second formula, $3x \approx 2.4981 \pm 2\pi n$, by 3, we obtain

$$x \approx \frac{2.4981}{3} \pm \frac{2\pi}{3} n \approx 0.8327 \pm \frac{2\pi}{3} n$$

which to the nearest 0.001 radian is $0.833 \pm \frac{2\pi}{3} n$. Some of the solutions are $x \approx 0.833$, 2.927, 5.022 and $x \approx -1.261, -3.356, -5.450$.   ∎

Summary of Differences: Radians vs. Degrees

Table 4.10 summarizes the differences between the properties of the trigonometric functions in degrees and in radians. Be sure you understand that not everything changes. In general you can see that when a number like $180°$ or $360°$ appears it should be replaced by π or 2π.

Table 4.10   Summary of differences between radians and degrees

Basic Definitions of Trigonometric Functions	Same in radians and degrees
Graphs of Trigonometric Functions	Degrees: Use x-interval like $-360° \leq x \leq 360°$ Radians: Use x-interval like $-2\pi \leq x \leq 2\pi$
The Tangent Function	Degrees: Approaches infinity at $x = 90°$ Radians: Approaches infinity at $x = \dfrac{\pi}{2}$
Period of Sine and Cosine	Degrees: $360°$ Radians: 2π
Amplitude of Sine and Cosine	Same in radians and degrees
Phase Shift between Sine and Cosine	Degrees: $90°$ Radians: $\dfrac{\pi}{2}$
Equation of a Sine Function with Frequency f	Degrees: $g(t) = \sin(360° f t)$ Radians: $g(t) = \sin(2\pi f t)$
Equation Representing a Musical Note with Frequency = 262 Hz	Degrees: $L(t) = L_0 \sin(94320t)$ Radians: $L(t) = L_0 \sin(524\pi t)$
Spacing of Roots of sin (Bx) or cos (Bx)	Degrees: $\dfrac{180}{\lvert B \rvert}$ Radians: $\dfrac{\pi}{\lvert B \rvert}$
Roots of cos x	Degrees: $90° \pm 180°(n)$, n an integer Radians: $\dfrac{\pi}{2} \pm n\pi$, n an integer
Roots of sin x	Degrees: $180°(n)$, n an integer Radians: $n\pi$, n an integer
Law of Sines	Same in degrees and radians
Law of Cosines	Same in degrees and radians
Pythagorean Trigonometric Identity	Same in degrees and radians
Circumference and Area of Circle	Same in degrees and radians
Size of Central Angle If Circle Is Divided into n Equal Sectors	Degrees: $\dfrac{360°}{n}$ Radians: $\dfrac{2\pi}{n}$

Moving in Circles

In Chapter 1 we discussed the average speed of an object in terms of the distance traveled divided by elapsed time. Suppose that an object moves around a circle of radius r at a constant speed. If s is the distance traveled around this circle in time t, then the average velocity of the object is

$$v = \frac{s}{t}$$

We will use angular speed to refer to average angular speed.

For this object traveling around the circle suppose that θ, measured in radians, is the central angle of the arc the object travels in time t. Then the average **angular speed** ω of this object is the angle, measured in radians, the object travels divided by time. Thus, the average angular speed is

$$\omega = \frac{\theta}{t}$$

Angular speed is usually expressed in revolutions per unit of time such as revolutions per minute (rpm) or revolutions per second (rps).

Why is it important that angular speed be measured in radians rather than in degrees? One of the major reasons is that radians express the ratio of two lengths—arc length and the length of the radius—and so radians are a "dimensionless" unit.

Example 4.36   Application

A flywheel rotates 5.25 radians in 0.750 seconds. Determine its angular speed in revolutions per minute.

Solution   Using the formula $\omega = \frac{\theta}{t}$, with $\theta = 5.25$ rad and $t = 0.750$ s, we have

$$\omega = \frac{5.25}{0.750} = 7 \text{ rad/s}$$

Converting this to revolutions per minute we obtain

$$\omega = \frac{7 \text{ rad}}{1 \text{ s}} \cdot \frac{60 \text{s}}{1 \text{ min}} \cdot \frac{1 \text{ rev}}{2\pi \text{ rad}}$$
$$\approx 66.85 \text{ rev/min}$$

This flywheel is rotating at 66.85 revolutions per minute. ∎

There is an important relationship between linear speed and angular speed. The distance s the object moves through an angle θ around the circle of radius r is given by $s = r\theta$ if θ is in radians. Substituting this value of s in the formula $v = \frac{s}{t}$ produces $v = \frac{r\theta}{t}$. But $\omega = \frac{\theta}{t}$, so we have the following relationship between linear speed and angular speed

$$v = r\omega$$

Example 4.37   Application

A wheel is rotating at 3675 rev/min. Find the linear speed in meters per second of a point 24.5 cm from the center of the wheel.

Solution   We begin by expressing the angular speed in terms of radians and use the fact that there are 2π radians in one revolution.

$$\omega = \frac{2\pi \text{ rad}}{\text{rev}} \cdot \frac{3675 \text{ rev}}{\text{min}}$$

$$\approx 23,090 \text{ rad/min}$$

Substituting this value of ω in the formula $v = r\omega$ we obtain

$$v = \frac{23,090 \text{ rad}}{\text{min}}(24.5 \text{ cm})$$

$$\approx 565,700 \text{ cm/min}$$

$$\approx 94.28 \text{ m/s}$$

The linear speed of a point on this wheel is about 94.28 meters per second. ■

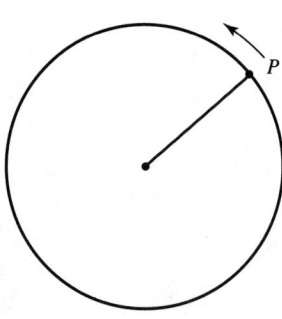

Figure 4.61

You might have found the previous example easier to work by thinking of the problem in the following manner.

In Figure 4.61, one revolution of P around the wheel means that P travels the circumference of the wheel, or $24.5(2\pi)$ cm $= 49\pi$ cm. Since P travels 3675 rev/min, it travels $3675 \times 49\pi$ cm/min. This is then converted to m/s

$$\frac{3675 \times 49\pi \text{ cm}}{1 \text{ min}} \cdot \frac{1 \text{ m}}{100 \text{ cm}} \cdot \frac{1 \text{ min}}{60 \text{ s}} \approx 94.24 \text{ m/s}$$

As you can see, 94.24 m/s is about the same answer as in Example 4.37.

A number of electrical formulas contain a multiplier of π, so it is often preferable to measure angles in radians rather than degrees. In a sinusoidal waveform the time required to complete one revolution is the period T of the waveform. In one complete revolution 2π radians are subtended. Thus the angular speed in radians per second of a sinusoidal waveform of period T seconds is

$$\omega = \frac{2\pi}{T}$$

Since the frequency is $f = 1/T$ we can express this formula as

$$\omega = 2\pi f$$

Example 4.38   Application

Determine the angular speed of the generator that creates a sine wave that has a frequency of 60 Hz.

Solution   Using the formula $\omega = 2\pi f$ we have

$$\omega = 2\pi f = (2\pi)(60) \approx 377$$

The angular speed is 377 rad/s. ■

Example 4.39

Find the angle through which a sine wave with a frequency of 60 Hz will pass in a period of 15 ms.

Solution   Rewriting the formula $\omega = \dfrac{\theta}{t}$ as $\theta = \omega t$ and using the formula $\omega = 2\pi f$, we have $\theta = 2\pi f t$. Here we are given $f = 60$ Hz and $t = 15$ ms $= 15 \times 10^{-3}$ s. Substituting these values we obtain

$$\theta = 2\pi f t = (2\pi)(60)\left(15 \times 10^{-3}\right) \approx 5.655$$

In 15 ms this sine wave covered an angle of 5.655 radians.   ■

Section 4.4 Exercises

In Exercises 1–6 convert each of the given angle measures from radians to degrees.

1. $\frac{2}{3}\pi$
2. 0.8π
3. -1.75π
4. $\frac{5}{8}\pi$
5. -3.58
6. 0.87

In Exercises 7–10 convert each of the given angle measures from degrees to radians.

7. $315°$
8. $-75°$
9. $-212°$
10. $15°$

In Exercises 11–18 solve each of the given equations to the nearest 0.001 rad in the interval $-\pi \le x \le 2\pi$.

11. $\cos 2x = 1$
12. $\tan 5x = 2$
13. $4\tan 3x = -2$
14. $3\sin 4x = -1.5$
15. $3 + \cos 2x = 2.4$
16. $2 - \sin 3x = 1.7$
17. $4 - 6\tan 2x = 1.8$
18. $2 + 5\sin 3x = 4.25$

Exercises 19 and 20 require a ramp much like the one you used in Exercise Set 2.5. In addition to the ramp you will need a stop watch or a watch that can be used to count seconds. However these exercises do not require a motion detector or a CBL.

19. Roll a toy car down the ramp and count (a) the number of revolutions of one of its tires and (b) the time it takes for the car to reach the bottom. (You might want to place a mark on one of the tires to help you count the number of revolutions.)
    (a) What was the average angular speed of this car in revolutions per second?
    (b) What was the average linear speed of this car? (You select the units—in., ft, cm, m—and report the speed in units per second.)

20. Roll a ball down the ramp and count (a) the number of revolutions and (b) the time it takes for the ball to reach the bottom. (Again, you might want to place a mark on the ball to help you count the number of revolutions.)
    (a) What was the average angular speed of this ball in revolutions per second?
    (b) What was the average linear speed of this ball?

21. *Computer science* A compact disc is a thin wafer of clear polycarbonate plastic and metal measuring 4.75 in. in diameter with a hole in its center.
    (a) When the disc is being read at a point 2.25 in. from the center the angular speed of the disc is 200 rpm. What is the linear speed of a point being read if that point is located 2.25 in. from the center of the disc?
    (b) When the disc is being read near the hub the angular speed of the disc is 530 rpm. What is the linear speed of a point being read if that point is located 0.85 in. from the center of the disc?
    (c) Your answers in (a) and (b) should have been about the same. Explain why you think the disc should spin at a constant linear speed.

22. *Automotive technology* A **skid pad** is a flat piece of pavement with a circle painted on it. A car is driven around the circle keeping the center of the car right on the line. Suppose that a car being tested on a skid pad with a radius of 150 feet turns a lap in 14.5 seconds.
    (a) What is the linear speed of the car?
    (b) What is the angular speed of the car?

23. *Sports technology* A bicycle has a wheel 66 cm in diameter.
    (a) How far in meters will the bicycle travel when the wheel makes one complete revolution?
    (b) How many revolutions of this wheel does it take for the bicycle to travel 1 km?
    (c) If the bicycle goes a distance of 1.0 km in 2.5 minutes, what is the angular speed in radians/sec of one of the wheels?

24. *Electronics* Find the angular speed of a waveform with a period of 2.5 s.

25. *Electronics* Find the angular speed of a waveform with a frequency of 50 Hz.

26. *Electronics* Find the angular speed of a waveform with a frequency of 0.045 MHz. (1 MHz = 1,000,000 Hz.)

27. *Electronics* Find the frequency and period of a sine wave with an angular speed of 865 rad/s.

28. *Electronics* Find the frequency and period of a sine wave with an angular speed of 9.6 rad/s.

29. *Electronics* Find the angle through which a sine wave with a frequency of 55 Hz will pass in a period of 12 ms.

30. *Aerospace technology* Engineers often need to know the tip speeds of rotating objects such as propellers or the rim speeds of turbines. Usually these devices are rated only by their angular velocity. The turbine on a certain jet engine has a diameter of 0.94 m. What is the rim speed in m/s when this turbine is rotating at 6500 rpm?

Prisms, Pyramids, and Other 3-D Figures

In this section we will introduce several three-dimensional figures and discuss their volumes and surface areas. The section will conclude with the solution of the second part of the Chapter Project involving finding the minimum area of a cylinder.

Prisms

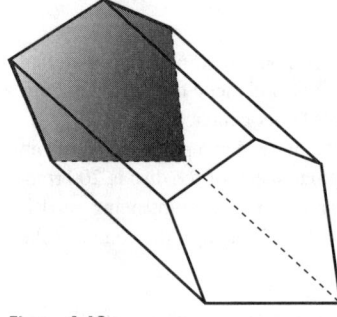

Figure 4.62

To imagine a **prism**, begin with a polygon drawn on a piece of paper. Think of drawing a line segment from one corner (called a **vertex**) of the polygon. Picture the line segment coming off the paper into space. Now imagine parallel line segments of the same length drawn from each of the other vertices. Then draw a copy of the polygon at the ends of the segments. The result will look something like Figure 4.62. Since the base of the prism in Figure 4.62 is a 5-sided polygon, called a **pentagon**, this prism is called a **pentagonal prism**. If the line segments are drawn at right angles to the polygon, then the prism is called a **right prism**.

At the beginning of this chapter you saw that the area of a rectangular solid, or box, was equal to the height times the area of the base. A box is a prism with a rectangular base. The volume of a prism is computed in the same way as the volume of a box.

Volume = Area of base × Height of prism

The total surface area of a prism is the sum of the areas of the top, bottom, and sides. You can see from Figure 4.62 that the sides of a right prism are rectangles, the areas of which you know how to compute. The lateral surface of a prism is the area of all the sides (but not the bases) of the prism.

Pyramids

A **pyramid** is defined similarly to the prism. Begin with a polygon and draw line segments from each vertex of the polygon. In a pyramid, however, the line segments all meet in a point (shown in Figure 4.63). If the point, called the **vertex of the pyramid**, is directly above the center of the base, then the pyramid is called a **right pyramid** (shown in Figure 4.64). The right square pyramid has a base in the shape of a square and is the type of pyramid used by the ancient Egyptians for the tombs of some of their pharaohs.

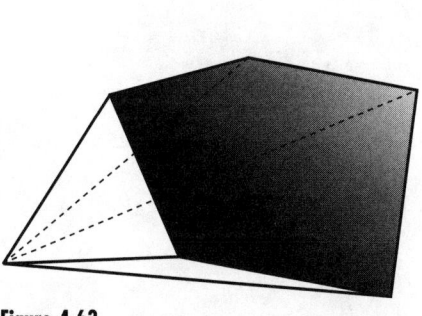

Figure 4.63

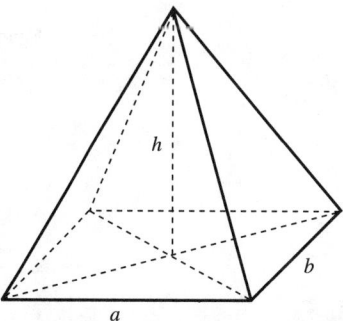

Figure 4.64

How to compute the volume of a pyramid is not at all obvious. It turns out that a volume equal to three prisms like the one in Figure 4.64 can fit in a prism with an identical base and the same height. For example, the pyramid in Figure 4.64 has $\frac{1}{3}$ the volume of the rectangular prism (a box) with dimensions a, b, and h shown in Figure 4.65. That is, the volume of a pyramid is

$$V = \frac{1}{3}\text{Area of base} \times \text{Height}$$

and the volume of the particular right square pyramid shown in Figure 4.64 is

$$V = \frac{1}{3}abh$$

All the sides of any pyramid are triangles, so the lateral surface area of a pyramid is the sum of the areas of the triangles on the sides. The total surface area of a pyramid is the lateral surface area plus the area of the base.

Suppose the top of a pyramid is removed by cutting the pyramid parallel to the base. The shape left after slicing away the top of the pyramid is called a **frustum of the pyramid**. The volume of the frustum is computed by subtracting the volumes of the two pyramids (the original and the sliced-off top).

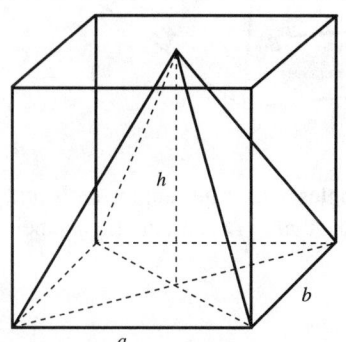

Figure 4.65

Example 4.40   Application

A concrete footing is designed to be a frustum of a pyramid, as shown in Figure 4.66. What is the volume of this footing in cubic inches?

Solution   This shape is a portion of a right pyramid, as shown in Figure 4.67. The missing information is the height of the two pyramids. It is possible to determine those heights from the given information by using trigonometry.

The key to finding the missing heights is found in the two triangles (ABE and DCE) in Figure 4.68.

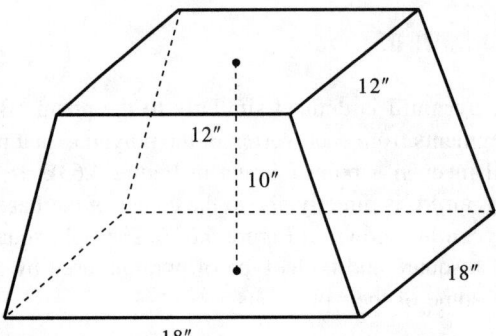

Figure 4.66

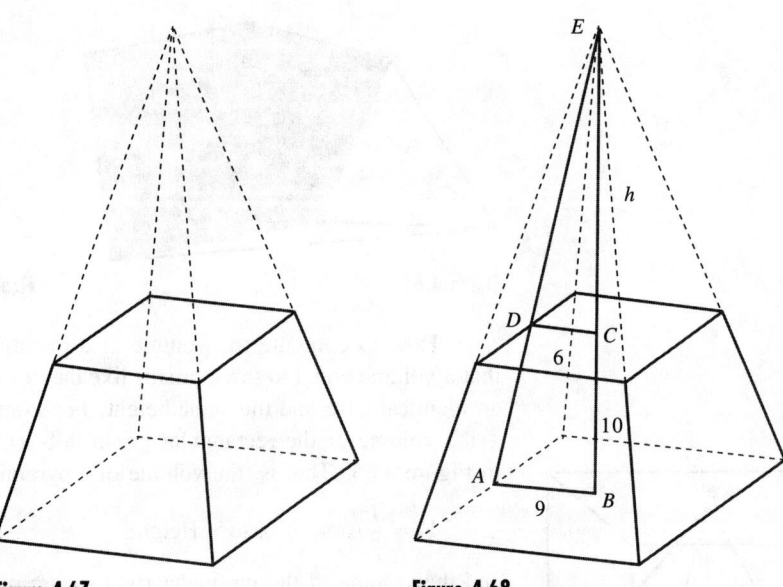

Figure 4.67   **Figure 4.68**

Triangles ABE and DCE are both right triangles with right angles at B and C because these are right pyramids. The angles at A and D are equal,* so their tangents must be equal:

$$\tan(\angle A) = \frac{10 + h}{9} \quad \text{and} \quad \tan(\angle D) = \frac{h}{6}$$

Therefore we must solve the equation

$$\frac{10 + h}{9} = \frac{h}{6}$$

To do that we first cross multiply and then proceed as indicated:

$$9h = 6(10 + h)$$
$$= 60 + 6h$$
$$3h = 60 \qquad \text{Subtract } 6h \text{ from both sides.}$$
$$h = 20 \qquad \text{Divide both sides by 3.}$$

---

*These angles are equal because both right triangles share an angle at E. Since two of the angles of the triangles are equal and the sum of the three angles of the triangle must add up to 180°, the third angles, in this case $\angle A$ and $\angle D$, must be equal.

So the height of the small pyramid is 20 in. and the height of the large pyramid is $h + 10 = 30$ in. The volume of the large pyramid is

$$V = \frac{1}{3}\text{Area of base} \times \text{Height} = \frac{1}{3} \times 18 \times 18 \times 30 = 3240 \text{ in.}^3$$

The volume of the small pyramid is

$$V = \frac{1}{3}\text{Area of base} \times \text{Height} = \frac{1}{3} \times 12 \times 12 \times 20 = 960 \text{ in.}^3$$

Subtracting we get $3240 - 960 = 2280$. Therefore the volume of this concrete footing is 2280 in.^{3}. ∎

Cylinders

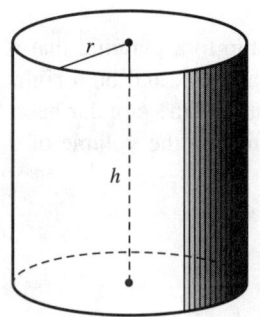

Figure 4.69

A **cylinder** is similar to a prism but its bases are not polygons. Normally the bases of a cylinder are circles, but they can be other curved figures such as ellipses. To imagine a circular cylinder, start with a circle and think of drawing equal and parallel line segments out from the paper at every point on the circle. If the segments are drawn at right angles to the base a **right circular cylinder** results, as shown in Figure 4.69. If the lines are not drawn at right angles to the circle a **slant circular cylinder**, like the one in Figure 4.70, is the result.

The volume of a cylinder is just like the volume of a prism:

$$V = \text{Area of base} \times \text{Height}$$

The volume of a circular cylinder is

$$V = \pi r^2 h$$

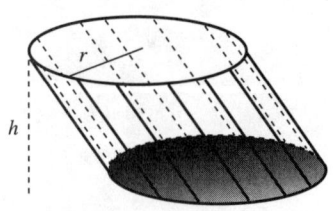

Figure 4.70

Finding a formula for the total surface area is more difficult. Let's look at it. The total surface area consists of two circular tops plus the lateral surface area. If it is a circular cylinder the two tops are circles, so we can write the total surface area S as

$$S = 2\pi r^2 + \text{lateral surface area}$$

Determining the lateral surface area of a right circular cylinder has a surprisingly simple solution. To see it, take a rectangular sheet of paper and roll it into a circular cylinder by bringing one edge together to meet the opposite edge, as shown in Figure 4.71. The top edge of the paper becomes the circumference of the circle, so

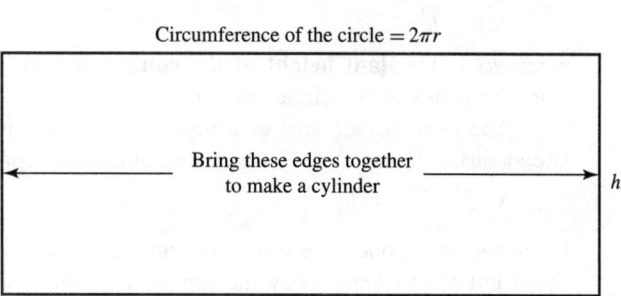

Figure 4.71

its length can be represented as $2\pi r$. The sides of the paper are equal to the height of the cylinder, as shown in the picture. If a right circular cylinder can be made from a rectangle, then the area of the lateral surface area of the cylinder is equal to the area of the rectangular sheet from which it was made. The lateral surface area of a cylinder is the length of the base × height. In the case of a right circular cylinder the lateral surface area L is

$$L = 2\pi rh$$

Now we can complete the formula for the total surface area S of a right circular cylinder:

$$S = 2\pi r^2 + 2\pi rh$$

Cones

A **cone** is pictured in Figure 4.72. It has the same relationship to a pyramid that a cylinder has to a prism. A **circular cone** has a base that is a circle, and on a **right circular cone** the vertex on the cone is directly over the center of the circular base. It should not be a surprise that the volume of a cone is similar to the volume of a pyramid:

$$V = \frac{1}{3}\text{Area of base} \times \text{Height}$$

The volume of a right circular cone is

$$V = \frac{1}{3}\pi r^2 h$$

This means that the volume of three cones can fit into a cylinder with the same base and height, as shown in Figure 4.73.

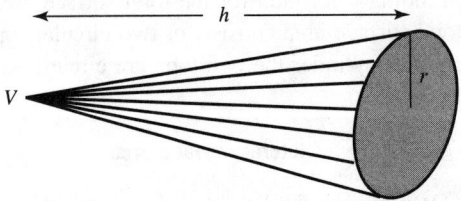

Figure 4.72

The lateral surface area of a cone is not as easy to explain as the lateral surface area of the cylinder. For a right circular cone the lateral surface area is

$$L = \pi rh_1$$

where h_1 is the **slant height of the cone**, which is the length of a segment drawn from the vertex to the circle (shown in Figure 4.73).

The total surface area of a right circular cone is the area of the base plus the lateral surface area. In the case of a right circular cone the total surface area S is

$$S = \pi r^2 + \pi rh_1$$

If the top of a cone is removed by cutting the cone parallel to the base, then the shape left after slicing away the top of the cone is called the **frustum of a cone**. The volume of the frustum of a cone is computed by subtracting the volumes of the two cones.

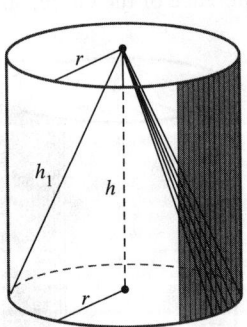

Figure 4.73

Solution of the Chapter Project's Cylinder Problem

In the opening portion of this chapter a problem about a circular cylinder was posed:

> One of our customers needs several cylindrical storage tanks with lids, each able to hold 10 000 liters of oil. Each tank will be the same size. In order to save money they want the tanks to use the least possible amount of metal. Please advise us on the dimensions of the tank we should construct.

The volume V of a circular cylinder is computed in a way similar to that of a rectangular box, using the area B of the base and the height of the cylinder h (see page 215 in Section 4.1 and Figure 4.69 on page 265). That is,

$$V = Bh = \pi r^2 h$$

We are given the volume and we must find the shape of the circular cylinder that gives the minimum surface area S which is

$$S = 2\pi r^2 + 2\pi rh$$

To restate the problem now, we must find the dimensions (r and h) of the cylinder that will produce the required volume with the smallest value of the surface area. This is another example of a max/min problem. In Section 4.1 (on page 219) we listed seven steps that will take you to the solution of a max/min problem. They are given again here.

Solving a Max/Min Problem

Step 1: Select a letter to represent what is being changed (in this case the size of the cut). This value is called the **independent variable**, because we have control over its possible values.

Step 2: Select a letter to represent what you want to maximize or minimize (in this case volume). This value is called the **dependent variable**, because its value depends on the value of the independent variable.

Step 3: Write other variables in the problem in terms of the independent variable.

Step 4: Write a formula that connects the dependent and the independent variables, and only these variables. This is the **mathematical model** of the problem.

Step 5: Decide what values of the independent variable are acceptable. The acceptable values for the independent variable make up what is called the **domain of the model**; so in this step you will determine the **practical domain** of the model.

Step 6: Use a table of values or a graph to locate the largest or smallest value of the dependent variable.

Step 7: Answer the question.

Here is our solution for the cylinder problem. We have two formulas to use:

$$V = \pi r^2 h$$
$$\text{and} \quad S = 2\pi r^2 + 2\pi rh$$

We will be studying the graph of S, and we must choose whether to keep r or h as the independent variable. Which would you choose? We'll show both options and then we'll select one of them.

Choose h as the independent variable. Then we must solve $V = \pi r^2 h$ for r. The answer is

$$r = \sqrt{\frac{V}{\pi h}}$$

We must substitute this expression for r into the equation for S. The result is

$$S = 2\pi \frac{V}{\pi h} + \sqrt{\frac{V}{\pi h}} h$$

$$= \frac{2V}{h} + 2\pi \sqrt{\frac{V}{\pi h}} h$$

More simplification can be done on this equation, but we'll leave it as is for now.

Choose r as the independent variable. Then we must solve $V = \pi r^2 h$ for h. The answer is

$$h = \frac{V}{\pi r^2}$$

We must substitute this expression for h into the equation for S. The result is

$$S = 2\pi r^2 + 2\pi r \frac{V}{\pi r^2}$$

$$= 2\pi r^2 + \frac{2V}{r}$$

Now compare the two different equations for S. The second one certainly seems simpler. Could we have predicted that in advance? Usually you could because you could have seen the square root coming since the variable r is squared in the volume formula. Also, there are two occurrences of r in the surface area and only one of h, so substitution for h would likely be easier.

This completes the first four steps of the solution. The fifth step, determining the domain of the model, is next. The radius must be positive, so the domain is $r > 0$.

Before we can use the formula for S we must substitute the given constant value for the volume V. Since $V = 10\,000$ liters $= 10^4$ liters, and one liter is 1000 cm$^3 = 10^3$ cm^3, we have $V = 10\,000\,000$ cm$^3 = 10^7$ cm^3. One m$^3 = 10^6$ cm$^3 = 1\,000\,000$ cm^3, so we have $V = \dfrac{10^7 \ \text{cm}^3}{10^6 \ \text{cm}^3/\text{m}^3} = 10$ m^3. Therefore the equation for S becomes

$$S = 2\pi r^2 + \frac{2 \cdot 10}{r}$$

Figures 4.74 and 4.75 show the table and graph for this function. A 1-decimal-place estimate of the value of r that minimizes the value of S is therefore 1.2 meters.

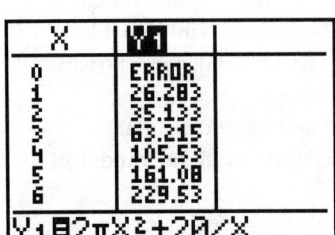

Figure 4.74

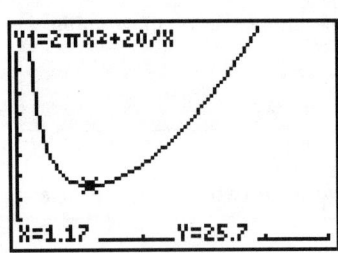

Figure 4.75

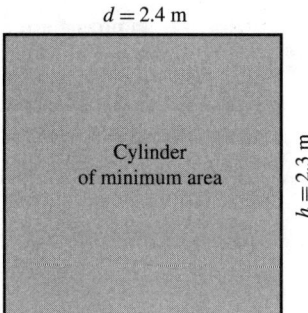

Figure 4.76

The value of h is determined by substituting $r = 1.17$ (*not 1.2—keep the precision!*) into the equation $h = \dfrac{V}{\pi r^2} = \dfrac{10}{\pi r^2}$. That answer is $h = \dfrac{10}{\pi \times 1.17^2} \approx$ 2.3 m.

Therefore the cylinder with minimum surface area has height about twice the radius. Looked at from the side (Figure 4.76), the shape of the cylinder is very close to a square.

Can you think of some products that are packaged in a cylinder with that shape? Why do you think that more cans are *not* in that shape if that shape is the most economical?

Section 4.5 Exercises

In Exercises 1–8 find the lateral surface area, total surface area, and volume of each of the solids pictured.

1.

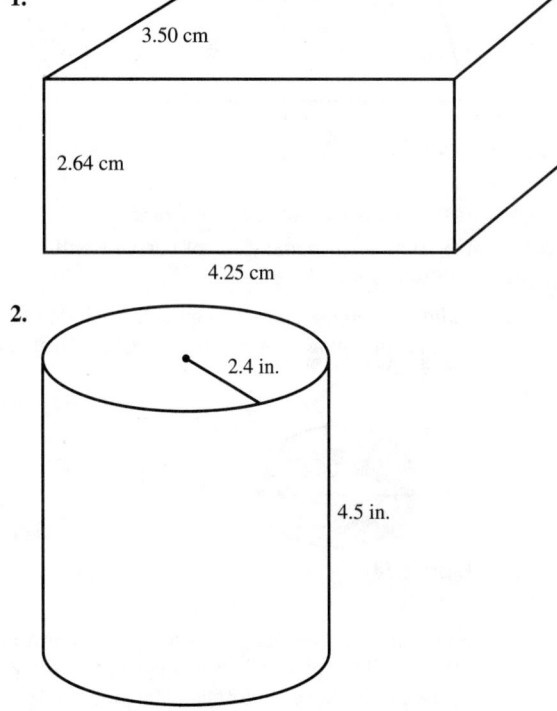

2.

3.

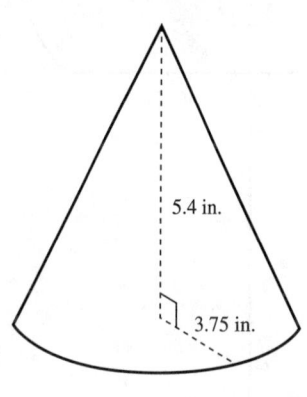

4.

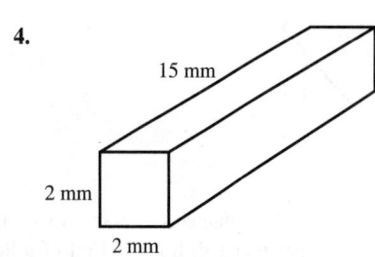

5.

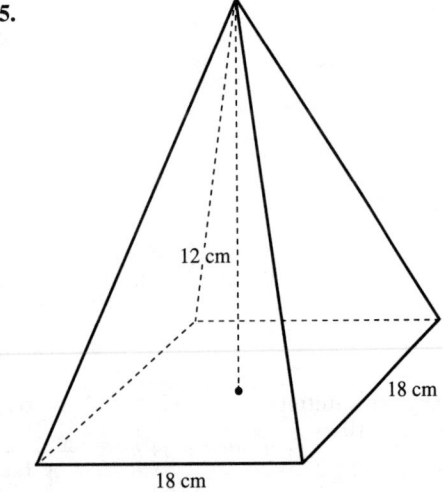

12 cm

18 cm

18 cm

6.

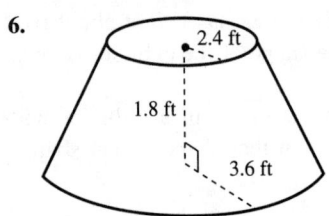

2.4 ft

1.8 ft

3.6 ft

7.

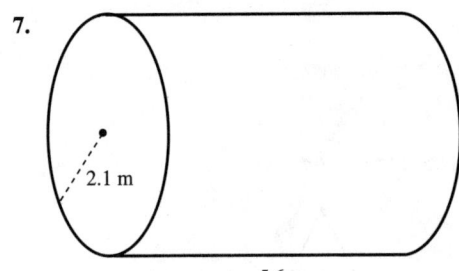

2.1 m

5.6 m

8.

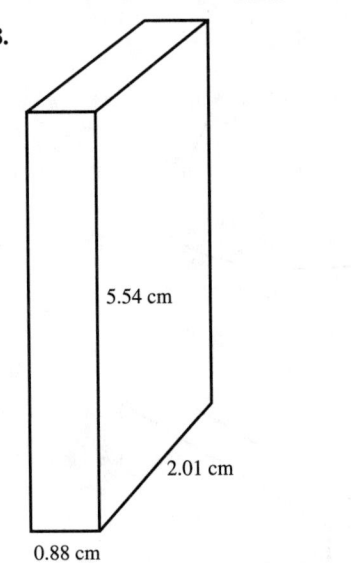

5.54 cm

2.01 cm

0.88 cm

9. *Industrial design*   An engineer is asked to design a cylindrical container that will hold 250 ml of a liquid.

In order for it to be easy to pick up with one hand, the diameter of the container should be 5.70 cm.

(a) Determine the height of the cylinder.

(b) A label will be placed around the outside of the container. The label will cover the entire lateral surface, but its ends need to overlap for sealing the label. If the overlap is 1.1 cm, what is the area of label?

(c) What is the total surface area of this container?

10. *Industrial design*   The container in Exercise 9 is to be made by stamping two circles and a rectangle from a rectangular piece of metal that measures 20 cm wide by 30 cm long.

(a) The metal for how many containers should be able to be stamped from this sheet of metal?

(b) Is it possible to get the metal for the number of containers you got in (a) from this sheet? If not, explain why it is not possible.

11. *Civil engineering*   The frustum of a right square pyramid (as shown in Figure 4.77) is used as the base for a light support. Each edge of the lower base is 30.0 in., each edge of the upper base is 18.0 in., and the slant height is 24.0 in.

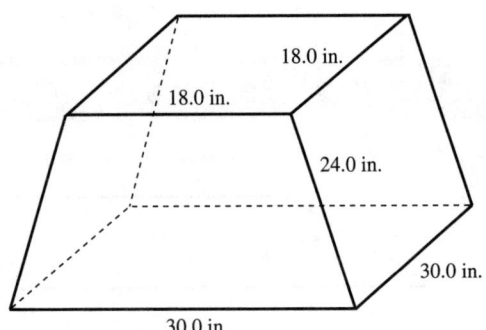

18.0 in.

18.0 in.

24.0 in.

30.0 in.

30.0 in.

Figure 4.77

(a) What is the volume of the base?

(b) How many cubic yards of concrete will be needed to make 50 bases?

12. *Industrial engineering*   Compute the weight of a steel washer (as shown in Figure 4.78), if steel weighs 0.283 lb/ft^3.

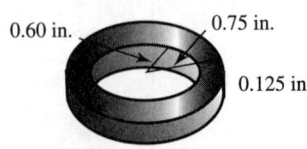

0.60 in.   0.75 in.

0.125 in.

Figure 4.78

13. *Mechanics*   A cross-section of a pipe is shown in Figure 4.79. If the pipe is 2.85 m long, what is the volume of the material needed to make the pipe?

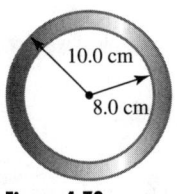

10.0 cm

8.0 cm

Figure 4.79

14. *Energy technology* A cylindrical gas tank has a radius of 64 ft. If the volume of the tank is 100,000 ft³, **(a)** determine the height of the tank and **(b)** find the amount of material that will be needed to construct the tank.

● ● ● ● ● ● ● ● ● ● Chapter 4 Summary and Review

Topics You Learned or Reviewed

▶ The volume of a cone or a pyramid is $V = \frac{1}{3}Bh$ where B is the area of the base and h is the height of the pyramid.

▶ The lateral surface area L and total surface area S of a right circular cone, respectively, are

$$L = \pi r h_1$$
$$S = \pi r^2 + \pi r h_1$$

where h_1 is the slant height of the cone.

▶ The Pythagorean theorem states that if a right triangle has legs a and b and hypotenuse c, then $a^2 + b^2 = c^2$.

▶ The law of sines states that for triangle $\triangle ABC$

$$\frac{\sin A}{a} = \frac{\sin B}{b} = \frac{\sin C}{c}$$

▶ The law of cosines states that for triangle $\triangle ABC$

$$c^2 = a^2 + b^2 - 2ab\cos C$$
$$\cos C = \frac{a^2 + b^2 - c^2}{2ab}$$

▶ You can solve a max/min problem by selecting independent and dependent variables, writing a mathematical model (formula) that connects the independent and dependent variables, determining the domain of the model, and using a table or graph to locate the largest or smallest value of the dependent variable.

▶ A radian is a measure of the angle that produces an arc of the circle equal in length to the radius. Thus π radians $= 180°$.

▶ If θ is the central angle of the distance an object moves around a circle of radius r, then $s = r\theta$ if θ is in radians. The angular speed of this object is $\omega = \dfrac{\theta}{t}$ where t is the time it takes the object to move this distance.

▶ Angular speed ω and linear speed v around a circle of radius r are related by the formula $v = r\omega$.

Review Exercises

In Exercises 1–2 write the formulas for the (a) volume and (b) total surface area of each rectangular solid.

1. The base is a square. Each side of the square has a length of x and the height of the solid is three times the length of one side of the square.

2. The height is 2″ shorter than the length, and the width is 10″ longer than the length.

In Exercises 3–5 determine the (a) lateral surface area L and (b) total surface area S of the rectangular solid with the given dimensions. Assume that the length and width are the dimensions of the base.

3. $l = 3.5$ cm, $w = 7.6$ cm, and $h = 12.0$ cm

4. $l = 3.25$ m, $w = 2.75$ m, and $h = 4.60$ m

5. $l = 6.2$ in., $w = 5.3$ in., and $h = 2.5$ in.

$w = 4.25$ cm, and $h = 5.40$ cm, determine l.

6. If the volume of a rectangular solid is 156.06 cm³,

In Exercises 7–8 find the distance between the given pairs of points.

7. $(-5, 2)$ and $(7, -4)$

8. $(-9, 12)$ and $(-1, -3)$

In Exercises 9–10 determine the area of the triangle.

9. $a = 9$ m, $\angle B = 65°$, $c = 12$ m

10. $\angle A = 36°$, $b = 14.2$ in., $c = 6.8$ in.

In Exercises 11–14 determine the measures of the three other parts of the triangle, if possible.

11. $\angle A = 36°$, $\angle B = 67°$, $a = 14.25$ cm

12. $\angle A = 117°$, $\angle B = 33°$, $c = 34.2$ ft

13. $a = 18$ in., $b = 24.5$ in., $\angle C = 125.2°$

14. $a = 21.35$ yd, $b = 8.96$ yd, $c = 16.87$ yd

In Exercises 15–16 find the equation for the circle with the given center C and radius r.

15. $C = (0, 0)$, $r = 4.5$

16. $C = (7.4, -3.1)$, $r = \sqrt{17}$

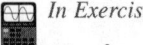 *In Exercises 17–18 (a) give the center and radius described by each equation and (b) sketch the circle on your calculator.*

17. $x^2 + y^2 = 121$

18. $(x + 4.38)^2 + (y - 5.12)^2 = 53.29$

19. What are the circumference and area of a circle with a radius of 6.25 cm?

20. What are the circumference and area of a circle with a diameter of 12.76 ft?

In Exercises 21–24 determine (a) arc length s and (b) area A_s of the sector.

21. The sector is $\frac{1}{12}$ of a circle with radius 5.4 cm.

22. The central angle is 148.5° and the radius is 9.75 in.

23. The central angle is $\frac{\pi}{5} \approx 0.6283$ radians and the radius is 77.46 mm.

24. The sector is $\frac{5}{16}$ of a circle with radius 61.8 mm.

In Exercises 25–26 convert each of the given angle measures from radians to degrees.

25. $\frac{3}{5}\pi$

26. 1.38

In Exercises 27–28 convert each of the given angle measures from degrees to radians.

27. 225°

28. 18°

In Exercises 29–32 solve each of the given equations to the nearest 0.001 rad in the interval $-\pi \le x \le 2\pi$.

29. $\tan 2x = 1$

30. $2\cos 3x = -1.5$

31. $3 - \sin x = 2.25$

32. $3 - 5\sin 2x = -1.75$

In Exercises 33–36 find the lateral surface area, total surface area, and volume of each of the solids pictured.

33.

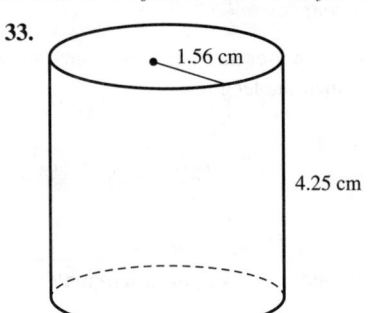

1.56 cm

4.25 cm

34.

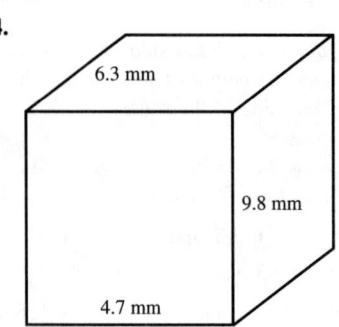

6.3 mm

9.8 mm

4.7 mm

35.

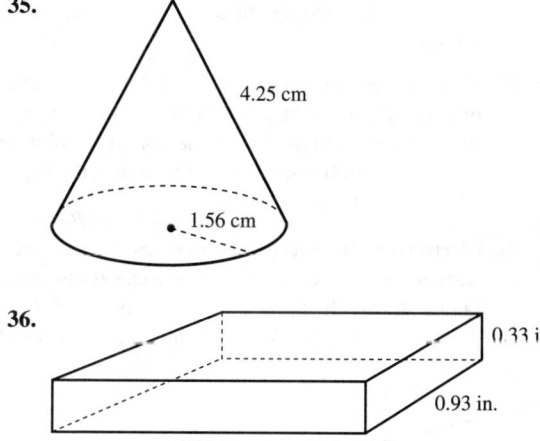

4.25 cm

1.56 cm

36.

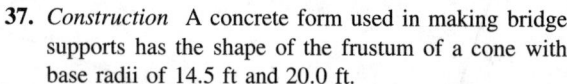

0.33 in

0.93 in.

2.07 in.

37. *Construction* A concrete form used in making bridge supports has the shape of the frustum of a cone with base radii of 14.5 ft and 20.0 ft.

(a) If the support is 42′3″ tall, determine the volume in ft³ of concrete needed to make one support.

(b) Concrete is usually purchased in cubic yards. Convert your answer in (a) to cubic yards.

38. *Construction* An engineer for a highway department has designed the divider in Figure 4.80 to be placed between two lanes of a highway. Find the number of cubic yards of concrete required to make this divider.

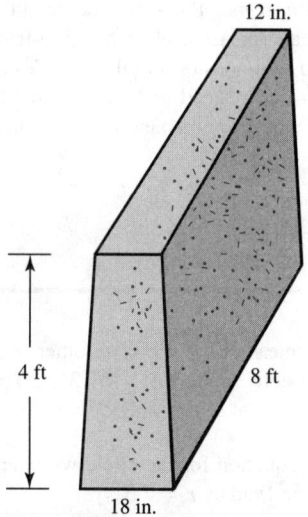

12 in.

4 ft

8 ft

18 in.

Figure 4.80

39. *Industrial design* Determine the volume of the gravity bin in Figure 4.81.

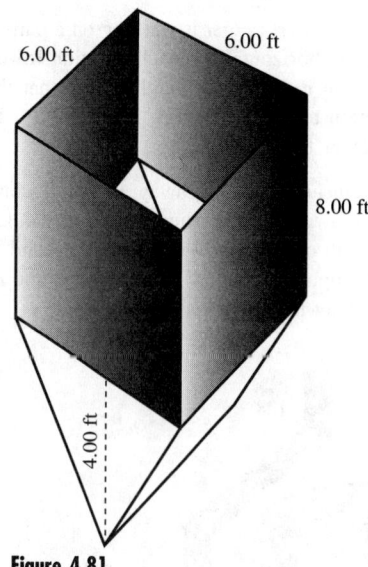

6.00 ft

6.00 ft

8.00 ft

4.00 ft

Figure 4.81

40. *Industrial design* How much sheet metal in ft² will be needed to make the gravity bin in Figure 4.81?

41. *Agriculture* A company has 15 000 m of available fencing. It wants to use the fencing to enclose a rectangular field. One side of the field is bordered by a warehouse and no fencing is needed along this side. If the fence is 1.5 m high, what are the length and width of the largest area that can be enclosed?

42. *Industrial design* A metal trash container is to have a capacity of 18 m³. To fit with the company's removal method, the length is 1.5 as long as the width. Because of environmental concerns, the bottom is 2 cm thick while the sides and the top are each 1 cm thick. You want to make this trash container using the least amount (volume) of material.

(a) Write a mathematical model for this problem.

(b) What is the practical domain for your model?

(c) What dimensions, to two decimal places, will require the least amount of material?

(d) Use your results in (c) to determine the least amount of material (volume) that will be needed to make this trash container.

(e) If this trash container is made from steel and steel weighs 7,721 kg/m³, what is the weight of this container?

43. *Civil engineering* A high-tension wire is to be strung across a river from tower A to tower C. To determine the distance between these two towers a third point B is located on the same side of the river as point A, 720.0 m from A. If $\angle ABC = 52.4°$ and $\angle BAC = 47.6°$, what is the distance between towers A and C?

44. *Forestry* A forest ranger is walking on a path inclined at 5° to the horizontal to an observation tower. At a rest area the ranger pauses and notices that the angle of elevation to the 125-ft-high tower is 38°. How far is the ranger from the tower?

45. *Civil engineering* A highway is being constructed around the wetland area shown in Figure 4.82. What is the length of highway needed to go around the wetland. (Compute the length of the side of the highway farthest away from the wetland.)

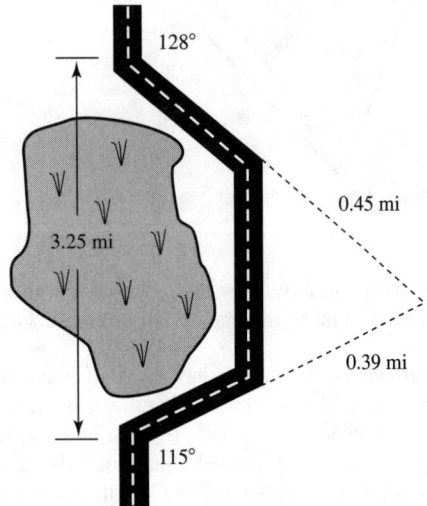

Figure 4.82

46. *Navigation* The navigator of a reconnaissance plane determines the distance from the plane to radar station *A* to be 4782.9 m. At the same time, the distance from the plane to radar station *B* is 6937.4 m. As viewed from the plane the angle between the two stations is

116.4°. Determine the distance between the two radar stations.

47. *Civil engineering* A guy wire runs 21.8 m from the top of a utility pole to the level ground. A second guy wire runs 32.6 m from the top of the pole to a point on the ground 12.7 m further from the pole than the first wire. What is the height of the pole?

48. *Electronics* Two electrical forces F_1 and F_2 act on a particle in an electric field at an angle θ (as shown in Figure 4.83). If $F_1 = 6.0 \times 10^{-6}$ N, $\theta = 75°$, and $F_2 = 8.5 \times 10^{-6}$ N, determine the magnitude of R.

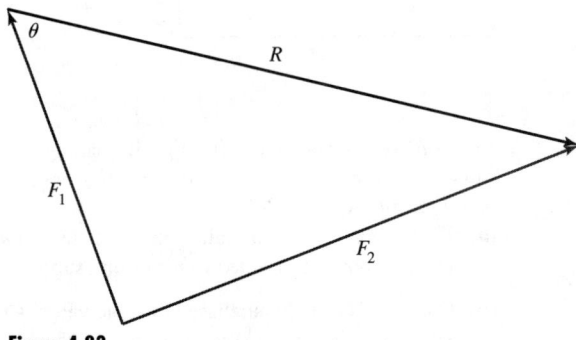

Figure 4.83

49. *Medical technology* A lab technician is centrifuging a blood sample at an angular speed of 3000 rpm. If the radius of the circular path followed by the sample is 0.15 m, find the linear speed of the sample in m/s.

50. *Industrial engineering* Plans for the world's largest Ferris wheel call for a wheel with a 500-ft diameter. The wheel will turn nonstop and the capsules for riders will move at a linear speed of 1 ft/sec. How long will it take for the wheel to complete one revolution?

●●●●●●●● Chapter 4 Test

1. Determine the **(a)** lateral surface area L and **(b)** total surface area S of a rectangular solid with dimensions $l = 7.5$ cm, $w = 4.3$ cm, and $h = 14.6$ cm. Assume that the length and width are the dimensions of the base.

2. If the volume of a rectangular solid is 309.96 cm³ with $l = 12.3$ cm and $h = 4.5$ cm, determine w.

3. Find the distance between the points $(-4, -3)$ and $(5, -4)$.

4. Determine the area of a triangle with sides of length $a = 9.2$ cm and $b = 5.3$ cm and $\angle C = 47°$.

5. Determine the measures of the three other parts of the triangle, if possible, when $\angle A = 67.3°$, $\angle B = 42.1°$, $a = 9.64$ cm.

6. Determine the equation for the circle with center $C = (-4.25, 9.61)$ and radius $r = 7.3$.

7. Give the center and radius of the circle described by the equation $x^2 + y^2 = 152$.

8. What are the circumference and area of a circle with a radius of 9.73 cm?

9. Determine the (a) arc length s and (b) area A_s of the sector with central angle 215° and radius 4.3 in.

10. Convert 2.54 from radians to degrees.

11. Convert the angle measures of 47° from degrees to radians.

12. Solve the equation $1 - 4\cos 2x = 3$ to the nearest 0.001 rad in the interval $-\pi \le x \le 2\pi$.

13. Find the lateral surface area, total surface area, and volume of the solid shown in Figure 4.84.

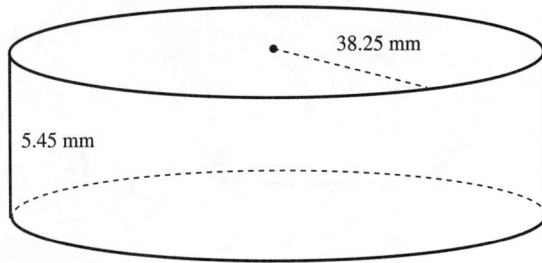

Figure 4.84

14. Determine the length of the brace BC of the framework in Figure 4.85.

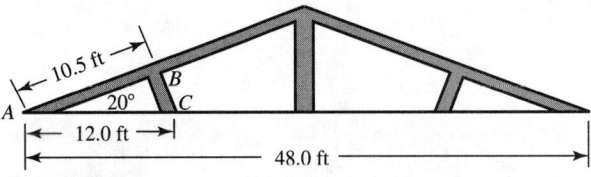

Figure 4.85

15. A race car is driven around a circular track at a constant speed of 196 mph. If the diameter of the track is 0.75 mile, what is the angular speed of the car?

16. Squares are to be cut from the corners of a piece of cardboard that measures 24.0 in. by 36.0 in. After the corners have been removed, the sides will be folded up to make an open box.
   (a) What size square will give a box with the largest volume?
   (b) What is the largest volume that a box made from this cardboard can hold?

CHAPTER

5

Modeling Motion in Two Dimensions
Vectors and Parametric Equations

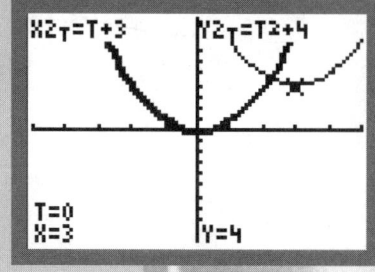

Topics You'll Need to Know

- ► Pythagorean theorem
- ► Trigonometry:
  - • circle definition
  - • applications to right triangles
  - • graphs
  - • law of cosines
- ► Use of formula for d vs. t for:
  - • constant velocity
  - • gravity-influenced motion
- ► Properties of a parallelogram
- ► Plotting graphs by hand
- ► Solving quadratic equations of the form $ax^2 - bx = 0$ by factoring
- ► Determining from the equation how a function is shifted horizontally or vertically from the basic function
- ► Converting mph to ft/sec and the reverse
- ► Use of the law of cosines

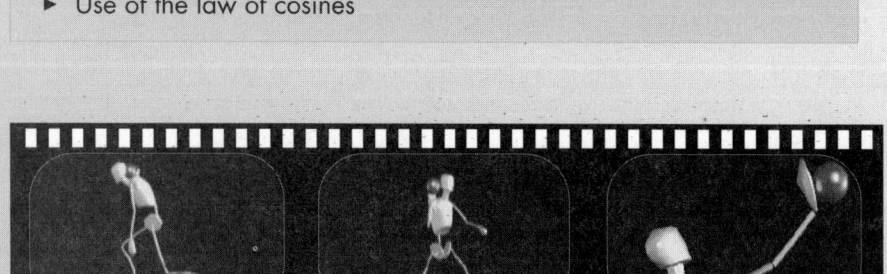

Calculator Skills You'll Use

- ► Path setting for graphs
- ► Simultaneous (Simul) mode for graphing
- ► Adjusting graph window for best display
- ► Use and interpretation of quadratic regression (QuadReg)
- ► Finding the maximum value of a function

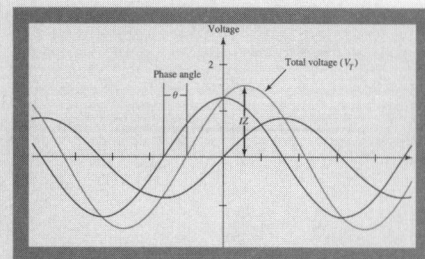

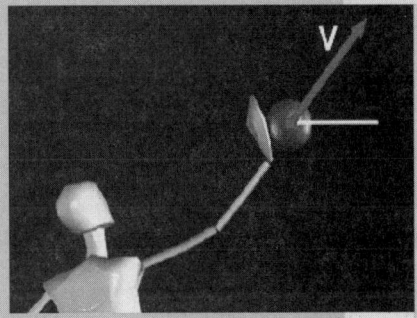

Topics You'll Learn or Review

- ► Vector notation
- ► Vector addition
- ► Parametric equations:
  - parabolas
  - circles
  - ellipses
  - plotting planetary orbits
- ► Using the discriminant to analyze roots of a quadratic equation
- ► Imaginary and complex numbers:
  - addition and multiplication
  - graphical representation
  - absolute value
  - the angle associated with a complex number

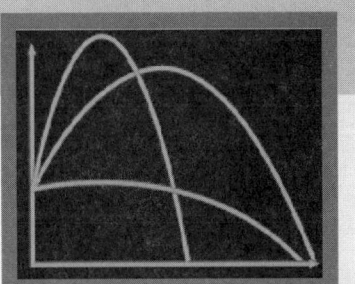

Calculator Skills You'll Learn

- ► Plotting parametric equations
- ► Working with complex numbers

What You'll Do in This Chapter

In this chapter you will learn how to analyze the paths of objects that are moving under the influence of gravity. To do so you'll learn two new mathematical modeling techniques: parametric equations and vectors. You'll see that parametric equations and vectors can be used to solve many different kinds of problems, from electronics to navigation.

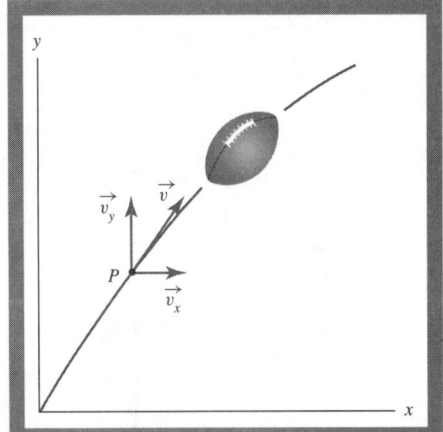

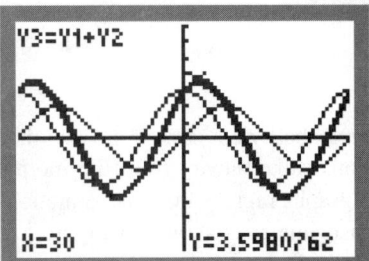

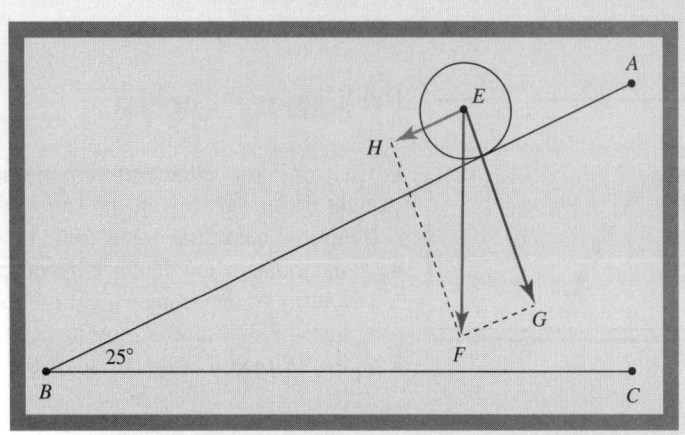

Chapter Project—What's the Best Angle?

Peak *Performance* Training

— "Our Tips Will Make You Tops" —

To: Research department

From: Management

Subject: Getting the longest distance

We have been getting requests from participants in many sports: tennis, soccer, baseball, softball, track, golf, football, even fishing. They all want to increase the distance of their hits, kicks, throws, or casts.

We've helped them to get the maximum power out of their bodies, but now they need our advice on how to get the longest distance.

A consultant has suggested that we look into the angle of release. She thinks there may be a best angle at which to throw, kick, hit, or cast.

Please study and report on the relation of the angle of release to the distance the object travels. Is there one angle that is better than all others?

Also, many of our clients, especially the baseball and football players, want to get distance but they also want to shorten the time that the ball is in flight. What can you recommend about getting the ball to its target as quickly as possible while still getting the maximum distance?

Please be prepared to defend your decision with pictures, graphs, and mathematical analysis. Our customers are depending on you, and so are we.

Remember, we're not the cheapest, but we *are* the best, and your mathematical knowledge is part of our competitive distinctiveness.

Preliminary Analysis

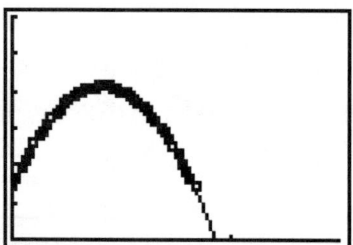

Figure 5.1

In Chapter 2 you studied motion in one dimension. A ball went straight up and then straight down, landing in the same spot. A person walked away from the motion detector and then turned and came toward it. The graphs representing these motions were parabolas when distance d was plotted against time t, as shown in Figure 5.1. In our study of one-dimensional motion we discussed the maximum height, the total time that the object was moving, and its changing velocity while it was in the air. You also learned that the slope of the tangent drawn at a point on the d vs. t graph

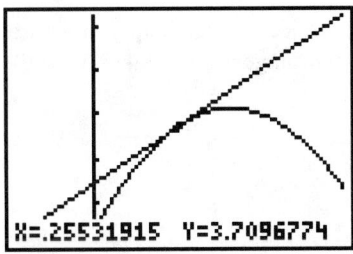

X=.25531915 Y=3.7096774

Figure 5.2

(see Figure 5.2) is equal to the velocity of the object at that particular time. The tangent line will continue to be important as our study of motion proceeds, but its meaning will be different.

The problem before us now is two-dimensional motion. A ball or other object is being thrown or kicked or hit into the air away from a person and the motion is in two dimensions: vertical (up) and horizontal (away). The kind of graph we have been using has only two dimensions, so we can't plot vertical distance, horizontal distance, and time on it; that would require a three-dimensional graph. Such graphing is possible and we'll study 3-D graphs in a later chapter, but our approach here will be to plot the vertical distance (y) against the horizontal distance (x), as shown in Figure 5.3.

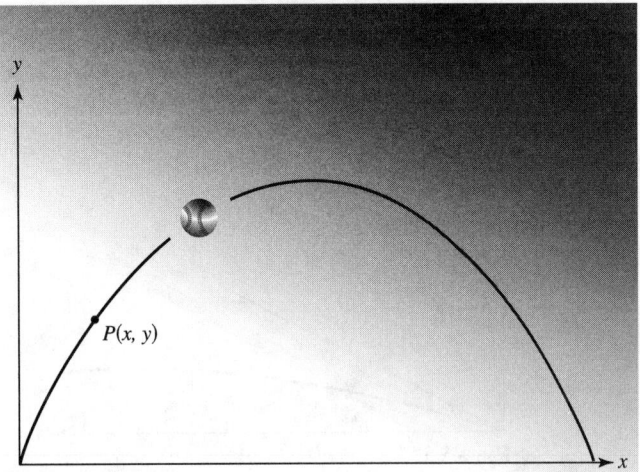

Figure 5.3

Think about the situation when you throw a ball straight up. It will come straight down and not go any horizontal distance at all. Let's refer to throwing a ball straight up as throwing it at an angle of 90°. If you throw it straight ahead (an angle of 0°) it goes away from you. How far it goes before touching the ground depends on the height of the ball above the ground when it is released and the initial velocity, v_0. Figure 5.4 shows three other angles at which a ball could be thrown, kicked, or batted with the same strength. The difference in the ball's path is caused only by the angle at which it is thrown. You can see that the 60° angle gives a longer horizontal distance than the others. The question is "Can you do better than that?"

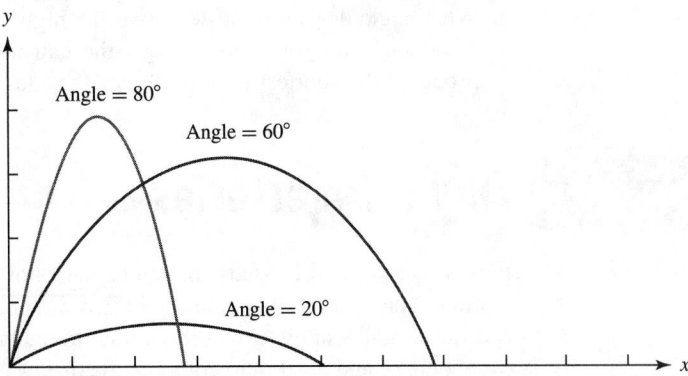

Figure 5.4

The tangent line drawn at the beginning of the trajectory can help you look at the angle at which the ball is released. The angle the tangent line makes with the positive *x*-axis is the angle of release (see Figure 5.5).

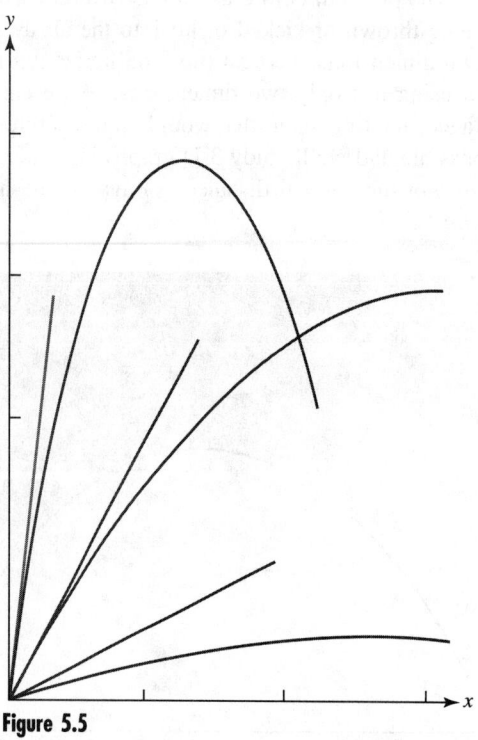

Figure 5.5

Activity 5.1
First Look at the Best-Angle Problem

Think about the problem in the Chapter Project and write answers to the following questions. You currently do not have enough information to get a correct answer to all the questions. For now, just write your best guesses. Assume that all tosses, hits, etc., begin with the same initial velocity.

1. What angle do you think will give the longest horizontal distance? Why?
2. Do you think there is only one best angle? Why?
3. What angle do you think will give the highest vertical distance? Why?
4. What angle do you think will get the ball to the ground fastest? That is, which produces the shortest time in the air (besides 0° of course)? Why?

5.1   Flight Trajectories

In this section we will study the mathematics of two-dimensional gravity-influenced motion. You will see that velocity in two dimensions has two components—vertical and horizontal—and you will learn how to analyze these components. You will use trigonometry and algebra to compute the distance a ball travels both horizontally and vertically after it has been thrown at an angle.

Looking at Trajectories

In the discussion following the Chapter Project you started thinking about the difference between a distance d vs. time t graph and a vertical distance y vs. horizontal distance x graph. It is easy to confuse them because both types have a parabolic shape. The y vs. x graph represents the actual path of the object in flight (see again Figure 5.3). This path is called the **trajectory** of the object.

In Chapter 2 you studied how gravity changes one-dimensional (up and down) motion. Gravity also affects motion in two dimensions, as we'll see next. First let's look at the role that the tangent to a curve plays in both kinds of motion.

Tangents to Curves

One-dimensional (up and down) motion. The slope of the tangent to the d vs. t curve at a particular point is the velocity at that point in time.
Two-dimensional motion. The tangent to the y vs. x curve at a particular point shows the direction of the motion at that point in space.

Vectors and the Components of Velocity

It is useful to think of velocity in two dimensions as being made up of two parts: left-right (horizontal) and up-down (vertical). These two parts of the velocity are called its **components**. The components are represented in the following way:

$\vec{v_x}$ = velocity in the horizontal direction

$\vec{v_y}$ = velocity in the vertical direction

$\vec{v}$ = velocity (sometimes called the **resultant** velocity)

Figure 5.6 shows the velocity and its components at a single point P. The following statements are true about the velocity at point P:

1. The ball is going up (the vertical velocity is positive).
2. The ball is going to the right (the horizontal velocity is positive).
3. The vertical velocity seems to be more than the horizontal velocity.

*Many texts use a boldface letter to represent a vector. Thus both **v** and $\vec{v}$ may represent the same vector.*

The symbols $\vec{v_x}$, $\vec{v_y}$, and $\vec{v}$ are used to remind us that these three velocities have direction as well as magnitude. Such quantities are called **vectors**. A vector is a quantity that has both direction and magnitude. Adding the components $\vec{v_x}$ and $\vec{v_y}$ produces the velocity $\vec{v}$. That is

$$\vec{v} = \vec{v_x} + \vec{v_y}$$

The sum of two perpendicular vectors can be represented by a right triangle or by a rectangle, as shown in Figure 5.7. The lengths of the sides of the triangle or the rectangle represent the relative magnitudes of the vectors. When we want to refer to the magnitude of the vectors we will use the symbols v_x, v_y, and v without the arrows.

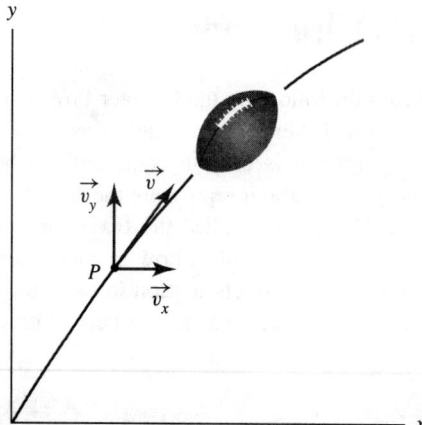

Figure 5.6

$$\vec{v}_x + \vec{v}_y = \vec{v}$$

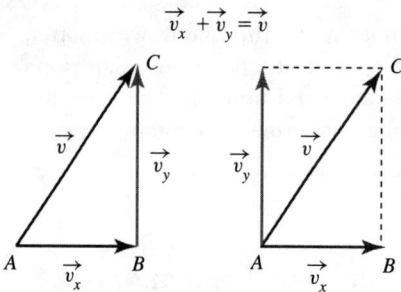

Figure 5.7

Example 5.1

Describe the motion of the ball in Figure 5.8 at points P_1, P_2, and P_3.

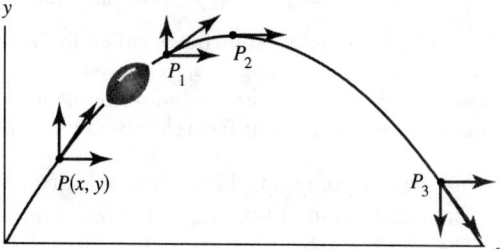

Figure 5.8

Solutions At Point P_1 the ball is still going up (the vertical velocity is positive) and to the right (the horizontal velocity is positive). At point P_1 the vertical velocity has decreased compared to when the ball was thrown. It is now a smaller number than the horizontal velocity.

At Point P_2 the ball has reached the top of its trajectory (the vertical velocity is zero) and the ball is still going to the right (the horizontal velocity is positive).

At Point P_3 the ball is now going down (the vertical velocity is negative), but the ball is still going to the right (the horizontal velocity is positive). In absolute value the vertical velocity is greater than the horizontal velocity. ■

Example 5.2

If an object is moving upward at 30 ft/sec and at an angle of 53° to the ground, what are the horizontal and vertical components of the velocity?

Solution   Using the notation of Figure 5.7 we can see that $\angle A = 53°$. Therefore for the horizontal component v_x we have

$$v_x = v \cos 53°$$
$$\approx 30 \times 0.6018 \approx 18$$

The horizontal velocity is about 18 ft/sec.

We determine the vertical component in a similar manner.

$$v_y = v \sin 53°$$
$$\approx 30 \times 0.7986 \approx 24$$

Thus the vertical velocity is about 24 ft/sec.

As a check we can use the Pythagorean theorem to determine the magnitude of the resultant velocity.

$$v = \sqrt{v_x^2 + v_y^2} \approx \sqrt{18^2 + 24^2}$$
$$= \sqrt{900} = 30$$

The magnitude of the resultant velocity of 30 ft/sec agrees with the given information.

■

Summary: Components of Velocity

If an object is moving at a speed of $|v|$ ft/sec and traveling at an angle of θ with the ground, then

$$\overrightarrow{v} = \overrightarrow{v_x} + \overrightarrow{v_y}$$

The velocity components can be computed as

$$v_x = v \cos \theta$$
$$v_y = v \sin \theta$$

and the magnitude of the resultant velocity is

$$v = \sqrt{v_x^2 + v_y^2}$$

It is important to remember that both the magnitude of the velocity v and angle θ are continually changing under the influence of gravity, as you can see in Figure 5.8.

The Influence of Gravity

In Chapter 2 you saw that vertical motion is influenced by gravity. Because of gravity the distance $d(t)$ of an object above the ground at any time t after the object is released is given by the function

$$d(t) = -\frac{1}{2}gt^2 + v_0 t + d_0$$

where

g = the force of gravity (technically this is the acceleration due to gravity)

v_0 = the initial velocity of the object

d_0 = the distance above ground when the object is released

Figure 5.9 summarizes motion in one dimension. There are no vectors in a d vs. t graph.

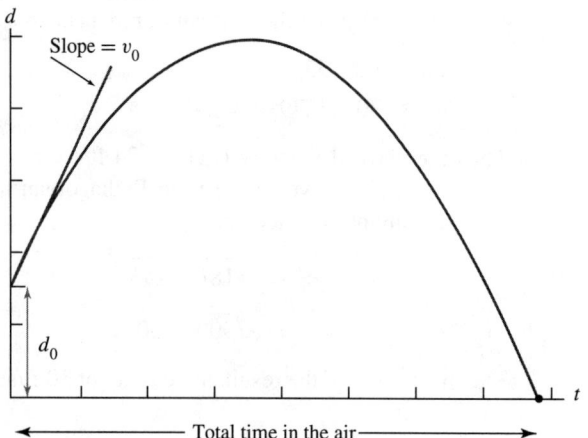

Figure 5.9

The force of gravity is different for different planets and satellites. On Earth g is approximately 9.8 m/s^2 (meters per second per second), which is how fast the velocity changes each second, or 32 ft/sec^2 (feet per second per second). The actual value of g on Earth varies from place to place. For example, at sea level near the equator it is about 9.780 327 m/s^2 (32.08730 ft/sec^2) while at the North Pole the value of g is about 9.832 186 m/s^2 (32.25744 ft/sec^2). Also there is a slight reduction in g at higher altitudes on any planet.

In two dimensions the vertical motion is affected by gravity in exactly the same way:

$$y(t) = -\frac{1}{2}gt^2 + v_{y_0} t + y_0$$

where

$y(t)$ = distance above ground at time t

v_{y_0} = initial vertical component of velocity

y_0 = height above ground at release

Figure 5.10 illustrates two-dimensional motion.

So far in our analysis of two-dimensional motion we have set $y_0 = 0$. Figure 5.3 shows that this can be done with any trajectory just by placing the origin of the coordinate system where the ball begins its flight. Therefore we have the slightly simpler form

$$y(t) = -\frac{1}{2}gt^2 + v_{y_0} t$$

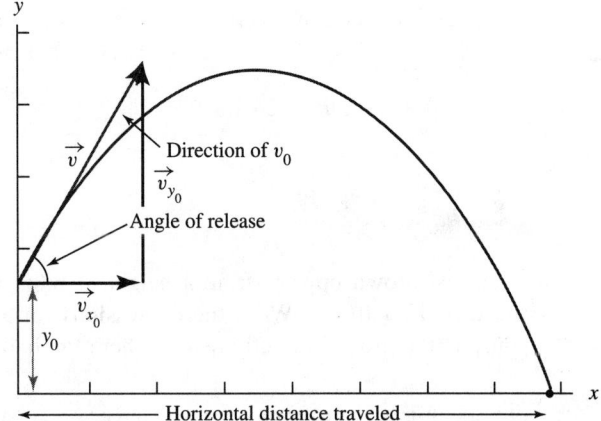

Figure 5.10

Horizontal motion is *not* affected by gravity. Gravity is a vertical force and it has no influence on horizontal motion. For example, a ball rolling on a level table slows down only because of friction, not because of gravity. Since v_x is not affected by gravity it remains constant during the flight. In Figure 5.8 all four of the vectors $\overrightarrow{v_x}$ are the same length. Because $\overrightarrow{v_x}$ is constant, the equations for $x(t)$ are identical to the equations of constant-velocity motion that you studied in Chapter 1. Therefore the horizontal distance at any time t is given by

$$x(t) = v_{x_0} t + x_0$$

If we place the origin of the coordinate system where the object begins its flight then $x_0 = 0$. This leads to the simplified form

$$x(t) = v_{x_0} t$$

Example 5.3

An object is fired upward from ground level at an angle of $57.2°$ with a speed of 28.7 m/s. Write the expressions for $x(t)$ and $y(t)$ as measured from the beginning position (that is, when $t = 0$).

Solution   Notice that the speed is given in terms of meters per second. Because of this we will use $g = 9.8$ m/s^2. Since the object is fired from ground level, at $t = 0$ we have $x_0 = 0$ and $y_0 = 0$.

For the horizontal component we have $v_{x_0} = 28.7 \cos 57.2°$. Using the formula for the horizontal component, $x(t) = v_{x_0} t + x_0$, we have

$$x(t) = (28.7 \cos 57.2°)\, t + 0$$
$$\approx 15.5 t + 0$$

The initial vertical velocity is $v_{y_0} = 28.7 \sin 57.2°$. Using the formula $y(t) = -\frac{1}{2} g t^2 + v_{y_0} t$ produces

$$y(t) = -\frac{1}{2}(9.8) t^2 + (28.7 \sin 57.2°)\, t + y_0$$

$$\approx -4.9 t^2 + 24.1 t + 0 = -4.9 t^2 + 24.1 t$$

Thus the horizontal and vertical components are, respectively,

$$x(t) \approx 15.5t$$
$$y(t) \approx -4.9t^2 + 24.1t$$

∎

Example 5.4

A ball is thrown upward from a height of 6.00 ft at an angle of 53.0° and with a speed of 73.4 ft/sec. Write the expressions for $x(t)$ and $y(t)$ as measured from a point on the ground directly below where the ball was released.

Solution For the horizontal component we have $x_0 = 0$ and $v_{x_0} = 73.4 \cos 53°$. Using the formula for the horizontal component, $x(t) = v_{x_0} t + x_0$, we have $x(t) = (73.4 \cos 53°) t \approx 44.2t$.

This ball was released from a height of 6.00 feet, so we have $y_0 = 6.0$. The initial vertical velocity is $v_{y_0} = 73.4 \sin 53° \approx 58.6$. Since the speed is given in terms of feet per second we use $g = 32$ ft/sec². Using the formula $y(t) = -\frac{1}{2}gt^2 + v_{y_0}t + y_0$ produces

$$y(t) = -\frac{1}{2}(32)t^2 + (73.4 \sin 53°) t + 6.0$$

$$\approx -16t^2 + 58.6t + 6.0$$

Thus the horizontal and vertical components are, respectively,

$$x(t) \approx 44.2t$$
$$y(t) \approx -16t^2 + 58.6t + 6.0$$

∎

Activity 5.2
Plotting the Trajectory

Plot the position of the ball in Example 5.4 at each of the times from $t = 0$ to $t = 4$ sec in intervals of 0.5 seconds.

First you need to complete Table 5.1 and then plot the points on a graph. Then connect the dots in a curve to produce the trajectory. We've filled in one pair of values and plotted that point on the graph (see Figure 5.11).

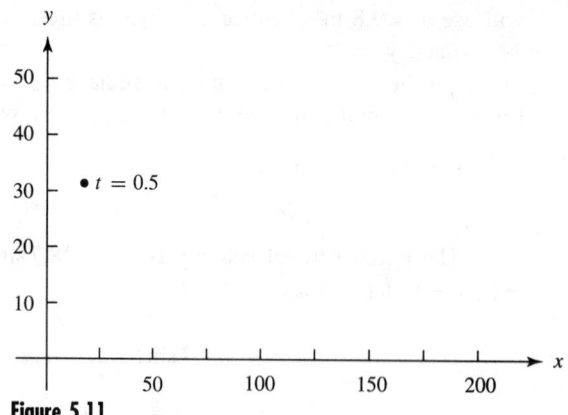

Figure 5.11

Table 5.1 Position of ball

t	$x(t) = 44.2t$	$y(t) = -16t^2 + 58.6t + 6$
0		
0.5	$44.2 \times 0.5 = 22.1$	$-16 \times (0.5)^2 + 58.6 \times 0.5 + 6 = 31.3$
1		
1.5		
2		
2.5		
3		
3.5		
4		

Earlier we said that it would be difficult to plot three variables—such as x, y, and t—on a single two-dimensional graph. Activity 5.2 shows one way to do that. We'll explore that kind of graphing in the next section of this chapter.

Example 5.5

How long will the ball in Activity 5.2 be in the air before it hits the ground?

Solution   You saw that the value of $y(4)$, which is the value of y when $t = 4$ seconds, was a negative number. This means that if the ball could continue it would be below ground level at 4 seconds. So the ball hits the ground some time between $t = 3.5$ and $t = 4$. One way to find a better estimate of the time of impact is by using algebra to solve an equation.

The ball hits the ground when the vertical distance $y(t)$ is equal to zero. Therefore we must solve the equation

$$y(t) = -16t^2 + 58.6t + 6.0 = 0$$

Using the quadratic formula we obtain

$$t = \frac{-58.6 \pm \sqrt{58.6^2 - 4 \times (-16) \times 6.0}}{2 \times (-16)}$$

$$= \frac{-58.6 \pm \sqrt{3817.96}}{-32}$$

There are now two solutions,

$$t = \frac{-58.6 - \sqrt{3817.96}}{-32} \approx 3.8 \quad \text{and} \quad t = \frac{-58.6 + \sqrt{3817.96}}{-32} \approx -0.1.$$

The first is when the ball hits the ground and is the answer we seek. The second seems to indicate that the ball hits the ground 0.1 sec before the ball was released. This of course does not make sense, so we reject it as a solution. ∎

Example 5.6

How far does the ball in Example 5.4 travel horizontally?

Solution   It is possible to estimate this distance from your graph. Go ahead and do that now before we compute a more exact number. We need to compute the horizontal distance at $t = 3.8$ sec. That is

$$x(3.8) \approx 44.2 \times 3.8 \approx 168.0$$

Therefore a ball thrown at 73.4 feet per second from a height of 6.0 feet and at an angle of 53° travels a horizontal distance of about 168 feet.   ■

One way to solve the best-angle problem is to repeat these calculations for other angles and estimate the angle that gives the greatest horizontal distance. We'll return to the best-angle problem after looking more closely at a new kind of graphing and at vectors.

Section 5.1 Exercises

1. Describe the motion of the ball in Figure 5.12 at point P_1.

2. Describe the motion of the ball in Figure 5.12 at point P_2.

3. Describe the motion of the ball in Figure 5.13 at point P_3.

4. Describe the motion of the ball in Figure 5.13 at point P_4.

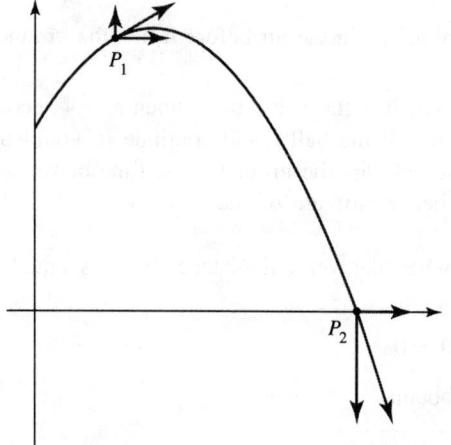

Figure 5.12

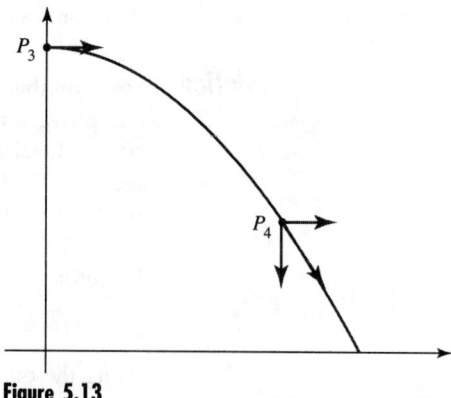

Figure 5.13

In Exercises 5–8 determine the horizontal v_x and vertical v_y components of each given vector $\vec{v}$ with magnitude $|\vec{v}|$ and direction θ.

5. $|\vec{v}| = 25.00, \theta = 35°$

6. $|\vec{v}| = 42.50, \theta = 123°$

7. $|\vec{v}| = 1.42, \theta = 312.4°$

8. $|\vec{v}| = 21.75, \theta = 222.2°$

9. If an object is moving upward at 42 ft/sec at an angle of 37.2° to the ground, what are the horizontal and vertical components of the velocity?

10. An object is moving upward at 21.9 m/s at an angle of 12.6° to the ground. What are the horizontal and vertical components of the velocity?

11. *Recreation* A golfer hits a ball off the ground with an initial velocity of 175 ft/sec and at an angle of elevation of 22.5°. Write expressions for $x(t)$ and $y(t)$ as measured from the instant the club head hits the ball.

12. *Recreation* For the golf shot in Exercise 11,
    (a) create a table, similar to Table 5.1, for the time the ball is in the air and
    (b) sketch the path of the ball while it is in the air.

13. *Recreation* For the golf shot in Exercise 11,
    (a) how long is the ball in the air and
    (b) how far does the ball travel with respect to the ground before it hits the ground?

14. *Recreation* A golf ball leaves the ground at a 25° angle with a speed of 95 ft/sec. Will it clear the top of a 30-ft tree 145 ft away?

15. *Recreation* A golfer hits a ball off the ground to the elevated green in Figure 5.14 with an initial velocity of 121 ft/sec at an angle of elevation of 50°. Write expressions for $x(t)$ and $y(t)$ as measured from the instant the club head hits the ball.

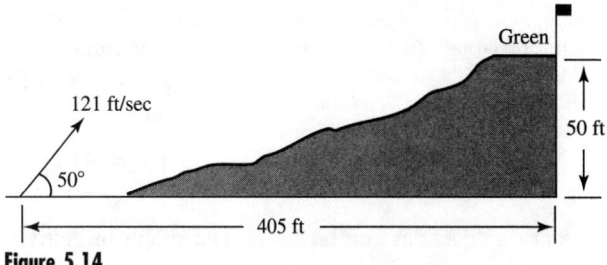

Figure 5.14

16. *Recreation* For the golf shot in Exercise 15,
    (a) create a table, similar to Table 5.1, for the time the ball is in the air, and
    (b) sketch the path of the ball while it is in the air.

17. *Recreation* For the golf shot in Exercise 15,
    (a) how long is the ball in the air and
    (b) does the ball land on the green?

18. *Navigation* An object is dropped from an airplane that is in level flight at an altitude of 12,000 feet. At the time the object is dropped the plane is flying at a speed of 165 mph. Write expressions for $x(t)$ and $y(t)$ in ft/sec as measured from the instant the object is dropped.

5.2

Parametric Equations

In Activity 5.2 (Section 5.1) you plotted a y vs. x graph but you used a third variable, t, to produce the graph. The equations for x and y in terms of t are called **parametric equations** and the third variable t is called the **parameter**.

When you plot an equation like $y = \sqrt{100 - x^2}$ you substitute values for the independent variable x and compute values of the dependent variable y. The graph is a set of points (x, y) whose values satisfy the equation $y = \sqrt{100 - x^2}$. You may remember that this equation produces a half-circle (like Figure 5.15). In this kind of graphing we say that y *is a function of* x.

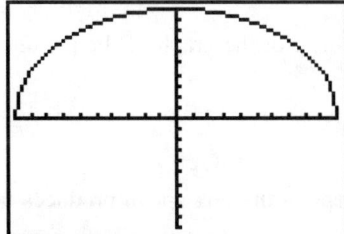

Figure 5.15

There are situations—like two-dimensional motion—in which the computations can be simplified if we introduce a third variable and describe x and y as functions of this new variable or parameter. It is normal to use the letter t to represent the

Usually you will see parametric equations written as shown here rather than with the notations $x(t)$ and $y(t)$.

parameter because many of the applications involve *time* as the independent variable. The set of values that the parameter can have is called the **domain of the parameter**.

Activity 5.2 of Section 5.1 gave you practice working with the set of parametric equations

$$\begin{cases} x = 44.2t \\ y = -16t^2 + 58.6t + 6.0 \end{cases}$$

The graph in Activity 5.2 was a parabola. In the next activities you will explore parametric equations on the calculator.

Parametric equations can be written without the left brace, as in

$$x = 44.2t$$
$$y = -16t^2 + 58.6t + 6.0$$

The brace is often used to show which equations belong together and to avoid possible confusion.

Graphing Parametric Equations

In order to graph a set of parametric equations on a calculator such as a *TI-83* you should follow the five steps in the following box.

Graphing Parametric Equations on a Calculator

1. In the MODE menu, change the graph mode from Func (Function) to Par (Parametric).
2. In the Y= menu, fill in the two parametric equations you want to plot. Use the [X,T,θ,n] key when you want to enter the parameter t, represented by the symbol T on the *TI-83*.
3. In the WINDOW menu, set the graph window settings to the values you need in order to show the graph well.* Notice that you have to set values for x, y, and t. Tstep refers to the increment in the T-values. We used the settings Tmin $= -5$, Tmax $= 5$, Tstep $= 0.5$, Xmin $= -5$, Xmax $= 5$, Xscl $= 1$, Ymin $= -10$, Ymax $= 10$, and Yscl $= 1$.
4. Press [GRAPH].
5. Repeat steps 3 and 4 until the graph looks the way you want. The calculator's Table feature also can help you figure out the best graph window settings.

Example 5.7

On your calculator, plot the graph of the parametric equations

$$x = t + 3$$
$$y = t^2 + 4$$

Solution   Following the steps in the box above produces the graph in Figure 5.16.

---

*Because the parameter t is often used for trigonometric graphs, the ZStandard (Zoom Standard) setting sets the domain of the parameter to $0° \leq t \leq 360°$ or $0 \leq t \leq 2\pi$. For nontrigonometric graphs you will need to change the domain of t and the value of Tstep.

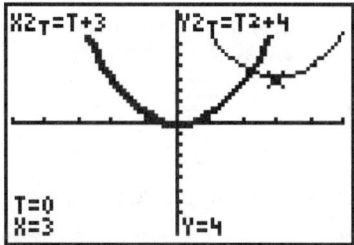

Figure 5.16

You can see in Figure 5.16 that the parabola you plotted in Example 5.7 is the same shape as the basic parabola $y = x^2$, but it is shifted. How can you predict the location of the parabola from its parametric equations? You'll find out in Calculator Lab 5.1.

Calculator Lab 5.1

Shifted Parabolas

In this calculator lab you will study the effect that a and b have on the graph of the parametric equations

$$\begin{cases} x = t + a \\ y = t^2 + b \end{cases}$$

Procedures

1. Systematically change the values of a and b in the parametric equations

$$\begin{cases} x = t + a \\ y = t^2 + b \end{cases}$$

Graph the parametric equations until you think you know how the values of a and b affect the graph of the parabola. Look at the vertex of the parabola you graphed and compare its location to the vertex of the basic parabola $y = x^2$, which is the point $(0, 0)$.

2. Write a paragraph explaining how a and b affect the graph of the parametric equations.

Example 5.8

In the second step of Calculator Lab 5.1 you wrote a paragraph explaining how a and b affect the graph of the parametric equations. Explain *why* the parametric equations in Calculator Lab 5.1 produce this effect on the graphs.

Solution   The values of a and b cause the parametric equations

$$\begin{cases} x = t + a \\ y = t^2 + b \end{cases}$$

to shift compared to the graphs of

$$\begin{cases} x = t \\ y = t^2 \end{cases}$$

We can understand the shifting by converting the parametric equations to a single y vs. x equation. First solve $x = t + a$ for t.

$$x = t + a$$
$$t = x - a$$

Next substitute $x - a$ for t in the equation for y.

$$y = t^2 + b$$
$$y = (x - a)^2 + b$$

You learned in Chapter 3 (see page 167) that $y = (x - a)^2 + b$ is the equation of the basic parabola $y = x^2$ shifted $|a|$ units to the right and $|b|$ units up. If a is negative, the shift is to the left, and if b is negative, the shift is down. For example, the parametric equations

$$\begin{cases} x = t - 2 \\ y = t^2 - 6 \end{cases}$$

produce the equation

$$y = (x + 2)^2 - 6$$

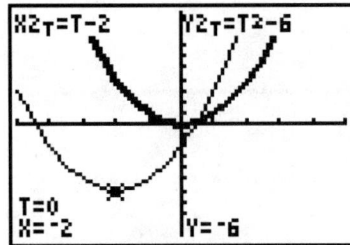

Figure 5.17

This is the graph of $y = x^2$ shifted 2 units to the left and 6 units down. Graphs of $y = x^2$ and $y = (x + 2)^2 - 6$ are shown in Figure 5.17. ■

Trigonometry, Circles, and Parametric Equations

When you were introduced to the equation of a circle you saw that it comes directly from the Pythagorean theorem and the distance formula. A version of Figure 4.43 (page 242) is reproduced in Figure 5.18. It illustrates the reasoning that produced the following equation for a circle of radius 6.

$$x^2 + y^2 = 36$$
$$y = \pm\sqrt{36 - x^2}$$

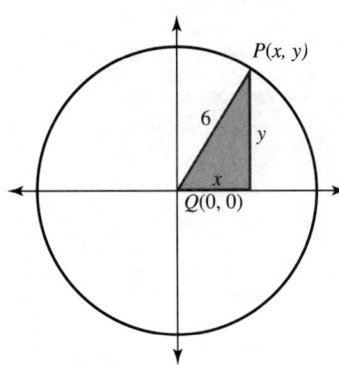

Figure 5.18

The parametric equations of a circle also come directly from that same figure. The third variable, on which x and y depend, is the angle θ. As $\angle\theta$ changes, the values of x and y change, but the point $P(x, y)$ remains on the circle. The equations that describe the dependence of x and y on $\angle\theta$ are

$$\begin{cases} x = 6\cos\theta \\ y = 6\sin\theta \end{cases}$$

Since the point (x, y) lies on a circle of radius 6, these are the parametric equations of the circle. In the next example we'll plot them.

Parametric Equations for a Circle Centered at the Origin

The parametric equations of a circle of radius $|r|$ and centered at the origin $(0, 0)$ are

$$\begin{cases} x = r\cos bt \\ y = r\sin bt \end{cases}$$

Example 5.9

Plot a circle of radius equal to 6 using parametric equations.

Solution   The two graphs are shown in Figure 5.19 (degrees) and 5.20 (radians). You can see that there is no apparent difference between them. The domain of the parameter is $0° \le t \le 360°$ in degrees or $0 \le t \le 2\pi$ in radians.

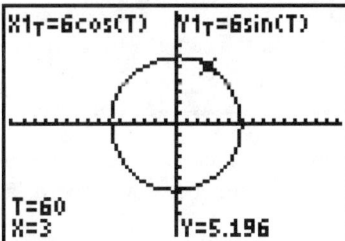 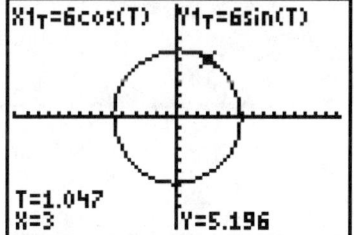

Figure 5.19
Parametric form of a circle (degrees)

Figure 5.20
Parametric form of a circle (radians)

Calculator Lab 5.2

Exploring Parametric Equations of a Circle

The three parts of this calculator lab will look at the parametric equations for a circle of radius $|r|$ and centered at the origin:

$$\begin{cases} x = r \cos bt \\ y = r \sin bt \end{cases}$$

In the first part you will explore the domain of t, the action of the calculator's Trace feature, and changing the radius. In the second part you will explore parametric equations such as

$$\begin{cases} x = a \cos t \\ y = b \sin t \end{cases} \quad \text{where } a \neq b.$$

In the last part you will look at the effect of replacing t with bt where b is a nonzero real number.

Procedures   1. Graph the parametric equations

$$\begin{cases} x = 6 \cos t \\ y = 6 \sin t \end{cases}$$

Experiment with the domain of the parameter and the equations themselves to be able to write the answers to the following questions about graphing circles with parametric equations. You might find it useful to change the graph style to the one named `path`, which allows you to see the points more clearly as they are being plotted.

Path is described on pages 3–9 of the TI-83 Graphing Calculator Guidebook.

(a) What change is there in the graph if the upper limit of t is changed to 720°?

(b) What change is there in the graph if the upper limit of t is changed to 180°?

(c) When you use the Trace feature and press the *right* arrow to move the trace cursor, why does the trace cursor move to the *left*?

(d) How can you make a circle with a different radius?

(e) What changes to the graph do you see if you enter the radius as a negative number? Why do you think that this change occurs?

2. What happens if the two parametric equations have different coefficients? For example, what happens if the equations are

$$\begin{cases} x = 4\cos t \\ y = 8\sin t \end{cases}$$

3. What happens if the coefficient of t is different from 1, as in

$$\begin{cases} x = 6\cos 2t \\ y = 6\sin 2t \end{cases}$$

For this question use the same coefficients of t in the expressions for both x and y. In a later lab you will explore the results of using different coefficients.

From what you have done in Calculator Lab 5.2 you should be able to conclude that the general set of parametric equations of a circle of radius $|r|$ and centered at the origin are

$$\begin{cases} x = r\cos bt \\ y = r\sin bt \end{cases}$$

The examples that follow will show you the importance of the discoveries you made in Calculator Lab 5.2.

Parametric Equations of Ellipses

An **ellipse** is a mathematical shape whose appearance is like a circle that has been stretched or shrunk in either the x- or the y-direction. An ellipse is shown in Figure 5.21. The **major axis** and the **minor axis** are like two different diameters of this stretched circle. The major axis is the longest segment through the center of the ellipse and the minor axis is the shortest segment through the center. The major and minor axes are perpendicular to each other. If you think of an ellipse as a stretched circle, then the center of the circle gets pulled in two directions along the major axis. These new "centers" are called the **foci** (singular **focus**) of the ellipse.

Ellipses are important in astronomy because all planets and satellites, and many comets, move in trajectories that are in the shape of ellipses.

The equation of an ellipse looks something like the equation of a circle but with a difference that accommodates the fact that it has a different "radius" in each

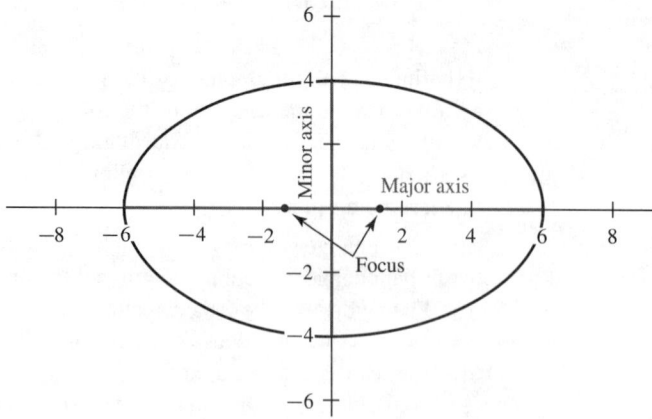

Figure 5.21

direction. An equation of the ellipse pictured in Figure 5.21 is

$$\frac{x^2}{6^2} + \frac{y^2}{4^2} = 1$$

In order to graph this ellipse on your calculator in function (Func) graphing mode, you need to solve this equation for y. Thus

$$\frac{x^2}{6^2} + \frac{y^2}{4^2} = 1$$

$$4^2 x^2 + 6^2 y^2 = 6^2 \cdot 4^2$$

$$16x^2 + 36y^2 = 576$$

$$36y^2 = 576 - 16x^2$$

$$y^2 = \frac{576 - 16x^2}{36}$$

$$y = \pm\frac{\sqrt{576 - 16x^2}}{6}$$

You can see that reaching the expression for y in terms of x can be quite a challenge and that calculating with it would be even harder. The parametric equations of this ellipse are

$$\begin{cases} x = 6\cos t \\ y = 4\sin t \end{cases}$$

These parametric equations are easier to use than the equations

$$y = \pm\frac{\sqrt{576 - 16x^2}}{6}.$$

Equations for an Ellipse Centered at the Origin

The general equation for an ellipse centered at the origin is

$$\frac{x^2}{a^2} + \frac{y^2}{b^2} = 1$$

One set of parametric equations for this ellipse are

$$\begin{cases} x = a\cos t \\ y = b\sin t \end{cases}$$

where $|2a|$ and $|2b|$ are the lengths of the major and minor axes. The major axis has length that is the larger of $|2a|$ and $|2b|$.

Plotting Planetary Trajectories

We began this chapter with the discussion of trajectories of objects under the influence of gravity. It turns out that even as you read this sentence you are moving on several enormous trajectories. You are moving at about 1000 mph as Earth spins around its center and you are racing over 65,000 mph as Earth moves in its orbit around the Sun. In addition, the Sun moves around the Milky Way galaxy at 480,000 mph.

The first trajectory model we will build is of the orbits of planets. Although technically an ellipse, Earth's orbit can be approximated with a circle. In fact the orbits of all the planets in our solar system except Pluto are very close to circular, as shown in Figure 5.22. The orbit of Pluto, which is the most uncircular orbit of the Sun's nine planets, can be modeled with the equations

$$\begin{cases} x = 4,580,000,000\cos t \\ y = 2,765,000,000\sin t \end{cases}$$

where x and y are in miles and t is in degrees.

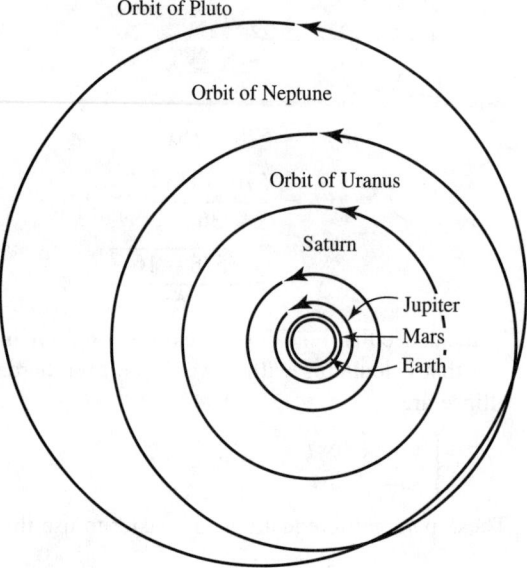

Orbit of Pluto

Orbit of Neptune

Orbit of Uranus

Saturn

Jupiter
Mars
Earth

Figure 5.22

Pluto's closest distance to the Sun is 1.8 billion miles less than its farthest distance! Comets, which often do not return within the view of people on Earth until after hundreds or thousands of years, have elliptical orbits that are very far from circular, as we shall see in Chapter 6.

Example 5.10

Make a parametric model of Earth's orbit, and plot the graph.

Solution   Earth's orbit around the Sun is an ellipse that is almost circular. We will assume it is a circle. The average distance from Earth to the Sun is 93 million miles, which equals the radius of the orbit, so the equations are

$$\begin{cases} x = 93,000,000\cos t \\ y = 93,000,000\sin t \end{cases}$$

To model this orbit on the *TI-83*, change the window settings to Xmin = −93000000, Xmax = 93000000, Xscl = 10000000, Ymin = −93000000, Ymax = 93000000, and Yscl = 10000000 and then press ⬛ZOOM⬛ ZSquare. The graph is shown in Figure 5.23.

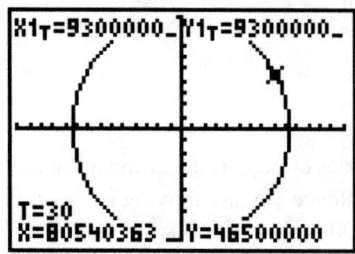

Figure 5.23

In the Exercises you will be asked to write the parametric equations for the ellipse that represents Earth's orbit around the Sun.

Degrees and Days

Remember that the functions $\cos t$ and $\sin t$ have periods of $360°$. Therefore, as the parameter t goes through $360°$, you will see the model of Earth making one complete orbit, which represents 365 days in time. This connection between degrees and days is not an accident. In fact ancient astronomers originally made the degree the size it is in order to make this connection. Like those ancient star-gazers, you will sometimes find it useful to think of a degree as corresponding approximately to a day in time. For example, in Figure 5.23 the trace cursor is on $t = 30$ and we can say that this point in the orbit is reached around the end of January.

Next we will add the orbit of the planet Mars to the graph. For convenience we will change the distance unit from miles to **astronomical units** (A.U.). One astronomical unit is equal to the average radius of Earth's orbit, or 93,000,000 miles. The radius of Mars's orbit is approximately 1.5 A.U.; its orbit also is very close to a circle.

Example 5.11

Model the orbits of Mars and Earth and plot them on the same coordinate system using A.U. as the unit of distance.

Solution Since the radius of Earth's orbit is 1 A.U. and the radius of Mars's orbit is 1.5 A.U., we have the following two sets of parametric equations.

$$\text{Earth} \quad \begin{cases} x = \cos t \\ y = \sin t \end{cases}$$

$$\text{Mars} \quad \begin{cases} x = 1.5\cos t \\ y = 1.5\sin t \end{cases}$$

The graphs are shown in Figure 5.24.

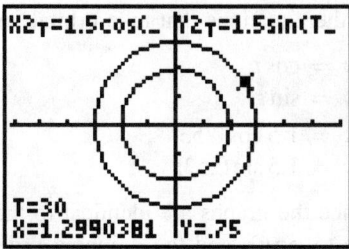

Figure 5.24

Modeling Planetary Velocity

The model you used in Example 5.11 produces a picture of the orbits of Mars and Earth. However a picture of orbits tells only part of the story. Planets travel at different speeds around the Sun and parametric equations are perfect for showing that difference.

If you divide the length of one of Earth's orbits $(2\pi \times 93{,}000{,}000$ miles) by the number of seconds in a year $(60 \times 60 \times 24 \times 365.25)$ you will get approximately 18.51 miles per second.

Mars travels at approximately 14.99 miles per second around the Sun; Earth travels at about 18.51 miles per second. Using its speed and the circumference of its orbit, you can show that Mars completes one orbit in approximately 687 Earth days. In the language of trigonometry the frequency of the Mars orbit is about $\frac{365}{687}$ times the frequency of the Earth orbit.

In Calculator Lab 5.2 you saw that the graph of

$$\begin{cases} x = \cos 2t \\ y = \sin 2t \end{cases}$$

is identical to the graph of

$$\begin{cases} x = \cos t \\ y = \sin t \end{cases}$$

but in the first graph the point moves twice as fast around the circle, completing two revolutions in the time of one revolution for the second set of equations. That is, the first set of equations represents a frequency that is twice as much as that of the second set. This is consistent with what you learned about frequency of trigonometric functions in Chapter 3.

We'll use this information to build a model of the orbits of Mars and Earth that takes into account both the sizes of their orbits and their velocities in those orbits.

Example 5.12

Model the orbits and velocities of Mars and Earth.

Solution   The general set of parametric equations of a circle are

$$\begin{cases} x = r \cos bt \\ y = r \sin bt \end{cases}$$

If we use the astronomical unit for the distance unit, then $r = 1$ for Earth and $r = 1.5$ for Mars. If we think about degrees as being like days, then it makes sense to use $b = 1$ to describe Earth's velocity because a frequency of $360°$ will mean one complete circuit in about a year. Since the frequency of Mars in its orbit is $\frac{365}{687}$ times the frequency of Earth in its orbit, then for Mars we should use $b = \frac{365}{687} \approx 0.53$.

Therefore the equations that take orbital velocity into account are

Earth $\begin{cases} x = \cos t \\ y = \sin t \end{cases}$

Mars $\begin{cases} x = 1.5 \cos 0.53t \\ y = 1.5 \sin 0.53t \end{cases}$

In final appearance the graphs are identical to the ones shown in Figure 5.24, but if you set the style to `path` and the graphing order (in the MODE menu) to `Simul` (Simultaneous) you will see a simulation of the two planets traveling around the Sun with different velocities.   ∎

Modeling Other Gravity-Influenced Trajectories

When a communications satellite or the Space Shuttle is launched it is given enough velocity to enable it to go into Earth's orbit instead of falling back to Earth. How much velocity is enough? The answer depends on the weight of the rocket and where you want the orbit to be. The Space Shuttle, for example, must be sent off at 17,500 mph in order to reach its orbit. If a projectile moves with an initial velocity that is below its **escape velocity**, it returns to Earth without going into orbit. We'll now return to motion that does not reach orbit and is therefore called **suborbital motion**.

In Activity 5.2 (page 286) you plotted a graph of the trajectory of a ball thrown in the air. Here is a review of the parametric equations of that kind of trajectory.

Summary of Suborbital Projectile Motion

The parametric equations that describe the position of an object thrown into the air with velocity less than the escape velocity for Earth are

$$\begin{cases} x(t) = v_{x_0}t + x_0 \\ y(t) = -\dfrac{1}{2}gt^2 + v_{y_0}t + y_0 \end{cases}$$

where t is the time after the toss; $x(t)$ the horizontal position; $y(t)$ the vertical position of the object, v_{x_0} the initial horizontal component of velocity (v_{x_0} is constant during flight); x_0 the horizontal distance from zero point at release; $g = 32$ if the units are feet and seconds and $g = 9.8$ if the units are meters and seconds; v_{y_0} the initial vertical component of velocity; and y_0 the height above ground at release.

Example 5.13

A ball is thrown upward at an angle of 50° from a height of 30 ft with a velocity of 92 ft/sec.

(a) Write the parametric equations of the position of the ball at any time t.
(b) Plot the trajectory on your calculator.
(c) Estimate to one decimal place the time when the ball reaches the ground.
(d) Estimate to the nearest foot the maximum height the ball reaches.
(e) Estimate to the nearest foot the horizontal distance the ball travels.

Solutions   We are given that $y_0 = 30$ and we can set $x_0 = 0$. Since $v_0 = 92$, we have $v_{x_0} = 92\cos 50° \approx 59$ ft/sec and $v_{y_0} = 92\sin 50° \approx 70$ ft/sec. Since y_0 and v_0 are both in terms of feet, we use $g = 32$.

(a) Using the above information the equations are

$$\begin{cases} x = 59t \\ y = -16t^2 + 70t + 30 \end{cases}$$

(b) The trajectory is shown in Figure 5.25.
(c) Using the Trace feature you can see that the ball hits the ground between $t = 4.5$ and $t = 5$ sec. To get a more precise estimate you can use Zoom or Table. We used Table with ΔTbl $= 0.1$ and got an estimate of $t = 4.8$ sec when $y = 0$, as shown in Figure 5.26.

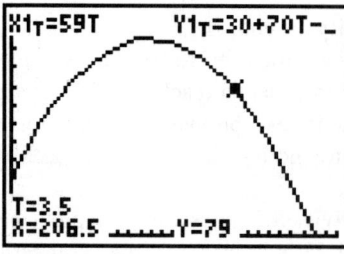

Figure 5.25   **Figure 5.26**

(d) Using the same Table settings we found that the maximum height was 107 ft.

(e) At the time it hits the ground (when $t = 4.8$) the ball has traveled $x(4.8)$ feet horizontally. Since $x(4.8) = 59 \times 4.8 \approx 283$, the ball lands 283 feet away from the point where it was thrown.   ∎

The next example uses the same ideas in a different setting.

Example 5.14   Application

A plane is attempting to drop a package of food near a campsite in the wilderness. The plane is moving horizontally in level flight at 100 mph and at a height of 1000 ft.

(a) Write the parametric equations that model the positions of the plane and the package.

(b) Plot the trajectories on your calculator.

(c) Estimate to the nearest tenth of a second the time when the food package reaches the ground.

(d) Estimate to the nearest foot the horizontal distance from where the food is dropped to where it lands.

Solutions   In this example $v_{y_0} = 0$ because the package is being dropped, not thrown. However $v_{x_0} = 100 \text{ mph} = \dfrac{100 \times 5280}{60 \times 60} \approx 147$ ft/sec. Finally, $x_0 = 0$ ft and $y_0 = 1000$ ft.

(a) The equations for the position of the package are

$$\begin{cases} x = 147t \\ y = -16t^2 + 1000 \end{cases}$$

The equations for the position of the plane are

$$\begin{cases} x = 147t \\ y = 1000 \end{cases}$$

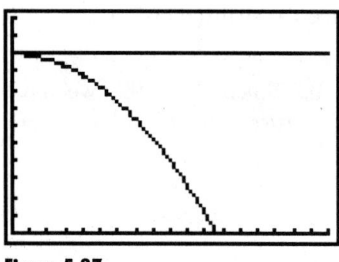

Figure 5.27

(b) The two trajectories are plotted in Figure 5.27 with the settings Tmin = 0, Tmax = 12, Tstep = 0.1, Xmin = 0, Xmax = 1800, Xscl = 100, Ymin = 0, Ymax = 1200, and Yscl = 100.

(c) Solving $y = -16t^2 + 1000 = 0$ we see that the package reaches the ground at about $t = 7.9$ sec.

(d) The horizontal distance the food package travels before it hits the ground is approximately $147 \times 7.9 \approx 1161$ ft.   ∎

If the plane had not been flying horizontally, then v_{y_0} would not be zero. We will discuss this situation in the next section. The next calculator lab gives you an opportunity to use your skills to drop the package on a target.

Calculator Lab 5.3

Landing on Target

In a contest, small planes attempt to drop a ball onto a target that is 20 feet in diameter. The planes can fly between 50 mph and 250 mph and at altitudes between 500 ft and 2500 ft. The rules also state that the drop must be made when the plane is directly above a point on the ground that is 750 ft from the center of the target. Your job is to find a combination of velocity and altitude that gets the ball onto the target (that is, within ten feet of the center of the target). The winner is the person who gets closest to the center of the target. Remember that the speed for the plane must be between 50 and 250 mph and that your parametric equations must use feet and seconds. The conversion between mph and ft/sec is

$$1 \text{ mph} \approx 1.467 \text{ ft/sec}$$
$$60 \text{ mph} = 88 \text{ ft/sec}$$

Procedures

1. Set up a coordinate system as shown in Figure 5.28.

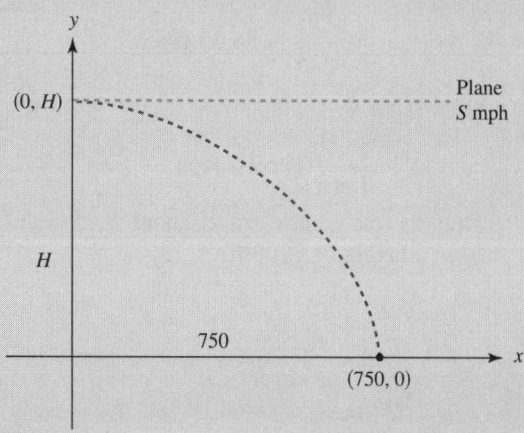

Figure 5.28

2. Write parametric equations for the trajectory of the falling ball, using S (the speed of the plane) and H (the height of the plane).
3. Experiment with values of S and H until you have a combination that makes the ball land within ten feet of the target. In mathematical terms the x-intercept of the trajectory must be between 740 and 760. Don't forget to convert S to feet per second for the experiments, and remember that there are limitations on the values of S and H.
4. Report a combination of S mph and H ft that produces your closest approach to the target.
5. Compare your best combinations with those of your classmates.

Example 5.15   Application

For the contest described in Calculator Lab 5.3, use algebra to find the plane's required speed in miles per hour if the plane is traveling at an altitude of 1200 ft. Report the answer to three significant figures.

Solution   Let S be the speed of the plane. Then the parametric equations for the dropped ball are

$$\begin{cases} x = St \\ y = -16t^2 + 1200 \end{cases}$$

We want to find the value of S that makes $x = 750$ when $y = 0$. One way to approach this is first to find the value of t that makes $y = 0$ and then use that time to solve for S. Therefore the first job is to solve the equation

$$-16t^2 + 1200 = 0$$

which is the same as

$$16t^2 = 1200$$

$$\text{or} \quad t = \sqrt{\frac{1200}{16}} = \sqrt{75} \text{ sec}$$

Here the calculations are done to four significant figures and the answer is reported to three significant figures. Note that we waited to compute numerical answers until the end. In general, answers will be more accurate if the calculations are done later in the problem.

We want x to be 750 ft when $t = \sqrt{75}$. Since $x = St$ we have the equation

$$750 = S \times \sqrt{75}$$

$$S = \frac{750}{\sqrt{75}} \approx 86.60 \text{ ft/sec}$$

Finally, we must convert 86.60 ft/sec to mph.

$$S \approx \frac{86.60}{1.467} \approx 59.0 \text{ mph}$$

Thus the plane must travel about 59.0 mph in order to have the ball hit the target from a height of 1200 ft. ∎

Section 5.2 Exercises

 In Exercises 1–4 use your calculator to sketch the graph of the given set of parametric equations for the indicated values of t.

1. $\begin{cases} x = t - 2 \\ y = t^2 - 2 \end{cases}$ where $-6 \le t \le 6$

2. $\begin{cases} x = t^2 - 5 \\ y = t^3 + 1 \end{cases}$ where $-4 \le t \le 4$

3. $\begin{cases} x = (t + 1)^2 \\ y = (t - 1)^2 \end{cases}$ where $-4 \le t \le 4$

4. $\begin{cases} x = t^2 - 2t - 5 \\ y = t^2 - 4t - 8 \end{cases}$ where $-6 \le t \le 6$

5. What are the parametric equations of a circle of radius 4 and centered at the origin?

6. **(a)** Graph the parametric equations

$$\begin{cases} x = -5 \sin t \\ y = 5 \cos t \end{cases}$$

(b) Describe the graph.

(c) How does the graph in (a) differ from the graph of

$$\begin{cases} x = 5 \sin t \\ y = 5 \cos t \end{cases}$$

7. The sets of parametric equations

$$\begin{cases} x = 5\cos t \\ y = 5\sin t \end{cases}$$

and $\quad \begin{cases} x = -4\cos t \\ y = -4\sin t \end{cases}$

both produce the graph of a circle. (The radii of the two circles are different so you can see how each circle is graphed.) Explain how the graphs are alike and how they are different.

8. The sets of parametric equations

$$\begin{cases} x = 5\cos t \\ y = 5\sin t \end{cases}$$

and $\quad \begin{cases} x = 4\sin t \\ y = 4\cos t \end{cases}$

both produce the graph of a circle. (The radii of the two circles are different so you can see how each circle is graphed.) Explain how the graphs are alike and how they are different.

In Exercises 9–12, (a) graph the ellipse described by the given parametric equations; (b) determine the x- and y-intercepts; (c) determine the lengths of the major and minor axes; and (d) describe any relationships you notice between the parametric equations and your answers to parts (a)–(c).

9. $\begin{cases} x = 5\cos t \\ y = 2\sin t \end{cases}$

10. $\begin{cases} x = 3\sin t \\ y = 5\cos t \end{cases}$

11. $\begin{cases} x = -4\sin t \\ y = \cos t \end{cases}$

12. $\begin{cases} x = 8\cos t \\ y = 2 - 3\sin t \end{cases}$

13. *Astronomy* The orbit of Venus has a radius of 0.72 A.U. and the frequency of Venus in its orbit is 1.63 times the frequency of Earth in its orbit. Write a set of parametric equations that models the orbits of Earth and Venus.

14. *Astronomy* The orbit of Jupiter has a radius of 5.20 A.U. and the frequency of Jupiter in its orbit is 0.084 times the frequency of Earth in its orbit.

(a) Write a set of parametric equations that models the orbits of Earth and Jupiter.

(b) What is Jupiter's average distance from the Sun?

(c) How many Earth days does it take for Jupiter to orbit the Sun?

15. *Automotive technology* The motion of a piece of gravel leaving a (rear) wheel at an angle α with speed V in ft/sec can be described by

$$\begin{cases} x = (V\cos\alpha)t \\ y = (V\sin\alpha)t - \frac{1}{2}gt^2 \end{cases}$$

where $g \approx 32$ ft/sec^2.

(a) Assume that a car is traveling 30 mph (44 ft/sec) and that three pieces of gravel leave its rear tire, one at $\alpha = 30°$, one at $\alpha = 45°$, and one at $\alpha = 50°$. Graph the path of each of these three pieces of gravel on the same set of axes.

(b) Rewrite these parametric equations as one equation in rectangular form.

(c) Use your answer to (b) to determine how far the gravel travels before hitting the road.

16. *Automotive technology* An automobile tire is driven over a nail. As a result the nail is embedded in the tire with only its head showing on the outside surface of the tire. As the tire moves down the highway the nail follows a path known as a **cycloid**, given by the parametric equations

$$\begin{cases} x = a(t - \sin t) \\ y = a(1 - \cos t) \end{cases}$$

where a is the radius of the tire. The tire has a radius of 0.9 feet and t is measured in seconds.

(a) Sketch the graph of the nail's path from the time it becomes stuck in the tire (when $t = 0$) until it hits the pavement three more times.

(b) What are the amplitude, period, and frequency of this periodic function?

17. *Sports technology* A baseball leaves a major-league pitcher's hand traveling horizontally with initial speed 140 ft/sec and initial height 7′6″. (Ignore wind resistance.)

(a) Write a set of parametric equations that describes the path of the ball.

(b) Graph your parametric equations from the instant the ball leaves the pitcher's hand until it hits the ground.

(c) If the ball is released 58.0 ft from home plate, how long does it take for the ball to cross the plate?

(d) If the strike zone is approximately from 1.5 to 4.0 ft above the ground, is this pitch a strike?

18. *Sports technology* A baseball leaves a major-league pitcher's hand traveling horizontally with initial speed 100 ft/sec and initial height 7′6″. (Ignore wind resistance.)

(a) Write a set of parametric equations that describes the path of the ball.

(b) Graph your parametric equations from the instant the ball leaves the pitcher's hand until it hits the ground.

(c) If the ball is released 58.0 ft from home plate, how long does it take for the ball to cross the plate?

(d) If the strike zone is approximately from 1.5 to 4.0 ft above the ground, is this pitch a strike?

19. *Navigation* A plane is attempting to drop a package of food near a campsite in the wilderness. The plane is in a 150.0-mph level flight. When the plane reaches a point that is 2500 ft above the ground the package is dropped. (Ignore the effects of air resistance.)

(a) Write the parametric equations that model the positions of the plane and the package.

(b) Plot the trajectories on your calculator.

(c) Estimate to the nearest tenth of a second the time when the food package reaches the ground.

(d) Estimate to the nearest foot the horizontal distance from where the food is dropped to where it lands.

5.3   Vectors

In this section we return to the study of vectors, including vectors that are not at right angles to one another.

You have seen that you can think of a velocity vector $\vec{v}$ in two dimensions as being made up of two component vectors: $\vec{v_x}$, the velocity in the *x*-direction, and $\vec{v_y}$, the velocity in the *y*-direction. In mathematical terms we say that

$$\vec{v_x} + \vec{v_y} = \vec{v}$$

You saw in Figure 5.7 (page 282) that the sum of two perpendicular vectors can be represented with a right triangle or a rectangle. However there are many applications of vectors in which the vectors are not perpendicular. What then? The following example shows you how to add vectors that are not perpendicular.

Vectors in Navigation

One science that makes great use of vectors is navigation. The following examples apply to the navigation of airplanes and ships. When discussing navigation we need to adjust the labeling of the coordinate system. In navigation due north is either 0° or 360°, east is 90°, south 180°, and west 270°, as shown in Figure 5.29. This is a clockwise rotation starting from the positive *y*-axis rather than the counterclockwise rotation starting from the positive *x*-axis that we have been using.

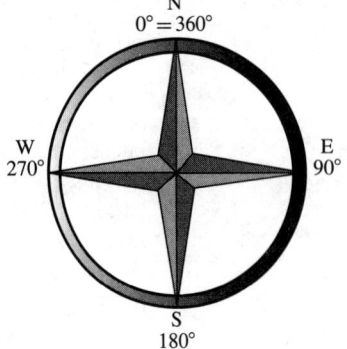

Figure 5.29

Example 5.16   *Application*

A ship heading 20 mph due north is being blown off course by a 30-mph wind coming from 315° (the northwest), as shown in Figure 5.30. Construct a vector diagram to show this situation.

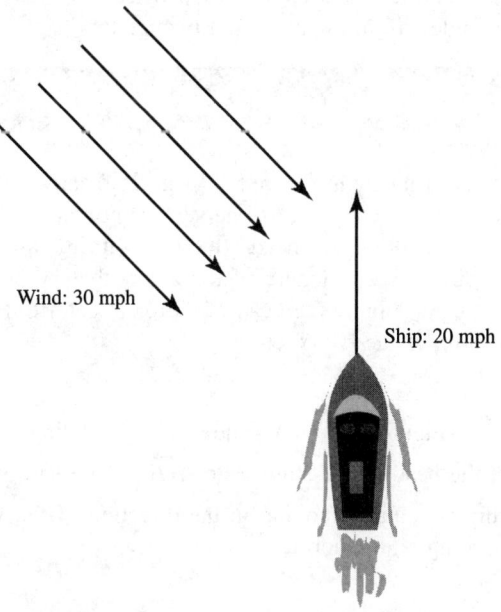

Wind: 30 mph

Ship: 20 mph

Figure 5.30

Solution   Vectors are useful in modeling a problem like this. The actual speed and direction of the ship is represented by the vector $\vec{v}$, which is the sum of the velocity due to the ship's motors and heading and the velocity due to the wind.

$$\vec{v} = \overrightarrow{v_{\text{ship}}} + \overrightarrow{v_{\text{wind}}}$$

The two components are shown in Figure 5.31. Since a 30-mph wind is 1.5 times the ship's 20-mph velocity, the wind vector is 1.5 times longer than the ship's motion vector. The wind is drawn coming from the compass direction 315°.

Ship heading north   Wind from the northwest

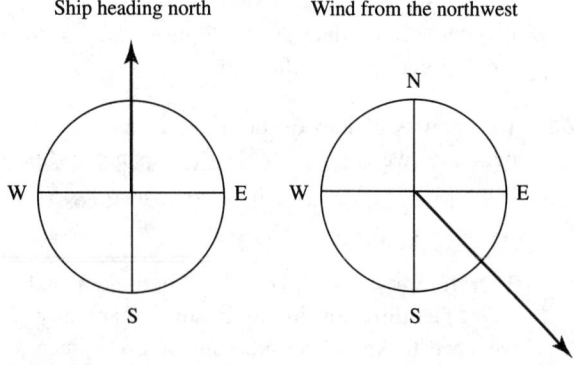

Figure 5.31

The actual direction and velocity of the ship is the resultant of these two vectors. Next we will see how we can add vectors.

Modeling Vector Sums with Triangles

In Figure 5.7 a triangle is shown as one way to model vector components. This triangle will allow us to add two vectors.

Addition of Vectors (Triangle Method)

The triangle in Figure 5.7 (page 282) shows one way to think about the addition of two vectors. Slide one vector until its "tail" (the end without the arrowhead) is touching the "head" (the end with the arrowhead) of the other vector. The direction and length of the arrow that is moved must remain the same. The vector joining the head of the first vector to the tail of the second vector is the sum of the two vectors.

Figure 5.32 shows the result of "sliding" the wind vector $\overrightarrow{BC}$ so that its tail is at the head of the ship vector $\overrightarrow{AB}$. Make sure you understand why $\angle B = 45°$. The ship is actually moving in the direction of the vector $\overrightarrow{AC}$ even though its compass heading is due north.

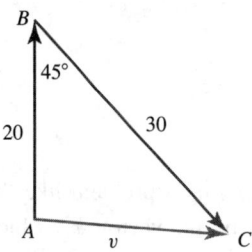

Figure 5.32

Now that we can see what direction the ship is moving, it is natural to inquire how fast it is moving and precisely in what direction.

Example 5.17

Use the vector diagram in Figure 5.32 to compute how fast the ship is actually moving and in what direction.

Solution   We know two sides of the triangle and one angle. Because the angle is between the two sides we can use the law of cosines to find the value of v.

$$v^2 = 20^2 + 30^2 - 2 \times 20 \times 30 \times \cos 45°$$
$$\approx 451.47$$

Therefore $v = \sqrt{451.47} \approx 21.2$, and the actual velocity of the ship is about 21 mph.

The direction in navigation is generally given in terms of degrees. Therefore we need to know the measure of $\angle A$. Since we know three sides and one angle, we could use either the law of sines or the law of cosines. However, as you saw in Activity 4.2 of Chapter 4, there are often two possible answers when you are computing an angle using the law of sines, because there are always two angles between 0° and 180° that produce the same value of the sine function. Therefore we'll use the law of cosines.

$$30^2 = 20^2 + 451.47 - 2 \times 20 \times \sqrt{451.47} \times \cos A$$

$$900 = 851.47 - 850 \cos A$$

$$\cos A = \frac{900 - 851.47}{-850} \approx -0.057$$

Therefore $\angle A \approx \cos^{-1}(-0.057) \approx 93°$ and the ship's heading is 93°. In navigation one might report this answer as either N93°E or S87°E. ■

When vectors are used to model straight-line constant velocities as in Example 5.17, the triangle vector diagram in Figure 5.32 has another use, as we'll see in the next example.

Example 5.18

After two hours of sailing under the conditions described in Example 5.16, what is the location of the ship?

Solution   The vector diagram in Figure 5.32 answers questions like this directly. Because the velocities are assumed to be constant, the formula for distance, $d = r \times t$, gives the distance traveled in t hours at r mph. The numbers in Figure 5.32 show actual distances covered after one hour. That is, if the ship starts at point A, then one hour later it is at point B, if there is no wind. Point C is the location to which it is forced by the wind after one hour.

To locate the ship after two hours, assuming that the location of point A is known, a navigator would draw a diagram like Figure 5.33, use the scale on the map, and lay the triangle over the map with point A placed on the known location of two hours before. Point C marks the location of the ship two hours after it leaves point A. Since the velocity of the ship is about 21 mph, this location is about 42 miles from where the ship started.

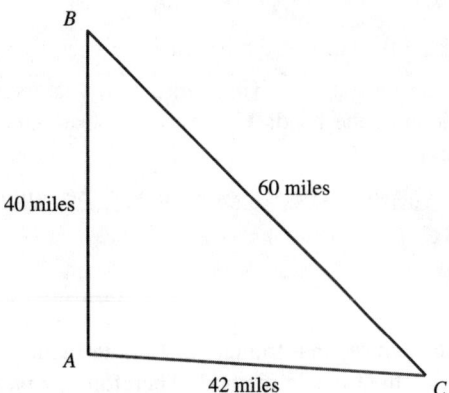

Figure 5.33

■

Modeling Vector Sums with Parallelograms

In Figure 5.7 (page 282) a rectangle is shown as one way to visualize addition of perpendicular vectors. If the components are not perpendicular a parallelogram is used, as shown in Figure 5.36. To make it easier to talk about adding vectors by using a parallelogram, we need to introduce the term **equal vectors**.

> **Definition: Equal Vectors**
> Two vectors that have the same direction and magnitude are said to be **equal vectors**.

Example 5.19

In Figure 5.34, $\vec{a} = \vec{b}$ but $\vec{a} \neq \vec{c}$ because they have different directions; $\vec{a} \neq \vec{d}$ because they have different magnitudes.

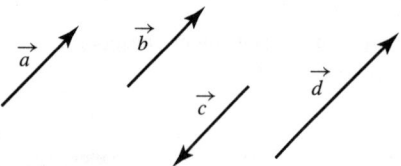

Figure 5.34

We now use the idea of equal vectors to explain a second method for adding vectors—the parallelogram method.

Adding Vectors (Parallelogram Method)

Follow these steps to use a parallelogram to show the vector sum of two vectors (here $\overrightarrow{AB}$ and $\overrightarrow{PQ}$, as in Figure 5.35).

1. From the tail and the head of one vector, say $\overrightarrow{AB}$, draw vectors equal to the other vector, $\overrightarrow{PQ}$. (In Figure 5.36 vectors $\overrightarrow{AD}$ and $\overrightarrow{BC}$ are equal to $\overrightarrow{PQ}$.)
2. Connect the heads C and D of these vectors. This should form parallelogram $ABCD$.
3. Vector $\overrightarrow{AC}$ represents the sum of the original vectors. Thus $\overrightarrow{AB} + \overrightarrow{PQ} = \overrightarrow{AC}$.

You can see that triangle $\triangle ABC$ in Figure 5.36 is the triangle that could result if you used the triangle method. Therefore the parallelogram can be used to compute angles and lengths in the same way that the triangle was used in Example 5.17.

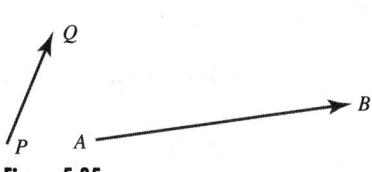

Figure 5.35   **Figure 5.36**

Example 5.20

Vector $\overrightarrow{A}$ joins the points $P(0,0)$ and $R(-6,7)$ on a coordinate system. Vector $\overrightarrow{B}$ joins the points $P(0,0)$ and $Q(7,1)$. Thus $\overrightarrow{A} = \overrightarrow{PR}$ and $\overrightarrow{B} = \overrightarrow{PQ}$.

(a) Draw on a sheet of graph paper the parallelogram that shows the vectors $\overrightarrow{A}$, $\overrightarrow{B}$, and their sum $\overrightarrow{C}$.

(b) What are the coordinates of the head of $\overrightarrow{C}$?

(c) Compute the magnitude of $\overrightarrow{C}$.

Solutions (a) Begin by drawing vectors $\overrightarrow{A} = \overrightarrow{PR}$ and $\overrightarrow{B} = \overrightarrow{PQ}$. Next draw $\overrightarrow{RS}$. Because parallel lines have the same slope, the best way to draw vector $\overrightarrow{RS}$ is to count spaces vertically and horizontally (rise and run) from point R. Where you stop will be point S. Connect points S and Q. Then $PRSQ$ should be a parallelogram, as in Figure 5.37, and $\overrightarrow{C} = \overrightarrow{PS}$.

(b) The point S is the head of $\overrightarrow{C}$. To get to S from R we emulate going from P to Q. Add 7 to the x-coordinate and 1 to the y-coordinate of R. This gives the coordinates of S as $(1, 8)$.

(c) The length of $\overline{PS}$ is the magnitude of the sum. From the distance formula
$$PS = |\overrightarrow{C}| = \sqrt{65} \approx 8.1.$$

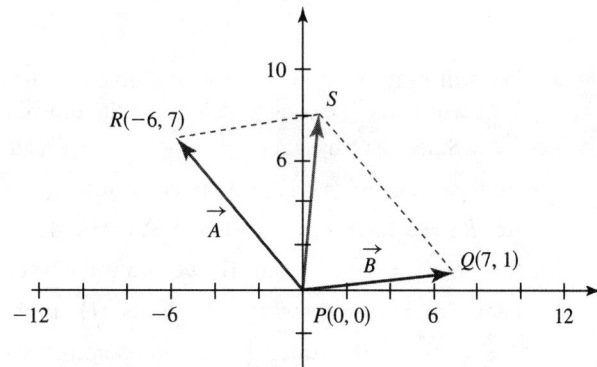

Figure 5.37

In Example 5.20 we computed the magnitude of $\overrightarrow{C} = \overrightarrow{PS}$. But a vector consists of magnitude and direction. How do we determine the direction of $\overrightarrow{PS}$? Mathematically, if the direction that $\overrightarrow{PS}$ makes with the positive x-axis is θ, then $\tan\theta$ is the direction of $\overrightarrow{PS}$.

Example 5.21

Determine the direction of $\overrightarrow{C} = \overrightarrow{PS}$ from Example 5.20.

Solution In Example 5.20, $m_{\overrightarrow{PS}} = \dfrac{8-0}{1-0} = 8$ and $\theta = \tan^{-1}8 \approx 82.9°$. Since $\overrightarrow{PS}$ is in the first quadrant, $\overrightarrow{PS}$ has magnitude $\sqrt{65}$ and direction $82.9°$. ∎

Example 5.22

As shown in Figure 5.38, vector $\overrightarrow{A}$ joins the points $P(3, 2)$ and $Q(-4, -1)$. Vector $\overrightarrow{B}$ joins the points $R(-1, 5)$ and $S(-2, 1)$. Thus $\overrightarrow{A} = \overrightarrow{PQ}$ and $\overrightarrow{B} = \overrightarrow{RS}$. If $\overrightarrow{C} = \overrightarrow{A} + \overrightarrow{B}$, compute the magnitude and direction of $\overrightarrow{C}$.

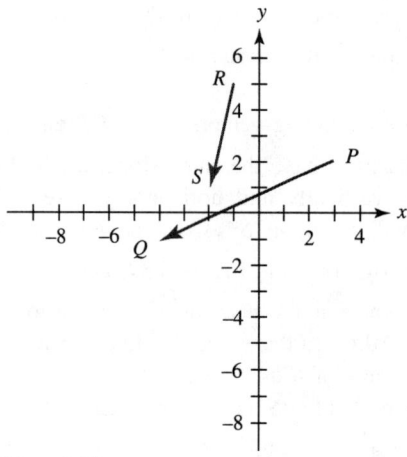

Figure 5.38

Solutions We will draw vectors equal to the given vectors but with their tails at the origin. This will make it easier to determine the direction of their sum.

Since $\overrightarrow{PQ}$ has $\Delta x = -4 - 3 = -7$ and $\Delta y = -1 - 2 = -3$, the vector from $O(0, 0)$ to $W(-7, -3)$ is equal to $\overrightarrow{PQ}$, as shown in Figure 5.39. Similarly, for $\overrightarrow{RS}$ we have $\Delta x = -1$ and $\Delta y = -4$. If X has the coordinates $(-1, -4)$, then $\overrightarrow{OX} = \overrightarrow{RS}$. From W we draw the vector $\overrightarrow{WY} = \overrightarrow{OX}$ and complete the parallelogram by drawing $\overline{YX}$. Thus $\overrightarrow{OY}$ is the desired vector sum $\overrightarrow{C}$. That is, $\overrightarrow{OY} = \overrightarrow{A} + \overrightarrow{B}$. Since Y has the coordinates $(-8, -7)$, the magnitude of $\overrightarrow{OY}$ is $\sqrt{(-8)^2 + (-7)^2} = \sqrt{113} \approx 10.63$.

The slope of $\overrightarrow{OY}$ is $m_{\overrightarrow{OY}} = \dfrac{-7}{-8} = \dfrac{7}{8}$ and $\theta = \tan^{-1}\left(\dfrac{7}{8}\right) \approx 41.2°$. But $\overrightarrow{OY}$ is in the third quadrant, so the direction of $\overrightarrow{OY}$ is $180° + 41.2° = 221.2°$. ∎

In the next example a parallelogram is used to solve a problem in which the sum of two vectors is known.

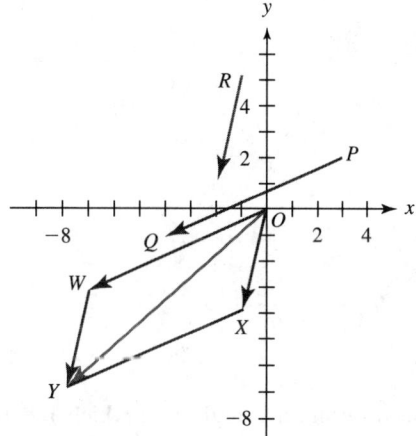

Figure 5.39

Example 5.23   Application

Jeff's small outboard motorboat can travel at 20 mph in still water. He wants to cross a river that flows from west to east, but the current is flowing at 10 mph. In order to go straight north across from point E to point R (Figure 5.40), (a) what direction should he point his boat and (b) what speed will he be able to make as he goes across the river?

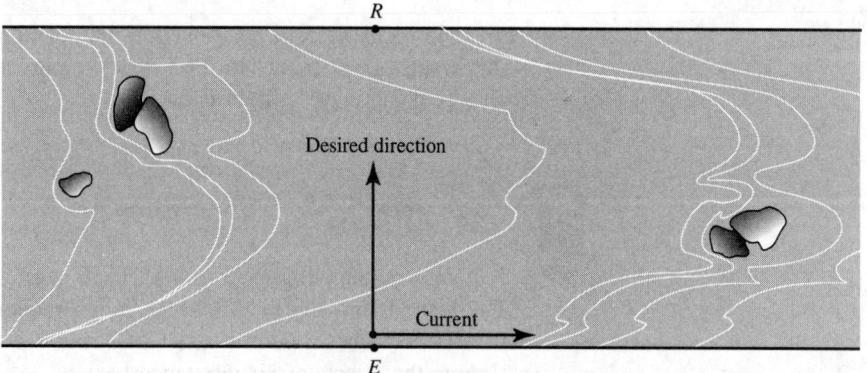

Figure 5.40

Solutions   We have information about three vectors. We are given the direction 90° (east) and magnitude (10 mph) of the vector that describes the current of the river. We are also given the magnitude (20 mph) but not the direction of the vector that describes the boat, and the direction (straight across the river, 90° to the current) but not the magnitude of the sum of the boat and current vectors.

Figure 5.41 shows all the important information within the parallelogram that shows all these vectors where $\overrightarrow{ES}$ represents the current, $\overrightarrow{ER}$ the direction Jeff wants to go, and $\overrightarrow{EQ}$ the direction he should head in order to go from E to F.

The vector sum is

$$\overrightarrow{EQ} \text{ (boat)} + \overrightarrow{ES} \text{ (current)} = \overrightarrow{ER} \text{ (actual speed and direction)}$$

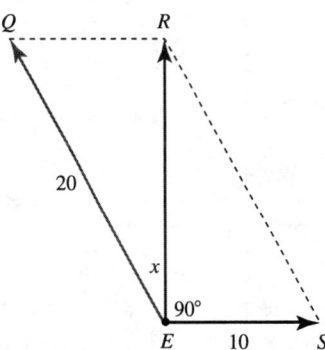

Figure 5.41

(a) We need to find the angle x at which Jeff must point his boat. The rest of the problem is trigonometry. Since the magnitude of $\overrightarrow{ES}$ is 10, the magnitude of $\overrightarrow{QR}$ is 10 also. We can find the angle x from trigonometry:

$$\sin x = \frac{10}{20} = 0.5$$

$$x = \sin^{-1}(0.5) = 30°$$

Therefore Jeff must point the boat in a direction 30° to the left of his destination and trust the current, his boat, and the mathematics to get him to his destination.

(b) His speed will be equal to the length of $\overrightarrow{PR}$, which is

$$PR(\text{actual speed}) = 20\cos 30° \approx 17 \text{ mph}$$ ■

In Example 5.14 we looked at the position of a package dropped from a plane that was in level flight. In the next example we will look at how the situation can change if the plane's flight is not level.

Example 5.24   Application

A plane is attempting to drop a package of food near a campsite in the wilderness. The plane is moving at 150.0 mph in a direction 12.5° below the horizontal. When the plane reaches a point that is 2500 ft above the ground, the package is dropped. (Ignore the effects of air resistance.)

(a) Write the parametric equations that model the position of the plane and the package.
(b) Plot the trajectories on your calculator.
(c) Estimate to the nearest tenth of a second the time when the food package reaches the ground.
(d) Estimate to the nearest foot the horizontal distance from where the food is dropped to where it lands.

Solutions   In this example the initial horizontal velocity v_{x_0} and the initial vertical velocity v_{y_0} of the package are determined by the components of the plane's velocity at the time the package is dropped. Now the plane's velocity v is 150.0 mph $= \dfrac{150 \times 5280}{60 \times 60} \approx$ 220 ft/sec.

Since the airplane is flying 12.5° *below* the horizontal, we will say it is flying vertically at $-12.5°$. Thus in ft/sec we have $v_{x_0} = 220\cos(-12.5°) \approx 214.8$ and $v_{y_0} = 220\sin(-12.5°) \approx -47.6$. Notice that v_{y_0} is negative since the plane is lowering its altitude.

(a) The equations for the position of the package are

$$\begin{cases} x = 214.8t \\ y = -16t^2 - 47.6t + 2500 \end{cases}$$

The equations for the position of the plane are

$$\begin{cases} x = 214.8t \\ y - -47.6t + 2500 \end{cases}$$

(b) With the settings Tmin = 0, Tmax = 12, Tstep = 0.1, Xmin = 0, Xmax = 2500, Xscl = 500, Ymin = 0, Ymax = 2500, and Yscl = 500, the two trajectories are plotted as shown in Figure 5.42.

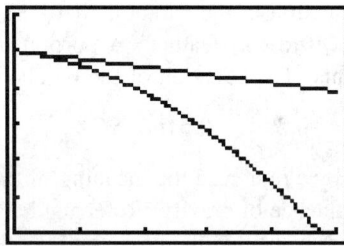

Figure 5.42

(c) Solving $y = -16t^2 - 47.6t + 2500 = 0$, we see that the package reaches the ground at about $t = 11.1$ sec.

(d) The horizontal distance the food package travels before it hits the ground is approximately $214.8 \times 11.1 \approx 2384$ ft.   ∎

Vectors and Gravity: Motion on a Ramp

Exercises 5 and 6 of Section 2.5 (page 103) described a way to use the Calculator-Based Laboratory (CBL) to study the motion of a ball as it rolls down a ramp. Figure 5.43 shows a way to set up that experiment.

Motion detector

Figure 5.43

The ramp is useful in an experiment with gravity-influenced motion because it slows down the ball and allows us to observe more data points. The ramp, however, exerts a frictional force on the rolling ball. The data can sometimes be confusing because gravity is not the only force acting on the rolling object.

Figure 5.44 shows our graph from an experiment of rolling a basketball down a ramp. The units of the graph are seconds and feet. You can see that the data includes measurements recorded before and after the ball begins rolling. In Section 2.5 you learned how to transform data like this to eliminate the unwanted points and to make the data begin at $t = 0$. We did that and Figure 5.45 shows the result.

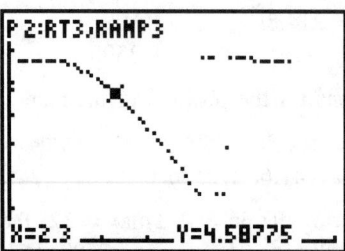

Figure 5.44

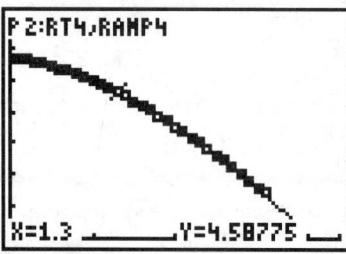

Figure 5.45

The equation of the parabola that best fits this set of data was found using the calculator's `QuadReg` feature. A portion of the parabola is visible to the right of the data points. The equation of the parabola of best fit is

$$d(t) = -0.29t^2 - 0.51t + 5.7$$

On page 283 we reviewed the meaning of the coefficients of this equation of motion under the influence of gravity. Referring back to Figure 5.44 and using the quadratic formula, we can see that the basketball is released 5.7 ft from the motion detector at an initial velocity of about -0.51 ft/sec. Of course we tried to release the ball with zero velocity, but this is what the model predicts. It is difficult to understand why the coefficient of t^2 is such a low number. For a falling ball in air the coefficient should be -16. To understand the coefficient of -0.29 for t^2 we will use vectors to look more closely at the ramp.

In Figure 5.46 vector $\overrightarrow{EF}$ shows the force of gravity directed vertically downward on the ball. The rectangle $EGFH$ shows two perpendicular vectors, $\overrightarrow{EH}$ and $\overrightarrow{EG}$, that add up to the gravity vector. That is, we have broken the force of gravity into two components; they are

▸ $\overrightarrow{EH}$, which pulls the ball down the ramp, and
▸ $\overrightarrow{EG}$, which is perpendicular to the motion and so does not affect it.

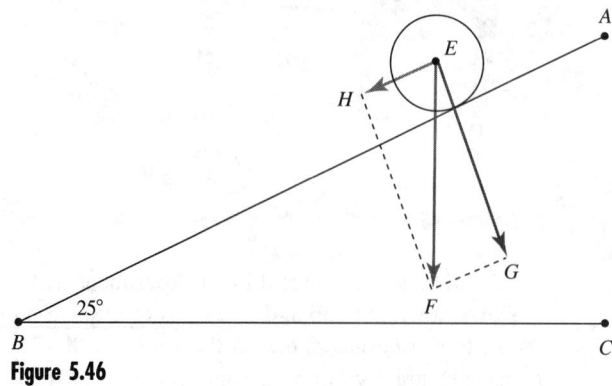

Figure 5.46

Can you see that $\triangle EFH$, if rotated around and flipped over, is a miniature version of the ramp? That's true no matter what the angle of the ramp because $\overline{HF}$ and $\overline{EF}$ will always be perpendicular to the ramp and the floor. Therefore $\angle EFH$ is also equal to the angle of the ramp, in this case 25°. So the magnitude of the vector along the ramp is equal to

$$EH = EF \sin \theta$$

where θ is the angle of the ramp. But EF is the magnitude of the force of gravity. So the force of gravity along a ramp is reduced because it is multiplied by the sine of θ, and the sine is always less than or equal to 1. If the angle is small, $\sin \theta$ is a small number and the force on the ball is small, so it moves more slowly. If the angle is close to 90°, then $\sin \theta$ is closer to 1 and the force of gravity along the ramp is close to the value without the ramp, which makes the ball move more quickly.

Figures 5.47 and 5.48 show the ramp at two other angles. Notice how the vector $\overrightarrow{EH}$ changes in size as the angle θ changes. This makes sense—the steeper the ramp, the larger the angle; the larger the value of $\sin \theta$, the greater the force applied to the ball and the faster it rolls.

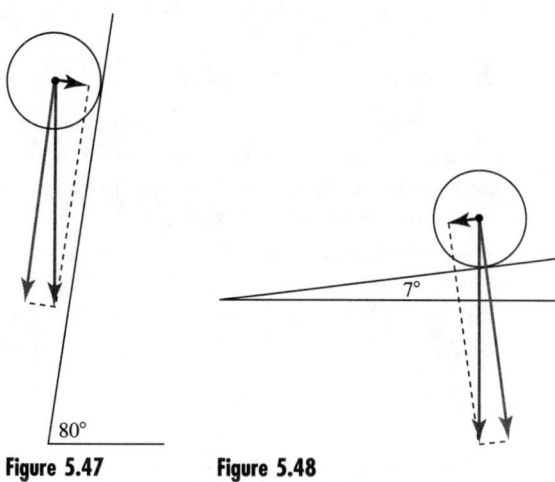

Figure 5.47   **Figure 5.48**

Example 5.25

Figures 5.44 and 5.45 show the data gathered from rolling a basketball down a ramp. Estimate the angle of the ramp from the data.

Solution   Remember that the force of gravity is reduced by the equation

$$EH \text{ (force along the ramp)} = EF\text{(vertical force of gravity)} \times \sin \theta$$

The coefficient of the t^2 term is $-\frac{1}{2}g$, where g is the vertical acceleration (change in velocity) caused by the force of gravity. When the units are feet and seconds the value of g is 32 ft per second per second. The parabola that fits the data for the basketball rolling down the ramp (Figure 5.45) is $d(t) = -0.29t^2 - 0.51t + 5.7$. The coefficient of the t^2 term for this parabola is -0.29. Therefore the magnitude of the vector component of gravity that is acting on the ball (vector $\overrightarrow{EH}$ in Figure 5.46) is 0.58. The angle of the ramp is then found from the following equation:

$$0.58 = 32 \sin \theta$$

$$\theta = \sin^{-1}\left(\frac{0.58}{32}\right) \approx 1°$$ ■

On most ramps the estimate of θ will be accurate only to within a few degrees, since friction is not accounted for in the model we are using.

Modeling Impedance in RC Circuits

Figure 5.49 shows an alternating current (ac) circuit with a voltage source, a resistor, and a capacitor. This is called an **RC circuit**. In this discussion we will be dealing with different measures of opposition to the flow of current in a circuit. They are all measured in ohms, but each has a different name, depending on the circuit component responsible:

▶ **Resistance** (R) is associated with a resistor
▶ **Capacitive reactance** (X_C) is associated with a capacitor
▶ **Impedance** (Z) is the total measure of opposition to current flow in a circuit.

A common problem in electronics is to compute Z for a circuit given values of R and X_C, and to compute the phase angle between the voltage source and the total resulting voltage in the circuit. Let's look at how trigonometric functions and vectors contribute to the calculation.

Figure 5.49

You have seen in Chapter 3 that the current flowing through an ac circuit can be modeled by a sine or cosine wave. Suppose the voltage source produces a current of I amps. Then the voltages across the resistor (V_R) and the voltage across the capacitor (V_C) can be computed by **Ohm's Law**.

$$V_R = IR$$

and $$V_C = IX_C$$

If these were two resistors, instead of a resistor and a capacitor, we would compute the total voltage in the circuit by adding $V_R + V_C$. But because of the nature of a capacitor, the two voltages are out of phase by 90°, so we will have to add the sine curves, not the voltages.

The two curves in Figure 5.50 show the effect of introducing a capacitor into a circuit. The capacitor causes the voltage across it (V_C) to be out of phase by 90° with the voltage across the resistor (V_R). In electronics they say that *the capacitor voltage lags the resistor voltage by 90°*. The vertical coordinate measures voltage, not current as was the case in Chapter 3. The amplitudes of the two curves are $V_R = IR$ and $V_C = IX_C$. (For Figure 5.50 we have let $V_R = 1.5V_C$.) The voltage V_R leads the voltage V_C by 90°. *Leads* means that the leading curve reaches its maximum point 90° *earlier* than the lagging curve, so the curve for V_C is shifted to the right by 90°.

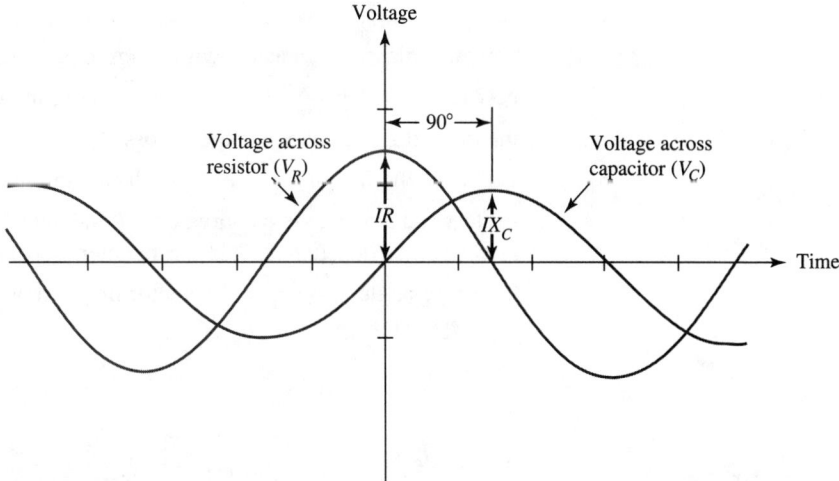

Figure 5.50

At any point in time along these curves the voltage of the circuit is the sum of the two voltages so the total voltage is another periodic function. That is, $V_T = V_R + V_C$. In Figure 5.51 we have three different curves, representing the three out-of-phase voltages. Since the voltage across the resistor is in phase with the source voltage, then the **phase angle of the circuit**, θ, is the phase difference between V_R and V_T. In an RC-circuit it is customary to report this as a negative angle.

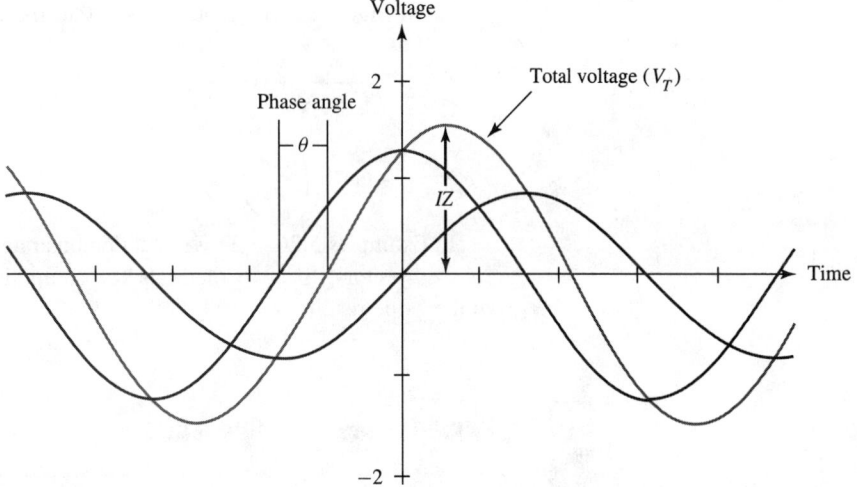

Figure 5.51

One way to compute the impedance is to measure the amplitude of the total voltage curve in Figure 5.51. You would do that by plotting that curve and finding the height at a maximum point.

There is another way to compute the impedance, without drawing any curves. Electrical engineers think of the voltages V_R and V_C as vectors, and V_T as the sum of these vectors. That is,

$$\vec{V_T} = \vec{V_R} + \vec{V_C}$$

Since the voltages V_R and V_C are out of phase by 90°, the vectors $\vec{V_R}$ and $\vec{V_C}$ are perpendicular vectors as shown in Figure 5.52(a). It is customary to show the angle between these vectors as −90°, and $\vec{V_C}$ pointing downward, since the voltage across the capacitor *lags* the voltage across the resistor. Notice that we have drawn $\vec{V_R}$ 1.5 times the length of $\vec{V_C}$, to match the amplitudes in Figure 5.50. Since all three vectors contain $\vec{I}$, we can divide by I and get the similar triangles in Figure 5.52(b), in which the impedance Z is the hypotenuse of the right triangle. Notice also that the phase angle appears in the vector diagram as the angle between $\vec{V_R}$ and $\vec{V_T}$, and is a negative angle.

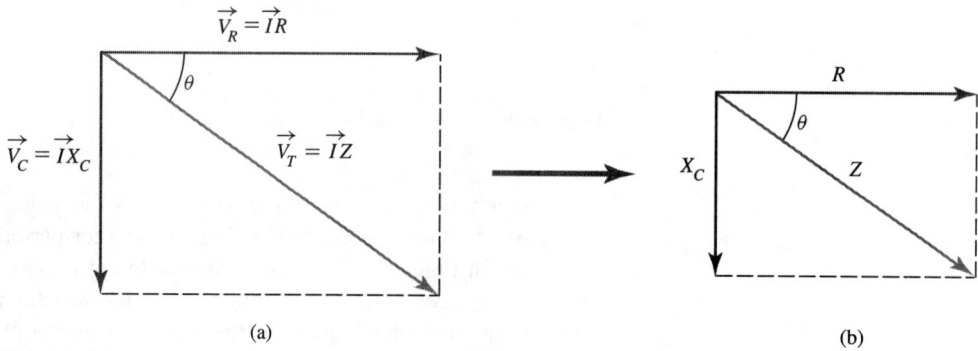

(a)   (b)

Figure 5.52

From Figure 5.52(b) you can see that the impedance and the phase angle are computed as

$$Z = \sqrt{R^2 + X_C^2}$$

$$\theta = -\tan^{-1}\left(\frac{X_C}{R}\right)$$

In Examples 5.26–5.27 we will demonstrate that whether you use trigonometric graphs or vectors, the two methods for computing the impedance and phase angle give the same result.

Example 5.26   **Application**

Show that the impedance of the circuit can be computed by measuring the amplitude of the periodic function that is the sum of the resistance curve and the reactance curve.

Solution   For simplicity we will use $R = 3$ and $X_C = 2$, and we will represent voltages V_R and V_C with simple trigonometric functions, in this case the cosine function. The equations will be entered into the calculator as follows:

Voltage across the resistor (V_R): $y = 3\cos(x)$

Voltage across the capacitor (V_C): $y = 2\cos(x - 90°)$

Figure 5.53, which looks quite similar to Figure 5.51, shows a calculator plot of the two voltages, and their sum, the total voltage. The `Trace` cursor is located at what appears to be the maximum point of the total voltage, and its y-coordinate, 3.598, is an estimate of the impedance in ohms.

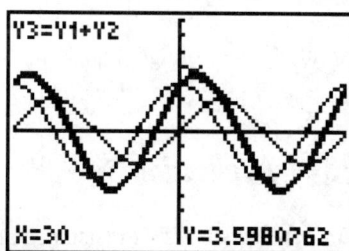

Figure 5.53   **Figure 5.54**

If a more precise estimate of the impedance is required, the calculator's `Table` feature is useful. Figure 5.54 shows that a better estimate is 3.6055 ohms. Note also that the angle at which the maximum occurs is 34°. The phase angle is the horizontal shift between the source voltage and the total current. Since the maximum of the source voltage occurs at 0°, and the maximum of the total current occurs at 34°, then the phase angle is the difference between them, or 34°. Since this is an RC-circuit we report this as −34°.   ■

Example 5.27   Application

Show that the impedance of the circuit can be computed by using the vector diagram.

Solution   The vector computations for the magnitude of Z are

$$Z = \sqrt{R^2 + X_C{}^2}$$

$$= \sqrt{3^2 + 2^2} \approx 3.606\ \Omega$$

Thus the magnitude of Z is about 3.606 Ω.

Next we determine the direction of Z.

$$\theta = -\tan^{-1}\left(\frac{X_C}{R}\right)$$

$$= -\tan^{-1}\left(\frac{2}{3}\right) \approx -34°$$

The phase angle of −34° reminds us that the total voltage lags the source voltage by 34°.

Thus, as in Examples 5.26–5.27, the impedance has a magnitude of 3.606 Ω and a phase angle of −34°.   ■

You can see from the results of Example 5.26 that a vector sum is equivalent to adding two out-of-phase cosine waves. You have also seen that vectors can be useful in the study of such different applications as trajectories of thrown objects, balls rolling on ramps, and navigation in winds or currents. In the next section we will study an area of mathematics which may seem completely different from anything you have studied before, and vectors will be useful there as well.

In an ac circuit, impedance takes the place of resistance and Ohm's law becomes

$$i = \frac{V}{Z}$$

where i is the current, V the voltage, and Z the impedance.

Example 5.28   Application

A 150-V RC circuit consists of 175-Ω resistance and a capacitive reactance of 265 Ω. Determine (a) the phase angle between the voltage and the current, (b) the impedance of the circuit, and (c) the current through the circuit.

Solutions   (a) The impedance diagram for this situation is shown in Figure 5.55. From the diagram we can see that $X_C = 265$ and $R = 175$, so

$$-\tan \theta = \frac{-X_C}{R} = \frac{-265}{175} \approx -1.51$$

$$\theta = -56.6°$$

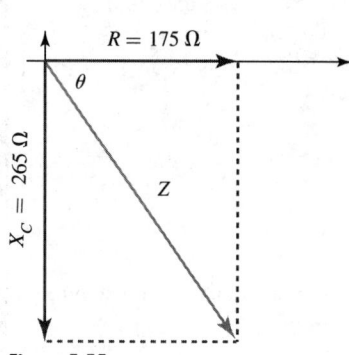

Figure 5.55

Thus the current is leading the voltage by 56.6°.

(b) In Figure 5.55 the impedance Z is the hypotenuse of a right triangle with legs $R = 175$ and $X_C = 265$. Thus we have

$$Z = \sqrt{R^2 + X_C{}^2}$$

$$Z = \sqrt{175^2 + 265^2} \approx 317.6$$

The impedance is about 317.6 Ω.

(c) Here we use the version of Ohm's law with $i = \frac{V}{Z}$ and obtain $i = \frac{150}{317.6} \approx 0.472$. The current is about 0.472 A. ■

In Example 5.28(b) we could have used trigonometry to determine the impedance. Referring to Figure 5.55 you can see that

$$Z = \frac{R}{\cos \theta} = \frac{175}{\cos 56.6°} \approx 317.9$$

$$\text{and} \quad Z = \frac{X_C}{\sin \theta} = \frac{265}{\sin 56.6°} \approx 317.4$$

You can see from the results of Example 5.26 that a vector sum is equivalent to adding two out-of-phase cosine waves. You have also seen that vectors can be useful in the study of such different applications as trajectories of thrown objects, balls rolling on ramps, and navigation in winds or currents. In the next section we will study an area of mathematics that may seem completely different from anything you have studied before, and vectors will be useful there as well.

Section 5.3 Exercises

In Exercises 1–4 use the triangle method to add the given vectors. Indicate the sum by drawing the resultant vector.

1.

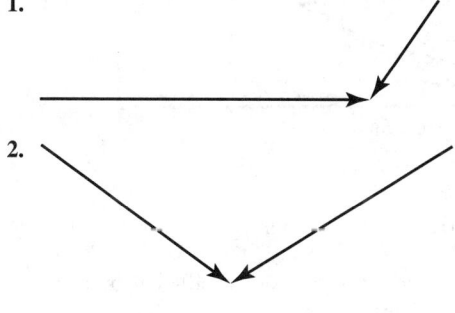

2.

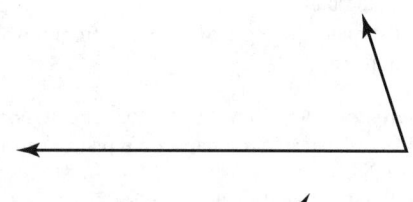

3.

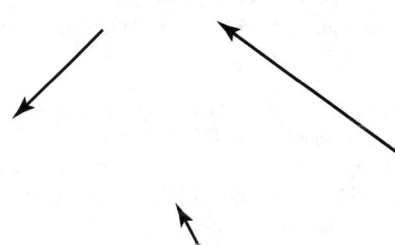

4.

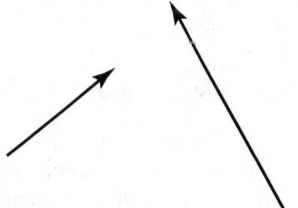

In Exercises 5–8 use the parallelogram method to add the given vectors. Indicate the sum by drawing the resultant vector.

5.

6.

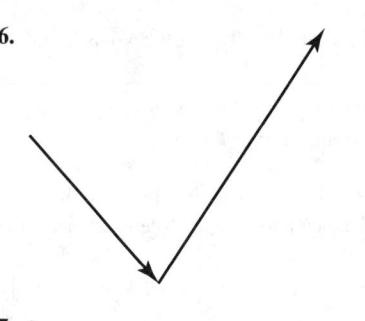

7.

8.

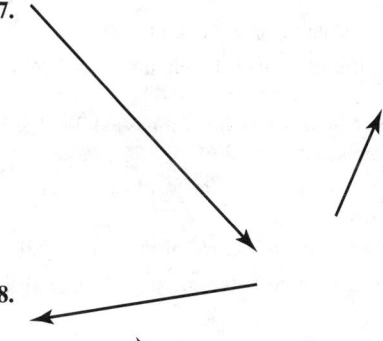

9. Given: $A(0, 0)$, $B(3, 4)$, and $C(5, -12)$.

   (a) Draw the parallelogram that shows vectors $\overrightarrow{AB}$, $\overrightarrow{AC}$, and $\overrightarrow{AB} + \overrightarrow{AC}$.

   (b) What are the coordinates of the head of $\overrightarrow{AB} + \overrightarrow{AC}$?

   (c) What is the magnitude of $\overrightarrow{AB} + \overrightarrow{AC}$?

   (d) What is the direction of $\overrightarrow{AB} + \overrightarrow{AC}$?

10. Given: $D(0, 0)$, $E(-4, 3)$, and $F(-2, -5)$.

   (a) Draw the parallelogram that shows vectors $\overrightarrow{DE}$, $\overrightarrow{DF}$, and $\overrightarrow{DE} + \overrightarrow{DF}$.

   (b) What are the coordinates of the head of $\overrightarrow{DE} + \overrightarrow{DF}$?

   (c) What is the magnitude of $\overrightarrow{DE} + \overrightarrow{DF}$?

   (d) What is the direction of $\overrightarrow{DE} + \overrightarrow{DF}$?

11. Given: $A(1, -2)$, $B(-3, 5)$, $C(4.2, 4.2)$ and $D(-3.5, 6.7)$.

   (a) What are the coordinates of the head of $\overrightarrow{AB} + \overrightarrow{CD}$?

   (b) What is the magnitude of $\overrightarrow{AB} + \overrightarrow{CD}$?

   (c) What is the direction of $\overrightarrow{AB} + \overrightarrow{CD}$?

12. Given: $E(-5.4, -2.1)$, $F(-6.3, 7.4)$, $G(2.8, -1.1)$ and $H(7.9, 6.8)$.

   (a) What are the coordinates of the head of $\overrightarrow{EF} + \overrightarrow{GH}$?

   (b) What is the magnitude of $\overrightarrow{EF} + \overrightarrow{GH}$?

   (c) What is the direction of $\overrightarrow{EF} + \overrightarrow{GH}$?

13. *Navigation* An airplane is flying in a wind blowing with a velocity of 55 mph from the north. The velocity of the plane is 475 mph and it has a compass heading of 250°. What are

   (a) the resultant velocity and

   (b) the actual direction of the plane?

14. *Navigation* An airplane is flying with a compass heading of 113° at an airspeed of 620 mph. If the wind blows at 85 mph from the direction of 168°, what are

    (a) the resultant velocity and

    (b) the actual direction of the plane with respect to the ground?

15. *Navigation* During a cruise, a ship travels 18 mi with a compass heading of 180°, turns and sails 24 mi with a heading of 106°, and makes a final turn and travels 11 mi with a heading of 121°. Assuming there was no current, what are the ship's

    (a) distance and

    (b) direction from the starting point?

16. *Navigation* During a cruise, a ship with a constant velocity of 8 mph travels 18 mi with a compass heading of 180°, turns and sails 24 mi with a heading of 106°, and makes a final turn and travels 11 mi with a heading of 121°. If the current during this time flowed at 6 mph with a direction of 247°, what is the ship's

    (a) distance and

    (b) direction from the starting point?

 17. *Navigation* A plane is attempting to drop a package of food near a campsite in the wilderness. The plane is moving at 150.0 mph in a direction 12.5° above the horizontal. When the plane reaches a point that is 2500 ft above the ground, the package is dropped. (Ignore the effects of air resistance.)

    (a) Write the parametric equations that model the positions of the plane and the package.

    (b) Plot the trajectories on your calculator.

    (c) Estimate to the nearest tenth of a second the time when the food package reaches the ground.

    (d) Estimate to the nearest foot the horizontal distance from where the food is dropped to where it lands.

 18. Roll a ball down a ramp using an arrangement like the one shown in Figure 5.43. Use a CBL to gather data, use the calculator to transform the data, and then use the calculator's `QuadReg` feature to determine the equation of the parabola that best fits this set of data.

    (a) Estimate the angle of the ramp from the data.

    (b) Determine the actual angle of the ramp.

 19. Increase the slope of the ramp in Exercise 18 and repeat the experiment.

 20. Figure 5.56 shows a graph of data gathered using a CBL and a ball rolling up a ramp. The data has been transformed as described in Section 2.5. The equation of the parabola that best fits this set of data, using the calculator's `QuadReg` feature, is $d(t) = 0.55t^2 + 0.56t + 1.3$. A portion of the parabola is visible to the right of the data points.

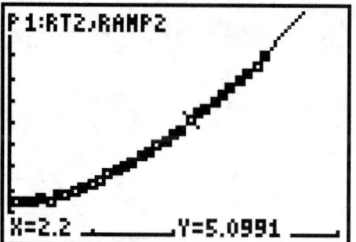

Figure 5.56

    (a) What basic change in the design of the experiment (compared with Figure 5.43) could have caused the graph to look so different from the graph in Figure 5.45?

    (b) The model predicts that the ball was how far away from the motion detector at the beginning of the experiment?

    (c) What does the model predict about the initial speed of the ball?

    (d) Estimate the angle of the ramp from this information.

21. *Electronics* A 120-V series RC circuit consists of 330-Ω resistance and a capacitive reactance of 121 Ω. Determine

    (a) the phase angle between the voltage and the current,

    (b) the impedance of the circuit, and

    (c) the current through the circuit.

22. *Electronics* A 120-V series RC circuit consists of 330-Ω resistance and a capacitive reactance of 60.3 Ω. Determine

    (a) the phase angle between the voltage and the current,

    (b) the impedance of the circuit, and

    (c) the current through the circuit.

23. *Electronics* A series RC circuit has a 3.0-kΩ resistance and an impedance of 3.2 kΩ.

    (a) Draw a vector diagram to illustrate this situation and determine

    (b) the capacitive reactance of the circuit and

    (c) the phase angle between the voltage and the current.

24. *Electronics* A series RC circuit has a 160-Ω resistance and an impedance of 215 Ω.

    (a) Draw a vector diagram to illustrate this situation and determine

    (b) the capacitive reactance of the circuit and

    (c) the phase angle between the voltage and the current.

5.4

Vectors and Complex Numbers

In this section you will meet two new kinds of numbers: the imaginary numbers and the complex numbers. You will see how these numbers can be added, multiplied, and graphed and how they are related to vectors.

Types of Numbers

You know many different kinds of numbers. Everyone is first exposed to the counting numbers, which are sometimes called the natural numbers. We will call them the

positive integers: 1, 2, 3, 4 ...

However, some problems require zero, so we added zero to the positive integers and named these the

whole numbers: 0, 1, 2, 3, 4 ...

Although fractions probably came next in your development, we'll finish the discussion of integers by including the negative integers. This gives the complete set of integers.

integers: ... − 4, −3, −2, −1, 0, 1, 2, 3, 4 ...

You started thinking about fractions when you needed to discuss dividing something, such as a cookie or a pizza, into halves, thirds, or quarters. The fractions and the integers make up a collection called the

rational numbers: any number that can be written as a fraction $\frac{p}{q}$ where p and q are integers and $q \neq 0$

An interesting property of all rational numbers is that their decimal representations either repeat or terminate.

Example 5.29

The numbers $\frac{2}{5} = 0.4$ and $\frac{286}{25} = 11.44$ are examples of terminating decimals. The numbers $\frac{4}{7} = 0.571428571428\ldots$ and $\frac{361}{15} = 24.06666666\ldots$ are examples of repeating decimals. In $\frac{4}{7}$ the group of numbers 571428 keeps repeating, and in $\frac{361}{15}$ the number 6 repeats. ∎

The set of rational numbers is sufficient for most people, but those who study mathematics and engineering soon encounter another kind of number:

irrational numbers: any decimal number that *cannot* be expressed as a fraction of the form $\frac{p}{q}$ where p and q are integers and $q \neq 0$

The word *irrational* means *not making sense*, and indeed the mathematicians who first studied these numbers were sure that something mysterious was happening. They noticed that it seemed impossible to measure exactly the length of the hypotenuse of a triangle with both legs of length 1. (The length of the hypotenuse is shown as the variable x in Figure 5.57.)

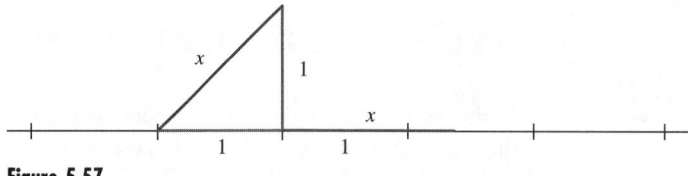

Figure 5.57

The length of the hypotenuse of a right triangle with both legs of length 1 is $\sqrt{2}$. The attempt to find an exact or repeating decimal for $\sqrt{2}$ went something like the following. You can see from Figure 5.57 that $\sqrt{2}$ is a little less than 1.5. If the length from 1 to 2 were marked off in tenths, the value could be estimated as between 1.4 and 1.5. Marking every hundredth reveals that the value is between 1.41 and 1.42. The early mathematicians could not draw enough markings on the ruler to locate the value of $\sqrt{2}$ accurately, and their effort never could be successful because it cannot be done.

Example 5.30

We now know that $\sqrt{2}$ and all roots that do not come out even (for example, $\sqrt{8}$, $\sqrt[3]{15}$, $\sqrt[7]{256}$, etc.) are irrational numbers. It turns out that the number π is irrational also. However $\sqrt{4} = 2$ is a rational number.   ∎

So we have the rational and the irrational numbers, which can all be represented as points on a number line. Together they make up the

real numbers: the collection of all rational and irrational numbers

Figure 5.58 illustrates the idea that all real numbers can be located on a number line.

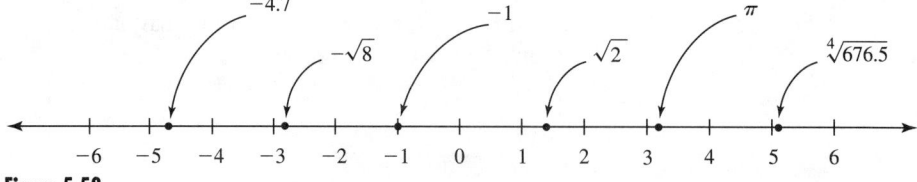

Figure 5.58

Before we meet imaginary and complex numbers, which are *not* located on the real number line, Calculator Lab 5.4 will review material from this chapter and help prepare you for imaginary numbers.

Calculator Lab 5.4	**The Human Cannonball**

The Human Cannonball is working on a new trick. He wants to be shot into the air and caught by a plane flying overhead. To ensure a gentle meeting between plane and person, he wants to meet the plane at the top of his trajectory, as shown in Figure 5.59.
    We need to determine the following two items:

1. How to produce a trajectory whose vertex is at the height of the plane
2. How to time the release of the Human Cannonball so that he and the plane meet at the vertex of the trajectory *at the same time*

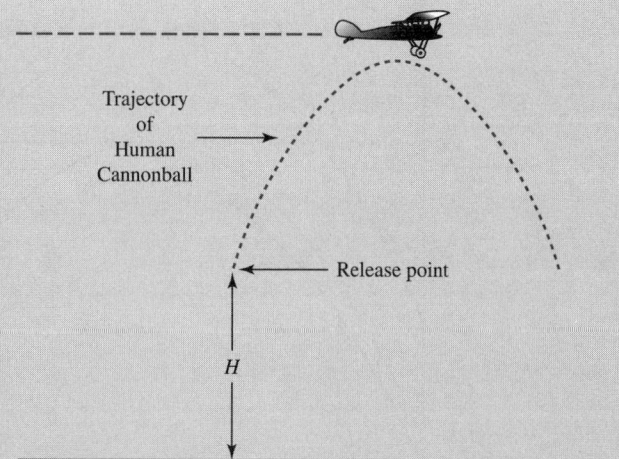

Figure 5.59

In this lab you will solve the first part of this problem by simulating the flights of the plane and the person with graphs of parametric equations. Here are the numbers you need:

▶ The plane travels 100 mph ≈ 147 ft/sec.
▶ The Human Cannonball's cannon will shoot him at an angle of 30°.
▶ The Human Cannonball is shot from the cannon at 200 ft/sec.
▶ The plane is flying 200 ft above the ground.
▶ The height H of the cannon can be varied to the nearest tenth of a foot.
▶ In order for the Human Cannonball to be picked up by the plane, the vertex of his trajectory must be no more than 3 ft below the plane. The vertex cannot be above the plane.

Your challenge is to find a value of the height H that brings the vertex of the Human Cannonball's trajectory to within 3 ft of the plane's trajectory. One way to accomplish this task is to try different values of H until you have produced a trajectory whose vertex comes within 3 ft of 200 ft.

Procedures

1. Write the parametric equations for the Human Cannonball's flight. Your equations should contain the variable H as well as the parameter t.
2. Write the parametric equations for the plane's flight.
3. Change the value of H until you have found a satisfactory trajectory.

Of course the Human Cannonball will pay more for a trajectory that gets him the closest to the plane.

In Examples 5.31–5.33 we will use algebra to analyze a problem similar to the Human Cannonball problem. Everything will be the same except that the height of the plane will be 300 ft. Figure 5.60 shows some trajectories that you might have produced in Calculator Lab 5.4. In the next examples we will focus on determining which trajectories have two points of intersection with the plane's trajectory, which have no points of intersection, and—the correct answer—which has one point of intersection.

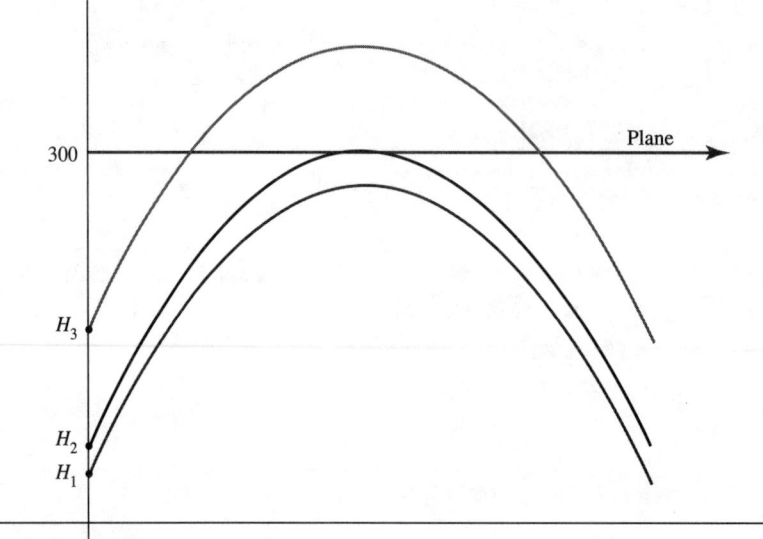

Figure 5.60

Example 5.31

A plane is flying at 300 ft above the ground with a speed of 100 mph $\approx$ 147 ft/sec. The Human Cannonball will be shot from the cannon at 200 ft/sec at an angle of 30°. If $H = 190$ ft, are there two, one, or zero points of intersection with the plane's trajectory?

Solution We care only about the vertical distance, so we will look at the parametric equation for y. We are given $v_0 = 200$, so $v_{y_0} = 200 \sin 30°$. We are also given $H = y_0 = 190$. Using the formula $y = -\frac{1}{2}gt^2 + v_0 t + y_0$ with $g = 32$ ft/sec^2, we have

$$y = -16t^2 + 200 \sin(30°)t + 190$$
$$= -16t^2 + 100t + 190$$

We want to know the values of t when $y = 300$. So the equation we want to solve is

$$-16t^2 + 100t + 190 = 300$$
$$-16t^2 + 100t - 110 = 0$$

The quadratic formula can be used to solve this equation, with $a = -16$, $b = 100$, and $c = -110$. The result is

$$t = \frac{-100 \pm \sqrt{100^2 - 4(-16)(-110)}}{2(-16)}$$
$$= \frac{-100 \pm \sqrt{10000 - 7040}}{-32}$$
$$= \frac{-100 \pm \sqrt{2960}}{-32}$$

We don't have to go any further to see that there are two solutions, because for one solution we add $\sqrt{2960}$ and for another solution we subtract $\sqrt{2960}$. However, just to finish up this problem, we can compute the two times, which are about $t = 1.4$ sec and $t = 4.8$ sec. ∎

For the quadratic equation to give only one solution, the number under the square root must be zero. If that should happen we would be adding and then subtracting zero, which would give two identical roots. Therefore the task becomes solving

$$b^2 - 4ac = 0$$

The expression within the square root, $b^2 - 4ac$, is called the **discriminant** of the quadratic equation because it helps you discriminate between equations with two solutions, one solution, and no solutions. Example 5.31 showed that if the discriminant is positive there will be two real number solutions.

Example 5.32

If the Human Cannonball is shot at 100 ft/sec and the plane is 300 ft high, find the value of H that gives only one solution.

Solution Here we are looking for the value of H that will make the discriminant equal to zero and produce only one point of intersection between the trajectory of the plane and the trajectory of the Human Cannonball.

The equation that must be solved for t is

$$-16t^2 + 100t + H = 300$$
$$-16t^2 + 100t + H - 300 = 0$$

This is a quadratic equation in which $a = -16$, $b = 100$, and $c = H - 300$. Therefore the discriminant is

$$b^2 - 4ac = 100^2 - 4(-16)(H - 300)$$
$$= 10000 + 64(H - 300)$$
$$= 64H - 9200$$

and so the discriminant is equal to zero when

$$64H - 9200 = 0$$
$$H = \frac{9200}{64} = 143.75$$

If the plane is 300 ft high and the original height of the Human Cannonball is 143.75 ft, then the trajectory of the person and the trajectory of the plane will touch in just one place. ∎

We have one more situation to study: the case when the Human Cannonball's trajectory does not intersect the trajectory of the plane.

Example 5.33

If the Human Cannonball is shot from 100 ft/sec at 100 ft/sec and the plane is 300 ft high, at what time will the two trajectories cross?

Solution The equation we need to solve is

$$-16t^2 + 100t + 100 = 300$$
$$-16t^2 + 100t - 200 = 0$$

The solutions are

$$t = \frac{-100 \pm \sqrt{100^2 - 4(-16)(-200)}}{2(-16)}$$

$$= \frac{-100 \pm \sqrt{10000 - 12800}}{-32}$$

$$= \frac{-100 \pm \sqrt{-2800}}{-32}$$

If you try to compute the value of one of these solutions, the calculator will say something like NONREAL ANSWER. Since only the discriminant has changed in this solution, you can guess that the calculator doesn't like taking the square root of this discriminant, -2800. And that makes sense because if $Q = \sqrt{-2800}$, then $Q^2 = -2800$. When you square a real number, whether that number is positive or negative, the answer must be positive. ∎

The number $\sqrt{-2800}$ is your first example of a new type of number, what mathematicians call **imaginary numbers**. We'll study imaginary numbers in more detail in the next subsection, but for now we can answer the question that was asked in this example by saying that there are no real values of t that solve the equation and therefore the Human Cannonball's trajectory does not meet the plane's trajectory for this starting height.

It turns out that there are many real problems in electronics and in other fields of science and engineering that are made simpler by using imaginary numbers.

Imaginary Numbers

In Example 5.33 you saw one example of an imaginary number, $\sqrt{-2800}$. Imaginary numbers use a new symbol: the letter i, called the **imaginary unit**, which is defined to be $\sqrt{-1}$. The imaginary number i has a property that no real number has: Its square is a negative number. That is, $i^2 = -1$.

Note: *i* vs. *j*

In electronics, the symbol i is used to represent current, so problems dealing with electricity use the symbol j for the imaginary unit. Thus $j = \sqrt{-1}$. We will follow the practice of using j for the imaginary unit when we are talking about electricity. The rest of the time we will use i.

Example 5.34

Determine the values of the imaginary numbers i^3, i^4, and i^5.

Solutions   Beginning with $i^2 = -1$, any other power of i can be found.

(a) $i^3 = i^2 \times i = (-1) \times i = -i$
(b) $i^4 = i^2 \times i^2 = (-1) \times (-1) = 1$
(c) $i^5 = i^4 \times i = i$ ∎

You can see from these results that as the exponent increases the answers cycle among the number i, -1, $-i$, and 1. We will use this idea in the next example.

Example 5.35

Determine the values of the imaginary numbers i^{100} and i^{103}.

Solutions   (a) $i^{100} = (i^4)^{25} = 1^{25} = 1$
(b) $i^{103} = i^{100} \times i^3 = -i$

■

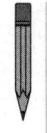

Activity 5.3
Large Powers of i

Review the results of Examples 5.34 and 5.35 and write an answer to the question, How can you predict the value of a large power of i, for example i^{5286}?

Example 5.36

State the solutions of Example 5.33 using imaginary numbers.

Solution   The two solutions were

$$t = \frac{-100 \pm \sqrt{-2800}}{-32}$$

We can rewrite this as

$$t = \frac{100}{32} \pm \frac{\sqrt{-2800}}{-32}$$

Since $\sqrt{-2800} = \sqrt{(-1) \times 2800} = \sqrt{-1} \times \sqrt{2800} = i \times \sqrt{2800}$, the answers are

$$\frac{100}{32} \pm \frac{\sqrt{2800}i}{-32} \approx 3.125 \pm (-1.654i)$$

Therefore, if the initial height of the Human Cannonball is 100 ft, the two solutions are

$$t = 3.125 + 1.654i \text{ and } t = 3.125 - 1.654i$$

■

Simplifying Numerical Expressions

Most students of algebra have learned that an answer like $\dfrac{\sqrt{2800}}{-32}$ should not be left in that form but should be simplified. This may or may not be the best thing to do, depending on the objective of the problem. First let's look at how the simplification is done. The first step is to rewrite the number inside the square root as the product of perfect squares, if possible.

$$\sqrt{2800} = \sqrt{4 \times 7 \times 100}$$
$$= \sqrt{4} \times \sqrt{7} \times \sqrt{100}$$
$$= 2 \times \sqrt{7} \times 10 = 20\sqrt{7}$$

Therefore the answer $\dfrac{100}{32} \pm \dfrac{20\sqrt{7}i}{-32}$ reduces to $\dfrac{25}{8} \pm \dfrac{5\sqrt{7}i}{-8}$.

Before calculators, people had to compute square roots by hand, by looking them up in a table, or by using something called logarithms. Because it makes the computation easier, the expression $20\sqrt{7}$ was thought to be a better way to write $\sqrt{2800}$. It was easier to find square roots of small numbers, so students were taught to simplify numbers like $\sqrt{2800}$. However, since your calculator can compute $20\sqrt{7}$ and $\sqrt{2800}$ in about the same amount of time and using about the same number of keystrokes, it really doesn't matter which form you use. Also, if numerical answers are required, it normally will not matter if fractions like $\frac{100}{32}$ are reduced to $\frac{25}{8}$.

Example 5.37

In Example 5.33 you found that $\dfrac{-100 \pm \sqrt{-2800}}{-32}$ are the solutions to the equation $-16t^2 + 100t - 200 = 0$. Use your calculator to write these solutions in decimal form.

Solution   In the *TI-83's* MODE menu, switch from Real to a+bi. Also switch to three decimal place accuracy because your answer will probably be too big for the screen in Float mode. Your answers should look like those in Figure 5.61. As you can see, they agree with the answers in Example 5.36.   ■

```
(-100+√(-2800))/
(-32)
        3.125-1.654i
(-100-√(-2800))/
(-32)
        3.125+1.654i
```

Figure 5.61

The answers to Example 5.36 are $3.125 + 1.654i$ and $3.125 - 1.654i$. Each of these numbers has a real part and an imaginary part. Such a number is called a **complex number**.

Definition: Complex Numbers
A **complex number** is any number of the form $a + bi$ where a and b are real numbers and $i = \sqrt{-1}$. The number a is called the **real part** of the complex number and b is the **imaginary part**.

Example 5.38

What are the real and imaginary parts of the numbers (a) $5 - \frac{1}{2}i$, (b) 14.7, and (c) $-\sqrt{12}\,i$?

Solutions   (a) $5 - \frac{1}{2}i$ is a complex number. The real part is 5 and the imaginary part is $-\frac{1}{2}$.

(b) 14.7 is both a real and a complex number. The real part is 14.7 and the complex part is 0.

(c) $-\sqrt{12}\,i$ is a complex number with real part 0 and imaginary part $-\sqrt{12}$. ■

Notice that any complex number $a + bi$ where $b = 0$ is also a real number. Thus every real number is a complex number. A complex number where $b \neq 0$ is *not* a real number and is called a **nonreal complex number**.

Summary: Roots of a Quadratic Equation

The roots of the quadratic equation $ax^2 + bx + c = 0$ with $a \neq 0$ are

$$x = \frac{-b \pm \sqrt{b^2 - 4ac}}{2a}$$

The discriminant $b^2 - 4ac$ determines the nature of the roots.

▶ If the discriminant is a positive number (that is, if $b^2 - 4ac > 0$) then there are two real roots and the parabola intersects the x-axis at two different points.

▶ If the discriminant is zero, then $b^2 - 4ac = 0$ and there is one root. This root is called a **double root** of the equation and we say that there are two equal roots. The parabola intersects the x-axis at one point, the vertex.

▶ If the discriminant is a negative number (that is, if $b^2 - 4ac < 0$) then there are two complex solutions and the parabola does not intersect the x-axis.

Figure 5.62 summarizes these three cases.

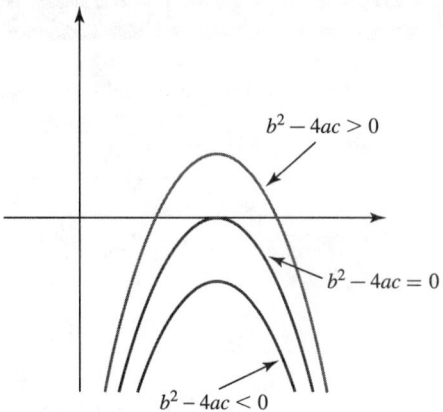

Figure 5.62

As you learned in Chapter 2, the sum of the roots is equal to $-\dfrac{b}{a}$ whether the roots are real or complex. In Example 5.36 the roots are $3.125 + 1.654i$ and $3.125 - 1.654i$. The value of $-\dfrac{b}{a}$ is $-\dfrac{100}{-16} = 6.25$. To add the two roots we create a new complex number by adding the real parts and the imaginary parts, with the result $6.25 \pm 0i = 6.25$.

Arithmetic of Complex Numbers

You can add, subtract, multiply, divide, and graph complex numbers. We will give the definitions for adding and multiplying complex numbers. We will also show how to multiply and divide complex numbers on your calculator. Those calculator instructions should enable you to use your calculator to add and subtract complex numbers.

Addition of Complex Numbers

To add two complex numbers, add the real parts and add the imaginary parts. Thus, if $a + bi$ and $c + di$ are two complex numbers, then

$$(a + bi) + (c + di) = (a + c) + (b + d)i$$

Example 5.39

Add $2 + 5i$ and $7 - 3i$.

Solution   Here $7 - 3i$ is the complex number $7 + {}^{-}3i$. So

$$(2 + 5i) + (7 - 3i) = (2 + 7) + (5 + {}^{-}3)i$$
$$= 9 + 2i$$

Two complex numbers can be multiplied in the same way that algebraic expressions like $(a + b)$ and $(c + d)$ are multiplied.

Multiplication of Complex Numbers

If $a + bi$ and $c + di$ are two complex numbers, then

$$(a + bi) \times (c + di) = (ac - bd) + (ad + bc)i$$

Example 5.40

Multiply $2 + 5i$ and $7 + 3i$.

Solution
$$(2 + 5i) \times (7 + 3i) = (2 \times 7 - 5 \times 3) + (2 \times 3 + 5 \times 7)i$$
$$= (14 - 15) + (6 + 35)i$$
$$= -1 + 41i$$

Example 5.41

Use your calculator to multiply $2 + 5i$ and $7 + 3i$.

Solution The calculator MODE setting should be in a+bi. To multiply these two complex numbers, write each number in parentheses. The i is accessed by pressing [2nd] [.] in the bottom row of the *TI-83* keypad. The product is computed by pressing the following sequence of keys:

The result, $-1 + 41i$, as shown in Figure 5.63, is the same as the answer obtained in Example 5.40.

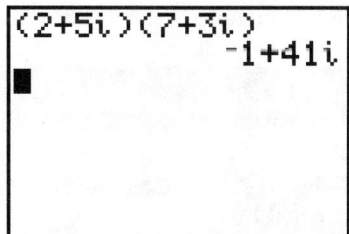

Figure 5.63

Example 5.42

Use your calculator to compute $(3.5 - 7.2i) \div (3 - 4i)$. Express your answer in both decimal and fraction forms.

Solution As in Example 5.41, each number should be placed in parentheses. The quotient is computed by pressing the following sequence of keys:

The result, $1.5272 - 0.304i$, is shown in Figure 5.64. To convert the real and imaginary parts of this answer to rational numbers press [MATH] [1] [ENTER]. The result, $\frac{393}{250} - \frac{38}{125}i$, is shown in Figure 5.65.

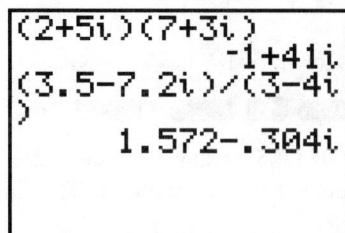

Figure 5.64

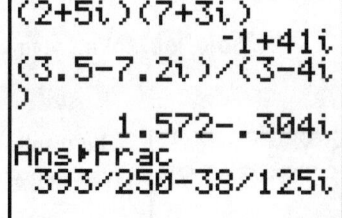

Figure 5.65

Example 5.43   Application

In an ac circuit the formula $V = ZI$ is used to relate the voltage V, impedance Z, and current I. Find the impedance in a given circuit if the voltage is $36 + 9j$ V and the current is $8 - 6j$ A.

Solution   Since we want to find the impedance Z, we rewrite the given formula as $Z = \dfrac{V}{I}$ and substitute the given values.

$$
\begin{aligned}
Z &= \frac{V}{I} \\[4pt]
&= \frac{36 + 9j}{8 - 6j} \\[4pt]
&= 2.34 + 2.88j
\end{aligned}
$$

The impedance in this circuit is $2.34 + 2.88j$ Ω.   ■

Absolute Value of Complex Numbers

The absolute value of a complex number $a + bi$ is represented by the symbol $|a + bi|$ and is equal to

$$|a + bi| = \sqrt{a^2 + b^2}$$

Example 5.44

Determine the absolute value of $3 - 5i$.

Solution   Since $3 - 5i = 3 + {}^-5i$, we have

$$
\begin{aligned}
|3 - 5i| &= \sqrt{3^2 + (-5)^2} \\[4pt]
&= \sqrt{9 + 25} = \sqrt{34}
\end{aligned}
$$
■

Example 5.45

Use your calculator to compute $|3 - 5i|$.

Solution   With your calculator in a+bi mode, press

The result, 5.830951895, is shown on the second line of Figure 5.66. Squaring this result on your calculator, you obtain 34. Thus, as in Example 5.44,

$$|3 - 5i| = \sqrt{34} \approx 5.830951895$$
■

```
abs(3-5i)
            5.830951895
Ans²
                     34
```

Figure 5.66

Geometry of Complex Numbers

To represent complex numbers on a graph requires the special kind of graph shown in Figure 5.67. The horizontal axis is a real number line, called the **real axis**. The vertical axis is something new. It is an imaginary number line, called the **imaginary axis**, on which any imaginary number can be located. The entire graph in Figure 5.67 represents the **complex plane**. A complex number is located in the complex plane by thinking of the complex number $a + bi$ as the ordered pair (a, b) and plotting the point with the coordinates (a, b) in the same way a point is plotted in the Cartesian coordinate system. Quadrants in the complex plane are numbered in the same way as in the Cartesian coordinate system.

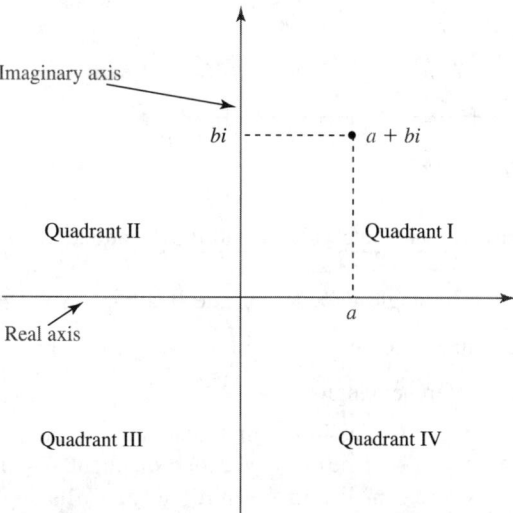

Figure 5.67

The absolute value of a complex number $a + bi$ is equal to the length of the line segment drawn from the point $0 + 0i$ to the point $a + bi$ (shown in Figure 5.68). Since the absolute value of the complex number is the hypotenuse of the right triangle with legs a and b, we can see that $|a + bi| = \sqrt{a^2 + b^2}$.

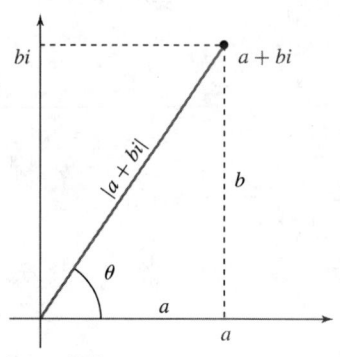

Figure 5.68

The Angle Associated with a Complex Number

In Figure 5.68 angle θ is the angle associated with the complex number $a + bi$. This angle is always measured from the positive real axis.

Example 5.46

Demonstrate the sum of two complex numbers $2 + 5i$ and $7 + 3i$ on the complex plane.

Solution   Figure 5.69 shows the complex numbers $2 + 5i$ and $7 + 3i$ and their sum, $9 + 8i$. Notice how the sum of complex numbers can be shown with a parallelogram in exactly the same way as the sum of vectors. You can verify that the figure is a parallelogram by showing that the opposite sides have the same slope.   ■

Example 5.47

What is (a) the absolute value and (b) the angle for each of the complex numbers (i) $2 + 6i$, (ii) $-2 + 6i$, (iii) $-2 - 6i$, and (iv) $2 - 6i$.

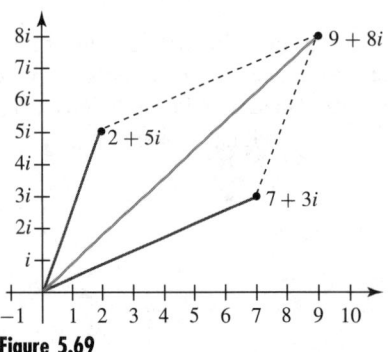

Figure 5.69

Solutions   (a) The absolute value of all these complex numbers is the same since $\sqrt{2^2 + 6^2} = \sqrt{(-2)^2 + 6^2} = \sqrt{(-2)^2 + (-6)^2} = \sqrt{2^2 + (-6)^2} = \sqrt{40} \approx 6.3246$.

(b) The angle depends on the quadrant in which the point lies. The angle in the first quadrant is $\tan^{-1}\left(\dfrac{6}{2}\right) \approx 71.6°$. The angles must be determined from this reference angle:

- ▶ $2 + 6i$ lies in the first quadrant. Its angle is $71.6°$.
- ▶ $-2 + 6i$ lies in the second quadrant. Its angle is $180° - 71.6° = 108.4°$.
- ▶ $-2 - 6i$ lies in the third quadrant. Its angle is $180° + 71.6° = 251.6°$.
- ▶ $2 - 6i$ lies in the fourth quadrant. Its angle is $360° - 71.6° = 288.4°$.

Figure 5.70 shows the location of all four of these complex numbers.

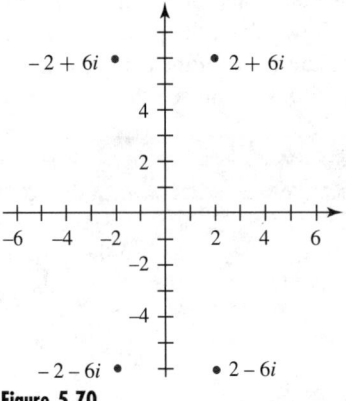

Figure 5.70

Calculator Lab 5.5

Patterns in Products of Complex Numbers

You saw in Example 5.46 that adding complex numbers is like adding vectors. When you multiply complex numbers you will notice that some interesting patterns emerge. This calculator lab gives you an opportunity to explore complex number multiplication in order to find those patterns.

Procedures   1. Table 5.2 contains complex numbers for you to multiply. The complex numbers are represented by z_1 and z_2, and $z_3 = z_1 z_2$. You are to complete the table. Figure 5.71 shows the symbols we'll use in this lab. The variables r_1, r_2, and r_3 are the absolute values of z_1, z_2, and z_3. The variables θ_1, θ_2, and θ_3 are the angles associated with z_1, z_2, and z_3. Remember to plot each complex number so that you can write the correct angle for the quadrant in which the number is located.

Table 5.2

z_1	r_1	θ_1	z_2	r_2	θ_2	z_3	r_3	θ_3
$2+5i$	$\sqrt{29}$		$7+3i$			$-1+41i$		
$3+4i$			$4+3i$					
i	1		i					
$4+3i$			i					
i			-1					
$3+4i$			-1					
$3+4i$			$-4+3i$					
$3+4i$			$-3-4i$					
$2+3i$			$2-3i$					

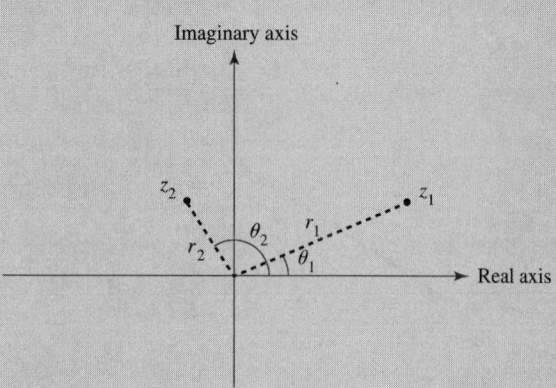

Figure 5.71

2. After you have completed the table and thought about the patterns in the answers, you should be able to answer some or all of these questions:

   (a) What is the effect on complex number z of multiplying z by the number -1?

   (b) What is the effect on complex number z of multiplying z by the number i?

   (c) What is the relationship between r_1, r_2, and r_3 in Table 5.2?

   (d) What is the relationship between θ_1, θ_2, and θ_3 in Table 5.2?

   (e) What is the effect on the complex number z of multiplying z by i? Compare the angle and the absolute value of a complex number z with the angle and absolute value of complex number $i \times z$. Another way of describing this relationship is to say that the effect of multiplying a complex number by i is the same as rotating the complex number through an angle of ____°.

   (f) Using the above answers, describe the complex number that would result from the computation $(3+4i)^{10}$?

Polar Form for a Complex Number

In Calculator Lab 5.5 you saw that useful patterns emerge if complex numbers are written in terms of their absolute values and their angles. When a complex number is written this way we are using the **polar form for a complex number**, which is defined formally below.

Polar Form for a Complex Number

Any complex number $z = a + bi$ can be expressed in the **polar form for the complex number** z where

$$z = r \underline{/\theta}$$

and where $r = |z| = \sqrt{a^2 + b^2}$ and $\tan \theta = \dfrac{b}{a}$ (as shown in Figure 5.72).
Using trigonometry you can see that $a = r \cos \theta$ and $b = r \sin \theta$.

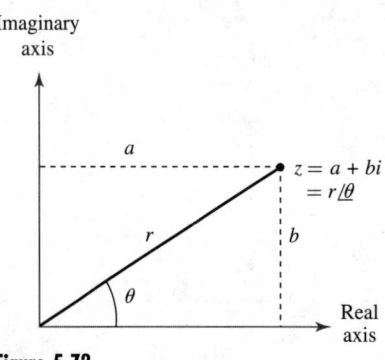

Imaginary
axis

a

$z = a + bi$
$= r \underline{/\theta}$

r

b

θ

Real
axis

Figure 5.72

Example 5.48

Convert the complex number $4.7 - 8.3i$ to polar form.

Solution   Here $a = 4.7$ and $b = -8.3$, so

$$r = \sqrt{a^2 + b^2}$$
$$= \sqrt{4.7^2 + (-8.3)^2}$$
$$\approx 9.54$$

This complex number is in Quadrant IV.

$$\tan \theta = \frac{b}{a}$$
$$= \frac{-8.3}{4.7}$$

so $\theta = \tan^{-1}\left(\frac{-8.3}{4.7}\right)$

$$\approx -60.5°$$

So $4.7 - 8.3i \approx 9.54 \underline{/-60.5°} = 9.54 \underline{/299.5°}$.

Example 5.49

Change the complex number $12\underline{/-30°}$ from polar form to rectangular form.

Solution The rectangular form of a complex number is written as $a+bi$. Here we have $r = 12$ and $\theta = -30°$. So

$$a = r\cos\theta$$
$$= 12\cos(-30°) \approx 10.392$$
$$b = r\sin\theta$$
$$= 12\sin(-30°) = -6$$

Thus the rectangular form of the complex number $12\underline{/-30°}$ is $10.392 - 6i$. ∎

One of the nice aspects of the polar form of complex numbers is that it makes it easy to multiply and divide complex numbers. Calculator Lab 5.5 laid the foundation for this by exploring the multiplication of complex numbers. This process can also be applied to the division of complex numbers when they are expressed in the polar form.

Multiplication and Division of Complex Numbers

If $z_1 = r_1\underline{/\theta_1}$ and $z_2 = r_2\underline{/\theta_2}$ are the polar forms of two complex numbers, then

$$z_1 \cdot z_2 = r_1 r_2\underline{/\theta_1 + \theta_2}$$
$$\frac{z_1}{z_2} = \frac{r_1}{r_2}\underline{/\theta_1 - \theta_2}, z_2 \neq 0$$

Example 5.50

If $z_1 = 3\underline{/40°}$ and $z_2 = 5\underline{/-15°}$, compute $z_1 z_2$ and $\frac{z_1}{z_2}$.

Solutions (a) We will use the formula $z_1 \cdot z_2 = r_1 r_2\underline{/\theta_1 + \theta_2}$ with $r_1 = 3$, $\theta_1 = 40°$, $r_2 = 5$, and $\theta_2 = -15°$. Thus

$$z_1 \cdot z_2 = r_1 r_2\underline{/\theta_1 + \theta_2}$$
$$= 3 \cdot 5\underline{/40° + (-15°)}$$
$$= 15\underline{/25°}$$

(b) Here we use the formula $\frac{z_1}{z_2} = \frac{r_1}{r_2}\underline{/\theta_1 - \theta_2}$ with $r_1 = 3$, $\theta_1 = 40°$, $r_2 = 5$, and $\theta_2 = -15°$. Thus

$$\frac{z_1}{z_2} = \frac{r_1}{r_2}\underline{/\theta_1 - \theta_2}$$
$$= \frac{3}{5}\underline{/40° - {}^-15°}$$
$$= 0.6\underline{/55°}$$
∎

Phasors and Complex Numbers

We have been using the term *vector* to identify a quantity that has direction. To completely describe a vector we need to provide both its magnitude and direction. The direction here refers to its direction in space—either two- or three-dimensional space.

A **phasor** is an electrical vector and is used to describe a quantity that varies in time. Thus the angle shown by a phasor arrow represents the differences in time between two quantities, such as voltage and current, in an ac circuit. Vectors and phasors are treated the same mathematically.

Phasors are drawn with their reference angles relative to some reference axis. Normally a horizontal line is used as the reference line for the other phasors in the same diagram.

In Section 5.3 we looked at vectors and RC circuits. In an RC circuit the voltage across the resistor V_R is in phase with the current I. The voltage across the capacitor V_C *lags* the current by 90°, hence V_C lags V_R by 90°. This situation is shown in Figure 5.73 where V_R and I are drawn in the same direction because they are in phase. Since V_C lags I by 90°, the V_C phasor is drawn at an angle of $-90°$.

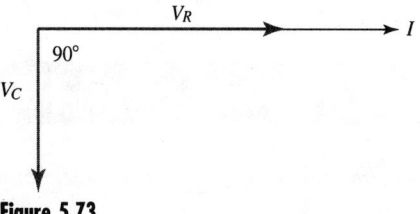

Figure 5.73

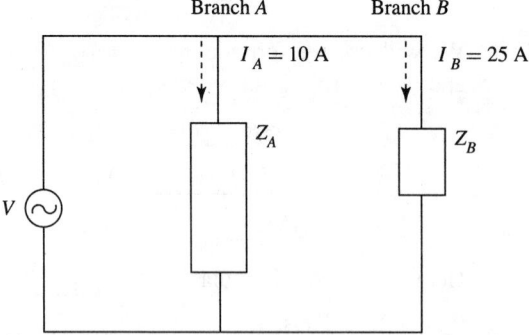

Example 5.51　Application

In a series-parallel circuit, like the one shown in Figure 5.74, a current of 10 A in branch A leads the total voltage by 25°. A current of 25 A in branch B leads the total voltage by 47°. Find (a) the total current and (b) the angle by which the total current leads the total voltage.

Figure 5.74

Solutions　The phasor diagram for the branch currents is drawn with the total voltage on the horizontal axis, as shown in Figure 5.75. The vector sum is shown in Figure 5.76. We need to determine the values of I_T and θ_T.

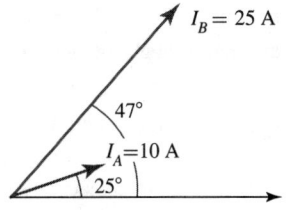

Figure 5.75

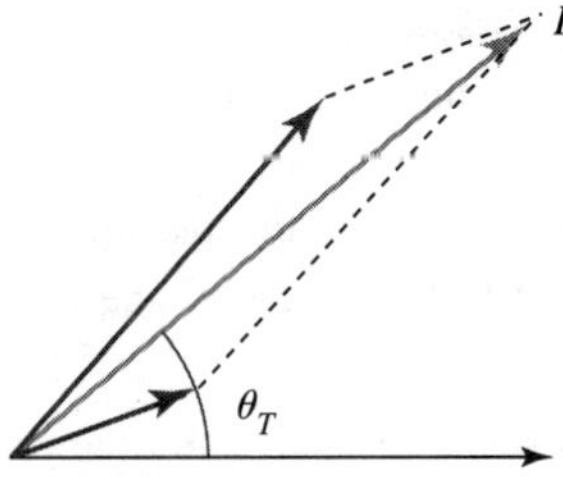

$I_B = 25$ A

$47°$

$I_A = 10$ A

$25°$

I_T

θ_T

Figure 5.76

(a) It helps to think of each of these phasors as a complex number. We are given $I_A = 10\underline{/25°} \approx 9.063 + 4.226j$ and $I_B = 25\underline{/47°} \approx 17.050 + 18.284j$. Thus we have

$$I_T = I_A + I_B$$
$$= (9.063 + 4.226j) + (17.050 + 18.284j)$$
$$= 26.113 + 22.510j$$

So the total current is the magnitude of $I_T = \sqrt{26.113^2 + 22.510^2} \approx 34.476$.

(b) The phase angle of I_T is $\tan^{-1}\frac{22.510}{26.113} \approx 40.8°$.

Thus we have found that $I_T = 26.113 + 22.510j \approx 34.476\underline{/40.8°}$. ∎

Section 5.4 Exercises

In Exercises 1–16 perform the indicated operation on the given complex numbers.

1. $(4 - 2i) + (3 + 7i)$

2. $(3 + 6i) - (-9 + 2i)$

3. $(-4 + 8i)(7 - 6i)$

4. $(3 + 2i) \div (8 - i)$

5. $|6 + 8i|$

6. $(5.3 - 6.1i) + (-0.4 + 8.9i)$

7. $(-4.25 + 7.5i)(1.4 - 3.2i)$

8. $(-3.4 + 6i) - (5.4 + 6i)$

9. $(7.6 - 0.5i) \div (-4.2 + 5.6i)$

10. $|-5.6 + 4.2i|$

11. $(-2i)^3$

12. i^{951}

13. $(4 + 3i)2i$

14. $10 \div (-3 + 4i)$

15. $12i(-2.5i)$

16. $(2 - 3.5i)(1.7i)$

In Exercises 17–20, (a) locate each of the given complex numbers on a graph of the complex plane and determine the (b) absolute value and (c) angle associated with each complex number.

17. $4 - 2i$

18. $3 + 5i$

19. $-5 + 6i$

20. $-7 - 3i$

In Exercises 21–24 solve each of the quadratic equations.

21. $4x^2 - 5x + 3 = 0$

22. $7x^2 + 6x = 9$

23. $2.1x^2 + 5.1 = 3.4x$

24. $6.7x^2 + 20.1x + 34.2 = 0$

25. *Electronics* In an ac circuit, if two sections are connected in series and have the same current in each section the voltage V is given by $V = V_1 + V_2$. Find the total voltage in a given circuit if the voltages in the individual sections are $12.57 - 4.82j$ and $6.43 + 2.32j$.

26. *Electronics* Find the total voltage in an ac series circuit that has the same current in each section if the voltages in the individual sections are $4.57 - 8.91j$ and $15.65 + 7.09j$.

27. *Electronics* In an ac circuit the voltage V, current I, and impedance Z are related by $V = IZ$. If $I = 5.8 + 3.5j$ A and $Z = 9.5 - 4.3j$ Ω, what is the voltage?

28. *Electronics* What is the impedance in an ac circuit when $V = 5.2 + 4.5j$ V and $I = 9 - 3.1$ A?

29. *Electronics* If an ac circuit contains two impedances Z_1 and Z_2 in parallel, then the total impedance Z is given by

$$Z = \frac{Z_1 Z_2}{Z_1 + Z_2}$$

What is Z when $Z_1 = 5 + 4j \ \Omega$ and $Z_2 = 8 - 4j \ \Omega$?

30. *Electronics* What is the total impedance in an ac circuit that contains two impedances Z_1 and Z_2 in parallel if $Z_1 = 16.4 - 6.9j$ and $Z_2 = 5.4 + 6.7j$?

31. *Electronics* An RL circuit consists of a resistance and an inductance connected in series with an ac source. In this situation the sine wave voltage across the inductance V_L is out of phase with the sine wave voltage across the resistance V_R.   Here V_L leads the current I and V_R by 90°. The total voltage is given by $V_T = V_R + V_L j$. Consider an RL circuit with $V_R = 12$ V and $V_L = 8.8$ V.

(a) Draw the voltage phasor diagram for this circuit.

(b) Find the magnitude and phase angle of the total voltage V_T.

32. *Electronics* Consider a series RL circuit with $V_R = 15.4$ V and $V_L = 7.6$ V.

(a) Draw the voltage phasor diagram for this circuit.

(b) Find the magnitude and phase angle of the total voltage V_T.

33. *Electronics* An induction motor draws a current of $24\underline{/-26°}$ and is in parallel with a motor that draws a current of $10\underline{/37°}$.

(a) Draw the voltage phasor diagram for this circuit.

(b) Find the magnitude and phase angle of the total current I_T.

34. Use algebra to solve the Human Cannonball problem as it is stated in Calculator Lab 5.4.

5.5   Solving the Best-Angle Problem

In this chapter you have used parametric equations to study the trajectories of thrown or propelled objects and to study trajectories of planets. The chapter opened with the question about the angle of release that gives the greatest horizontal distance. We'll begin our analysis of the best-angle problem with a calculator lab and then complete the solution with algebra and trigonometry.

Calculator Lab 5.6

Looking for the Best Angle

In Examples 5.4 and 5.6, beginning on page 286, you saw how to find the distance traveled if a ball were tossed with an angle of release of 53° and an initial velocity of 73.4 ft/sec. In this calculator lab carry out the same studies for the angles of release in Table 5.3 below. Then plot the distance traveled vs. the angle of release and use the graph to estimate the angle that gives the longest distance.

Table 5.3

Angle of Release	Parametric Equations	Time T (sec) When Object Hits Ground	Horizontal Distance $x(T)$ (in feet) Object Travels
20°			
30°			
40°			
50°			
53°		3.6	158
60°			
70°			

Here are the characteristics of the trajectory for this lab:

▶ The initial velocity is 73.4 ft/sec.
▶ The initial horizontal distance is zero.
▶ The initial height (vertical distance) is zero.
▶ The final height (the height of the object when it lands) is the same as the initial height.

The second and third assumptions mean that we are setting the origin of the coordinate system, $(0, 0)$, at the point of release. This is different from the Human Cannonball problem in which the initial height H was an important variable. Figure 5.77 shows a typical trajectory.

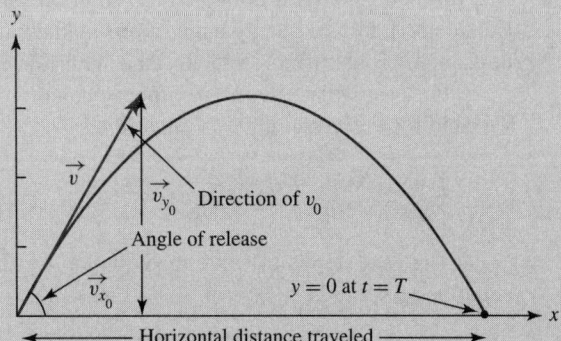

Figure 5.77

Procedures

1. Write the parametric equations for each trajectory in Table 5.3, and then find the horizontal distance each angle produces.
2. Determine the time T in seconds when the object returns to the ground (that is, what is the value of t when $y = 0$?).
3. Determine the horizontal distance $x(T)$ in feet that the object traveled. You can choose to estimate $x(T)$ from the graph of the parametric equations, from a table of values, or from the equations themselves. All three methods are shown in Examples 5.4 and 5.6.
4. When you have completed Table 5.3, plot the graph of distance traveled vs. angle on a graph like the one shown in Figure 5.78.

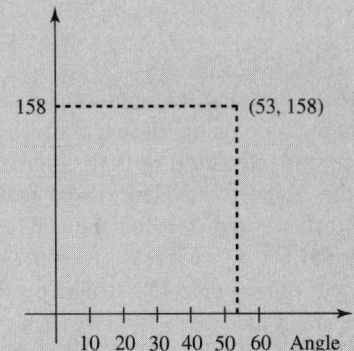

Figure 5.78

5. After you have completed the graph, respond to the following:
   (a) Estimate the angle that produces the longest horizontal distance.
   (b) Estimate the longest horizontal distance for an initial velocity of 73.4 ft/sec

(c) If the initial velocity were doubled to 146.8 ft/sec,
   i. what changes would you expect in the graph of horizontal distance vs. angle?
   ii. What change would you expect in the estimate of the best angle?

Solution That Uses Algebra and Trigonometry

The following solution builds on the solution shown in Examples 5.4 and 5.6 and in Calculator Lab 5.6, but by using more mathematics we will reach the answer to the best-angle problem more quickly and with more accuracy.

First we will write the parametric equations that model the trajectory using the variable for θ for the angle of release:

$$\begin{cases} x = 73.4t \cos\theta \\ y = -16t^2 + 73.4t \sin\theta \end{cases}$$

The next step is to solve the equation for the value of t that makes $y = 0$. The equation to solve is

$$-16t^2 + 73.4t \sin\theta = 0$$

This can be factored as

$$t(-16t + 73.4 \sin\theta) = 0$$

The two solutions of this equation are $t = 0$ (the starting time) and $t = \dfrac{73.4 \sin\theta}{16}$ (the time when the object reaches the ground the second time). This value was referred to as T in Calculator Lab 5.6. Thus $T = \dfrac{73.4 \sin\theta}{16}$.

The third step is to write the expression for $x(T)$, the horizontal distance traveled in T seconds:

$$\text{Horizontal distance} = 73.4T \cos\theta = 73.4 \times \frac{73.4 \sin\theta}{16} \times \cos(\theta)$$

$$= \frac{73.4^2}{16} \sin(\theta) \cos(\theta)$$

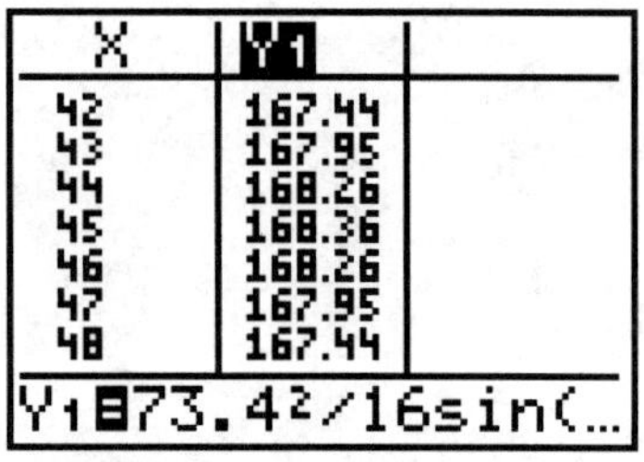

Figure 5.79

The final step is to study this last equation and find the value of θ that produces the maximum value of the distance traveled. You have a choice of using the graph or table on your calculator, or both. Figure 5.79 shows the table near the maximum value of the distance traveled. Note that Y1 in the figure is the expression for horizontal distance and X is the angle θ. The maximum horizontal distance appears to occur at 45°.

Did you guess that 45° would be the best angle? Do you think that it is always the angle that gives the largest horizontal distance? We'll begin to answer that question in Example 5.52.

There are two parts of the model for horizontal distance that could take on other values. They are the initial velocity v_0, which was 73.4 ft/sec in the above equation, and the acceleration due to gravity g, which is 32 ft/sec² (the value for Earth) in the above equation. Do you think there would be a different answer to the best-angle problem if either of these variables had a different value?

Example 5.52

Write the necessary equations for solving the best-angle problem using the variables v_0 and g.

Solution   The parametric equations are

$$\begin{cases} x = v_0 t \cos \theta \\ y = \dfrac{-gt^2}{2} + v_0 t \sin \theta \end{cases}$$

The first equation to solve is $-gt^2 + v_0 t \sin \theta = 0$. Factoring this as $t\left(-\dfrac{gt}{2} + v_0 \sin \theta\right) = 0$ produces the two solutions $t = 0$ (the starting time) and $t = \dfrac{2v_0 \sin \theta}{g}$ (the time when the object reaches the ground the second time). Substituting the second solution for t into the equation for the horizontal distance x gives the following for the horizontal distance:

$$\text{Horizontal distance} = v_0 T \cos \theta$$

$$= v_0 \times \frac{2v_0 \sin \theta}{g} \times \cos \theta$$

$$= \frac{2v_0{}^2}{g} \sin \theta \cos \theta$$

There is a trigonometric identity that states $2 \sin \theta \cos \theta = \sin 2\theta$. Using this identity we can rewrite this as

$$\text{Horizontal distance} = \frac{v_0{}^2}{g} \sin 2\theta$$

The equation in Example 5.52 looks very similar to the equation whose table is shown in Figure 5.79. The question is, "Does the angle that gives the maximum distance change?" You can get a better idea of the answer to that question by trying different values of v and g. We'll return to this question in one of the Section Exercises.

What If the Beginning and Ending Heights Are Different

The solution we found to the best-angle problem was based on the assumption that $h_0 = 0$. What happens if $h_0 \neq 0$? The next calculator lab and the examples that follow will begin to explore this possibility. Exercises 6–11 will complete this exploration.

In Calculator Lab 5.7 you will explore finding the best angle of release if the initial height and ending height are different. In particular we are interested in the following characteristics of the trajectory:

► The initial horizontal distance is zero.
► The initial height h_0 is *not* zero.
► The height of the object when it lands h_T is not h_0.

The first two assumptions suggest that the point of release is at the point $(0, h_0)$. The last assumption means that $y(T) = h_T$; the height of the object when it lands is not h_0. In Figure 5.80 the object lands at a point higher than h_0. In Figure 5.81 the object lands at a point lower than h_0. The situation in Figure 5.81 happens when a shotputter releases the shot about 8 ft above the ground. The length of the put is determined by measuring the horizontal distance from a ring on the ground to the point where the shot hits the ground. At what angle should the shot be put in order to cover the greatest distance?

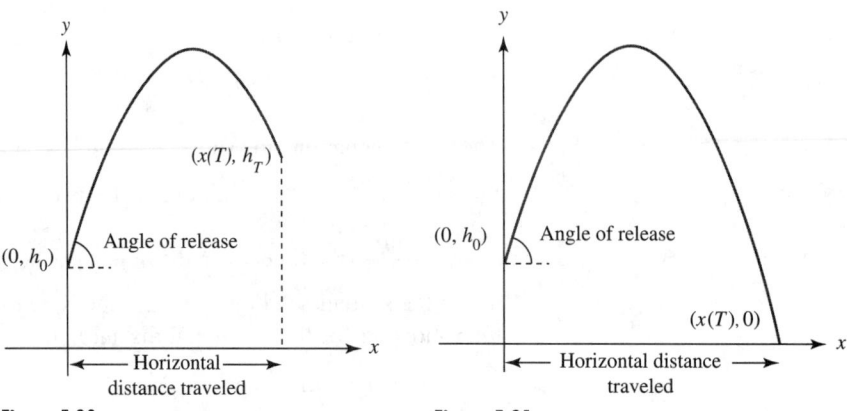

Figure 5.80   **Figure 5.81**

Calculator Lab 5.7

Getting Off the Ground

In this calculator lab you will explore the angle at which a projectile should be released for the projectile to travel the greatest distance. This lab uses the following assumptions:

▶ The initial velocity is 73.4 ft/sec.
▶ The initial horizontal distance is zero.
▶ The initial height (vertical distance) is 8 ft.
▶ The height of the object when it lands is 0 ft.

Figure 5.81 matches these assumptions.
    The last two assumptions suggest that we set the origin so the point of release is $(0, 8)$ and the landing point at time $t = T$ is $(x(T), 0)$.

Procedures
1. Write the parametric equations for each trajectory in Table 5.4.
2. Determine the time T (in seconds) when the object returns to the ground; that is, what is the value of t when $y = 0$? You can choose to estimate T from the graph of the parametric equations, from a table of values, or from the equations themselves.
3. Determine the horizontal distance $x(T)$ (in feet) that the object travels. You can choose to estimate $x(T)$ from the graph of the parametric equations, from a table of values, or from the equations themselves.
4. When you have completed Table 5.4, plot the graph of distance traveled vs. angle on a graph like the one shown in Figure 5.78.
5. After you have completed the graph, do the following:

    (a) Estimate the angle that produces the greatest horizontal distance.
    (b) Estimate the greatest horizontal distance for an initial velocity of 73.4 ft/sec.

Table 5.4

Angle of Release	Parametric Equations	Time T (sec) When Object Hits Ground	Horizontal Distance $x(T)$ (in feet) Object Travels
30°			
35°			
37°			
40°			
42.5°			
45°			
47.5°			
50°			
53°			
55°			
60°			

Section 5.5 Exercises

1. Find the angle that gives the largest value of time in the air (T) for the model described in Calculator Lab 5.6.

2. Suppose the experiment in Calculator Lab 5.6 was performed on the Moon, where the acceleration due to gravity is about $\frac{1}{6}$ of its value on Earth.

   (a) What angle would produce the greatest horizontal distance?

   (b) What is the value of the maximum horizontal distance?

   (c) How do the answers to (a) and (b) compare with the answers in Calculator Lab 5.6?

3. Suppose the experiment in Calculator Lab 5.6 was performed on Earth, but the initial velocity v was 100 times as fast ($v = 7340$ ft/sec).

   (a) What angle would produce the greatest horizontal distance?

   (b) What is the value of the maximum horizontal distance?

   (c) How do the answers to (a) and (b) compare with the answers in Calculator Lab 5.6?

4. Explain why the answers to Exercises 2 and 3 are the same as the answer found in Calculator Lab 5.6?

5. Find the angle that produces the largest horizontal distance if a ball is thrown from a height of 200 feet with an initial velocity of 96.4 ft/sec. (The situation would be as pictured in Figure 5.10 on page 285.)

6. Suppose that the experiment in Calculator Lab 5.7 is performed on the Moon, where the acceleration due to gravity is about $\frac{1}{6}$ of its value on Earth.

   (a) What angle produces the greatest horizontal distance?

   (b) What is the value of the maximum horizontal distance?

7. Suppose that the experiment in Calculator Lab 5.7 was performed on Earth, but that the initial velocity v is 100 times as fast ($v = 7340$ ft/sec).

   (a) What angle would produce the greatest horizontal distance?

   (b) What is the value of the maximum horizontal distance?

   (c) How do the answers to (a) and (b) compare with the answers in Calculator Lab 5.7?

8. Predict the maximum horizontal distance if the experiments in Calculator Labs 5.6 and 5.7 are performed on a planet that has gravity $\frac{1}{n}$ the gravity of Earth.

9. Predict the maximum horizontal distance if the experiments in Calculator Labs 5.6 and 5.7 are performed with an initial velocity of $73.4n$ ft/sec.

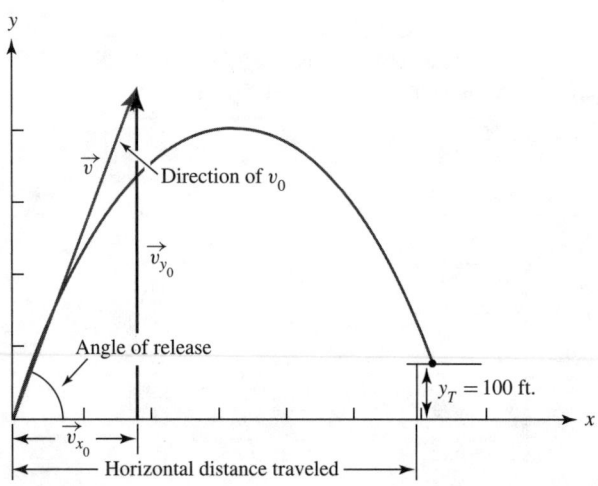

Figure 5.82

10. Find the angle that produces the greatest horizontal distance when a ball is thrown from the ground with an initial velocity of 96.4 ft/sec to an elevated platform 100 feet above the ground. The situation is as pictured in Figure 5.82.

11. Suppose the experiment in Exercise 10 is performed on Earth, but with an initial velocity v that is 100 times as fast ($v = 9640$ ft/sec).

    (a) What angle would produce the greatest horizontal distance?

    (b) What is the value of the maximum horizontal distance?

● ● ● ● ● ● ● ● ● Chapter 5 Summary and Review

Topics You Learned or Reviewed

▶ A vector can be written either with an arrow over its name, as in $\overrightarrow{v}$, or in boldface type, as in **v**.

▶ Two vectors can be added graphically by either the triangle or parallelogram methods. Algebraically, if $A(x_1, y_1)$, $B(x_2, y_2)$, $C(x_3, y_3)$, and $D(x_4, y_4)$ are points, then

$$\overrightarrow{AB} + \overrightarrow{CD} = ([x_2 - x_1] + [x_4 - x_3], [y_2 - y_1] + [y_4 - y_3])$$

▶ Parametric equations are used to express equations for x and y in terms of a third variable, usually t. The third variable t is called the **parameter**.

    • The parametric equations for a parabola are

$$\begin{cases} x = t + a \\ y = t^2 + b \end{cases}$$

    • The parametric equations of a circle of radius $|r|$ and centered at the origin $(0, 0)$ are

$$\begin{cases} x = r \cos bt \\ y = r \sin bt \end{cases}$$

    • One set of parametric equations for an ellipse are

$$\begin{cases} x = a \cos t \\ y = b \sin t \end{cases}$$

    where $|2a|$ and $|2b|$ are the lengths of the major and minor axes. The major axis has length that is the larger of $|2a|$ and $|2b|$.

    • Parametric equations can be used to plot planetary orbits.

▶ The expression $b^2 - 4ac$ is called the **discriminant** of the quadratic equation because it helps you discriminate between equations with two real number solutions (if the discriminant is positive), one solution (the discriminant is zero), and no real solutions, but two nonreal complex solutions (if the discriminant is negative).

▶ An imaginary number is a number of the form $\sqrt{-b} = \sqrt{b}i$ where $b > 0$. The letter i, called the **imaginary unit**, is defined to be $\sqrt{-1}$.

▶ A **complex number** is any number of the form $a + bi$ where a and b are real numbers and $i = \sqrt{-1}$. The number a is called the **real part** of the complex number and b is an **imaginary part** of the complex number.

- To add two complex numbers, add the real parts and add the imaginary parts. Thus, if $a + bi$ and $c + di$ are two complex numbers, then $(a + bi) + (c + di) = (a + c) + (b + d)i$.
- To multiply two complex numbers $a + bi$ and $c + di$, write the product as $(a + bi) \times (c + di) = (ac - bd) + (ad + bc)i$.
- In representing complex numbers on a graph, the horizontal axis is a real number line, called the **real axis**, and the vertical axis is an imaginary number line, called the **imaginary axis**. A complex number is located in the complex plane by thinking of the complex number $a + bi$ as the ordered pair (a, b) and plotting the point with the coordinates (a, b) the same way a point is plotted in the Cartesian coordinate system.
- The absolute value of a complex number $a + bi$ is represented by the symbol $|a + bi|$ and is equal to $|a + bi| = \sqrt{a^2 + b^2}$.
- The angle associated with a complex number $a + bi$ is the angle θ measured from the positive real axis to the line from the origin to the point (a, b).
- Parametric equations are used to plot planetary motion and suborbital trajectories.
- Complex numbers are especially useful in applications to electronics.

Review Exercises

1. Describe the motion of the ball in Figure 5.83 at point P_1.

2. Describe the motion of the ball in Figure 5.83 at point P_2.

3. Describe the motion of the ball in Figure 5.83 at point P_3.

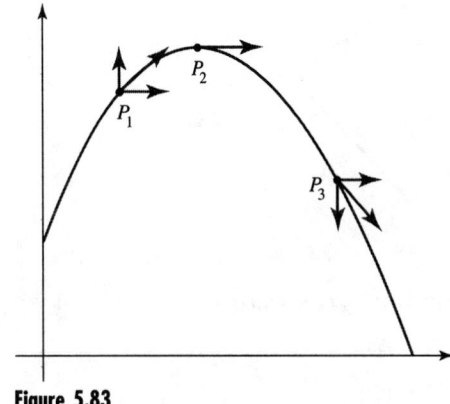

Figure 5.83

In Exercises 4–7 determine the horizontal $\overrightarrow{v_x}$ and vertical $\overrightarrow{v_y}$ components of each given vector $\overrightarrow{v}$ with magnitude $|\overrightarrow{v}|$ and direction θ.

4. $|\overrightarrow{v}| = 15.00$, $\theta = 72°$

5. $|\overrightarrow{v}| = 35.84$, $\theta = 321°$

6. $|\overrightarrow{v}| = 5.92$, $\theta = 247.3°$

7. $|\overrightarrow{v}| = 84.73$, $\theta = 152.8°$

In Exercises 8–9 use your calculator to sketch the graph of the given set of parametric equations for the indicated values of t.

8. $\begin{cases} x = t + 5 \\ y = t^2 - 6 \end{cases}$ where $-6 \le t \le 6$

9. $\begin{cases} x = t^2 - 6t + 9 \\ y = t^2 - 5t \end{cases}$ where $-6 \le t \le 6$

10. What are the parametric equations of a circle of radius 5 centered at the origin?

11. (a) Graph the parametric equations

$$\begin{cases} x = -4 \sin t \\ y = 8 \cos t \end{cases}$$

$0 \le t \le 2\pi$ or $0 \le t \le 360°$

(b) Describe the graph.

(c) How does the graph in (a) differ from the graph of

$$\begin{cases} x = 4 \sin t \\ y = -8 \cos t \end{cases}$$

In Exercises 12–13, (a) graph the ellipse described by the given parametric equations, (b) determine the x- and y-intercepts, (c) determine the lengths of the major and minor axes, and (d) describe any relationship you notice between the parametric equations and your answers to parts (a)–(c).

12. $\begin{cases} x = 6 \cos t \\ y = 2 \sin t \end{cases}$

13. $\begin{cases} x = -2 \sin t \\ y = 6 \cos t \end{cases}$

In Exercises 14–15 use the triangle method to add the given vectors. Indicate the sum by drawing the resultant vector.

14.

15.

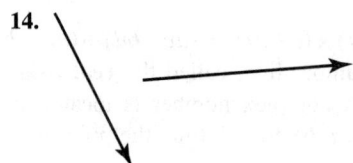

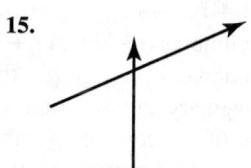

In Exercises 16–17 use the parallelogram method to add the given vectors. Indicate the sum by drawing the resultant vector.

16.

17.

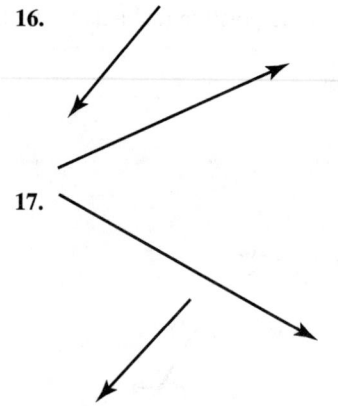

18. Given: $A(0, 0)$, $B(-3, 5)$, and $C(2, -7)$.

  (a) Draw the parallelogram that shows vectors $\overrightarrow{AB}$, $\overrightarrow{AC}$, and $\overrightarrow{AB} + \overrightarrow{AC}$.

  (b) What are the coordinates of the head of $\overrightarrow{AB} + \overrightarrow{AC}$?

  (c) What is the magnitude of $\overrightarrow{AB} + \overrightarrow{AC}$?

  (d) What is the direction of $\overrightarrow{AB} + \overrightarrow{AC}$?

19. Given: $A(-2, 3)$, $B(-4, 5)$, $C(1.6, -6.2)$, and $D(-4.8, 3.7)$.

  (a) What are the coordinates of the head of $\overrightarrow{AB} + \overrightarrow{CD}$?

  (b) What is the magnitude of $\overrightarrow{AB} + \overrightarrow{CD}$?

  (c) What is the direction of $\overrightarrow{AB} + \overrightarrow{CD}$?

In Exercises 20–28 perform the indicated operation on the given complex numbers.

20. $(7 - 3i) + (-5 + 4i)$

21. $(2 - i) - (4 - 6i)$

22. $|5 - 12i|$

23. $(3.25 + 8.5i)(6.8 - 9.2i)$

24. $(-6.2 + 4.5i) \div (2.5 - 6i)$

25. $(-3i)^4$

26. i^{2001}

27. $(-3 + 4i)5i$

28. $(5.2 - 3.8i)(9.5i)$

In Exercises 29–30, (a) locate each of the given complex numbers on a graph of the complex plane and determine the (b) absolute value and (c) angle associated with each complex number.

29. $-3 + 8i$

30. $5 - 12i$

In Exercises 31–32 solve each of the quadratic equations.

31. $5x^2 - 4x + 2 = 0$

32. $9.1x^2 + 5.4x + 12.8 = 0$

33. *Recreation* A golfer hits a ball off the ground with an initial velocity of 115 ft/sec and at an angle of elevation of 32.5°. Write expressions for $x(t)$ and $y(t)$ as measured from the instant the club head hits the ball.

34. *Recreation* For the golf shot in Exercise 33,

  (a) create a table for the time the ball is in the air.

  (b) sketch the flight of the ball while it is in the air.

  (c) how long is the ball in the air before it hits the ground?

  (d) how far, with respect to the ground, does the ball travel before it hits the ground?

35. *Recreation* A golfer hits a ball from an elevated tee to the green (as shown in Figure 5.84) with an initial velocity of 97 ft/sec at an angle of elevation of 45°. Write expressions for $x(t)$ and $y(t)$ as measured from the instant the club head hits the ball.

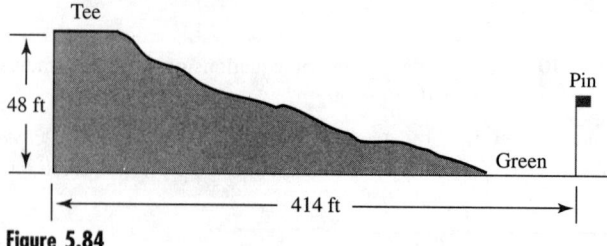

Figure 5.84

36. *Recreation* For the golf shot in Exercise 35,

    (a) create a table for the time the ball is in the air.

    (b) sketch the flight of the ball while it is in the air.

    (c) how long is the ball in the air before it hits the ground?

    (d) if the front of the green is 10 yd from the pin, determine whether the ball lands on the green?

37. *Navigation* An object is dropped from an airplane that is in level flight at an altitude of 15,500 ft. At the time the object is dropped the plane is flying at a speed of 135 mph. Write expressions for $x(t)$ and $y(t)$ in ft/sec as measured from the instant the object is dropped.

38. *Astronomy* The orbit of the planet Mercury has a radius of 0.39 A.U. and the frequency of Mercury in its orbit is 6.23 times the frequency of Earth in its orbit. Write a set of parametric equations that models the orbits of Earth and Venus.

39. *Sports technology* A baseball leaves a major-league pitcher's hand traveling downward at 1° above the horizontal with initial speed 120 ft/sec and initial height 7'6". (Ignore wind resistance.)

    (a) Write a set of parametric equations that describes the path of the ball.

    (b) Graph your parametric equations from the instant the ball leaves the pitcher's hand until it hits the ground.

    (c) If the ball is released 58.0 ft from home plate, how long does it take for the ball to cross the plate?

    (d) If the strike zone is approximately approximately from 1.5 to 4.0 ft above the ground, is this pitch a strike?

40. *Navigation* An airplane is flying in a wind blowing with a velocity of 47 mph from the south. The velocity of the plane is 385 mph and it has a compass heading of 130°. What are (a) the resultant velocity and (b) the actual direction of the plane?

41. *Navigation* An airplane is flying with a compass heading of 317° at an airspeed of 580 mph. The wind blows at 62 mph from the direction of 227°. What are (a) the resultant velocity and (b) the actual direction of the plane with respect to the ground?

 42. *Navigation* A plane is attempting to drop a package of food near a campsite in the wilderness. The plane is moving at 135.0 mph in a direction 8.5° above the horizontal. When the plane reaches a point that is 2750 ft above the ground the package is dropped. (Ignore the effects of air resistance.)

    (a) Write the parametric equations that model the positions of the plane and the package.

    (b) Plot the trajectories on your calculator.

    (c) Estimate to the nearest tenth of a second the time when the food package reaches the ground.

    (d) Estimate to the nearest foot the horizontal distance from where the food is dropped to where it lands.

43. *Electronics* A 120-V series RC circuit consists of 360-Ω resistance and a capacitive reactance of 132 Ω. Determine

    (a) the phase angle between the voltage and the current,

    (b) the impedance of the circuit, and

    (c) the current through the circuit.

44. *Electronics* A 120-V series RC circuit consists of 360-Ω resistance and a capacitive reactance of 66 Ω. Determine (a) the phase angle between the voltage and the current, (b) the impedance of the circuit, and (c) the current through the circuit.

45. *Electronics* In an ac circuit, if two sections are connected in series and have the same current in each section the voltage V is given by $V = V_1 + V_2$. Find the total voltage in a given circuit if the voltages in the individual sections are $14.32 - 6.91j$ and $8.68 + 3.41j$.

46. *Electronics* Find the total voltage in an ac series circuit that has the same current in each section if the voltages in the individual sections are $6.97 - 5.31j$ and $12.43 + 8.42j$.

47. *Electronics* In an ac circuit the voltage V, current I, and impedance Z are related by $V = IZ$. If $I = 3.5 + 8.5j$ A and $Z = 4.2 - 6.4j$ Ω, what is the voltage?

48. *Electronics* If an ac circuit contains two impedances Z_1 and Z_2 in parallel, then the total impedance Z is given by

$$Z = \frac{Z_1 Z_2}{Z_1 + Z_2}$$

What is Z when $Z_1 = 8 - 10j$ Ω and $Z_2 = 12 + 5j$ Ω?

49. *Electronics* Consider a series RL circuit with $V_R = 12.4$ V and $V_L = 9.2$ V.

    (a) Draw the voltage phasor diagram for this circuit.

    (b) Find the magnitude and phase angle of the total voltage V_T.

50. *Electronics* An induction motor draws a current of $15\underline{/-13°}$ and is in parallel with a motor that draws a current of $12\underline{/26°}$.

    (a) Draw the voltage phasor diagram for this circuit.

    (b) Find the magnitude and phase angle of the total current I_T.

●●●●●●●● Chapter 5 Test

1. Describe the motion of the ball in Figure 5.85 at point P_1.

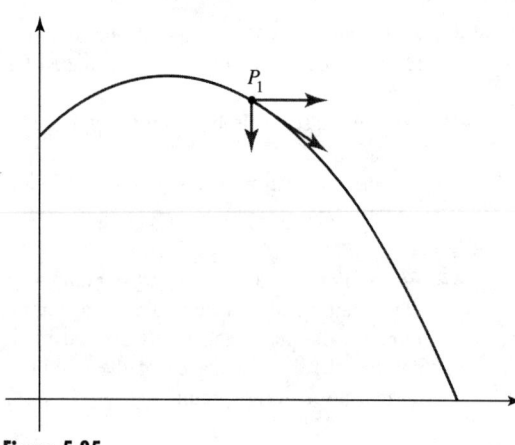

Figure 5.85

2. Determine the horizontal $\vec{v_x}$ and vertical $\vec{v_y}$ components of $\vec{v}$ with magnitude $|\vec{v}| = 27.5$ and direction $\theta = 257°$.

3. Use your calculator to sketch the graph of the parametric equations

$$\begin{cases} x = 2t - 3 \\ y = t^2 + 2t - 1 \end{cases}$$

for $-6 \le t \le 6$.

4. What are the parametric equations of a circle of radius 7.5 centered at the origin?

In Exercises 9–14 perform the indicated operation on the given complex numbers.

9. $(9 - 6i) + (4.5 + 2i)$

10. $(8.6 - 3.2i)(9.1 - 7.5i)$

11. $(4 + 6.2i) \div (3 - 4i)$

12. $(2 - 3i)^5$

13. $|-8.2 + 7.3i|$

14. $(7 - 4i) - (6 - 8i)$

15. (a) Locate each of the complex numbers $-5 + 2i$ on a graph of the complex plane and determine the (b) absolute value and (c) angle associated with them.

16. Solve the quadratic formula $3.1x^2 - 6.5x + 7.5 = 0$.

17. An archer shoots at a target 90 m away. When the arrow is released it has an initial velocity of 150 mph. The wind is blowing at a speed of 15 mph into the archer's face from an angle of 27° to the right of a line from the arrow to the center of the target (as shown in

5. (a) Graph the parametric equations

$$\begin{cases} x = -3 \sin t \\ y = 5 \cos t \end{cases}$$

for $0 \le t \le 2\pi$ or $0 \le t \le 360°$.

(b) Describe the graph. Be sure to include a description of its shape and any dimensions of the graph that can be determined from the parametric equations.

6. Use the triangle method to add the vectors given in Figure 5.86. Indicate the sum by drawing the resultant vector.

7. Use the parallelogram method to add the vectors given in Figure 5.87. Indicate the sum by drawing the resultant vector.

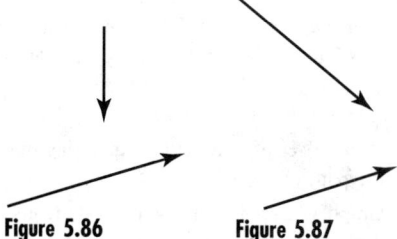

Figure 5.86   **Figure 5.87**

8. Given: $A(-3, 5)$, $B(2, -1)$, $C(4, 3)$, and $D(-2, 1)$.
   (a) What are the coordinates of the head of $\overrightarrow{AB} + \overrightarrow{CD}$?
   (b) What is the magnitude of $\overrightarrow{AB} + \overrightarrow{CD}$?
   (c) What is the direction of $\overrightarrow{AB} + \overrightarrow{CD}$?

Figure 5.88). By how much will the arrow miss the center of the target?

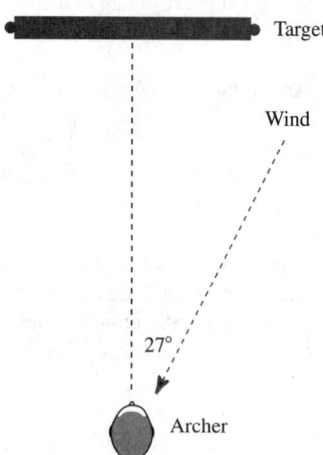

Figure 5.88

18. An archer shoots at a target 90 m away. The target is 122 cm in diameter and the bottom of the target is 130 cm above the ground. When the arrow is released it has an initial velocity of 150 mph (about 67.06 m/s) and is released at an angle of elevation of 5.5° and from a height of 180 cm.

(a) Write expressions for $x(t)$ and $y(t)$ as measured from the instant the arrow is released.

(b) Create a table for the time the arrow is in the air.

(c) How long is the arrow in the air before it hits the target?

(d) If the bullseye (center of the target) has a diameter of 12.2 cm, does the arrow hit the target?

19. A 120-V series RC circuit consists of 380-Ω resistance and a capacitive reactance of 152 Ω. Determine

(a) the phase angle between the voltage and the current,

(b) the impedance of the circuit, and

(c) the current through the circuit.

20. In an ac circuit the voltage V, current I, and impedance Z are related by $V = IZ$. If $V = 7.2 - 3.5j$ A and $Z = 4.2 + 8.6j$ Ω, what is the current?

21. *Electronics* If an ac circuit contains two impedances Z_1 and Z_2 in parallel, then the total impedance Z is given by

$$Z - \frac{Z_1 Z_2}{Z_1 + Z_2}$$

What is Z when $Z_1 = 9.5 - 7.3j$ Ω and $Z_2 = 8 + 6j$ Ω?

CHAPTER

6

Polar Graphing and Elementary Programming

Topics You'll Need to Know

▶ Parametric equations of motion
▶ Parametric form of an ellipse
▶ Solving an equation for one of the variables

Topics You'll Learn or Review

▶ Converting parametric equations to a rectangular equation
▶ Describing symmetry in graphs
▶ Using parametric equations to model an oscilloscope
  • analyzing Lissajous curves
▶ Plotting points and equations in polar coordinates
▶ Using properties of an ellipse
  • polar form of the equation
  • rectangular form of the equation
  • the role of the focus in astronomy and reflection
  • effect of eccentricity on the shape
  • Two reflective properties

Calculator Skills You'll Use

▶ Using the Table feature
▶ Plotting parametric equations
▶ Using the path graph style

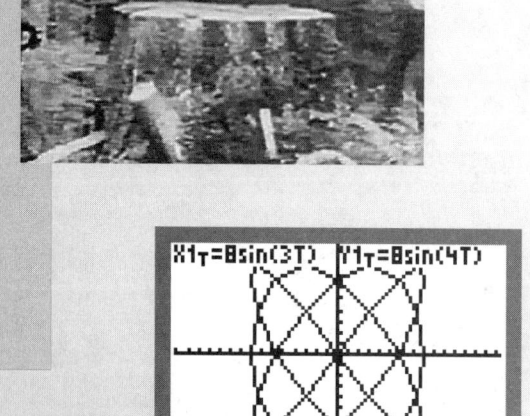

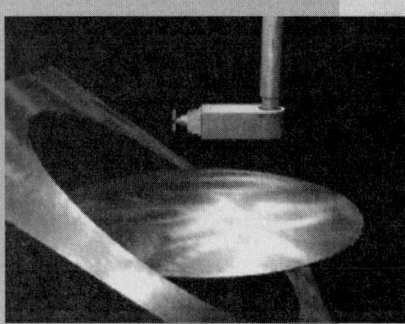

Calculator Skills You'll Learn

▶ Plotting polar equations
▶ Programming, running, and editing programs
▶ Entering new programs
▶ Understanding the programming language
  • setting the initial environment
  • input instructions
  • output instructions
  • control instructions
  • variables and functions

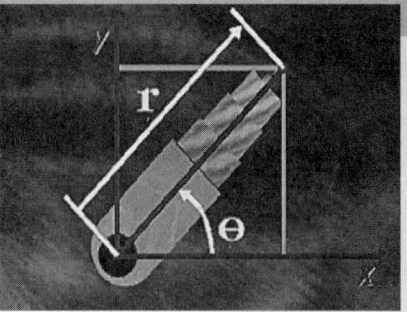

What You'll Do in This Chapter

This chapter builds on what you have learned in previous chapters about trigonometry and graphing. You will return to parametric equations and see how they are related to oscilloscope displays. You will learn more about ellipses and see their applications in the orbits and planets and in modern medical technology.

We will introduce a new coordinate system based on trigonometry, called the polar coordinate system, and you'll see some of the attractive designs that can be produced with polar equations. You will also see why astronomers use only polar coordinates to plot orbits of planets.

Throughout the chapter you will be introduced to the logic of programming a computer or a calculator. You will learn how to run programs, change the instructions in programs written by others, and write your own programs on the *TI-83*.

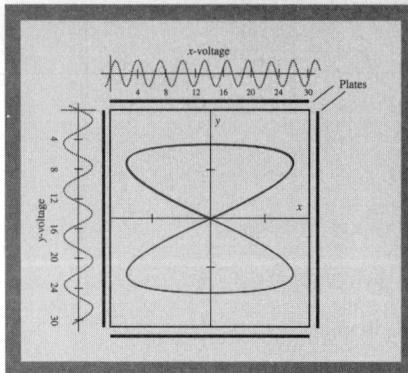

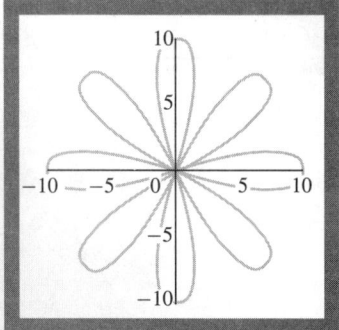

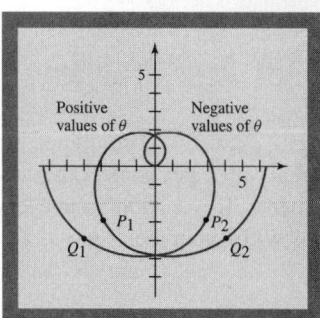

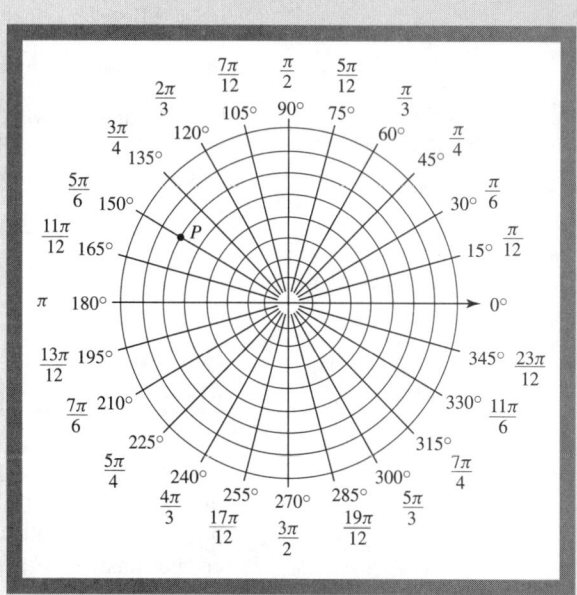

Chapter Project—Programming a Robot Arm

Robots-R-Us

"If humans can't do the job, ask us"

To: Research Department
From: Management
Subject: Automatic shape-cutting

Shapes Galore, Inc. has asked us to redesign our *X-Y Grabber* robot to fit their new needs. We are to reprogram the *X-Y Grabber*'s arm, so their new diamond-edged blade or a pen can be attached. The diamond-edged blade will let the robot cut shapes out of sheet metal; the pen will draw shapes.

Currently the *X-Y Grabber*'s hand moves horizontally or vertically—up and down or left and right. When it receives the polar *r-θ* coordinates of an object, it first converts them to *x-y* coordinates; then it extends or contracts the horizontal and vertical arms to reach the object, which it picks up with our *Grabbing Hand* mechanism. With the new project, which we are calling *Shape Cutter*, the blade or pen will move around a shape defined by a polar equation.

The old project and the new project are shown here.

They want the *Shape Cutter* to trace many different shapes. So far they have mentioned a circle, square, hexagon, pentagram, cardioid, *n*-leafed rose, and (something new to us) a trisectrix.

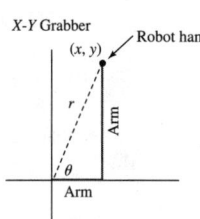
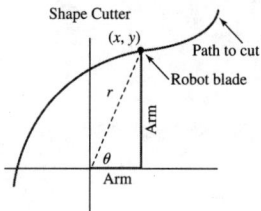

Your tasks are the following:

• Review the mathematics and programming that make the *X-Y Grabber* work.

• Determine how to adapt its programming for tracing out points on a curve.

• Develop a single equation that will draw a square or a hexagon.

• Learn the definitions of a pentagram and a trisectrix and determine their equations.

Engineering will take care of the blade design and operation.

Activity 6.1
First Look at the Robot Arm Problem

The mathematics of graphing is used in the computer programs that control movements of robots. In this section you'll get some experience in working with simple programs that control movement. If you don't have enough information to get a correct answer to a question, write your best guess; later you should get enough information to fully answer the question. You can assume that the point where the robot arm swivels is at the origin of the coordinate system.

1. Let's review the mathematics that makes the *X-Y Grabber* work. You saw in Chapter 5 how x- and y-coordinates can be converted to θ and r, where θ is the angle the arm makes with the positive x-axis and r is the length of the arm. A program that could control the *X-Y Grabber* would look like the following set of instructions.* As you read this and other programs in this chapter, try to think like the robot and imagine following each instruction exactly.

```
1  Wait for input
2  Receive polar position data r and θ
3  Compute x-coordinate (x)
4  Compute y-coordinate (y)
5  Move arm to rectangular point (x, y)
6  Perform task
7  Go to step 1
```

► Refer to the figure in the memo. What equations will be used in steps 3 and 4?
► Step 2 may be unclear. You will learn about polar coordinates in this chapter.

2. Here is a another program written in pseudocode. Read it carefully and try to figure out what would happen if a person or a robot followed the instructions precisely.

```
1  Begin at point A facing north
2  Walk ten paces
3  Turn right 90 degrees
4  Are you back at point A? If yes: go to step 5; If no: go to step 2
5  Stop
```

► If you followed these instructions, what shape would you trace?
► If step 3 were changed to Turn right 60 degrees, what kind of shape would be traced?
► How would you change the program to trace a triangle whose sides were 6 paces on each side?

3. Here is another program:

```
1  Set the value of x to −5
2  Compute the value of y if y = 2x − 5
3  Plot the point (x, y)
4  Increase x by 1
5  Is x < 5? If Yes, go to step 2; If no, go to step 6
6  Stop
```

► What are the first and last points that are plotted?
► How many points are plotted?
► What kind of path do the points trace?
► How would you change the program in order to produce points that are closer together?
► How would you change the program to connect all the points with straight lines?

---

*The programs in this chapter will be written not in an actual programming language, but in a language that programmers call *pseudocode*. When the programs are actually written for a computer, the pseudocode must be translated into a language the computer understands, such as Assembler, BASIC, FORTRAN, PASCAL, or C++.

6.1   Review of Two-Dimensional Graphing

The polar coordinate system is widely used in robotic operations. Before we introduce the polar coordinate system, let's review the coordinate system and equations that you know how to plot already.

Connecting Cartesian and Parametric Graphing

Up to this point you have used two kinds of coordinate systems: the Cartesian coordinate system and the system for plotting complex numbers. Both of these are based on distances from two perpendicular axes (as shown in Figure 6.1).

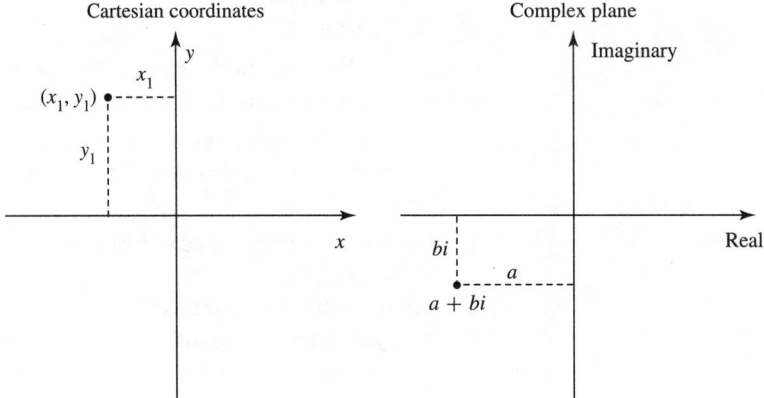

Figure 6.1

You have used two kinds of equations to produce graphs in the Cartesian coordinate system:

▶ Write y as a function of x. For example, $y = x^2$ produces a graph in the shape of a parabola. These equations can come in two forms.

  • If the equation is written in a form like $y = \dfrac{2x^3}{4x + 7}$, you can compute values of y from any value of x except $x = -\frac{7}{4}$ and then plot the pairs of values of x and y that fit the equation. If the equation is in this form, we say that **y is given in terms of x** or that **y is a function of x**.

  • If the equation is not written in the function form, it can sometimes be transformed to the form you need. For example, the equations $x + y = 15$ or $x^2 - 2y = 3x$ can be rewritten as $y = 15 - x$ and $y = \dfrac{3x - x^2}{-2}$ so that they can be treated as a function. However not all equations in x and y can be rewritten this way. For example, the equation $y = x \cos y$ cannot be transformed so that y is a function of x. (The equation can be written as $x = \dfrac{y}{\cos y}$ and in this case x is a function of y.) Even though there is a graph for such an equation, it cannot be drawn by hand or by calculator without great difficulty.

▶ Write x and y as functions of a third variable t. For example, the equations

$$\begin{cases} x = 5\cos t \\ y = 5\sin t \end{cases}$$

produce a graph in the shape of a circle of radius 5. You have seen that parametric equations have practical uses when t represents time or angle measure.

Equations and Cartesian Coordinates

The Cartesian coordinate system can be used to graph equations of two forms:

▶ Single equations using two variables
▶ Pairs of parametric equations in which two variables are written in terms of a third variable called a parameter

To distinguish between these two types of equations, we'll use the term **rectangular equations** when we mean an equation written in two variables.*

In robotics and fields such as navigation it is often necessary to translate from one kind of equation or coordinate system to another. Therefore we'll begin with an example in which a pair of parametric equations is translated to a form whereby y is expressed as a function of x.

Example 6.1

A ball is thrown upward from a height of 40 ft in such a way that the initial velocity in the x-direction is 20 fps and the initial velocity in the y-direction is 50 fps. Write the rectangular equation of the trajectory of the ball giving the height y in terms of the horizontal distance x.

Solution   You have learned how to write the parametric equations that describe this motion. They are

$$\begin{cases} x = 20t \\ y = 40 + 50t - 16t^2 \end{cases}$$

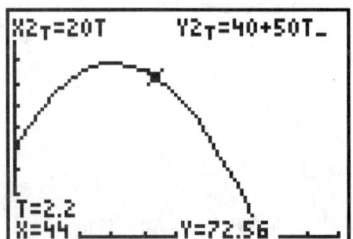

Figure 6.2

The resulting graph represents the trajectory of the ball (Figure 6.2). Remember that the trajectory is a picture of the actual parabola along which the ball travels.

But we have been asked to write the equation of the trajectory in terms of y and x only. To do that you need to look closely at the parametric equations and do some algebra.

(a) We want to write y in terms of x, but we have y in terms of t. We ask, "What is the connection between t and x?"

(b) Since $x = 20t$ we can write $t = \dfrac{x}{20}$ and then substitute $\dfrac{x}{20}$ for t in the equation for y.

(c) Substituting $\dfrac{x}{20}$ for t in the equation $y = 40 + 50t - 16t^2$, we obtain $y = 40 + 50\left(\dfrac{x}{20}\right) - 16\left(\dfrac{x}{20}\right)^2$.

---

*We will not use Func, the term that is used to refer to rectangular equations on the TI graphing calculators, because functions can be written with other kinds of equations as well.

The last equation is an answer to the question. It can be simplified to

$$y = 40 + 2.5x - 0.04x^2$$

The decimals in the last form of the equation are exact; but since they may not be exact, you will more often see an equation like this written with fractions instead of decimals so that the equation is exactly equivalent to the parametric equations. One form of the equation that uses fractions is

$$y = 40 + \frac{5x}{2} - \frac{x^2}{25}$$

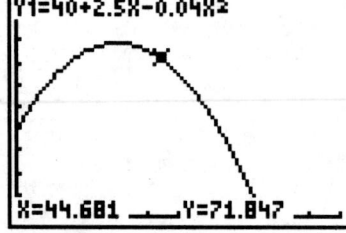

Figure 6.3

When the last equation is graphed in the calculator's Func mode you can see that the result shown in Figure 6.3 is identical to the graph in Figure 6.2. Both forms describe the trajectory of the ball. An important difference is that the parametric equations can be written directly from a description of the motion. It is not easy to write the equation in x and y from the same description of the motion. Can you see how to do it?

A summary of the technique of converting from two parametric equations to one rectangular equation follows.

Converting Parametric Equations to x-y Equations

The goal is to eliminate the parameter t and to express y in terms of x.

▶ Write t in terms of x.
▶ Substitute that expression for t into the equation of y in terms of t.
▶ Simplify if necessary or desired.

Some parametric equations are not as easy to change into a rectangular equation, as we will see in Examples 6.2 and 6.3.

Example 6.2

A circle centered at the origin and with radius 6 has the parametric equations

$$\begin{cases} x = 6\cos t \\ y = 6\sin t \end{cases}$$

Rewrite these parametric equations as a single rectangular equation in y and x.

Solution Unlike the equations in Example 6.1, it is not as easy to write t in terms of x. However the Pythagorean theorem will help.

We begin by thinking about the equation $x = 6\cos t$. Dividing both sides of the equation by 6 we have $\cos t = \frac{x}{6}$, so t can be represented as an angle in a right triangle, like $\triangle ABC$ in Figure 6.4. The side $\overline{AC}$ has length x and the hypotenuse $\overline{AB}$ has length 6. Since $y = 6\sin t$, you can see that the third side of the triangle has length y.

Finally, to express y in terms of x we use the Pythagorean theorem to get $x^2 + y^2 = 36$, a familiar equation for a circle of radius 6 and centered at the origin.

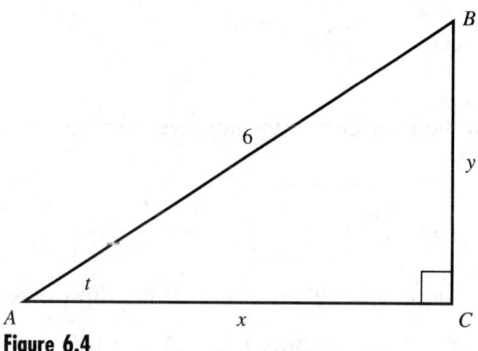

Figure 6.4

If you wanted to write the equation $x^2 + y^2 = 36$ where y is a function of x, then you would have to rewrite this equation as $y = \sqrt{36 - x^2}$. But x and y both can be negative as well as positive, so we need to write two equations:

$$y = +\sqrt{36 - x^2} \quad \text{and} \quad y = -\sqrt{36 - x^2}$$

Therefore the original parametric equations are equivalent to two equations in y and x.

Example 6.3

An ellipse centered at the origin with major axis in the x-direction of 12 and minor axis in the y-direction of 8 has the parametric equations

$$\begin{cases} x = 6\cos t \\ y = 4\sin t \end{cases}$$

Rewrite these parametric equations as a single rectangular equation in y and x.

Solution   This solution begins with the diagram in Figure 6.4 except that y is not the length of the side $\overline{BC}$. Figure 6.5 shows a way to picture y and x in $\triangle ADE$.

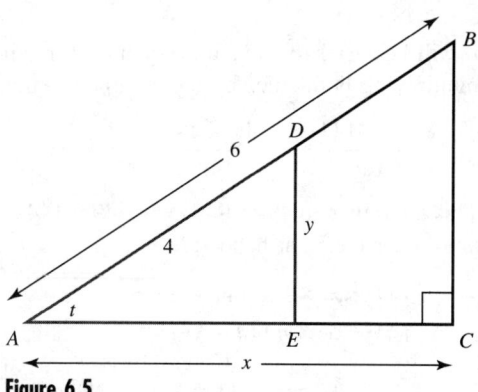

Figure 6.5

The angle t remains fixed and the hypotenuse is changed to 4 because $y = 4\sin t$. $AE = 4\cos t$, and since $\cos t = \dfrac{x}{6}$, then

$$AE = 4\left(\frac{x}{6}\right) = \frac{2x}{3}$$

From the Pythagorean theorem we then have $\left(\dfrac{2x}{3}\right)^2 + y^2 = 16$, or

$$y = +\sqrt{16 - \left(\frac{2x}{3}\right)^2} \quad \text{and} \quad y = -\sqrt{16 - \left(\frac{2x}{3}\right)^2}$$

These are the equations of an ellipse written in rectangular form. You will usually see the equation of an ellipse written in the simpler form $\dfrac{x^2}{6^2} + \dfrac{y^2}{4^2} = 1$. From this equation you can see immediately that the length of the major axis in the x-direction is 12 and the length of the minor axis in the y-direction is 8. ∎

Notice that in both Examples 6.2 and 6.3 the coordinates of a point on the graph can be determined either from the parametric equations or from the rectangular equation in x and y, but that the parametric equations are simpler to use.

Although the equation $\dfrac{x^2}{6^2} + \dfrac{y^2}{4^2} = 1$ quickly tells you important information about the ellipse, in order to be useful in computations we must convert it to a form that expresses y in terms of x. Example 6.4 shows how this can be done.

Example 6.4

An ellipse is centered at the origin and has an axis in the x-direction of 24 and an axis in the y-direction of 14. Write the rectangular equations of this ellipse as (a) a single equation in x^2 and y^2 and (b) two equations that express y in terms of x.

Solution   The solutions follow directly from the solution to the second part of Example 6.2.

(a) Go through the second part of Example 6.2 with 12 and 7 replacing 6 and 4. The results are

$$\frac{x^2}{144} + \frac{y^2}{49} = 1$$

(b) It will be easier to solve this equation for y if we first eliminate the fractions by multiplying both sides by the common denominator 49×144:

$$\frac{49 \times 144}{144}x^2 + \frac{49 \times 144}{49}y^2 = 49 \times 144$$

This simplifies to $49x^2 + 144y^2 = 49 \times 144$. Now subtract $49x^2$ from both sides and factor the right-hand side

$$144y^2 = 49 \times 144 - 49x^2$$
$$144y^2 = 49(144 - x^2) \qquad \text{Factor.}$$
$$y^2 = \frac{49}{144}(144 - x^2) \qquad \text{Divide by 144.}$$

Taking the square root of both sides produces the two answers.

$$y = +\frac{7}{12}\sqrt{144 - x^2}$$

$$y = -\frac{7}{12}\sqrt{144 - x^2}$$

■

Example 6.5

An ellipse is centered at the origin and has an axis in the x-direction of $2a$ and an axis in the y-direction of $2b$. Write the rectangular equations of this ellipse as (a) a single equation in x^2 and y^2 and (b) two equations that express y in terms of x.

Solution (a) Use the solution of Example 6.4(a) with a replacing 12 and b replacing 7. The results are

$$\frac{x^2}{a^2} + \frac{y^2}{b^2} = 1$$

(b) Use the solution of Example 6.4(b) with a replacing 12 and b replacing 7. This produces

$$y = +\frac{b}{a}\sqrt{a^2 - x^2}$$

$$y = -\frac{b}{a}\sqrt{a^2 - x^2}$$

■

These last two equations are equivalent to the more complicated pair that resulted from Example 6.3. They reveal more clearly how the ellipse is related to a circle. The equations $y = \pm\sqrt{a^2 - x^2}$ graphed together form a circle centered at the origin with radius a. If you multiply the y-values by $\frac{b}{a}$, then the circle is stretched or shrunk in the y-direction. If $b > a$ it is stretched vertically, and if $b < a$ it is stretched horizontally.

Summary: Equations of Ellipses

(Refer to Figure 6.6.)

► Parametric equations

$$\begin{cases} x = a\cos t \\ y = b\sin t \end{cases}$$

► Rectangular equations

$$\frac{x^2}{a^2} + \frac{y^2}{b^2} = 1$$

or $\quad y = \pm\frac{b}{a}\sqrt{a^2 - x^2}$

► If $b = a$ the ellipse becomes a circle.

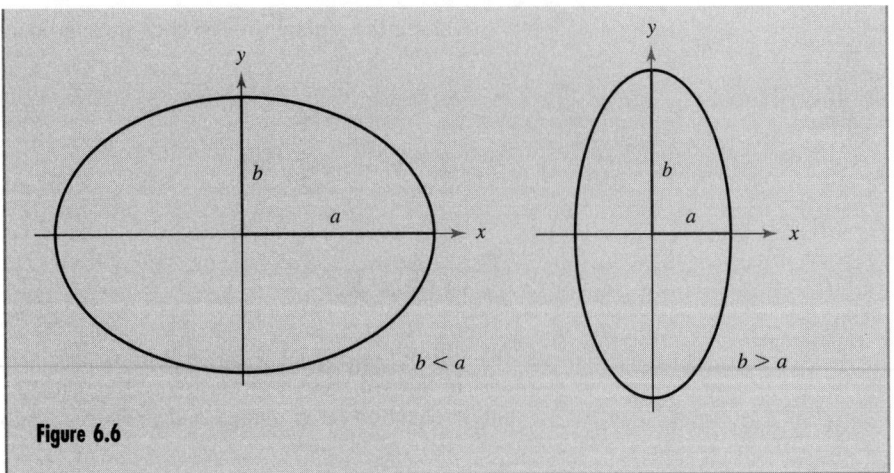

Figure 6.6

Now that you have learned some general information about parametric and rectangular equations for an ellipse, Example 6.6 shows you how to use that information to complete a part of the Chapter Project.

Example 6.6

Write a pseudocode program that would direct a robot to plot the points of an ellipse with $a = 24$ and $b = 18$. Write one program to cover the equation for an ellipse in (a) parametric equations and the other for (b) two rectangular equations in x and y.

Solutions   The programs will be similar to those presented in Activity 6.1.

(a) 1 Comment: Parametric Equations
    2 Set mode to degrees
    3 Let $t = 0$
    4 Compute $x = 24 \cos(t)$
    5 Compute $y = 18 \sin(t)$
    6 Plot the point (x, y)
    7 Increase t by 10
    8 Is $t > 360$? If yes: go to step 9; If no: go to step 4
    9 Stop

Make sure you understand this program before going on to the next part of the solution. What would happen in step 7 if t were increased by 5 or 15 instead of 10?

(b) 1 Comment: Rectangular Equation
    2 Let $x = -24$
    3 Compute $y = +\frac{3}{4}\sqrt{24^2 - x^2}$
    4 Plot the point (x, y)
    5 Compute $y = -\frac{3}{4}\sqrt{24^2 - x^2}$
    6 Plot the point (x, y)
    7 Increase x by 1
    8 Is $x > 24$? If yes: go to step 9; If no: go to step 3
    9 Stop

Notice that two points are plotted for each value of x, just as you would expect. Also note that this program plots points only; a curve is not drawn between the points.   ■

Before we move on to Section 6.2 and polar coordinates, the final part of this section introduces an important application of parametric equations to electronics.

Parametric Equations and the Oscilloscope

We now turn to an important application of parametric equations, one that will allow us, among other things, to study the underlying mathematics of the oscilloscope and to create some unusual shapes on the graphing calculator. In this example of parametric equations you will see how to model the voltages that are applied to the horizontal and vertical plates of an oscilloscope. An **oscilloscope** is an electronics instrument designed to display periodic functions that represent voltage combinations. In the simplest use of the oscilloscope an alternating current is passed across one set of plates and the oscilloscope displays a graph of the current vs. time.

A more complex experiment involves alternating currents flowing across two pairs of parallel plates. Each pair of plates is perpendicular to the other (see Figure 6.7).

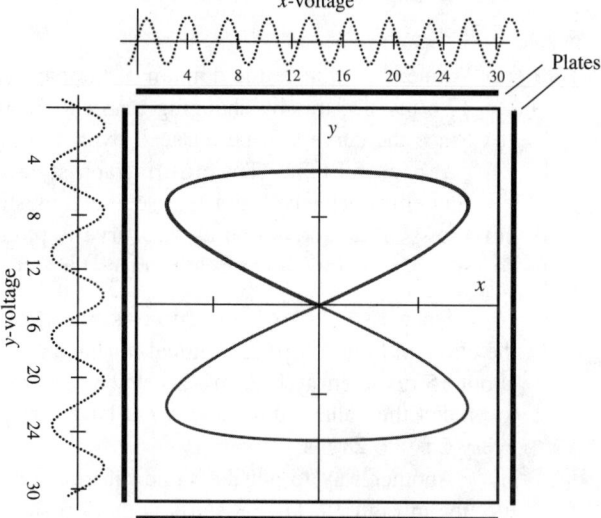

Figure 6.7

The curve displayed on the oscilloscope represents a two-dimensional view of the two currents. Any point (x, y) on the oscilloscope curve gives the two voltages at some time. The figure-eight curve shown in Figure 6.7 is characteristic of the voltage graphs shown on the top and side. That is, the x-voltage has twice the frequency of the y-voltage.

The experiment pictured in Figure 6.7 can be modeled with the parametric equations

$$\begin{cases} x = V_0 \sin 2t \\ y = V_0 \sin t \end{cases}$$

Example 6.7

Plot on your calculator the parametric equations $\begin{cases} x = V_0 \sin 2t \\ y = V_0 \sin t \end{cases}$ with $V_0 = 8$.

Solution   Follow the same steps you used in creating the ellipse and the circles. Your settings can be in degrees or radians. The resulting graph, plotted in radians, is shown in Figure 6.8.   ■

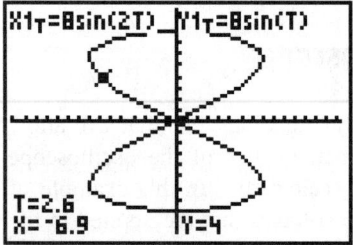

Figure 6.8

Look at the coordinates given in Figure 6.8. First let's suppose that the unit of time is milliseconds (ms) and the unit of voltage is volts (V). The meaning of the coordinates is that at $t = 2.6$ ms the voltage in the x-direction (call it V_x) is -6.9 V, and in the y-direction $V_y = 4.0$ V. What do you think the voltages would be 15 milliseconds later? As long as the voltages continue to be applied, the oscilloscope continues to plot the curve, whose basic shape does not change. We'll explore this idea more in Example 6.8.

Example 6.8

Plot the graph in Figure 6.8 and determine the voltage about 15 ms after $t = 2.6$ ms.

Solution   (a) We need to change the domain of the parameter so that t will reach $15 + 2.6 = 17.6$ ms. Do this by changing `Tmax` to 18 in the WINDOW menu.
(b) Since the curve will be retraced over itself, you will find the graph style called `path` useful here. The `path` graph style is indicated on the Y= screen by a tiny ellipse attached to a line segment, as shown in Figure 6.9.
(c) Press `GRAPH` and watch as the curve is plotted. After $t = 2\pi \approx 6.28$ you will see that the traveling circle retraces the curve.

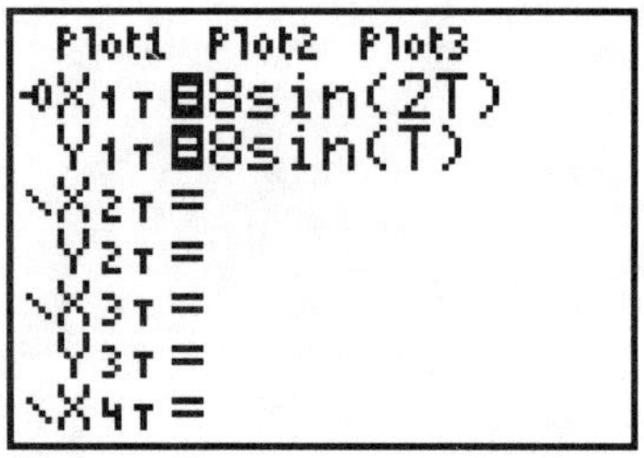

Figure 6.9

Since $18/6.28 \approx 3$, the curve is traced almost three times. You can see that the curve in Figure 6.10 is identical in shape to the curve in Figure 6.8. The voltages about 15 ms later, at 17.6 ms, are $V_x = -4.8$ V and $V_y = -7.6$ V. You can begin to predict the values of V_x and V_y at later times by realizing that the graph repeats every $2\pi \approx 6.28$ ms.

Another way to get the same information is to look at a table. For example, the line in Figure 6.11 that starts with 17.6 shows $V_x = -4.8$ V and $V_y = -7.6$ V. You need to remember here that the column headed X_{1T} represents V_x and the Y_{1T} column represents V_y.

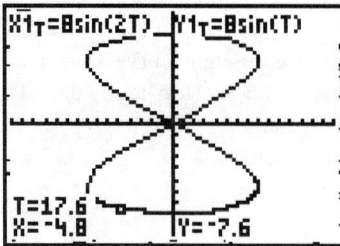

Figure 6.10

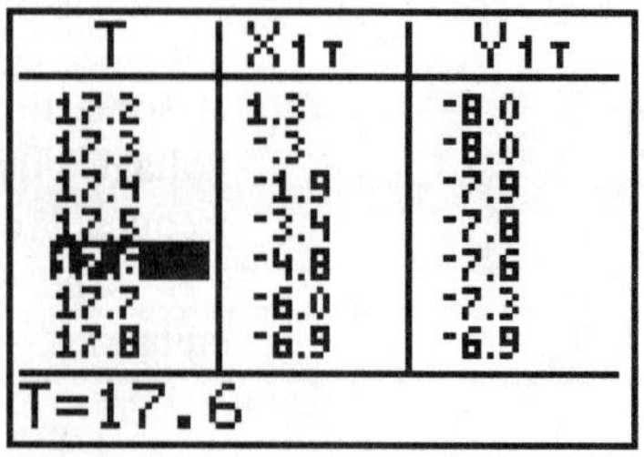

T	X_{1T}	Y_{1T}
17.2	1.3	-8.0
17.3	-.3	-8.0
17.4	-1.9	-7.9
17.5	-3.4	-7.8
17.6	-4.8	-7.6
17.7	-6.0	-7.3
17.8	-6.9	-6.9
T=17.6		

Figure 6.11

Lissajous Curves

You can produce interesting graphs on your own by varying the coefficients in the parametric equations

$$\begin{cases} x = A_x \sin B_x t \\ y = A_y \sin B_y t \end{cases}$$

Curves of this type are called **Lissajous curves**.

For example, Figures 6.12 and 6.13 show two of the many Lissajous curves that are possible.

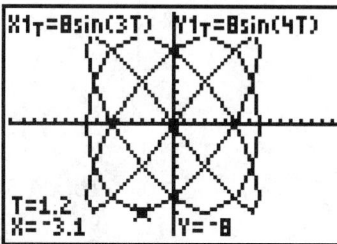

Figure 6.12

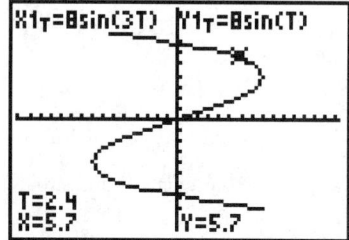

Figure 6.13

Calculator Lab 6.1

Patterns in Lissajous Curves

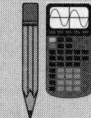

Experiment with values of A_x, A_y, B_x, and B_y to produce a series of Lissajous curves and determine the effects each of these four variables has on the curves.

Procedures

1. Keep B_x and B_y fixed; change A_x and A_y, but keep them equal in value.
2. Keep A_x and A_y fixed; change B_x and B_y, but keep them equal in value.
3. Fix B_x, B_y, and A_x. Increase and decrease A_y.
4. Fix B_x, A_x, and A_y. Increase and decrease B_y.

Answer the following questions.

1. Which combinations of B_x and B_y produce curves like Figure 6.13, whose ends do not connect? Can you make a general statement about these combinations?
2. In the following two pairs of equations B_x is twice the value of B_y. How do these two pairs compare in appearance?

$$\begin{cases} x = 8 \sin 2t \\ y = 8 \sin t \end{cases} \quad \text{and} \quad \begin{cases} x = 8 \sin 4t \\ y = 8 \sin 2t \end{cases}$$

Make a general statement about pairs like these.

3. What is the connection between the coefficients and the number of "loops" each curve has at the *top* of the graph (Figure 6.12 has four) and the number of "loops" each has on the *side* of the graph (Figure 6.12 has three).
4. How are any of the results of your explorations different if the voltages are out of phase by 90° or $\frac{\pi}{2}$? That is, what if the equations are

$$\begin{cases} x = A_x \cos B_x t \\ y = A_y \sin(B_y t + 90°) \end{cases}$$

Section 6.1 Exercises

In Exercises 1–6, (a) graph the given parametric equations, (b) write the rectangular equations for these parametric equations, and (c) if possible, graph the rectangular equation from (b) and check that it matches the graph in (a).

1. $\begin{cases} x = 5t \\ y = 12 - 10t + 15t^2 \end{cases}$

2. $\begin{cases} x = 4t^2 + 6t - 8 \\ y = 2t \end{cases}$

3. $\begin{cases} x = 4\cos t \\ y = 9\sin t \end{cases}$

4. $\begin{cases} x = t \\ y = \dfrac{4}{t} \end{cases}$

5. $\begin{cases} x = \dfrac{1}{\sin t} \\ y = \dfrac{1}{\cos t} \end{cases}$

6. $\begin{cases} x = \dfrac{1}{\sin t} \\ y = \tan t \end{cases}$

In Exercises 7–10 sketch each of the parametric equations.

7. This will produce the graph of an astroid.

$$\begin{cases} x = \cos^3 t \\ y = \sin^3 t \end{cases}$$

[Enter $\cos^3 t$ in your calculator as $(\cos(T))^3$.]

8. This will produce the graph of a prolate cycloid:

$$\begin{cases} x = t + 2\sin t \\ y = 1 - 2\cos t \end{cases}$$

9. $\begin{cases} x = 3\cos^4 t \\ y = 3\sin^3 t \end{cases}$

10. $\begin{cases} x = 3\cos^3 2t \\ y = 3\cos^3 t \end{cases}$

11. Read the following program written in pseudocode and describe what would happen if a person or robot followed the code exactly.

```
1   Begin at point A facing north
2   Walk ten paces
3   Turn right 45 degrees
4   Are you back at point A? If yes: go to
    step 5; If no: go to step 2
5   Stop
```

12. Read the following program written in pseudocode and answer the questions at the end of the program.

```
1   Set the value of x to -3
2   Compute the value of y if y = x² - 4
3   Plot the point (x, y)
4   Increase x by 1
```

```
5   Is x < 5? If yes, go to step 2; If no,
    go to step 6
6   Stop
```

(a) What would happen if a person or robot followed the code exactly?

(b) What are the first and last points that are plotted?

(c) How many points are plotted?

(d) What kind of path do the points trace?

13. (a) Write a program in pseudocode that will result in a rectangle being traced.

   (b) Ask a classmate to follow your program and hand you the result. Did he or she draw a rectangle? If not, work with your classmate to revise your program.

14. Reconsider the program in Exercise 12.

   (a) How can you change the program so that twice as many points are plotted?

   (b) How can you change the original program so that each point is connected to the one(s) before and after it.

15. Write a program in pseudocode that will draw six different rectangles that each have an area of 16 in.^{2}.

16. Write a program in pseudocode that begins with Exercise 15. Select a classmate to test—and, if necessary, help revise—the program. Write the program so that new classmates will be added to test and help revise the program until it is correct.

In Exercises 17–22, (a) predict the ratio of the number of loops along the top to the number of loops along the side and (b) graph the Lissajous curve.

17. $x = \sin 3t, \; y = \cos 5t$

18. $x = \sin 4t, \; y = \cos 2t$

19. $x = \sin 6t, \; y = \cos 2t$

20. $x = \sin t, \; y = \cos 2t$

21. $x = \sin 4t, \; y = \cos 3t$

22. $x = \sin 2t, \; y = \cos 3t$

23. *Electronics*   Two voltages $V_1 = 110 \sin 120\pi t$ and $V_1 = 110 \sin\left(120\pi t - \frac{\pi}{2}\right)$ are applied respectively to the horizontal and vertical plates of an oscilloscope.

(a) Sketch the resulting Lissajous curve in radian mode for $0 \le t \le \dfrac{7\pi}{24}$.

(b) Sketch the figure again for $0 \le t \le 2\pi$.

24. *Electronics*

(a) Double the frequencies of the voltages in Exercise 23 and sketch the curve.

(b) How did changing the frequency affect the graph?

(c) Cut the frequencies of the voltages in Exercise 23 in half and sketch the curve.

(d) How did changing the frequency affect this graph?

25. Consider the parametric equations

$$\begin{cases} x = 3\sin(2t + D_x) \\ y = 2\cos 4t \end{cases}$$

Use the `path` graph style on your calculator and experiment with the values of D_x to produce a series of Lissajous curves and determine the effect that D_x has on these curves.

26. Consider the parametric equations

$$\begin{cases} x = 3\sin(2t + D_x) \\ y = 2\cos(4t + D_y) \end{cases}$$

Use the `path` graph style on your calculator and experiment with the values of D_x and D_y to produce a series of Lissajous curves and determine the effect that D_x and D_y have on these curves.

6.2   Polar Coordinates

In this section we introduce a new kind of coordinate system, **polar coordinates**, and a new kind of equation, the **polar equation**. Polar equations can produce some spectacular graphs, they will help you review your trigonometry, and they have several applications that we will see later in the chapter.

Plotting Points in Polar Coordinates

In polar coordinates the location of points is based on the distance from a single point—called the **pole** of the graph—and a line that passes through the pole—the **polar axis**. The polar axis is located in the same position as the positive x-axis in a Cartesian coordinate system and the pole is located at the origin; the $\frac{\pi}{2}$-axis is located in the same position as the positive y-axis. Figure 6.14 shows that a point, point P for example, is located a directed distance $r = 6$ from the pole and at an angle $\theta = 40°$ from the polar axis.

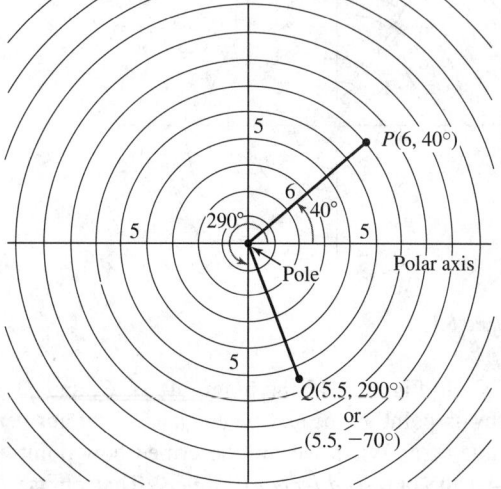

Figure 6.14

In Figure 6.14 point Q is located by two different coordinate pairs, $(5.5, 290°)$ and $(5.5, -70°)$. In fact you can probably write several other different pairs that locate this same point. Unlike Cartesian coordinates, where every point has only one pair of coordinates, in polar coordinates any point can be represented by an infinite number of coordinate pairs. For example, point Q could also be located by the coordinates $(5.5, 650°)$ or $(5.5, -430°)$. We arrived at the angles 650° and $-430°$ by adding and subtracting 360°. Thus $650° = 290° + 360°$ and $-430° = -70° - 360°$.

Next we'll get more practice at locating polar points by plotting a few polar equations.

Plotting Polar Equations

A polar equation is an equation in which r and θ are the variables *and* the graph is plotted in a polar coordinate system. Just as you would make a table of values for x and y in order to plot the graph of $y = x^2$, in polar coordinates you make a table of the values of r and θ that satisfy the polar equation.

Example 6.9

Plot by hand the five points of the polar equation $r = 10\cos\theta$ when $\theta = 0°$, 30°, 60°, 90°, and 120°.

Solution   To plot these five points by hand, we make a table and then locate the points as shown in Figure 6.15.

θ	0°	30°	60°	90°	120°
r	10	8.7	5	0	-5
Label	A	B	C	D	E

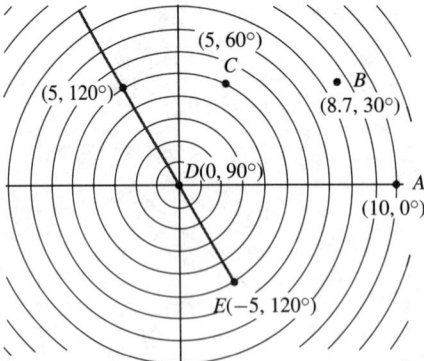

Figure 6.15

In Figure 6.15 the points A, B, C, and D are located as described above. But why is point E located where it is? The answer is in how we interpret a negative value of r. When we first described how points are plotted in polar coordinates we used the phrase *directed distance*. That phrase means that the value of r could be either positive or negative depending on its direction.

How is the number -5 located on a number line? It is 5 units from 0 but in the opposite direction from the location of 5. Negative values of r are handled in a similar way. To plot a point in polar coordinates with a negative value of r, do the following:

▶ Locate the angle.
▶ Measure the distance from the pole equal to the absolute value of r but in the opposite direction along the same line of the angle. That is, the point $E(-5, 120°)$ will be on the line that makes a 120° angle with the polar axis, but it will be on the opposite side of the pole from the point $(5, 120°)$.

You may be wondering how to connect the dots in the graph. We'll consider that in Example 6.10, which follows.

Example 6.10

Plot $r = \theta$ by hand on polar graph paper. Use θ measured in radians from 0 to 2π.

Solution   Make a table like the one below and fill in the values of r. Two values are included in the table and all the points are plotted on the graph in Figure 6.16. For example, if $\theta = \dfrac{5\pi}{4}$ then $r = \dfrac{5\pi}{4} \approx 3.9$.

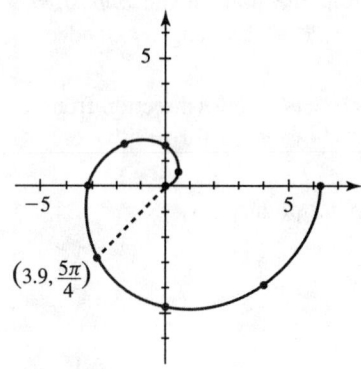

$(3.9, \frac{5\pi}{4})$

Figure 6.16

θ	0	$\frac{\pi}{4}$	$\frac{\pi}{2}$	$\frac{3\pi}{4}$	π	$\frac{5\pi}{4}$	$\frac{3\pi}{2}$	$\frac{7\pi}{4}$	2π
r	0					3.9			

The points should be connected in the order of increasing values of the angle θ. In this example the points appear to follow the pattern of a spiral curve. As you do more of these plots you will gain more experience that will help you connect the points correctly. ■

Why did we work Example 6.10 in radians? If we had worked the problem in degrees, then, for example, we would have $r = 45°$. But r represents a length from the pole and 45° is a direction, not a length. Therefore, we had to switch to the "dimensionless" radian in order to get meaningful values for r. In Example 6.11 we will get more practice with negative angles.

Example 6.11

Plot $r = \theta$ by hand using negative values of θ measured in radians from $-2\pi \le \theta \le 0$.

Solution   The negative angles give values of r that are the same in absolute value, but negative. For example, if $\theta = -\dfrac{5\pi}{4}$ then $r = -\dfrac{5\pi}{4} \approx -3.9$. Negative angles are measured in the clockwise direction and negative values of r are handled as described following Example 6.9. Study Figure 6.17 to see how the point $\left(-3.9, -\dfrac{5\pi}{4}\right)$ is plotted.

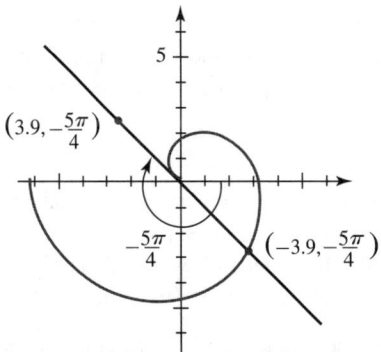

Figure 6.17

Plotting Polar Points

▶ If θ is positive, measure θ degrees or radians in a counterclockwise direction from the polar axis (the positive x-axis). One side of the angle θ is the polar axis and the other side is called the **terminal side** of the angle.

- If r is positive, plot the point r units from the pole along the terminal side of the angle.
- If r is negative, plot the point $|r|$ units from the pole in the *opposite* direction on the extension of the terminal side of the angle extended through the pole.

▶ If θ is negative, measure $|\theta|$ degrees or radians in a *clockwise* direction from the polar axis (the positive x-axis). Repeat the above steps for positive and negative values of r.

Figure 6.18 shows each of these types of polar points.

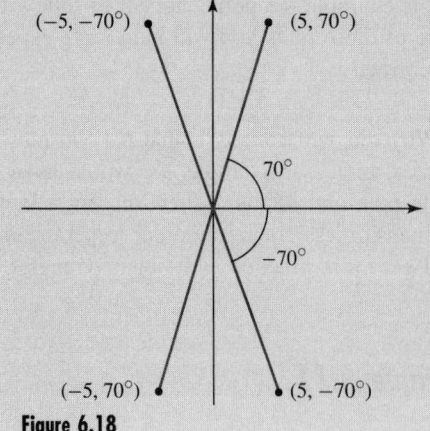

Figure 6.18

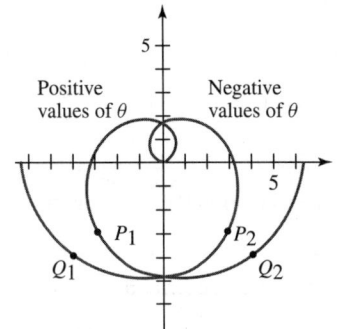

Figure 6.19

The graph in Figure 6.17 is similar to the one in Figure 6.16. How would you describe the similarities and differences between the two graphs?

Figure 6.19 shows the two graphs drawn on the same set of axes. Look at the location of the two points $P_1\left(3.9, \dfrac{5\pi}{4}\right)$ and $P_2\left(-3.9, -\dfrac{5\pi}{4}\right)$. One way to describe these two points is to say that they are mirror images of each other. The mirror is the y-axis. Now pick another pair of points on the graph, for example,

$Q_1\left(5.5, \dfrac{7\pi}{4}\right)$ and $Q_2\left(-5.5, -\dfrac{7\pi}{4}\right)$. You can see that they too are mirror images of each other and that the mirror is once again the y-axis. In fact each of the points with negative values of θ is a mirror image of a point with positive value of θ. A graph in which all points have a mirror image on the other side of the y-axis is said to have **symmetry about the y-axis**.

Example 6.12

Plot $r = 6$ by hand on polar graph paper. Use θ measured in degrees where $-360° \le \theta \le 360°$.

Solution   This table is very easy to complete. No matter what the value of θ, the value of r is 6. Therefore all the points are a distance of 6 units from the pole. Some of the points are shown in Figure 6.20. The graph of $r = 6$ is a circle of radius 6 centered at the origin.

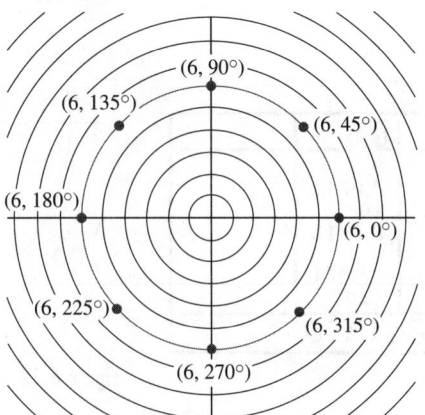

Figure 6.20

Example 6.12 shows a third way to graph a circle centered at the origin. As a summary, a list of the three methods we have discussed follows.

Three Equations for a Circle Centered at the Origin

The equation for a circle centered at the origin $(0, 0)$ with a radius equal to a units has the following forms:

► Cartesian coordinates: $x^2 + y^2 = a^2$ or $y = \pm\sqrt{a^2 - x^2}$
► Parametric equations plotted on a Cartesian coordinate system: $x = a\cos t$ and $y = a\sin t$
► Polar coordinates: $r = a$

Example 6.13

Plot $r = 10\sin\theta$ (a) by hand and (b) by graphing calculator. Use θ measured in degrees for $0° \le \theta \le 360°$.

Solution (a) First make a table of values. To let the calculator produce a table, first set the calculator in both polar and degree modes by using the MODE menu and setting the graph mode to `Pol` (Polar). To enter the polar equation $r = 10 \sin\theta$, press `Y=` 10 `SIN` `X.T.θ.n` `)`. Next press TBLSET and set `TblStart` to 0 and Δ `Tbl` to 45. Finally press `2nd` [TABLE]. The result in Figure 6.21 shows seven of the points. Additional points are seen by scrolling to the bottom of the table.

When the calculator is in polar mode, pressing the `X.T.θ.n` key puts a θ on the screen.

When you plot the points shown in Figure 6.21, as well as the points when $\theta = 315°$ and $\theta = 360°$, you will get only the four distinct points shown in Figure 6.22. The nine entries in the table of the calculator make only four different points. In one case, that is because $(10, 90°) = (-10, 270°)$. The other cases are similar.

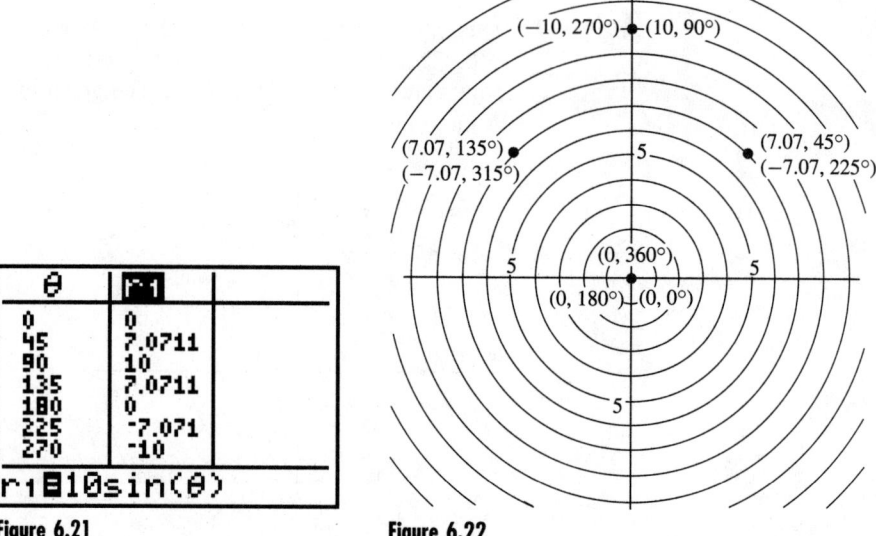

Figure 6.21

Figure 6.22

What does the graph in this example actually look like? One way to figure this out is to plot more points. If you change the value of ΔTbl to 5 and plot these points, the graph looks like the one shown in Figure 6.23. The four largest points remind you that, for this equation, angles separated by 45° give only a small portion of the graph. It looks as if the graph of $r = 10 \sin\theta$ is a circle of radius 5 and centered at $(5, 90°) = \left(5, \dfrac{\pi}{2}\right)$.

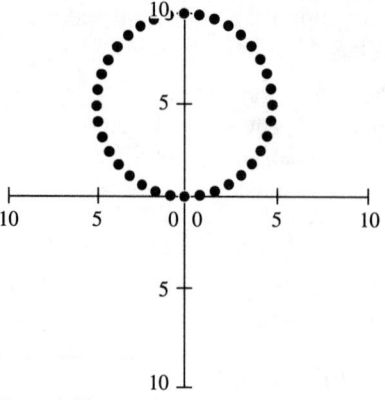

Figure 6.23

(b) To graph the curve on your calculator press ⬛ZOOM⬛ 6 (ZStandard) ⬛ZOOM⬛ 5 (ZSquare). Next press ⬛WINDOW⬛ and set θstep= 2 and then press ⬛GRAPH⬛. The result is shown in Figure 6.24.

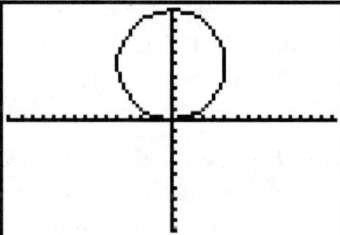

Figure 6.24   ◼

If you trace this curve you will get something like Figure 6.25. Notice that this shows the value of θ where the cursor is located and the x- and y-coordinates of that point. You need to calculate $r = \sqrt{x^2 + y^2} \approx \sqrt{4.33^2 + 2.5^2} \approx 5$.

If you would like the polar coordinates of the point instead of the Cartesian coordinates, press ⬛2nd⬛ [FORMAT] and select PolarGC. Pressing ⬛GRAPH⬛ ⬛TRACE⬛ produces Figure 6.26. Notice that the cursor is at the point with polar coordinates $(5, 30°)$.

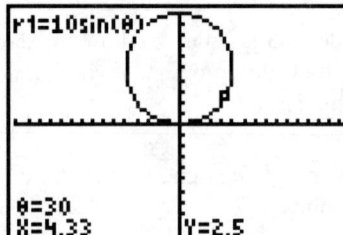   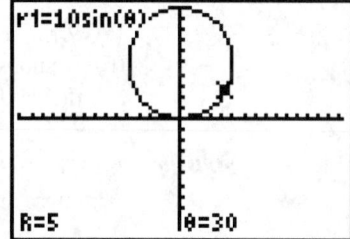

Figure 6.25   **Figure 6.26**

In order to graph a polar equation on a calculator such as a *TI-83*, you should follow the six steps in the following box. (Step 3 is optional.)

Graphing Polar Equations on a Calculator

1. In the MODE menu, set the graph mode to Pol (Polar).
2. In the Y= menu fill in the polar equation you want to plot. Use the ⬛X,T,θ,n⬛ key when you want to enter the parameter θ.
3. If you want the values of r and θ to appear when the curve is traced, press ⬛2nd⬛ [FORMAT] to select PolarGC.
4. In the WINDOW menu, set the graph window settings to the values you need in order to show the graph well. Notice that you have to set values for x, y, and θ. Here θstep refers to the increment in the θ-values.
5. Press ⬛GRAPH⬛.
6. Repeat steps 4 and 5 until the graph looks the way you want. The calculator's Table feature can also help you figure out the best graph window settings.

Activity 6.2
Can You Predict the Result?

You just saw that the graph of the polar equation $r = 10\sin\theta$ is a circle of radius 5. Try to make a guess about the shape and location of the graph produced by each of the following polar equations. Verify your guess by graphing each equation on your calculator.

1. $r = 10\cos\theta$ (*Hint:* Look at the points that were plotted in Example 6.9.)
2. $r = -10\sin\theta$ (*Hint:* Each point has a value of r with the opposite sign as the corresponding point in Figure 6.23.)
3. $r = -10\cos\theta$

To complete our study of the polar form of a circle, Example 6.14 shows how to program a robot to trace out the curve. This example will be different from Example 6.6 (on page 364), which only plotted points. The program must connect the points in order to design a program that is able to guide a cutting arm.

Example 6.14

(a) Write a pseudocode program to draw the circle shown in Figure 6.24.
(b) Demonstrate how this program is similar to the way the graphing calculator draws the same circle.

Solution (a) The following program is one way to draw a curve. It draws segments between consecutive points.

```
1   Comment: Polar equation of circle
2   Set mode to degrees; set graph mode to polar
3   Let θmin = 0; Let θmax = 360; Let θstep = 7.5
4   Let θ1 = θmin
5   Compute r1 = 10sin(θ1)
6   Let θ2 = θ1 + θstep
7   Compute r2 = 10sin(θ2)
8   Draw a segment between the point (r1, θ1) and the point (r2, θ2)
9   Increase θ1 by θstep
10  Is θ2 ≥ θmax? If yes: go to step 12; If no: go to step 5
11  Stop
```

Study the program before going on to the second part of the solution (designed to increase your understanding of the program).

(b) The program in part (a) is very similar to the program that the graphing calculator runs when it plots a polar curve:

▶ Step 2 sets the calculator's MODE menu.
▶ Step 3 is similar to what you see in the WINDOW menu (Figure 6.27). These are the settings that result from selecting ZStandard (Zoom Standard) and then ZSquare (Zoom Square, which makes circles look round).
▶ The calculator's WINDOW menu also includes the boundaries of the x-axis and the y-axis. We'll assume that we don't need to include that information in the curve-drawing program. ■

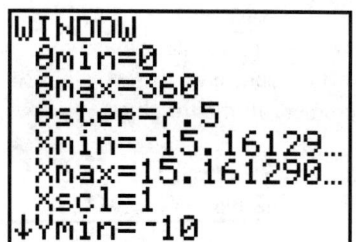

Figure 6.27

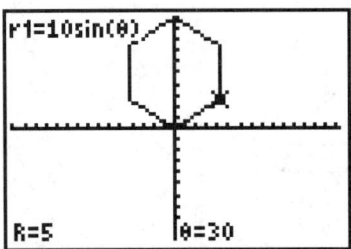

Figure 6.28

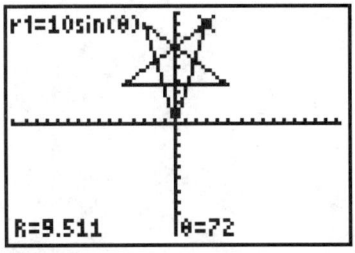

Figure 6.29

Think about what difference the value of θstep makes in the picture that the robot or the calculator would produce. θstep is the angular difference between two consecutive points. Therefore if θstep is smaller, the points would be closer together and the graph would look smoother. You'll see that θstep will be important when we need to get a more accurate picture of more complicated polar curves. If θstep is a larger number, the points are farther apart and the curve looks very different. For example, Figure 6.22 shows the points that are produced if θstep $= 45°$; Figures 6.28 and 6.29 show the graphs produced if θstep $= 30°$ and θstep $= 72°$. The five-pointed star shown in Figure 6.29 is called a pentagram. This looks something like a five-pointed star you may have drawn as a child, with the sides going through the "middle" and forming a pentagon inside the pentagram.

Try this yourself with different values of θstep. See if you can find other values of θstep that produce graphs identical to the ones shown here. These examples show that θstep makes a big difference in the appearance of a graph. You wouldn't want your robot to produce a hexagon or a five-pointed star if the customer has asked for a circle!

It would seem natural that for greatest accuracy one should use the smallest possible value of θstep. However, if you try something like θstep $= 0.05$, you will see that the plot is painfully slow. Any program that cuts out shapes must have a value of θstep that makes a balance between speed and accuracy.

Patterns in Polar Graphs

Polar equations can produce some unusual and attractive graphs that would be very difficult to plot in Cartesian coordinates. Some examples follow.

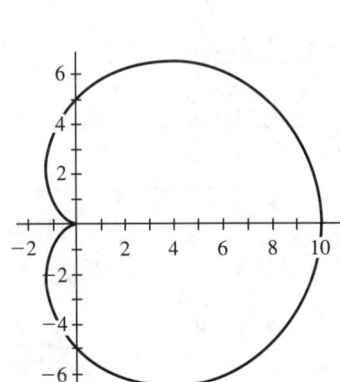

Figure 6.30

1. The **cardioid** is shaped somewhat like a heart. The cardioid in Figure 6.30 is the graph of $r = 5\cos\theta + 5$.

   Figure 6.31 shows the graph of another cardioid, this one with the equation $r = 5\sin\theta + 5$.
2. A variation of the cardioid is the **trisectrix** (shown in Figure 6.32). Its equation is $r = 10\cos\theta + 5$.

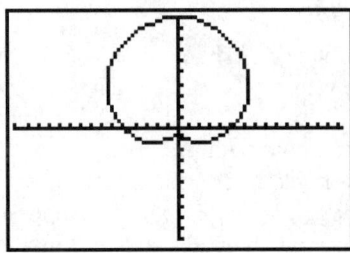

Figure 6.31

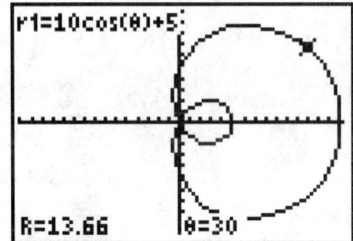

Figure 6.32

3. An **8-leafed rose** is generated by the equation $r = 10\cos 4\theta$. Its graph when done by a computer is shown in Figure 6.33, and when done on a *TI-83* in Figure 6.34.
4. The curve produced by the equation $r = 10\cos\left(\dfrac{\theta}{4}\right)$ does not have a name— perhaps you will think of a good one. The graph is shown in Figure 6.35.

   To plot the complete curve on your calculator you need to change the upper limit of θ to $4 \times 360° = 1440°$, or in radians to $4 \times 2\pi = 8\pi \approx 25.1$.

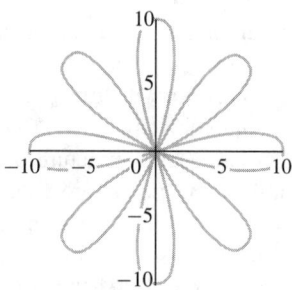

Figure 6.33

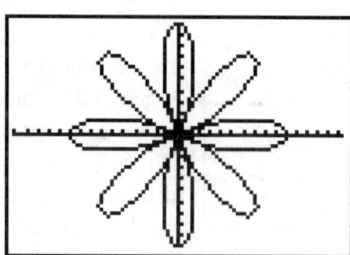

Figure 6.34

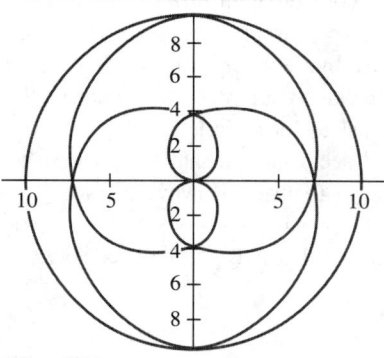

Figure 6.35

Patterns in *n*-Leafed Roses and Their Relatives

As you can see, there are many interesting patterns produced by polar equations. In the calculator labs that follow you will explore a few of them.

Calculator Lab 6.2	**Exploring the *n*-Leafed Rose**

The general equation for the *n*-leafed rose is

$$r = a \sin n\theta \qquad \text{or} \qquad r = a \cos n\theta$$

where *n* is an integer. The goal of your exploration in this lab is to understand how different values of *a* and *n* change the shape of the graph.

Procedures   Using positive whole numbers for *a* and *n*, change their values to determine the effects each variable has on the graph.

(*Graphing calculator tips:* To better understand how the graph is being plotted, use the `path` graph style. You can improve the appearance of your graph by reducing the value of θ`step` to something like 2 or 3.)

Answer the following questions.

1. How does the value of *a* change the graph?
2. How does the value of *n* change the graph?
3. What value of *n* would produce a 24-leafed rose?
4. What value of *n* would produce a 51-leafed rose?
5. What is the difference in the graph when sine is used as compared to when cosine is used?

Calculator Lab 6.3

The n-Leafed Rose with Negative Coefficients

Use the same basic equations

$$r = a \sin n\theta \quad \text{or} \quad r = a \cos n\theta$$

and experiment with values of n and a that are negative integers.

Procedures Using negative integers for a and n, change their values to determine the effect each variable has on the graph. Write your answers to each of the following.

1. How does the graph for $a = 10$ compare with the graph for $a = -10$?
2. Knowing what you do about graphing negative values of r, explain your answer to question 1.
3. Is the answer to question 1 the same for sine and cosine?
4. How does the graph for $n = 2$ compare with the graph for $n = -2$?
5. Using what you have learned about trigonometry, explain your answer to question 4.
6. Is the answer to question 4 the same for sine and cosine?

Calculator Lab 6.4

The n-Leafed Rose with Fractional Coefficients

Use the same basic equations

$$r = a \sin n\theta \quad \text{or} \quad r = a \cos n\theta$$

and experiment with values of n and a that are not integers.

Procedures Explore and describe the graphs that result from using values like the following for n: $\frac{1}{2}, \frac{1}{3}, \frac{1}{4}, \frac{1}{5}, \ldots$. Using $n = \frac{1}{4}$ is the same as plotting the equation $r = a \sin\left(\frac{\theta}{4}\right)$, which is the nameless curve plotted in Figure 6.35. To get the complete graph for these values of n you must increase θmax. For example, if $n = \frac{1}{5}$ you must set θmax to $5 \times 360° = 1800°$, or in radians to $5 \times 2\pi = 10\pi \approx 31.4$.

Calculator Lab 6.5

Exploring the Cardioid Family

In Figures 6.31 and 6.32 you saw two of the variations of the equation $r = a \cos\theta + b$. In this lab you will determine the effects of a and b on the graph of $r = a \cos\theta + b$.

Procedures Experiment with the coefficients a and b and answer the following questions. Use the `path` graph style so you can see any changes in the way the graphs are plotted.

1. Describe the effect of keeping a and b equal but changing their values.
2. How does your answer to part 1 change if a and b are both negative numbers?
3. Describe the shape if $a > b$ and both are positive.
4. Describe the shape if $a < b$ and both are negative.
5. How do the answers to all the previous questions change if a and b have different signs?

Connecting Polar and Rectangular Equations

Now that you have worked with polar equations you can appreciate the differences among polar, parametric, and rectangular equations. Figure 6.36 reminds you how different the graphs of similar equations are, depending on the type of equation and the type of coordinate system. (The plots in Figure 6.36 are in radians.)

▶ Rectangular equation on a Cartesian coordinate system:

$$y = \cos x \quad \text{(Figure 6.36(a))}$$

▶ Parametric equations on a Cartesian coordinate system:

$$\begin{cases} x = \cos t \\ y = \sin t \end{cases} \quad \text{(Figure 6.36(b))}$$

▶ Polar equation on a polar coordinate system:

$$r = \cos \theta \quad \text{(Figure 6.36(c))}$$

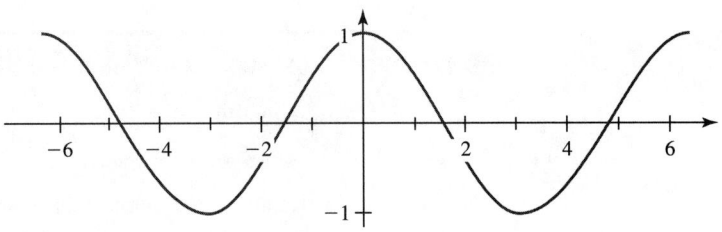

(a) Rectangular: $y = \cos x$

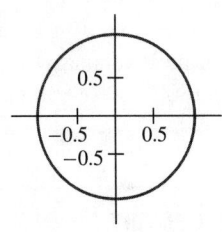
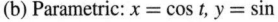

(b) Parametric: $x = \cos t, \ y = \sin t$

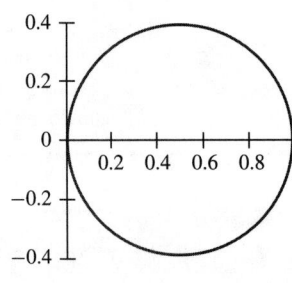

(c) Polar: $r = \cos \theta$

Figure 6.36

Converting between Polar and Rectangular Equations

In Section 6.1 you saw how to convert from parametric to rectangular equations. Now let's see how to convert a polar equation to a rectangular equation. You saw how to convert the polar form of a complex number to the rectangular form in Example 5.49 (page 339). In general, to convert a polar point (r, θ) to (x, y), use the conversions

$$x = r \cos \theta$$
$$y = r \sin \theta$$

Example 6.15

Convert the polar points $P(6, 40°)$ and $Q(5.5, 290°)$ to points in Cartesian coordinates.

Solution   For $P(6, 40°)$ we have $x = 6 \cos 40°$ and $y = 6 \sin 40°$. Thus the Cartesian coordinates of the polar point $(6, 40°)$ are $(6 \cos 40°, 6 \sin 40°) \approx (4.6, 3.9)$. In the same way, for $Q(5.5, 290°)$ we have $x = 5.5 \cos 290°$ and $y = 5.5 \sin 290°$. Hence $(5.5 \cos 290°, 5.5 \sin 290°) = (1.9, -5.2)$.

These two points are plotted in Figure 6.14 (page 369). Refer to that graph and you will see that the locations of points P and Q agree with these results. ∎

To convert an entire equation you must convert *all* the points at once. It is not as difficult as it sounds; however the conversion is the opposite of the one used in Example 6.15. We must express r and θ in terms of x and y, so we need the conversions that were used in Chapter 5 to find the polar form of a complex number.

Rectangular Form of Polar Equations

Any polar equation can be converted to a rectangular equation by replacing r and θ with the expressions

$$r = \sqrt{x^2 + y^2}$$

$$\theta = \tan^{-1}\left(\frac{y}{x}\right)$$

Example 6.16

Convert the polar equation $r = 5 \cos \theta$ to a rectangular equation in x and y.

Solution   Replacing r and θ we have

$$\sqrt{x^2 + y^2} = 5 \cos\left(\tan^{-1}\left(\frac{y}{x}\right)\right)$$

This is certainly an equation in x and y. It can be simplified, however, as you saw in Example 6.2 (page 360). The key to the simplification is to think about the triangle that includes x, y, r, and θ (Figure 6.37).

We started with $\cos \theta$, and the triangle reminds us that $\cos \theta = \dfrac{x}{r}$. But we know that $r = \sqrt{x^2 + y^2}$, so $\cos \theta = \dfrac{x}{\sqrt{x^2 + y^2}}$. Therefore the equation $r = 5 \cos \theta$ becomes

$$\sqrt{x^2 + y^2} = 5 \frac{x}{\sqrt{x^2 + y^2}}$$

which can be simplified by multiplying both sides of the equation by $\sqrt{x^2 + y^2}$ to give

$$x^2 + y^2 = 5x$$

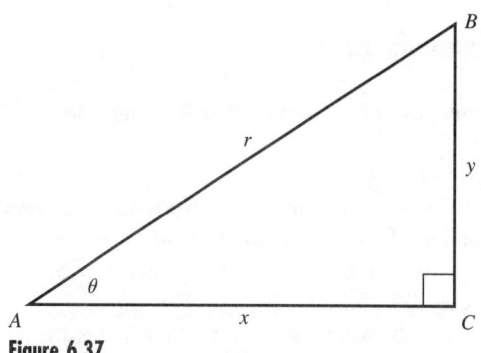

Figure 6.37

This last equation is equivalent to the first equation of this solution, but it is easier to work with. In order to plot an equation like this on a graphing calculator, however, we would have to solve for y, which gives two equations

$$y = +\sqrt{5x - x^2} \qquad \text{and} \qquad y = -\sqrt{5x - x^2}$$

If these were plotted on the calculator, the combination would be the same circle that is produced by $r = 5 \cos \theta$ in polar coordinates.   ■

From Example 6.16 you can see once again that some equations are much easier to work with in polar coordinates. As you'll see in Example 6.17, there are also many that are easier to work with in rectangular equations.

Example 6.17

Graph in polar coordinates the line whose equation is $y = 2x + 5$.

Solution   We know that $x = r \cos \theta$ and $y = r \sin \theta$. Substituting these expressions we have

$$r \sin \theta = 2r \cos \theta + 5$$

To graph such an equation we must solve it for r. This requires a few steps, including factoring out a common factor:

$$r \sin \theta - 2r \cos \theta = 5$$
$$r(\sin \theta - 2 \cos \theta) = 5$$
$$r = \frac{5}{\sin \theta - 2 \cos \theta}$$

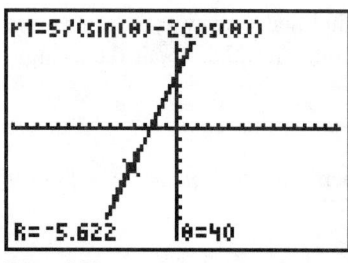

Figure 6.38

This last equation produces the exact straight line you would expect, with a slope of 2 and a y-intercept of 5 (as shown in Figure 6.38). If you plot the graph with the `path` graphing style you will see the unusual order in which the points are plotted. For example, it's quite a surprise that the point shown in the figure is for $\theta = 40°$.   ■

In the following calculator lab you'll have a chance to explore this version of the straight line equation.

Calculator Lab 6.6

Polar Version of the Straight Line

You saw in Example 6.17 that the equation

$$r = \frac{5}{\sin\theta - 2\cos\theta}$$

produces a straight line with slope equal to 2 and y-intercept equal to 5. In this calculator lab you will explore how to write the slope and y-intercept just by looking at the above equation. You learned in Chapter 1 that the rectangular equation of such a line is $y = 2x + 5$, and that in general if the equation of the line is $y = mx + b$, then the slope is m and the y-intercept is b. The purpose of this lab is for you to see how you can read the slope and y-intercept from the general polar equation of a line

$$r = \frac{a}{b\sin\theta + c\cos\theta}$$

Procedures   Experiment with values of a, b, and c to make lines of different slope and y-intercept until you can write a complete answer to the problem posed above.

Section 6.2 Exercises

In Exercises 1 and 2, for each of the points shown in the respective figures,

(a) write polar coordinates with r and θ both positive.

(b) write polar coordinates with r and θ both negative.

(c) write polar coordinates with r positive and θ negative.

(d) write polar coordinates with r negative and θ positive.

1.

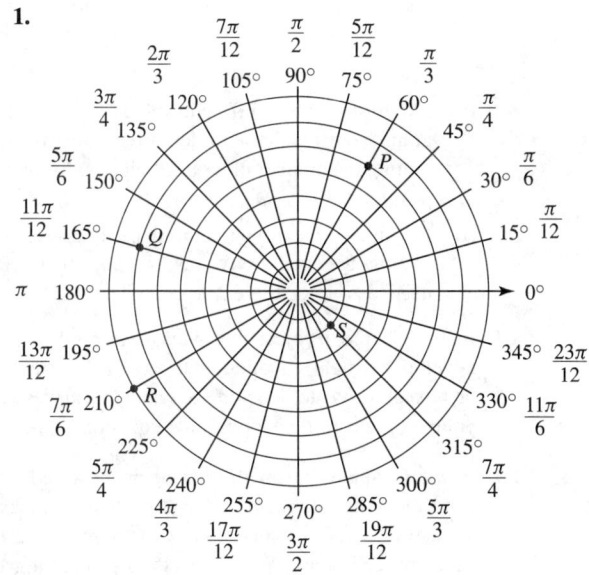

2.

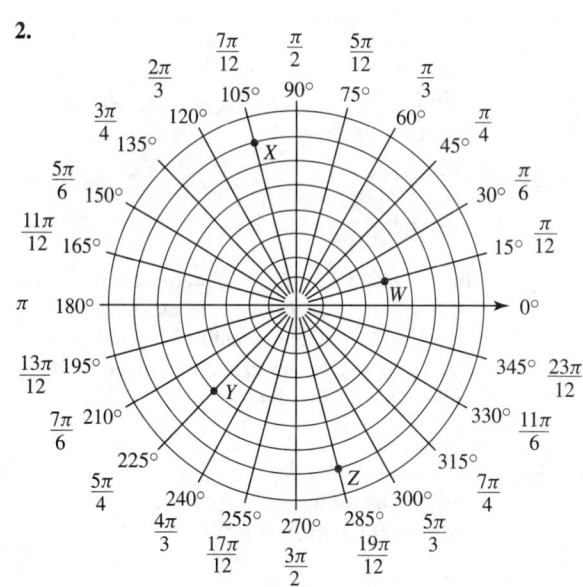

In Exercises 3 and 4 plot each of the given polar points.

3. $A(5, 90°)$, $B(-3, 45°)$, $C(4, -195°)$, and $D(-6, -105°)$

4. $A\left(-2, \frac{\pi}{6}°\right)$, $B\left(3, -\frac{7\pi}{6}\right)$, $C\left(-5, -\frac{5\pi}{12}\right)$, and $D\left(4, \frac{13\pi}{6}°\right)$

5. *Air traffic control*   An air traffic controller sees two flights on the radar screen. One, Quantus Flight 1234, is 19 miles from the tower on a heading of 194°; the other, Delta Flight 444, is 16 miles from the airport on a heading of 284°.

    (a) Plot the locations of these flights. Remember to use the navigation coordinate system shown in Figure 5.29 on page 304.

    (b) How far apart are the airplanes?

6. *Air traffic control*   Later in the day the air traffic controller in Exercise 5 sees two other flights on the radar screen. One, United Flight 4321, is 28 miles from the tower on a heading of 73°; the other, American Flight 111, is 32 miles from the airport on a heading of 343°.

    (a) Plot the locations of these flights.

    (b) How far apart are the airplanes?

In Exercises 7–14 graph each of the polar curves. (If the curve has a name, it is given after each equation.)

7. $r = 1 + 3\cos\theta$ (Limaçon)

8. $r = 2\cos 3\theta$ (3-leafed rose)

9. $r = 1 + 4\cos 3\theta$

10. $r = 1 + 4\cos 4\theta$

11. $r^2 = 5\cos 2\theta$

12. $r = \sin^7\theta + 9\sin\theta\cos^5\theta$

13. $r\theta = 4$ (Hyperbolic spiral)
(You may want to put your calculator in Dot mode in order to remove the false lines.)

14. $r = \dfrac{6}{\cos 2\theta}$ (Cruciform)

(You may want to put your calculator in Dot mode in order to remove the false lines.)

15. Write a polar equation for a circle of radius 4 centered at the polar point $(4, 0)$.

16. Write a polar equation for a circle of radius 3 centered at the polar point $\left(3, \frac{3\pi}{2}\right)$.

17. Write a polar equation for a 5-leafed rose that has symmetry about the polar axis.

18. Write a polar equation for a trisectrix that has symmetry about the $\frac{\pi}{2}$-axis.

19. Graph $r = 10\sin 2\theta$ with your calculator in degree mode and with θstep set at (a) 30 and (b) 60.

20. Graph $r = 10\sin 3\theta$ with your calculator in degree mode and with θstep set at (a) 45 and (b) 60.

21. Write a polar equation for a line that has a slope of $\frac{2}{3}$ and a y-intercept of 5.

22. Write a polar equation for a line that has a slope of $-\frac{4}{3}$ and a y-intercept of 2.

In Exercises 23–26, (a) convert the given polar equation into an equation involving rectangular coordinates and (b) identify the graph of the equation.

23. $r\sin\theta = 4$

24. $2r\sin\theta - 3r\cos\theta = -5$

25. $r^2 - 2r\cos\theta = 8$

26. $r = \dfrac{2}{1 - \cos\theta}$

27. In Calculator Lab 6.6 you were asked to experiment with values of a, b, and c to make lines of different slopes and y-intercepts until you could determine the slope and y-intercept of the line given by the general polar equation of a line

$$r = \frac{a}{b\sin\theta + c\cos\theta}$$

Write your results and explain why you think this is the correct interpretation.

28. Consider the family of "hybrid rose curves" given by $r = a + b\sin(c\theta)$ where a, b, and c are whole numbers.

    (a) Experiment with different values of a, b, and c, graph your results, and determine rules for determining the number of loops, their lengths, and their positions.

    (b) If θmin $= 0$, can you determine any rules for predicting the smallest value of θmax that will be needed to trace the entire curve?

29. In the Chapter Project the management asked you to develop a single equation that would draw a square or hexagon. Write a report to management describing your progress in developing this equation.

30. In the Chapter Project the management asked you to learn the definitions of a pentagram and a trisectrix and to determine their equations. Write a report to management describing your progress in completing these tasks.

6.3 Applications of Polar Graphing: Ellipses

In this section you'll see that polar graphing is an ideal way of plotting the elliptical orbits of planets and comets. We'll also explore several other applications of ellipses in nature, engineering, architecture, and medicine.

In Chapter 5 you saw the following rectangular and parametric forms for the equation of an ellipse centered at the origin.

Rectangular and Parametric Equations for an Ellipse

An ellipse centered at the origin, with major axis of length $|2a|$, and minor axis of length $|2b|$ has the following equations.

Rectangular equation: $\dfrac{x^2}{a^2} + \dfrac{y^2}{b^2} = 1$

Parametric equations: $\begin{cases} x = a \cos t \\ y = b \sin t \end{cases}$

The major axis has length that is the larger of $|2a|$ and $|2b|$.

Ellipses and Eccentricity

We saw in Chapter 5 that the orbits of planets around the Sun can be plotted with parametric equations and that these orbits are ellipses that are very nearly circular.

In this section we'll see that the orbits of comets are ellipses that are quite far from being circular. In mathematical language we say that the orbits of comets have a high **eccentricity**. Before we define eccentricity we need to look again at another characteristic of an ellipse, the focus, which we introduced in Chapter 5. If a planet or comet orbits the Sun, the Sun is located at a focus of the ellipse.

Figure 6.39 shows a model of a comet orbiting the Sun. The lengths a and b in the figure are familiar to you because they are used in the parametric equations of the ellipse:

$$\begin{cases} x = a \cos t \\ y = b \sin t \end{cases}$$

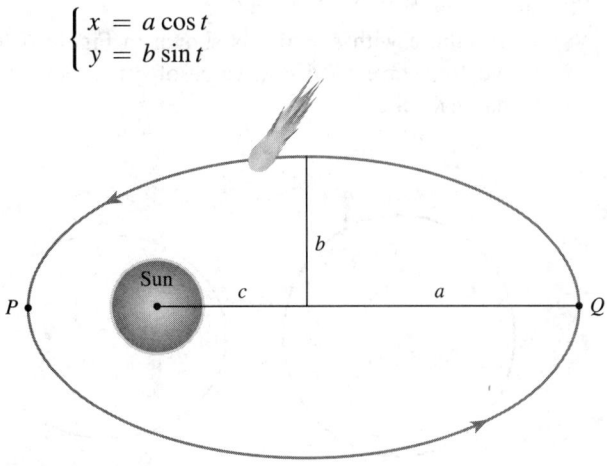

Figure 6.39

The length c is the distance from the center of the orbit to the focus at the center of the Sun.

There are various ways to define the eccentricity of an ellipse. We'll use two definitions that at first seem quite different; but they turn out to be equivalent because of mathematical properties of the ellipse.

Definition: Eccentricity (e) of an Ellipse

The eccentricity e of an ellipse whose major axis has length $2a$, minor axis length $2b$, and whose foci are located c units from the center is

$$e = \sqrt{1 - \frac{b^2}{a^2}} \qquad \text{or} \qquad e = \frac{c}{a}$$

Notice that this first definition requires that e be nonnegative, since a number like $\sqrt{9}$ has only a positive result. In Examples 6.18 and 6.19 you'll see how the value of e affects the shape of the ellipse.

Example 6.18

If $e = 0.9$ and $a = 10$, what are the values of b and c and what is the shape of the ellipse?

Solution   The second definition of eccentricity gives

$$\frac{c}{10} = 0.9 \text{ or } c = 9$$

The first definition gives

$$0.9 = \sqrt{1 - \frac{b^2}{100}}$$

$$0.81 = 1 - \frac{b^2}{100} \qquad \text{Square both sides.}$$

$$\frac{b^2}{100} = 1 - 0.81 = 0.19 \qquad \text{Solve for } b.$$

$$b = \sqrt{19} \approx 4.36$$

An ellipse with $e = 0.9$ is shown in Figure 6.40. Notice that the locations of the two foci correspond to a value of c that is close to a and a value of b that is less than half of a.

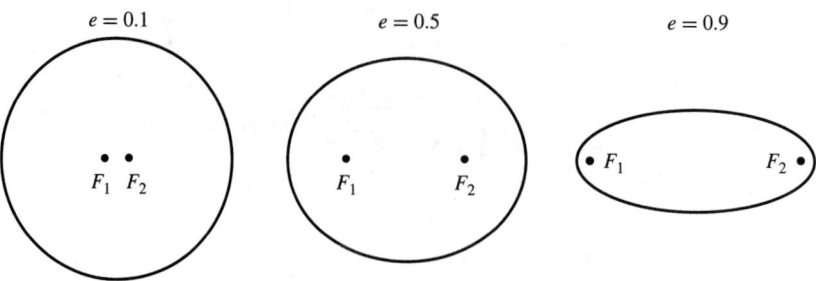

Figure 6.40

Putting the two definitions of eccentricity together we have $\dfrac{c}{a} = \sqrt{1 - \dfrac{b^2}{a^2}}$.
Squaring both sides we obtain

$$\frac{c^2}{a^2} = 1 - \frac{b^2}{a^2} = \frac{a^2 - b^2}{a^2}$$

Thus $c^2 = a^2 - b^2$, and so $c = \sqrt{a^2 - b^2}$.

Example 6.19

Discuss how the value of the eccentricity affects the graph of the ellipse. Discuss (a) $e = 0$ and (b) $e \approx 1$. (c) What are the possible values of e?

Solutions   (a) If $e = 0$ then $\dfrac{b^2}{a^2} = 1$, so $b = a$. Thus the radii in each direction have the same length. Also, if $e = 0$ then $\dfrac{c}{a} = 0$, so $c = 0$. Therefore the distance between the foci is zero.

Both of these statements mean that the ellipse comes closer to being a circle the closer that e gets to 0. (See Figure 6.40.)

(b) If e is close to 1 then $\sqrt{1 - \dfrac{b^2}{a^2}}$ is close to 1, so $\dfrac{b^2}{a^2}$ is close to 0. Thus b is small compared to a. Also, if e is close to 1 then $\dfrac{c}{a}$ is close to 1, so c is close to a.

Both of these statements imply that when $e \approx 1$ the ellipse is very far from being circular. See Figure 6.40.

(c) If you look at the definitions of e you can see that for an ellipse the value of e is in the range

$$0 < e < 1$$

Technically a circle is an ellipse, so it is possible for e to be equal to 0. However if $e = 1$ then $b = 0$, so we cannot have an ellipse.   ■

Next we'll see why eccentricity is important to people who need to plot accurate orbits of planets and comets.

Plotting Orbits of Comets

Table 6.1 gives information about the orbits of the planets and a few comets. In the table you can see that the orbits of Pluto and Mercury are less circular than the orbits of other planets. In Figure 5.20 (in Chapter 5) you can see that, even with this higher eccentricity, Pluto's orbit is quite circular.

Many astronomy handbooks describe orbits of planets and comets with two numbers, a and e. This is because the length a, in addition to being equal to the radius in the x-direction, has the additional property of being the average of the minimum and maximum distances of the planet or comet from the Sun. To see that this is true, look at Figure 6.39, which shows that the closest approach to the Sun (at point P) is $a - c$ and that the length of the farthest distance of the planet or

Table 6.1   Eccentricities of planets and some comets

Body	a (AU)	b (AU)	e
Mercury	0.387	0.379	0.206
Venus	0.723	0.723	0.007
Earth	1.000	1.000	0.017
Mars	1.524	1.517	0.093
Jupiter	5.203	5.197	0.049
Saturn	9.539	9.524	0.056
Uranus	19.182	19.161	0.047
Neptune	30.058	30.057	0.009
Pluto	39.439	38.186	0.250
Comet Halley (1986 and 2061)	17.8	4.535	0.967
Comet Hyakutake (1996)	1150.5	23.009	0.9998
Comet Hale-Bopp (1997)	186.551	18.445	0.9951

comet from the Sun (when the planet or comet is at point Q) is $a + c$. Therefore the average of the minimum and the maximum distances is

$$\frac{(a-c)+(a+c)}{2} = \frac{2a}{2} = a$$

Example 6.20   Application

The Moon has an orbit that is an ellipse with Earth at one focus. If the major axis of the orbit is 477,710 miles and $e \approx 0.0549$, determine a, b, and c.

Solution   We are given $2a = 477{,}710$, so $a = 238{,}855$ miles. From the first definition of eccentricity we have $e = \dfrac{c}{a}$, then $0.0549 = \dfrac{c}{238{,}855}$ and $c \approx 13{,}113$ miles.

From the statement between Examples 6.18 and 6.19 we have $c^2 = a^2 - b^2$. Using the values we determined for c and a, we have

$$13113^2 = 238855^2 - b^2$$
$$b^2 = 238855^2 - 13113^2$$
$$b \approx 238495$$

Thus we have determined that $a = 238{,}855$ miles, $b \approx 238{,}495$ miles, and $c \approx 13{,}113$ miles.   ∎

Polar Equations of Ellipses

It turns out that polar equations are perfect for plotting orbits of comets and planets and that this is the method used by astronomers. To see why polar equations are so suitable we start with Example 6.21, which shows why parametric equations are *not* suitable for plotting orbits of planets and comets.

Example 6.21   Application

Using the parametric plots of the orbits of Jupiter, Earth, and Comet Halley, explain why the method you have learned for plotting ellipses with parametric equations is not suitable for plotting the orbits of comets.

Solution   We saw in Chapter 5 (page 295) that the values of a and b are used in the parametric equations of an ellipse:

$$\begin{cases} x = a \cos \omega t \\ y = b \sin \omega t \end{cases}$$

where ω is equal to the angular velocity of the planet's motion. Parametric equations are ideal for comparing the orbiting speed of planets because speed of movement is included directly in the modeling equations. In the present example, however, we are only interested in the shape of the orbits, not the different velocities, so we will use the following equations.

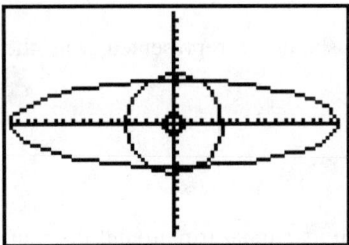

Figure 6.41

Jupiter $\begin{cases} x = 5.203 \cos t \\ y = 5.197 \sin t \end{cases}$

Earth $\begin{cases} x = 1.0 \cos t \\ y = 1.0 \sin t \end{cases}$

Comet Halley $\begin{cases} x = 17.8 \cos t \\ y = 4.535 \sin t \end{cases}$

Their graphs are shown in Figure 6.41.

These orbits are the correct shape. The problem with these parametric models is that the orbits are all centered at the origin of the graph. Where is the Sun? The Sun should be at a focus of *each* of these ellipses. For the two planets Jupiter and Earth the Sun is located approximately at the origin of the graph. However the focus of an elliptical orbit like Comet Halley would be very far from the origin of the graph. Therefore if the Sun is at the origin it cannot be at the focus of the comet's orbit. In Example 6.22 we'll see how you can use polar coordinates to take care of this problem and place the Sun correctly at the focus.

Polar Equation for an Ellipse

The polar equation of an ellipse with eccentricity e and length of major axis $|2a|$ is

$$r = \frac{a(1 - e^2)}{1 - e \cos \theta}$$

If this equation is used to model the orbit of a comet around the Sun, the Sun will be located at the origin of the coordinate system.

Example 6.22   Application

Use polar coordinates to plot the orbits of Jupiter, Earth, and Comet Halley.

Solution   Substituting the values of a and e from the table into the polar equation of the orbit gives the following three equations:

$$\text{Jupiter} \qquad r = \frac{5.203(1 - 0.049^2)}{1 - 0.049\cos\theta}$$

$$= \frac{5.191}{1 - 0.049\cos\theta}$$

$$\text{Earth} \qquad r = \frac{1(1 - 0.017^2)}{1 - 0.017\cos\theta}$$

$$= \frac{0.9997}{1 - 0.017\cos\theta}$$

$$\text{Comet Halley} \quad r = \frac{17.8(1 - 0.967^2)}{1 - 0.967\cos\theta}$$

$$= \frac{1.155}{1 - 0.967\cos\theta}$$

The graphs in Figure 6.42 model the orbits as they should be represented, with the Sun at the focus of each ellipse. ■

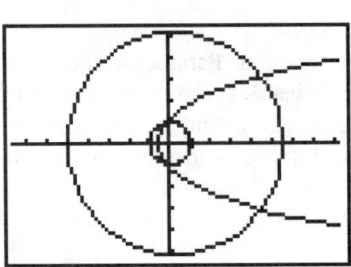

Figure 6.42

Other Applications of Ellipses

In addition to being the pattern our planet follows in its annual trip around the Sun, the ellipse is used in many applications. We'll mention two of them: gears that are made in the shapes of ellipses and a machine used to destroy kidney stones in patients.

Elliptical Gears

Because elliptical gears produce motion that is not constant velocity and not circular, they are used for special purposes in some machinery to provide a quick-return mechanism or a slow power stroke (for heavy cutting, for example) in each revolution. Figure 6.43 shows two identical gears that remain in contact as each rotates around one of its foci.

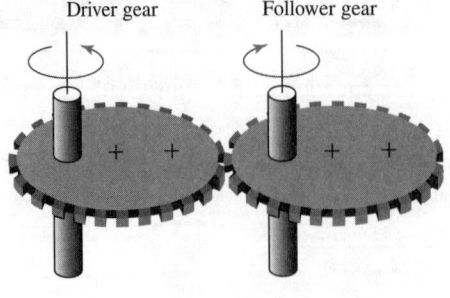

Elliptic gears

Figure 6.43

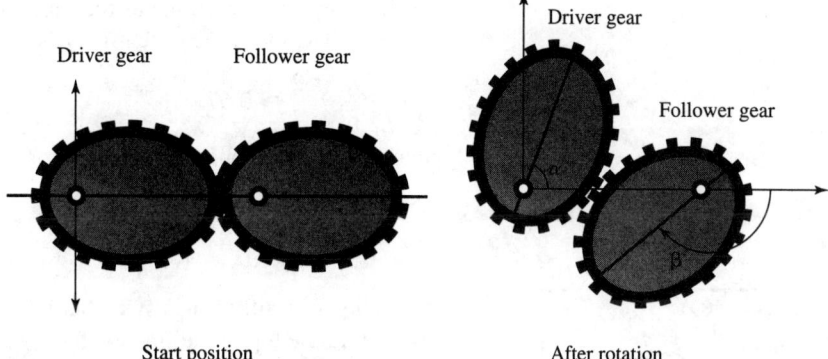

Figure 6.44

Figure 6.44 shows that when the driver gear has rotated through an angle α, the follower gear has rotated in the opposite direction through a larger angle β. That is, during that period of rotation the follower gear rotates faster than the driver gear. At other parts of the rotation cycle the follower gear rotates more slowly than the driver. The variable rotation speeds of the follower gear are designed into equipment because the slower speeds produce a higher force.

Example 6.23   Application

(a) What is the eccentricity of the identical gears shown in Figure 6.45?
(b) Write a polar equation for one of the gears.

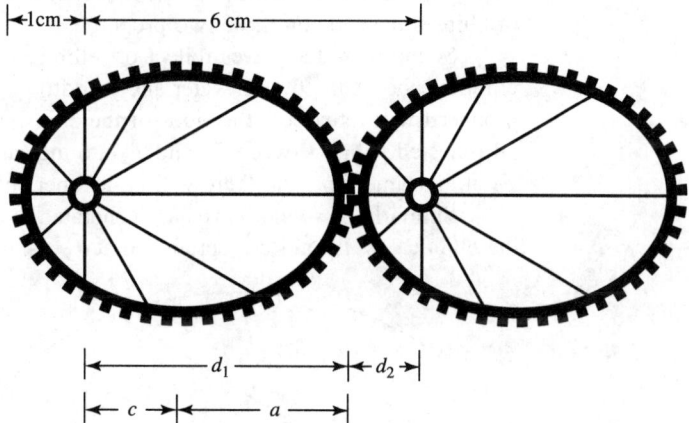

Figure 6.45

Solution   The dimensions given in the figure are sufficient to answer these questions. We need to know the value of a, half the length of the major axis, and c, the distance from the center of the ellipse to one focus.

(a) Since the gears are identical, the distance $d_2 = 1$ cm. Therefore the distance $d_1 = 5$ cm and the length of the major axis, which equals $d_1 + 1$, is 6 cm; so $a = 3$ cm.

Next we need to compute the value of c. Since $c + a = d_1$, then $c = d_1 - a = 2$. Therefore the eccentricity is

$$e = \frac{c}{a} = \frac{2}{3} \approx 0.67$$

(b) A polar equation for one of these gears is

$$r = \frac{a(1 - e^2)}{1 - e\cos\theta} = \frac{3\left(1 - \left(\frac{2}{3}\right)^2\right)}{1 - \frac{2}{3}\cos\theta}$$

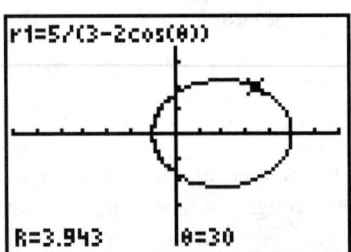

r1=5/(3-2cos(θ))

R=3.943   θ=30

Figure 6.46

This equation is a sufficient answer, but it is also possible to simplify it somewhat by computing the top of the fraction $3\left(1 - \frac{4}{9}\right) = 3\left(\frac{5}{9}\right) = \frac{5}{3}$ and then multiplying top and bottom by 3. These two steps produce the result

$$r = \frac{\frac{5}{3}}{1 - \frac{2}{3}\cos\theta} = \frac{5}{3 - 2\cos\theta}$$

Any of these three forms of the same polar equation can be graphed, as shown in Figure 6.46. Remember that this form of the polar equation places one focus of the ellipse at the origin of the coordinate system. ∎

Using the Reflective Properties of the Ellipse: The Lithotripter

Every ellipse has two unusual properties that have surprising applications. Imagine a pool table in the shape of an ellipse. If a ball with no "english" on it rolls across one focus and bounces off the side of the ellipse, it will rebound in a path that takes it to the other focus. Figure 6.47 shows two paths that begin at one focus and end at the other. A second property of the ellipse guarantees that the lengths of all paths from F_1 to F_2 that bounce off the elliptical border will be the same. Our next application of ellipses depends on these two properties.

Sound or water waves reflect off elliptical boundaries in the same way. If an elliptical pool is filled with water and you drop a stone at the focus F_1, all the waves produced will reflect off the sides of the ellipse and meet at the focus F_2. Because all reflected paths between F_1 and F_2 are the same length, all the waves will reach F_2 at the same time and there will be another splash at F_2.

Similarly, if sound waves are generated at F_1, the sounds will be reflected off the elliptical walls and be reproduced at F_2. A room with smooth walls built in the

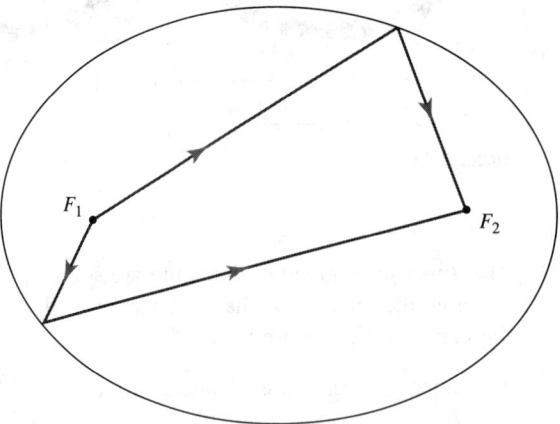

Figure 6.47

shape of an ellipse will have the unusual property that a person sitting at one focus can hear a person at the other focus, even though both are whispering across a distance of more than 50 feet. Examples of such "whispering galleries" include St. Paul's Cathedral in London, England, and the Rotunda of the U.S. Capitol Building in Washington, D.C.

Example 6.24   Application

Fred and Flora sit at the foci of a whispering gallery (Figure 6.48), each a distance of 5 feet from the nearest wall and facing away from each other.

(a) What are the values of a, b, c, and e for this ellipse?
(b) Write a rectangular equation of this ellipse.
(c) Write the parametric equation of this ellipse.
(d) Write the polar equation of this ellipse.

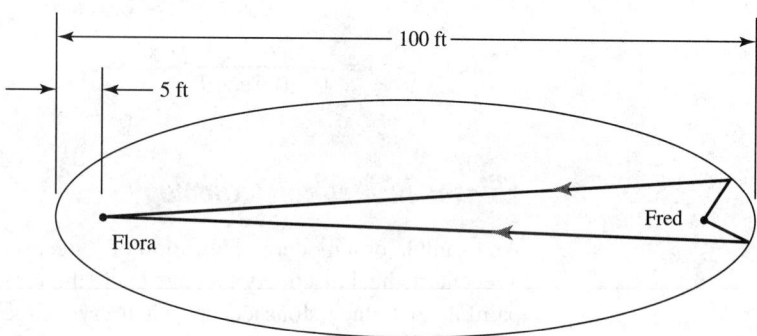

Figure 6.48

Solution (a) The four values are computed as follows:

▶ a is half the length of the major axis, so $a = 50$.
▶ $c + 5 = a$, so $c = 45$.
▶ Eccentricity is $\dfrac{c}{a}$, so $e = \dfrac{45}{50} = 0.9$.
▶ To compute b we use the second definition of e.

$$e = \sqrt{1 - \frac{b^2}{a^2}}$$

$$0.9 = \sqrt{1 - \frac{b^2}{50^2}}$$

$$0.81 = 1 - \frac{b^2}{50^2} \qquad \text{Square both sides.}$$

Now we have to isolate b.

$$\frac{b^2}{50^2} = 1 - 0.81 = 0.19$$

$$b^2 = 50^2 \times 0.19 = 475$$

$$b = \sqrt{475} \approx 21.8 \text{ ft}$$

(b) There are two forms of the rectangular equation. The standard form is

$$\frac{x^2}{a^2} + \frac{y^2}{b^2} = 1$$

$$\frac{x^2}{2500} + \frac{y^2}{475} = 1$$

Solving for y produces two other equations

$$y \approx \pm\frac{21.8}{50}\sqrt{2500 - x^2}$$

(c) Parametric equations

$$\begin{cases} x = 50\cos t \\ y = 21.8\sin t \end{cases}$$

(d) Polar equation

▶ Since $a = 50$ and $e = 0.9$ we have

$$r = \frac{a(1 - e^2)}{1 - e\cos\theta} = \frac{50(1 - 0.81)}{1 - 0.9\cos\theta}$$

$$= \frac{9.5}{1 - 0.9\cos\theta}$$

Ellipses in Medical Technology

An example of a modern application of reflection in ellipses is **medical lithotripsy**,* a recent medical discovery that has made the removal of kidney stones safer and less painful. A kidney stone occurs when crystalline deposits form a roughly spherical hard shape within the kidney. A kidney stone can cause great pain and was previously treated by surgery or the introduction of chemicals to dissolve the stone.

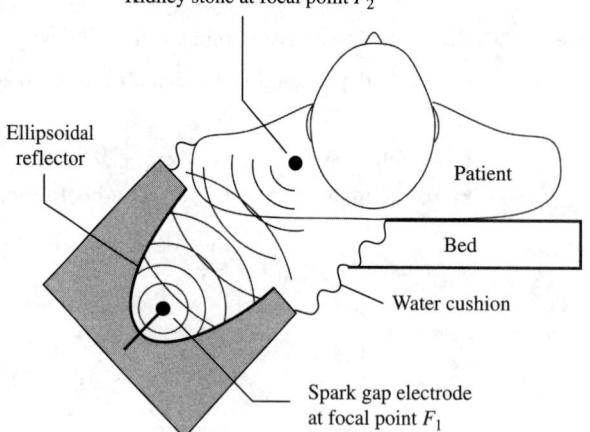

Figure 6.49

Experimentation determined that concentrated sound waves could break up kidney stones. In 1980, German doctors and mathematicians teamed up to produce

---

*Pictures and text adapted from an article entitled *Medical Lithotripsy*, by Marc Frantz (World Wide Web site http://www.math.iupui.edu/m261vis/litho.html).

a device called a **lithotripter**, in which high intensity sound waves are produced at one focus of an ellipse. The patient is positioned so that the kidney stone is at the other focus. Figure 6.49 shows the concentric circles of the sound waves leaving one focus (F_1) and arriving together at the other (F_2). Notice that there is only half an ellipse inside the machine, but that the sound waves still arrive together at the other focus outside the machine.

The phrase *ellipsoidal reflector* is used in Figure 6.49. An **ellipsoid** is a three-dimensional version of an ellipse. You can imagine an ellipsoid resulting if an ellipse is rotated around its major axis (similar in shape to an egg or a football). For example, if the ellipse plotted in Figure 6.50 were rotated around the x-axis, a three-dimensional figure like Figure 6.51 would result. The figure shows only half of the ellipsoid in order to give you an idea of what the ellipsoid inside the lithotripter looks like. One version of an actual lithotripter is shown in Figure 6.52. At the right side of the machine you can see the open end of its ellipsoid, where sound waves are directed at the patient.

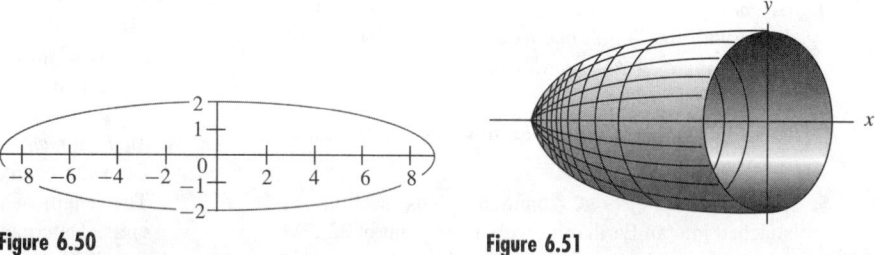

Figure 6.50

Figure 6.51

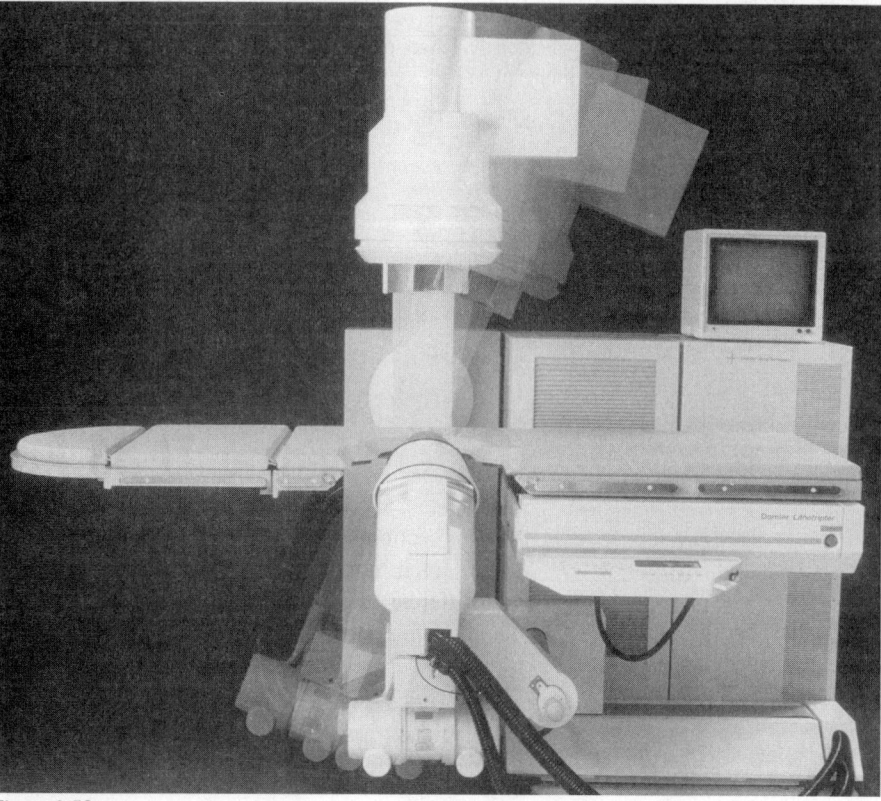

Figure 6.52
(Photo courtesy of Dornier Medical Systems)

Section 6.3 Exercises

1. *Astronomy* The comet Giacobini-Zinner has an orbit that is an ellipse with the Sun at one focus. If the major axis of the orbit is 7.04 AU and $e \approx 0.706$,
   (a) determine a, b, and c and
   (b) write the polar equation for this comet.

2. *Astronomy* The comet Chiron has an orbit that is an ellipse with the Sun at one focus. If the major axis of the orbit is 27.4 AU and $e \approx 0.38$,
   (a) determine a, b, and c and
   (b) write the polar equation for this comet.

 3. Graph the comets Halley, Giacobini-Zinner, and Chiron on the same axes.

4. *Astronomy* The asteroid Ceres has an orbit that is an ellipse with the Sun at one focus. If the major axis of the orbit is 5.534 AU and $e \approx 0.079$,
   (a) write the polar equation for this asteroid and
   (b) graph the asteroid and Earth's orbit on the same axes.

5. *Space technology* A communications satellite is launched into an Earth orbit with a low point of 22,500 miles and a high point of 24,000 miles.

In Exercises 9–14 write an answer to the questions about lithotripters.

9. What would be a good value for the length of the major axis of the ellipse in a lithotripter?

10. Explain your answer to Exercise 9.

11. What would be a good value for the eccentricity of the ellipse in a lithotripter?

12. Explain your answer to Exercise 11.

13. Besides the possibility of providing comfort to the patient, why do you suppose there is a water cushion near the patient?

(a) Determine the eccentricity of the orbit.

(b) Determine the polar equation for this satellite.

6. *Radio technology* The pattern of strongest radiation of a certain bidirectional radio antenna is given by the equation $r = 1 + \sin 2\theta$.
   (a) Graph this radiation pattern.
   (b) Use the identity $\sin 2\theta = 2 \sin \theta \cos \theta$ to convert this equation into rectangular form.

7. *Radio technology* The pattern of strongest radiation of a certain bidirectional radio antenna is given by the equation $r = 1 + 2 \sin 2\theta$. Notice that this pattern has *sidelobes*.
   (a) Graph this radiation pattern.
   (b) Use the identity $\sin 2\theta = 2 \sin \theta \cos \theta$ to convert this equation into rectangular form.

8. *Medical technology* A patient is placed 12 units away from the source of the shock waves of a lithotripter. The length of the minor axis of the lithotripter is 16 units. Determine an equation for the ellipse that satisfies these conditions.

14. Why would an ellipsoid (three dimensions) be better than an ellipse (two dimensions) for a lithotripter?

15. In the Chapter Project the management asked you to determine how to adapt the programming of the *X-Y Grabber* for tracing out points on a curve. While you have not had a chance to see the programming of the *X-Y Grabber*, write a report to management explaining your understanding of drawing curves using polar equations.

6.4   Understanding *TI-83* Programs

In this section you will learn how to write programs for the *TI-83 Graphing Calculator*.* Even though programming languages on computers and other calculators are all different, they share many basic concepts. You will be able to apply what you learn in this section to other technology.

---

*The programming commands and procedures in this section are designed to work on the *TI-83*. If your calculator is different, consult your user's guide.

Introduction to *TI-83* Programming

Up until now the programs you have seen in this book have been written in pseudo-code. We'll begin with a comparison of the pseudocode program for the *X-Y Grabber* and the same program translated for the calculator. The program for the *X-Y Grabber* was introduced in Activity 6.1 (page 356) and is reproduced below with four new lines numbered 1, 2, 7, and 8. Make sure you understand what the statements in these new lines are trying to do.

1   Set mode to degrees or radians, as desired
2   Define the limits of the robot's movement
3   Wait for input
4   Receive polar position data *r* and θ
5   Compute *x*-coordinate (*x*)
6   Compute *y*-coordinate (*y*)
7   Is (*x*, *y*) outside the limits of the robot's movement?
8   If yes, inform user and go to step 3; If no, continue
9   Move arm to rectangular point (*x*, *y*)
10   Perform task
11   Go to step 1

The above program was translated into the programming language that is understood by the *TI-83* and stored as the program XYGRABBR. First you can run the program to get some experience with it, and then we'll look at the instructions that make the program perform as it does. Finally, you'll learn how to write programs of your own.

Calculator Lab 6.7

Running a Program

Your task in this lab is to run the program XYGRABBR until you have produced points in each of the four quadrants, on each of the four parts of the axes, and at the origin.

Procedures

1. Selecting and starting the program

   ▶ List of programs: Press **PRGM**. You will see a list of programs that are stored on your calculator.

   ▶ Finding the program you want: Press either ▼ or ▲ until you see the name XYGRABBR.

   ▶ Selecting the program: Press **ENTER** or, if it has a number, the number of the program to select it; this brings the command prgmXYGRABBR to the home screen.

   ▶ Starting the program: Press **ENTER** to execute the instruction prgmXYGRABBR, which starts the program.

2. Running the program

   ▶ On the screen you'll see the word RADIUS. Type the number 5 and press **ENTER**.

   ▶ Next you'll see the word ANGLE. Type the number 60 and press **ENTER**.

   ▶ A graph window like Figure 6.53 will appear. The polar point $(5, 60°)$ has been plotted.

   ▶ In Figure 6.53 you can see a faint set of dots forming a vertical line segment in the upper right-hand corner. This means the program is waiting for you to press **ENTER** before going to the next instruction.   *(continued)*

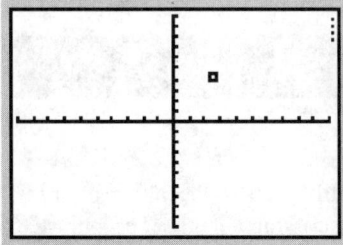

Figure 6.53

▶ Press **ENTER** and then type the numbers 15 for radius and 522 for angle.

▶ You will see the message   OFF THE SCALE   because the polar point $(15, 522°)$ is outside the standard window. In this case, the x-coordinate of $(15, 522°)$ is $15 \cos 522° \approx -14.3$. The standard viewing window has x- and y-values from -10 to 10.

▶ Enter a few more numbers. For example, enter the numbers 6 and 522 and see what happens. Then try some negative entries and try to predict where the point will be.

3. Stopping the program

▶ Press the calculator's **ON** key; then press **ENTER** to select QUIT on the next screen. It may be difficult to tell by looking at the calculator's screen, but the program has stopped and your calculator is ready to do something else.

4. Restarting the program

▶ Clearing the screen: This program does not clear the screen, so you have to do it yourself. Press **2nd** [DRAW] and select ClrDraw, the first item on the DRAW menu. This brings the instruction ClrDraw to the home screen. Press **ENTER** to execute this instruction.

▶ Restarting the program: You can repeat the steps outlined above, or if you have recently run the program, you can press **2nd** [ENTRY] until the instruction prgmXYGRABBR reappears on the screen. Press **ENTER** to start the program. If you have done nothing since you last ran the program, just pressing **ENTER** will restart it, since **ENTER** always executes the most recent command.

Run the program until you have produced points in each of the four quadrants, on each of the four parts of the axes, and at the origin—something like what you see in Figure 6.54. When you have finished this lab, stop the program.

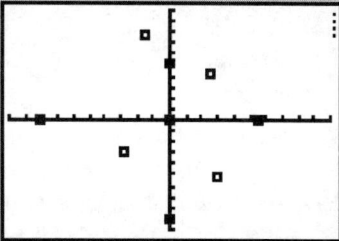

Figure 6.54

Here follows a summary of what you have learned in Calculator Lab 6.7 about running a program.

Summary: Running Programs

1. Selecting and starting the program

▶ Press **PRGM**, then the **▼** or **▲** key until you see the name of the program you want. Select the program by pressing **ENTER**. The phrase prgm⟨*Name*⟩ will appear on the home screen. Press **ENTER** to start the program.

2. Continuing a program that has paused

   ▶ Press **ENTER** to continue a program that has paused.

3. Stopping the program

   ▶ Press the calculator's **ON** key, then press **ENTER** or 1 to select QUIT on the next screen.

4. Restarting the program

   ▶ Press **2nd** [ENTRY] until you see the command prgm⟨*Name*⟩ reappear on your screen. Press **ENTER** to start the program. If you have done nothing since you quit the program, just press **ENTER** to restart it.

5. Related commands

   ▶ Clear the home screen.
   - Press **CLEAR** once or twice.
   ▶ Clear the graphing window.
   - Select ClrDraw from the DRAW menu (by pressing **2nd** [DRAW]) and press **ENTER** twice.

How the Program XYGRABBR Works

Now let's look at the programming that makes XYGRABBR perform the way it does. Like the pseudocode examples you have seen, a program on a calculator or computer consists of instructions written in a language the machine can understand. Since these machines are in fact non-intelligent servants, they can only do as they are told. Therefore we must be careful as we write instructions. Figure 6.55 shows the *TI-83* program and Figure 6.56 describes each of its steps.

```
Degree
FnOff:PlotsOff
-10→Xmin:10→Xmax:-10→Ymin:10→Ymax
Lbl 1
Input "RADIUS",R
Input "ANGLE",θ
Rcos(θ)→X
Rsin(θ)→Y
If (X>Xmax) or (Y>Ymax) or (X<Xmin) or (Y<Ymin)
Then
Disp "OFF THE SCALE"
Goto 1
Else
Pt-On(X,Y,2)
Pause
Goto 1
```
Figure 6.55

Study Figure 6.56 to begin to understand the language and logic used by the *TI-83*. The point was plotted with the Pt-On command. Even though the calculator was plotting a point in polar coordinates, the Pt-On command plots points in rectangular coordinates only. Thus we use the command Pt-On(X,Y,2). The 2 describes the type of point we want displayed. The following boxed text is a summary of the programming instructions used in the program XYGRABBR.

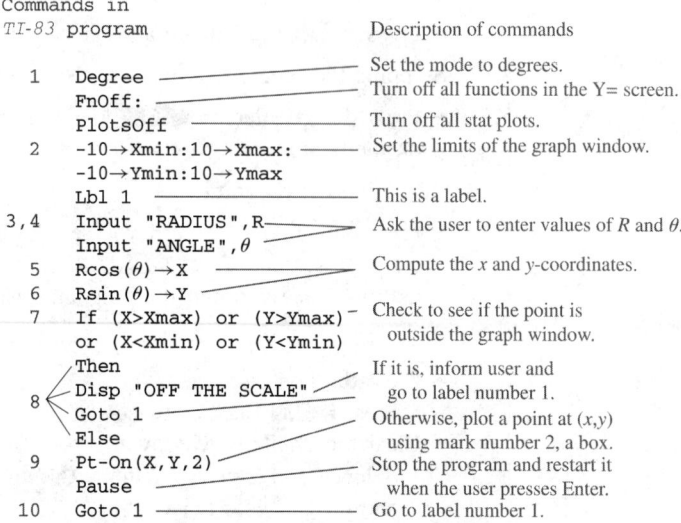

```
Commands in
TI-83 program                              Description of commands

  1    Degree ————————————————————— Set the mode to degrees.
       FnOff: ————————————————————— Turn off all functions in the Y= screen.
       PlotsOff ——————————————————— Turn off all stat plots.
  2    -10→Xmin:10→Xmax: ——————————— Set the limits of the graph window.
       -10→Ymin:10→Ymax
       Lbl 1 ——————————————————————— This is a label.
  3,4  Input "RADIUS",R—————————————— Ask the user to enter values of R and θ.
       Input "ANGLE",θ
  5    Rcos(θ)→X ——————————————————— Compute the x and y-coordinates.
  6    Rsin(θ)→Y
  7    If (X>Xmax) or (Y>Ymax)———— Check to see if the point is
       or (X<Xmin) or (Y<Ymin)        outside the graph window.
       Then ————————————————————————— If it is, inform user and
       Disp "OFF THE SCALE"—————————     go to label number 1.
  8    Goto 1 ——————————————————————— Otherwise, plot a point at (x,y)
       Else                              using mark number 2, a box.
  9    Pt-On(X,Y,2)————————————————— Stop the program and restart it
       Pause ———————————————————————     when the user presses Enter.
  10   Goto 1 ——————————————————————— Go to label number 1.
```

Figure 6.56

Summary: Programming Commands

1. Input instructions
   ▸ **Input** "⟨*text in quotes*⟩", ⟨*variable*⟩
     • The calculator waits for the user to type a number and press ⌷ENTER⌷. That value is then stored in the *variable* that is named. The text is optional. The name of the variable is not printed.
   ▸ **Prompt** ⟨*variable(s)*⟩
     • **Prompt** is an alternative to **Input**. Only the variable name is printed. More than one variable can be used.

2. Output instructions
   ▸ **Disp** "⟨*text in quotes*⟩", ⟨*variable(s)*⟩
     • Prints text on the home screen, and the value of the variables if any are listed.
   ▸ **Pt-on**(⟨*x-coordinate, y-coordinate, type of mark*⟩)
     • Places a mark at the given coordinates in the graphing window.
     • Type of mark: 1:dot, 2:box, 3:cross.

3. Changing the mode
   ▸ **Degree** changes to degree mode.
   ▸ **Radian** changes to radian mode.

4. Computations
   ▸ ⟨*number*⟩→⟨*variable*⟩
     • Stores a number in the variable that is named.
   ▸ ⟨*computation*⟩→⟨*variable*⟩
     • Stores the number produced by the computation in the variable that is named.

5. Controlling movement within the program
   ▸ **Pause**
     • Stops the program and waits for the user to press ⌷ENTER⌷.

> ► `Lbl ⟨nn⟩`
>   • Creates a place within the program to jump to when directed by the `Goto` instruction.
> ► `Goto ⟨nn⟩`
>   • Causes the program to move to the line defined by `Lbl ⟨nn⟩`.
> ► `If...Then...Else`
>   • Sets up a test. If the test is passed, the program advances to the `Then` line. If the test is failed, the program advances to the `Else` line.

In the next section you will learn how to make changes to existing programs and to enter new programs into the calculator.

6.5   Writing *TI-83* Programs

In this section you will learn how to enter and run your own programs on the *TI-83*. Then you will learn how to make changes to an existing program. The process of changing a program is called **editing**.

Entering a Program

The program you will write makes the calculator ask for the radius and height of a cone and then displays the volume of the cone. Here is a listing of the program

```
Prompt R,H
πR²H/3→V
Disp "VOLUME IS ",V
```

To enter this program follow these steps:

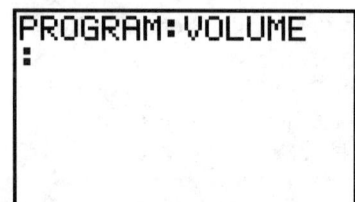

Figure 6.57

1. Reaching the Program Editor
   - ► Press `PRGM`, then select NEW and press `ENTER`.
   - ► Type the name VOLUME and press `ENTER`. Notice that the cursor is in `Alpha Lock` mode, so you don't have to press the `ALPHA` key each time. Just press the keys according to their green letters.
   - ► You are now in the Program Editor (Figure 6.57).

2. Entering the program
   - ► Notes and cautions:
     - • No spaces are required in programs except within quotes, as described below.
     - • If you find yourself looking at a menu you didn't intend, press `CLEAR` to return to the Program Editor.
     - • Do not press `2nd` [QUIT] unless you are ready to leave the Program Editor. If you do this unintentionally, follow the instructions for editing a program (beginning with Example 6.25).
   - ► `Prompt R,H`
     - • Press `PRGM`, then select the I/O (Input/Output) menu. Select number `2:Prompt`. The instruction `Prompt` will appear in the first line of your Editor.
     - • Press `ALPHA` [R] `,` `ALPHA` [H]. The instruction `Prompt R,H` should be in the first line of your program.

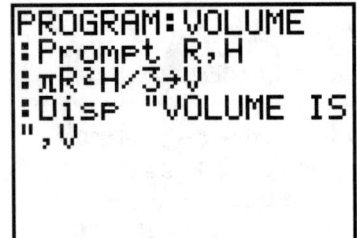

Figure 6.58

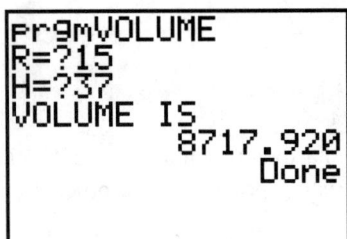

Figure 6.59

- Press **ENTER** to move to the next line.
▶ $\pi R^2 H/3 \to V$
  - Press **2nd** [π], then **ALPHA** [R] **x^2** **ALPHA** [H], then **÷** 3, then **STO ➡**
    ALPHA [V].
  - Press **ENTER** to move to the next line.
▶ Disp "VOLUME IS ",V
  - Press **PRGM**, then select the I/O menu. Select number 3: Disp. The
    instruction Disp will appear in the third line of the Editor.
  - Because you have a lot of letters in a row, press **2nd** [A-LOCK].
  - Press the keys for "VOLUME IS ". The space symbol is located above the
    0 key and the quotation marks are above the **+** key.
  - Press **ALPHA** to turn off A-LOCK.
  - Press **,** **ALPHA** [V].
  Your screen should now look like Figure 6.58.
▶ If you need to change a part of the program, move around with the cursor keys
  and then follow the boxed directions in the next subsection of this chapter—
  Editing an Existing Program—for adding or deleting characters or lines.
▶ When your program is complete, press **2nd** [QUIT] to return to the home
  screen.

Now run your program and enter the two numbers 15 and 37. Your screen
should look like Figure 6.59. To run the program again, press **ENTER**.

The following is a summary of the steps for writing a new program.

Summary: Entering a New Program

1. Reaching the Program Editor
   ▶ Press **PRGM**, select NEW, and press **ENTER**.
   ▶ Type the name of the program and press **ENTER**.
2. Entering the program—things to remember
   ▶ Do not press **2nd** [QUIT] until you are ready to leave the Program Editor.
   ▶ Press **ENTER** after each line is complete.
   ▶ Except for names of variables and the numbers and keys on the keypad,
     all instructions are entered from menus. For example, you cannot enter
     the instruction Prompt using the letters on the calculator keypad; it must
     be entered from the PRGM I/O menu.
   ▶ Press **CLEAR** to return to the Program Editor if you find yourself looking
     at a menu you don't want.
3. Press **2nd** [QUIT] to leave the Program Editor when you are ready to test
   your program.
4. The program is saved every time you leave the Program Editor.

Editing an Existing Program

Example 6.25

Change the VOLUME program to make it compute the volume of a pyramid with a
rectangular base. We'll use L and W for the length and width of the base, and the

formula $V = \dfrac{LWH}{3}$ for the volume. The new program will look like this:

```
Prompt L,W,H
LWH/3 →V
Disp "VOLUME IS ",V
```

Solution Here are the steps for making these changes:

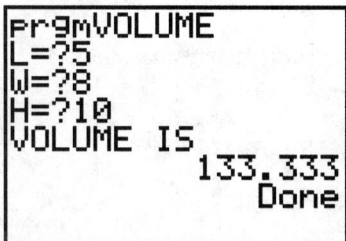

Figure 6.60

Figure 6.61

(a) Enter the Program Editor.

> ▸ Press (PRGM), then select EDIT and select the program name, VOLUME.

> What you see will look like Figure 6.58.

(b) `Prompt L,W,H`

> ▸ Delete the R with the (DEL) key, then press (2nd) [INS] (to insert), then type L,W.

(c) `LWH/3→V`

> ▸ Press (ENTER) or (▼) to move the cursor to the second line.

> ▸ Delete the πR^2 and insert LW. You can also write directly over the characters πR^2.

> ▸ The third line need not be changed.

> Your screen should look like Figure 6.60.

(d) Press (2nd) [QUIT] to leave the Program Editor.

Run the program a few times to see that it does what it is supposed to do. Check your results by comparing your output with Figure 6.61. ■

Example 6.26

Change the VOLUME program so that it automatically goes back to the beginning each time. The new version will look like this:

```
Lbl 1
Prompt L,W,H
LWH/3→V
Disp "VOLUME IS ",V
Goto 1
```

Solution We must insert two lines. Here are the steps:

(a) Enter the Program Editor. What you will see will look like Figure 6.60.

(b) `Lbl 1`

> ▸ Move to the beginning of the first line. Your cursor should be blinking on the P of Prompt.

> ▸ Press (2nd) [INS] to enter Insert mode.

> ▸ Press (ENTER) to insert a new line.

> ▸ On the new line, press (PRGM), then move your cursor down the CTL (Control) menu until you see 9:Lbl. Select Lbl.

> ▸ Press number 1.

(c) `Goto 1`

> ▸ Move the cursor to any position in the last line of the program.

> ▸ Press (ENTER). A new line opens up.

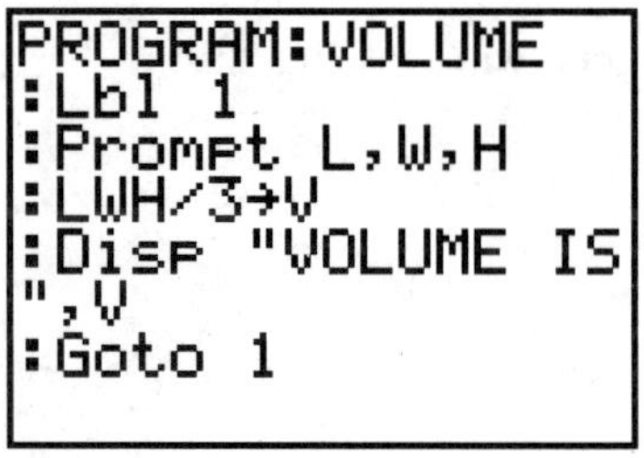

```
PROGRAM:VOLUME
:Lbl 1
:Prompt L,W,H
:LWH/3→V
:Disp "VOLUME IS
",V
:Goto 1
```

Figure 6.62

▶ Press **PRGM** and select 0:Goto from the CTL menu.
▶ Press number 1.

(d) Your program should look like Figure 6.62.

(e) Press **2nd** [Quit] to leave the Program Editor and save the program.

Run the program a few times and see how it has changed. Press **ON** when you want to stop the program. ▪

Summary: Inserting and Deleting in the Editor

1. To insert characters
   ▶ Move the cursor to the location where you want the new characters to appear.
   ▶ Press **2nd** [INS].
   ▶ Enter the characters or commands you want.

2. To delete characters
   ▶ Move the cursor to the location where you want to delete characters.
   ▶ Press **DEL**.

3. To insert a line
   ▶ At the end of the program:
     • Move your cursor to the last line and press **ENTER**.
   ▶ Within the program:
     • Move the cursor to the beginning of the line *below which* you want a new line.
     • Press **2nd** [INS] **ENTER**.

4. To delete a line
   ▶ Move the cursor into the line you want to delete.
   ▶ Press **CLEAR** to erase the characters in the line, then press **DEL** to delete the blank line.

In Example 6.27 you will have one more opportunity to follow the step-by-step process of entering a program. This program will show you how to graph a function.

Example 6.27

Write a program that asks for the values of M and B and then plots the equation $Y = MX + B$.

Solution   Here is a program to accomplish the task:

```
FnOff
PlotsOff
Lbl 1
Prompt M,B
"MX+B"→ Y1
DispGraph
Pause
Goto 1
```

The steps for entering this program follow.

(a) Create a new program called GRAPH.

(b) FnOff

[This instruction turns off the functions in the Y= screen so they will not be graphed.]

- ▶ Press **VARS**, then select 4:On/Off from the Y-VARS menu.
- ▶ Select 2:FnOff.

(c) PlotsOff

[This instruction turns off the stat plots.]

- ▶ Press **2nd** [STAT PLOT], then select 4:PlotsOff from the STAT PLOTS menu.

(d) Lbl 1

- ▶ See Example 6.26.

(e) Prompt M,B

- ▶ See Example 6.25.

(f) "MX+B" **STO ➡** Y1

- ▶ "MX+B"
  (The quotation marks (") are required.)
- ▶ Press **STO ➡**.
- ▶ Press **VARS**, then select 1:Function from the Y-VARS menu and select 1:Y1.

(g) DispGraph

(This instruction plots all equations in the Y= menu.)

- ▶ Press **PRGM** and select 4:DispGraph from the I/O menu.

(h) Pause

(This instruction stops the program until the user presses **ENTER**).

- ▶ Press **PRGM** and select 8:Pause from the CTL menu.

(i) Goto 1

- ▶ See Example 6.26.

The program now looks like Figure 6.63.

(j) Press **2nd** [Quit] to leave the Program Editor.

Run the program a few times to see how it works. ■

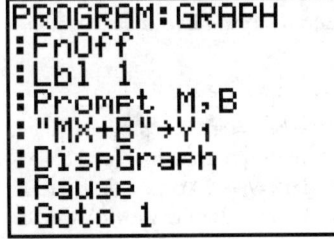

```
PROGRAM:GRAPH
:FnOff
:Lbl 1
:Prompt M,B
:"MX+B"→Y1
:DispGraph
:Pause
:Goto 1
```

Figure 6.63

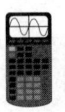

Activity 6.3
Changing the GRAPH Program

Make the following changes to the program named GRAPH:

1. Add a line at the beginning to set the graph window to the standard setting.

- ▶ Insert a blank line, then press **ZOOM** and select 6:ZStandard from the ZOOM menu. This accomplishes the same thing as setting the dimensions of the graphing window with instructions, as was done in the XYGRABBR program (shown in Figure 6.55 on page 399).

2. Change the function to $MX^2 + B$. Remember that you must use quotation marks around the expression.

In this section you have learned how to write and change programs. In the last section of this chapter you will see how to enter the programs that will simulate the running of a robot designed to cut out shapes.

Section 6.5 Exercises

 In Exercises 1–14, (a) make the indicated changes to the function in the program named GRAPH *and (b) test your program with the given values.*

1. $MX^2 + BX$; $M = 4$ and $B = -7$

2. $MX^3 + B$; $M = 0.5$ and $B = -3$

3. $MX^3 + BX$; $M = 0.1$ and $B = -4$

4. $MX^3 + BX^2$; $M = -0.1$ and $B = 0.75$

5. $M + X^B$; $M = -4$ and $B = \frac{2}{3}$

6. $M^{|X|+B}$; $M = 1.25$ and $B = -3$

7. $M\cos(BX)$; $M = 4$ and $B = -2$
   [Use the █COS█ key. *Do not type* COS. For a better view of the graph, change the Zoom instruction in the program to ZTrig.]

8. $\cos(MX + B)$; $M = 1$ and $B = 45$

9. $ax^2 + bx + c$; $a = 2$, $b = 3$, and $c = -5$
   [Make sure to change the Prompt line to ask for values of a, b, and c. Since the *TI-83* does not use lowercase letters you will have to use A for a, B for b, etc. Set the viewing window back to ZStandard.]

10. $a(x + b)^2 + bx + c$; $a = 2$, $b = 3$, and $c = -5$

11. $a\sin(bx + c)$; $a = 3$, $b = 0.75$, and $c = 45$

12. $a\tan(bx + c)$; $a = -0.5$, $b = 1$, and $c = 30$

For Exercises 13 and 14, modify your GRAPH *program so it operates in radian mode. To do this, insert a blank a line at the beginning, press* █MODE█, *and select* Radian.

13. $a\cos(bx^2 + c)$; $a = 2$, $b = .5$, $c = \frac{\pi}{3}$

14. $a\cos(bx) + \sin(cx)$; $a = -\frac{1}{3}$, $b = 2$, $c = -1$

 15. Write a program QUADFORM to use the quadratic formula to find the roots of the quadratic equation $ax^2 + bx + c = 0$. Test your program on the following quadratic equations.
   (a) $2x^2 + 5x - 3 = 0$
   (b) $4x^2 + 12x + 9 = 0$
   (c) $3x^2 + 4x + 5 = 0$

 16. Write a program CIRCLE to draw a circle with the equation $(x - h)^2 + (y - k)^2 = r^2$. The program should ask for h, k, and r. Have the program first set the viewing window so that Xmin $= h-r$, Xmax $= h+r$, Ymin $= k-r$, and Ymax $= k+r$. After the viewing window is set, use ZSquare so that the circles look circular. Test your program on the following circles.
   (a) $(x + 2)^2 + (y - 3)^2 = 5^2$
   (b) $(x - 4)^2 + (y + 10)^2 = 49$

6.6   Solving the Chapter Project

We began this chapter by posing the problem of controlling the operation of a robot programmed to cut out shapes. This is not a simple matter of telling the calculator to draw a graph. The robot must be programmed to draw one point at a time or to draw the curve one small segment at a time. In this section you'll see how to do that.

Plotting Curves One Point at a Time

Example 6.28

Write a calculator program for graphing a pair of parametric equations one point at a time. Model the program after the pseudocode program in Example 6.6(a) (page 364), reproduced here:

```
1  Comment: Parametric Equations
2  Set mode to degrees
3  Let t = 0
4  Compute x = 24cos(t)
5  Compute y = 18sin(t)
6  Plot the point (x, y)
7  Increase t by 10
8  Is t > 360? If yes: go to step 9; If no: go to step 4
9  Stop
```

Solution Most of the *TI-83* instructions for this program have been discussed in Section 6.4, especially in Figure 6.56 (page 400). Here is a *TI-83* program that implements this pseudocode (notice the new instruction used in step 7):

```
Degree
FnOff
PlotsOff
-25→Xmin:25→Xmax:-20→Ymin:20→Ymax
0 → T
Lbl 1
24cos(T)→X
18sin(T)→Y
Pt-On(X,Y,2)
T+10 → T
IF T>360
Then
Pause
Else
Goto 1
```

Here is a description showing how to enter the program POINTS in your calculator.

(a) Enter the Program Editor with a new program named POINTS.

(b) `Degree`

  ▶ Press **MODE** and move the cursor to `Degree`. Press **ENTER** **ENTER**.

(c) `FnOff`

  ▶ See Example 6.27.

(d) `PlotsOff`

  ▶ See Example 6.27.

(e) `-25→Xmin:25→Xmax:-20→Ymin:20→Ymax`
  [These instructions set the viewing window.]

  ▶ `Xmin`, `Xmax`, `Ymin`, and `Ymax` are accessed by pressing **VARS** and selecting the appropriate command from the X/Y menu.

(f) 0 [STO ►] [ALPHA] T

[This instruction stores the value 0 in the variable T.]

(g) Lbl 1

▶ See Example 6.26.

(h) 24cos(T) [STO ►] [ALPHA] X
     18cos(T) [STO ►] [ALPHA] Y

▶ Again, you must use the [COS] key. You cannot enter cosine by typing the letters COS. Quotes are not needed here because you are not assigning the entire formula to X or Y, only the numerical value that results for a specific value of T.

(i) Pt-On (X,Y,2)

[This instruction places a mark of type 2 at the point (X,Y). Note that the Pt-On command will only plot a point using the *x*- and *y*-coordinates of that point. If you want to plot a point using its polar coordinates, you *must* convert them to Cartesian coordinates where $x = r\cos\theta$ and $y = r\sin\theta$. This is what we did above in (g).]

▶ Press [2nd] [DRAW] and select 1:Pt-On(from the POINTS menu.
▶ Enter the remainder of the instruction. Don't forget the final parenthesis.
▶ *Note:* The 2 is a code to plot each point with a square. Use code 1 for a dot and code 3 for a cross.

(j) T+10 → T

[This instruction replaces the number stored in T with a number that is 10 greater than T. The effect is to increase the value of T by 10.]

(k) IF T> 360

▶ Press [PRGM] and select 1:If from the CTL menu.
▶ Press [ALPHA] [T].
▶ Press [2nd] [TEST] and select 3:> from the TEST menu.
▶ Enter 360.

(l) Then

▶ Press [PRGM] and select 2:Then from the CTL menu.

(m) Pause

▶ See Example 6.27.

(n) Else

▶ Press [PRGM] and select 3:Else from the CTL menu.

(o) Goto 1

▶ See Example 6.26.

The resulting program is shown in Figure 6.64. We put the last four instructions of the program into two lines, separated by colons. Figure 6.65 shows the program's result.

```
PROGRAM:POINTS
:Degree
:FnOff
:PlotsOff
:-25→Xmin:25→Xma
x:-20→Ymin:20→Ym
ax
:0→T
:Lbl 1
:Acos(T)→X
:Bsin(T)→Y
:Pt-On(X,Y,2)
:T+10→T
:If T>360
:Then:Pause
:Else:Goto 1
```

Figure 6.64

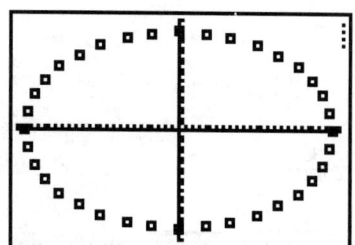

Figure 6.65

The POINTS program is rather complex. Study it and make sure you understand each of its steps before you move on to Activity 6.4.

Activity 6.4
Changing the Points Program

Caution: Some of the changes to the program POINTS described here will change it considerably and the original version will be gone. To avoid losing the original program, you need to copy all the lines of POINTS to a program with a different name. The procedure for copying a program uses the Rcl instruction and is described in the *TI-83 Graphing Calculator Guidebook*, page 16-7.

1. Plot the points with a mark other than a square.
2. The points in Figure 6.65 are spaced approximately 10° apart. How would you change the program so that it plots points that are closer together?
3. Plot the points of the ellipse in Example 6.6(b) (page 364) in its rectangular equation form. The pseudocode program is reproduced here:

```
1   Comment: Rectangular Equation
2   Let x = -24
3   Compute y = +3/4 √(24² - x²)
4   Plot the point (x, y)
5   Compute y = -3/4 √(24² - x²)
6   Plot the point (x, y)
7   Increase x by 1
8   Is x > 24? If yes: go to step 9; If no: go to step 3
9   Stop
```

Plotting Connected Curves

Now that you have learned how to plot curves one point at a time, we'll move to the final goal of this chapter, which is to plot curves in which the individual points are connected by lines resulting in a smooth curve. This will be similar to the program that a robot would need to cut out a shape.

Also you will learn how to use polar coordinates in the programs that follow so that a new set of curves can be plotted, such as the cardioid, the trisectrix, and the *n*-leafed rose.

Example 6.29

Design a calculator program for graphing the polar equation $r = 10 \sin \theta$ two points at a time; connect each pair of points with a straight line. Model the program after the pseudocode program shown below:

```
1   Comment: Polar equation of circle
2   Set mode to degrees; set graph mode to polar
3   Let θmin = 0; Let θmax = 360; Let θstep = 7.5
4   Let θ = θmin
```

5   Compute r1 = 10sin(θ)
6   Convert the point (r1,θ) to Cartesian coordinates (x1, y1)
7   Compute r2 = 10sin($\theta + \theta$step)
8   Convert the point (r2,$\theta + \theta$step) to Cartesian coordinates (x2, y2)
9   Draw a segment between the point (x1, y1) and the point (x2, y2)
10   Increase θ by θstep
11   Is $\theta > \theta$max? If yes: go to step 12; If no: go to step 5
12   Stop

Solution The calculator program has two new ideas. First, since the original equation is in polar coordinates, each point must be translated into Cartesian coordinates with the usual equations

$$x = r\cos\theta$$
$$y = r\sin\theta$$

The second new aspect of this program is the idea of plotting two points at a time and then drawing a line segment between them. Here is a way to write this program for the *TI-83*:

(a) The first lines set up the initial conditions. These lines are executed only once.

```
FnOff
PlotsOff
Degree
ZStandard
ZSquare
0→ θmin
7.5→ θstep
360→ θmax
θmin→ θ
```

(b) The next lines compute the coordinates of the first point (x, y).

```
Lbl 1
10sin(θ)→R
Rcos(θ)→ X
Rsin(θ)→ Y
```

(c) Then we increase θ and compute the coordinates of the second point (p, q).

```
θ+θstep→ T
10sin(T)→ R
Rcos(T)→ P
Rsin(T)→ Q
```

(d) Here is a new instruction. It draws a line segment from (x, y) to (p, q).

```
Line(X,Y,P,Q)
```

(e) The last two lines increase θ and check to see if it is still under the maximum. If it is, go back and do it again with the increased value of θ.

```
θ+θstep→ θ
If θ < θmax:Then:Goto 1
```

Figure 6.66 gives another way to look at this program.

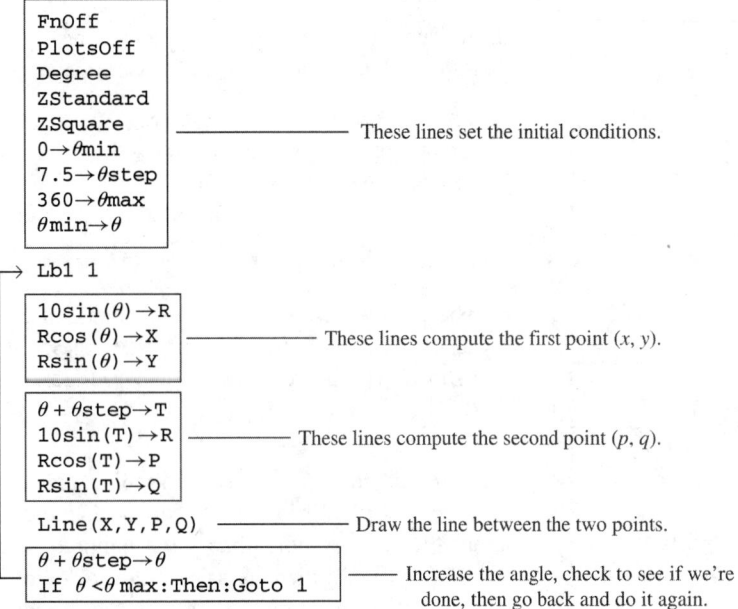

```
FnOff
PlotsOff
Degree
ZStandard
ZSquare
0→θmin
7.5→θstep
360→θmax
θmin→θ
```
———————— These lines set the initial conditions.

→ Lbl 1

```
10sin(θ)→R
Rcos(θ)→X
Rsin(θ)→Y
```
———————— These lines compute the first point (x, y).

```
θ+θstep→T
10sin(T)→R
Rcos(T)→P
Rsin(T)→Q
```
———————— These lines compute the second point (p, q).

Line(X,Y,P,Q) ———————— Draw the line between the two points.

```
θ+θstep→θ
If θ<θmax:Then:Goto 1
```
———————— Increase the angle, check to see if we're
done, then go back and do it again.

Figure 6.66

Activity 6.5
Entering the Robot-Draw Program

Enter the program from Example 6.29 into the calculator. Give it the name ROBOTDRW. Follow previous instructions about creating a new program and entering instructions. Each of the instructions has been covered in earlier examples, with the following exceptions:

▶ θmin, θmax, and θstep   You can see how to enter these by looking at page A-44 of the *TI-83 Graphing Calculator Guidebook*. There you will see all of the calculator's menus displayed. For example, θmin would be entered by pressing 【VARS】, then selecting 1:Window..., and then selecting 4:θmin from the T/θ menu.

▶ θ can be entered directly from the keyboard using 【ALPHA】. You'll find θ above the 【3】 key.

▶ Line(X,Y,P,Q)   The way to enter this instruction is described on page A-44 of the *TI-83 Graphing Calculator Guidebook* under the DRAW menu. It is also described on page A-13. You'll see that the path to this instruction is to press 【2nd】 [DRAW] and then select 2:Line(from the DRAW menu.

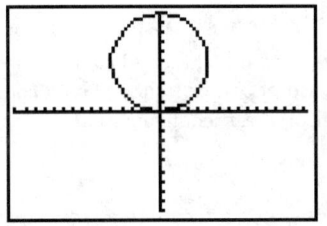

Figure 6.67

Your program should produce a graph like the one shown in Figure 6.67.

Activity 6.6
Changing the Robot-Draw Program

Change the ROBOTDRW program to do each of the following tasks. When you change the polar equation, remember that you must change it in *two* different command lines. Remember that you can keep the original version of ROBOTDRW by copying

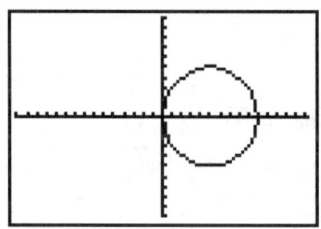

Figure 6.68

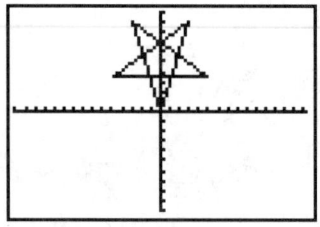

Figure 6.69

its instructions into a new program using the `Rcl` instruction as described in the *TI-83 Graphing Calculator Guidebook*, page 16-7.

1. Create a graph like Figure 6.68.
2. Graph the cardioid $r = 5\cos\theta + 5$.
3. The program accepts A and B as input and then plots a member of the cardioid-limaçon family, the polar equation of which is

$$r = A\cos\theta + B$$

4. The program is like the original version of ROBOTDRW, but it replaces the instruction $7.5 \rightarrow \theta$`step` with a `Prompt` instruction that asks the user to enter θ`step`. Then the program can produce pictures like Figure 6.69, which used θ`step` = 72.

In this chapter we have written programs that have simulated those that might be used to control robots or other mechanical or electronic objects. You have used about fifty of the more than three hundred instructions and functions available on the *TI-83*. While there is a lot more to learn, both about the calculator and about programming, the programs of this section have covered some important concepts upon which you can build.

Section 6.6 Exercises

1. In the ROBOTDRW program, outlined in Example 6.29, there are two lines that begin $\theta + \theta$`step`. One line is $\theta + \theta$`step`$\rightarrow$`T` and the other is $\theta + \theta$`step`$\rightarrow \theta$. (*Caution:* You may want to work on a copy of the program ROBOTDRW.)

   (a) How would the program be affected if you deleted the line that now reads $\theta + \theta$`step`$\rightarrow \theta$ and changed the $\theta + \theta$`step`$\rightarrow$`T` line to $\theta + \theta$`step`$\rightarrow \theta$?

   (b) Make the changes suggested in (a), run this new version of ROBOTDRW, and sketch its graph.

   (c) In the lines `10sin(T)` $\rightarrow$`R`, `Rcos(T)` $\rightarrow$`P`, and `Rsin(T)` $\rightarrow$`Q`, change the `T` to θ. How do you think these changes will affect the program?

   (d) After you have made the changes given in (c), run this version of ROBOTDRW and sketch its graph.

2. (a) Change the `POINTS` program to graph the general ellipse

$$\begin{cases} x = a\cos t \\ y = b\sin t \end{cases}$$

   You will need to make the following changes to the program:
   ▶ Use the statement `Prompt A,B` to ask the user for the values of a and b.
   ▶ Use a and b in defining the values of `Xmin`, etc. For example, one of the instructions might be `-(A+2)`$\rightarrow$`Xmin`.

   (b) Use your program to graph the ellipse

$$\frac{x^2}{25} + \frac{y^2}{9} = 1.$$

3. (a) Change the `POINTS` program to graph the Lissajous patterns $\begin{cases} x = a\sin(bt) \\ y = c\sin(dt) \end{cases}$

   (b) Test your program with $a = c = 8$, $b = 3$, $d = 4$, and with the points spread $10°$ apart.

   (c) Change to points spread $2°$ apart.

4. *Electronics* The low-pass, constant-$k\pi$ filter shown in Figure 6.70 is composed of two capacitors with a coil connected between them in a π configuration.

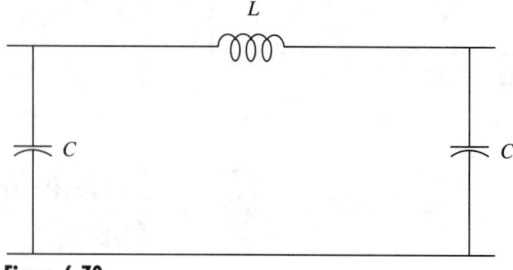

Figure 6.70

   (a) Write a program that will ask you to enter the terminating resistance R of the filter in ohms (Ω) and the cutoff frequency F in hertz (Hz). The program will then calculate the values of the inductance L in millihenries (mH) and the two capacitors C, which are equal in value, in microfarads (μF).

The necessary formulas are $L = \dfrac{10^3 R}{\pi F}$ and $C = \dfrac{10^6}{2\pi R F}$. The program should round your answers to two decimal places.

(b) Test your program with $R = 75\ \Omega$ and $F = 4000$ Hz. The answers should be $L = 5.97$ mH and $C = 0.53\ \mu$F.

 5. *Electronics* The band-elimination filter can be thought of as a combination of a low-pass filter and a high-pass filter. (Of course the cutoff frequency of the low-pass filter is lower than the cutoff frequency of the high-pass filter.) A gap is left between the two cutoff frequencies where the frequency is attenuated, allowing the band-elimination filter to attenuate a specific frequency selectively. A band-elimination constant-k T filter is shown in Figure 6.71.

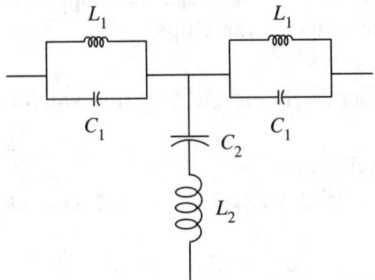

Figure 6.71

(a) Write a program that will ask you to enter the terminating resistance R of the filter in ohms (Ω) and the upper U and lower L bandpass frequencies in hertz (Hz). The program will then calculate the values of the two inductances L_1 and L_2 in millihenries (mH) and the two capacitors C_1 and C_2 in microfarads (μF). The necessary formulas

are

$$L_1 = \frac{10^3 R(U - L)}{2\pi LU} \qquad L_2 = \frac{10^3 R}{4\pi (U - L)}$$

$$C_1 = \frac{10^6}{2\pi R(U - L)} \qquad C_2 = \frac{10^6 (U - L)}{\pi LUR}$$

The program should round your answers to two decimal places. (*Hint:* You may want to put a PAUSE in the program to allow you to read the first answers.)

(b) Test your program with $R = 60\ \Omega$, $L = 45$ Hz, and $U = 65$ Hz. The answers should be $L_1 = 65.29$ mH, $L_2 = 238.73$ mH, $C_1 = 132.63\ \mu$F, and $C_2 = 36.27\ \mu$F.

 6. *Computer science*
(a) Write a program to determine the polar coordinates (r, θ) given the Cartesian coordinates (x, y) of the point. Round answers to three decimal places and report answers in both degrees and radians.

(b) Test your program on the following points: $(5, -5)$ (answers should be $r = 7.017$, $\theta = -45°$, $\theta = -0.785$ rad) and $(-8, 4)$ (answers: $r = 8.944$, $\theta = 153.435°$, $\theta = 2.678$ rad).

 7. *Computer science* Write a program to convert the polar coordinates of a point to its Cartesian coordinates. The program should you give the option of entering θ in either radians or degrees.

8. In the Chapter Project the management asked you to
(a) Review the mathematics and programming that make the *X-Y Grabber* work.
(b) Determine how to adapt its programming to trace out points on a curve.
(c) Develop a single equation that will draw a square or hexagon.
(d) Learn the definitions of a pentagram and a trisectrix and determine their equations.

Write a report to management responding to these four requests.

9. How do a limaçon and a trisectrix differ? How are they alike?

● ● ● ● ● ● ● ● ● ● **Chapter 6 Summary and Review**

Topics You Learned or Reviewed

▶ Converting parametric equations to a rectangular equations. In some cases you need only solve one equation for t and substitute that answer for t in the other equation.

▶ A graph where all points have a mirror image on the other side of the y-axis is said to have symmetry about the y-axis.

▶ Parametric equations can be used to model the voltages that are applied to the horizontal and vertical plates of an oscilloscope. Any point (x, y) on the oscilloscope curve gives the two voltages at some time. This point can be modeled with the parametric equations

$$\begin{cases} x = V_0 \sin B_x t \\ y = V_0 \sin B_y t \end{cases}$$

Curves of this type are called Lissajous curves.

▶ Points and equations in polar coordinates can be plotted by hand and by calculator.

▶ An ellipse centered at the origin, with major axis of length $2a$, minor axis of length $2b$, and eccentricity e has the following equations.

Rectangular equation: $\dfrac{x^2}{a^2} + \dfrac{y^2}{b^2} = 1$ or $y = \pm\dfrac{b}{a}\sqrt{a^2 - x^2}$

Parametric equations: $\begin{cases} x = a\cos t \\ y = b\sin t \end{cases}$

Polar equation: $r = \dfrac{a(1 - e^2)}{1 - e\cos\theta}$

▶ In astronomy, the Sun is at the focus of an ellipse that forms the path of an object, such as a planet or comet, that is orbiting the Sun.

▶ Anything, unless it is acted upon by another force, that passes through one focus of an ellipse will be reflected through the other focus. Examples are sound waves in a lithotripter.

▶ The eccentricity of an ellipse is between 0 and 1. The closer the eccentricity is to 0, the closer the shape of the ellipse is to a circle. The closer the eccentricity is to 1, the more elongated is the shape of the ellipse.

▶ Running, editing, and entering new programs in a *TI-83*.

▶ Using the programming language to set the initial environment of a program, defining the graphing window, setting the MODEs, clearing the graphing window, and turning off function.

▶ A program will ask for instructions with either the `Input` or `Prompt` commands.

▶ The output from a program is produced by using the `Disp`, `DispGraph`, `DispTable`, `Pt-On`, and `Line` commands.

▶ Programming control instructions include `Pause`, `If...Then...Else`, `Goto`, and `Lbl`.

▶ A value can be assigned to a variable and the program can be directed to increase the value of a variable.

▶ An expression can be assigned to a function that is to be graphed, provided the function is of the form $y = \ldots$.

Review Exercises

In Exercises 1 and 2, (a) graph the given parametric equations, (b) write the rectangular equation for the parametric equations, and (c) if possible, graph the rectangular equation from (b) and check that it matches the graph in (a).

1. $\begin{cases} x = 4t + 3 \\ y = 6t^2 + \dfrac{1}{t} \end{cases}$

2. $\begin{cases} x = \dfrac{1}{\cos t} \\ y = 5\sin t \end{cases}$

In Exercises 3 and 4 sketch each of the parametric equations.

3. $\begin{cases} x = \dfrac{1}{\sin t} \\ y = 3t^2 + 1 \end{cases}$

4. $\begin{cases} x = 6\cos^3 t \\ y = 10\sin^4 t \end{cases}$

In Exercises 5–8 sketch each of the polar equations.

5. $r = 3 - 8\sin 3\theta$

6. $r = 3 + 9\cos 1.5\theta$

7. $r = 2\sin\theta + \cos 2\theta$

8. $r = 3\sin\theta + \cos 3\theta$

In Exercises 9–12, (a) predict the ratio of the number of loops along the top to the number of loops along the side and (b) graph the Lissajous curve.

9. $x = \sin 3t$, $y = \cos 4t$

10. $x = \sin 2t$, $y = \cos 5t$

11. $x = \sin 6t$, $y = \cos 3t$

12. $x = \sin 5t$, $y = \cos 3t$

13. Read the following program written in pseudocode and describe what would happen if a person or robot followed the code exactly.

```
1   Begin at point A facing north
2   Walk ten paces
3   Are you back at point A? If yes: go to
    step 9; If no: go to step 4
4   Turn right 45 degrees
5   Walk 20 paces
6   Are you back at point A? If yes: go to
    step 9; If no: go to step 7
7   Turn right 90 degrees
8   Go to step 2
9   Stop
```

14. Write a program in pseudocode that will result in two triangles being traced so that the triangles touch at only one vertex.

15. Write a program in pseudocode that will draw six different triangles, each having an area of 8 cm².

16. For each of the five points shown in Figure 6.72,
   (a) write polar coordinates with r and θ both positive.
   (b) write polar coordinates with r and θ both negative.
   (c) write polar coordinates with r positive and θ negative.
   (d) write polar coordinates with r negative and θ positive.

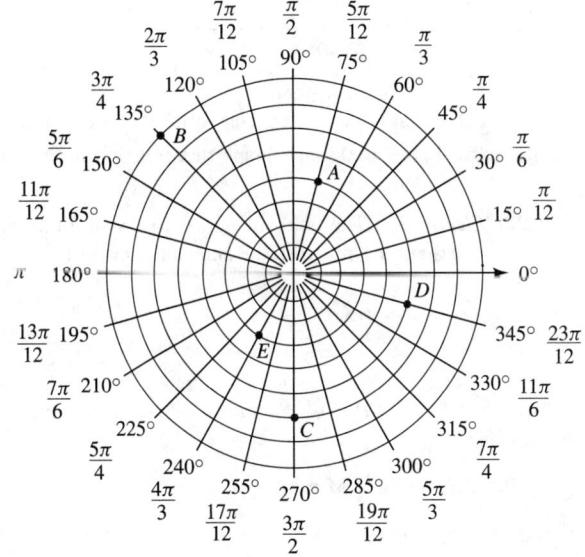

Figure 6.72

17. Plot each of the following polar points and give their x- and y-coordinates: $A(-3, 90°)$, $B(2, 30°)$, $C(-5, -215°)$, $D(8, -315°)$, and $E\left(4, \frac{7\pi}{6}\right)$.

18. Write the polar equation for a circle of radius 6 centered at the polar point $(6, 270°)$.

19. Write a polar equation for a 5-leafed rose that has symmetry about the polar axis.

20. Write a polar equation for a line that has a slope of $\frac{2}{5}$ and a y-intercept of 5.

21. Write a polar equation for a line that has a slope of -2 and a y-intercept of 4.

In Exercises 22–23, (a) convert the given polar equation into an equation involving rectangular coordinates and (b) identify the graph of the equation.

22. $r \cos \theta = -3$

23. $r = 3 \sin \theta$

24. *Astronomy* The comet Temple-Tuttle has an orbit that is an ellipse with the Sun at one focus. If the major axis of the orbit is 20.66 AU and $e \approx 0.906$, **(a)** determine a, b, and c and **(b)** write the polar equation for this comet.

25. *Astronomy* The asteroid Icarus has an orbit that is an ellipse with the Sun at one focus. If the major axis of the orbit is 2.156 AU and $e \approx 0.827$, **(a)** write the polar equation for this asteroid and **(b)** graph the asteroid and Earth's orbit on the same axes.

26. *Business* After a new consumer electronics product is introduced, sales rise quickly and the price gradually decreases. Let t be the number of years since a product was introduced (that means the product was introduced at $t = 0$). Suppose the unit price at time t, in hundreds of dollars, is $p(t) = \dfrac{t^2 + 20}{t^2 + 5}$, and the monthly sales, in 100,000 units, are $s(t) = \dfrac{t^2 + 3t}{t^2 + 1}$.

   (a) Graph this pair of parametric equations for 5 years with $s(t)$ on the horizontal axis and $p(t)$ on the vertical axis.

   (b) What was the price when the product was introduced? after 1 year? after 5 years?

   (c) What were the monthly sales during the 12th month? after 5 years?

27. *Architecture*   In the design of a geodesic dome, an architect uses the equation

$$r^2 = \frac{E^2}{E^2 \cos^2 \theta + \sin^2 \theta}$$

where E is a constant.

(a) What is the rectangular form of this equation?

(b) Let $E = 0.5$ and graph this function on your calculator.

28. *Electronics*   The field strength r, in μV/m, of a broadcast station 1 mi from the antenna is given by

$$r = 2 + 5 \cos 2\theta + \sin \theta.$$

Graph this antenna pattern.

29. *Electronics*   Two voltages $V_1 = 110 \sin 120\pi t$ and $V_2 = 110 \sin \left(120\pi t + \frac{\pi}{2}\right)$ are applied respectively to the horizontal and vertical plates of an oscilloscope.

(a) Sketch the resulting Lissajous curve in radian mode for $0 \le t \le \pi$ with θstep $= 0.5$.

(b) Sketch the figure again with θstep $= 0.51$.

(c) Sketch the figure again with θstep $= 0.051$.

(d) Explain why the three figures are so different.

In Exercises 30–31, (a) make the indicated changes to the function in the program named GRAPH *and (b) test your program with the given values.*

30. $MX^3 - BX^2$; $M = 0.5$ and $B = 2$

31. $a^{bx} + c$; $a = 0.9$, $b = 3$, and $c = -4$

32. Write a program HYPERBLA to draw a hyperbola with the equation

$$\frac{(x - h)^2}{a^2} - \frac{(y - k)^2}{b^2} = 1$$

The program should ask for a, b, h, and k. Have the program first set the viewing window so that Xmin $= h - |2a|$, Xmax $= h + |2a|$, Ymin $= k - |2b|$, and Ymax $= k + |2b|$. After the viewing window is set, use ZSquare so that the circles look circular. Test your program on the hyperbola

$$\frac{(x - 3)^2}{4^2} - \frac{(y + 2)^2}{5^2} = 1$$

33. *Electronics*   The high-pass, series, M-derived filter shown in Figure 6.73 is composed of three capacitors, two of which have the same capacitance, with one inductor.

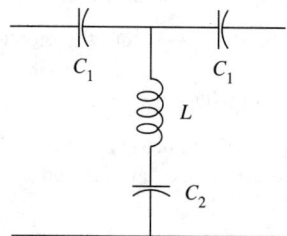

Figure 6.73

(a) Write a program that will ask you to enter the terminating resistance R of the filter in ohms (Ω), the cutoff frequency F in hertz (Hz), and the maximum attenuation frequency M. The program will then calculate the values of the inductance L in millihenries (mH) and the capacitors C_1 and C_2 in microfarads (μF). The necessary formulas are

$$L = \frac{10^3 R}{4\pi F \sqrt{1 - \left(\frac{M}{F}\right)^2}}$$

$$C_1 = \frac{2 \cdot 10^6}{4\pi R F \sqrt{1 - \left(\frac{M}{F}\right)^2}}$$

$$C_2 = \frac{10^6 \sqrt{1 - \left(\frac{M}{F}\right)^2}}{\pi R F \left(\frac{M}{F}\right)^2}$$

The program should round your answers to two decimal places. (*Hint:* To save some work you might want to store

$$\sqrt{1 - \left(\frac{M}{F}\right)^2}$$

as a variable and use that variable three times.)

(b) Test your program with $R = 400 \ \Omega$, $F = 1000$ Hz, and $M = 950$. The answers should be $L = 101.94$ mH, $C_1 = 1.27 \ \mu$F, and $C_2 = 0.28 \ \mu$F.

Chapter 6 Test

1. **(a)** Graph the parametric equations $\begin{cases} x = 3t - 5 \\ y = \sqrt{t} - \dfrac{3}{t} \end{cases}$

   for $0 < t \le 6$.

   (b) Write the rectangular equation for these parametric equations.

2. **(a)** Sketch the parametric equations $\begin{cases} x = \dfrac{\cos t}{2 + \sin t} \\ y = 1 + \sin t \end{cases}$

   (b) What are the value(s) of y when $x = 1$?

3. Sketch the polar equation $r = \theta + \sin 3\theta$.

4. Write the parametric equations that will produce a Lissajous curve with three loops along the top and two loops along the side.

5. Describe what would happen and the shape that would be traced if a person or robot enacted the following pseudocode exactly.
   ```
   1   Begin at point A facing north
   2   Walk ten paces
   3   Are you back at point A? If yes: go to
       step 9; If no: go to step 4
   4   Turn right 45 degrees
   5   Walk 20 paces
   6   Are you back at point A? If yes: go to
       step 9; If no: go to step 7
   7   Turn right 135 degrees
   8   Go to step 2
   9   Stop
   ```

6. Write a program in pseudocode that will result in a triangle being traced on top of a square.

7. For point P shown in Figure 6.74,
   (a) write polar coordinates with r and θ both positive.
   (b) write polar coordinates with r negative and θ positive.

8. Give the x- and y-coordinates of the polar points $A(5, -120°)$ and $B\left(-2, \frac{7\pi}{4}\right)$.

9. Write the equation for a circle of radius 5 centered at the polar point $(5, 180°)$.

10. Write an equation for a 4-leafed rose that has symmetry about the polar axis.

11. Write an equation for a line that has a slope of $\frac{1}{3}$ and a y-intercept of 4.

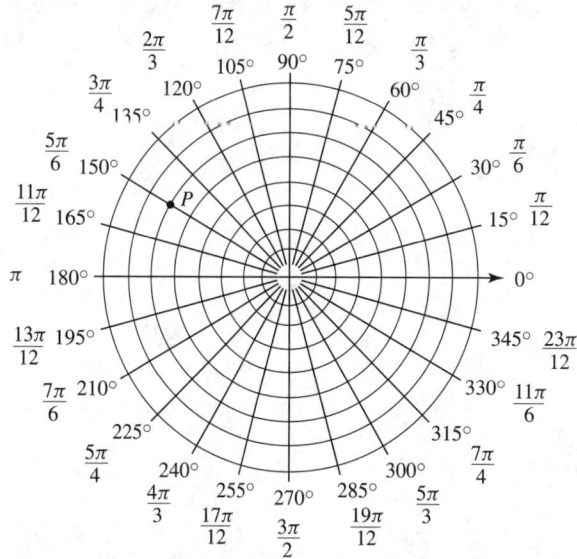

Figure 6.74

12. Convert the polar equation $r \cos\theta + 2r \sin\theta = 4$ into an equation involving rectangular coordinates and identify the graph of the equation.

13. The comet Kahoutek has an orbit that is an ellipse with the Sun at one focus. If the major axis of the orbit is 6.8 AU and $e \approx 0.537$, **(a)** determine the values of a, b, and c and **(b)** write the polar equation for this comet.

 14. Write a program CONICS that will draw a rotated circle, ellipse, parabola, or hyperbola with the equation

   $$r = \frac{ek}{1 + e\cos(\theta - \alpha)}$$ where e is the eccentricity, k

   is a constant that will enlarge or shrink the conic, and α is the angle the axis of the conic is rotated from the x-axis. The program should ask for e, k, and α. (For your information: $e = 0$ produces a circle, $0 < e < 1$ an ellipse, $e = 1$ a parabola, and $e > 1$ a hyperbola.)

CHAPTER

7

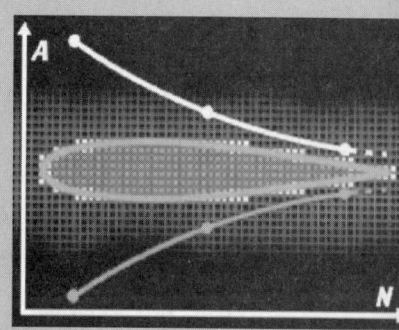

Modeling with Sequences and Series

Topics You'll Need to Know

▸ Definitions and uses
  • percents, ratios, exponents, scientific notation, nth root
  • subscripts: $a_n = n$th term in a sequence
▸ Properties
  • $a^n \times a^m = a^{n+m}$
  • $(a^n)^m = a^{nm}$
  • $(\sqrt[n]{b})^n = b$ for $b \geq 0$
  • factoring: $ar^n + a = a(r^n + 1)$
▸ Making and using graphs

Calculator Skills You'll Use

▸ Graph, Table, Trace
▸ Solving an equation in one variable with a graph or table
▸ Running a calculator program

Topics You'll Learn or Review

▸ Meaning and use of fractional, negative, and zero exponents
▸ Geometric and arithmetic sequences
  • definite and recursive definitions
  • finding the formula for nth term from patterns
  • using the recursive definition in fixed points and chaos theory
  • limits as $n \to \infty$
▸ Geometric and arithmetic series
  • definition
  • formula for adding first n terms
  • limits
▸ Limits
  • guessing limits from a table
  • guessing limits from a graph
  • limits of sequences defined by $\frac{1}{n}$, $\left(\frac{1}{2}\right)^n$, or r^n
  • using algebraic properties of limits to guess limits of sequences
▸ Estimating areas with Riemann sums

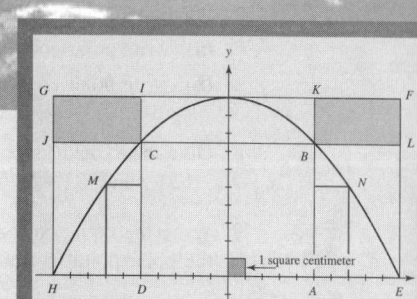

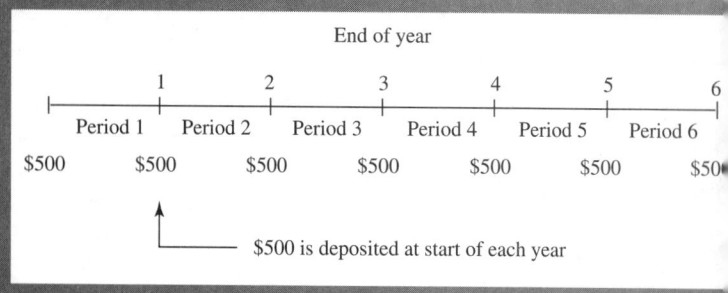

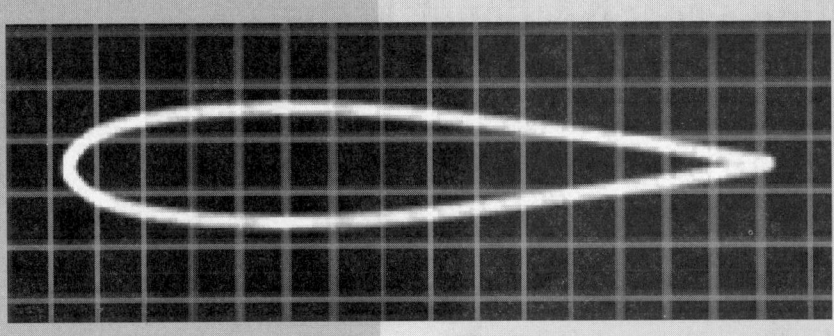

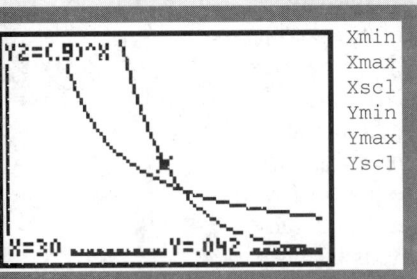

Xmin
Xmax
Xscl
Ymin
Ymax
Yscl

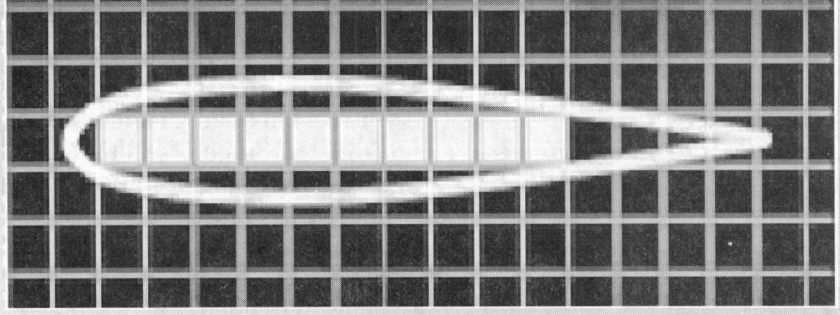

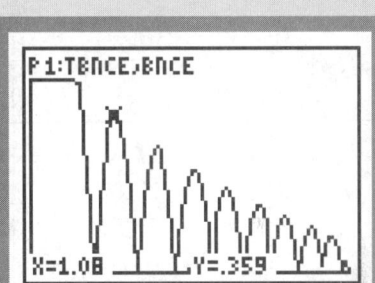

Calculator Skills You'll Learn

▶ Sequences
  • seq function
  • seq mode
    • defining a sequence with the
      nth term
    • defining a sequence recursively
    • guessing a limit from a graph or table
▶ Programming
  • For instruction
  • Menu instruction

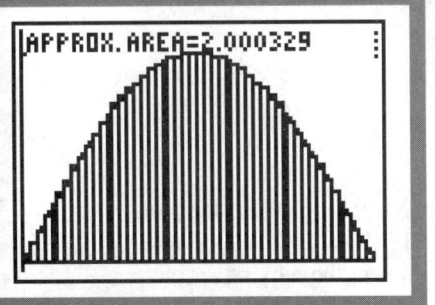

What You'll Do in This Chapter

The chapter begins with the problem of estimating the area under a curve. You'll invent your own method and later learn one of the methods that mathematicians have developed for the solution. You'll learn to look at your estimates as a sequence of numbers that follow a pattern, and you'll learn to look at the pattern of numbers in a sequence and guess the limit of the sequence.

The study of mathematical sequences leads to applications including interest on investments, charging a capacitor in an RC circuit, and estimating the age of ancient objects using the mathematics of radioactive decay and half-life. You will also have occasion to find the sum of the terms of a sequence. Finally, you will use and study a calculator program that can estimate the area under a curve using a mathematical sequence.

Chapter Project—Estimating Cross-Sectional Areas

Streamline Testing Labs
Customized
Wind Tunnel Testing

To:	Research Department
From:	Management
Subject:	Estimating Airfoil Areas

If you're a fan of drag racing you may have heard of the new cars that Jeff Smith of Smith Racing Team, Inc. is developing. He's working on interesting new designs for airplane-like airfoils for his next generation of dragsters. Jeff has asked us to test his designs in our high-speed wind tunnel. That series of tests will be nothing new for us. However, he has also asked us to figure a way to estimate the cross-sectional area of each design we test. If we knew the equations of the curves that form the exterior of the airfoil we could use the math you learned in school to produce estimates to whatever accuracy is required. However, we'll have to use other methods because Jeff and his engineers believe in hand-crafting every shape. They think of themselves as artists and they don't like to use equations.

Here's what we need to do in order to crack this area estimation problem:

• Develop a method to estimate the area of a two-dimensional curved shape.

• Develop a method to estimate areas if the equations of the boundary curves are known.

Preliminary Analysis

When a structure is placed in a wind tunnel, instruments help engineers study the way that air flows around the structure. Figures 7.1 and 7.2 show that the cross-section of the same wing, placed at different angles, will deflect the moving air quite differently. Notice that when the wing is tilted there is turbulence at the back of the wing. Experts can look at this turbulence and determine if the angle or the shape of the wing needs to be changed to achieve a desired result. Some desired results may be to provide lift for the plane or to slow it down.

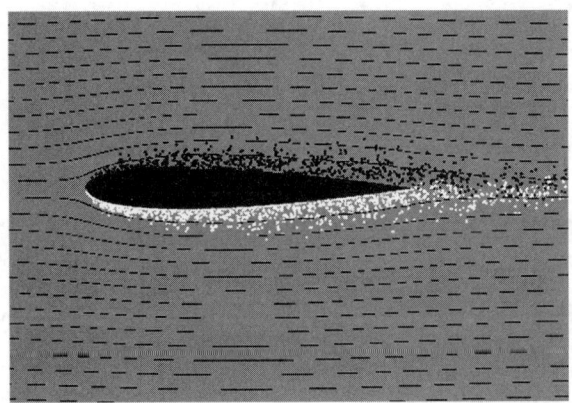

Figure 7.1

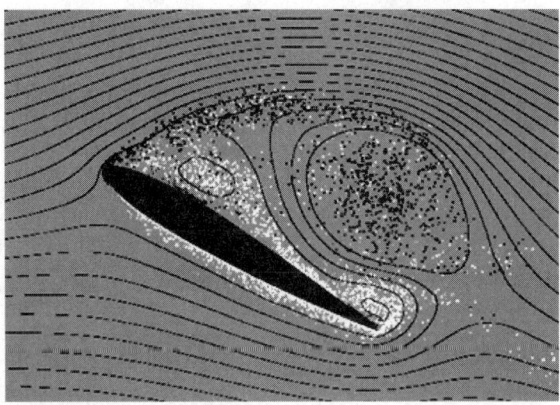

Figure 7.2

Courtesy Leon van Dommelen, FAMU-FSU College of Engineering. Work supported by the U.S. Air Force Office of Scientific Research. Originally found at http://www.scri.fsu.edu/~dommelen/research/airfoil/airfoil.html

Analysis of these wing shapes often requires an estimate of the area of the cross-section of the wing.

Activity 7.1
First Look at the Area Estimation Problem

None of Figures 7.3–7.6 are actual size.

A parabola is shown in Figure 7.3. What is an estimate of the area under the curve and above the x-axis? Compare your methods and results with other students in the class. Answer these questions:

1. Make a quick estimate of the area. Is it more than 50 cm^2? Less than 100 cm^2? How do you know?
2. Without doing any measuring, give two numbers that you're sure the area is between. For example, can you be sure that the area A is bigger than 0 and smaller than 1 000 000 cm^2? That is, do you know that $0 < A < 1\ 000\ 000$?

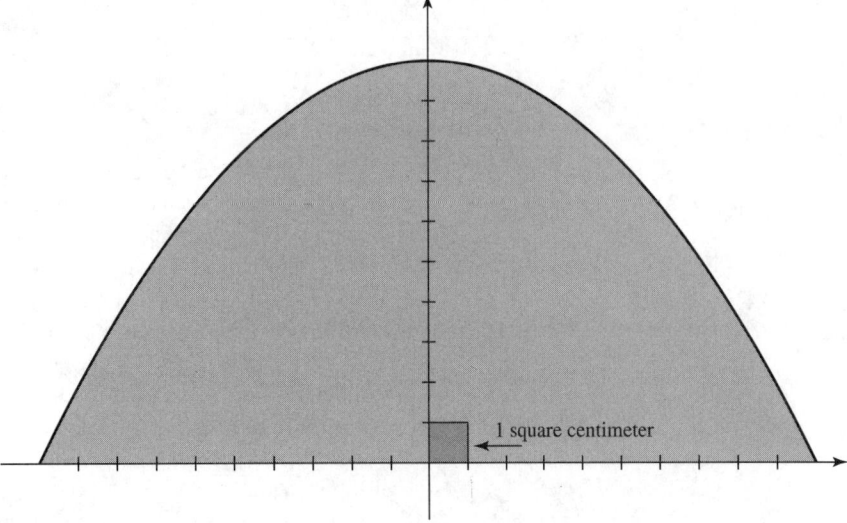

1 square centimeter

Figure 7.3

Can you do better than that? Find a smaller interval and explain your reasoning. For example, you can be sure that the area is between the areas of rectangles $ABCD$ and $EFGH$ in Figure 7.4.

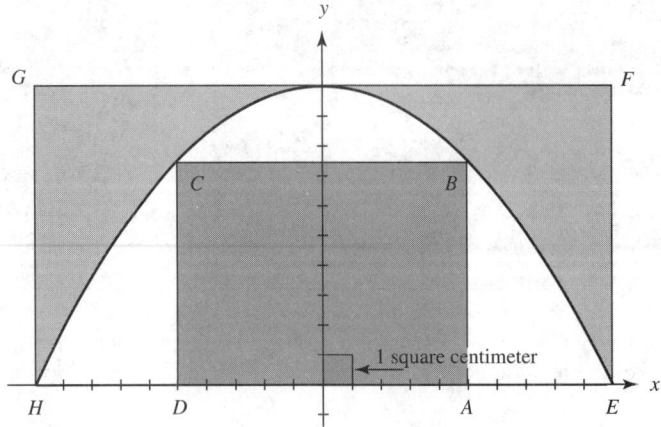

Figure 7.4

3. Use the average of the highest and lowest numbers in your interval to give you another estimate of the true area. Record your answers in a table like Table 7.1. Use the symbols L_1 for the first lower estimate, U_1 for the first upper estimate, and M_1 for the first average where $M_1 = \dfrac{L_1 + U_1}{2}$. For example, if you use Figure 7.4, L_1 would be the area of rectangle $ABCD$ and U_1 the area of rectangle $EFGH$.

Table 7.1   Upper and lower area approximations

Estimate #	Lower Estimate	Upper Estimate	Average
1	$L_1 =$	$U_1 =$	$M_1 =$
2	$L_2 =$	$U_2 =$	$M_2 =$
3	$L_3 =$	$U_3 =$	$M_3 =$
4	$L_4 =$	$U_4 =$	$M_4 =$

4. Do you think the true area is more or less than the average you calculated for M_1?

5. See if you can make a better estimate. Continue to use the idea of an interval that you can be sure contains the area. Give the lower and upper endpoints of your interval, and report the average of these two numbers. The estimates for your second interval are L_2, U_2, and M_2.

   To determine the next upper estimate, U_2, extend $\overline{CB}$, $\overline{AB}$, and $\overline{CD}$ as shown in Figure 7.5. You can exclude the areas of the shaded rectangles $CIGJ$ and $BKFL$ from U_1 to get U_2. Notice that $U_2 < U_1$.

   To determine the next lower estimate, L_2, you might want to draw vertical lines from the x-axis at $x = -7$ to the curve at M and from $x = 7$ to the curve at N. Complete the rectangles by drawing horizontal lines from M and N to the side of the previous rectangles. Thus, L_2 would be the area of the rectangle used for L_1 and the two new areas. Use L_2 and U_2 to determine M_2 and enter these three numbers in Table 7.1. Notice that $L_2 > L_1$.

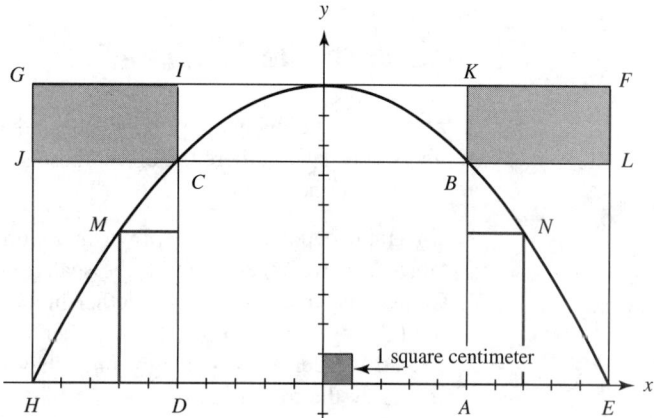

Figure 7.5

6. Let's try something different to find L_3. We will use some triangles. Label the top of the curve P and draw segments $\overline{HC}$, $\overline{CP}$, $\overline{PB}$, and $\overline{BE}$. This forms the three triangles $\triangle HDC$, $\triangle CBP$, and $\triangle BEA$. If we add the areas of each of these triangles to the area of rectangle $ABCD$, we will get L_3. From the diagram in Figure 7.6 you can see that $L_3 > L_2$ and that L_3 is a better approximation of the desired area.

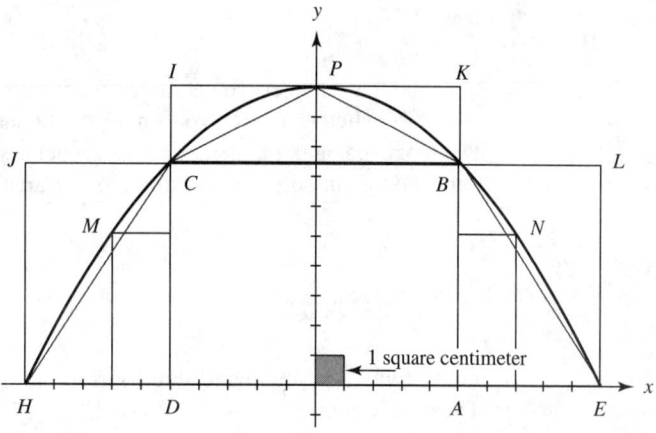

Figure 7.6

7. Continue and determine U_3 and M_3. Decide the approach you will take. You might draw more triangles and rectangles or you might draw some other figures. Enter your results in a table like Table 7.1. Remember, the estimate is the average of the upper and lower limits

$$M_n = \frac{L_n + U_n}{2}$$

8. Determine L_4 and U_4 and use these values to calculate M_4. You decide on the approach to take.

Now that you have made the estimates, let's take a look at the numbers.

Activity 7.2
Looking at the Estimates

A picture of data always improves your understanding of the relationship among the numbers you have gathered. Follow these steps for another way to estimate the true area under the parabola.

1. On graph paper like that shown in Figure 7.7, plot the 12 estimates from your table. You will need to mark the scale on the vertical axis.
2. Connect each set of points together in a curve. That is, connect the L's together, the U's together, and the A's together.
3. Extend the curves to the right until they meet. Do you think the point of intersection is the true area?

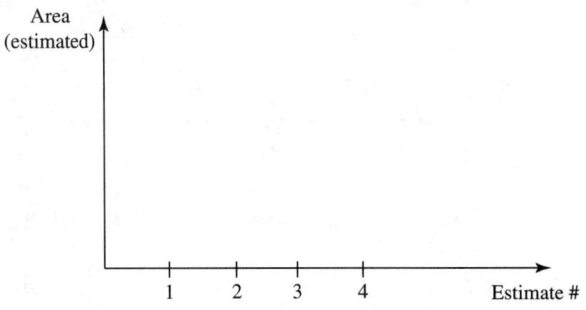

Figure 7.7

In this chapter you will learn mathematical techniques for estimating areas like this, whether or not you know the equation of the curve. You will also learn the language and mathematics of sequences and series, which have many other applications, in addition to estimation of area.

7.1   Sequences

In Activity 7.1 you made three sets of estimates of the area under the parabola. These sets were $\{L_1, L_2, L_3, L_4\}$, $\{U_1, U_2, U_3, U_4\}$, and $\{A_1, A_2, A_3, A_4\}$. Each of these sets is an example of a mathematical **sequence**. Mathematical sequences are sets of numbers that follow a pattern. The numbers that make up the sequence are called **terms of the sequence**. For example, the sets

$$A = \{1, 2, 3, 4, 5, \ldots\}$$

$$B = \{15, 11, 7, 3, \ldots\}$$

are sequences with patterns that involve addition or subtraction of the same number. Sequences like this are called **arithmetic sequences**,* and we say that there is a **common difference** between consecutive terms. The common difference in sequence A is 1 and the common difference in sequence B is -4.

---

*The word *arithmetic* is used here as an adjective and is pronounced with the accent on "met."

C and D are sequences with patterns that involve multiplication or division of the same number.

$$C = \{1, 2, 4, 8, 16, \ldots\}$$

$$D = \{1000, 200, 40, 8, 1.6, \ldots\}$$

Sequences like this are called **geometric sequences**, and we say that there is a **common ratio** between consecutive terms. The common ratio in sequence C is 2, and the common ratio in sequence D is $\frac{1}{5}$.

Sequences of numbers occur quite commonly in the world. For example, regular deposits of $100 per month into a bank account produce the arithmetic sequence $\{100, 200, 300, 400, 500, \ldots\}$, where each number is the amount in the account each month, excluding interest. The amount that a single deposit of $1000 becomes each year, at 10% interest per year, turns out to be a geometric sequence, as we will see in Section 7.4. Population growth and radioactive decay can both be modeled with geometric sequences.

There are sequences that are neither arithmetic or geometric. A very simple example of one of these is a sequence whose rule is "double the previous number and add 10." One sequence that can be built from that pattern is $\{5, 20, 50, 110, 230 \ldots\}$. You can see that there is neither a common difference nor a common ratio between consecutive terms.

Defining Sequences Directly

One way to define a sequence is give a formula for the nth term. For example, the sequence $\{4, 5, 6, 7, \ldots\}$ can be defined by the formula

$$a_n = n + 3$$

where a_n represents the nth term in the sequence, and n is a positive integer $1, 2, 3, 4, \ldots$.

Throughout this chapter, the variable n will always be a positive integer. When the sequence begins with a_1, then a_n will represent the nth term of the sequence.

Example 7.1

For the sequence $a_n = \dfrac{1}{n}$

(a) Write the first five terms of the sequence.
(b) Write the 100th and 101st terms.
(c) Write the difference and ratio between any two consecutive terms that you have written.
(d) Tell whether the sequence is an arithmetic sequence, a geometric sequence, or neither of these types of sequences.

Solutions (a) The sequence is $a_n = \dfrac{1}{n}$. If we let $n = 1, 2, 3, 4$, and 5 we will get the first five terms of the sequence. If $n = 1$, then the first term of the sequence is $a_1 = \frac{1}{1} = 1$. If $n = 2$, the second term of the sequence is $a_2 = \frac{1}{2}$. $n = 3$ produces $\frac{1}{3}$, the third term of the sequence. The fourth and fifth terms of the sequence are $a_4 = \frac{1}{4}$ and $a_5 = \frac{1}{5}$. Thus, the first five terms are $\left\{1, \frac{1}{2}, \frac{1}{3}, \frac{1}{4}, \frac{1}{5}\right\}$.

(b) For the 100th and 101st terms we let $n = 100$ and $n = 101$, respectively. This gives the terms $a_{100} = \frac{1}{100}$ and $a_{101} = \frac{1}{101}$. So, $\{a_{100}, a_{101}\} = \left\{\frac{1}{100}, \frac{1}{101}\right\}$.

(c) To find the difference between two consecutive terms we will subtract the earlier term from the later term. Thus, for the difference between the second and third terms we will find $a_3 - a_2 = \frac{1}{3} - \frac{1}{2} = -\frac{1}{6}$. Similarly, for the ratio between two consecutive terms we will divide the later term by the earlier. Thus, for the ratio between the second and third terms we compute $\frac{a_3}{a_2} = \frac{1/3}{1/2} = \frac{2}{3}$. The table below shows the difference and ratio between any two consecutive terms we found in (a) or (b). The table looks a little strange because the terms are listed in every other box, which allows us to list the difference and ratio between two consecutive terms in the column between the two terms.

n (term)	1		2		3		4		5	...	100		101
$a_n = \dfrac{1}{n}$	1		$\frac{1}{2}$		$\frac{1}{3}$		$\frac{1}{4}$		$\frac{1}{5}$	...	$\frac{1}{100}$		$\frac{1}{101}$
Difference		$-\frac{1}{2}$		$-\frac{1}{6}$		$-\frac{1}{12}$		$-\frac{1}{20}$				$-\frac{1}{10100}$	
Ratio		$\frac{1}{2}$		$\frac{2}{3}$		$\frac{3}{4}$		$\frac{4}{5}$				$\frac{100}{101}$	

(d) This sequence, $a_n = \dfrac{1}{n}$, is not an arithmetic sequence because all the differences are different. It is also not a geometric sequence because all the ratios are different. Thus, the sequence $a_n = \dfrac{1}{n}$ is neither an arithmetic nor a geometric sequence.   ∎

Example 7.2

For the sequence $b_n = 2n + 4$

(a) Write the first five terms of the sequence.
(b) Write the 100th and 101st terms.
(c) Write the difference and ratio between any two consecutive terms that you have written.
(d) Tell whether the sequence is an arithmetic sequence, a geometric sequence, or neither of these types of sequences.

Solutions (a) The sequence is $b_n = 2n + 4$. If $n = 1$, then the first term of the sequence is $b_1 = 2(1) + 4 = 6$. If $n = 2$, the second term of the sequence is $b_2 = 2(2) + 4 = 8$. The next three terms of the sequence are $b_3 = 2(3) + 4 = 10$, $b_4 = 2(4) + 4 = 12$, and $b_5 = 2(5) + 4 = 14$. As a result, the first five terms of $\{b_n\}$ are $\{6, 8, 10, 12, 14\}$.

(b) For the 100th and 101st terms we let $n = 100$ and $n = 101$, respectively. This gives the terms $b_{100} = 2(100) + 4 = 204$ and $b_{101} = 2(101) + 4 = 206$, and so $\{b_{100}, b_{101}\} = \{204, 206\}$.

(c) As in Example 7.1, the table below lists the first five, 100th, and 101st terms of $\{b_n\}$ and the difference and ratio between two consecutive terms.

n (term)	1		2		3		4		5	...	100		101
$b_n = 2n + 4$	6		8		10		12		14	...	204		206
Difference		2		2		2		2				2	
Ratio		$\frac{8}{6}$		$\frac{10}{8}$		$\frac{12}{10}$		$\frac{14}{12}$				$\frac{206}{204}$	

(d) This sequence, $b_n = 2n + 4$, is an arithmetic sequence because all the differences are the same: 2. It is not a geometric sequence because all the ratios are not the same. ∎

Example 7.3

For the sequence $c_n = 3 \times 2^n$

(a) Write the first five terms of the sequence.
(b) Write the 100th and 101st terms.
(c) Write the difference and ratio between any two consecutive terms that you have written.
(d) Tell whether the sequence is an arithmetic sequence, a geometric sequence, or neither of these types of sequences.

Solutions   (a) The sequence is $c_n = 3 \times 2^n$. The first term of the sequence is $c_1 = 3 \times 2^1 = 3 \times 2 = 6$. The second term of the sequence is $c_2 = 3 \times 2^2 = 3 \times 4 = 12$. Continuing in this manner produces the first five terms of the sequence $\{c_n\}$ as $\{6, 12, 24, 48, 96\}$.

(b) The 100th term is $c_{100} = 3 \times 2^{100}$ and the 101st term is $c_{101} = 3 \times 2^{101}$, so $\{c_{100}, c_{101}\} = \{3 \times 2^{100}, 3 \times 2^{101}\}$.

(c) As in Examples 7.1 and 7.2 the table below lists the first five, 100th, and 101st terms of $\{c_n\}$ and the difference and ratio between two consecutive terms. You might guess the difference $c_{101} - c_{100}$ by looking at the pattern formed by the first four differences.

n (term)	1		2		3		4		5	...	100		101
$c_n = 3 \times 2^n$	6		12		24		48		96	...	3×2^{100}		3×2^{101}
Difference		6		12		24		48				3×2^{100}	
Ratio		2		2		2		2				2	

(d) This sequence, $c_n = 3 \times 2^n$, is a geometric sequence because all the ratios are the same: 2. It is not an arithmetic sequence because all the differences are not the same. ∎

Now that you have had some experience with mathematical definitions of sequences, the following activity gives you an opportunity to think about the connection between the definition and the type of sequence that results.

Activity 7.3
Identifying Arithmetic and Geometric Sequences

Review the answers to Examples 7.1–7.3 and try to answer the following questions:

1. How can you tell from the expression for the nth term of a sequence if it is an arithmetic sequence?
2. If the sequence is arithmetic, how can you determine the common difference from the expression for the nth term?
3. How can you tell from the expression for the nth term of a sequence if it is a geometric sequence?
4. If the sequence is geometric, how can you determine the common ratio from the expression for the nth term?

Example 7.4

Write a formula in terms of n for the difference between the $(n + 1)$st term and the nth term for each of the sequences in Examples 7.1–7.3. Substitute $n = 100$ into each result and check your answers.

Solutions (a) The sequence of Example 7.1 is $a_n = \dfrac{1}{n}$.

The $(n + 1)$st term is $a_{n+1} = \dfrac{1}{n + 1}$ and the difference is $\dfrac{1}{n + 1} - \dfrac{1}{n} = \dfrac{n - (n + 1)}{n(n + 1)} = \dfrac{-1}{n(n + 1)}$.

If $n = 100$, then the difference is $\dfrac{-1}{100(101)}$.

(b) The sequence of Example 7.2 is $b_n = 2n + 4$.

The $(n + 1)$st term is $b_{n+1} = 2(n + 1) + 4 = 2n + 6$. The difference is $(2n + 6) - (2n + 4) = 2$.

If $n = 100$, then the difference is 2.

(c) The sequence from Example 7.3 is $c_n = 3 \times 2^n$.

Here the $(n + 1)$st term is $c_{n+1} = 3 \times 2^{n+1}$. The difference is

$$\left(3 \times 2^{n+1}\right) - \left(3 \times 2^n\right) = \left(3 \times 2^n \cdot 2\right) - \left(3 \times 2^n\right)$$

Factoring out 3×2^n produces

$$\left(3 \times 2^n \cdot 2\right) - \left(3 \times 2^n\right) = 3 \times 2^n (2 - 1) = 3 \times 2^n$$

If $n = 100$, then the difference is $3 \times 2^{100} \approx 3.8 \times 10^{30}$.   ∎

Example 7.5   Application

You deposit $500 in the bank and then plan to deposit $20 from your paycheck at the end of each of the following weeks.

(a) Ignoring interest, how much have you deposited after n weeks?
(b) How much have you deposited after 5 years (again, without interest)?

Solutions (a) The amount in the account each week can be expressed by the sequence that begins $\{\$500, \$520, \$540, \$560, \$580, \ldots\}$.

If you can't write the nth term immediately, don't worry. A table can help you organize the numbers to see the pattern. Notice the relationship between the number of 20s and the value of n.

Week n	Amount in the bank
1	500
2	$500 + 20$
3	$500 + 20 + 20 = 500 + 20 \times 2$
4	$500 + 20 + 20 + 20 = 500 + 20 \times 3$
5	$500 + 20 + 20 + 20 + 20 = 500 + 20 \times 4$
⋮	⋮
20	$500 + 20 \times 19$

Following the pattern, after n weeks the amount in the bank is $500 + 20(n - 1)$.

(b) 5 years has $5 \times 52 = 260$ weeks. Therefore, at the end of five years $n = 260$. The amount, without interest, is

$$500 + 20 \times (260 - 1) = 500 + 20 \times 259$$
$$= 500 + 5180 = 5680$$

Incidentally, if the bank gives 5% interest per year, compounded quarterly, the total amount after five years would be over $7000. You will learn how to do interest computations like that in Section 7.4.

Example 7.6   Application

A store is holding a "Going out of Business" sale. On the first day all items are marked 10% off the regular price. On each day afterwards, each item is reduced by another 10%.

(a) What is the price of a $200 jacket after 10 days?
(b) What is the price after n days?
(c) When is the price equal to zero?

Solutions   (a) It helps to begin looking at this problem with a table. Pay close attention to what happens in day 2, where we factor out the common factor 200×0.9.

In general, if the price is P one day, then the next day, after taking a discount of 10%, the price is

$$P - 0.1P = P(1 - 0.1)$$
$$= 0.9P$$

Therefore each day the price is multiplied again by 0.9 and the sequence of prices can be written as

$$\{200 \times 0.9^1, 200 \times 0.9^2, 200 \times 0.9^3, 200 \times 0.9^4, 200 \times 0.9^5\}$$

or $\{\$180, \$162, \$145.80, \$131.22, \$118.10, \dots\}$. Continuing this sequence we see that on the 10th day the price is $200 \times 0.9^{10} \approx \$69.74$.

n (Day)	Price Calculation	Price (in $)
0	200 (beginning price, before the sale)	200
1	$200 - 200 \times 0.1 = 200 \times 0.9$	180
2	$\begin{aligned}(200 \times 0.9) - (200 \times 0.9) \times 0.1 &= (200 \times 0.9) \times (1 - 0.1)\\ &= (200 \times 0.9) \times 0.9\\ &= 200 \times 0.9^2\end{aligned}$	162
3	200×0.9^3	145.80
4	200×0.9^4	131.22
20	200×0.9^{20}	24.32

(b) From what we did in (a) we can see that on the nth day the price is

$$200 \times 0.9^n$$

(c) Trying higher values of n, we have the following table

n	Price
20	$24.32
40	2.96
60	0.36
80	0.04
100	0.01

Under this rule of discounting, some time after 100 days the price of the jacket would drop below one cent. Theoretically, the value never reaches zero, as you'll see later in this chapter. ■

Direct Definition of Arithmetic Sequences

An **arithmetic sequence** is a sequence in which there is a common difference between consecutive terms. The formula for the nth term of an arithmetic sequence is

$$a_n = a_1 + d \times (n - 1)$$

where d is the common difference.

Example 7.7

The sequence $\{5, 9, 13, 17, 21, \ldots\}$ is an example of an arithmetic sequence. It has a common difference of 4. The nth term of this sequence is

$$a_n = 5 + 4(n - 1)$$

■

Direct Definition of Geometric Sequences

A **geometric sequence** is a sequence where there is a common ratio between consecutive terms. The formula for the nth term of a geometric sequence is

$$a_n = a_1 \times r^{(n-1)}$$

where r is the common ratio.

Example 7.8

The sequence $\{4, 12, 36, 108, 324, \ldots\}$ is an example of a geometric sequence. It has a common ratio of 3. The nth term of this sequence is

$$a_n = 4 \times 3^{n-1}$$ ■

Most graphing calculators can be used to calculate the terms of sequences, as you'll see in the following calculator lab.

Calculator Lab 7.1

Sequences on the Calculators

In this calculator lab you will use a calculator to check your answers to Examples 7.1–7.3.

The *TI-83* has several built-in ways to compute sequences. The first one we'll look at is the `seq` function, which is selection 5 on the OPS submenu of the LIST menu (see Figure 7.8).

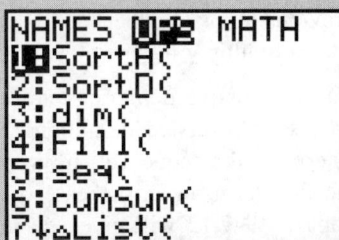

```
NAMES OPS MATH
1:SortA(
2:SortD(
3:dim(
4:Fill(
5:seq(
6:cumSum(
7↓ΔList(
```

Figure 7.8

Procedures

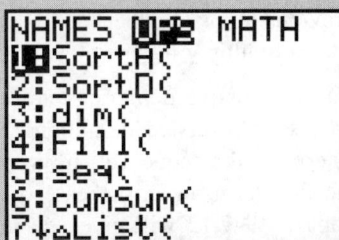

```
seq(1/X,X,1,5)
{1 .5 .33333333...
```

Figure 7.9

1. To compute the first five terms of the sequence defined by $a_n = \dfrac{1}{n}$, use the command: `seq(1/x,x,1,5)`.

    There are four items within the parentheses. The first is the expression for the nth term. The second is the name of the variable being used. Note that any variable can be used. We used x because it is easier to enter into the *TI-83* than other letters. The third entry is the first value of the variable, and the final entry is the last value of the variable. The result is shown in Figure 7.9. Notice that the sequence goes off the screen.

2. Press the right arrow key to move the cursor to the hidden portions of the sequence .

3. The decimals can be converted to fractions using the ▶Frac feature of the *TI-83*. (The "convert to fraction" function is selection 1 on the MATH menu.) Select that function and then press ⟨ENTER⟩ and you will see the decimals in the list converted to fractions (Figure 7.10).

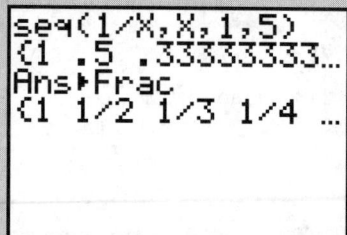

```
seq(1/X,X,1,5)
{1 .5 .33333333...
Ans▶Frac
{1 1/2 1/3 1/4 ...
```

Figure 7.10

4. To compute the 100th and 101st term of a sequence, enter `seq(1/x,x,100, 101)`.

5. Continue with the remaining sequences from Examples 7.1–7.3.

Defining Sequences Recursively

Another common way to describe a sequence is to tell what you do to one term to get the next term. For example, the sequence $\{2, 4, 8, 16, 32, \ldots\}$, can be described by giving the nth term, $a_n = 2^n$. It can also be defined with the phrase "multiply any term by 2 to get the next term." When each term of a sequence is defined using the preceding terms, the sequence is called a **recursive sequence**. Other examples of recursive definitions are:

▶ Multiply the previous term by 2 and add 7.
▶ Divide the previous term by 3.
▶ Add the two previous terms together.
▶ To get the nth term, multiply the previous term by n.

However, these recursive definitions are incomplete. If you think carefully about these examples of recursive definitions, you will see that it is impossible to write a single term in the sequence. In order to start writing the sequence we must be given the first term, or in the case of the third example above, the first two terms. Examples 7.9–7.12 will give you more experience with definitions of recursive sequences.

Example 7.9

Consider the recursive sequence where you multiply the previous term by 2 and add 7 and where $a_1 = -5$. (a) Write the first six terms of this sequence and (b) identify the sequence as arithmetic or geometric, or neither.

Solutions (a) Here $a_1 = -5$. From the definition $a_2 = 2a_1 + 7 = -3$, $a_3 = 2a_2 + 7 = 1$, $a_4 = 2a_3 + 7 = 9$, and so on. Thus, $a_1 = -5$, $a_2 = -3$, $a_3 = 1$, $a_4 = 9$, $a_5 = 25$, and $a_6 = 57$. The first six terms of this sequence are $\{-5, -3, 1, 9, 25, 57\}$.

(b) You can check to see that this is neither an arithmetic nor geometric sequence.

Example 7.10

Consider the recursive sequence where you subtract 4 from the previous term and where $a_1 = 92$. (a) Write the first six terms of this sequence and (b) identify the sequence as arithmetic or geometric, or neither.

Solutions (a) In symbols, we have $a_1 = 92$, $a_2 = a_1 - 4 = 88$, $a_3 = a_2 - 4 = 84$, $a_4 = a_3 - 4 = 80, \ldots$. The first six terms of this sequence are $\{92, 88, 84, 80, 76, 72\}$.
(b) This is an arithmetic sequence with a common difference of -4. ∎

Example 7.11

Consider the recursive sequence where you divide the previous term by 3 and where $a_1 = 27$. (a) Write the first six terms of this sequence and (b) identify the sequence as arithmetic or geometric, or neither.

Solutions (a) In symbols, we have $a_1 = 27$, $a_2 = a_1 \div 3 = 9$, $a_3 = a_2 \div 3 = 3$, $a_4 = a_3 \div 3 = 1, \ldots$. The first six terms of this sequence are $\left\{27, 9, 3, 1, \frac{1}{3}, \frac{1}{9}\right\}$.
(b) This is a geometric sequence with common ratio of $\frac{1}{3}$. ∎

The recursive definition in Example 7.11 stated that a term is obtained by dividing the previous term by 3. Each term of a geometric sequence is multiplied by a common ratio. Since dividing by 3 is the same as multiplying by $\frac{1}{3}$, that is, $1 \div 3 = 1 \times \frac{1}{3}$, the common ratio in Example 7.11 is $\frac{1}{3}$.

Example 7.12

Consider the recursive sequence where you add the previous two terms of the sequence and with $a_1 = 1$ and $a_2 = 1$. (a) Write the first six terms of this sequence and (b) identify the sequence as arithmetic, geometric, or neither.

Solutions (a) In symbols, we have $a_1 = 1$, $a_2 = 1$, $a_3 = a_1 + a_2 = 2$, $a_4 = a_2 + a_3 = 3, \ldots$. The first six terms of this sequence are $\{1, 1, 2, 3, 5, 8\}$.
(b) This is neither an arithmetic nor a geometric sequence. This particular sequence is called the **Fibonacci sequence** after Leonardo of Pisa, better known as Fibonacci or "son of Bonaccio," an Italian merchant. ∎

Recursive sequences are normally written as formulas for a_n, the nth term, as a function of the previous term a_{n-1}. The value of a_1 must also be given. Example 7.13 gives you practice in using the correct mathematical notation for a recursive sequence.

Example 7.13

Write each of the recursive definitions in Examples 7.9–7.12 as a formula.

Solutions   (a) For the sequence in Example 7.9 you multiply the previous term by 2 and then add 7. In symbols this is $a_n = 2a_{n-1} + 7$ with $a_1 = -5$.

(b) In the sequence in Example 7.10 you subtract 4 from the previous term, or $a_n = a_{n-1} - 4$ and $a_1 = 92$.

(c) For the sequence in Example 7.11 the previous term is divided by 3 or $a_n = \dfrac{a_{n-1}}{3}$. This could also be written as $a_n = \frac{1}{3}a_{n-1}$. In both cases, $a_1 = 27$.

(d) With the sequence in Example 7.12 you add the previous *two* terms of the sequence or, in symbols, $a_n = a_{n-1} + a_{n-2}$ with $a_1 = 1$ and $a_2 = 1$.   ∎

Now that you have had some experience with recursive definitions of sequences, the following activity gives you an opportunity to think about the connection between the recursive definition and the type of sequence that results.

Activity 7.4
Identifying Arithmetic and Geometric Sequences

Review the solutions to Examples 7.9–7.12 and write answers to the following questions.

1. How can you tell from the recursive definition of a sequence if it is an arithmetic sequence?
2. If the sequence is arithmetic, how can you determine the common difference from the recursive definition?
3. How can you tell from the recursive definition of a sequence if it is a geometric sequence?
4. If the sequence is geometric, how can you determine the common ratio from the recursive definition?

Recursive Definitions of Sequences

▶ The recursive definition of an **arithmetic sequence** is

$$a_n = a_{n-1} + d$$

where d is the common difference.

▶ The recursive definition of a **geometric sequence** is

$$a_n = ra_{n-1}$$

where r is the common ratio.

Example 7.14

An example of a recursively defined arithmetic sequence is $a_n = a_{n-1} + 4$ with $a_1 = 5$. This produces the sequence $\{5, 9, 13, 17, \ldots\}$.   ∎

Example 7.15

An example of a recursively defined geometric sequence is $a_n = 3a_{n-1}$ with $a_1 = 4$. This produces the sequence $\{4, 12, 36, 108, \ldots\}$. ■

The *TI-83* will compute and plot sequences using the `Seq` mode. This mode is particularly useful for recursively defined sequences. The following calculator lab gives you practice defining and graphing sequences.

Calculator Lab 7.2

Graphing Sequences

Use the `Seq` mode of your calculator (not the `seq` function) to produce terms of the following sequences and to plot the sequences.

(a) $a_n = 0.6a_{n-1}$; $a_1 = 9$

(b) $a_n = -(n-4)^2$

Procedures

1. Set the graph window and the table:

   (a) Select `ZStandard` from the ZOOM menu.
   (b) In the `TBLSET` menu set `TblStart=1` and `ΔTbl=1`.

2. Select the `Seq` mode, which is on the fourth line of the MODE menu.

3. Press [Y=]. You will see a very different screen (see Figure 7.11).

   Three different sequences, $\{u_n\}$, $\{v_n\}$, and $\{w_n\}$ can be defined on this screen. The calculator uses function notation $u(n)$ instead of the subscript notation u_n that we have been using. They are equivalent. You are given the choice of changing the first value of n. For the two parts of this lab this value will always be 1, so leave the nMin set to 1.

4. Enter `0.6u(n-1)` for $u(n)$ and 9 for $u(n$Min$)$. You must use the lowercase u, which is above the [7] key. Use the [X.T.θ.n] key to produce the lowercase n. Press [2nd] [TABLE] to see a portion of the table (Figure 7.12).

```
Plot1  Plot2  Plot3
 nMin=1
·.u(n)=
 u(nMin)=
·.v(n)=
 v(nMin)=
·.w(n)=
 w(nMin)=
```

Figure 7.11

n	u(n)
1	9
2	5.4
3	3.24
4	1.944
5	1.1664
6	.69984
7	.4199

$n=1$

Figure 7.12

5. Press [GRAPH] to see Figure 7.13. The graph plots the value of $u(n)$ on the vertical axis, and the value of n on the horizontal axis. There are actually ten points plotted but five of them are so close to the x-axis that they are not visible. You can evaluate the term of a sequence by using the table or by tracing the points on the graph. For example, in Figure 7.13, $u(3) = u_3 = 3.24$.

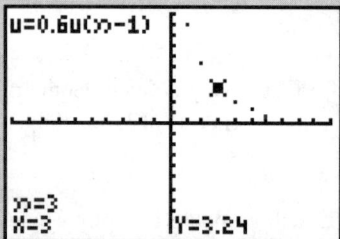

Figure 7.13

6. The second sequence is defined directly in terms of n, so just enter the expression $-(n-4)^2$ for the second sequence, $v(n)$. For a directly defined sequence, press **CLEAR** to leave the $v(nMin)$ field blank.

   The tabular result is shown in Figure 7.14 and the graph is in Figure 7.15. There are six points visible below the x-axis. The fourth term in the sequence, which is zero, is plotted on the x-axis, and the eighth through tenth terms are outside the graphing window. For both Figures 7.14 and 7.15 we see that $v(4) = v_4 = 0$.

n	$u(n)$	$v(n)$
1	9	-9
2	5.4	-4
3	3.24	-1
4	1.944	0
5	1.1664	-1
6	.69984	-4
7	.4199	-9

$v(n)=0$

Figure 7.14

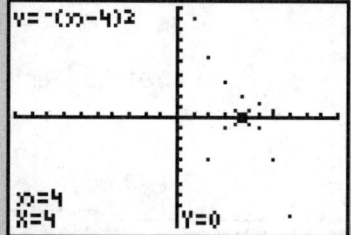

Figure 7.15

Section 7.1 Exercises

For each of the sequences in Exercises 1–6 (a) write the first five terms of each sequence, (b) write the 100th and 101st terms, (c) write the difference between any two consecutive terms you have written, (d) write the ratio between any two consecutive terms you have written, and (e) tell whether the sequence is an arithmetic sequence, a geometric sequence, or neither of these types of sequences. Explain your answer to (e).

1. $a_n = \dfrac{1}{n^2}$

2. $b_n = 3n - 5$

3. $c_n = \frac{1}{2} \times 2^n$

4. $d_n = 100 - 3n$

5. $e_n = 3n^2$

6. $f_n = 2 \times 10^n$

In Exercises 7–12 write a formula in terms of n for the difference and ratio between the $(n + 1)$st and nth term for each sequence.

7. $a_n = \dfrac{1}{n}$

8. $b_n = 4n - 2$

9. $c_n = 2 \cdot 5^n$

10. $d_n = 4(-3)^n$

11. $e_n = 7 - 3n$

12. $f_n = 2n^2 - n + 1$

In Exercises 13–20 the first four terms of a sequence are shown. (a) Write the most likely next four terms, (b) write a formula for the nth term of the sequence, and (c) tell whether the sequence is an arithmetic sequence, a geometric sequence, or neither of these types of sequences. Explain your answer.

13. $-4, -1, 2, 5, \ldots$

14. $200, 100, 50, 25, \ldots$

15. $2, -2, 2, -2, \ldots$

16. $\frac{1}{3}, \frac{2}{9}, \frac{4}{27}, \frac{8}{81}, \ldots$

17. $6, 4, 2, 0, \ldots$

18. $1, 0, 1, 1 \ldots$

19. $-1, 2, 1, 3, \ldots$

20. $7, 3, -1, -5, \ldots$

In Exercises 21–28 a sequence is defined recursively. (a) Write the first five terms of the sequence and (b) graph each sequence on your calculator

21. $a_n = a_{n-1} - 2$ with $a_1 = 100$

22. $b_n = \frac{1}{2}b_{n-1} + \frac{1}{2}$ with $b_1 = 5$

23. $c_n = 2c_{n-1} + 1$ with $c_0 = -0.75$

24. $d_n = (-2)^{n-1}d_{n-1}$ with $d_1 = 1$

25. $e_n = \dfrac{1}{e_{n-1}} - 1$ with $e_1 = \frac{1}{3}$

26. $f_n = f_{n-1} - 2f_{n-2}$ with $f_1 = f_2 = 1$

27. $g_n = \sqrt{g_{n-1}}$ with $g_1 = 65,536$

28. $h_n = (h_{n-1} - 1)^2$ with $h_1 = 3$

In Exercises 29–34 a recursive sequence is given in words. (a) Write the first five terms of the sequence, (b) write a recursive formula for the sequence, and (c) tell whether the sequence is an arithmetic sequence, a geometric sequence, or neither of these types of sequences. Explain your answer.

29. To get the nth term, multiply the previous term by n. $a_1 = 1$

30. Multiply the previous term by -2. $a_1 = 1$

31. Find the cosine of the previous term in radians. $a_1 = \dfrac{\pi}{2}$

32. Multiply each term by 2 and subtract 10 to get the next term. $a_1 = 10$

33. Multiply each term by 2 and subtract 10 to get the next term. $a_1 = 9$

34. Multiply each term by 2 and subtract 10 to get the next term. $a_1 = 11$

35. *Finance* You deposit $750 in the bank and then plan to deposit $25 from your paycheck at the end of each of the following weeks.

(a) Ignoring interest, how much have you deposited after n weeks?

(b) How much have you deposited after 5 years (again, without interest)?

36. *Finance* Your savings account has $12,500 in it. You plan to withdraw $35 from your account at the end of each of the following weeks.

(a) How much have you withdrawn after n weeks?

(b) How much have you withdrawn after 5 years?

(c) After 5 years, ignoring interest, how much money do you have left in your account?

37. *Business* A store is holding a "Going out of Business" sale. On the first day all items are marked 5% off the regular price. On each day afterward, each item is reduced by another 5%.

(a) What is the price of a $200 jacket after 10 days?

(b) What is the price after n days?

(c) When is the price equal to zero?

(d) Compare your answer in (c) with the answer to Example 7.5(c).

38. *Business* The company you work for has the following retirement plan. The first year they put 10% of your salary into a retirement fund. Every year after that they add 1% to the amount they deposit in your retirement account. Thus, the second year they put 11% into your account; the third year 12%, and so on. This continues until they are putting 25% of your salary into your retirement. From then on they continue to deposit 25% of your salary into your retirement account. To make the problem easier assume that your starting salary at age 20 is $20,000 and that your salary remains the same until you retire.

(a) How much does the company deposit in your retirement account at the end of the first year?

(b) How much does the company deposit in your retirement account at the end of the second year?

(c) How much does the company deposit in your retirement account at the end of the fifth year?

(d) If you continue to work until you are 65, what is the total amount the company has deposited in your retirement account?

39. *Environmental Science* An ocean beach is eroding at the rate of 3 in. per year. If the beach is currently 25 ft wide, how wide will the beach be in 25 years? (*Hint:* $a_1 = 25$ ft -3 in.; find a_{25}.)

40. *Construction* A contractor is preparing a bid for constructing an office building. The foundation and basement will cost $750,000. The first floor will cost $320,000. The second floor will cost $240,000. Each floor above the second will cost $12,000 more than the floor below it. **(a)** How much will the 10th floor cost? **(b)** How much will the 20th floor cost?

41. *Medical technology* A group of people plan an exercise program. The first day they will jog $\frac{1}{2}$ mi. After that they will jog a certain amount more each day until on the 61st day, 8 mi are jogged. How much was the distance increased each day?

42. *Physics* A ball is dropped to the ground from a height of 80 m. Each time it bounces it goes $\frac{7}{8}$ as high as the previous bounce. How high does it go on the sixth bounce?

43. *Physics* A pendulum swings 15 cm on its first swing. Each subsequent swing is reduced by 0.3 cm. How far is the 10th swing? How many times will the pendulum swing before it comes to rest?

44. *Aeronautical engineering* The atmospheric pressure at sea level is approximately 100 kPa and decreases 12.5% for each km increase in altitude. What is the atmospheric pressure at the top of Mt. Everest, which is about 8.8 km high?

45. *Environmental science* A chemical spill pollutes a river. A monitor located 1 mi downstream from the spill measures 940 parts of the chemical for every million parts of water. (This is written as 940 ppm.) The readings decrease 16.2% for each mile farther downstream. **(a)** How far downstream from the spill will it be before the concentration is reduced to 100 ppm? **(b)** The water is considered safe for human consumption at 1.5 ppm. How far downstream from the spill will you have to go before the water is considered safe for humans to drink?

46. *Environmental science* The level of a particular metal pollutant in a certain lake was found to be increasing geometrically. In January, there were 3.50 parts per billion (ppb) and in April there were 18.76 ppb.

(a) By what ratio is the level increasing each month?

(b) What will the pollution level be in December?

47. Explain how an arithmetic sequence and a geometric sequence are alike and how they are different.

48. In Exercises 1–20 you were asked to determine if each given sequence is an arithmetic sequence, a geometric sequence, or neither. Describe how you made your decision.

7.2   Limits of Sequences

In this section we will look closely at how the terms of sequences change as the value of n increases. You have seen that sequences, like functions, can have very different behavior for large values of n.

1. In the sequence $\{1, 2, 6, 24, 120, 720, \ldots\}$ the terms get larger and larger.
2. The sequence from Example 7.11 is $\left\{27, 9, 3, 1, \frac{1}{3}, \frac{1}{9}, \ldots\right\}$. Here the terms get smaller and smaller but are all positive.
3. The sequence in Exercise 30 of Exercise Set 7.1 is $\{1, -2, 4, -8, 16, -32, \ldots\}$. Here the terms alternate between large (positive) and small (negative).
4. The sequence in Exercise 32 of Exercise Set 7.1 is $\{10, 10, 10, 10, 10, 10, \ldots\}$. In this sequence the terms are constant even though the recursive definition, $a_n = 2a_{n-1} - 10$, does not look like any constant function, such as $y = 10$, that you have seen before.

Actually, we touched on limits earlier in this text when we considered the question of finding the slope of a tangent at a particular point.

The key idea here is that some sequences in the above list approach a specific number as n gets larger and some do not. Those sequences that do approach a specific number as n gets larger are said to have a **limit**. The limit of a sequence, if it exists, is a real number. Limits of sequences and functions is a topic that is at the heart of calculus. We'll introduce the concept in this chapter and revisit it in chapters that follow.

Estimating Limits

To estimate the limit of a sequence, write the first five or six terms, or until you can make a guess for the limit. Then carry the sequence out to a large number of terms such as twenty or a hundred, and see if your guess still looks good. If it doesn't, guess again and repeat the process.

Example 7.16

Guess a limit of each of the following sequences.

(a) $a_n = 2a_{n-1} + 7$; $a_1 = -5$
(b) $a_n = a_{n-1} - 4$; $a_1 = -5$
(c) $a_n = \dfrac{a_{n-1}}{3}$; $a_1 = 27$
(d) $a_n = a_{n-1} + a_{n-2}$; $a_1 = 1$ and $a_2 = 1$

Solutions   These are the same sequences that we studied in Examples 7.9–7.12. Although we won't get into the mathematical theory of limits right now, we can look at sequences and see whether or not they have a limit.

(a) The first six terms are $\{-5, -3, 1, 9, 25, 57\}$. There does not appear to be a limit, because a_n gets larger and larger, without bound, as n continues to increase.
(b) The sequence begins with $\{-5, -9, -13, -17, -21\}$. There does not appear to be a limit, because a_n is a negative number with larger and larger absolute value, without bound, as n continues to increase.
(c) The first six terms are $\left\{27, 9, 3, 1, \frac{1}{3}, \frac{1}{9}\right\}$. The limit seems to be 0 because as n gets larger, a_n gets closer and closer to 0. For example, the 10th term is $\frac{1}{9}\left(\frac{1}{3}\right)\left(\frac{1}{3}\right)\left(\frac{1}{3}\right)\left(\frac{1}{3}\right) = \frac{1}{729} \approx 0.00137$. Continued division by 3 will continue to reduce a_n, but it will never go below 0.
(d) The first six terms are $\{1, 1, 2, 3, 5, 8\}$. There doesn't seem to be a limit because continued addition of positive numbers will make the sequence increase without bound. ∎

Sequences and Infinity

Sequences that do not have a limit can be classified into three different groups:

▶ The terms for high values of n might be larger and larger positive numbers, for example $\{1, 2, 4, 8, 16, 32, \ldots\}$. In this case the sequence approaches **positive infinity**. Since infinity is the concept of a number that is larger than any real number, it is not itself a real number. However, by saying the sequence approaches positive infinity, we are at least saying in what direction the numbers are going.

> ▶ The terms for high values of n might be negative numbers that are larger and larger in absolute value, for example $\{5, 1, -3, -7, -11, -15, \ldots\}$. In this case the sequence approaches **negative infinity**.
> ▶ The terms for high values of n might oscillate back and forth between positive and negative or between one number and another, for example $\{1, -2, 4, -8, 16, -32, \ldots\}$ or $\{10, 5, 10, 5, 10, 5, \ldots\}$. In this case the sequence is said to **oscillate without limit**.
>
> In cases where a sequence approaches positive or negative infinity, the terms of the sequence increase beyond any positive number or decrease beyond any negative number. Such sequences are said to increase or decrease **without bound**.

Example 7.17 gives you more practice with several kinds of sequences that have, or don't have, a limit. You will also see a sequence that has different limits depending on the value of a_1.

Example 7.17

Write the first five terms of each sequence and guess its limit. If you need more terms, enter the expression into the calculator's Y= screen in the Seq mode and look at the table for the sequence.

(a) $a_n = -n$

(b) $a_n = \dfrac{10}{n}$

(c) $a_n = \dfrac{9n - 4}{n}$

(d) $a_n = 2 + 4\left(\dfrac{1}{2}\right)^n$

Solutions
(a) The first five terms are $\{-1, -2, -3, -4, -5\}$. This sequence will continue to decrease without bound. The sequence approaches negative infinity.

(b) The first five terms of the sequence $a_n = \dfrac{10}{n}$ are $\{10, 5, \frac{10}{3} \approx 3.333, 2.5, 2\}$. This sequence remains positive but continues to decrease. We need more terms to guess the limit. The next five terms are

$$\left\{ \ldots, \frac{10}{6} \approx 1.667, \frac{10}{7} \approx 1.429, 1.25, \frac{10}{9} \approx 1.111, 1 \right\}$$

It's still decreasing. The 20th and the 100th terms are $\frac{10}{20} = 0.5$ and $\frac{10}{100} = 0.1$, respectively. We could go further but a good guess for the limit is 0.

(c) The sequence is $a_n = \dfrac{9n - 4}{n}$. The first five terms are $\{5, 7, \frac{23}{3} \approx 7.667, 8, \frac{41}{5} = 8.2\}$. The sequence is increasing, but there is not enough evidence to guess the limit. The next five terms of the sequence are

$$\{\ldots, 8.333, 8.429, 8.5, 8.556, 8.6\}$$

It's still hard to see the limit. Looking at the 20th we see $\dfrac{9(20) - 4}{20} = 8.8$, and the 100th term is $\dfrac{9(100) - 4}{100} = 8.96$. Although the sequence is increasing, the limit may not be clear. The 10,000th term, 8.9996, seems to indicate that the sequence is approaching 9. This is correct, and we shall see later how to solve a problem like this with algebra.

(d) The first five terms of the sequence $a_n = 2 + 4 \left(\frac{1}{2} \right)^n$ are $\{4, 3, 2.5, 2.25, 2.125\}$. The sequence is decreasing, but there is not enough evidence to guess the limit. Let's look at the next five terms:

$$\{ \ldots, 2.063, 2.031, 2.016, 2.008, 2.004 \}$$

The evidence indicates that the limit is 2. However, as a check, the 20th term is about 2.000004. For now this is enough evidence to say with some confidence that the limit is equal to 2. Again, we shall see later how to tackle a problem like this with algebra.   ■

Example 7.18

Consider the recursive sequence $a_n = 2a_{n-1} - 7$. How does changing the first term affect the sequence? To help see the effect of the first term, use each of the first terms $a_1 = 7.1$, $a_1 = 7$, and $a_1 = 6.9$ to produce a sequence. Write the first five terms of each sequence and guess its limit. If you need more terms, enter the expression into the calculator's Y= screen in the Seq mode, and look at the table for the sequence.

Solutions (a) If $a_1 = 7.1$, the first five terms are $\{7.1, 7.2, 7.4, 7.8, 8.6\}$. This does not give enough information. The next five terms are

$$\{ \ldots, 10.2, 13.4, 19.8, 32.6, 58.2 \}$$

This seems to be enough evidence that the sequence is increasing without bound, but let's peek at the 20th term: $a_{20} = 52,435.8$. This is good evidence that the sequence approaches positive infinity.

(b) If $a_1 = 7$, then the first five terms are $\{7, 7, 7, 7, 7\}$. This is good evidence that the sequence continues to be constant with a limit of 7.

(c) If $a_1 = 6.9$, then the first five terms are $\{6.9, 6.8, 6.6, 6.2, 5.4\}$. This does not give enough information so we list the next five terms:

$$\{ \ldots, 3.8, 0.6, -5.8, -18.6, -44.2 \}$$

This seems to indicate that the sequence is decreasing without bound, but let's peek at the 20th term: $a_{20} \approx -52,421.8$. We can guess that the limit is negative infinity.   ■

    Next we will look at how the graphs of sequences give a visual image of whether or not there is a limit to a sequence.

Activity 7.5
Graphs of Sequences

This activity will give you experience in interpreting graphs of sequences. You will be asked to match each of the seven sequences of Examples 7.17 and 7.18 to one of seven graphs.

1. The seven sequences of Examples 7.17 and 7.18 are

(a) $a_n = -n$

(b) $b_n = \dfrac{10}{n}$

(c) $c_n = \dfrac{9n - 4}{n}$

(d) $d_n = 2 + 4\left(\dfrac{1}{2}\right)^n$

(e) $e_n = 2e_{n-1} - 7$ with $e_1 = 7.1$

(f) $f_n = 2f_{n-1} - 7$ with $f_1 = 7.0$

(g) $g_n = 2g_{n-1} - 7$ with $g_1 = 6.9$

Match each sequence to one of the graphs in Figure 7.16.

2. Describe in words and pictures the graph of a sequence that converges to a real number limit.

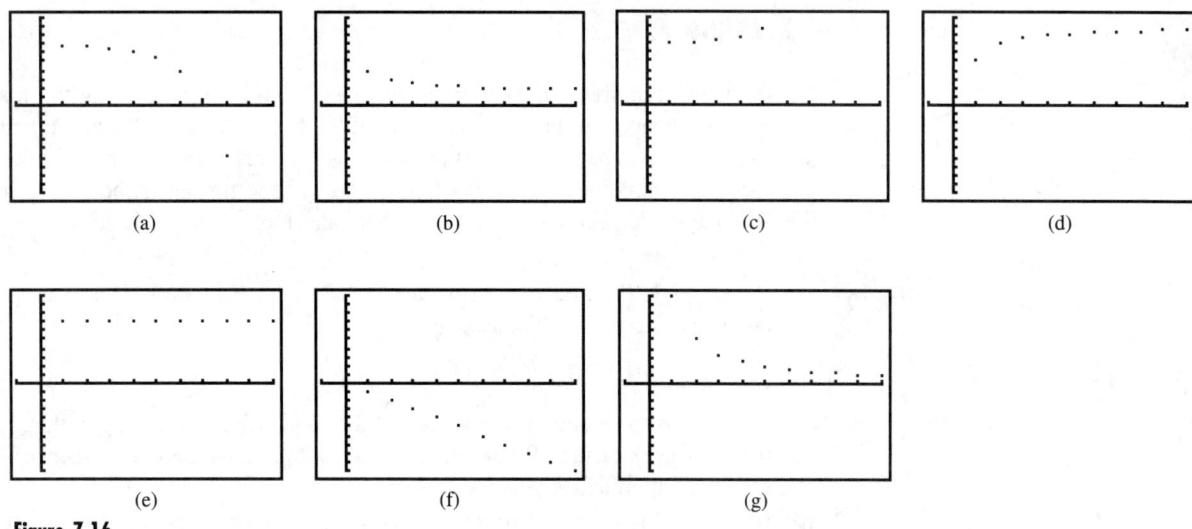

(a) (b) (c) (d)

(e) (f) (g)

Figure 7.16

Sequences and Chaos Theory

Something strange happened in Example 7.18. The recursion formula for each of the three sequences is identical, and the values of a_1 differ only by a small amount. In fact you would find a similar result if the three values of a_1 were 7.00001, 7, and 6.99999. That is, a tiny change of only 0.00002, the difference between 7.00001 and 6.99999, can determine whether the sequence approaches negative infinity or positive infinity.

For the sequence $a_n = 2a_{n-1} - 7$, the number 7 is called the **fixed point of the sequence**, because if $a_1 = 7$ then the terms of the sequence are constant. But if a_1 slips a very small amount to either side of 7, then the sequence veers off to positive or negative infinity.

The mathematics of **chaos theory**, which was invented in the 1970s, studies situations in which a small change in the inputs makes a huge change in the outputs. Chaos theory has been defined as "... a field of study devoted to processes that exhibit complex, apparently random behavior, such as cloud formation or fluctuations of biological populations."[*] Both cloud formation and changes in populations are affected quite dramatically by small changes in initial conditions, like the effects that a small change in a_1 has on the direction of this sequence.

Activity 7.6
Chaos and Fixed Points

In this activity your goal is to experiment to find the fixed point for several sequences.

1. Experiment with different values of a_1 until you find the fixed point for the following sequences:

   (a) $a_n = 2a_{n-1} - 10$
   (b) $a_n = 2a_{n-1} + 10$
   (c) $a_n = 3a_{n-1} - 12$

2. Determine a general way to find the fixed point for the general sequence

$$a_n = C \times a_{n-1} - D$$

Next we look at the part of the Chapter Project that concerns estimating the area under a curve. You will see that our work with sequences and their limits will be very important.

A Sequence of Areas

In Activity 7.1 you made several estimates of the area under a curve. The values of the lower estimate, upper estimate, and average formed the terms of three sequences. We are interested in the limits of these sequences because these limits are related to the area under the curve. Perhaps you can see from the graph you drew in Activity 7.2 that your three sequences are all getting closer together. Do you think the curves are approaching the same limit?

Now we'll take another look at the problem of estimating the area under a curve, using a systematic approach to finding the sequence of upper and lower estimates of the area.

In Activity 7.1 you made one initial guess at the area and then three estimates that were based on intervals within which you were sure the area would be located. This process of guessing and then guessing again is very powerful, but only if it is carried out in a systematic manner. Next you'll see one way to make systematic approximations that get more and more accurate. You may have thought of a different way to make the estimates from the one we'll show. That's all right because there is more than one good approach to this problem.

First, we'll focus entirely on the right half of the area under the parabola shown in Figure 7.3 on page 421. Since the graph is symmetric about the y-axis, the two halves on either side of the y-axis have the same area. Thus, double the area under the right half of the parabola to find the total area.

Our process will be to draw rectangles that are under the curve and rectangles that extend above the curve. The sum of the areas of the rectangles under the curve will become our lower estimate L_n of the area, and the sum of the areas of the

---

[*]*The Concise Columbia Encyclopedia.* Copyright © 1995 by Columbia University Press.

rectangles that extend above the curve will become our upper estimate U_n of the area. For an example look at Figure 7.17. The shaded rectangle is the start of our first lower estimate of the area. The first lower estimate L_1 will be twice the area of this rectangle.

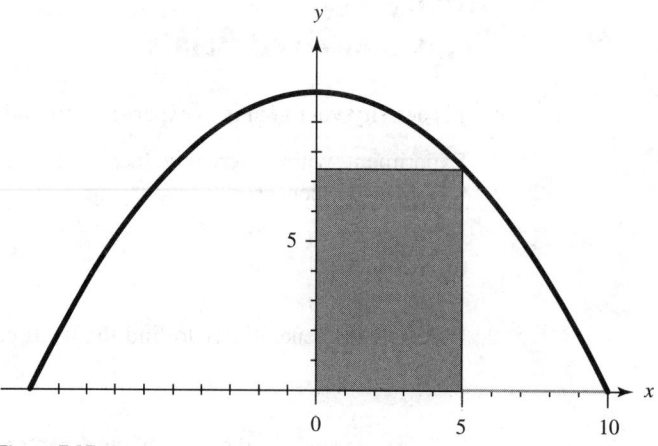

Figure 7.17

In order to make these estimates systematically, we have to describe how these rectangles were formed. The base of the rectangle in Figure 7.17 is half the distance from the origin to the x-intercept. That is, its base is 5 cm. Its height could be determined from the equation of the parabola, if that were known, or it can be estimated from the graph. The height looks like 7.5 cm. Therefore the area of the first lower rectangle is $5 \times 7.5 = 37.5$ cm^2, and $L_1 = 2 \times 37.5 = 75$ cm^2.

For U_1 we will use the two upper rectangles in Figure 7.18. The two upper rectangles in Figure 7.18 extend from the x-axis. Their bases are both 5 cm and their heights are 10 cm and 7.5 cm, respectively. Therefore the sum of the areas of these two rectangles is $5 \times 10 + 5 \times 7.5 = 87.5$ cm^2 and $U_1 = 2 \times 87.5 = 175$ cm^2. Therefore the average is $M_1 = \frac{75+175}{2} = 125$ cm^2. These numbers have been entered on the first line of Table 7.2.

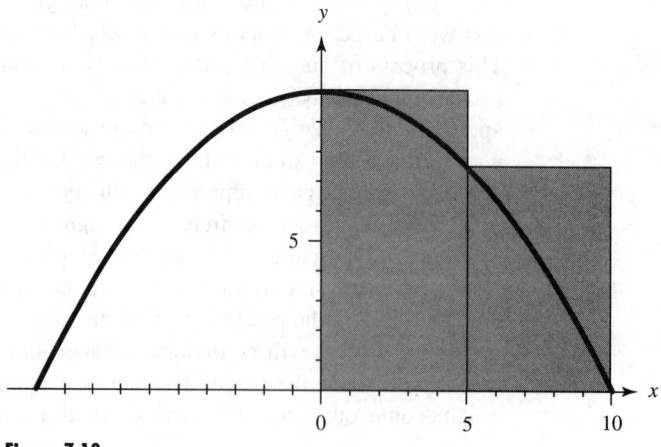

Figure 7.18

Table 7.2

Estimate #	Lower Estimate	Upper Estimate	Average $= \dfrac{L_n + U_n}{2}$
1	$L_1 = 75$	$U_1 = 175$	$M_1 = 125$
2	$L_2 =$	$U_2 =$	$M_2 =$
3	$L_3 =$	$U_3 =$	$M_3 =$

Activity 7.7
Systematic Estimation

The estimation process continues with Figures 7.19 and 7.20, which show the construction of rectangles below and above the curve. The rectangles have been formed by dividing the distance from 0 to 10 into four equal parts. Therefore all the rectangles have the same base, which is $\frac{10}{4} = 2.5$ cm. Use the graphs to estimate the heights of the rectangles. Calculate the lower and upper estimated areas and then enter L_2, U_2, and M_2 in the table. Remember to multiply the estimated areas by 2 so that you get an estimate for the entire area.

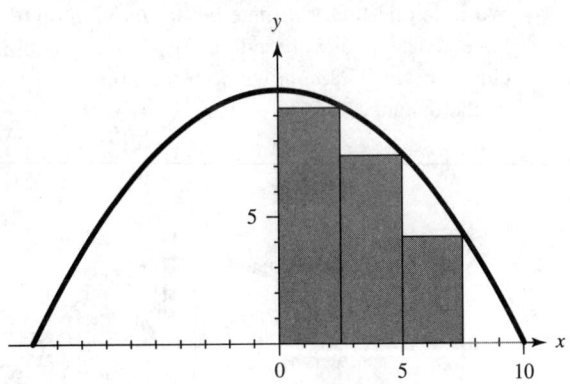

Figure 7.19

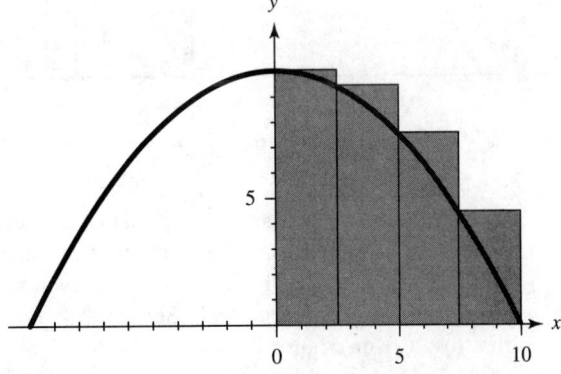

Figure 7.20

Figures 7.21 and 7.22 show the third stage in the estimation. The rectangles have been formed by dividing the distance from 0 to 10 into eight equal parts.

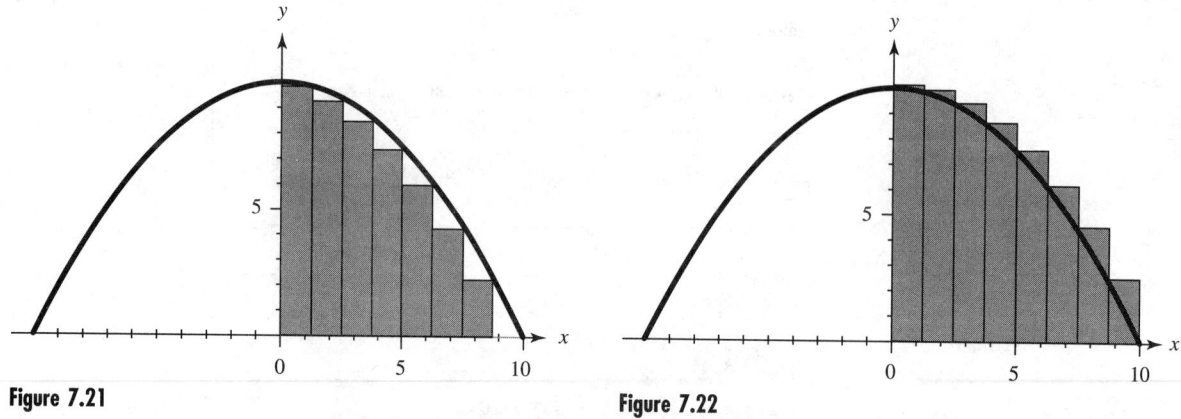

Figure 7.21   **Figure 7.22**

Therefore all the rectangles have the same base, which is $\frac{10}{8} = 1.25$ cm. Use the graphs to estimate the heights of the rectangles. Calculate the lower and upper estimated areas and then enter L_3, U_3, and M_3 in the table. Remember to multiply the estimated areas by 2 so that you get an estimate for the entire area.

Finally, plot these nine estimates on a graph like the one in Figure 7.7.

You have now computed three sequences of estimates, each with three terms. We'll do one final estimate before moving on to other applications of sequences. We have repeated the procedure for drawing additional rectangles under and over the curve. Figure 7.23 shows both the upper and lower rectangles from 16 subdivisions of the distance from 0 to 10.

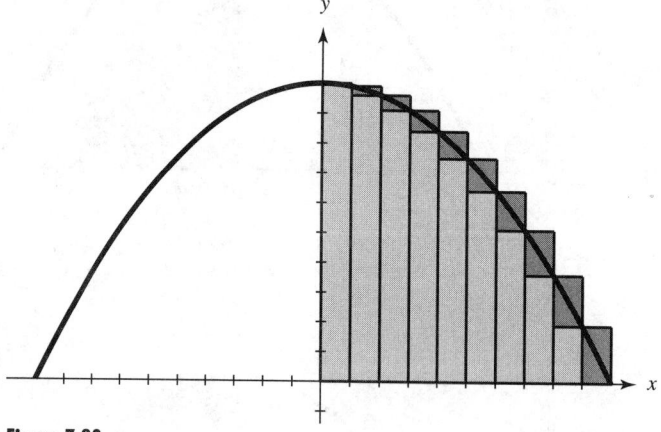

Figure 7.23

You can guess that the estimate of the area from this set of rectangles would be quite a bit better than the previous three estimates, and you can probably guess that the estimate after that one would be even better. The idea of an estimate getting better and better is discussed in Section 7.3. For now, we'll just look at the estimates that result from Figure 7.23.

Our values of the estimates were $L_4 = 126.95$, $U_4 = 139.45$, and $M_4 = 133.20$. Figure 7.24 shows our plot of the four estimates in the lower, upper, and average sequences. Your graph should look something like this figure, but without the last three points on the right, which represent L_4, U_4, and M_4.

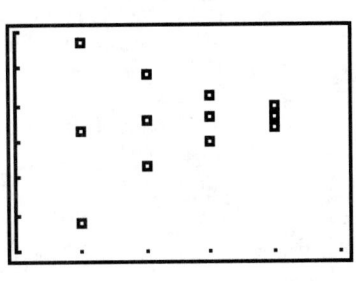
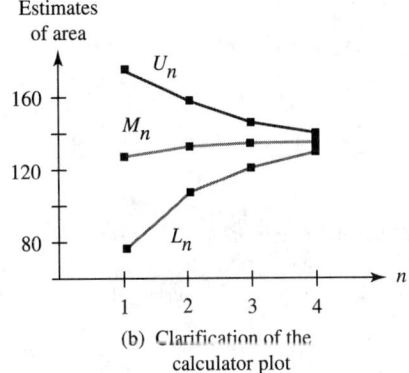

(a) Three sequences plotted
on the calculator

(b) Clarification of the
calculator plot

Figure 7.24

Activity 7.8
Is There a Better Estimate of the Area?

Here are the values of the lower and upper estimates for the area that we got from the four sets of rectangles. We have rounded off the estimates to the nearest hundredth. Check your answers to Activity 7.7 with the first three terms in each sequence.

$$\{L_n\} = \{75, 106.25, 120.31, 126.95\}$$

$$\{U_n\} = \{175, 156.25, 145.31, 139.45\}$$

1. Compute the terms of the sequence $\{M_n\}$, the average of the lower and upper estimates.
2. Use all these estimates to produce one estimate of the area under the curve. That is, if you had just one guess right now, what would it be? There are many ways that you might make this guess. You might use the graphs, or guess the next values of the lower and upper estimates, or look for a pattern in the values of the estimates, or do something else.
3. Write an explanation of your estimate.

Section 7.2 Exercises

In Exercises 1–18 write the first five terms of each sequence and guess its limit. If you need more terms, enter the expression into the calculator's Y= screen in the Seq *mode, and look at the table for the sequence.*

1. $a_n = \dfrac{5n + 8}{n}$

2. $a_n = \dfrac{n^2 - n}{n + 1}$

3. $a_n = \dfrac{2n + 1}{3n - 4}$

4. $a_n = \sqrt{n - 1}$

5. $a_n = \left(1 + \dfrac{1}{n}\right)^n$

6. $a_n = \left(1 - \dfrac{1}{n}\right)^{1/n}$

7. $a_n = na_{n-1}; a_1 = 1$

8. $a_n = -2a_{n-1}; a_1 = 1$

9. $a_n = \dfrac{1}{n} a_{n-1}; a_1 = 1$

10. $a_n = \dfrac{-1}{2n(2n - 1)} a_{n-1}; a_0 = 1$

11. $a_n = 1.8(1 - 0.001a_{n-1})a_{n-1}; a_0 = 10$

12. $a_n = 3.2(1 - 0.001a_{n-1})a_{n-1}; a_0 = 10$

13. $a_n = 2a_{n-1} - 10; a_1 = 9$

14. $a_n = 2a_{n-1} - 10$; $a_1 = 10$

15. $a_n = 2a_{n-1} - 10$; $a_1 = 11$

16. $a_n = \frac{1}{2}a_{n-1} + 5$; $a_1 = 9$

17. $a_n = \frac{1}{2}a_{n-1} + 5$; $a_1 = 10$

18. $a_n = \frac{1}{2}a_{n-1} + 5$; $a_1 = 11$

19. Sketch the graph of $f(x) = \cos x$ where x is in radians. Approximate the lower L_1, upper U_1, and middle M_1 area under the curve, above the x-axis, from

$x = -0.5\pi$ to $x = 0.5\pi$. Record your results in a table like Table 7.2. Refine your results by making second and third approximations for the lower and upper areas.

20. Sketch the graph of $f(x) = \sqrt{x}$. Approximate the lower L_1, upper U_1, and middle M_1 area under the curve, above the x-axis, from $x = 0$ to $x = 4$. Record your results in a table like Table 7.2. Refine your results by making second and third approximations for the lower and upper areas.

In Exercises 21–26 use the Seq *mode of your calculator to graph each of the given sequences.*

21. Sketch the graph of the first 20 terms of the sequence $u_n = \sqrt{u_{n-1}}$ with **(a)** $u_0 = 16$ and **(b)** $u_0 = 0.0016$. Set your graphing window to Xmin=0, Xmax=20, Ymin=0. You may want to change the value of Ymax for each graph.

22. Consider the sequence $u_n = 3u_{n-1} - 8$. **(a)** Determine the fixed point F of this sequence, **(b)** sketch the graph of the first 20 terms of the sequence with $u_0 = F$, and **(c)** sketch the graph of the first 20 terms of the sequence with $u_0 = F - 0.001$. Set your graphing window to Xmin=0, Xmax=20, Ymin=0, and Ymax=10.

23. Match the graphs in Figure 7.25(a)–(d) with each of the following sequences. The window settings in each graph are Xmin = 0, Xmax = 12, Xscl = 1, Ymin = -4, Ymax = 8, and Yscl = 1. (Since there are more sequences than graphs, at least one of the sequences cannot be matched with a graph.)

(a) $a_n = \left(1 + \dfrac{2}{n}\right)^n$

(b) $b_n = \dfrac{6 - 5n}{2n}$

(c) $c_n = 2.5c_{n-1} + 3.6$ with $c_1 = -2.3999$

(d) $d_n = 2.5d_{n-1} + 3.6$ with $d_1 = -2.4$

(e) $e_n = 2.5e_{n-1} + 3.6$ with $e_1 = -2.4001$

24. Sketch the graph of the first 100 terms of the sequence $u_n = 3.25u_{n-1}(1 - u_{n-1})$ with $u_0 = 0.9$. Describe what seems to be happening.

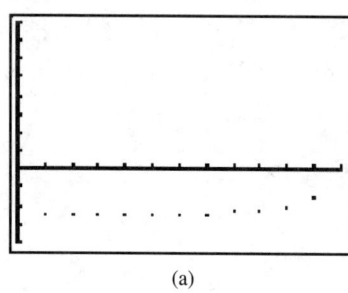

(a)

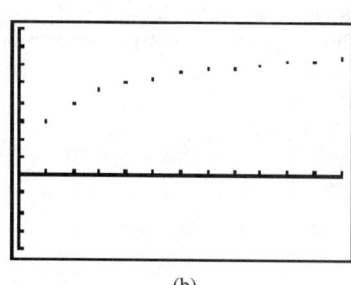

(b)

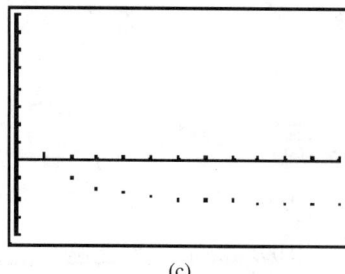

(c)

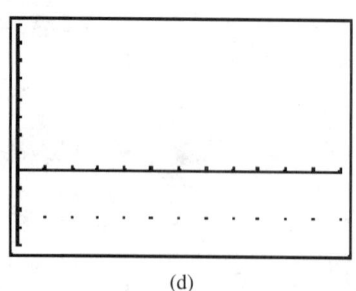

(d)

Figure 7.25

In Exercises 25–26 use the Seq *mode of your calculator to graph each of the given sequences. Set your graphing window to* Xmin=0, Xmax=100, Ymin=0, *and* Ymax=1.

25. Sketch the graph of the first 100 terms of the sequence $u_n = 3.5u_{n-1}(1 - u_{n-1})$ with $u_0 = 0.9$. Describe what seems to be happening.

26. Attempt to sketch the graphs of the first 100 terms of the sequence $u_n = 3.5u_{n-1}(1 - u_{n-1})$ with **(a)** $u_0 = 0.9999$ and **(b)** $u_0 = 1.0001$. If your calculator gives an error message (for example, ERR:OVERFLOW) then use the table feature of the calculator to look at values for the sequence. Describe what seems to be happening and try to explain why.

27. *Automotive technology* The cross-section of an airfoil for some race cars can be described by the graphs of $f(x) = \sqrt{x}$ and $g(x) = -\sqrt{x} + \frac{1}{32}x^2$. Approximate the lower L_1, upper U_1, and middle M_1 area between the curves from $x = 0$ to $x = 16$. Record your results in a table like Table 7.2. Refine your results by making second and third approximations for the lower and upper areas.

28. *Automotive technology* For this exercise you will need either the used-car advertising section of a large city newspaper or a book of used-car prices. Select a model of automobile that has been in production for at least seven years.
(a) Complete a table like the one below for the price of this model according to its age. The age at year 0 would be the original price of the car:

Age of car (yr)	0	1	2	3	4	5	6	7
Price ($)								

(b) Make a graph of the data you entered in the table.
(c) Does this data seem to form an arithmetic sequence, a geometric sequence, or neither?
(d) Use your sequence to estimate the price two years from now of the car in your sequence. Explain how you arrived at your answer.

29. In Activity 7.6 you determined a formula in terms of C and D that gives the fixed point for the general sequence

$$a_n = C \times a_{n-1} - D$$

Use that formula to explain why there is no fixed point for the sequence $a_n = a_{n-1} - 7$, or for any recursive definition in the form

$$a_n = a_{n-1} - D$$

30. Write a memo to the management in which you describe what you have learned about finding area and how you can use this information to determine the area of the cross-section of the wing in the Chapter Project.

7.3

Looking More Closely at Limits

Now that you have had some experience guessing limits of sequences, we will look at how you can use algebra to help prove that your guess is correct.

Three Important Sequences

We will begin by discussing in detail the limits of three sequences that are quite common in mathematics. They are simpler than the ones we have been analyzing so far, but they are like building blocks. Many other important sequences can be analyzed beginning with these three:

▸ $\{1, 2, 3, 4, 5, \ldots\}$
▸ $\{1, \frac{1}{2}, \frac{1}{3}, \frac{1}{4}, \frac{1}{5}, \ldots\}$
▸ $\{\frac{1}{2}, \frac{1}{4}, \frac{1}{8}, \frac{1}{16}, \frac{1}{32}, \ldots\}$

Example 7.19

Discuss the sequence described by $a_n = a_{n-1} + 1$ with $a_1 = 1$ as n gets larger and larger.

Solution   The sequence looks like $\{1, 2, 3, 4, 5, \ldots\}$. The terms get bigger because each term is one more than the term before it. The nth term will eventually reach and pass any number you can name because if someone claimed to have thought of the largest term, we would just add one to that term and get the next term, which would be even larger. Here are two different ways of summarizing these thoughts about this sequence:

► This sequence has no limit as n gets larger and larger.
► The nth term of this sequence approaches positive infinity as n gets larger and larger.   ∎

When a sequence like $\{1, 2, 3, 4, 5, \ldots\}$ has no limit, we say that it **diverges**. When a sequence gets closer and closer to a limiting value, like the one in the following example, we say that it **converges**.

Example 7.20

Discuss the sequence described by $a_n = \dfrac{1}{n}$ as n gets larger and larger.

Solution   The sequence looks like this: $\left\{1, \frac{1}{2}, \frac{1}{3}, \frac{1}{4}, \frac{1}{5}, \ldots\right\}$. The table shows the decimal form for these terms and for three of the later terms.

n	a_n
1	1
2	0.5
3	0.33333
4	0.25
5	0.2
⋮	⋮
2,000	0.0005
⋮	⋮
2,000,000	0.0000005
⋮	⋮
200,000,000	0.000000005

You can see what is happening. All the terms are positive, each term is smaller than the one before it, and there is no end to the process of getting smaller. Just as there was no largest number in Example 7.19, there is no smallest number in this sequence. However, although the terms are decreasing, they are not decreasing without limit. In fact, the terms are getting closer and closer to zero. Here are three different ways of summarizing these thoughts about this sequence:

- This sequence has a limit of 0 as n gets larger and larger.
- This sequence converges to a limit of 0 as n gets larger and larger.
- As n gets larger and larger, the nth term of this sequence approaches zero. ■

Calculator graphs can give another view of how the sequence in Example 7.20 behaves when n gets larger and larger. We'll use the calculator's `Func` mode to plot the function $y = \frac{1}{x}$ instead of the `Seq` mode to plot the sequence $u(n) = \frac{1}{n}$, because if you use the `Seq` mode with large values of n the calculator will be very slow. Figures 7.26–7.28 show the result of plotting $y = \frac{1}{x}$ and changing the graph window to capture higher and higher values of x. You can see that the graphs continue to show the same shape in all the graph windows. We could continue to look at smaller and smaller graph windows until we reach the limit of the smallest numbers the calculator can handle, and there would still be numbers smaller than the ones you could see on the graph.

These graphs give additional evidence that the limit of the sequence is zero.

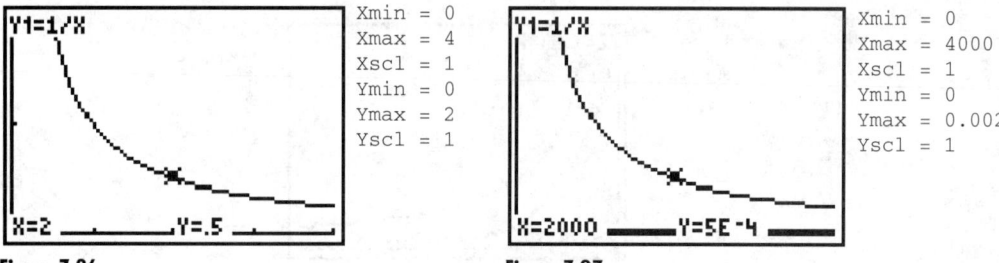

Xmin = 0
Xmax = 4
Xscl = 1
Ymin = 0
Ymax = 2
Yscl = 1

Figure 7.26

Xmin = 0
Xmax = 4000
Xscl = 1
Ymin = 0
Ymax = 0.002
Yscl = 1

Figure 7.27

Xmin = 0
Xmax = 4000000
Xscl = 1
Ymin = 0
Ymax = 0.000002
Yscl = 1

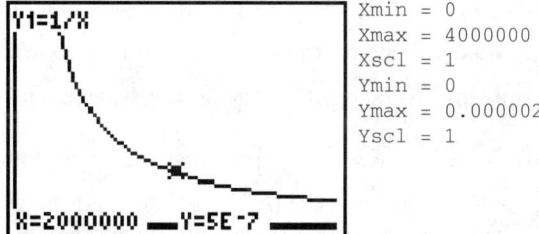

Figure 7.28

In Example 7.21 we will discuss the sequence $\left\{\frac{1}{2}, \frac{1}{4}, \frac{1}{8}, \frac{1}{16}, \ldots\right\}$. There are three ways of describing this sequence:

- In this sequence each term is computed by multiplying the previous term by $\frac{1}{2}$. Therefore it can be written recursively as $a_n = \frac{1}{2}a_{n-1}$ and $a_1 = \frac{1}{2}$.
- In this sequence each term is computed by dividing the previous term by 2. Therefore it can be written recursively as $a_n = \dfrac{a_{n-1}}{2}$ and $a_1 = \dfrac{1}{2}$.

▶ Since the second term is $\frac{1}{2} \times \frac{1}{2} = \frac{1}{4}$ and the third term is $\frac{1}{2} \times \frac{1}{2} \times \frac{1}{2} = \frac{1}{8}$, then the nth term can be represented by the direct expression $a_n = \left(\frac{1}{2}\right)^n$.

Example 7.21

Discuss the sequence $\left\{\frac{1}{2}, \frac{1}{4}, \frac{1}{8}, \frac{1}{16}, \ldots\right\}$ as n gets larger and larger.

Solution   Let's compare this sequence with the sequence $\left\{1, \frac{1}{2}, \frac{1}{3}, \frac{1}{4}, \frac{1}{5}, \ldots\right\}$. First we'll look at the sequences term by term.

n	1	2	3	4	5
$a_n = \dfrac{1}{n}$	1	$\frac{1}{2}$	$\frac{1}{3}$	$\frac{1}{4}$	$\frac{1}{5}$
$b_n = \frac{1}{2} b_{n-1}$	$\frac{1}{2}$	$\frac{1}{4}$	$\frac{1}{8}$	$\frac{1}{16}$	$\frac{1}{32}$

You can see that each term of the sequence $\{b_n\}$ is smaller than the sequence $\{a_n\}$. Graphs can show us many more terms. Let's see what they can tell us about the two sequences. Figure 7.29 shows the graphs of $y = \left(\frac{1}{2}\right)^x$ and $y = \frac{1}{x}$.

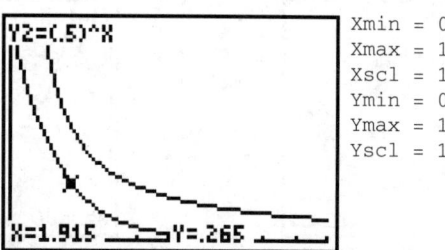

```
Xmin = 0
Xmax = 10
Xscl = 1
Ymin = 0
Ymax = 1
Yscl = 1
```

Figure 7.29

The graph of $y = \left(\frac{1}{2}\right)^x$ is below the graph of $y = \frac{1}{x}$ and always above the x-axis. Since we know that the upper curve, which represents the sequence described by $a_n = \frac{1}{n}$, approaches zero as n gets larger and larger, then we can be quite sure that the sequence $b_n = \left(\frac{1}{2}\right)^n$ will also get closer and closer to zero as n gets larger and larger. The sequence $\{b_n\}$ gets "squeezed" between the sequence $\{a_n\}$ and the x-axis, so the sequence $\{b_n\}$ must approach the same limit as the sequence $\{a_n\}$. ∎

Here are three different ways of summarizing these thoughts:

▶ This sequence has a lower limit of 0 as n gets larger and larger.
▶ This sequence converges to a limit of 0 as n gets larger and larger.
▶ As n gets larger and larger the nth term of this sequence approaches zero.

Now that we have studied how three important mathematical sequences act when we look at more and more terms, we'll introduce some mathematical symbols that are used to describe the limits of sequences.

Symbolism for Limits

If the sequence $\{a_n\}$ has a limit L as n gets larger and larger then we write

$$\lim_{n \to \infty} a_n = L$$

Let's take these symbols one at a time.

▶ lim means "the limit."
▶ $n \to \infty$ means "as n gets larger and larger" or "as n approaches infinity."
▶ L is the number to which the nth term a_n gets closer and closer.

Putting the meaning of these symbols together along with the name of the sequence we say, "The limit of a_n as n approaches infinity is equal to L."

We'll use this limit symbolism to describe the three sequences we have studied in this section. There is nothing new in the following summary; there is just a new way to write what you found out about the three sequences.

Three Important Limits

▶ $\{1, 2, 3, 4, 5, \ldots\}$ is defined by $a_n = a_{n-1} + 1$ and $a_1 = 1$. The limit of the sequence is described by

$$\lim_{n \to \infty} n = +\infty$$

or you could say *this sequence has no limit* because the terms approach positive infinity.

▶ $\left\{1, \frac{1}{2}, \frac{1}{3}, \frac{1}{4}, \frac{1}{5}, \ldots\right\}$ is defined by $a_n = \dfrac{1}{n}$. The limit of the sequence is described by

$$\lim_{n \to \infty} \frac{1}{n} = 0$$

▶ $\left\{\frac{1}{2}, \frac{1}{4}, \frac{1}{8}, \frac{1}{16}, \frac{1}{32}, \ldots\right\}$ is defined by $a_n = \left(\frac{1}{2}\right)^n$. The limit of the sequence is described by

$$\lim_{n \to \infty} \left(\frac{1}{2}\right)^n = 0$$

The Algebra of Limits

We have looked closely at three limits. Next you will see how to use knowledge of these three limits in order to estimate the limits of other sequences.

Example 7.22

If $a_n = \dfrac{3 + 10n}{n}$, what is $\lim_{n \to \infty} a_n$?

Solution   The first 5 terms are {13, 11.5, 11, 10.75, 10.6} and a graph of the sequence is shown in Figure 7.30.

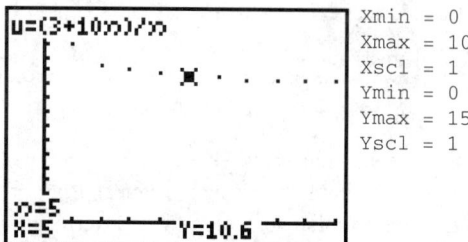

Xmin = 0
Xmax = 10
Xscl = 1
Ymin = 0
Ymax = 15
Yscl = 1

Figure 7.30

The graph seems to show that the sequence is approaching a limit, but what is that limit? We'll use algebra to find out. First, since the fraction means to divide by n, we can rewrite $\dfrac{3 + 10n}{n}$ as $\dfrac{3}{n} + \dfrac{10n}{n} = \dfrac{3}{n} + 10$. Now we want to know the value of

$$\lim_{n \to \infty} \left(\frac{3}{n} + 10 \right)$$

There are two parts to this limit. The fraction $\dfrac{3}{n}$ depends on n and the number 10 does not depend on n. Since 10 does not depend on n it will remain 10 no matter what n does. However, $\dfrac{3}{n}$ will change as n increases. What limit will it approach? Now think about $\dfrac{3}{n}$ as $3 \times \dfrac{1}{n}$. Find

$$\lim_{n \to \infty} \left(3 \times \frac{1}{n} \right)$$

Again there are two parts to this limit. The first part, 3, does not depend on n, and the second part, $\dfrac{1}{n}$, does depend on n. Therefore the limit of $3 \times \dfrac{1}{n}$ is $3 \lim\limits_{n \to \infty} \dfrac{1}{n}$. But you know that $\lim\limits_{n \to \infty} \dfrac{1}{n} = 0$, and since $3 \times 0 = 0$, then

$$\lim_{n \to \infty} \frac{3}{n} = \lim_{n \to \infty} \left(3 \times \frac{1}{n} \right)$$

$$= 3 \lim_{n \to \infty} \frac{1}{n} = 0$$

But we wanted to know the limit of $\dfrac{3}{n} + 10$. It is $0 + 10 = 10$. Therefore

$$\lim_{n \to \infty} \frac{3 + 10n}{n} = \lim_{n \to \infty} \left(\frac{3}{n} + 10 \right)$$

$$= 0 + 10 = 10 \qquad \blacksquare$$

In a similar way you can find the limits of many sequences. Here is another example.

Example 7.23

Find the value of the limit

$$\lim_{n \to \infty} \left[4 + 3 \left(\frac{1}{2} \right)^n \right]$$

Solution   The first 5 terms are {5.5, 4.75, 4.375, 4.188, 4.094} and a graph of the sequence is shown in Figure 7.31.

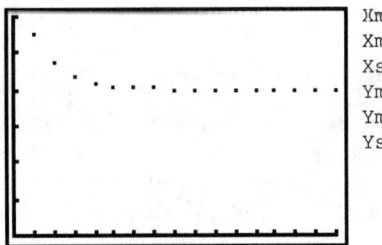

```
Xmin = 0
Xmax = 16
Xscl = 1
Ymin = 0
Ymax = 6
Yscl = 1
```

Figure 7.31

You might be able to guess the limit now. Let's look at algebra for the solution. There are two limits:

$$\lim_{n \to \infty} 4 \quad \text{and} \quad \lim_{n \to \infty} 3 \left(\frac{1}{2} \right)^n$$

The first limit is 4. That is, $\lim_{n \to \infty} 4 = 4$.

The second limit can be broken into the two parts:

$$\lim_{n \to \infty} 3 \times \lim_{n \to \infty} \left(\frac{1}{2} \right)^n$$

But you know that the number 3 does not depend on n, so its limit is 3. And the limit of $\left(\frac{1}{2} \right)^n$ is 0, so we have:

$$\lim_{n \to \infty} 4 + 3 \left(\frac{1}{2} \right)^n = \lim_{n \to \infty} 4 + \lim_{n \to \infty} 3 \times \lim_{n \to \infty} \left(\frac{1}{2} \right)^n$$

$$= 4 + 3 \times 0 = 4 \qquad \blacksquare$$

You now can apply algebra to prove to yourself that your guess of a limit is correct. The method used in the previous two examples depends on the following three properties of limits.

Three Properties of Limits

▶ The limit of a constant is that constant. In symbols, if k is a constant then

$$\lim_{n \to \infty} k = k$$

▶ The limit of a sum is equal to the sum of the limits. That is, if you want to know the limit of the sum of two or more expressions just look at the limit of each of the expressions separately and add the answers. In symbols this statement is

$$\lim_{n \to \infty} (A + B) = \lim_{n \to \infty} A + \lim_{n \to \infty} B$$

▶ The limit of a product is equal to the product of the limits. That is, if you want to know the limit of the product of two or more expressions just look at the limit of each of the expressions separately and multiply the answers. In symbols this statement is

$$\lim_{n \to \infty} (AB) = \lim_{n \to \infty} A \times \lim_{n \to \infty} B$$

Be Careful

Although the second property of limits seems sensible, it is important to realize that it is not always true for other kinds of algebraic operations. For instance, consider these two statements:

▶ The square root of a sum is not equal to the sum of the square roots. For example $\sqrt{4+9} \neq \sqrt{4} + \sqrt{9}$ since $\sqrt{4+9} = \sqrt{13} \approx 3.61$ is not the same as $\sqrt{4} + \sqrt{9}$, which is $2 + 3 = 5$.
▶ The square of a sum is not equal to the sum of the squares. For example, $(4+9)^2 \neq 4^2 + 9^2$ because $(4+9)^2 = 13^2 = 169$ is not equal to $4^2 + 9^2 = 16 + 81 = 97$.

Example 7.24

Discuss the sequence described by $a_n = 0.9^n$, as n gets larger and larger. Use graphs to help you estimate the limit the sequence is approaching.

Solution   We will use the same strategy that was used in Example 7.21, by comparing the graph with one whose limit we know, the sequence described by $a_n = \dfrac{1}{n}$. Figure 7.32 shows the graphs of two functions $y = \dfrac{1}{x}$ and $y = 0.9^x$.

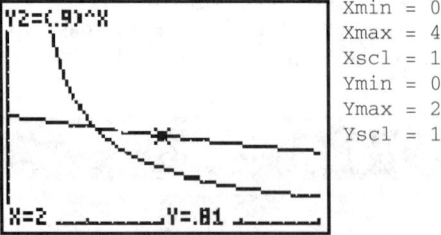

```
Y2=(.9)^X


              Xmin = 0
              Xmax = 4
              Xscl = 1
              Ymin = 0
              Ymax = 2
              Yscl = 1

X=2          Y=.81
```

Figure 7.32

For x larger than about 1, you can see in the figure that the graph of $y = 0.9^x$ is above the graph of $y = \dfrac{1}{x}$. Therefore we can conclude that $0.9^x > \dfrac{1}{x}$ for some values of x. But the graph only shows values of x from 0 to 4. If we want x to approach infinity we've got to use numbers larger than 4.

Figure 7.33 shows a different interval and the graphs seem to be coming back together, but it's not clear what is happening. One more view is shown in Figure 7.34. Now it is quite clear that 0.9^x is less than $\dfrac{1}{x}$ if x is larger than about 35.

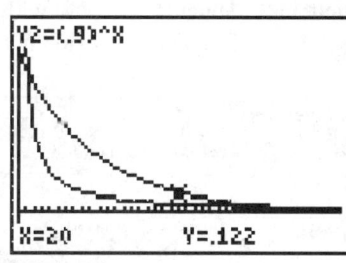

```
Xmin = 0
Xmax = 40
Xscl = 1
Ymin = -0.2
Ymax = 1.2
Yscl = 1
```

Figure 7.33

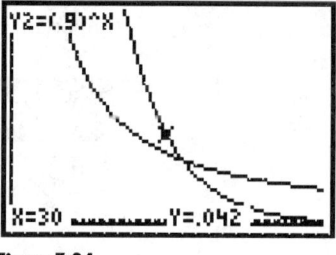

```
Xmin = 0
Xmax = 60
Xscl = 1
Ymin = 0
Ymax = 0.1
Yscl = 1
```

Figure 7.34

If this process were carried out for larger and larger intervals the graphs would continue to show that $0.9^x < \dfrac{1}{x}$. We also know that 0.9^x is always positive, because a positive number multiplied x times will always be positive. From these two statements we can conclude that if $\dfrac{1}{x}$ approaches zero then 0.9^x must have the same limit.

Therefore we can conclude that

$$\lim_{n \to \infty} (0.9)^n = 0$$

Calculator Lab 7.3

Chaos in Geometric Sequences

In this calculator lab you will study the limits of the following sequences, as $n \to \infty$: (a) $a_n = 0.99^n$, (b) $a_n = 1.01^n$, (c) $a_n = (-0.99)^n$, and (d) $a_n = (-1.01)^n$.

We have studied previous limits with graphs. To provide a different experience we will use tables to study the sequences in this lab. After this lab you can decide whether to use graphs, tables, or both. Once again, we will use the calculator's `Func` mode because if you use the `Seq` mode with large values of n the calculator will be very slow. For example if you set `TblStart=1000` and `Tbl=1000` the calculator will compute each and every value from $n = 1000$ to $n = 2000$, although only the entries for $n = 1000, 2000, \ldots$ will be shown in the table.

Procedures

1. Set the calculator mode to `Func`.
2. Enter $y = 0.99^x$ in the Y= screen.
3. Set `TblStart=0` and `Tbl=10`.
4. Look at the resulting table and see if you can guess what the function will approach for large values of x.
5. If necessary, change the values of `TblStart` and `Tbl` and look at the table again.
6. Continue until you can be sure what the sequence is approaching as $n \to \infty$.
7. Continue with the other functions.

8. For each sequence write your answer to the two questions that follow. Explain each answer.

(a) What limit does each sequence approach as $n \to \infty$?

(b) How are these sequences related to chaos theory?

The sequences you studied in Calculator Lab 7.3 will turn out to be very important in the applications of Section 7.4. Here is a summary of their properties.

Summary: The Sequences Described by $a_n = r^n$

▶ If $-1 < r < 1$ (for example, $r = \dfrac{1}{2}$ or $r = -0.9$) then

$$\lim_{n\to\infty} r^n = 0$$

▶ If $r > 1$ (for example, $r = 2$ or $r = 1.1$), then

$$\lim_{n\to\infty} r^n = +\infty$$

▶ If $r < -1$ (for example, $r = -2$ or $r = -1.1$), then the sequence oscillates without limit

Section 7.3 Exercises

In Exercises 1–10, (a) graph the sequence and estimate its limit and (b) use algebra to determine the limit of the sequence.

1. $a_n = \dfrac{2}{3}$

2. $b_n = 2n - 1$

3. $c_n = \dfrac{3n - 8}{8}$

4. $d_n = \dfrac{5 - 2n}{n}$

5. $e_n = \dfrac{2n^2 + 4n - 1}{n^2}$

6. $f_n = \dfrac{5 - 4n - 6n^2}{2n^2}$

7. $g_n = 5 - 4\left(\dfrac{1}{2}\right)^n$

8. $h_n = -6 + 3\left(\dfrac{1}{4}\right)^n$

9. $j_n = \left(\dfrac{3}{4}\right)^n$

10. $k_n = \left(-\dfrac{3}{4}\right)^n$

11. Consider the recursive sequence $s_n = \dfrac{1}{2}\left(s_{n-1} + \dfrac{4}{s_{n-1}}\right)$ with $s_1 = 4$. **(a)** Graph the sequence and **(b)** estimate its limit.

12. Consider the recursive sequence $s_n = \sqrt{2 + \sqrt{s_{n-1}}}$ with $s_1 = \sqrt{2}$. **(a)** Graph the sequence and **(b)** estimate its limit.

13. *Medical technology* Some strains of bacteria are *histidine auxotrophs*; that is, they cannot produce their own histidine, an amino acid needed to construct proteins and make possible cell reproduction. When the number of cells becomes large, competition for the limited amount of histidine ensues, and the bacteria get an inadequate supply to sustain division at the same frequency. If C_n is the total mass of cells after n intervals, then when the population is large we have

$$C_{n+1} = \dfrac{2}{1 + \frac{C_n}{3}}\, C_n$$

Graph the sequence and estimate its limit for **(a)** $C_1 = 2$, **(b)** $C_1 = 3$, and **(c)** $C_1 = 4$.

14. *Environmental science* The level of cadmium metal pollution present in a lake following an accidental discharge at an industrial plant was 693 parts of cadmium metal per billion parts of water (ppb) in January and 462 ppb in February. If there is no further pollution the level of contamination will follow a geometric sequence. How long will it take for the amount of cadmium metal pollution to go below 0.200 ppb?

15. *Finance* Phillipe has $120,000 in a retirement fund that pays 5% interest per year.

   (a) If he plans to withdraw $10,000 per year, write a sequence that describes the amount left in his account each year.

   (b) At this rate how long will it take Phillipe to take all the money out of his retirement fund?

   (c) Determine the amount that Phillipe should withdraw each year if he does not wish to reduce the balance of the account.

16. *Environmental science* The level of pollution present in a lake following an accidental discharge at an industrial plant was 820 parts of cadmium metal per billion parts of water (ppb) in January. The lake washes out 2% of the pollution each month. However, a manufacturing plant discharges an average of 45 ppb of pollution in the lake each month.

   (a) Write a sequence that describes the amount of pollution in the lake each month.

   (b) Determine the amount of pollution the plant can discharge into the lake without raising the level above the present 820 ppb.

17. *Medical technology* A person is given an initial dose of 42 ml of a certain medication. The kidneys remove 18% of the medication from the bloodstream each day.

   (a) Write a sequence that describes the amount of medication in the bloodstream each day.

   (b) Determine the amount of medication that should be given each day in order to maintain a level of 35 ml in the bloodstream.

18. *Medical technology* A bacterial culture grows according to the sequence

$$a_n = 1.3a_{n-1} - 0.00006a_{n-1}^2$$

with $a_1 = 750$.

   (a) Graph the first 30 terms of this sequence.

   (b) Estimate the limit of the growth of this bacterial culture.

7.4   Applications of Geometric Sequences

In Section 7.2 we introduced the idea that a geometric sequence is a sequence that has a constant multiplier between consecutive terms, for example, $\{1, 2, 4, 8, 16, 32, \ldots\}$ (multiply by 2) or $\{9, 3, 1, \frac{1}{3}, \frac{1}{9}, \ldots\}$ (multiply by $\frac{1}{3}$). In this section you will study some of the applications of geometric sequences. Here are some of the applications we will study:

▸ Interest earned by a savings account or an investment
▸ Bouncing ball data from Chapter 2
▸ Discharging and charging a capacitor in an electric circuit
▸ Radioactive decay and age measurement using half-life
▸ Frequency of notes on a musical scale

Interest on a Savings Account

Most accounts compute interest quarterly (four times per year), monthly (twelve times per year), or daily, but we'll begin with annual compounding (once per year) to make things simpler.

Savings accounts accumulate interest according to a geometric sequence. Let's see how it works.

Imagine a deposit of $100 that earns 5% per year, **compounded annually**, which means that at the end of each year 5% of the account balance is added to the account.

Table 7.3 shows the balance at the beginning of each year. The interest is computed at the end of the year.

Table 7.3

Year	Balance at Beginning of Year	Interest Computation	Balance at End of Year
0	100	$100 + 0.05 \times 100 = 100 + 5$	105
1	105	$105 + 0.05 \times 105 = 105 + 5.25$	110.25
2	110.25	$110.25 + .005 \times 110.25 \approx 110.25 + 5.51$	115.76

Activity 7.9
Computing Interest

1. Complete Table 7.4 for the next four years (from $n = 3$ through $n = 6$).

Table 7.4

Year	Balance at Beginning of Year	Interest Computation at 5%	Balance at End of Year
0	100	$100 + 0.05 \times 100 = 100 + 5$	105
1	105	$105 + 0.05 \times 105 = 105 + 5.25$	110.25
2	110.25	$110.25 + 0.05 \times 110.25 \approx 110.25 + 5.51$	115.76
3			
4			
5			
6			

2. Plot the annual balance for six years, plus $n = 0$, in a graph like the one in Figure 7.35.

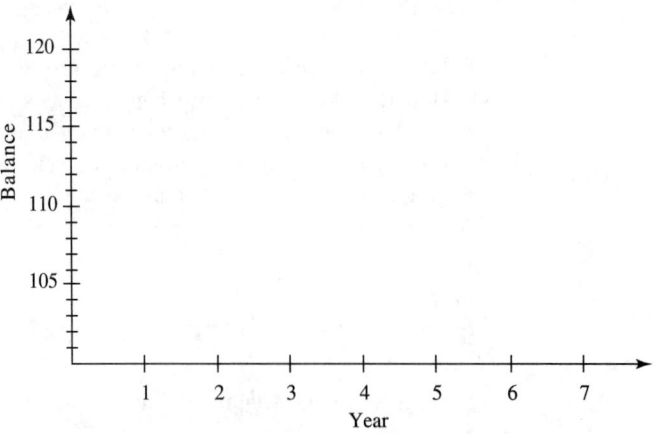

Figure 7.35

3. Change the interest rate to 10% per year and repeat the interest computations.
   (a) Fill in a table like Table 7.5 up through year 6.
   (b) Plot the annual balance on a graph like the one in Figure 7.35.

Table 7.5

Year	Balance at Beginning of Year	Interest Computation at 10%	Balance at End of Year
0	100		
1			
2			
3			
4			
5			
6			

In Example 7.25 we will consider the interest computation as an example of a sequence.

Example 7.25   Application

$1000 is deposited in an account that earns 8% per year, compounded annually.

(a) Write the recursive definition of the sequence.
(b) Write the formula for the amount in the account after n years.
(c) Repeat steps 1 and 2 for a deposit of $D and an annual interest rate of r, compounded annually.

Solution (a) The recursive definition of the sequence is

$$a_n = a_{n-1} + 0.08a_{n-1} \text{ and } a_0 = 1000$$

We use a_0 and not a_1 because a_1 will include the first interest payment at the end of year one. This way we match a_n with the nth interest payment.

You always have the choice of subscript for the first term in the sequence. Watch how we make that choice in each of the examples that follow.

(b) If we factor a_{n-1} out of the recursive expression, then it becomes

$$a_n = a_{n-1}(1 + 0.08)$$
$$= a_{n-1} \times 1.08 = 1.08a_{n-1}$$

This is the form of a geometric sequence with a common ratio equal to 1.08. You saw in Section 7.2 how to write the formula for the nth term. Although you probably remember how to write the expression for the nth term, we'll go through the explanation with a table.

n	a_n
0	1000
1	1000×1.08
2	$1000 \times 1.08 \times 1.08 = 1000 \times 1.08^2$
3	$1000 \times 1.08 \times 1.08 \times 1.08 = 1000 \times 1.08^3$

The pattern of exponents leads to the equation

$$a_n = 1000 \times 1.08^n$$

(c) For a deposit of D and interest rate r the entire process is the same. We won't show the table, but the results are summarized after the conclusion of this example. ∎

Growth of Investments—Annual Compounding

The interest rate r must be written as a decimal. Thus, an interest rate of 5% would have $r = 0.05$.

If a deposit of D is invested at annual interest rate r, compounded annually, the amount after n years, A_n, can be described either recursively or directly.

▶ Recursive:

$$A_n = A_{n-1}(1+r) \text{ and } A_0 = D$$

▶ Direct:

$$A_n = D(1+r)^n$$

Example 7.26   Application

What is the amount that a deposit of $500 becomes after 10 years at an interest rate of 7% compounded annually?

Solution   Since $7\% = 0.07$, we have $a_{10} = 500(1.07)^{10} \approx \$983.58$ ∎

Compounding More than Once per Year

If interest is computed quarterly (four times per year) or monthly (twelve times per year) the interest rate used in each computation is less than the annual rate. The rate used is the annual rate divided by the number of times the interest is computed in one year. That is, if the interest rate is r then each quarter the rate would be $\frac{r}{4}$ and each month the rate would be $\frac{r}{12}$. Table 7.6 summarizes how $100 invested at a annual rate of 12% will accumulate interest in one year under three different methods of compounding.

Example 7.27   Application

What is the amount that a deposit of $500 becomes after one year, at an interest rate of 7% compounded (a) annually, (b) quarterly, (c) monthly, and (d) daily.

Solutions   (a) After one year with annual compounding, the deposit is worth $500 \times 1.07 = \$535$
(b) Each time the interest is computed the rate is $\frac{0.07}{4} = 0.0175$. The problem then becomes just like compounding annually at a rate of 0.0175 for four years:

$$A = 500(1.0175)^4 \approx \$535.93$$

Table 7.6

Month	Annual	Quarterly	Monthly
1			1% on Jan 31
2			1% on Feb 28
3		3% on Mar 31	1% on Mar 31
4			1% on Apr 30
5			1% on May 31
6		3% on Jun 30	1% on Jun 30
7			1% on July 31
8			1% on Aug 31
9		3% on Sep 30	1% on Sep 30
10			1% on Oct 31
11			1% on Nov 30
12	12% on Dec 31	3% on Dec 31	1% on Dec 31
Result	$100(1.12) = \$112$	$100(1.03)^4 \approx \$112.55$	$100(1.01)^{12} \approx \$112.68$

(c) Each time the interest is computed, the rate is $\frac{0.07}{12} \approx 0.0058$. For better accuracy, use the fraction $\frac{0.07}{12}$. The problem then becomes just like compounding annually at a rate of $\frac{0.07}{12}$ for twelve years:

$$A = 500\left(1 + \frac{0.07}{12}\right)^{12} \approx \$536.15$$

(d) Daily compounding means 365 times per year:

$$A = 500\left(1 + \frac{0.07}{365}\right)^{365} \approx \$536.25$$

Example 7.28   Application

What is the amount that a deposit of \$500 becomes after 10 years, at an interest rate of 7% compounded (a) annually, (b) quarterly, (c) monthly, and (d) daily.

Solutions   (a) $500 \times (1.07)^{10} \approx \$983.58$

(b) Each time the interest is computed, the rate is $\frac{0.07}{4} = 0.0175$. In 10 years the interest is computed $4 \times 10 = 40$ times. Therefore the problem is the same as compounding annually at the rate of $\frac{0.07}{4}$ for 40 years. The amount A in the account grows to

$$A = 500(1.0175)^{40} \approx \$1000.80$$

(c) Each time the interest is computed, the rate is $\frac{0.07}{12} \approx 0.0058$. For better accuracy, use the fraction $\frac{0.07}{12}$. In 10 years the interest is computed $12 \times 10 = 120$ times.

Therefore the problem is the same as compounding annually at the rate of $\frac{0.07}{12}$ for 120 years. The amount A in the account grows to

$$A = 500 \left(1 + \frac{0.07}{12}\right)^{120} \approx \$1004.83$$

(d) Daily compounding means 365 times per year. In 10 years the interest is computed $365 \times 10 = 3,650$ times. The amount A in the account grows to

$$A = 500 \left(1 + \frac{0.07}{365}\right)^{3650} \approx \$1006.81 \qquad \blacksquare$$

Compounding More than Once per Year

If interest is computed k times per year on a deposit of $\$D$ and the annual interest rate is r, then the amount A in the account after one year is a function of D, r, and k:

$$A = D \left(1 + \frac{r}{k}\right)^{k}$$

If the interest computation is continued for n years, then the amount A in the account is a function of D, r, k, and n:

$$A = D \left(1 + \frac{r}{k}\right)^{kn}$$

The original $500 in Examples 7.26 through 7.28 approximately doubled to about $1000 in 10 years at a 7% annual interest rate. Let's see how long it takes to double your money at other interest rates.

Example 7.29   Application

How long would it take to double $500 if it is invested at 9% compounded annually?

Solution   For what value of n does the expression $500(1.09)^n$ equal 1000? To answer a question like this one, you need to solve the equation

$$500(1.09)^n = 1000$$

Equations that have the unknown in the exponent are called **exponential equations** and we'll meet them in a later chapter. For now you can use your calculator to solve the equation. We only seek the approximate year that the amount goes over $1000 as outlined by the following steps.

(a) Enter the equation $y = 500 \times (1.09)^x$ in the Y= screen.
(b) Change the values of Tbl Set and ΔTbl to suit your needs.
(c) Look through the table (Figure 7.36) to find the year in which the amount is closest to $1000.

For this problem the original amount is approximately doubled between the 8th and 9th years. Since 996.28 is closer to 1000 than is 1085.90 we would say that the original $500 doubles in about eight years. $\qquad \blacksquare$

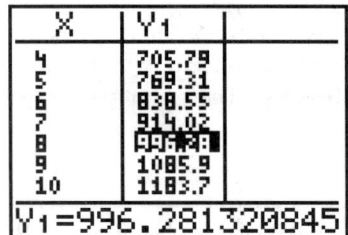

X	Y1
4	705.79
5	769.31
6	838.55
7	914.02
8	996.28
9	1085.9
10	1183.7

Y1=996.281320845

Figure 7.36

In the next calculator lab you will explore the relationship between doubling time for a savings account and the annual interest rate that is given by the bank.

Calculator Lab 7.4

Procedures

Double Your Money

Investigate the number of years required to double your money at each of the following annual rates of interest: 2%, 3%, 4%, 5%, 6%, 7%, 8%, 9%, 10%, and 12%.

1. Follow the steps in Example 7.29 for each of the interest rates.
2. Complete a table like the one below. Give the number of years to the nearest whole number. Two values that were computed in previous examples have been entered for you.

Interest Rate (r)	2	3	4	5	6	7	8	9	10	12
Doubling Time (t years)						10		8		

3. Plot the data on a graph of t vs. r like the one shown in Figure 7.37.

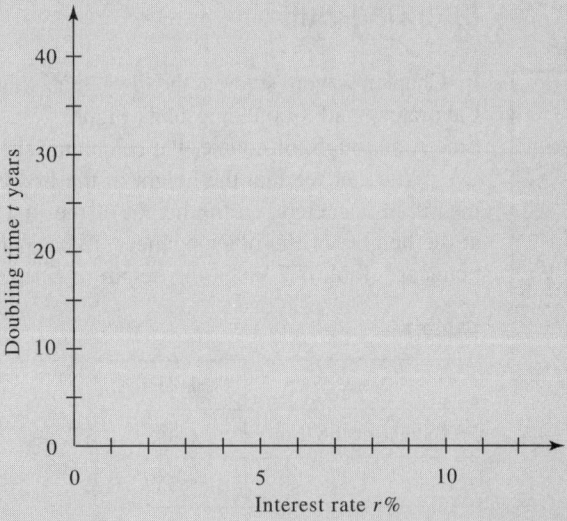

Figure 7.37

4. Connect the data points on the graph with a smooth curve.
5. Use the graph to estimate the doubling time for the following interest rates (a) 3.5%, (B) 11%, and (c) 15%.

Calculator Lab 7.5

Does Compounding Make You Rich?

Think about the sequence of numbers $\{A_k\}$ where $A_k =$ the amount earned in ten years from a deposit of $500 at 7% annual interest rate, compounded k times per year. The table below shows the part of this sequence that was computed in Example 7.28. As k continues to increase, what does the value of A_k do? Does

A_k increase without limit, or does it approach a limit? That is, find the value of

$$\lim_{k \to \infty} A_k$$

k	1	4	12	365			
A_k	983.58	1000.80	1004.83	1006.81			

Procedures
1. First, make a guess at how big A_k will get for extremely large values of k.
2. Write the formula for A_k.
3. Check your formula by using it to verify the values in the table.
4. Compute more values of A_k. For example, compounding every hour would mean that $k = 365 \times 24 = 8760$. How about every minute or every second?
5. Use a graph or a table or a lot of calculations to study the change in A_k as k gets larger and larger. It is not necessary to be able to describe the value of k, as in "compounding every hour," to study this question.
6. Based on your computations answer the questions again as to $\lim_{k \to \infty} A_k$.

Bouncing Ball

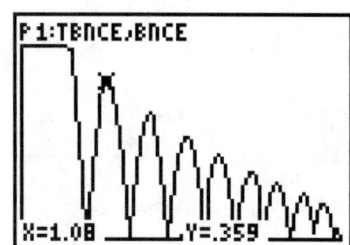

Figure 7.38

In Chapter 2 you studied the data that was produced with the Calculator-Based Laboratory and a bouncing ball. Figure 7.38 shows the height h of the ball for any time t, although, of course, the calculator shows these variables as x and y.

You can see that the height of the first bounce is about 0.359. This distance is measured in meters, so the height of the first bounce is a little over one foot. Look at the heights of the other bounces. What pattern do you think there is between the bounces? Table 7.7 gives the height attained after each of the eight bounces.

Table 7.7

Bounce	Height (meters)
0 (height of the drop)	0.443
1	0.359
2	0.289
3	0.233
4	0.191
5	0.155
6	0.128
7	0.104
8	0.083

In Activity 7.10 you will have an opportunity to explore the sequence of heights of the bouncing ball. First let's look at a summary of arithmetic and geometric sequences.

Arithmetic and Geometric Sequences

▶ If a sequence is arithmetic then either of the equations

$$a_n = d + a_{n-1}$$

or $$a_n = a_1 + (n - 1)d$$

describes the sequence, where d is the constant difference between consecutive terms $d = a_n - a_{n-1}$, and a_1 is the first term.

▶ If a sequence of nonzero terms is geometric then either of the following equations

$$a_n = r \times a_{n-1}$$

or $$a_n = a_1 r^{n-1}$$

describes the sequence, where r is the constant ratio between consecutive terms $r = \dfrac{a_n}{a_{n-1}}$, and a_1 is the first term.

Activity 7.10
Exploring the Bouncing Ball Sequence

The heights of a bouncing ball form a sequence, but what kind? Do the heights form an arithmetic sequence, a geometric sequence, or neither? Complete Table 7.8 with the differences and ratios between the heights of successive bounces. Give your answer to two significant figures, as shown for bounce #1. Since the heights are decreasing, the values of d will be negative and the values of r will be less than 1.

Table 7.8

Bounce No.	Height (meters)	Difference d	Ratio r
0 (height of the drop)	0.443	–	–
1	0.359	−0.084	0.81
2	0.289		
3	0.233		
4	0.191		
5	0.155		
6	0.128		
7	0.104		
8	0.083		

If you were expecting a constant difference or constant ratio you might have been disappointed. This is real data, which includes errors of measurement. Therefore we have to be reminded that the question is not "Which kind of sequence *is formed* by the heights of the bounces?" but "Which kind of mathematical sequence is the *best model* for these heights?" Your differences should be quite far from being constant, but the numbers in your ratio column of Table 7.8 should all be between 0.80 and 0.83. Another thing to observe about the ratios is that they do not form a trend of increasing or decreasing numbers. That is another sign that the sequence of ratios can be successfully modeled by a constant number. The average of the ratios is 0.813, so we'll use the number 0.81 as our estimate of the common ratio.

Figure 7.39 shows a graph of two sequences:

▸ The heights of consecutive bounces
▸ A sequence described by the recursive definition $a_n = 0.81 \times a_{n-1}$ with $a_0 = 0.443$.

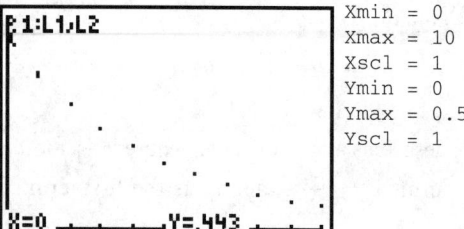

Xmin	= 0
Xmax	= 10
Xscl	= 1
Ymin	= 0
Ymax	= 0.5
Yscl	= 1

Figure 7.39

Note that we have used a_0 for the first term. This is consistent with the first value of n in the table but different from what is normally used in sequences. Either a_0 or a_1 could have been used for the first term, but the important point is to use the same starting point for both sequences in the graph.

The two sequences differ so little that all the points overlap. Even at the only observable difference, when $n = 1$, the points are still so close that they touch each other. The closeness of the fit between the data and the model allows us to say that the geometric sequence is a very good model for the successive heights of a bouncing ball.

Example 7.30   Application

Suppose a ball is dropped from a height of 3.4 meters, and the ratio of consecutive heights reached by the ball is approximately 0.7.

(a) Write the mathematical model for this bouncing ball as a recursive expression and as a direct expression for the nth term.
(b) What percent is each height of the bounce preceding it?
(c) What percent of the height is lost with each bounce?
(d) What is the height reached by (i) the first bounce, (ii) the 10th bounce, and (iii) the 100th bounce?
(e) Which bounce first produces a height less than 0.001 m (one millimeter)?
(f) Does the ball ever stop bouncing?

Solution   (a) The model is a geometric sequence. It can be can be described as the recursive sequence $a_n = 0.7 \times a_{n-1}$ with $a_0 = 3.4$. Here we use a_0 to represent the first term (the height from which the ball was dropped) because we are talking about bounces, and the height of the first bounce is actually the second term, a_1, in the sequence. This sequence can be described by its nth term as $a_n = 3.4 \times 0.7^n$.

(b) Since each height is 0.7 times the height before it, then each height is 70% of the height of the one preceding it.

(c) If you think about the loss of height each time, then the percent that is lost is 30%.

(d) The height reached after each bounce is modeled by the direct expression $a_n = 3.4 \times 0.7^n$. Note that the recursive expression would require us to compute each term up to the one we want. Therefore the direct expression is used for these computations:

   i. $a_1 = 0.7 \times 3.4 \approx 2.4$ m
   ii. $a_{10} = 0.7^{10} \times 3.4 \approx 0.096041585 \approx 0.096$ m
   iii. $a_{100} = 0.7^{100} \times 3.4 \approx 0.0000000000000011 \approx 0.00$ m

(e) To determine the bounce that first produces a height less than 0.001 m, we have to solve the equation $0.7^n \times 3.4 = 0.001$. You will learn how to use algebra to solve equations like that in Chapter 9 For now just guess a value of n and calculate the value of a_n. Then increase and decrease n until you reach an answer that is approximately 0.001 m. Our results were $a_{22} \approx 0.0013$ and $a_{23} \approx 0.0009$. Therefore the 23rd bounce is the first one with height below one millimeter.

(f) You know from experience that a ball in an experiment like this does stop bouncing eventually. However, the mathematical model predicts that the hundredth bounce will still have some positive nonzero height. The model's prediction of the height of the hundredth bounce could not be measured by any human equipment; it is about a millionth of the diameter of a hydrogen atom. If we had a measuring device that could measure to the nearest millimeter (0.001 m) then all bounces after the 22nd could not be detected, and as far as our measurements are concerned the ball has stopped bouncing. █

Charging and Discharging a Capacitor

In Chapter 3 we studied an RC circuit, which is an electrical circuit that contains a voltage source, capacitor, and resistor (see Figure 3.114 on page 191). You saw then that the capacitor changes the flow of current so that the current becomes 90° out of phase with the voltage source. For current to flow through a capacitor, one plate of the capacitor must assume a positive charge and the other must assume a negative charge. The process of building up the charge is called **charging** the capacitor. It takes time for this charge to reach its maximum, and it is this rate of charging that we will study with a geometric sequence. See Figure 7.40.

While the capacitor is charging (Figure 7.40(b)) current flows through the capacitor but at a reduced level. When the capacitor is fully charged the voltage across the capacitor equals that of the voltage source.

Likewise, when the voltage source is switched off (Figure 7.40(d)) it takes time for the charge on the plates to return to their original state. This process, called **discharging**, also takes time, and it is accompanied by a gradual decrease in voltage in the circuit as shown in Figure 7.41(a)–7.41(c).

Electrical engineers need to know how fast the capacitor charges and discharges and how the voltage varies with time. Using sequences and knowledge of how a capacitor works we can build mathematical models for the voltage variations when a capacitor is charging or discharging.

A geometric sequence describes how the voltage increases in the circuit as the capacitor is being charged and how the voltage decreases as the capacitor is being discharged. The **time constant** of a capacitor is defined as the time required for the voltage across a discharging capacitor to be reduced by 63% to 37% of the value

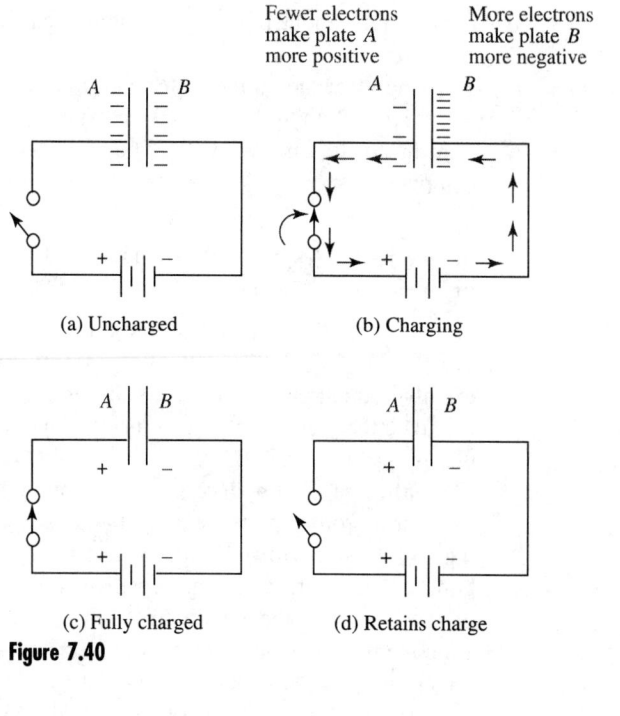

Fewer electrons
make plate *A*
more positive

More electrons
make plate *B*
more negative

(a) Uncharged   (b) Charging

(c) Fully charged   (d) Retains charge

Figure 7.40

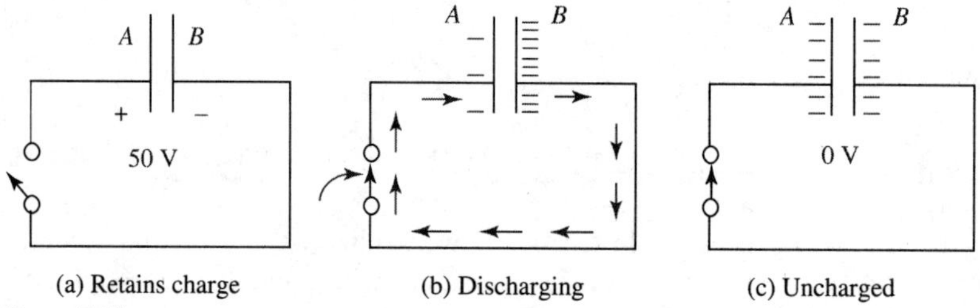

(a) Retains charge   (b) Discharging   (c) Uncharged

Figure 7.41

it had at the beginning of the time period. For example if the time constant of a capacitor is 5 seconds, then after 5 seconds the voltage is reduced by 63% to 37% of the original amount. After another 5 seconds the voltage is reduced again to 37% of the previous amount, which results in a current equal to $0.37 \times 0.37 \approx 14\%$ of the original voltage.

Similar mathematics works in reverse when the same capacitor is being charged. First we will look at discharging.

Example 7.31   *Application*

A fully charged capacitor with a voltage of 70 volts has a time constant of 6 seconds. Estimate the voltage remaining in the capacitor circuit after 18 seconds. Use both the recursive definition of the geometric sequence and the direct expression for the *n*th term.

Solution   The voltage follows the geometric sequence described recursively as

$$V_n = 0.37V_{n-1} \text{ with } V_0 = 70$$

V_n is the voltage after n periods of 6 seconds each or $6n$ seconds. Table 7.9 shows the computations for the first three seconds.

Table 7.9

Number of Time Constants (n)	Elapsed Time (s)	V_n
0	0	70 V
1	6	$70 \times 0.37 \approx 26$ V
2	12	$(70 \times 0.37) \times 37 \approx 9.6$ V
3	18	$(70 \times 0.37 \times 0.37) \times 37 \approx 3.5$ V

From the table you can see a pattern, which is to multiply by 0.37 for every time constant. The general form for the nth term is $V_n = 70(0.37)^n$.

A graph of the equation $y = 70(0.37)^x$ in Figure 7.42 shows that after three time constants the voltage has dropped to about 3.55 V.

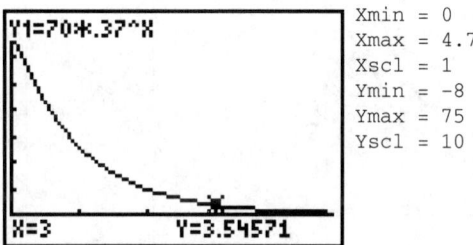

```
Y1=70*.37^X                     Xmin = 0
                                Xmax = 4.7
                                Xscl = 1
                                Ymin = -8
                                Ymax = 75
                                Yscl = 10



X=3              Y=3.54571
```

Figure 7.42

Example 7.32   Application

If the voltage in the capacitor circuit in Example 7.31 were measured with a voltmeter that could measure to the nearest 0.1 volt, about how long would it be before the voltage reading had dropped to zero volts?

Solution   A similar question came up in Example 7.30 where we determined which bounce produced a height less than 0.001 meter. Here we need to estimate the value of n (the number of time constants) before the voltage drops below 0.1 volts. To do so, find the value of n by trial and error that makes the voltage drop below 0.1 volts. That is, find an approximate solution for the equation

$$70(0.37)^n = 0.1$$

A good strategy is to form a table for the function

$$y = 70(0.37)^x$$

```
 X      Y1
 4     1.3119
 5     .48541
 6     .1796
 7     .06644
 8     .02459
 9     .0091
10     .00337

Y1=.066452313993
```

Figure 7.43

then look in the table for the number of time constants (x) when the voltage is about 0.1 V. Figure 7.43 shows that $V_6 \approx 0.18$ V and $V_7 \approx 0.07$ V. Therefore the voltage drops to zero as measured on our voltmeter somewhere between 6 time constants (36 seconds) and 7 time constants (42 seconds).

Charging a Capacitor

The question of charging is a little more complex, but 63% still plays a role. As the charge builds up during each time constant the charge in the capacitor increases by 63% of the amount remaining to charge to full capacity, which we'll call 100%. After one time constant the charge increases to 63% of capacity. There remains 37% to charge. In the next time constant 63% of that 37% or 23% will be charged. That means that the total charge after two time constants is 63% + 23% = 86%. This leaves 14% to charge. 63% of that 14% will be charged during the next time constant. This is demonstrated in Figure 7.44.

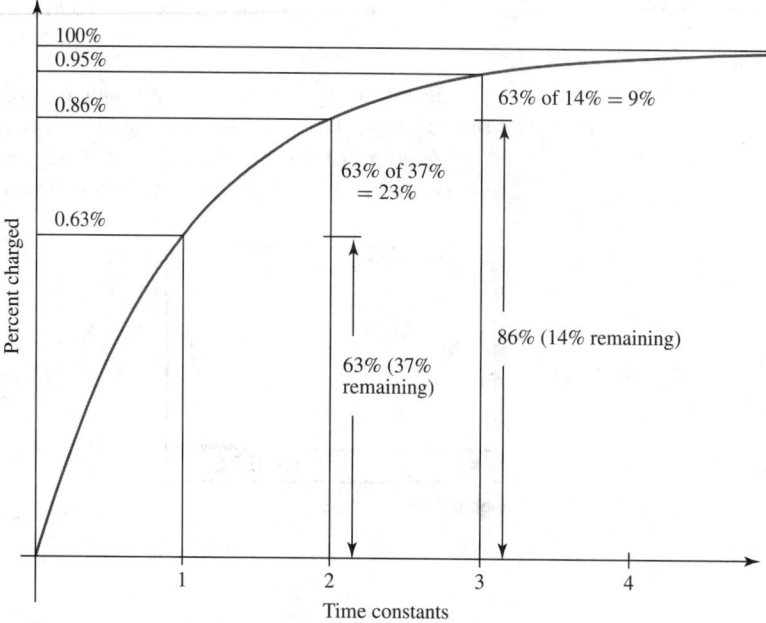

Figure 7.44

Example 7.33   *Application*

For the process of charging a capacitor, (a) write the first five terms of the sequence of percents of charge (the sequence begins {0, 63, ...}) and (b) write the recursive definition for the percent charged during the nth time constant.

Solutions   The description of the way a capacitor charges is: *the charge in the capacitor increases by 63% of the amount remaining to charge to full capacity, which we'll call 100%.*

(a) The first four terms, as given in Figure 7.44, are {0, 63, 86, 95}. The fifth term is determined through the following steps:

▶ The amount remaining to be charged is 100% − 95% = 5%.
▶ 63% of the amount remaining to be charged is 0.63 × 5% ≈ 3%.
▶ The new amount of the charge is 95% + 3% = 98%.
▶ The first five terms of the sequence of percents of full capacity are {0, 63, 86, 95, 98}

(b) To find the recursive definition we let a_{n-1} represent the previous amount of charge. That makes $100 - a_{n-1}$ the amount remaining to be charged. 63% of the amount remaining to be charged is $0.63(100 - a_{n-1})$. This is added to the previous charge. This recursive expression can be simplified if we first multiply by 0.63 and then factor out a_{n-1}.

$$a_n = a_{n-1} + 0.63(100 - a_{n-1})$$
$$= a_{n-1} + 63 - 0.63a_{n-1}$$
$$= 63 + (1 - 0.63)a_{n-1}$$
$$= 63 + 0.37a_{n-1}$$

Thus the recursive definition for the percent of a capacitor charged during the nth time constant is $a_n = 63 + 0.37a_{n-1}$. ∎

The recursive expression $a_n = 63 + 0.37a_{n-1}$ will lead to a direct expression for a_n when you study geometric series in Section 7.5. The recursive expression will be fine for the next example because you will only need to compute a few terms.

Example 7.34   Application

If a voltage of 36 V is applied to a capacitor with a time constant of 3 s

(a) Determine the voltage on the capacitor 12 seconds after the voltage source is switched on.
(b) Approximately how many seconds does it take for the voltage across the capacitor to reach 35.9 V?
(c) Approximately how long will it take for the voltage across the capacitor to reach 36 V?

Solutions   We will use the percents computed in Example 7.33.

(a) 12 s is equal to four time constants of 3 s each, and $a_4 \approx 98.1\%$. Therefore the voltage is $0.981 \times 36 \approx 35.32$ V.
(b) Here 35.9 V is $\frac{35.9}{36} \approx 0.997 = 99.7\%$ of a full charge. We need to continue the sequence of percents until a term exceeds 99.7. Set your calculator to Seq mode and enter the sequence $63 + 0.37a_{n-1}$. The table of values for that sequence is shown in Figure 7.45. It will take almost 6 time constants or 18 seconds for the voltage across the capacitor to reach 35.9 V.

n	$u(n)$
0.000	0.000
1.000	63.000
2.000	86.310
3.000	94.935
4.000	98.126
5.000	99.307
6.000	99.743

$u(n)=99.74342736$

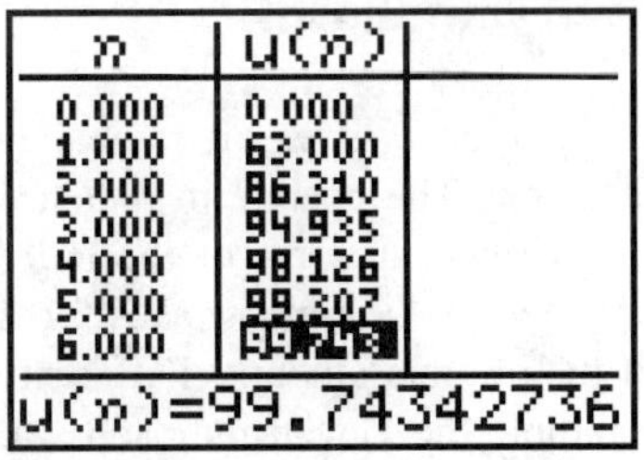

Figure 7.45

(c) Figure 7.46(a) shows the percent charged after the next six time constants. The mathematical model predicts that there will always be a positive amount of charge remaining, and the charge will never reach 100%.

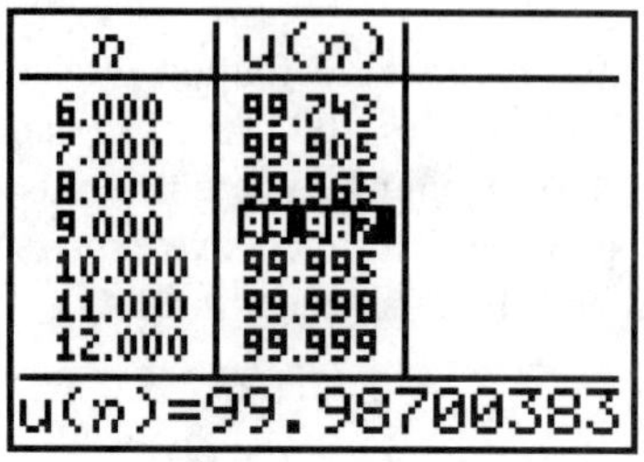

n	$u(n)$
6.000	99.743
7.000	99.905
8.000	99.965
9.000	99.987
10.000	99.995
11.000	99.998
12.000	99.999

$u(n)=99.98700383$

(a) Percents

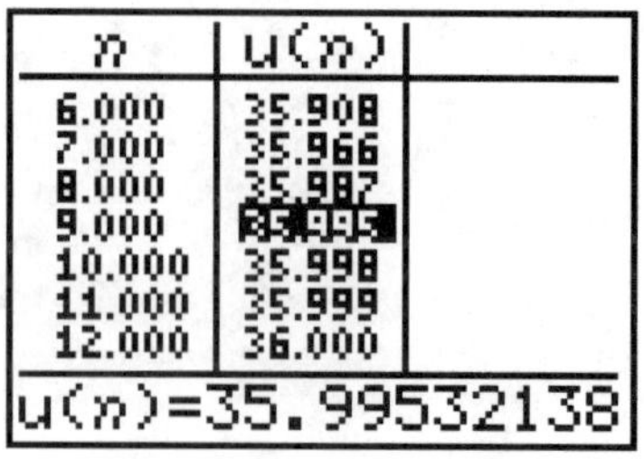

n	$u(n)$
6.000	35.908
7.000	35.966
8.000	35.987
9.000	35.995
10.000	35.998
11.000	35.999
12.000	36.000

$u(n)=35.99532138$

(b) Based on maximum of 36 V

Figure 7.46

Practically speaking, the charge reaches 100%, or 36 V in this example, when the measured voltage is 36 V. That is, when the voltmeter gives a reading of 36 V then the charge has reached its maximum. Different voltmeters will give different answers to the question. Figure 7.46(b) shows that a voltmeter accurate to the nearest hundredth of a volt will show 36 V after the 8th time constant. ∎

Look Ahead: Continuous vs. Discrete Variation

Money earning interest in the bank and a bouncing ball are examples of a **discrete sequence**. Interest is computed at regular intervals, and no interest is earned between computations. There is a first bounce and a second bounce, and no bounce in between. In discrete variation the values of the variable are integers, for example 1, 2, 3, ... As we have seen in Figure 7.39, a geometric sequence models discrete variation very well.

On the other hand the decrease in voltage in a capacitor circuit is not discrete. This variation is **continuous** because voltage drops steadily. That is, if the time constant is 6 seconds the decrease in voltage doesn't wait until 6 seconds but begins immediately. The voltage reaches 37% of the initial voltage after 6 seconds, but it has been decreasing continuously from the beginning of the time interval. We will see in a later chapter that exponential functions make the best models for continuous variation but that the geometric sequence does a fairly good job.

Radioactivity and Half-Life

A radioactive substance gives off radioactive emissions in the form of electrons or x-rays or neutrons. The change in radioactivity follows a geometric sequence that is generally described in terms of the **half-life** of the substance. The half-life of a radioactive element is the time it takes for the amount of radioactivity in the sample to reach half the original amount. Different radioactive elements have different half-lives. For example the eleven radioactive isotopes* of the element plutonium have

---

***Isotopes** of an element differ from each other in the number of neutrons in the nucleus. One isotope can change into another by losing a neutron by radioactive emission. If an isotope loses electrons and protons through radioactive emission it will change into another element.

half-lives that range from 26 minutes to 500,000 years. Half-life is like a capacitor's time constant, but the percent lost during a half-life is 50%, and the percent of voltage decrease during a time constant is 63%.

Although radioactivity, like the discharging of a capacitor, is a continuous process, a geometric sequence can give a good model of the process and it allows for quick computations. If A_0 is the amount of the radioactive substance when it is formed, or at the time when we start the clock, the modeled geometric sequence is described by the recursive equation $A_n = \frac{1}{2}A_{n-1}$ where A_0 is the original amount. This produces the sequence $\{A_0, \frac{1}{2}A_0, \frac{1}{4}A_0, \frac{1}{8}A_0, \ldots\}$. The direct expression $A_n = \left(\frac{1}{2}\right)^n A_0$ describes the same sequence. Remember, each term in the sequence represents another half life. That is, after one half-life the amount remaining is $\frac{1}{2}A_0$ and after three half-lives the amount remaining is $\frac{1}{8}A_0$.

Example 7.35 Application

If 3.5 grams of plutonium 235 with a half-life of 26 minutes is produced by atomic fission, (a) write the geometric sequence that models the situation and (b) estimate how much of it remains after 24 hours?

Solutions (a) This sequence can be described by the equation

$$A_n = \left(\frac{1}{2}\right)^n A_0$$

where n is the number of half-lives and A_0 is the amount of the radioactive substance when it is formed, or at the time when we start the clock. So $A_n = (\frac{1}{2})^n \times 3.5$.

(b) The half-life of plutonium 235 is in minutes, so we have to convert 24 hours to minutes. Thus 24 hours $= 24 \times 60 = 1440$ minutes which equals $\frac{1440}{26} \approx 55$ half-lives. Therefore the amount that remains after 24 hours is $(\frac{1}{2})^{55} \times 3.5 \approx 10^{-16}$ g. This amount is zero in the real world of actual weights, since it is less than the weight of one atom of plutonium 235. ∎

Example 7.36 Application

If 0.08 grams (80 milligrams) of plutonium 242 with a half-life of 500,000 years is produced by atomic fission, (a) write the geometric sequence that models the situation and (b) approximate how much of it remains after 2 million years?

Solutions (a) This sequence can be described by the equation

$$A_n = \left(\frac{1}{2}\right)^n A_0$$

where n is the number of half-lives and A_0 is the amount of the radioactive substance when it is formed, or at the time when we start the clock. So $A_n = (\frac{1}{2})^n \times 0.08$.

(b) Here the half-life is in years so 2 million years $= \frac{2,000,000}{500,000} = 4$ half-lives. Therefore the amount that remains is $\left(\frac{1}{2}\right)^4 \times 0.08 \approx 0.005$ g, or 5 milligrams. ∎

Using Half-life to Find the Age of Ancient Objects

Most elements contain a mixture of isotopes; that is, different atoms have different numbers of neutrons. In the carbon dioxide in the air there is a mixture of the isotopes of carbon. Although 99% of the carbon is carbon 12 (with 12 neutrons in the nucleus) there is always a tiny but fixed percent (less than 0.001%) of its atoms that are the radioactive isotope called carbon 14, which has 14 neutrons in its nucleus. During a plant's life carbon 14 decays to carbon 12 by giving off neutrons, but the plant continues to bring in new carbon atoms from the air through photosynthesis (see Figure 7.47). Thus the ratio of carbon 14 to carbon 12 atoms in living plants remains constant. When a plant dies photosynthesis stops, so while the carbon 14 continues to decay no new carbon 14 atoms are brought into the plant. Therefore by looking at a sample of the carbon atoms from a dead tree, scientists can tell how many years have passed since the tree was cut down.

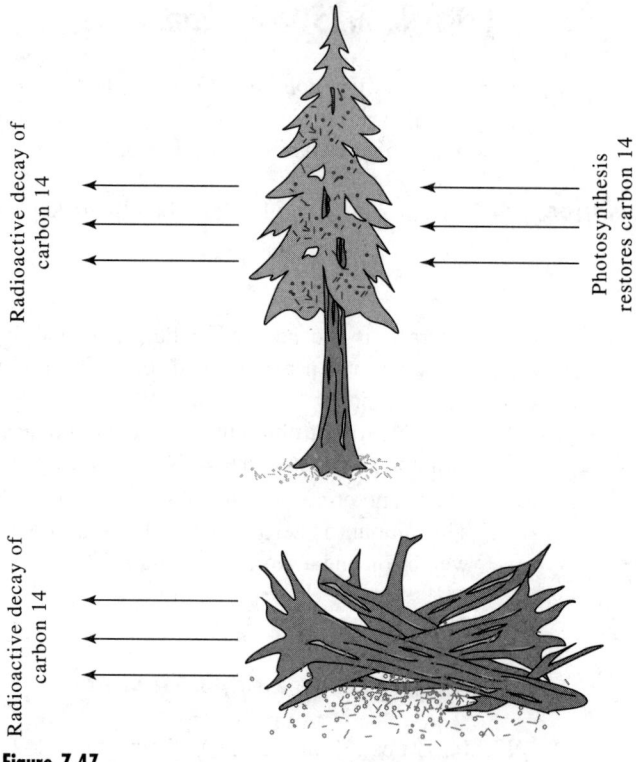

Figure 7.47

The half-life of carbon 14 is 5700 years. Suppose a piece of wood is found from an ancient building or fire pit. If a Geiger counter reveals that the amount of carbon 14 in the wood is half the amount you would expect in a living tree, we can estimate that the tree was cut about 5700 years before.

Example 7.37   Application

If the amount of carbon 14 in a piece of charcoal found in a Native American fire pit is measured to be approximately $\frac{1}{4}$ of the amount that would have been there in the living tree, what is the approximate age of the settlement?

Solution Since $\frac{1}{4} = \frac{1}{2} \times \frac{1}{2}$, two half-lives have passed since the tree was cut down. We can estimate that the settlement is $2 \times 5700 = 11,400$ years old. ∎

Example 7.38 Application

A sample of wood was taken from the beam of a building in the ruins of a Mayan city in Mexico. If the amount of carbon 14 in the sample was 83% of what it would be in a living tree, what is the estimate of the time that has passed since the tree was cut down?

Solution The decay of carbon 14 can be modeled with the same equation you saw for plutonium in Example 7.35. That is,

$$A_n = \left(\frac{1}{2}\right)^n A_0$$

We are given that $\dfrac{A_n}{A_0} = 0.83$, or $A_n = 0.83 \times A_0$. Therefore we must solve the equation

$$0.83 A_0 = \left(\frac{1}{2}\right)^n A_0$$

which is the same as $0.83 = \left(\frac{1}{2}\right)^n$. Again, we will see how to solve an equation like this in Chapter 9. For now we will try an approach that uses the Table feature of the calculator. Define the function $y = \left(\frac{1}{2}\right)^x$ and explore the table until 0.83 appears in the Y1 column.

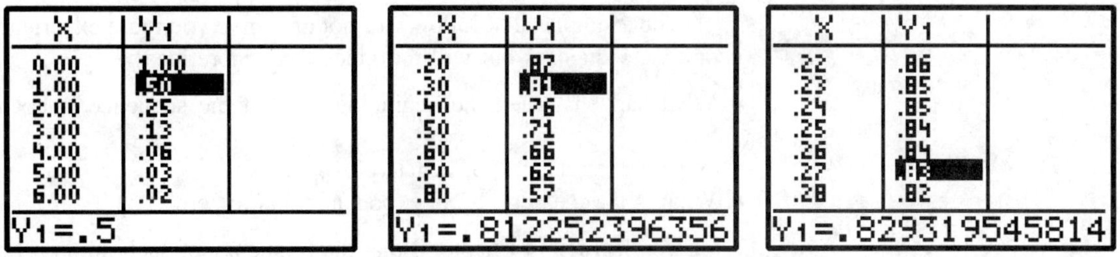

Figure 7.48

Figure 7.48 shows the three tables we used, beginning with TblStart=0 and Tbl=1. To produce the second table we could see that our value of x was somewhere between 0 and 1, so we chose TblStart=0 and Tbl=0.1. The third table, which used TblStart=0.20 and Tbl=0.01, gives an estimate of x which is 0.27. This is the number of half-lives since the tree was cut down. Therefore the estimated age of the building is $0.27 \times 5700 \approx 1539$ years. ∎

Exponents and Notes in a Musical Scale

In Chapter 3 we studied the sine waves produced by musical notes of different frequency. Table 7.10 shows the frequencies of the twelve notes in one octave, plus the first note in the next octave. Although music traditions differ around the world, most music composed in Europe and America in the past five hundred years has been based on this twelve-note scale. However, in many parts of Asia, including

China, Indonesia, and Vietnam, traditional music is based on a five- or seven-note scale.

Table 7.10   Frequencies of the notes in one octave

Note	Frequency (hertz)
C (middle C)	262
C-sharp	277
D	294
D-sharp	311
E	330
F	349
F-sharp	370
G	392
G-sharp	415
A	440
A-sharp	466
B	494
C	523

The examples that follow will not only give you more experience working with geometric sequences but will introduce three more ideas:

▶ What happens when the beginning term of the sequence is not in fact the first term?
▶ What is the meaning of negative exponents?
▶ What is the meaning of an exponent equal to zero?

You learned in Chapter 3 that the frequency of each note is computed by multiplying the frequency of the previous note by the number $\sqrt[12]{2} \approx 1.0595$. Therefore the notes in this musical scale form a geometric sequence that is described recursively by the expression

$$a_n = \sqrt[12]{2}a_{n-1}$$

What value should be assigned to a_0? Since the notes do not start at middle C there is no particular reason for the sequence to start there. It turns out that for musical reasons the notes are all computed based on note A above middle C, which has a frequency of 440 Hz. Therefore we will assign $a_0 = 440$.

Example 7.39   Application

What is the frequency of the 12th note above the note designated by a_0?

Solution   Since $a_0 = 440$ Hz then a_0 is note A. The first note above A is a_1 and the second note above A is a_2. Therefore the 12th note is a_{12}. To compute a_1 we multiply a_0

once by $\sqrt[12]{2}$. To compute a_2 we multiply a_0 twice by $\sqrt[12]{2}$. To get a_n we multiply a_0 by $\sqrt[12]{2}$ a total of n times. Therefore to compute a_{12} we multiply a_0 by $\sqrt[12]{2}$ a total of 12 times or $a_{12} = \left(\sqrt[12]{2}\right)^{12} 440$. Table 7.11 summarizes these statements.

What is the value of $\left(\sqrt[12]{2}\right)^{12} 440$? This calculation can be done without a calculator because of the meaning of $\sqrt[12]{2}$, which is "the number of 12 identical numbers you multiply to get 2." Therefore $(\sqrt[12]{2})^{12} = 2$, so $a_{12} = 440 \times 2 = 880$ Hz.

Table 7.11   Computing notes of higher frequency

Note	n	a_n
A	0	440
A-sharp	1	$\sqrt[12]{2} \times 440$
B	2	$\sqrt[12]{2} \times \sqrt[12]{2} \times 440 = \left(\sqrt[12]{2}\right)^2 440$
C	3	$\sqrt[12]{2} \times \sqrt[12]{2} \times \sqrt[12]{2} \times 440 = \left(\sqrt[12]{2}\right)^3 440$
$\vdots$	$\vdots$	$\vdots$
A	12	$(\sqrt[12]{2})^{12} 440$

Example 7.40

Write $\sqrt[12]{2}$ in the exponent form 2^x and generalize the result to other roots.

Solution   The problem asks, if $\sqrt[12]{2} = 2^x$ what is the value of x?

Begin with the idea that was expressed in Example 7.39, that the meaning of $\sqrt[12]{2}$ is "the 12 identical numbers you multiply to get 2," and the accompanying equation,

$$(\sqrt[12]{2})^{12} = 2$$

If $\sqrt[12]{2} = 2^x$, then $(2^x)^{12} = 2$. But 2 is the same as 2^1 and raising a power to a power allows us to multiply the exponents, so we have the equation

$$2^{12x} = 2^1$$

Therefore $12x = 1$, or $x = \frac{1}{12}$. That means that $\sqrt[12]{2}$ in exponential form is $2^{1/12}$. In general,

$$a^{1/n} = \sqrt[n]{a}$$

The result in Example 7.39 would be the same for any chosen note. That is, the 12th note above any particular note x will have doubled the frequency of note x. The last note in the table (523 Hz) is not quite double the first note (262 Hz) because of round-off error. In fact the frequencies of the two notes are about 261.625565301 and 523.251130601, which have the ratio 1 : 2 up to the last decimal place (again, there's round-off error).

Example 7.41   Application

Compute the frequency of the 8th note below a_0. What symbolism should be used for this note?

Solution   Since we multiply by $\sqrt[12]{2}$ to get the next higher note we would divide by $\sqrt[12]{2}$ to get the next lower note. How many $\sqrt[12]{2}$'s should we divide by to get the 8th note below A? We need to follow the same logic as in Example 7.39. Read Table 7.12 from the bottom up. You can see that it would be natural to designate the eighth note below a_0 as a_{-8}.

Table 7.12 Computing notes of lower frequency (read from the bottom)

Note	n	a_n
C-sharp	-8	$\dfrac{440}{\left(\sqrt[12]{2}\right)^8}$
$\vdots$	$\vdots$	$\vdots$
F-sharp	-3	$\dfrac{440}{\left(\sqrt[12]{2}\right)^2} \div \sqrt[12]{2} = \dfrac{440}{\left(\sqrt[12]{2}\right)^3}$
G	-2	$\dfrac{440}{\sqrt[12]{2}} \div \sqrt[12]{2} = \dfrac{440}{\left(\sqrt[12]{2}\right)^2}$
G-sharp	-1	$440 \div \sqrt[12]{2} = \dfrac{440}{\sqrt[12]{2}}$
A	0	440

The value of $a_{-8} = \dfrac{440}{\left(\sqrt[12]{2}\right)^8} \approx 277$. So, the frequency of the eighth note below a_0 is 277 Hz. ∎

Finally we will look for the general formula for any note whether it is above or below the starting point, a_0.

You can see from Example 7.39 that the 12th note above a_0 is calculated by the formula

$$a_{12} = a_0 \times \left(\sqrt[12]{2}\right)^{12}$$

and in general,

$$a_n = a_0 \times \left(\sqrt[12]{2}\right)^{n}$$

If we apply this formula to the 8th note below a_0 we get

$$a_{-8} = a_0 \times \left(\sqrt[12]{2}\right)^{-8}$$

But what is the meaning of the negative exponent in $\left(\sqrt[12]{2}\right)^{-8}$?
There are two expressions for a_{-8}.

1. a_{-8} was calculated in Table 7.12 to be

$$a_{-8} = a_0 \div (\sqrt[12]{2})^8 = \frac{a_0}{(\sqrt[12]{2})^8}$$

2. Earlier we found that the result was

$$a_{-8} = a_0 \times (\sqrt[12]{2})^{-8}$$

We want the two expressions for a_{-8} to be equal. Therefore we have the following explanation for negative exponents:

$$a_0 \times (\sqrt[12]{2})^{-8} = \frac{a_0}{(\sqrt[12]{2})^8}$$

The negative exponent, which seems to be a strange idea, means to multiply not by the number but by the reciprocal of the number. For example,

$$5^{-1} = \frac{1}{5}$$

$$5^{-2} = \frac{1}{5^2} = \frac{1}{5} \times \frac{1}{5}$$

$$5^{-3} = \frac{1}{5^3} = \frac{1}{5} \times \frac{1}{5} \times \frac{1}{5}$$

$$\vdots$$

What is the meaning of an exponent equal to 0? We can use the general formula for a_n to come to an understanding of the meaning of a zero exponent. We know that $a_n = a_0 \times (\sqrt[12]{2})^n$. Substituting 0 for n we have

$$a_0 = a_0 \times (\sqrt[12]{2})^0$$

We can conclude from this that $(\sqrt[12]{2})^0$ must equal 1, which is indeed true for all numbers except 0. That is $x^0 = 1$, except when $x = 0$, because 0^0 is not defined.

These conclusions about exponents are summarized below.

Exponents that Are Fractions, Negative, or Zero

If n and m are positive whole numbers, then

▶ Positive exponents:

$$x^n = x \times x \times x \times x \times \cdots \times x \ (n \text{ factors})$$

▶ Fractional exponents: For $x \geq 0$

$$x^{1/n} = \sqrt[n]{x}$$

and, since $(x^m)^{1/n} = x^{m/n}$ and $(x^{1/n})^m = x^{m/n}$, then

$$x^{m/n} = \sqrt[n]{x^m} \ \text{ if } x^m \geq 0 \text{ and } x^{m/n} = \left(\sqrt[n]{x}\right)^m \text{ if } x \geq 0$$

▶ Negative exponents:

$$x^{-n} = \frac{1}{x} \times \frac{1}{x} \times \frac{1}{x} \times \cdots \times \frac{1}{x} \ (n \text{ factors})$$

or $$x^{-n} = \frac{1}{x^n}$$

unless $x = 0$, since $\frac{1}{0}$ is undefined.

▶ Zero exponents:

$$x^0 = 1$$

unless $x = 0$, because 0^0 is undefined.

Section 7.4 Exercises

1. *Finance* $1000 is deposited in an account that earns 6% per year compounded annually.
   (a) Write the recursive definition of the sequence.
   (b) Write the formula for the amount in the account after n years.

2. *Finance* The amount of $2,500 is placed in a savings account at 6% interest. If interest is compounded (a) annually, (b) semiannually, (c) quarterly, and (d) monthly, what is the total after 10 years?

3. *Finance* One bank offers 7% interest compounded semiannually. A second bank offers the same interest but compounded monthly. How much more income will result by depositing $1,000 in the second account for five years than by depositing $1,000 in the first account for five years?

4. *Finance* One bank offers 9.25% interest compounded annually. A second bank offers 9.00% compounded quarterly.
   (a) Write the recursive definition of each sequence.
   (b) Write the formula for the amount in each account after n years.
   (c) Determine the value of $1,000 if it is deposited in each account and allowed to collect interest for 10 years.
   (d) Which account will be worth more at the end of the 10-year period?

5. *Finance* Compare the answers to Calculator Lab 7.4 with monthly compounding. That is, how much does monthly compounding decrease the doubling times?

6. *Finance* Think about doubling time and total interest earned, and explain how important monthly compounding is compared to annual compounding.

7. *Physics* A ball is dropped to a level piece of ground from a height of 60 m. Each time it bounces it goes $\frac{7}{8}$ as high as the previous bounce.
   (a) Write a recursive sequence that describes the height of the nth bounce.
   (b) How high does the ball bounce on the sixth bounce?
   (c) How many bounces does it take before the ball bounces less than 1 cm?

(d) If we consider that the ball has come to rest when it bounces less than 0.1 mm, how long does it take for the ball to come to rest?

(e) What is the total vertical distance the ball travels during all its bounces? (The answer is not 60 m but the total of all the distances the ball fell and the heights it rebounded.)

8. *Physics* Suppose a ball is dropped from a height of 13 meters and the ratio of consecutive heights reached by the ball is approximately 0.75.
   (a) Write the mathematical model for this bouncing ball as a recursive expression and as an expression for the nth term.
   (b) What percent is each height of the bounce preceding it?
   (c) What percent of the height is lost with each bounce?
   (d) What is the height reached by (i) the first bounce, (ii) the 10th bounce, and (iii) the 100th bounce?
   (e) Which bounce first produces a height less than 0.001 m (one millimeter)?

9. *Machine technology* When a motor is turned off, a flywheel attached to the motor coasts to a stop. In the first second the flywheel makes 250 revolutions. Each of the following seconds it makes $\frac{13}{20} = 0.65$ of the number of revolutions it made the previous second.
   (a) Write a recursive sequence for the number of revolutions the flywheel makes in the nth second.
   (b) How many seconds will it take before the flywheel makes less than 1 revolution?
   (c) How many seconds will it take before the flywheel makes less than 0.1 revolution?
   (d) How long will it take before the flywheel comes to rest? What do you consider coming to rest? According to your formula for this geometric sequence the flywheel will never stop turning. Explain why you think your sequence is not a good description for the number of revolutions when the flywheel is turning slowly.

10. *Electronics* A fully charged capacitor with a voltage of 120 volts has a time constant of 8 seconds. Estimate the voltage remaining in the capacitor circuit after (a) 24 and (b) 48 seconds. Use both the recursive definition of the geometric sequence and the direct expression for the nth term.

11. *Electronics* If the voltage on the capacitor in Exercise 10 were measured with a voltmeter that could measure to the nearest 0.1 volt, about how long would it be before the voltage reading had dropped to zero volts?

12. *Electronics* A fully charged capacitor with a voltage of 220 volts has a time constant of 10 seconds. Estimate the voltage remaining in the capacitor circuit after **(a)** 30 and **(b)** 50 seconds. Use both the recursive definition of the geometric sequence and the direct expression for the nth term.

13. *Electronics* If the voltage in the capacitor circuit in Exercise 12 were measured with a voltmeter that could measure to the nearest 0.1 volt, about how long would it be before the voltage reading had dropped to zero volts?

14. *Nuclear technology* Tritium has a half-life of 12.5 years. **(a)** Write the geometric sequence that models the situation and **(b)** estimate how much of 100 grams of tritium will remain after 50 years?

15. *Nuclear technology* The half-life of polonium 210 is 140 days. **(a)** Write the geometric sequence that models the situation, and **(b)** estimate how much of 10 grams of tritium will remain after 1 year?

16. *Archaeology* A fossil originally contained 100 g of carbon 14 and now has 10 g. What is the approximate age of the fossil?

17. *Archaeology* Paint from the Lascaux caves of France contains 15% of the normal amount of carbon 14. Estimate the age of the caves' paintings.

18. *Medical technology* An epidemic is spreading through the student body of a certain school with an enrollment of 9,000. Past experience with similar epidemics shows that the number of students that have been infected by the disease can be approximated by the sequence

$$s_n = \frac{427500}{45 + 9455 \times 2^{-n}}$$

where s_n is the number of students infected at the end of week n and s_0 is the number of students initially infected.

  (a) How many students were initially infected?

  (b) How many students were infected at the end of the 5th week?

  (c) How long will it take until 95% of the students are infected?

  (d) How long will it take to infect all students?

19. *Music* If the frequency of $a_0 = 440$ Hz, what is the frequency of the 10th note above the note designated by a_0?

20. *Music* If the frequency of $a_0 = 440$ Hz, what is the frequency of the 6th note below the note designated by a_0?

21. Use a CBL set up like the one shown in Figure 7.49. (You used this earlier in Section 3.5, Exercise 11, page 186.) Attach a spring, such as a Slinky or a bungee cord, directly over the motion detector. To make it easier for the motion detector to "see" the spring, attach a square piece of cardboard about 2″–3″ on a side to the bottom of the spring. The object should be light enough so that it does not noticeably stretch the spring. As an alternative, attach a smaller mass but tape a piece of cardboard to the bottom of the mass. The cardboard will serve as a target for the motion detector. Pull the spring straight down, turn on the CBL, and start the program SLINKY on the calculator. Release the spring and allow it to oscillate. As soon as the motion of the spring is only up and down (with no side motion) press TRIGGER on the CBL.

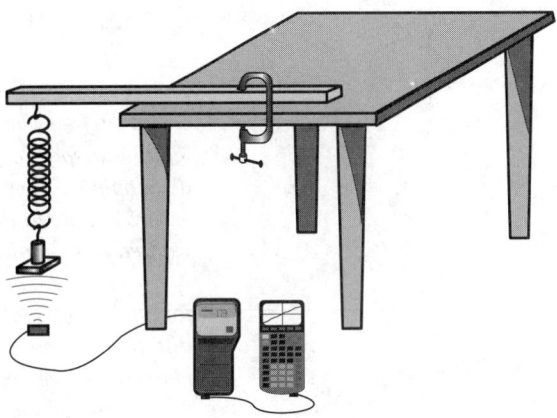

Figure 7.49

  (a) Use the list of the heights to create a sequence formed by the high and low position of each oscillation.

  (b) Examine this sequence. Does it appear to be an arithmetic sequence, a geometric sequence, or neither? Write an explanation that supports your answer.

  (c) What is the limit of this sequence?

  (d) Approximate the sequence with either a direct or recursive formula.

22. Use the sequence from Exercise 21(a) and form a new sequence by subtracting the limit of the sequence from each of the heights.

  (a) What is the limit of this sequence?

  (b) Approximate the sequence with either a direct or recursive formula.

  (c) How many bounces does it take before one bounce of the Slinky is less than 1 mm?

7.5   Arithmetic and Geometric Series

A mathematical **series** is the sum of the terms of a sequence. If the sequence is

$$\{1, 2, 3, 4, 5, \ldots, n\}$$

then the series that is produced from that sequence is

$$S_n = 1 + 2 + 3 + 4 + 5 + \cdots + n$$

A **geometric series** is a series that is produced from a geometric sequence, and an **arithmetic series** is a series that is produced from an arithmetic sequence.

In this section you'll see how to write formulas for arithmetic and geometric series, and we'll explore applications of these series.*

We'll introduce this topic with a modern version of a legend that has been told in many countries for many centuries.

A Legendary Bonus

The year is 2010 A.D. In the world's largest computer company an old and experienced engineer has worked on many interesting projects during his long career. After leading a special project team for several years the group was finally successful in developing a new kind of computer that could be miniaturized down to the size of several hundred atoms. The company had long hoped for such a breakthrough and wanted to reward each member of the team.

The old engineer made sure the members of his team were given good bonuses for their excellent work and devoted service. But for himself he wanted something different. He told the vice president, "I don't want much for myself, but I really like chess, and I've been thinking about having a bonus that relates to the chessboard. As you know the chessboard has eight rows of eight squares, with a total of 64 squares in all. I'd like one penny on the first square, two pennies on the second square, four pennies on the third square, eight pennies on the fourth square, and so on, doubling the amount each time until you reach the 64th square."

The boss replied, "Your work has been incredibly valuable to us; it will eventually bring the company hundreds of millions or even a billion dollars in revenue. This talk about pennies can't be serious."

"But I said I don't want much for myself. Will you agree to this bonus?"

"Yes, if that's what you want."

"Thanks. And would you mind signing this agreement?" He unfolded the agreement, which his lawyer had prepared, and the boss signed it.

"Let's get started," said the boss. "I've got some pennies in a jar back in my office."

We interrupt this story to investigate the beginning of the payment.

---

*The plural of the word series is series. One series; two series.

Example 7.42

(a) Write the first eight terms of the geometric sequence that describes how the engineer's bonus would be paid.
(b) Write the recursive description of this sequence.
(c) Write the expression for the nth term in the sequence.
(d) What is the total amount (the sum) of the payments on the first row (8 squares) of the chessboard?

Solution (a) $\{1, 2, 4, 8, 16, 32, 64, 128\}$

(b) $a_n = 2a_{n-1}$ and $a_1 = 1$

(c) We'll build up the sequence with the recursive definition and you'll see the pattern of the exponents:

$$a_2 = 2a_1 = 2 = 2^1$$
$$a_3 = 2a_2 = 2 \times 2 = 2^2$$
$$a_4 = 2a_3 = 2 \times 2 \times 2 = 2^3$$
$$a_5 = 2a_4 = 2 \times 2 \times 2 \times 2 = 2^4$$

You can see that on the fifth square we multiply four 2's together. That pattern will continue so

$$a_8 = 2^7 \text{ and in general, } a_n = 2^{n-1}$$

(d) The sum of the pennies on the first eight squares is $1 + 2 + 4 + 8 + 16 + 32 + 64 + 128 = 255$ pennies, or \$2.55. ∎

There were still plenty of pennies in the jar, and the boss began to count out what he needed for the ninth square, when he stopped. "I've paid you \$2.55 for the first row, but this counting is getting tiresome. Let's do some math to figure out the sum for each row, so I don't have to count all these pennies out."

Example 7.43

Figure out a quick way to add up the pennies needed for the first row and for the second row.

Solution Here's a way to add up $1 + 2 + 4 + 8 + 16 + 32 + 64 + 128$ without doing any addition. Let $S =$ the sum of the first eight terms and then write $2S$ above it. Each term of $2S$ is twice as much as a term of S:

$$2S = 2 + 4 + 8 + 16 + 32 + 64 + 128 + 256$$
$$S = 1 + 2 + 4 + 8 + 16 + 32 + 64 + 128$$

First, notice that each equation contains seven terms in common. The identical parts are

$$2 + 4 + 8 + 16 + 32 + 64 + 128$$

Now we will apply a simple method. We will subtract $2S - S$, which of course equals S. It will help you see what's happening if we first rewrite the two equations, with the identical parts in parentheses lined up over each other.

$$2S = \quad (2 + 4 + 8 + 16 + 32 + 64 + 128) + 256$$
$$S = 1 + (2 + 4 + 8 + 16 + 32 + 64 + 128)$$

When we subtract, the identical parts subtract to zero, leaving only the last term of $2S$ and the first term S:

$$2S - S = 256 - 1$$
$$S = 255$$

This method shows that if you multiply by the common ratio of the geometric sequence (in this case 2) then you will produce many of the same terms in the second equation. When you subtract the equations you end up with only the first term of one series and the last term of the second series. The rest of the terms subtract themselves out.

Let's apply the method to the second row. We must know the 9th term, which is 2^8, and the 16th term, which is 2^{15}.

$$2S = \qquad (2^9 + 2^{10} + \ldots + 2^{15}) + 2^{16}$$
$$S = 2^8 + (2^9 + \ldots + 2^{14} + 2^{15})$$

Subtracting, we have

$$2S - S = 2^{16} - 2^8$$
$$S = 65536 - 256 = 65280$$

The manager pulled out his calculator and computed the sum for the second row. He didn't have enough pennies, but he knew that compared to the bonus that he himself was going to get this was very little, so he wrote a personal check for $652.80. The engineer initialed on the contract that the first two rows had been paid.

The boss was getting impatient. "Let's get this over with. Let's figure out the total that you need for the rest of the rows. I've got a golf date this afternoon."

Activity 7.11
Computing the Bonus

Answer the following questions about this bonus payment plan:

1. What is the sum required for the third row, that is, from a_{17} to a_{24}?
2. Compare the amount placed on the 25th square with the sum of all the terms from a_1 to a_{24}.
3. Complete this sentence: The amount that is placed on a square is _____ compared with the total amount that has been placed up to, but not including, that square.
4. What is the sum required for the fourth row?
5. What is the sum required for *all* the rows, that is, from a_1 to a_{64}?
6. The boss thought that the invention of the new computer could bring in a billion dollars to the company. On what square would that total amount have been reached? Remember that a billion dollars is equal to 100 billion pennies.
7. What would be the approximate amount placed on the *next* square after a billion dollars had been reached on all the previous squares?
8. IBM, the world's largest computer company today, has a net worth of approximately 60 billion dollars. On which square would that amount be placed?
9. Now that you know something about the numbers involved, write an ending to the story.

In solving the problem of the legendary bonus, you learned how to find the sum of the terms of a geometric sequence. This result, and the way to reach it, is summarized below.

Geometric Series

A geometric series is the sum of the terms of a geometric sequence. If the sum of n terms is represented by S_n then

$$S_n = a + ar + ar^2 + ar^3 + \cdots + ar^{n-1}$$

where a is the first term and r is the ratio between consecutive terms.

To produce a formula for S_n we multiply each term of S_n by the ratio r.

$$rS_n = ar + ar^2 + ar^3 + \ldots + ar^{n-1} + ar^n$$

Again we have many identical terms in the two series. Before subtracting the two equations we'll group the identical parts.

$$rS_n = ar^n + (ar^{n-1} + ar^{n-2} + \cdots + ar^3 + ar^2 + ar)$$
$$S_n = \qquad (ar^{n-1} + ar^{n-2} + \cdots + ar^3 + ar^2 + ar) + a$$

Subtracting S_n from rS_n we have

$$rS_n - S_n = ar^n - a$$

Both sides can be factored and then we can solve for S_n.

$$S_n(r - 1) = a(r^n - 1)$$
$$S_n = \frac{a(r^n - 1)}{(r - 1)}$$

Sum of First n Terms of a Geometric Series

The sum of the first n terms of the geometric series

$$S_n = a + ar + ar^2 + ar^3 + \cdots + ar^{n-1}$$

is

$$S_n = \frac{a(r^n - 1)}{(r - 1)}$$

where a is the first term and r is the ratio between consecutive terms.

Example 7.44

What is the sum of the first 24 terms of the geometric sequence $\{3, 6, 12, 24, \ldots\}$?

Solution   We will use the formula $S_n = \dfrac{a(r^n - 1)}{(r - 1)}$ with $a = 3$, $r = 2$, and $n = 24$. $S_n = \dfrac{3(2^{24} - 1)}{1} = 50331645$.

The sum of the first 24 terms of the geometric sequence $\{3, 6, 12, 24, \ldots\}$ is $50,331,645$.

Example 7.45

What is the sum of the first 10 terms of the geometric sequence $\{9, 3, 1, \frac{1}{3}, \frac{1}{9}, \ldots\}$?

Solution   We will use the formula $S_n = \dfrac{a(r^n - 1)}{(r - 1)}$ with $a = 9$, $r = \frac{1}{3}$, and $n = 10$. Thus,

$$S_n = \frac{9\left(\left(\frac{1}{3}\right)^{10} - 1\right)}{\frac{1}{3} - 1} \approx \frac{-8.9998}{-\frac{2}{3}} \approx 13.4997$$

The sum of the first 10 terms of the geometric sequence $\{9, 3, 1, \frac{1}{3}, \frac{1}{9}, \ldots\}$ is about 13.4997.   ■

Example 7.46

What is the sum of the first 10 terms of the geometric sequence $\{100, -50, 25, -12.5, \ldots\}$?

Solution   Again we use the formula $S_n = \dfrac{a(r^n - 1)}{(r - 1)}$. Here $a = 100$, $r = -\frac{1}{2}$, and $n = 10$. So,

$$S_n = \frac{100\left(\left(\frac{-1}{2}\right)^{10} - 1\right)}{-\frac{1}{2} - 1} \approx \frac{-99.9}{-\frac{3}{2}} \approx 66.6$$

Thus, the sum of the first 10 terms of the geometric sequence $\{100, -50, 25, -12.5, \ldots\}$ is about 66.6.   ■

Example 7.47

What is the sum of the first 11 terms of the geometric sequence $\{100, -50, 25, -12.5, \ldots\}$?

Solution   Notice that this is the same sequence we summed in Example 7.46 with one additional term. Again we use the formula $S_n = \dfrac{a(r^n - 1)}{(r - 1)}$. Here $a = 100$, $r = -\frac{1}{2}$, and $n = 11$. So,

$$S_n = \frac{100\left(\left(\frac{-1}{2}\right)^{11} - 1\right)}{-\frac{1}{2} - 1} \approx \frac{-100.05}{-\frac{3}{2}} \approx 66.7$$

Thus the sum of the first 11 terms of the geometric sequence $\{100, -50, 25, -12.5, \ldots\}$ is about 66.7. This is what you would expect since the additional term in this example compared to Example 7.46 was $100\left(-\frac{1}{2}\right)^{10} \approx 0.098$.   ■

Limits of Geometric Series

Some of the answers to Examples 7.44–7.47 might give you the idea that some geometric series (sums of geometric sequences) might converge to a finite limit as you add more terms.

The idea of adding an infinite number of terms and still getting a finite number for the answer seems quite strange to most people when they first hear of it. To help you think of this finite sum of an infinite sequence let's look at cutting up a square birthday cake as shown in Figure 7.50.

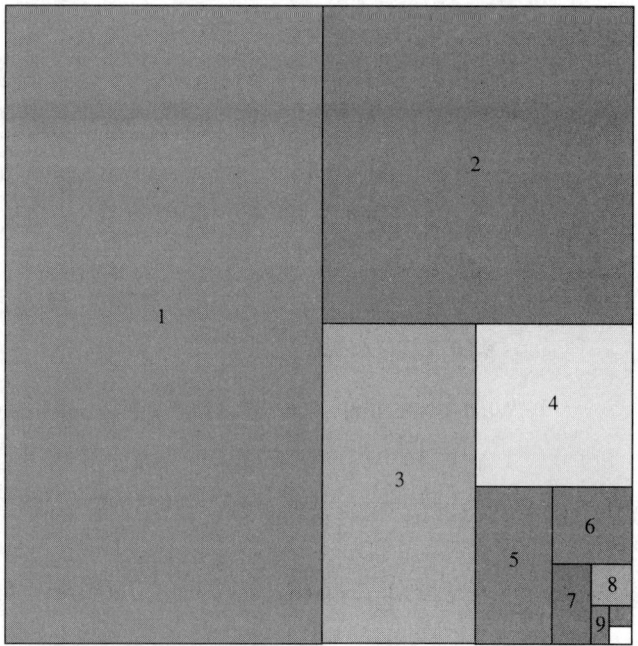

Figure 7.50

Cut the cake in half and cut one of those two pieces in half. From then on, each time a cut is made, one of the smallest two pieces is cut in half. Imagine these cuts going on forever. You can see that the pieces of cake are in a geometric sequence

$$\left\{ \frac{1}{2}, \frac{1}{4}, \frac{1}{8}, \frac{1}{16}, \cdots \right\}$$

Let S_n be the sum of the first n of these pieces. Then

$$S_n = \frac{1}{2} + \frac{1}{4} + \frac{1}{8} + \frac{1}{16} + \cdots + \frac{1}{2^n}$$

What is the sum of these pieces as $n \to \infty$? Well, they're all pieces of the same one cake, so the sum of all these pieces must approach 1 as $n \to \infty$. That is,

$$\lim_{n \to \infty} S_n = 1$$

We can prove that last statement mathematically. We know that the sum S_n is a geometric series with $a = \frac{1}{2}$ and $r = \frac{1}{2}$. Therefore

$$S_n = \frac{\frac{1}{2}\left(\left(\frac{1}{2}\right)^n - 1\right)}{\frac{1}{2} - 1}$$

To figure out the limit as $n \to \infty$, let's look at the parts of S_n separately. The only variable part is $\left(\frac{1}{2}\right)^n$. But you saw in Section 7.3 that

$$\lim_{n \to \infty} \left(\frac{1}{2}\right)^n = 0$$

Therefore

$$\lim_{n \to \infty} S_n = \frac{\frac{1}{2}(0-1)}{\frac{1}{2}-1} = \frac{-\frac{1}{2}}{-\frac{1}{2}} = 1$$

just as we figured out by thinking of all those pieces adding up to just one birthday cake.

Convergence and Divergence

For a sequence S_n, if $\lim\limits_{n \to \infty} S_n = L$ for some finite number L then we say that S_n **converges** to the limit L. If S_n does not converge we say it **diverges**.

Example 7.48

What is the limit as $n \to \infty$ of the geometric series from the sequence $\{3, 6, 12, 24, \ldots\}$ in Example 7.44?

Solution   We will use the formula $S_n = \dfrac{a(r^n - 1)}{(r - 1)}$ with $a = 3$ and $r = 2$.

$$S_n = \frac{3(2^n - 1)}{1}$$

Since $\lim\limits_{n \to \infty} 2^n = +\infty$ then the limit of $S_n = +\infty$. Thus this geometric series diverges.   ∎

Example 7.49

What is the limit as $n \to \infty$ of the geometric series from the sequence $\{9, 3, 1, \frac{1}{3}, \frac{1}{9}, \ldots\}$ in Example 7.45?

Solution   We will use the formula $S_n = \dfrac{a(r^n - 1)}{(r - 1)}$ with $a = 9$ and $r = \dfrac{1}{3}$. Thus

$$S_n = \frac{9\left(\left(\frac{1}{3}\right)^n - 1\right)}{\frac{1}{3} - 1}$$

Since $\lim\limits_{n \to \infty} \left(\frac{1}{3}\right)^n = 0$, then

$$\lim_{n \to \infty} S_n = \frac{9(0 - 1)}{\frac{1}{3} - 1} = \frac{-9}{-\frac{2}{3}} = \frac{27}{2} = 13.5$$

This geometric series converges to 13.5.   ∎

Example 7.50

What is the limit as $n \to \infty$ of the geometric series from the sequence $\{100, -50, 25, -12.5, \ldots\}$ in Examples 7.46–7.47?

Solution Again we use the formula $S_n = \dfrac{a\,(r^n - 1)}{(r - 1)}$. Here $a = 100$ and $r = -\frac{1}{2}$ so $S_n = \dfrac{100\left(\left(\frac{-1}{2}\right)^n - 1\right)}{-\frac{1}{2} - 1}$.

The sequence represented by $a_n = \left(\frac{-1}{2}\right)^n$ needs some discussion. The first few terms are $\left\{-\frac{1}{2}, \frac{1}{4}, -\frac{1}{8}, \frac{1}{16}, \ldots\right\}$. The terms alternate between positive and negative, getting smaller and smaller in absolute value, and the limit, like the sequence defined by $a_n = \left(\frac{1}{2}\right)^n$, is zero. Therefore

$$\lim_{n\to\infty} S_n = \frac{100\,(0 - 1)}{-\frac{1}{2} - 1} = \frac{-100}{-\frac{3}{2}} = \frac{200}{3} \approx 66.667$$

This geometric series converges to $\frac{200}{3} \approx 66.667$. ∎

Charging a Capacitor

In Section 7.4 you saw that all capacitors charge and discharge in a similar way. In a given period of time, called the time constant for the capacitor, 63% of the existing charge is lost or 63% of the remaining capacity is charged.

In Section 7.4 we were able to model the process of discharging a capacitor with a geometric sequence, but we delayed the analysis of charging until you learned about series. Well, here we are. We showed in Example 7.33 on page 472 that if a_n represents the percent of the maximum voltage that is on the capacitor after n time constants, then

$$a_n = 63 + 0.37 a_{n-1} \text{ and } a_0 = 0$$

Table 7.13 develops a pattern involving geometric series.

Table 7.13

n	a_n
0	0
1	$63 + 0.37 \times 0 = 63$
2	$63 + 0.37 \times 63 = 63\,(1 + 0.37)$ (Factor out 63)
3	$63 + 0.37 \times 63(1 + 0.37)$ Factoring out a 63 produces $63\,[1 + 0.37\,(1 + 0.37)] = 63\left(1 + 0.37 + 0.37^2\right)$
4	$63 + 0.37 \times 63(1 + 0.37 + 0.37^2)$ Factoring out a 63 we get: $63\left[1 + 0.37\left(1 + 0.37 + 0.37^2\right)\right] = 63(1 + 0.37 + 0.37^2 + 0.37^3)$

The pattern is a surprising one. The nth term of the sequence contains a geometric series of n terms. That is,

$$a_n = 63 S_n$$

where

$$S_n = 1 + 0.37 + 0.37^2 + 0.37^3 + \cdots + 0.37^{n-1}$$

But you know the formula for the sum S_n of a geometric series with $a = 1$ and $r = 0.37$. Hence

$$S_n = \frac{a\,(r^n - 1)}{(r - 1)}$$

$$= \frac{1\,(0.37^n - 1)}{0.37 - 1}$$

$$= \frac{1 - 0.37^n}{0.63}$$

Therefore the expression for a_n becomes

$$a_n = 63 \left(\frac{1 - 0.37^n}{0.63} \right)$$

$$= 100 \left(1 - 0.37^n \right)$$

To verify this formula let's compare its graph with the one shown in Figure 7.44 on page 472. The graph of

$$y = 100 \left(1 - 0.37^x \right)$$

is shown in Figure 7.51. You can see that after two time constants the charge reaches 86% as we had computed with the recursive formula in Example 7.33. Also notice how the curve seems to approach a limit as $n \to \infty$. What do you think that limit is?

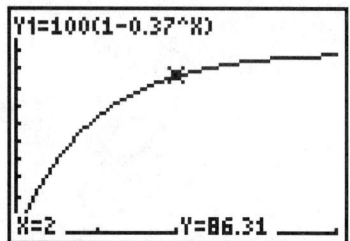

Y1=100(1-0.37^X)

X=2   Y=86.31

Xmin = 0
Xmax = 4
Xscl = 1
Ymin = 0
Ymax = 120
Yscl = 10

Figure 7.51

Example 7.51

What is the limit approached in Figure 7.51? That is, what is

$$\lim_{n \to \infty} 100 \left(1 - 0.37^n \right)$$

Solution   Look at the parts of $100\,(1 - 0.37^n)$. Only 0.37^n changes as n changes, and we know its limit. Remember that

$$\lim_{n \to \infty} r^n = 0$$

if $-1 < r < 1$. Since 0.37 fits that inequality then

$$\lim_{n \to \infty} 0.37^n = 0$$

Therefore the limit we seek is $100\,(1 - 0) = 100$. It makes sense that the limit is 100 because that means the charge approaches 100%. It is also interesting that the mathematical model predicts that the charge never actually reaches 100%. As we've said before the charge reaches 100% in reality when our voltmeter says it has. The time it takes to be fully charged depends on the accuracy of the voltmeter in use and the accuracy required. ∎

Annuities

Earlier we looked at how money accumulates interest if it is deposited and left untouched for a certain period of time. A sequence of equal payments made at equal periods of time is called an **annuity**. The sum of the payments and interest on the payments is the **future value** of the annuity.

Example 7.52   Application

Suppose $500 is deposited and left in an account paying 7% interest compounded annually. Another $500 is deposited at the end of each year after the first deposit. How much will be in the account after six years?

Solution   Figure 7.52 shows the annuity in this example.

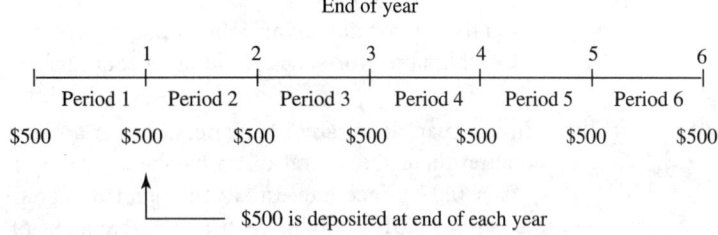

Figure 7.52

The first of these payments will produce a compound amount of

$$500(1 + 0.07)^6 \approx 750.37$$

The second payment of $500 will produce a compound amount of $500(1 + 0.07)^5 \approx 701.28$.

As shown in Table 7.14 the future value of the annuity is

$$500(1.07)^6 + 500(1.07)^5 + 500(1.07)^4 + 500(1.07)^3$$
$$+ 500(1.07)^2 + 500(1.07)^1 + 500$$

Table 7.14

Year	6	5	4	3	2	1	0
Deposit	$500	$500	$500	$500	$500	$500	$500
Compound amount	$500(1.07)^6$	$500(1.07)^5$	$500(1.07)^4$	$500(1.07)^3$	$500(1.07)^2$	$500(1.07)^1$	500

Notice that the last deposit does not earn any interest. But this is a geometric series with $a = 500$, $r = 1.07$, and $n = 7$, so its sum is

$$S_7 = \frac{500(1.07^7 - 1)}{1.07 - 1} \approx 4327.01$$

The future value of this annuity is $4327.01.

Example 7.52 leads to the following conclusion about the future value of an annuity.

Future Value of an Annuity

The future value S of an annuity of n payments of R dollars each at the beginning of each consecutive interest period, with interest compounded at a rate of i per period, is

$$S = R\left[\frac{(1+i)^n - 1}{i}\right]$$

Example 7.53   Application

Find the future value of an annuity due if payments of \$125 are made at the beginning of each quarter for six years in an account paying 7% compounded quarterly.

Solution   In six years there are 24 quarters. Since a payment is made at the end of the last quarter there are a total of 25 payments and so $n = 25$. The annual interest rate is $7\% = 0.07$. Since interest is compounded quarterly, the interest rate for each period is $\frac{0.07}{4} = 0.0175$. With each annual payment of \$500 we have $R = 125$ and the future value is

$$S = R\left[\frac{(1+i)^n - 1}{i}\right]$$

$$= 125\left[\frac{(1 + 0.0175)^{25} - 1}{0.0175}\right]$$

$$\approx 3878.43$$

Thus, the future value of this annuity is \$3878.43.   ∎

Compare the result in Example 7.53 to the one in Example 7.52. In each example a total of \$500 was deposited each year (except for the final payment) and the annual interest rates were the same. Yet the annuity in Example 7.52 produced \$4327.01 while the one in Example 7.53 produced \$3878.43.

Arithmetic Series

An arithmetic series is the sum of the terms of an arithmetic sequence. For example

$$S_n = 2 + 5 + 8 + 11 + 14 + \cdots$$

is an arithmetic series because the numbers being added are in an arithmetic sequence.

Example 7.54

Find the sum of the first 100 terms of the arithmetic series

$$S_n = 2 + 5 + 8 + 11 + 14 + \cdots$$

Solution   One way to do this is to add 3 to each successive term until we reach the 100th term. Then add up all 100 numbers. That's a lot of work. There must be a simpler way. Let's first do this for 10 terms and see if we can find a way that will simplify the work needed for 100 terms.

$$S_{10} = 2 + 5 + 8 + 11 + 14 + 17 + 20 + 23 + 26 + 29$$

You learned a method to use when analyzing geometric series. The method with arithmetic series is similar. We write S_{10} again, but in reverse order.

$$S_{10} = 29 + 26 + 23 + 20 + 17 + 14 + 11 + 8 + 5 + 2$$

Now add the two series. On the left-hand side of the equal sign we have $2S_{10}$, and on the right-hand side we will add the terms two at a time

$$2S_{10} = 31 + 31 + 31 + 31 + 31 + 31 + 31 + 31 + 31 + 31$$

Do you see where all those 31's came from? There are ten 31s, so the right-hand side equals $10 \times 31 = 310$:

$$2S_{10} = 310$$

$$S_{10} = \frac{310}{2} = 155$$

What about S_{100}? As you can see from our work with S_{10} we need to know the 100th term and we'll be in business. The nth term of an arithmetic sequence is

$$a_n = a + d(n - 1)$$

So for our sequence, with $a = 2$, $d = 3$, and $n = 100$, the 100th term is $2 + 3 \cdot 99 = 299$. Therefore if we follow the same procedure we used for 10 terms we get

$$S_{100} = 2 + 5 + 8 + \cdots + 293 + 296 + 299$$
$$S_{100} = 299 + 296 + 293 + \cdots + 8 + 5 + 2$$
$$2S_{100} = 301 + 301 + 301 + \cdots + 301 + 301 + 301$$

There are one hundred 301's, so

$$2S_{100} = 100 \cdot 301$$

and

$$S_{100} = \frac{100 \times 301}{2} = 15050$$

■

What is the sum of n terms of the arithmetic series that is based on the sequence $a_n = a_{n-1} + d$ and $a_1 = a$?

This is the general form of the arithmetic sequence

$$\{a, a + d, a + 2d, a + 3d, \ldots, a + d(n - 1)\}$$

You should be able to follow the same steps that we used in Example 7.51 to derive the formula. We'll take a shortcut to the result. Figure 7.53 explains where each part of the result of Example 7.54 came from.

First let's write the first term plus the nth term in our sequence. Here a is the first term and $a + d(n - 1)$ is the nth term. Therefore their sum is

$$a + a + d(n - 1) = 2a + d(n - 1)$$

and the formula for the sum of the series is

$$S_n = \frac{n[2a + d(n - 1)]}{2}$$

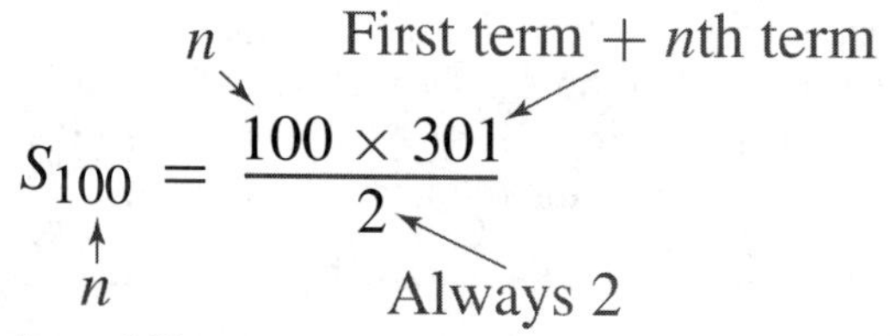

$$S_{100} = \frac{\overbrace{100}^{n} \times \overbrace{301}^{\text{First term} + n\text{th term}}}{\underbrace{2}_{\text{Always 2}}}$$

$$\underset{n}{\uparrow}$$

Figure 7.53

A good way to remember the formula for S_n is to think about Figure 7.53. The formula $\frac{100 \times 301}{2}$ can be rewritten as $100 \times \frac{301}{2}$. But $\frac{301}{2}$ is the average of the first term and the last term, and 100 is the number of terms. So the sum of n terms of an arithmetic sequence can be remembered as: *Take the average of the first term and the last term and multiply by the number of terms.*

Arithmetic Series

An arithmetic series is the sum of the terms of an arithmetic sequence. If the sum of the first n terms is represented by S_n then

$$S_n = a + (a + d) + (a + 2d) + (a + 3d) + \cdots + a + d(n - 1)$$

where a is the first term and d is the difference between consecutive terms. There are two ways to think of the formula for S_n:

$$S_n = n \times (\text{average of first and last term})$$

and $$S_n = \frac{n[2a + d(n - 1)]}{2}$$

Section 7.5 Exercises

In Exercises 1–6 determine the sum of the first 10 terms for the given geometric sequence.

1. $\{3, 6, 12, 24, \ldots\}$

2. $\{5, 30, 180, 1080, \ldots\}$

3. $\{16, -4, 1, -\frac{1}{4}, \ldots\}$

4. $\{25, 5, 1, \frac{1}{5}, \ldots\}$

5. $\{36, 12, 4, \frac{4}{3}, \ldots\}$

6. $\{\frac{2}{9}, -\frac{2}{3}, 2, -6, \ldots\}$

In Exercises 7–12 determine the sum of the first 10 terms for the given arithmetic sequence.

7. $\{3, 7, 11, 15, \ldots\}$

8. $\{5, 30, 55, 80, \ldots\}$

9. $\{16, 4, -8, -20, \ldots\}$

10. $\{25, 5, -15, -35, \ldots\}$

11. $\left\{-\frac{2}{3}, \frac{1}{3}, \frac{4}{3}, \frac{7}{3} \ldots\right\}$

12. $\left\{-\frac{13}{3}, -\frac{10}{3}, -\frac{7}{3}, -\frac{4}{3}, \ldots\right\}$

13. Determine (a) S_{12} and (b) S_{13} for the sequence $\{1, -2, 4, -8, 16, \ldots\}$.

14. What is the limit as $n \to \infty$ of the geometric series, from the sequence $\{1, -2, 4, -8, 16, \ldots\}$?

In Exercises 15–20 determine (a) whether the sum of the terms of each geometric series converges or diverges and (b) the limit of the series if it converges.

15. $270 - 90 + 30 - 10 + \cdots$

16. $256 - 64 + 16 - 1 + \cdots$

17. $\frac{3}{4} + \frac{3}{8} + \frac{3}{16} + \frac{3}{32} + \cdots$

18. $1 - \frac{1}{1.01} + \frac{1}{(-1.01)^2} + \frac{1}{(-1.01)^3} + \cdots$

19. $5 - 5 \cdot 2.1 + 5 \cdot 2.1^2 + 5(-2.1)^3 \cdots$

20. $3 + 3 \cdot 1.5 + 3 \cdot 1.5^2 + 3 \cdot 1.5^3 \cdots$

21. *Physics* A ball is dropped to a level piece of ground from a height of 60 m. Each time it bounces it goes $\frac{7}{8}$ as high as the previous bounce. What is the total vertical distance it has traveled when it hits the ground the fourth time?

22. *Physics* A ball is dropped to a level piece of ground from a height of 60 m. Each time it bounces it goes $\frac{7}{8}$ as high as the previous bounce. About how far (vertically) will the ball travel before it comes to rest. (*Hint:* Consider the sum of two sequences.)

23. *Finance* At the beginning of each year, the owner of an automotive repair shop deposits $3,000 into a retirement account paying 5.25% compounded annually. How much money will be in the account when the owner retires 25 years later?

24. *Electronics* A transistor's leakage current increases 1% for each increase of one degree Celsius. If a certain transistor's leakage is 1.40 nA at 20°C, how much will it be at 30°C? (*Hint:* An increase of 1% means that $r = 1.01$.)

25. *Medical technology* Certain medical conditions are treated with a fixed dose of a drug administered at regular intervals. Suppose that a person is given 2.5 mg of a drug each day and that during each 24-hour period, the body uses 40% of the amount of the drug that was in the system at the beginning of the period.

(a) How much of the drug is present in the body at the end of n days?

(b) Is there a maximum amount of drug that will be in the body at any one time? If so, how much?

26. *Construction* A contractor was preparing a bid for the construction of an office building. The foundation and basement will cost a total of $875,000 and the first floor will cost $380,000. The second floor will cost $255,000 and each floor above the second will cost $12,500 more than the floor below it.

(a) Write the cost of each floor as a sequence.

(b) Write a direct definition of this sequence.

(c) What is the cost of the 15th floor?

(d) What is the total cost of a 15-story building?

27. *Finance* The comment at the end of Example 7.5 on page 429 states "Incidentally, if the bank gives 5% interest per year compounded quarterly the total amount after five years would be over $6500." Determine the total amount in this annuity after five years. (*Caution:* Even though $20 is deposited in the account each week, interest is compounded quarterly, not weekly. How do you handle the initial deposit of $500?)

28. Describe the difference between a series and a sequence.

7.6 Estimating Areas under Curves

In this section we'll show a widely used method for estimating area under a curve. The method, which is one kind of a more general approach to area estimation called **Riemann sums**, is widely used in computer and calculator programs. We'll show you how to use the method and then you can run a *TI-83* program that does the computations for you.

One important new programming instruction will be introduced in this section, the FOR instruction.

Riemann Sums

The general problem is to estimate the area bounded by a curve whose equation is known, the x-axis, and two vertical lines drawn at $x = a$ and $x = b$. Figure 7.54 shows an example of this kind of area.

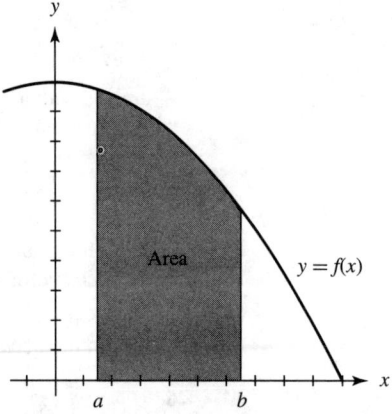

Figure 7.54

The area estimation method is similar to the one you used in Section 7.2 because it also divides the area into rectangles. All rectangles will have one side (the bottom) on the x-axis and the opposite side will touch the curve in one or more points. Figure 7.55(a) and 7.55(b) show two possibilities for lower rectangles. With an upper rectangle, the entire top side of the rectangle is above or touching the curve. The top side of an upper rectangle must touch the curve in at least one point as shown in Figure 7.56(a) and 7.56(b). As you can see from Figures 7.55(b) and 7.56(b) finding the x-value at a higher or lower point is not always easy.

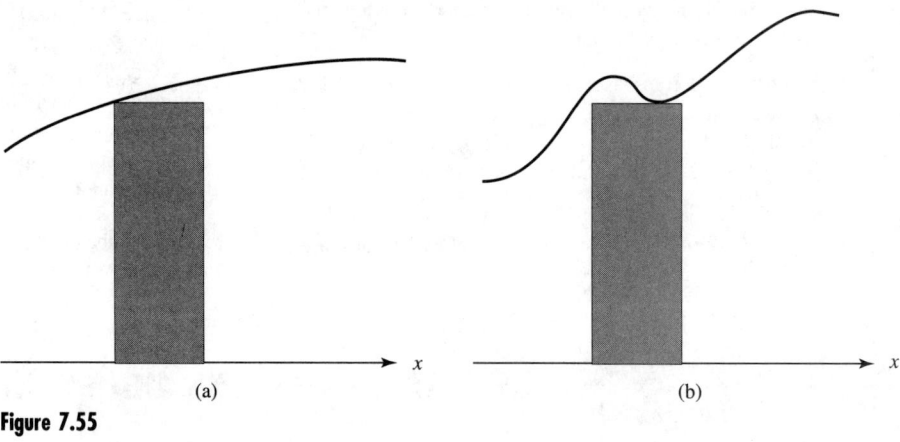

(a)   (b)

Figure 7.55

(a)   (b)

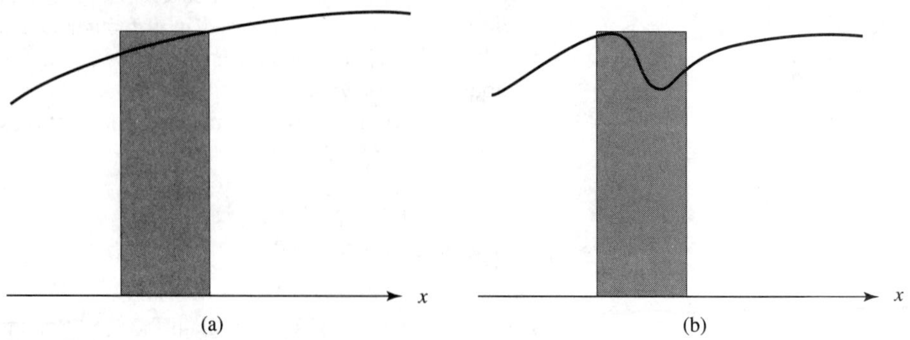

Figure 7.56

The method you used divided the area into lower and upper rectangles by systematically cutting the width of the rectangles in half for each level of estimation. The method we will introduce now uses the following steps:

1. Decide how many rectangles to use. Let n equal the number of rectangles.
2. Divide the horizontal distance between $x = a$ and $x = b$ into n equal parts called subdivisions. The width of each rectangle will be $\dfrac{|b - a|}{n}$.
3. Each rectangle stretches from the x-axis to the curve. The height of each rectangle depends on the type of estimate you are making: lower, upper, or other.
4. Add all the areas.

Example 7.55

Estimate the area under the curve $y = \sin x$, above the x-axis, and between $x = 0$ and $x = \pi$, where x is in radians. Use four rectangles and use (a) lower rectangles, (b) upper rectangles, and (c) the average of upper and lower.

Solution   The area to be estimated is shaded in Figure 7.57.

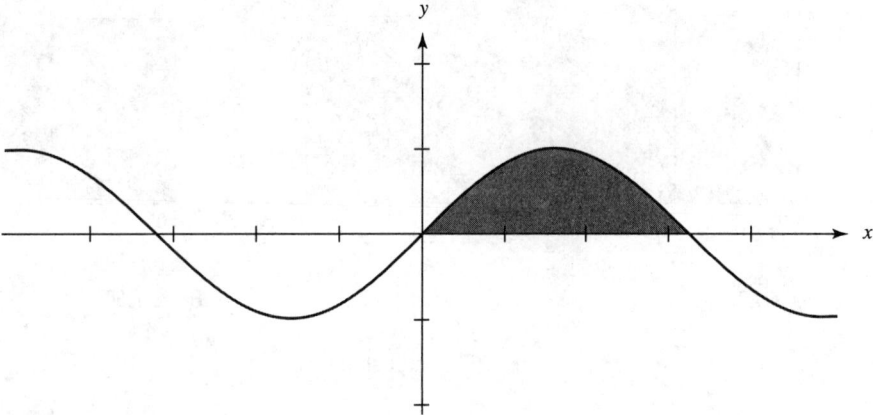

Figure 7.57

The lower rectangles A_1 and A_2 are shown in Figure 7.58. Notice that although we expected four rectangles there are only two because the first and last lower rectangles have height equal to zero. The figure also shows the error of the lower estimate. The errors are all negative errors, which means that the estimate will be less than the actual area.

Figure 7.59 shows the four upper rectangles, A_1, A_2, A_3, and A_4, and Figure 7.60 indicates the error of the upper estimate. The errors are all positive errors, so the estimate will be greater than the actual area.

The computation requires that we know the values of the function $y = \sin x$ at three points: $\sin \frac{\pi}{4}$, $\sin \frac{\pi}{2}$, and $\sin \frac{3\pi}{4}$. The width of each rectangle is $\frac{\pi}{4}$.

(a) Lower rectangles:

$$A_1 = \frac{\pi}{4} \sin\left(\frac{\pi}{4}\right) \approx 0.5554$$

$$A_2 = \frac{\pi}{4} \sin\left(\frac{3\pi}{4}\right) \approx 0.5554$$

$$A_1 + A_2 \approx 1.111$$

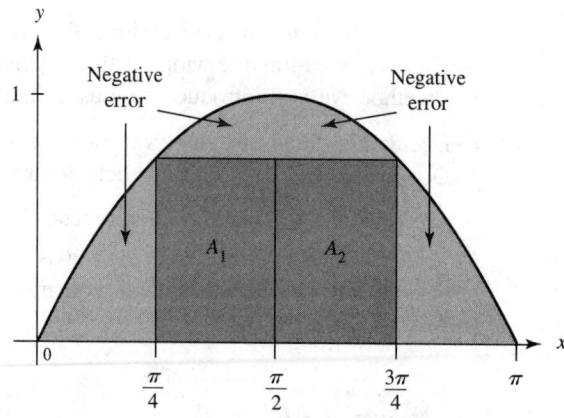

Figure 7.58

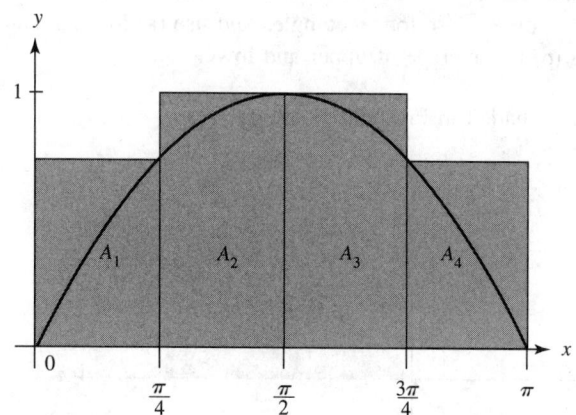

Figure 7.59

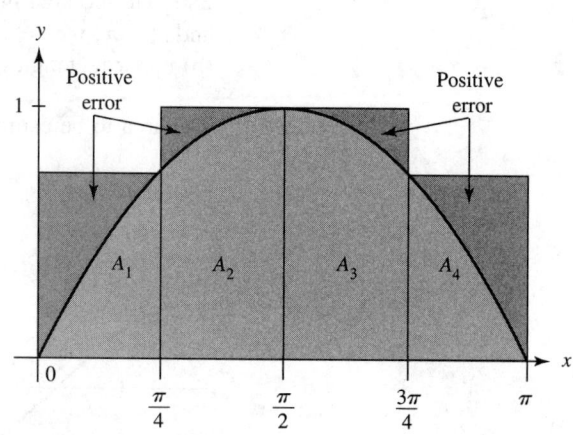

Figure 7.60

(b) Upper rectangles:

$$A_1 = \frac{\pi}{4} \sin\left(\frac{\pi}{4}\right) \approx 0.5554$$

$$A_2 = \frac{\pi}{4} \sin\left(\frac{\pi}{2}\right) \approx 0.7854$$

$$A_2 = \frac{\pi}{4} \sin\left(\frac{\pi}{2}\right) \approx 0.7854$$

$$A_4 = \frac{\pi}{4} \sin\left(\frac{3\pi}{4}\right) \approx 0.5554$$

$$A_1 + A_2 + A_3 + A_4 \approx 2.682$$

(c) Average of upper and lower:

$$\frac{1.111 + 2.682}{2} = 1.897$$

Estimation with Middle Rectangles

When you look at the errors that both the lower and upper rectangles introduce into the computation you might wonder if there is a better way to draw the rectangles. If we use "middle rectangles," as are shown in Figure 7.61, there will be less error.

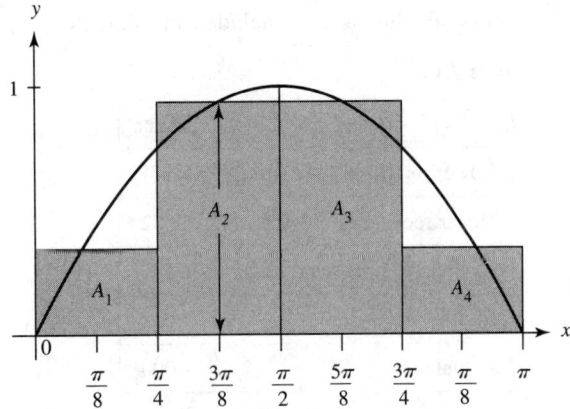

Figure 7.61

The procedure for constructing middle rectangles is as follows:

1. Decide how many rectangles to use. Let n equal the number of rectangles.
2. Divide the horizontal distance between $x = a$ and $x = b$ into n equal parts called subdivisions. The width of each rectangle will be $\dfrac{|b - a|}{n}$.
3. Mark each of the n equal subdivisions along the x-axis.
4. Mark the midpoint of each subdivision. The height of each rectangle extends from the midpoints to the curve.
5. Add up all the areas.

Example 7.56

Estimate the area above the x-axis under the curve $y = \sin x$ in radians between $x = 0$ and $x = \pi$. Use four "middle" rectangles.

Solution   The middle rectangles are shown in Figure 7.61.

The midpoints of each subdivision are at $x = \frac{\pi}{8}$, $\frac{3\pi}{8}$, $\frac{5\pi}{8}$, and $\frac{7\pi}{8}$. The heights of the four middle rectangles are $\sin\frac{\pi}{8}$, $\sin\frac{3\pi}{8}$, $\sin\frac{5\pi}{8}$, and $\sin\frac{7\pi}{8}$. The width of the rectangles are all $\frac{\pi}{4}$. Therefore the areas are computed as follows:

$$A_1 = \frac{\pi}{4}\sin\left(\frac{\pi}{8}\right) \approx 0.3006$$

$$A_2 = \frac{\pi}{4}\sin\left(\frac{3\pi}{8}\right) \approx 0.7256$$

$$A_2 = \frac{\pi}{4}\sin\left(\frac{5\pi}{8}\right) \approx 0.7256$$

$$A_4 = \frac{\pi}{4}\sin\left(\frac{7\pi}{8}\right) \approx 0.3006$$

$$A_1 + A_2 + A_3 + A_4 \approx 2.052$$

Finally, we will compare the four methods of estimation we have used in Examples 7.55 and 7.56.

The results are shown in Table 7.15. The actual area, which was computed with calculus, is also included in the table.

Table 7.15

Method	Estimate
Lower rectangles	1.111
Upper rectangles	2.682
Average of upper and lower	1.897
Middle rectangles	2.052
Actual area	2.000

After seeing the errors shown in Figures 7.58 and 7.60 you could guess that the estimates based on lower and upper rectangles would be less accurate. The average of the two is much better because it balances the positive and the negative errors. The middle rectangles gives an even better estimate than the average, although it will not always be superior. Both the average of the upper and lower rectangles and the middle rectangle are within about 5% of the actual area.

The errors of the middle rectangles are shown in Figure 7.62. Notice that some errors are positive and some are negative. As you can see the sum of the positive errors is approximately the same as the sum of the negative errors, which is the reason that the estimate is quite good.

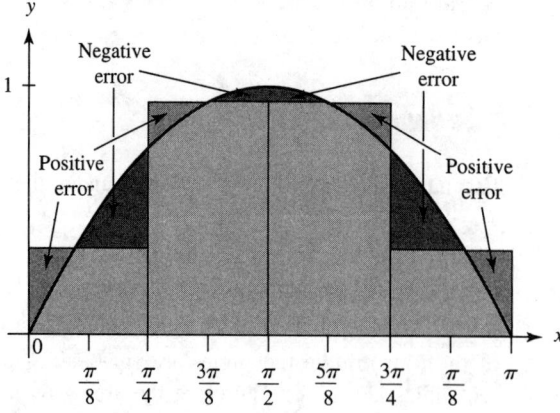

Figure 7.62

You saw in Section 7.2 that the three methods of estimation used there seem to converge to the same limit. The method of middle rectangles also converges to that same limit. However, the method of middle rectangles requires fewer calculations to reach a very good estimate, so it is the one we will continue to study with a program for the *TI-83*.

A Program to Estimate Areas

Because estimation of areas is an important need in many fields of technology, there are many calculator and computer programs that have been written for this purpose.

The *TI-83* program called AREAMID estimates the area bounded by the x-axis and a curve, between two values of x. It also draws the middle rectangles so you can see what it is doing.

Example 7.57

Use the program AREAMID to estimate the area above the x-axis and under the curve $y = \sin x$ between $x = 0$ and $x = \pi$. Use the following number of rectangles: (a) 4, (b) 10, (c) 20, (d) 50.

Solutions Before you run the program, enter the function $y = \sin x$ as Y1 in the Y= screen. Change your calculator mode to Radians if necessary. When you run the program you will see screens similar to those shown in Figure 7.63. Answer the questions in the way shown there. Use 2nd [π] to enter π. Press ENTER after each entry.

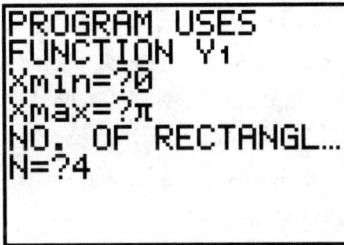

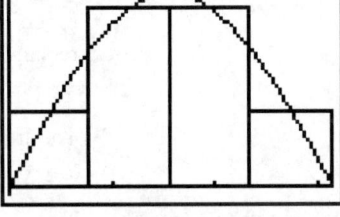

Input screen   Results screen

Figure 7.63

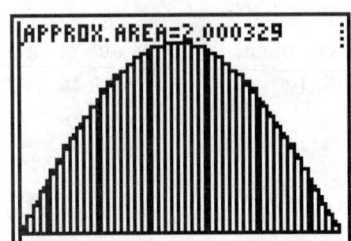

Figure 7.64

When you are looking at the graph there are four blinking dots arranged vertically in the upper right of the screen. This means that the program has come to a Pause command and is waiting for you to press ENTER. After you press ENTER the screen invites you to do another approximation with the same function and interval (0 to π). Press 1 for YES and then enter 10 for the number of rectangles. Continue through the other values of n required for this example. Notice that 50 rectangles gives a very good approximation of the area (Figure 7.64). You see black vertical lines on the calculator screen because the program is trying to plot more points than the resolution of the screen can handle.   ◼

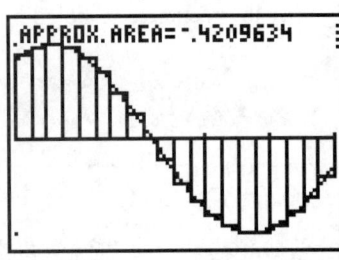

Figure 7.65

Example 7.58

Use the program AREAMID to estimate the area between the curve $y = \sin x$ and the x-axis, between $x = 1$ and $x = 6$. Use 20 rectangles.

Solution Figure 7.65 shows the result. The answer is negative because the rectangles *below* the x-axis have negative height where the value of the function is negative. If you wanted to know the actual area, for example if you were going to cover the area with cloth, then you would have to do separate intervals, one between $x = 1$ and $x = \frac{\pi}{2}$ and another between $x = \frac{\pi}{2}$ and $x = 6$. The second interval would produce a negative result. Use its absolute value and add it to the area from the first interval.   ◼

How the Program Works

The program AREAMID is much more complicated than the programs you used in Chapter 6. However, we have added comments within the program that tell you what the program is doing at that point. A comment in a programming language is a line of text that is ignored by the program. Its only purpose is to tell the reader what different parts of the program are doing. In the *TI-83* programming language comments are written within quotation marks.

You will see the following comments. Below each comment is a brief explanation of how this is accomplished in the program.

▶ "THE NEXT PART SETS THE GRAPH WINDOW"

- The program gets Xmin and Xmax from the user and computes Ymin, Ymax, Xscl, and Yscl.

▶ "ASK FOR THE NUMBER OF RECTANGLES"

- The program uses a Disp instruction and a Prompt instruction.

▶ "COMPUTE WIDTH OF RECTANGLE"

- Since $a = $ Xmin and $b = $ Xmax, then (Xmax–Xmin)/n is the width of the rectangle. In the program the variable D is this width.

▶ "BEGIN THE COMPUTATION"

- The first instruction is 0→S, which sets the sum to zero.

▶ "DRAW N RECTANGLES"

- The instruction FOR(I,0,N-1,1) tells the calculator to carry out all the instructions that follow, until the End command, for every value of I from 0 to $N - 1$ in steps of 1. That is, $I = 0, 1, 2, 3, \ldots, N - 1$. For each value of I, one rectangle is drawn.

▶ "M IS THE HEIGHT OF ONE RECTANGLE"

- The instruction is Y1(L+(I+.5)*D)→M
- Y1 is the function. L+(I+.5)*D is the x-coordinate of the midpoint of this interval. M is the value of Y1 at this value of x and therefore M is the height of this rectangle.
- L is Xmin. The factor (I+.5) causes the height of the rectangle to be computed from the midpoint of the interval.

▶ "DRAW ONE RECTANGLE"

- You saw the Line instruction in Chapter 6.

▶ "D*M IS THE AREA OF ONE RECTANGLE. ADD IT TO S THEN GO BACK AND DRAW THE NEXT RECTANGLE"

- This step adds the area of this one rectangle to all the others that have come before it.

▶ "ALL DONE WITH RECTANGLES"

- The End instruction marks the end of the block of instructions that the For instruction controls.

▶ "DRAW THE FUNCTION AND REPORT THE AREA"

- The instruction Text(0,2,"APPROX. AREA=",S) puts a message and the value of the estimated area S on the graph at screen coordinates $(0, 2)$. As with most computers and calculators, screen coordinates are measured down and to the right from the upper-left corner of the screen. (0, 2) means begin the text 0 pixels* down and 2 pixels to the right of the corner.

▶ "ASK ABOUT ANOTHER TRIAL"

- The `Menu` instruction offers a choice of going to `Lbl A` (start again with a different number of rectangles) or `Lbl B` (the end of the program).

The AREAMID Program

```
Func
FnOff
FnOn 1
ClrHome
Disp "PROGRAM USES"
Disp "FUNCTION Y1"
Prompt Xmin
Prompt Xmax
"THE NEXT PART SETS THE GRAPH WINDOW"
Xmin→L
Xmax→U
seq(Y1,X,L,U,(U-L)/20)→L1
augment({0},L1)→L1
"INCLUDE ZERO IN THE LIST IN CASE Y1 IS ALL POSITIVE OR ALL
NEGATIVE. THIS PUTS X-AXIS IN PICTURE"
min(L1)→Ymin
max(L1)→Ymax
Ymax-Ymin→H
max(1,iPart(((U-L)/10))→Xscl
max(1,iPart((H/10))→Yscl
(H/54)*6+Ymax→Ymax
-(H/54)*2+Ymin→Ymin
Lbl A
"ASK FOR THE NUMBER OF RECTANGLES"
ClrDraw
Disp "NO. OF RECTANGLES"
Prompt N
"COMPUTE WIDTH OF RECTANGLE"
(Xmax-Xmin)/N→D
"BEGIN THE COMPUTATION"
0→S
"DRAW N RECTANGLES"
For(I,0,N-1,1)
"M IS THE HEIGHT OF ONE RECTANGLE"
Y1(L+(I+.5)*D)→M
"DRAW ONE RECTANGLE"
Line(L+I*D,0,L+I*D,M)
Line(L+I*D,M,L+(I+1)*D,M)
Line(L+(I+1)*D,0,L+(I+1)*D,M)
"D*M IS THE AREA OF ONE RECTANGLE. ADD IT TO S THEN GO BACK AND
DRAW THE NEXT RECTANGLE"
S+D*M→S
End
```

(continues on next page)

---

*A pixel is the smallest area that can be used on the screen. The word *pixel* is an abbreviation of *picture element*. On the *TI-83* there are 95 pixels horizontally and 58 pixels vertically.

```
"ALL DONE WITH RECTANGLES"
"DRAW THE FUNCTION AND REPORT THE AREA"
DrawF Y1
Text(0,2,"APPROX. AREA=",S)
Pause
"ASK ABOUT ANOTHER TRIAL"
ClrHome
Menu("ANOTHER APPROX?","YES",A,"NO",B)
Lbl B
```

Calculator Lab 7.6

Solving the Original Area Problem

This lab has two parts. First you will estimate an area using your calculator for the computations of the areas. Then you will use the program AREAMID to estimate the area. The program can check your computation for the first part and, with a large number of rectangles, can give you a good estimate of the actual area.

Part 1: Estimate the area between the curve $y = 10 - \dfrac{x^2}{10}$ and the x-axis from $x = -10$ and $x = 10$. This function is the one pictured in Figure 7.3 on page 421, the one whose area we estimated in various ways. Use 10 rectangles in this part.

Procedures
1. Enter the function as Y1 in the Y= screen
2. Use a TABLE to find the values of Y1 for the values of x that you need.
3. Add the areas of all the rectangles. The answers given in Section 7.2 will give you an idea if your answer is approximately correct.

Part 2: Use the program AREAMID to estimate the area between the curve $y = 10 - \dfrac{x^2}{10}$ and the x-axis from $x = -10$ and $x = 10$.

Procedures
1. Enter the function as Y1 in the Y= screen
2. Run the program with 10 rectangles to check your answer to part 1.
3. Run the program with a large number of rectangles until you have a good estimate of the area.
4. Compare your best estimate with the best estimate you made in Section 7.2.

Section 7.6 Exercises

1. Estimate the area under the graph of $f(x) = \cos x$, above the x-axis, from $x = -0.5\pi$ to $x = 0.5\pi$ where x is in radians. Use four rectangles and use **(a)** lower, **(b)** upper, and **(c)** middle rectangles. **(d)** Use the program AREAMID to estimate the area with 20 rectangles.

2. Sketch the graph of $f(x) = \sqrt{x}$. Approximate the lower L_1, upper U_1, and middle M_1 area under the curve, above the x-axis, from $x = 0$ to $x = 4$. Use four rectangles and use **(a)** lower, **(b)** upper, and **(c)** middle rectangles. **(d)** Use the program AREAMID to estimate the area with 20 rectangles.

3. Sketch the graph of $f(x) = x^2 + 1$. Approximate the lower L_1, upper U_1, and middle M_1 area under the curve, above the x-axis, from $x = -1$ to $x = 2$. Use four rectangles and use **(a)** lower, **(b)** upper, and **(c)** middle rectangles. **(d)** Use the program AREAMID to estimate the area with 20 rectangles.

4. Sketch the graph of $f(x) = \sin 2x + \cos x + 2$. Approximate the lower L_1, upper U_1, and middle M_1 area under the curve, above the x-axis, from $x = 0$ to $x = 2\pi$. Use eight rectangles and use (a) lower, (b) upper, and (c) middle rectangles. (d) Use the program AREAMID to estimate the area with 30 rectangles.

5. Sketch the graph of $f(x) = x \sin x + \sin x$. Approximate the lower L_1, upper U_1, and middle M_1 area under the curve, above the x-axis, from $x = -\pi$ to $x = \pi$. Use eight rectangles and use (a) lower, (b) upper, and (c) middle rectangles. (d) Use the program AREAMID to estimate the area with 30 rectangles.

6. *Ecology* A town wants to drain and fill the swamp shown in Figure 7.66.

   (a) What is the surface area of the swamp?

   (b) If the swamp has an average depth of 6 ft how many cubic yards of dirt will it take to fill the "hole" that is left after the swamp is drained?

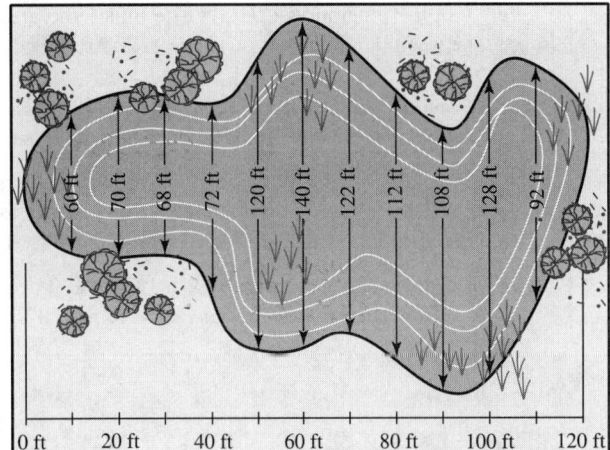

Figure 7.66

7. *Construction* Table 7.16 gives the results of a series of drillings to determine the depth of the bedrock at a drilling site. These drillings were taken along a straight line down the middle of the lot where the building will be placed. In the table x is the distance from the front of the lot and y is the corresponding depth of the bedrock. Both x and y are measured in feet. Approximate the area of this cross-section.

Table 7.16

x	0	20	40	60	80	100	120	140	160	180
y	38	42	53	42	37	25	25	37	42	48

8. *Automobile technology* The cross-section of an airfoil for some racing cars can be described by the graphs of $f(x) = \sqrt{x}$ and $g(x) = -\sqrt{x} + \frac{1}{32}x^2$. Use eight rectangles and use (a) lower, (b) upper, and (c) middle rectangles. (d) Use the program AREAMID to estimate the area with 30 rectangles.

9. *Environmental science* The level of pollution in San Juan Pedro Bay due to a sewage spill is estimated to be $f(t) = \dfrac{1200t}{\sqrt{t^2 + 10}}$ parts per million, where t is the time in days since the spill occurred. Find the total amount of pollution during the first five days of the oil spill by using the method described in this section and 10 divisions.

10. *Electronics* The charge on a capacitor in millicoulombs can be estimated by finding the area under the graph of $f(t) = 0.6t^2 - 0.2t^3$ from $t = 1$ to $t = 3$. Estimate the charge on this capacitor by using the method described in this section and 8 divisions.

11. Sketch the graph of $f(x) = 0.5x$.

    (a) Without using a calculator, determine the area between the graph of f and the x-axis from $x = 0$ to $x = 6$.

 (b) Use the program AREAMID to estimate the area in (a) using 24 rectangles.

   (c) Without using a calculator, determine the area between the graph of f and the x-axis from $x = -6$ to $x = 0$.

   (d) Use the program AREAMID to estimate the area in (c) using 24 rectangles.

   (e) Without using a calculator, determine the area between the graph of f and the x-axis from $x = -6$ to $x = 6$. Is this the same as the sum of your answers to (a) and (c)?

   (f) Use the program AREAMID to estimate the area in (e) using 24 rectangles. Is this the same as the sum of your answers to (b) and (d)?

   (g) Are your answers in (e) and (f) the same? If not, discuss why they are different.

 12. Sketch the graph of $g(x) = x^3 - x$.

    (a) Use the program AREAMID to estimate the area between the graph of g and the x-axis from $x = 0$ to $x = 1.5$ using 24 rectangles.

    (b) Use the program AREAMID to estimate the area between the graph of g and the x-axis from $x = -1.5$ to $x = 0$ using 24 rectangles.

(c) Use the program `AREAMID` to estimate the area between the graph of g and the x-axis from $x = -1.5$ to $x = 1.5$ using 24 rectangles.

(d) Compare the answer in **(c)** with the sum of the answers to **(a)** and **(b)**. How are they alike and how are they different? Explain your response.

13. Describe how to use the procedures of this chapter to find the area of an irregular-shaped region.

14. What changes would you have to make in the procedures of this chapter if the rectangles were not all the same width?

15. Write a memo to the management in which you describe what you have learned; respond to the memo on page 420.

(a) Develop a method to estimate the area of a two-dimensional curved shape.

(b) Develop a method to estimate areas if the equations of the boundary curves are known.

(c) Develop a method for finding an approximate area that fits a shape for which there is no equation such as the airfoil in the Chapter Project.

●●●●●●●● Chapter 7 Summary and Review

Topics You Learned or Reviewed

▶ Fractional exponents are used to indicate roots. Thus, $b^{m/n} = \sqrt[n]{b^m}$.

▶ Any number, except zero, raised to a negative exponent indicates the reciprocal of the number to that power. In symbols, this is $b^{-n} = \dfrac{1}{b^n}$.

▶ Any number, except zero, raised to a zero exponent has a value of 1. Thus, $b^0 = 1$, if $b \neq 0$.

▶ A sequence can be defined directly by giving a formula for the nth term of the sequence.

▶ Geometric sequence

• A geometric sequence has a common ratio between consecutive terms.
• The nth term of a geometric sequence is ar^{n-1}, where r is the common ratio.

▶ Arithmetic sequence

• An arithmetic sequence has a common difference between consecutive terms.
• The nth term of an arithmetic sequence is $a + (n-1)d$, where d is the common difference.
• An arithmetic sequence diverges as $n \to \infty$.

▶ A recursive sequence has each term of the sequence defined in terms of the preceding terms of the sequence.

▶ The `seq` function and `Seq` mode of a graphing calculator can be used to sketch the graph of a sequence.

▶ Geometric and arithmetic series

• A mathematical series is the sum of the terms of a sequence. A geometric series is a series that is produced from a geometric sequence, and an arithmetic series is a series that is produced from an arithmetic sequence.
• The sum of the first n terms of the geometric series

$$S_n = a + ar + ar^2 + ar^3 + \cdots + ar^{n-1}$$

is

$$S_n = \frac{a(r^n - 1)}{(r - 1)}$$

where a is the first term and r is the ratio between consecutive terms.
• The sum of the first n terms of an arithmetic series is

$$S_n = a + (a + d) + (a + 2d) + (a + 3d) + \cdots + a + d(n - 1)$$

where a is the first term and d is the difference between consecutive terms.
• A geometric series converges to a finite number L as $n \to \infty$ if $-1 < r < 1$.

▶ Tables and graphs can be used to approximate limits of sequences and series.

▶ The algebraic properties of limits can be use to determine or approximate the limits of sequences.
▶ The area under a curve can be approximated using Riemann sums and limits.
▶ The `For` and `Menu` instructions can be used in programming.

Review Exercises

In Exercises 1–4, (a) write the first six terms of the given sequence, (b) graph the first 10 terms of the sequence, and (c) guess the limit of the sequence.

1. $\{-2\}$

2. $\left\{\dfrac{n-6}{n-2}\right\}$ beginning with $n = 3$

3. $\left\{\sin\dfrac{n\pi}{6}\right\}$ in radians

4. $\left\{\left(2+\dfrac{2}{n}\right)^n\right\}$

For each of the sequences in Exercises 5–8, (a) write the first five terms of each sequence, (b) write the 100th and 101st terms, (c) write the difference between any two consecutive terms you have written, (d) write the ratio between any two consecutive terms you have written, and (e) tell whether the sequence is an arithmetic sequence, a geometric sequence, or neither of these types of sequences. Explain your answer.

5. $a_n = \dfrac{1}{n+1}$

6. $b_n = 5n - 3$

7. $c_n = 4\left(\frac{1}{3}\right)^n$

8. $d_n = \frac{1}{5}n^3$

In Exercises 9–12 write (a) the nth term, (b) the $(n+1)$st term, and (c) a formula in terms of n for the difference and ratio between the $(n+1)$st and nth terms for each sequence.

9. $a_n = \dfrac{1}{n-1}$

10. $b_n = 5n + 2$

11. $c_n = \frac{2}{3}(-5)^n$

12. $d_n = 6 - 3n$

In Exercises 13–16 the first four terms of a sequence are shown. (a) Write the most likely next four terms, (b) write a formula for the nth term of the sequence, and (c) tell whether the sequence is an arithmetic sequence, a geometric sequence, or neither of these types of sequences. Explain your answer.

13. $\frac{1}{2}, -\frac{2}{3}, \frac{3}{4}, -\frac{4}{5}, \ldots$

14. $1, 3, 6, 10, 15, \ldots$

15. $\frac{2}{5}, \frac{4}{15}, \frac{6}{45}, \frac{8}{135}, \ldots$

16. $4, 1, -2, -5, \ldots$

In Exercises 17–20 a sequence is defined recursively. Write the first five terms of the sequence.

17. $s_n = \frac{1}{2}n + 1 + s_{n-1}$ with $s_1 = 1$

18. $s_n = \dfrac{4}{s_{n-1}} + 2$ with $s_1 = 2$

19. $a_n = a_{n-1} + 2a_{n-2}$ with $a_1 = 1$ and $a_2 = 2$

20. $b_n = 2^{b_{n-1}} - 1$ with $b_1 = 1$

In Exercises 21–24 a recursive sequence is in words. (a) Write the first five terms of the sequence, (b) write a recursive formula for the sequence, and (c) tell whether the sequence is an arithmetic sequence, a geometric sequence, or neither of these types of sequences. Explain your answer.

21. Multiply the previous term by $n + 5$. $a_1 = 2$

22. Multiply the previous term by -2 and then add 4. $s_1 = 1$

23. Square each term and subtract 1 to get the next term. $s_1 = 2$

24. Divide 1 by the sine in radians of each term to get the next term. $s_1 = \frac{\pi}{2}$

In Exercises 25–26 the first four or five terms of a sequence are shown. (a) Write the most likely next four terms and then (b) write a recursive formula for the sequence. (Don't forget to define the first term.)

25. $5, 8, 17, 33, 58, \ldots$

26. $10, 2, 0.4, 0.08, \ldots$

In Exercises 27–30 write the first five terms of each sequence and guess its limit. If you need more terms, enter the expression into the calculator's Y= screen in the Seq *mode and look at the table for the sequence.*

27. $a_n = \dfrac{8 - 2n}{4n}$

28. $a_n = \dfrac{n^3 - n}{n - 1}, \ n \neq 1$

29. $a_n = \sqrt{4 - n}$

30. $a_n = \dfrac{1}{n^2} + \dfrac{3}{n} - 5$

In Exercises 31–36, (a) graph the sequence and estimate its limit and (b) use algebra to determine the limit of the sequence.

31. $a_n = -\frac{2}{3}n$

32. $b_n = (-1)^n$

33. $c_n = \dfrac{8n^2 - 3n + 5}{4n^2 - 1}$

34. $d_n = \left(-\frac{7}{8}\right)^n$

35. $e_n = \cos\left(\dfrac{1}{n}\right)$ in radians

36. $f_n = \sin^2 n + \cos^2 n$

In Exercises 37–40 determine the sum of the first 10 terms for the given sequence.

37. $\{1, 3, 5, 7, 9, \ldots\}$

38. $\{1, -3, 9, -27, \ldots\}$

39. $\{50, 10, 5, 1, \ldots\}$

40. $\{2, 9, 16, 23, \ldots\}$

41. Determine (a) S_{12} and (b) S_{13} for the sequence $\{15, -5, \frac{5}{3}, -\frac{5}{9}, \ldots\}$.

In Exercises 42–45 determine (a) whether each geometric series converges or diverges and (b) the limit of the series if it converges.

42. $160 - 40 + 10 - 2.5 + \cdots$

43. $0.6 + 0.3 + 0.15 + 0.075 + \cdots$

44. $5 + 5 \cdot (-0.2) + 5 \cdot (0.2)^2 + 5(-0.2)^3 \cdots$

45. $3 + \frac{3}{1.5} + \frac{3}{1.5^2} + \frac{3}{1.5^3} \cdots$

46. Sketch the graph of $f(x) = 4 - x^2$. Approximate the lower L_1, upper U_1, and middle M_1 area under the curve, above the x-axis, from $x = -2$ to $x = 2$. Record your results in a table like Table 7.1 on page 422. Refine your results by making second and third approximations for the lower and upper areas.

47. Sketch the graph of $f(x) = \sin x + 2\cos^2 x$ where x is in radians. Approximate the lower L_1, upper U_1, and middle M_1 area under the curve, above the x-axis, from $x = -0.5\pi$ to $x = 0.5\pi$. Record your results in a table like Table 7.2. Refine your results by making second and third approximations for the lower and upper areas.

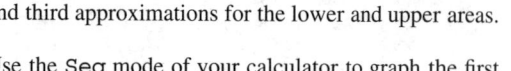 **48.** Use the Seq mode of your calculator to graph the first 20 terms of the sequence $u_n = 2u_{n-1} - 5$.

    (a) Determine the fixed point F of this sequence,

    (b) sketch the graph of the first 20 terms of the sequence with $u_0 = F$, and

    (c) sketch the graph of the first 20 terms of the sequence with $u_0 = F - 0.001$. Set your graphing window to Xmin=0, Xmax=20, Ymin=0, and Ymax=10.

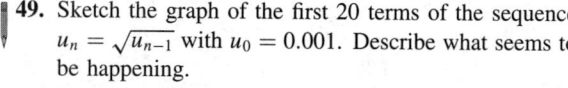 **49.** Sketch the graph of the first 20 terms of the sequence $u_n = \sqrt{u_{n-1}}$ with $u_0 = 0.001$. Describe what seems to be happening.

50. *Computer technology*

    (a) What is the maximum number of computer dies $0.35'' \times 0.35''$ that can be cut from a circular silicon wafer $4''$ in diameter? Note that the entire die must fit on the wafer.

    (b) How many dies will fit on a wafer if the diameter is doubled to $8''$?

    (c) Suppose the size of a die could be reduced to $0.30'' \times 0.30''$. How does this change your answers to (a) and (b)?

51. Consider the recursive sequence $s_n = 1 + \dfrac{1}{3}s_{n-1} - s_{n-1}$ with $s_1 = 9$. (a) Graph the sequence and (b) estimate its limit.

 52. Match the graphs in Figures 7.67(a)–(c) with each of the following sequences. The window settings in each graph are Xmin = 0, Xmax = 12, Xscl = 1, Ymin = -4, Ymax = 8, and Yscl = 1. (Since there are more sequences than graphs, at least one of the sequences cannot be matched with a graph.)

    (a) $a_n = -3 + 2\left(\frac{1}{3}\right)^n$

    (b) $b_n = \dfrac{3 - n + n^2}{n}$

    (c) $c_n = 8c_{n-1} - 14$ with $c_1 = 2$

    (d) $d_n = \left(-\dfrac{2}{n}\right)^2 + 1$

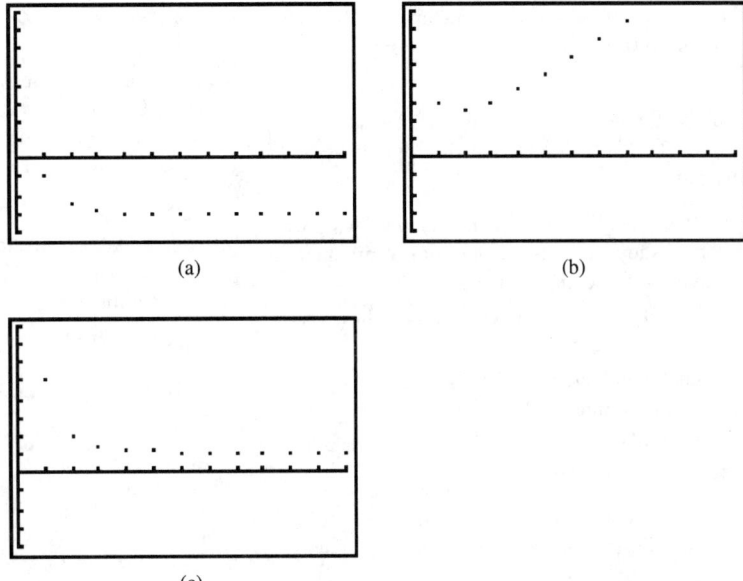

Figure 7.67

53. *Environmental science* The level of pollution present in a lake following an accidental discharge at an industrial plant was 942 parts of pollution per billion parts of water (ppb) in January and 725 ppb in February. If there is no further pollution the level of contamination will follow a geometric sequence. How long will it take for the amount of pollution to go below 0.150 ppb?

54. *Medical technology* A person is given an initial dose of 40 ml of a certain medication. The kidneys remove 22% of the medication from the bloodstream each day.

    (a) Write a sequence that describes the amount of medication in the bloodstream each day.

    (b) Determine the amount of medication that should be given each day in order to maintain a level of 40 ml in the bloodstream.

55. *Medical technology* A bacterial culture grows according to the sequence

$$a_n = 1.2a_{n-1} - 0.00008a_{n-1}^2$$

with $a_1 = 950$.

    (a) Graph the first 30 terms of this sequence.

    (b) Estimate the limit of the growth of this bacterial culture.

56. *Finance* $1,500 is deposited in an account that earns 6.5% per year compounded annually.

    (a) Write the recursive definition of the sequence.

    (b) Write the formula for the amount in the account after n years.

57. *Finance* The amount of $1 500 is placed in a savings account at 6.5% interest. If interest is compounded

    (a) annually, **(b)** semiannually, **(c)** quarterly, and **(d)** monthly, what is the total after 10 years?

58. *Physics* A new tennis ball is dropped to a level piece of ground from a height of 2.5 m. Each time it bounces it goes $\frac{3}{5}$ as high as the previous bounce.

    (a) Write a recursive sequence that describes the height of the nth bounce.

    (b) How high does the ball bounce on the sixth bounce?

    (c) How many bounces does it take before the ball bounces less than 1 cm?

    (d) If we consider that the ball has come to rest when it bounces less than 0.1 mm, how long does it take for the ball to come to rest?

    (e) What is the total vertical distance the ball travels during all its bounces?

59. *Electronics* A fully charged capacitor with a voltage of 220 volts has a time constant of 9 seconds. Estimate the voltage remaining in the capacitor circuit after **(a)** 27 and **(b)** 56 seconds. Use both the recursive definition of the geometric sequence and the direct expression for the nth term.

60. *Electronics* If the voltage on the capacitor in Exercise 15 were measured with a voltmeter that could measure to the nearest 0.1 volt, about how long would it be before the voltage reading had dropped to zero volts?

61. *Waste technology* A certain amount of radioactive waste has a half-life of 38.2 years. **(a)** Write the geometric sequence that models the situation and **(b)** estimate how much of 100 grams of the material will remain after 200 years?

62. *Archaeology* A fossil originally contained 150 g of carbon 14 and now has 10 g. What is the approximate age of the fossil?

63. *Music* If the frequency of $a_0 = 440$ Hz, what is the frequency of the **(a)** 12th note and **(b)** 15th above the note designated by a_0?

64. *Finance* At the beginning of each year the owner of an electronics shop deposits \$4,000 into a retirement account paying 6.5% compounded annually.

  (a) Ignoring interest, how much will she have deposited after n years?

  (b) How much will have been deposited after 25 years (again, without interest)?

  (c) How much money will be in the account when the owner retires 25 years later?

65. *Construction* If a contractor does not complete a multi-million-dollar construction project on time, a penalty must be paid of \$500 for the first day the project is late, \$750 for the second day, \$1,000 for the third day, and so on. Each day the penalty is \$250 larger than the previous day.

  (a) Write a formula for the penalty on the nth day.

  (b) What is the penalty for the 15th day?

  (c) Write a formula for the total penalty for n days.

  (d) What is the total penalty if the project is 15 days late?

66. *Recreation* Marcia jumps from a platform with a bungee cord tied to her legs. She falls 180 feet before being pulled back upward by the bungee cord. She always rebounds $\frac{1}{3}$ of the distance she fell and then falls $\frac{2}{3}$ of the distance of her last rebound. Approximately how far does she travel before coming to rest?

67. *Business* A sales manager sets a team goal of \$1 million in sales during the next quarter. The team plans to get $\frac{1}{2}$ of the goal the first week of the quarter, and every week after they estimate that they will sell only $\frac{1}{2}$ as much as they sold the previous week because the market will become saturated. How close will they come to their sales goal after the 13-week quarter?

68. Sketch the graph of $f(x) = 9 - x^2$. Approximate the lower L_1, upper U_1, and middle M_1 area under the curve, above the x-axis, from $x = -2$ to $x = 3$. Use four rectangles and use **(a)** lower, **(b)** upper, and **(c)** middle rectangles. **(d)** Use the program AREAMID to estimate the area with 20 rectangles.

69. *Construction* Table 7.17 gives the results of a series of laser measurements across a small lake. These measurements were taken along the one side of the lake. In the table, x is the distance from the shoreline across the lake to the other shore. Both x and y are measured in meters.

  (a) Approximate the surface area of this lake.

Table 7.17

x	0	50	100	150	200	250	300	350
y	138	252	523	432	346	265	225	137

  (b) This lake is to be drained and filled with earth in order to build a shopping center. If the average depth of the lake is 42 m how much fill will it take to replace the water?

70. Lay the palm of your hand on a sheet of paper and use your pencil to draw around your hand. Use 10 rectangles and use **(a)** lower, **(b)** upper, and **(c)** middle rectangles to estimate the area of your hand.

71. *Environmental science* The level of pollution in Ivan Petrovich Bay due to a sewage spill is estimated to be $f(t) = \dfrac{1500t}{\sqrt{t^2 + 50}}$ parts per million, where t is the time in days since the spill occurred. Find the total amount of pollution during the first five days of the oil spill.

72. *Electronics* The charge on a capacitor in millicoulombs can be estimated by finding the area under the graph of $f(t) = 1.5t^2 - 0.5t^3$ from $t = 0$ to $t = 2$. Estimate the charge on this capacitor by using the program AREAMID with 20 rectangles.

●●●●●●●● Chapter 7 Test

1. Consider the sequence $\left\{\dfrac{n+2}{n+5}\right\}$. **(a)** Write the first 6 terms of the given sequence, **(b)** graph the first 10 terms of the sequence, and **(c)** guess the limit of the sequence.

2. For the sequence $s_n = \dfrac{3}{n+1}$

  (a) Write the first five terms of each sequence.

  (b) Write the 100th and 101st terms.

  (c) Write the difference between any two consecutive terms you have written.

  (d) Write the ratio between any two consecutive terms you have written.

  (e) Tell whether the sequence is an arithmetic sequence, a geometric sequence, or neither of these types of sequences. Explain your answer.

3. For the sequence $s_n = 2 - 3n$ write **(a)** the nth term, **(b)** the $(n + 1)$st term, and **(c)** a formula in terms of n for the difference and ratio between the $(n + 1)$st and nth terms for each sequence.

4. $\{\frac{1}{3}, -\frac{1}{12}, \frac{1}{48}, -\frac{1}{192}, \}$ are the first four terms of a sequence. **(a)** Write the most likely next four terms, **(b)** write a formula for the nth term of the sequence, and **(c)** tell whether the sequence is an arithmetic sequence, a geometric sequence, or neither of these types of sequences. Explain your answer.

5. Write the first five terms of the recursive sequence $s_n = \frac{2}{3}n + 1 - s_{n-1}$ with $s_1 = 2$.

6. "Subtract the previous answer and multiply by 5" describes a recursive sequence in words. **(a)** Write the first five terms of the sequence. **(b)** Write a recursive formula for the sequence and **(c)** tell whether the sequence is an arithmetic sequence, a geometric sequence, or neither of these types of sequences. Explain your answer.

7. Consider the sequence whose first four terms are $\{3, 17, 31, 45, \ldots\}$.
   (a) Write the most likely next four terms.
   (b) Write a recursive formula for the sequence.

8. For the sequence $a_n = \frac{1}{5}n + 1$, **(a)** graph the sequence and estimate its limit and **(b)** use algebra to determine the limit of the sequence.

9. What is the sum of the first 10 terms for the sequence that begins $\{1, 5, 9, 13, 17, \ldots\}$?

10. For the sequence $2(0.9)^n$
   (a) Determine whether the sum of the terms of each geometric series converges or diverges.
   (b) Determine the limit of the series if it converges.

11. Sketch the graph of $f(x) = 4 - 3x + x^2$. Use four subdivisions to approximate the lower L_1, upper U_1, and middle M_1 area under the curve, above the x-axis, from $x = 0$ to $x = 2$.

 12. Match the graphs in Figure 7.68(a)–(b) with each of the following sequences. The window settings in each graph are Xmin = 0, Xmax = 12, Xscl = 1, Ymin = -4, Ymax = 8, and Yscl = 1. (Since there are more sequences than graphs, at least one of the sequences cannot be matched with a graph.)
   (a) $a_n = 6 - 0.5n$
   (b) $b_n = 6 - \dfrac{1}{n}$
   (c) $c_n = 6 + \left(\dfrac{-1}{n}\right)^n$

13. Consider the recursive sequence $s_n = 4 - \dfrac{2}{3}s_{n-1} - s_{n-1}$ with $s_1 = 9$. **(a)** Graph the sequence and **(b)** estimate its limit.

14. $\$1,200$ is deposited in an account that earns 5.8% per year compounded annually.
   (a) Write the recursive definition of the sequence.
   (b) Write the formula for the amount in the account after n years.
   (c) If interest is compounded quarterly, what is the total after 10 years?

15. A fully charged capacitor with a voltage of 110 volts has a time constant of 5 seconds. Estimate the voltage remaining in the capacitor circuit after **(a)** 15 and **(b)** 25 seconds. Use both the recursive definition of the geometric sequence and the direct expression for the nth term.

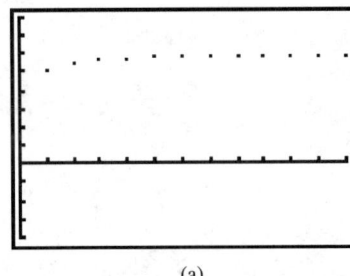

(a)

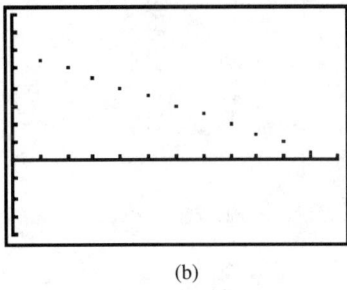

(b)

Figure 7.68

Modeling with Algebraic Functions

Topics You'll Need to Know

▶ Properties of exponents
  • negative, zero, and fractional exponents
  • raising a fraction to a power
▶ How to solve proportions
▶ Point slope equation of the straight line
▶ Quadratic formula
▶ Horizontal and vertical shifts of a function $y = f(x)$
  • $y = f(x - a)$
  • $y = f(x) + b$

Calculator Skills You'll Use

▶ Using `Table` and `TBLSET` to estimate roots of equations
▶ Entering numbers into a list
▶ Running a program
▶ Using `Seq` function to generate sequences in lists
▶ Regression from data in lists
▶ Plotting lists
▶ Using `Zoom In`, `Zoom Out`, and `ZBox` (Zoom Box)

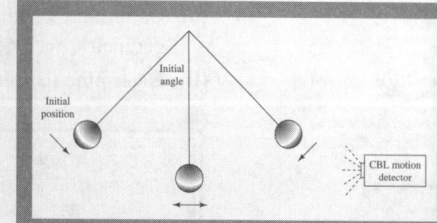

Topics You'll Learn or Review

▶ Types of functions and their graphs
  • power, polynomial, rational, algebraic, odd and even
▶ Properties of graphs
  • predict shape of graph from the function
  • domain and range
  • roots and factored form
  • asymptotes
  • limits as $x \to +\infty$ and $x \to -\infty$
  • symmetry
▶ Direct and inverse variation
▶ Maclaurin series approximation for sine and cosine
▶ Consecutive differences method
▶ Tangents to curves
  • writing the equation of the tangent to a curve at a point
  • sketching the slope function of a function

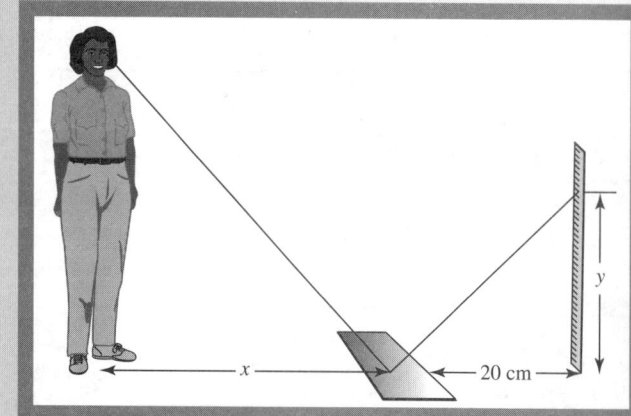

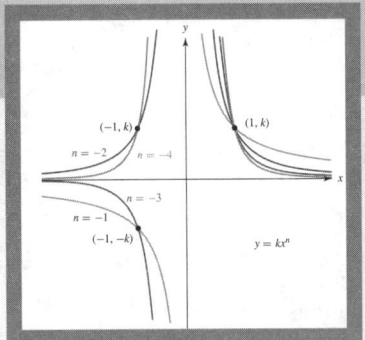

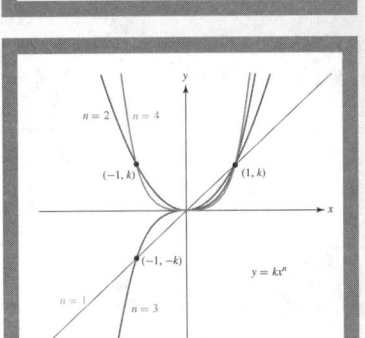

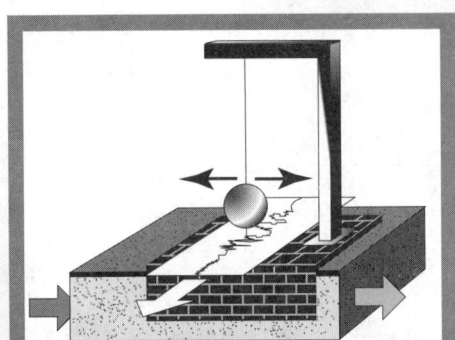

Calculator Skills You'll Learn

▶ `CBL` with light sensor
▶ Δ`List` function
▶ Use `Table` to find the best dimensions of the viewing window
▶ Use `SetFactors` to change the zoom ratio
▶ Entering the regression equation into the `Y=` screen

What You'll Do in This Chapter

The chapter begins with the problem of finding the mathematical model for a set of data from an experimental setting. Within the chapter you will study a wide variety of algebraic functions and their applications, including gravitational attraction, variation of light intensity with the distance from the light source, and direct and inverse variation.

As you learn about the properties of the power and polynomial functions you will also learn a number of useful techniques for fitting these functions to data, including the wise use of regression, the Consecutive Differences method, and how to guess a function by studying the shape of the graph produced by the data. You will apply these methods to experimental data in calculator labs and exploration activities.

You will revisit the idea of drawing tangents to curves and get your first introduction to a powerful calculus idea, the slope function of a function. Finally, you will have learned enough about functions and fitting them to data so that you will be able to tackle the Chapter Project and find the function that fits the experimental data.

Chapter Project—Seismographs and Pendulums

Science Toolkits, Inc.
Why buy when you can build?

To: Research Department
From: Management
Subject: Pendulums and Seismographs

I've just returned from California, and the people there are always talking about earthquakes. A few distributors wanted to know if we have a kit for a seismograph so they can measure earthquakes. I saw an article from the World Wide Web (http:psn.quake.net/lehmntxt.html) showing how to build one. It seems like a great opportunity for a new kit. We just have to add our special touches to make our seismograph kit stand out.

Because a seismograph is basically a pendulum that is sensitive to very small vibrations, we need to learn something about pendulums:

▶ The article mentions the *natural period* of a pendulum. What is that, and what are the factors that influence it? Please test how the following variables change the period of a pendulum:

  • The mass on the end of the pendulum (the *bob*).
  • How far back the pendulum is pulled before it starts swinging \
      (initial displacement).
  • The length of the pendulum.

▶ What is a good way to damp (reduce) the pendulum's motion?

▶ What is the mathematics of the motion of a damped pendulum?

Finally, we're having a contest to name the product. One of the entries is *Great Shakes*. What's your suggestion?

Preliminary Analysis

A **seismograph** is a device that records movement on the earth's surface. Vibrations produced by large trucks, underground nuclear tests, and earthquakes all cause the ground to shake, and a record of this motion is transferred to the paper or screen of a seismograph.

Earthquakes have influenced the present location and shape of continents, and they are responsible for all the mountains of the earth. The surface of the earth, which seems so stable to us, in fact consists of separate but interconnected land masses, called **plates**, that float on a sea of hot fluid rock. The boundaries between plates are called **faults**. Figure 8.1 shows that earthquakes tend to occur along the boundaries of major geologic plates.

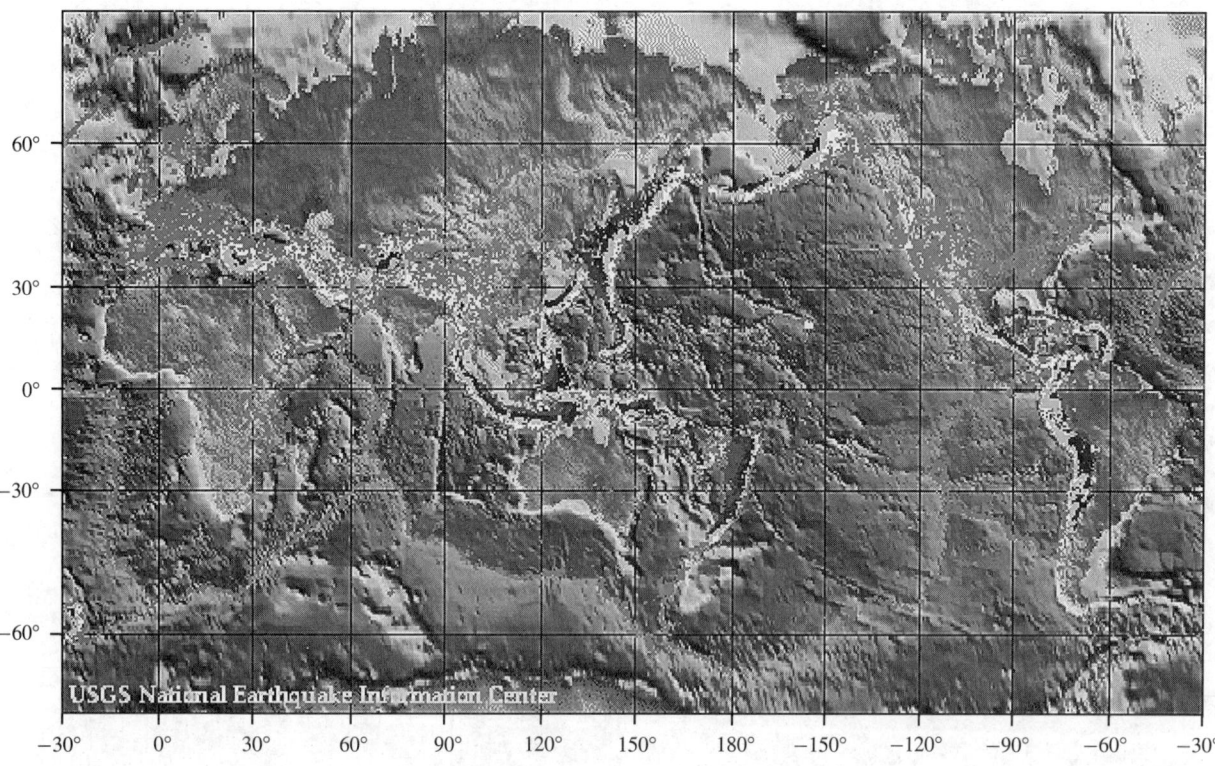

Figure 8.1
World Seismicity: 1975–1995. Courtesy of U.S. Geological Survey. This map was taken from the world wide web site: http://wwwneic.cr.usgs.gov/neis/general/seismicity/world.html

Movement between plates along the faults is not continuous like a flowing stream, but stops and starts like a stretched rubber band that suddenly breaks, making your hands fly apart. The breaking rock between the plates causes what we call an earthquake. Large earthquakes can release power equivalent to many nuclear bombs.

An earthquake produces waves within the earth. The ground vibrates as the waves spread out from the point on the fault where the break occurred (the **epicenter**). When the earthquake's waves reach the seismograph the motion of the vibrating ground is captured by a pen attached to a kind of pendulum built into the seismograph. Figure 8.2 shows a conceptual diagram of how the earth's motion is transferred to marks on a moving paper tape.

Figure 8.3 shows a seismograph record of the famous 1906 San Francisco earthquake that led to the destruction of most of the city. Large earthquakes can be recorded on seismographs anywhere in the world; the seismograph that drew these curves was located in Germany.

In Figure 8.3 the vertical line labeled P, located at about 750 seconds on the horizontal scale, marks the first wave received from the 1906 San Francisco earthquake. The P and the S represent the two different kinds of waves produced by

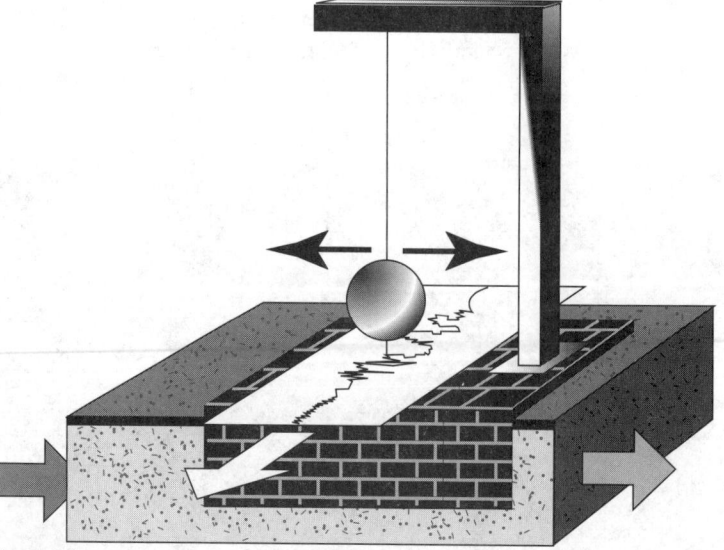

Figure 8.2
Courtesy of John Louie, University of Nevada, Reno. This image is adapted
from the world wide web site:
http://www.seismo.unr.edu/ftp/pub/louie/class/100/seismic-waves.html

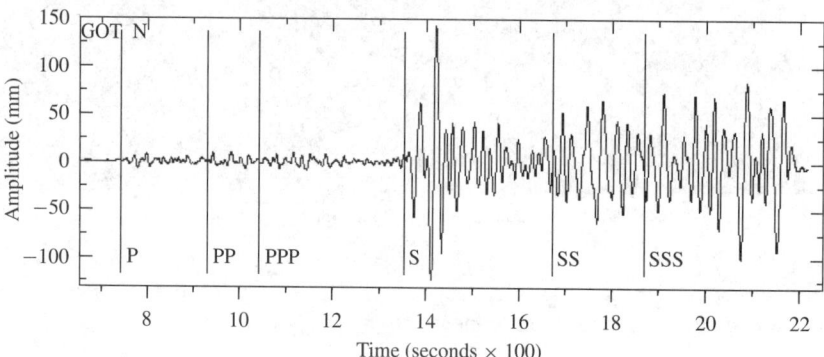

Figure 8.3
David Wald, U.S. Geological Survey. This image was adapted from an image
found at the world wide web site:
http://www-socal.wr.usgs.gov/wald/1906/fig7.html

earthquakes. These waves travel at different speeds through the earth, so they arrive
at different times at the seismograph. Seismologists use the difference in arrival time
to estimate the distance between the seismograph and the earthquake's epicenter.

A pendulum (Figure 8.4) consists of a weight suspended from a fixed point
by a string or rod. Until recently, most clocks used pendulums to keep time (see
Figure 8.5). In Activity 8.1 you will learn why a pendulum can be used to keep
accurate time.

The natural period or natural frequency of the pendulum used in a seismograph
determines the frequency of detected earthquake waves. The **natural period** is the
time required for the pendulum to make one complete swing if it is swinging freely
under the influence of gravity. The **natural frequency** is the number of swings per
unit of time. We'll focus on the period in this chapter, and we'll use the term *period*
instead of *natural period*, since in our activities the pendulum will be swinging freely

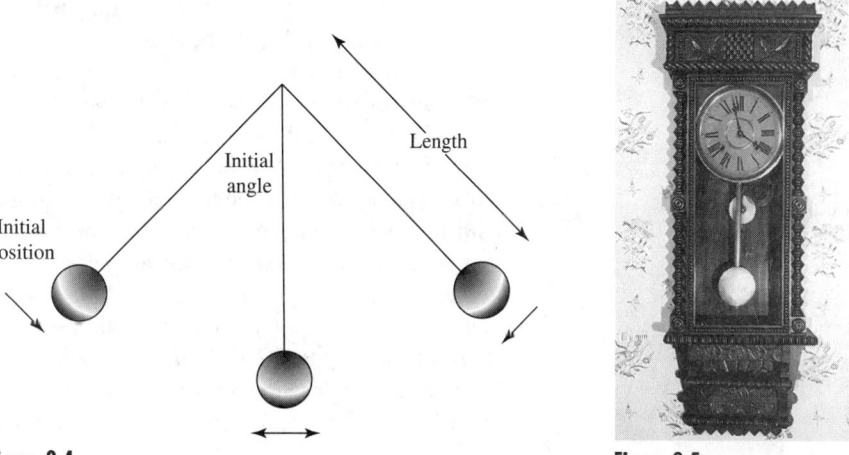

Figure 8.4

Figure 8.5
Courtesy Arthur Weeks.

and not under the influence of an earthquake. In Activity 8.1 you will see which variables affect the period of the pendulum.

Figure 8.6 represents a setup to study the motion of a pendulum with a CBL device connected to a calculator. The back and forth motion is translated into a picture like the one shown in Figure 8.7. Like other situations you have studied—musical notes, vibrations of a Slinky, and the number of hours of daylight—the motion of the pendulum appears to be modeled by a periodic function. Since the time scale on the horizontal axis is marked off in seconds from 0 to 6, you can see that the period is a little less than 2 seconds.

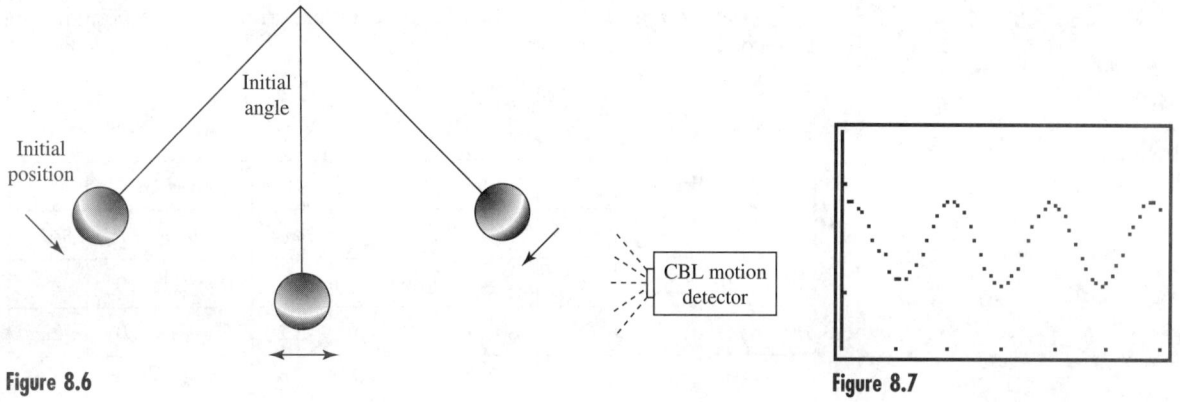

Figure 8.6

Figure 8.7

Activity 8.1
What Affects the Period of a Pendulum?

We will consider three variables and see what effect each has on the period of the pendulum. The variables are:

▶ the mass of the pendulum **bob** (the weight on the end of the string)
▶ the length of the string to the center of the bob
▶ the size of the initial angle of the pendulum

The masses we used were a racquet ball (50 g), a baseball (140 g) and a croquet ball (275 g). We will call these masses M_1, M_2, and M_3. The lengths of the string were 67 cm, 122 cm, and 155 cm. We will call these lengths L_1, L_2, and L_3. The initial angles were 10°, 20°, and 30° which we will call θ_1, θ_2, and θ_3.

To test the influence of each variable on the period we must change its value while keeping the other variables constant. That is, we must test the three masses with the same L and θ; the three lengths with the same M and θ; and the three angles with the same values of M and L. The tests we will run are shown in Table 8.1. How many different experiments are necessary? In this design the three experiments in the middle row of the table all have $M = M_2$, $L = L_2$, and $\theta = \theta_2$, and so they are identical. The three experiments in the first row have the following configurations: Cell 1: L_2, M_1, θ_2; Cell 2: L_1, M_2, θ_2; and Cell 3: L_2, M_2, θ_1. These are all different as are the three experiments in the last row. Since each experiment in the middle row is identical there are a total of seven different experiments to be run.

Table 8.1

Testing Mass	Testing Length	Testing Displacement
$L = L_2, \theta = \theta_2$	$M = M_2, \theta = \theta_2$	$L = L_2, M = M_2$
$M = M_1$	$L = L_1$	$\theta = \theta_1$
$M = M_2$	$L = L_2$	$\theta = \theta_2$
$M = M_3$	$L = L_3$	$\theta = \theta_3$

Table 8.2 gives the results of our seven experiments. The period is measured in seconds.

Table 8.2

Testing Mass		Testing Length		Testing Angle	
$L = L_2, \theta = \theta_2$	Period	$M = M_2, \theta = \theta_2$	Period	$L = L_2, M = M_2$	Period
$M = M_1 = 50$ g	2.2	$L = L_1 = 67$ cm	1.7	$\theta = \theta_1 = 10°$	2.1
$M = M_2 = 140$ g	2.2	$L = L_2 = 122$ cm	2.2	$\theta = \theta_2 = 20°$	2.2
$M = M_3 = 275$ g	2.2	$L = L_3 = 155$ cm	2.7	$\theta = \theta_3 = 30°$	2.2

1. Which of the three variables has the biggest effect on the period of the pendulum?
2. Complete each of the following sentences:

   ▶ As the mass increases, the period of the pendulum seems to _____.
   ▶ As the length increases, the period of the pendulum seems to _____.
   ▶ As the angle increases, the period of the pendulum seems to _____.

Activity 8.1 shows that length of the pendulum has more effect on the period of the pendulum than either the mass of the bob or the initial angle. Despite one measurement that is different, it turns out that changing the initial angle of the pendulum's bob does not change the period. Therefore the period remains constant even as the angle the pendulum swings decreases. The discovery that a pendulum always takes the same time to move back and forth made it possible to build clocks based on the periods of pendulums.

In this chapter you will learn how to find a mathematical model that describes the relationship between period and length of a pendulum. The data you will use comes from an experiment in which we changed the length of a pendulum and measured the time for ten swings. Dividing by 10 then gave us an estimate for the period of that pendulum. In Activity 8.2 you will work with that data.

Activity 8.2
The Effect of the Length of the String on the Period

Table 8.3 contains the results of 12 different pendulum experiments.

Table 8.3

Length (cm)	Period (seconds)
12.7	0.7
20.3	0.9
26.7	1.0
38.1	1.2
53.3	1.5
65.4	1.6
76.2	1.8
90.2	1.9
108.0	2.1
142.2	2.4
165.1	2.6
181.6	2.7

You can either enter these data in your calculator or get them from lists named PENLG (length) and PENPD (period).

1. Plot the data on your calculator.
2. What mathematical equation could model this equation?

You will study many kinds of functions in this chapter. Among them you will find the function that best matches the graph of the period of a pendulum vs. its length. As you learn about each function, think about whether a form of it could fit the data in your graph from Activity 8.2.

8.1

Power Functions: $f(x) = kx^n$

In this section you will learn about the properties and applications of functions like the following:

- $f(x) = 4.5x$
- $g(x) = -x^4$
- $h(x) = 0.7x^{-2}$
- $j(x) = \dfrac{3x^{1/2}}{4}$

- $k(x) = \dfrac{2}{x}$
- $m(x) = 5\sqrt[3]{2x}$
- $n(x) = \pi x^{3.14}$

Each of these is an example of a **power function**, $f(x) = kx^n$, and they are all said to be in the family of power functions. In this section you will study the effect of the values of k and n on the shape and symmetry of the graphs. In Section 8.2 you will see some of the applications of power functions. The function that models the relationship between the length of a pendulum and its period is a member of the power function family.

Power Functions

A **power function** is any function of the form $f(x) = kx^n$ where k and n are real numbers.

Example 8.1

The seven functions above, (a) $f(x) = 4.5x$, (b) $g(x) = -x^4$, (c) $h(x) = 0.7x^{-2}$, (d) $j(x) = \dfrac{3x^{1/2}}{4}$, (e) $k(x) = \dfrac{2}{x}$, (f) $m(x) = 5\sqrt[3]{2x}$, and (g) $n(x) = \pi x^{3.14}$, are all power functions of the form $f(x) = kx^n$. Determine the value of k and n for each function.

Solutions   (a) Even though the exponent in $f(x) = 4.5x$ is not written, it is understood that $x = x^1$. Thus, $k = 4.5$ and $n = 1$.

(b) Since $-x^4 = -1 \cdot x^4$, for the function $g(x) = -x^4$ we see that $k = -1$ and $n = 4$.

(c) For $h(x) = 0.7x^{-2}$ we have $k = 0.7$ and $n = -2$.

(d) Rewriting $j(x) = \dfrac{3x^{1/2}}{4}$ as $j(x) = \frac{3}{4}x^{1/2}$, we see that $k = \frac{3}{4}$ and $n = \frac{1}{2}$.

(e) We can rewrite the function $k(x) = \dfrac{2}{x}$ as $k(x) = 2 \cdot \dfrac{1}{x} = 2x^{-1}$. Thus, $k = 2$ and $n = -1$.

(f) The function $m(x) = 5\sqrt[3]{2x}$ can be rewritten as $m(x) = 5\sqrt[3]{2x} = 5\sqrt[3]{2}\sqrt[3]{x}$. Since $x^{1/n} = \sqrt[n]{x}$ or the nth root of x, then $\sqrt[3]{x} = x^{1/3}$ and so $k = 5\sqrt[3]{2}$ and $n = \frac{1}{3}$.

(g) For $n(x) = \pi x^{3.14}$ we have $k = \pi$ and $n = 3.14$.   ∎

Power functions are an example of a type of functions called **algebraic functions** that we will study next.

Algebraic Functions

As you look at the following definition of algebraic functions, think about which of the functions you have studied are algebraic and which are not.

> **Definition: Algebraic Functions**
> An **algebraic function** is a function whose values can be computed with a finite number of elementary operations of mathematics. These operations are addition, subtraction, multiplication, division, and roots, such as square root, cube root, 12th root, etc.

Example 8.2

Here are some functions you have studied. Which of them are algebraic functions?

(a) Linear functions, for example, distance vs. time if you are walking at a constant rate of 5 feet per second, beginning at a point 8 feet from the measuring device:

$$d(t) = 5t + 8$$

(b) Quadratic functions, for example the distance vs. time graph of a ball thrown upwards with an initial velocity of 30 feet per second from a distance 20 feet above the ground:

$$h(t) = -16t^2 + 30t + 20$$

(c) Functions involving roots, for example the frequency of the nth note on the scale, beginning with middle C:

$$f(t) = 262 \sqrt[12]{n}$$

(d) Trigonometric functions, for example the mathematical model of the loudness of the tone of middle C, as a function of time:

$$L(t) = L_0 \sin(524\pi t)$$

(e) Exponential functions, for example the percent of full charge of a capacitor in terms of the number of time constants since it began charging:

$$C = 100(1 - 0.37^t)$$

Solutions   (a)–(c) From the definition of algebraic function you can see that the first three are algebraic because any value can be computed with a small number of the computations listed in the definition. What about the last three?

(d) Trigonometric functions are not algebraic. Even though there is an easy-to-understand definition of how to compute the sine of an angle from the ratio of two sides of a right triangle or the location of a point on a circle, you cannot compute the value of, for example, $\sin 52°$ with a finite number of elementary operations.

(e) Exponential functions are not algebraic. They seem like they would be, because raising a number to a power is just a certain number of multiplications. For example,

$$C(3) = 100(1 - 0.37^3)$$
$$= 100(1 - 0.37 \cdot 0.37 \cdot 0.37)$$
$$= 100(1 - 0.050653)$$
$$= 94.9347$$

However, the variable in a function can be any number. What is the value of $C(\pi)$? Since $\pi = 3.141592653\ldots$ a decimal number which never stops and never repeats, the value of 0.37^π could not be computed in a finite number of steps. Of course it could be approximated by $0.37^{3.14}$, which the calculator says is about 0.0441, but the exact value could not be computed in a finite number of steps.   ∎

Exploring the Graphs of $f(x) = kx^n$

In the following calculator labs you will study the graphs of different kinds of power functions. You will see how the shapes change as the exponent changes, and how the shapes of the graphs with negative exponents compare with those for positive exponents. After you have had some experience with the graphs of the functions we will study the symmetry that each graph shows.

There will be three parts to the study:

▶ the positive values of x
▶ A closer look at the values of x from 0 to 1, that is $0 \le x \le 1$
▶ the negative values of x

At the end of each lab you should be able to answer the following questions:

1. In the interval $0 \le x \le 1$, how does the exponent of the function determine which function has the largest value?
2. In the interval $x \ge 1$, how does the exponent of the function determine which function has the largest value?
3. Answer questions 1 and 2 for the negative values of x, that is, for the intervals (a) $-1 \le x \le 0$ and (b) $x \le -1$.
4. What point or points do all these power functions have in common?

There are three labs that follow. One is for power functions with positive integer exponents, one with negative integer exponents, and one with fractional exponents. After you do the explorations in each lab, you will be asked to answer the above questions about the particular functions you have studied. A discussion on symmetry of power functions follows the first two labs.

These labs also include the task of providing a sketch of the function's graph. A sketch shows the shape of a graph and the main features, without plotting individual points.

Calculator Lab 8.1   Power Functions—Positive Exponents

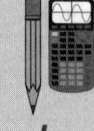

In this lab you will study the four functions defined by $f(x) = kx^n$ where $n = 1$, 2, 3, and 4.

Procedures   1. For each function complete a table of values like the one shown below. Let k have the same value in each function. (You might want to enter these functions in the Y= screen by letting $T_1=KX$, $T_2=KX^2$, etc. Then select and store a value for K.) Use the Table feature of your calculator to complete the table.

x	0	0.5	1	1.5	2	2.5	3	3.5	4
$f_1(x) = kx$									
$f_2(x) = kx^2$									
$f_3(x) = kx^3$									
$f_4(x) = kx^4$									

Now let's take a look at the functions when x is a small nonnegative number.

2. For each function, fill in a table of values like the one shown below.

x	0	0.2	0.4	0.6	0.8	1.0
$f_1(x) = kx$						
$f_2(x) = kx^2$						
$f_3(x) = kx^3$						
$f_4(x) = kx^4$						

Next we'll explore negative values of x.

3. For each function, fill in a table of values like the one shown below.

x	0	-0.5	-1	-1.5	-2	-2.5	-3	-3.5	-4
$f_1(x) = kx$									
$f_2(x) = kx^2$									
$f_3(x) = kx^3$									
$f_4(x) = kx^4$									

Now let's take a look at the functions when x is a negative number close to 0.

4. For each function, fill in a table of values like the one shown below.

x	0	-0.2	-0.4	-0.6	-0.8	-1.0
$f_1(x) = kx$						
$f_2(x) = kx^2$						
$f_3(x) = kx^3$						
$f_4(x) = kx^4$						

5. Plot all four functions on your calculator. Sketch each of the graphs on paper and label each graph according to its equation.

6. Answer each of the following questions:

   (a) In the interval $0 \leq x \leq 1$, how does the exponent of the function determine which function has the largest value?

   (b) In the interval $x \geq 1$, how does the exponent of the function determine which function has the largest value?

   (c) In the interval $-1 \leq x \leq 0$, how does the exponent of the function determine which function has the largest value?

(d) In the interval $x \leq -1$, how does the exponent of the function determine which function has the largest value?

(e) What point or points do all these power functions have in common?

Example 8.3

Explain how the function $f(x) = 1$ fits into the power function family.

Solution   The function $f(x) = 1$ is the same as the power function $g(x) = x^0$, with one exception: $g(0)$ is undefined because 0^0 is undefined.   ■

Calculator Lab 8.2

Power Functions—Negative Exponents

In this calculator lab you will study the four functions defined by $f(x) = kx^n$, where $n = -1, -2, -3,$ and -4. Again, you may want to store these functions by using the Y= menu of your calculator.

Procedures   1. Complete the table below.

x	0	0.5	1	1.5	2	2.5	3	3.5	4
$f_1(x) = kx^{-1}$									
$f_2(x) = kx^{-2}$									
$f_3(x) = kx^{-3}$									
$f_4(x) = kx^{-4}$									

Now let's take a look at the functions when x is a positive number close to 0.

2. For each function, fill in a table of values like the one shown below.

x	0	0.2	0.4	0.6	0.8	1.0
$f_1(x) = kx^{-1}$						
$f_2(x) = kx^{-2}$						
$f_3(x) = kx^{-3}$						
$f_4(x) = kx^{-4}$						

Next we'll explore negative values of x.

3. For each function, fill in a table of values like the one shown below.

x	0	−0.5	−1	−1.5	−2	−2.5	−3	−3.5	−4
$f_1(x) = kx^{-1}$									
$f_2(x) = kx^{-2}$									
$f_3(x) = kx^{-3}$									
$f_4(x) = kx^{-4}$									

Finally let's take a look at the functions when x is a negative number close to 0.

4. For each function, fill in a table of values like the one shown below.

x	0	−0.2	−0.4	−0.6	−0.8	−1.0
$f_1(x) = kx^{-1}$						
$f_2(x) = kx^{-2}$						
$f_3(x) = kx^{-3}$						
$f_4(x) = kx^{-4}$						

5. Plot all four functions on your calculator. Sketch each of the graphs on paper and label each graph according to its equation.

6. Answer each of the following questions.

   (a) In the interval $0 \leq x \leq 1$, how does the exponent of the function determine which function has the largest value?

   (b) In the interval $x \geq 1$, how does the exponent of the function determine which function has the largest value?

   (c) In the interval $-1 \leq x \leq 0$, how does the exponent of the function determine which function has the largest value?

   (d) In the interval $x \leq -1$, how does the exponent of the function determine which function has the largest value?

   (e) What point or points do all these power functions have in common?

Example 8.4

What symmetry is shown by the graphs of the power functions $f(x) = x^4$ and $f(x) = x^3$?

Solutions   We begin with the function $f(x) = x^4$, because it is an example of a kind of symmetry that we have seen before. Figure 8.8 shows the graph, and the coordinates of four points, A, B, C, and D. Notice that each of the points A and C has a mirror image across the y-axis. Therefore we say that the graph of the function $f(x) = x^4$ has symmetry with respect to the y-axis. That is, the part of the curve to the left of the y-axis is mirrored in the part to the right of the y-axis.

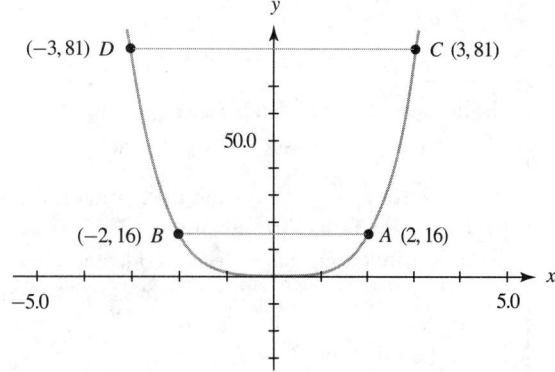

Figure 8.8

Now compare the coordinates of these mirror points B and D with the coordinates of A and C. You can see that $f(3)$ and $f(-3)$ are both equal to 81, and $f(2)$ and $f(-2)$ are both equal to 16. In general for a function whose graph is symmetric about the y-axis we can say that

$$f(a) = f(-a)$$

Next we consider a different shape, but one that also has symmetry. The function $f(x) = x^3$ is plotted in Figure 8.9 with the coordinates of four points, A, B, C, and D.

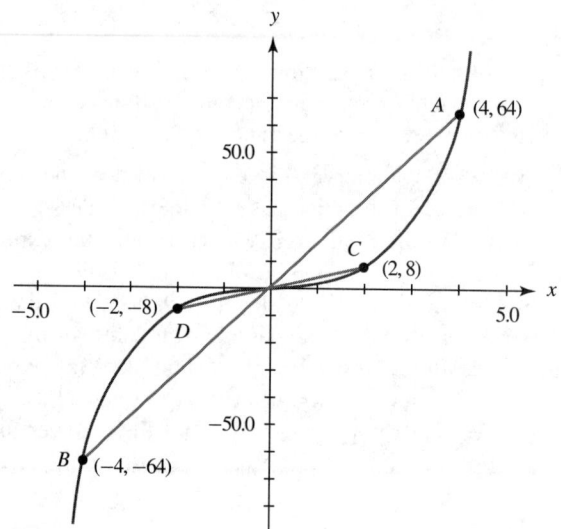

Figure 8.9

Notice that each of the points A and C has a mate with similar coordinates and location. Notice also that the lines joining each pair of points passes through the origin of the coordinate system. In fact, any point you could pick in the first quadrant would have a mate in the third quadrant, and the line joining them would go through the origin. For graphs having this property we say that the function has **symmetry with respect to the origin**.

Now compare the coordinates of the points B and D with the coordinates of A and C. You can see that $f(2)$ and $f(-2)$ have opposite signs, as do $f(4)$ and $f(-4)$. In general, for a function whose graph is symmetric with respect to the origin, we can say that

$$f(a) = -f(-a)$$

■

Definition: Odd and Even Functions

The definitions of odd and even functions are connected to their symmetry.

▶ A function is said to be an **odd function** if it has symmetry with respect to the origin. This also means that if $f(a) = -f(-a)$ for all values of a that fit the function, then f is an odd function.

▶ A function is said to be an **even function** if it has symmetry with respect to the y-axis. This also means that if $f(a) = f(-a)$ for all values of a that fit the function, then f is an even function.

Example 8.5

Tell whether the following functions are odd, even, or neither.

(a) $f(x) = \cos x$

(b) $g(\theta) = \tan \theta$

(c) $h(t) = \sin t$

(d) $j(t) = \sin\left(t - \dfrac{\pi}{4}\right)$, or $\sin(\theta - 45°)$

Solutions   The first three functions will probably be familiar to you. If you need a reminder of the shape and location of any of the functions, plot it with your calculator. You can see from the graphs that cosine is an even function and sine and tangent are odd functions. On the other hand the function $j(t) = \sin\left(t - \dfrac{\pi}{4}\right)$ or $\sin(\theta - 45°)$ is different. Its graph is shown in Figure 8.10.

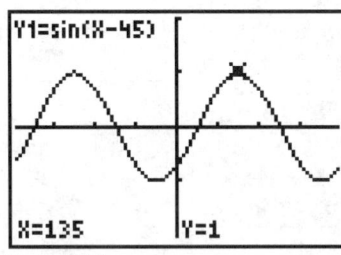

```
Xmin = -352.5
Xmax = 352.5
Xscl = 90
Ymin = -2
Ymax = 2
Yscl = 1
```

Figure 8.10

If a line is drawn from the marked point $(135°, 1)$ through the origin to the point $(-135°, -1)$, you can see that $(135°, -1)$ is not on the graph. So the function is not odd. The mirror image across the y-axis of the point $(135°, 1)$ is the point $(-135°, 1)$. This point is not on the graph of j either, so the function $j(x) = \sin\left(t - \dfrac{\pi}{4}\right) = \sin(t - 45°)$ is not even. Combining our results we conclude that j is neither even nor odd.   ∎

Finally we turn to the power functions that relate to square roots, cube roots, etc.

Calculator Lab 8.3

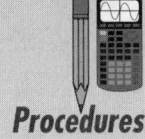

Power Functions—Fractional Exponents

In this lab you will study the four functions defined by $f(x) = kx^n$ where $n = 1$, $\frac{1}{2}$, $\frac{1}{3}$, and $\frac{1}{4}$.

Procedures   1. For each function, complete a table of values like the one shown below.

x	0	5	10	15	20	25	30
$f_1(x) = kx$							
$f_2(x) = kx^{1/2}$							
$f_3(x) = kx^{1/3}$							
$f_4(x) = kx^{1/4}$							

Now let's take a look at the functions when x is a small positive number.

2. Complete a table of values like the one shown below.

x	0	0.2	0.4	0.6	0.8	1.0
$f_1(x) = kx$						
$f_2(x) = kx^{1/2}$						
$f_3(x) = kx^{1/3}$						
$f_4(x) = kx^{1/4}$						

Next we'll explore negative values of x. Remember that the square root of a negative number is not a real number. For example you have learned that $\sqrt{-4} = 2i$, which is an imaginary number. However, the cube root of a negative number is a real number. For example $\sqrt[3]{-8} = -2$, because $(-2)(-2)(-2) = -8$.

3. For each function, fill in a table of values like the one shown below.

x	0	−0.5	−1	−1.5	−2	−2.5	−3	−3.5	−4
$f_1(x) = kx$									
$f_2(x) = kx^{1/2}$									
$f_3(x) = kx^{1/3}$									
$f_4(x) = kx^{1/4}$									

Now let's take a look at the functions when x is a negative number close to 0.

4. For each function, fill in a table of values like the one shown below.

x	0	−0.2	−0.4	−0.6	−0.8	−1.0
$f_1(x) = kx$						
$f_2(x) = kx^{1/2}$						
$f_3(x) = kx^{1/3}$						
$f_4(x) = kx^{1/4}$						

5. Plot all four functions on your calculator. Sketch each of the graphs on paper and label each graph according to its equation.

6. Answer each of the following questions.

   (a) In the interval $0 \leq x \leq 1$, how does the exponent of the function determine which function has the largest value?

   (b) In the interval $x \geq 1$, how does the exponent of the function determine which function has the largest value?

   (c) In the interval $-1 \leq x \leq 0$, how does the exponent of the function determine which function has the largest value?

   (d) In the interval $x \leq -1$, how does the exponent of the function determine which function has the largest value?

   (e) What point or points do all these power functions have in common?

Summary: Shapes of Power Function Graphs

Positive Integer Values of n

Figure 8.11 summarizes the shapes of power functions $f(x) = kx^n$ for positive integer values of n.

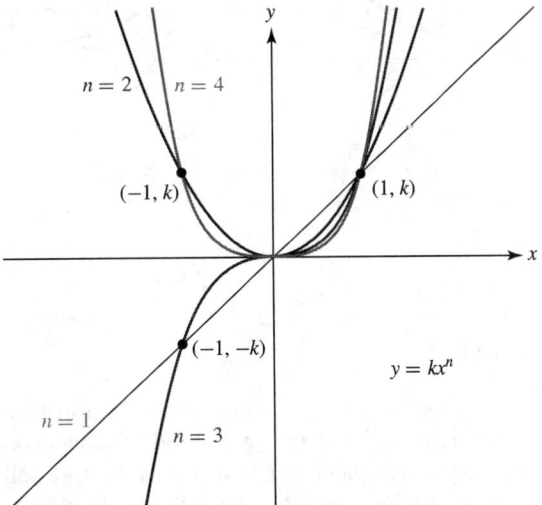

Figure 8.11

Make sure you understand the following main points about these graphs:

▶ The curve for $n = 1$ is a straight line and $n = 2$ produces a parabola.
▶ The curve for $n = 4$ is curved in a similar way to a parabola
▶ The curves for $n = 2$ and $n = 4$ are symmetric with respect to the y-axis and represent even functions.
▶ The curves for $n = 1$ and $n = 3$ are symmetric with respect to the origin and represent odd functions.
▶ All curves share the points $(1, k)$ and $(0, 0)$.
▶ In addition, even functions (those with even exponents) share the point $(-1, k)$, and odd functions (those with odd exponents) share the point $(-1, -k)$.
▶ For positive values of k and x all the curves are rising; that is, they go up as you move to the right. Another way to say this is that a tangent drawn at any point where $x > 0$ has a positive slope.

Negative Integer Values of n

Figure 8.12 summarizes the shapes of power functions $f(x) = kx^n$ for negative integer values of n.

Make sure you understand the following main points about these graphs:

▶ All functions have no result if $x = 0$.
▶ The curves for $n = -2$ and $n = -4$ are symmetric with respect to the y-axis and represent even functions.
▶ The curves for $n = -1$ and $n = -3$ are symmetric with respect to the origin and represent odd functions.
▶ All curves share the point $(1, k)$.

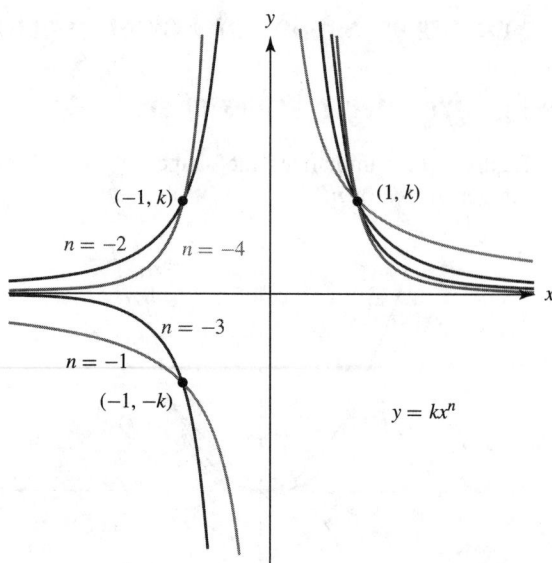

Figure 8.12

▶ In addition, even functions (those with even exponents) share the point $(-1, k)$, and odd functions (those with odd exponents) share the point $(-1, -k)$.
▶ For positive values of k all the curves are falling; that is, they go down as you move to the right. Another way to say this is that a tangent drawn at any point where $x > 0$ has a negative slope.
▶ For positive values of k the values of y increase indefinitely as x gets close to zero. That is

$$\text{if } x > 0 \text{ then } \lim_{x \to 0} kx^n = +\infty$$

▶ For negative values of x, with $k > 0$, the even functions also approach positive infinity as x approaches zero.
▶ For negative values of x, with $k > 0$, the odd functions approach negative infinity as x approaches zero. That is

$$\text{if } k > 0 \text{ and } n \text{ is a negative integer then } \lim_{x \to 0} kx^n = -\infty$$

▶ All functions approach zero as x approaches either positive or negative infinity. That is

$$\lim_{x \to \infty} f(x) = 0$$
$$\text{and} \quad \lim_{x \to -\infty} f(x) = 0$$

Notice how the curves get very close to the x- and y-axes. This is typical of functions in which x appears in the denominator, as in $f(x) = kx^{-2} = \dfrac{k}{x^2}$. This observation leads to two definitions.

Definition: Asymptotes
The graphs of all power functions with negative exponents have the following two properties:

1. The x-axis (or the line whose equation is $y = 0$) is a line that all curves approach (but never reach) as $x \to \pm\infty$. Such a line is called a **horizontal asymptote**.

2. The y-axis (or the line whose equation is $x = 0$) is a line that all curves approach (but never reach) as $x \rightarrow 0$. Such a line is called a **vertical asymptote**.

Other functions we will study have horizontal and vertical asymptotes that are different from the x-axis and the y-axis. We will later give examples of functions whose graphs cross a horizontal asymptote. However, these crossings are relatively close to zero. The important concept is that a graph approaches, but does not cross, a horizontal asymptote as $x \rightarrow \infty$ or $x \rightarrow -\infty$. That statement does not forbid the graph from crossing the horizontal asymptote "near" the y-axis.

Positive Fractional Values of n

Figure 8.13 summarizes the shapes of power functions $f(x) = kx^{1/n}$, where $n = 1, 2, 3, 4, \ldots$ and $k > 0$. We have the following main points about these graphs:

▸ If n is even the function has no value if x is negative.
▸ If n is odd the function is an odd function.
▸ All curves share the points $(1, k)$ and $(0, 0)$.
▸ If n is odd the functions share the point $(-1, -k)$.
▸ All the curves are rising; that is, they go up as you move to the right. Another way to say this is that a tangent drawn at any point has a positive slope.

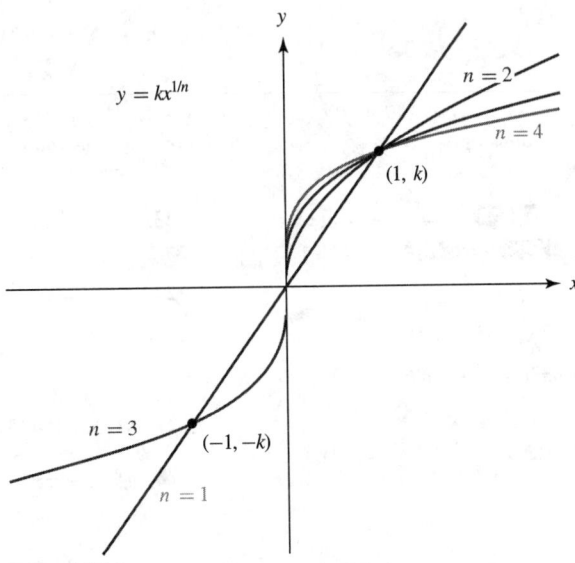

Figure 8.13

Section 8.1 Exercises

Each function in Exercises 1–4 is a power function of the form $y = kx^n$. Determine the value of k and n for each function.

1. $f(x) = -7\pi x^{4.2}$

2. $g(x) = 3\sqrt{5x^7}$

3. $h(x) = \dfrac{\sin 4}{x^{2\pi}}$

4. $j(x) = -\dfrac{3}{\pi} x^{\cos 1.2\pi}$

In Exercises 5–8, (a) determine whether each function is an algebraic function, and (b) explain why you made that decision.

5. $f(x) = 14.2$

6. $g(x) = \sin^2 x + \cos x$

7. $h(x) = 3^x$

8. $j(x) = 9x^7 - 3x^2 + \sqrt{4x^3} + 9\sin 3x$

In Exercises 9–12, (a) describe the symmetry of the given function, and (b) determine whether the function is odd, even, or neither odd nor even.

9. $f(x) = 17x^4$

10. $g(x) = -0.2x^5$

11. $h(x) = \sin x + \tan x$

12. $j(x) = \cos x - \dfrac{1}{\tan x}$

In Exercises 13–16, (a) describe the symmetry of the given function, (b) determine and justify whether the function is odd, even, or neither odd nor even, and (c) describe the values of y as x gets close to zero.

13. $f(x) = 3x^{-4}$

14. $g(x) = -0.2x^{-5}$

15. $h(x) = -\dfrac{1}{\pi x^3}$

16. $j(x) = \dfrac{\pi^3}{x^7}$

In Exercises 17–20, (a) describe the symmetry of the given function, (b) determine and justify whether the function is odd, even, or neither odd nor even, and (c) describe the slope of a tangent to the function's graph.

17. $f(x) = 5x^{1/4}$

18. $g(x) = 0.08x^{1/7}$

19. $h(x) = -\frac{2}{\pi} x^{1/5}$

20. $j(x) = -1.2x^{1/8}$

Exercises 21–24, each contain two functions f and g with $g(x) = f(x) + 1$. That is, the graph of g is the graph of f translated 1 unit vertically to help you tell the graphs apart. (a) Graph each function on your calculator and (b) describe how the two graphs are alike and how they are different.

21. $f(x) = x^{2/4}$ [Enter as x ▬ ^ ▬ (2/4).] and $g(x) = \left(x^2\right)^{1/4} + 1$ [Enter as (x ▬ ^ ▬ 2) ▬ ^ ▬ (1/4)+1.]

22. $f(x) = x^{2/6}$ and $g(x) = \left(x^2\right)^{1/6} + 1$

23. $f(x) = x^{-2/6}$ and $g(x) = \left(x^{-2}\right)^{1/6} + 1$

24. $f(x) = x^{-1/2}$ and $g(x) = \left(x^{-1}\right)^{1/2} + 1$

In Exercises 25–28 give all horizontal and vertical asymptotes of the given function.

25. $f(x) = -3x^{-2}$

26. $g(x) = -0.275x^{-3}$

27. $h(x) = 0.435x^{-4}$

28. $j(x) = \dfrac{\pi}{25} x^{-5}$

Exercises 29–32 show the graph of a power function $y = kx^n$ graphed on a TI-83 using the indicated viewing window. One point on the curve is shown. Based on your knowledge of power functions, symmetry, and asymptotes, (a) estimate the values of k and n, and (b) justify your estimates.

29.

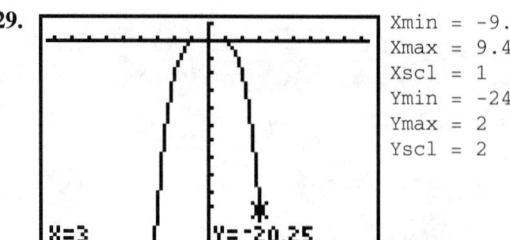

```
Xmin = -9.4
Xmax = 9.4
Xscl = 1
Ymin = -24
Ymax = 2
Yscl = 2
```

30.

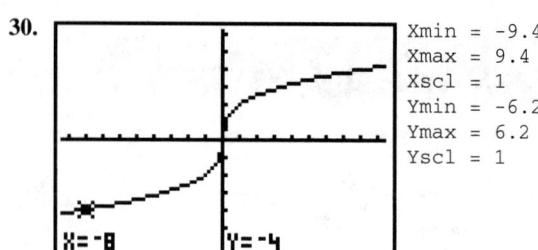

```
Xmin = -9.4
Xmax = 9.4
Xscl = 1
Ymin = -6.2
Ymax = 6.2
Yscl = 1
```

31.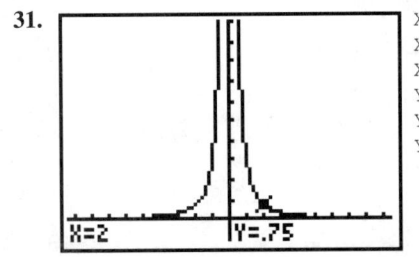

```
Xmin = -9.4
Xmax = 9.4
Xscl = 1
Ymin = -1
Ymax = 10
Yscl = 1
```

32.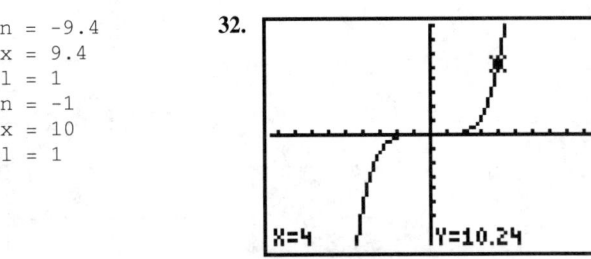

```
Xmin = -9.4
Xmax = 9.4
Xscl = 1
Ymin = -16
Ymax = 16
Yscl = 2
```

8.2

Variation and Power Functions

Power functions are important because they are the mathematical models of all relationships that can be described by direct or inverse variation. We'll introduce this topic with a story.

A Legend about Gifts and Gold

The king was angry. He had received a gift of a cube of gold 2 cm on a side from the ambassador from Transylvania, and he had angrily rejected it. "Too small," he raged. "I expect a cube of gold that weighs four times as much as this." Later the ambassador returned to offer a larger cube of gold. This one was 4 cm on a side.

The king screamed, "I asked for four times as much and you're offering me twice as much. Be gone!"

"But sire," said the ambassador "let me explain with mathematics. You see, volume is computed with a power function..."

But the king interrupted him. "You think I don't know mathematics? Tell your king this means war!"

Who is right? How can you get a cube that is four times the volume of another? This is an example of direct variation. We'll take a look at it with a picture, with numbers, with formulas, and with a graph.

Direct Variation

Figure 8.14 shows two cubes like the ones that the ambassador offered. What is the ratio of their volumes? You can see that the cube on the left is identical to one part of the cube on the right. Can you see that eight identical cubes make up the cube on the right? Therefore the ambassador's second gift has **eight** times the volume as the first gift.

The volume of a cube is expressed by the power function $V(x) = x^3$. The volume of the first gift $V(2) = 2^3 = 8$ cm^3 and the volume of the second gift is $V(4) = 4^3 = 64$ cm^3. The ratio of the volumes of the two gifts is $\dfrac{V_2}{V_1} = \dfrac{V(4)}{V(2)} = \dfrac{64}{8} = 8$. The second gift has eight times the volume of the first.

Finally, look at the graph of the power function $V(x) = x^3$ (Figure 8.15). The ratio of the x values is $\frac{4}{2} = 2$ and the ratio of the V values is $\frac{64}{8} = 8$. The second gift has eight times the volume of the first.

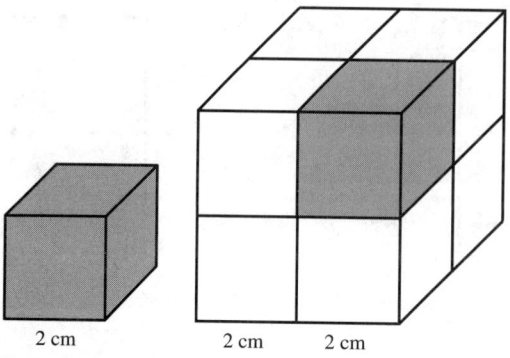

2 cm   2 cm   2 cm

Figure 8.14

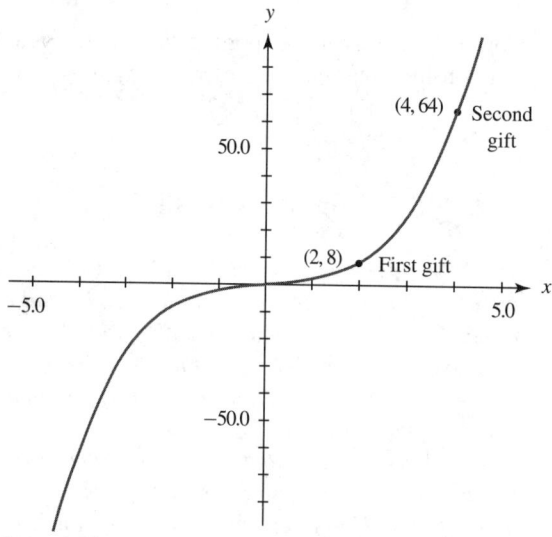

Figure 8.15

Example 8.6

How long should the side of the cube be so that its volume is four times the volume of a cube 2 cm on a side?

Solution We'll solve this problem using the formula and the graph. The volume of the first gift is 8 cm³, so the volume of a cube with four times the volume must be $4 \times 8 = 32$ cm³. The graph in Figure 8.16 shows how to estimate the side of the cube whose volume is 32 cm³. Begin at 32 on the V-axis and draw a horizontal line to the curve and then a vertical line down. The value of x is approximately 3.2 cm.

The formula will give a more exact answer. We know that $V(x) = 32$. What is x?

$$V(x) = x^3 = 32$$
$$x = \sqrt[3]{32}$$

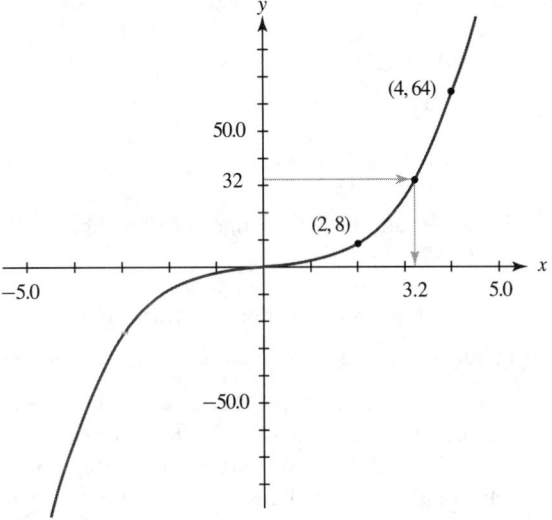

Figure 8.16

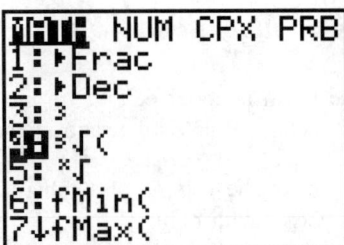

Figure 8.17

The value of $\sqrt[3]{32}$ can be computed in two ways with the *TI-83*. First look at the MATH menu (press **MATH**). As shown in Figure 8.17, selection number 4 is the cube root. Or you can remember that the cube root of a number x is the same as $x^{1/3}$. The following entry will produce the cube root of 32:

32 **∧** **(** 1 **÷** 3 **)** **ENTER**

Either way the result is the same:

$$\sqrt[3]{32} = 32^{1/3} \approx 3.175$$

Each side of the cube must be about 3.175 cm in order for the cube to have a volume that is four times the volume of the first gift. ■

In general if the side of a cube is multiplied by the ratio r, what is the volume multiplied by? This question is restated and answered in Example 8.7.

Example 8.7

Two cubes have sides with lengths of x_1 and x_2. What is the ratio of the volumes of the cubes?

Solution   If $\dfrac{x_2}{x_1} = r$, what is $\dfrac{V_2}{V_1}$? Since $V_2 = x_2{}^3$ and $V_1 = x_1{}^3$, then

$$\frac{V_2}{V_1} = \frac{x_2{}^3}{x_1{}^3}$$

$$= \left(\frac{x_2}{x_1}\right)^3 = r^3$$

Thus the ratio of the volumes of two cubes is the cube of the ratio of the lengths of their sides.   ■

The ratio r is called the **ratio of the linear dimensions** since x_1 and x_2 are in measurements in one dimension, such as centimeters or feet. The ratio $\dfrac{V_2}{V_1}$ is called the **ratio of cubic dimensions** because V_2 and V_1 measure quantities in three dimensions, such as cubic centimeters or cubic feet.

What if the king had gotten gold spheres instead of gold cubes? Or gold ellipsoids? Would the same rule of ratios be true? Example 8.8 answers these questions.

Example 8.8   Application

The radius of the Earth is about 6.38×10^6 meters and the radius of the Moon is about 1.74×10^6 meters. What is the ratio of their volumes?

Solution   The power function for the volume of a sphere is $V(R) = \frac{4}{3}\pi R^3$. The ratio of the volumes of two spheres with radius R_1 and R_2 is

$$\frac{V_1}{V_2} = \frac{\frac{4}{3}\pi R_1{}^3}{\frac{4}{3}\pi R_2{}^3}$$

Since the factor $\frac{4}{3}\pi$ appears in both the numerator and denominator then

$$\frac{V_1}{V_2} = \frac{R_1{}^3}{R_2{}^3} = \left(\frac{R_1}{R_2}\right)^3$$

Notice that the ratio of the volumes is the cube of the ratio of the linear dimensions.

For the Earth and Moon the ratio of their radii, in kilometers, is

$$\frac{\text{Earth}}{\text{Moon}} = \frac{R_1}{R_2} \approx \frac{6378}{1738} \approx 3.67$$

Cubing this answer we get the ratio of their volumes: $3.67^3 \approx 49.4$. That is, the volume of the Earth is about 50 times the volume of the Moon.   ■

We have seen that for two power functions, $V(x) = x^3$ and $V(R) = \frac{4}{3}\pi R^3$, the ratio $\dfrac{V_2}{V_1}$ is equal to the cube of the ratio of the linear dimensions. This is generally true for the function $f(x) = kx^3$ because the k factor, like the factor $\frac{4}{3}\pi$, will be canceled out of the top and bottom parts of the fraction that represents the ratio. Also, if you look back at Examples 8.7 and 8.8 you can see that the power was 3 but it could have been any positive power and the result would be similar. We will consider power functions with negative powers separately. Therefore we have the following summary of direct variation:

Direct Variation as the nth Power

If $f(x) = kx^n$ with $n > 0$, and if $\dfrac{x_2}{x_1} = r$, then

$$\frac{f(x_2)}{f(x_1)} = r^n$$

Two sentences are often used to describe this relationship in words:

▶ $f(x)$ **varies directly** as the nth power of x
▶ If the value of x is multiplied by r, then the value of $f(x)$ would be multiplied by r^n

The next example returns to a familiar topic, gravity influenced motion, from the point of view of power functions.

Example 8.9   Application

If air resistance is disregarded, the distance d that a ball falls varies directly as the square of the time t since it was dropped.

(a) What is the power function that models this situation?
(b) If a ball falls 64 feet in the first two seconds, how far does it fall in the first four seconds?
(c) How far would it fall in the first nine seconds?
(d) What is the value of k in the formula for the power function?

Solutions (a) The phrase *varies directly as* tells us that this is a power function of the form $d = kt^n$. The words *the square* means that $n = 2$. Therefore the formula is $d = kt^2$.

(b) Since the ratio of times is $\frac{4}{2} = 2$, then the ratio of distances is $2^2 = 4$. So a ball that falls 64 ft the first two seconds will fall four times as far in the first four seconds. Therefore the distance traveled in the first four seconds is $4 \times 64 = 256$ ft.

(c) The ratio of times is $\frac{9}{2} = 4.5$, so the ratio of distances is $4.5^2 = 20.25$. Therefore the distance traveled in the first nine seconds is $20.25 \times 64 = 1296$ ft.

(d) Since $d = kt^2$, we can find the value of k by substituting one pair of values for d and t, except $d = 0$ and $t = 0$. Let's take the first given pair of values, $d = 64$ and $t = 2$. Substituting these values into the equation $d = kt^2$, we have the equation $64 = k \cdot 2^2 = k \cdot 4$, which leads to $k = 16$. Therefore the power function for this example is $d = 16t^2$. This is the familiar equation for gravity influenced motion with initial velocity zero and initial distance zero. The coefficient of t^2 is 16 if d is measured in feet and t is measured in seconds. ■

The computation of k is very important in determining a good model for a set of data. We will need that technique when we find the model for light intensity as a function of distance.

Variation of Light Intensity with Distance

You know that the intensity, or brightness, of a light changes as you change your distance from the light. Take a moment to think about this relationship. Does the graph go up or down as the distance increases? If the distance doubles, how does the intensity change?

Draw a set of coordinate axes like the ones shown in Figure 8.18 and sketch a graph of how you think the intensity of a light bulb changes as your distance from it changes. Think about the intensity you see when you are one mile away from the light, compared with when you are one foot away.

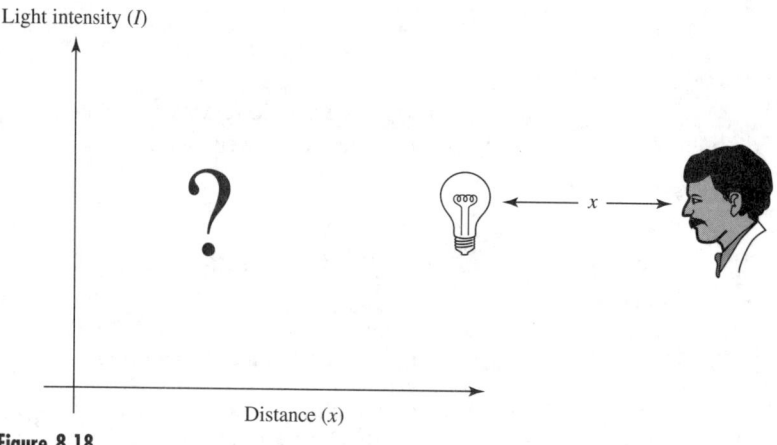

Figure 8.18

Calculator Lab 8.4 provides an opportunity to investigate the relationship between light intensity and distance.

Calculator Lab 8.4

Measuring Light Intensity

The Calculator-Based Laboratory System (CBL) is shipped with a light sensor that measures the intensity of a light. Light bulbs are rated in watts, which are a measure of power, or energy used per second. Brightness or intensity of light represents the power observed in a given area, for example, watts per square foot. The light sensor of the CBL unit records light intensity in milliwatts per square centimeter. One milliwatt (mW) = 0.001 W.

Using the light sensor, the CBL unit, and the calculator program named LIGHT, we did an experiment to measure the changing intensity of a light bulb as the distance between the sensor and the bulb was changed. Figure 8.19 shows the experimental setup we used.*

The data we collected are in Table 8.4. You can see that the intensity decreases as the distance increases.

---

*This experiment is described in the publication, *CBL System Experiment Workbook*, 1994, published by Texas Instruments, Inc., pages 54–56.

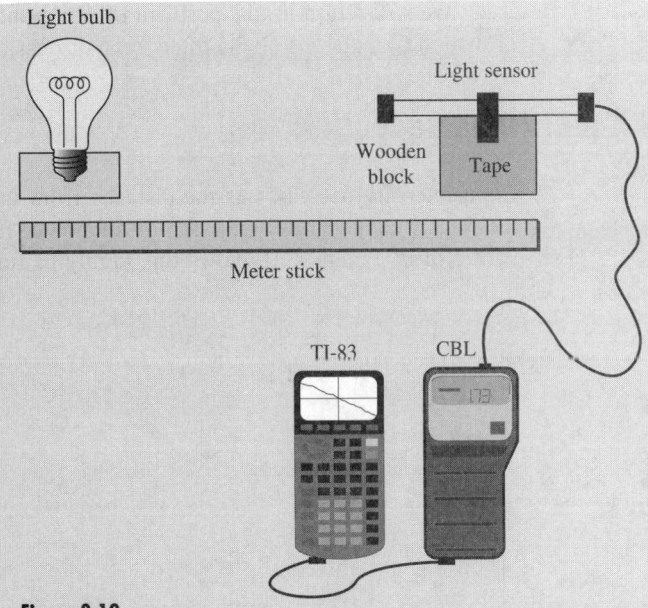

Figure 8.19

Table 8.4

Distance (cm)	30	35	40	45	50	55	60
Light intensity (mW/cm²)	0.825	0.655	0.522	0.412	0.334	0.278	0.230

Distance (cm)	65	70	75	80	85	90	95	100
Light intensity (mW/cm²)	0.193	0.163	0.146	0.129	0.117	0.104	0.096	0.087

Procedures

1. Now that you see the data in Table 8.4, how would you change your graph in Figure 8.18?

2. Plot the data from the table in a graph, either by hand or with the calculator. Use *distance* on the horizontal axis and *light intensity* on the vertical axis. To use the calculator, enter the data into two lists or use the lists named LTDIS (distance) and LTINT (intensity). The graph should look like Figure 8.20.

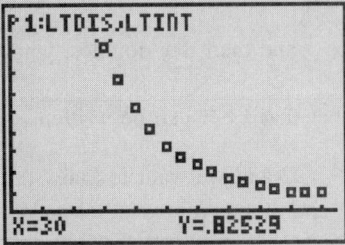

Figure 8.20

3. Which of the graphs that you sketched in Calculator Labs 8.1–8.3 look like the graph of light intensity vs. distance? It is close to one of them; however, without additional analysis, it is not possible to decide which it is.

We will return to this problem of fitting the graph of a power function to a set of data in Calculator Lab 8.5 after some discussion of inverse variation mathematics.

Inverse Variation

Light intensity decreases as the distance from the light source increases. Therefore the model could be one of a family of power functions that have negative integer values of n. Figure 8.21 shows this family of functions for positive values of x.

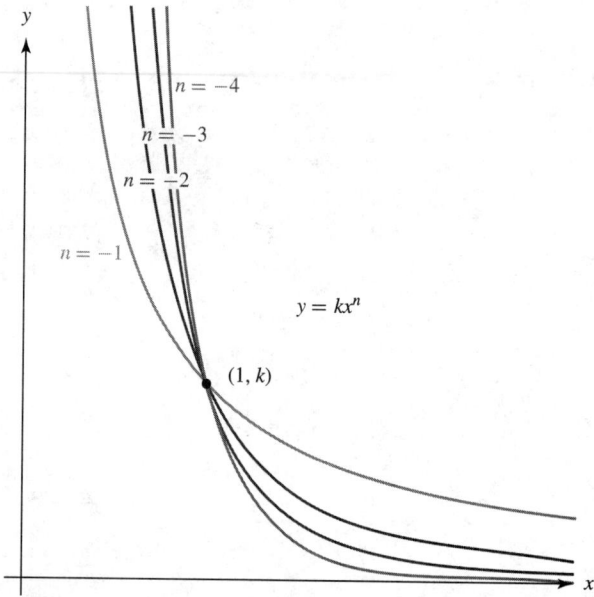

Figure 8.21

One of these curves could be the correct model for light intensity. It will take more analysis to identify which one. For now just take a guess about which curve it *could* be and which curve it could *not* be.

First we'll study inverse variation, which is modeled by power functions with negative exponents.

Example 8.10

If $f(x) = kx^{-1}$ and then if x doubles, what is the effect on $f(x)$?

Solution   First remember that kx^{-1} can be written as $\dfrac{k}{x}$ because $kx^{-1} = k \cdot \dfrac{1}{x} = \dfrac{k}{x}$. We are given that $\dfrac{x_2}{x_1} = 2$ and we want to know the ratio $\dfrac{f(x_2)}{f(x_1)}$, which we'll call $\dfrac{f_2}{f_1}$.

$$\frac{f_2}{f_1} = \frac{f(x_2)}{f(x_1)} = \frac{\frac{k}{x_2}}{\frac{k}{x_1}}$$

$$= \frac{k}{x_2} \div \frac{k}{x_1} = \frac{k}{x_2} \cdot \frac{x_1}{k}$$

$$= \frac{x_1}{x_2}$$

That is, if $\dfrac{x_2}{x_1} = 2$, then $\dfrac{f_2}{f_1} = \dfrac{x_1}{x_2} = \dfrac{1}{2}$. In words, if x doubles then $f(x)$ is cut in half. ∎

Just as we did for power functions that have positive exponents, we'll look at the general case of multiplying x by a factor r. Remember that r is any positive value.

Example 8.11

If $f(x) = kx^{-2} = \dfrac{k}{x^2}$, and x is multiplied by a factor r, what is the effect on $f(x)$?

Solution   We are given $x_2 = rx_1$, which is the same as $\dfrac{x_2}{x_1} = r$, and we want to know $\dfrac{f(x_2)}{f(x_1)}$, which we'll call $\dfrac{f_2}{f_1}$. The analysis is similar to the previous example:

$$\frac{f_2}{f_1} = \frac{f(x_2)}{f(x_1)} = \frac{\dfrac{k}{x_2{}^2}}{\dfrac{k}{x_1{}^2}}$$

$$= \frac{k}{x_2{}^2} \div \frac{k}{x_1{}^2} = \frac{k}{x_2{}^2} \cdot \frac{x_1{}^2}{k}$$

$$= \frac{x_1{}^2}{x_2{}^2} = \left(\frac{x_1}{x_2}\right)^2$$

That is, if $f(x) = kx^{-2}$ and $\dfrac{x_2}{x_1} = r$, then $\dfrac{f_2}{f_1} = \left(\dfrac{x_1}{x_2}\right)^2 = \left(\dfrac{1}{r}\right)^2$, or $\dfrac{1}{r^2}$. In words, if x is multiplied by r, then $f(x)$ is divided by r^2. Here are two applications of this result:

▶ If x is multiplied by 2, then $f(x)$ is divided by 4
▶ If x is multiplied by 0.2 (or $\frac{1}{5}$), then $f(x)$ is divided by $0.2^2 = 0.04$ or $\frac{1}{25}$. To say it another way, if x is multiplied by $\frac{1}{5}$, then $f(x)$ is multiplied by 25   ∎

In summary, with negative exponents and positive values of x the power functions produce a decrease in $f(x)$ with every increase in x. Here is a summary of what is called inverse variation:

Inverse Variation as the nth Power

If $f(x) = kx^{-n}$ and n is a positive integer, then $f(x)$ can be written as $\dfrac{k}{x^n}$.
If $\dfrac{x_2}{x_1} = r$, then

$$\frac{f(x_2)}{f(x_1)} = \frac{f_2}{f_1} = \left(\frac{1}{r}\right)^n = \frac{1}{r^n}$$

Two sentences are often used to describe this relationship in words.

▶ $f(x)$ **varies inversely** as the nth power of x.
▶ If the value of x is multiplied by r, then the value of $f(x)$ would be divided by r^n.

The gravitational attraction felt by a satellite orbiting the Earth is inversely proportional to the square of the distance to the center of the Earth. The next two examples will look at these situations.

Example 8.12 Application

Compare the gravitational attraction of a satellite in an orbit 200 kilometers above the Earth's surface with the gravitational attraction of an orbit 800 kilometers above the Earth's surface.

Solution First we need to find the ratio of the distances, $\dfrac{d_2}{d_1}$, then we will find the ratios of the gravitational attractions, $\dfrac{G_2}{G_1}$. The radius of the Earth is about 6400 kilometers, so $d_1 = 6400 + 200 = 6600$ km and $d_2 = 6400 + 800 = 7200$ km. The ratio of distances is $\dfrac{7200}{6600} \approx 1.091$. Therefore the ratio $\dfrac{G_2}{G_1} = \dfrac{1}{1.091^2} \approx 0.840$. The attraction felt in the outer orbit is about 84% of the attraction felt in the inner orbit. ■

Next we will find the value of k for the situation described in Example 8.12.

Example 8.13 Application

The gravitational attraction felt by a satellite weighing 2000 kg in an orbit 800 kilometers above the surface of the Earth is about 15 000 newtons.

(a) What is the power function that describes this situation?
(b) What is the value of k in the power function?
(c) Use the equation of the power function to compute the gravitational force felt by the same satellite in an orbit 200 km above the surface of the Earth.
(d) Check your answer to (c) with the answer to Example 8.12.

Solutions (a) The gravitational attraction felt by a satellite orbiting Earth *varies inversely* as the *square* of the distance from the satellite to the center of Earth. Let G represent the gravitational attraction and d the distance from the satellite to the Earth's center. *Varies inversely* means that $G = \dfrac{1}{d^n}$ and *square* means that $n = 2$. The power function is $G = \dfrac{k}{d^2}$.

(b) Substituting the known values of $G = 15000$ N and $d = 6400 + 800 = 7200$ km, we have $15000 = \dfrac{k}{7200^2}$, which gives $k \approx 7.8 \times 10^{11}$.*

*This constant, called the Universal Gravitational Constant, is normally reported using meters instead of kilometers. Its value here is therefore different from what you might see in a physics text.

(c) If the satellite is 200 km above Earth's center, then $d = 6400 + 200 = 6600$ km.

So, $G = \dfrac{7.8 \times 10^{11}}{d^2} = \dfrac{7.8 \times 10^{11}}{6600^2} \approx 17900$ newtons.

(d) 84% of 17900 is about 15000, so the answer to part (c) is approximately correct. ∎

In the final example, before we return to the discussion of light intensity, we will study some data and try to match it with a power function.

Example 8.14

The points shown in Table 8.5 come from a power function. The values of y were rounded to the nearest whole number. Which power function best fits these data?

Table 8.5

x	3	4	6	8	10
y	99	42	12	5	3

Solution   First we observe that as x increases y decreases. That is, x and y vary inversely. Only one family of power functions behaves like this when x and y are both positive: a function of the form $y = kx^{-n} = k \cdot \dfrac{1}{x^n}$ for some values of k and n where $n > 0$. Furthermore, Figure 8.22 shows that the points follow the characteristic shape of the power function with negative exponent. You can also see the horizontal asymptote, and you can guess that there might be a vertical asymptote.

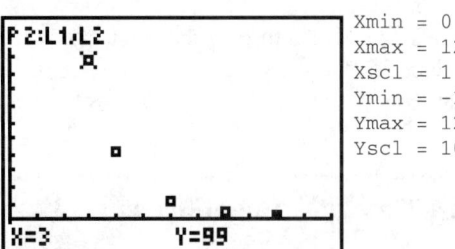

```
Xmin = 0
Xmax = 12
Xscl = 1
Ymin = -15
Ymax = 120
Yscl = 10
```

Figure 8.22

To find the mathematical model we must compute both k and n. There are several ways to approach this problem. For example, you will learn how to solve this problem using a system of two equations in Chapter 10. The strategy we will adopt here is to compute k for several values of n and see how the different curves match the data.

We could choose any of the data points. Let's take the first one, $x = 3$ and $y = 99$. The equation of the power function then becomes $99 = \dfrac{k}{3^n}$, which produces $k = 99 \cdot 3^n$. Table 8.6 shows the computed values of k for each value of n.

Table 8.6

n	k
1	$99 \cdot 3 = 297$
2	$99 \cdot 3^2 = 891$
3	$99 \cdot 3^3 = 2673$
4	$99 \cdot 3^4 = 8019$

The four curves with these four values of k and n are shown in Figure 8.23. You can see that all of them pass through the point $(3, 99)$ because we used that point to calculate the values of k. The third curve, $y = \dfrac{2673}{x^3}$, is the one that best fits the data. Therefore we conclude that this is the best power function with integer exponents to select as a model for this data. An isolated view of the five points and the graph of $y = \dfrac{2673}{x^3}$ is shown in Figure 8.24.

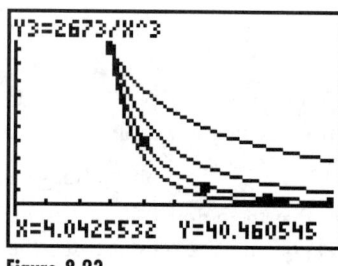

```
Xmin = 0
Xmax = 10
Xscl = 1
Ymin = -15
Ymax = 120
Yscl = 10
```

Figure 8.23

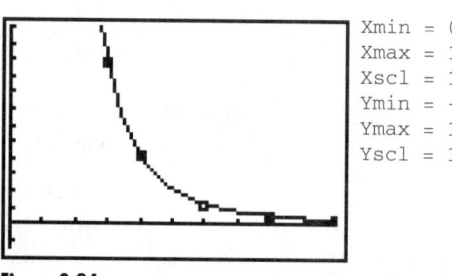

```
Xmin = 0
Xmax = 10
Xscl = 1
Ymin = -15
Ymax = 120
Yscl = 10
```

Figure 8.24

Next we will return to the light intensity data in the Calculator Lab and use the techniques shown in the previous examples to find the best mathematical model for that data.

Calculator Lab 8.5

Finding the Mathematical Model for Light Intensity

Look again at the data in Calculator Lab 8.4, and at the graph of the data in Figure 8.20 on page 541. You have learned that the curve shown in the figure is characteristic of a power function of the form $y = \dfrac{k}{x^n}$. We have two jobs: find the best value of n and find the best value of k, so that the curve matches the data as close as possible.

Procedures

1. Use what you have learned in this section to calculate values of k and n that produce a power function that models this data.
2. After you have decided on the value of n, change the value of k to improve the fit of your curve to the data.
3. Compare your results to those found with the *TI-83*'s power regression feature, which is selection A on the STAT Calc menu. On the calculator's home screen, the command will look like this:

```
PwrReg LTDIS,LTINT
```

4. The result of the power regression computation will be the best possible fit for the data. Discuss the difference between the regression result and the best model you were able to find without regression.

Section 8.2 Exercises

1. **(a)** What is the volume of a cube that measures 3 cm on each side?
   (b) How long should each side be for a cube that has twice the volume of the cube in (a)?

2. **(a)** What is the volume of a cube that measures 1.5 cm on each side?
   (b) How long should each side be for a cube that has three times the volume of the cube in (a)?

3. Each side of cube A is 2.5 in. long and each side of cube B is 3.75 in. long. What is the ratio of the volume of cube B to the volume of cube A?

4. Each side of cube C is 5 in. long and each side of cube D is 2 in. long. What is the ratio of the volume of cube D to the volume of cube C?

5. *Recreation* A golf ball has a diameter of 1.68 in. and a tennis ball has a diameter of 2.5 in.
   (a) What is the ratio of their radii?
   (b) What is the ratio of their volumes?

6. *Recreation* What is the ratio of the surface areas of the golf and tennis balls in Exercise 5?

7. *Space technology* In Example 8.8 we showed that the ratio of the volume of the Moon to the volume of the Earth is approximately equal to $0.020 = \frac{1}{50}$. What do you expect the ratio of their masses to be? Think about this problem before moving on to Exercise 8.

8. *Space technology* The actual volume of the Moon is about 7.3×10^{22} kg, and the actual volume of the Earth is about 6.0×10^{24} kg. Give possible reasons why there is a difference between the ratio you predicted and the actual ratio.

9. If a ball falls 19.6 meters in the first two seconds, how far does it fall in the first three seconds?

10. How far would the ball in Exercise 9 fall in the first five seconds?

11. If $f(x) = kx^{-4}$ and x is multiplied by a factor r, what is the effect on $f(x)$?

12. If $f(x) = kx^{-5}$ and x is multiplied by a factor r, what is the effect on $f(x)$?

13. *Space technology* A communications satellite is in orbit 22,300 mi above the Earth's surface. The space shuttle is in orbit 245 mi above the Earth's surface. What is the ratio of the gravitational attraction of the communication satellite and the space shuttle?

14. *Space technology* What is the gravitational attraction felt by a 25,000 kg communications satellite in an orbit 35,680 km above the Earth's surface. (See Example 8.13.)

15. The points shown in the table below come from a power function. Which power function best fits these data?

x	-6	-4	3	5	7
y	-1555.2	-307.2	-97.2	-750	-2881.2

16. The points shown in the table below come from a power function. Which power function best fits these data?

x	-2	0	2	4	5	8
y	-8	0	8	256	781.25	8192

Exercises 17–20 show the graph of a power function $y = kx^n$ graphed on a TI-83 using the indicated viewing window. One point on the curve is shown. Based on your knowledge of power functions, symmetry, and asymptotes, (a) estimate the values of k and n, and (b) justify your estimates.

17.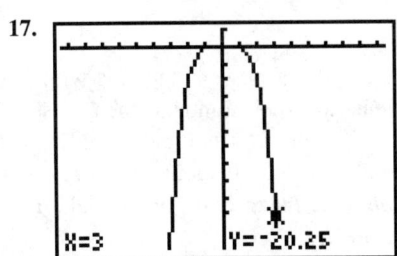

```
Xmin = -9.4
Xmax = 9.4
Xscl = 1
Ymin = -24
Ymax = 2
Yscl = 2
```

18.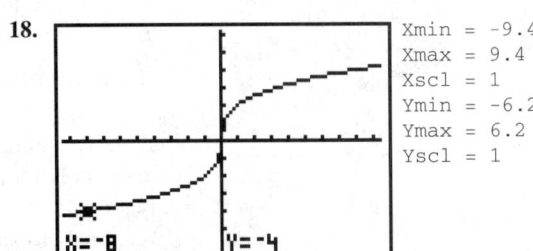

```
Xmin = -9.4
Xmax = 9.4
Xscl = 1
Ymin = -6.2
Ymax = 6.2
Yscl = 1
```

19.

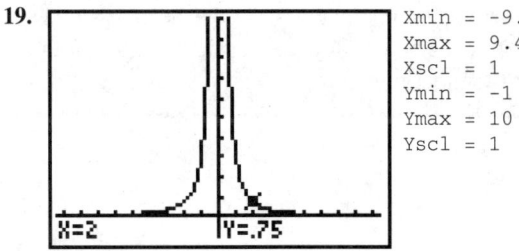

```
Xmin = -9.4
Xmax = 9.4
Xscl = 1
Ymin = -1
Ymax = 10
Yscl = 1
```

20.

```
Xmin = -9.4
Xmax = 9.4
Xscl = 1
Ymin = -16
Ymax = 16
Yscl = 2
```

21. Tape a meter stick to the wall and place a mirror on the floor 20 cm from the base of the meter stick. Have a student stand so the mirror is directly between him or her and the meter stick. Record the highest point on the meter stick the student can see reflected in the mirror, y, and the distance from the student to the reflection of this highest point in the mirror, x, as shown in Figure 8.25. Ask the student to repeat this for four other values of x.

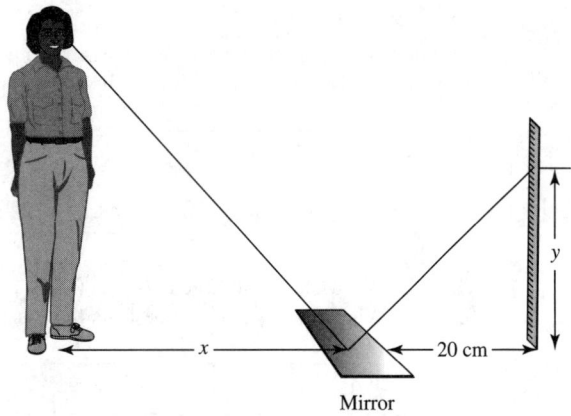

Mirror

Figure 8.25

(a) Determine the power function that best fits these values.

(b) Describe the type of variation (direct or inverse) these values satisfy.

(c) Write a sentence describing the relationship between the distance a person is from the mirror and the highest point on the meter stick that can be observed in the mirror.

22. Lay a pencil on top of a table or other level surface and place a ruler across the pencil at or near the midpoint of the ruler so the ruler is balanced on the pencil. The pencil acts as a fulcrum. Stack 2 to 5 dimes on top of each other at one end of the ruler. Place weights at a point 1 in. from the pencil until the ruler is again balanced on the pencil. Record the distance of the weight from the fulcrum and the weight used. Remove the weights and repeat by placing weights 2 in. from the pencil as shown in Figure 8.26. Again record the distance and the weight. Repeat for 3, 4, and 5 in. from the pencil.

Figure 8.26

(a) Determine the power function that best fits these values.

(b) Describe the type of variation (direct or inverse) these values satisfy.

(c) Write a sentence describing the relationship between the amount of weight and its distance from the fulcrum.

8.3 Polynomial Functions

You can think of a **polynomial function** as a sum of power functions with nonnegative integer exponents. If all the coefficients are real numbers, then it is called a **real polynomial**. If one or more of the coefficients is a nonreal complex number, then the polynomial is a **complex polynomial**. We shall be concerned only with real polynomials and will use the word *polynomial* to mean real polynomial. You used polynomials to study motion in Chapters 1 and 2.

Examples of Polynomial Functions

▶ The distance d from home of a car moving with a constant velocity of 45 mph after t hours, beginning at a point 10 miles from home:

$$d(t) = 45t + 10$$

▶ The height h above the ground of a ball t seconds after it is thrown upward at 30 feet per second from a point 20 feet above the ground:

$$h(t) = -16t^2 + 30t + 20$$

If you remember that numbers like 20 and 10 can be written as $20t^0$ and $10t^0$, then you can see that each of the above polynomials is the sum of power functions. The largest exponent in a polynomial is called the **degree of the polynomial**. The degree of the polynomial function h is 2 and the degree of the polynomial function d is 1. You can say that $h(t) = -16t^2 + 30t + 20$ is a second degree polynomial function (also called a **quadratic function**), and that $d(t) = 45t + 10$ is a first degree polynomial function (also called a **linear function**). Likewise, a third degree polynomial function is called a **cubic function**.

Roots of Polynomials

Roots of a function f are solutions to the equation $f(x) = 0$. Graphically the real roots are the x-coordinates of the points where the graph of the function crosses the x-axis. In Chapter 2 you learned a way to write the equation of a parabola in factored form

$$f(x) = a(x - r_1)(x - r_2)$$

where $x = r_1$ and $x = r_2$ are the roots and $a \neq 0$.

When you studied the quadratic formula, which is used to calculate the roots of a quadratic equation, you learned that there can be two real roots or two nonreal complex roots. These ideas are summarized in Figure 8.27. Note that when the parabola is tangent to the x-axis, we describe the one point of intersection with the phrase *two equal real roots* or simply *two equal roots*. This makes sense when you think of the factored form for such a function as $f(x) = a(x - r_1)(x - r_1)$ and $a \neq 0$. There are two identical factors, so we say that there are two equal roots.

We know that a first degree polynomial (the linear function $f(x) = kx + b$ with $k \neq 0$) has one root (see Figure 8.28).

If a linear function has one real root and a quadratic function has either two real roots or two nonreal complex roots, what about a cubic function? All the cubic functions you have seen in this chapter have had only one root, at $x = 0$, because they have all been power functions, which always pass through the point $(0, 0)$ if the exponent is positive. However there are more possibilities with cubic polynomial functions, as we will see in Calculator Lab 8.6.

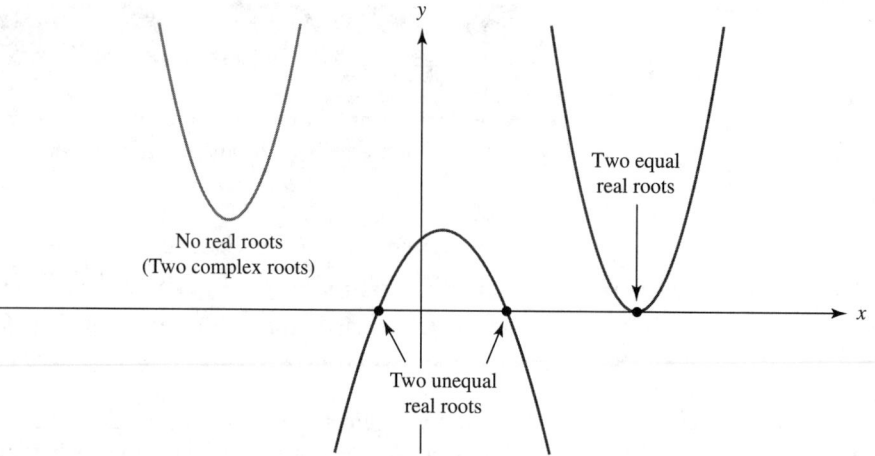

Figure 8.27

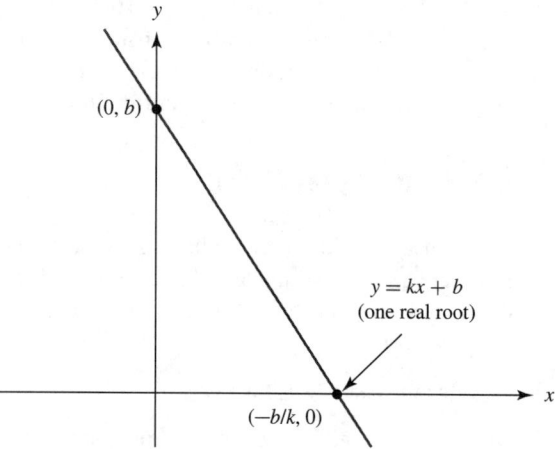

Figure 8.28

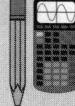

Calculator Lab 8.6

Exploring Cubic Functions

The cubic function $f(x) = x^3 + x^2 - 4x$ can be graphed in the ZStandard viewing window as shown in Figure 8.29.

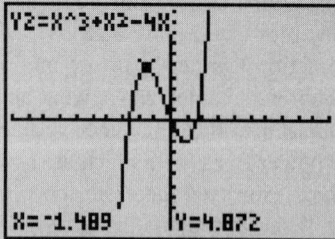

Figure 8.29

This function has three different roots. The aim of this lab is to change this function in order to see the different numbers of roots that a cubic function can have.

Procedures

1. Estimate the numerical value of each root of f by looking at Figure 8.29.
2. Estimate the roots to the nearest tenth with a table.

   (a) Enter the function in the calculator's Y= screen.
   (b) Use the calculator's `Table` and `TblSet` features to estimate the roots. For example Figure 8.30 shows that there is a root at approximately $x = -2.6$. We chose -2.6 and not -2.5 because $f(-2.6)$ is closer to zero than $f(-2.5)$.

3. Change the function by adding or subtracting a number, for example, plot $y = x^3 + x^2 - 4x + 1$ or $y = x^3 + x^2 - 4x - 3$
4. Estimate the roots for the new function.
5. Find one positive number that can be added to $f(x)$ so that the resulting function has only one real root. What is the number and what is the root?
6. Find one positive number that can be subtracted from $f(x)$ so that the resulting function has only one real root. What is the number and what is the root?
7. Take one of your new functions and change every sign in it. For example change $f(x) = x^3 + x^2 - 4x + 1$ to $g(x) = -x^3 - x^2 + 4x - 1$. [Thus $g(x) = -f(x)$.] Plot the new function and compare the roots of $f(x)$ and $g(x)$, describe what seems to be true about the roots, and explain why you think this is true.
8. Make some other changes to the coefficients of $f(x)$ and try to produce a cubic function that has only two real roots.
9. Draw a rough sketch of the following cubic functions, *if possible*:

   (a) A cubic with only one root, and the real root is negative.
   (b) A cubic with only one root, and the real root is positive.
   (c) A cubic with three real roots, all positive.
   (d) A cubic with three real roots, one positive and two negative.
   (e) A cubic with three real roots, and symmetry with respect to the origin.
   (f) A cubic whose graph has symmetry with respect to the y-axis.
   (g) A cubic with three real roots, but two of them are equal. For example, part of it will look something like the parabola that has equal roots in Figure 8.27.
   (h) A cubic with two and only two different roots.

10. Explain why it is impossible for a cubic function to have two and only two roots.
11. Explain why it is impossible for a cubic function to have a graph that has symmetry with respect to the y-axis.

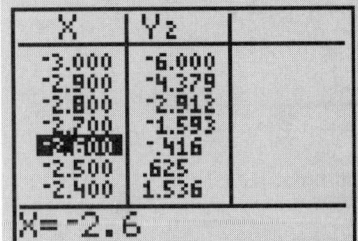

X	Y2
-3.000	-6.000
-2.900	-4.379
-2.800	-2.912
-2.700	-1.593
-2.600	-.416
-2.500	.625
-2.400	1.536

X=-2.6

Figure 8.30

Factored Form of a Polynomial

It turns out that every polynomial can be written as a product of factors in much the same way a quadratic function can be written in its factored form. That is, a cubic polynomial function can be written as

$$f(x) = a(x - r_1)(x - r_2)(x - r_3)$$

But the three factors $a(x - r_1)(x - r_2)$ form a quadratic polynomial, so we could write the cubic function $f(x)$ as

$$f(x) = (\text{quadratic polynomial})(x - r_3)$$

In the above equation if f is a real polynomial then the root r_3 must be a real number because if it was complex then $f(x)$ would have a complex number as part of its expression. But then $f(x)$ wouldn't be a real polynomial. Therefore a cubic real polynomial must have at least one real root. Since we know that a quadratic expression can have two real roots or two nonreal complex roots, we know that a cubic can have either one real root or three real roots. It cannot have exactly two real roots. This is the answer to one of the questions in Calculator Lab 8.6.

Example 8.15

How many real roots can a fourth degree real polynomial have?

Solution   If $g(x)$ is a fourth degree or *quartic* real polynomial function then we can write $g(x)$ in factored form as

$$g(x) = a(x - r_1)(x - r_2)(x - r_3)(x - r_4)$$

where r_1, r_2, r_3, and r_4 are the roots and $a \neq 0$.

In a similar way that we could rewrite the cubic polynomial function f, we can rewrite the quartic function as

$$g(x) = a(\text{quadratic polynomial})(\text{quadratic polynomial})$$

We have seen that a quadratic expression can have no real roots or two real roots. The table in Figure 8.31 helps organize all the possibilities for multiplying two quadratic polynomials. Study the table until you can explain why a quartic function can have no real roots, two real roots, or four real roots.

	Quadratic polynomial 1	
	r_1 and r_2 are real	r_1 and r_2 are complex
Quadratic polynomial 2 — r_3 and r_4 are complex	Quartic has 2 real roots	Quartic has no real roots
Quadratic polynomial 2 — r_3 and r_4 are real	Quartic has 4 real roots	Quartic has 2 real roots

Figure 8.31

Activity 8.3
Sketching Quartic Functions

1. Use the result of Example 8.15 to sketch the possible different appearances of a quartic polynomial function. One of your sketches will have no real roots, one will have two different real roots, and one will have four different real roots.
2. Make a fourth graph that has two different real roots and two equal real roots.

3. Make a fifth graph of a quartic function that has four identical real roots.
4. There are other possibilities. Describe each of them and sketch a graph that shows how each one can happen.

Limits and Graphs of Polynomial Functions

To get an idea of what polynomial graphs look like, it helps to think about the limits of the function as $x \rightarrow +\infty$ (moving to the right along the x-axis) and $x \rightarrow -\infty$ (moving to the left along the x-axis). That is, what does the graph look like way out to the right and way out to the left?

Example 8.16

Figure 8.32 shows the graph of the cubic function $y = 0.1x^3 + 2x^2 - 5$ using the ZStandard viewing window. This appears to have only two real roots. Is this possible?

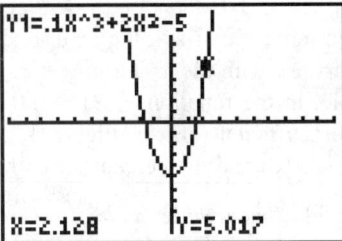

Figure 8.32

Solution In the paragraph just before Example 8.15 we stated that a cubic function can have either one real root or three real roots. Since we can see that this function has two real roots we know that there must be another real root that we cannot see. One easy way to try to find the third root is to zoom out by pressing ⬚ZOOM⬚ 3 ⬚ENTER⬚. The authors' result is shown in Figure 8.33. By tracing we can see that the highest point to the left of the origin is near $(-13.6, 113.4)$ and the lowest point near the origin is $(0, -5)$. This can help us set the viewing window. Figure 8.34 is the same graph with the viewing window adjusted so you can see the top and bottom of each "hump" and "valley" in the graph.

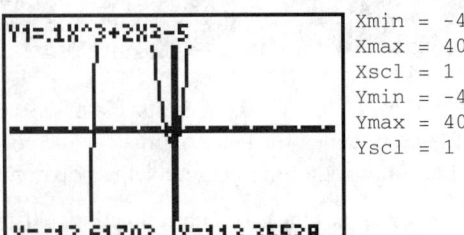

```
Xmin = -40
Xmax = 40
Xscl = 1
Ymin = -40
Ymax = 40
Yscl = 1
```

Figure 8.33

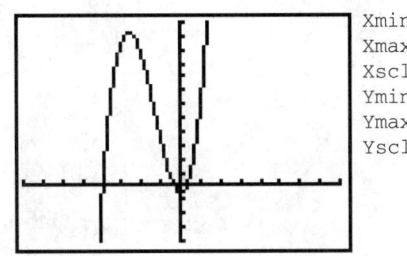

```
Xmin = -40
Xmax = 40
Xscl = 5
Ymin = -40
Ymax = 120
Yscl = 10
```

Figure 8.34

What happens to the values of $f(x)$ for $x > 0$? A glance at the graph in Figure 8.34 seems to indicate that the y-values will continue to increase. That is, from this graph we can guess that

$$\lim_{x \to \infty} f(x) = \infty$$

What happens to the y-values when the values of $x < -13.7$ decrease? As you trace the graph and the cursor moves to smaller values of x the values of y start to decrease. This is evidence that

$$\lim_{x \to -\infty} f(x) = -\infty$$ ■

What happened in Example 8.16 is quite common. In our first look at the graph, important parts were outside the boundaries of the viewing window. In addition to zooming out on a graph, the calculator's Table feature can help you see what's going on out to the right and left. The table will also help you see how big to make the viewing window.

The same function $f(x) = 0.1x^3 + 2x^2 - 5$ can teach another lesson. The coefficient of x^3 is $\frac{1}{20}$th the size of the coefficient of x^2. Our first graph looked like a parabola for small values of x because the term $2x^2$ is larger than the term $0.1x^3$. But eventually, no matter how small the coefficient of x^3, as you move farther to the right or left on the x-axis the x^3 term will get much larger in absolute value than the x^2 term. Even more important, the x^2 term and all the others eventually become *insignificant* compared with the x^3 term.

For example, in the function $y_1(x) = 0.01x^3 + 100x^2$ the coefficient of x^2 is $10,000$ times the coefficient of x^3. Figure 8.35 shows the values of the function $y_1(x) = 0.01x^3 + 100x^2$ and the percent difference between y_1 and $0.01x^3$. To compute this percent difference we let $y_2 = \dfrac{(y_1 - 0.01x^3)}{y_1} \times 100$. After $x = 1,000,000$ the percent error is under 1%. That is, when x is far to the left or right, only the x^3 term makes any difference.

X	Y1	Y2
500000	1.3E15	1.9608
600000	2.2E15	1.6393
700000	3.5E15	1.4085
800000	5.2E15	1.2346
900000	7.4E15	1.0989
1E6	1E16	.9901
1.1E6	1.3E16	.9009

Y₂=.99009009901

Figure 8.35

Summary: Limits and Graphs of Polynomials

Thinking about the shape of the graph at its far ends means thinking about the size of the function values for large or small values of x. The behavior is different depending on whether the degree of the polynomial is odd or even.

▶ **Odd degree,** for example linear or cubic functions. If the coefficient of the term with the highest degree is positive, then the function will look like the power function x^3 for large positive or small negative values of x. (See Figure 8.36).

> ► **Even degree,** for example quadratic or quartic functions. If the coefficient of the term with the highest degree is positive then the function will look like $5x^4$ for large positive or small negative values of x. (See Figure 8.37).
> ► If the coefficient of the term with the highest degree is negative the shape of the graph will not change but the graph will be flipped over the x-axis.

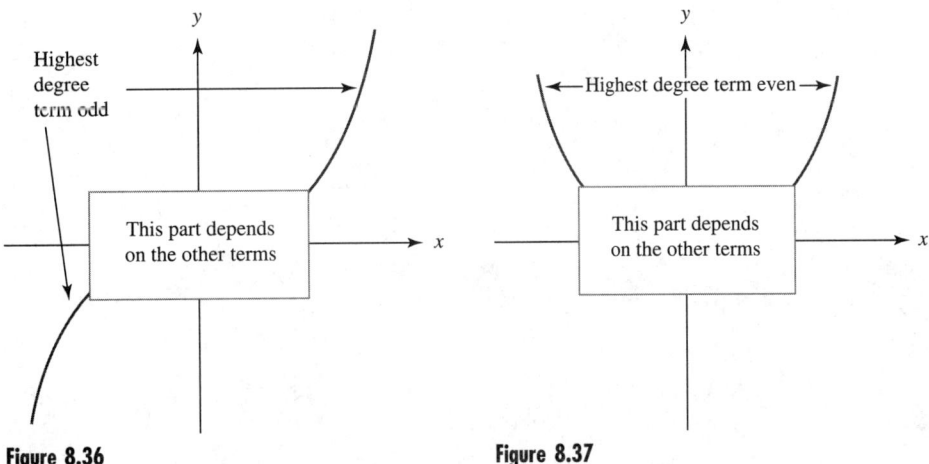

Figure 8.36   **Figure 8.37**

Example 8.17

Describe the shape of $y = 2x^7 - 3x^5 + 4$ for large positive or small negative values of x.

Solution   The graph of $y = 2x^7 - 3x^5 + 4$ with a ZStandard viewing window is shown in Figure 8.38. The largest exponent is 7, an odd number, and the coefficient of the x^7-term is positive. Therefore when you zoom out the curve should look much like the graph in Figure 8.36. This can partially be seen by the graph of $y = 2x^7 - 3x^5 + 4$ shown in Figure 8.39.

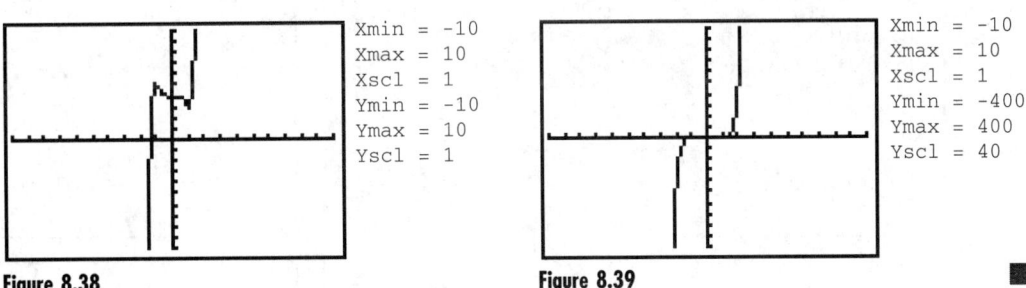

```
Xmin = -10
Xmax = 10
Xscl = 1
Ymin = -10
Ymax = 10
Yscl = 1
```

```
Xmin = -10
Xmax = 10
Xscl = 1
Ymin = -400
Ymax = 400
Yscl = 40
```

Figure 8.38   **Figure 8.39**

Example 8.18

Describe the shape of $y = -0.1x^6 + 2x^3$ for large positive or small negative values of x.

Solution　The graph of $y = -0.1x^6 + 2x^3$ with a ZStandard viewing window is shown in Figure 8.40. The largest exponent is 6 which is even and the coefficient of the x^6-term is -0.1 a negative. Therefore when you zoom out the curve should look much like the graph in Figure 8.37 that has been flipped over the x-axis. This can be partially seen by the graph of $y = -0.1x^6 + 2x^3$ shown in Figure 8.41.

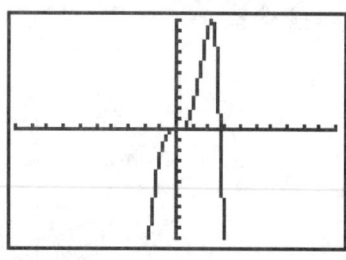

```
Xmin = -10
Xmax = 10
Xscl = 1
Ymin = -10
Ymax = 10
Yscl = 1
```

Figure 8.40

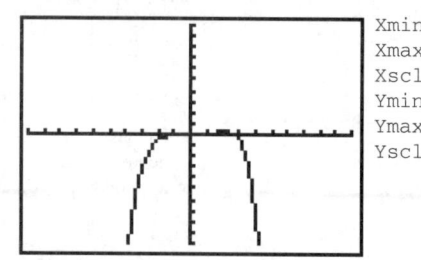

```
Xmin = -10
Xmax = 10
Xscl = 1
Ymin = -400
Ymax = 400
Yscl = 40
```

Figure 8.41

Section 8.3 Exercises

In Exercises 1–2 give the degree of the given polynomial.

1. $f(x) = 7.2x^7 - 3.1x^2 + 4x^3 + 7x^{10} - 2$

2. $g(x) = 9x^3 + 6x^{12} - 3 + 2x^2$

In Exercises 3–8 give the roots of the given polynomial functions.

3. $f(x) = 3.1(x - 6.4)\left(x + \sqrt{2}\right)(x - 2\pi)(x + 1)$

4. $g(x) = -4\left(x - \sqrt{5}\right)(x + 3.1)\left(x + 4\sqrt{2}\right)(x - 2)(x + 1)$

5. $h(x) = -5x\,(x + 5)^2\left(x - 3\pi^2\right)(5x + 1)$

6. $j(x) = 6x^2\,(x + 7)^3\,(2x - 1)$

7. $k(x) = 7x\left(x^2 - 4\right)\left(x^2 + 1\right)$

8. $m(x) = 8x^2\left(x^2 + 3x + 2\right)\left(x^2 - 5x - 6\right)\left(x^2 + 4\right)$

In Exercises 9–12 estimate the real roots of the given polynomial function to three decimal places.

9. $f(x) = 7x^3 - 4x^2 + 1$

10. $g(x) = 2.1x^4 - 3.4x^2 + 2x - 5$

11. $h(x) = 0.01\,x^3 - 0.022\,x^2 - 0.032\,x + 0.072$

12. $j(x) = -0.02\,x^4 + 0.0806\,x^3 - 0.119100\,x^2 + 0.0763360\,x - 0.01792000$

In Exercises 13–16 examine the graphs and estimate (a) the degree of the polynomial, (b) the number of real roots, and (c) the number of nonreal roots.

13.

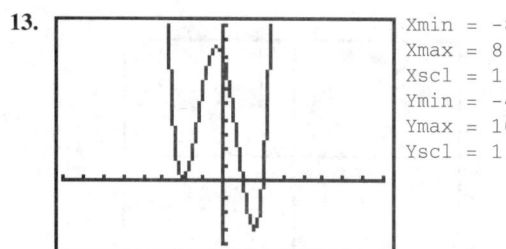

```
Xmin = -8
Xmax = 8
Xscl = 1
Ymin = -4
Ymax = 10
Yscl = 1
```

15.

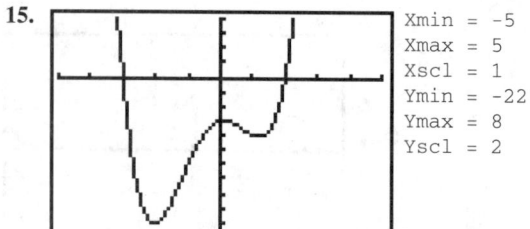

```
Xmin = -5
Xmax = 5
Xscl = 1
Ymin = -22
Ymax = 8
Yscl = 2
```

14.

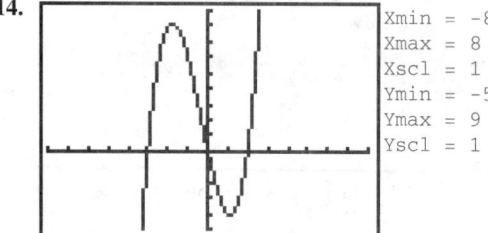

```
Xmin = -8
Xmax = 8
Xscl = 1
Ymin = -5
Ymax = 9
Yscl = 1
```

16.

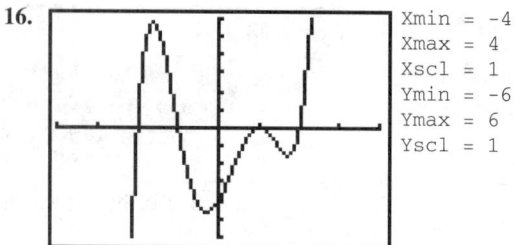

```
Xmin = -4
Xmax = 4
Xscl = 1
Ymin = -6
Ymax = 6
Yscl = 1
```

In Exercises 17–18 estimate the root from the given table.

17.

X	Y₁
-.43	-.1371
-.42	-.076
-.41	-.017
-.4	.04
-.39	.09501
-.38	.14804
-.37	.19914

X=-.43

18.

X	Y₁
.79	-.2412
.8	-.2
.81	-.1572
.82	-.1128
.83	-.0667
.84	-.0189
.85	.03063

X=.79

In Exercises 19–22, (a) graph the given polynomial function on your calculator using the ZStandard *window setting, (b) use Figures 8.36 and 8.37 to describe the shape of the given function for large positive or small negative values of x, (c) zoom out on your calculator to confirm your response to (b).*

19. $f(x) = \frac{1}{240}(x^3 - 24x^2 - 240x)$

20. $g(x) = \frac{1}{60}(x^4 - 24x^3 - 60x^2 - 120x - 60)$

21. $h(x) = 0.01\,x^4 - 0.09\,x^3 - 0.45\,x^2 + 1.13\,x - 0.60$

22. $j(x) = 0.00005\,x^5 + 0.00220\,x^4 + 0.01290\,x^3 - 0.35800\,x^2 - 2.24375\,x - 1.87500$

23. Design a table like the one in Figure 8.31 to show how many roots are possible for a 5th degree polynomial.

24. Sketch graphs that show three of the possibilities for the roots of 5th degree polynomial functions. Include one sketch that shows the possibility that there are two equal real roots.

25. Make a table like the one in Figure 8.31 to show how many roots are possible for a 6th degree polynomial.

26. Sketch graphs that show three of the possibilities for the roots of 6th degree polynomial functions. Include one sketch that shows the possibility that there are two equal real roots.

27. Extend your table to show the number of combinations for a 7th degree polynomial.

28. What are the possible number of roots of an 8th degree polynomial? 100th degree?

29. What are the possible number of roots of an 9th degree polynomial? 99th degree?

30. Write a memo to the management in which you describe what you have learned about the properties of the power and polynomial functions and how you can apply these methods to the Chapter Project and the memo on page 516.

8.4   Application of Polynomials

In this section we will look at several applications of polynomials. We will begin with a look at how calculators compute trigonometric values. Then we will examine a method for determining the nth term of a sequence.

How do Calculators Compute Trigonometric Values?

When your calculator needs to graph a polynomial like $f(x) = 0.1x^3 + 2x - 5$, it divides the width of the viewing window into 94 equal parts and computes $f(x)$ by substituting each of the resulting 95 values of x into the polynomial expression. It then plots the 95 points.* How does it calculate the values of a function like $\sin x$? One of the most important but hidden uses of polynomials is to make approximations

---

*The *TI-83* uses 95 points for plotting the graph of a function. Other calculators and computer programs may use a different number of points.

for sine, cosine, and other non-algebraic functions. A form of these polynomial approximations is programmed into every scientific calculator and every mathematics software program.

Polynomial Approximations for Sine and Cosine

The following polynomial functions are theoretically infinite. If you want more accuracy use more terms in the polynomial; that is, take it to a higher degree. The value of x is assumed to be in radians.

$$\sin x = s(x) \approx x - \frac{x^3}{3!} + \frac{x^5}{5!} - \frac{x^7}{7!} + \frac{x^9}{9!} - \cdots$$

$$\cos x = c(x) \approx 1 - \frac{x^2}{2!} + \frac{x^4}{4!} - \frac{x^6}{6!} + \frac{x^8}{8!} - \cdots$$

The symbol 8! means the product $8 \cdot 7 \cdot 6 \cdot 5 \cdot 4 \cdot 3 \cdot 2 \cdot 1$, which equals 40320. 8! is pronounced 8 *factorial*. Factorials can be calculated using your calculator. Some calculators have an $\boxed{n!}$ key or an n! command. The *TI-83* has a ! command at ⬛MATH ◀ 4. To calculate 8! press

8 ⬛MATH ◀ 4 ⬛ENTER

The result is 40320.

The polynomials described in the above box are named for the Scottish mathematician Colin Maclaurin (1698–1746) who formulated them. Each is an example of the **Maclaurin polynomial** or **Maclaurin series**, which is itself an example of the more general **Taylor series**. You might see these polynomials in other books under either of these names.

Calculator Lab 8.7

Exploring Maclaurin Series Approximations

Procedure   Plot the 9th degree Maclaurin polynomial approximation $s(x)$ together with the function $\sin x$. Set your calculator to Radian mode and use the ZTrig setting. Answer the following questions.

1. How many real roots does this polynomial have? Approximately what are they? How do its roots compare with the roots of $\sin x$?
2. In what interval does the approximation of $s(x)$ for $\sin x$ seem close?
3. How could you expand that interval?
4. In what intervals does the approximation seem loose?
5. Compare the value of $\sin 0.5$ with the value of the polynomial evaluated at $x = 0.5$. How many decimal places of accuracy does the polynomial give? Figure 8.42 shows how to use the Table feature to look at all the decimal places in a calculation. The number at the bottom of the screen is the computed value of $s(0.5)$. Compare it with the calculator's value for $\sin(0.5)$, which is in the next column. How many decimal places does the estimate $s(0.5)$ agree with the actual value $\sin(0.5)$?

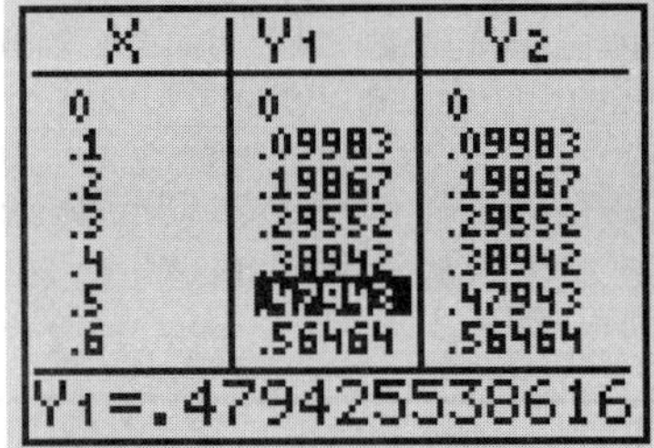

X	Y1	Y2
0	0	0
.1	.09983	.09983
.2	.19867	.19867
.3	.29552	.29552
.4	.38942	.38942
.5	.47943	.47943
.6	.56464	.56464

Y1=.47942553616

Figure 8.42

6. Make the same comparison for $x = 1.5$. Is the approximation for $\sin(1.5)$ more accurate or less accurate than the approximation of $\sin(0.5)$?

   ▶ Write an explanation of why you think one estimate was better than the other.

7. Add one term to the polynomial to make it 11th degree, and report how many more digits of accuracy are gained for the estimation of $\sin 1.5$.

8. Explain how to use the periodic nature of the sine function to estimate the value of $\sin 25$. Carry out this procedure and report how many decimal places of accuracy there are in your estimate.

The *n*th Term of a Sequence

Looking for patterns in the world is a major activity of mathematicians and scientists. The following puzzle has a solution that reveals a pattern and, believe it or not, leads back to polynomials. The question is simple. How many squares are there in Figure 8.43?

Figure 8.43

Spend some time thinking about this question. Your first answer might be 16, but of course if that were the answer it wouldn't be much of a puzzle. Did you count the big square as one? There are others! When you are satisfied with your answer move on to Activity 8.4 where you will explore the mathematical patterns in this puzzle.

Activity 8.4
Counting the Squares

The secret to solving this problem lies in looking carefully for geometric and numerical patterns. It turns out that there are four kinds of squares in Figure 8.43. We pointed out the little ones and the big one. The trick is to see the others. Figure 8.44 shows the other kinds of squares.

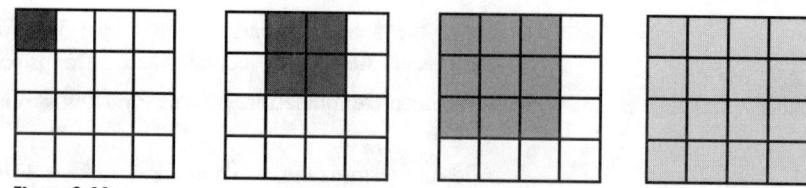

Figure 8.44

Count the number of each kind of square and you will have the answer. But organize your answer into parts as shown in Figure 8.45.

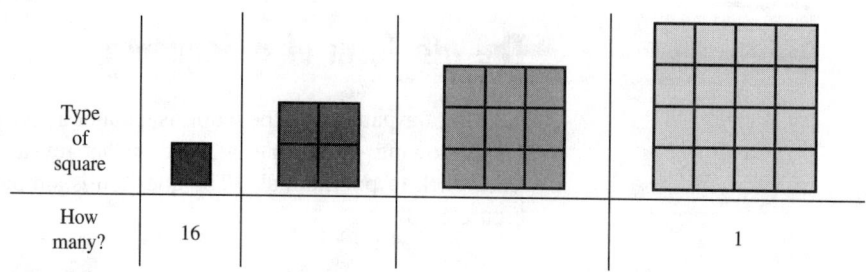

Type of square				
How many?	16			1

Figure 8.45

There is a pattern in the numbers that you have written in the table. If you see the pattern you are ready to try it on a similar problem: How many squares are there in the 5×5 square shown in Figure 8.46? To solve this problem organize your work in the way we suggested for the 4×4 square.

Figure 8.46

Example 8.19

How many squares are there on an 8×8 checkerboard?

Solution   In Chapter 7 the old engineer was counting the 64 little squares. But now you know that there are squares of many other sizes. The pattern that you may have seen in Activity 8.4 makes it possible to solve this puzzle with a little computation. The answer is

$$1^2 + 2^2 + 3^2 + 4^2 + 5^2 + 6^2 + 7^2 + 8^2$$
$$= 1 + 4 + 9 + 16 + 25 + 36 + 49 + 64 = 204$$ ■

The next question is not one that anyone will ever ask you on the job. But the approach to the solution requires the kind of skills that are used regularly in technical workplaces. The question is: *How many squares are there on a* 1000×1000 *checkerboard?* Of course if you had enough time you could solve the problem with a calculator or a computer because you can write out an expression that represents the solution

$$S_{1000} = 1^2 + 2^2 + 3^2 + 4^2 + \cdots + 999^2 + 1000^2$$

However, the important thing is not the answer; it is the mathematical method for finding the answer. We are not going to find one answer; we are going to find *all* the answers for *any* sized checkerboard. Next you'll see how to find the general formula for S_n in terms of n, where

$$S_n = 1^2 + 2^2 + 3^2 + 4^2 + \cdots + (n-1)^2 + n^2$$

In Chapter 7 you worked with sequences like this and you saw how to find formulas for the nth term and the sum of n terms of arithmetic and geometric sequences. Some of that work began with a table of values like the one shown in Table 8.7.

Table 8.7

n	How to Compute S_n	S_n
1	1	1
2	$1 + 4$	5
3	$1 + 4 + 9$	14
4	$1 + 4 + 9 + 16$	30
5	$1 + 4 + 9 + 16 + 25$	55
6	$1 + 4 + 9 + 16 + 25 + 36$	91
7	$1 + 4 + 9 + 16 + 25 + 36 + 49$	140
8	$1 + 4 + 9 + 16 + 25 + 36 + 49 + 64$	204

We want to know the 1000th entry in the table; that is, we're looking for the value of S_{1000}. If you could extend this table to 1000 rows, you would have the solution for S_{1000}. But we're not after one answer; we're after all of them. We seek a formula for S_n in terms of n. To find this formula we'll use an approach to patterns that often gives very good results. In this approach, called the **consecutive differences**

method, we write a table of values for the information we know, for example the values of S_n, and write the differences between consecutive entries. Then do it again, taking the difference of the differences. Repeat the process until the consecutive differences are constant. Table 8.8 shows that for S_n the 3rd differences are constant.

Table 8.8

n	S_n	$S_n - S_{n-1}$	1st Difference	2nd Difference	3rd Difference
1	1				
2	5	$5 - 1$	4		
3	14	$14 - 5$	9	5	
4	30	$30 - 14$	16	7	2
5	55	$55 - 30$	25	9	2
6	91	$91 - 55$	36	11	2
7	140	$140 - 91$	49	13	2
8	204	$204 - 140$	64	15	2

The constant differences are important because of the result in the box below.

Consecutive Differences Method

1. In a sequence $a_1, a_2, a_3, \ldots, a_n$ if the kth differences between consecutive terms of a sequence are constant then the nth term, a_n, of the sequence can be represented by a polynomial of degree k.
2. In general you need $k+1$ data points from the table to produce a polynomial of degree k.
3. The method only works if the numbers in the n column (the independent variable) form an arithmetic sequence, as in Table 8.8.

Example 8.20

Can the number of squares on a checkerboard be described by a polynomial?

Solution   In the sequence described by S_n the 3rd difference is constant, so $S_n = an^3 + bn^2 + cn + d$. You can compute the values of a, b, c, and d using any four data points from the table. ■

The problem now is to find the values of the coefficients a, b, c, and d that define the polynomial function $S_n = an^3 + bn^2 + cn + d$. There is an algebraic way to do this, using four equations. You'll see how to do that in Chapter 11. At this point, however, knowing that it is a cubic polynomial leads to a calculator solution. We have the data in Table 8.7. To find a formula for this data we need to know that a cubic polynomial fits the data and to have a program that will fit a cubic polynomial to a set of data points. That mathematical feature is called **cubic regression**, which is selection 6 in the *TI-83* STAT CALC menu (Figure 8.47).

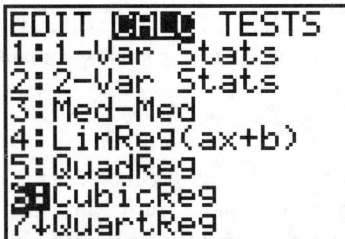

Figure 8.47

You have done regression before. You need two lists, one for the values of n and one for the values of S_n. Figure 8.48 shows how to use the ⬚STO⬚ key to enter the two lists and how to enter the instruction for cubic regression. Don't forget to use the curly braces { } when you enter a list. Why have we used only four entries in these lists? A cubic polynomial has 4 coefficients, $a, b, c,$ and d, and it turns out that four data points are sufficient to find four coefficients.

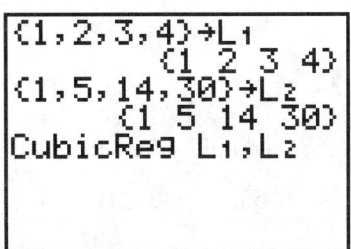

Figure 8.48

To get the same results change your calculator's precision to `Float`

When you press ⬚ENTER⬚ to execute the cubic regression instruction, the result is shown in Figure 8.49.

The regression equation predicts that the formula for S_n is

$$S_n = 0.3333333333n^3 + 0.5n^2 + 0.1666666667n - 8.2 \times 10^{-12}$$

To prove this is the formula we seek enter it into the calculator's Y= screen. A quick way is to put the cursor in an empty line in the Y= screen and press the following key sequence.

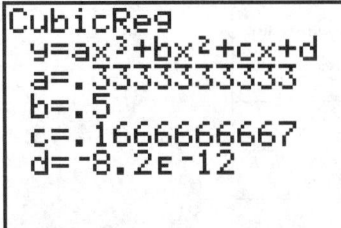

Figure 8.49

The cubic regression equation shown in Figure 8.49 will almost fill your Y= screen. Now use the Table feature to see the results. After setting both `TblStart` and `ΔTbl` to 1 your table will look like Figure 8.50, which looks exactly like our values for S_n in Tables 8.7 and 8.8. We now have a formula for S_n that can be used for any value of n. For example if you set the table to show $n = 1000$, you'll see that $S_n \approx 3.34 \times 10^8$ (the actual value is 333833500). But of course this result is not as important as the fact that we can get *any value* using the formula provided by cubic regression.

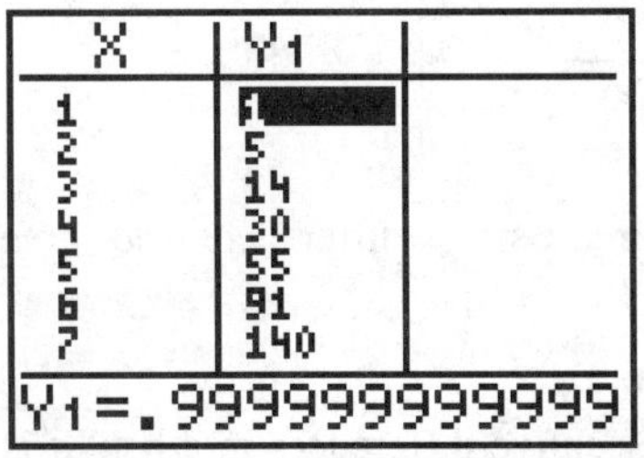

Let's take one more look at the formula for S_n. Look closely at Figure 8.50. The first entry is actually not 1 but 0.999999999999. All calculators and computers will occasionally produce errors like this. The error itself is too small to be really important, but it reminds us that there is a small error in the formula. Look at the coefficients in the formula. The value of a is probably $\frac{1}{3}$. And d is so small it's probably 0. The value of c is less obvious, but it is a decimal approximation of the fraction $\frac{1}{6}$. Calculators can never be absolutely precise with fractions (or, for that

Figure 8.50

matter, with numbers like $\sqrt{2}$ or π) but that's where human intelligence comes in. We can guess from the calculator's answer that the actual answer could be

$$S_n = \frac{1}{3}n^3 + \frac{1}{2}n^2 + \frac{1}{6}n$$

Remember that the main problem for this chapter is to find a formula for the relationship between the natural period of a pendulum and the length of the pendulum. That data is real and every real world measurement has some error, so you shouldn't expect the differences to be actually constant, like the differences for S_n.

We'll close this section with a calculator lab that gives you an opportunity to apply the method of consecutive differences to real data: the heights of the bouncing ball from Chapter 2. However, first we'll look at the consecutive differences graphically because although real data will not produce absolutely constant differences, the graph of the points will be approximately horizontal. In Example 8.21 we will demonstrate this method with the data for the number of squares before turning to real data in Calculator Lab 8.8.

Example 8.21

Plot the first, second, and third differences for the list

$$S = \{1, 5, 14, 30, 55, 91, 140, 204\}$$

Solution   The differences will be plotted on the vertical axis, and the horizontal axis will be related to the variable n shown in the left column of Table 8.8, repeated below as Table 8.9. Notice that in the table the first differences correspond to the values of n in the list $\{2, 3, 4, 5, 6, 7, 8\}$, the second differences to the list $\{3, 4, 5, 6, 7, 8\}$, and the third differences correspond to the list $\{4, 5, 6, 7, 8\}$.

Table 8.9

n	S_n	$S_n - S_{n-1}$	1st Difference	2nd Difference	3rd Difference
1	1				
2	5	$5 - 1$	4		
3	14	$14 - 5$	9	5	
4	30	$30 - 14$	16	7	2
5	55	$55 - 30$	25	9	2
6	91	$91 - 55$	36	11	2
7	140	$140 - 91$	49	13	2
8	204	$204 - 140$	64	15	2

To plot three differences we need to form three lists of differences and three lists based on n.

(a) First enter the list named S.
(b) The *TI-83* has a function designed to take consecutive differences in a list. It is the ΔList function, which is selection 7 on the LIST OPS menu. For example if

a list named A contains the items $\{1, 5, 14, 30, 55\}$ then ΔList(A) will produce the list $\{4, 9, 16, 25\}$, the differences between consecutive items in list A.

Figure 8.51 shows how we produced the three sets of differences. To enter the name of the list inside the parentheses, select the name you want from the lists you see when you press [2nd] [LIST].

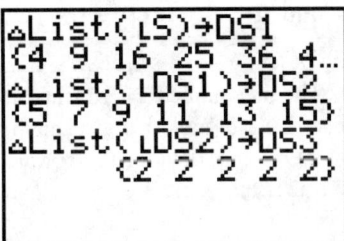

Figure 8.51

Figure 8.52

(c) Forming the three lists based on n would be easy for these short lists, but to prepare you for the need to make longer lists we will remind you of the calculator function named **seq** which you used in Chapter 7 to make lists from arithmetic sequences. Figure 8.52 shows how we formed the three independent variable lists.

To access seq press [2nd] [LIST]
▶ 5

(d) Figure 8.53 shows how we set one of the **StatPlot** definitions. Set Plot2 with **Xlist: N2, Ylist: DS2,** and **Mark: +** and set Plot3 with **Xlist: N3, Ylist: DS3,** and **Mark: □**. Figure 8.54 shows the graphs of all three plots. The squares mark the plot of the third differences. A closer view is shown in Figure 8.55. You can tell from the horizontal line formed by the squares that the third differences are constant. Notice that the second differences form a straight line with positive slope, which you would expect if the third differences are a positive constant.

Figure 8.53

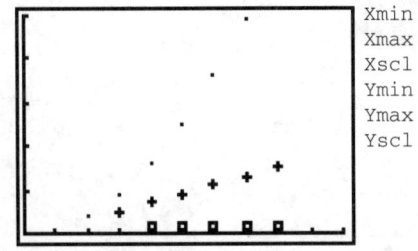

Figure 8.54

```
Xmin = 0
Xmax = 10
Xscl = 1
Ymin = 0
Ymax = 50
Yscl = 10
```

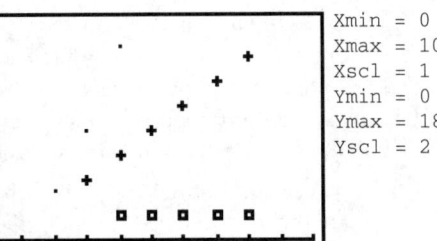

Figure 8.55

```
Xmin = 0
Xmax = 10
Xscl = 1
Ymin = 0
Ymax = 18
Yscl = 2
```

The plots in Figures 8.54 and 8.55 show how a graph can reveal which difference is constant. The graphical method of studying differences will be more important when we look at real data, which never has actual constant differences, because of measurement errors.

Calculator Lab 8.8

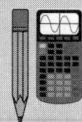

Differences and the Bouncing Ball

The data we will study next is from the bouncing ball CBL experiment you used in Chapter 2. The lists named TBNC1 and BNCE1 are the time and the height for the first bounce. You learned that the equation that fits the data best is a quadratic function (that is, a 2nd degree polynomial). So if we look at the differences we should find that the second differences are constant. Let's do that, using the graphical method of Example 8.21.

Procedures

1. Use ΔList as shown in Figure 8.56 to make the two difference lists L1 and L2.

Figure 8.56

2. Count the items in list BNCE1. One way to do this is by using 1-Var Stats on the STAT CALC menu, the same menu that contains the regressions you have used, and issuing the instruction

    1-Var Stats BNCE1

   On the resulting screen look for the line that begins n=.
3. Use the seq function to generate lists. One list begins with 2 and goes to the number of items in BNCE1, and the other begins with 3 and continues to the same last number.

   Store the integer list that goes with L1 in L3 and the integer list that goes with L2 in L4. You now have lists whose sizes match so you can graph them.
4. Set the dimensions of the viewing window to match the values in the lists.
5. Plot L1 and L3 in one stat plot and L2 and L4 in another stat plot. Use L3 and L4 for the Xlist, as shown in Figure 8.57. Look at the values of the differences to figure out the dimensions of the viewing window.
6. Compare your plots with the ones we got with our data. They are shown in Figure 8.58. In our plots the crosses represent the first differences (L1) and the squares represent the second differences (L2). With the exception of the first and last points the approximately horizontal line of squares is good evidence that the second differences (L2) are constant.

Figure 8.57

7. Write an answer to the following questions.

   (a) If the second differences are approximately constant, what would you expect the first differences to look like?
   (b) Do the data points marked with crosses agree with your answer?
   (c) Why do the first differences in height tend to go down throughout the experiment?

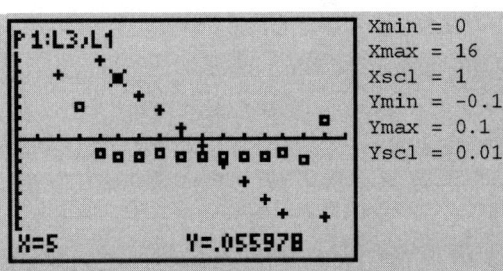

Figure 8.58

(d) In this part of the experiment the ball bounced up and then down. The first differences are positive and then negative. What would the first difference in height be at the top of the bounce? Why?

Section 8.4 Exercises

For Exercises 1–6 plot the 8th degree Maclaurin polynomial approximation

$$c(x) \approx 1 - \frac{x^2}{2!} + \frac{x^4}{4!} - \frac{x^6}{6!} + \frac{x^8}{8!} - \cdots$$

together with the function $\cos x$. *Set your calculator to Radian mode and use the* ZTrig *setting.*

1. (a) How many real roots does this polynomial have?

   (b) Approximately what are they?

   (c) How do its roots compare with the roots of $\cos x$?

2. (a) In what interval does the approximation of $c(x)$ for $\cos x$ seem good?

   (b) How could you expand that interval?

3. In what intervals does the approximation seem too rough?

4. Compare the value of $\cos 0.5$ with the value of the polynomial evaluated at $x = 0.5$. How many decimal places of accuracy does the polynomial give?

   (a) Make the same comparison for $x = 1.5$.

   (b) Is the approximation for $\cos 1.5$ more accurate or less accurate than the approximation of $\cos 0.5$?

   (c) Write an explanation of why you think one estimate was better than the other.

5. Add one term to the polynomial to make it 10th degree and report how many more digits of accuracy are gained for the estimation of $\cos 1.5$.

6. Explain how to use the periodic nature of the cosine function to compute an estimate for the value of $\cos 25$. Carry out this procedure and report how many decimal places of accuracy there are in your estimate.

For Exercises 7a–12 plot the 4th degree Maclaurin polynomial approximation $f(x) \approx 1 + x + x^2 + x^3 + x^4 + \cdots$ *together with the function* $g(x) = \dfrac{1}{1-x}$.

7. (a) In what interval does the approximation of $f(x)$ for $g(x)$ seem close?

   (b) How could you expand that interval?

8. In what intervals does the approximation seem too rough?

9. Compare the value of $g(0.5)$ with the value of the polynomial evaluated at $x = 0.5$. How many decimal places of accuracy does the polynomial give?

10. (a) Compare the value of $g(1.5)$ with the value of the polynomial evaluated at $x = 1.5$. How many decimal places of accuracy does the polynomial give?

    (b) Is the approximation for $g(1.5)$ more accurate or less accurate than the approximation of $g(0.5)$?

    (c) Write an explanation of why you think one estimate was better than the other.

11. (a) Compare the value of $g(-1.5)$ with the value of the polynomial evaluated at $x = -1.5$. How many decimal places of accuracy does the polynomial give?

    (b) Is the approximation for $g(-1.5)$ more accurate or less accurate than the approximation of $g(0.5)$?

    (c) Write an explanation of why you think one estimate was better than the other.

12. Add one term to the polynomial to make it 5th degree, and report how many more digits of accuracy are gained for the estimation of $g(0.5)$.

13. Consider the following table of values:

x	1	2	3	4	5	6	7
y	1	9	31	62	117	188	275

(a) Make a table of the first, second, and third differences (see Table 8.9).

(b) Based on your table in (a), what is the degree of the polynomial that fits your data?

 (c) Use your calculator to determine the regression equation for your data.

(d) Graph the original data and the regression equation on the same graph. How well does the regression curve fit your actual data? How can you account for any differences?

14. Consider the following table of values:

x	1	2	3	4	5	6	7
y	5.45	5.6	6.25	7.3	12.95	20.5	38.95

(a) Make a table of the first, second, and third differences.

(b) Based on your table in (a), what is the degree of the polynomial that fits your data?

 (c) Use your calculator to determine the regression equation for your data.

(d) Graph the original data and the regression equation on the same graph. How well does the regression curve fit your actual data? How can you account for any differences?

15. Consider the following table of values:

x	1	2	3	4	5	6	7	8
y	5	7	14	29	63	128	234	396

(a) Make a table of the first, second, third, and fourth differences.

(b) Based on your table in (a), what is the degree of the polynomial that fits your data?

 (c) Use your calculator to determine the regression equation for your data.

(d) Graph the original data and the regression equation on the same graph. How well does the regression curve fit your actual data? How can you account for any differences?

16. Consider the following table of values:

x	1	2	3	4	5	6	7	8
y	−5	−2.5	6.5	29	74	160	308	544

(a) Make a table of the first, second, third, and fourth differences.

(b) Based on your table in (a), what is the degree of the polynomial that fits your data?

 (c) Use your calculator to determine the regression equation for your data.

(d) Graph the original data and the regression equation on the same graph. How well does the regression curve fit your actual data? How can you account for any differences?

17. For this exercise you will need a large coffee urn that has a spigot at the bottom and a large measuring container. Fill the urn with water. Open the spigot for 10 seconds and let the water drain into the container. Record the volume of water that was released during that 10-second interval. Repeat for a second 10-second interval. Record the total volume of water that was released in the first 20 seconds. Repeat until all the water has been drained from the urn.

(a) Make a graph of the volume as a function of time.

(b) Make a table of the first, second, and third differences (see Table 8.9).

(c) Based on your table in (b), what is the degree of the polynomial that fits your data?

 (d) Use your calculator to determine the regression equation for your data.

(e) Graph the regression equation on the same graph from (a). How well does the regression curve fit your actual data? How can you account for any differences?

18. For this exercise you will need to draw six different polygons with from three to eight sides.

(a) Draw all the diagonals of each polygon and make a table of the number of sides and the number of diagonals.

(b) Make a graph of the number of diagonals as a function of number of sides.

(c) Make a table of the first, second, and third differences.

(d) Based on your table in (c), what is the degree of the polynomial that fits your data?

 (e) Use your calculator to determine the regression equation for your data.

(f) Graph the regression equation on the same graph from (b). How well does the regression curve fit your actual data? How can you account for any differences?

19. For this exercise you will need several objects with different masses to serve as the bobs of a pendulum. Weigh each bob. Start with the lightest object and fasten it to one end of the string. Fasten the other end of the string to a solid object so that the pendulum can swing freely. Measure the distance from one end of the string to the center of the bob. Pull back the bob to a predetermined initial position (make sure the string remains taut). Let go and measure how long it takes for the pendulum to complete ten periods. Repeat with this same bob and average the two results. Change the bob to the next heaviest object and repeat the experiment. Make sure to keep the length of the string and the initial position the same. Continue until you have done this twice with each object.

   (a) Determine the regression equation that seems to fit the relationship between the weight of the bob and the time it takes to complete ten swings.

   (b) Write a memo to the management reporting your results.

20. For this exercise you will need one object to serve as the bob of a pendulum. Fasten the bob to one end of the string. Fasten the other end of the string to a solid object so that the pendulum can swing freely. Measure the distance from one end of the string to the center of the bob. Pull back the bob to a predetermined initial position (make sure the string remains taut). Let go and measure how long it takes for the pendulum to complete ten periods. Repeat with this same initial position and average the two results. Change the initial position and repeat the experiment. Make sure to use the same string and bob. Continue until you have done this with five different initial positions.

   (a) Determine the regression equation that seems to fit the relationship between the initial position and the time it takes to complete ten swings.

   (b) Write a memo to the management reporting your results.

8.5   Patterns in Slopes of Power Functions

You have seen previously that the slope of the tangent to a distance vs. time curve is the velocity of the object at that time. Slopes of curves are valuable tools in science and engineering because they tell how a function changes at a particular point. In Chapter 2 you saw how to construct the tangent to a curve. In this section we'll study how the slope of the tangent to a curve varies depending on its location on the curve. Through experimentation and analysis you will see how to find the slope function for some power functions.

Review: Tangents to Curves

▶ A tangent line to a smooth curve at a point, called the **point of tangency**, goes through the point of tangency. For our purposes, a tangent line does not cross the graph at the point of tangency and lies completely on one side of the graph near that point. Three examples of tangent lines are shown in Figures 8.59(a)–(c). In Figure 8.59(c) the tangent line intersects the curve at one point, but near the point of tangency it only touches the curve. The line in Figure 8.59(d) is not a tangent to the curve at this point because it crosses the curve.

▶ The slope of a tangent line is approximately equal to the slope of the line segment connecting the point of tangency and a nearby point on the curve.

▶ The closer the nearby point, the better estimate of the slope of the tangent line we can get as shown in Figure 8.60. Using the language of limits, we can say that

$$\lim_{P \to T} \text{slope of } PT = \text{Slope of tangent line at } T$$

▶ The slope of the curve at the point of tangency is defined as the slope of the tangent line.

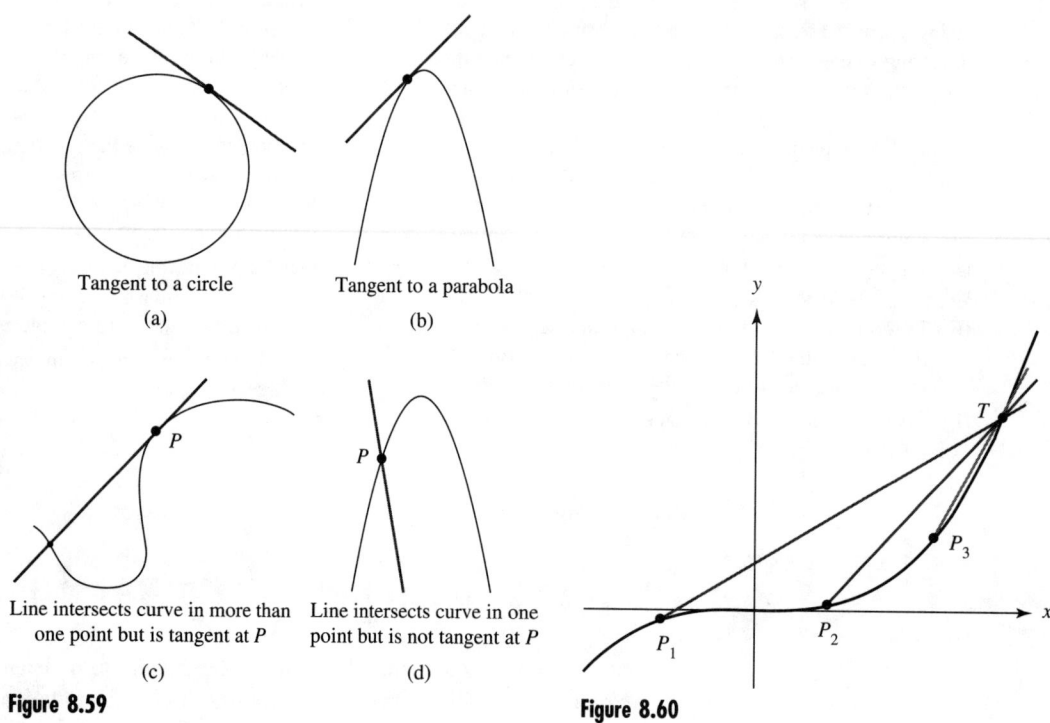

Tangent to a circle
(a)

Tangent to a parabola
(b)

Line intersects curve in more than
one point but is tangent at P
(c)

Line intersects curve in one
point but is not tangent at P
(d)

Figure 8.59

Figure 8.60

Plotting a Tangent to a Curve

Example 8.22

Estimate the slope of the tangent line for the power function $f(x) = 2x^3$ if the point of tangency is the point $(2, 16)$. Report your estimate with a precision of one decimal place.

Solution   The method is to locate a point P_1 near the point $(2, 16)$ and calculate the slope of the line through those two points. Then move to a point P_2 closer to $(2, 16)$ and compute the slope of the line through P_2 and T. Continue this process until two consecutive computations of the slope give the same result up to the desired number of decimal places. Figure 8.60 shows an example of the sequence of points P_1, P_2, P_3 moving closer to the point of tangency, T. The process of doing the calculation is described in a **flow chart** in Figure 8.61. Begin with the instruction at the top of the flow chart and continue until you have reached the desired degree of precision. Because the flow chart *loops* back near the beginning and repeats several instructions, the points P_1, P_2, P_3, etc. are all called P.

   Your results will depend on the size of your starting view window. For greatest accuracy you should change your MODE setting to Float. Figure 8.62 shows the graph with the first position of point P, the nearest point on the right of $(2, 16)$ that the Trace function identifies.

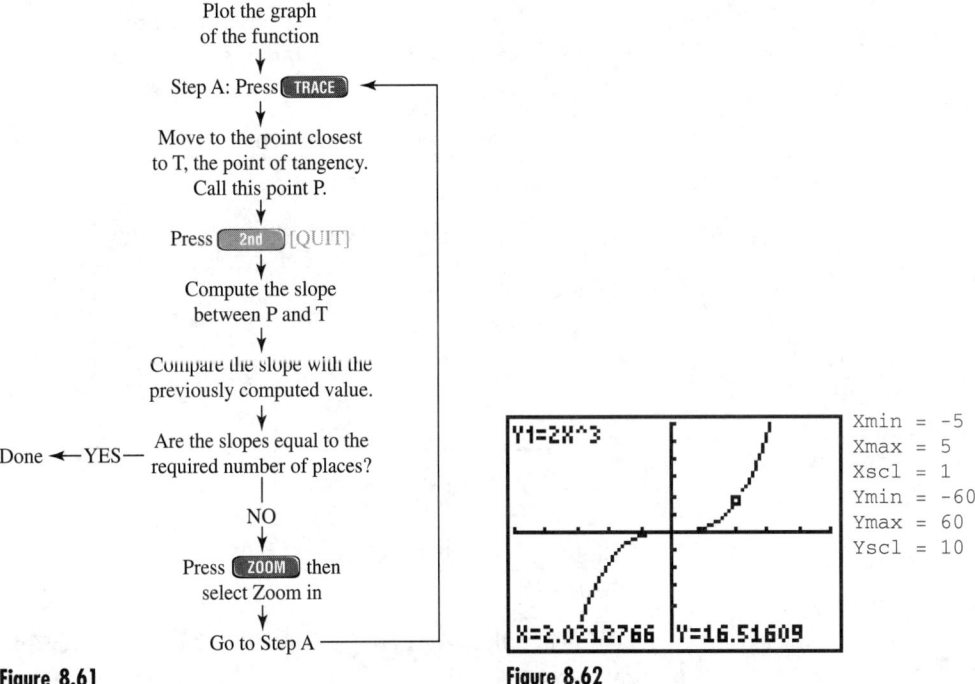

Plot the graph
of the function
↓
Step A: Press [TRACE] ←
↓
Move to the point closest
to T, the point of tangency.
Call this point P.
↓
Press [2nd] [QUIT]
↓
Compute the slope
between P and T
↓
Compare the slope with the
previously computed value.
↓
Done ← YES — Are the slopes equal to the
required number of places?
|
NO
↓
Press [ZOOM] then
select Zoom in
↓
Go to Step A

Figure 8.61

Figure 8.62

The first estimate of the slope of the tangent line is

$$\frac{16.51609 - 16}{2.0212766 - 2} = \frac{0.51609}{0.0212766} \approx 24.256$$

Table 8.10 shows two additional locations of point P and the estimates of the slope of the tangent that each one produced. The value of n in Table 8.10 refers to the number of zoom-ins it took to obtain the coordinates of P_n. We chose to keep point P on the right of (2, 16). In general, it is more convenient to stick to one side or the other.

Sometimes when you zoom in, the point you used in the previous step is still the closest one. If this happens, just zoom in again.

Table 8.10

n	Coordinates of Point P_n	Slope
1	(2.0212766, 16.51609)	24.256
2	(2.0013298, 16.031936)	24.0156
3	(2.0000831, 16.001995)	24.0072

An estimate of the slope of the tangent to one decimal place is 24.0. ■

Now that we have an estimate of the slope of the tangent, we will draw the line that passes through the point of tangency (2, 16) with that slope. In Chapter 1 you learned how to find the equation of a line if you know its slope and one point on the line. Example 8.23 reminds you how to do that.

Example 8.23

Write the equation of the line that passes through the point (2, 16) with a slope equal to 24.0. Plot this line on the same graph with the function $f(x) = 2x^3$.

Solution   The point-slope form of the equation of a straight line is $y - y_1 = m(x - x_1)$ where m is the slope and (x_1, y_1) is one point on the line. So we have the equation

$$y - 16 = 24(x - 2)$$

In order to use this equation with the *TI-83*, we must solve for y

$$y = 24(x - 2) + 16$$

This equation could be simplified further if you want to enter fewer symbols in the Y= screen of the calculator. The graph of the function and the tangent are shown in Figure 8.63.

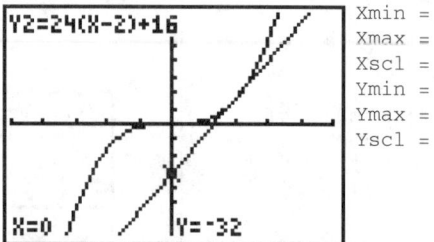

Xmin = -5
Xmax = 5
Xscl = 1
Ymin = -60
Ymax = 60
Yscl = 10

Figure 8.63

In Calculator Lab 8.9 you will calculate the slope of the tangent at other points of the same curve $y = 2x^3$.

Calculator Lab 8.9

Plotting the Slopes

In this lab you will use a flow chart to find the slope of the tangent to the curve at other points on the same curve. Then you will plot the graph of these slopes. From the results of this lab you will be able to find a formula that can give the slope at *any* point on the curve.

Procedures 1. For each of the points of tangency in Table 8.11, estimate the slope of the tangent line for the function $f(x) = 2x^3$. Report your estimate to one decimal place. As you make these estimates, look for ways that the symmetry of the graph can help you make some of the estimates without any calculations.

Table 8.11

Point of Tangency	Estimate of the Slope of the Tangent
$(-3, -54)$	
$(-2, -16)$	
$(-1, -2)$	
$(0, 0)$	
$(1, 2)$	
$(2, 16)$	24.0
$(3, 54)$	

2. Plot the slope values in the blank graph in Figure 8.64. The y-coordinate of each point on the lower graph is the slope of the tangent to the curve shown above at the same value of x.

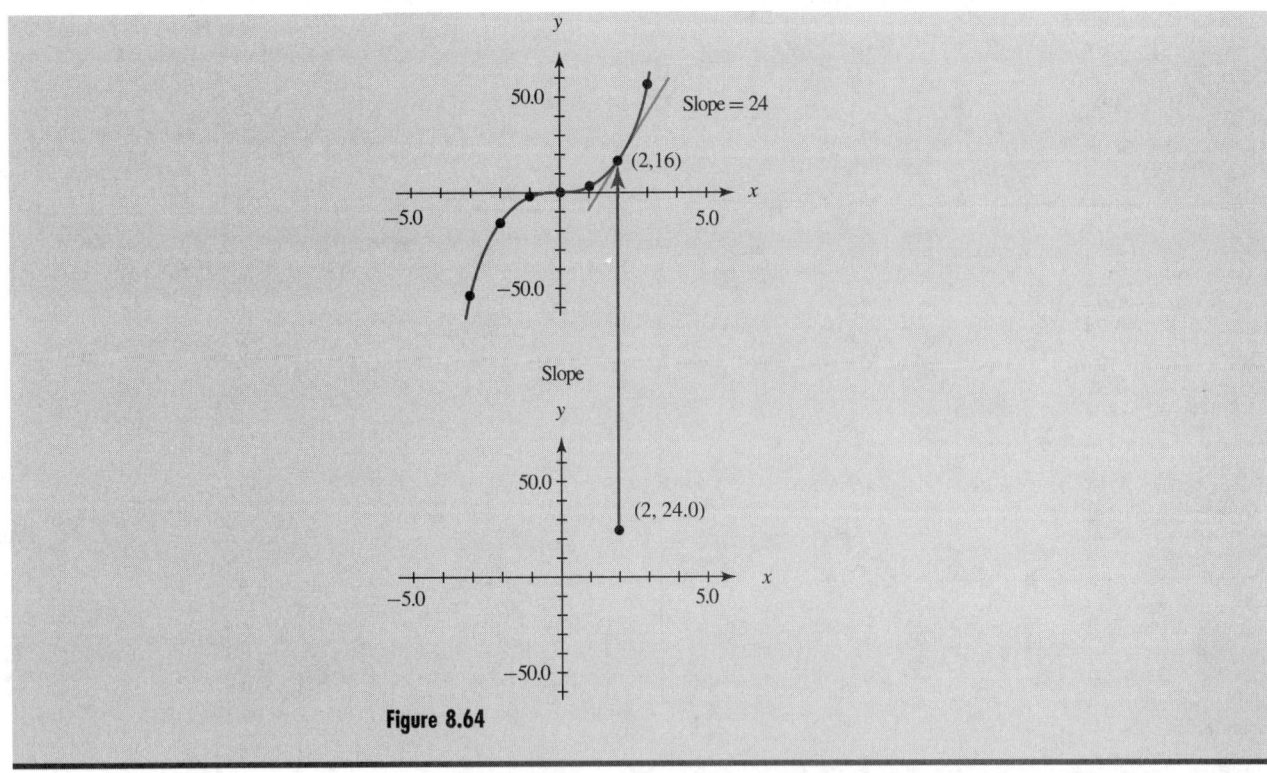

Figure 8.64

The points you plotted in the bottom graph of Figure 8.64 are part of a new kind of function, one that gives the slope of $f(x) = 2x^3$ for any value of x. This new function is called the **slope function**.

> **Definition: Slope Function**
> The slope of the graph of $y = f(x)$ at any point where a tangent can be drawn is given by the curve's **slope function**. To find the slope of the tangent at some point (x, y) evaluate the slope function for that value of x.

Procedures for finding a function's slope function is studied in detail in calculus. In this section we will find a few slope functions by experimenting and looking at patterns.

The points you plotted in Calculator Lab 8.9 were part of the slope function of the function $f(x) = 2x^3$. In Section 8.4 you studied the Consecutive differences method, which can help you use those points to write an equation for a slope function. In Example 8.24 you will find that equation. But first in Activity 8.5 you will get more experience thinking about the shape of the graphs of other slope functions.

Activity 8.5
Plotting Slopes

In Figure 8.65 the top graph is a power function and the bottom graph is for you to make a *sketch* of the slope function. For this activity, do not calculate estimates of the slope; just think about how the slope is changing in magnitude and in sign, and draw a rough graph. Later you will see the correct shapes of the graphs of these slope functions.

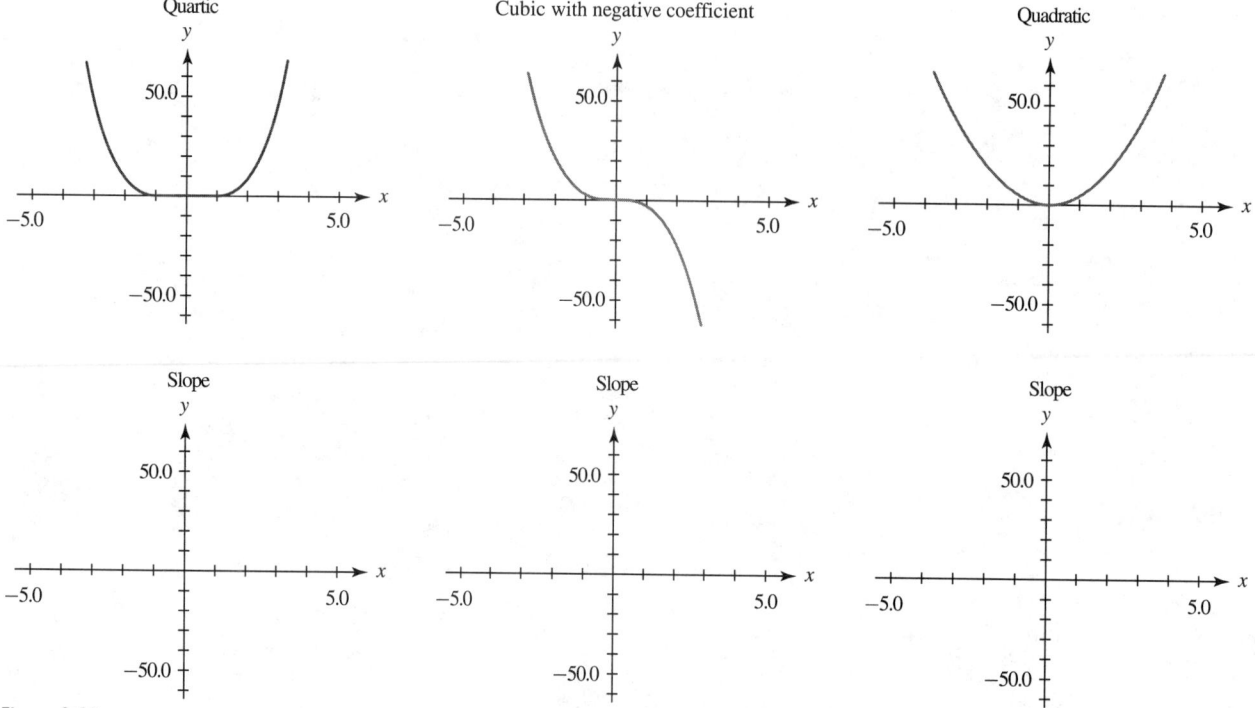

Figure 8.65

A Program that Draws Tangents

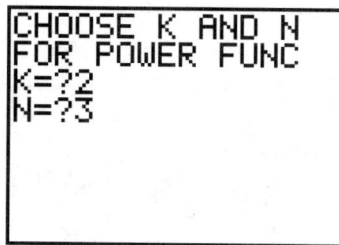

Figure 8.66

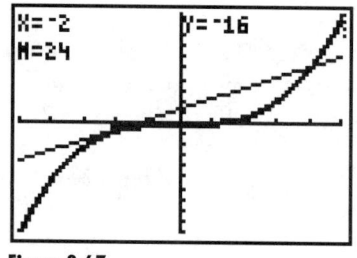

Figure 8.67

The *TI-83* program named POWERFUN was written to provide you with experience looking at the slopes of tangents to power functions. When you run the program, the first screen looks like Figure 8.66. It invites you to enter values of k and n for the power function $f(x) = kx^n$. Our values, $k = 2$ and $n = 3$, produce the function $f(x) = 2x^3$, which we worked with in the previous labs and examples.

The next part of the program draws a tangent at ten points with integer x-values from -5 to 4. The tangent drawn at $x = -2$ is shown in Figure 8.67. On this screen you also see the x and y coordinates and the value of the slope of the tangent, written as M=24. Press ENTER whenever you see the four blinking dots in the upper right corner of the screen and the next tangent will be drawn.

Figure 8.68 shows that after the last tangent is drawn, all the tangents are shown on one screen. In the background you can see ten square dots. The y-value at each dot is the slope of the tangent to the power function (Figure 8.69) for that value of x. The square dots lie on the slope function of $f(x) = 2x^3$. Pressing ENTER again shows the graph of the slope function for the given power function and pressing ENTER again superimposes the graph of the original power function as shown in Figure 8.70. You can compare the curve these dots make with your answer to Calculator Lab 8.9.

When you press ENTER again, you should see Figure 8.71. Select option 1, 2, or 4; we will use option 3 in a later calculator lab.

Calculator Lab 8.10

Drawing Tangents with POWERFUN

In this lab you will gather more data about the slopes of tangents for different power functions. The purpose of the lab is to give you data so that you can guess the slope function for each power function.

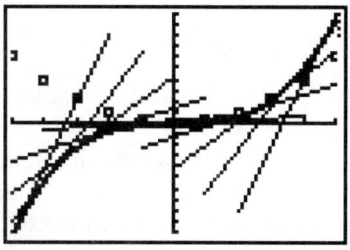

Figure 8.68

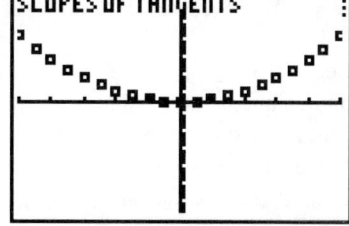

Figure 8.69

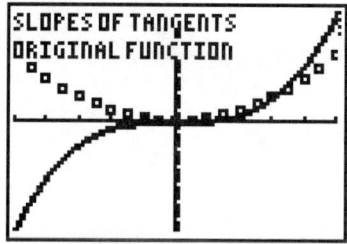

Figure 8.70

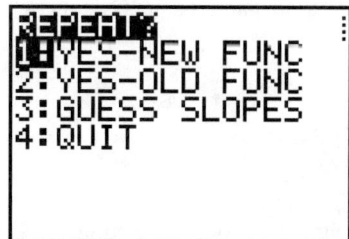

Figure 8.71

Procedures

1. Run the program POWERFUN for the function $f(x) = 2x^3$. As each tangent is drawn, check it against the values you computed in Calculator Lab 8.9.

2. Run the program POWERFUN for five other power functions. You may choose to use functions with different values of k and the same value of n. For example you might use $n = 3$ with other values of k to see if you can detect a pattern in the slope functions. Or you may experiment with other values of n.

3. For each power function you study fill in a table like Table 8.12. Complete the difference columns until the difference is constant.

Table 8.12

$k =$_____ $n =$_____				
Value of x	Slope of Tangent	Difference 1	Difference 2	Difference 3
−5				
−4				
−3				
−2				
−1				
0				
1				
2				
3				
4				

Modeling the Slope Function

In Section 8.4 you learned how to use the Consecutive Differences method to find a mathematical model that will fit certain data. Remember that there are two restrictions on this method: first the x-values must be in an arithmetic sequence, as they are in Table 8.12. Next the differences must be constant at some point. If both of these restrictions occur then we know that the mathematical model will be a polynomial whose degree is equal to the number of differences taken until the differences are constant. Example 8.24 will show you how to apply that method to the data you gathered in Calculator Labs 8.9 and 8.10 for the function $f(x) = 2x^3$.

Example 8.24

Estimate the slope function for the function $f(x) = 2x^3$ from data you have collected.

Solution   We will use the Consecutive Differences method. Table 8.13 summarizes the results you found in Calculator Labs 8.9 and 8.10. We have included the difference columns. There are some negative numbers in the difference column because we always subtract one number from the one below it. Negative values of the difference mean that the slopes are decreasing.

Table 8.13

k = 2	n = 3			
Value of x	Slope of Tangent	Difference 1	Difference 2	Difference 3
−5	150			
−4	96	−54		
−3	54	−42	12	
−2	24	−30	12	
−1	6	−18	12	
0	0	−6	12	
1	6	6	12	
2	24	18	12	
3	54	30	12	
4	96	42	12	

The second difference is constant. Therefore the function is a quadratic function $S(x) = ax^2 + bx + c$, where $S(x)$ is the slope function for $y = 2x^3$ at any x value and $a \neq 0$. To estimate the three coefficients a, b, and c we need to select any three points on the graph of $y = 2x^3$ and then use quadratic regression. We'll select the points $(1, 6)$, $(2, 24)$, and $(3, 54)$, although *any* three points could be used. Figure 8.72 shows how we entered the x-list, y-list, and regression instruction.

```
{1,2,3}→L₁
            {1 2 3}
{6,24,54}→L₂
            {6 24 54}
QuadReg L₁,L₂
```

Figure 8.72

The resulting regression equation is

$$S(x) = 6x^2$$

This is the slope function for the function $f(x) = 2x^3$. With it, the slope can be computed for any value of x. For example, at the point on the curve where the x-coordinate is 0.3 the slope is $6(0.3^2) = 0.54$.   ■

In Calculator Lab 8.11 you will use the results of Calculator Lab 8.10 to write the slope functions for other power functions. Try to find a pattern in the answers. That is, you might be able to see how to predict that the slope function for $f(x) = 2x^3$ is $S(x) = 6x^2$.

Calculator Lab 8.11

Patterns in Slope Functions

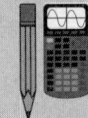

In Calculator Lab 8.10 you made tables and computed differences. Now use the Consecutive Differences method to determine the slope function for each of the five power functions you studied in Calculator Lab 8.10.

Procedures

1. For each table you made in Calculator Lab 8.10 find the equation for the slope function.
2. Complete a table like Table 8.14 with your results.

Table 8.14

Power Function	Slope Function
$2x^3$	$6x^2$

3. Write about the patterns you see in the table. Can you predict the slope function for some power functions?

Calculator Lab 8.12 returns you to the calculator program POWERFUN to let you use what you have figured out so far about patterns in the slope functions of power functions.

Calculator Lab 8.12

Guessing Slope Functions

Now you've learned enough to guess some slope functions. Using the same program you can see which of the patterns you have discovered are correct. You can also use it to help you think of other patterns in the slopes and slope functions.

Procedures 1. Run the program POWERFUN with the values $k = 2$ and $n = 3$. When you reach the menu in Figure 8.71 select option 3, GUESS SLOPES. Figure 8.73 shows the program reminder for the current values of k and n, and invites you to enter the values of k and n for your guess of the slope function. We entered the incorrect values $k = 4$ and $n = 2$. Figure 8.74 shows the graph of the slopes of the tangents and displays the graph of the incorrect function we guessed. Finally, if you enter $k = 6$ and $n = 2$, the slope function exactly matches the slopes of the tangents as shown in Figure 8.75.

Figure 8.73

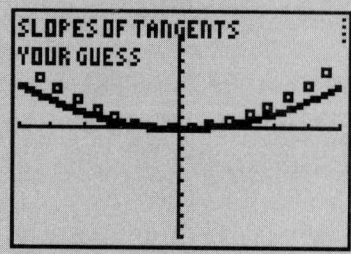

Figure 8.74

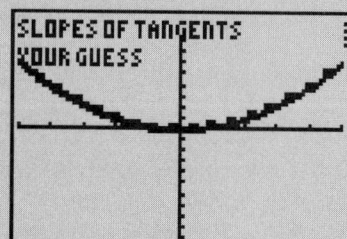

Figure 8.75

2. Run the program POWERFUN with other values of k and n that you have studied in this section, and check to see that your slope functions are correct.
3. Revise what you wrote in Calculator Lab 8.11 about patterns in slope functions. Are there any values of n in the power functions $f(x) = kx^n$ for which you know the rule for writing the slope function?

Section 8.5 Exercises

1. Write the equation of the line that passes through the point $(-2, 20)$ with a slope of -20.0. Plot this line on the same graph with the function $f(x) = 5x^2$.

2. Write the equation of the line that passes through the point $(-1, 3)$ with a slope of -12.0. Plot this line on the same graph with the function $g(x) = 3x^4$.

3. Write the equation of the line that passes through the point $(3, -8.1)$ with a slope of -10.8. Plot this line on the same graph with the function $h(x) = -0.1x^4$.

4. Write the equation of the line that passes through the point $(-4, -16)$ with a slope of 12.0. Plot this line on the same graph with the function $j(x) = 0.25x^3$.

5. (a) Run the program POWERFUN for the function $f(x) = 5x^2$ and complete a table like Table 8.15.

    (b) Use the data in (a) to estimate the slope function for the function $f(x) = 5x^2$.

6. (a) Run the program POWERFUN for the function $g(x) = 3x^4$ and complete a table like Table 8.13.

    (b) Use the data in (a) to estimate the slope function for the function $g(x) = 3x^4$.

7. (a) Run the program POWERFUN for the function $h(x) = -0.1x^4$ and complete a table like Table 8.13.

    (b) Use the data in (a) to estimate the slope function for the function $h(x) = -0.1x^4$.

8. (a) Run the program POWERFUN for the function $j(x) = 0.25x^3$ and complete a table like Table 8.13.

    (b) Use the data in (a) to estimate the slope function for the function $j(x) = 0.25x^3$.

Table 8.15

k =____ n =____				
Value of x	Slope of Tangent	Diff. 1	Diff. 2	Diff. 3
−5				
−4				
−3				
−2				
−1				
0				
1				
2				
3				
4				

8.6   Analyzing Rational Functions

Definition and Examples

In this chapter on algebraic functions there is one more type of function to consider, the rational function. A **rational function** is formed from polynomials and ratios of polynomials. Here are examples of rational functions that we will study in this section:

▶ Shifted forms of power functions:

- $f(x) = \dfrac{3}{x-4}$
- $g(x) = \dfrac{3}{x+4} + 2$

▶ Adding a variable term to a familiar function makes it look quite different:

- $f(x) = x^2 + \dfrac{5}{x-3}$

▶ Application to light intensity or gravitation problem:

- $f(x) = \dfrac{1}{x^2} - \dfrac{2}{(10-x)^2}$

Shifted Forms of Power Functions

When we studied trigonometric functions, the horizontal and vertical shift of the graph were important in electrical and other applications. Here is a summary of the algebra behind these shifts.

Review: Functions Shifted Horizontally

The graph of the function $f(x - h)$ is identical in shape to the graph of $f(x)$ but $f(x - h)$ is shifted *to the right* h units, if h is a positive number. If h is negative, the shift is *to the left* the same number of units.

Example 8.25

Plot the graphs of $f(x) = \sin\left(x - \dfrac{\pi}{2}\right)$ and $g(x) = \sin(x)$ on the same axes and discuss the similarities and differences between them.

Solution   The graphs of $f(x) = \sin\left(x - \dfrac{\pi}{2}\right)$ and $g(x) = \sin(x)$ are shown in Figure 8.76 with g drawn as the thicker curve. The graph of $f(x) = \sin\left(x - \dfrac{\pi}{2}\right)$ is the same shape as $g(x) = \sin(x)$, but the graph of f is shifted $\dfrac{\pi}{2}$ units to the right of the graph of g.

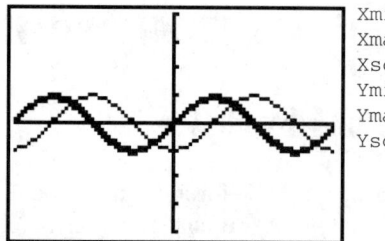

```
Xmin = -2π
Xmax = 2π
Xscl = π/2
Ymin = -4
Ymax = 4
Yscl = 1
```

Figure 8.76

Review: Functions Shifted Vertically

The graph of the function $f(x) + k$ adds a vertical shift *up* k units to the graph of $f(x)$ if $k > 0$. On the other hand if $k < 0$ then the shift is *down* the same number of units.

Example 8.26

Plot the graphs of $y = 3x - 7$ and $y = 3x$ on the same axes and discuss the similarities and differences among them.

Solution   The graphs of $y = 3x - 7$ and $y = 3x$ are shown in Figure 8.77 with $y = 3x$ drawn as the thicker curve. The graph of $y = 3x - 7$ is the same line as $y = 3x$ but it is shifted 7 units down.

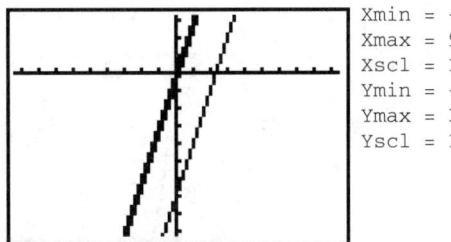

```
Xmin = -9.4
Xmax = 9.4
Xscl = 1
Ymin = -9.2
Ymax = 3.2
Yscl = 1
```

Figure 8.77

■

Example 8.27

Plot the graph of each function and discuss the similarities and differences among them:

(a) $f_1(x) = \dfrac{3}{x}$

(b) $f_2(x) = \dfrac{3}{x-4}$

(c) $f_3(x) = \dfrac{3}{x+4} + 2$

Solutions   We begin by comparing the first two functions, $f_1(x) = \dfrac{3}{x}$ with $f_2(x) = \dfrac{3}{x-4}$. Notice that $f_2(x) = f_1(x-4)$. Your calculator graph, if plotted using the `ZStandard` setting, will look something like Figure 8.78. We used the *thick* graph style for f_1 so that you can more easily identify the two different curves. You can see from this figure that the curve for f_2 is identical to f_1, but shifted four units to the right, as expected from the term $x - 4$ in $f_2(x)$. As you saw in the tangent function graph, the *TI-83* adds a vertical line to the second curve. That vertical line approximates an asymptote of the function, but it is an error to display it as part of the graph of the function. A slight change in the window setting, like those shown in Figure 8.79, can sometimes eliminate the graphing errors.

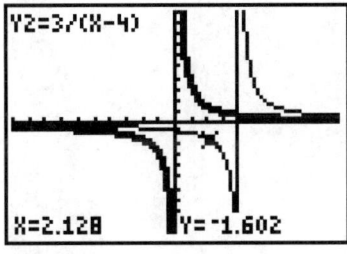

Figure 8.78

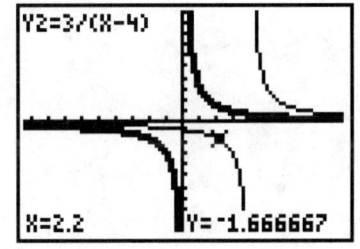

Figure 8.79

```
Xmin = -9.4
Xmax = 9.4
Xscl = 1
Ymin = -10
Ymax = 10
Yscl = 1
```

Figure 8.80 shows the same plot using a computer graphing program. Many computer graphing programs correctly draw a curve and use color to help you distinguish two or more graphs.

Figure 8.81 shows the computer plot of $f_1(x) = \dfrac{3}{x}$ and $f_3(x) = \dfrac{3}{x+4} + 2$. We have added dashed lines representing the asymptotes of the third function. The arrow indicates that f_3 is just the first function shifted to the left 4 units and up 2 units.

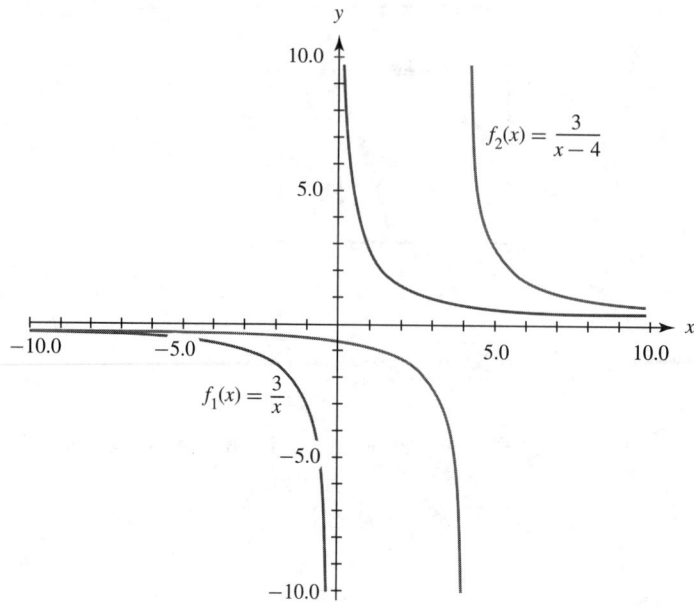

Figure 8.80

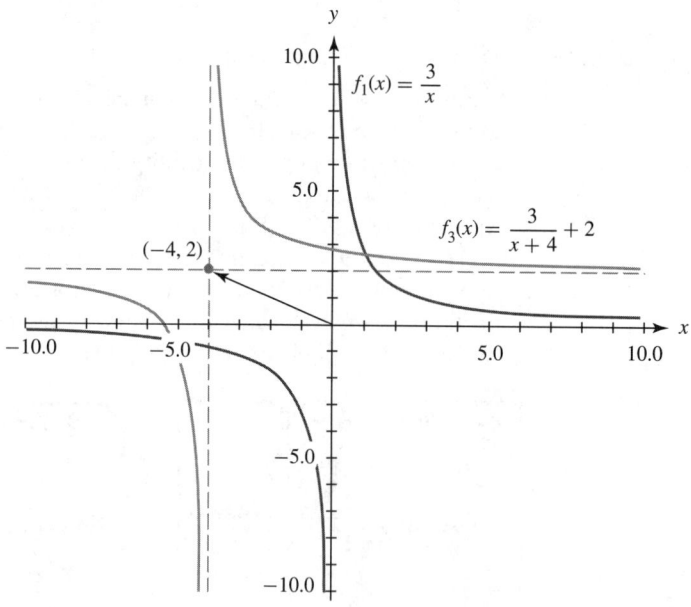

Figure 8.81

Example 8.27 gives us reason to revise the definition of horizontal and vertical asymptotes that we gave in Section 8.1 on page 532.

Definition: Asymptotes

1. A **horizontal asymptote** is a line that a curve approaches (but never reaches) as $x \to \pm\infty$. Algebraically, if there is a real number L such that $f(x) \to L$ as $x \to -\infty$ or $x \to \infty$ then the line $y = L$ is a horizontal asymptote of the function f.

2. A **vertical asymptote** is a vertical line that a curve approaches (but never reaches) as $x \to c$ where c is a real number. Algebraically, if $f(x) \to \pm\infty$ as $x \to c$ then the line $x = c$ is a vertical asymptote of the function f.

Example 8.27 reviewed three characteristics that we have used to analyze graphs of functions: intercepts, asymptotes, and symmetry. We will also include a new characteristic of every function, the values not included in the domain and range of the function.

Definition: Domain of a Function

The **domain** of a function $f(x)$ is the set of values of x that can be successfully substituted into the expression for f to produce values for $f(x)$.

Example 8.28

What is the domain of $f(x) = \dfrac{1}{x-3}$?

Solution   The domain of the function $f(x) = \dfrac{1}{x-3}$ consists of all real numbers except 3 because $f(3) = \frac{1}{3-3} = \frac{1}{0}$ is not a real number. This domain is described in symbols by writing $x \neq 3$, which is spoken as "x is not equal to 3." ∎

Example 8.29

What is the domain of $f(x) = \sqrt{x}$?

Solution   The domain of the function $f(x) = x^{1/2} = \sqrt{x}$ is all real numbers greater than or equal to zero, since the square root of a negative number is not a real number. The domain of this function is $x \geq 0$. ∎

Definition: Range of a Function

The **range** of a real function is the set of values of $f(x)$ that result from substituting each number in the domain.

Example 8.30

What is the range of $f(x) = x^2$?

Solution   The range of the function $f(x) = x^2$ is $f(x) \geq 0$, since negative numbers cannot result from the expression x^2 if x is a real number. ∎

Example 8.31

What is the range of $f(x) = \sqrt{x}$?

Solution   The range of the function $f(x) = x^{1/2} = \sqrt{x}$ is $f(x) \geq 0$, since negative numbers cannot result from the expression $\sqrt{x}$ if x is a real number. ∎

Example 8.32

For the function $f_1(x) = \dfrac{3}{x}$ from Example 8.27, discuss each of the following characteristics of the graph: (a) The domain and range, (b) the intercepts, (c) the asymptotes, and (d) the symmetry. It is possible that the graph may not have any intercepts, asymptotes, or symmetry.

Solutions
(a) Find the domain by asking yourself what values of x could not be used in the function. For this function any number except $x = 0$ can be used, so we describe the domain as $x \neq 0$. The range is not as easy to determine from the equation, but by looking at the graph we can see that any number except zero is produced, so the range is $f(x) \neq 0$, or $y \neq 0$.

(b) There are no intercepts because the function does not cross either the x-axis or the y-axis. This is more obvious when you look at the domain and range, which state that $x \neq 0$ and $y \neq 0$.

(c) The y-axis (whose equation is $x = 0$) is a vertical asymptote, and the x-axis (whose equation is $y = 0$) is a horizontal asymptote.

(d) As we have seen in Section 8.1, a power function whose exponent is an odd negative integer has symmetry with respect to the origin $(0, 0)$. ∎

Example 8.33

For the function $f_2(x) = \dfrac{3}{x - 4}$ from Example 8.27, discuss each of the following characteristics of the graph: (a) The domain and range, (b) the intercepts, (c) the asymptotes, and (d) the symmetry. It is possible that the graph may not have any intercepts, asymptotes, or symmetry.

Solutions
(a) For this function $x = 4$ makes the denominator zero. But, a denominator of zero is not allowed. Hence, the domain is $x \neq 4$. Again the range is not obvious from the expression, but the graph in Figure 8.78 or 8.80 indicates that the function reaches all values except zero. Therefore the range is $y \neq 0$.

(b) The graph reveals that there is no x-intercept, but the function does cross the y-axis between $y = 0$ and $y = -1$. If you wanted to get a better estimate of this intercept you could zoom in on the graph, use the calculator's Table feature, or substitute $x = 0$ into the expression for $f_2(x)$. Letting $x = 0$ produces

$$y = \frac{3}{0 - 4} = -\frac{3}{4} = -0.75$$

The y-intercept is $(0, -0.75)$.

(c) If there are asymptotes they come directly from the domain and range. The vertical asymptote is the line whose equation is $x = 4$, and the horizontal asymptote is the x-axis.

(d) f_2 has symmetry with respect to the point $(4, 0)$ since this point plays a role similar to that of the origin when a curve is symmetric with respect to the origin. Test it by drawing a line from any point on the curve through $(4, 0)$ and see where your line intersects the other branch of the curve. ∎

Example 8.34

For the function $f_3(x) = \dfrac{3}{x+4} + 2$ from Example 8.27, discuss each of the following characteristics of the graph: (a) The domain and range, (b) the intercepts, (c) the asymptotes, and (d) the symmetry. It is possible that the graph may not have any intercepts, asymptotes, or symmetry.

Solutions (a) Since $x = -4$ makes the denominator zero the domain is $x \neq -4$. Again the range is not obvious from the expression, but the graph in Figure 8.81 indicates that the function reaches all values except 2. Therefore the range is $y \neq 2$.

(b) The graph reveals that there is one x-intercept, and one y-intercept. The y-intercept can be calculated in the same way as we did with the function f_2, by substituting 0 for x:

$$y = \frac{3}{0+4} + 2 = \frac{3}{4} + 2 = 2.75$$

The x-intercept is computed by substituting $y = 0$ into the equation and solving for x:

$$0 = \frac{3}{x+4} + 2$$
$$-2 = \frac{3}{x+4}$$
$$-2(x+4) = 3 \qquad\qquad \text{Multiply both sides by } x + 4$$
$$-2x - 8 = 3$$
$$-2x = 11$$
$$x = \frac{11}{-2} = -5.5$$

Check Figure 8.81 to see that these intercepts of $(0, 2.75)$ and $(-5.5, 0)$ appear to be correct.

(c) A look at the graph in Figure 8.81 reveals the vertical asymptote as the line whose equation is $x = -4$, and the horizontal asymptote is the line $y = 2$.

(d) f_3 has symmetry with respect to the point $(-4, 2)$. ∎

Example 8.35

For the function $g(x) = \dfrac{x}{x^2 - 4} - 0.5$ discuss each of the following characteristics of the graph: (a) The domain and range, (b) the intercepts, (c) the asymptotes, and (d) the symmetry. It is possible that the graph may not have any intercepts, asymptotes, or symmetry.

Solutions (a) Domain: Since the denominator $x^2 - 4 = 0$ at $x = \pm 2$, the domain is all real numbers except $x = -2$ and $x = 2$. Again the range is not obvious from the expression, but the graph in Figure 8.82 indicates that the function reaches all values. Therefore the range is all real numbers.

(b) The graph reveals that there are two x-intercepts and one y-intercept. The y-intercept can be calculated in the same way as we did with the function f_2, by substituting 0 for x:

$$y = \frac{0}{0^2 - 4} - 0.5 = 0 - 0.5 = -0.5$$

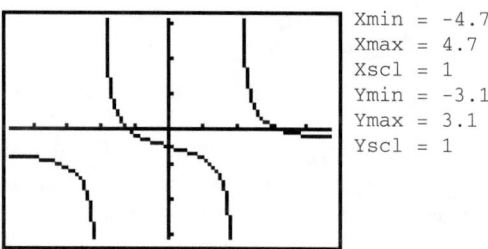

Figure 8.82

The x-intercept is computed by substituting $y = 0$ into the equation and solving for x:

$$0 = \frac{x}{x^2 - 4} - 0.5$$
$$0.5 = \frac{x}{x^2 - 4}$$
$$0.5(x^2 - 4) = x \qquad \text{Multiply both sides by } x^2 - 4$$
$$0.5x^2 - 2 = x$$
$$0.5x^2 - x - 2 = 0$$
$$x = 1 \pm \sqrt{5} \approx -1.236, 3.236$$

Check Figure 8.82 to see that these x-intercepts of $(-1.236, 0)$ and $(3.236, 0)$ are approximately correct.

(c) A look at the graph reveals that there are two vertical asymptotes—one for each point where the denominator is zero. These vertical asymptotes are the lines $x = -2$ and $x = 2$. The horizontal asymptote is the line $y = -0.5$. Notice that the graph *does* cross the horizontal asymptote at the point $(0, -0.5)$. It is even possible for a graph to cross a horizontal asymptote more than once.

(d) f_3 has symmetry with respect to the point $(0, -0.5)$.   ■

To summarize, the graph of $g(x) = \frac{x}{x^2 - 4} - 0.5$ in Example 8.35 *does* cross the horizontal asymptote at the point $(0, -0.5)$. It is possible for a graph to cross a horizontal asymptote more than once and for a graph to have two horizontal asymptotes. A horizontal asymptote describes the behavior of a graph as $x \to \pm\infty$ and does not say anything about the behavior of the graph "near" $x = 0$.

Variations on Familiar Functions

We have seen that the shape of the graph of a power function with negative integer exponent is very distinctive. A typical graph for odd exponents is shown in Figure 8.83 and a typical graph for even exponents is shown in Figure 8.84. In either

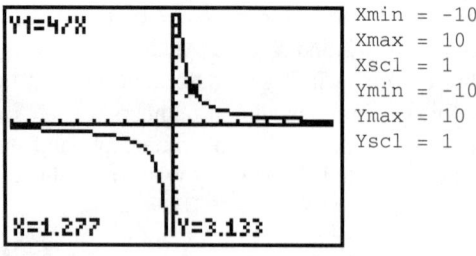

Figure 8.83   **Figure 8.84**

case the values of the function go from very large to very small as x increases for $x > 0$. That property leads us to a single function that will look like one of its components in some portions of the domain and like the other components in other portions of its domain.

First you will get some experience plotting a graph when the standard window does not give a complete picture.

Example 8.36

Find a suitable viewing window to plot the graph of the function $g(x) = x^2 + \dfrac{50}{x - 3}$.

Solution   Figure 8.85 shows the graph plotted on the calculator with the ZStandard viewing window setting.

The figure tells you almost nothing about the graph, except that there is some part of the curve within the window, and there is a vertical asymptote at $x = 3$. Therefore the viewing window must be expanded, but in which direction? Earlier, in Example 8.16, we zoomed out to get a better view. Figure 8.86 indicates that zooming out on this figure does not tell us more about the graph. We need a different approach.

Figure 8.85

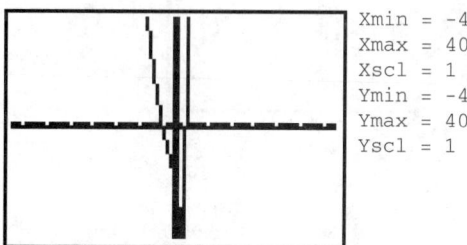

```
Xmin = -40
Xmax = 40
Xscl = 1
Ymin = -40
Ymax = 40
Yscl = 1
```

Figure 8.86

You can get a better idea of what the graph looks like by using the calculator's Table feature. Figures 8.87 and 8.88 show the values for negative and positive values of x.

X	Y₁
-6.000	30.444
-5.000	18.750
-4.000	8.857
-3.000	.667
-2.000	-6.000
-1.000	-11.50
0.000	-16.67

Y1=30.444444444

Figure 8.87

X	Y₁
1.000	-24.00
2.000	-46.00
3.000	ERROR
4.000	66.000
5.000	50.000
6.000	52.667
7.000	61.500

Y1=ERROR

Figure 8.88

First notice that $g(3)$ is undefined, which fits with the observation of a vertical asymptote at $x = 3$. From the tables you can see that the values of y range from -46 to 66, so the next window we should try is $-7 \leq x \leq 7$ and $-50 \leq y \leq 70$. Figure 8.89 shows this graph.

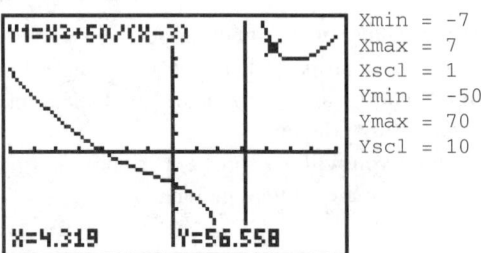

Figure 8.89

For some purposes the graph in Figure 8.89 would be sufficient. However, we'll take one more step to make sure we're not losing some portions of the curve. We'll next expand the viewing window in both directions by multiplying the limits by the same factor. The Zoom Out feature does exactly what we want; it multiplies the limits of the window by a factor of 4.

One use of Zoom Out produces the graph in Figure 8.90 which tells us more about the shape of the graph. Note that the horizontal and vertical dimensions have been multiplied by a factor of 4, but the actual y-values are not multiplied by 4 because the screen was expanded from its center, not from the origin.

The multiplying factor of Zoom Out and Zoom In can be changed by selecting SetFactors from the MEMORY submenu of the ZOOM menu.

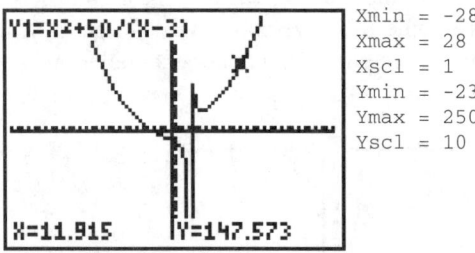

Figure 8.90

If you think about the shape of the parabola whose graph is $y = x^2$ you might recognize it in Figure 8.90, which looks like a parabola with something unusual going on to the right of the origin. One more zoom out as shown in Figure 8.91 and we see almost nothing but a parabola. (In Figure 8.91 we changed Xscl and Yscl after we zoomed out.)

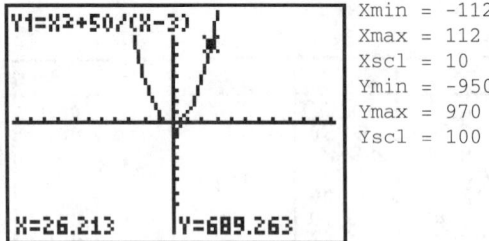

Figure 8.91

Except for an almost invisible dot in the fourth quadrant near the origin, this graph looks exactly like a parabola. The vertical asymptote has vanished. Any calculator or computer graphs can be made to play tricks on you in this way because they can jump right over a very important part of the graph. Figure 8.92 summarizes our previous attempts to graph $y = x^2 + \dfrac{5}{x - 3}$. Each gives a different picture of the graph. Sometimes one view will be useful, sometimes another. As you gain experience graphing complicated functions like this one, you will get better at using Table and Zoom to find the view that you need.

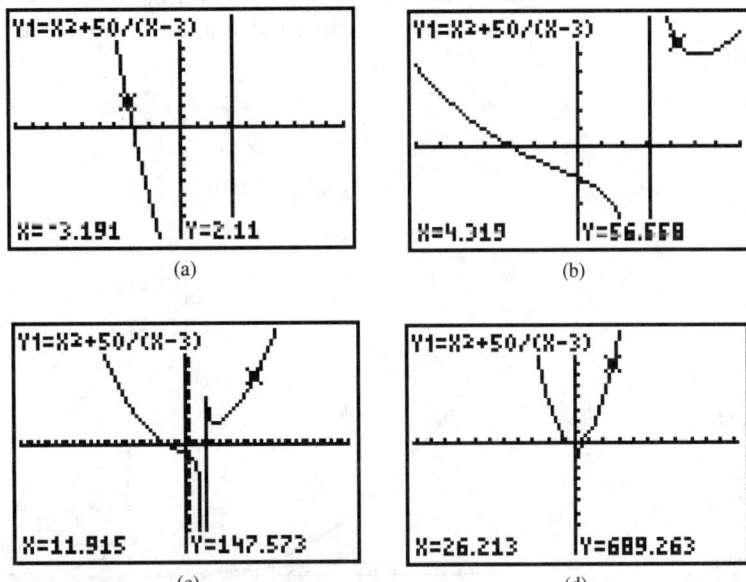

(a)

(b)

(c)

(d)

Figure 8.92

Example 8.37

Plot the graph of the function $g(x) = x^2 + \dfrac{5}{x-3}$ and compare it with the graphs of its components, $g_1(x) = x^2$ and $g_2(x) = \dfrac{5}{x-3}$.

Solution   Figures 8.93 through 8.95 show the graphs of the function g and its two components. Compare these graphs with the calculator versions of the same graph in Figure 8.92. You can see what a challenge it is to get a good idea of what a graph actually looks like.

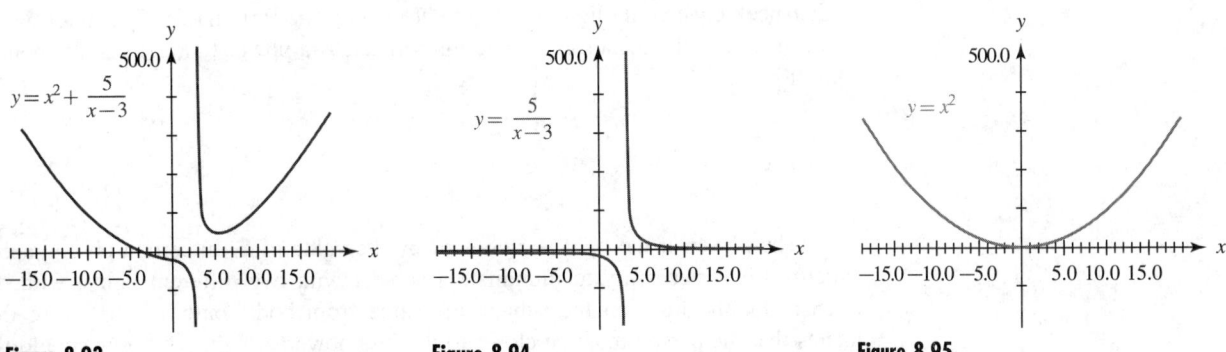

Figure 8.93

Figure 8.94

Figure 8.95

Finally look at Figure 8.96. The function g and its two components are shown on the same graph. It shows that for values of x large in absolute value, g is very close to x^2 and the curve is very close to the parabola. For values of x close to 3, the graph of $y = g(x)$ doesn't look anything like a parabola; now it looks like a shifted power function with negative exponent.

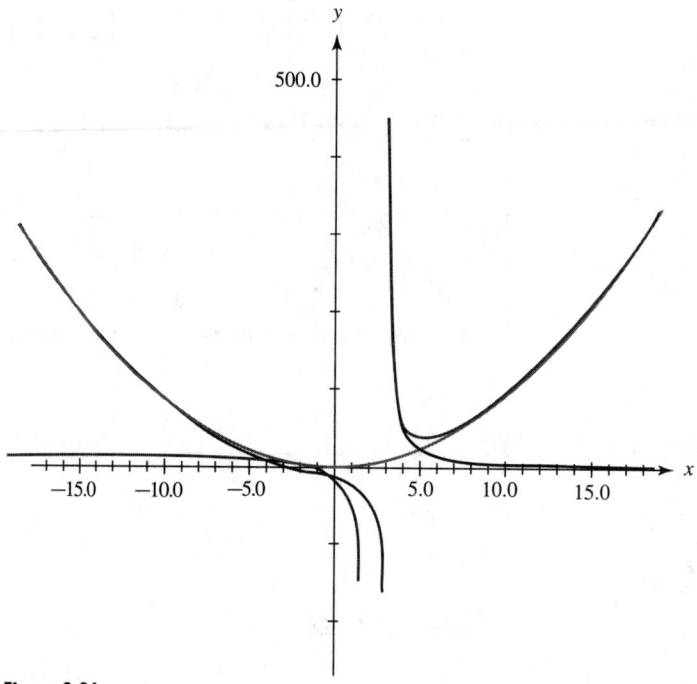

Figure 8.96

Solving a Light Intensity Problem with Graph, Table, and Algebra

In Section 8.2 you learned that the intensity of light varies inversely as the square of distance between the light and the light meter. The formula for light intensity I, as a function of the distance x to the light meter, is a power function with exponent equal to -2:

$$I(x) = \frac{k}{x^2}$$

The power of the light is measured by the value of k. In certain photographic situations it is necessary to illuminate a subject with two different lights, with the intensity of the light on the subject the same from both sources. Of course that means that the person must be closer to the less powerful light. But how should the two distances compare? Example 8.38 shows how to solve this problem when the person is between the two lights.

Example 8.38   Application

A person sits between two different lights that are 10 feet apart. The intensity of the more powerful light is twice that of the less powerful light. What is the equation that must be solved to find the point between the two lights where the person should sit so that the light intensity from each light is the same?

Solution   The situation is shown in Figure 8.97 where x represents the distance from the light on the left to the person.

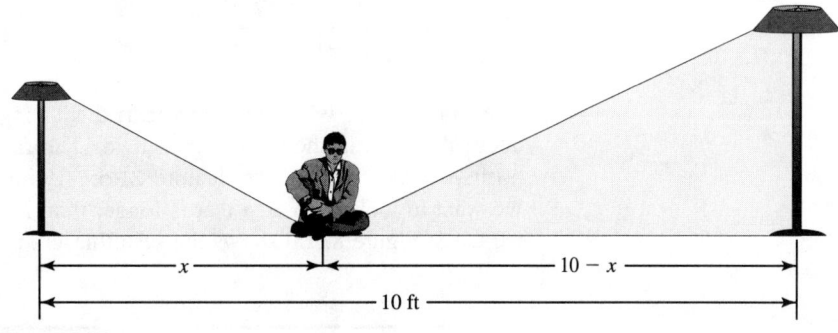

Figure 8.97

The illumination that comes from the two lights are given by the two equations

$$I_1 = \frac{k_1}{x^2}$$

$$I_2 = \frac{k_2}{(10-x)^2}$$

These two intensities are equal when x is the solution of the equation

$$\frac{k_1}{x^2} = \frac{k_2}{(10-x)^2}$$

Since we are given the ratio of the intensities of the two lights as $\frac{2}{1}$ and from the figure we can see that $k_2 > k_1$, then $k_2 = 2k_1$. If we replace k_2 with $2k_1$ the equation becomes

$$\frac{k_1}{x^2} = \frac{2k_1}{(10-x)^2}$$

Now we can divide both sides of the equation by k_1 with the result

$$\frac{1}{x^2} = \frac{2}{(10-x)^2}$$

The solution to this equation will tell us where a person should sit so the light intensity from each light is the same.   ∎

The equation developed in Example 8.38 can be solved with a graph, with tables, and with algebra. We'll show all these methods in Examples 8.39–8.41.

Example 8.39   Application

Solve the light intensity equation $\dfrac{1}{x^2} = \dfrac{2}{(10-x)^2}$ with a graph.

Solution   We'll form a new function from the difference between the two sides of the equation, and look for the value of x that makes the difference equal to zero. The function is

$$f(x) = \frac{1}{x^2} - \frac{2}{(10-x)^2}$$

The graph of f, with the `ZStandard` viewing window and MODE set to three decimal points, is shown in Figure 8.98. The function is zero somewhere along the positive x-axis. The Zoom feature `ZBox` (Zoom Box) is most useful here because we want to look at an area that is longer than it is wide. Figure 8.99 shows the box we used. Figure 8.100 shows the resulting graph.

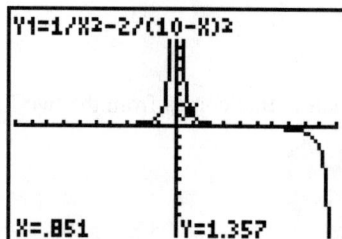

Figure 8.98
Graphed using the `ZStandard` window setting.

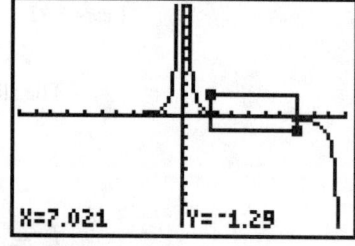

Figure 8.99
Box drawn on Figure 8.98 using `ZBox`.

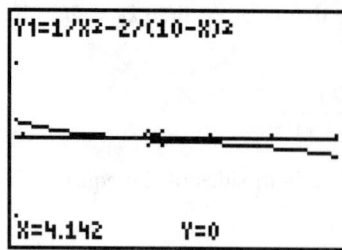

Figure 8.100
Result after using `ZBox`.

Now Trace along the curve to find the value of x where $y = 0$. The decimal precision was set to 3 places, so the result, $x = 4.142$, produces a y-value of 0 to three places. ∎

Example 8.40   Application

Solve the light intensity equation $\dfrac{1}{x^2} = \dfrac{2}{(10-x)^2}$ with a table.

Solution   Figure 8.101 shows that the solution is between 4 and 5. We changed the values of `TblStart` to 4 and `Tbl` to 0.1 to make the table in Figure 8.102. If we continue the process to get the desired accuracy for the value of x that makes the difference equal to 0; we see again that $x = 4.142$.

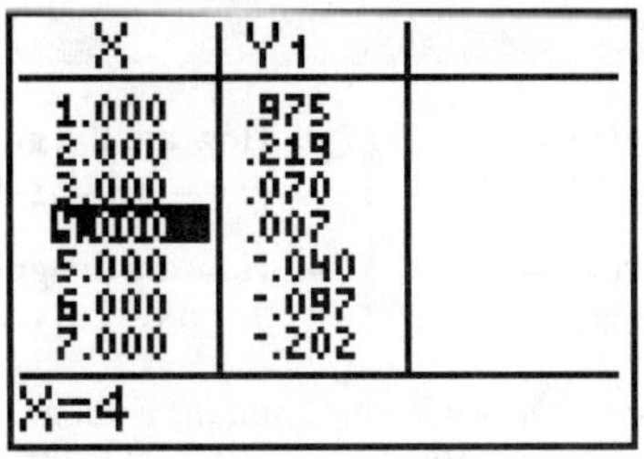

Figure 8.101

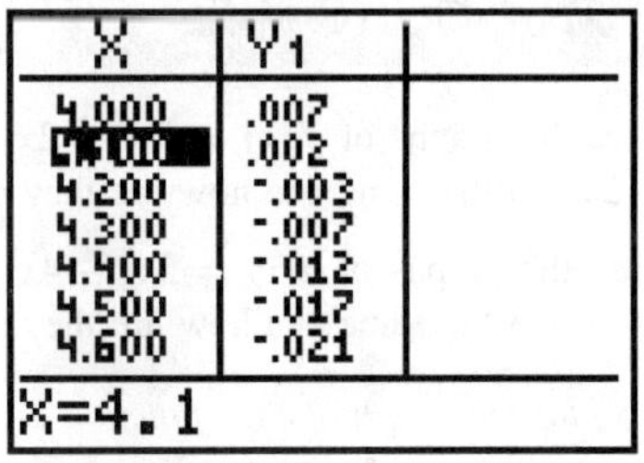

Figure 8.102

Example 8.41   Application

Solve the light intensity equation $\dfrac{1}{x^2} = \dfrac{2}{(10-x)^2}$ with algebra.

Solution   The equation is

$$\frac{1}{x^2} = \frac{2}{(10-x)^2}$$

First remember that an equation in which two fractions are equal can be solved by the method of cross-multiplication. This produces

$$(10-x)^2 = 2x^2$$

Since

$$(10-x)^2 = 100 - 20x + x^2$$

the equation can be written as

$$100 - 20x + x^2 = 2x^2$$

This is a quadratic equation solved by getting zero on one side of the equation and then using the quadratic formula.

$$0 = x^2 + 20x - 100$$

Therefore

$$x = \frac{-b \pm \sqrt{b^2 - 4ac}}{2a}$$

with $a = 1$, $b = 20$, and $c = -100$, which gives

$$x = \frac{-20 \pm \sqrt{20^2 - 4 \cdot 1 \cdot (-100)}}{2}$$

There are two values for x: approximately 4.142 and -24.142. The negative value of x does not make sense to this problem, so the one answer is $x \approx 4.142$. One advantage of the algebraic method is that it can easily be computed to as many decimal places as desired.

In reality of course, the problem of determining a location of a person between two lights need not be solved with more accuracy than one decimal place. However, it is quite possible that a technical experiment with lasers would have to be set up with a higher degree of accuracy. ■

Section 8.6 Exercises

1. How are the graphs of $f(x) = 3x^2 + 2x$ and $g(x) = 3x^2 + 2x - 5$ the same and how are they different?

2. How are the graphs of $h(x) = 2x^3 - 4x$ and $g(x) = 2x^3 - 4x + 3$ the same and how are they different?

3. How are the graphs of $f(x) = 4x^2 - 2$ and $f(x-3) = 4(x-3)^2 - 2$ the same and how are they different?

4. How are the graphs of $g(x) = 7x^3 + 4x$ and $g(x+5) = 7(x+5)^3 + 4(x+5)$ the same and how are they different?

In Exercises 5–8 plot the graph of each function and discuss the similarities and differences among them. You might want to use the viewing windows $Xmin = -9.4$, $Xmax = 9.4$, $Xscl = 1$, $Ymin = -10$, $Ymax = 10$, and $Yscl = 1$.

5. $f_1(x) = \dfrac{5}{x}$, $f_2(x) = \dfrac{5}{x+3}$, $f_3(x) = \dfrac{5}{x+3} - 2$

6. $g_1(x) = \dfrac{x+4}{x}$, $g_2(x) = \dfrac{x+4}{x+3}$, $g_3(x) = \dfrac{x+4}{x+3} - 2$

7. $h_1(x) = \dfrac{x^2 - 4}{x}$, $h_2(x) = \dfrac{x^2 - 4}{x-2}$, $h_3(x) = \dfrac{x^2 - 4}{x-2} - 5$

8. $j_1(x) = \dfrac{x^2 - 9}{x^2 - 3x}$, $j_2(x) = \dfrac{x^2 - 9}{x^2 - 3x - 4}$, $j_3(x) = \dfrac{x^2 - 9}{x^2 - 3x - 4} - 5$

In Exercises 9–22 discuss each of the following characteristics of the graph of the given function: (a) The domain and range, (b) the intercepts, (c) the asymptotes, and (d) the symmetry. It is possible that the graph may not have any intercepts, asymptotes, or symmetry.

9. $f(x) = \dfrac{5}{x}$

10. $g(x) = \dfrac{5}{x+3}$

11. $h(x) = \dfrac{5}{x+3} - 2$

12. $j(x) = \dfrac{x+4}{x}$

13. $k(x) = \dfrac{x+4}{x+3}$

14. $m(x) = \dfrac{x+4}{x+3} - 2$

15. $n(x) = \dfrac{x^2 - 4}{x}$

16. $p(x) = \dfrac{x^2 - 4}{x-2}$

17. $q_3(x) = \dfrac{x^2 - 4}{x-2} - 5$

18. $r(x) = \dfrac{x^2 - 9}{x^2 - 3x}$

19. $s(x) = \dfrac{x^2 - 9}{x^2 - 3x - 4}$

20. $t(x) = \dfrac{x^2 - 9}{x^2 - 3x - 4} - 5$

21. $v(x) = \dfrac{\sin x}{x}$. Give x-intercepts from $-2\pi \le x \le 2\pi$. (Make sure your calculator is in Radian mode.)

22. $v(x) = \dfrac{\sin x}{x - 1}$. Give x-intercepts from $-2\pi \le x \le 2\pi$. (Make sure your calculator is in Radian mode.)

In Exercises 23–26, (a) find a suitable viewing window for the function f, (b) plot the graph of the function f, and (c) compare the plotted function to the graphs of its components f_1 and f_2.

23. $f(x) = x^2 + \dfrac{4}{x-5}$, $f_1(x) = x^2$, $f_2(x) = \dfrac{4}{x-5}$

24. $f(x) = 0.25x^2 + \dfrac{x^2}{x-5}$, $f_1(x) = 0.25x^2$, $f_2(x) = \dfrac{x^2}{x-5}$

25. $f(x) = x^3 - \dfrac{2x^2}{x^2 - 1}$, $f_1(x) = x^3$, $f_2(x) = -\dfrac{2x^2}{x^2 - 1}$

26. $f(x) = x^3 + \dfrac{x^2 - 2}{x-2}$, $f_1(x) = x^3$, $f_2(x) = \dfrac{x^2 - 2}{x-2}$

In Exercises 27–30 sketch the slope function for the given function.

27. $f(x) = 3x^2 + 2x$

28. $g(x) = \dfrac{5}{x}$

29. $h(x) = 7x^3 + 4x$

30. $h(x + 5) = 7(x + 5)^3 + 4(x + 5)$

31. *Medical technology* When the distance from an x-ray film (the film-focus distance) is increased, the intensity of the x-ray beam must be increased as well in order to maintain an exposure of similar quality. Conversely, if the distance between the film and the x-ray source is decreased, the intensity must be decreased. The time of exposure to the x-ray beam is directly related to the square of the distance to the film by the formula

$$\frac{T_n}{T_0} = \frac{D_n^2}{D_0^2}$$

where T_0 is the original exposure time, T_n is the new exposure time, D_0 is the original film-focus distance, and D_n is the new film-focus distance.

A good quality x-ray image may be made with an exposure time of 2.0 seconds at a distance of 87.5 cm. What time is required if the distance is shortened to 75 cm?

32. *Police science* The polynomial function $A(x) = -0.015x^3 + 1.058x$ gives the approximate alcohol concentration (in tenths of a percent) in an average person's bloodstream x hours after drinking 8 oz of 100 proof whiskey. This function is approximately valid for $0 \le x \le 8$.

(a) Graph A over the interval $0 \le x \le 8$.

(b) Estimate the time of maximum alcohol concentration.

(c) Graph the slope function for A over the interval $0 \le x \le 8$.

8.7 Solving the Chapter Project

Two questions were posed in the beginning of this chapter:

▶ What variable affects the natural period of a pendulum?

• The answer is the length of the pendulum, not the mass or the initial angle.

▶ What mathematical function could model the relationship between the length and the period of the pendulum?

• We will show how to determine the mathematical function in this section.

The measurements are shown in Table 8.16 and are stored in the lists named PENLG (length) and PENPD (period). To get this data we hung a croquet ball attached to a string and set the ball swinging. We measured the time for 10 swings (periods), and then divided by 10 to get an estimate of the time for one period.

The graph of length vs. period is shown in Figure 8.103.

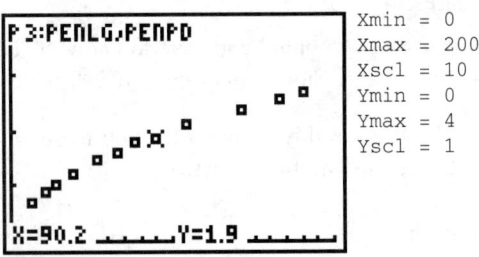

```
Xmin = 0
Xmax = 200
Xscl = 10
Ymin = 0
Ymax = 4
Yscl = 1
```

Figure 8.103

You have studied power functions and polynomials in this chapter. A function of one of those types will model this data. You have studied several tools that can

Table 8.16

Length (cm)	Period (seconds)
12.7	0.7
20.3	0.9
26.7	1.0
38.1	1.2
53.3	1.5
65.4	1.6
76.2	1.8
90.2	1.9
108.0	2.1
142.2	2.4
165.1	2.6
181.6	2.7

help you figure out the mathematical model. Here are some of them:

▶ The shape of the graph

- Does it look like a portion of a power function or a polynomial function?
- Power functions with positive exponents all pass through the point $(0, 0)$. Does it make sense that when the length is near zero the period is also near zero? Think about it.
- If it is a power function does it have a positive exponent or a negative exponent?
- If it is a power function does it have an integer or fractional exponent?

▶ Ratios

- There are some pairs of lengths that are in the ratio of $\frac{1}{2}$. What are the ratios of the corresponding periods? If they seem to be constant, you might have a power function, and the ratio can help you figure out what the exponent is.

▶ Consecutive differences method

- Why can't this method be used to analyze the data?

▶ Regression

- If it's a polynomial you have to know the degree before you can use regression.
- Power regression is only good if you have evidence that it's a power function.

We'll leave this problem for you to solve, but first Example 8.42 shows how you can be mislead by regression.

Example 8.42

Fit a quadratic function to the pendulum data. Discuss the fit.

Solution

To review how to do this see Calculator Lab 2.7 on page 97.

We used the command QuadReg PENLG, PENPD to get the regression equation, and then put that equation into the Y= screen with the following commands: **VARS** 5:Statistics EQ RegEQ. The graph of the regression parabola and the data are shown in Figure 8.104.

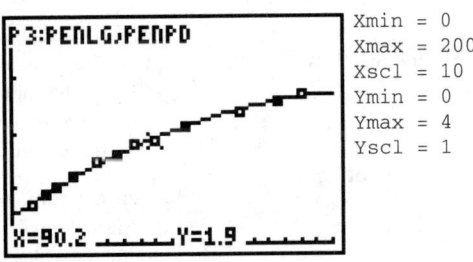

```
Xmin = 0
Xmax = 200
Xscl = 10
Ymin = 0
Ymax = 4
Yscl = 1
```

Figure 8.104

The fit looks very good. But wait! It's too good to be true. Figure 8.105 shows what happens if we zoom out on this graph. The parabola fits the data that we have, but it predicts that the period will eventually decrease as the length gets longer. That shouldn't fit with your intuition about how pendulums work.

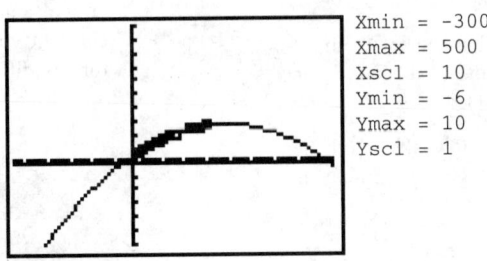

```
Xmin = -300
Xmax = 500
Xscl = 10
Ymin = -6
Ymax = 10
Yscl = 1
```

Figure 8.105

We got a very good fit, but the function doesn't make sense. In addition to fitting the data, the function you find must also make sense. ■

You have learned a lot about functions in this chapter. Now you have the opportunity to exercise your knowledge.

Activity 8.6
What Function Fits the Data?

Physicists have developed a theoretical formula for the relationship between the length of a pendulum and its period. The function is one that we have studied in this chapter.

1. Use the various tools you have learned about to write a function that models the relationship between the length of a pendulum and its period.
2. Defend your choice by showing that it makes sense.

Section 8.7 Exercises

For Exercises 1–6 plot the 4th degree Maclaurin polynomial approximation

$$sq(x) \approx 1 + \frac{1}{2}x - \frac{1}{2^2} \cdot \frac{x^2}{2!} + \frac{1 \cdot 3}{2^3} \cdot \frac{x^3}{3!} - \frac{1 \cdot 3 \cdot 5}{2^4} \cdot \frac{x^4}{4!} + \cdots$$

together with the function $f(x) = \sqrt{1+x}$.

1. (a) How many real roots does this polynomial have?
   (b) Approximately what are they?
   (c) How do its roots compare with the roots of f?

2. (a) In what interval does the approximation of $sq(x)$ for $f(x)$ seem good?
   (b) How could you expand that interval?

3. In what intervals does the approximation seem terrible?

4. Compare the value of $f(0.5)$ with the value of the polynomial evaluated at $x = 0.5$. How many decimal places of accuracy does the polynomial give?
   (a) Make the same comparison for $x = 1.1$.
   (b) Is the approximation for $f(1.1)$ more accurate or less accurate than the approximation of $f(0.5)$?
   (c) Write an explanation of why you think one estimate was better than the other.

5. Add one term to the polynomial to make it 5th degree, and report how many more digits of accuracy are gained for the estimation of $f(1.1)$.

6. Add one term to the polynomial to make it 6th degree, and report how many more digits of accuracy are gained for the estimation of $f(1.1)$.

7. Consider the following table of values:

x	1	2	3	4	5	6	7
y	1	1.4	1.8	2.1	2.3	2.4	2.5

   (a) Use your calculator to determine the regression equation of the power function for your data.

   (b) Graph the original data and the regression equation on the same graph. How well does the regression curve fit your actual data? How can you account for any differences?

8. Consider the following table of values:

x	1	2	3	4	5	6	7
y	2.8	11.2	31.8	70.0	130.5	227.1	355.6

   (a) Use your calculator to determine the regression equation of the power function for your data.

   (b) Graph the original data and the regression equation on the same graph. How well does the regression curve fit your actual data? How can you account for any differences?

9. Set up a ramp with books about 20 cm from the edge of a table as shown in Figure 8.106. Cover a marble with chalk dust to mark the spot where it lands on the floor. Let the marble roll down the ramp, across the table, and hit the floor. Measure the height of the ramp at the point the marble was released and the distance the marble landed from the table. Change the steepness of the ramp and repeat the experiment until you have at least six measurements.
   (a) Use your calculator to determine the power regression that fits your data.
   (b) Graph the original data and the regression equation on the same graph. How well does the regression curve fit your actual data? How can you account for any differences?

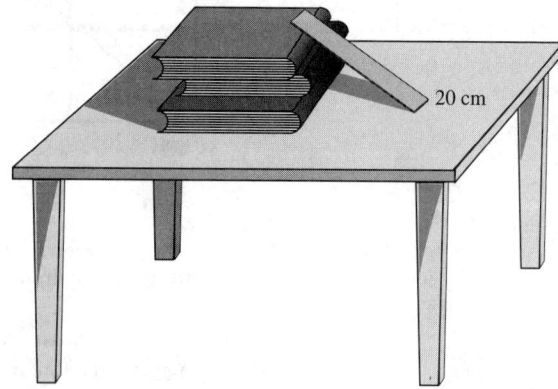

Figure 8.106

10. Set up a ramp with books on the floor. Roll a ball down the ramp and across the floor until it stops. Measure the height of the ramp at the point the ball was released and the distance the ball rolled from the base of the ramp. Change the steepness of the ramp and repeat the experiment until you have at least six measurements.
    (a) Make a table of the first, second, third, and fourth differences.
    (b) Based on your table in (a), what is the degree of the polynomial that fits your data?

(c) Use your calculator to determine the power regression equation for your data.

(d) Graph the original data and the regression equation on the same graph. How well does the regression curve fit your actual data? How can you account for any differences?

11. Write a memo to the management in which you describe what you have learned; respond to the memo on page 516 and explain how the following variables change the period of a pendulum.

(a) The mass on the end of the pendulum (the bob).

(b) How far back the pendulum is pulled back before it starts swinging (initial displacement).

(c) The length of the pendulum.

(d) What is a good way to damp (reduce) the pendulum's motion?

(e) What is the mathematics of the motion of a damped pendulum?

12. At the end of the memo on page 516 the management said they were having a contest to name the product. One of the entries is *Great Shakes*.

(a) What's your suggestion?

(b) Explain why you think it is a good name for this product.

Chapter 8 Summary and Review

Topics You Learned or Reviewed

► You learned about six different types of functions.

- A **power function** is any function of the form $f(x) = kx^n$ where k and n are real numbers.
- A **polynomial function** is a sum of power functions with nonnegative integer exponents.
- An **algebraic function** is a function whose values can be computed with a finite number of elementary operations of mathematics.
- A **rational function** is formed from polynomials and ratios of polynomials.
- A function is said to be an **odd function** if it has symmetry with respect to the origin. This also means if $f(a) = -f(-a)$ for all values of a that fit the function, then f is an odd function.
- A function is said to be an **even function** if it has symmetry with respect to the y-axis. This also means if $f(a) = f(-a)$ for all values of a that fit the function, then f is an even function.

► You learned the following properties of functions and their graphs:

- How to predict the shape of the graph of a power or polynomial function.
- The **domain** of a function f is the set of values of x that can be successfully substituted into the expression for f to produce values for $f(x)$.
- The **range** of a real function is the set of values of $f(x)$ that result from substituting each number in the domain.
- Any polynomial function of degree n can be written in the form $a(x - r_1)(x - r_2)(x - r_3) \cdots (x - r_n)$ where r_1, r_2, r_3, ..., r_n are the n roots of the polynomial.
- A **horizontal asymptote** is a line that a curve approaches (but never reaches) as $x \to \pm\infty$. Algebraically, if there is a real number L such that $f(x) \to \pm L$ as $x \to -\infty$ or $x \to \infty$ then the line $y = L$ is a horizontal asymptote of the function f.
- A **vertical asymptote** is a vertical line that a curve approaches (but never reaches) as $x \to c$ where c is a real number. Algebraically, if $f(x) \to \pm\infty$ as $x \to c$ then the line $x = c$ is a vertical asymptote of the function f.
- Limits as $x \to +\infty$ and $x \to -\infty$.
- Even functions are symmetric with respect to the y-axis and odd functions are symmetric with respect to the origin.

► If $f(x) = kx^n$ with $n > 0$, and if $\dfrac{x_2}{x_1} = r$, then $\dfrac{f(x_2)}{f(x_1)} = r^n$ says that $f(x)$ **varies directly** as the nth power of x.

► If $f(x) = kx^{-n}$ and $n > 0$, then $f(x)$ can be written as $\dfrac{k}{x^n}$. If $\dfrac{x_2}{x_1} = r$, then $\dfrac{f(x_2)}{f(x_1)} = \dfrac{f_2}{f_1} = \dfrac{1}{r^n} = \left(\dfrac{1}{r}\right)^n$ says that $f(x)$ **varies inversely** as the nth power of x.

▶ The Maclaurin series approximation for the sine and cosine functions are

$$\sin x \approx x - \frac{x^3}{3!} + \frac{x^5}{5!} - \frac{x^7}{7!} + \frac{x^9}{9!} - \cdots$$

$$\cos x \approx 1 - \frac{x^2}{2!} + \frac{x^4}{4!} - \frac{x^6}{6!} + \frac{x^8}{8!} - \cdots$$

▶ If the **Consecutive differences method** is used in a sequence $a_1, a_2, a_3, \ldots, a_n$ and if the kth differences between consecutive terms of a sequence is constant, then the nth term, a_n, of the sequence can be represented by a polynomial of degree k. In general you need $k + 1$ data points from the table to produce a polynomial of degree k.

▶ The equation of the tangent to a curve at a point can be written using the coordinates of the point and the slope of the tangent to the curve at that point.

▶ How to sketch the slope function of a function.

▶ How to use the CBL with a light sensor.

▶ How to use the ΔList function to compute the differences between consecutive items in a list.

▶ How to use Table to find the best dimensions of the viewing window on the calculator.

▶ SetFactors can be used to change the zoom ratio on the calculator.

▶ How to enter a regression equation into the Y= screen.

Review Exercises

Each function in Exercises 1–2 is a power function of the form $y = kx^n$. Determine the value of k and n for each function.

1. $f(x) = -3.2\pi^4 x^{-7.1}$

2. $g(x) = 2\sqrt[7]{12}x^{3.8}$

In Exercises 3–4, (a) determine if each function is an algebraic function, and (b) explain why you made that decision.

3. $f(x) = -3.4x$

4. $j(x) = 9x^7 - \sin(3x + 1)$

In Exercises 5–6, (a) describe the symmetry of the given function, and (b) tell whether the function is odd, even, or neither odd nor even.

5. $f(x) = -5x^6 - 3x^2$

6. $j(x) = \sin x - \frac{1}{4}x^3$

In Exercises 7–8, (a) describe the symmetry of the given function, (b) determine and justify whether the function is odd, even, or neither odd nor even, and (c) describe the values of y as x gets close to zero.

7. $f(x) = 4x^{-3}$

8. $j(x) = \frac{5\pi}{x^6}$

In Exercises 9–10, (a) describe the symmetry of the given function, (b) determine and justify whether the function is odd, even, or neither odd nor even, and (c) describe the slope of a tangent to the function's graph.

9. $f(x) = -4x^{1/5}$

10. $j(x) = 2.7x^{1/6}$

In Exercises 11–12 give all horizontal and vertical asymptotes of the given function.

11. $f(x) = -2x^{-4}$

12. $j(x) = 0.25\pi x^{-3}$

Exercises 13–14 show the graph of a power function $y = kx^n$ graphed on a TI-83 using the indicated viewing window. One point on the curve is shown. Based on your knowledge of power functions, symmetry, and asymptotes, (a) estimate the values of k and n, and (b) justify your estimates.

13.

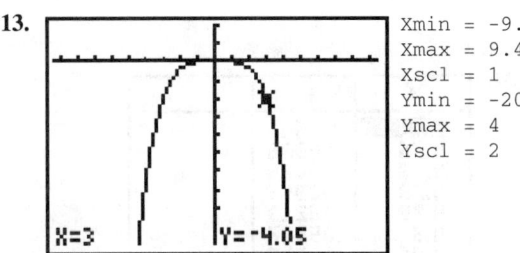

Xmin = -9.4
Xmax = 9.4
Xscl = 1
Ymin = -20
Ymax = 4
Yscl = 2

14.

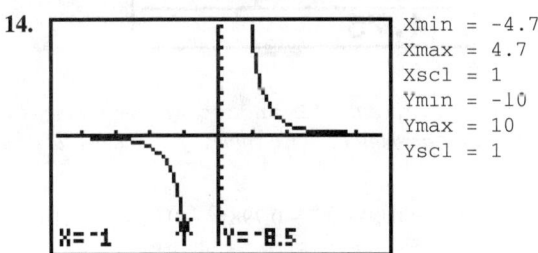

Xmin = -4.7
Xmax = 4.7
Xscl = 1
Ymin = -10
Ymax = 10
Yscl = 1

15. (a) What is the volume of a cube that measures 2.5 cm on each side?

  (b) How long should each side be for a cube that has twice the volume of the cube in (a)?

16. Each side of cube A is 2.25 in. long and each side of cube B is 3.50 in. long. What is the ratio of the volume of cube B to the volume of cube A?

17. *Recreation* A golf ball has a diameter of 1.68 in. and a croquet ball has a diameter of $3\frac{5}{8}$ in.

  (a) What is the ratio of their radii?

  (b) What is the ratio of their volumes?

18. If a ball falls 19.6 meters in the first two seconds, how far does it fall in the first six seconds?

19. *Space technology* What is the gravitational attraction felt by a 250,000 lb space shuttle in an orbit 250 mi above the Earth's surface?

20. The points shown in the table below come from a power function. Which power function best fits these data?

x	-5	-3	2	5	7
y	31.25	6.75	-2	-31.25	-85.75

21. The points shown in the table below come from a power function. Which power function best fits these data?

x	-6	0	2	4	5	8
y	-3.180	0	2.205	2.778	2.992	3.500

22. The points shown in the table below come from a power function. Which power function best fits these data?

x	-5	-2	3	5	7	8
y	-23.796	-6.598	11.639	23.796	38.113	45.948

23. Give the degree of the polynomial

$$f(x) = -3.9x^5 - 2.5x^{11} + 5x^7 - 9x^{10} + 3.75$$

In Exercises 24–25 give the roots of the given polynomial functions.

24. $f(x) = -7.3(x+7.25)\left(x + \sqrt{37}\right)(x - 7\pi)(x - 8.75)$

25. $g(x) = 2.75x\left(x^2 + 5x - 6\right)\left(x^2 + 7x + 12\right)\left(x^2 + 25\right)$

In Exercises 26–27 estimate the real roots of the given polynomial function to three decimal places.

26. $h(x) = 2x^3 - 5x^2 - 4x + 3.5$

27. $j(x) = -3.1x^5 - 3.1x^4 + 43.4x^3 + 43.4x^2 - 139.5x - 139.5$

In Exercises 28–29 examine the graphs and estimate (a) the degree of the polynomial, (b) the number of real roots, and (c) the number of nonreal roots.

28.

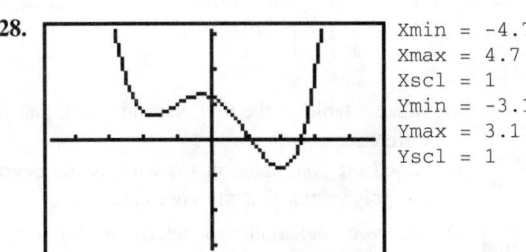

Xmin = -4.7
Xmax = 4.7
Xscl = 1
Ymin = -3.1
Ymax = 3.1
Yscl = 1

29.

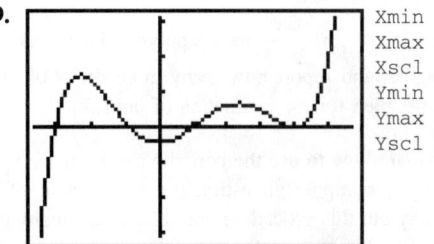

Xmin = -3.7
Xmax = 5.7
Xscl = 1
Ymin = -3.1
Ymax = 3.1
Yscl = 1

In Exercises 30–31 estimate the root from the given table.

30.

X	Y₁
-1.77	.05787
-1.76	.0461
-1.75	.03424
-1.74	.02229
-1.73	.01026
-1.72	-.0018
-1.71	-.014

X=-1.77

31.

X	Y₁
4.75	-.0642
4.76	-.0336
4.77	-.0024
4.78	.02956
4.79	.06213
4.8	.09537
4.81	.12929

X=4.75

In Exercises 32–33, (a) graph the given polynomial function on your calculator using the ZStandard *window setting, (b) use Figures 8.36 and 8.37 to describe the shape of the given function for large positive or small negative values of x, and (c) zoom out on your calculator to confirm your response to (b).*

32. $f(x) = \frac{1}{240}(x^4 - 96x^2 - 360x + 480)$

33. $g(x) = 0.0005x^5 - 0.798x^2 - 0.075x + 0.16$

For Exercises 34–39 plot the 7th degree Maclaurin polynomial approximation

$$t(x) \approx x + \frac{x^3}{3} + \frac{2x^5}{15} + \frac{17x^7}{315} + \cdots$$

together with the function tan x. *Set your calculator to Radian mode and use the* ZTrig *setting.*

34. (a) How many real roots does this polynomial have?

(b) Approximately what are they?

(c) How do its roots compare with the roots of tan x?

35. (a) In what interval does the approximation of $t(x)$ for tan x seem good?

(b) How could you expand that interval?

36. In what intervals does the approximation seem too rough?

37. Change the window settings so that Xmin = -1.57, Xmax = 1.57, and Xscl = 0.25. Compare the value of tan 0.5 with the value of the polynomial evaluated at x = 0.5. How many decimal places of accuracy does the polynomial give?

(a) Make the same comparison for x = 1.5.

(b) Is the approximation for tan 1.5 more accurate or less accurate than the approximation of tan 0.5?

(c) Write an explanation of why you think one estimate was better than the other.

38. Add the term $\frac{62x^9}{2835}$ to the polynomial to make it 9th degree, and report how many more digits of accuracy are gained for the estimation of tan 1.5.

39. Explain how to use the periodic nature of the tan function to compute an estimate for the value of tan 25. Carry out this procedure and report how many decimal places of accuracy there are in your estimate.

40. Consider the following table of values:

x	-2	-1	0	1	2	3	4
y	-0.5	1.5	1.0	-0.5	-1.5	-0.5	4

(a) Make a table of the first, second, and third differences.

(b) Based on your table in (a) what is the degree of the polynomial that fits your data?

 (c) Use your calculator to determine the regression equation for your data.

(d) Graph the original data and the regression equation on the same graph. How well does the regression curve fit your actual data? How can you account for any differences?

41. Consider the following table of values:

x	-2	-1	0	1	2	3	4	5
y	-0.4	2	2.6	-0.4	-4	-0.4	23	83.6

(a) Make a table of the first, second, third, and fourth differences.

(b) Based on your table in (a) what is the degree of the polynomial that fits your data?

 (c) Use your calculator to determine the regression equation for your data.

(d) Graph the original data and the regression equation on the same graph. How well does the regression curve fit your actual data? How can you account for any differences?

42. Write the equation of the line that passes through the point $(-1, -2)$ with a slope of -6.0. Plot this line on the same graph with the function $f(x) = 2x^{-3}$.

43. Write the equation of the line that passes through the point $(8, 12)$ with a slope of 0.5. Plot this line on the same graph with the function $g(x) = 6x^{1/3}$.

44. (a) Run the program POWERFUN for the function $f(x) = -3x^2$ and complete a table like Table 8.17.
(b) Use the data in (a) to estimate the slope function for the function $f(x) = -3x^2$.

45. How are the graphs of $f(x) = -2x^3 + 3x$ and $g(x) = -2x^3 + 3x + 4$ the same and how are they different?

46. How are the graphs of $f(x) = x^{-2} + x^3 - 7$ and $f(x - 1) = (x - 1)^{-2} + (x - 1)^3 - 7$ the same and how are they different?

Table 8.17

$k =$____ $n =$____				
Value of x	Slope of Tangent	Diff. 1	Diff. 2	Diff. 3
-5				
-4				
-3				
-2				
-1				
0				
1				
2				
3				
4				

In Exercises 47–51 discuss each of the following characteristics of the graph of the given function: (a) the domain and range, (b) the intercepts, (c) the asymptotes, and (d) the symmetry. It is possible that the graph may not have any intercepts, asymptotes, or symmetry.

47. $f(x) = \dfrac{7}{x - 2}$

48. $g(x) = \dfrac{7}{x - 2} + 4$

49. $h(x) = \dfrac{x + 5}{x^2 - 4}$

50. $j(x) = \dfrac{x + 2}{x^2 - 4} + 5$

51. $k(x) = \dfrac{x^2 + 5x - 6}{x^2 + 3x - 4}$

In Exercises 52–53, (a) find a suitable viewing window for the function f, (b) plot the graph of the function f, and (c) compare it with the graphs of its components f_1 and f_2.

52. $f(x) = 3x^2 - \dfrac{7}{x - 4}$, $f_1(x) = 3x^2$, $f_2(x) = -\dfrac{7}{x - 4}$

53. $f(x) = x^3 + \dfrac{x^2 - 4}{x - 2}$, $f_1(x) = x^3$, $f_2(x) = \dfrac{x^2 - 4}{x - 2}$

In Exercises 54–55 sketch the slope function for the given function.

54. $f(x) = 2x^2 - 5x$

55. $g(x - 5) = (x - 5)^3 + 3(x - 5)^2$

56. *Medical technology* The time of exposure to the x-ray beam is directly related to the square of the distance to the film by the formula

$$\frac{T_n}{T_0} = \frac{D_n^2}{D_0^2}$$

where T_0 is the original exposure time, T_n is the new exposure time, D_0 is the original film-focus distance, and D_n is the new film-focus distance.

A good quality x-ray image may be made with an exposure time of 2.0 seconds at a distance of 87.5 cm. What time is required if the distance is lengthened to 1.25 meters?

57. Consider the following table of values:

x	1	2	3	4	5	6	7
y	3.2	3.7	4.0	4.6	4.6	4.8	5.2

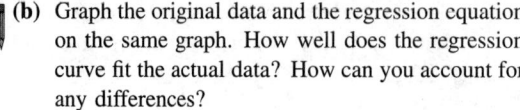 **(a)** Use your calculator to determine the regression equation of the power function for the data.

 (b) Graph the original data and the regression equation on the same graph. How well does the regression curve fit the actual data? How can you account for any differences?

58. Consider the following table of values:

x	1	2	3	4	5	6	7
y	2.2	1.7	1.6	1.2	1.2	1.1	1.1

 (a) Use your calculator to determine the regression equation of the power function for the data.

(b) Graph the original data and the regression equation on the same graph. How well does the regression curve fit the actual data? How can you account for any differences?

● ● ● ● ● ● ● ● Chapter 8 Test

1. Determine the value of k and n for the power function $f(x) = -3\sqrt{17}x^{-2.5}$.

2. **(a)** Is the function $g(x) = \tan^2 x - \dfrac{9x}{x^2 + 1}$ an algebraic function? **(b)** Explain your answer.

3. **(a)** Describe the symmetry of the function $h(x) = -9x^4 + 3x^2 - 5$ and **(b)** determine whether the function is odd, even, or neither odd nor even.

4. **(a)** Describe the symmetry of the function $j(x) = -3.1x^{1/4}$, **(b)** determine and justify whether the function is odd, even, or neither odd nor even, and **(c)** describe the slope of a tangent to the function's graph.

5. Give all horizontal and vertical asymptotes of the function $k(x) = 1.5\pi x^{1/3}$.

6. Figure 8.107 shows the graph of a power function $y = kx^n$ graphed on a *TI-83* using the indicated viewing window. One point on the curve is shown. Based on your knowledge of power functions, symmetry, and asymptotes **(a)** estimate the values of k and n and **(b)** justify your estimates.

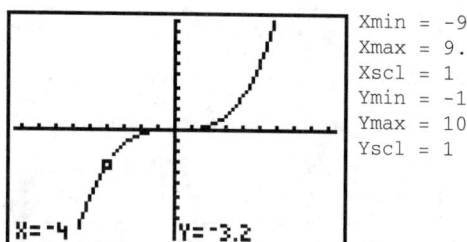

```
Xmin = -9.4
Xmax = 9.4
Xscl = 1
Ymin = -10
Ymax = 10
Yscl = 1
```

X=-4   Y=-3.2

Figure 8.107

7. If a ball falls 29.4 meters in the first three seconds, how far does it fall in the first seven seconds?

8. The points shown in the table below come from a power function. Which power function best fits these data?

x	−5	−3	−1	0	4
y	1.313	0.967	0.5	0	−1.149

9. Give the degree of the polynomial

$$f(x) = -\sqrt{2}x^4 - 2.5x^9 + 21.4x^6 - 9x^7 - 2\pi$$

10. Give the roots of the polynomial function

$$g(x) = -1.25x \left(x^2 + 6x + 5\right) \left(x^2 - 12\right) \left(x^2 + 16\right)$$

11. Estimate to three decimal places the real roots of the polynomial function $h(x) = -x^3 + 4x^2 - 2x - 1.7$.

12. Examine the graph in Figure 8.108 and estimate **(a)** the degree of the polynomial, **(b)** the number of real roots, and **(c)** the number of nonreal roots.

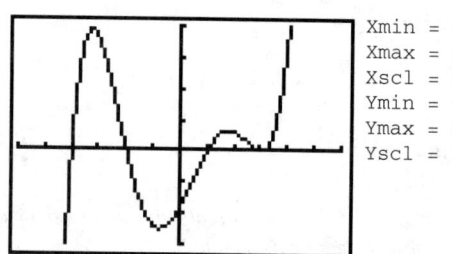

```
Xmin = -6
Xmax = 6
Xscl = 1
Ymin = -3
Ymax = 4
Yscl = 1
```

Figure 8.108

13. Estimate the root from the table in Figure 8.109.

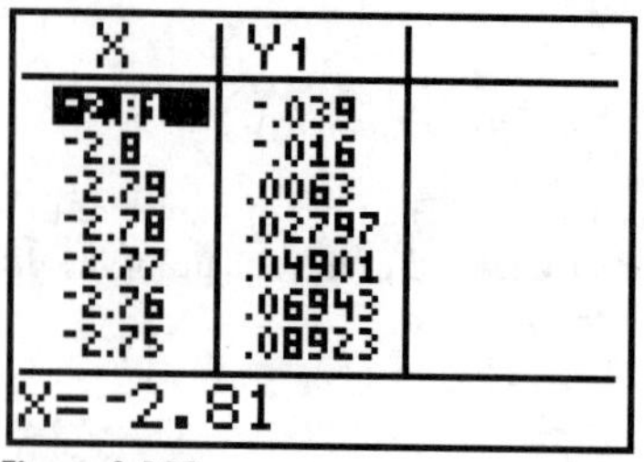

X	Y1
-2.81	-.039
-2.8	-.016
-2.79	.0063
-2.78	.02797
-2.77	.04901
-2.76	.06943
-2.75	.08923

X=-2.81

Figure 8.109

14. Consider the following table of values:

x	−2	−1	0	1	2	3	4
y	8.85	4.65	1.25	−1.35	−3.15	−4.15	-4.35

(a) Make a table of the first, second, and third differences.

(b) Based on your table in (a) what is the degree of the polynomial that fits your data?

 (c) Use your calculator to determine the regression equation for your data.

15. Write the equation of the line that passes through the point $(-2, -2.5)$ with a slope of -1.25.

16. (a) Run the program POWERFUN for the function $f(x) = -2x^4$ and complete a table like Table 8.18.

Table 8.18

k =_____ n =_____				
Value of x	**Slope of Tangent**	**Diff. 1**	**Diff. 2**	**Diff. 3**
−5				
−4				
−3				
−2				
−1				
0				
1				
2				
3				
4				

(b) Use the data in (a) to estimate the slope function for the function $f(x) = -2x^4$.

17. Discuss each of the following graph characteristics of the function $f(x) = \dfrac{-3}{x-1} + 2$ (a) The domain and range, (b) the intercepts, (c) the asymptotes, and (d) the symmetry. It is possible that the graph may not have any intercepts, asymptotes, or symmetry.

18. Consider the following table of values:

x	1	2	3	4	5	6	7
y	1.8	1.5	1.3	1.2	1.2	1.1	1.1

 (a) Use your calculator to determine the regression equation of the power function for your data.

(b) Graph the original data and the regression equation on the same graph. How well does the regression curve fit your actual data? How can you account for any differences?

19. Plot the 4th degree Maclaurin polynomial approximation $f(x) \approx 1 + 2x + 4x^2 + 8x^3 + 16x^4$ together with the function $g(x) = \dfrac{1}{1 - 2x}$.

(a) In what interval does the approximation of $f(x)$ for $g(x)$ seem good?

(b) How could you expand that interval?

(c) In what intervals does the approximation seem too rough?

(d) Compare the value of $g(0.75)$ with the value of the polynomial evaluated at $x = 0.75$. How many decimal places of accuracy does the polynomial give?

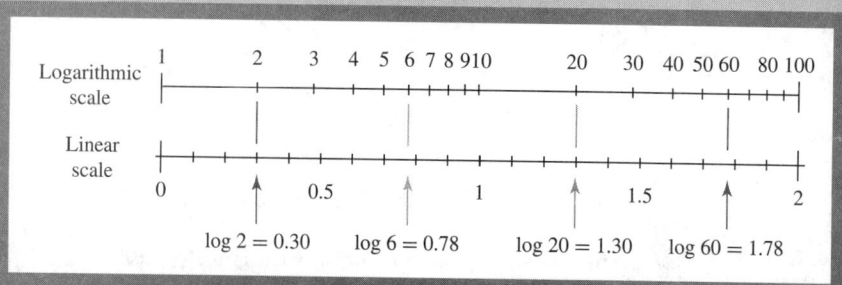

CHAPTER
9

Exponential and Logarithmic Functions

Topics You'll Need to Know

▶ Applications of geometric sequences
 • charging and discharging a capacitor
 • raidoactive decay
 • interest on a bank account
▶ Domain and range
 • of a function
 • of a model
▶ Limit notation; for example, $\lim_{x \to +\infty} f(x)$
▶ Laws of exponents
▶ How to sketch a slope function

Calculator Skills You'll Use

▶ Copying the contents of one list to another list with a new name
▶ Linear regression
▶ Subtracting the same number from every entry in a list and storing the result in a new list
▶ Running and stopping calculator programs
▶ STAT PLOT
▶ The calculator keys **LOG**, **LN**, $[e^x]$, $[10^x]$, $[\sin^{-1}]$, $[\cos^{-1}]$, **VARS**
▶ The calculator functions Δlist, seq, min, max, abs

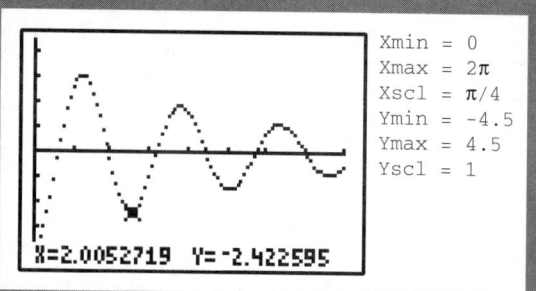

What You'll Do in This Chapter

Run a CBL cooling experiment, gather data, and generate a unique graph shape relating to exponential and logarithmic functions and their applications. The applications include radioactive decay, population growth, interest on a bank account, charging a capacitor, and measuring concentrations of acids.

You will learn how to use two new kinds of graphing scales, semilog and log-log, and you'll see that the graphs of some familiar functions are a straight line on these new coordinate systems.

The chapter concludes by applying the new material to discover and evaluate the mathematical model for the cooling data. This model follows a physical law discovered in the 17th century by the same genius who discovered calculus and gravity-influenced motion.

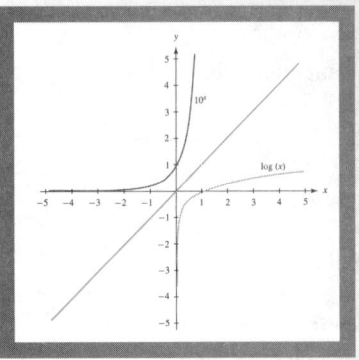

Calculator Skills You'll Learn

- ▶ Using CBL with the temperature probe
- ▶ Exponential regression
- ▶ Semilog plots of functions and data
- ▶ Plotting the slope function for a stat plot
- ▶ Computing the average percent error by comparing the predictions of the model with the data

Topics You'll Learn or Review

- ▶ Discrete and continuous functions
- ▶ Exponential functions
  - • sketch and recognize graphs
  - • applications to population growth, radioactive decay, interest in a bank account, charging a capacitor
- ▶ Logarithms
  - • definition; common and natural logarithms
  - • properties: addition, subtraction, multiplication
  - • how to write the inverse function for a linear function or power function
  - • symmetry of the graphs of a function and its inverse function
  - • relationship between the domain and range of a function and its inverse function
- ▶ Solve exponential equations
- ▶ Logarithmic scales: semilog and log-log
  - • plotting points
  - • straightening exponential and power functions
- ▶ Computing average percent error by comparing a model and the data
- ▶ Newton's law of cooling

Chapter Project—Modeling Temperature Changes

Advanced Temperature Control, Inc.
You name the temperature, we'll keep it there

To: Research Department
From: Management
Subject: Pleasing a new customer

We have the possibility of a big sale to Exotic Metals, Ltd. They are interested in our high-temperature ovens and thermostats to control the cooling of metal rods. They have devised a way to change the hardness and tensile strength by heating the rods to a high temperature and then transferring them to a lower temperature oven.

I think our high-temperature ovens and thermostats are just what they need.

We must prove two things to them:

 That our ovens can maintain a specified temperature between 100° and 400°C to within an accuracy of 5°C.

 That we can accurately predict how long it will take for a metal rod to cool from over 500°C to a temperature in the desired range. They want us to be able to estimate the cooling time given the starting temperature and the final temperature.

Engineering and Marketing are working on the first item. Please find out what you can about the mathematics of cooling, and meet with me next Wednesday at 2 P.M.

Preliminary Analysis

The process of heating and cooling metals to change physical properties such as hardness and tensile strength is very old. Over two thousand years ago the process was used with iron and other metals throughout Europe and Asia. In this chapter we will study one important part of this process: the way that an object cools down from a high temperature. We will begin with an experiment you can do with the Calculator-Based Laboratory system, but first let's think about how an object cools in air.

Activity 9.1
Think about Cooling and Heating: Part 1

Imagine you have a device that can measure temperature every second for about two minutes. Use three charts like the one in Figure 9.1 to sketch graphs of how temperature changes with time in each situation described below.

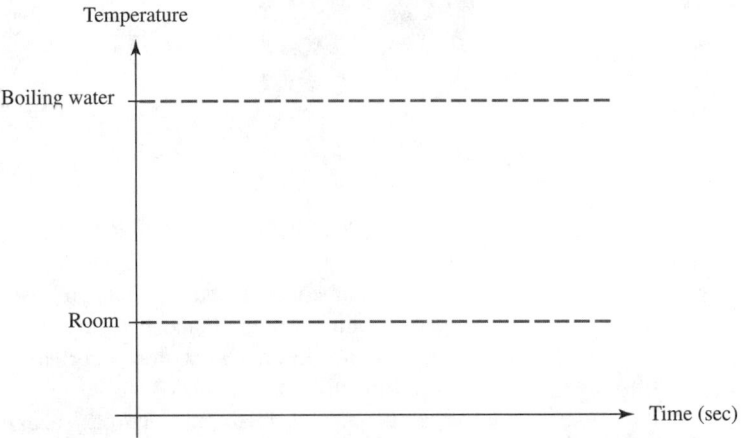

Figure 9.1

▶ Sketch a graph that represents how you think the temperature of water will change with time as it is heated to boiling. Begin your graph ($t = 0$) when the heat is first turned on.
▶ Sketch a graph that represents how you think the temperature will change with time when an object is removed from boiling water and allowed to cool in air at room temperature. Begin your graph ($t = 0$) when the object is removed from the boiling water.
▶ Sketch a graph that represents how you think the temperature will change with time when an object is removed from boiling water and allowed to cool in *water* at room temperature. Begin your graph ($t = 0$) when the object is removed from the boiling water.

Calculator Lab 9.1

A Cooling Experiment

In this lab you will use the CBL to gather temperature data as an object is cooled in air from the boiling point of water to near room temperature.

Procedures

1. Connect the temperature probe to the CBL.
2. Start the program named COOLTEMP on the *TI-83*.
3. Measure the room temperature by running the program once with the temperature probe in the air.

   The temperature is recorded once per second for about a minute and a half. After the experiment you will see a horizontal line representing the constant room temperature drawn on your calculator screen. Use Trace to determine the temperature of the air in the room.
4. Using the setup shown in Figure 9.2, place the temperature probe in boiling water for about thirty seconds.

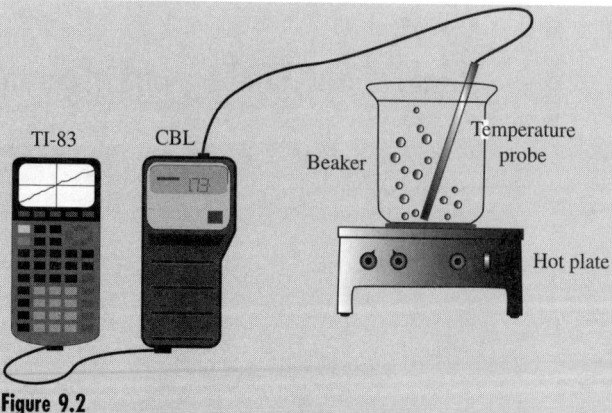

Figure 9.2

5. Restart the COOLTEMP program and press (ENTER) to begin temperature measurements.
6. At the same time as you start the data collection, remove the temperature probe from the boiling water. Place it gently on a flat surface and ensure that the probe tip is not in contact with the surface. The probe tip should only be in contact with still air.
7. The temperature is recorded once per second for about a minute and a half. After the experiment you will see a graph drawn on your calculator similar to the one in Figure 9.3.

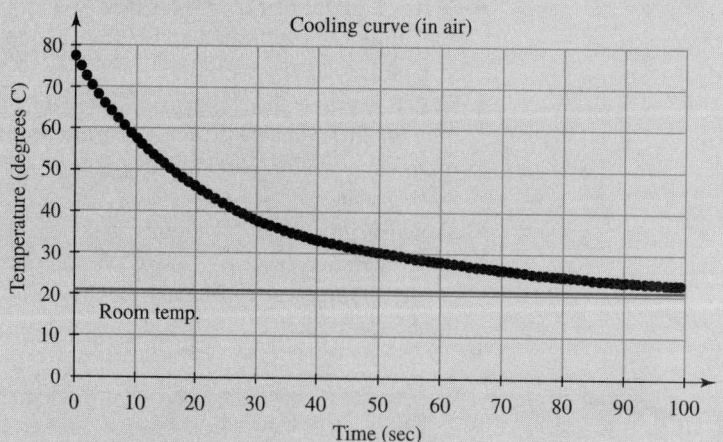

Figure 9.3

8. If you are satisfied with the success of the experiment, store the data with new names. The calculator stored the temperature data in list L4, and the time is stored in list L2. Use the STO instruction to store these lists with different names. For example, the following instruction stores the time data in the list named TTIME:

$$\text{L2} \quad \boxed{\text{STO} \blacktriangleright} \quad \text{TTIME.}$$

Be careful in naming new lists because if the name you choose is already in use on your calculator, the old data will be erased. For example there may already be a list named TIME stored on your calculator.

The experiment is described in detail in Texas Instruments' *CBL System Experiment Handbook*.

Our results are stored in the lists named TMPT (time from 0 to 98 seconds) and TMP (temperature in degrees Celsius). The graph of our data is shown in Figure 9.3.

Activity 9.2
Thinking about Cooling and Heating: Part 2

Now that you have seen one set of data for cooling to room temperature, think about the graphs you sketched in Activity 9.1; change your graphs, if necessary, and answer each of the following questions.

1. The temperature of boiling water is 100°C. Why do you think the first point in Figure 9.3 is about 80°?
2. Describe how the *rate* of cooling changes with time. That is, compare how fast the temperature changes at first with how fast it is changing after 70 seconds.
3. Sketch a graph of the *slope* of the temperature vs. time graph as a function of time. Use a graph like Figure 9.4.

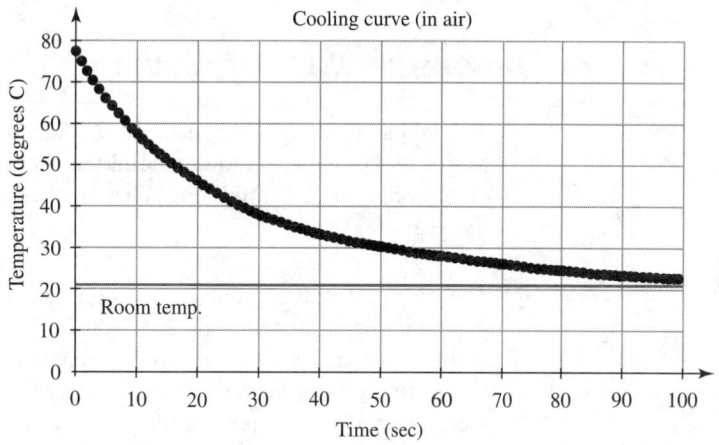

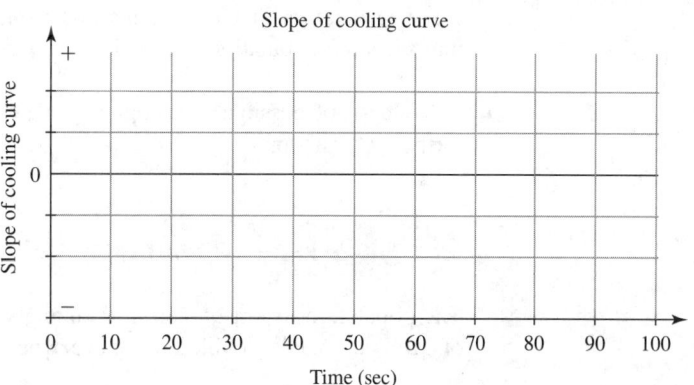

Figure 9.4

9.1

Exponential Function

An **exponential function** is any function of the form

$$f(x) = b^x$$

where $b > 0$, $b \neq 1$, and x is any real number. The number b is called the base.

Exponential Functions

An **exponential function** is one in which the exponent contains the variable. In Chapter 7 you saw that exponential functions are related to finding the nth term and the sum of n terms of geometric sequences. Here are some examples of exponential functions from Chapter 7:

▶ $S(n) = 2^n - 1$, where $S(n)$ is the sum of the first n terms of the geometric sequence

$$1, 2, 4, 8, 16, \ldots, 2^{n-1}$$

$S(n)$ equals the total number of pennies placed on the nth square of a checkerboard in our legend of the compensation requested by the old engineer.

▶ $A(n) = 50(\frac{1}{2})^n$, where $A(n)$ is the amount that remains of 50 grams of a radioactive substance after n half-lives have elapsed.

▶ $B(n) = 1000 \cdot 1.02^{4n}$, where $B(n)$ is the number of dollars in a bank account n years after a deposit of $1000 if the annual rate of interest is 8%, and the interest is compounded quarterly (four times per year).

▶ $C(n) = 100(1 - 0.37^n)$, where $C(n)$ is the percent of the maximum charge of a capacitor after n time constants have elapsed.

Example 9.1 introduces another application of exponential functions: the study of population growth.

Example 9.1   Application

According to an estimate made in 1992,* the population of the African country Morocco was expected to double in the following 35 years. In 1992 the population of Morocco was 25.8 million. What is an estimate of the population of Morocco in the year 2000?

Solution The given information only tells us that in the year 2027 (35 years from 1992) the population is estimated to be $2 \times 25.8 = 51.6$ million. To estimate the population in 2000, we need to know what kind of function the population will follow. Figure 9.5 shows four curves that connect the points (1992, 25.8) and (2027, 51.6). Of course there are many other possibilities. The vertical line at the year 2000 intersects the curves in points A, B, C, and D, respectively. To estimate the population that each curve predicts for that year, draw a horizontal line from each point and see where it intersects the population (vertical) axis. ∎

It is not possible to completely answer the question in Example 9.1 without further information.

Example 9.2   Application

The population expert quoted in Example 9.1 actually estimated that the population of Morocco would double during *every* period of 35 years beginning in 1992.

---

*The estimate is located at http://www.macroint.com/dhs/factsht.html

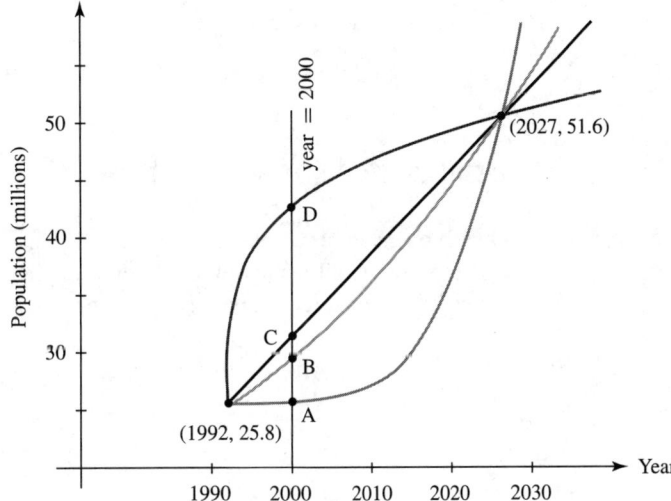

Figure 9.5

(a) What is the geometric sequence that describes this growth?
(b) What function describes this growth?
(c) What is the estimate of the population in the year 2000?

Solution   We are told that the estimated population changes according to the following table.

Year	Population (millions)
1992	25.8
2027	$25.8 \cdot 2 = 51.6$
2062	$51.6 \cdot 2 = 25.8 \cdot 2^2 = 103.2$

(a) The sequence is $25.8, 51.6, 103.2, \ldots$. The nth term of this sequence is $25.8 \cdot 2^{n-1}$.

(b) One possible function similar to the one for radioactive decay, except there is increase, not decrease is

$$P(n) = 25.8 \times 2^n$$

where $P(n)$ = the population after n doubling times, beginning in 1992.

Sometimes there will be a difference between the function and the formula for the nth term of the sequence because of different interpretations of the meaning of the variable n. In our function, $n = 1$ represents the *first* doubling time and corresponds to the *second* term of the sequence.

However, since population does not jump suddenly every 35 years but increases steadily with time, a more appropriate model for this growth is the function P in terms of the year y

$$P(y) = 25.8 \times 2^{(y-1992)/35}$$

We use $\dfrac{y - 1992}{35}$ as the exponent because when $y = 1992$ the exponent is zero, so $2^0 = 1$ and the population is 25.8. When $y = 2027$, we have 2^1 and the population is 51.6.

(c) Substituting $y = 2000$ we have

$$P(2000) = 25.8 \times 2^{(2000-1992)/35} \approx 30.2 \text{ million}$$

Graphs of Exponential Functions

Once you have some experience with exponential functions you will begin to see that the graphs all share some common characteristics and shapes. Figures 9.6–9.8 show graphs of three typical exponential functions from page 612.

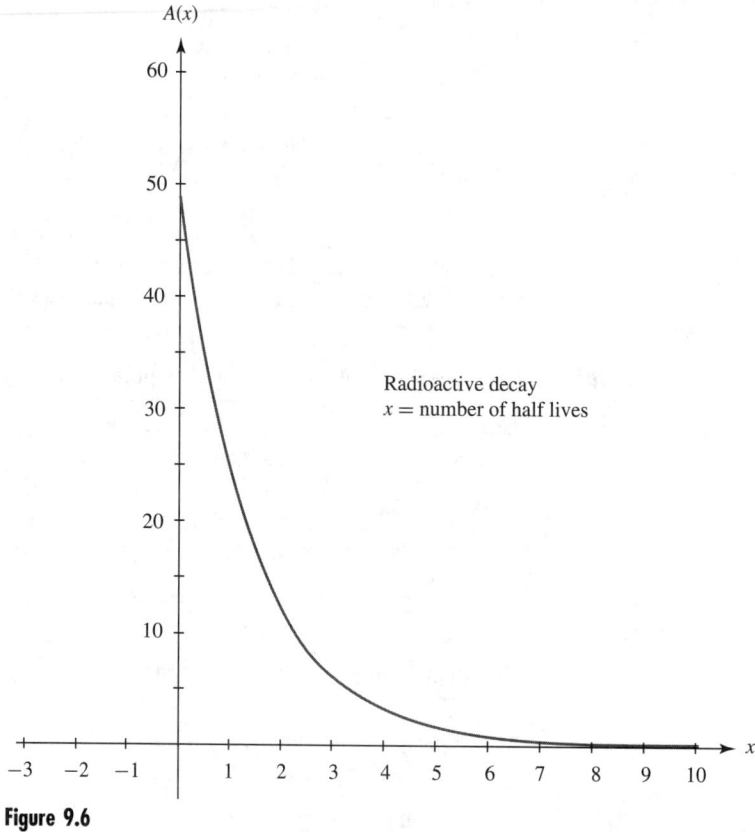

Radioactive decay
$x =$ number of half lives

Figure 9.6

The three exponential graphs are quite similar in shape. Calculator lab 9.2 and Activity 9.3 will help you explore the similarities and differences among these graphs.

Calculator Lab 9.2   **Variations on Exponential Graphs**

Procedures Use your calculator to explore variations of each of the three graphs in Figures 9.6–9.8. On the graphs provided, sketch and label each of the following functions:

▶ Variations on $A(x) = 50(\frac{1}{2})^x$. Use Figure 9.6.

1. $A_1(x) = 40(\frac{1}{2})^x$ (decay of 40 grams)

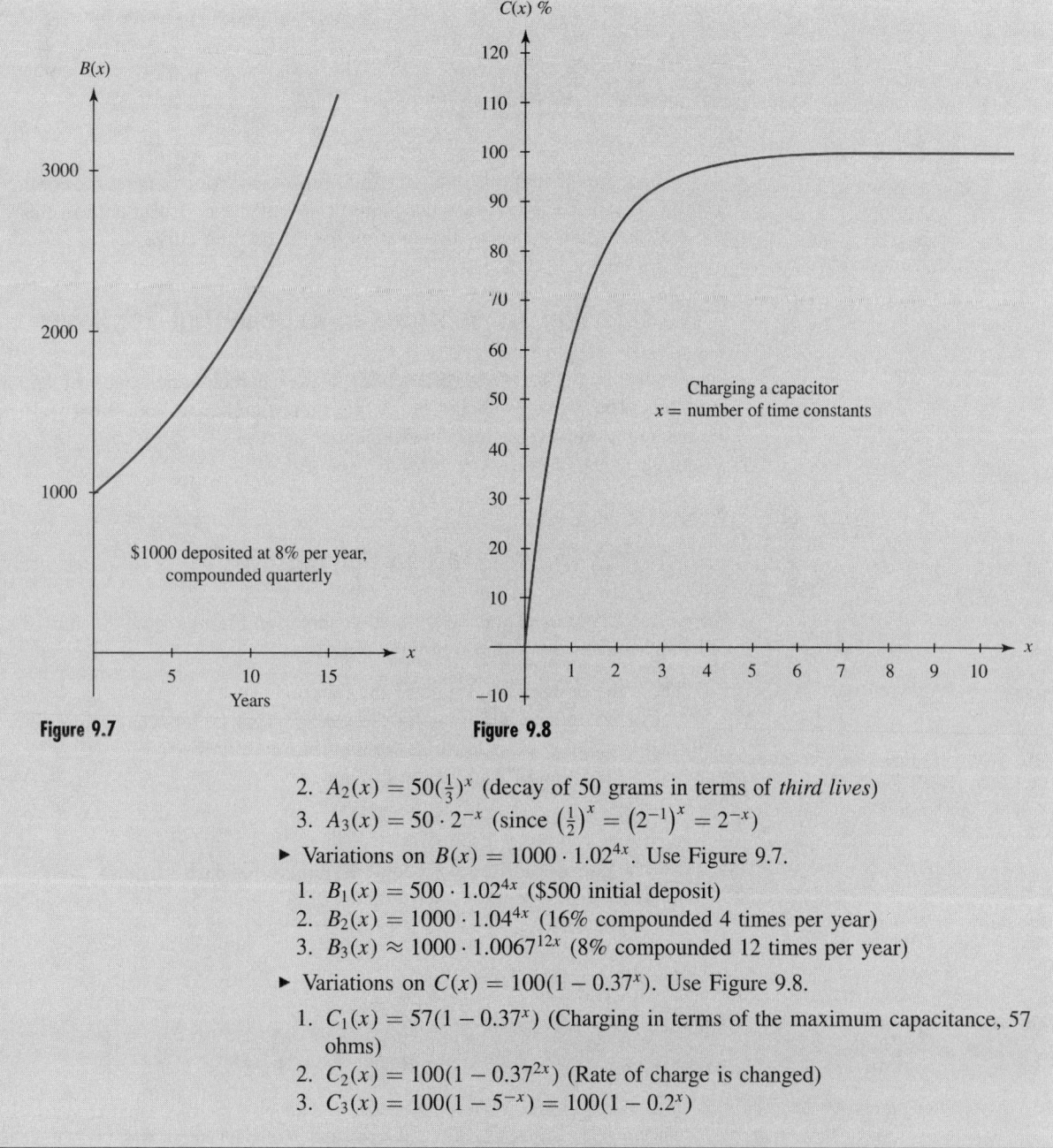

Figure 9.7

Figure 9.8

2. $A_2(x) = 50(\frac{1}{3})^x$ (decay of 50 grams in terms of *third lives*)
3. $A_3(x) = 50 \cdot 2^{-x}$ (since $(\frac{1}{2})^x = (2^{-1})^x = 2^{-x}$)

▶ Variations on $B(x) = 1000 \cdot 1.02^{4x}$. Use Figure 9.7.

1. $B_1(x) = 500 \cdot 1.02^{4x}$ ($500 initial deposit)
2. $B_2(x) = 1000 \cdot 1.04^{4x}$ (16% compounded 4 times per year)
3. $B_3(x) \approx 1000 \cdot 1.0067^{12x}$ (8% compounded 12 times per year)

▶ Variations on $C(x) = 100(1 - 0.37^x)$. Use Figure 9.8.

1. $C_1(x) = 57(1 - 0.37^x)$ (Charging in terms of the maximum capacitance, 57 ohms)
2. $C_2(x) = 100(1 - 0.37^{2x})$ (Rate of charge is changed)
3. $C_3(x) = 100(1 - 5^{-x}) = 100(1 - 0.2^x)$

The three different forms of the exponential function are summarized as follows:

Summary: Three Types of Exponential Curves

For values of the constants $a, b,$ and c with $b > 0$ and $b \neq 1$

▶ The function $f(x) = a \cdot b^{cx}$, with $b > 1$, represents exponential growth with a value equal to a when $x = 0$.

> ▸ The function $f(x) = a \cdot b^{-cx}$, with $b > 1$, represents exponential decay with a value equal to a when $x = 0$.
> ▸ The function $f(x) = a(1 - b^x)$, where $0 < b < 1$, represents exponential growth with an upper bound equal to a.

There are other forms of exponential functions. For example, the cooling curve from the Chapter Project does not quite fit any of these. Eventually in this chapter you will learn how to write the equation for the cooling curve.

Intercepts and Asymptotes in Exponential Functions

The graphs in Figures 9.6–9.8 each have the characteristic shape of exponential functions. These shapes should be getting more familiar to you. Next we'll look at the intercepts and asymptotes of exponential functions.

Activity 9.3
Identifying Intercepts and Asymptotes

For each example of exponential function shown in Figures 9.6–9.8, study the graph and write the answer to the following questions.

1. What is the domain and range of the function?
2. What is the domain and range of the *model*? That is, for what values of x and y do the equations make sense in the application described?
3. What is the value of each of the limits

$$\lim_{x \to \infty} f(x) \text{ and } \lim_{x \to -\infty} f(x)$$

4. Write the equation of the horizontal asymptote for each function.
5. What are the intercepts of each function?

Section 9.1 Exercises

1. *Ecology* Estimate the population of Morocco in the year 2000, given that: **(a)** the population in 1992 is 25.8 million, **(b)** the population in 2027 is 51.6 million, and **(c)** the changing population between those two years follows a straight line. Compare your answer with the vertical coordinate of point C in Figure 9.5.

2. *Ecology* Estimate the population of Morocco in the year 2015, given that: **(a)** the population in 1992 is 25.8 million, **(b)** the population in 2027 is 51.6 million, and **(c)** the changing population between those two years follows a straight line.

3. *Ecology* According to an estimate made in 1990 the population of the Asian country Bangladesh was ex-

pected to double in the following 32 years. In 1990 the population of Bangladesh was 111.4 million.
(a) What is your estimate of the population of Bangladesh in the year 2005?
(b) What assumptions and procedures did you use to get your answer in (a)?

4. *Ecology* According to an estimate made in 1990 the population of the South American country Paraguay was expected to double in the following 24 years. In 1990 the population of Paraguay was 4.3 million.
(a) What is your estimate of the population of Paraguay in the year 2015?

(b) What assumptions and procedures did you use to get your answer in (a)?

5. *Sports management* Table 9.1 gives the average major league baseball player's salary on opening day for 1989 and 1997.

Table 9.1

Year	1989	1997
Average salary ($)	512,804	1,383,578

Source: http://xenocide.nando.net/newsroom/ap/bbo/
1997/mlb/mlb/feat/archive/040297/mlb54801.html

(a) What is your estimate of the average baseball salary on opening day for 2005?

(b) What assumptions and procedures did you use to get your answer in (a)?

(c) Consider the additional information that the average salary on opening day in 1993 was $1,120,254. Use a quadratic regression to predict the average baseball salary on opening day for 2005.

6. *Agriculture* Table 9.2 gives the total number of farms in the United States for selected years.

Table 9.2

Year	1978	1982	1987	1992
Total farms	2258			1925

Source: 1996 Statistical Abstracts of the United States,
Table No. 1074, p. 661

(a) What are your estimates of the number of farms in 1982 and 1987?

(b) What assumptions and procedures did you use to get your answer in (a)?

(c) Use the additional information that there were 2241 farms in 1982 to revise your prediction for the number of farms in 1987.

(d) Use the additional information that there were 2088 farms in 1987 to predict the number of farms in 2002.

7. *Medical technology* A pharmaceutical company is growing an organism to be used in a vaccine. The organism's growth is given by the equation $Q = Q_0 2.5^t$, where Q_0 is the initial number of bacteria (that is, the number of bacteria when $t = 0$) and t is the time in hours since the initial count was taken. When $t = 1$, we know that $Q = 3,150$.

(a) Find Q_0.

(b) What are the number of organisms at the end of 4 hours?

(c) How long will it take for Q to become 5 times as large as Q_0?

8. *Lighting technology* Each 1 mm thickness of a certain translucent material reduces the intensity of a light beam passing through it by 14%. This means that the intensity I is a function of the thickness T of the material as given by $I = 0.86^T$.

(a) What is the intensity when the thickness is 1.75 mm?

(b) Graph I as a function of T.

(c) Use the graph from (b) to approximate the thickness that produces an intensity of 0.50?

9. (a) Graph the two functions $f(x) = x^2$ and $g(x) = 2^x$ on your calculator.

(b) Explain how the two graphs are similar and how they are different.

(c) Describe how you would help someone learn which was the graph of $f(x) = x^2$ and which was the graph of $g(x) = 2^x$.

10. What variables would affect the accuracy of a population model used to predict the population in 5, 10, or 50 years?

9.2 Logarithms and Inverse Functions

What Are Logarithms?

The calculator keys ⬛ **LOG** and ⬛ **LN** represent two kinds of logarithmic functions. After a little exploration with the ⬛ **LOG** key in the following activity, you will learn a lot about logarithms and the function log.

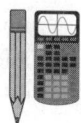

Activity 9.4
Exploring Logarithms: Part 1

Make sure your calculator is set to Real mode.

1. Use your calculator to evaluate each of these values of the log function. The patterns in the answers will help you understand the meaning of this function.

   (a) $\log(1)$ (d) $\log(1000)$
   (b) $\log(10)$ (e) $\log(10^{12})$
   (c) $\log(100)$ (f) $\log(10^{351})$

2. What do you think should be the answer to question 1(f)? Why doesn't the calculator show that?

3. Write two other logarithm problems like those in question 1 that you can do without the calculator. Use the calculator to check your answers.

4. Write a sentence describing how the log function seems to work.

5. Use your calculator to evaluate each of these log function values:

   (a) $\log(0.1)$ (d) $\log(0.00001)$
   (b) $\log(0.01)$ (e) $\log(10^{-5})$
   (c) $\log(\frac{1}{10})$ (f) $\log(10^{-500})$

   The patterns in the answers will help you understand the meaning of the log function.

6. What do you think should be the answer to question 5(f)? Why doesn't the calculator show that?

7. Write two logarithm problems like those in question 5 that you can do without the calculator. Use the calculator to check your answers.

8. (a) Write a sentence describing how the log function works for values of x between 0 and 1.
   (b) Is this answer consistent with your answer to question 4?

9. Use your calculator to evaluate each of the values

   (a) $\log(0)$
   (b) $\log(-1)$
   (c) $\log(-100)$

10. Write a sentence that explains why you think problems 9(a)–(c) have no answer.

11. Think about the graph of the function $y = \log(x)$

    (a) What appears to be the domain of the function $y = \log(x)$?
    (b) What appears to be the range of the function $y = \log(x)$?
    (c) What values of x give a negative value of $\log(x)$?
    (d) Use some of the values you have computed above to *sketch* a graph of the function $y = f(x)$ on a graph like the one in Figure 9.9. On your graph show the coordinates of any x- or y-intercepts of the curve. Label any horizontal or vertical asymptotes of the function.

12. Think about the values that you have calculated and complete the following by filling in each blank with a number:

    (a) If $\log(x) > 2$ then $x >$ _____.
    (b) If $10 < \log(x) < 11$ (that is, $\log(x)$ is between 10 and 11), then _____ $< x <$ _____ (that is, x is between what two numbers?).

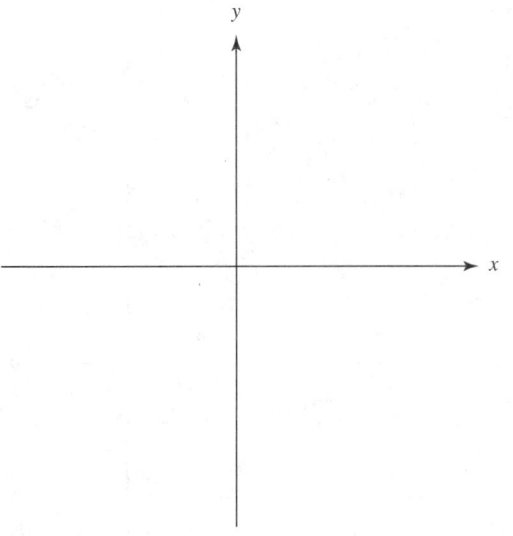

Figure 9.9

(c) If $0 < \log(x) < 1$ then _____ $< x <$ _____.
(d) If $\log(x) < 0$, then $x <$ _____.

13. Write a definition of the log function based on what you know so far.

Example 9.3

Draw the graph of $y = \log(x)$ and discuss its intercepts, asymptotes, domain, and range.

Solution A graph of $y = \log(x)$ appears in Figure 9.10. As you may have noticed, the values of the logarithm (the y-values) are much smaller than the x-values. There is only one intercept, at $x = 1$. Even though the graph gets very close to the y-axis, they never touch each other, so the y-axis is the vertical asymptote of the graph. It should be clear from the graph that the domain of the function consists only of positive numbers.

The range of the function is more difficult to guess. Does the graph eventually reach any large y-value? We saw in Activity 9.4 that $y = 12$ is reached at $x = 10^{12}$. Similarly the value $y = 2000$ will be reached at $x = 10^{2000}$. Using that same logic, *any value* of y, positive, negative, or zero, can be reached. Therefore the range of the function is all real numbers, and since the function value can exceed any number there is no horizontal asymptote. ■

The graph in Figure 9.10 reminds us that there are values of $\log(x)$ when x is not a power of 10. For example $\log(3)$ seems to be approximately equal to 0.5. The next activity will help you understand the meanings of logarithms of numbers that are not powers of 10.

If $\log(10) = 1$ and $\log(100) = 2$ what can you say about the logarithm of 55, the number halfway between 10 and 100? That is, what estimate can you make of the value of $\log(55)$? Activity 9.5 begins our exploration of the values of the logarithm function between the powers of 10.

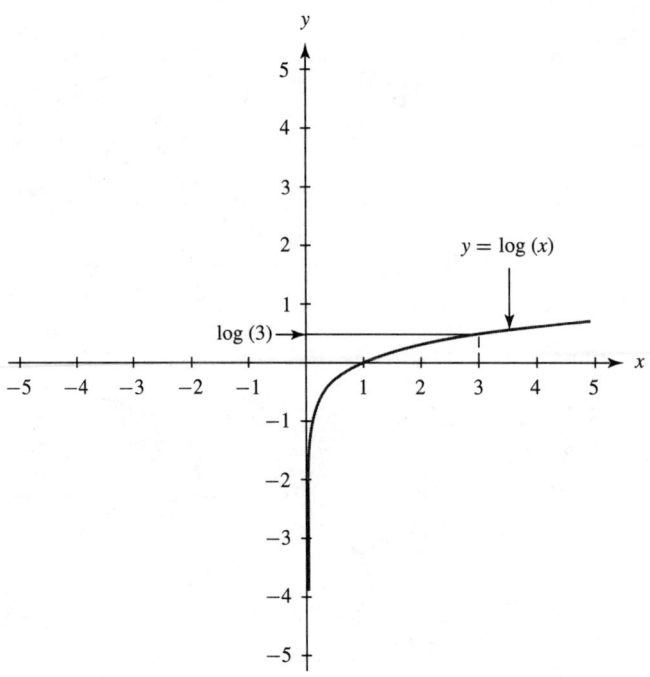

Figure 9.10

Activity 9.5
Exploring Logarithms: Part 2

1. Compute each of the following logarithms and think about the answers.
   (a) log(55)   (d) $\log(5.5 \times 10^{23})$
   (b) log(550)   (e) $\log(5.5 \times 10^{-4})$
   (c) log(55000)
2. Write a sentence that describes how to compute $\log(a \times 10^{b})$ if a is positive.
3. Test your answer to question 2 with the following activity: Start with the fact that $\log(3.56) \approx 0.551$, and using the pattern of the answers in question 1 guess the value of each of the numbers below and then check your answers with the calculator.
   (a) log(35.6)   (c) $\log(3.56 \times 10^{20})$
   (b) log(3560)   (d) log(0.356)
4. Change your answer to question 2 if necessary.
5. For each computation tell what two integers each answer is between.
   (a) log(2)   (d) log(0.12345)
   (b) log(543)   (e) log(0.000023456)
   (c) log(432198)

Properties of Logarithms

The explorations in the previous activities have prepared you for the following definition of logarithmic functions.

Definitions of Logarithmic Functions

▶ Logarithms are exponents of a base.
- The base of the function **log** is the number 10. The function log, that is $y = \log x$, is called the **common logarithmic** function.
- The base of the function **ln** is the number called e. The function ln is called the **natural logarithmic** function.

▶ The value of log is the power of 10 that equals x. That is, $\log(x) = a$ means that $10^a = x$. Similarly the value of ln is the power of e that equals x. That is, $\ln(x) = a$ means that $e^a = x$.

From now on we will drop the parentheses unless they are necessary for clarity. The symbols *log*(a) and *log a* have the same meaning. The *TI-83*, however, always requires that you use the parentheses.

We'll look at some important properties of logarithms in the next activity.

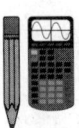

Activity 9.6
Properties of Logarithms

1. Compute each of the following values:

   (a) $\log 48$ (d) $4\log 2 + \log 3$
   (b) $\log 8 + \log 6$ (e) $\log 96 - \log 2$
   (c) $\log 2 + \log 4 + \log 6$ (f) $\log 1440 - \log 30$

2. Write two more expressions like those in question 1 that have the same answer.

3. (a) Based on your answers to question 1, write some mathematical rules that you think might be true.
   (b) Thinking about the definition of logarithms, why do you think these rules are true?

4. Use the rules you wrote in answer to question 2 to fill in the blanks, then check your answer with the calculator.

   (a) $\log \underline{\hspace{1.5cm}} = \log 7 + \log 4$
   (b) $\log 560 = \log 2 + \log 5 + \log \underline{\hspace{1.5cm}}$
   (c) $\log 8 = \log 64 - \log \underline{\hspace{1.5cm}}$
   (d) $\log 240 = \log 14400 - \log \underline{\hspace{1.5cm}}$
   (e) $\log 270 = \log \underline{\hspace{1cm}} + \log \underline{\hspace{1cm}} + \log \underline{\hspace{1cm}} + \log \underline{\hspace{1cm}}$

Some of the mathematical rules you wrote and used in Activity 9.6 probably were related to addition and subtraction of logarithms. In Example 9.4 the addition property of logarithms will be developed in more detail.

Example 9.4

The addition property of logarithms is:

$$\log a + \log b = \log ab$$

Explain why it is true.

Solution

All the properties of logarithms that we will study are true for any logarithm with any base.

The properties of logarithms are based on the properties of exponents. Make sure you know the reason for each of these statements:

▶ The definition of the common logarithm tells us that $x = \log(10^x)$ and $y = \log(10^y)$

- Therefore $x + y = \log(10^x) + \log(10^y)$
- And also, $x + y = \log(10^{x+y})$ because of the definition of the common logarithm
- Therefore $\log(10^x) + \log(10^y) = \log(10^{x+y})$ because both are equal to $x + y$

▶ But we know that $10^{x+y} = 10^x \cdot 10^y$

- Therefore $\log(10^x) + \log(10^y) = \log(10^{x+y}) = \log(10^x \cdot 10^y)$

Therefore if $a = 10^x$ and $b = 10^y$, then $\log a + \log b = \log ab$ ■

Properties of Logarithms

The following properties are true for logarithms with any base, for positive values of a and b, and for *any* real value of n:

▶ Addition: $\log a + \log b = \log ab$

▶ Subtraction: $\log a - \log b = \log \dfrac{a}{b}$

▶ Multiplication: $n \cdot \log a = \log a^n$

Example 9.5

Use the addition property of logarithms to rewrite each of the following: **(a)** $\log 39$, **(b)** $\log 7x$, **(c)** $\log 7 + \log 5$, and **(d)** $\log 3 + \log t$.

Solutions
(a) $\log 39 = \log(3 \cdot 13) = \log 3 + \log 13$
(b) $\log 7x = \log(7 \cdot x) = \log 7 + \log x$
(c) $\log 7 + \log 5 = \log(7 \cdot 5) = \log 35$
(d) $\log 3 + \log t = \log(3 \cdot t) = \log 3t$ ■

Example 9.6

Use the subtraction property of logarithms to rewrite each of the following: **(a)** $\log \frac{3}{7}$, **(b)** $\log \dfrac{7}{5x}$, **(c)** $\log 7 - \log 5$, and **(d)** $\log 3x^2 - \log 11$.

Solutions
(a) $\log \frac{3}{7} = \log 3 - \log 7$
(b) $\log \dfrac{7}{5x} = \log 7 - \log 5x$
(c) $\log 7 - \log 5 = \log \frac{7}{5}$
(d) $\log 3x^2 - \log 11 = \log \dfrac{3x^2}{11}$ ■

Example 9.7

Use the multiplication property of logarithms to rewrite each of the following: **(a)** $\log 5^3$, **(b)** $\log \sqrt[3]{25}$, **(c)** $\log 15^4$, and **(d)** $2\log 5 + 3\log 7 - 6\log 2$.

Solutions   (a) $\log 5^3 = 3\log 5$
(b) $\log \sqrt[3]{25} = \log 25^{1/3} = \frac{1}{3}\log 25$
(c) $\log 15^4 = 4\log 15 = 4\log(3 \cdot 5) = 4(\log 3 + \log 5)$
(d) $2\log 5 + 3\log 7 - 6\log 2 = \log 5^2 + \log 7^3 - \log 2^6$
$$= \log\left(5^2 7^3\right) - \log 2^6$$
$$= \log\left(\frac{5^2 7^3}{2^6}\right)$$

∎

Notice that Example 9.7 required the use of the addition and subtraction properties of logarithms as well as the multiplication property.

Why Study Logarithms?

There are two basic reasons that people have studied logarithms—to assist in calculations and because of the applications of the logarithm function in science and engineering. Since scientific calculators became generally available in the early 1970's, the first reason is no longer important. However, you might be interested to see how logarithms assisted engineers and scientists before calculators and computers were available to everyone, so we will look at how logarithms simplified computations that were almost impossible without them.

Before calculators were generally available all engineers used either a slide rule or a table of logarithm values whenever they had to do certain computations. Many high school and college mathematics books included a table of logarithms. With a table of logarithms it was possible to obtain results accurate to three or four significant figures; the slide rule was less accurate but faster to use.

The slide rule has been completely replaced by scientific calculators, but before 1970 every engineer and engineering student owned a slide rule. The design of a slide rule is based on logarithms.

Example 9.8

Show how logarithms can provide an estimate for the computation of $\sqrt[3]{438.6^{11}}$. Assume that you only have a table of logarithms for the numbers between 1 and 10 and that the values in the table are given to four decimal places.

Solution   The steps are as follows:

(a) Recognize that $\sqrt[3]{438.6^{11}} = 438.6^{11/3}$
(b) Use the properties of logarithms to rewrite the problem like this:

$$\log(438.6^{11/3}) = \frac{11}{3}\log 438.6$$
$$= \frac{11}{3}\log(4.386 \times 10^2)$$
$$= \frac{11}{3}((\log 4.386) + 2)$$

(c) Look up log 4.386 in the table. The answer is 0.6421

(d) Add 2 and multiply by $\frac{11}{3}$. The answer is approximately 9.6877

(e) This is the logarithm of the answer. That is, the answer is $10^{9.6877} = 10^9 \times 10^{0.6877}$

(f) To find the value of $10^{0.6877}$ we find what number has a logarithm equal to 0.6877. The answer from the table is approximately 4.872.

(g) Therefore the answer to the problem is approximately 4.872×10^9.

(h) Compare this with the answer your calculator gives, and you will see that the table derived answer is incorrect in the fourth significant figure. The result should be about 4.871×10^9.   ∎

Example 9.8 shows how logarithms change a multiplication problem to an addition problem and a power problem to multiplication. These properties of logarithms simplified computations for over 300 years prior to the 1970's. They were essential and students spent a lot of time learning how to do problems like the one in Example 9.8.

Since calculators quickly do any computation that would have been done using logarithms we will turn to the *applications* of logarithms. Here are two:

▶ One common way to report the concentration of an acid in a water solution is with the **pH** value, which is the *negative logarithm of the concentration* of the hydrogen ions in moles per liter. Thus, if H^+ represents the concentration of the hydrogen ions in moles per liter then

$$\text{pH} = -\log H^+$$

If the pH is 3, then the concentration of hydrogen ions is 10^{-3} moles per liter. A mole is a measure of weight of chemical compounds. More specifically a mole is the molecular mass of a substance in grams.

▶ Plotting some functions on **logarithmic coordinate axes** makes their graphs easier to read. Some kinds of data will lie on a straight line on logarithmic graph paper and indicates an equation that could model the data. We will study logarithmic graphing techniques in Section 9.4.

Examples 9.9–9.10 apply logarithms to the concentration of an acid whose pH value is known.

Example 9.9

Calculate the pH of a solution of nitric acid whose concentration of hydrogen ions is 0.005 moles per liter.

Solution   The pH is the negative of the logarithm of the concentration of hydrogen ions. Thus $-\log(0.005) = 2.30$. Therefore the pH of the solution is 2.3.   ∎

Example 9.10   Application

If normal blood has $H^+ = 3.4 \times 10^{-8}$, find the pH value of normal blood.

Solution Since pH $= -\log H^+$ and $H^+ = 3.4 \times 10^{-8}$ we have

$$\text{pH} = -\log H^+$$

$$= -\log \left(3.4 \times 10^{-8}\right)$$

$$\approx 7.47$$

The pH of normal blood is about 7.47. ■

The main reason that logarithmic functions are important is because of their relationship to exponential functions. Logarithms can make the large numbers encountered in exponential functions more manageable, as in $\log 100000 = 5$. The relationship between logarithmic functions and exponential functions is a special one because they are **inverse functions**.

Inverse Functions

Here are three ways to think about inverse functions:

▶ An inverse function reverses the process of another function.

- For example, to solve the equation $\tan\theta = 0.2345$ we have seen that the answer is $\theta = \tan^{-1}(0.2345) \approx 13.2°$. The function symbolized by $\tan^{-1}$ is the inverse of the tangent function.

▶ The inverse of a function can be formed by switching x and y in the equation that defines the function and solving for y. If this result is a function then it is the inverse function of the original function. If it is not a function then it is called an **inverse relation** of the original function.

- Examples 9.11–9.13 illustrate this process.

▶ The *TI-83* calculator has some inverse functions printed in yellow above the corresponding function key.

- Look at the [SIN], [COS], [TAN], [LOG], [LN], and [x^2] keys. The function in yellow and the function on the key are inverses of each other. They each reverse the process of the other. To make $\sin^{-1}, \cos^{-1}, \tan^{-1}$, and $\sqrt{x}$ inverse functions, the domains of sin, cos, tan, and x^2 have to be restricted.

Example 9.11

Determine the inverse of $y = 3x + 2$. Is this an inverse function?

Solution If $y = 3x + 2$ then the inverse is found by first switching x and y in the original equation. This produces $x = 3y + 2$ and we next solve this equation for y. This gives $y = \dfrac{x-2}{3}$ which is a function. You can think of the function

$$f(x) = \frac{x-2}{3} \qquad \text{(subtracting 2 and then dividing by 3)}$$

as reversing the process of its inverse function

$$g(x) = 3x + 2 \qquad \text{(multiplying by 3 and then adding 2)} \qquad ■$$

Example 9.12

Determine the inverse of $y = 4x^2 - 8$. Is this an inverse function?

Solution   Switching each x and y in the original equation produces $x = 4y^2 - 8$. Solving this equation for y we get $y = \pm\sqrt{\dfrac{x+8}{4}}$. This is not a function since $x = 1$ produces two values for y, namely $y = \sqrt{\dfrac{x+8}{4}} = \dfrac{3}{2}$ and $y = -\sqrt{\dfrac{x+8}{4}} = -\dfrac{3}{2}$. Thus $y = \pm\sqrt{\dfrac{x+8}{4}}$ is the inverse relation of $y = 4x^2 - 8$.   ∎

Example 9.13

Determine the inverse of $y = \log x$. Is this an inverse function?

Solution   Switching each x and y in the original equation produces $x = \log y$. In the boxed definition of logarithms on page 621 we see that $x = \log y$ means $y = 10^x$. This is a function (an exponential function) and hence $y = 10^x$ is the inverse function of $y = \log x$.   ∎

Example 9.14

Draw the graphs of $y = \log x$ and $y = 10^x$ and discuss the shapes and symmetry of the graphs

Solution   Figure 9.11 shows the graphs, along with the graph of the equation $y = x$. You can see that the curves have the same basic shape. Furthermore, the curves are symmetric with respect to the line $y = x$. That is, they are mirror images if the mirror is placed on the line $y = x$. This symmetry can be seen in all pairs of inverse functions.   ∎

Examination of the graphs of the inverse functions $y = \log x$ and $y = 10^x$ in Figure 9.11 shows one important relationship between the domain and range of inverse functions. The domain of a function is the range of its inverse function and the range of a function is the domain of its inverse function. For example, the domain of $y = \log x$ is $x > 0$; the range of $y = 10^x$ is $y > 0$. The range of $y = \log x$ is all real numbers; the domain of $y = 10^x$ is all real numbers. This relationship is not true for a function and its inverse relation if that relation is not an inverse function.

A *TI-83* calculator will draw the inverse relation of an expression placed in the Y= screen. To access `DrawInv` press `2nd` [DRAW] and select `8:DrawInv`. Notice that you have to hold down the `▼` to see `8:DrawInv`. You must be on the home screen when you access `DrawInv` and it requires the name of the expression in the Y= screen.

Example 9.15

Use a *TI-83* to draw $y = \log x$ and its inverse function.

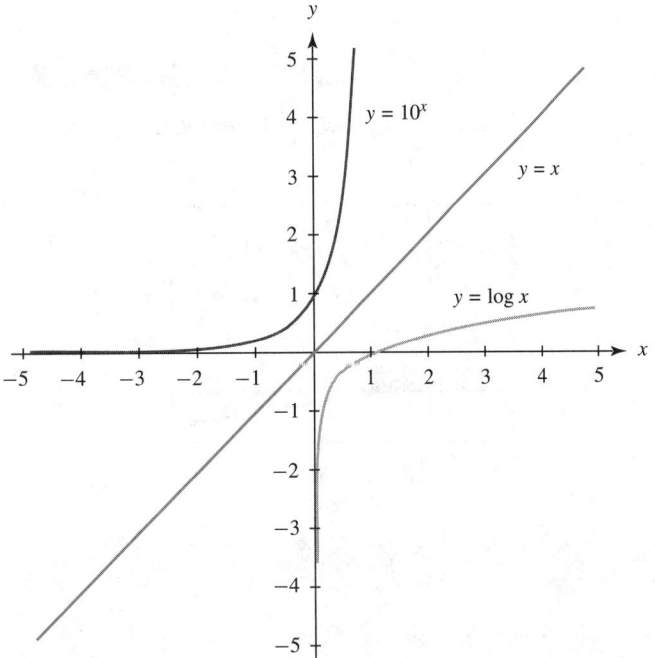

Figure 9.11

Solution   We know that the inverse function of $y = \log x$ is $y = 10^x$ so we could graph these directly. However we want to use the calculator command `DrawInv`.

Caution: You must enter the window settings before you use `DrawInv`.

Begin by entering $y = \log x$ in the Y= screen by pressing ⬛Y=⬛ ⬛LOG⬛ ⬛X,T,θ,n⬛. This should enter $y = \log x$ on the Y1 line. Exit the Y= screen by pressing ⬛2nd⬛ [QUIT]. Now draw the inverse of $y = \log x$ by pressing the following sequence of keys:

⬛2nd⬛ [DRAW] ⬛8⬛ ⬛VARS⬛ ⬛▶⬛ 1(Function) 1(Y1) ⬛ENTER⬛

The result is shown in Figure 9.12. Notice that this graph also has the line $y = x$ so it should correspond to Figure 9.11.

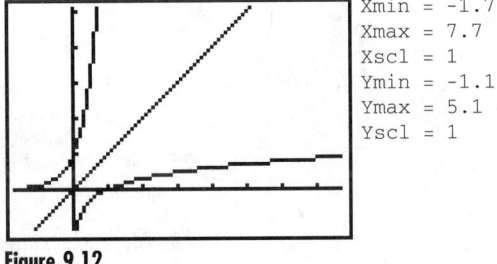

```
Xmin = -1.7
Xmax = 7.7
Xscl = 1
Ymin = -1.1
Ymax = 5.1
Yscl = 1
```

Figure 9.12

Example 9.16

Use a *TI-83* to draw $y = (x - 2)^2$ and its inverse relation.

Solution   Begin by entering $y = (x - 2)^2$ on the Y1 line in the Y= screen. To help distinguish its graph from the inverse relation graph, use the thick drawing mode. Exit the Y= screen by pressing ⬛2nd⬛ [QUIT]. Now draw the inverse of $y = (x - 2)^2$ by pressing

628 Chapter 9 EXPONENTIAL AND LOGARITHMIC FUNCTIONS

the following sequence of keys:

[2nd] [DRAW] [8] [VARS] [▶] 1(Function) 1(Y1) [ENTER]

The result is shown in Figure 9.13. Notice that this graph also has the graph of the line $y = x$.

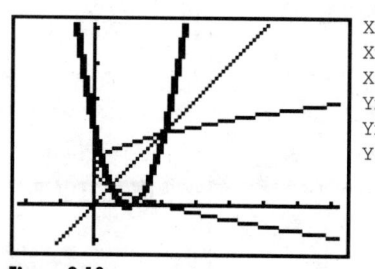

```
Xmin = -4.4
Xmax = 14.4
Xscl = 2
Ymin = -2.2
Ymax = 10.2
Yscl = 2
```

Figure 9.13

Section 9.2 Exercises

In Exercises 1–4 evaluate each of the following logarithms without *the use of a calculator.*

1. $\log 10000$

2. $\log 0.001$

3. $\log 10^{23}$

4. $\log 10^{-67}$

In Exercises 5–8 use the fact that $\log 7.34 \approx 0.8657$ *to evaluate each of the given logarithms without* the use of a calculator.

5. $\log 734$

6. $\log 73,400$

7. $\log 0.734$

8. $\log 0.0000734$

In Exercises 9–16 use one of the properties of logarithms to rewrite each of the following.

9. $\log 18$

10. $\log \frac{3}{5}$

11. $\log 5^7$

12. $\log 24$

13. $\log 12 - \log 6$

14. $\log 2 + \log 5 - \log 4$

15. $7 \log 3 + 9 \log x - \log z$

16. $\log x - 4 \log 2 + 2 \log 5$

In Exercises 17–22 (a) determine the inverse of the given function, (b) determine if the answer in (a) is an inverse function or an inverse relation, and (c) graph the given function and its inverse.

17. $y = 4x - 8$

18. $y = (x + 1)^3$

19. $y = x^2 - 4$

20. $y = \sqrt{x - 5}$

21. $y = \sqrt[5]{x + 2}$

22. $y = \frac{1}{2}x + 3$

23. *Environmental science* The pH of some acid rain is 5.3. What is the concentration of hydrogen ions in acid rain?

24. *Food technology* If the pH of eggs is 7.8 what is the concentration of hydrogen ions in an egg?

25. *Food technology* If beer has $H^+ \approx 3.98 \times 10^{-5}$ hydrogen ions what is the pH of beer?

26. *Medical technology* In normal human intestinal contents we find that $H^+ \approx 3.16 \times 10^{-7}$ hydrogen ions. Find the pH of normal human intestinal contents.

27. *Sound technology* The loudness L of a sound, as measured in decibels (dB), is given by the equation

$$L = 10 \log \left(\frac{I}{I_0} \right)$$

where I is the intensity of the sound and $I_0 = 10^{-12}$ W/m². If the intensity of a lawn mower is 10^{-2} W/m² calculate the loudness of the lawn mower.

28. *Sound technology* If the intensity of a jet aircraft is 10^2 W/m^2 calculate the loudness of the jet plane.

29. *Seismology* The Richter scale magnitude R of an earthquake is given by the equation

$$R = \log\left(\frac{I}{I_0}\right)$$

where I is the intensity of the quake and I_0 is a minimum perceptible intensity. What is the value of an earthquake whose intensity is $3{,}000{,}000 I_0$?

30. *Seismology* One earthquake has a magnitude 100 times as intense as another. How much higher is its magnitude on the Richter scale?

31. *Seismology* One earthquake has a magnitude $10{,}000$ times as intense as another. How much higher is its magnitude on the Richter scale?

32. *Astronomy* The magnitude M of a star is a measure of its brightness. The larger the magnitude the fainter the star. The magnitude of stars made visible through a telescope is given by

$$M(a) = 9 + \log a$$

where a is the aperture in inches of the telescope. Calculate the magnitude of the faintest star made visible by a telescope with an aperture of

(a) 2.5 inches

(b) 50 inches

For Exercises 33–34 put your calculator in radian mode and set the viewing window to `ZTrig`.

 33. In the Y= screen of your calculator let $y_1 = \sin x$ and $y_2 = \sin^{-1} x$. Set the calculator so it draws y_2 in the "thick" mode. Use the procedure outlined in Example 9.16 to draw the inverse relation of y_1.
  (a) Sketch the graphs of $y_2 = \sin^{-1} x$ and `DrawInv(Y1)`.
  (b) How are the graphs similar and how are they different?
  (c) Explain your answer in (b).

 34. In the Y= screen of your calculator let $y_1 = \tan x$ and $y_2 = \tan^{-1} x$. Set the calculator so it draws y_2 in the "thick" mode. Use the procedure outlined in Example 9.16 to draw the inverse relation of y_1.
  (a) Sketch the graphs of $y_2 = \tan^{-1} x$ and `DrawInv(Y1)`.
  (b) How are the graphs similar and how are they different?
  (c) Explain your answer in (b).

9.3   Natural Logarithms and the Number *e*

In Section 9.2 you learned that common logarithms are exponents of the number 10 and natural logarithms are exponents of a number called e. That is, if

$$a = \log b \text{ then } b = 10^a$$
$$\text{and if} \quad a = \ln b \text{ then } b = e^a$$

But what is e and why is it important? On the *TI-83* you can see that e is important enough to have its own key (look above the ▭ key). You can use that key to see that an approximate value is

$$e \approx 2.718281828$$

You will also find that e^x is above the `LN` key.

This value of $e \approx 2.718281828$ looks like it might repeat. However, the next two digits are 4 and 6, and in fact e has been proven to be an irrational number (a number in which the decimal form will never repeat).

Before exploring the number e in detail we'll study the natural logarithmic function.

Calculator Lab 9.3

Exploring the Natural Logarithmic Function

Study the two functions $y = \ln x$ and $y = \log x$ with graph, trace, table, and any other calculator feature you think will help you to answer the following questions:

Procedures   1. Complete the following table with the properties of the two functions:

	$y = \log x$	$y = \ln x$
Domain		
Range		
x-intercept		
y-intercept		
Horizontal asymptote		
Vertical asymptote		

2. On a set of axes like the one shown in Figure 9.14, sketch and label the graphs of both functions. Make sure you show which one is above the other.

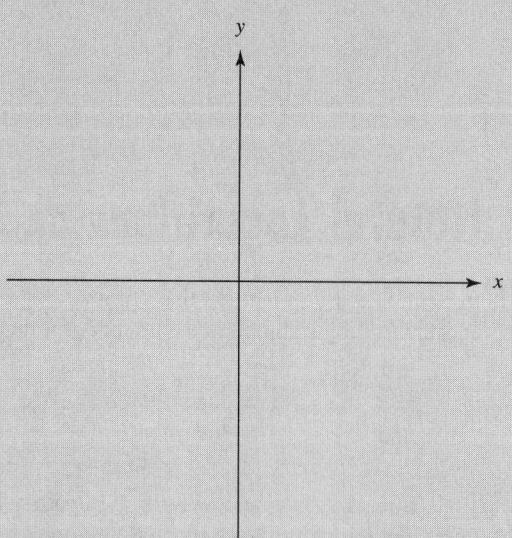

Figure 9.14

3. Fill in the blanks with one of these symbols: $=$, $<$, $>$, $\leq$, or $\geq$.

    (a) If $x > 1$, then $\ln x$ _____ $\log x$

    (b) If $x = 1$, then $\ln x$ _____ $\log x$

    (c) If $0 < x < 1$, then $\ln x$ _____ $\log x$

Continuous Growth and the Number *e*

Our study of *e* begins, surprisingly, with compound interest. You learned in Chapter 7 that if $P is invested at the rate r for n years, compounded k times per year, then the amount of money in the account is expressed by the discrete function A where

$$A(P, k, r, n) = P\left(1 + \frac{r}{k}\right)^{kn}$$

The rate r is often given as a percent and must be changed to a decimal. For example, $3.5\% = 0.035$.

The symbol $A(P, k, r, n)$ means that A is a function of four variables, P, k, r, and n.

Banks sometimes advertise the importance of compounding daily ($k = 365$) compared with compounding monthly ($k = 12$). How much difference do you think that makes? Example 9.17 shows how to make that comparison.

Example 9.17   Application

Compare monthly and daily compounding for $100 invested at 8% per year for 5 years.

Solution   We are given $P = 100$, $r = 0.08$, and $t = 5$, so the formula is now a function of k only:

$$A(k) = 100\left(1 + \frac{0.08}{k}\right)^{5k}$$

Substituting $k = 12$ and $k = 365$ we find that

$$A(12) \approx 148.9845708 \approx \$148.98$$
$$\text{and} \qquad A(365) \approx 149.1759314 \approx \$149.18$$

You are not going to get rich on that 20 cent difference!   ∎

Even though there is not much financial difference between compounding daily and compounding monthly, the difference is very interesting mathematically, as we will see. In Activity 9.7 we'll look more closely at the function $A(k)$.

Activity 9.7
Compounding Many Times per Year

Suppose that $200 is invested at 7% per year and compounded k times per year.

1. Write the formula for the function $A(k)$, the amount of money in the account after 12 years.
2. Set your calculator's precision to Float.
3. Complete the table. One row is done for you.
4. In dollars and cents, determine

$$\lim_{k \to \infty} A(k)$$

Type of Compounding	k	A(k)	A(k)
		dollars and cents	calculator accuracy
Yearly			
Quarterly			
Monthly	12	$462.14	462.1441488
Daily			
Hourly			
Every minute			
Every second			

5. In terms of the calculator's greatest accuracy, determine

$$\lim_{k \to \infty} A(k)$$

One result of Activity 9.7 is that the amount produced in a savings account appears to approach a limit as $k \to \infty$. Next we'll investigate that limit.

Example 9.18

Rewrite the compound interest function $A(k)$ to study its value as $k \to \infty$.

Solution   Any version of $A(k)$ will be satisfactory for this exercise. We'll choose the one used in Activity 9.7:

$$A(k) = 200 \left(1 + \frac{0.07}{k}\right)^{12k}$$

We'll introduce a new variable $c = \dfrac{k}{0.07}$. Since $k = 0.07c$, the formula becomes

$$A(c) = 200 \left(1 + \frac{1}{c}\right)^{12 \cdot 0.07c}$$

$$= 200 \left(1 + \frac{1}{c}\right)^{0.84c}$$

We have $k = 0.07c$. That means that as $c \to \infty$ then it is also true that $k \to \infty$, and the limits of $A(c)$ and $A(k)$ are equal. We introduce the variable A to represent these limits:

$$A = \lim_{k \to \infty} A(k) = \lim_{c \to \infty} A(c)$$

The variable A is an estimate of the amount that a bank account will become for large values of k. We can use one of the properties of exponents to rewrite A. Remember that, for example, $(5^2)^4 = 5^8$, and in general $a^{xy} = (a^x)^y$. Our exponent is $0.84 \cdot c$. Therefore we can rewrite A as

$$A = 200 \left(\lim_{c \to \infty} \left(1 + \frac{1}{c}\right)^c\right)^{0.84}$$

To get the formula for A we need the value of the limit which we will call L:

$$L = \lim_{c \to \infty} \left(1 + \frac{1}{c}\right)^c$$

Therefore the formula for A is

$$A = 200L^{0.84}$$

If we could find the value of the limit L then we could write a formula for savings account growth if the compounding is done a very large number of times per year.

Look at the last formula in Example 9.18. Remember that for this problem the original amount invested was $200, the number of years was 12, and the interest rate was 7% = 0.07. You can see those numbers in the formula for the amount A. Therefore we get a new formula for A by substituting the variables for these numbers and arrive at

$$A = PL^{nr}$$

The formula for A describes continuous rather than discrete growth. Many growth processes in nature are continuous because growth is happening all the time, or at least it is happening so often that continuous growth is a good mathematical model. If we knew the value of L then we could use a simple formula to estimate the amount of growth present at any time after the growth began. Calculator Lab 9.4 will help you estimate the value of the limit L.

Calculator Lab 9.4

Estimating the Limit L

In this lab you will use graphs and tables to estimate the value of the limit L when

$$L = \lim_{c \to \infty} \left(1 + \frac{1}{c}\right)^c$$

Procedures 1. Enter the formula $y = \left(1 + \frac{1}{x}\right)^x$ into the calculator.

2. (a) Use GRAPH and TRACE to look at the graph for large values of x.
   (b) Describe the graph and what you can conclude from it.

3. Use TABLE with ΔTbl set to a large number to investigate the value of y for large values of x Remember that if you move the cursor into the y column you can see the full value of the entry as shown in Figure 9.15.

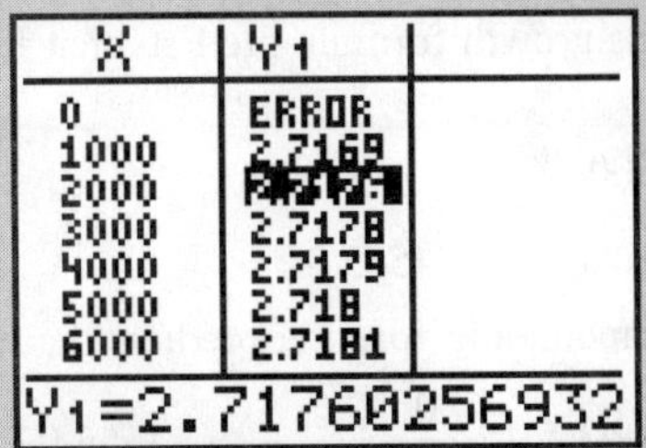

X	Y1	
0	ERROR	
1000	2.7169	
2000	2.7174	
3000	2.7178	
4000	2.7179	
5000	2.718	
6000	2.7181	
Y1=2.7176025693		

Figure 9.15

4. What are your conclusions from looking at the table?

Calculator Lab 9.4 should convince you that the limit L is a number around 2.718. You probably have reached an estimate that has more decimal places than that. However the calculator can only give you a *guess* about such a limit because when numbers get too large any computer or calculator will stop giving accurate answers. For example the *TI-83* reports that if $x = 1,000,000$ (one million) then $\left(1 + \dfrac{1}{x}\right)^x \approx 2.718280469$ and if $x = 5,000,000$ (five million) the estimate increases a bit to 2.718281557. If you increase x beyond ten million you will see that the values stop increasing and instead they oscillate. For example, when $x = 12,000,000$ we get $\left(1 + \dfrac{1}{x}\right)^x \approx 2.718280628$, for $x = 16,000,000$ we have $\left(1 + \dfrac{1}{x}\right)^x \approx 2.718281744$, and for $x = 17,000,000$ we see that $\left(1 + \dfrac{1}{x}\right)^x \approx 2.718280389$. This oscillation indicates that the limit of the calculator's accuracy has been reached and we can't get a better estimate than

$$L \approx 2.71828$$

As you may have guessed already the limit L turns out to be the number e. Therefore we have the following summary.

A Formula for Continuous Growth

If interest on a savings account is compounded a very large number of times per year the value A of the account after t years can be estimated by the formula

$$A(P, r, t) = Pe^{rt}$$

where P = original amount deposited, r = annual interest rate expressed as a decimal, and t = the time in years.

Example 9.19   Application

Estimate the amount that $2500 will become in 10 years if it is invested in an account paying 6% compounded daily. Compare the continuous growth formula with the compounding formula.

Solution   The continuous growth formula predicts that for daily compounding the amount will be

$$A = 2500e^{10 \cdot 0.06}$$

$$= 2500e^{0.6} \approx \$4555.30$$

The compounding formula predicts that the amount will be

$$A = 2500\left(1 + \frac{0.06}{365}\right)^{365 \cdot 10} \approx \$4555.07$$

The small difference, 23 cents over a period of 10 years, tells you why many people use the simpler continuous growth formula to get an estimate. Banks, however, use the compounding formula because they need to predict the exact amount in the account.

Continuous growth is also an accurate description of growth in nature. Look at the following example.

Example 9.20   Application

Some bacteria will increase at the rate of about 2% per minute in a laboratory dish. Suppose 1000 bacteria are introduced into the dish.

(a) Estimate the number of bacteria after 2 hours.
(b) Estimate the number of bacteria after 8 hours.
(c) Estimate the number after 3 days.

Solutions   Here $P = 1000$ and $r = 0.02$ so the formula for the bacteria present after t minutes is

$$A(t) = 1000e^{0.02t}$$

(a) We are given a time of 2 hours. Two hours is $2 \cdot 60 = 120$ minutes so $t = 120$ and $A(120) \approx 11023$. There will be about $11,023$ bacteria in the dish after two hours.
(b) Eight hours is $8 \cdot 60 = 480$ minutes so $t = 480$ and we obtain $A(480) \approx 14,764,781$. In eight hours the original 1000 bacteria will have grown to almost $14,765,000$ bacteria.
(c) Three days has $3 \cdot 24 \cdot 60 = 4320$ minutes and with $t = 4320$ we get $A(4320) \approx 3.3 \times 10^{40}$. ■

The answer to Example 9.20(c) is mathematically correct but we hope physically impossible. For example, one *E.coli* bacterium has a mass of 10^{-12} grams, and 3.3×10^{40} *E.coli* bacteria would have a mass of 3.3×10^{28} g $= 3.3 \times 10^{25}$ kg. This is about five times the Earth's mass of 6×10^{24} kg.

Escherichia coli (E.coli) is a dangerous bacteria most frequently acquired through consumption of undercooked ground beef and unpasteurized milk. According to the Center for Disease Control approximately 20,000 people are infected by E.coli each year. Approximately 200 deaths are reported annually. Source: http://www.sddt.com/ keeney_waite/files/ec-prim.html

Solving Exponential Equations

Logarithms simplify the process of solving equations in which the unknown is in the exponent. To demonstrate the process, we return to the bacteria of Example 9.20.

Example 9.21   Application

What is the doubling time of the bacteria in Example 9.20? The doubling time is the time required for twice as many bacteria to be present.

Solutions   There are several approaches that can be taken to this problem. We'll mention two of them.

(a) Find the value of t that satisfies the equation $1000e^{0.02t} = 2000$. This equation is the same as $e^{0.02t} = 2$ and can be solved by either of the following methods that we have used previously:
  ▶ Plot a graph of $y = e^{0.02x}$ and use TRACE to find the point where $y = 2$.
  ▶ Use a TABLE to find the value of x that makes $y = 2$.

(b) We can also use our new knowledge of logarithms to solve the exponential equation. The first step in solving exponential equations is to take a logarithm of both sides of the equation. Because the base is e we'll take the natural logarithm of both sides.

$$\ln\left(e^{0.02t}\right) = \ln 2$$

But natural logarithms are the exponent of e, so $\ln\left(e^{0.02t}\right) = 0.02t$. Therefore we have

$$0.02t = \ln 2$$

$$0.02t \approx 0.693$$

$$t = \frac{0.693}{0.02} \approx 35$$

It will take about 35 minutes for the number of bacteria to double.   ∎

Example 9.21 demonstrates how we can use logarithms to more easily solve exponential equations. Example 9.22 takes this one step further.

Example 9.22

Solve the equation $0.83 = \left(\frac{1}{2}\right)^n$ in Example 7.38 where $n =$ the number of half-lives of carbon 14 since the tree was cut down to construct the building in ancient Mexico.

Solution We can take either the natural logarithm or the common logarithm of both sides. We'll choose the common logarithm.

$$\log 0.83 = \log\left(\frac{1}{2}\right)^n$$

The multiplication law of logarithms states that $n \log a = \log a^n$. Therefore $\log\left(\frac{1}{2}\right)^n = n \log\left(\frac{1}{2}\right)$. So the equation becomes

$$\log 0.83 = n \log\left(\frac{1}{2}\right)$$

$$= n \log 0.5$$

$$n = \frac{\log 0.83}{\log 0.5} \approx 0.27$$

The half life of carbon 14 is 5700 years, so the estimated age of the building is

$$0.27 \times 5700 \approx 1539 \text{ years}$$

In Example 7.38 we found that the estimated age of the building is $0.27 \times 5700 \approx 1539$ years—the same answer we got using logarithms.   ∎

The final example of solving exponential equations involves the effective annual interest rate of a stock market investment.

Example 9.23 Application

In a magazine advertisement for the Kemper Technology Fund, a mutual fund that invests in various stocks in the technology field, the statement was made that an investment of $10,000 in 1948 would have become $3.9 million in 1997. What is the average effective annual interest rate of this investment?

Solution   The effective interest rate is the annual rate that would have produced the growth if it had been in effect every year. Obviously the stock market has its ups and downs, and there is never a promise of a specific interest rate. However, the effective rate can be used to compare the history of two different investments. If we assume continuous compounding the growth of the original investment can be approximated by the formula

$$A = Pe^{rt}$$

where $A = 3,900,000$, $P = 10,000$, and $t = 1997 - 1948 = 49$. We need to solve for r. The equation is

$$3,900,000 = 10,000e^{49r}$$

or $$390 = e^{49r}$$

Take the natural log of both sides of the equation $390 = e^{49r}$. Using the addition and multiplication properties of logarithms we have

$$\ln 390 = \ln e^{49r}$$
$$\ln 390 = 49r$$
$$\frac{\ln 390}{49} = r$$
$$0.12 \approx r$$

Thus $r \approx 12\%$ and since 1948 the average effective annual interest rate of the Kemper Technology Fund has been 12%.   ∎

Charging a Capacitor and the Number *e*

The exponential function that represents the charging of a capacitor was given at the beginning of Section 9.1. It is

$$C(n) = 100\left(1 - 0.37^n\right)$$

where $C(n)$ is the percent of the maximum charge of a capacitor after n time constants have elapsed.

The formula is the same for all capacitors although different capacitors have different time constants. What is so special about the number 0.37? It turns out that $0.37 \approx \frac{1}{e}$. Therefore the charging formula is actually

$$C(n) = 100\left(1 - \left(\frac{1}{e}\right)^n\right)$$
$$= 100\left(1 - e^{-n}\right)$$

Once again the number *e* has appeared in a real application.

Electrical engineers and technicians write the formula for $C(n)$ in terms of the elapsed time t since the charging began. They introduce the product RC, where R is the resistance and C is the capacitance. RC stands for a time constant for the capacitor. Then the elapsed time $t = nRC$, so $n = \dfrac{t}{RC}$. We have the following summary of the charging and discharging functions for a capacitor:

Charging and Discharging a Capacitor

The percent of a capacitor's maximum charge is given by $C(t)$.

▶ Discharging (See Example 7.31 for this derivation.)

$$C(t) = 100 \left(e^{-t/RC}\right)$$

▶ Charging

$$C(t) = 100 \left(1 - e^{-t/RC}\right)$$

where $t =$ elapsed time and $RC =$ time constant for the capacitor (R is the resistance and C is the capacitance) and where t and RC are measured in the same units of time.

Example 9.24   Application

Consider a capacitor with a 100% charge, $R = 400\ \Omega$, and $C = 12.5\ \mu\text{F}$,

(a) What percent has been discharged after 1.5 ms?
(b) How long does it take for the capacitor to discharge 75% of its original charge?

Solutions   We begin by evaluating RC. Since $R = 400\ \Omega$ and $C = 12.5\ \mu\text{F} = 12.5 \times 10^{-6}\ \text{F}$, we have

$$RC = (400)(12.5 \times 10^{-6}) = 0.005\ \text{s} = 5\ \text{ms}$$

Thus, we will be using the function $C(t) = 100e^{-t/5}$ where t is in milliseconds.

(a) After 1.5 ms the percent of the maximum charge is

$$C(t) = 100e^{-t/5}$$

$$C(1.5) = 100e^{-1.5/5}$$

$$\approx 74.08$$

After 1.5 ms the capacitor retains about 74% of its original charge.

(b) In order for the capacitor to discharge 75% of its original charge, we want to determine the value of t when $C(t) = 25$.

$$C(t) = 100e^{-t/5}$$

$$25 = 100e^{-t/5}$$

$$0.25 = e^{-t/5}$$

$$\ln 0.25 = \frac{-t}{5}$$

$$-5 \ln 0.25 = t$$

$$\approx 6.93$$

It takes about 6.93 milliseconds for the capacitor to lose 75% of its initial charge.

■

Converting any Exponential Function to an Exponential Function with Base *e*

The number *e* is widely used in mathematics and science, as we have seen. Sometimes it is useful to have a way to convert any exponential function to an exponential function with base *e*. We examine this process next.

Example 9.25

Convert the function $y = a \cdot b^{cx}$ to a function base *e*.

Solution   The constant *a* will remain the same. We will convert the exponential factor b^{cx} to e^{kx} by solving the equation $e^{kx} = b^{cx}$ for *k*. We have seen that to solve an equation that has the variable in the exponent, we must take a logarithm of both sides of the equation. But which logarithm (log or ln) should we use? The natural logarithmic function ln is usually the better choice when *e* is involved. Therefore we have the equation

$$\ln \left(e^{kx} \right) = \ln \left(b^{cx} \right)$$

The definition of the logarithm means that $\ln \left(e^{kx} \right) = kx$. The multiplication property of logarithms means that $\ln \left(b^{cx} \right) = cx \ln b$. Thus we have

$$kx = cx \ln b$$

which means that

$$k = c \ln b$$

Therefore the function $y = a \cdot b^{cx}$ can be restated as $y = a \cdot e^{(c \ln b)x}$ ■

Example 9.26 shows how to apply this result to radioactive decay.

Example 9.26   Application

The radioactive isotope strontium 90 (Sr^{90}) was found in increased amounts around the world after the accidental explosion in the Russian nuclear reactor at Chernobyl in 1986. The half-life of Sr^{90} is 25 years. For this problem suppose that 4 milligrams of Sr^{90} was estimated to have fallen on an area in Sweden.

(a) Write the function that shows how much remains of 4 mg of Sr^{90} *t* years after the accident.
(b) Convert this function to one that uses *e* as the base.
(c) Calculate how much of the 4 mg will remain in the year 2000.
(d) What percent of all the Sr^{90} that was released in 1986 will still remain in the year 2100?

Solutions   (a) The function $A(t)$ represents the amount remaining of 4 mg in t years. In t years a total of $\dfrac{t}{25}$ half-lives will have elapsed. We have seen this formula in Chapter 7 and again earlier in this chapter: $A(t) = 4 \cdot \left(\frac{1}{2}\right)^{t/25}$

(b) If $A(t) = a \cdot b^{ct}$, then $a = 4$, $b = \frac{1}{2}$, and $c = \frac{1}{25}$. Therefore the value of k in the expression e^{kt} is

$$k = c \ln b = \frac{1}{25} \ln \frac{1}{2} \approx -0.028$$

So the equation is

$$A(t) = 4 \left(\frac{1}{2}\right)^{t/25} = 4 \cdot e^{-0.028t}$$

(c) The year 2000 will be 14 years after the Chernobyl accident so we let $t = 14$. After 14 years the estimated amount remaining is

$$A(14) = 4 \cdot e^{-0.028 \cdot 14} \approx 2.70 \text{ mg}$$

(d) The year 2100 is 114 years after the accident. After 114 years the estimated amount remaining of the 4 mg is

$$A(14) = 4 \cdot e^{-0.028 \cdot 114} \approx 0.164 \text{ mg}$$

The percent remaining can be estimated as $\frac{0.164}{4} = 0.041 = 4.1\%$ ■

Exponential Damping

In Section 3.5 we discussed the motion of two Slinkies. The graph of the larger Slinky completed about five cycles in 10 seconds. The graph from that experiment was shown in Figure 3.95 and is reproduced in Figure 9.16. You can see in Figure 9.16 that the amplitude decreases between $t = 0$ and $t = 10$. The vibration of the Slinky is slowing down during that time interval. All vibrations, whether from an oscillating spring (or a Slinky), a swinging pendulum, the height of a bouncing ball, or a plucked guitar string slow down eventually. These are all examples of **exponential damping** and the models are known as **damped simple harmonic motion**.

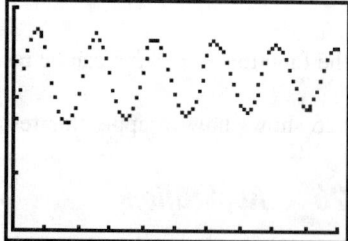

Figure 9.16

Damped Simple Harmonic Motion

A mathematical model for damped simple harmonic motion is

$$y = ae^{-mt} \sin (2\pi f t + D)$$

or $$y = ae^{-mt} \cos (2\pi f t + D)$$

where the positive constant m is called the **damping factor** and $\dfrac{1}{f}$ is the **quasiperiod** or, simply, the period.

Example 9.27

A weight is attached to a spring, pulled below the equilibrium position $y = 0$, and then released at $t = 0$. The position $s(t)$ of the weight at time t is given by the equation $s(t) = -4e^{-0.25t} \cos \pi t$ cm. Determine (a) the maximum displacement from $y = 0$, (b) the frequency, and (c) the displacement from equilibrium after one cycle.

Solutions   The graph of s shown in Figure 9.17 lies between the graphs of $y = -4e^{-0.25t}$ and $y = 4e^{-0.25t}$. These two bounding curves are shown in Figure 9.18 along with the graph of s between them.

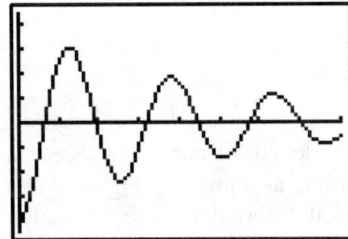

```
Xmin = 0
Xmax = 2π
Xscl = π/4
Ymin = -4.5
Ymax = 4.5
Yscl = 1
```

Figure 9.17

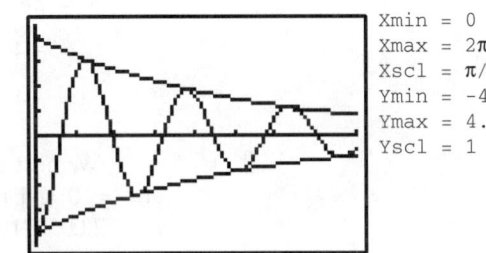

```
Xmin = 0
Xmax = 2π
Xscl = π/4
Ymin = -4.5
Ymax = 4.5
Yscl = 1
```

Figure 9.18

(a) The maximum displacement occurs at $t = 0$, the time the object is released. Since $s(0) = -4$ the maximum displacement is 4 cm below the equilibrium position.

(b) Since $2\pi f = \pi$ then $f = \frac{1}{2} = 0.5$ oscillation per second and the period is $\dfrac{1}{f} = 2$ seconds.

(c) Since the period is 2 seconds the displacement after one cycle is found at $t = 2$. It is $s(2) = -4e^{-0.5} \cos 2\pi = -4e^{-0.5} \approx -2.426$ cm. or about 2.4 cm below equilibrium.   ∎

Now let's examine another situation and assume we have data from a Slinky. We want to determine an approximate equation for such a damped simple harmonic function.

Example 9.28

Use the data from a previous experiment to approximate the damped simple harmonic function represented by the two points on the dotted curve in Figures 9.19 and 9.20.

Solution   By using the TRACE function of the calculator we find the two data points shown in Figures 9.19 and 9.20. This appears to be a damped simple harmonic function so we

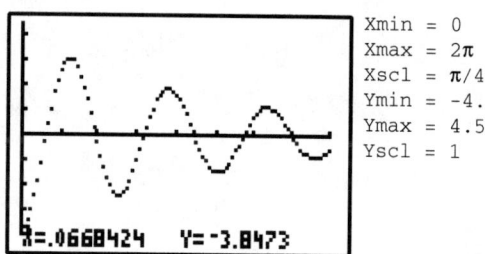

Xmin = 0
Xmax = 2π
Xscl = π/4
Ymin = -4.5
Ymax = 4.5
Yscl = 1

X=.0668424   Y=-3.8473

Figure 9.19

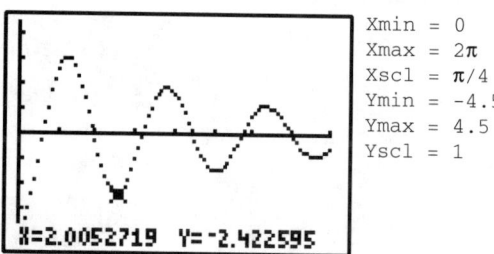

Xmin = 0
Xmax = 2π
Xscl = π/4
Ymin = -4.5
Ymax = 4.5
Yscl = 1

X=2.0052719   Y=-2.422595

Figure 9.20

know that it is of the form $y = ae^{-mt} \sin(2\pi ft + D)$ or $y = ae^{-mt} \cos(2\pi ft + D)$. Looking at the curves, we select the cosine function with $D = 0$.

For simplicity's sake we have to make some approximations. We do not have the value when $t = 0$. However, when $t = 0.0668424$ we have $y = -3.8473$. The graph is increasing at this point so it would seem reasonable that when $t = 0$ $y < -3.8473$. A reasonable choice for y at $t = 0$ would be -4. This is the value of a in our equation. So far we have

$$y = ae^{-mt} \cos(2\pi ft + D)$$
$$= -4e^{-mt} \cos(2\pi ft + 0)$$
$$= -4e^{-mt} \cos(2\pi ft)$$

We next try to determine the length of one cycle. Since we have a low point at $t = 0$ and the next lowest point, as shown in Figure 9.20, seems to be when $t = 2.0052719$, it appears as if the period is approximately $2.0052719 \approx 2.00$ seconds. Thus $\dfrac{1}{f} = 2$ and $f = \frac{1}{2} = 0.5$. So now the equation approximates

$$y = -4e^{-mt} \cos(2\pi 0.5t)$$
$$= -4e^{-mt} \cos \pi t$$

Finally we need to approximate the value of m. Using Figure 9.20 we solve $y = -4e^{-mt} \cos \pi t$ when $t = 2.0052719$ and $y = -2.422595$.

$$y = -4e^{-mt} \cos \pi t$$
$$-2.422595 = -4e^{-2.0052719m} \cos(\pi \cdot 2.0052719)$$
$$-2.422595 = -4e^{-2.0052719m}(0.9998628505)$$
$$0.6057318258 = e^{-2.0052719m}$$
$$\ln 0.6057318258 = -2.0052719m$$
$$\frac{\ln 0.6057318258}{-2.0052719} = m$$
$$0.250 \approx m$$

Although we show all the work, use your calculator to perform the calculations.

Thus if we let $m = 0.25$ we have an approximation for the damped simple harmonic function of $y = -4e^{-0.25t} \cos \pi t$. ∎

The Slope Function for Exponential and Logarithmic Functions

We now return to the slope function, which plays a major role in engineering and science. The slope function is important because it measures how fast the value of a function is changing at a particular point. In this section we will extend your experience with slope functions to include exponential and logarithmic functions.

You have studied five kinds of functions in this chapter. There are three kinds of exponential functions

$$f_1(x) = a^{bx}$$
$$f_2(x) = a^{-bx}$$
$$f_3(x) = 1 - a^{-bx}$$

with $a > 1$ and $b > 0$. We have looked at two logarithmic functions

$$f_4(x) = a \log bx$$
$$f_5(x) = a \ln bx$$

with $b > 0$.

You have worked with several variations of exponential functions, including those with $a = 10$, $a = 2$, $a = \frac{1}{2}$, and $a = e \approx 2.71828$. The logarithmic functions you saw had $a = 1$ and $b = 1$. You should be able to sketch graphs of any of the functions.

Activity 9.8
Sketching Slope Functions

For each of the functions f_1 through f_5, sketch a graph of the function and of the slope function. Use a graph like the one shown in Figure 9.21. To produce your sketches of the slope function, think about the answer to these questions:

▶ For what values of x is the slope positive?
▶ For what values of x is the slope negative?
▶ Does the slope equal or approach zero anywhere?

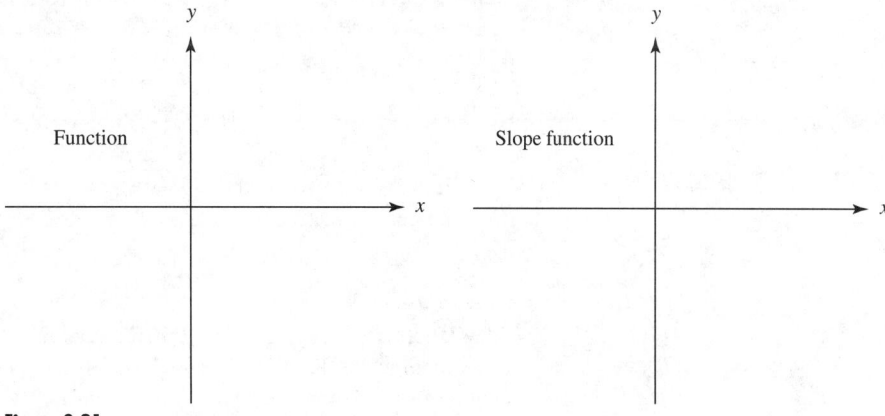

Figure 9.21

In Figure 9.22 we completed this task for f_2. The sketch of f_2 shows the characteristic shape of the negative exponential function and gives the coordinates of the y-intercept. The slope function shows that the slope of f_2 is always negative and gets closer and closer to zero as x gets larger. The slope at $x = 0$ is unclear so we did not show the y-intercept coordinates of the slope function graph. What shape does the slope function appear to have?

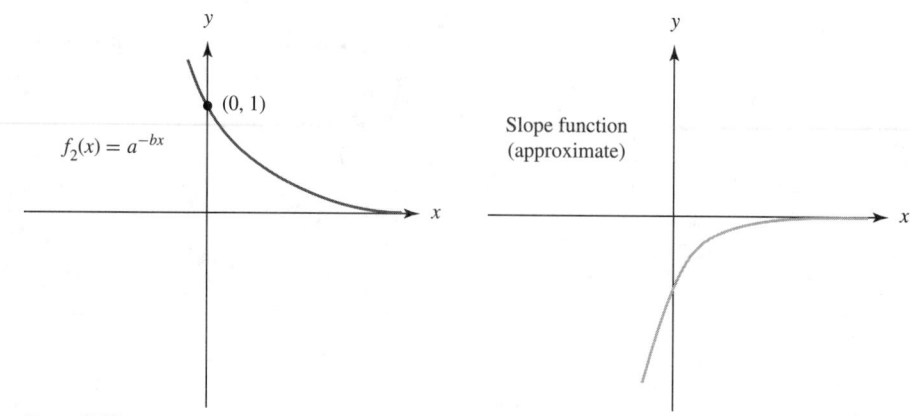

Figure 9.22

From the sketches in Activity 9.8 you can begin to see patterns in slope functions for exponential and logarithmic functions. We have developed a calculator program that can help you explore these relationships in more detail. In Calculator Lab 9.5 you will use that program.

Calculator Lab 9.5

Exploring Slope Functions

The program OTHERFUN is similar to the program POWERFUN, which you used to study power functions in Chapter 8. Your goal in this lab is to find relationships between the exponential and logarithmic functions and their slope functions. You should be able to answer the following questions:

1. What kind of function is the slope function for exponential functions?
2. What kind of function is the slope function for logarithmic functions?
3. How does the slope function for a^{-x} compare with the slope function for a^x?
4. How does the slope function for $\log x$ compare with the slope function for $\log 2x$?
5. The simplest functions based on the number e are $\ln x$ and e^x. Can you write the equation for their slope functions? Your answer gives a hint at why e is so important in math and science.
6. What can you say about the slope functions of the sine and cosine functions?

Procedures

1. Run the program called OTHERFUN. The menu screen shown in Figure 9.23 will appear. Notice that in addition to exponential and logarithm functions, two trigonometric functions are included for your experimentation. Also, option 4 is a form of the capacitor charging curve.
2. Select the function type that you want to investigate. If you select #1, a screen like Figure 9.24 will appear.

Figure 9.23

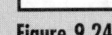

Figure 9.24

Figure 9.25

3. Type the values of the coefficients *A* and *B*, and press ⟨ENTER⟩ after each entry. Remember that the number *e* is listed above the ⟨÷⟩ key.

4. The next part is similar to what you saw in the program POWERFUN. The program draws tangents to the curve and then plots the slope function. Press ⟨ENTER⟩ whenever you see the four vertical dots in the upper right corner of the screen.

5. When you are ready to guess an expression for the slope function, select option 2 on the menu that follows the graphing. You will see the menu of function types shown in Figure 9.25. Notice that in addition to the trig, log, and exponential functions, option 4 is a power function which is the slope function for several of the functions studied in this program.

Section 9.3 Exercises

1. *Finance* $5000 is placed in a savings account at $6\frac{1}{2}\%$ interest. What is the total after 5 years if the interest is compounded (a) annually, (b) monthly, and (c) continuously?

2. *Finance* If $7500 is invested in an account that pays 5.75% interest compounded continuously, how much can we expect to have after 10 years?

3. *Biology* A bacteria culture originally numbers 800. It is known that this particular bacteria increases at the rate of about 0.5% per hour.
   (a) Estimate the number of bacteria after 2 hours.
   (b) Estimate the number of bacteria after 8 hours.
   (c) Estimate the number after 3 days.

4. *Biology* The number of bacteria in a certain culture increases from 5,000 to 15,000 in 20 hours. If we assume these bacteria grow exponentially:
   (a) How many will be there after 10 hours?
   (b) How many can we expect after 30 hours?
   (c) How many can we expect after 3 days?

5. *Ecology* The population of a certain city is increasing at the rate of 7% per year. The present population is 200,000.
   (a) What will be the population in 5 years?
   (b) What can the population be expected to reach in 10 years?

6. *Ecology* In 1980 the population of Canada was about 24,400,000 and by 1996 it had increased to about 28,821,000. If we assume that the population of Canada is growing exponentially
   (a) What will the population be in the year 2010?
   (b) Compute the constant *k* in this equation for exponential growth?
   (c) How long will it take for the population of Canada to double its 1980 population?

7. *Nuclear technology* The number of milligrams of a radioactive substance present after *t* years is given by $Q = 125e^{-0.375t}$.
   (a) How many milligrams are present after 1 year?
   (b) How many milligrams are present after 16 years?

8. *Nuclear technology* Radium decays exponentially and has a half-life of 1,600 years. How much of 100 mg will be left after 2,000 years?

9. *Medical technology* A pharmaceutical company is growing an organism to be used in a vaccine. The organism grows at a rate of 4.5% per hour. How many units of this organism must they begin with in order to have 2,000 units at the end of 7 days?

10. *Electricity* A circuit contains a resistance *R*, a voltage *V*, and an inductance *L*. The current *I* at *t* seconds (s)

after the switch is closed is given by

$$I = I_0 \left(1 - e^{-t/T}\right)$$

where I_0 is the steady state current $\dfrac{V}{R}$ and $T = \dfrac{L}{R}$. If a circuit has $V = 120\,\text{V}$, $R = 40\,\Omega$, and $L = 3.0\,\text{H}$ (henrys), determine the current in the circuit **(a)** 0.01 s, **(b)** 0.1 s, and **(c)** 1.0 s after the connection is made.

11. *Electricity* A 130-μF capacitor is charged by being connected to a 120-V circuit. The resistance is $4\,500\,\Omega$. What is the charge on the capacitor 0.5 s after the circuit is disconnected?

12. *Nuclear technology* The half-life of tritium is 12.5 years. How much of 100 g will remain after 40 years?

13. *Medical technology* A radioactive material used in radiation therapy has a half-life of 5.4 days. This means that the radioactivity decreases by one-half each 5.4 days. A hospital gets a new supply that measures 1200 microcuries (μCi). How much of this material will still be radioactive after 30 days?

14. *Medical technology* The average doubling time for a breast cancer cell is 100 days. It takes about 9 years before a lump can be seen on a mammogram and 10 years before a breast cancer is large enough to be felt as a lump. (Use 365.25 days = 1 year.)
    1. How many cancer cells are in a lump that is just large enough to be seen on a mammogram?
    2. How many cancer cells are in a lump that is just large enough to be felt?
    3. If a lump must be approximately 1 cm in diameter before it can be felt, how large is each cancer cell? (Assume the lump is a perfect sphere.)

15. *Electronics* In a capacitive circuit the equation for the current in amperes is given by

$$i = \frac{V}{R}e^{-t/RC}$$

where t is elapsed time in seconds after the switch is closed, V is the impressed voltage, C is the capacitance of the circuit in F, and R is the circuit resistance in Ω. A capacitance of 500 μF in series with 1 kΩ is connected across a 50-V generator.
    1. What is the value of the current at the instant the switch is closed?
    2. What is the value of the current 0.02 s after the switch is closed?
    3. What is the value of the current 0.04 s after the switch is closed?

16. *Lighting technology* When a beam of light with an initial intensity I_0 enters a medium such as water or smoky air, its intensity decreases based on the thickness or concentration of the medium. The intensity I at a

depth of x units is given by

$$I = I_0 e^{-\mu x}$$

where μ, the **coefficient of absorption**, varies with the medium. For example, light through a certain density of fog has $\mu \approx 1.004$ if x is in meters.
    (a) What percentage of I_0 remains at a distance of 3.5 meters in this type of fog?
    (b) If 50% of I_0 remains in this fog, what is the value of x?

17. *Business* A business estimates that the salvage value V in dollars of a desktop computer after t years is given by $V(t) = 3500e^{-0.75t}$.
    (a) What was the original cost of the computer?
    (b) What is the value of the computer after three years?

18. *Meteorology* The atmospheric pressure P, in pounds per square inch (psi or lb/in.^{2}), at an altitude h, in miles above sea level, can be approximated from the formula

$$P = 14.7e^{-0.21h}$$

for $0 \le h \le 12$.
    (a) What is the atmospheric pressure on the top of Mount Everest, which is 29,028 feet high?
    (b) The atmospheric pressure on the top of a certain mountain is about 8.25 lb/in.^{2}. How high is the mountain?

19. *Ecology* A reservoir has become polluted due to an industrial waste spill, causing a buildup of algae. The number of algae $N(t)$ present per 1000 gallons of water t days after the spill is given by the formula

$$N(t) = 100e^{1.97t}$$

How long will it take before the algae count reaches 25,000?

 20. The number e can be approximated by the Maclaurin series

$$e = 2 + \frac{1}{2} + \frac{1}{2 \cdot 3} + \frac{1}{2 \cdot 3 \cdot 4} + \frac{1}{2 \cdot 3 \cdot 4 \cdot 5} \cdots$$

$$= 2 + \frac{1}{2!} + \frac{1}{3!} + \frac{1}{4!} + \frac{1}{5!}$$

(Remember that $3! = 3{\cdot}2{\cdot}1 = 6$ and $4! = 4{\cdot}3{\cdot}2{\cdot}1 = 24$. In general, $n! = n \cdot (n-1) \cdot (n-2) \cdots 4 \cdot 3 \cdot 2 \cdot 1$.) Calculate e to four figures by adding the first six terms of this series.

21. A weight is attached to a spring, pulled below the equilibrium position $y = 0$, and then released at $t = 0$. The position $s(t)$ of this object at time t is given by the equation $s(t) = -3.5e^{-0.15t}\cos 0.5\pi t$ cm. Determine **(a)** the maximum displacement from $y = 0$, **(b)** the frequency, and **(c)** the displacement from equilibrium after one cycle.

22. A technician attached a weight to a spring, pulled the weight *above* the equilibrium position $y = 0$, and then released it at $t = 0$. Using a CBL the following data was collected:

t	0	0.1	0.2	0.3	0.4	0.5	0.6	0.7	0.8	0.9	1.0	1.1	1.2	1.3	1.4	1.5
y	2.40	1.90	0.70	−0.70	−1.80	−2.15	−1.70	−0.65	0.65	1.60	1.95	1.55	0.60	−0.55	−1.45	−1.80

t	1.6	1.7	1.8	1.9	2.0	2.1	2.2	2.3	2.4	2.5	2.6	2.7	2.8	2.9	3.0	3.1
y	−1.40	−0.55	0.50	1.35	1.60	1.30	0.50	−0.45	−1.20	−1.45	−1.15	−0.45	0.40	1.10	1.30	1.05

(a) Plot the given data on your calculator.

(b) Determine the damped simple harmonic function that approximates these data.

(c) Graph your answer to (b) over your graph from (a).

 23. Use a CBL set up like the one shown in Figure 9.26. Attach a spring, such as a Slinky or a bungee cord, directly over the motion detector. To make it easier for the motion detector to "see" the spring, attach a piece of cardboard about 2″–3″ square to the bottom of the spring. The object should be light enough so that it does not noticeably stretch the spring. As an alternative, attach a smaller mass but tape a piece of cardboard to the bottom of the mass. The cardboard will serve as a target for the motion detector. Pull the spring straight down, turn on the CBL, and start the program SLINKY on the calculator. Release the spring and allow it to oscillate. As soon as the motion of the spring is only up and down (with no side motion), press **TRIGGER** on the CBL. Use the resulting graph and data to approximate the damped simple harmonic function.

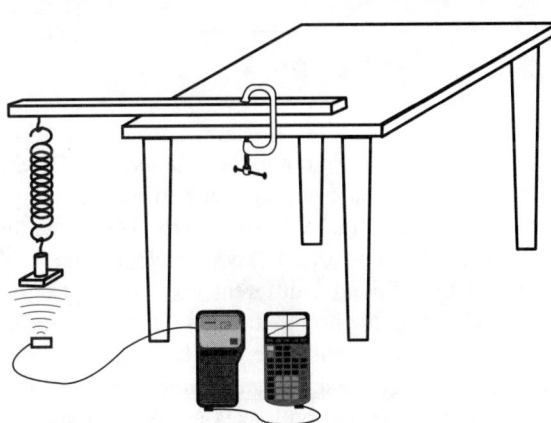

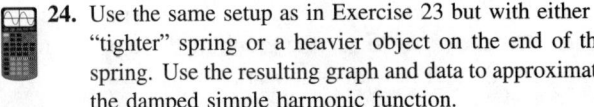

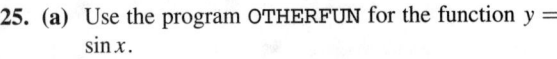

Figure 9.26

24. Use the same setup as in Exercise 23 but with either a "tighter" spring or a heavier object on the end of the spring. Use the resulting graph and data to approximate the damped simple harmonic function.

25. (a) Use the program OTHERFUN for the function $y = \sin x$.

(b) Based on the plots of the slopes of the tangents to $y = \sin x$, what function do you think de-

scribes the slopes of the tangent of the sine function?

(c) Check your answer to (b) with the program OTHERFUN.

26. (a) Use the program OTHERFUN for the functions $f_1(x) = \sin 2x$ and $f_2(x) = \sin 0.5x$.

(b) What functions do you think describe the slopes of the tangent to f_1 and f_2?

(c) Check your answer to (b) with the program OTHERFUN.

27. (a) Use the program OTHERFUN for the function $g(x) = \cos x$.

(b) What functions do you think describe the slopes of the tangent to g?

(c) Check your answer to (b) with the program OTHERFUN.

28. (a) Use the program OTHERFUN for the functions $g_1(x) = \cos 2x$ and $g_2(x) = \cos 0.5x$.

(b) What functions do you think describe the slopes of the tangent to g_1 and g_2?

(c) Check your answer to (b) with the program OTHERFUN.

29. Make a copy of OTHERFUN and save it under the name OTHRFUN2. Modify the program OTHRFUN2 so it will graph the slopes of the tangents to $A \sin(Bx - C)$ and $A \cos(Bx - C)$.

30. Suppose that you are shown the graphs of $y = e^x$, $y = 2^x$, $y = x^2$, and $y = x^3$, but the graphs are not labeled. Explain how you would distinguish among the graphs.

31. Write a report to management in which you describe the progress you have made toward the request in their memo on page 608. In your report you should describe how you would prove that you can accurately predict the time needed for a metal rod to cool from over 500° to a temperature in the desired range. In particular tell how you would be able to estimate the cooling time given the starting temperature and the final temperature.

9.4

Using Logarithmic Scales

If you want to graph a set of numbers that covers a wide range of values a base 10 **logarithmic scale** is often useful. For example let's look at two ways to plot the population of the United States over a period of 350 years, since the earliest colonial times.

The population estimates range from a low of 4000 in 1630* to a total of about 250 million in 1990. Figure 9.27 shows the data displayed in a normal Cartesian coordinate system.

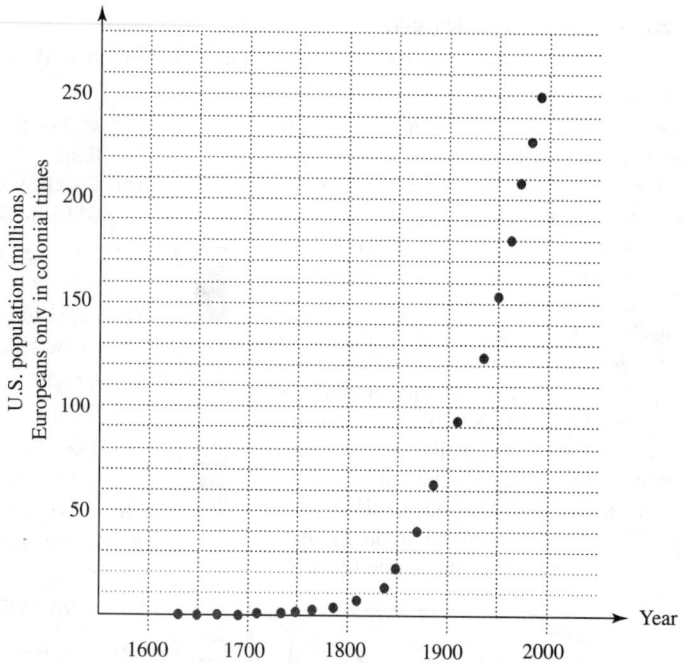

Figure 9.27

You can see that it is difficult to see any pattern in the data until 1800, when the population first exceeded 5 million. On a scale large enough to include the present population, the early numbers are not distinguishable from zero. The curve through these points does have the characteristic shape of an exponential function.

Figure 9.28 shows the population plotted a different way. In Figure 9.28 the vertical axis uses a logarithmic scale while the horizontal axis retains the linear scale of the Cartesian coordinate system. A graph with one axis based on a logarithmic scale and the other on a linear scale is called a **semilogarithmic** or **semi-log graph**. On a Cartesian or rectangular coordinate system both axes use a linear scale.

The graph in Figure 9.28 reveals more about the population growth before 1800 than does Figure 9.27. Although population in this century has increased quite rapidly, the *rate* of the increase (the *slope function*!) appears to be slowing. The use of a logarithmic scale in Figure 9.28 allows us to see the graph of these points more clearly. What is this new kind of scale and how does it work?

---

*The early population estimates do not include Native Americans.

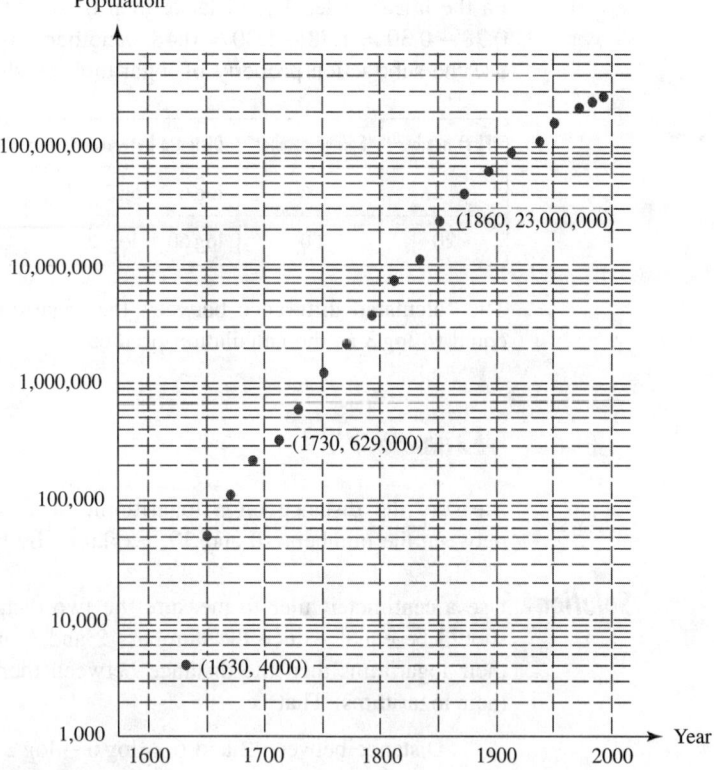

Figure 9.28

The Mathematics of the Logarithmic Scale

The scale used in the Cartesian coordinate system is called linear because there is constant spacing between consecutive whole numbers. In the logarithmic scale used for the vertical axis in Figure 9.28 there is constant spacing between consecutive powers of 10. Since the logarithm is only defined for positive numbers a logarithmic scale has no zero point as shown in Figure 9.29. It is therefore not possible to plot positive and negative numbers on the same graph.

Let's look at how the numbers between the powers of 10 are spaced on the logarithmic scale. Figure 9.30 shows a logarithmic scale and a linear scale below it.

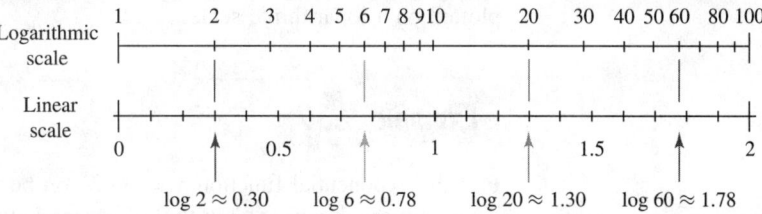

Figure 9.30

First notice that in Figure 9.30 each number (such as 2) on the logarithmic scale is directly above its logarithm ($\log 2 \approx 0.30$) on the linear scale. On the logarithmic scale, look at the distance between two points, 2 and 6, and compare it to the distance between 20 and 60. They are the same distance apart. Why? Each number on the logarithmic scale is located according to the position of its logarithm

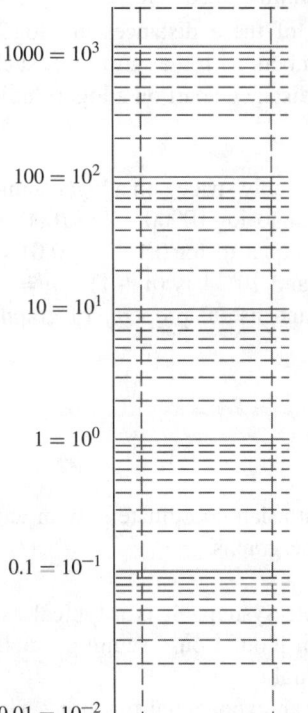

Figure 9.29

on the linear scale. The distance then is the difference between the logarithms, and $0.78 - 0.30 = 1.78 - 1.30 = 0.48$. Another way to talk about these differences is to use the subtraction property of logarithms as shown in the table below.

First number	Second number	Distance between First and Second Numbers on Logarithmic Scale
2	6	$\log 6 - \log 2 = \log \frac{6}{2} = \log 3$
20	60	$\log 60 - \log 20 = \log \frac{60}{20} = \log \frac{6}{2} = \log 3$

The calculated difference between the logarithms is 0.48, which is approximately equal to $\log 3$ as the calculation predicts.

Example 9.29

Measure the distance on a logarithmic scale between the numbers 2 and 6 and between the numbers 10 and 30. Explain why the results are equal.

Solution   Use a centimeter ruler to measure the two distances in Figure 9.30. The distances should be equal. Since the numbers 2 and 6 are plotted according to the value of their logarithms then the distance between them should be the difference between their logarithms. That is

Distance between 2 and 6 $= \log 6 - \log 2$

and

Distance between 10 and 30 $= \log 30 - \log 10$

But because of the subtraction property of logarithms, $\log 6 - \log 2 = \log \frac{6}{2} = \log 3$, and $\log 30 - \log 10 = \log \frac{30}{10} = \log 3$. Both of these distances are $\log 3$. Therefore the distances are equal. In fact any two numbers with the ratio of 3, such as 75 and 25 or 36000 and 12000, would be the same distance apart on a logarithmic scale. ∎

Example 9.29 explains why the powers of 10 are evenly spaced on a logarithmic scale. The distance between 1000 and 10000 is equal to $\log 10000 - \log 1000 = 4 - 3 = 1$, and the distance between 0.01 and 0.1 is equal to $\log 0.1 - \log 0.01 = -1 - (-2) = 1$. In general the distance between 10^n and 10^{n+1} is $(n + 1) - n = 1$.

In the next example you will see a surprising and useful property of graphs plotted on a logarithmic scale.

Example 9.30

Plot the exponential function $y = 3 \cdot 2^x$ on both a Cartesian coordinate system and a semilog graph and explain the difference between the graphs.

Solution   Figure 9.31 shows the graph on a Cartesian coordinate system. It is a typical exponential curve. Figure 9.32 shows the same function plotted on a semilog graph. This time the graph of the function $y = 3 \cdot 2^x$ is a straight line.

Why does the logarithmic scale "straighten out" an exponential function? The secret was stated earlier: plotting numbers on a logarithmic scale is the same as

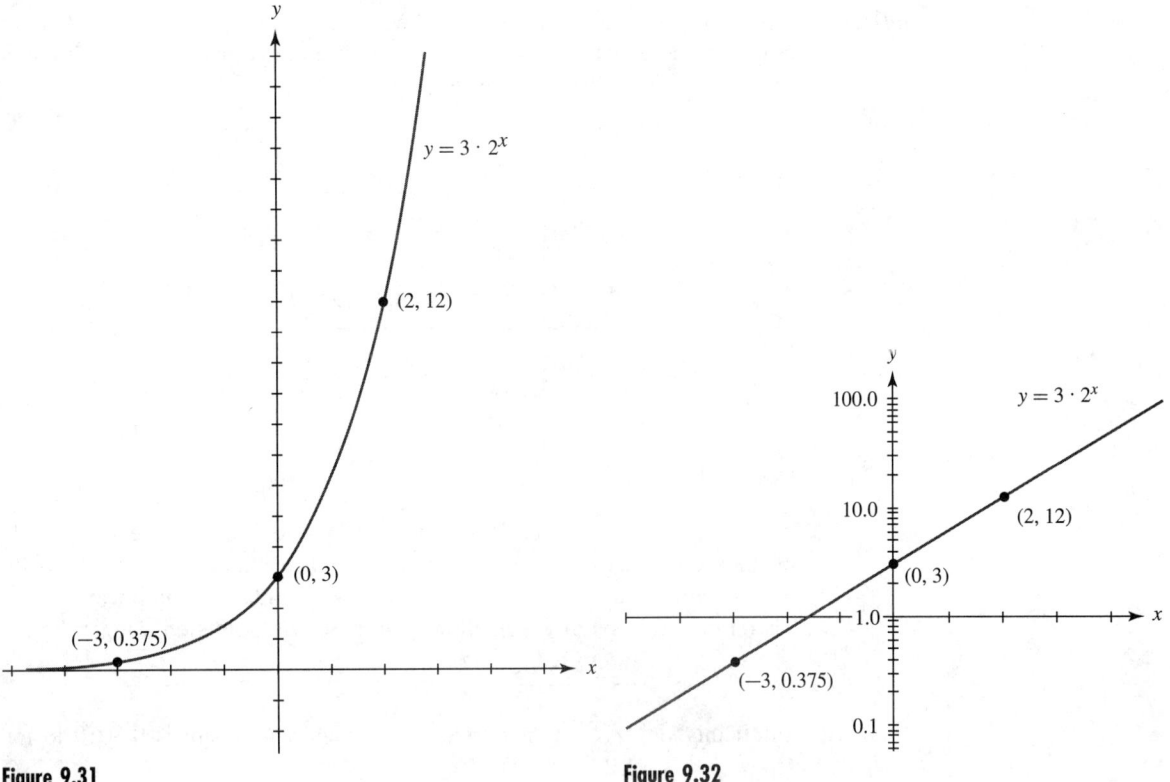

Figure 9.31

Figure 9.32

plotting the logarithms of those numbers on a linear scale. If we transform the function by taking the logarithm of both sides of the equation, we can see why a straight line results:

$$y = 3 \cdot 2^x$$

$$\log y = \log(3 \cdot 2^x)$$

And using the addition and multiplication laws of logarithms we have

$$\log y = \log 3 + \log 2^x$$

$$\log y = \log 3 + x \log 2$$

The last equation is a form of the straight line equation $y = b + mx$. It indicates that if we plot $\log y$ vs. x then a straight line results, with slope of $\log 2$ and y-intercept at $\log y = \log 3$ or $y = 3$. ∎

In Figure 9.33 you can see that the y-intercept is $y = 3$. (What seems to be the x-axis in Figure 9.33 is the line $y = 1$ not $y = 0$.) It is not so obvious that the slope is $\log 2$. We will study the slope in Example 9.31.

Example 9.31

Show that the slope of the straight line in Figure 9.32 is $\log 2$. Do this by (a) using the calculator to compute the slope and (b) using properties of logarithms to compute the slope.

Solutions   To compute the slope of a straight line we need two points. Figure 9.33 shows two points on the graph of $y = 3 \cdot 2^x$. We will use the points $P(-3, 0.375)$ and $Q(2, 12)$ to calculate the slope of the line.

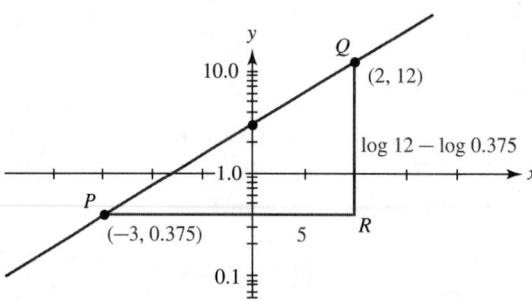

Figure 9.33

The vertical distance RQ between the points is the difference of the logarithms of the y-coordinates $\log 12 - \log 0.375$. The horizontal distance is the difference in the x-coordinates $2 - (-3) = 5$ since the horizontal axis is linear not logarithmic. Therefore the slope of the line between these two points on a logarithmic scale is

$$\text{Slope} = \frac{\log 12 - \log 0.375}{5}$$

(a) If this slope is computed with a calculator the result is 0.3010 to four places, which is $\log 2$.

(b) Using a calculator only for division we can apply the laws of logarithms to the vertical change as follows:

$$\log 12 - \log 0.375 = \log \frac{12}{0.375} = \log 32$$
$$= \log(2 \cdot 2 \cdot 2 \cdot 2 \cdot 2)$$
$$= 5 \log 2$$

Therefore the slope is

$$\text{Slope} = \frac{5 \log 2}{5} = \log 2 \qquad \blacksquare$$

To prove that the slope is the same for *any two points* on the line is a little more difficult. Example 9.32 shows you how to do that.

Example 9.32

Show that if the function $y = 3 \cdot 2^x$ is plotted on a semilog graph the slope between any two points is $\log 2$.

Solution   We choose two points P and Q with x-coordinates equal to a and b. Therefore the y-coordinates are $3 \cdot 2^a$ and $3 \cdot 2^b$. The two points are shown on the graph in Figure 9.34.

The slope between these two points when plotted on a semi-logarithmic scale is

$$\text{slope} = \frac{\log(3 \cdot 2^b) - \log(3 \cdot 2^a)}{b - a}$$

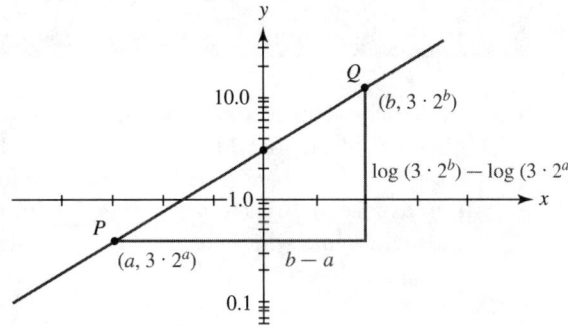

Figure 9.34

But $\log(3 \cdot 2^b) = \log 3 + \log 2^b = \log 3 + b \log 2$. Therefore we have

$$\text{slope} = \frac{\log 3 + b \log 2 - (\log 3 + a \log 2)}{b - a}$$

$$= \frac{b \log 2 - a \log 2}{b - a} = \frac{(b - a)(\log 2)}{b - a}$$

$$= \log 2$$

Since the slope does not depend on the choice of a or b the slope is the same everywhere on the line. ■

Finally we look at the general exponential function $y = a \cdot e^{kx}$ and ask what the slope of the line will be when this function is plotted on a logarithmic scale.

Example 9.33

Show that if an exponential function $y = a \cdot e^{kx}$ is plotted on a logarithmic scale the result is a straight line with y-intercept $(0, a)$ and slope equal to $k \log e$.

Solution

We need to use the common logarithm here, even though we are dealing with an equation with base e, because almost all logarithmic scales use base 10.

We take two points, (x_1, y_1) and (x_2, y_2), on the curve. The slope of the line between the two points is:

$$\text{Slope} = \frac{\log y_2 - \log y_1}{x_2 - x_1}$$

We will use the same approach you saw in Example 9.32.

$$\log y_2 = \log(a \cdot e^{kx_2}) = \log a + kx_2 \log e$$

$$\log y_1 = \log(a \cdot e^{kx_1}) = \log a + kx_1 \log e$$

$$\log y_2 - \log y_1 = \log a + kx_2 \log e - (\log a + kx_1 \log e) = (x_2 - x_1)k \log e$$

$$\text{Slope} = \frac{(x_2 - x_1)k \cdot \log e}{x_2 - x_1} = k \log e$$

If $x = 0$ you obtain $y = ae^{k \cdot 0} = ae^0 = 1$. Thus the y-intercept is $(0, a)$. ■

We have been working with a graph that uses the logarithmic scale on the vertical axis and a linear scale on the horizontal axis. A graph set up this way is said to have **semilogarithmic** or **semilog** axes. If a logarithmic scale is used on both axes, then the graph is said to have **log-log** axes and the graph is called a logarithmic or **log-log** graph. The following is a summary of the properties of semilog axes.

Plotting on Semilog Axes

▶ The spacing between two numbers on a logarithmic scale is equal to the difference of the logarithms of the two numbers.
▶ The powers of 10 are equally spaced on a logarithmic scale.
▶ If an exponential function $y = a \cdot e^{kx}$ is plotted on semilog axes the result is a straight line with intercept $(0, a)$ and slope equal to $k \log e \approx 0.434k$.

Before we turn to graphing with semilog axes on the calculator we'll explore log-log axes, which cannot be demonstrated on the *TI-83*. In the next activity you will explore the patterns in the graphs of power functions on log-log axes.

Activity 9.9
Graphing on Log-log Axes

You need a single sheet of log-log graph paper like the one shown in Figure 9.35. Set up the coordinates from 1 to 100 on the *x*-axis and 1 to 1000 on the *y*-axis. Make a table of four points for each of the following power functions. Plot each set of four points and draw a line connecting them. Label your graphs. Check your calculations if any of your lines is not straight.

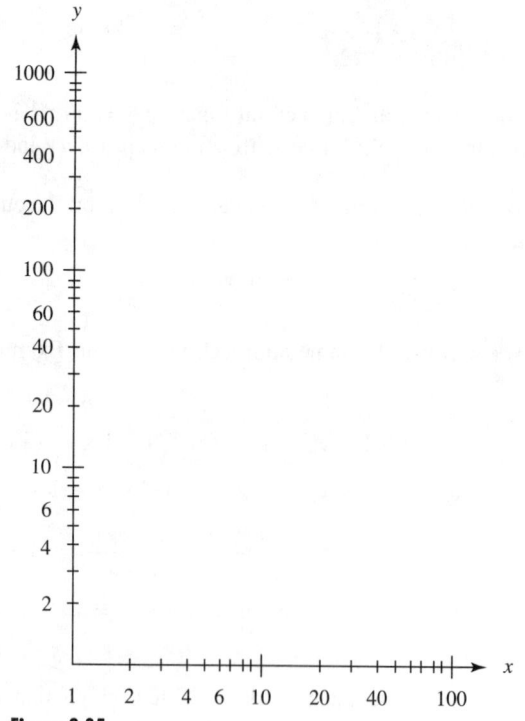

Figure 9.35

The functions are (a) $y = x$, (b) $y = 2x$, (c) $y = x^2$, (d) $y = 3x^2$, (e) $y = x^3$, and (f) $y = 2x^3$. Answer these questions about the graphs:

1. Describe the differences and similarities between the functions with the same exponent and different coefficients. For example how is $y = x^2$ the same and different compared with $y = 3x^2$?
2. Describe the differences and similarities between the functions with the same coefficient and different exponents. For example how is $y = 2x$ the same and different compared with $y = 2x^3$?
3. Why should power functions have straight line graphs on log-log axes? *Hint:* take the logarithm of both sides of the equation and simplify with the properties of exponents.
4. What can you predict about the graph of the function $y = 10x^6$ on the log-log paper? Draw the graph to check your answer.

Activity 9.9 should help you see that log-log graphs are especially useful in graphing power functions, that is, functions of the form $y = ax^b$. The graph of a power function appears as a straight line when graphed on log-log axes.

Using Graphs to Help Determine an Equation

If the graph of data is a straight line on semilog paper then the data are related by an exponential function of the form $y = ab^x$.

If the graph of data is a straight line with slope b on log-log paper then the data are related by a power function of the form $y = ax^b$.

Example 9.34   Application

Are the population estimates in Figures 9.27 and 9.28 related by an exponential function?

Solution A curve through the points in Figure 9.27 looks like an exponential curve. However, looking at the graph in Figure 9.28 we can see that these points are not connected by a straight line. Thus these points are not related by an exponential function. ∎

Example 9.35   Application

In a tungsten lamp, data for the current i in milliamps (mA) for various voltages v in volts (V) are shown in the table below. Determine an equation that relates i as a function of v.

v (V)	1	2	8	25	50	100	150	200
i (mA)	16.5	24.6	56.9	113	172	261	330	387

Solution These points are plotted on semilog graph paper in Figure 9.36 and on log-log graph paper in Figure 9.37. When the points are connected we get the curves shown in the two figures.

The curve in Figure 9.37 is a straight line. This curve was drawn on log-log graph paper, so we conclude that these data satisfy a power function of the form $i = av^b$.

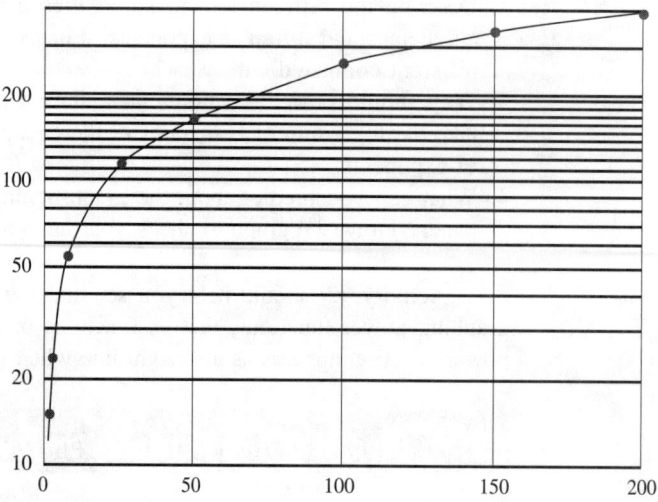

Figure 9.36

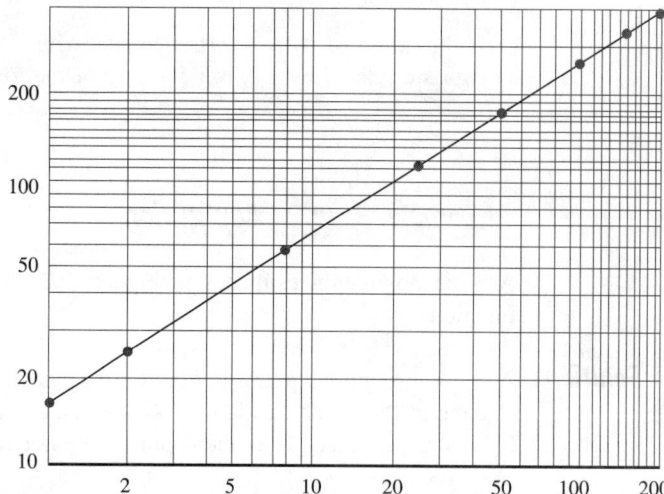

Figure 9.37

The slope of this straight line is b, and if (x_1, y_1) and (x_2, y_2) are two points on the curve we have

$$b = \frac{\log y_2 - \log y_1}{\log x_2 - \log x_1}$$

If we use the points $(1, 16.5)$ and $(2, 24.6)$, we will get

$$b = \frac{\log 24.6 - \log 16.5}{\log 2 - \log 1}$$

$$\approx \frac{1.390935107 - 1.217483944}{0.301029996 - 0}$$

$$= \frac{0.1734511629}{0.301029996}$$

$$\approx 0.576$$

Thus we know that these data are of the form $i = av^{0.576}$.

To determine a we use the point $(1, 16.5)$ and get $i = a \cdot 1^{0.576} = a$, so $a = 16.5$. Thus the data in the table satisfy the equation $i = 16.5v^{0.576}$. ∎

Semilog Graphing on the Graphing Calculator

Many computer graphing programs offer the options of using semilog and log-log axes. The *TI-83*, like most graphing calculators, does not offer a logarithmic scale. However there is a way around this, as shown in Example 9.36.

Example 9.36

Plot the line $y = 2 \cdot 5^x$ on your graphing calculator.

Solution Enter expressions for Y₁ and Y₂ as shown in Figure 9.38. Plot the graph using the ZStandard (Zoom Standard) setting, and then use Zoom In once and Trace to reach Figure 9.39.

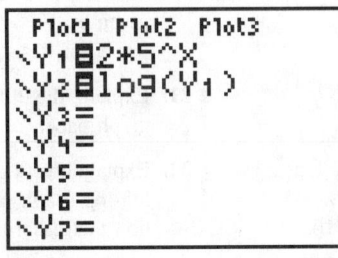

Figure 9.38

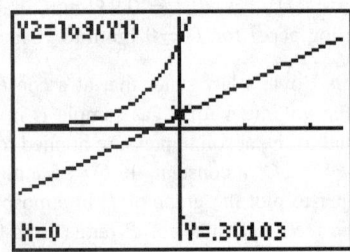

Figure 9.39

The y-intercept is shown in Figure 9.39 with coordinates $(0, 0.30103)$, which is the same as $(0, \log 2)$. If the plot were done on a true semilog graph the intercept would be $(0, 2)$, as you would expect for this exponential function. We cannot change the coordinates shown by the calculator, but by plotting Y2=log(Y1) we know that the point $(0, 0.30103)$ on the calculator is a good approximation to the point $(0, 2)$ on a semilog graph. ∎

When we study the cooling data in Section 9.5 we will use the calculator to plot the data on a semilog graph to determine the equation that models the data.

Section 9.4 Exercises

In Exercises 1–6 sketch the graphs of the given functions on semilog graph paper.

1. $y = 3^x$

2. $y = 3(5^x)$

3. $y = 3^{-x}$

4. $y = x^3$

5. $y = 5x^3$

6. $y = 5x^2 + 3x$

In Exercises 7–12 sketch the graphs of the given functions on log-log graph paper.

7. $y = x^{1/3}$

8. $y = 2x^3$

9. $y = x^3 + x$

10. $y = 5x^{-3}$

11. $y^2 = 4x^3$

12. $x^2 y^3 = 27$

13. *Meteorology* The atmospheric pressure P in atmospheres is given by $P = e^{-kh}$ where k is a constant and h is the altitude, in feet. Graph P vs. h on semilog paper for $k = 1.25 \times 10^{-4}$ at 0°C.

14. *Meteorology* The atmospheric pressure P in kilopascals (kPa) is given approximately by $P = 100e^{-0.3h}$, where h is the altitude in kilometers. Graph P vs. h on semilog paper.

15. *Electricity* The resistance R of a copper wire varies inversely as the square of its cross-sectional diameter D. Thus $RD^2 = k$. If $k = 0.9\,\Omega\cdot$mm, graph R vs. D on log-log paper for $D = 0.1$ mm to 10 mm.

16. *Physics* Boyle's law states that at a constant temperature the volume V of a gas sample is inversely proportional to the absolute pressure applied to the gas P. Thus $PV = C$, a constant. If $C = 5$ atm·ft³ use log-log paper to plot the graph of P in atmospheres vs. V in cubic feet. Let values of P range from 0.1 to 10.

17. A technician gathered the data in the following table. Plot the data on semilog and log-log papers and determine the equation that relates y as a function of x.

x	1	2	5	10
y	1,000	250	40	10

18. A technician gathered the data in the following table. Plot the data on semilog and log-log papers and determine the equation that relates y as a function of x.

x	1	2	3	4	5
y	16	80	400	2,000	10,000

19. *Electronics* For a certain electric circuit the voltage is given by $V = e^{-0.25t}$. Graph V vs. t on semilog graph paper for $0 \le t \le 5$.

20. *Electronics* For a certain electric circuit, the voltage is given by $V = e^{-0.35t}$. Graph V vs. t on semilog graph paper for $0 \le t \le 10$.

21. Explain the difference between semilog and log-log graph paper.

22. Explain how a graph of a function drawn on semilog or log-log graph paper can help you determine the equation that describes that function.

9.5   Analysis of the Cooling Data

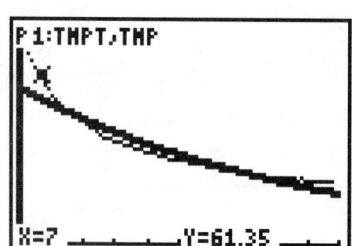

P 2:THPT,THP

X=20   Y=44.55

Figure 9.40

P 1:THPT,THP

X=7   Y=61.35

Figure 9.41

Our Chapter Project was to find a mathematical model for the decreasing temperature of an object cooling in air after being removed from boiling water. One graph of temperature vs. time appears in Figure 9.3 on page 610. We stored the temperatures (degrees Celsius) in the list named TMP and the times (0 to 98 seconds) in the list named TMPT. A calculator plot of lists TMP vs. TMPT is shown in Figure 9.40.

Exponential Regression

The curve in Figure 9.40 has the look of an exponential function. Using the calculator's exponential regression feature (Selection 0 on the STAT CALC menu) we produced the exponential function that is plotted with the cooling data in Figure 9.41. The calculator's exponential regression feature produces an equation of the form $y = a \cdot b^x$. The regression equation plotted in the calculator's "thick" style is $y = 57.7 \cdot 0.989^x$.

This regression equation is unsatisfactory for two reasons. First, it does not fit the data very well. More importantly, like all exponential functions of the form $y = a \cdot b^x$, it will have a horizontal asymptote at $y = 0$, the x-axis. The horizontal asymptote at $y = 0$ means that this regression equation predicts the final temperature will be zero degrees—not likely in an experiment conducted inside on a day in May. In fact we measured the room temperature to be $21.6°$ C. More work is needed to find the correct model for this set of data.

Slope Function of Data

Let's move on to a tool that we have used in recent chapters, the slope function. You studied slope functions of exponential functions in your work with the calculator program OTHERFUN. Perhaps you noticed that for exponential functions the slope function also has an exponential shape.

The slope between two points of a temperature vs. time graph is computed with the formula

$$\text{slope} = \frac{\text{change in temperature}}{\text{change in time}}$$

We have 99 measurements from TMPT(1) $= 0$ seconds to TMPT(99) $= 98$ seconds, so we can compute the value of the slope for each second of the experiment with the formula

$$\text{slope(n)} = \frac{\text{TMP}(n) - \text{TMP}(n-1)}{\text{TMPT}(n) - \text{TMPT}(n-1)}$$

There are 98 different values of the slope that can be computed, from $n = 2$ to $n = 99$. The temperatures were collected once each second, so TMPT$(n) -$ TMPT$(n-1) = 1$ for all values of n. Since the denominator of the slope formula is equal to 1, each of the 98 slopes can be computed from the formula

$$\text{slope(n)} = \text{TMP}(n) - \text{TMP}(n-1)$$

Example 9.37

Compute and plot the 98 values of the slope function from $n = 2$ to $n = 99$.

Solution   In Chapter 8 you learned how to use the calculator's ∆List feature, which is selection 7 on the LIST OPS menu, and the seq function to generate a new list of the integers 1 through 98. Our calculator instructions are shown in Figure 9.42. List L2 contains the temperature differences TMP$(n)-$TMP$(n-1)$ and List L3 contains the integers 1 through 98. The plot of L2 vs. L3 is shown in Figure 9.43. The dark horizontal line is the x-axis.

Remember: If the data points were not collected every one second, the slope calculation would require division by the time between data points.

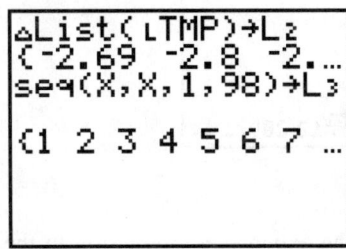

Figure 9.42   **Figure 9.43**

Although the plot of the slope function is not as orderly as other plots we have seen, its general shape is like one of the exponential functions we have studied. Compare it, for example, to the slope function shown in Figure 9.22 on page 644. The horizontal asymptote at $y = 0$ is consistent with other slope functions you have seen in this chapter. The shape of the slope function is further evidence that the cooling data can be modeled with an exponential function. But which one?

Vertical Transformation of the Data

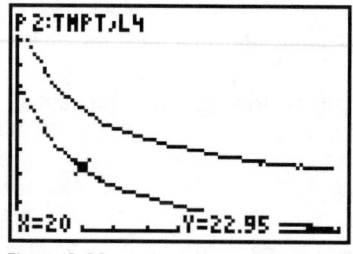

Figure 9.44

Figure 9.3 on page 610 shows that the plot of the cooling data has a horizontal asymptote at room temperature, 21.6°C, and it makes sense that the temperature of the cooling object will approach room temperature. To use an exponential function of the form $y = a \cdot b^{cx}$, which has an asymptote at $0°$, we will need to subtract 21.6 from each of the temperatures.

The command we used is TMP$-21.6 \rightarrow$L4. With this command 21.6 is subtracted from all 99 temperatures and the result is stored in list L4. The plot of list L4 vs. TMPT is shown, along with the original data, as the lower curve in Figure 9.44.

Now the graph has the horizontal asymptote that we want. These transformed temperatures are the ones we will try to fit with an exponential function. We will compare two ways of producing the model for the data:

1. Exponential regression on L4 vs. TMPT.
2. Semilog plot of L4 vs. TMPT and using linear regression on the resulting line.

Example 9.38

Compute the exponential regression equation for the adjusted temperatures in L4 vs. the times in TMPT.

Solution As noted above, exponential regression on the *TI-83* is Selection 0 on the STAT CALC menu. The calculator instruction is

    ExpReg TMPT,TMP

Our result was the equation $y = 49.72 \cdot 0.966^x$. The plot of the data and the regression equation is shown in Figure 9.45.

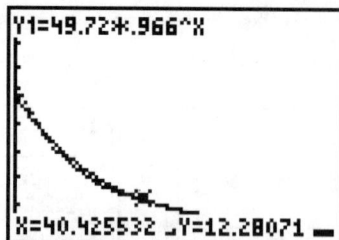

Figure 9.45   ■

The fit of the exponential regression to the transformed data looks extremely good. Our next task is to add back in the 21.6 degrees that we subtracted out so we can have the model of the actual data.

Example 9.39

In Calculator lab 9.1 you used the CBL to gather temperature data of an object that cooled in air from the boiling point of water to near room temperature. In Example 9.38 you computed an exponential regression equation for the adjusted temperatures from this lab.

(a) Form the model of the original data.
(b) Restate the equation with e as the base.
(c) Plot the resulting equation along with the original data.

Solution   Since we subtracted 21.6 from the data and then found an exponential equation that modeled that data we must add the 21.6 back in now.

(a) The resulting equation is $y = 21.6 + 49.72 \cdot 0.966^x$.
(b) Using the relationship between $y = a \cdot b^{cx}$ and $y = e^{kx}$ that was developed in Example 9.25 on page 639, we have $y = 21.6 + 49.72 \cdot e^{-0.0346x}$.
(c) The graph in Figure 9.46 shows that the model is quite good. Next we will study just how well the model fits the data.

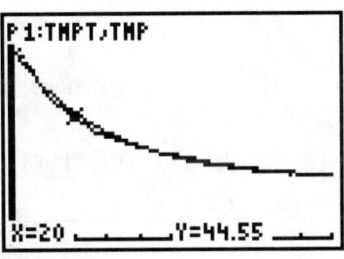

Figure 9.46

The Accuracy of the Model

There are many ways to estimate the accuracy of a model. The easiest is just to look at the graph of the model and the data, as in Figure 9.46. The calculator regression feature found the exponential function that had the lowest total of the *squared* errors, but how good is the calculator's best equation? Next we'll see another way to look at accuracy: the average absolute percent error. First we need to define a few terms.

The **error** is the difference between a measurement produced by prediction and the actual measurement. A value of the error is produced for each point in the data by computing

$$\text{Error} = \text{Predicted} - \text{Actual}$$

for each measurement. Notice that the error is positive if the prediction is too high and negative if it is too low.

The **percent error** is the error expressed as a percent of the actual measured temperature. Thus,

$$\text{Percent error} = \frac{\text{Error}}{\text{Actual}} \cdot 100$$

In the future when we are discussing temperature we will use $T_{\text{predicted}}$ for the predicted temperature and T_{actual} for the measured, or actual, temperature.

Example 9.40

If a model predicts 72 degrees when the measured temperature is 70 degrees, what is (a) the error and (b) the percent error?

Solutions (a) The error is $T_{predicted} - T_{actual}$. Since $T_{predicted} = 72°$ and $T_{actual} = 70°$ the error is $72° - 70° = 2°$.

(b) The percent error is $\dfrac{\text{Error}}{T_{actual}} \cdot 100$. For this example we have $\dfrac{2°}{70°} \cdot 100 \approx 2.9\%$. ∎

The **absolute error** is the absolute value of the error, and the **absolute percent error** is the absolute value of the percent error.

Example 9.41

If a model predicts 34 degrees when the measured temperature is 35 degrees, what is (a) the error, (b) the percent error, (c) the absolute error, and (d) the absolute percent error?

Solutions (a) The error is $T_{predicted} - T_{actual}$. Since $T_{predicted} = 34°$ and $T_{actual} = 35°$ the error is $34° - 35° = -1°$.

(b) The percent error is $\dfrac{\text{Error}}{T_{actual}} \cdot 100$. For this example we have $\dfrac{-1°}{35°} \cdot 100 \approx -2.9\%$.

(c) The absolute error is $|-1°| = 1°$.

(d) The absolute percent error is $|-2.9\%| = 2.9\%$. ∎

The absolute error is twice as much in Example 9.40 as in Example 9.41 but the absolute percent errors in the two examples are the same. In general the percent error is computed by the formula

$$\frac{T_{predicted} - T_{actual}}{T_{actual}} \cdot 100$$

In judging the overall accuracy of a model compared with a set of data it is useful to compute the average of all the individual errors. For this purpose we'll use the calculator list feature named mean, which computes the average value of the numbers in a list. **Mean** is another name for average. The calculator's mean feature is selection 3 on the LIST MATH menu.

Example 9.42

Compute the average of each of the above errors for the cooling data and its exponential model.

Solution We have 99 different errors, one for each of the data points. The *TI-83* can calculate each of the percent errors all at once with one calculator formula:

$$100\frac{Y_1(\text{TMPT}) - \text{TMP}}{\text{TMP}}$$

For our data this formula produces 99 calculations, one for each member of the lists. The expression $Y_1(\text{TMPT})$ means to evaluate the function Y_1 for each time value in the list TMPT. For each of those calculations the corresponding temperature in TMP is subtracted, and the result divided by that same value of TMP. Multiplication by 100 converts to a percent.

The actual calculator command is shown in the top two lines of Figure 9.47, which we produced with the decimal mode setting changed to 1. The three resulting percent errors shown are negative because early in the cooling process the model predicts temperatures that are lower than the measured temperatures.

The computation of the average percent error and the average absolute percent error are shown in Figure 9.48.

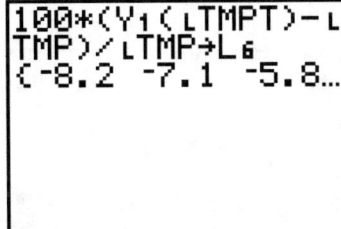

Figure 9.47   **Figure 9.48**

The mean (average) percent error is 0.4%, a low number. However the average percent error can be misleading. For example two huge errors of 80% and -70% would average to be 5% and would give an incorrect view of the overall accuracy of the estimates.

The average absolute percent error is 1.9%, which assures us that after the first few larger errors the model gives better predictions. ∎

The low average absolute errors can make us quite confident in the model of the cooling data. In fact, the physics of heat transfer predicts that an exponential function will model cooling data like ours. The law was discovered by the 17th-century Englishman Isaac Newton who also discovered the mathematics of gravity-influenced motion we have used in this book. **Newton's Law of Cooling** states that the temperature $T(t)$ at any time t is

$$T(t) = T_{\text{Final}} + (T_{\text{Begin}} - T_{\text{Final}})e^{kt}$$

where T_{Final} is the temperature of the surroundings, or the 'room' temperature in our experiment, and T_{Begin} is the first measured temperature. In our experiment T_{Begin} should be close to the boiling point of water. The value of k depends on the experiment. If the object is cooled in air the value of k will be smaller than if the object is cooled in water. In cooling, k is a negative number and for heating (for example taking an object out of a freezer and letting it warm up to room temperature) k is a positive number.

Example 9.43

Estimate the beginning temperature T_{Begin} from the mathematical model for the cooling data. Compare this with the first temperature measured.

Solution   There are two ways to answer this question. First we could substitute $t = 0$ into the model and get the beginning temperature:

$$T(t) = y = 21.6 + 49.72 \cdot e^{-0.0346t}$$

and

$$T(0) = 21.6 + 49.72 = 71.32°C$$

Another way is to work from Newton's Law of Cooling: $T_{Begin} - T_{Final}$ is the coefficient of the exponential part of the function. Therefore

$$T_{Begin} - T_{Final} = 49.72$$

and since T_{Final} is room temperature, or 21.6, then

$$T_{Begin} = 49.72 + 21.6 = 71.32°C$$

The actual measured beginning temperature is $TMP(1) = 77.65°C$ so the model measures low at the beginning, as the first three calculated percent errors in Figure 9.48 told us.   ■

Using a Semilog Plot of the Data

We have seen that an exponential function plotted on semilog axes will become a straight line. The slope and intercept of the straight line can be used to predict the exponential function that produced the straight line. In the final activity of this chapter we will guide you in finding the exponential function using this technique.

Calculator Lab 9.6

Procedures

Analyzing the Cooling Data with a Semilog Plot

For this lab you can use your own cooling data or the data from our experiment, which is stored in lists TMP (temperature) and TMPT (time).

1. Make a list of the common logarithm of the temperatures with room temperature subtracted:

    LOG(TMP-21.6) → L1

2. Look at the values of L1 and change the dimensions of your graph window to match them. One way to do this is to use:

    max(L1) and min(L1)

    max and min are in the LIST MATH menu.

3. Plot L1 vs. TMPT to determine the straightness of the semilog line.
4. Fit a regression line to the plot of L1 vs. TMPT, using:

    LinReg L1, TMPT

5. Compute the coefficients of the estimated exponential function $y = c \cdot e^{kx}$ from the estimated linear equation.

    (a) The equation of the line is $y = ax + b$. According to Example 9.33 on page 653 the coefficient a, the slope of the line, is equal to $k \log e$.
    (b) According to Example 9.33 the value b, the intercept of the line, is equal to the intercept c in the exponential function equation. However since our regression is based on logarithms, not the numbers themselves, b is equal to $\log c$, so $c = 10^b$.

6. Compute the average percent error for the exponential model that results. To assign the regression equation to Y₁, position your cursor on line Y₁ in the Y= screen, and follow these steps:

VARS 5 EQ 1

7. How does your model compare with the exponential model function that was found in the text?

Section 9.5 Exercises

In Exercises 1–4 determine the (a) error, (b) percent error, (c) absolute error, and (d) absolute percent error.

1. *Construction*   An electrical contractor measures out 25.2 meters of wire for a house that is known to require 24.7 meters.

2. *Construction*   The actual length of an I-beam is 12.445 m. An engineer measures the beam as 12.45 m.

3. *Computer science*   A computer is designed to run at 450 MHz. The first machines off the production line test at 462 MHz.

4. *Computer science*   A microcomputer chip is supposed to measure 24 mm long, 8 mm wide, and 3 mm thick. One chip, when measured with a micrometer, is 23.72 mm long, 8.35 mm wide, and 2.98 mm thick. What are the absolute, relative, and percent errors of each of these measurements?

5. *Medical technology*   Suppose that a small amount of *e. coli* bacteria is placed on a microscope slide and maintained under conditions that allow growth. Every 20 minutes the slide is examined and the number of bacteria cells are estimated. The results for the first two hours are shown in the table below.
   (a) What is the exponential regression equation that satisfies this information?
   (b) Express the regression equation in (a) as an exponential equation of the form $n = a \cdot e^{bt}$.
   (c) Graph the original data and your regression equation.

Time, t, (min)	0	20	40	60	80	100	120
No. of cells, n	20	40	79	157	314	624	1241

6. *Biology*   The life expectancy at birth of females from 1970–1994 is shown in the table below.
   (a) What is the exponential regression equation that satisfies this information?
   (b) What is the natural logarithmic regression equation that satisfies this information?
   (c) A female born in what year could expect to have a life expectancy of 85 years?

Year, t	1970	1975	1980	1985	1990	1992	1994
Life expectancy L	74.7	76.6	77.4	78.2	78.8	79.1	79.0

7. *Thermodynamics*   According to **Newton's Law of Cooling**, the rate at which a hot object cools is proportional to the difference between its temperature and the temperature of its surroundings. The temperature T of the object after a period of time t is

$$T = T_m + (T_0 - T_m)e^{-kt}$$

where T_0 is the initial temperature and T_m is the temperature of the surrounding medium. An object cools from 180°F to 150°F in 20 min when surrounded by air at 60°F. What is the temperature at the end of 1 hour of cooling?

8. *Thermodynamics*   A piece of metal is heated to 150°C and is then placed in the outside air, which is 30°C. After 15 min the temperature of the metal is 90°C. What will its temperature be in another 15 min?

9. *Thermodynamics*   You like your drinks at 45°F. When you arrive home from the store, the cans of drink you bought are 87°F. You place the cans in a refrigerator. The thermostat is set at 37°F. When you open the refrigerator 25 min later, the drinks are at 70°F. How long will it take for the drinks to get to 45°F?

10. *Thermodynamics* A cup of coffee is poured from a pot whose contents are 95°C into a noninsulated cup in a room at 20°C. One minute later the coffee in the cup has cooled to 90°C. How long will it take before the coffee is cool enough to drink at 65°C?

11. *Medical technology* About 11:30 A.M. the body of an apparent homicide victim is found in a room that is kept at a constant temperature of 72°F. At noon (12:00 P.M.) the temperature of the body is 82°F and at 1:00 P.M. it is 76°F. Assume that the body at time of death was 98.6°F.
  (a) What is the exponential equation that satisfies this information?
  (b) When was the time of death?

12. *Medical technology* An early morning jogger notices a body in a shaded alley. Upon her arrival at 7:30 A.M., the medical examiner takes the temperature of the the the body and air in the alley. She finds that the body is 71.2°F while the air is 57°F. One hour later she takes the two temperatures again. The air is still at 57°F but the body has cooled off to 68.5°F. Finally, one hour and 15 minutes later, just before the body is removed, she takes the two temperatures again and finds that the body is 66.7°F and the air temperature is still 57°F. Assume that the body at time of death was 98.6°F.
  (a) What is the exponential equation that satisfies this information?
  (b) When was the time of death?

13. In Chapter 1 you entered the Olympic men's freestyle 100-meter swimming records into your *TI-83* and saved them using the names YEAR and TIME for the two lists.
  (a) Produce and plot an exponential regression line for the men's 100-meter freestyle swimming event in the Olympics.
  (b) Use your regression curve in **(a)** to predict the times for 1920, 1960, 1980, and 1996.
  (c) What are the error and percent error for your prediction for 1996?
  (d) Compare your results to those you got using linear regression in Example 1.45 and the actual results.*

Year	1896	1904	1908	1912	1920	1924	1928	1932	1936	1948	1952
Time (sec.)	82.2	62.8	65.6	63.4		59.0	58.6	58.2	57.6	57.3	57.4

Year	1956	1960	1964	1968	1972	1976	1980	1984	1988	1992	1996
Time (sec.)	55.4		53.4	52.2	51.2	50.0		49.8	48.6	49	

14. Write a memo to management in which you would prove that you can accurately predict the time needed for a metal rod to cool from over 500° to a temperature in the desired range. Be sure to include how you would be able to estimate the cooling time given the starting temperature and the final temperature.

●●●●●●●● Chapter 9 Summary and Review

Topics You Learned or Reviewed

▶ An **exponential function** is one in which the exponent contains the variable. An exponential function is of the form $y = a \cdot b^x$ or $y = ae^x$ where $e \approx 2.71828$.

  • You know how to sketch and recognize the graph of an exponential function.

  • exponential functions can be used to model population growth, radioactive decay, interest in a bank account, and charging a capacitor.

▶ A logarithmic function is an inverse function of an exponential function.

  • The **common logarithmic** function, written $y = \log x$, is the inverse of the exponential function $y = 10^x$. The **natural logarithmic** function, written $y = \ln x$, is the inverse of the exponential function $y = e^x$.

---

*The linear regression results were (1920, 63.5), (1960, 54.1), (1980, 49.5), and (1996, 45.7). The actual times were (1920, 60.4), (1960, 55.2), (1980, 50.4), and (1996, 48.74).

- The following properties for logarithms with any base are true for positive values of a and b, and for *any* real value of n:
  - Addition: $\log a + \log b = \log ab$
  - Subtraction: $\log a - \log b = \log \dfrac{a}{b}$
  - Multiplication: $n \cdot \log a = \log a^n$
- You know how to sketch and recognize the graph of a logarithmic function.

▶ An inverse function can be thought of in three ways:
  - An inverse function reverses the process of another function.
  - The inverse of a function can be formed by switching x and y in the function and solving for y.
  - The *TI-83* calculator has some inverse functions printed in yellow above the corresponding function key. Look at the ❨SIN❩, ❨COS❩, ❨TAN❩, ❨LOG❩, ❨LN❩, and ❨x^2❩ keys. The function in yellow and the function on the key are inverses of each other. They each reverse the process of the other.

▶ If the inverse of a function can be formed by switching x and y in the function, and if solving for y results in a function, then it is the inverse function of the original function. If it is not a function then it is called an **inverse relation** of the original function. This process of finding an inverse function is easiest for a linear function or a power function.

▶ The graphs of a function and its inverse function are mirror images of each other, if the mirror is placed on the line $y = x$.

▶ The domain of a function is the range of its inverse function and the range of a function is the domain of its inverse function. This relationship is not true for a function and its inverse relation if that relation is not an inverse function.

▶ An exponential equation can be solved by using tables, graphs, or logarithms.

▶ A graph with one axis based on a logarithmic scale and the other on a linear scale is called a **semilogarithmic** or **semi-log graph**. On a Cartesian or rectangular coordinate system both axes use a linear scale. If a logarithmic scale is used on both axes, then the graph is called a **logarithmic** or **log-log graph**.

▶ If the graph of data is a straight line on semi-log paper, then the data are related by an exponential function of the form $y = ab^x$.

▶ If the graph of data is a straight line with slope m on log-log paper, then the data are related by a power function of the form $y = ax^m$.

▶ The **error** is the difference between a measurement produced by prediction and the actual measurement. A value of the error is produced for each point in the data by computing

$$\text{Error} = \text{Predicted} - \text{Actual}$$

for each measurement. Notice that the error is positive if the prediction is too high and negative if it is too low.

▶ **Newton's Law of Cooling** states that the temperature $T(t)$ at any time t is

$$T(t) = T_{\text{Final}} + (T_{\text{Begin}} - T_{\text{Final}})e^{kt}$$

where T_{Final} is the temperature of the surroundings and T_{Begin} is the first measured temperature.

Review Exercises

1. *Ecology* Estimate the population of China in the year 2010 given that the population in 1980 is 985 million and the population in 1996 is 1210 million and that the changing population between those two years follows a straight line.

2. *Ecology* Estimate the population of China in the year 2010 given the population data in the Table 9.3 and that the population change between those years follows a quadratic curve.

Table 9.3

Year	1980	1990	1996
Population $\times 1000$	$984,736$	$1,133,710$	$1,210,005$

Source: http://census.gov/statab/freq/96s1325.txt

3. *Ecology* According to an estimate made in 1991 the population of the Asian country Bangladesh was expected to double in the following 32 years. In 1991 the population of Bangladesh was 111.4 million.

(a) What is your estimate that the population of Bangladesh will be in the year 2005?

(b) What assumptions and procedures did you use to get your answer in (a)?

In Exercises 4–7 do not use a calculator to evaluate each of the following logarithms.

4. $\log 100$

5. $\log 0.0001$

6. $\log 10^{17.5}$

7. $\log 10^{-38}$

In Exercises 8–11 use the fact that $\log 3.87 \approx 0.5877$ to evaluate each of the given logarithms without the use of a calculator.

8. $\log 38.7$

9. $\log 3,870$

10. $\log 0.0387$

11. $\log 0.000387$

In Exercises 12–17 use the properties of logarithms to rewrite each of the following.

12. $\log 25$

13. $\log \frac{4}{7}$

14. $\log 7^4$

15. $\log 75$

16. $\log 18 - \log 3$

17. $\log x + 4\log 3 - x\log 5$

In Exercises 18–21 (a) determine the inverse of the given function, (b) determine if the answer in (a) is an inverse function or an inverse relation, and (c) graph the given function and its inverse.

18. $y = 7x - 5$

19. $y = (x - 3)^2$

20. $y = \sqrt{x - 5}$

21. $y = \dfrac{2}{x} - 5$

22. *Environmental science* The pH of some eggs is 7.8. What is the concentration of hydrogen ions in these eggs?

23. *Sound technology* The loudness L of a sound, as measured in decibels (dB), is given by the equation

$$L = 10\log\left(\frac{I}{I_0}\right)$$

where I is the intensity of the sound and $I_0 = 10^{-12}$ W/m^2. If the intensity of a live rock concert is 1 W/m^2 calculate the loudness of the concert.

24. *Seismology* The Richter scale magnitude R of an earthquake is given by the equation

$$R = \log\left(\frac{I}{I_0}\right)$$

where I is the intensity of the quake and I_0 is a minimum perceptible intensity. What is the value of an earthquake whose intensity is $12,500,000I_0$?

25. *Astronomy* The magnitude M of a star is a measure of its brightness. The larger the magnitude the fainter the star. The magnitude of stars made visible through a telescope is given by

$$M(a) = 9 + \log a$$

where a is the aperture in inches of the telescope. Calculate the magnitude of the faintest star made visible by a telescope with an aperture of 35 inches.

26. *Finance* If \$7,500 is placed in a savings account at 5.75% interest, what is the total after 5 years if the interest is compounded (a) annually, (b) monthly, (c) daily, and (d) continuously?

27. *Biology* A bacteria culture originally numbers 1,200. It is known that this particular bacteria increase at the rate of about 0.25% per hour in a laboratory dish.

(a) Estimate the number of bacteria after 2 hours.

(b) Estimate the number of bacteria after 10 hours.

(c) Estimate the number after 4 days.

28. *Ecology* The population of a certain city is increasing at the rate of 6.7% per year. The present population is 200,000.

(a) What will be the population in 5 years?

(b) What can the population be expected to reach in 10 years?

29. *Medical technology* A pharmaceutical company is growing an organism to be used in a vaccine. The organism grows at a rate of 4.25% per day. How many units of this organism must they begin with in order to have 1,000 units at the end of 4 weeks?

30. *Electricity* A 130-μF capacitor is charged by being connected to a 120-V circuit. The resistance is 3 800 Ω. What is the charge on the capacitor 0.5 seconds after the circuit is disconnected?

31. *Electronics* In a capacitive circuit the equation for the current in amperes is given by

$$i = \frac{V}{R}e^{-t/RC}$$

where t is any elapsed time in seconds after the switch is closed, V is the impressed voltage, C is the capacitance of the circuit in F, and R is the circuit resistance in Ω. A capacitance of 700 μF in series with 1 kΩ is connected across a 50-V generator.

   1. What is the value of the current at the instant the switch is closed?

   2. What is the value of the current 0.02 s after the switch is closed?

   3. What is the value of the current 0.04 s after the switch is closed?

32. *Business* A business estimates that the salvage value V in dollars of a desktop computer after t years is given by $V(t) = 4750e^{-0.65t}$.

   (a) What was the original cost of the computer?

   (b) What is the value of the computer after three years?

33. A reservoir was polluted due to an industrial waste spill. The pollution created an algae bloom that caused many of the fish to die. The number of algae $N(t)$ present per 1000 gallons of water t days after the spill is given by the formula

$$N(t) = 100e^{1.62t}$$

How long will it take before the algae count reaches 20,000?

In Exercises 34–36 sketch the graphs of the given functions on semilog graph paper.

34. $y = 5^x$

35. $y = 5(3^x)$

36. $y = 5^{-x}$

In Exercises 37–39 sketch the graphs of the given functions on log-log graph paper.

37. $y = x^{1/5}$

38. $y = 2x^5$

39. $y^2 = 4x^5$

40. *Electronics* If a charged capacitor is connected to a coil by closing a switch, energy is transferred to the coil and then back to the capacitor in an oscillatory motion. The voltage V, in volts, across the capacitor will gradually diminish to 0 with time t.

   (a) Graph the function $V(t) = e^{-1.25t}\cos \pi t$ for $0 \leq t \leq 3$.

   (b) Determine the maximum displacement from $V = 0$.

   (c) Determine the frequency.

   (d) Determine the displacement from equilibrium after one cycle.

41. A technician attached a weight to a spring, pulled the weight below the equilibrium position $y = 0$, and then released it at $t = 0$. The following data was collected using a CBL:

t	0	0.1	0.2	0.3	0.4	0.5	0.6	0.7	0.8	0.9	1.0	1.1	1.2	1.3	1.4	1.5
y	−1.90	−1.60	−0.85	0.15	1.05	1.65	1.70	1.30	0.50	−0.40	−1.15	−1.55	−1.50	−1.00	−0.25	0.60

t	1.6	1.7	1.8	1.9	2.0	2.1	2.2	2.3	2.4	2.5	2.6	2.7	2.8	2.9	3.0	3.1
y	1.20	1.45	1.30	0.75	0	−0.7	−1.20	−1.35	−1.05	−0.50	−0.20	0.80	1.20	1.20	0.85	0 .30

   (a) Plot the given data on your calculator.

   (b) Determine the damped simple harmonic function that approximates these data.

   (c) Graph your answer to **(b)** over your graph to **(a)**.

42. A technician gathered the data in the following table. Plot the data on semilog and log-log papers and determine the equation that relates y as a function of x.

x	1	2	5	10	20
y	2,000	250	16	2	0.25

43. *Construction* The actual length of an I-beam is 45.625 ft. An engineer measures the beam as 45.875 ft. What are the (a) error, (b) percent error, (c) absolute error, and (d) absolute percent error?

44. *Biology* The life expectancy at birth of males from 1970–1994 is shown in the table below.
    (a) What is the exponential regression equation that satisfies this information?
    (b) What is the natural logarithmic regression equation that satisfies this information?
    (c) A male born in what year could expect to have a life expectancy of 85 years?

Year, t	1970	1975	1980	1985	1990	1992	1994
Life expectancy L	67.1	68.8	70.0	71.1	71.8	72.3	72.3

45. *Thermodynamics* According to **Newton's Law of Cooling**, the rate at which a hot object cools is proportional to the difference between its temperature and the temperature of its surroundings. An object cools from 212°F to 172°F in 20 min when surrounded by air at 72°F. What is the object's temperature at the end of 2 hours of cooling?

Chapter 9 Test

1. Estimate the population of Mexico in the year 2010 given the population data in the Table 9.4 and that the changing population follows an exponential curve.

Table 9.4

Year	1980	1990	1996
Population $\times 1000$	68,686	85,121	95,772

Source: http://census.gov/statab/freq/96s1325.txt

2. According to an estimate made in 1991 the population of the Philippines was expected to double in the following 28 years. In 1993 the population of the Philippines was 64.6 million.
    (a) What is your estimate of the population of the Philippines in the year 2005?
    (b) What assumptions and procedures did you use to get your answer in (a)?

3. Use the fact that $\log 7.52 \approx 0.8762$ to evaluate each of the given logarithms *without* the use of a calculator.
    (a) $\log 7520$
    (b) $\log 0.0752$

4. Use the properties of logarithms to rewrite each of the following.
    (a) $\log 4x^3$
    (b) $\log x - y \log 7$

5. Consider the function $y = 2x^3 - 4$.
    (a) Determine the inverse of the given function.
    (b) Is your answer in (a) an inverse function or an inverse relation?
    (c) Sketch the graph of the given function and its inverse.

6. The pH of some bread is 5.5. What is the concentration of hydrogen ions in this type of bread?

7. The loudness L of a sound, as measured in decibels (dB), is given by the equation

$$L = 10 \log \left(\frac{I}{I_0} \right)$$

where I is the intensity of the sound and $I_0 = 10^{-12}$ W/m^2. At a point 100 feet away the intensity of the sound of a jet airplane taking off is 100 W/m^2. Calculate the loudness of the airplane.

8. If $9,000 is placed in a savings account at 6.25% interest, what is the total after 10 years if the interest is compounded continuously?

9. A reservoir has become polluted due to an industrial waste spill. The pollution caused a large fish kill. A count of one species of fish estimates that there are 150 fish in the reservoir 10 days days after the spill. The wildlife biologist thinks this fish reproduces at a rate of 175% each year. Before they can allow people to fish in the reservoir they believe there should be at least 3000 of this fish in the reservoir. How long will it take before the fish count reaches 3000?

10. Sketch the graph of $y = 2.5^x$ on semilog graph paper.

11. Sketch the graph of $= 4x^{2.5}$ on log-log graph paper.

12. The actual length of a beam for a house is suppose to be 24.5 ft. A supervisor checks a carpenter's work and measures the beam at 24.875 ft. What are the (a) error, (b) percent error, (c) absolute error, and (d) absolute percent error?

13. The number of infant deaths per 1000 live births (infant mortality rate) for the United States from 1970–1993 is shown in the table below.

(a) What is the exponential regression equation that satisfies this information?

(b) What is the natural logarithmic regression equation that satisfies this information?

(c) In what year would you expect that infant mortality would first reach 3.5 per 1000 live births?

Year, t	1970	1980	1984	1987	1990	1993
Infant mortality rate M	20.0	12.6	10.8	10.1	9.2	8.4

14. Data from an experiment produced a straight line on semi-log paper. Two of the data points were $(2, 15.25)$ and $(4, 596.5)$. Find an equation that satisfies these data.

15. Newton's Law of Cooling states that the temperature $T(t)$ at any time t is

$$T(t) = T_{Final} + (T_{Begin} - T_{Final})e^{kt}$$

where T_{Final} is the temperature of the surroundings and T_{Begin} is the first measured temperature. A package of food at $-10°F$ is removed from a freezer and placed in a room whose temperature is $72°F$. After 15 minutes the temperature of the package is $5°F$. Use Newton's Law of Cooling to determine its temperature after 2 hours.

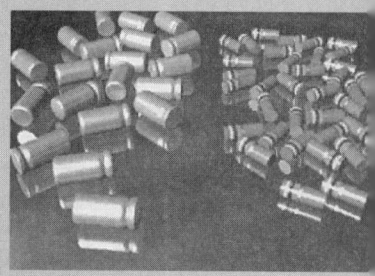

Systems of Equations and Inequalities

Topics You'll Need to Know

▶ Solving linear and quadratic equations in one variable

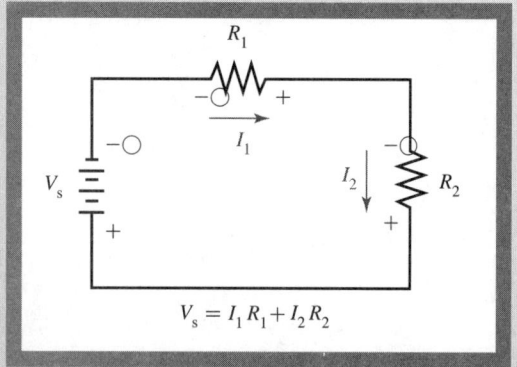
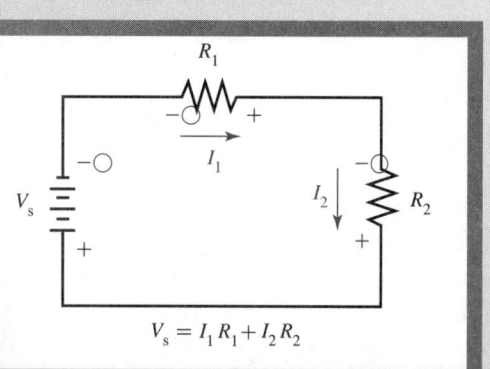

$$V_s = I_1 R_1 + I_2 R_2$$

Calculator Skills You'll Use

▶ Using inverse function to solve certain equations
▶ Linear regression to fit straight lines to data
▶ Solve equations in one variable:
  • using graph
  • using table

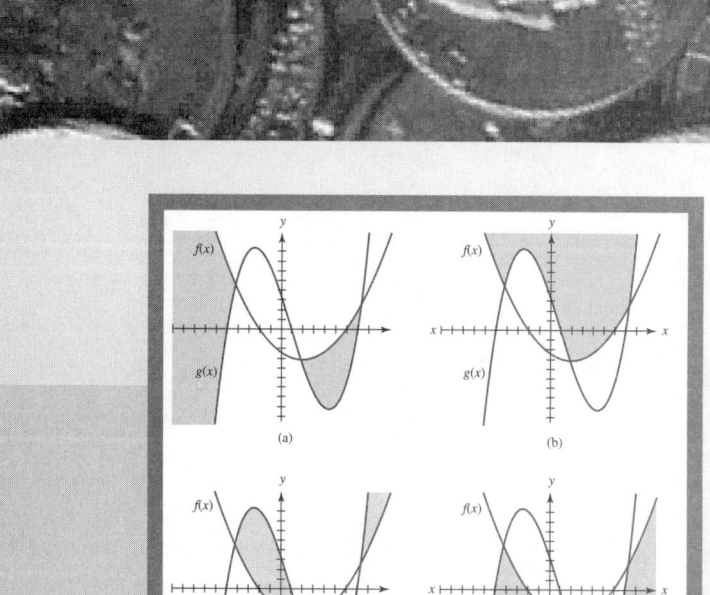

Calculator Skills You'll Learn

▶ Solving a 2 × 2 linear or nonlinear system
  • graph, using the `intersect` feature
  • display individual solutions using the free-moving cursor
▶ Using the `Test` and `Logic` menus to solve inequalites by means of binary logic
▶ Displaying solutions of inequalities with `above` and `below` graph styles

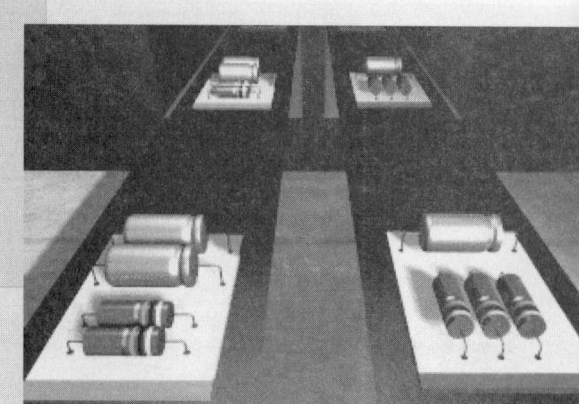

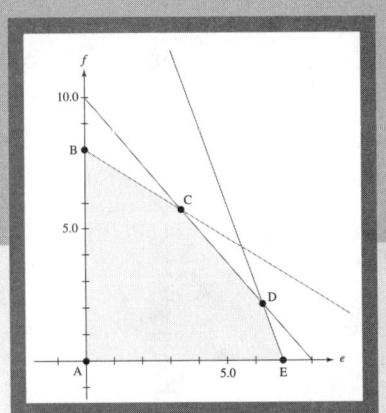

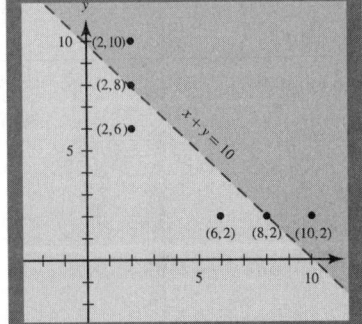

Topics You'll Learn or Review

▶ Deciding when to use a calculator to solve an equation
▶ Extraneous roots of an equation
▶ Solving linear and nonlinear systems of equations
  • addition method
  • graphical method
  • substitution method
  • tabular method
▶ Solving and graphing inequalities
  • one variable: solution shown on a number line
  • two variables: solution shown on a Cartesian coordinate system
▶ Binary logic (Boolean algebra)
▶ Using the simplex method of linear programming to solve a system of inequalites

What You'll Do in This Chapter

This chapter provides practice in solving several kinds of equations, including systems of two or more equations. You will practice deciding when to use the calculator and when to use pencil and paper algebraic methods. The solutions to two or more equations or inequalities will be applied to electronics and to a new kind of problem involving a manufacturing situation that is limited by the supply of parts. You will learn how to use a relatively new field of mathematics, linear programming, to solve these problems.

Chapter Project—Maximizing the Profit

 Custom Circuits Corporation

"Your design is our command"

To: Research Department
From: Management
Subject: Can Linear Programming Help Us?

Since we shifted to our "just in time" parts supply system, we have saved a lot of money, but I've just heard from a consultant that we could save even more if we apply linear programming to our manufacturing.

I know we have the right people to figure out linear programming. Please research this field, and use it to solve the following simplified problems based on our manufacturing processes.

Let me know when you have solved the two problems below and we can begin to streamline our actual process.

▶ **Resistors and capacitors on boards A and B**

   • PARTS REQUIREMENTS:
      • Board A requires three resistors and one capacitor
      • Board B requires two resistors and two capacitors
   • PARTS SUPPLY (daily):
      • 36 resistors and 20 capacitors
   • PROFIT:
      • $1.00 per board

   How many of each board should be made in order to make the most profit?

▶ **Resistors, capacitors, and transistors on boards E and F**

   • PARTS REQUIREMENTS:
      • Board E requires 30 resistors, 15 capacitors, and 6 transistors
      • Board F requires 10 resistors, 12 capacitors, and 9 transistors
   • PARTS SUPPLY (daily):
      • 210 resistors, 120 capacitors, and 72 transistors
   • PROFIT:
      • $6.00 for Board E and $5.00 for board F

   How many of each board should be made in order to make the most profit?

Preliminary Analysis

In order to solve the problems posed in the memo, it is necessary that you understand the statement of the problem. The following two activities are designed to help you make sure that you understand the relationships between the products, the parts, and the profit. Although you can make many different combinations of the two products, the goal is to find the combination that gives you the largest profit.

Activity 10.1
First Look at the Manufacturing Problem—Boards A and B

1. If you only manufactured board A, how many could you make with the number of parts available?
2. If you only manufactured board B, how many could you make with the number of parts available?
3. Can you make 9 A's and 4 B's?
4. Can you make 5 A's and 7 B's?
5. If you made 5 A's, what is the maximum number of B's that can be made?
6. If you made 5 B's, what is the maximum number of A's that can be made?
7. On a graph like the one in Figure 10.1, mark with a dot those points that represent combinations of boards A and B that are possible to make.

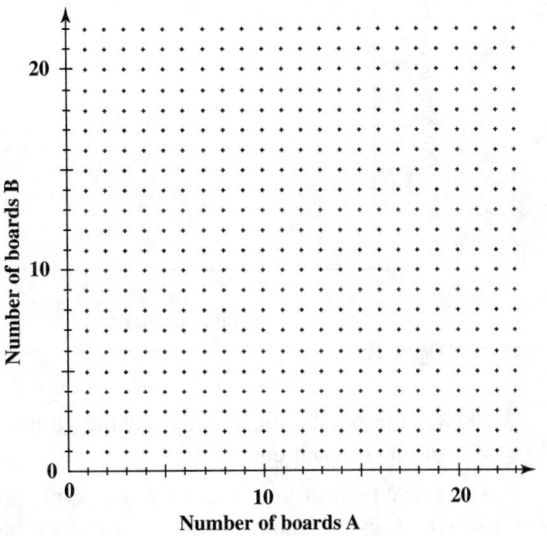

Figure 10.1

8. Of the points you marked on the graph, choose four of them and compute the profit for each one.

   (a) Which of your four selected points gives the largest profit?
   (b) Can you locate a point that gives a higher profit?

9. If a represents the number of boards A that are manufactured and b the number of boards B, write the following two functions:

   (a) $R(a, b)$, the number of resistors used.
   ▸ For example, if $a = 2$ and $b = 5$, then $R(a, b) = R(2, 5) = 3 \cdot 2 + 2 \cdot 5 = 16$
   (b) $C(a, b)$, the number of capacitors used.

Activity 10.2
First Look at the Manufacturing Problem—Boards E and F

1. If you only manufactured board E, how many could you make with the number of parts available?

2. If you only manufactured board F, how many could you make with the number of parts available?

3. Can you make 3 E's and 4 F's?

4. Can you make 5 E's and 5 F's?

5. If you made 3 E's, what is the maximum number of F's that can be made?

6. If you made 2 F's, what is the maximum number of E's that can be made?

7. On a graph like the one in Figure 10.2, mark with a dot those points that represent combinations of boards E and F that are possible to make.

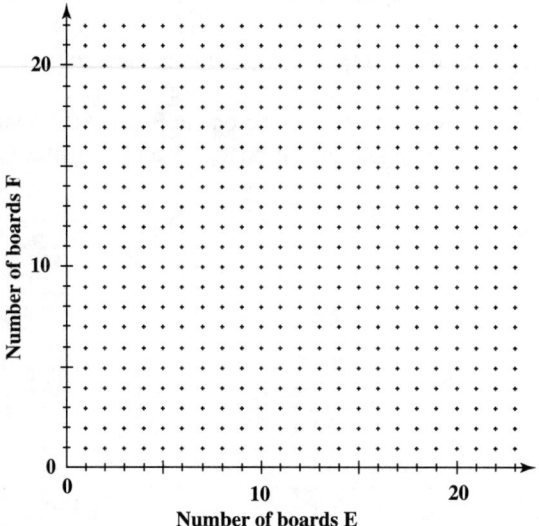

Figure 10.2

8. Of the points you marked on the graph, choose four of them and compute the profit for each one.

   (a) Which of your four selected points gives the largest profit?

   (b) Can you locate a point that gives a higher profit?

9. If e represents the number of boards E made and f is the number of boards F made, write the function $P(e, f)$ for the profit from the total number of boards.

10. If e shows the number of boards E made and f is the number of boards F made, write the following three functions:

   (a) $R(e, f)$, the number of resistors used.

      ▶ For example, if $e = 2$ and $f = 5$ then $R(e, f) = R(2, 5) = 30 \cdot 2 + 10 \cdot 5 = 110$.

   (b) $C(e, f)$, the number of capacitors used.

   (c) $T(e, f)$, the number of transistors used.

10.1   Review: Equations in One Variable

Throughout this book you have solved equations to find the value of the variables that make the equation true. In general, the solution has been the number or numbers that turn the equation into a true statement. For example, the equation $x + 5 = 7$ is true for $x = 2$ since $2 + 5 = 7$. The equation $x + 5 = 7$ is not true for any value of x other than 2.

An equation may have only one solution, others may have several, and some equations have no solutions. You have learned pencil and paper methods and calculator methods to solve equations. In this section we will review the different kinds of equations you can solve and remind you that although a calculator can be helpful, there are many equations that can be solved without a calculator.

Just as you do not reach for the calculator to multiply 5×6, you can solve some simpler equations quicker without a calculator. In this section you'll get more experience deciding which equations require a calculator and which you can do on your own.

Solving Equations Without a Calculator

Although all of the equations listed below can be solved with a graph or a table, you should be able to solve each of them with pencil and paper up until the point that the calculator is needed to do a division, a square root, an inverse sine, or a logarithm. Doing these equations by hand will actually save time, as the graph or table solution always takes longer.

Equations you can solve without a calculator come in several categories:

▶ Those that can be transformed into an easily solved equation:

1. $2a + 3 = 17$
2. $3(2b - 5) = 17$
3. $\dfrac{c}{3} = 24$
4. $2d - 8 = 4d + 5$
5. $(f - 3)(f + 6) = 0$
6. $(2g + 5)(3g - 4) = 0$
7. $\dfrac{2}{h - 3} = \dfrac{5}{4}$

▶ Equations that require the calculator for computation after you have done some pencil work on them.

 • Quadratic equations like the following may require the quadratic formula:
8. $k^2 - 6k + 8 = 0$
9. $m^2 + 7m = 15$
10. $10n + 5 = 3n(n - 5)$

▶ Equations that require one of the calculator's inverse functions after you have done some pencil work on them:

11. $p^2 = 19$
12. $\dfrac{100}{q^2 - 5} = 5$
13. $\sin r = 0.4$ (use degrees and give three answers to the nearest degree)
14. $\cos 2s = 1$ (use radians and give three answers to the nearest tenth)
15. $10^t = 43.25$
16. $e^{2u} = 10$
17. $5 \cdot 10^{4v} = 403$
18. $\sqrt[4]{w} = 2.5$
19. $5\sqrt[3]{x + 3} = 20$

Activity 10.3
Pencil and Paper Equation Solving

Working alone or with a partner, solve each of the 19 equations listed above. Give your non-integer answers to three significant figures. The answers are:

1.	$a = 7$	2.	$b = 5.33$	3.	$c = 72$
4.	$d = -6.50$	5.	$f = 3$ and -6	6.	$g = -2.5$ and 1.33
7.	$h = 4.60$	8.	$k = 2$ and 4	9.	$m = 1.72$ and -8.72
10.	$n = -0.195$ and 8.529	11.	$p = \pm 4.36$	12.	$q = \pm 5$
13.	$r = 24°, 156°,$ and $384°$	14.	$s = 0, \pi, 2\pi$	15.	$t = 1.64$
16.	$u = 1.15$	17.	$v = 0.477$	18.	$w = 39.1$
19.	$x = 61$				

In the next activity you will practice writing directions for solving equations. This activity not only will sharpen your communications skills, but will also improve your understanding of the equation solving process.

Activity 10.4
Equation Solving—Describing Solutions

Working with a team of three to five members, divide up the 19 equations listed above among the group. Ask each person to do the following for each of their equations:

1. Write a description of how to solve the equation with pencil and paper.
2. Create a similar problem.
3. Give the description of the solution and the new problem to another member of the group.
4. Each person will then judge how well the written directions lead to the solution of the equation.

When to Use the Calculator?

Equations for which we recommend calculator use fall into one of the following two categories:

To solve an equation **exactly** means to write the solution as a real number (for example 2.5, $\frac{4}{3}$, π, or $\sqrt[3]{23}$) or as a function of a real number (for example log 5 or $\sin^{-1}(0.34)$).

▶ An equation that cannot be solved exactly with pencil and paper. These include the simple equations that follow, for which there is no known method for getting an exact answer. Yes, that's right; no matter how much mathematics you study or which experts you ask, you will never learn how to find an exact solution to simple looking equations like the four below:

1. $\cos a = a$ (use radians whenever the angle appears outside the trig function)
2. $b^6 + b = 1$
3. $10^c + c = 15$
4. $\ln d = \dfrac{d}{5}$

▶ Equations that can be solved with algebraic pencil and paper methods are complicated enough (and even the most careful person may make an error) that the time spent setting up the calculator solution will be worthwhile. For example, there are methods for finding exact solutions to the following equations, but you would probably be better off using a graphing calculator to get an approximation of the answer.

1. $\dfrac{2}{\sqrt{f+2}} = 2f + 5$

2. $\dfrac{5}{g-3} - \dfrac{4}{g+4} = 2$

3. $\sqrt{h+5} = \sqrt{4+5h^2}$

4. $\sqrt{3k-5} = \sqrt{4+5k}$

Solving Equations With a Calculator

The basic method for a calculator solution involves plotting a graph and either looking for where your graph crosses the x-axis or where the graph crosses another graph. We'll call these two methods the *roots method* and the *equal functions* method.

▶ **Roots method: $f(x) = 0$.** If an equation can easily be changed into a form which has zero on one side of the equal sign you can graph that function and find its roots. That is, you can find the x-coordinate of the points where its graph crosses the x-axis.

▶ **Equal functions method: $f(x) = g(x)$.** Any equation can be approached this way. Think of the equation as being an equal sign separating two functions, $f(x)$ and $g(x)$. Then plot both functions and find where their graphs intersect; that is, find what values of x make $f(x) = g(x)$.

You can also use these methods with a table. The goal is still the same: Either find the value of x that makes the function $f(x)$ equal zero, or find the value of x that makes the functions $f(x)$ and $g(x)$ have equal values. The following examples show the two methods both with graphing and with tables.

Calculator Tips—Solving Equations Using Graphs

▶ Begin with the graph window set to ZStandard (Zoom Standard) or ZTrig (Zoom Trig) for the first look at the graph.

▶ When you see a solution within the graph window, use TRACE to estimate the solution.

▶ To get a more precise estimate of the answer:

• Use Zoom In to get a more accurate estimate of the answer(s).

• Repeat Zoom In and TRACE until two answers in a row are accurate to the required number of decimal places.

▶ Use ZStandard and then Zoom Out to expand your view of the graph and make sure you have all the solutions.

▶ Trigonometric equations:

• If the variable appears outside the trig function, set the mode to radians.

• Select ZTrig.

Calculator Tips—Solving Equations Using Tables

▶ For the first look at the table, begin with `TBLStart = 0` and `ΔTbl = 0`.
▶ To get a more precise estimate of the answer:

   • Reset `TBLStart` to just below the current estimate and decrease the value of `ΔTbl` until two answers in a row are accurate to the required number of decimal places.

▶ Trigonometric equations:

   • If the variable appears outside the trig function, set the mode to radians.

Caution:

With the calculator methods you are searching for solutions. There is no guarantee that you have found *all* the solutions. A combination of tables, graphs, and understanding the basic shapes of curves will help, but it still is possible to miss a solution. Fortunately, when equations come from real applications, you usually know approximately where to look for solutions.

Example 10.1

Solve the equation $\cos a = a$ with a graph using the roots method. Report your answer to two decimal places.

Solution   First we transform the equation so that zero is on one side of the equal sign. This produces the function $f(a) = \cos a - a$. We'll find the roots of the equation by plotting $y = f(a)$ in radian mode and looking for the roots. Figure 10.3 shows the first graph. The solution is approximately $a = 0.79$.

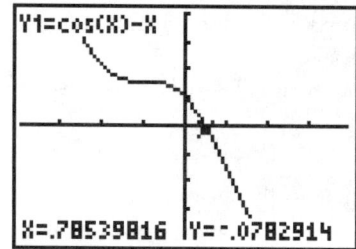

Xmin = -6.152285613
Xmax = 6.152285613
Xscl = π/2
Ymin = -4
Ymax = 4
Yscl = 1

Figure 10.3

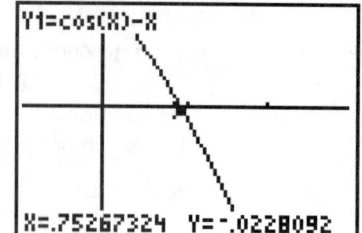

Xmin = -0.7526732403
Xmax = -2.323469566
Xscl = π/2
Ymin = -1.129032258
Ymax = 0.870967742
Yscl = 1

Figure 10.4

Next we zoom in to get a closer look at the root. Figures 10.4–10.6 show that the next three results are approximately 0.75, 0.74, and 0.74. Therefore we report the answer as $a = 0.74$.

To look for more solutions, we return to `ZTrig` and then `Zoom Out`. The form of the graph in Figure 10.7 leads us to believe that there is only one root.

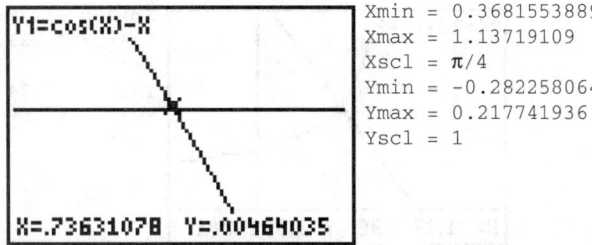

```
Y1=cos(X)-X
X=.73631078   Y=.00464035
```

```
Xmin = 0.3681553889
Xmax = 1.13719109
Xscl = π/4
Ymin = -0.282258064
Ymax = 0.217741936
Yscl = 1
```

Figure 10.5

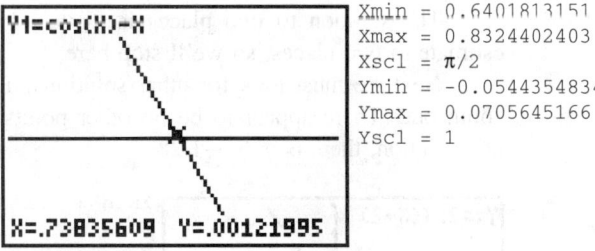

```
Y1=cos(X)-X
X=.73835609   Y=.00121995
```

```
Xmin = 0.6401813151
Xmax = 0.8324402403
Xscl = π/2
Ymin = -0.0544354834
Ymax = 0.0705645166
Yscl = 1
```

Figure 10.6

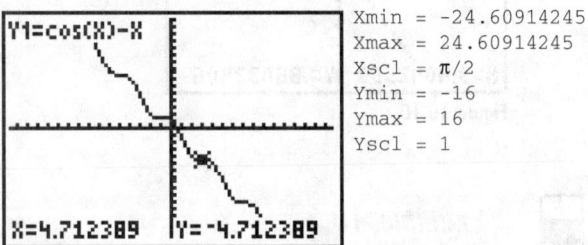

```
Y1=cos(X)-X
X=4.712389   Y=-4.712389
```

```
Xmin = -24.60914245
Xmax = 24.60914245
Xscl = π/2
Ymin = -16
Ymax = 16
Yscl = 1
```

Figure 10.7

 Example 10.2

Solve the equation $\dfrac{2}{\sqrt{f+2}} = 2f + 5$ with a graph using the Equal functions method. Report the solution to two decimal places.

Solution   It is possible to subtract $2f + 5$ from both sides of the equation and produce the equation

$$\frac{2}{\sqrt{f+2}} - (2f + 5) = 0$$

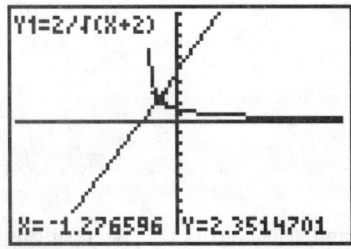

```
Y1=2/√(X+2)
X=-1.276596   Y=2.3514701
```

Figure 10.8

which could be solved with the roots method. However, the equal functions method will also work, and by allowing us to focus on each side of the equation as a separate function, the problem can be reduced in difficulty. The plot of the two functions

$$y_1 = \frac{2}{\sqrt{f+2}}$$
$$y_2 = 2f + 5$$

are shown in Figure 10.8. The first estimated solution is $f \approx -1.28$.

Xmin = -3.776595745
Xmax = 1.223404255
Xscl = 1
Ymin = -0.24193548
Ymax = 4.75806452
Yscl = 1

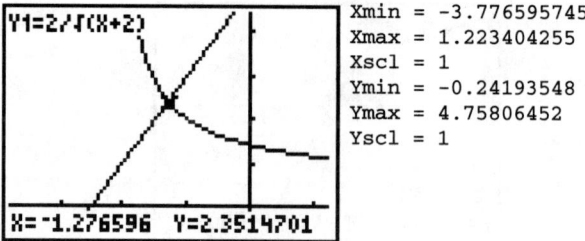

Figure 10.9

Figure 10.9 shows the result after the first zoom in.

The solution to two places is $f \approx -1.28$, which is equal to the previous estimate to two places, so we'll stop here.

Next we must look for other solutions. Figure 10.10 shows the effect of one zoom out. There appear to be no other points of intersection for the graphs. The one solution, then, is $f = -1.28$.

Xmin = -40
Xmax = 40
Xscl = 10
Ymin = -40
Ymax = 40
Yscl = 10

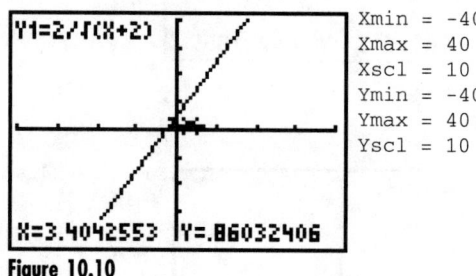

Figure 10.10

■

Example 10.3

Solve the equation $\ln d = \frac{d}{5}$, with a table using the roots method. Report the solution to two decimal places.

Solution First we transform the equation so that zero is on one side of the equal sign. This produces the function $f(d) = \ln d - \frac{d}{5}$. We'll find the roots of the equation in the table for $y = f(x)$ and look for sign changes in the y column. A change in sign usually indicates that the value $y = 0$ is somewhere nearby. Figure 10.11 shows the first table. A solution seems to be between $x = 1$ and $x = 2$.

X	Y1
0	ERROR
1	-.2
2	.29315
3	.49861
4	.58629
5	.60944
6	.59176

Y1=ln(X)-X/5

Figure 10.11

We'll change the value of `TBLStart` to 1 and `ΔTBL` to 0.1. The resulting table in Figure 10.12 shows that the root seems to be between $x = 1.2$ and $x = 1.3$.

The next step is to change TBLStart to 1.2 and ΔTBL to 0.01. Figure 10.13 shows that the root is between $x = 1.29$ and $x = 1.30$, and seems to be closer to $x = 1.30$. Therefore a solution is $d = 1.30$. Are there others?

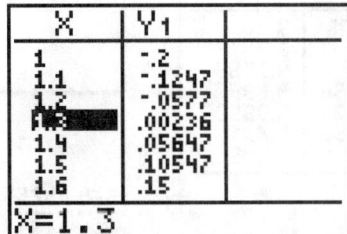

Figure 10.12

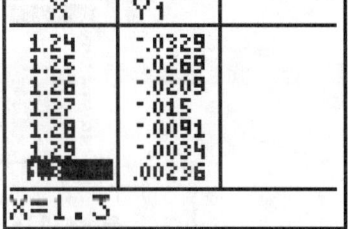

Figure 10.13

To check for additional roots, we must go back to the first table and do the equivalent of zooming out. Try changing TBLStart to -10 and ΔTBL to 5. The table in Figure 10.14 shows that there is another root between $x = 10$ and $x = 15$. The word "ERROR" reminds us that logarithm functions are not defined for $x \leq 0$.

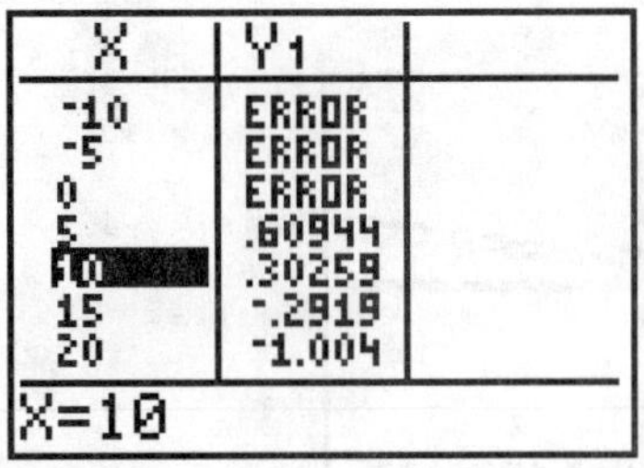

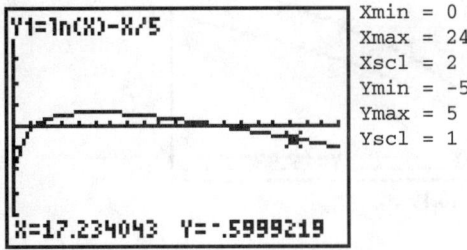

Figure 10.14

Following the same process of changing TBLStart and ΔTBL we arrive at an estimate of the root as $x = 12.71$. Are there more roots? Let's look at the graph. We set the graph window in Figure 10.15 at $0 \leq x \leq 20$ because $\ln x$ is undefined for $x \leq 0$ and $-5 \leq y \leq 5$. The graph indicates that the equation might intersect the x-axis only twice. Additional zooms out confirm that guess. The solutions are therefore $d = 1.30$ and $d = 12.71$.

Y1=ln(X)-X/5

X=17.234043   Y=-.5999219

Xmin = 0
Xmax = 24
Xscl = 2
Ymin = -5
Ymax = 5
Yscl = 1

Figure 10.15

Example 10.4

Solve the equation $\sqrt{3k - 5} = \sqrt{4 + 5k}$ with a table using the equal functions method. Report the solution to two decimal places.

Solution First define Y1 $= \sqrt{3x-5}$ and Y2 $= \sqrt{4+5x}$. The table is shown in Figure 10.16. There is no obvious place in the table where the values of Y1 and Y2 are equal.

X	Y1	Y2
1	ERROR	3
2	1	3.7417
3	2	4.3589
4	2.6458	4.899
5	3.1623	5.3852
6	3.6056	5.831
7	4	6.245

X=1

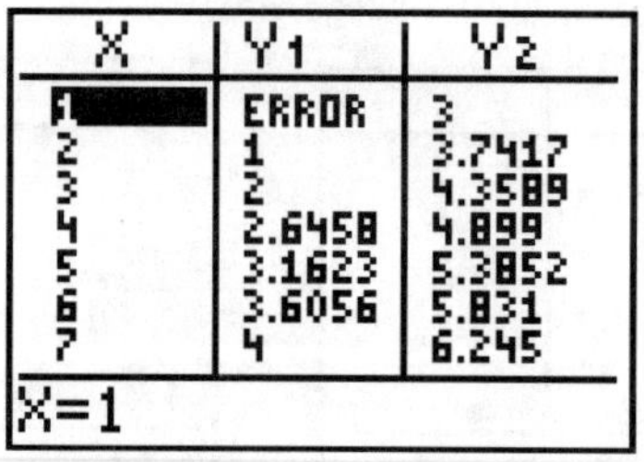

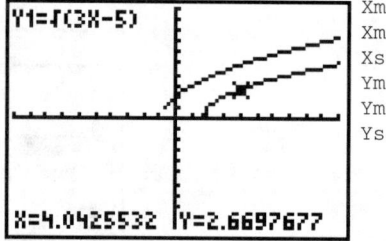

Figure 10.16

Figure 10.17

The best next step is to use a graph to give a visual perspective. Figure 10.17 shows that the graphs of the two functions seem to be on parallel paths. Are they parallel?

The only way to find out if these graphs are likely to meet is to zoom out. Zooming out (Figure 10.18) doesn't give enough information, so we need to examine only the first quadrant.

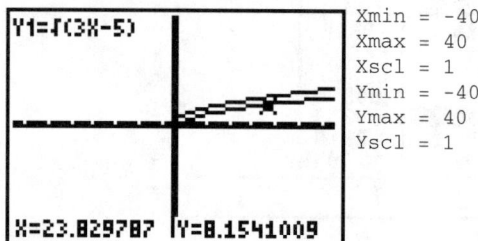

Figure 10.18

Figure 10.19, which sets the window dimension to $0 \leq x \leq 100$ and $0 \leq y \leq 25$, shows that the curves are actually moving apart. Therefore there is no solution to this equation.

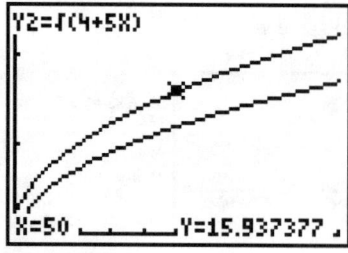

Figure 10.19

■

Notice in Example 10.4 that graphing played an important role in what started out as a table solution. The graphs gave visual feedback about how the function changed. It's usually best to use a combination of graphs and tables, carefully observing the shape of the curves.

The solution (that there is no solution) to Example 10.4 is not as uncommon as you might think. It often happens that there is no solution to an equation. When you are solving an equation with algebra only, it may not be as obvious that there is no solution. Example 10.5 shows an algebraic reason why there is no solution.

Example 10.5

Explain algebraically why there is no solution to the equation $\sqrt{3k-5} = \sqrt{4+5k}$.

Solution   We have seen in Chapter 9 that to solve exponential equations you can take the logarithm of both sides of the equation. In this equation we will apply another function to both sides of the equation: We will *square* both sides. The result of squaring both sides is

$$3k - 5 = 4 + 5k$$

which reduces to $k = -4.5$. This would appear to be a solution until you try to check the solution by substituting -4.5 for k into the original equation. Since $\sqrt{3(-4.5)-5} = \sqrt{-18.5}$, then $k = -4.5$ is not a solution because it is not possible to take the square root of a negative real number such as -18.5. So $k = -4.5$ is not a solution of the original equation.   ■

Squaring both sides of the equation, as we did in Example 10.5, will sometimes introduce numbers that are in fact *not* roots. Roots introduced in this way are called **extraneous roots**.

Section 10.1 Exercises

In Exercises 1–4 use the roots method to graphically solve the given equation.

1. $b^6 + b = 1$

2. $10^c + c = 15$

3. $0.1x^4 + 1.9x^3 - 2.1x^2 = 1.9x - 2$

4. $\sin\left(x^2\right) = x - x^3$

In Exercises 5–8 use the equal functions method to graphically solve the given equation.

5. $\log x^2 = x^2 - 1$

6. $\sin t = \ln(t + 1)$

7. $\cos x = \ln x + 1$

8. $\dfrac{5}{g-3} = 2 + \dfrac{4}{g+4}$

In Exercises 9–12 use the roots method with a table to solve the given equation.

9. $\sqrt{h+5} = \sqrt{4+5h^2}$

10. $\dfrac{2}{m+5} = \sqrt{m}$

11. $\dfrac{\sin x}{x} = 1$

12. $\dfrac{1}{x-5} = \sqrt{x+1}$

In Exercises 13–16 use the equal functions method with a table to solve the given equation.

13. $(\cos x)e^x = \ln x$

14. $\dfrac{1}{x+1} = \dfrac{1}{x - \sqrt{5}}$

15. $\sqrt{x+5} = \log x$

16. $\sin\left(\sqrt{x^2+1}\right) = e^x$

17. *Electricity* Three electric resistors are connected in parallel. The second resistor is $4\,\Omega$ more than the first and the third resistor is $1\,\Omega$ larger than the first. The total resistance is $1\,\Omega$. In order to find the first resistance R, we must solve the equation

$$\frac{1}{R} + \frac{1}{R+4} + \frac{1}{R+1} = 1$$

What are the values of the resistances?

18. *Architecture* The bending moment of a beam is given by $M(d) = 0.1d^4 - 2.2d^3 + 15.2d^2 - 32d$, where d is the distance in meters from one end. Find the values of d, where the bending moment is zero. (*Hint:* First multiply by 10 to eliminate the decimals.)

19. *Mechanical engineering* The characteristic polynomial for a certain material is

$$S^3 - 6S^2 - 78S + 108 = 0$$

Find the approximate value(s) of the stress between 1 and 2 psi.

20. *Medical technology* One of Poiseuille's laws states that the resistance R encountered by blood flowing through a blood vessel is given by $R(r) = \dfrac{CL}{r^4}$, where C is a positive constant determined by the viscosity of the blood, L is the length of the blood vessel, and r is the radius of the blood vessel. If $C = 1$, $L = 100\,\text{mm}$, and $R = 6.25/\text{mm}^3$, determine r.

21. *Chemistry* The reference polynomial for $C_{10}H_8$ (naphthalene) is given by

$$R(x) = x^{10} - 11x^8 + 41x^6 - 61x^4 + 31x^2 - 3$$

Solve this reference polynomial.

22. *Automotive technology* The displacement d of a piston is given by

$$d = \sin \omega t + \frac{1}{2} \sin 2\omega t$$

For what primary solutions of ωt being less than 2π is $d = 0$?

23. *Forestry* The yield, Y, in total ft³/acre of thinned stands of yellow poplar can be predicted by the equation

$$\ln Y = 5.36437 - 101.16296 S^{-1}$$
$$-22.00048 A^{-1} + 0.97116 \ln \text{BA}$$

where S is the site index, A is the current age of the trees, and BA is the basal area. What is the basal area if the predicted yield of a 60-year-old stand growing on a site with an index of 110 is $4{,}720\,\text{ft}^3/\text{acre}$?

10.2   Solving Two Equations in Two Variables

Up to this point, when you solved equations you have usually been looking for the value of one variable or the expression that one variable is equal to in terms of the other variables. However, in fields as different as engineering and economics, it is often necessary to solve more than one equation for more than one variable. When there is more than one equation and we seek the value of more than one variable, we are dealing with a **system of equations**. When there are two equations and two variables, we have a **2 × 2 system**. Three equations in two variables would be a 3×2 system. If the equations are all linear like $2x + 3y = 10$, but not like $2x^2 + 3y = 10$, the system is called a **linear system**. A solution to the system must satisfy all the equations in the system, as shown in Example 10.6.

2 × 2 is pronounced "2 by 2."

Example 10.6

Which of the points [a] (5, 2), [b] (2, 8), [c] (9, 7), [d] (6, 4), and [e] (−4, −6) are solutions of the 3 × 2 system

$$\begin{cases} x - y = 2 \\ x + y = 10 \\ \quad\quad y = 2x - 8 \end{cases}$$

Solution A solution of the system must make each of the equations true. If you substitute the values of x and y represented by each of the points, you will see that only one of the points satisfies *all three* of the equations. For example, the point $(5, 2)$ represents the values $x = 5$ and $y = 2$, which satisfy the third equation but not the others, so it is not a solution to the system. The only point that satisfies all three equations is $(6, 4)$. ∎

Some situations require a system of ten or even one hundred equations to be solved. For example, to find the maximum profit for our manufacturing situations in the Chapter Project, we will be required to work with systems of up to five equations and two variables: a 5×2 system.

In this chapter and the next we will explore different methods for solving systems of equations. As usual, there will be sets of equations that you can solve by pencil and paper methods and others by calculator methods.

We'll begin with a problem that you may want to solve in your head. We'll then show you two algebraic methods to solve 2×2 systems and two graphing calculator methods. With these tools you will be able to solve even the most difficult 2×2 system.

In this chapter we will deal mostly with 2×2 linear systems, and in the next chapter we will discuss systems of more than two equations and more than two variables.

Activity 10.5
Guess My Numbers

Try to answer the following questions without doing any algebra or using your calculator. After this activity, you'll learn how to solve this type of problem by algebra and with your calculator.

In each of the problems, the person speaking claims to be thinking of two numbers. See if you can figure out what the two numbers are. If you think any of the problems are impossible, write "No solution exists."

1. I am thinking of two numbers. If you add my numbers you get 10 and if you subtract them you get 2. What are the numbers?
2. I am thinking of two numbers. If you add my numbers you get 100 and if you subtract them you get 30. What are the numbers?
3. I am thinking of two numbers. If you add my numbers you get 25 and if you subtract them you get 4. What are the numbers?
4. I am thinking of two numbers. If you add my numbers you get 10 and if you subtract them you get -2. What are the numbers?
5. I am thinking of two numbers. If you add my numbers you get 10 and if you subtract them you get 10. What are the numbers?
6. I am thinking of two numbers. If you add my numbers you get 10 and if you subtract them you get 20. What are the numbers?
7. I am thinking of two numbers. If you add my numbers you get 10.2 and if you subtract them you get 3.7. What are the numbers?

Solving Two Equations: The Addition Method

Even if you were able to solve all of the problems in Activity 10.5 in your head, there are steps you can use to solve these problems that will help in more difficult ones.

The first step in solving problems like the ones in Activity 10.5 is to write an equation for each of the statements about the numbers. There will be two equations for each problem. In the addition method the two equations are added to produce a new equation with just one variable. In the next example you will learn how to solve these problems with algebra using the addition method.

Example 10.7

Solve question 7 from Activity 10.5 using the addition method. The statement was, "I am thinking of two numbers. If you add my numbers you get 10.2 and if you subtract them you get 3.7. What are the numbers?"

Solution   The first step is to write an equation for each statement. If we let x represent the larger of the two numbers and y the other number, then the two equations for question 7 are

$$\begin{cases} x + y = 10.2 \\ x - y = 3.7 \end{cases}$$

The addition method requires you to add these two equations to make a new equation with only one variable. For this pair of equations, the sum is

$$2x = 13.9$$

whose solution is $x = 6.95$. We can then substitute 6.95 for x into either of the first two equations and determine the value of y. We'll choose the first equation, since the positive coefficient of y will make it easier. Therefore we have

$$6.95 + y = 10.2$$

$$y = 10.2 - 6.95 = 3.25$$

Thus we have $x = 6.95$ and $y = 3.25$.

As a check, we substitute the values of x and y into the second equation, $x - y = 3.7$, producing

$$6.95 - 3.25 = 3.7$$

$$3.7 = 3.7$$

This is a true statement and so the answers are correct.   ■

The addition method for solving two equations is extremely powerful. However, there are two questions about it that many students ask:

▶ Why does the addition method work?
▶ What if I add the equations and the resulting equation still contains two variables?

We'll return to these two questions after we look at the second of our methods for solving two equations, the graphical method.

Solving Two Equations: The Graphical Method

Let's look at the first question in Activity 10.5. The statement was: "I am thinking of two numbers. If you add my numbers you get 10 and if you subtract them you get 2. What are the numbers?" Translated into equations we have

$$\begin{cases} x + y = 10 \\ x - y = 2 \end{cases}$$

Figure 10.20 shows the graphs of the two equations. There are many pairs of numbers that satisfy the equation $x + y = 10$, and many that satisfy $x - y = 2$, but only one point, $(6, 4)$ lies on both the lines. Therefore the pair of numbers $x = 6$ and $y = 4$ is the solution to the system.

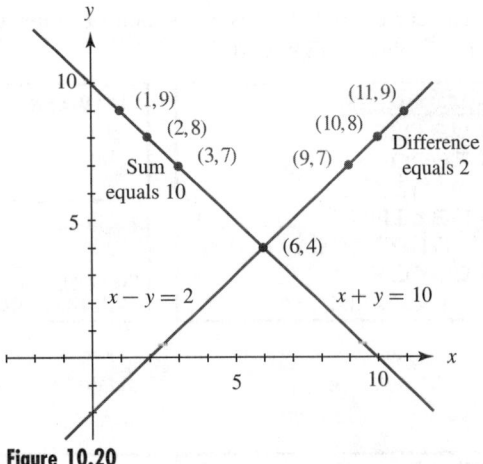

Figure 10.20

To solve two equations using the graphical method, plot the two equations and find any point or points of intersection.

Example 10.8

Solve question 7 of Activity 10.5 using the graphical method on your calculator. Use the `intersect` operation of the *TI-83*.

Solution   The equations are

$$\begin{cases} x + y = 10.2 \\ x - y = 3.7 \end{cases}$$

To graph these equations with the calculator, we must solve them for y.

$$\begin{cases} y = 10.2 - x \\ y = x - 3.7 \end{cases}$$

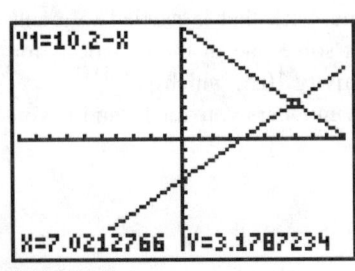

Figure 10.21

The graphs of the two equations are shown in Figure 10.21 with the `ZStandard` window setting. Using TRACE, we see that the approximate coordinates of the intersection point are (7.0, 3.2) which add up to 10.2, but they do not subtract to give 3.7.

You know that we can get better accuracy with ZOOM. However, we'll use instead a feature of the *TI-83* that can give a good estimate of the point of intersection. The feature is called `intersect` and is found on the [CALC] menu, which is above [TRACE].

(a) Press [2nd] [CALC] and you see the CALCULATE menu as shown in Figure 10.22(a).

(b) Select 5:intersect and you will see Figure 10.22(b), which requires you to identify the two curves you want to study. In general, press [ENTER] if you want to use the equation that is displayed or press either [▲] or [▼] to move to another equation without selecting the displayed equation.

(c) After you have selected the two equations you will see Figure 10.22(c) requiring your guess for the intersection. You can type a value of x, or press the right or left arrows to move the cursor closer to the point of intersection. Once you have the cursor moved near the point of intersection, press [ENTER] to accept the guess that is displayed.

(d) The correct coordinates of the point of intersection, $x = 6.95$ and $y = 3.25$, will be displayed (Figure 10.22(d)).

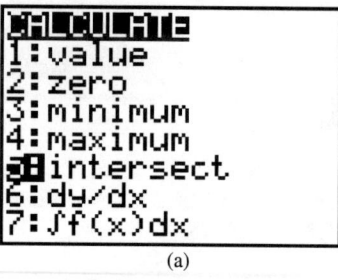

(a)

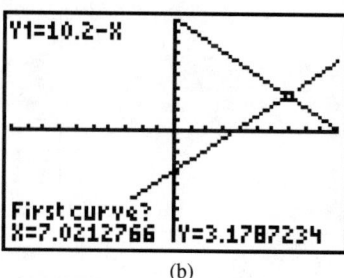

(b)

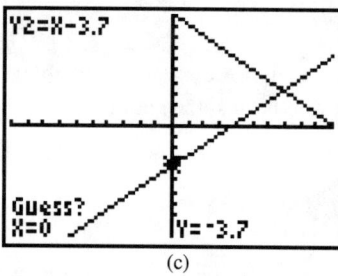

(c)

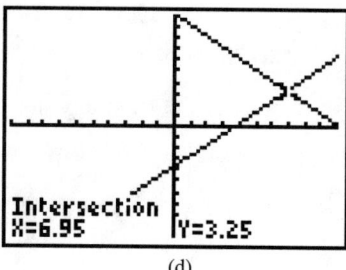

(d)

Figure 10.22

Solving Two Equations: The Substitution Method

Another algebraic method requires some of the same work that is needed to graph the equations on your calculator. That is, you must solve one of the equations for one of the variables. Let's look at question 6 of Activity 10.5, which is:

"I am thinking of two numbers. If you add my numbers you get 10 and if you subtract them you get 20. What are the numbers?"

The equations are

$$\begin{cases} x + y = 10 \\ x - y = 20 \end{cases}$$

Many students are suspicious of this problem because they think it does not have a solution. Of course we could solve it with the addition method, as you may have already done. But we will use this problem to demonstrate the substitution method.

Example 10.9

Solve question 6 of Activity 10.5 using the method of substitution.

Solution   Look at the two equations and choose the equation and variable you think would be easiest to solve. We'll choose the first equation and solve for x. The two equations are

$$\begin{cases} x = -y + 10 \\ x - y = 20 \end{cases}$$

The idea is to substitute $-y + 10$ for x in the other equation. This results in

$$-y + 10 - y = 20$$

We now solve this new equation for y.

$$-2y + 10 = 20$$
$$-2y = 10$$
$$y = -5$$

If $y = -5$ then by substituting -5 for y in the equation $x = -y + 10$ we have $x = -(-5) + 10 = 15$. So the answers are $x = 15$ and $y = -5$.

Check by substituting into the two equations:

$$\begin{cases} x + y = 10 \\ 15 + (-5) = 10 \\ 10 = 10 \end{cases} \quad \text{and} \quad \begin{cases} x - y = 20 \\ 15 - (-5) = 20 \\ 20 = 20 \end{cases}$$

Since both the statements $10 = 10$ and $20 = 20$ are true the answers of $x = 15$ and $y = -5$ are correct.   ■

In the following example we address a nonlinear system. The addition method will generally not work for a nonlinear system, so we will use substitution and graphing. You will see that there can be more than one solution to a nonlinear 2×2 system.

Example 10.10

Find all the solutions of the system

$$\begin{cases} y = -0.5x^2 + 3x + 10 \\ y = 0.7x^2 - 8 \end{cases}$$

Solution   We'll show two methods of solution. First, you could solve this problem with substitution. Since in both equations y is expressed in terms of x, we can put them together in one equation:

$$-0.5x^2 + 3x + 10 = 0.7x^2 - 8$$

This is a quadratic equation and it can be solved with the quadratic formula. You have also learned that a quadratic equation can have one, two, or zero real number solutions, depending on the coefficients of the equation. Therefore in this example we must be prepared for these possibilities.

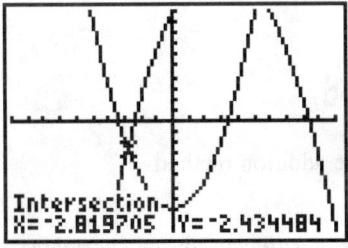

Intersection
X=-2.819705   Y=-2.434484

Figure 10.23

We can also solve this system using the calculator's `intersect` operation. The graph with one point of intersection $(-2.8, -2.4)$ is shown in Figure 10.23.

You can see that there appears to be another point of intersection off the screen in the upper right. The `intersect` operation will find the closest point to the guess that you give. To find the second point of intersection, make your guess closer to the right hand intersection. The second solution is approximately $x = 5.3$ and $y = 11.8$.   ■

There is one more type of 2×2 system for which the substitution method is the best tool. In Activity 10.6 we return to the Olympic swimming data that you studied in Chapter 1. We'll compare the men's records with the women's records over the years.

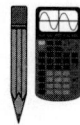

Activity 10.6
Will Women Swimmers Catch the Men?

Figure 10.24 shows the graph of the Olympic 100 meter swimming records for men (squares) and women (crosses).

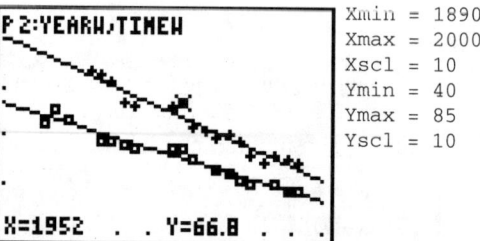

```
Xmin = 1890
Xmax = 2000
Xscl = 10
Ymin = 40
Ymax = 85
Yscl = 10
```

Figure 10.24

You studied the men's data in Chapter 1. The women's records (stored in lists named `YEARW` and `TIMEW`) show that their times have been higher than the men's, but the difference between the times has been decreasing. The two lines in Figure 10.24 are the linear regression lines for each set of data. In each case the time for the first race was omitted because it was quite a bit higher than the other times. The two regression lines are

Men: $y = -0.180x + 406$
Women: $y = -0.268x + 586$

1. What do the coordinates at the intersection point of the two lines in Figure 10.24 mean?
2. Use substitution to solve the 2 × 2 system and calculate the coordinates of the point of intersection.
3. Check your result by using the calculator's `intersect` operation.
4. Which method do you prefer, and why?
5. Explain why the prediction is or is not a realistic one.

Questions About the Addition Method

Let's return to the two questions we raised about the addition method.

▶ *Why does the addition method work?*
▶ *What if I add the equations and the resulting equation still contains two variables?*

The addition method works because whenever you add two equations that have at least one point of intersection, the equation that results from the addition passes through the same point(s) of intersection. This is illustrated in Figure 10.25, which shows the two equations

$$\begin{cases} 2x + 3y = 12 \\ 3x - 2y = 18 \end{cases}$$

and their sum $5x + y = 30$. The third line passes through the point of intersection of the other two. If we can force the new equation to have only one variable, like $3y = 15$ or $2x = -20$, then we'll have the solution to the system. However, a sum

The equation $5x + y = 30$ is not completely useless, as it could be easily solved for y to give $y = -5x + 30$, and this equation could be used in the substitution method.

like $5x + y = 30$ gives nothing new about the first two equations and so is almost useless.

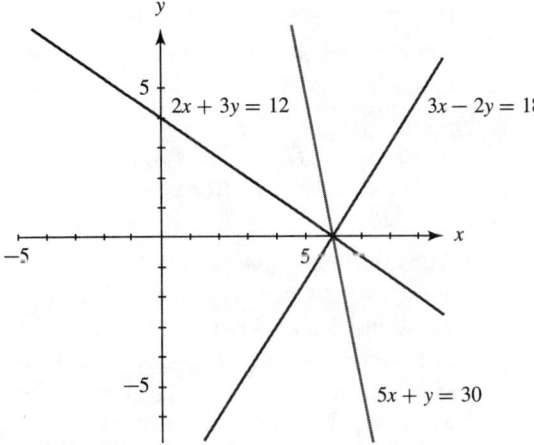

Figure 10.25

So we now come to the second question about the addition method: *What if addition doesn't give the answer directly?* The answer is that we change the equations so that addition (or subtraction) will eliminate one variable in the sum. Example 10.11 shows how by using the equations in Figure 10.25. The method is a variation of the addition method and requires some multiplication or division first.

Example 10.11

Use the addition method to solve the equations

$$\begin{cases} 2x + 3y = 12 \\ 3x - 2y = 18 \end{cases}$$

Solution Look at the coefficients of y in the two equations; they are 3 and -2. If you multiply the first equation by 2 and the second equation by 3, then the coefficients of y will be 6 and -6. Now addition of the two equations will eliminate y.

$$2(2x + 3y = 12) \quad \text{or} \quad 4x + 6y = 24$$

$$3(3x - 2y = 18) \quad \text{or} \quad 9x - 6y = 54$$

These new equations are in fact not new—their graphs are identical to the original equations. When you add them you get

$$13x = 78$$

$$x = 6$$

So we have the value of x. To find the value of y we must substitute $x = 6$ into either of the *original* equations. We'll take the first one because it has positive coefficients.

$$2(6) + 3y = 12$$

$$3y = 0$$

$$y = 0$$

The solution, $x = 6$ and $y = 0$, is the same one you can see in Figure 10.25. ∎

Summary: Methods for 2 × 2 Systems

All 2×2 systems can be solved by each of these methods, except that nonlinear systems are not usually solved with the addition method.

▶ **Graphical:** Graph the equations and find the point of intersection

- To use the *TI-83*, you will have to solve the equations for the variable y.
- The calculator's `intersect` operation will give a good estimate of the solution.

▶ **Substitution:** Solve one of the equations for y and substitute that expression for y in the other equation. You could also solve one equation for x and substitute that expression for x in the other equation.

▶ **Addition:** Multiply or divide each equation by numbers that will produce equal in absolute value (but opposite signed) coefficients for one of the variables. Then add the equations to produce a new equation that does not contain that variable. Solve for the remaining variable, then substitute that value back into one of the *original* equations to solve for the second variable.

Example 10.12 *Application*

Two different gasohol mixtures are available. One mixture contains 6% ethonol and the other 12%. State regulations require that gasohol contain 10% ethonol. How much of each mixture should be used in order to get 25 000 liters of gasohol that satisfies state regulations?

Solution We will let x represent the number of liters of 6% ethonol mixture it will take and y the number of liters for the 12% mixture. We mix these two mixtures to get a total of 25 000 liters, so $x+y = 25\,000$. Of this 25 000 liters, 10%, or $(0.10)(25\,000) = 2500$ liters, will be ethonol. This will consist of $0.06x$ liters of the 6% mixtures and $0.12y$ liters of the 12% mixture. Thus, $0.06x + 0.12y = 2500$. We now have the system of two equations with two variables

$$\begin{cases} x + y = 25000 \\ 0.06x + 0.12y = 2500 \end{cases}$$

If we solve each of these equations for y we can use the calculator to solve this system. The resulting equations are

$$\begin{cases} y = 25000 - x \\ y = \dfrac{2500 - 0.06x}{0.12} \end{cases}$$

We cannot have a negative number of liters and 25 000 is the most number of liters of each type that are possible. Thus, for any solution $0 < x < 25\,000$ and $0 < y < 25\,000$. As you can see in Figure 10.26, the solution is $x \approx 8333.33$ and $y \approx 16\,666.67$.

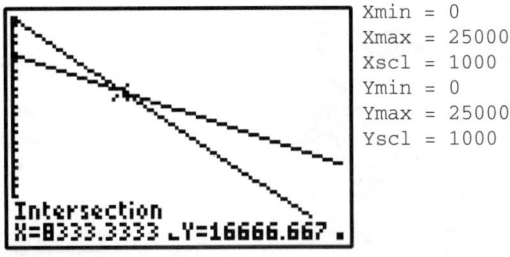

```
Xmin = 0
Xmax = 25000
Xscl = 1000
Ymin = 0
Ymax = 25000
Yscl = 1000
```

Figure 10.26

Application: Kirchhoff's Laws of Circuit Analysis

Two of the most useful relationships in electronics were discovered in the 19th century by a German physicist, Gustav Kirchhoff, who also was a co-discoverer of two chemical elements, cesium and rubidium. With Kirchhoff's laws you can produce systems of equations that are used to analyze the current flow in series and parallel direct current circuits. Now that we know something about solving systems, we can test our skills on a real application.

Kirchhoff's Laws of Electric Circuits

1. **Current through a point in a circuit (Kirchhoff's current law)**

   The sum of the currents entering a point in a circuit is equal to the sum of the currents leaving that point.

   For example, in Figure 10.27 two wires meet at point A and the current that leaves point A is equal to their sum.

2. **Voltages within a circuit (Kirchhoff's voltage law)**

   In any closed circuit, the applied voltage is equal to the sum of the voltages through the components in the loop. Sometimes this is rephrased as *the applied voltage equals the sum of the voltage drops across each of the components in the circuit.*

   For example, Figure 10.28 shows a circuit with a voltage source and two resistors. Since Ohm's law states that $V = I \cdot R$, then Kirchhoff's voltage law gives the relationship shown in the figure.

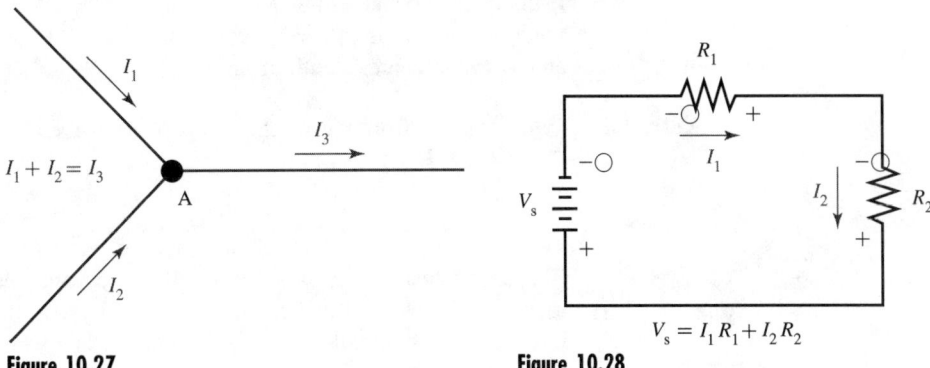

Figure 10.27

Figure 10.28

Example 10.13

Answer the following questions about the parallel circuit shown in Figure 10.29. There are two voltage sources, V_a and V_b, and three resistors, R_1, R_2, and R_3.

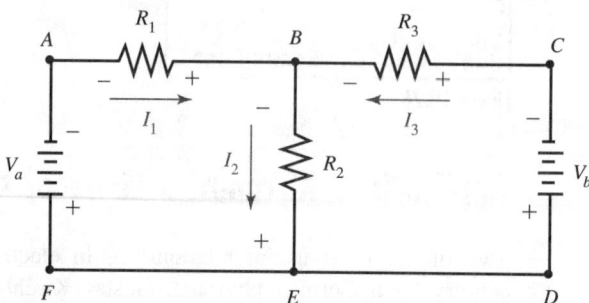

Figure 10.29

(a) What does Kirchhoff's current law predict about the three currents I_1, I_2, and I_3 at point B in the circuit?

(b) What does Kirchhoff's voltage law predict about each of these closed circuits?

    i. The left circuit A–B–E–F

    ii. The right circuit B–C–D–E

Solution (a) Currents I_1 and I_3 are flowing in to point B, and current I_2 is flowing out. Therefore

$$I_1 + I_3 = I_2$$

(b) Each circuit is analyzed separately

$$\begin{cases} V_a = I_1 R_1 + I_2 R_2 \\ V_b = I_3 R_3 + I_2 R_2 \end{cases}$$

■

Now that you have seen how to write the voltage and current equations, you can use those equations to predict the currents through the three resistors if you know the values of the resistances and the voltage sources.

Example 10.14   *Application*

If the three resistances are $R_1 = 6$ ohms, $R_2 = 12$ ohms, and $R_3 = 4$ ohms, and the voltages sources are $V_a = 80$ volts and $V_b = 100$ volts, what are the currents I_1, I_2, and I_3 to the nearest tenth of an amp?

Solution Two equations from the solution in part (b) of Example 10.13 produce the following:

$$\begin{cases} 80 = 6I_1 + 12I_2 \\ 100 = 4I_3 + 12I_2 \end{cases}$$

This is a 2×3 system, which we have not yet learned how to solve. You'll see in Chapter 11 that solving a system with three variables requires a three-dimensional graph. For now we'll just change this 2×3 system into a 2×2 system by using the information provided by Kirchhoff's current law where

$$I_1 + I_3 = I_2$$

We can use this equation to replace any of the three currents. We'll choose to make the substitution $I_1 = I_2 - I_3$ into the two voltage equations:

$$\begin{cases} 80 = 6(I_2 - I_3) + 12I_2 \\ 100 = 4I_3 + 12I_2 \end{cases}$$

Simplifying and rearranging the equations so that the variables line up, we have

$$\begin{cases} 80 = 18I_2 - 6I_3 \\ 100 = 12I_2 + 4I_3 \end{cases}$$

You could solve these with a graph or with the addition or substitution method. We'll show addition and graphing methods.

Addition Method: We can eliminate I_3 by multiplying the first equation by 2 and the second equation by 3. This will change the coefficients of I_3 to -12 and 12, respectively. That means that I_3 will be eliminated if we add the equations.

$$80 = 18I_2 - 6I_3$$
$$160 = 36I_2 - 12I_3 \text{ Multiply by 2}$$
$$100 = 12I_2 + 4I_3$$
$$300 = 36I_2 + 12I_3 \text{ Multiply by 3}$$

Adding the second and fourth equations produces a single equation with one variable

$$460 = 72I_2$$

$$I_2 \approx 6.39$$

Thus, we see that $I_2 \approx 6.39$ amps.

Substituting $I_2 = 6.39$ into the *original* equation $80 = 18I_2 - 6I_3$ we get

$$80 = 18 \cdot 6.39 - 6I_3$$

$$6I_3 = 18 \cdot 6.39 - 80 = 35.02$$

$$I_3 = \frac{35.02}{6} \approx 5.84$$

Thus, $I_3 \approx 5.84$ amps. Finally $I_1 = I_2 - I_3 = 0.55$ amp. The answers, to the nearest tenth, are $I_1 = 0.6$ A, $I_2 = 6.4$ A, and $I_3 = 5.8$ A.

Graphical Method: To solve this system with a calculator graph, we must choose to replace one of the variables with x and the other with y, then solve the two equations for y. The choice could go either way. We'll choose to replace I_2 with x and I_3 with y. The two equations are

$$80 = 18x - 6y$$

$$y = \frac{80 - 18x}{-6}$$

$$100 = 12x + 4y$$

$$y = \frac{100 - 12x}{4}$$

Both of the fractions $\dfrac{80 - 18x}{-6}$ and $\dfrac{100 - 12x}{4}$ could be reduced by dividing top and bottom by common factors, but that simplification is not usually necessary when you are doing calculator work. The graphs of the two equations, and their point of intersection, are shown in Figure 10.30. Make sure you understand how the values of x and y in the figure relate to the answers of the problem.

Keep two decimal places now because the answer is requested to one decimal place. Remember not to round off too soon.

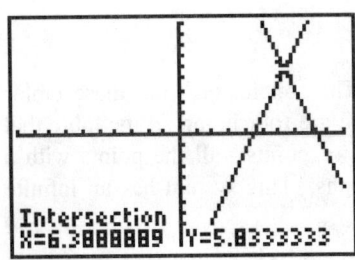

Intersection
X=6.3888889 Y=5.8333333

Figure 10.30

How Many Solutions are Possible?

Each of the linear systems we have studied had only one solution: the point where the two lines crossed. The one nonlinear 2×2 system we studied had two solutions because the parabolas in Figure 10.23 on page 691 intersected in two points. What are the different possibilities for a 2×2 system of quadratic equations? Three of the possibilities are shown in Figure 10.31. Example 10.15 explores a third possibility for a 2×2 system of quadratic equations.

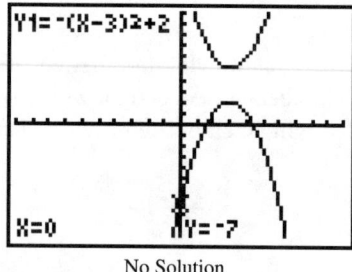
No Solution

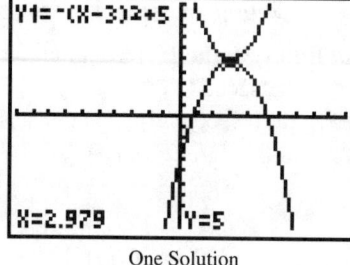
One Solution

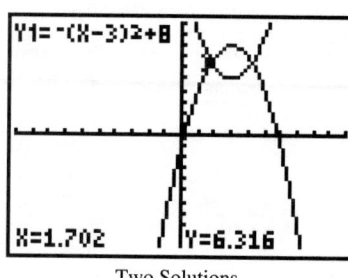
Two Solutions

Figure 10.31

Example 10.15

Complete the table of values for each equation in the system and discuss how many solutions there are to the system.

$$\begin{cases} A : y = x^2 + 3 \\ B : x = \sqrt{y - 3} \end{cases}$$

A:

x	−4	−2	0	2	4	6
y						

B:

x	−4	−2	0	2	4	6
y						

Solution Table A is easy to complete from the formula for y. The results are $19, 7, 3, 7, 19$, and 39. Table B is not as easy. If $x = -4$ there is no value of y because there is no real number whose square root is equal to -7. If $x = 4$ we substitute, square both sides, and solve for y

$$4 = \sqrt{y - 3}$$
$$16 = y - 3$$
$$y = 19$$

You can now complete the table for equation B. The conclusion from these tables is that even though the equations are different, there are four points in the table that are identical. In fact, there are an infinite number of points—all the points with a positive or zero value for x—that are on both graphs. This system has an infinite number of solutions. ∎

What are the possibilities for a linear 2×2 system? It is always possible to make up a situation in which two straight lines are identical as we did for the equations in Example 10.15. The other possibilities are that the lines would be parallel (no solution) or that they would intersect (one solution).

In Activity 10.7 you will see how many solutions there could be in different kinds of 2 × 2 systems.

Activity 10.7
How Many Solutions for Two Polynomials?

The table below shows different combinations of equations that could appear in a 2 × 2 system. For example, we have filled in the correct responses for $n = 1$ and $m = 1$ (two straight lines) and $n = 2$ and $m = 2$ (two parabolas). Other entries in the table will require you to sketch different kinds of curves to see the different intersection possibilities. You studied the shapes of these polynomials in Chapter 8. Figure 10.32 shows one of the possibilities for $n = 1$ (straight line) and $m = 4$ (fourth degree polynomial). In this graph there are four solutions to the system. What other possibilities are there for these two curves? Is it possible to get five solutions? Is it possible to get zero solutions?

Degree of first polynomial (n)	Degree of second polynomial (m)	Possible number of solutions
1	1	0, 1, infinite
1	2	
2	2	0, 1, 2, infinite
1	3	
2	3	
3	3	
1	4	
2	4	
3	4	
4	4	

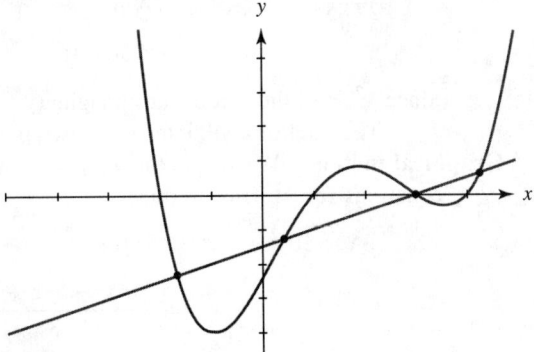

Figure 10.32

Example 10.16   *Application*

Two satellites are in orbit in the same plane. One orbit is described by $5(x + 10)^2 + 3y^2 = 3\,000$. The other is described by $5x^2 + 4y^2 = 4\,000$. What are the points where the two orbits intersect?

Solution   We will first solve this algebraically by using the substitution method and then the graphical method.

Substitution method: We want to solve the system that contains the equations of two ellipses.

$$\begin{cases} 5(x + 10)^2 + 3y^2 = 3\,000 \\ 5x^2 + 4y^2 = 4\,000 \end{cases}$$

We will multiply the first equation by 4 and the second equation by -3. This will give the y^2-terms the coefficient of 12 and -12, respectively.

$$\begin{cases} 20(x + 10)^2 + 12y^2 = 12\,000 \\ -15x^2 - 12y^2 = -12\,000 \end{cases}$$

Adding, we get

$$20(x + 10)^2 - 15x^2 = 0$$

or $$5x^2 + 400x + 2\,000 = 0$$

$$x^2 + 80x + 400 = 0$$

Using the quadratic formula we get

$$x = \frac{-80 \pm \sqrt{80^2 - 4(1)(400)}}{2} = \frac{-80 \pm \sqrt{4800}}{2}$$

$$= -40 \pm 20\sqrt{3}$$

Substituting $x = -40 + 20\sqrt{3} \approx -5.36$ into the second of the original two equations we get

$$143.65 + 4y^2 = 4\,000$$

$$y^2 = 964.09$$

$$y \approx \pm 31.05$$

Substituting $x = -40 - 20\sqrt{3} \approx -74.64$ into the second of the original two equations we get

$$27\,855.65 + 4y^2 = 4\,000$$

$$y^2 \approx -5\,963.91$$

Since $y^2 < 0$, these roots are imaginary.

The satellites will intersect at two points: $(-5.36, 31.05)$ and $(-5.36, -31.05)$.

Graphical method: We begin the graphical method by solving the two original equations for y. This produces

$$5(x + 10)^2 + 3y^2 = 3\,000$$

$$y^2 = \frac{3\,000 - 5(x + 10)^2}{3}$$

$$y = \pm \sqrt{\frac{3\,000 - 5(x + 10)^2}{3}} \tag{1}$$

$$5x^2 + 4y^2 = 4\,000$$

$$y^2 = \frac{4\,000 - 5x^2}{4}$$

$$y = \pm\sqrt{\frac{4\,000 - 5x^2}{4}} \tag{2}$$

In Figure 10.33 the two equations in (1) are graphed in the thick graph style and the two equations in (2) are graphed in the normal graph style. By using `intersect` we see in Figure 10.33 that one point of intersection is about $(-5.36, 31.05)$. By using `intersect` again we obtain the second point of intersection as $(-5.36, -31.05)$.

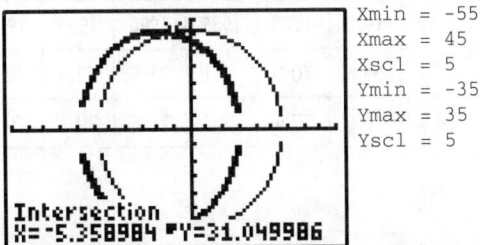

```
Xmin = -55
Xmax = 45
Xscl = 5
Ymin = -35
Ymax = 35
Yscl = 5
```

Figure 10.33

Section 10.2 Exercises

In Exercises 1–8 use the addition method to solve each system of equations.

1.
$$\begin{cases} x + y = 17 \\ x - y = -5 \end{cases}$$

2.
$$\begin{cases} x + 2y = 9 \\ 3x - 2y = 7 \end{cases}$$

3.
$$\begin{cases} 5x + 2y = 17 \\ 3x - 2y = 75 \end{cases}$$

4.
$$\begin{cases} 3x - 4y = 1 \\ 3x + 2y = 13 \end{cases}$$

5.
$$\begin{cases} x - 5y = 1 \\ 2x + y = 2 \end{cases}$$

6.
$$\begin{cases} 4x + y = 7 \\ 2x - 2y = 9 \end{cases}$$

7.
$$\begin{cases} 0.2x + 0.5y = 6 \\ 1.2x + 0.1y = 4.1 \end{cases}$$

8.
$$\begin{cases} 3.2x - 0.5y = -6.4 \\ 0.4x + 1.5y = 4.2 \end{cases}$$

In Exercises 9–12 use the substitution method to solve each system of equations.

9.
$$\begin{cases} x + 2y = 9 \\ 2x + y = 12 \end{cases}$$

10.
$$\begin{cases} 3x - y = 6 \\ x + 5y = -14 \end{cases}$$

11.
$$\begin{cases} 1.2x + 2.5y = -14 \\ 2.4x - y = 20 \end{cases}$$

12.
$$\begin{cases} 1.5x + 7.5y = 41.4 \\ x + 2y = 9 \end{cases}$$

In Exercises 13–20 use the graphical method to solve each system of equations.

13.
$$\begin{cases} 2.4x - 5.4y = 9.222 \\ 6.1x + 1.2y = 0.007 \end{cases}$$

14.
$$\begin{cases} 6.5x - 4.6y = 44.09 \\ 2.5x + 1.2y = 9.98 \end{cases}$$

15.
$$\begin{cases} 9.4x - 3.6y = -2.2 \\ 2.7x - 5.3y = 31.96 \end{cases}$$

16.
$$\begin{cases} 6.3x + 24.5y = -6.58 \\ 1.5x + 12.5y = 12.8 \end{cases}$$

17.
$$\begin{cases} x^2 + y = 3 \\ x + y^2 = 3 \end{cases}$$

18.
$$\begin{cases} 8x^2 + 4y^2 = 36 \\ y = x^2 - 3 \end{cases}$$

19. $\begin{cases} y = x^3 + 2x^2 - x - 2 \\ y^2 = x + 2 \end{cases}$

20. $\begin{cases} y = \dfrac{1}{x+5} - 2 \\ x + y = 6 \end{cases}$

21. Use regression to produce a straight line that fits the men's Olympic swimming records stored in the lists YEAR and TIME (men's records), and a straight line that fits the lists YEARW and TIMEW (women's records). Find the point of intersection and compare your results with the answer you got to Activity 10.6, in which the regression lines were computed without the first records in each list. Explain why there is a difference in the results.

22. *Biology* The life expectancies at birth of males and females from 1970–1994 are shown in the table below.

Year, t	1970	1975	1980	1985	1990	1992	1994
Life expectancy (males), L_1	67.1	68.8	70.0	71.1	71.8	72.3	72.3
Life expectancy (females), L_2	74.7	76.6	77.4	78.2	78.8	79.1	79.0

(a) What are the linear regression equations that satisfy this information? You will need a linear regression for male life expectancy and one for female life expectancy.

(b) Using your results from **(a)**, in what year will the life expectancies be the same and what will the average life expectancy be of a child born in that year?

(c) What are the natural logarithmic regression equations that satisfy this information? You will need a logarithmic regression for male life expectancy and one for female life expectancy.

(d) Using your results from **(c)**, in what year will the life expectancies be the same and what will the average life expectancy be of a child born in that year?

(e) Discuss the results. Are these answers reasonable? What other techniques could you use? Why do you think these other techniques would give more reasonable results?

23. *Petroleum technology* Two different gasohol mixtures are available. One mixture contains 4% alcohol and the other, 12% alcohol. How much of each mixture should be used to get 20 000 L of gasohol containing 10% alcohol?

24. *Automotive technology* A 12-liter cooling system is filled with 25% antifreeze. How many liters must be replaced with 100% antifreeze to raise the strength to 45% antifreeze?

25. *Electronics* Find the currents in the three resistors of the circuit shown in Figure 10.34, given that $E_1 = 8$ V, $E_2 = 5$ V, $R_1 = 3\,\Omega$, $R_2 = 5\,\Omega$, and $R_3 = 6\,\Omega$. [Use $E_1 = I_1R_1 + I_2R_2$ and $E_2 = R_2I_2 - R_3(I_1 - I_2)$.]

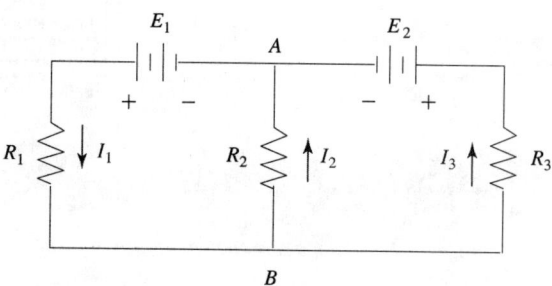

Figure 10.34

26. *Electronics* In Figure 10.34, if $E_1 = 10$ V, $E_2 = 15$ V, $R_1 = 2\,\Omega$, $R_2 = 4\,\Omega$, and $R_3 = 8\,\Omega$, find I_1, I_2, and I_3.

27. *Electronics* In Figure 10.35, we have $I_2 = I_1 + I_3$, $E_1 = R_1I_1 + R_2I_2$, and $E_2 = R_3I_3 + R_2I_2$. If $E_1 = 6$ V, $E_2 = 10$ V, $R_1 = 8\,\Omega$, $R_2 = 4\,\Omega$, and $R_3 = 7\,\Omega$, find I_1, I_2, and I_3.

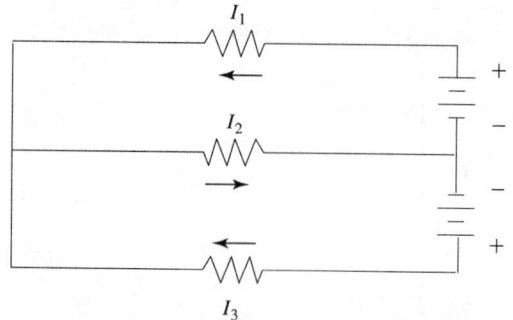

Figure 10.35

28. *Electronics* In Figure 10.36, we have a series–parallel circuit with $V_1 = 9$ V, $R_1 = 3\,\Omega$, $R_1 = 5\,\Omega$, and $R_3 = 6\,\Omega$. Find I_1 and I_2.

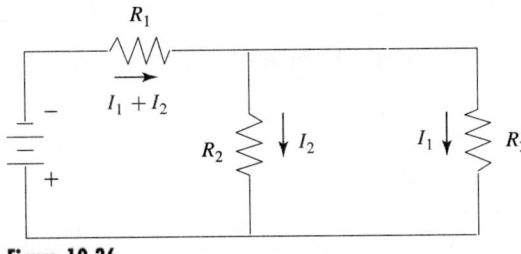

Figure 10.36

29. *Electrical engineering*   In determining the Thomson electromotive force of a certain thermocouple, it is necessary to solve the system of equations

$$0.02\left(T_2^2 - T_1^2\right) = 4280$$

$$T_2 - T_1 = 250$$

Solve this system of equations for T_1 and T_2 in degrees Kelvin.

30. *Business* In business, the breakeven point for a product is where the revenue from selling the product equals the cost of producing the item. A company produces and sells x units of product A and y units of product B. The company has determined that the cost of producing the two products is given by $x^2 + 1.5y^2 - xy = 2,891,189$. Meanwhile, the revenue is given by $80x + 100y = 252,640$. What is the breakeven point?

31. Compare the advantages and disadvantages of the addition, substitution, and graphical methods.

32. Explain how you decide whether to use the substitution method or the addition and subtraction method to solve a system of nonlinear equations.

10.3

Inequalities in One Variable

Inequalities are like equations, with the $=$ sign replaced with one of the inequality symbols $<$, $>$, $\leq$, $\geq$, or $\neq$. Each of the symbols is described in the chart below.

Symbol	Meaning	Example
$<$	is less than	The inequality $4 < 8$ is true because 4 is less than 8.
$>$	is greater than	The inequality $4 > 4$ is false because 4 is not greater than 4.
$\leq$	is less than or equal to	The inequality $4 \leq 4$ is true.
$\geq$	is greater than or equal to	The inequality $x + 2 \geq 5$ is true if x is any real number, 3 or greater.
$\neq$	is not equal to	The statement $3 \neq 5$ is true because 3 is not equal to 5.

The $\neq$ symbol is formed by drawing a slash through the $=$ symbol. You can negate any of the other inequality symbols by putting a / through it. Thus, $\not>$ means *is not greater than* and is logically the same as $\leq$.

When we work with inequalities we usually want the answer to the question "What value or values make this inequality true?" In this section we will study inequalities that contain only one variable, and in Section 10.4 we will study inequalities that contain two variables. In both sections we will return to the manufacturing problems that began this chapter.

Activity 10.8
Exploring Inequalities

For each of the following inequalities, try different values for x until you can write four values that will make the inequality true. In each set of four answers, include one negative number, if that is possible.

1. $2x \geq 7$

   ▶ What are four values of x that make this true?
   ▶ Is 3.5 a solution of this inequality?

2. $3x - 2 < 6$

   ▶ What are four values of x that make this true?
   ▶ Is $\frac{8}{3}$ a solution of this inequality?

3. $10^x \leq 100$

   ▶ What are four values of x that make this true?
   ▶ Is 2 a solution of this inequality?

4. $\log x > 5$

   ▶ What are four values of x that make this true?
   ▶ Is 10,000 a solution of this inequality?

5. $x^2 \geq 50$

   ▶ What are four values of x that make this true?
   ▶ Is $\sqrt{50}$ a solution of this inequality?

6. $x^2 < 4$

   ▶ What are four values of x that make this true?
   ▶ Is 2 a solution of this inequality?

7. $-2 < x < 2$

   ▶ What are four values of x that make this true?
   ▶ Is -2 a solution of this inequality?

8. $|x| < 2$

   ▶ What are four values of x that make this true?
   ▶ Is 2 a solution of this inequality?

9. $-x < 5$

   ▶ What are four values of x that make this true?
   ▶ Is -5 a solution of this inequality?

10. $-2x \geq -10$

    ▶ What are four values of x that make this true?
    ▶ Is 5 a solution of this inequality?

Representing the Solution Of an Inequality With a Graph

You have plotted an equation in two variables on a two dimensional graph. However, an equation or inequality in one variable requires only a one-dimensional graph. It is customary to show the solutions of one-variable inequalities using a one-dimensional number line.

Figure 10.37 shows the customary way to graph five different inequalities. The shaded parts of the number line represent values that are solutions to the inequality. When the endpoint of a number line graph is filled in, the endpoint is included in the solution. When the endpoint of the graph is hollow, then the endpoint is not included in the solution.

The inequalities that are graphed in Figure 10.37 represent five types of solutions you should be able to use without algebra or a calculator. Table 10.1 contains a description of the solutions graphed in Figure 10.37.

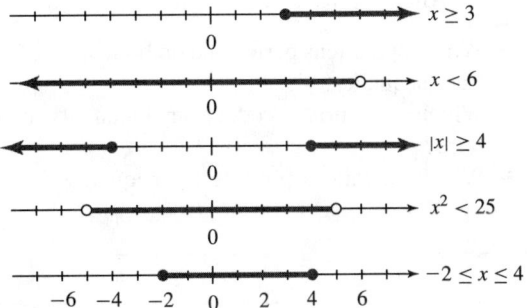

Figure 10.37

Table 10.1

Inequality	Description of the solution
$x \geq -3$	All real numbers greater than or equal to -3.
$x < 6$	All real numbers less than 6.
$|x| \geq 4$	All real numbers greater than 4 or less than -4, including 4 and -4.
$x^2 < 25$	All real numbers between, but not including, -5 and 5.
$-2 \leq x \leq 4$	All real numbers between, and including, -2 and 4.

A trial-and-error approach can be useful to find a few solutions to an inequality. However we are usually interested in finding *all* the values of the variable that solve the inequality. To do that we now look at some algebraic and graphical methods for solving an inequality.

Solving an Inequality with Algebra

An inequality is solved like an equation, with some important exceptions that you will meet soon. Just as with equations, you will simplify inequalities by adding, subtracting, multiplying, and dividing both sides by the same number. Not all of these operations follow the same rules you use with equations, so watch carefully for the differences. In Activity 10.9 you will experiment with different operations to see which produce inequalities that remain true.

Activity 10.9
Transforming Inequalties

The four inequalities listed on the top of the following table are true numerical statements. Each line begins with an operation. For example, the first begins with "add 5." Use that operation on each of the given inequalities. So for the inequality $2 < 8$ you are to add 5 to both sides of the inequality. This produces $2 + 5 < 8 + 5$ or $7 < 13$. This is entered in the column headed "Result." Finally, state whether this is a true inequality. Some of the entries have been completed. The entries marked **N.A.** stand for "Not Applicable," because, for example, you can't take the square root or logarithm of a negative number.

After you have completed the table, answer the following questions:

▶ Which operations performed on both sides of the inequality *always* produce another true inequality?
▶ Which operations produce an inequality that is *sometimes* true and *sometimes* false?
▶ Which operations produce an inequality that is *never* true?

Start with:→	2 < 8		2 > −8		−2 < 8		−2 > −8	
Operation ↓	Result	Still True?	Result	Still True?	Result	Still True?	Result	Still True?
add 5	7 < 13	Yes						
add −5					−7 < 3	Yes		
subtract 3								
subtract −3								
multiply by 4			8 > −32	Yes				
multiply by −4								
divide by 2								
divide by −2							1 > 4	No
square root	1.4 < 2.8	Yes					N.A.	N.A.
square			4 > 64	No				
log								
raise 10 to a power	$10^2 < 10^8$	Yes						

As a result of Activity 10.9 we can make the following statements about operations performed on both sides of the inequality:

Solving Inequalities, Part 1

▶ To solve an inequality you can do either of the following and the resulting inequality will still be true:
 • Add or subtract any number
 • Multiply or divide by a positive number
▶ If you multiply or divide by a negative number, then you must change the direction of the inequality sign to make the resulting inequality true.
▶ The following operations do not give predictable results. None of them should be applied to both sides of an inequality:
 • Square, square root, logarithm
 • Raise to any power (square and square root are examples of raising to the power 2 and to the power $\frac{1}{2}$. Other powers can also give unpredictable results.)

Examples 10.17–10.19 show how to use the rules stated above to solve four types of inequalities. Express each solution in symbols and in words.

Example 10.17

Solve the inequality $3x - 2 < 6$. Write the solution in symbols, with a number line graph, and in words.

Solution   Our job is to convert the inequality to one whose solution we can state quickly, like the ones in Figure 10.37. We can add, subtract, multiply, or divide on both sides of the inequality sign, but if we multiply or divide by a negative number we must change the direction of the inequality sign. And, of course, we cannot multiply or divide by zero.

For the inequality $3x - 2 < 6$ we will add 2 and then divide by 3.

$$3x - 2 < 6$$
$$3x < 8 \qquad \text{Add 2 to both sides.}$$
$$x < \frac{8}{3} \approx 2.67 \quad \text{Divide both sides by 3.}$$

Thus, the symbolic solution is $x < \frac{8}{3} \approx 2.67$. The graphical solution is shown in Figure 10.38. The solution can be expressed as "all real numbers less than $\frac{8}{3}$" or "all real numbers less than approximately 2.67." Notice that the number $\frac{8}{3}$ is *not* a solution of this inequality.

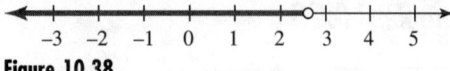

Figure 10.38

Example 10.18

Solve the inequality $-3x - 2 < 6$. Write the solution in symbols, with a number line graph, and in words.

Solution   Once again, our job is to convert the inequality to one whose solution we can state quickly, like the ones in Figure 10.37. We can add, subtract, multiply, or divide on both sides, but if we multiply or divide by a negative number we must change the direction of the inequality.

We will add 2 to both sides and then divide by -3.

$$-3x - 2 < 6$$
$$-3x < 8 \qquad \text{Add 2 to both sides.}$$
$$x > \frac{8}{-3} \approx -2.67 \quad \begin{array}{l}\text{Divide both sides by } -3. \text{ Notice that the}\\ \text{direction of the inequality changes because}\\ \text{we are dividing by a negative number.}\end{array}$$

Thus, the symbolic solution is $x > -\frac{8}{3} \approx -2.67$. The graphical solution is shown in Figure 10.39. The solution can be expressed as "all real numbers greater than $-\frac{8}{3}$" or "all real numbers greater than approximately -2.67."

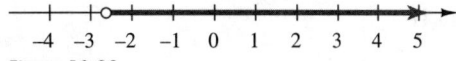

Figure 10.39

Example 10.19

Solve the inequality $3\,|x| - 2 > 6$. Write the solution in symbols, with a number line graph, and in words.

Solution For the inequality $3\,|x| - 2 > 6$ we will add 2 and then divide by 3.

$$3\,|x| - 2 > 6$$
$$3\,|x| > 8$$
$$|x| > \frac{8}{3} \approx 2.67$$

The symbolic solution is $|x| > \frac{8}{3} \approx 2.67$. The graphical solution is shown in Figure 10.40. The solution can be expressed as "all real numbers greater than $\frac{8}{3}$ or less than $-\frac{8}{3}$" or "all real numbers greater than approximately 2.67 or less than approximately -2.67."

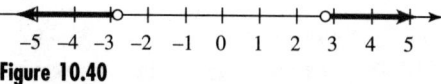

−5 −4 −3 −2 −1 0 1 2 3 4 5
Figure 10.40

Example 10.20

Solve the inequality $-3x^2 - 2 > -6$. Write the solution in symbols, with a number line graph, and in words.

Solution Once again, we will add 2 and then divide by -3

$$\begin{aligned} -3x^2 - 2 &> -6 \\ -3x^2 &> -4 \\ x^2 &< \frac{-4}{-3} = \frac{4}{3} \approx 1.33 \end{aligned}$$

Change direction of inequality sign because of division by -3.

Now, if $x^2 = \frac{4}{3}$ then $x = -\sqrt{\frac{4}{3}}$ or $x = \sqrt{\frac{4}{3}}$. These two equations can be combined into the one statement $|x| = \sqrt{\frac{4}{3}}$. In a similar way, if $x^2 < \frac{4}{3}$ then taking the square root of both sides produces $|x| < \sqrt{\frac{4}{3}} \approx 1.15$. Thus, $x < \sqrt{\frac{4}{3}} \approx 1.15$ **and** $x > -\sqrt{\frac{4}{3}} \approx -1.15$. The graphical solution is shown in Figure 10.41. The solution can be expressed as "all real numbers between, but not including, $-\sqrt{\frac{4}{3}}$ and $\sqrt{\frac{4}{3}}$" or "all real numbers between, but not including, approximately -1.15 and 1.15."

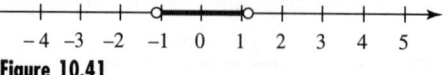

−4 −3 −2 −1 0 1 2 3 4 5
Figure 10.41

An inequality like $2x + 5 < -4x - 8$ would be solved in much the same way: add $4x$ to both sides, then subtract 5, and finally divide by 6. This is shown in Example 10.21.

Example 10.21

Solve the inequality $2x + 5 < -4x - 8$.

Solution

$$
\begin{aligned}
2x + 5 &< -4x - 8 \\
6x + 5 &< -8 \qquad \text{Add } 4x \text{ to both sides.} \\
6x &< -13 \qquad \text{Subtract 5 from both sides.} \\
x &< \frac{-13}{6} \approx -2.17 \quad \text{Divide both sides by 6.}
\end{aligned}
$$

The solution is all real numbers less than $\frac{-13}{6}$ and is shown graphically in Figure 10.42.

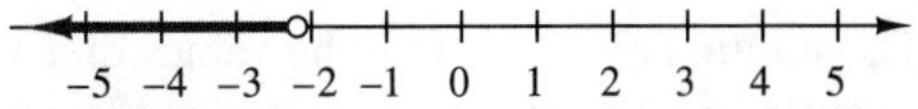

Figure 10.42

Solving an Inequality With a Calculator

Examples 10.22–10.23 show how the calculator can give you a picture of the inequality, and how to use `intersect` to get a good idea of the description of the solution. We'll show calculator solutions for two of the inequalities presented above.

Example 10.22

Use a graphing calculator to solve the inequality $2x + 5 < -4x - 8$. This is the same inequality you solved algebraically in Example 10.21.

Solution

Our strategy is to plot the graphs of each side of the inequality and then study the graph, while remembering the following

- if $f(x) > g(x)$ then the graph of f lies *above* the graph of g
- if $f(x) < g(x)$ then the graph of f lies *below* the graph of g

For the inequalty $2x + 5 < -4x - 8$, we let $f(x) = 2x + 5$ and $g(x) = -4x - 8$. Figure 10.43 shows the plots of $y = 2x + 5$ (plotted in the thick style), and $y = -4x - 8$ with the `ZStandard` window setting.

The thick line, the graph of f, lies below the other line when x is to the *left* of the point of intersection. The calculator's `intersect` feature estimates the point of intersection to be $(-2.17, 0.67)$. Therefore $x < -2.17$ is the solution to two decimal places, which agrees with the answer we got algebraically.

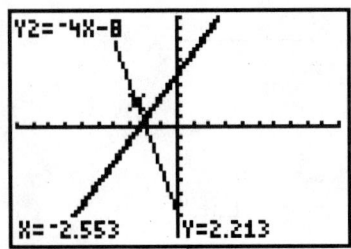

Figure 10.43

Example 10.23

Use a graphing calculator to solve the inequality $-3x^2 - 2 > -6$. This is the same inequality you solved algebraically in Example 10.20.

Solution   For the inequality $-3x^2 - 2 > -6$ we let $f(x) = -3x^2 - 2$ and $g(x) = -6$. Figure 10.44 shows the graph of $y = -3x^2 - 2$ and $y = -6$, with one point of intersection near $(1.15, -6)$ displayed.

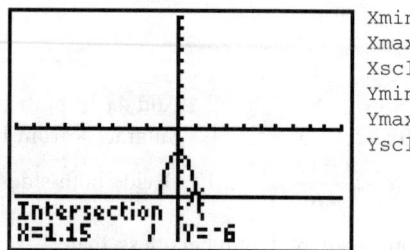

```
Xmin = -9.4
Xmax = 9.4
Xscl = 1
Ymin = -9.2
Ymax = 3.2
Yscl = 1
```

Intersection
X=1.15   Y=-6

Figure 10.44

The solution consists of all the values of x where the parabola is above the line (all the values of x between the two points of intersection). The `intersect` feature reports that the left hand point of intersection is $(-1.15, -6)$, so our solution is $-1.15 < x < 1.15$, which is the same answer shown in Example 10.20. ■

Example 10.24   Application

A technician determines that an electronic circuit (Figure 10.45) fails to operate because the resistance between A and B, $600\,\Omega$, exceeds the specifications. The specifications state that the resistance must be between $200\,\Omega$ and $500\,\Omega$. If a shunt resistor of $R\,\Omega$, $R > 0$, is added, the circuit will satisfy the specifications. The equivalent resistance for the parallel connection is $\dfrac{1}{\dfrac{1}{600} + \dfrac{1}{R}}$ or $\dfrac{600R}{600 + R}\,\Omega$. What are the possible values for R?

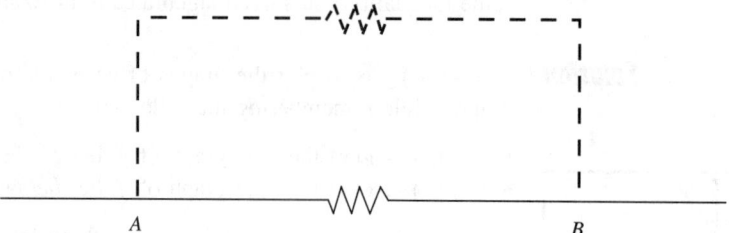

A B

Figure 10.45

Solution   To satisfy the specifications, the following inequality must be true.

$$200 < \frac{600R}{600 + R} < 500$$

Since $R > 0$, then $600 + R > 0$, and we can multiply the inequality by $600 + R$ without changing the direction of the inequality signs.

$$200(600 + R) < 600R < 500(600 + R)$$

$$120\,000 + 200R < 600R < 300\,000 + 500R$$

We will now write this compound inequality as two simple inequalities:

$$
\begin{array}{rcl}
120\,000 + 200R < 600R & \text{and} & 600R < 300\,000 + 500R \\
120\,000 < 400R & \text{and} & 100R < 300\,000 \\
300 < R & \text{and} & R < 3\,000
\end{array}
$$

Thus, $300 < R < 3\,000$ and the shunt resistance must be between 300 Ω and 3 000 Ω.

■

Solving Inequalities: Binary Logic and the Step Function

Binary means to have two possible results. Profit/loss, true/false, yes/no are binary choices. In their basic operations, computers and calculators use the **binary number system**, in which the two states of an electrical off/on switch are represented by the numbers 0 and 1. In the binary number system all numbers are written entirely with 0's and 1's. For example, the number 2345 in our usual base ten number system is equal to 100100101001 in the binary number system.*

In the **binary logic system**, calculators and computers use the numbers 0 and 1 to represent false and true statements. If you enter the expression 5=5 into a calculator or computer, the machine would reply in a rather unexpected way: it says that the answer is 1. It answers with the number 1 because the expression is true, and 1 is the value of true in the binary logic system. If 5=10 were entered, the response would be 0, meaning that $5 = 10$ is a false statement.

For some more complicated logical expressions, look at Figure 10.46. The first line, $5 = 2 + 3$, is true so the answer is 1. The second and third lines contain the two logical connections *and* and *or*. They are defined as follows

▶ If two statements are joined by **and**, the result is true only if *both* statements are true, and false otherwise.

▶ If two statements are joined by **or**, the result is true if *either* or *both* statements are true, and false otherwise.

When we solved a system of equations, we were actually using the logical *and* without stating it directly. To solve a system of two equations, we are looking for the point that solves the first equation *and* the second equation. The point of intersection is the only point that solves both equations, so it is the solution of a logical *and* statement like

$$x + y = 10 \text{ and } x - y = 2$$

If we made the same statement but with *or* then the solution would be all the points of the first graph plus all the points of the second graph.

Example 10.25 shows how to use binary logic to solve the inequality from Example 10.19. First, however, you need to know how to enter the symbols in Figure 10.46 on your *TI-83*. The logic symbols are in the **TEST** menu. [TEST] is above the ⬛ MATH key on the left side. If you press ⬛ 2nd [TEST] you'll see

Binary logic is often referred to as **Boolean algebra**, after the 19th century British mathematician George Boole.

```
5=2+3
                    1
5=2+3 and 4=6
                    0
5=2+3 or 4=6
                    1
```

Figure 10.46

---

*Just as 2345 is equal to $2 \cdot 10^3 + 3 \cdot 10^2 + 4 \cdot 10^1 + 5 \cdot 10^0$, so 100100101001 is based on powers of 2. It is equal to $1 \cdot 2^{11} + 1 \cdot 2^8 + 1 \cdot 2^5 + 1 \cdot 2^3 + 1 \cdot 2^0$.

Figure 10.47, and if you then press the right arrow key to select the LOGIC menu, you'll see Figure 10.48. Notice that the first two entries are the logical connectives and and or. When you want to enter any of the symbols in Figures 10.47 and 10.48, make the selection from the TEST or LOGIC menu.

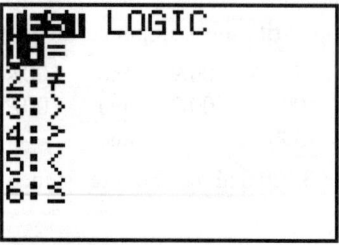

Figure 10.47

Figure 10.48

Example 10.25

Solve the inequality $3\,|x| - 2 > 6$ using the binary logic of your calculator.

Solution The first step is to enter the entire inequality into the calculator's Y= screen, as shown in Figure 10.49.

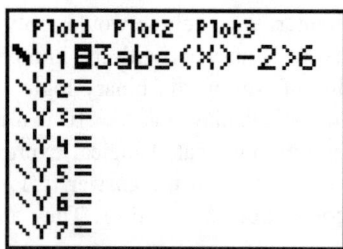

Figure 10.49

The absolute value function is selection 1 on the NUM submenu of the MATH menu. The graph, with the ZStandard (Zoom Standard) setting, is shown in Figure 10.50 and again in Figure 10.51, in which the vertical scale is changed to $-2 \le y \le 2$.

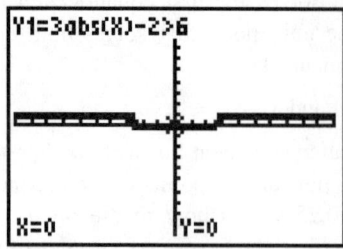

Figure 10.50

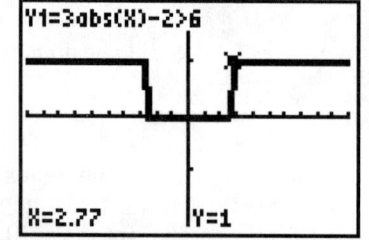

Figure 10.51

The graph shows a **step function**, which gets its name from the step up it takes just before $x = 2.77$ and the step down it takes just after $x = -2.77$. The graph shows where the inequality is true ($y = 1$) and where it is false ($y = 0$). Like number line graphs, the step function graph is a direct visual display of our solutions.

A step function could have one step or it could have many. The two almost vertical lines in Figure 10.51 are not part of the graph. They, like the vertical lines in a calculator graph of the tangent function or the rational function in Figure 8.78 on page 581, can be eliminated by putting the calculator in Dot mode as shown in Figure 10.52.

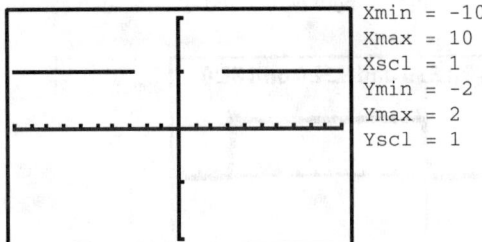

```
Xmin = -10
Xmax = 10
Xscl = 1
Ymin = -2
Ymax = 2
Yscl = 1
```

Figure 10.52

Next we must estimate the value of x at which the function changes from true to false. While the calculator cannot plot points that are directly above one another, we can zoom in to get an estimate of the point where the inequality changes from true to false. Figure 10.53 shows the graph after two uses of Zoom In.

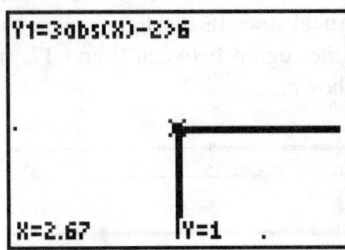

Figure 10.53

The step graph is now more pronounced and we estimate the point where the step takes place at $x = 2.67$. Similarly, we can see that $3|x| - 2 > 6$ is true for $x < -2.67$. Thus, the entire solution is $x < -2.67$ or $x > 2.67$. This corresponds to the solution given in Example 10.19.   ■

Binary Logic and the Manufacturing Problem

In the first of the Chapter Project questions we are told that 36 resistors and 20 capacitors are received daily for production. Also we know that board A requires three resistors and one capacitor, and board B requires two resistors and two capacitors. In Activity 10.1 you answered this question:

"If you only manufactured board A, how many could you make with the number of parts available?"

Although you may have already answered this question correctly, we'll derive the answer again using inequalities and binary logic. Suppose we made x copies of board A and zero copies of board B. Then the supply of resistors and capacitors limits the value of x according to the inequalities.

$$3x \leq 36 \quad \text{(3 resistors for every board and 36 are available)}$$
$$x \leq 20 \quad \text{(1 capacitor for every board and 20 are available)}$$

Both of these inequalities must be true, and furthermore, the number of boards cannot be negative. Therefore x must satisfy the following system of inequalities:

$$3x \leq 36 \text{ and } x \leq 20 \text{ and } x \geq 0$$

Figure 10.54 shows the graph of this logical statement. From the figure we conclude that x must be between 0 and 12, inclusive.

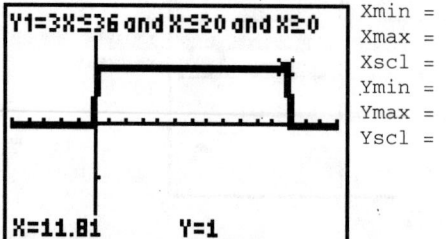

Xmin = -5
Xmax = 15
Xscl = 1
Ymin = -2
Ymax = 2
Yscl = 1

Figure 10.54

Another way to show the solution of this system of inequalities is to look at the graphs of the individual inequalities, which are shown in Figure 10.55. Looking at all three graphs at once, we need to find the region where *all* the inequalities are true. The vertical lines in the figure mark the only region that is shaded on all the graphs; it is the region between 0 and 12, inclusive. Therefore the solution is $0 \leq x \leq 12$, as shown.

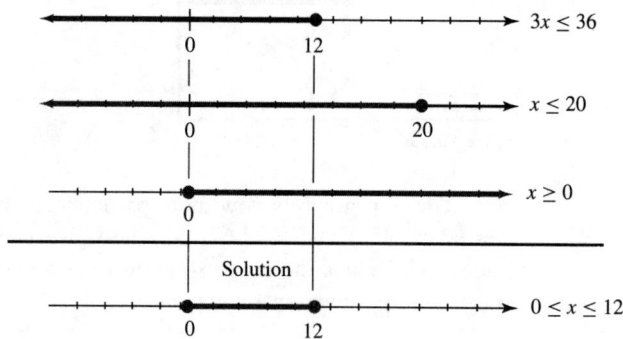

Figure 10.55

Section 10.3 Exercises

In Exercises 1–6, (a) determine four values of x that make the given inequality true and (b) determine if -2π is a solution of the inequality.

1. $4x + 6 < 10$

2. $3 - 5x \geq 38$

3. $-2 < x \leq 1$

4. $\dfrac{1}{x} < \dfrac{1}{2}$

5. $\cos x > 0.5$

6. $\dfrac{x + 6}{x - 2} \leq -1$

In Exercises 7–12 use the inequality $x < 7$ to (a) state the inequality that results when the given operation is performed on each side of the inequality and (b) tell whether the resulting inequality is always, sometimes, or never true.

7. Add -7

8. Subtract 4

9. Multiply by -2.5

10. Divide by 2

11. Square

12. Take the common logarithm

In Exercises 13–20 write the solution (a) in symbols, (b) with a number line graph, and (c) in words.

13. $4 - 3x \leq 19$

14. $2x + 1 > 11$

15. $2.5|x| + 1 < 14.75$

16. $-2x + 1 \geq 4 + 3x$

17. $x^2 - 3.7 \geq 9.1$

18. $-27 \leq 4r < 19$

19. $x - 3x^2 > x^2 + 5 + x$

20. $1.5|x| - 6 \geq 7.65$

In Exercises 21–30 use a graphing calculator to solve the given inequality.

21. $3x + 7 \leq 15 + 5x$

22. $7x + 5 > 12 - 2x$

23. $12x - 3 < 6x + 9$

24. $2x^2 - 3x > -1$

25. $5x^2 - 3x > 6x - 3$

26. $(x^2 - 1)(x - 99) \geq 0$

27. $|6x - 2| \leq 2x + 1$

28. $(x^2 - 4)(5x + 1) \geq 0$

29. $\dfrac{1}{x^2 - 4} + \dfrac{1}{x} \leq 0$

30. $\sqrt{x^2 - 4} < \sqrt{28 - x^2}$

31. *Business* The weekly cost of manufacturing x microcomputers of a certain type is given by $C = 1500 + 25x$. The revenue from selling these is given by $R = 60x$. In order to have a profit the revenue must be larger than the cost. At least how many microcomputers must be made and sold each week to produce a profit?

32. *Energy technology* A rectangular solar collector is to have a height of 1.5 m. The collector will supply 600 W per m^2 and is to supply a total wattage between 2 400 and 4 000 W. What range should the width of the collector have to provide this voltage?

33. *Medical technology* The medical dosage for a very young child is sometimes calculated by the formula $c = \dfrac{Ad}{150}$, where d is the adult dose, c is the child's dose, and A is the child's age in months. For what ages is the child's dosage between 25% and 50% of an adult's dose?

34. *Computer science* A certain computer diskette must be kept within a temperature range given by $|x - 29.4°C| \leq 3.7°C$. What are the minimum and maximum temperatures of the range?

35. *Lighting technology* The intensity I in candelas (cd) of a certain light is $I = 75d^2$, where d is the distance in meters from the source. For what range of distances will the intensity be between 75 and 450 cd?

36. *Architecture* The deflection of a beam d is given by $x^2 - 1.1x + 0.2$, where x is the distance from one end. For what values of x is $d > 0.08$?

37. *Physics* A ball is thrown vertically upward into the air from the roof of a 452-ft building. If the initial velocity of the ball is 64 ft/sec, the height of the ball above the ground is given by the function $h(t) = -16t^2 + 64t + 452$.
(a) For what times t will the height be greater than 500 feet?
(b) For what times t will the height be less than 395 feet?

38. *Forestry* Volume estimates V, in cubic feet, for shortleaf pine trees are based on D, the d.b.h. (diameter at breast height) in inches; top d.i.b., the diameter, in inches, inside the bark at the top of the tree; and H, the height of the tree in feet. One formula for trees with a 3-in. top d.i.b. is

$$V = 0.002837D^2H - 0.127248.$$

Determine D for a 75 ft tree that has a volume estimate greater than 47.75 ft^3.

39. *Forestry* The Scribner log-rule equation for 16 ft logs is $V = 0.79D^2 - 2D - 4$, where V is the volume, in board feet, of the log and D is the diameter, in inches, of the small end of a log inside the bark. What diameter of the small end of the log inside the bark is needed for a 16 ft log to have a volume of at least 926.0 board ft?

40. *Material science* The range of temperatures, T (in degrees Kelvin), for 100.00 g of silver when the silver receives 2.00×10^4 cal of heat is given by $|-2.8T^2 + 820T - 44,000| < 10,000$. Solve for T to the nearest 0.1 K.

10.4   Inequalities in Two Variables

In Section 10.2 you learned how to find the solution to the 2×2 system of equations:

$$\begin{cases} x + y = 10 \\ x - y = 2 \end{cases}$$

The answer was the single point $x = 6$ and $y = 4$. To solve the manufacturing problems presented at the beginning of this chapter, you will need to be able to solve systems of *inequalities*. We'll begin by exploring a 2×2 system of inequalities based on these same equations.

Activity 10.10
Exploring a System of Linear Inequalities

Write ten pairs of numbers that could be answers to the question that follows. Try to find two answers in each of the four quadrants and one answer on each axis. If you think that it is impossible to find answers in one of these areas, explain why.

▶ I am thinking of two numbers. If you add my numbers you get less than 10 and if you subtract them you get more than 2. What are the numbers?

As with inequalities in one variable, we usually want to know *all* the points that satisfy a system, not just a few of them. Of course it is not possible to write all of the pairs because there are an infinite number of them. What we need is a way to display the solutions on a graph.

The Graph of a System of Inequalities

The process for graphing an inequality is described in the box below. Example 10.26 shows how inequalities in two variables are usually graphed.

How to Graph a System of Inequalities

▶ Draw the graph of the equation that results from replacing the inequality sign with the equals sign.
▶ The points that satisfy the inequality will be on one side of the line. Find out which side by either of the following methods:
  • Solve the inequality for y. If the result begins $y < \ldots$ then the points that satisfy the inequality are below the line. If the result begins $y > \ldots$ then the points that satisfy the inequality are above the line.
  • Substitute the coordinates of the one point into the inequality and see if the result is true. If yes, then the solution is all the points on the same side as that point. If not, the solution is all the points on the other side of the line. *Any* point not on the line can be used but the origin is often chosen because its zeroes make the calculations easier.

Example 10.26

Draw the graph of the inequality $x + y < 10$.

Solution   Figure 10.56 shows the graph of the line $x + y = 10$. Because this line is not part of the solution it has been dashed. The regions on either side of the line are shaded and a few points are indicated to help you decide which side of the line will represent the solution to the inequality.

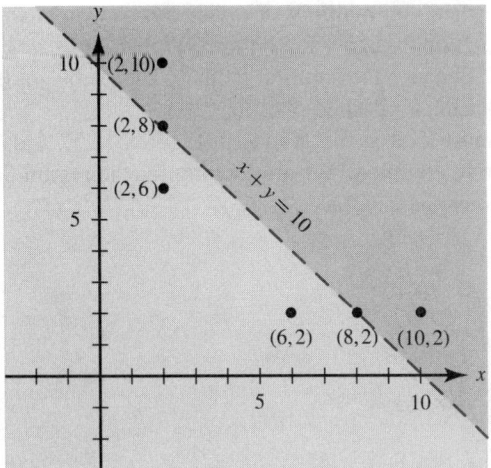

Figure 10.56

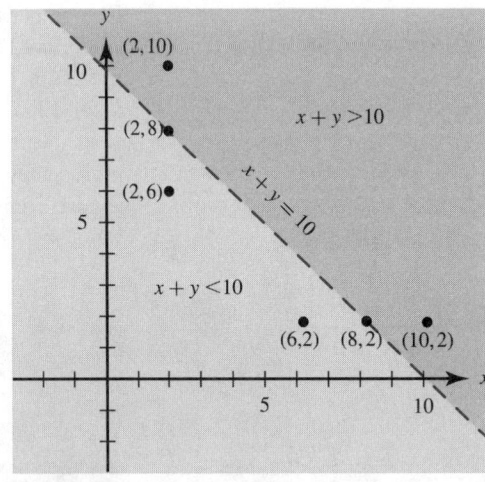

Figure 10.57

To decide which side of the line satisfies the inequality $x + y < 10$ we'll demonstrate both of the methods described above.

▶ Solving the inequality $x + y < 10$ for y produces $y < 10 - x$. Therefore the solutions of the inequality are all points *below* the line.
▶ The origin $(0, 0)$ is a solution of the inequality because $0 + 0 < 10$. Therefore the solution of the inequality is all the points on the same side of line as the origin (below the line).

Therefore the solutions of $x + y < 10$ are all the points in the lighter colored region (*below* the line) shown in Figure 10.57.   ■

It is important to remember that the line itself is *not* a solution of $x + y < 10$, because the points that satisfy $x + y = 10$ do not satisfy $x + y < 10$.

The solution of an inequality in two variables consists of all the points in a region that is bounded by the equation that corresponds to the inequality. This region is called the **graph of the inequality**. To solve a system of two or more inequalities, you must find the region that is common to the graphs of both the inequalities. Example 10.27 shows how to represent the solutions of a 2×2 system of inequalities.

Example 10.27

Graph the solution of the 2×2 system

$$\begin{cases} x + y < 10 \\ x - y \geq 2 \end{cases}$$

Solution   Remember that the solution to this system is the set of points that satisfy *both* of the inequalities. The graph of the first inequality was produced in Example 10.26.

Let's look at how to graph the solution of the second inequality $x - y \geq 2$. Solving this inequality for y we get

$$x - y \geq 2$$

$$-y \geq -x + 2$$

$$y \leq x - 2$$

Remember that when you multiply or divide by -1 you must reverse the inequality. Therefore the solution is the region below *and including* the line in the graph shown in Figure 10.58. Because the solution includes the points on the line we have drawn a solid line rather than a dashed one.

Next we must look at the graphs in Figures 10.57 and 10.58 to find out what region they have in common. We are looking for the region that is *below* both lines. Figure 10.59 shows that region.

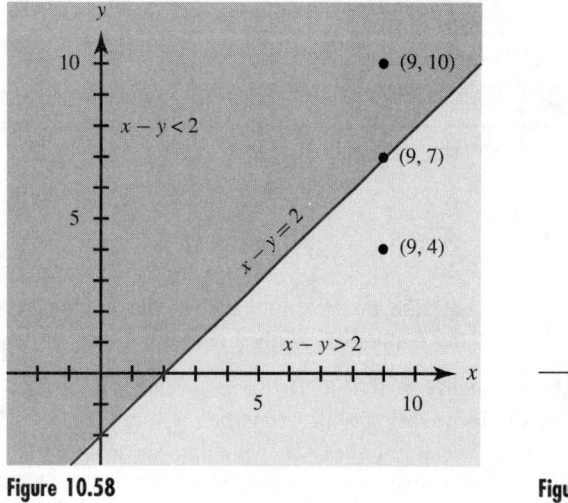

Figure 10.58

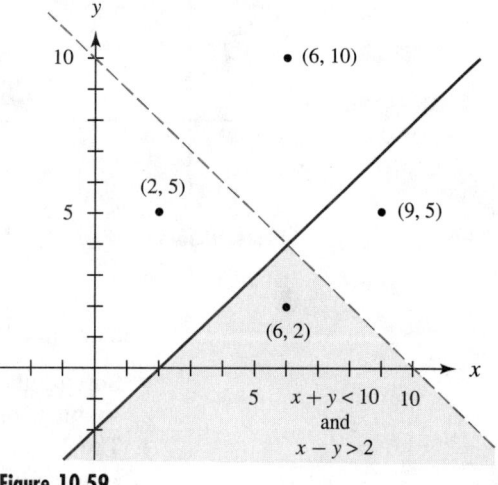

Figure 10.59

As you look at the graph of the solution of the system of inequalities in Figure 10.59 you can see why it was difficult to find some points in quadrant II of the graph that satisfied both inequalities. There aren't any!

The next two activities are designed to increase your understanding of the graphs associated with systems of inequalities.

Activity 10.11
Exploring Systems of Nonlinear Inequalities—1

The graph in Figure 10.60 shows the equations $y = x \sin x$ and $2x - 5y = 5$. Using four copies of the figure, shade the graph of each of these inequalities

1. $y > x \sin x$ and $2x - 5y > 5$
2. $y > x \sin x$ and $2x - 5y < 5$
3. $y < x \sin x$ and $2x - 5y > 5$
4. $y < x \sin x$ and $2x - 5y < 5$

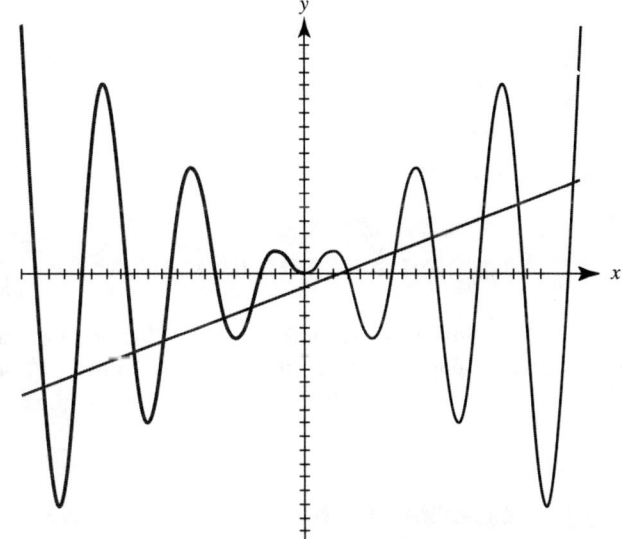

Figure 10.60

Activity 10.12
Exploring Systems of Nonlinear Inequalities—2

Figure 10.61 shows the graph of four different systems of 2×2 inequalities. Each system contains a quadratic function $y = f(x)$ and a cubic function $y = g(x)$.

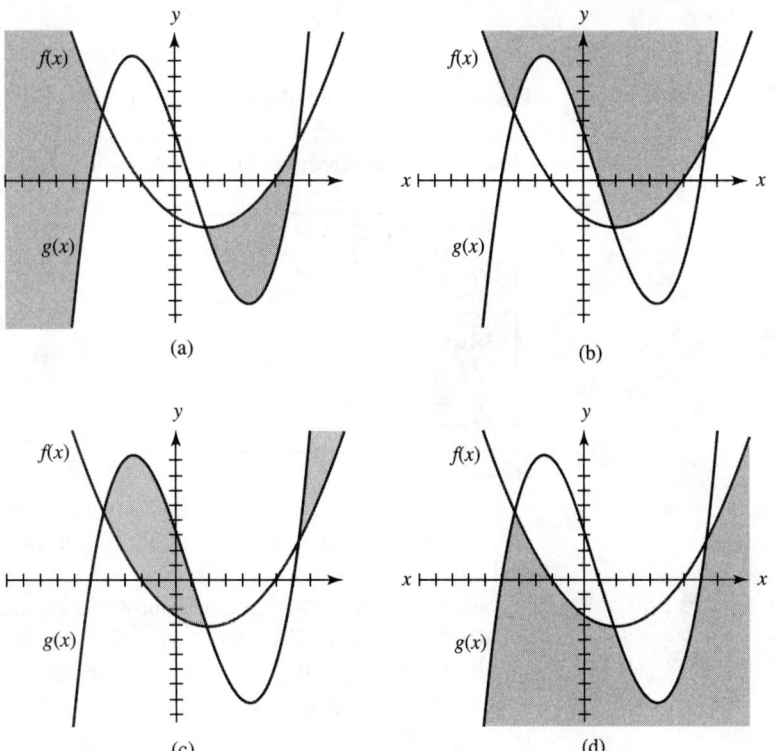

Figure 10.61

Match each of the graphs in Figure 10.61 with one of the systems given below.

1. $y \geq f(x)$ and $y \leq g(x)$
2. $y \geq f(x)$ and $y \geq g(x)$
3. $y \leq f(x)$ and $y \leq g(x)$
4. $y \leq f(x)$ and $y \geq g(x)$

Graphing Systems of Inequalities on the Calculator

You must do some thinking about which side of the line contains the solution before attempting to graph an inequality. Example 10.28 explains how to use the calculator to display the graph for a system of inequalities.

Example 10.28

Use the *TI-83* to graph the system of inequalities

$$\begin{cases} x + y < 10 \\ x - y \geq 2 \end{cases}$$

These are the same inequalities in Example 10.27.

Solution Indicate on the *TI-83* which side of the line contains the solution, and solve the equation for y if it is not already in that form. For this system we obtain

$$\begin{cases} y < 10 - x \\ y \leq x - 2 \end{cases}$$

After you have solved each inequality for y, the calculator can be used to display the solution. Figure 10.62 shows the Y= screen for this system. You cannot tell by looking at this screen whether the edge of the region is part of the solution.

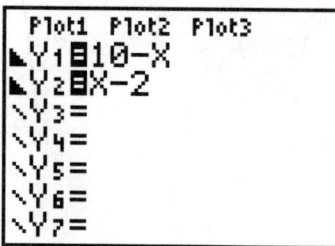

Figure 10.62

The key part is shown to the left of each equation. There we instruct the calculator to shade the region *under* each line, by indicating the `below` graph style for each. The calculator graph is shown in Figure 10.63. You can see that two areas of the figure are shaded once, one area is shaded twice, and one area is not shaded at all. The area that is shaded twice is the solution to *both* of the inequalities, so it is the graph of the system. Compare Figure 10.63 to Figure 10.59. They communicate the same answer.

Just as the calculator shaded the region under the curve it can shade the region above a curve by using the `above` graph style.

The dark band along part of the x-axis in Figure 10.63 is three horizontal lines drawn in a small space and merged into one line.

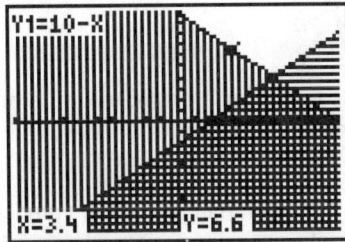

Figure 10.63

Figure 10.63 shows the customary way to show the graph of an inequality. However, some people prefer to shade the opposite way so that the graph of the inequality can be seen more clearly. In Figure 10.64 you can see the graph of the region a little more clearly. Since this method is different from what most people expect for the graph of a system of inequalities, you must always be sure to state that the graph is the *white* area.

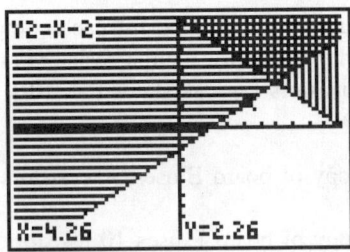

Figure 10.64

Graphing the Systems for the Chapter Project

The question in the Chapter Project asks how many of each board should be made so that the total profit is maximized. There are two main parts to this kind of problem:

► Find the combinations of boards that are possible, given the parts each requires and the total number of each part that is available.
► Find the choice among all the possible choices that gives the highest profit.

We'll solve the first part by graphing a system of inequalities. The second part will be solved by the mathematical technique called linear programming, which you'll learn about in Section 10.5.

We will use the second question in the Chapter Project (boards E and F) for the following examples and calculator lab, and leave the problem of boards A and B for the exercises.

The supply and usage of parts for boards E and F are described as follows:

► **Parts requirements:**

  • Board E requires 30 resistors, 15 capacitors, and 6 transistors
  • Board F requires 10 resistors, 12 capacitors, and 9 transistors

► **Parts supply (daily):**

  • 210 resistors, 120 capacitors, and 72 transistors

Example 10.29

Write three of the five inequalities needed to represent the number of each board that can be manufactured.

Solution   First we need to assign variables for the number of each kind of board that can be made. Although when we use the calculator we will have to change these to x and y, for now we'll use e and f:

Let e = the number of boards E made

Let f = the number of boards F made

Since e and f must be positive or zero, we have the first two inequalities:

$e \geq 0$

$f \geq 0$

One inequality will be used for each component to indicate the supply limits. We will show the inequality for the resistors and leave the inequalities for the capacitors and transistors for Calculator Lab 10.1. First, look at the usage of resistors on each board:

▶ Since each copy of board E uses 30 resistors, then e copies of board E will use $30e$ resistors.
▶ Since each copy of board F uses 10 resistors, then f copies of board F will use $10f$ resistors.
▶ Therefore the number of resistors used is $30e + 10f$.

Finally, since you can't use more than 210 resistors, we have the third inequality

$30e + 10f \leq 210$ ■

Before we graph these inequalities, look at Table 10.2 which presents a useful way to organize your work. There is a column for each component of the boards.

Table 10.2

	Resistors	Capacitors	Transistors
Daily Supply	210		
1 Board E requires	30		
e Board E's require	$30e$		
1 Board F requires	10		
f Board F's require	$10f$		
Total used	$30e + 10f$		
Inequality	$30e + 10f \leq 210$		

You'll fill in the remaining blanks of Table 10.2 in Calculator Lab 10.1.

Example 10.30

Graph the three inequalities from Example 10.29 and list four possible manufacturing choices.

Solution   We'll replace x for the variable e and f with the variable y. The three inequalities are therefore

$$x \geq 0$$

$$y \geq 0$$

$$30x + 10y \leq 210$$

To graph the last inequality we must solve it for y.

$$30x + 10y \leq 210$$

$$10y \leq 210 - 30x$$

$$y \leq 21 - 3x$$

Figure 10.65 shows the Y= screen. We are using the technique of shading the areas that are *not* part of the solution by reversing the shading pattern associated with each line. This means that the solution will be the unshaded part of the graph. It is not possible to graph the solution of $x \geq 0$ on the *TI-83*, so we must remember that the solution should not include negative values of x. Thus we will ignore all values to the left of the y-axis. The graph is shown in Figure 10.66.

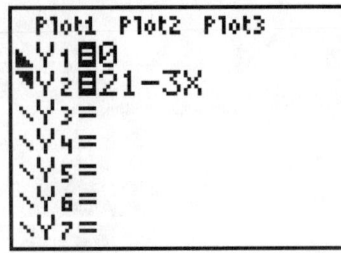

Figure 10.65

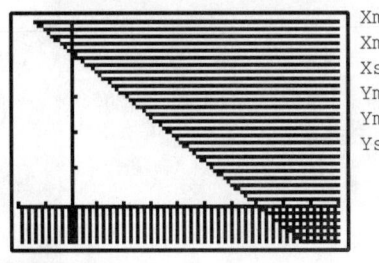

Xmin = -2
Xmax = 10
Xscl = 1
Ymin = -5
Ymax = 25
Yscl = 5

Figure 10.66

To locate points within the solution, we use the calculator's free-moving cursor, which can go anywhere on the screen. Press any of the arrow keys to display the free-moving cursor and move it away from the center of the screen. Although you cannot force the free-moving cursor to land on points with integer coordinates, Figure 10.67 shows that the point $(2, 10)$ is inside the region and therefore is a possible manufacturing choice.

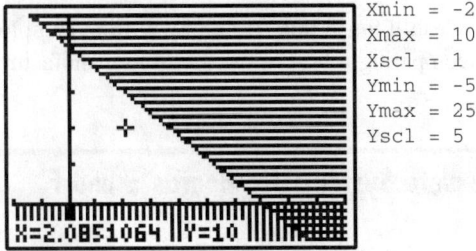

Xmin = -2
Xmax = 10
Xscl = 1
Ymin = -5
Ymax = 25
Yscl = 5

Figure 10.67

Note: to see the free-moving cursor it is essential to leave the solution blank and shade the areas that are not part of the solution.

Three other points that we found were $(0, 20)$, $(5, 3)$, and $(4, 9)$. The last point is on the line.

The profit is \$6 for every E board and \$5 for every F board, so the profit function is a function of x and y

$$P(x, y) = 6x + 5y$$

The profit for each of our points is

$$P(2, 10) = \$62$$
$$P(0, 20) = \$100$$
$$P(5, 3) = \$45$$
$$P(4, 9) = \$69$$

Of these choices, manufacturing no E boards and 20 F boards is the most profitable.

■

When the graph in Figure 10.66 is plotted on axes with the same scale on the x-axis and y-axis, the result is shown in Figure 10.68. The shaded region is the set of points that satisfy all three inequalities from Example 10.30.

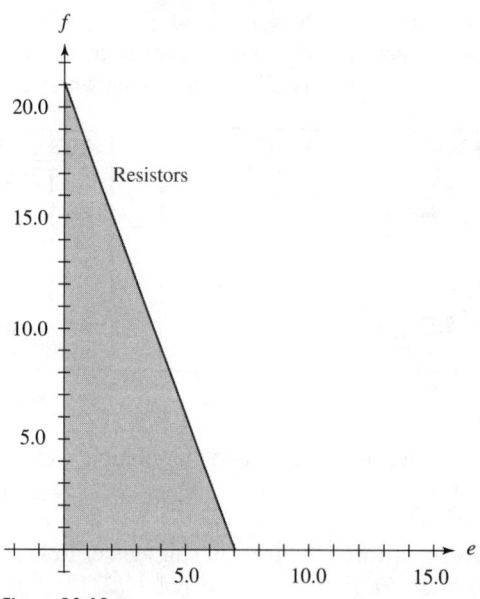

Figure 10.68

You'll finish this section by writing and plotting the two remaining inequalities and comparing a few of the possible manufacturing choices.

Calculator Lab 10.1

Complete Systems for Boards E and F

In this lab you will complete the analysis of the manufacturing choices for boards E and F. Although you may not yet be able to find the most profitable choice, the graphs you draw will lead to that solution in Section 10.5.

Procedures   1. Complete the capacitor and transistor columns in Table 10.2. The table is reproduced below as Table 10.3.

Table 10.3

	Resistors	Capacitors	Transistors
Daily Supply	210		
1 Board E requires	30		
e Board E's require	$30e$		
1 Board F requires	10		
f Board F's require	$10f$		
Total used	$30e + 10f$		
Inequality	$30e + 10f \leq 210$		

2. Add the two inequalities to the plot shown in Figure 10.66. Don't forget that the *white* area represents the solution of the system of inequalities.
3. Check to see if the four points listed in Example 10.30 are still within the solution.
4. Find the coordinates of four additional points within the solution.
5. Of the points you now have, find the one that gives the greatest profit.
6. How many points are within the solution?
7. Make a guess at the point that gives the highest profit of all.

While it is possible to check all the integer-valued points within the solution area, the method of linear programming that we will use in Section 10.5 gives a proven method for finding the greatest profit quickly.

Section 10.4 Exercises

In Exercises 1–4, (a) graph the given linear inequality and (b) identify any points of intersections.

1. $x + y \leq -2$

2. $0.5x - y \leq 4$

3. $y \leq -2$

4. $x \geq 3$

In Exercises 5–18 sketch the graphical solution to each system of inequalities.

5. $\begin{cases} x - y > 0 \\ y - 2x < 6 \end{cases}$

6. $\begin{cases} x + y \geq 4 \\ y - 3x < -2 \end{cases}$

7. $\begin{cases} x + y \geq 3 \\ y < 1 \end{cases}$

8. $\begin{cases} x + 2y \leq 6 \\ y < -2 \end{cases}$

9. $\begin{cases} 2x - 3y \leq 7 \\ 3x + y \leq -2 \end{cases}$

10. $\begin{cases} y > 6 - 0.5x \\ 4y - x \leq 8 \end{cases}$

11. $\begin{cases} x - 2y + 3 \leq 0 \\ 2x + y - 6 > 0 \end{cases}$

12. $\begin{cases} 2x - 3y < 6 \\ 4x - 3y \leq 12 \end{cases}$

13. $\begin{cases} xy \geq 2 \\ 2y + x \leq 8 \end{cases}$ for $x > 0$

14. $\begin{cases} 2y + 4 \geq x^2 \\ x^4 + 4y \leq 1 \end{cases}$

15. $\begin{cases} \cos x - y < -1 \\ 3\sin(2x - \pi) + y \le 0.5 \end{cases}$ from $x = 0$ to $x = 2\pi$

16. $\begin{cases} \cos\left(x - \frac{\pi}{4}\right)x + y < 1.25 \\ y + 0.1x^2 \le 0.5x\cos x + 2 \end{cases}$ from $x = -2\pi$ to $x = 3\pi$

17. $\begin{cases} x + 2y < 4 \\ x - 2y < 4 \\ y \ge 2 \end{cases}$

18. $\begin{cases} x > 1 \\ y < 2 \\ 8x + 3y \ge 24 \end{cases}$

19. *Business* A company manufactures two products. Each product must pass two inspection points, A and B. Each unit of product X requires 30 minutes at A and 45 minutes at B. Product Y requires 15 minutes at A and 10 minutes at B. There are enough trained people to provide 100 hours at A and 80 hours at B.

(a) Write the inequalities needed to represent the number of each product that can be manufactured.

(b) Graph the inequalities from (a) and list four possible manufacturing choices.

20. *Business* An electronics company manufactures two models of computer chips. Model A requires 1 unit of labor and 5 units of parts. Model B requires 1 unit of labor and 3 units of parts. There are 120 units of labor, and 390 units of parts are available.

(a) Write the inequalities needed to represent the number of each model chip that can be manufactured.

(b) Graph the inequalities from (a) and list four possible manufacturing choices.

21. *Business* A computer company manufactures a personal computer (PC) and a business computer (BC). Each computer uses two types of chips, an AB chip and an EP chip. The number of chips needed for each computer is given in the following table.

Computer	AB	EP
PC	2	3
BC	3	8

The company has 200 AB chips and 450 EP chips in stock.

(a) Write the inequalities needed to represent the number of each computer that can be manufactured.

(b) Graph the inequalities from (a) and list four possible manufacturing choices.

22. *Business* The loan department of a bank has set aside $50 million for commercial and home loans. The bank's policy is to allocate at least five times as much money

to home loans as to commercial loans. The policy also states that it must make some commercial loans. This bank's return on a home loan is 8.5% and 6.75% on a commercial loan. The manager of the loan department wants to earn a return of at least $3.6 million on these loans.

(a) Write the inequalities needed to represent the number of each each type of loan that can be made.

(b) Graph the inequalities from (a) and list four possible choices for the types of loans.

23. *Nutrition* A brand-X multivitamin tablet contains 15 mg of iron and 10 mg of zinc. One brand-Y multivitamin tablet contains 9 mg of iron and 12 mg of zinc. A nutritionist advises a patient to take at least 90 mg of iron and 90 mg of zinc each day.

(a) Write the inequalities needed to represent the number of each each type of multivitamin the patient should take.

(b) Graph the inequalities from (a) and list four possible choices for the types of multivitamins.

24. *Industrial management* The chemistry department of a company decides to stock at least 8,000 small test tubes and 5,000 large test tubes. It can take advantage of a special price if it buys at least 15,000 test tubes. The small tubes are broken twice as often as the large tubes, so the company plans to order at least twice as many small tubes as large ones.

(a) Write the inequalities needed to represent the number of each each type of test tube the company can order.

(b) Graph the inequalities from (a) and list four possible choices for the number of test tubes the company can order.

25. (a) Write the inequalities that describe the possible numbers of A and B circuit boards in the Chapter Project.

(b) Show the solution in a graph.

26. *Industrial management* Two types of industrial machines are produced by a certain manufacturer. Machine A requires 3 hours of labor for the body and 1 hour for wiring. Machine B requires 2 hours of labor for the body and 2 hours for wiring. The body shop can provide 120 hours of time per week and the wiring area can furnish 80 hours.

(a) Write the inequalities needed to represent the number of each each type of machine the company can manufacture in an eight hour day.

(b) Graph the inequalities from (a) and list four possible choices for the number of each type of machine the company can make.

10.5   Linear Programming

In this section you will learn how to apply the mathematical technique called *linear programming* to problems like the ones that opened this chapter. Let's review the major ideas behind the questions in the Chapter Project.

▶ Choices need to be made in the number of two or more quantities.
  • In the Chapter Project the products are circuit boards to be manufactured.
  • We assigned variables e for the number of E boards that would be made and f for the number of F boards to be made.

▶ Each product is made up of two or more components.
  • The E and F boards contain resistors, capacitors, and transistors in different combinations.

▶ There are limited supplies of some components.
  • The limitations on the process are called **constraints**.
  • In our Chapter Project the constraints are caused by the limited supply of parts.
  • The straight lines in Figure 10.69, including the e- and f-axes, are graphical representations of the constraints. The inequalities that express the constraints are

$$\begin{cases} e \geq 0 \\ f \geq 0 \\ 30e + 10f \leq 210 \\ 15e + 12f \leq 120 \\ 6e + 9f \leq 72 \end{cases}$$

▶ There are many pairs of values that satisfy the description and constraints of the system.
  • Each of the possible pairs of values is called a **feasible point**. The part of the graph where the feasible points lie is called the **feasible region**.
  • The feasible region for the problem of boards E and F is shaded in Figure 10.69.

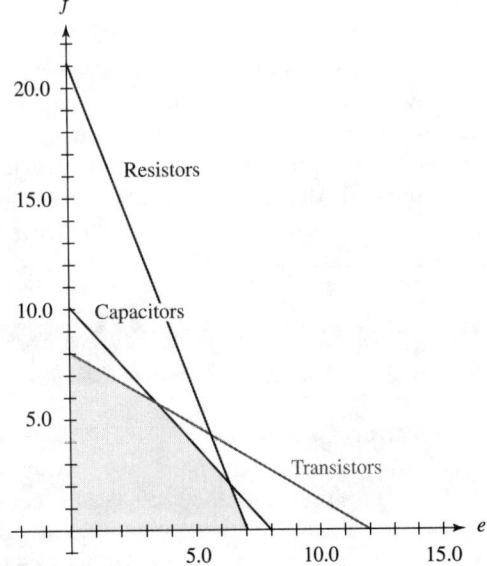

Figure 10.69

▶ There is a measure of the relative value of each product.

- The function that describes the value is called the **objective function**.
- In the Chapter Project the objective function measures the profit that can be made from each type of circuit board. For boards E and F the objective function is

$$P(e, f) = 6e + 5f$$

▶ We must consider whether the coordinates of the points within the feasible region must be integers or can be any real number.

- In the case of circuit boards, the number manufactured must be an integer greater than or equal to zero. If we were producing refined petroleum products, the values might be represented in fractions of a gallon. However, if the petroleum products were measured in barrels, then the values would have to be nonnegative integers.

Activity 10.13
Thinking About the Assumptions

Problems like the ones we considered in the Chapter Project contain many simplifying assumptions. For example there is the assumption that the supply constraints cannot be changed by adding to the cost. Other assumptions are that the same manufacturing line can make either board, and there is no significant down time when the product is changed.

1. Discuss the assumptions listed above and give your opinion of whether it is valid to ignore any of them.
2. Identify any other assumptions in the Chapter Project and discuss their importance.

What is Linear Programming?

The word *linear* is used because the constraints and the objective function must be linear functions. The word *programming* is used possibly because this field was one of the first applications of computer programming.

During the American effort in World War II (1941–1945), there was a great need to improve systems of supply, manufacturing, and shipping under time, budget, and manpower constraints. This need to streamline large systems created a need for some way to simplify and automate the evaluation of a huge number of choices. An effort was made to use electronic computers, which were just becoming available to large institutions, for the calculations. In 1947, the new field of **linear programming** began with the discovery by mathematician George Danzig of a way to simplify many of the calculations. His discovery is known as the **simplex method**.

The Simplex Method and Linear Programming

Linear programming is the name for the field of mathematics that is used to solve problems like the ones described in this section.

The **simplex method** reduces the calculations necessary to maximize the objective function. For situations that involve only two products, such as those in the Chapter Project, the simplex method says that when a set of points (x, y) is bounded by straight lines, forming a polygon, the maximum and minimum values of a linear function $P(x, y)$ occur at the vertices of the polygon.

> If the greatest value of the objective function occurs at a point whose coordinates are not integers then the simplex method says to look at acceptable nearby points.

Because of the simplex method, it is only necessary to compute the value of the objective function at each vertex, not at the interior points. Therefore to find the maximum value of $P(e, f) = 6e + 5f$, we only need to look at points A, B, C, D, and E in Figure 10.70.

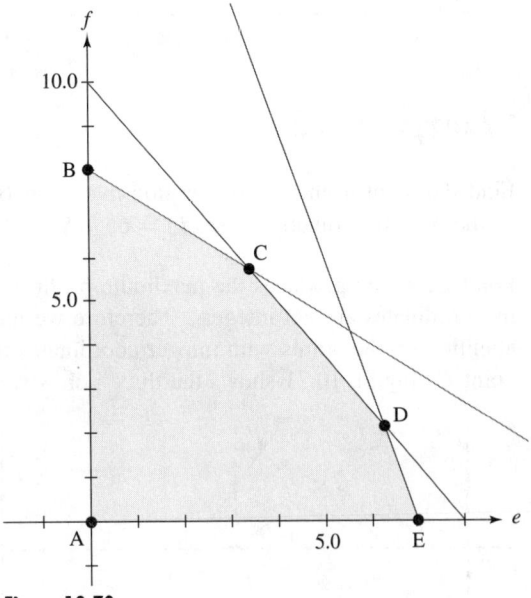

Figure 10.70

Example 10.31

Find the maximum value achieved by the objective function $P(e, f) = 6e + 5f$ across all the points within the feasible region shaded in Figure 10.70.

Solution The simplex method states that we only need to compute the value of $P(e, f)$ for the five points at the vertices. Complete Table 10.4. Use the calculator's `intersect` feature to find the coordinates of points C and D. You can read the coordinates of points A, B, and E off the graph. Two of the points are completed for you.

When the Maximum Occurs at an Unacceptable Point

The simplex method tells you where to look to find the maximum value of the objective function. However, it doesn't consider that we require the number of circuit boards to be integers. When the maximum occurs at a point whose coordinates are not integers, we must look at acceptable points that are located close to the maximum point. Example 10.32 shows how to do that.

Table 10.4

Point	Coordinates (e, f)	Objective Function $P(e, f) = 6e + 5f$
A		
B	$(0, 8)$	$40
C	$(3.43, 5.71)$	$49.13
D		
E		

Example 10.32

Find the point in the feasible region in Figure 10.70 that gives the maximum value of the objective function $P(e, f) = 6e + 5f$.

Solution   Point C, which produces the maximum profit, is not an acceptable solution because its coordinates are not integers. Therefore we must continue our search by focusing attention on the points with integer coordinates that are in the immediate vicinity of point C. Figure 10.71 shows the three points we need to evaluate.

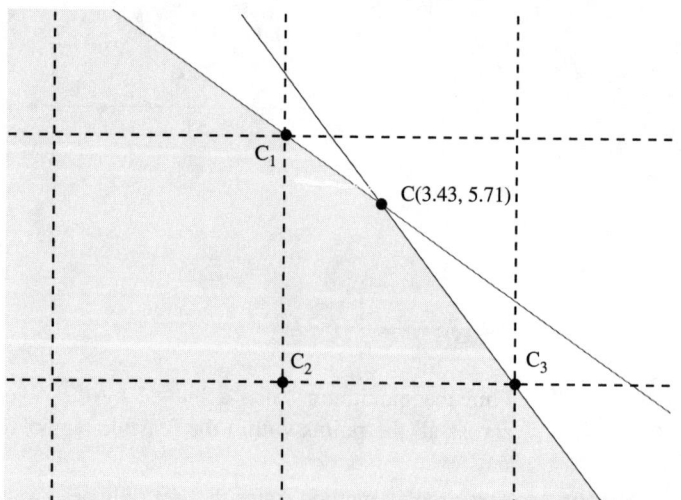

Figure 10.71

There are three parts to our evaluation of the points C_1, C_2, and C_3.

▶ What are the coordinates of the points?
▶ Are the points within the feasible region? While C_2 is clearly in the shaded region, we have to make sure that points C_1 and C_3 lie on or below the boundary lines.

  • For example, to see if point C_1 is inside the feasible region, we see if its coordinates $(3, 6)$ satisfy the inequality represented by the line it is nearest to, $6e + 9f \leq 72$.

▶ What is the value of $P(e, f) = 6e + 5f$ at each of the acceptable points?   ■

We leave the actual analysis of these points to Calculator Lab 10.2.

Calculator Lab 10.2

Finding the Best Point

In this calculator lab you will answer the three questions from Example 10.32.

Procedures   1. For each of the points C_1, C_2, and C_3 in Example 10.32, answer the three questions listed in the example and repeated below.

▶ What are the coordinates of the points?
▶ Are the points within the feasible region? While C_2 is clearly in the shaded region, we have to make sure that points C_1 and C_3 lie on or below the boundary lines.
▶ What is the value of $P(e, f) = 6e + 5f$ at each of the acceptable points?

2. Compare the maximum value of $P(e, f)$ with the other values you found in Example 10.31 to make sure that none of the other points at the vertices of the feasible region are better than the best of C_1, C_2, and C_3.
3. What is the maximum profit and what combination of boards E and F produces that profit?

Adding a New Constraint

There are other kinds of constraints that are often introduced into the manufacturing process. Suppose that the marketing department knows that the E board is more popular than the F board. Then they might insist that the number of F boards must not be greater than the number of E boards. That is, $f \le e$ becomes an additional constraint. If this constraint is added to Figure 10.70, the feasible region changes to the shaded region in Figure 10.72.

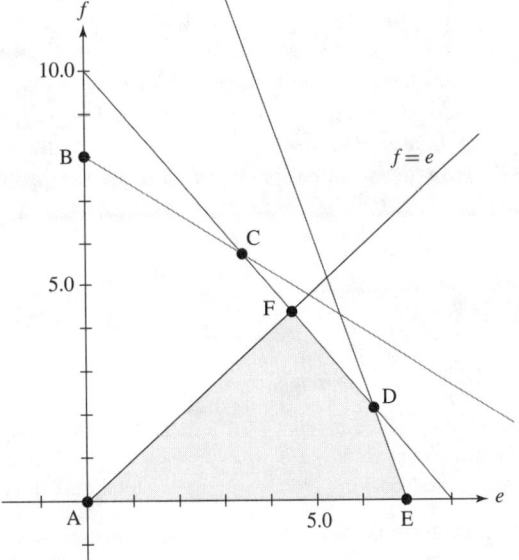

Figure 10.72

The new constraint shown in Figure 10.72 reduces the size of the feasible region. Only point *F* needs to be evaluated and compared with the profits for points *A*, *D*, and *E*. In Calculator Lab 10.3 you'll find the impact that the marketing requirement has on the total profit from these circuit boards.

Calculator Lab 10.3

Handling the Marketing Constraint

In this calculator lab you will examine the effect of requiring that the F boards must not be greater than the number of E boards that are made.

Procedures

1. What are the coordinates of point *F*?
2. What is the maximum profit in the new feasible region?
3. How did the marketing requirement affect the total profit from these boards?
4. Do you think the marketing department should be allowed to add constraints to the manufacturing process? Why or why not?

Finally, we return to the first problem of the chapter, the one involving A and B circuit boards.

Calculator Lab 10.4

Analysis of A and B Circuit Boards

In this calculator lab you will use linear programming to determine the number of A and B boards the company should make.

Procedures

1. Write the inequalities that describe the constraints of the problem.
2. Graph the inequalities and find the feasible region.
3. Write the objective function $P(a, b)$.
4. Find the coordinates of the vertices of the feasible region.
5. Find the point with integer coordinates that maximizes the profit.
6. Add an additional constraint from the marketing department: The number of A boards must be less than or equal to the number of B boards.
7. How does the new constraint affect the profit from these circuit boards?

Section 10.5 Exercises

In Exercises 1–4 write the objective function for the given data.

1. A company makes a profit of $15 on each of product *A* and a profit of $8.95 on each of product *B*. What is the company's total profit on these products?

2. A company makes a profit of $76.25 on each of product *C* and a profit of $27.32 on each of product *D*. What is the company's total profit on these products?

3. It costs a company $57.19 to make each product *E* and $6.49 to make each product *F*. What is the company's total cost to make these two products?

4. A company only makes two items. Its fixed costs average $753.84 each day. It costs $117.19 to make one item *G* and $12.76 to make one item *H*. What is the company's total daily cost to make these two products?

In Exercises 5–10 find the maximum or minimum value (as specified) of the objective function that is subject to the given constraints.

5. Maximize $P = 2x + y$ given the constraints
$$\begin{cases} 3x + 4y \leq 24 \\ x \geq 2 \\ y \geq 3 \end{cases}$$

6. Minimize $C = 9x + 5y$ given the constraints
$$\begin{cases} 3x + 4y \geq 25 \\ x + 3y \geq 15 \\ x \geq 0 \\ y \geq 0 \end{cases}$$

7. Maximize $F - 15x + 10y$ given the constraints
$$\begin{cases} 3x + 2y \leq 80 \\ 2x + 3y \leq 70 \\ x \geq 0 \\ y \geq 0 \end{cases}$$

8. Minimize $F = 6x + 9y$ given the constraints
$$\begin{cases} 2x + 5y \leq 50 \\ x + y \leq 12 \\ 2x + y \leq 20 \\ y - x \geq 2 \\ x \geq 0 \end{cases}$$

9. Maximize $P = 8x + 10y$ given the constraints
$$\begin{cases} 5x + 10y \leq 180 \\ 10x + 5y \leq 180 \\ x \geq 0 \\ y \geq 3 \end{cases}$$

10. Minimize $M = 10x - 8y + 15$ given the constraints
$$\begin{cases} y \geq -3 \\ x - y \geq -5 \\ x + y \geq -5 \\ x \leq 2 \end{cases}$$

11. *Business* A company manufactures two products. Each product must pass two inspection points, A and B. Each unit of product X requires 30 min at A and 45 min at B. Product Y requires 15 min at A and 10 min at B. There are enough trained people to provide 100 hours at A and 80 hours at B. The company makes a profit of \$10 and \$8 on each of X and Y, respectively.

   (a) How many of each should be manufactured to make the most profit?

   (b) How much is the most profit?

   (c) Which one of the inspection points would you add people to increase profits, A or B (but not both)?

12. *Business* An electronics company manufactures two models of computer chips. Model A requires 1 unit of labor and 4 units of parts. Model B requires 1 unit of labor and 3 units of parts. 120 units of labor and 390 units of parts are available, and the company makes a profit of \$7.00 on each model A chip and \$5.50 on each model B. How many of each chip should it manufacture to maximize its profits?

13. *Business* The company in Exercise 12 raises its prices so that it makes a profit of \$8.00 on each model A chip and \$9.50 on each model B. Now, how many should it manufacture to maximize profits?

14. *Business* A computer company manufactures a personal computer (PC) and a business computer (BC). Each computer uses two types of chips, an AB chip and an EP chip. The number of chips needed for each computer and the profit for each are given in the following table.

Computer	AB	EP	Profit
PC	2	3	145
BC	3	8	230

The company has 200 AB chips and 450 EP chips in stock.

   (a) How many of each computer should be made in order to maximize profits?

   (b) What is that maximum profit?

15. *Business* The loan department of a bank has set aside \$50 million for commercial and home loans. The bank's policy is to allocate at least five times as much money to home loans as to commercial loans. The policy also states that it must make some commercial loans. The bank's return on a home loan is 8.5% and 6.75% on a commercial loan. The manager of the loan department wants to earn a return of at least \$3.6 million on these loans. What is the minimum of each kind of loan the bank should make in order to reach its goal of at least \$3.6 million dollars?

16. *Nutrition* A brand-X multivitamin tablet contains 15 mg of iron and 10 mg of zinc. One brand-Y multivitamin tablet contains 9 mg of iron and 12 mg of zinc. A nutritionist advises a patient to take at least 90 mg of iron and 90 mg of zinc each day. If brand-X pills cost 5 cents each and brand-Y pills cost 3.5 cents each, how many of each should the patient take per day to fulfill the prescription at minimum cost?

17. *Industrial management* The chemistry department of a company decides to stock at least 8,000 small test tubes and 5,000 large test tubes. It can take advantage of a special price if it buys at least 15,000 test tubes. The small tubes are broken twice as often as the large tubes, so the company plans to order at least twice as many small tubes as large ones. If the small test tubes cost 15 cents each and the large ones, made of cheaper glass, cost 12 cents each, how many of each size should they order to minimize cost?

18. *Industrial management* Two types of industrial machines are produced by a certain manufacturer. Machine A requires 3 hours of labor for the body and 1 hour for wiring. Machine B requires 2 hours of labor for the body and 2 hours for wiring. The profit on machine A is \$32, and the profit on machine B is \$45. The body shop can provide 120 hours of time per week, and the wiring area can furnish 80 hours. How many of each type of machine should be manufactured each week in order to maximize profit?

19. Explain what is meant by a feasible region and a feasible point? How are they alike? How are they different?

20. Describe how to use the objective function to determine the solution to a linear system.

●●●●●●●●● **Chapter 10 Summary and Review**

Topics You Learned or Reviewed

► You should decide when to use your calculator to solve equations:

• Use a calculator for equations that cannot be solved exactly with pencil and paper.
• Equations that can be solved with algebraic pencil and paper methods but are complicated enough (and likely to cause even the most careful person to make an error) that the time spent setting up the calculator solution will be worthwhile.

► Extraneous roots are false roots that are often introduced when both sides of an equation are raised to a power.

► Linear and nonlinear systems of equations can be solved using several methods:

Addition method: Add two equations to produce a new equation with just one variable.
Graphical method: Plot the equations and find any point or points of intersection.
Substitution method: Solve one equation for one variable and substitute that solution into the other equations.
Tabular method: Use a table of values to either find the value of x that makes the function $f(x)$ equal zero, or find the value of x that makes the functions $f(x)$ and $g(x)$ have equal values.

► Solve and graph inequalities.

• The graphical solution to an inequality with a single variable is shown on a number line.
• The graphical solution to an inequality in two variables is shown on a Cartesian coordinate system.

► In binary logic, calculators and computers use the numbers 0 and 1 to represent false and true statements. A calculator or computer gives the answer 1 if the statement is true and 0 if the statement is false.

► A 2×2 linear or nonlinear system can be solved using one of several methods:

• Graphically by using the *TI-83* feature called `intersect` (found on the [CALC] menu above the `TRACE` key).
• An individual solution can be determined by using the calculator's free-moving cursor.

► Use the `Test` and `Logic` menus to solve inequalities using binary logic.
► Display solutions of inequalities with the *TI-83*'s `above` and `below` graph styles.
► The simplex method of linear programming says that when a set of points (x, y) is bounded by straight lines forming a polygon, the maximum and minimum values of a linear function $P(x, y)$ occur at the vertices of the polygon. If the greatest or smallest value of the objective function occurs at a point whose coordinates are not integers then the simplex method says to look at acceptable nearby points.

Review Exercises

In Exercises 1–4 use the roots method to graphically solve the given equation.

1. $5a - 3 = 12$

2. $3b^2 - 4b = 3b - 2$

3. $c^3 = 9c^2 - c - 9$

4. $\ln(x^2 - 1) = 0.5x$

In Exercises 5–8 use the equal functions method to graphically solve the given equation.

5. $\log(x^2 + 1) = x^2 - 1$

6. $\log(x^2 + 1) = \cos(x - 1)$

7. $\log(x^2 + 1) = e^{\sin 0.5x}$ from $x = -2\pi$ to $x = 2\pi$

8. $\log(x^2 + 1) = |x| \sin 0.25x$ from $x = -\pi$ to $x = 4\pi$

In Exercises 9–12 use the roots method with a table to solve the given equation.

9. $\sqrt{a + 2} = \dfrac{a - 1}{a - 4}$

10. $\dfrac{2}{b - 5} = 2\cos b + 3$

11. $\dfrac{\cos x}{x - 1} = 2\cos x + 3$

12. $\ln(5d - 1) + 2 = 3\cos d + 2.5$

In Exercises 13–14 use the equal functions method with a table to solve the given equation.

13. $\dfrac{e^x}{x} = 3x + 1$

14. $\sin\left(\cos\left(x - \tfrac{\pi}{4}\right)\right) = 0.2x - 1$

In Exercises 15–18 use the addition method to solve the given system.

15. $\begin{cases} 2x + 3y = 1 \\ -3x + 6y = 16 \end{cases}$

16. $\begin{cases} 2x + y = 5 \\ 4x - y = 1 \end{cases}$

17. $\begin{cases} -2x + 2y = 5 \\ x + 6y = 1 \end{cases}$

18. $\begin{cases} 3x + 2y = 2 \\ 2x - 6y = -39 \end{cases}$

In Exercises 19–22 use the substitution method to solve each system of equations.

19. $\begin{cases} y = 3x - 4 \\ x + y = 8 \end{cases}$

20. $\begin{cases} 2x + 5y = 6 \\ x - y = 10 \end{cases}$

21. $\begin{cases} 2x + 3y = 3 \\ 6x + 4y = 15 \end{cases}$

22. $\begin{cases} 4.8x - 1.3y = 16.9 \\ -7.2x - 2.8y = -9.2 \end{cases}$

In Exercises 23–26 use the graphical method to solve each system of equations.

23. $\begin{cases} 3x - 2y = 5 \\ -7x + 4y = -7 \end{cases}$

24. $\begin{cases} 2x + 5y = 6 \\ x - y = 10 \end{cases}$

25. $\begin{cases} 2x + 2y = -3 \\ 4x + 9y = 5 \end{cases}$

26. $\begin{cases} 2.3x + 1.7y = 8.5 \\ -6.7x + 3.7y = 38.4 \end{cases}$

In Exercises 27–36 solve each inequality both (a) algebracially and (b) graphically.

27. $3x < -18$

28. $4x + 7 \geq 10$

29. $3x - 7 \leq 19$

30. $8 - 3x > 4x - 5$

31. $\dfrac{2x - 5}{4} < \dfrac{4x - 1}{3}$

32. $|2x - 1| \leq 9$

33. $|5x - 1| > 4$

34. $-6 > 2x^2 - 5$

35. $\sqrt{x + 3} \geq 0.5\sqrt{x + 16}$

36. $5 \leq x^2 - 6x + 9$

In Exercises 37–44, (a) graph each system of equations or inequalities and (b) identify all points of intersection.

37. $\begin{cases} x + y = 8 \\ x^2 + y^2 = 49 \end{cases}$

38. $\begin{cases} x^2 + y^2 = 9 \\ x^2 = 5y \end{cases}$

39. $\begin{cases} 2x^2 - y^2 = 8 \\ x^2 + 2y^2 = 4 \end{cases}$

40. $\begin{cases} 2x + 4y^2 = 16 \\ xy = 10 \end{cases}$

41. $\begin{cases} x + 3y < 5 \\ x - 4y \leq 8 \end{cases}$

42. $\begin{cases} 4x - y \leq 4 \\ y + x > -2 \\ y - x < 1 \end{cases}$

43. $\begin{cases} 2x + y - 3 < 0 \\ x - 2y - 4 \geq 0 \end{cases}$

44. $\begin{cases} 3x - y \leq -2 \\ 4 - x - y \geq 0 \\ x > -3 \\ 2x + y > -4 \end{cases}$

In Exercises 45–46, find the maximum or minimum value (as specified) of the objective function that is subject to the given constraints.

45. Maximize $P = 3x + 5y$ given the constraints
$$\begin{cases} x + y \leq 5 \\ y - x \geq -2 \\ x \geq -1 \\ x \leq 3 \end{cases}$$

46. Minimize $F = 4x - 3y$ given the constraints
$$\begin{cases} 2x + y \leq 6 \\ y \geq \dfrac{x}{2} - 3 \\ x \geq -1 \\ y \leq 4x - 1 \end{cases}$$

47. *Business* A store owner makes a special blend of coffee from Colombian Supreme costing $4.99/lb and Mocha Java costing $5.99/lb. The mixture costs $5.39/lb. If this mixture is made in 50-lb batches, how many pounds of each type should be used?

48. *Business* A computer company makes two kinds of computers. One, a personal computer (PC), uses 4 type A chips and 11 type B chips. The other, a business computer (BC), uses 9 type A chips and 6 type B chips. The company has 670 type A chips and 1,055 type B chips in stock. How many PC and BC computers can the company make so that all the chips are used?

49. *Business* An office building has 146 rooms made into 66 offices. The smallest offices each have 1 room and rent for $300 per month. The middle-sized offices have 2 rooms each and rent for $520 per month. The largest offices have 3 rooms each and rent for $730 each. If the rental income is $37,160 per month, how many of each type of offices are there?

50. *Electricity* Two resistances R_1 and R_2 connected in parallel must have a total resistance R of at least $4\,\Omega$. If $R_1 R_2 = 20\,\Omega$, what are acceptable values for $R_1 + R_2$? Remember $R = \dfrac{R_1 R_2}{R_1 + R_2}$.

51. *Transportation* A trucker was delayed 30 minutes because of trouble loading the shipment. To make up for the delay, the trucker increased his speed on the interstate highways. The speed was increased by an average of 4 mph, and as a result the trucker was on time, 7 hours and 30 minutes after leaving. Find the usual average speed of the trucker and the distance the truck traveled.

52. *Business* A robotics manufacturing company makes two types of robots: a cylindrical coordinate robot (Model C) and a spherical (polar) coordinate robot (Model S). Each robot has two assembly stations. The number of hours needed at each station and the profit of each robot are given in the following table.

Robot	Station 1 (hours)	Station 2 (hours)	Profit ($)
C	12	15	1,250
S	18	10	1,620

During the next month, the workers will take their vacations. As a result, the company will only have enough workers for 220 hours at station 1 and 180 hours at station 2. How many of each robot should be manufactured in order to make the most profit?

●●●●●●●●● Chapter 10 Test

1. Use the addition method to solve the following system.
$$\begin{cases} 2x + 3y = 1 \\ -3x + 6y = 16 \end{cases}$$

2. Solve the following system graphically.
$$\begin{cases} 2x + 3y = 5 \\ x - 4y = -14 \end{cases}$$

3. Solve the following system by the substitution method.
$$\begin{cases} 4x + 3y = 9 \\ 2x + y = 2 \end{cases}$$

4. Find the currents of the system shown in Figure 10.73.

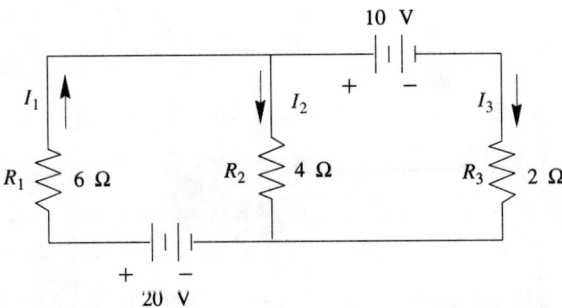

Figure 10.73

5. Solve the inequality $5x > 30$ both algebraically and graphically.

6. Solve $|4x + 5| < 17$ both algebraically and graphically.

7. Solve $\dfrac{4x - 5}{2} \leq \dfrac{7x - 2}{3}$ both algebraically and graphically.

8. Solve $(x - 2)(x - 5) > 0$ both algebraically and graphically.

9. Solve $5x^2 + 3x \leq 2x^2 - x - 1$ both algebraically and graphically.

10. Graph this system of inequalities.

$$\begin{cases} x + 2y > 4 \\ 3x - y \leq 3 \end{cases}$$

11. A brand X multivitamin tablet contains 12 mg of iron and 10 mg of zinc. Each brand Y tablet contains 5 mg of iron and 8 mg of zinc. A nutritionist suggests that a patient take at least 80 mg of iron and 90 mg of zinc each day. The brand X tablets cost 4 cents each and brand Y pills are 6 cents each.
   (a) How many of each pill should the patient take per day to satisfy the suggestion at the minimum cost?
   (b) What is the minimum cost?

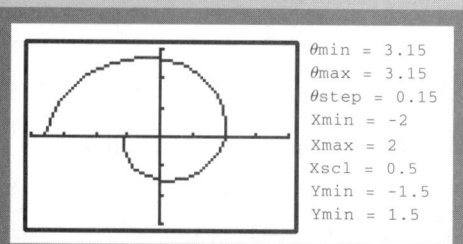

θmin = 3.15
θmax = 3.15
θstep = 0.15
Xmin = -2
Xmax = 2
Xscl = 0.5
Ymin = -1.5
Ymin = 1.5

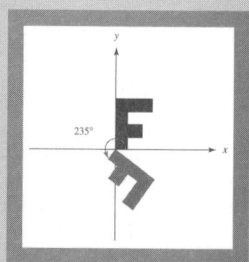

Matrices and 3-D Graphing

Topics You'll Need to Know

▶ Coordinate geometry
▶ Definition and use of sine and cosine
▶ Graphing in two dimensions with
  • Cartesian coordinates
  • parametric equations
  • polar coordinates

Topics You'll Learn or Review

▶ Graphing in three dimensions
  • plotting points and visualizing graphs
  • Cartesian coordinates
  • distance between two points in 3-D
  • parametric equations
  • cylindrical coordinates
  • spherical coordinates
▶ Expressions for sine and cosine of angle sums
▶ Matrix algebra
  • definition of a matrix
  • multiplication
  • identity matrix
  • inverse matrix
  • solving systems of 2, 3, and more equations using matrices
  • representation of linear algebra
  • representation of rotations and scaling geometric transformations
▶ Reflective properties and uses of the paraboloid

Calculator Skills You'll Use

▶ Graphing in Cartesian, polar coordinates, and parametric equations

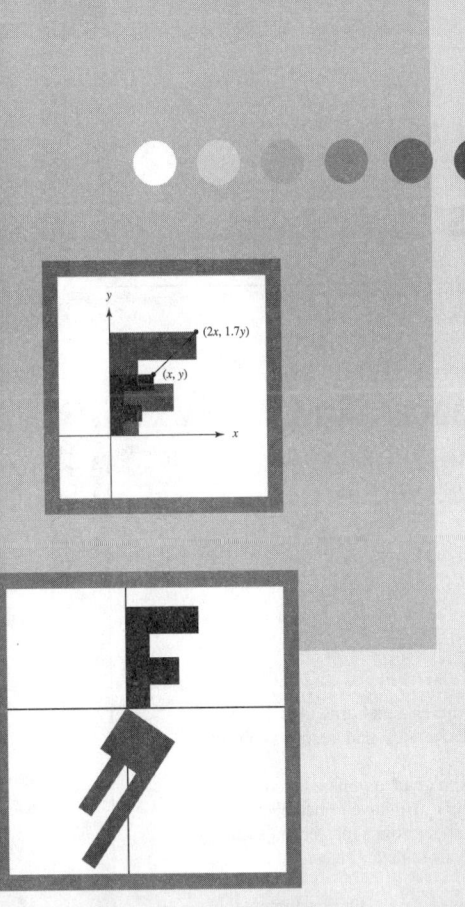

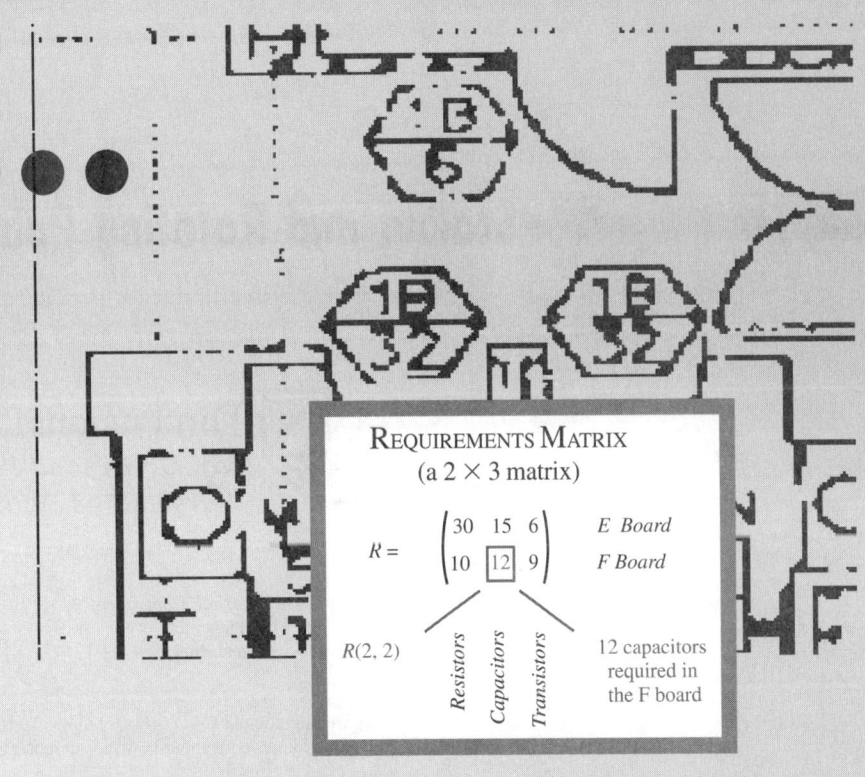

REQUIREMENTS MATRIX
(a 2 × 3 matrix)

$$R = \begin{pmatrix} 30 & 15 & 6 \\ 10 & \boxed{12} & 9 \end{pmatrix} \quad \begin{matrix} E \; Board \\ F \; Board \end{matrix}$$

$R(2, 2)$ — Resistors, Capacitors, Transistors

12 capacitors required in the F board

What You'll Do in This Chapter

This chapter begins with the problem of rotating and scaling figures in two dimensions and later extends the problem to three dimensions. After exploring these transformations on graph paper you'll see that trigonometry gives a quick way to find the new location of a set of points. You'll meet matrix algebra and see how it can be used to represent both geometric transformations and linear programming problems. You'll also see how matrix algebra can be used to solve systems of two or more equations. When the topic turns to three dimensional graphing you'll meet new coordinate systems and some very unusual mathematical curves and surfaces.

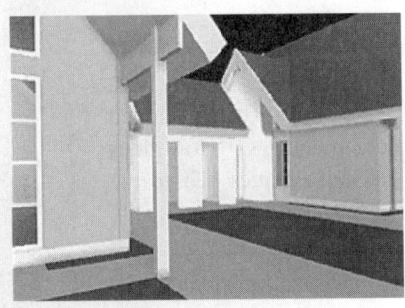

Chapter Project—Scaling and Rotating Points

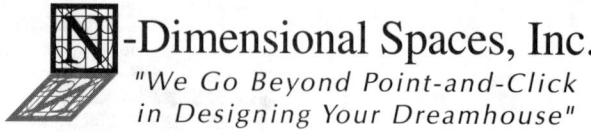

-Dimensional Spaces, Inc.
"We Go Beyond Point-and-Click in Designing Your Dreamhouse"

To:	Research Department
From:	Management
Subject:	Transformations in 3-D

As we upgrade our CAD software to handle three-dimensional drawings, I want us all to know more about how the computer moves, rotates, and reshapes objects.

Since Chris retired and settled near Cancun we haven't had anyone who could explain this to our mathematically inclined customers. We have to be able to answer their questions. I recall Chris talking a lot about how a matrix can simplify things when the computer needs to rotate, translate, or scale an image.

But that was two dimensions — now we're going to use three. Our customers rely on us not to be mere mouse clickers and button pushers, but to understand what's going on inside the computer. We've got to rebuild our expertise, particularly in the 3-D area.

Please prepare a report on the following:

- What is the mathematics behind rotation and scaling objects in two dimensions?

- How does a matrix contribute to the solution?

- How does this work in three dimensions?

I know you can do it. Don't ask me to get Chris out of retirement to explain all this. It takes an enormous consulting rate to budge her from that dream house on the beach.

Preliminary Analysis

Architects and others doing design work use computers to build and change images. For example, the house shown in Figures 11.1 and 11.2 has been rotated, moved around the screen, and scaled differently in the two screen images. How does the computer move figures around like that? What is the mathematics behind the programs that designers use?

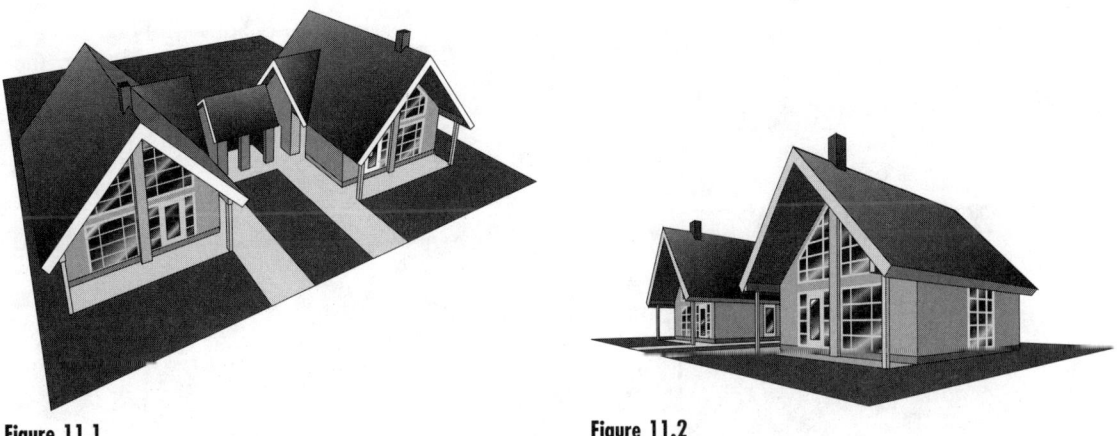

Figure 11.1

Figure 11.2

Activity 11.1
A First Look at Rotation and Scaling

Rotations, scaling, and translations are the transformations that are used to move drawings around on paper or on a computer screen. Earlier in the text you used horizontal and vertical translations. The present activity will give you an idea of what is meant by rotation and scaling. As we progress through the chapter we will mathematically explore these two types of transformations.

Figure 11.3 shows a side-drawing of a saltbox house.

1. Place a blank sheet of paper over Figure 11.3. The paper should be thin enough that you can see the figure through the paper. On the blank sheet of paper trace the *x*- and *y*-axes and the outline of the house.
2. Before you move the paper, place your pencil point at the origin and then rotate the paper 90° in a counterclockwise direction. Stop when the *x*-axis on the paper rests on top of the *y*-axis of Figure 11.3.
3. Hold the paper still and, pressing hard with your pencil, trace the house. Remove the paper. You should see an impression of the rotated house on Figure 11.3.
4. Lightly trace over the rotated house and give the coordinates of its five vertices.

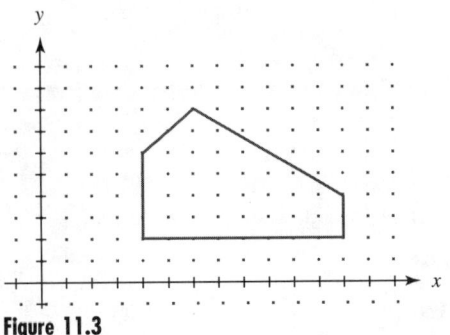

Figure 11.3

Figure 11.4 shows a different side-drawing of a saltbox house.

1. Scale this drawing by multiplying the *x*-coordinate of each vertex by 2 and the *y*-coordinate by 0.5. Scaling $A(6, 2)$ in this way produces the point $A'(2 \times 6, 0.5 \times 2) = A'(12, 1)$.

2. Scale each of the other four vertices of the house in Figure 11.4.
3. Plot each of the scaled points A', B', C', D', and E' and draw the scaled saltbox house $A'B'C'D'E'$.

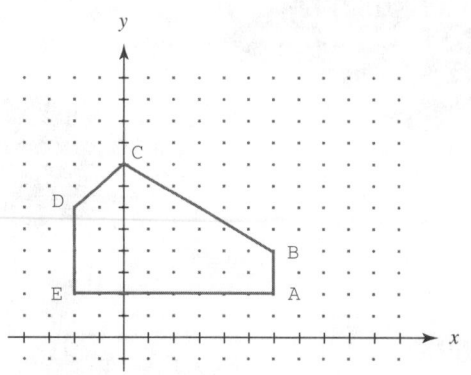

Figure 11.4

11.1

Introduction to the Transformation of Points

Figures are transformed on a computer screen when all the points are moved according to one of the **transformation** processes called **scaling**, **rotation**, and **translation**. These concepts are summarized as follows.

Transforming Points

▶ **Scaling** transforms every point of the figure as follows.
  - Each x-coordinate is multiplied by the same number, called the **x-scaling factor**.
  - Each y-coordinate is multiplied by the same number, called the **y-scaling factor**.

If m is the x-scaling factor and n is the y-scaling factor, then the point (x, y) becomes (mx, ny). The scaling factors m and n can be positive or negative.
▶ **Rotation** transforms every point of the figure as if an imaginary line segment was drawn from each point to the origin of the coordinate system, and that segment was rotated through an angle called the **angle of rotation**.

   If the angle of rotation is equal to α, then in polar coordinates the point (r, θ) becomes $(r, \theta + \alpha)$.
▶ **Translation** transforms every point by moving it horizontally a distance c and vertically a distance d. The point (x, y) becomes $(x + c, y + d)$.

Alpha, α, is the Greek letter a.

In this chapter we will focus on rotation and scaling. Because computer graphics systems use Cartesian coordinates, we will work on a way to express rotation in terms of Cartesian, not polar coordinates. For example, the letter F shown in black in Figure 11.5 has been transformed by rotating it through an angle of 235°. If the rotated letter in Figure 11.5 is then scaled with an x-scaling factor of 2 and a y-scaling factor of 1.7, the result is shown in Figure 11.6.

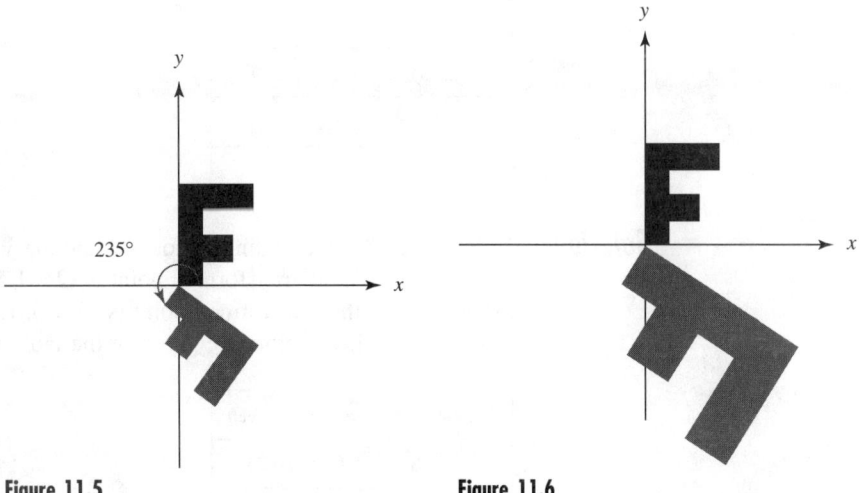

Figure 11.5   **Figure 11.6**

Scaling

Figure 11.7 shows how a smaller letter F can be transformed into a larger one, using an x-scaling factor of 2 and a y-scaling factor equal to 1.7. Every point (x, y) in the old figure corresponds to a new point $(2x, 1.7y)$ in the transformed figure.

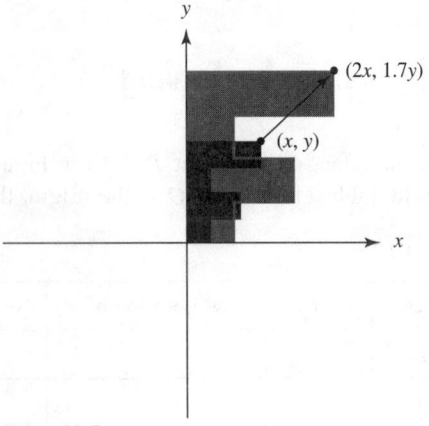

Figure 11.7

Example 11.1

Using the scaling factors shown in Figure 11.7, calculate the missing coordinates of the points in the table that follows. (The first entry has been completed.)

Original Point	Transformed Point
$(-2.5, 6)$	$(-5, 10.2)$
$(1.5, 4)$	$(\_\_, \_\_)$
$(0, 3)$	$(\_\_, \_\_)$
$(1.2, 0)$	$(\_\_, \_\_)$
$(0, 0)$	$(\_\_, \_\_)$
$(\_\_, \_\_)$	$(3, 5)$

Solutions In Figure 11.7 the x-scaling factor is 2 and the y-scaling factor is 1.7. If the original point is $(1.5, 4)$, the transformed point is $(2 \times 1.5, 1.7 \times 4) = (3, 6.8)$. If the original point is $(0, 3)$ the transformed point is $(2 \times 0, 1.7 \times 3) = (0, 5.1)$. The approximate answers for all six points are given in the table below.

Original Point	Transformed Point
$(-2.5, 6)$	$(-5, 10.2)$
$(1.5, 4)$	$(3.0, 6.8)$
$(0, 3)$	$(0, 5.1)$
$(1.2, 0)$	$(2.4, 0)$
$(0, 0)$	$(0, 0)$
$(1.5, 2.9)$	$(3, 5)$

■

Activity 11.2 gives you practice locating points that are transformed by scaling.

Activity 11.2
Transformation by Scaling

1. Show the effect on the point $P(8, 6)$ in Figure 11.8 for each of the scale factors given in Table 11.1. Since O is the origin, the length $OP = \sqrt{8^2 + 6^2} = 10$.

Table 11.1

x-factor	y-factor	New Location of P	New Length of $\overline{OP}$
1	2		
0.5	-1.5		
1.5	0.5		
0	3		
-1	-1		
1	-1		

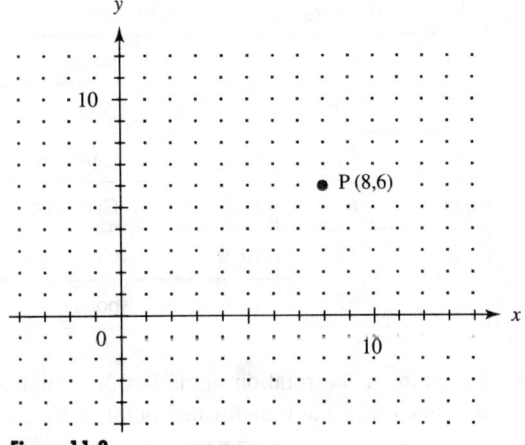

Figure 11.8

2. Plot each new point on a graph like the one in Figure 11.8.
3. Give an example of two sets of scale factors that produce a rotation of 180°.
4. Give the x and y scale factors that produce a reflection across the x-axis.
5. Give the x and y scale factors that produce a reflection across the y-axis.
6. Give the x and y scale factors that produce a point that is symmetric to P with respect to the origin.
7. Give the x and y scale factors that leave point P unchanged.

Rotation

Figure 11.5 on page 743 shows how the letter F can be transformed by rotation. As a review, Figure 11.9 shows the relationship between the Cartesian coordinates (x, y) and the polar coordinates (r, θ).

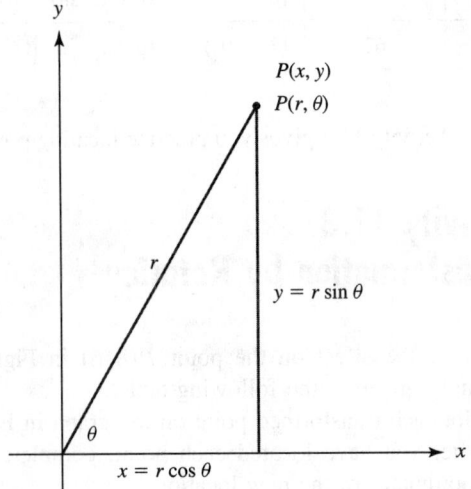

Figure 11.9

Example 11.2

Calculate the missing coordinates of the points in the following table.

Point	Rotation Angle	Original Point		Transformed Point	
A	62°	(6, 32°)	(polar)	(__, __°)	(polar)
B	129°	(1.5, 240°)	(polar)	(__, __°)	(polar)
C	62°	(1.2, 0)	(Cartesian)	(__, __)	(polar)
D	62°	(1.2, 0)	(Cartesian)	(__, __)	(Cartesian)
E	285°	(0, 0)	(Cartesian)	(__, __)	(Cartesian)
F	62°	(__, __°)	(polar)	(3, 52°)	(polar)

Solutions   For point A the rotation angle is 62°. Point A has polar coordinates (6, 32°). If A is rotated 62°, the transformed point is (6, 32° + 62°) = (6, 94°).

Point D has Cartesian coordinates (1.2, 0) and is to be rotated 62°. The polar coordinates of D are $r = \sqrt{1.2^2 + 0^2} = 1.2$ and $\theta = 0°$ or (1.2, 0°). The transformed point D has polar coordinates (1.2, 0° + 62°) = (1.2, 62°), which has Cartesian coordinates $x = 1.2\cos 62° \approx 0.563$ and $y = 1.2\sin 62° \approx 1.060$ or, to two decimal places, (0.56, 1.06).

The answers, to two decimal places, for all six points are given in the table below.

Point	Rotation Angle	Original Point		Transformed Point	
A	62°	(6, 32°)	(polar)	(6, 94°)	(polar)
B	129°	(1.5, 240°)	(polar)	(1.5, 9°)	(polar)
C	62°	(1.2, 0)	(Cartesian)	(1.2, 62°)	(polar)
D	62°	(1.2, 0)	(Cartesian)	(0.56, 1.06)	(Cartesian)
E	285°	(0, 0)	(Cartesian)	(0, 0)	(Cartesian)
F	62°	(3, 350°)	(polar)	(3, 52°)	(polar)

Activity 11.3 gives you practice locating points that are transformed by rotation.

Activity 11.3
Transformation by Rotation

1. Show the effect on the point $P(8, 6)$ in Figure 11.10 for each of the rotation angles given in the following table.
2. Plot each transformed point on the graph in Figure 11.10.
3. After you have located each point, complete the table with an estimate of the coordinates of the new location.
4. Use the equations $x = r\cos\theta$ and $y = r\sin\theta$ to check that your polar coordinates match your Cartesian coordinates.
5. For the polar point $P(6, 32°)$, give two rotations that produce the reflection of P across the x-axis.
6. For the polar point $P(6, 32°)$, give two rotations that produce the point that is symmetric with P with respect to the origin.

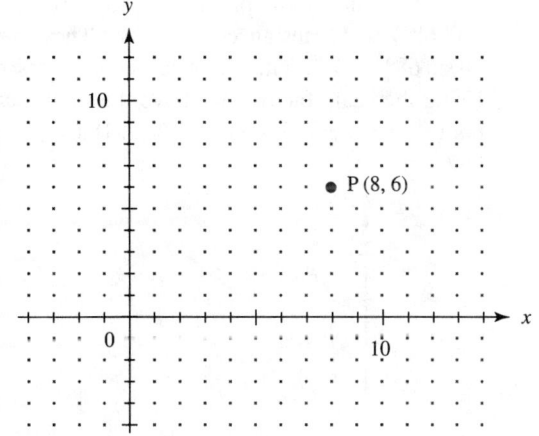

Figure 11.10

Rotation Angle α	New Location of P (polar coordinates)	New Location of P (Cartesian coordinates)
50°		
90°		
180°		
220°		
−71°		
−220°		

Rotation and Scaling Together

We'll approach the combination of rotation and scaling by looking at the change in a single point as we did in Activities 11.2 and 11.3. Example 11.3 gives some review of necessary trigonometry.

Example 11.3

Give the final location in Cartesian coordinates of the point $P(8, 6)$ after each of the following combinations of transformations.

(a) Rotate by 31° and then use an x-scaling factor of 0.5 and a y-scaling factor of 2.
(b) Use an x-scaling factor of -1.5 and a y-scaling factor of 0.5 followed by a rotation of 50°.

Solutions   (a) The transformations are shown in Figure 11.11. The rotation of 31° results in the point P_1 and the scaling of P_1 results in point P_2.

The original angle that segment $\overline{OP}$ makes with the positive x-axis is $\tan^{-1}\left(\frac{6}{8}\right) \approx 37°$ and the length of OP is $\sqrt{6^2 + 8^2} = 10$. Therefore $P(8, 6)$ has polar coordinates $P(10, 37°)$. The polar coordinates of the point P_1 after a rotation of 31° are $(10, 68°)$.

We can only apply scaling to Cartesian coordinates, so we must convert $(10, 68°)$ to Cartesian coordinates. They are $x = 10\cos 68° \approx 3.7$ and $y = 10\sin 68° \approx 9.3$. The Cartesian coordinates of the rotated point are $(3.7, 9.3)$. Using the scale factors, we have the final location of P_2, the rotated and scaled point: $(0.5 \times 3.7, 2 \times 9.3) \approx (1.9, 18.6)$.

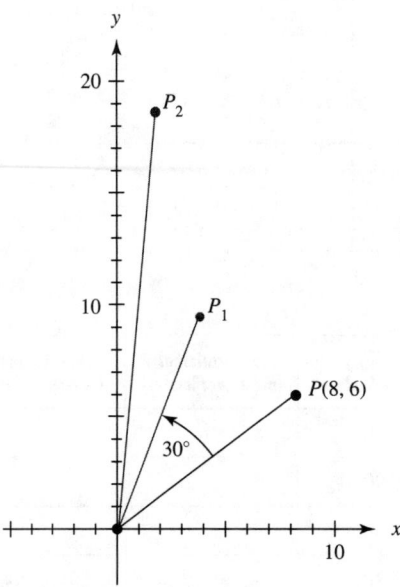

Figure 11.11

(b) The transformations are shown in Figure 11.12. The scaling of P results in P_1 and the rotation of $50°$ provides point P_2.

The Cartesian coordinates of the point P_1 after scaling are $(8 \times -1.5, 6 \times 0.5) = (-12, 3)$. At present we can only apply rotation to polar coordinates, so we must convert $(-12, 3)$ to polar coordinates. The length of $\overline{OP_1} = \sqrt{(-12)^2 + 3^2} \approx 12.4$. The angle is in the second quadrant, so it is equal to $180° - \angle P_1 O Q = 180° - \tan^{-1}\left(\frac{3}{12}\right) \approx 166°$. The polar coordinates of P_1 are $(12.4, 166°)$.

Now we can apply the rotation, producing the new point $(12.4, 166+50) = (12.4, 216°)$. The Cartesian coordinates for this point are $x = 12.4\cos 216° \approx -10.0$ and $y = 12.4\sin 216° \approx -7.3$. Therefore the Cartesian coordinates of the final location are $(-10.0, -7.3)$.

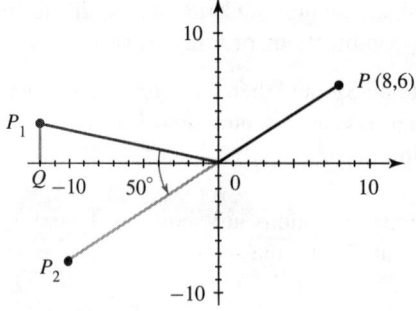

Figure 11.12

Activity 11.4
Rotation and Scaling

As in the previous examples, begin with the point $P(8, 6)$ shown in Figure 11.8. In part 1, use rotation first and then scaling. In part 2, use scaling first and then rotation. Compare your answers.

1. Fill in the coordinates of point P after a rotation (these will be the same as those worked out in Activity 11.3). Then apply the scaling factors to that point and enter the coordinates in the table.

Rotation Angle α	New Location (Cartesian)	Scale Factors	New Location (Cartesian)
50°		$x : 1, \ y : 2$	
90°		$x : 1.5, \ y : 0.5$	
180°		$x : 0.5, \ y : -1.5$	
220°		$x : -1, \ y : -1$	

2. Next you will locate the points using the same transformations, but with scaling first, followed by rotation. You can complete the fourth column with an estimate of the coordinates from your graph and use the method shown in Example 11.3 to check your result. Compare the final location in part one with the final location in part two.

Scale Factors	New Location (Cartesian)	Rotation Angle α	New Location (Cartesian)
$x : 1, \ y : 2$		50°	
$x : 1.5, \ y : 0.5$		90°	
$x : 0.5, \ y : -1.5$		180°	
$x : -1, \ y : -1$		220°	

3. Based on your results from parts 1 and 2, does it matter in which order you do rotation and scaling?

To demonstrate that the transformation resulting from rotation and scaling depends on the order in which you do the individual transformations, look at Figures 11.13 and 11.14. The results are quite different, but the rotation angles and scaling factors were identical; only the order was switched. In Figure 11.13 the black F was rotated through 235° and then an x-scaling factor of 2 and a y-scaling factor of 0.6 were applied. In Figure 11.13 we first applied the x- and y-scaling and then the same rotation.

The Trigonometry of Rotation

Now that you have had some experience with rotation, we'll do the mathematics necessary for rotation to be carried out by a computer program, which deals only in

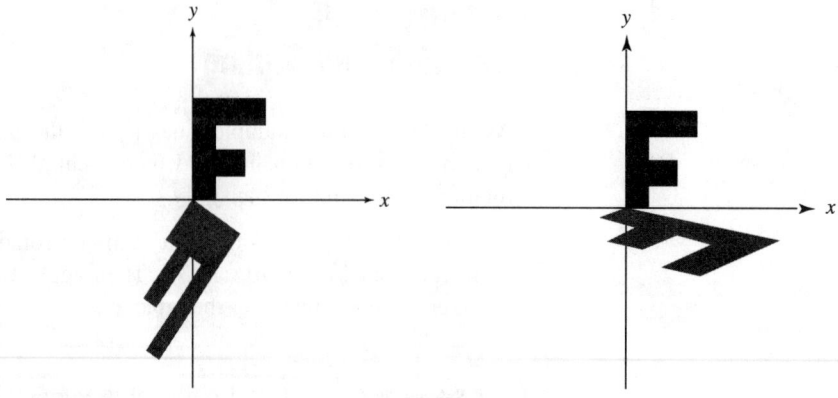

Figure 11.13   **Figure 11.14**

Cartesian coordinates. As you have seen, with polar coordinates it is easy to write the new coordinates of a rotated point. If we rotate the point $P(x, y)$ through an angle α to produce the point $P_1(a, b)$, we have to do some of the mathematics you have seen in Example 11.3.

Since we seek a general result that can be used by the computer, we will seek formulas for the new coordinates a and b in terms of the original coordinates x and y and the angle of rotation α. Figure 11.15 shows a picture that represents this problem. Although P is shown in the first quadrant and α is an angle less than $90°$ in the figure, the mathematics we will use will be good for any location of P and any size and direction of the rotation.

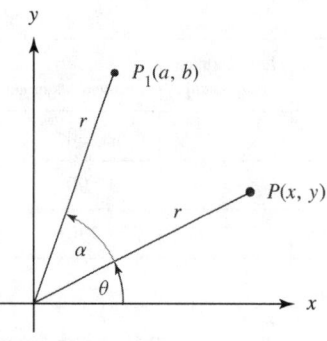

Figure 11.15

Example 11.4

Write the coordinates a and b of point P_1 in Figure 11.15 in terms of x, y, and α only.

Solution   First notice that the value of r for both points is the same since $\overline{OP}$ is rotated but not changed in length. We will use r in the formulas that follow, but we can always replace r with $\sqrt{x^2 + y^2}$ to satisfy the statement of the problem. Next notice that the polar coordinates for point P_1 are $(r, \theta + \alpha)$. Finally we can write the Cartesian coordinates from the polar coordinates using trigonometry

$$a = r\cos(\theta + \alpha)$$

$$b = r\sin(\theta + \alpha)$$

We can now substitute $r = \sqrt{x^2 + y^2}$ and $\theta = \tan^{-1}\left(\dfrac{y}{x}\right)$ to produce formulas that are only in terms of x, y, and α. However for computer and other engineering applications there is a more useful way to rewrite these formulas, as we will see next. ∎

To simplify the formulas in Example 11.4 we will need trigonometric formulas for $\sin(\theta + \alpha)$ and $\cos(\theta + \alpha)$. They are:

Trigonometric Functions of Angle Sums

$$\sin(\theta + \alpha) = \sin\theta\cos\alpha + \sin\alpha\cos\theta$$

$$\cos(\theta + \alpha) = \cos\theta\cos\alpha - \sin\theta\sin\alpha$$

Activity 11.5
Computing with the Angle Sum Formulas

Show that the formulas for the trigonometric functions of angle sums are true for the following specific angles.

1. Show that the formula works for $\sin 70°$ when we write $70°$ as $25° + 45°$. That is, show that

$$\sin 70° = \sin 25°\cos 45° + \sin 45°\cos 25°$$

2. If there is a difference between $\sin 70°$ and $\sin 25°\cos 45° + \sin 45°\cos 25°$, explain why you think that might have happened.
3. Repeat parts 1 and 2 for $\cos 180°$ when we write $180° = 63° + 117°$.

Example 11.5

Use the formulas for the trig functions of the sum of angles to rewrite the answer to Example 11.4.

Solution   The equations from Example 11.4 are

$$a = r\cos(\theta + \alpha)$$
$$b = r\sin(\theta + \alpha)$$

which the angle sum formulas convert into

$$a = r\cos\theta\cos\alpha - r\sin\theta\sin\alpha$$
$$b = r\sin\alpha\cos\theta + r\sin\theta\cos\alpha$$

But the Cartesian coordinates of point P are $x = r\cos\theta$ and $y = r\sin\theta$. These expressions for x and y are hidden in the above equations, as shown by the parentheses in the equations below.

$$a = (r\cos\theta)\cos\alpha - (r\sin\theta)\sin\alpha$$
$$b = \sin\alpha(r\cos\theta) + (r\sin\theta)\cos\alpha$$

Substituting the appropriate values of x and y for the parenthetical expressions produces the final expressions in terms of only x, y, and α:

$$a = x \cos\alpha - y \sin\alpha$$

$$b = x \sin\alpha + y \cos\alpha$$

These equations have many applications because of their connection to the field of matrix algebra, which we turn to in the next section.

Section 11.1 Exercises

1. Using an x-scaling factor of 3 and a y-scaling factor of -0.5, calculate the missing coordinates in the table below.

Original Point	Transformed Point
$(-2.5, 6)$	$(\_\_, \_\_)$
$(1.5, 4)$	$(\_\_, \_\_)$
$(0, 3)$	$(\_\_, \_\_)$
$(1.25, -2)$	$(\_\_, \_\_)$
$(\_\_, \_\_)$	$(4, 3)$
$(\_\_, \_\_)$	$(6, -2)$

2. Using an x-scaling factor of -0.25 and a y-scaling factor of 1.5, calculate the missing coordinates in the table below.

Original Point	Transformed Point
$(4, 4)$	$(\_\_, \_\_)$
$(8, -6)$	$(\_\_, \_\_)$
$(6, 2)$	$(\_\_, \_\_)$
$(-5, 4.2)$	$(\_\_, \_\_)$
$(\_\_, \_\_)$	$(1, 1.5)$
$(\_\_, \_\_)$	$(1.5, -2)$

In Exercises 3–6 use Figure 11.16.

3. (a) What is the area of $ABCD$?

   (b) Use an x-scaling factor of 2 and a y-scaling factor of 0.5 to transform $ABCD$ into rectangle $A'B'C'D'$.

   (c) What is the area of $A'B'C'D'$?

4. (a) Use an x-scaling factor of 0.5 and a y-scaling factor of 2 to transform $ABCD$ into rectangle $A''B''C''D''$.

   (b) What is the area of $A''B''C''D''$?

5. (a) Use an x-scaling factor of 2 and a y-scaling factor of 2 to transform $ABCD$ into rectangle $EFGH$.

   (b) What is the area of $EFGH$?

6. (a) Use an x-scaling factor of -1 and a y-scaling factor of -1 to transform $ABCD$ into rectangle $E''F''G''H''$.

   (b) What is the area of $E''F''G''H''$?

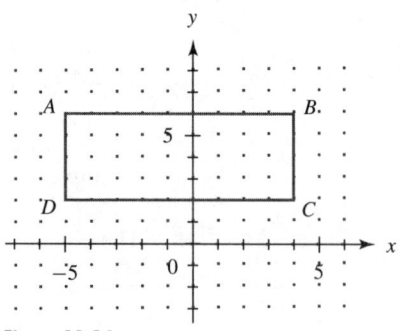

Figure 11.16

In Exercises 7–10 use Figure 11.17.

7. (a) What is the area of $ABCD$?

   (b) Use an x-scaling factor of 3 and a y-scaling factor of $\frac{1}{3}$ to transform $ABCD$ into rectangle $A'B'C'D'$.

   (c) What is the area of $A'B'C'D'$?

8. (a) Use an x-scaling factor of 1.5 and a y-scaling factor of $\frac{2}{3}$ to transform $ABCD$ into rectangle $A''B''C''D''$.

   (b) What is the area of rectangle $A''B''C''D''$?

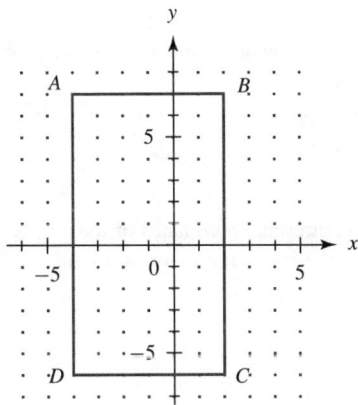

Figure 11.17

In Exercises 11–14 use Figure 11.18.

11. (a) What is the area of triangle $\triangle ABC$?

    (b) Use an x-scaling factor of -2 and a y-scaling factor of 0.5 to transform triangle $\triangle ABC$ into triangle $\triangle A'B'C'$.

    (c) Describe the effect of these scaling factors on triangle $\triangle ABC$.

    (d) What is the area of $A'B'C'$?

12. (a) Use an x-scaling factor of -0.5 and a y-scaling factor of 2 to transform triangle $\triangle ABC$ into triangle $\triangle A''B''C''$.

    (b) Describe the effect of these scaling factors on triangle $\triangle ABC$.

    (c) What is the area of triangle $\triangle A''B''C''$?

13. (a) Use an x-scaling factor of -2 and a y-scaling factor of 2 to transform triangle $\triangle ABC$ into triangle $\triangle DEF$.

    (b) Describe the effect of these scaling factors on triangle $\triangle ABC$.

    (c) What is the area of $\triangle DEF$?

In Exercises 15–18 use Figure 11.19.

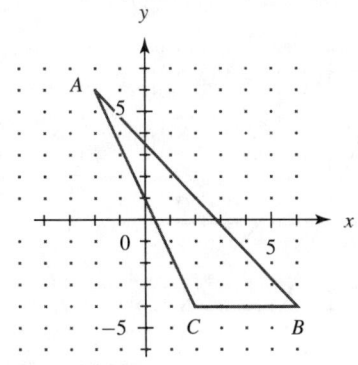

Figure 11.19

9. (a) Use an x-scaling factor of -2 and a y-scaling factor of -2 to transform $ABCD$ into rectangle $EFGH$.

    (b) What is the area of $EFGH$?

10. (a) Use an x-scaling factor of -1 and a y-scaling factor of -1 to transform $ABCD$ into rectangle $E''F''G''H''$.

    (b) What is the area of $E''F''G''H''$?

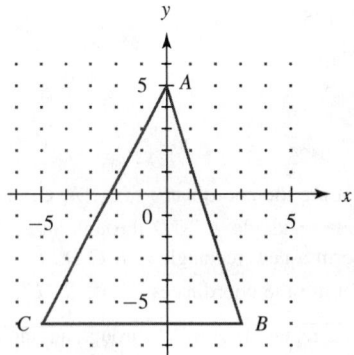

Figure 11.18

14. (a) Use an x-scaling factor of -2 and a y-scaling factor of -2 to transform triangle $\triangle ABC$ into triangle $\triangle D'E'F'$.

    (b) Describe the effect of these scaling factors on triangle $\triangle ABC$.

    (c) What is the area of $\triangle D'E'F'$?

15. (a) What is the area of triangle $\triangle ABC$?

    (b) Use an x-scaling factor of -2 and a y-scaling factor of -0.5 to transform triangle $\triangle ABC$ into triangle $\triangle A'B'C'$.

    (c) Describe the effect of these scaling factors on triangle $\triangle ABC$.

    (d) What is the area of $A'B'C'$?

16. (a) Use an x-scaling factor of -1 and a y-scaling factor of -2 to transform triangle $\triangle ABC$ into triangle $\triangle A''B''C''$.

    (b) Describe the effect of these scaling factors on triangle $\triangle ABC$.

    (c) What is the area of $\triangle A''B''C''$?

17. (a) Use an x-scaling factor of -1 and a y-scaling factor of -1 to transform triangle $\triangle ABC$ into triangle $\triangle DEF$.

   (b) Describe the effect of these scaling factors on triangle $\triangle ABC$.

   (c) What is the area of triangle $\triangle DEF$?

18. (a) Use an x-scaling factor of -0.5 and a y-scaling factor of -2 to transform triangle $\triangle ABC$ into triangle $\triangle D'E'F'$.

   (b) Describe the effect of these scaling factors on triangle $\triangle ABC$.

   (c) What is the area of triangle $\triangle D'E'F'$?

19. Give the polar coordinates of the points that result if each of the following points are rotated 142°.

   (a) $A(9, 37°)$

   (b) $B(2, 135°)$

   (c) $C(5, 216°)$

   (d) $D(7, 345°)$

In Exercises 23–26 use Figure 11.20.

23. (a) What are the coordinates of A, B, C, and D?

   (b) Rotate rectangle $ABCD$ through an angle of 90° to form a new rectangle $A'B'C'D'$.

   (c) What are the coordinates of A', B', C', and D'?

24. (a) Rotate rectangle $ABCD$ through an angle of 180° to form a new rectangle $EFGH$.

   (b) What are the coordinates of E, F, G, and H?

25. (a) Rotate rectangle $ABCD$ through an angle of 270° to form a new rectangle $E'F'G'H'$.

   (b) What are the coordinates of E', F', G', and H'?

26. (a) Rotate rectangle $ABCD$ through an angle of 135° to form a new rectangle $IJKL$.

   (b) What are the coordinates of I, J, K, and L?

In Exercises 27–30 use Figure 11.21.

27. (a) What are the coordinates of A, B, and C?

   (b) Rotate triangle ABC through an angle of 90° to form a new triangle $A'B'C'$.

   (c) What are the coordinates of A', B', and C'?

28. (a) Rotate triangle ABC through an angle of 180° to form a new triangle DEF.

   (b) What are the coordinates of D, E, and F?

29. (a) Rotate triangle ABC through an angle of 270° to form a new triangle $D'E'F'$.

   (b) What are the coordinates of D', E', and F'?

30. (a) Rotate triangle ABC through an angle of 135° to form a new triangle GHI.

   (b) What are the coordinates of G, H, and I?

20. Give the polar coordinates of the points that result if each of the following points are rotated 237°.

   (a) $E(6.5, 57°)$

   (b) $F(3.7, 153°)$

   (c) $G(2.5, 261°)$

   (d) $H(4.7, 297°)$

21. Give the Cartesian coordinates of the points that result if each of the following points are rotated 142°.

   (a) $A(9, 5)$

   (b) $B(2, -6)$

   (c) $C(-5, 7)$

   (d) $D(-7, -3)$

22. Give the Cartesian coordinates of the points that result if each of the following points are rotated 237°.

   (a) $E(-6.5, 5.7)$

   (b) $F(-3.7, -1.5)$

   (c) $G(2.5, 2.6)$

   (d) $H(4.7, -7.9)$

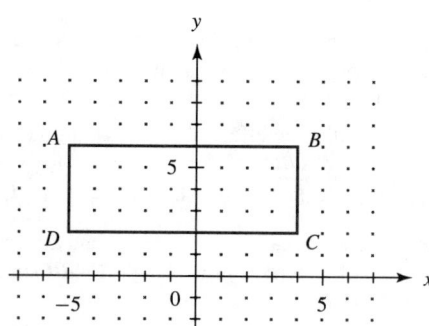

Figure 11.20

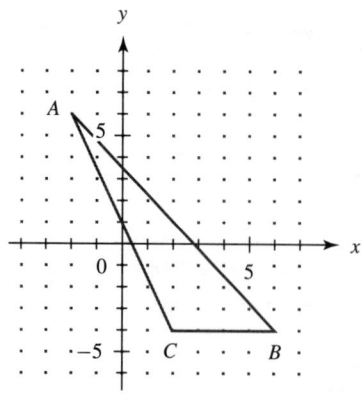

Figure 11.21

31. For this exercise use Figure 11.22.

    (a) Rotate rectangle $ABCD$ through an angle of 90° to form a new rectangle $A'B'C'D'$.

    (b) Use an x-scaling factor of -2 and a y-scaling factor of 0.5 to transform $A'B'C'D'$ into rectangle $A''B''C''D''$.

    (c) Use an x-scaling factor of -2 and a y-scaling factor of 0.5 to transform $ABCD$ into rectangle $EFGH$.

    (d) Rotate rectangle $EFGH$ through an angle of 90° to form a new rectangle $E'F'G'H'$.

    (e) Compare rectangles $A''B''C''D''$ and $E'F'G'H'$. How are they alike? How are they different? How do they compare to rectangle $ABCD$?

32. For this exercise use Figure 11.23.

    (a) Rotate triangle ABC through an angle of 180° to form a new triangle $A'B'C'$.

    (b) Use an x-scaling factor of -0.5 and a y-scaling factor of -1.5 to transform $A'B'C'$ into triangle $A''B''C''$.

    (c) Use an x-scaling factor of -0.5 and a y-scaling factor of -1.5 to transform ABC into triangle DEF.

    (d) Rotate triangle DEF through an angle of 180° to form a new triangle $D'E'F'$.

    (e) Compare triangles $A''B''C''$ and $D'E'F'$. How are they alike? How are they different? How do they compare to triangle ABC?

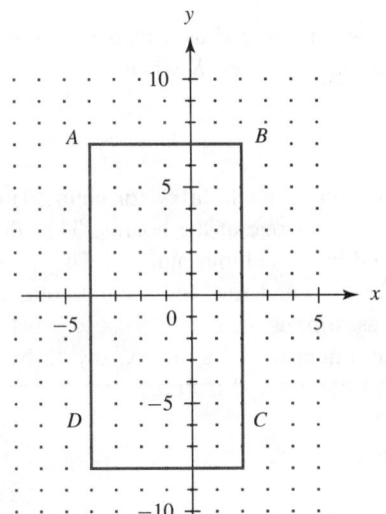

Figure 11.22

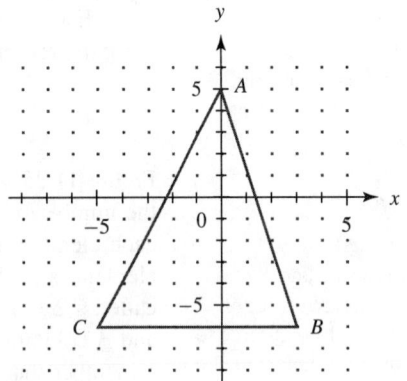

Figure 11.23

11.2 Introduction to Matrix Algebra

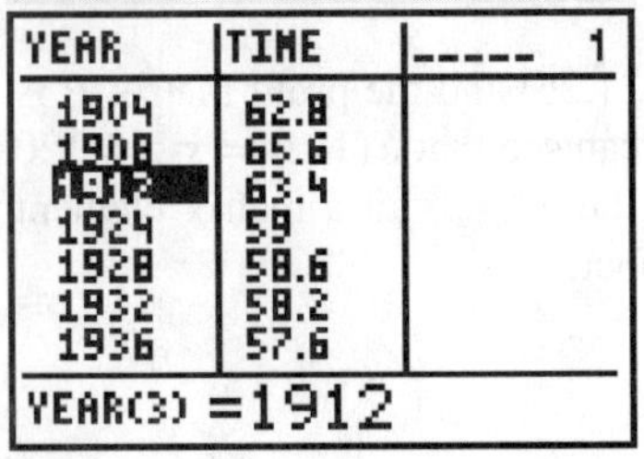

Figure 11.24

Actually a matrix can have more than two dimensions, but in this book we will stop at two.

Throughout this book you have studied data that is stored in calculator lists. For example, Figure 11.24 shows a portion of two lists that we used in Chapter 1. The lists are *YEAR*, the list of years in which the men's 100-yard freestyle swimming event was held in the Olympics, and *TIME*, the list of the winning times in seconds for each year. These lists are each written as a column of numbers, so in mathematics we think of them as one-dimensional lists. Therefore to identify the address of any of the entries, we need to say just one number. Figure 11.24 shows how the calculator identifies the third number in the list as YEAR(3) = 1912.

    A **matrix** is a two-dimensional list of data that contains numbers or variables in rows and columns. For an example of the application of matrices we will return to the second opening problem in Chapter 10 and present the same data in matrix form. First let's look again at the statement of the problem.

Building the "E" and "F" Circuit Boards

▶ **Parts Requirements**
- board E requires 30 resistors, 15 capacitors, and 6 transistors
- board F requires 10 resistors, 12 capacitors, and 9 transistors

▶ **Parts supply (daily)**
- 210 resistors, 120 capacitors, and 72 transistors

▶ **Profit**
- $6.00 for board E and $5.00 for board F

How many of each board should be made in order to make the most profit?

Some mathematicians use brackets to enclose a matrix. They would write matrix R as $\begin{bmatrix} 30 & 15 & 6 \\ 10 & 12 & 9 \end{bmatrix}$.

Many books name a matrix with a capital, or uppercase, letter and use a subscripted lowercase version of that same letter for the entries.

Each of these sets of numbers can be represented as a matrix. For example, the parts requirements can be written in a matrix named R where

$$R = \begin{pmatrix} 30 & 15 & 6 \\ 10 & 12 & 9 \end{pmatrix}$$

Figure 11.25 gives an explanation of this matrix. Each **entry** (or **element**) of R is the number of one type of part that is needed for one of the boards. The address of each element uses its row number followed by its column number. The highlighted element is $R(2, 2)$ or r_{22}, the number in the second row and second column. R is called a 2×3 matrix, or we say that R has **dimension** 2×3, because it has 2 rows and 3 columns. When you pronounce the dimension of a matrix, say "2 by 3" and remember that a 2×3 matrix is different from a 3×2 matrix.

REQUIREMENTS MATRIX
(a 2×3 matrix)

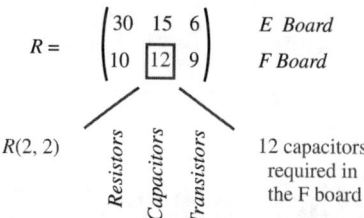

Figure 11.25

The supply matrix S is a 1×3 matrix (Figure 11.26) and the profit matrix P is a 2×1 matrix (Figure 11.27). From the figures you can see that $S(1, 2) = s_{12} = 120$ and $D(2, 1) = d_{21} = 5$. Remember to always give the address of a matrix element using the row number followed by the column number.

Matrix Multiplication

Now that we have represented the given information with matrices, we are ready to make some use of the effort. Remember that the question was to find the number of each board to manufacture so that the available supplies were not exceeded and the profit was maximized. Therefore we need a matrix that shows the number of each

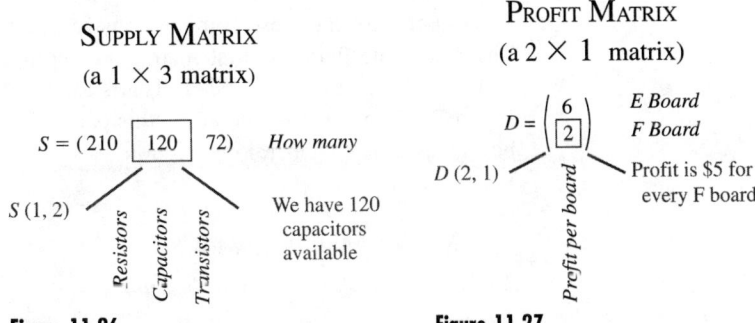

Figure 11.26

Figure 11.27

board to make. Figure 11.28 shows a production matrix with values chosen for e and f, the number of "E" boards and "F" boards we might want to produce.

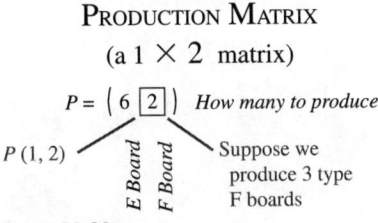

Figure 11.28

You might wonder why the production matrix is written in one row and the profit matrix is written in one column. As will be explained later, the answer is that matrix algebra requires them to be different.

The total profit from 6 E boards and 2 F boards is computed as

$$\text{Profit} = 6 \times 6 + 2 \times 5 = 46$$

Thus the total profit in this case is $46.

To compute the total number of each part that is used we would do three computations:

$$\text{Number of resistors} = 6 \times 30 + 2 \times 10 = 200$$

$$\text{Number of capacitors} = 6 \times 15 + 2 \times 12 = 114$$

$$\text{Number of transistors} = 6 \times 6 + 2 \times 9 = 54$$

Computations like this, in which numbers from one table are multiplied by numbers in another table and the results are added, are common in many types of applications. Matrix algebra was invented just for this purpose. For example, when the production matrix P is multiplied by the profit matrix D we get

$$PD = (6 \quad 2)\begin{pmatrix} 6 \\ 5 \end{pmatrix} = (46)$$

and when the production matrix P is multiplied by the requirements matrix R, the result is

$$PR = (6 \quad 2)\begin{pmatrix} 30 & 15 & 6 \\ 10 & 12 & 9 \end{pmatrix} = (200 \quad 114 \quad 54)$$

You can see that the results of these matrix multiplications match the hand computations of profit and parts supplies. How does matrix multiplication work? Look at each of the these computations for the total number of parts and the results. The 6 and the 2 in the first matrix are multiplied by each column of the second matrix and the results are added. That's exactly how matrix multiplication is defined. Figure 11.29 gives more detail on this particular matrix multiplication; the following box gives a more general description.

MATRIX MULTIPLICATION

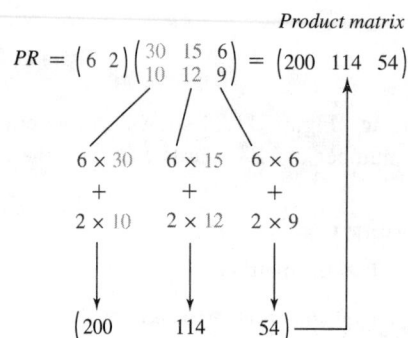

Figure 11.29

Summary: Matrix Multiplication

Let A be an $m \times n$ matrix and B an $n \times p$ matrix. The product AB is an $m \times p$ matrix where:

▶ The numbers in each row of matrix A are paired up with the numbers in each column of matrix B.
▶ Each pair of numbers is multiplied and the products are added.
▶ The result obtained from multiplying row r of matrix A and column c in matrix B is placed in row r, column c of the product matrix.

Example 11.6

If $C = AB$ where $A = \begin{pmatrix} 5 & 2 \\ 4 & 1 \\ 3 & 0 \end{pmatrix}$ and $B = \begin{pmatrix} 1 & 2 & 3 & 4 \\ 5 & 6 & 7 & 8 \end{pmatrix}$, what is c_{23}, the element in row 2, column 3 of C?

Solution   The element c_{23} comes from multiplying the second row of A and the third column of B. The result of that combination is $4 \times 3 + 1 \times 7 = 19$. Thus $c_{23} = 19$.

Example 11.7

What are the dimensions of C if $C = AB$ where

$$A = \begin{pmatrix} 5 & 2 \\ 4 & 1 \\ 3 & 0 \end{pmatrix} \quad \text{and} \quad B = \begin{pmatrix} 1 & 2 & 3 & 4 \\ 5 & 6 & 7 & 8 \end{pmatrix}$$

Solution   Since there are three rows in A and four columns in B, there will be three rows and four columns of C. We say that C is a 3×4 matrix.   ■

Example 11.8

Compute C, the product of matrices A and B where

$$A = \begin{pmatrix} 5 & 2 \\ 4 & 1 \\ 3 & 0 \end{pmatrix} \quad \text{and} \quad B = \begin{pmatrix} 1 & 2 & 3 & 4 \\ 5 & 6 & 7 & 8 \end{pmatrix}$$

Solution   Each of the 12 elements of C is computed using the process shown in Example 11.6. The result is

$$C = AB = \begin{pmatrix} 15 & 22 & 29 & 36 \\ 9 & 14 & 19 & 24 \\ 3 & 6 & 9 & 12 \end{pmatrix}$$ ■

Matrix Algebra on the *TI-83*

The *TI-83* handles all the matrix computations that you will need to do in this course. However there are some computations that it does not do, so check your calculator manual when you need to use matrix algebra in later courses or jobs.

Example 11.9

Enter the matrices A and B from Example 11.6 into your calculator. Here $A = \begin{pmatrix} 5 & 2 \\ 4 & 1 \\ 3 & 0 \end{pmatrix}$ and $B = \begin{pmatrix} 1 & 2 & 3 & 4 \\ 5 & 6 & 7 & 8 \end{pmatrix}$.

Solution   Press **MATRX** and select the **MATRIX EDIT** menu. If no matrices have been previously entered, you will see Figure 11.30. Select matrix **A** and you will see the **MATRIX EDIT** screen (Figure 11.31). Enter the dimensions 3 and 2 in place of the 1×1 that appears at the top of the screen, pressing **ENTER** after each number. When you have finished you will see a 3×2 matrix containing all zeros. Replace each zero with the desired numbers. Make sure to press the **ENTER** key after typing each number. When you have finished your calculator screen should look like Figure 11.32. Press **2nd** [QUIT] when you have entered all the elements.

If you wish to delete existing matrices, press **2nd** [MEM]; then select option 2:Delete. From the DELETE FROM menu, select option 5:Matrix. You will see a list of the existing matrices. Press **ENTER** to delete the matrices one at a time.

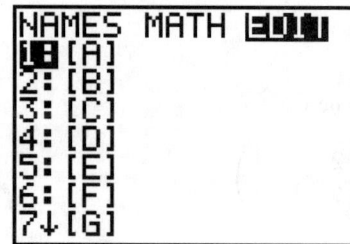
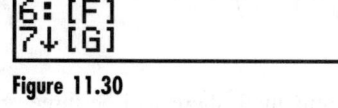

Figure 11.30

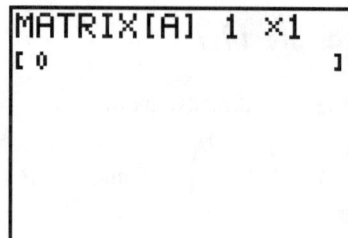

Figure 11.31

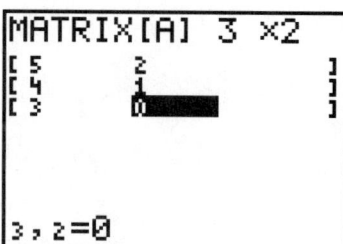

Figure 11.32

Repeat the process for matrix *B*. You will not be able to see the entire matrix all at once on the edit screen. One way to check that you have entered the matrix correctly is to view it on the home screen. After you have finished entering the elements of *B*, press (2nd) [QUIT] (MATRX) 2 (ENTER). The result should look like Figure 11.33.

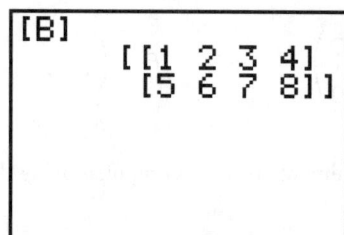

Figure 11.33

Example 11.10

Multiply the matrices *A* and *B* on your calculator.

Solution   Like all calculations, matrix multiplication is done on the home screen, and you need to enter the multiplication command there. You do this in the same way that you would enter a list: Press (MATRX), then select A; press (MATRX) again and select B. The symbols [A] [B] appear on the screen. It is not necessary to enter the multiplication sign between the matrices; just press (ENTER) to execute the multiplication and you will see the product shown in Figure 11.34.

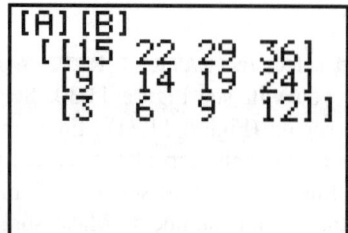

Figure 11.34

Which Matrices Can Be Multiplied?

If you enter the product BA into the calculator, you will see Figure 11.35. Why? The reason will be clear if you return to the explanation of matrix multiplication.

$$BA = \begin{pmatrix} 1 & 2 & 3 & 4 \\ 5 & 6 & 7 & 8 \end{pmatrix} \begin{pmatrix} 5 & 2 \\ 4 & 1 \\ 3 & 0 \end{pmatrix}$$

Each row of B contains four elements and each column of A contains three elements. There is truly a "dimension mismatch," as shown on the calculator screen.

```
ERR:DIM MISMATCH
1▮Quit
2:Goto
```

Figure 11.35

Summary: Matching Matrices for Multiplication

In order for the matrix product AB to make sense, the number of rows of A must equal the number of columns of B. This can be summarized as follows.

▶ Write the dimensions of the two matrices in the order in which you want to multiply them. For A and B as defined in Example 11.6, that would be $(3 \times 2)(2 \times 4)$. If the "inner" numbers are equal (in this case 2 and 2) the matrices can be multiplied.

▶ The "outer" numbers in the two dimensions tell you the dimensions of the product matrix. In the case of AB the product matrix has dimension 3×4.

Activity 11.6
Matrix Match-Making

Begin with the matrix A as defined in Example 11.6. That is,

$$A = \begin{pmatrix} 5 & 2 \\ 4 & 1 \\ 3 & 0 \end{pmatrix}$$

You saw in Example 11.8 that the product AB makes sense if B is a 2×4 matrix.

1. Write the dimensions of four other forms that B could take for the multiplication AB to be successful. Give the dimensions of the product matrix in each case.
2. Write the dimensions of four forms that B could take for the multiplication BA to be successful. Give the dimensions of the product matrix in each case.

Choosing the Dimensions of a Matrix

You have seen in this section that the dimensions of matrices are very important in matrix multiplication. In setting up the matrix solution to the linear programming problem, why did we choose to give the requirements and production matrices the dimensions they have? Did we have any other choices?

We began with defining the requirements for matrix R as follows:

$$R = \begin{pmatrix} 30 & 15 & 6 \\ 10 & 12 & 9 \end{pmatrix}$$

This shape was chosen because of the way the problem was stated, which was:

▶ Board E requires 30 resistors, 15 capacitors, and 6 transistors.
▶ Board F requires 10 resistors, 12 capacitors, and 9 transistors.

However, if the problem had been stated differently, for example,

▶ Resistors: 30 for board E and 10 for board F
▶ Capacitors: 15 for board E and 12 for board F
▶ Transistors: 6 for board E and 9 for board F

then it would have made perfect sense to write

$$R_1 = \begin{pmatrix} 30 & 10 \\ 15 & 12 \\ 6 & 9 \end{pmatrix}$$

The production matrix P was determined by the dimensions of R, which are 2×3. In order for the matrix multiplication PR to be possible, P needed to be a 1×2 matrix, and it was defined as

$$P = (6 \quad 2)$$

This allows us to compute the product PR.

To match R_1, a 3×2 matrix, the production matrix would have to be defined as a 2×1 matrix,

$$P_1 = \begin{pmatrix} 6 \\ 2 \end{pmatrix}$$

and we could compute the product $R_1 P_1$.

Note also that the definitions also determined the way each product was written. In the first case the product had to be PR, and in the second case the product would have to be written $R_1 P_1$. Both methods of writing the matrices would lead to the same answers.

Matrices and Transformation of Points in Two Dimensions

In two dimensions, the scaling and rotation of points can be expressed in terms of matrix multiplication. This is useful because matrix algebra is built into advanced calculators and many computer languages. We begin with scaling.

Scaling

We defined scaling as the multiplication of each coordinate by a constant called the scaling factor. The scaling factor for the x-coordinate can be different from the scaling factor for the y-coordinate. Let's call these scaling factors m and n. Then if the point (x, y) is scaled using these factors, the new point (a, b) is equal to (mx, ny). The transformation equations are

$$a = mx$$

$$b = ny$$

Example 11.11 shows how to write these equations using matrix multiplication.

Example 11.11

Express the scaling transformation equations in matrix form.

Solution   Check that the following matrix equation does the job.

$$\begin{pmatrix} m & 0 \\ 0 & n \end{pmatrix} \begin{pmatrix} x \\ y \end{pmatrix} = \begin{pmatrix} a \\ b \end{pmatrix}$$

Row 1, column 1 of the result is formed by row 1 of the first matrix and column 1 of the second. Therefore that element is equal to $mx + 0y = mx$, which corresponds to the first scaling equation. Row 2, column 1 of the result is formed by row 2 of the first matrix and column 1 of the second. Therefore that element is equal to $0x + ny = ny$, corresponding to the second scaling equation. ∎

Example 11.12   Application

Use matrices to determine the coordinates of rectangle $ABCD$—in Figure 11.36— with an x-scaling factor of 2.5 and a y-scaling factor of -1.7.

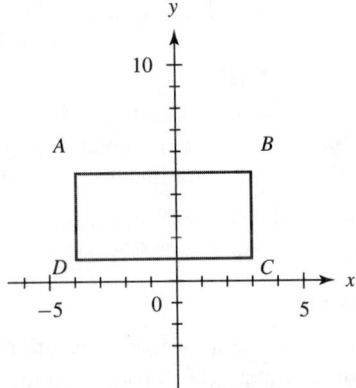

Figure 11.36

Solution   The vertices of the rectangle in Figure 11.36 have the coordinates $A(-4, 5)$, $B(3, 5)$, $C(3, 1)$, and $D(-4, 1)$. These coordinates can be represented by matrix U where

$$U = \begin{matrix} & \begin{matrix} A & B & C & D \end{matrix} \\ \begin{matrix} x \\ y \end{matrix} & \begin{pmatrix} -4 & 3 & 3 & -4 \\ 5 & 5 & 1 & 1 \end{pmatrix} \end{matrix}$$

(The rows and columns have been labeled to show where each entry comes from.)

The scaling factors are $m = 2.5$ and $n = -1.7$, which can be represented by matrix

$$S = \begin{pmatrix} 2.5 & 0 \\ 0 & -1.7 \end{pmatrix}$$

The transformed matrix is $T = SR$ or

$$T = \begin{pmatrix} 2.5 & 0 \\ 0 & -1.7 \end{pmatrix} \begin{pmatrix} -4 & 3 & 3 & -4 \\ 5 & 5 & 1 & 1 \end{pmatrix}$$

$$= \begin{pmatrix} -10 & 7.5 & 7.5 & -10 \\ -8.5 & -8.5 & -1.7 & -1.7 \end{pmatrix}$$

The coordinates of the scaled rectangle are $A'(-10, -8.5)$, $B'(7.5, -8.5)$, $C'(7.5, -1.7)$, and $D'(-10, -1.7)$, as shown in Figure 11.37.

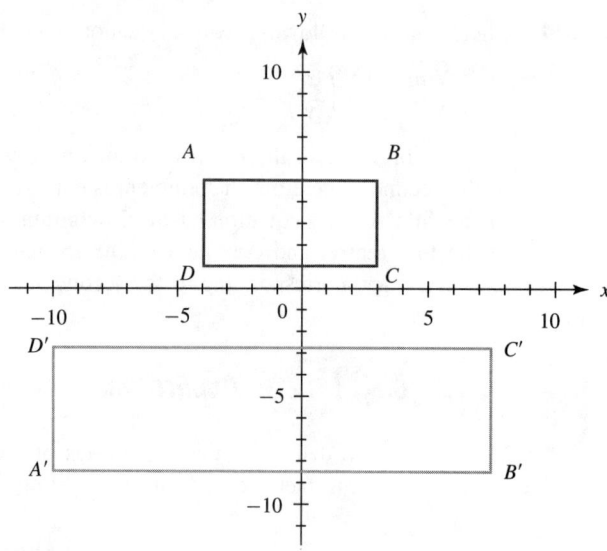

Figure 11.37

Rotation

In Section 11.1 you saw that if the point $P(x, y)$ is rotated through an angle α with respect to the origin of the coordinate system, the new location $P_1(a, b)$ is given by two equations:

$$a = x \cos\alpha - y \sin\alpha$$

$$b = x \sin\alpha + y \cos\alpha$$

This turns out to be a perfect application for matrix multiplication. First we form a 2×2 rotation matrix R for the α terms

$$R = \begin{pmatrix} \cos\alpha & -\sin\alpha \\ \sin\alpha & \cos\alpha \end{pmatrix}$$

We could form the matrix multiplication in two ways. The variables x and y could be in a 1×2 matrix or a 2×1 matrix because the multiplication would work either way. However, because of how we are going to use matrices in equation solving in the next section, we'll choose a 2×1 matrix for x and y:

$$P = \begin{pmatrix} x \\ y \end{pmatrix}$$

Then the multiplication RP will produce the new coordinates. That is,

$$RP = \begin{pmatrix} \cos\alpha & -\sin\alpha \\ \sin\alpha & \cos\alpha \end{pmatrix} \begin{pmatrix} x \\ y \end{pmatrix}$$

The matrix of new coordinates is equal to the product of these matrices:

$$\begin{pmatrix} a \\ b \end{pmatrix} = RP = \begin{pmatrix} x\cos\alpha - y\sin\alpha \\ x\sin\alpha + y\cos\alpha \end{pmatrix}$$

The result is a matrix form of the two equations.

Because many calculators and most computer languages now have built-in matrix multiplication and other matrix operations, the matrix form for problems like linear programming or geometric transformation has become more common than the approach that uses systems of equations. Later in this chapter we will show how to form the matrices for rotation in three dimensions.

Example 11.13

Write the rotation matrix for a rotation of $75°$ and apply it to the point $P(8, 6)$ to determine the final position of the rotated point. Plot the two points on a graph. Check that the angle of rotation is approximately $75°$ and that the distance of the new point to the origin is unchanged.

Solution   The rotation matrix is

$$\begin{pmatrix} \cos 75° & -\sin 75° \\ \sin 75° & \cos 75° \end{pmatrix}$$

The new point P_1 is formed by the product

$$\begin{pmatrix} \cos 75° & -\sin 75° \\ \sin 75° & \cos 75° \end{pmatrix} \begin{pmatrix} 8 \\ 6 \end{pmatrix} = \begin{pmatrix} -3.73 \\ 9.28 \end{pmatrix}$$

Thus P_1 has the coordinates $(-3.73, 9.28)$. (P_1 is plotted in Figure 11.38.) The angle $\angle POP_1$ appears to be approximately $75°$. The length OP_1 is equal to $\sqrt{(-3.73)^2 + 9.28^2} \approx 10.00$, which is the length of $\overline{OP}$.

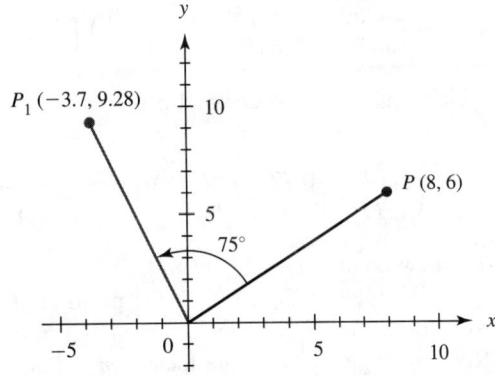

Figure 11.38

Activity 11.7
Practice with Rotations

Show on a graph like Figure 11.8 the effect of each of the following matrices if they were used as a rotation matrix for the point $(8, 6)$. Estimate the angle of rotation caused by each matrix. By rotating other points with the same matrices, determine whether the angle of rotation would be the same for any original point.

1. $\begin{pmatrix} 1 & 0 \\ 0 & 1 \end{pmatrix}$ 3. $\begin{pmatrix} 0 & -1 \\ 1 & 0 \end{pmatrix}$ 5. $\begin{pmatrix} 0 & 1 \\ 1 & 0 \end{pmatrix}$

2. $\begin{pmatrix} -1 & 0 \\ 0 & -1 \end{pmatrix}$ 4. $\begin{pmatrix} 0 & 1 \\ -1 & 0 \end{pmatrix}$

Rotation and Scaling

In the opening section you noticed that the order in which you do rotation and scaling usually makes a difference. Rotation followed by scaling reshapes a geometric figure quite differently from scaling followed by rotation. This can be demonstrated with matrix multiplication.

The scaling and rotation matrices are

$$\begin{pmatrix} m & 0 \\ 0 & n \end{pmatrix} \quad \text{and} \quad \begin{pmatrix} \cos\alpha & -\sin\alpha \\ \sin\alpha & \cos\alpha \end{pmatrix}$$

Activity 11.8
Using Matrices to Scale and Rotate

In Activity 11.4 (page 749) you tested the results of different orderings of scaling and rotation. In this activity, use the same beginning point $(8, 6)$ and transformations as you used there, but do the computation using matrices. The first one is done for you.

1. Rotation angle: $50°$, scale factors $m = 1$, $n = 2$.

   For scaling first, the coordinates of the new point are given by the matrix equation

$$\begin{pmatrix} \cos 50° & -\sin 50° \\ \sin 50° & \cos 50° \end{pmatrix} \begin{pmatrix} 1 & 0 \\ 0 & 2 \end{pmatrix} \begin{pmatrix} 8 \\ 6 \end{pmatrix}$$

   which equals, to two decimal places,

$$\begin{pmatrix} 0.64 & -0.77 \\ 0.77 & 0.64 \end{pmatrix} \begin{pmatrix} 1 & 0 \\ 0 & 2 \end{pmatrix} \begin{pmatrix} 8 \\ 6 \end{pmatrix} = \begin{pmatrix} -4.12 \\ 13.84 \end{pmatrix}$$

   The new point is $(-4.12, 13.84)$.

   For rotation first, switch the position of the rotation and scaling matrices.

2. Rotation angle: $90°$, scale factors $m = 1.5$, $n = 0.5$.
3. Rotation angle: $180°$, scale factors $m = 0.5$, $n = -1.5$.
4. Rotation angle: $220°$, scale factors $m = -1$, $n = -1$.

Example 11.14 continues the process begun in Example 11.12 to use matrices to transform a rectangle. This time the transformation will include both rotation and scaling transformations.

Example 11.14   Application

Use matrices to determine the coordinates of quadrilateral $WXYZ$ in Figure 11.39.

(a) First with a rotation of 125°, then with an x-scaling factor of 2.5 and a y-scaling factor of -1.7.
(b) First with an x-scaling factor of 2.5 and a y-scaling factor of -1.7, then with a rotation of 125°.

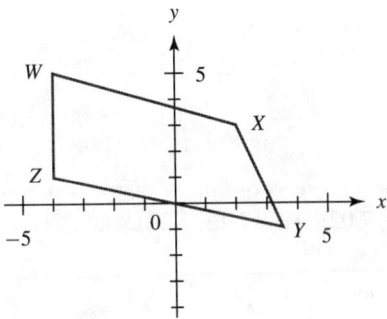

Figure 11.39

Solution   The vertices of the quadrilateral in Figure 11.39 have the coordinates $W(-4, 5)$, $X(2, 3)$, $Y(4, -1)$, and $Z(-4, 1)$. These coordinates can be represented by matrix C where

$$C = \begin{pmatrix} -4 & 2 & 4 & -4 \\ 5 & 3 & -1 & 1 \end{pmatrix}$$

The rotation factors for a rotation of $\alpha = 125°$ are

$$R = \begin{pmatrix} \cos\alpha & -\sin\alpha \\ \sin\alpha & \cos\alpha \end{pmatrix}$$

$$= \begin{pmatrix} \cos 125° & -\sin 125° \\ \sin 125° & \cos 125° \end{pmatrix}$$

When this matrix is entered in a *TI-83* as matrix A, you should obtain a screen like the one in Figure 11.40. While the entries were made using the trigonometric functions, the display shows decimal approximations for these values. Thus, in Figure 11.40, a_{22} has a value of -0.5736 in the matrix and a longer decimal value at the bottom of the screen. This longer value is the one the calculator will use in its calculations.

Figure 11.40

The scaling factors are $m = 2.5$ and $n = -1.7$, which can be represented by matrix

$$S = (2.5 \quad 00 \quad -1.7)$$

This is entered as matrix B in the *TI-83*.

(a) If T_{RS} denotes the matrix that results by first rotating quadrilateral $WXYZ$ and then scaling, we obtain

$$T_{RS} = BAC$$

$$= \begin{pmatrix} 2.5 & 0 \\ 0 & -1.7 \end{pmatrix} \begin{pmatrix} \cos 125° & -\sin 125° \\ \sin 125° & \cos 125° \end{pmatrix} \begin{pmatrix} -4 & 2 & 4 & -4 \\ 5 & 3 & -1 & 1 \end{pmatrix}$$

$$\approx \begin{pmatrix} -4.5 & -9.01 & -3.69 & 3.69 \\ 10.45 & 0.14 & -6.55 & 6.55 \end{pmatrix}$$

The approximate coordinates of the scaled quadrilateral are $W_1(12.70, -3.32)$, $X_1(1.31, 7.02)$, $Y_1(-7.13, 7.22)$, and $Z_1(7.13, -7.22)$, as shown in Figure 11.41.

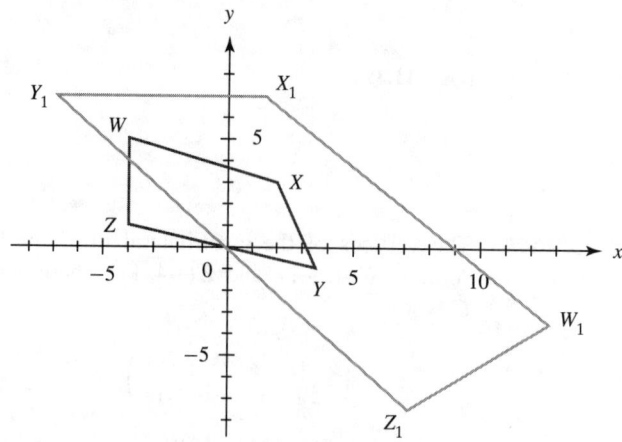

Figure 11.41

(b) If T_{SR} denotes the matrix that results by first scaling quadrilateral $WXYZ$ and then rotating, we obtain

$$T_{SR} = ABC$$

$$= \begin{pmatrix} \cos 125° & -\sin 125° \\ \sin 125° & \cos 125° \end{pmatrix} \begin{pmatrix} 2.5 & 0 \\ 0 & -1.7 \end{pmatrix} \begin{pmatrix} -4 & 2 & 4 & -4 \\ 5 & 3 & -1 & 1 \end{pmatrix}$$

$$\approx \begin{pmatrix} 12.70 & 1.31 & -7.13 & 7.13 \\ -3.32 & 7.02 & 7.22 & -7.22 \end{pmatrix}$$

The approximate coordinates of the scaled quadrilateral are $W_2(-4.5, 10.45)$, $X_2(-9.01, 0.14)$, $Y_2(-3.69, -6.55)$, and $Z_2(3.69, 6.55)$, as shown in Figure 11.42.

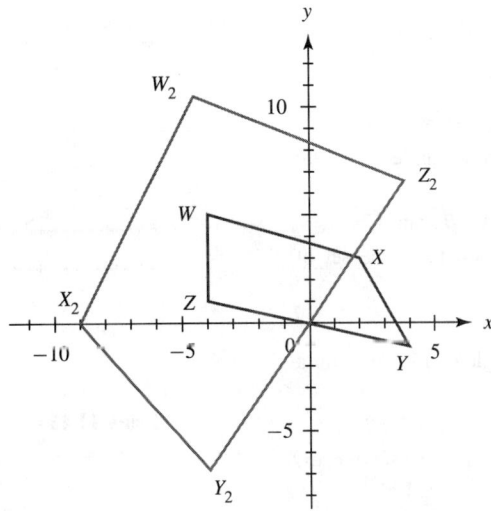

Figure 11.42

Section 11.2 Exercises

In Exercises 1–4 use matrix $A = \begin{pmatrix} -2 & 5 & 17 \\ 0 & 2 & 1 \end{pmatrix}$.

1. What are the dimensions of A?

2. $A(2, 3) =$

3. $a_{21} =$

4. $a_{12} =$

In Exercises 5–8 use matrix $B = \begin{pmatrix} 3 & -4 & 1 \\ 5 & 0 & 2 \\ 0 & -1 & 14 \\ 5 & 1 & 23 \\ -13 & -5 & 21 \end{pmatrix}$.

5. What are the dimensions of B?

6. $B(5, 2) =$

7. $b_{23} =$

8. $b_{32} =$

9. *Business* At one storage location a company has 27 of computer chip A, 14 keyboards, and 17 motherboards. At a second storage location the company has 43 of computer chip A, 31 keyboards, and 19 motherboards. Express this information as a matrix and indicate what each row and column represents.

10. *Business* A computer retailer sells three types of computer: the personal computer (PC), the minicomputer (MC), and the Web computer (WC). The retailer has two stores. Store 1 has 57 PC, 31 MC, and 92 WC computers. Store 2 has 42 PC, 16 MC, and 53 WC computers. Express this information as a matrix and indicate what each row and column represents.

11. What is the dimension of AB if A is a 5×2 matrix and B is a 2×3 matrix?

12. What is the dimension of XY if X is a 4×6 matrix and Y is a 6×1 matrix?

In Exercises 13–18 use matrix $C = \begin{pmatrix} 4 & 1 & 12 \\ 5 & 7 & -2 \\ -3 & 0 & 1 \end{pmatrix}$ *and* $D = \begin{pmatrix} -2 & 3 & 8 \\ -5 & 6 & -7 \\ 1 & -3 & 2 \end{pmatrix}$.

13. What are the dimensions of $E = CD$?

14. $e_{12} =$

15. $e_{23} =$

16. $e_{32} =$

17. $CD =$

18. $DC =$

In Exercises 19 and 20 use Figure 11.43.

19. *Computer Aided Design (CAD)*

    (a) What are the coordinates of A, B, and C?

    (b) Rotate triangle ABC through an angle of 45° to form a new triangle $A'B'C'$.

    (c) What are the coordinates of A', B', and C'?

    (d) Graph $\triangle ABC$ and $\triangle A'B'C'$ on the same set of axes.

20. *Computer Aided Design (CAD)*

    (a) Rotate $\triangle ABC$ through an angle of 150° to form a new triangle DEF.

    (b) What are the coordinates of D, E, and F?

    (c) Graph $\triangle ABC$ and $\triangle DEF$ on the same set of axes.

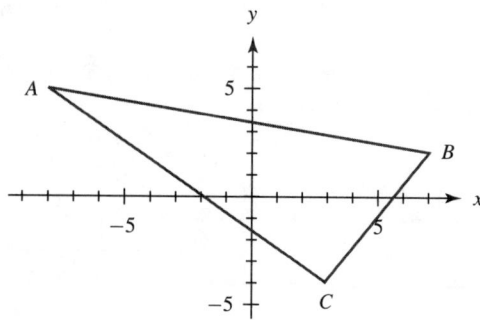

Figure 11.43

In Exercises 21 and 22 use Figure 11.44.

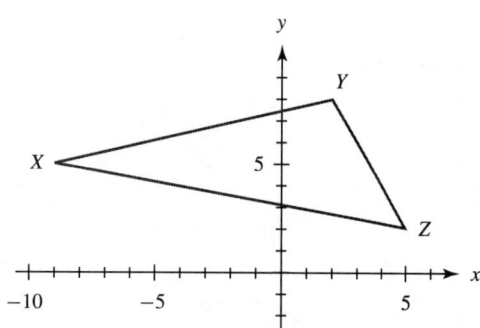

Figure 11.44

21. *Computer Aided Design (CAD)*

    (a) What are the coordinates of X, Y, and Z?

    (b) Rotate triangle XYZ through an angle of 245° to form a new triangle $X'Y'Z'$.

    (c) What are the coordinates of X', Y', and Z'?

    (d) Graph $\triangle XYZ$ and $\triangle X'Y'Z'$ on the same set of axes.

22. *Computer Aided Design (CAD)*

    (a) Rotate $\triangle XYZ$ through an angle of $-115°$ to form a new triangle PQR.

    (b) What are the coordinates of P, Q, and R?

    (c) Graph $\triangle XYZ$ and $\triangle PQR$ on the same set of axes.

In Exercises 23–26 use Figure 11.45.

23. *Computer Aided Design (CAD)*

    (a) What are the coordinates of D, E, and F?

    (b) Rotate triangle DEF through an angle of 105° and then use an x-scaling factor of 0.25 and a y-scaling factor of -0.5 to form a new triangle $\triangle D_1 E_1 F_1$.

    (c) What are the coordinates of D_1, E_1, and F_1?

    (d) Graph $\triangle DEF$ and $\triangle D_1 E_1 F_1$ on the same set of axes.

24. *Computer Aided Design (CAD)*

    (a) First apply an x-scaling factor of 0.25 and a y-scaling factor of -0.5 to $\triangle DEF$ and then rotate the result through an angle of 105° to form a new triangle $\triangle D_2 E_2 F_2$.

    (b) What are the coordinates of D_2, E_2, and F_2?

    (c) Graph $\triangle DEF$ and $\triangle D_2 E_2 F_2$ on the same set of axes.

    (d) Compare $\triangle D_2 E_2 F_2$ with $\triangle D_1 E_1 F_1$ from Exercise 23.

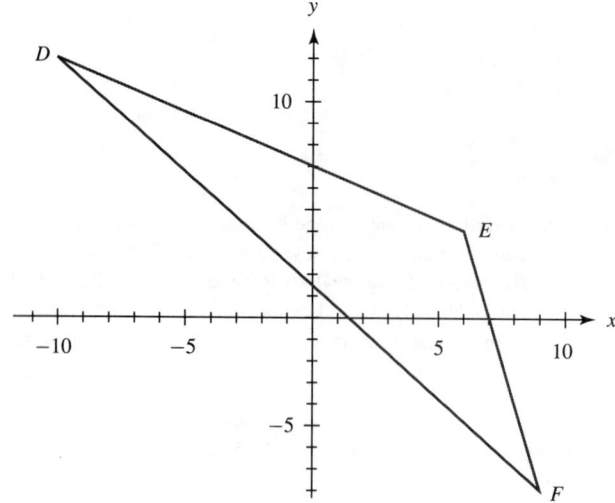

Figure 11.45

25. *Computer Aided Design (CAD)*

    (a) What are the coordinates of D, E, and F?

    (b) Rotate triangle DEF through an angle of $215°$ and then use an x-scaling factor of -0.5 and a y-scaling factor of 0.25 to form a new triangle $\triangle D_1 E_1 F_1$.

    (c) What are the coordinates of D_1, E_1, and F_1?

    (d) Graph $\triangle DEF$ and $\triangle D_1 E_1 F_1$ on the same set of axes.

In Exercises 27–30 use Figure 11.46.

27. *Computer Aided Design (CAD)*

    (a) What are the coordinates of A, B, C, and D?

    (b) Rotate quadrilateral $ABCD$ through an angle of $145°$ and then use an x-scaling factor of 1.5 and a y-scaling factor of -2 to form a new quadrilateral $A_1 B_1 C_1 D_1$.

    (c) What are the coordinates of A_1, B_1, C_1, and D_1?

    (d) Graph $ABCD$ and $A_1 B_1 C_1 D_1$ on the same set of axes.

28. *Computer Aided Design (CAD)*

    (a) First apply an x-scaling factor of 1.5 and a y-scaling factor of -2 to quadrilateral $ABCD$ and then rotate the result through an angle of $145°$ to form a new quadrilateral $A_2 B_2 C_2 D_2$.

    (b) What are the coordinates of A_2, B_2, C, and D_2?

    (c) Graph $ABCD$ and $A_2 B_2 C_2 D_2$ on the same set of axes.

    (d) Compare $A_2 B_2 C_2 D_2$ with $A_1 B_1 C_1 D_1$ from Exercise 27.

29. *Computer Aided Design (CAD)*

    (a) What are the coordinates of A, B, C, and D?

    (b) Rotate quadrilateral $ABCD$ through an angle of $-65°$ and then use an x-scaling factor of -1.5 and a y-scaling factor of 2 to form a new quadrilateral $A_3 B_3 C_3 D_3$.

    (c) What are the coordinates of A_3, B_3, C_3 and D_3?

    (d) Graph $ABCD$ and $A_3 B_3 C_3 D_3$ on the same set of axes.

30. *Computer Aided Design (CAD)*

    (a) What are the coordinates of A, B, C, and D?

    (b) First apply an x-scaling factor of 2 and a y-scaling factor of -1.5 to $ABCD$ and then rotate the result

26. *Computer Aided Design (CAD)*

    (a) What are the coordinates of D, E, and F?

    (b) First apply an x-scaling factor of -0.5 and a y-scaling factor of 0.25 to $\triangle DEF$ and then rotate the result through an angle of $215°$ to form a new triangle $\triangle D_2 E_2 F_2$.

    (c) What are the coordinates of D_2, E_2, and F_2?

    (d) Graph $\triangle DEF$ and $\triangle D_2 E_2 F_2$ on the same set of axes.

    (e) Compare $\triangle D_2 E_2 F_2$ with $\triangle D_1 E_1 F_1$ from Exercise 25.

through an angle of $-65°$ to form a new quadrilateral $A_4 B_4 C_4 D_4$.

    (c) What are the coordinates of A_4, B_4, C_4, and D_4?

    (d) Graph $ABCD$ and $A_4 B_4 C_4 D_4$ on the same set of axes.

    (e) Compare $A_4 B_4 C_4 D_4$ with $A_3 B_3 C_3 D_3$ from Exercise 29.

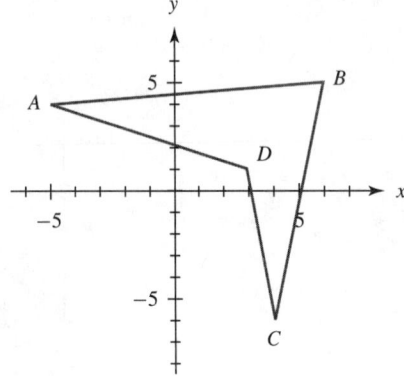

Figure 11.46

31. In the Chapter Project memo from management you were asked to prepare a report that answers the following three questions.

    ▶ What is the mathematics behind rotation and scaling objects in two dimensions?

    ▶ How does a matrix contribute to the solution?

    ▶ How does this work in three dimensions?

Write a memo to the management describing the progress you have made toward answering these questions.

11.3 Matrices and Equation Solving

In Chapter 10 you learned several methods for solving systems of two equations in two variables. Then you applied those methods to the solution of linear programming problems. In this section you will see how matrices are used to solve systems of equations. The matrix approach is in fact the method most widely used in engineering and social science to solve systems, especially those that have more than two equations.

Representing Systems with Matrices

To solve the circuit board manufacturing problems of Chapter 10 you needed to solve several pairs of equations in two variables. You learned how to solve these by hand and with the calculator's graph and trace features. The matrix method has not usually been taught in college math classes because some of the calculations are quite difficult to do by hand. With the *TI-83* you will see that the matrix solutions come quite easily.

One system of equations from Chapter 10 is

$$30e + 10f = 210 \quad \text{the resistor supply equation}$$

$$15e + 12f = 120 \quad \text{the capacitor supply equation}$$

where e is the number of E-type circuit boards to be manufactured and f is the number of F-type boards. The graphs of these equations form two of the boundaries of the feasible region in the linear programming problem. Their solution is represented by point D in Figure 11.47 (which you saw in Chapter 10 as Figure 10.70, on page 729).

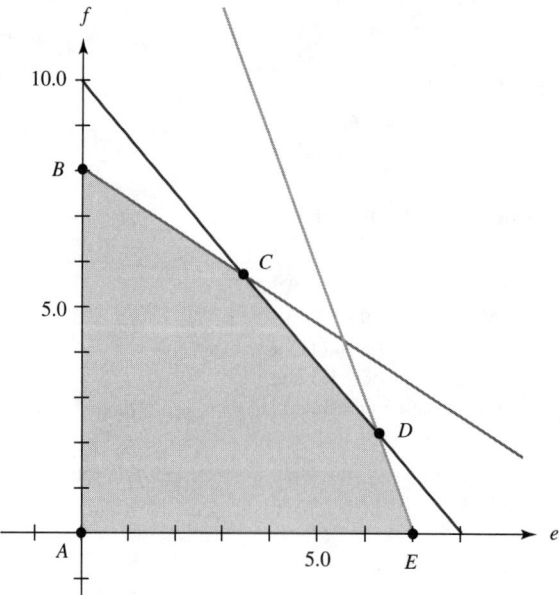

Figure 11.47

On the left-hand sides of the equal signs we have the expressions

$$30e + 10f$$

$$15e + 12f$$

These two expressions can be written as the product of two matrices

$$\begin{pmatrix} 30 & 10 \\ 15 & 12 \end{pmatrix} \begin{pmatrix} e \\ f \end{pmatrix} = \begin{pmatrix} 30e + 10f \\ 15e + 12f \end{pmatrix}$$

The entire system of two equations can be written as the matrix equation

$$\begin{pmatrix} 30 & 10 \\ 15 & 12 \end{pmatrix} (ef) = \begin{pmatrix} 210 \\ 120 \end{pmatrix}$$

If you could solve this equation for the matrix $\begin{pmatrix} e \\ f \end{pmatrix}$ you would know the values of e and f. For example, if it turned out that $\begin{pmatrix} e \\ f \end{pmatrix} = \begin{pmatrix} 7 \\ 3 \end{pmatrix}$ then we would know that the solutions are $e = 7$ and $f = 3$. Soon you'll know how to solve that equation for the matrix $\begin{pmatrix} e \\ f \end{pmatrix}$.

Example 11.15

Another of the points in Figure 11.47 is the intersection of the lines

$$15e + 12f = 120 \quad \text{the capacitor supply equation}$$
$$6e + 9f = 72 \quad \text{the transistor supply equation}$$

(a) Which lines in Figure 11.47 represent these equations?
(b) What is the point of intersection?
(c) Write the equations as a single matrix equation.
(d) What are the unknowns in the matrix equation?

Solution (a) The easiest way to identify the lines is to look at the intercepts. For the first equation the f-intercept (where $e = 0$) is the point $(0, 10)$, so this is line $\overleftrightarrow{C D}$. For the second equation the f-intercept is the point $(0, 8)$, so this is the line $\overleftrightarrow{B C}$.

(b) Since $\overleftrightarrow{C D}$ and $\overleftrightarrow{B C}$ meet at C, the intersection is point C.
(c) The matrix equation is

$$\begin{pmatrix} 15 & 12 \\ 6 & 9 \end{pmatrix} \begin{pmatrix} e \\ f \end{pmatrix} = \begin{pmatrix} 120 \\ 72 \end{pmatrix}$$

(d) The unknowns in the matrix equation are represented by the matrix $\begin{pmatrix} e \\ f \end{pmatrix}$. ■

Solving a Matrix Equation

The matrix equations written above are of the form

$$AB = C$$

where matrix B contains the unknowns. To think about how to solve this equation, let's look closely at how we solve an equation like

$$3x = 5$$

You know that the solution to this equation is $x = \frac{5}{3} \approx 1.67$. But we need to look at the steps used to get this solution in order to apply similar steps to a matrix equation.

$$3x = 5$$
$$\frac{1}{3} \times 3x = \frac{1}{3} \times 5 \qquad \text{Multiply both sides by } \frac{1}{3}$$
$$1x = \frac{5}{3} \qquad \text{because } \frac{1}{3} \times 3 = 1 \text{ and } \frac{1}{3} \times 5 = \frac{5}{3}$$
$$x = \frac{5}{3} \qquad \text{because } 1x = x$$

We multiplied by $\frac{1}{3}$ because it produces the number 1 when multiplied by 3. The technical names for these special numbers are the multiplicative identity and multiplicative inverse.

The Multiplicative Identity and Multiplicative Inverses

▶ The number 1 is called the **multiplicative identity** because

$$1n = n$$

for all values of n.

▶ One name for $\frac{1}{3}$ is the **multiplicative inverse of** 3; $\frac{1}{n}$ is the multiplicative inverse of n. A number times its multiplicative inverse produces the number 1.

$$\frac{1}{3} \times 3 = 1 \qquad \text{and} \qquad \frac{1}{n} \cdot n = 1$$

for all values of n except $n = 0$. From our discussion of exponents we know that $\frac{1}{n}$ can be written as n^{-1}, and that $n^1 n^{-1} = n^0 = 1$ if $n \neq 0$.

▶ The multiplicative inverse is also called the **reciprocal**. The reciprocal of 5 is $\frac{1}{5}$ and the reciprocal of -0.025 is $\frac{1}{-0.025} = -40$.

We will use the concepts of multiplicative identity and multiplicative inverse to solve a matrix equation. Here are the same ideas in matrix form.

The $n \times n$ Identity Matrix

An $n \times n$ **identity matrix** is an $n \times n$ matrix I_n usually written as I, that when multiplied by another matrix A produces the same matrix A. That is,

$$IA = A \qquad \text{and} \qquad AI = A$$

Each element along the main diagonal from i_{11} to i_{nn} is 1. All other elements are 0.

Example 11.16

Show that $I = I_4 = \begin{pmatrix} 1 & 0 & 0 & 0 \\ 0 & 1 & 0 & 0 \\ 0 & 0 & 1 & 0 \\ 0 & 0 & 0 & 1 \end{pmatrix}$ is a 4×4 identity matrix of

$P = \begin{pmatrix} 1 & 2 & 5 & -6 \\ 7 & -8 & 0 & 2 \\ 9 & 0 & -12 & 1 \end{pmatrix}$.

Solution If I_4 is an identity matrix for P, then either $I_4 P = P$ or $P I_4 = P$. Since P is a 3×4 matrix and I_4 is a 4×4 matrix, we can only multiply $P I_4$. Multiplying $P I_4$, we get

$$\begin{pmatrix} 1 & 2 & 5 & -6 \\ 7 & -8 & 0 & 2 \\ 9 & 0 & 012 & 1 \end{pmatrix} \begin{pmatrix} 1 & 0 & 0 & 0 \\ 0 & 1 & 0 & 0 \\ 0 & 0 & 1 & 0 \\ 0 & 0 & 0 & 1 \end{pmatrix} = \begin{pmatrix} 1 & 2 & 5 & -6 \\ 7 & -8 & 0 & 2 \\ 9 & 0 & -12 & 1 \end{pmatrix}$$

Thus I_4 is an identity matrix for P. ■

Because the matrix P in Example 11.16 is not a square matrix, I_4 is called a **right identity** for P (because $P I_4 = P$). Since P is a 3×4 matrix, P has a **left identity** I_3.

The Inverse Matrix

An **inverse matrix** for a square matrix A produces the identity matrix I when multiplied by A. The symbol for the inverse of matrix A is A^{-1}. That is,

$$A A^{-1} = I \qquad \text{and} \qquad A^{-1} A = I$$

If we apply the same principles from the simple equation solving example, we get the following steps:

$$
\begin{array}{ll}
AB = C & \text{the matrix equation} \\
A^{-1} A B = A^{-1} C & \text{multiply both sides of the equation by } A^{-1} \\
I B = A^{-1} C & \text{because } A^{-1} A = I \\
B = A^{-1} C & \text{because } I B = B
\end{array}
$$

Therefore if we can find the inverse of A and multiply it by C we would have the solution to the equation. Before we can do that we'll explore the identity matrix. As you will see, there are one or two identity matrices for a given matrix.

Example 11.17

What are the dimensions of the identity matrices for matrix A where

$$A = \begin{pmatrix} 1 & 2 \\ 3 & 4 \\ 5 & 6 \end{pmatrix}$$

776 Chapter 11 MATRICES AND 3-D GRAPHING

Solution A is a 3×2 matrix. We seek a matrix I that has the following property:

$$AI = A \quad \text{or} \quad IA = A$$

We first look at $AI = A$. Suppose the dimensions of I are $p \times q$. First write the dimensions of the two matrices to be multiplied:

$$(3 \times 2)(p \times q)$$

In matrix multiplication the inner dimensions (2 and p) must be equal, so $p = 2$. Also, the outer dimensions (3 and q) tell us the dimensions of the product matrix, which is 3×2, because the product matrix is A itself. Therefore $q = 2$ and I is a 2×2 matrix.

Using a similar process we see that I_3, a 3×3 matrix, is the identity for $IA = A$. ∎

You can see from Example 11.17 that if A had different dimensions, then the dimensions of I would be different from 2×2. In Example 11.18 you'll see how to determine I.

Example 11.18

If $A = \begin{pmatrix} 1 & 2 \\ 3 & 4 \\ 5 & 6 \end{pmatrix}$, the same matrix as in Example 11.17, what is I_2?

Solution We know that I is a 2×2 matrix, so we can write it as

$$I = \begin{pmatrix} a & b \\ c & d \end{pmatrix}$$

We know that

$$\begin{pmatrix} 1 & 2 \\ 3 & 4 \\ 5 & 6 \end{pmatrix}\begin{pmatrix} a & b \\ c & d \end{pmatrix} = \begin{pmatrix} 1 & 2 \\ 3 & 4 \\ 5 & 6 \end{pmatrix}$$

What are the values of a, b, c, and d? Figure 11.48 partially illustrates our answer to this question.

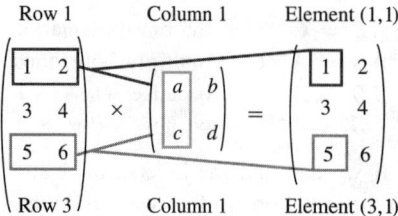

Figure 11.48

Remember that the first row of A and the first column of I produce the element in the first row, first column of the product, which is 1. Therefore $1a + 2c = 1$ or $a + 2c = 1$. There are many solutions to this equation. Let's turn to the third row of A and the first column of I. This produces the element in the third row, first column of I, which is 5. Therefore $5a + 6c = 5$. We have two equations for a and c:

$$a + 2c = 1$$

$$5a + 6c = 5$$

Multiplying the top equation by -5 we get

$$-5a - 10c = -5$$

$$5a + 6c = 5$$

Adding these two equations produces

$$-4c = 0 \qquad \text{or} \qquad c = 0$$

Substituting $c = 0$ in the first equation, we find $a = 1$. Working with the second column of I and using the same method, we would find that $b = 0$ and $d = 1$, and the matrix I is

$$I = \begin{pmatrix} 1 & 0 \\ 0 & 1 \end{pmatrix}$$

The identity $I = \begin{pmatrix} 1 & 0 \\ 0 & 1 \end{pmatrix}$ is called an **identity of order 2** to indicate that it is a 2×2 matrix. Notice that in each of the previous examples the identity matrix I was a **square matrix**; that is, each had the same number of rows and columns. Do you think that all identity matrices are square?

In Calculator Lab 11.1 you explore matrices on the calculator and return to matrix A to discover its *other* identity matrix.

Calculator Lab 11.1

Exploring Matrices on the Calculator

In this calculator lab you will explore five of the matrix features on a *TI-83* calculator. We'll skip the others, which are used in more advanced study of matrix algebra.

First switch to `Float` on the MODE menu. Then enter matrix $A = \begin{pmatrix} 1 & 2 \\ 3 & 4 \\ 5 & 6 \end{pmatrix}$ from Example 11.17 into your calculator. You will explore the matrix instructions 2 through 6 on the MATRIX `Math` menu (Figure 11.49). For each of these five selections, enter the instructions that we suggest and any others that you make up until you can write a sentence or two for each instruction that describes what it does.

Figure 11.49

Procedures 1. Selection `2:T`. Enter the instruction A^T. The keys to press are

▶ `MATRX` 1 (to enter matrix A on the home screen)

▶ `MATRX` ▶ 2 (to enter the second instruction, T. after A)

▶ `ENTER` (to execute this instruction)

2. Selection 3:dim(. Enter the instruction dim([A]). The keys to press are

▶ **MATRX** **▶** 3 (to enter dim(on the home screen)
▶ **MATRX** (1 to enter matrix A)
▶ **)** (for the closing parenthesis)
▶ **ENTER** (to execute this instruction)

Try the dim(instruction on other matrices; for example, you could compute dim(A^T).

3. Selection 4:Fill(. Enter the instruction Fill(5,[A]). We will leave it for you to figure out the keys to press.

4. Selection 5:identity(. Enter the instruction identity(6).

▶ What happens when you enter a number larger than 6?
▶ What is the largest number that your calculator will accept in this instruction? The answer depends on how much available memory your calculator has.

5. Selection 6:randM(. Enter the instruction randM(4,3). Press **2nd** [ENTRY] to enter the same instruction again. Repeat a few more times.

Calculator Lab 11.2

Exploring Identities on the Calculator

In this lab you will use what you learned in Example 11.17 and Calculator Lab 11.1 to find the other identity matrix for matrix A. That is, you will answer the question, "What is the matrix I that satisfies the equation

$$IA = A$$

where A is any 3×2 matrix?" Your answer will be the other identity matrix for A. Use your calculator to explore and to check your ideas.

First switch to Float on the MODE menu. Then enter matrix $A = \begin{pmatrix} 1 & 2 \\ 3 & 4 \\ 5 & 6 \end{pmatrix}$ from Example 11.17 into your calculator.

Procedure Use selection 5:identity(from the MATRIX Math menu to select an identity. What number should you enter to produce $IA = A$? What is the order of this identity? Is the solution to $IA = A$ the same as the answer to $AI = A$ in Example 11.18?

The Inverse Matrix

Finding the inverse of a matrix is the second part of the matrix solution of a system of equations. Before computers and powerful calculators were available, this was the most difficult part of matrix algebra. Now you'll see that technology makes finding inverses quite simple. However, although the computation will be easy, you still need to understand the ideas involved in order to use them.

You may have seen that the calculator uses the **x^{-1}** key to give the multiplicative inverse (reciprocal) of a number. For example, $7^{-1} \approx 0.1428571429$. This same key is used to produce the inverse of a matrix.

Example 11.19

Enter the 2×2 matrix $A = \begin{pmatrix} -1 & -6 \\ -5 & 8 \end{pmatrix}$ into the calculator and find its inverse. Convert the elements of the inverse matrix to fractions.

Solution You can partially see A^{-1} on the third and fourth lines of Figure 11.50. The instruction in line 5 of Figure 11.50 will convert the elements of A^{-1} to fractions. The ▶Frac feature is the first entry on the Math menu of the ◼MATH◼ key. The results for our matrix are shown in Figure 11.50. Thus

$$\begin{pmatrix} -1 & -6 \\ -5 & 8 \end{pmatrix}^{-1} = \begin{pmatrix} -\frac{4}{19} & -\frac{3}{19} \\ -\frac{5}{38} & \frac{1}{38} \end{pmatrix}$$

```
               [-5  8  ]]
[A]-1
[[ -.2105263158 ...
 [ -.1315789474 ...
[A]-1▶Frac
 [[ -4/19  -3/19]
  [ -5/38  1/38 ]]
```

Figure 11.50   ◼

In Example 11.20 we'll use the results of Example 11.19 to review matrix multiplication and computations with fractions and signed numbers. The solution shows the computation for our matrix A. Repeat them for your matrix A.

Example 11.20

For the matrix you produced in Example 11.19, compute AA^{-1} by hand and with a calculator. Our matrices were

$$A = \begin{pmatrix} -1 & -6 \\ -5 & 8 \end{pmatrix}$$

$$A^{-1} = \begin{pmatrix} -\frac{4}{19} & -\frac{3}{19} \\ -\frac{5}{38} & \frac{1}{38} \end{pmatrix}$$

Solution There are four parts to the pencil and paper multiplication of AA^{-1}. We'll do the first two.

▶ Row 1 of A and column 1 of A^{-1}:

$$-1\left(-\frac{4}{19}\right) + -6\left(-\frac{5}{38}\right) = \frac{4}{19} + \frac{30}{38}$$

$$= \frac{4}{19} + \frac{15}{19} = \frac{19}{19} = 1$$

▶ Row 2 of A and column 1 of A^{-1}:

$$-5\left(-\frac{4}{19}\right) + 8\left(-\frac{5}{38}\right) = \frac{20}{19} + \frac{-40}{38}$$

$$= \frac{20}{19} + \frac{-20}{19} = \frac{0}{19} = 0$$

Figure 11.51 shows the same computation done by calculator.

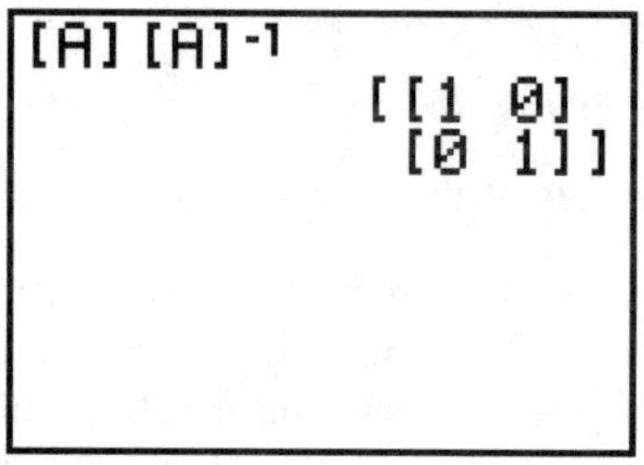

```
[A] [A]⁻¹
              [[1  0]
               [0  1]]
```

Figure 11.51   ■

As you can see, $\left(\begin{array}{cc} -\frac{4}{19} & -\frac{3}{19} \\ -\frac{5}{38} & \frac{1}{38} \end{array} \right)$ is the inverse of $A = \left(\begin{array}{cc} -1 & -6 \\ -5 & 8 \end{array} \right)$ because $AA^{-1} = I$. You can check that $A^{-1}A = I$.

Activity 11.9
Thinking about Square Matrices

Write an answer to each of the following questions.

1. The matrix we studied in Example 11.20 was a square matrix. Is it possible for a matrix that is not square to have an inverse? If yes, show an example. If no, explain.
2. The inverse matrix we studied in Example 11.20 was a square matrix. Is it possible for an inverse matrix not to be square? If yes, show an example. If no, explain.
3. For the matrix we studied in Example 11.20, is it true that $AA^{-1} = A^{-1}A$?
4. If two matrices A and B are square, would $AB = BA$ be always true, sometimes true, or never true? Explain.

The Calculator Solution of a Matrix Equation

After exploring identity and inverse matrices we're ready to show the solution of a matrix equation. Here is a summary of the important points about matrix equations.

Summary: Solving Systems with Matrices

Here are the steps for solving the general 2×2 system of equations

$$\begin{cases} ax + by = c \\ dx + ey = f \end{cases}$$

1. If the equations are not presented in this form, they must be manipulated until they are in this form. For example, the equation $3y - 45 = 2x$ would have to be changed to $-2x + 3y = 45$.
2. Set up three matrices:

$$A = \left(\begin{array}{cc} a & b \\ d & e \end{array} \right) \qquad B = \left(\begin{array}{c} x \\ y \end{array} \right) \qquad C = \left(\begin{array}{c} c \\ f \end{array} \right)$$

3. Rewrite the system of equations as the matrix equation

$$AB = C$$

Remember that order is important. BA is not correct.
4. The solution is

$$B = A^{-1}C$$

Again, order is important. CA^{-1} is not correct.
5. Use the calculator to enter the matrices and to compute the value of $A^{-1}C$. The values of x and y are found in the product $A^{-1}C$.

Example 11.21   Application

For the circuit board linear programming problem discussed in this section, compute the coordinates of point D, the intersection of the graphs of the following two equations.

$$30e + 10f = 210 \qquad \text{the resistor supply equation}$$
$$15e + 12f = 120 \qquad \text{the capacitor supply equation}$$

Solution   Although you might be able to solve this system by hand in a shorter time than we will take with the calculator, the point here is to learn the calculator solution. Here are the steps in the solution.

The matrices are

$$A = \begin{pmatrix} 30 & 10 \\ 15 & 12 \end{pmatrix} \qquad B = \begin{pmatrix} e \\ f \end{pmatrix} \qquad C = \begin{pmatrix} 210 \\ 120 \end{pmatrix}$$

Enter A and C into your calculator.

The equation is $AB = C$.

The solution is $B = A^{-1}C$.

Figure 11.52 shows the solution in both decimal form and fraction form. The solutions are

$$e = \frac{44}{7} \approx 6.3$$

$$f = \frac{15}{7} \approx 2.1$$

```
[A]⁻¹[C]
  [[6.285714286]
   [2.142857143]]
[A]⁻¹[C]▶Frac
        [[44/7]
         [15/7]]
```

Figure 11.52

Check the solutions. Do the values of e and f seem right for point D in Figure 11.47? ■

No Solution and Many Solutions

A 2×2 system of equations in x and y can have no solution if the two lines are parallel and therefore never intersect. Parallel lines have the same slope but different y-intercepts. For example, the following equations would have parallel graphs that would not intersect.

$$2x + 3y = 10$$

$$2x + 3y = 5$$

Another possibility is that the equations would actually have the same graph; that is, they could have the same slope and same y-intercept. In this case there would be an infinite number of solutions: all the points on the line. For example, the following equations are actually the same.

$$2x + 3y = 10$$

$$4x + 6y = 20$$

If we were to follow the steps in Example 11.21 to solve either of the above systems, the calculator would report the message shown in Figure 11.53.

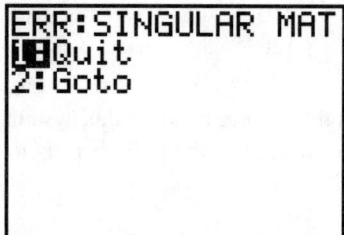

Figure 11.53

A **singular matrix** is a matrix that does not have an inverse. Just as the number zero does not have a multiplicative inverse, so there are many matrices that do not have inverses. The matrices that would be used to solve the above systems are

$$\begin{pmatrix} 2 & 3 \\ 2 & 3 \end{pmatrix} \text{ and } \begin{pmatrix} 2 & 3 \\ 4 & 6 \end{pmatrix}$$

Neither of these matrices has an inverse. Systems that use these matrices cannot be solved no matter what numbers are to the right of the equals sign in the original equations. Since the calculator does not distinguish between a system with no solution and a system with infinite solutions, you would have to look carefully at the given system of equations to figure out which is the case.

Determinant of a Matrix

In the previous lab we skipped over $\texttt{det(}$, the first entry in the MATRIX $\texttt{Math}$ menu. $\texttt{det(}$ stands for **determinant**. If you enter the command $\texttt{det([A])}$, the calculator will respond with the value of the determinant of matrix A.

A determinant is a number. For the 2×2 matrix $\begin{pmatrix} a & b \\ c & d \end{pmatrix}$ the value of the determinant is $ad - bc$. For example, the determinant of the matrix $\begin{pmatrix} 3 & -4 \\ 2 & 5 \end{pmatrix}$ is $3(5) - 2(-4) = 15 + 8 = 23$. Only square matrices have determinants. The determinant of a 3×3 matrix is calculated by a much more complicated formula.

Determinants were formerly used in the solution of systems of equations. They are less important now that calculators can do the matrix solution of systems that you will soon learn.

Determinants are often used to compute an inverse matrix. One way to tell if a matrix does not have an inverse is to look at its determinant. If the determinant of a matrix is zero then the matrix does not have an inverse. You can use the definition of determinant to see for yourself that the two matrices $\begin{pmatrix} 2 & 3 \\ 2 & 3 \end{pmatrix}$ and $\begin{pmatrix} 2 & 3 \\ 4 & 6 \end{pmatrix}$ each have a determinant of zero. A singular matrix has a determinant of zero.

Calculator Lab 11.3

Exploring Random Systems

This calculator lab seeks to answer the question, "How likely is it that a randomly produced pair of equations in two variables would produce a pair of lines that intersect in one point?" To do this you will randomly produce twenty different versions of the 2×2 matrix A in the equation $AB = C$ and count how many times the matrix A has an inverse.

Procedures

1. Enter the two instructions shown in Figure 11.54. (We used ▶Frac to show the patterns in the answers, but that is not required in this lab.)

```
randM(2,2)→[A]
          [[5  -2]
           [0  -4]]
[A]⁻¹▶Frac
    [[1/5  -1/10]
     [0    -1/4 ]]
```

Figure 11.54

2. Then use ⬚ 2nd ⬚ [ENTRY] to repeat the two instructions a total of 20 times.
3. Record how many of the matrices turn out to be singular (no inverse).
4. Compare your results with class members and record what percent of all the tests produced matrices that do not have inverses.

We end this section with a construction application that uses matrices in its solution.

Example 11.22   Application

A contractor mixes some aggregate with cement to make concrete. Two mixtures are on hand. One of the mixtures, which we will call mixture A, is 40% sand and 60% aggregate. The other mixture, mixture B, is 70% sand and 30% aggregate. How much of each should be used to get a 500-lb mixture that is 46.2% sand and 53.8% aggregate?

Solution Let a represent the amount of mixture A and b the amount of mixture B used in the final mixture. The final 500-lb mixture will contain 46.2% sand (231 lb) and 53.8%

aggregate (269 lb). Thus we get two equations. The first

$$0.4a + 0.7b = 231$$

represents the amounts of sand from each of mixtures A and B that are needed to make the 231 lb in the final mixture.

The second equation

$$0.6a + 0.3b = 269$$

represents the amount of aggregate from mixtures A and B needed to make the aggregate in the final mixture.

Thus we have the system

$$\begin{cases} 0.4a + 0.7b = 231 \\ 0.6a + 0.3b = 269 \end{cases}$$

Here $A = \begin{pmatrix} 0.4 & 0.7 \\ 0.6 & 0.3 \end{pmatrix}$, $X = \begin{pmatrix} a \\ b \end{pmatrix}$, and $C = \begin{pmatrix} 231 \\ 269 \end{pmatrix}$. As a result we have the matrix equation $AX = C$ and the solution is

$$X = A^{-1}C$$

$$= \begin{pmatrix} 0.4 & 0.7 \\ 0.6 & 0.3 \end{pmatrix}^{-1} \begin{pmatrix} 231 \\ 269 \end{pmatrix}$$

$$\approx \begin{pmatrix} 396.7 \\ 103.3 \end{pmatrix}$$

The final mixture should contain about 396.7 lb of mixture A and 103.3 lb of mixture B.   ∎

Section 11.3 Exercises

1. What is the multiplicative inverse of -4?

2. What is the multiplicative inverse of $\frac{1}{5}$?

3. What is the multiplicative inverse of $\frac{2}{3}$?

4. What is the multiplicative inverse of $-7\frac{1}{2}$?

 5. If $A_1 = \begin{pmatrix} -2 & 1 \\ 5 & 2.5 \end{pmatrix}$, calculate the determinant of A_1.

6. If $A_2 = \begin{pmatrix} 1 & -2 \\ 2.5 & 5 \end{pmatrix}$, calculate the determinant of A_2.

 7. If $B = \begin{pmatrix} 2 & -4 & 3.4 \\ 1 & 4.1 & 2 \\ 0 & 8 & -7 \end{pmatrix}$, calculate the determinant of B.

8. If $C = \begin{pmatrix} 6 & -2 & 1 & 7 \\ 4 & 0 & 2 & 1 \\ -2 & 1 & 0 & 4 \\ 3 & 0 & 1 & 1 \end{pmatrix}$, calculate the determinant of C.

9. If $A = \begin{pmatrix} 2 & 5 & 1 \\ 0 & -6 & 3 \end{pmatrix}$, determine (a) A^T, (b) $\left(A^T\right)^T$, and (c) $\left(\left(A^T\right)^T\right)^T$.

10. If $B = \begin{pmatrix} -3 & 1 \\ 4 & 2 \\ 0 & 7 \\ 9 & 2.4 \end{pmatrix}$, determine (a) B^T, (b) $\left(B^T\right)^T$, and (c) $\left(\left(B^T\right)^T\right)^T$.

11. If C is a 4×3 matrix, what is the dimension of C^T?

12. If D is a 1×7 matrix, what is the dimension of D^T?

13. Explain the function of the selection `4:Fill(` in the MATRIX Math menu.

14. Explain the function of the selection `5:identity(` in the MATRIX Math menu.

In Exercises 15–22 use matrices to solve each of the following systems of linear equations.

15. $\begin{cases} 9x - 4y = 30 \\ 4x + 9y = -19 \end{cases}$

16. $\begin{cases} 2x + 3y = 11 \\ 5x - 4y = 16 \end{cases}$

17. $\begin{cases} 2.1x + 5.3y = 2.4 \\ -4.6x + 8.7y = 31.3 \end{cases}$

18. $\begin{cases} 1.5x + 2.5y = 14.65 \\ 3.2x + 2.6y = 21.1 \end{cases}$

19. $\begin{cases} 5x + 6y - 2z = -16 \\ 3x - y + 4x = 23 \\ x + y + 6z = 26 \end{cases}$

20. $\begin{cases} -2x + 4y = -26 \\ 4x - y + 5z = 49 \\ x - 12y - 10z = -1 \end{cases}$

21. $\begin{cases} 3.2x + 7.3y + 2.5z = -19 \\ 4.7x + 2.3y + 4z = 22.1 \\ 2.6x - 4.7y + 1.7z = 43.14 \end{cases}$

22. $\begin{cases} 9.7x - 4.3y + 6.2z = -29.8 \\ 2.5x + 6.5y - 7z = 4.4 \\ 6.1x - 1.2y - 2.1z = -26.67 \end{cases}$

23. If $A = \begin{pmatrix} 1 & 2 & -3 \\ 5 & 7 & -1 \\ 0 & 2 & 1 \end{pmatrix}$, calculate A^2.

24. If $B = \begin{pmatrix} 4 & 0 & -3 \\ 2 & 1 & -1 \\ 0 & -5 & 0.5 \end{pmatrix}$, calculate B^3.

25. *Computer Aided Design (CAD)* Suppose we rotate some points through an angle $\alpha = 60°$. The effect of this rotation on these points can be determined by multiplying a matrix of the x- and y-coordinates of the points by the matrix

$$R = \begin{pmatrix} \cos\alpha & -\sin\alpha \\ \sin\alpha & \cos\alpha \end{pmatrix}$$

Evaluate R^6 and explain the result.

26. *Computer Aided Design (CAD)* Suppose we have a rotation of θ followed by a rotation of $-\theta$. The result should leave all points unchanged. Algebraically this means

$$\begin{pmatrix} \cos\theta & -\sin\theta \\ \sin\theta & \cos\theta \end{pmatrix}\begin{pmatrix} \cos(-\theta) & -\sin(-\theta) \\ \sin(-\theta) & \cos(-\theta) \end{pmatrix} = \begin{pmatrix} 1 & 0 \\ 0 & 1 \end{pmatrix}$$

(a) Try this for $\theta = 90°$.

(b) Try this for $\theta = 52°$.

(c) Multiply the two matrices with θ in them and see if you can make the product simplify to $\begin{pmatrix} 1 & 0 \\ 0 & 1 \end{pmatrix}$.
Hint: $\cos(-\theta) = \cos(\theta)$ and $\sin(-\theta) = -\sin(\theta)$.

27. *Computer Aided Design (CAD)* Suppose that a rotation of $82°$ produced the new point $(5, 4)$. What was the original point? Solve this in two ways:

(a) Rotate the new point *backward* by the same rotation.

(b) If R is the rotation matrix, solve the following matrix equation

$$R\begin{pmatrix} x \\ y \end{pmatrix} = \begin{pmatrix} 5 \\ 4 \end{pmatrix}$$

(c) Check your results with each other and with a graph.

28. *Computer Aided Design (CAD)* Triangle ABC in Figure 11.55 is the result transforming $\triangle A_1 B_1 C_1$ through a rotation of $145°$ and then applying an x-scaling factor of 1.5 and a y-scaling factor of -2.

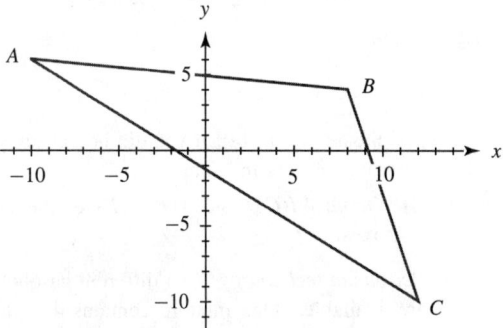

Figure 11.55

(a) What are the coordinates of A, B, and C?

(b) Write the matrices that were used to transform $\triangle A_1 B_1 C_1$ into $\triangle ABC$.

(c) Write the matrices that you would use to transform $\triangle ABC$ back into $\triangle A_1 B_1 C_1$.

(d) Use your matrices from (c) on the coordinates of $\triangle ABC$ to determine the coordinates of $\triangle A_1 B_1 C_1$. What are the coordinates of $\triangle A_1 B_1 C_1$?

(e) Check your answer in (d) by using your matrices from (b) on your answer to (d). You should get the answer of (a); if you did not, check your work and answers in (c) and (d).

(f) Graph ABC and $A_1 B_1 C_1$ on the same set of axes.

29. *Computer Aided Design (CAD)* Triangle ABC in Figure 11.56 is the result of transforming quadrilateral $A_1 B_1 C_1 D_1$ through a rotation of $215°$ and then applying an x-scaling factor of 2.5 and a y-scaling factor of -1.

(a) What are the coordinates of A, B, C, and D?

(b) Write the matrices that were used to transform $A_1 B_1 C_1 D_1$ into $ABCD$.

(c) Write the matrices that you would use to transform $ABCD$ back into $A_1 B_1 C_1 D_1$.

(d) Use your matrices from (c) on the coordinates of $ABCD$ to determine the coordinates of $A_1 B_1 C_1 D_1$. What are the coordinates of $A_1 B_1 C_1 D_1$?

(e) Check your answer in (d) by using your matrices from (b) on your answer to (d). You should get

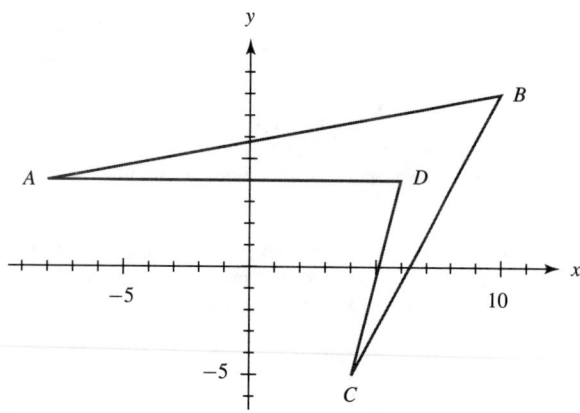

Figure 11.56

the answer of (a); if you did not, check your work and answers in (c) and (d).

(f)   Graph $ABCD$ and $A_1B_1C_1D_1$ on the same set of axes.

30. *Petroleum technology* Two different gasohol mixtures are available. One mixture contains 4% alcohol and the other contains 12% alcohol. How much of each mixture should be used to get 20 000 L of gasohol containing 10% alcohol?

31. *Automotive technology* A 12-liter cooling system is filled with 25% antifreeze. How many liters must be replaced with 100% antifreeze to raise the strength to 45% antifreeze?

32. *Electrical engineering* In determining the Thomson electromotive force of a certain thermocouple, it is necessary to solve the system of equations

$$0.02 \left(T_2{}^2 - T_1{}^2 \right) = 4280$$

$$T_2 - T_1 = 250$$

Solve this system of equations for T_1 and T_2 in degrees Kelvin.

33. *Business* In business the break-even point for a product is where the revenue from selling the product equals the cost of producing the item. A company produces and sells x units of product A and y units of product B. The company has determined that the cost of producing the two products is given by $x^2 + 1.5y^2 - xy = 2,891,189$. Meanwhile the revenue is given by $80x + 100y = 252,640$. What is the break-even point?

34. What conditions are necessary for the use of matrices to solve a system of equations?

35. In the Chapter Project memo from management you were asked to prepare a report that answers the following three questions.

▶ What is the mathematics behind the rotation and scaling of objects in two dimensions?
▶ How does a matrix contribute to the solution?
▶ How does this work in three dimensions?

Write a memo to the management describing the progress you have made toward answering these questions. This could be a continuation of the memo you wrote for Exercise 31 in Section 11.2.

11.4   Graphing in Three Dimensions

We live in a three-dimensional world, and we all find it difficult to represent that world two-dimensionally on paper or on the screens of our calculators or computers. In this section we'll demonstrate how three-dimensional locations, graphs, and motion can be represented in two dimensions.

Cartesian Coordinates in Three Dimensions

To represent two-dimensional locations we use two perpendicular axes, so you might guess that we will represent three dimensions with three perpendicular axes. There's probably a good model of a three-dimensional Cartesian coordinate system near you. Look at the corner of the room you are in. Unless you're in an A-frame cabin or a room with a floor that is not rectangular, you'll see three perpendicular lines meeting in the corner. We'll call the two lines along the floor the x-axis and the y-axis, and we'll call the vertical line between the two walls the z-axis. Figures 11.57 and 11.58 show two ways to represent the three axes on paper. Figure 11.57 shows how a three-dimensional coordinate system is pictured in many mathematics books. This is not always the best angle for viewing a figure; a better view can often be obtained if the axes are rotated as in Figure 11.58.

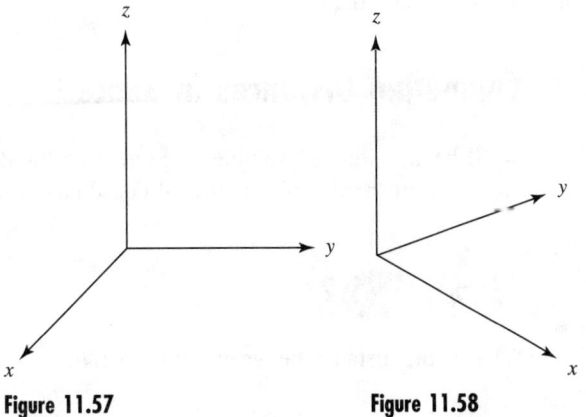

Figure 11.57   **Figure 11.58**

Three **planes** intersect at the corner of the room. The walls form the **xz-plane** and the **yz-plane**, and the floor forms the **xy-plane**. A graph using only two of the variables x, y, and z could be plotted entirely in one of those planes.

The location of any point in the room you are in could be described by giving its three coordinates. The x- and y-coordinates locate the point somewhere on the floor, and the z-coordinate describes the distance up from the floor. When we write the coordinates of a point in three dimensions, we give the three coordinates in the order (x, y, z). Point P in Figure 11.59 is shown with three positive coordinates, a, b, and c (the negative portions of the three axes are shown by dotted lines).

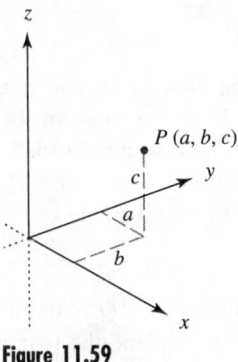

Figure 11.59

Activity 11.10
Describing Points in 3-D

If point $P(a, b, c)$ describes the location of a point somewhere in the room you are in, for example the point halfway between your head and the ceiling, then how would you describe the locations around you that are represented by the following eight points? (Assume that the origin is in one corner of the room, and describe the locations using phrases like "in the next room," "upstairs," "outside," "downstairs," "directly above," "directly below," "on the wall," "on the floor," etc.)

1. Point $P_2(a, b, 0)$
2. Point $P_3(0, b, c)$
3. Point $P_4(a, b, -c)$
4. Point $P_5(a, 0, c)$

5. Point $P_6(0, 0, c)$
6. Point $P_7(a, -b, c)$
7. Point $P_8(a, \frac{b}{2}, 0)$

Computing Distances in Space

We'll begin with the example of computing the distance between a single point and the origin of the three-dimensional coordinate system.

Example 11.23

What is the distance between point $P(a, b, c)$ and point $R(0, 0, 0)$ in Figure 11.60?

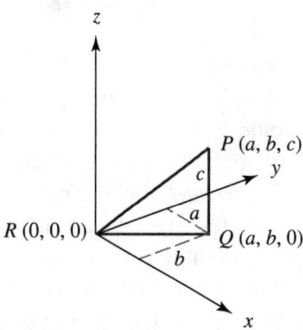

Figure 11.60

Solution   For convenience we assume that a, b, and c are all positive (the argument that follows would also be true if one or more of them was not positive). We want the distance PR. Our strategy is to look at right triangle $\triangle RQP$ and use the Pythagorean theorem to calculate the length of hypotenuse PR from the distances RQ and PQ. The vertical distance up from the xy-plane is $PQ = c$. Therefore we have

$$PR = \sqrt{RQ^2 + c^2}$$

To compute RQ we recognize that RQ is the distance between $Q(a, b, 0)$ and the origin in the xy-plane. In the xy-plane the points have the coordinates $Q(a, b)$ and $R(0, 0)$. Therefore

$$(RQ)^2 = a^2 + b^2$$

Substituting this expression for $(RQ)^2$ into the equation $PR = \sqrt{RQ^2 + c^2}$, we have

$$PR = \sqrt{a^2 + b^2 + c^2}\qquad\blacksquare$$

This answer seems right. In two dimensions the distance is $\sqrt{a^2 + b^2}$, and in three dimensions there are three terms inside the square root. Do you wonder what the distance between two points might be in four dimensions?

It may help you to understand the next examples if you think of the distance PR as the length of the diagonal distance between opposite corners of a rectangular box like the one shown in Figure 11.61.

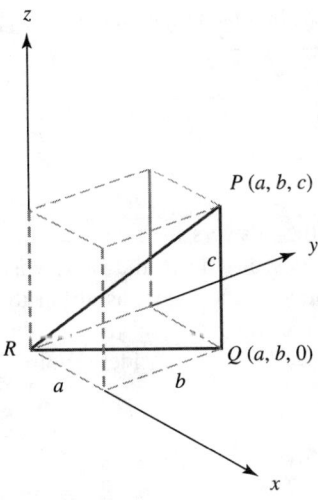

Figure 11.61

In Examples 11.24–11.26 we'll try some similar problems with different values of a, b, and c.

Example 11.24

Compute the distance between $R(0, 0, 0)$ and $P(a, b, c)$ if $a = 2$, $b = 3$, and $c = 4$.

Solution   Substituting the given values for a, b, and c in the distance formula, we obtain

$$PR = \sqrt{a^2 + b^2 + c^2}$$
$$= \sqrt{2^2 + 3^2 + 4^2}$$
$$= \sqrt{29} \approx 5.385$$

Example 11.25

Compute the distance between $R(0, 0, 0)$ and $P(2, -3, 4)$.

Solution   Here we have $a = 2$, $b = -3$, and $c = 4$. Substituting these values for a, b, and c in the distance formula, we obtain

$$PR = \sqrt{a^2 + b^2 + c^2}$$
$$= \sqrt{2^2 + (-3)^2 + 4^2}$$
$$= \sqrt{29} \approx 5.385$$

Example 11.26

Compute the distance between $R(0, 0, 0)$ and $P(a, b, c)$ if $a = -2$, $b = 3$, and $c = -4$.

Solution   Once again we use the distance formula with the given values of a, b, and c.

$$PR = \sqrt{a^2 + b^2 + c^2}$$
$$= \sqrt{(-2)^2 + 3^2 + (-4)^2}$$
$$= \sqrt{29} \approx 5.385$$

Notice that the answers in Examples 11.24–11.26 were all the same. That is because all the values of a have the same absolute value. Similarly, all the values of b have the same absolute value, as do all the values of c.

To move to the more general case of the distance between *any* two points in three dimensions, we have to consider a point S, whose coordinates are (d, e, f).

Example 11.27

What is the formula for the distance between the point $P(a, b, c)$ and $S(d, e, f)$ as shown in Figure 11.62?

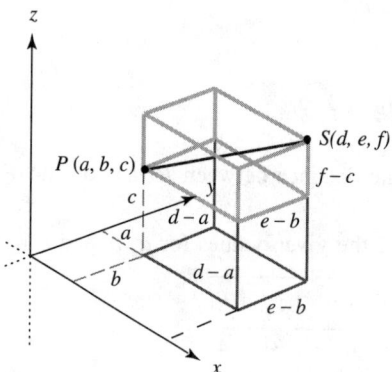

Figure 11.62

Solution   To make this problem a little easier to visualize, in Figure 11.62 we have drawn the rectangular box and shown the distance PS as the length of the diagonal line connecting opposite corners. Also, in order to make the notation easier we assume that a, b, c, d, e, and f are all positive. Reasoning from the result of Example 11.23, we can say that the distance is

$$PS = \sqrt{(d - a)^2 + (e - b)^2 + (f - c)^2}$$

Before we present an example using this new formula, we give a summary of distance in two and three dimensions.

Summary: Distance in 2-D and 3-D

▶ *Two dimensions:* The distance d between the points $P_1(x_1, y_1)$ and $P_2(x_2, y_2)$ is

$$d = \sqrt{(x_2 - x_1)^2 + (y_2 - y_1)^2}$$

▶ *Three dimensions*: The distance d between the points $P_1(x_1, y_1, z_1)$ and $P_2(x_2, y_2, z_2)$ is

$$d = \sqrt{(x_2 - x_1)^2 + (y_2 - y_1)^2 + (z_2 - z_1)^2}$$

Example 11.28

Determine the distance between $A(-2, 5, 7)$ and $B(4, -3, 1)$.

Solution   If A has the coordinates $(-2, 5, 7)$, then we can let $x_1 = -2$, $y_1 = 5$, and $z_1 = 7$. B has the coordinates $(4, -3, 1)$, so $x_2 = 4$, $y_2 = -3$, and $z_2 = 1$. Thus the distance between A and B, AB, is

$$AB = \sqrt{(x_2 - x_1)^2 + (y_2 - y_1)^2 + (z_2 - z_1)^2}$$
$$= \sqrt{[4 - (-2)]^2 + (-3 - 5)^2 + (1 - 7)^2}$$
$$= \sqrt{6^2 + (-8)^2 + (-6)^2} \approx 11.66$$

The distance from A to B is about 11.66 units.   ■

Example 11.29   Application

A builder wants to place a temporary brace from one corner of a rectangular room to the diagonally opposite corner in the ceiling. The dimensions of the room are $18'6'' \times 12'9'' \times 10'0''$. How long should the brace be?

Solution   We will let one corner of the room, point P, be the origin of a three-dimensional coordinate system. Since it is the origin, this corner has coordinates $P(0, 0, 0)$. For convenience we will assume that the distances to the other corner, point Q, are all positive. So Q has the coordinates $Q(18'6'', 12'9'', 10'0'')$. We will convert each of these distances to feet: $18'6'' = 18.5$ ft, $12'9'' = 12.75$ ft, and $10'0'' = 10.0$ ft. Thus we can also write the coordinates of Q as $Q(18.5, 12.75, 10.0)$.

Using the distance formula, we have

$$PQ = \sqrt{(18.5 - 0)^2 + (12.75 - 0)^2 + (10.0 - 0)^2}$$
$$\approx 24.59$$

Thus the brace should be about $24.59 = 24'7''$ long.   ■

Motion in 3-D

As you recall from Chapter 5, motion in two dimensions can be described by vectors $\overrightarrow{v_x}$ and $\overrightarrow{v_y}$, the components of the velocity in the x-direction and the y-direction. In three dimensions we just add a third component, $\overrightarrow{v_z}$, to the velocity, as shown in Figure 11.63. In two dimensions we compute the combined speed by using the

Pythagorean theorem. We can compute the speed in three dimensions by applying the formula for distance in three dimensions. Therefore the speed is equal to

$$|v| = \sqrt{v_x^2 + v_y^2 + v_z^2}$$

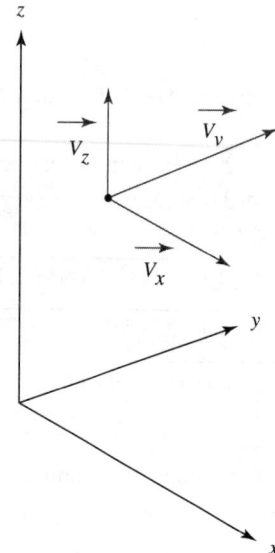

Figure 11.63

Example 11.30

An airplane bound for Las Vegas is flying with a constant velocity at 30,000 feet and is 120 miles east and 60 miles south of its destination. The components of its velocity are 300 mph west, 100 mph north, and 3 mph down.

(a) How far is the plane from Las Vegas?
(b) After flying for ten minutes at this rate what are the coordinates of the plane and how far is it from Las Vegas?
(c) How far has it traveled?
(d) What was the airplane's actual speed in miles per hour over the ten minutes?

Solution   The first thing we should do is establish a coordinate system. It makes sense to use the location of Las Vegas as the origin, since we are given distances relative to that point. Then we must decide which way is positive and which is negative. We selected the x-axis for east and the y-axis for north to conform with the earlier navigation material in Section 5.3. This also conforms to the notation in Figure 11.64.

The airplane's location is shown in Figure 11.64. In order to give all three coordinates in the same units of measure, we will change the altitude from feet to miles by dividing: $\frac{30000}{5280} \approx 5.68$. Thus 30,000 feet is about 5.68 miles. Since east, north, and up are positive directions, the airplane's location at 60 miles south of Las Vegas is recorded as -60.

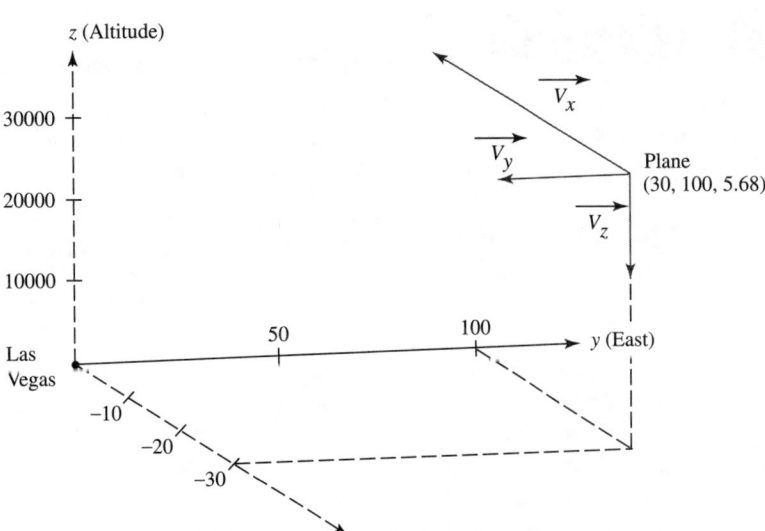

Figure 11.64

(a) The plane's initial straight line distance from Las Vegas is $\sqrt{120^2 + (-60)^2 + 5.68^2} \approx 134.3$ miles.

(b) After ten minutes ($\frac{1}{6}$ of an hour), the plane has traveled $\frac{-300}{6} = -50$ mi west, $\frac{100}{6} \approx 16.7$ mi north, and $\frac{-30}{6} \approx -0.5$ mi down.

   The new coordinates of the airplane (after ten minutes) are found by adding the distances flown to the original coordinates. The new coordinates are $120 + (-50) = 70$ east, $-60 + 16.7 = -43.3$ north, and $5.68 + (-0.5) = 5.18$ miles higher than Las Vegas. Notice that -43.3 north is the same as 43.3 south. The coordinates after ten minutes are therefore $(70, -43.3, 5.18)$. The new distance from the airplane to Las Vegas is $\sqrt{70^2 + (-43.3)^2 + 5.18^2} \approx 82.5$ miles.

(c) In ten minutes the plane has traveled from a point with coordinates $(120, -60, 5.68)$ to the point with coordinates $(70, -43.3, 5.18)$. This distance is

$$\sqrt{(120 - 70)^2 + [-60 - (-43.3)]^2 + (5.68 - 5.18)^2}$$
$$= \sqrt{50^2 + (-16.7)^2 + 0.5^2} \approx 52.7$$

The airplane traveled about 52.7 miles during those ten minutes.

(d) The speed during that ten-minute period was therefore $\dfrac{52.7}{\frac{1}{6}} \approx 316.2$ mph.

   The speed could also have been computed from the velocity components.

$$|v| = \sqrt{v_x^2 + v_y^2 + v_z^2}$$
$$= \sqrt{(-300)^2 + 100^2 + (-3)^2}$$
$$\approx 316.2$$

Again we see that the speed during this ten-minute period was about 316.2 mph.

Section 11.4 Exercises

In Exercises 1–6 let the point $(0, 0, 0)$ represent the location of your eyes in a three-dimensional coordinate system. If you are looking straight ahead then you are looking along the positive x-axis; the positive y-axis is to your left and the positive z-axis is above. Describe the location of each of the following points if all units are in meters.

1. $P(2, 0, 0)$

2. $Q(0, 2, 0)$

3. $R(0, 0, -2)$

4. $S(-2, 2, 0)$

5. $T(2, -2, 2)$

6. $U(0, 2, 1)$

In Exercises 7–10 compute the distance between the given point and the origin.

7. $P(-2, 5, -1)$

8. $Q(3.2, -0.4, 6)$

9. $R(1.25, 2.51, -5.12)$

10. $P(12.06, -6.12, 0.61)$

In Exercises 11–16 determine the distance between the two given points.

11. $P(-2, 0, 4)$ and $Q(4, 0, -2)$

12. $R(4, -1, 5)$ and $S(3, 2, -3)$

13. $T(2.5, -3.1, 4.6)$ and $U(5.1, -4.1, -4.6)$

14. $V(12.6, -1.3, 2.4)$ and $W(-8.4, 9.7, -7.6)$

15. $A(-4.1, 2.5, -6.3)$ and $B(4.7, -8.5, -6.9)$

16. $C(4.3, -9.4, 0)$ and $D(-6.2, 0, 5.7)$

17. *Transportation* An airplane taking off from Atlanta, Georgia is flying with a constant velocity at an altitude of 22,500 feet and is 25 miles west and 30 miles north of the Atlanta airport. The components of its velocity are 325 mph west, 37 mph north, and 8 mph up.

(a) How far is the plane from Atlanta?

(b) After flying for ten minutes at this rate, what is the plane's location and distance from Atlanta?

(c) How far did the plane travel during this ten-minute period?

(d) What was the airplane's actual speed in miles per hour over this ten-minute period?

18. *Transportation* An airplane bound for Zurich, Switzerland is flying with a constant velocity at an altitude of 10 100 meters and is 120 km west and 90 km north of its destination. The components of its velocity are 600 km/h east, 100 km/h south, and 5 km/h down.

(a) How far is the plane from Zurich?

(b) After flying for 15 minutes at this rate, what is the plane's location and distance from Zurich?

(c) How far did the plane travel during this 15-minute period?

(d) What was the airplane's actual speed in km/h over this 15-minute period?

11.5 Plotting Equations in 3-D

Plotting equations in 3-D is somewhat like plotting equations in 2-D. The major difference is that there are three variables and three axes to consider instead of two. We must find coordinates of points that satisfy the equation and then connect the points to show the pattern created by the points. Connecting the points in 3-D is more difficult to visualize, of course, but with experience it should become easier.

We touched on the concept of the graph of an equation in three dimensions when we discussed the coordinate planes referred to as xy, xz, and yz. In the example of the Cartesian system represented by the corner of your room, we said that the floor was the xy-plane. The z-coordinate of every point in the xy-plane is zero, so we can say that the graph of the equation $z = 0$ is the xy-plane. Example 11.31 covers a slightly different equation.

Example 11.31

Analyze the equation $z = 10$ in a three-dimensional Cartesian coordinate system.

(a) Give five points whose coordinates fit the equation.
(b) At what point(s) will the graph cross the x-axis?
(c) At what point(s) will the graph cross the y-axis?
(d) At what point(s) will the graph cross the z-axis?
(e) On what shape do the points lie?

Solution

(a) Any point with a z-coordinate of 10 will satisfy the equation $z = 10$. Five of these points are $(2, 3, 10)$, $(-2, -3, 10)$, $(0, 5, 10)$, $(10, -10, 10)$, and $(-5000, 4000, 10)$.

(b) If a point is on the x-axis, then both its y- and the z-coordinates are zero. Since every point that satisfies the equation $z = 10$ has a z-coordinate of 10, this graph cannot cross the x-axis.

(c) For a similar reason as in (b), the graph cannot cross the y-axis.

(d) Any point on the z-axis has x- and y-coordinates of zero. The only point where this occurs is the point $(0, 0, 10)$.

(e) The points are all 10 units above the xy-plane. A graph of this equation appears in Figure 11.65.

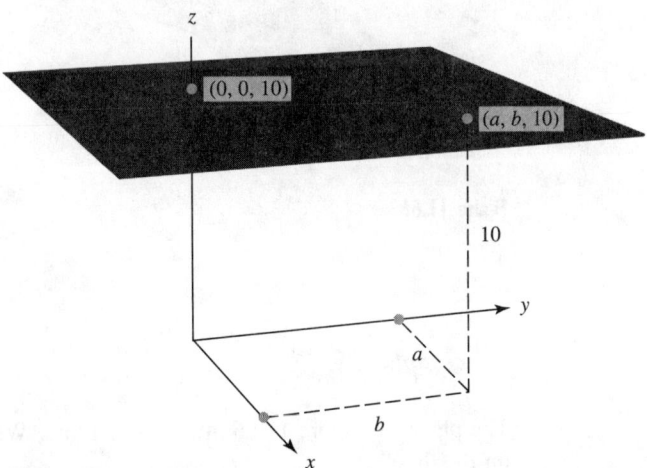

Figure 11.65

The plane in Figure 11.65 will extend indefinitely in the x- and y-directions, always remaining parallel to the xy-plane.

Example 11.32

Analyze the equation $z = x$ in a three-dimensional Cartesian coordinate system.

(a) Give five points whose coordinates fit the equation.
(b) At what point(s) will the graph cross the x-axis?
(c) At what point(s) will the graph cross the y-axis?
(d) At what point(s) will the graph cross the z-axis?
(e) On what shape do the points lie?

Solution   (a) All points that satisfy the equation $z = x$ must have the same x- and z-coordinates. Five such points are $(10, 3, 10)$, $(-2, -3, -2)$, $(0, 5, 0)$, $(3, -10, 3)$, and $(-5000, 4, -5000)$.

(b) All points on the x-axis have y- and z-coordinates of zero. Since this equation is $z = x$, the x- and z-coordinates must be the same. The only point on the x-axis that satisfies these two conditions is the origin $(0, 0, 0)$.

(c) On the y-axis the x- and z-coordinates are zero. Therefore this graph crosses the y-axis at any point $(0, y, 0)$. This graph contains the entire y-axis.

(d) On the z-axis are the x- and y-coordinates zero, and since $x = z$, then $z = 0$. This graph crosses the z-axis only at the origin $(0, 0, 0)$.

(e) A graph of this equation appears in Figure 11.66, which also includes the graph of $z = 10$. Notice that the graph of $z = x$ contains the origin and the entire y-axis.

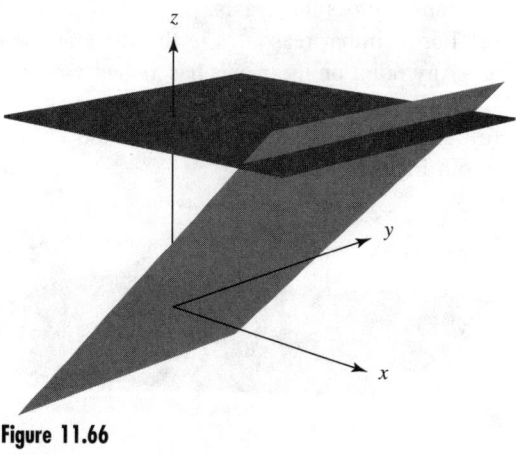

Figure 11.66

Example 11.33

The planes in Figure 11.66 intersect in a line. What are the coordinates of five points on this line?

Solution   The points on the line of intersection must satisfy both equations $z = 10$ and $z = x$, which means $x = 10$. Five points that accomplish that are $(10, 1, 10)$, $(10, -5, 10)$, $(10, \pi, 10)$, $(10, -2000, 10)$, and $(10, 3.4, 10)$. There is no simple single equation of this line. ■

The planes of Examples 11.32 and 11.33 intersect in a line. Two planes that are not parallel will always intersect in a line. You can verify this statement for yourself by modeling two planes with your hands flattened or with two sheets of stiff paper. Any way that you place your planes you will see that they are either parallel or (if extended to meet) they intersect in a line.

The next several examples discuss how three planes could intersect. You can model three planes by adding the table top to the two planes you can hold.

Example 11.34

Show the graph of $z = y$ on the same axes with $z = 10$ and $z = x$.

Solution   This graph is similar to the graph of $z = x$. The graph of $z = y$ will intersect the y-axis and the z-axis at the origin, $(0, 0, 0)$. It will contain the entire x-axis. Two views of these three graphs are shown in Figures 11.67 and 11.68.

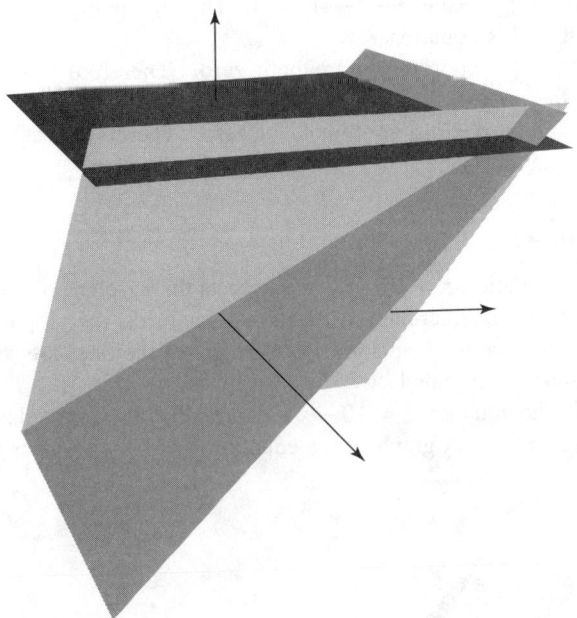

Figure 11.67
One view of three planes intersecting.

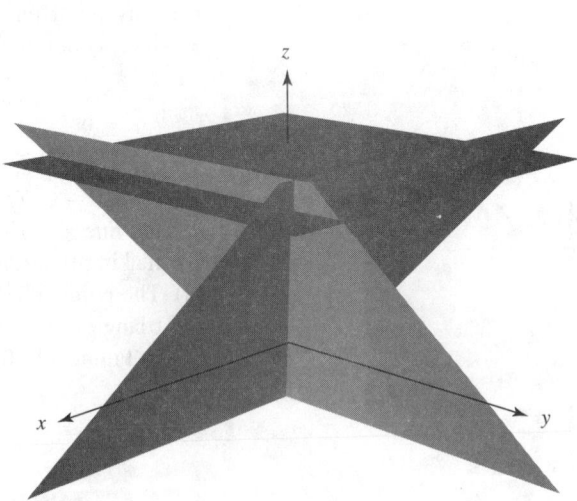

Figure 11.68
Another view of three planes intersecting.

Example 11.35

What are the coordinates of the point where the three planes intersect in Figures 11.67 and 11.68?

Solution   The planes intersect at the point where $z = 10$, $z = x$, and $z = y$. Since $z = x$ and $z = y$, then all three coordinates must be equal. And since $z = 10$, the answer must be the point $(10, 10, 10)$.

In Example 11.36 we turn to an equation that is more challenging.

Example 11.36

Analyze the graph of the equation $z = 10 - x - y$ in a three-dimensional Cartesian coordinate system.

(a) Give five points whose coordinates fit the equation.
(b) At what point(s) will the graph cross the x-axis?

(c) Describe the intersection between the graph of this equation and the xy-plane.

(d) What shape is described by the points that satisfy this equation?

Solution   (a) The equation might be easier to understand if we restate it as $x + y + z = 10$. Then we can think of the equation with the phrase, "The sum of the coordinates is 10." The points $(2, 3, 5)$, $(-2, -3, 15)$, $(0, 5, 5)$, $(20, -10, 0)$, $(10, 0, 0)$, and $(-5000, 4000, 1010)$ all satisfy that description. Another way to solve this is to take any two numbers for x and y and solve the equation for z. For example, if $x = 3$ and $y = -4$, then $z = 10 - 3 - (-4) = 11$. Thus the point $(3, -4, 11)$ also lies on the graph of this equation.

(b) On the x-axis the y- and z-coordinates are both zero. Therefore this graph crosses the x-axis at the point $(10, 0, 0)$.

(c) Any point on the xy-plane has a z-coordinate of zero. Therefore the line of intersection has the equation

$$0 = 10 - x - y$$
or $$y = 10 - x$$

The graph of the equation $z = 10 - x - y$ intersects the xz-plane in the line $z = 10 - y$. The graph of the equation $z = 10 - x - y$ intersects the yz-plane in the line $z = 10 - x$. All three intercepts are 10. Portions of the lines are shown making the triangle ABC in Figure 11.69.

(d) The points that satisfy the equation $z = 10 - x - y$ are all in the same plane as triangle ABC in Figure 11.69. A graph of the equation $z = 10 - x - y$ appears in Figure 11.70.

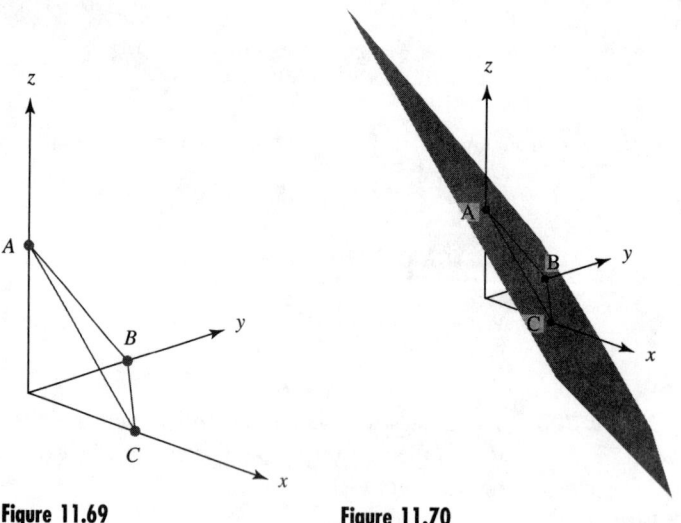

Figure 11.69

Figure 11.70
The graph of $z = 10 - x - y$.

Activity 11.11
Intersections of Three Planes

We have seen that two planes can intersect in a line or, if they are parallel, not intersect. We should also add that two planes could have equations that are in fact identical, so that the graphs would be indistinguishable from one another. Those two sentences cover all the possibilities for two planes. As you can imagine, there

It is not always obvious from the equation if two planes are identical. For example, the equations $2x + 2y = 2(10 - z)$ and $z = 10 - x - y$ are identical and their graphs would produce only one plane.

are more possibilities with three planes. We have seen one example of three planes intersecting in a point. Of course three planes could be parallel and not intersect at all. There are five other possibilities, including the ones in which two or three of the planes are identical.

Using stiff paper, 4×6 file cards, or your hands and the table, explore the possibilities for the intersection of three planes. You might even look around the room for examples of three planes. Write or draw a description of the seven possibilities for the intersection of three planes.

Solving Systems in Three or More Dimensions

Now that you have learned to visualize graphs in three dimensions, let's look at how to solve systems of three equations and three variables. Figure 11.68 shows the solution of one system of three equations. The equations are

$$z = 10$$

$$z = x$$

$$z = y$$

and the solution is $(10, 10, 10)$, which you could probably guess from looking at the equations for a while. Example 11.37 shows the matrix solution of these equations.

Example 11.37

Solve the preceding system by using matrices.

Solution   The matrix solution is similar to the method for systems of two equations. First we must write the equations with all the variables on the left (it helps to include all the variables, even if their coefficients are zero).

$$0x + 0y + z = 10$$

$$-x + 0y + z = 0$$

$$0x - y + z = 0$$

The matrix equation is

$$\begin{pmatrix} 0 & 0 & 1 \\ -1 & 0 & 1 \\ 0 & -1 & 1 \end{pmatrix} \begin{pmatrix} x \\ y \\ z \end{pmatrix} = \begin{pmatrix} 10 \\ 0 \\ 0 \end{pmatrix}$$

and the solution is

$$\begin{pmatrix} x \\ y \\ z \end{pmatrix} = \begin{pmatrix} 0 & 0 & 1 \\ -1 & 0 & 1 \\ 0 & -1 & 1 \end{pmatrix}^{-1} \begin{pmatrix} 10 \\ 0 \\ 0 \end{pmatrix}$$

To get the solution on the calculator, we defined $A = \begin{pmatrix} 0 & 0 & 1 \\ -1 & 0 & 1 \\ 0 & -1 & 1 \end{pmatrix}$ and $C = \begin{pmatrix} 10 \\ 0 \\ 0 \end{pmatrix}$. Figure 11.71 shows the calculator solution.

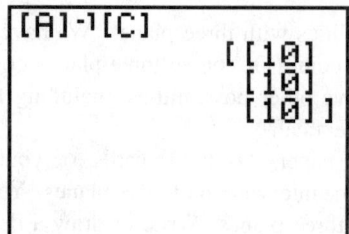

Figure 11.71

To build a more interesting example we will expand the circuit board linear programming problem. Suppose there is one more board to be made, board G, with the other boards and the parts supply limits remaining the same. A statement of the problem follows (the text in *italics* shows what is new).

Adding the "G" Circuit Board

▶ **Parts requirements**
- Board E requires 30 resistors, 15 capacitors, and 6 transistors
- Board F requires 10 resistors, 12 capacitors, and 9 transistors
- *Board G requires 20 resistors, 8 capacitors, and 10 transistors*

▶ **Parts supply (daily)**
- 210 resistors, 120 capacitors, and 72 transistors

▶ **Profit**
- $6.00 for board E, $5.00 for board F, *and $7.00 for board G*

Find the daily level of production that stays within the available supply and maximizes the profit.

The three equations that we wrote for this problem now have three variables, e, f, and g, the numbers of each board to be manufactured:

$$30e + 10f + 20g = 210 \quad \text{the resistor supply equation}$$

$$15e + 12f + 8g = 120 \quad \text{the capacitor supply equation}$$

$$6e + 9f + 10g = 72 \quad \text{the transistor supply equation}$$

The equations are graphed in Figure 11.72, which shows point P, the intersection of the three planes.

Figure 11.73 shows the graph viewed from a different angle. The feasible region is now a three-dimensional space between the origin of the graph and the three planes, bounded also by the ef-plane on the bottom, the eg-plane on the right, and the fg-plane on the left.

The problem of finding the maximum profit within a three-dimensional feasible region is also solved with the simplex method that we introduced in Chapter 10. We need only look at the vertices of the region. Point P is one of these vertices, and its coordinates are found from the solution of three equations written above. To solve these three equations, we use matrices.

Example 11.38

Solve the three circuit board equations for the point of intersection P.

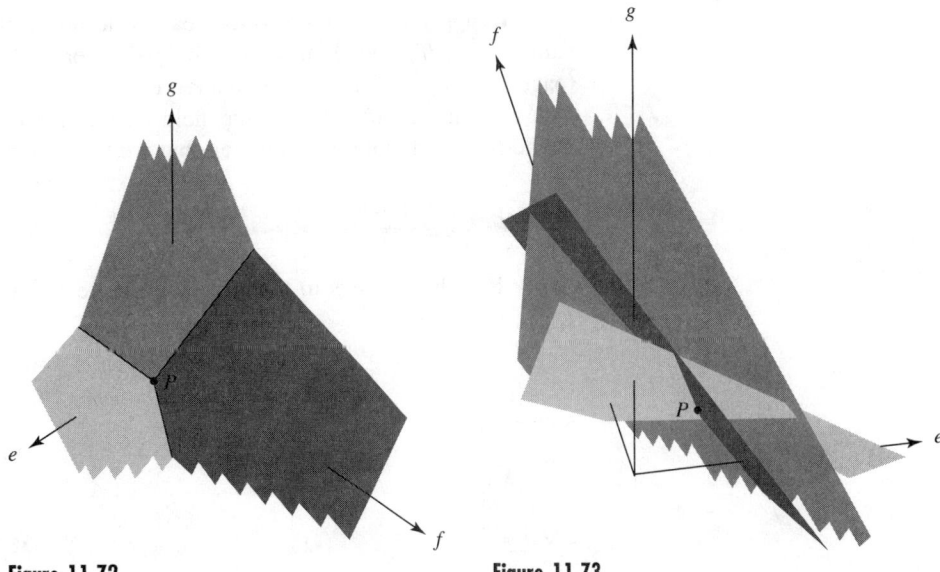

Figure 11.72   **Figure 11.73**

Solution The solution is similar to the simpler one in the previous example. The matrix equation is

$$\begin{pmatrix} 30 & 10 & 20 \\ 15 & 12 & 8 \\ 6 & 9 & 10 \end{pmatrix} \begin{pmatrix} e \\ f \\ g \end{pmatrix} = \begin{pmatrix} 210 \\ 120 \\ 72 \end{pmatrix}$$

and the solution is

$$\begin{pmatrix} e \\ f \\ g \end{pmatrix} = \begin{pmatrix} 30 & 10 & 20 \\ 15 & 12 & 8 \\ 6 & 9 & 10 \end{pmatrix}^{-1} \begin{pmatrix} 210 \\ 120 \\ 72 \end{pmatrix}$$

To get the solution on the calculator, we defined $A = \begin{pmatrix} 30 & 10 & 20 \\ 15 & 12 & 8 \\ 6 & 9 & 10 \end{pmatrix}$ and $C = \begin{pmatrix} 210 \\ 120 \\ 72 \end{pmatrix}$. As you can see in Figure 11.74, the calculator solution is $X \approx \begin{pmatrix} 4.9 \\ 2.7 \\ 1.9 \end{pmatrix}$.

Therefore the coordinates of point P are approximately (4.9, 2.7, 1.9).

```
[A]-¹[C]
  [[4.857142857]
   [2.678571429]
   [1.875      ]]
```

Figure 11.74

To apply the solution to the real-world problem, we must remember that the values of e, f, and g must be whole numbers. As you saw in Example 10.32 (page 730), this is not a simple matter, even in two dimensions. In three dimensions it is even more difficult. In the field of linear programming there are advanced methods for handling this situation, but we will not get into them in this course.

Example 11.39   Application

Apply Kirchhoff's laws to the circuit in Figure 11.75 and determine the currents in I_1, I_2, and I_3.

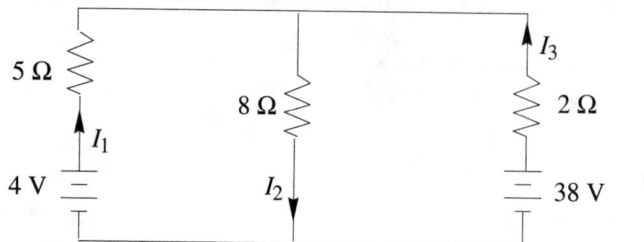

Figure 11.75

We can write this system as the matrix equation $AX = C$ where $A = \begin{pmatrix} 1 & -1 & 1 \\ 5 & 8 & 0 \\ 0 & 8 & 2 \end{pmatrix}$, $X = \begin{pmatrix} I_1 \\ I_2 \\ I_3 \end{pmatrix}$, and $C = \begin{pmatrix} 0 \\ 4 \\ 38 \end{pmatrix}$. (Notice that the coefficients of the missing variables are entered as 0 in matrix A.) The solution of this equation is

$$X = A^{-1}B$$
$$= \begin{pmatrix} -4 \\ 3 \\ 7 \end{pmatrix}$$

Thus the three currents are $I_1 = -4\,\text{A}$, $I_2 = 3\,\text{A}$, and $I_3 = 7\,\text{A}$. ■

Into the Fourth Dimension and Beyond

It is easy to imagine that there could be a choice between manufacturing four, five, or even 20 products, each with 20 components in limited supply. While easy to imagine the problem, it is hopeless to try to imagine a picture of the solution in 20 dimensions. However one thing remains the same: No matter how many products and components, a system of n equations and n variables can be solved using an $n \times n$ matrix A for the variable terms on the left side of the equal sign and an $n \times 1$ matrix C for the constant terms on the right side. And whether there are two equations or 20 equations, the solution is still given by a simple matrix equation:

$$X = A^{-1}C$$

Section 11.5 Exercises

In Exercises 1–4 analyze the equation $y = 8$ in a three-dimensional coordinate system.

1. Give four points that satisfy the equation.

2. At what point(s) does the graph cross the z-axis?

3. At what point(s) does the graph cross the y-axis?

4. Describe the graph of $y = 8$.

In Exercises 5–8 analyze the equation $x = -4$ in a three-dimensional coordinate system.

5. Give four points that satisfy the equation.

6. At what point(s) does the graph cross the z-axis?

7. At what point(s) does the graph cross the y-axis?

8. Describe the graph of $x = -4$.

In Exercises 11–14 analyze the graph of $z = 10 - x - y$ from Example 11.35.

11. At what point(s) will the graph cross the y-axis?

12. At what point(s) will the graph cross the z-axis?

13. Describe the intersection between the graph of this equation and the xz-plane.

14. Describe the intersection between the graph of this equation and the yz-plane.

15. (a) Describe the intersection of $x = 10 - x - y$ and $y = 8$.

   (b) Give the coordinates of four points that lie on the intersection of $x = 10 - x - y$ and $y = 8$.

16. (a) Describe the intersection of $x = 10 - x - y$ and $x = -4$.

   (b) Give the coordinates of four points that lie on the intersection of $x = 10 - x - y$ and $x = -4$.

17. *Electronics* If Kirchhoff's laws are applied to the circuit in Figure 11.76, the following equations are obtained. Determine the indicated currents.

$$I_1 - I_2 - I_3 = 0$$

$$26I_1 + 10I_2 = 100$$

$$-10I_2 + 35I_3 = 80$$

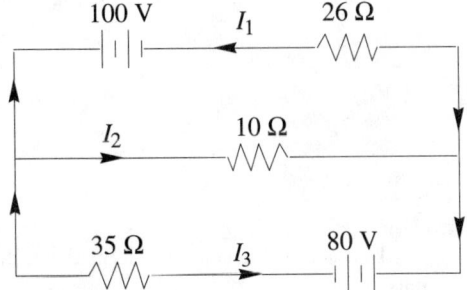

Figure 11.76

18. *Electronics* Applying Kirchhoff's laws to the circuit in Figure 11.77 results in the following equations. Determine the indicated currents.

$$I_3 - I_1 - I_2 = 0$$

$$20I_1 + 0.5I_1 - 15I_2 - 0.4I_2 = 120 - 80$$

$$15I_2 + 0.4I_2 + 10I_3 + 0.6I_3 = 140$$

9. Describe the intersection of $y = 8$ and $x = -4$ in a three-dimensional coordinate system.

10. Give the coordinates of four points that lie on the intersection of $y = 8$ and $x = -4$ in a three-dimensional coordinate system.

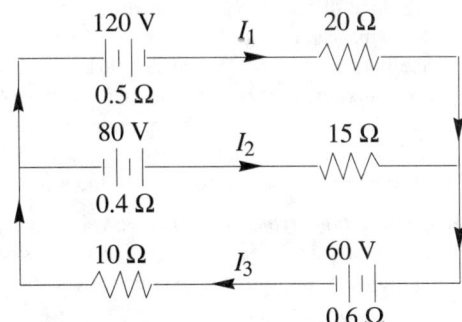

Figure 11.77

19. *Electronics* The currents through the resistors in Figure 11.78 produce the following equations. Determine the currents.

$$11.6I_1 - 3I_2 - 2.5I_3 = 10$$

$$-3I_1 + 10.3I_2 + 2.6I_3 = 15$$

$$2.5I_1 + 2.6I_2 + 8.4I_3 = 20$$

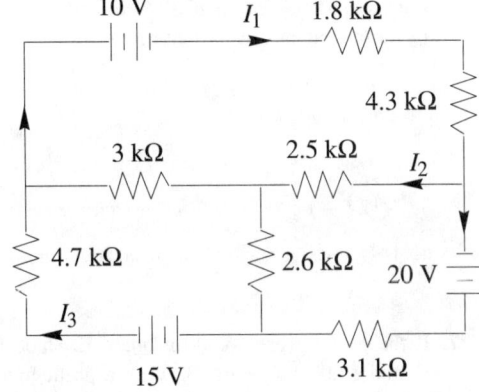

Figure 11.78

20. *Business* A company manufactures robotic controls. Their current models are the RC-1 and RC-2. Each RC-1 unit requires 8 transistors and 4 integrated circuits. Each RC-2 unit uses 9 transistors and 5 integrated circuits. Each day the company receives 1,596 transistors and 860 integrated circuits. How many units of each model can be made if all parts are used?

21. *Metallurgy* An alloy is composed of three metals, A, B, and C. The percentages of each metal are indicated by the following system of equations:

$$\begin{cases} A + B + C = 100 \\ A - 2B = 0 \\ -4A + C = 0 \end{cases}$$

Determine the percentage of each metal in the alloy.

22. *Automotive technology* A petroleum engineer was testing three different gasoline mixtures, A, B, and C, in the same car and under the same driving conditions. She noticed that the car traveled $90\,\text{km}$ farther when it used mixture B than when it used mixture A. Using fuel C, the car traveled $130\,\text{km}$ more than when it used fuel B. The total distance traveled was $1\,900\,\text{km}$. Find the travel distances of each of the three fuels.

23. *Construction technology* If three cables are joined at a point and three forces are applied so the system is in equilibrium, the following system of equations results.

$$\begin{cases} \dfrac{6}{7}F_B - \dfrac{2}{3}F_C = 2{,}000 \\[2mm] -F_A - \dfrac{3}{7}F_B + \dfrac{1}{3}F_C = 0 \\[2mm] \dfrac{2}{7}F_B + \dfrac{2}{3}F_C = 1{,}200 \end{cases}$$

Determine the three forces, F_A, F_B, and F_C, measured in newtons (N).

24. *Environmental science* To control ice and protect the environment, a certain city determines that the best mixture to be spread on roads consists of 5 units of salt, 6 units of sand, and 4 units of a chemical inhibiting agent. Three companies, A, B, and C, sell mixtures of these elements according to the following table

	Salt	Sand	Inhibiting Agent
Company A	2	1	1
Company B	2	2	2
Company C	1	5	1

(a) In what proportion should the city purchase from each company in order to spread the best mixture? (Assume that the city must buy complete truckloads.)

(b) If the city expects to need $3{,}630{,}000$ units for the winter, how many units should be bought from each company.

25. *Automotive technology* The relationship between the velocity v of a car (in mph) and the distance d (in ft) required to bring it to a complete stop is known to be of the form $d = av^2 + bv + c$, where a, b, and c are constants. Use the following data to determine the values of a, b, and c. When $v = 20$, then $d = 40$; when $v = 55$, then $d = 206.25$; and when $v = 65$, then $d = 276.25$.

26. *Metallurgy* An alloy is composed of four metals, A, B, C, and D. The percentages of each metal are indicated by the following system of equations:

$$\begin{cases} A + B + C + D = 100 \\ A + B - C = 0 \\ -1.64A + D = 0 \\ 3A - 2C + 2D = -1 \end{cases}$$

Determine the percentage of each metal in the alloy.

Exercises 27–29 continue Example 11.38 to find the maximum profit for the company making E, F, and G circuit boards. The number of boards to be manufactured were

$$30e + 10f + 20g = 210 \quad \text{the resistor supply equation}$$
$$15e + 12f + 8g = 120 \quad \text{the capacitor supply equation}$$
$$6e + 9f + 10g = 72 \quad \text{the transistor supply equation}$$

27. If the profits were \$6.00 for board E, \$5.00 for board F, and \$7.00 for board G, write a profit function that will determine the profit from making e of board E, f of board F, and g of board G.

28. The solution in Example 11.38 did not give whole number answers. Round each of the whole number solutions in Example 11.38 either up or down to get all the possible whole number solutions around P that satisfy the restrictions on the number of resistors, capacitors, and transistors. For example, is $e = 5$, $f = 3$, and

$g = 1$ an acceptable solution? Remember, the restrictions are

$$30e + 10f + 20g = 210 \quad \text{the resistor supply equation}$$
$$15e + 12f + 8g = 120 \quad \text{the capacitor supply equation}$$
$$6e + 9f + 10g = 72 \quad \text{the transistor supply equation}$$

29. Which of your answers from Exercise 28 gives the largest profit?

11.6   Curves and Surfaces in 3-D

In this section we'll study a few examples of three-dimensional equations that lead to some interesting applications. First we'll look at the parametric equations that can be used to model trajectories in three dimensions.

Parametric Equations in 3-D

In Chapter 5 you saw that parametric equations provide a good way to look at mathematical models of trajectories of objects in flight. At that time we were modeling motion in two dimensions; now we will turn to motion in three dimensions, again with parametric equations. Let's take a look at a modification of the familiar parametric equations for a circle:

$$\begin{cases} x = \cos t \\ y = \sin t \end{cases}$$

As you remember, these parametric equations trace a point around a circle of radius 1. If t starts at 0 and the upper limit of t is larger than 2π or $360°$, then the curve will be traced more than once. By adding a third parametric equation, $z = t$, we extend the curve to three dimensions. This gives the new set of parametric equations

$$\begin{cases} x = \cos t \\ y = \sin t \\ z = t \end{cases}$$

Every point that is plotted on this curve is raised a distance t above the xy-plane. The graph is called a **helix** and looks like a coiled spring. The graph is shown from four perspectives in Figure 11.79. The curve is plotted in radians over the interval $0 \le t \le 14\pi$, so there are seven complete revolutions with the curve ending its motion directly above where it started. Notice that the curve appears like a circle from the top and like a sine or cosine curve from the side. Viewed from the top you see only a circle because you cannot distinguish the different heights of the curve. Viewed from the side you see only the cosine curve because we are looking straight down the y-axis and we cannot see any of the motion in the y-direction.

Let's imagine that the curve is the trajectory of a very mathematically inclined fly buzzing around a room, or the trajectory of a robot arm under control of a program. Example 11.40 looks at positions on the curve at two different times.

Example 11.40

For the graph shown in Figure 11.80, the two points P and Q correspond to the times $t = 4$ and $t = 20$, respectively.

(a) What are the coordinates of P and Q?
(b) Draw P and Q on the different views in Figure 11.79.
(c) What is the straight line distance between P and Q?
(d) What is the average speed along the straight line path between P and Q?
(e) What is the actual distance along the curve between P and Q?
(f) Do you think the average speed along the curve is more or less than the average speed in the straight line path?

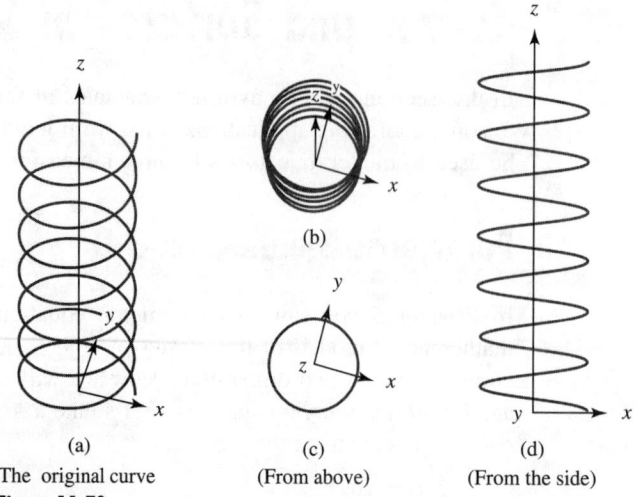

(a)   (c)   (d)
The original curve   (From above)   (From the side)
Figure 11.79
Four views of a parametric curve in space.

Solution Because $z = t$, we will have to keep in mind that t measures vertical distance along the z-axis, but also that it measures an angle in radians. Furthermore, for the purpose of this example we will also assume that t measures time elapsed in seconds.

(a) The coordinates of the points are $P(\cos 4, \sin 4, 4)$ and $Q(\cos 20, \sin 20, 20)$, or $P(-0.65, -0.76, 4)$ and $Q(0.41, 0.91, 20)$.

(b) The points are shown in Figure 11.80. We can see now that the scale on the z-axis is different from the scales on the x-axis and y-axis.

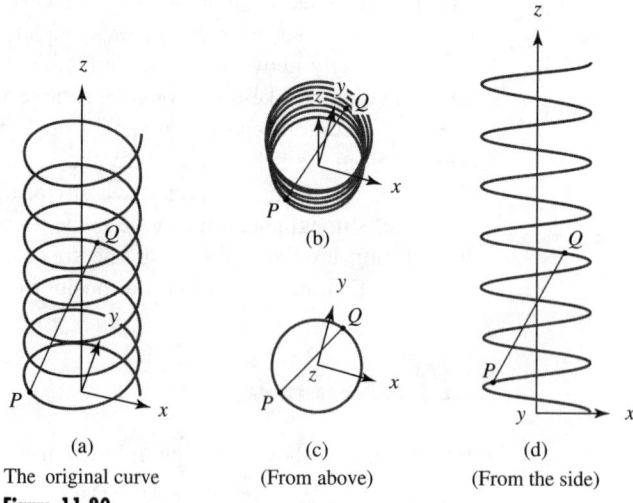

(a)   (c)   (d)
The original curve   (From above)   (From the side)
Figure 11.80
Point P corresponds to $t = 4$ and point Q corresponds to $t = 20$.

(c) The straight line distance is indicated by the line segments joining P and Q in Figure 11.80(a)–(d). The straight line distance between the points is

$$\sqrt{(0.41 - (-0.65))^2 + (0.91 - (-0.76))^2 + (20 - 4)^2} \approx 16.122$$

Notice that the distance is close to 16—the vertical distance between the two points—because the differences between the x-coordinates and the y-coordinates contribute very little to the computation compared with 16, whose square is much larger than the squares of the other two terms.

(d) The average speed in a straight line is $\dfrac{16.122}{20 - 4} \approx 1.01$ units per second.

(e) The distance along the curve is not simply a matter of counting the number of times around the circle and multiplying by 2π, because the curve is also rising and this increases the distance. The distance along a curve requires calculus, so we'll postpone discussion of this matter for now.

(f) We do know that the distance along the curve is more than the straight line distance, so the average speed along the curve is greater. However, until we have answered (e) we cannot determine the distance along the curve.   ■

Spherical Coordinates

Two commonly used coordinate systems in three dimensions are variations of polar coordinates, which use distance r and angle θ to locate the point $P(r, \theta)$ in two dimensions. The two systems, called spherical coordinates and cylindrical coordinates, each have their applications in robot control, mapping, and graphing certain images.

In the **spherical coordinate system** we locate points in space with two angles and a distance from a fixed point called the pole, as in two-dimensional polar coordinates. Figure 11.81 shows a point P plotted in spherical coordinates. The angle θ is in the xy-plane and is measured from the positive x-axis (the polar axis). The angle ϕ is measured down from the positive z-axis. Note that if $\phi = 90°$ the point is in the xy-plane.

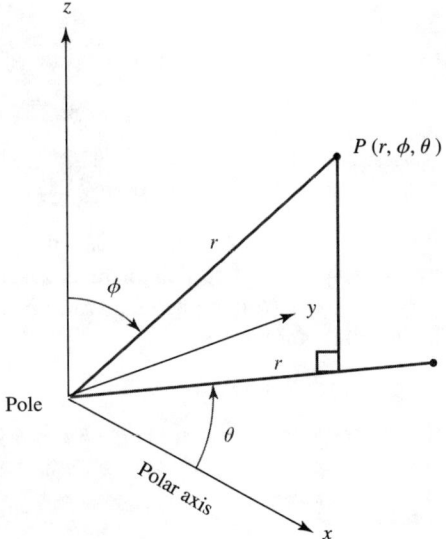

Figure 11.81

Locating a Point in Spherical Coordinates

To locate the point (r, θ, ϕ) in spherical coordinates,

1. Move a distance $|r|$ vertically along the z-axis from the pole. If r is positive, move up; if r is negative, move down.
2. Rotate an angle ϕ down from the z-axis toward the x-axis (the polar axis).
3. Without changing the angle ϕ and the distance r, rotate along an angle θ as measured in the xy-plane.

Figure 11.82 shows a graph of one of the simplest formulas in spherical coordinates. The equation is $r = 10$. Just as that equation in polar coordinates produces a circle in two dimensions, in spherical coordinates it produces a sphere in three dimensions.

Figure 11.83 shows the graph of the formula $r = 1.2^\theta \sin \phi$ plotted in spherical coordinates. To get a better understanding of this graph, let's look at it from two different perspectives. First we'll look down on it from above, as shown in Figure 11.84. There you can see that the image is made up of many spirals of the form $r = k(1.2^\theta)$ where $k = \sin \phi$. For each value of ϕ there is a spiral.

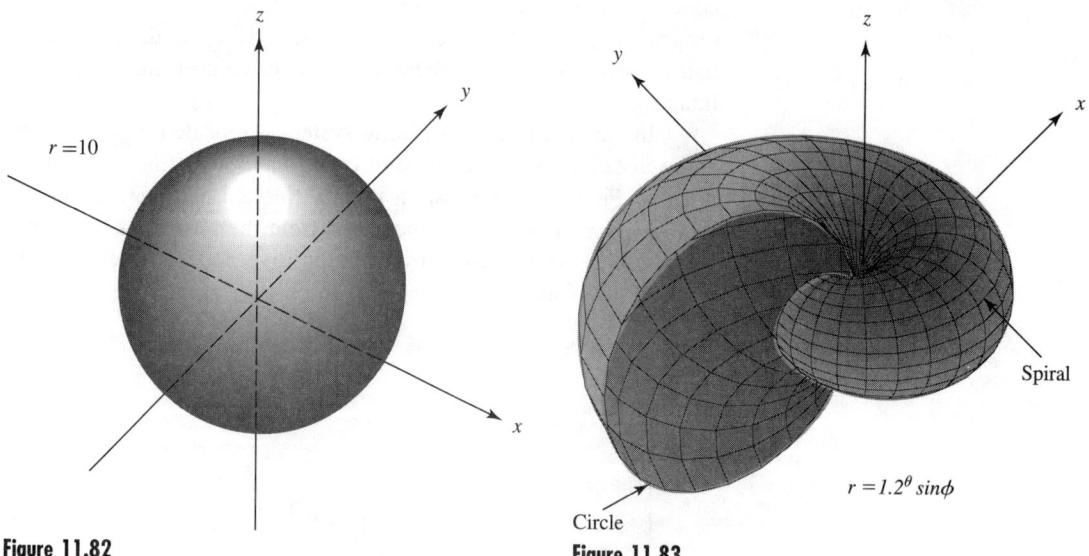

Figure 11.82   **Figure 11.83**

Figure 11.85 shows the view along the y-axis. The horizontal lines in this figure are the spirals. Notice that there is a series of circles of increasing radius. You learned in Chapter 6 that one polar form of a circle is $r = a \sin \theta$. The equation of the largest circle on the left is

$$r = 1.2^\pi \sin \phi \approx 1.77 \sin \phi$$

Its location corresponds to the angle $\theta = \pi = 180°$ in Figure 11.84.

Example 11.41

Plot the outer spiral in Figure 11.84 on your calculator and then describe how to plot the figure's family of spirals.

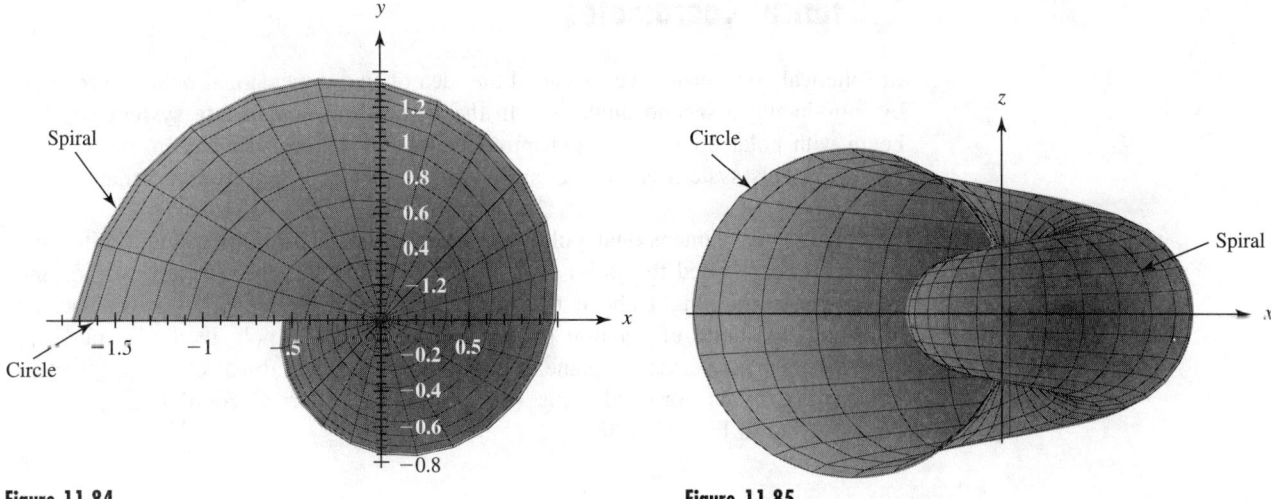

Figure 11.84

Figure 11.85

Solution   Your calculator can plot these spirals one at a time. The largest spiral occurs for the largest value of $\sin\phi$. Since the sine function reaches its maximum value at π radians, or $90°$, the outer spiral in the figure has the equation $r = 1.2^\theta \sin\pi = 1.2^\theta$. The graph of the polar equation $r = 1.2^\theta$ is shown in Figure 11.86.

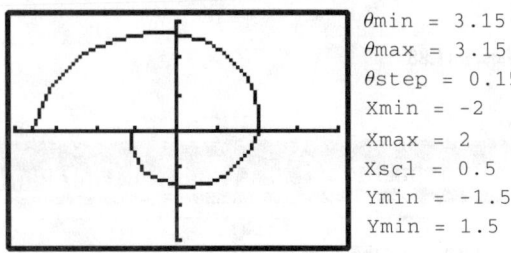

θmin = 3.15
θmax = 3.15
θstep = 0.15
Xmin = -2
Xmax = 2
Xscl = 0.5
Ymin = -1.5
Ymin = 1.5

Figure 11.86

One way to plot a family of spirals with different coefficients is to put the coefficients in a list and then use that list to define an equation in the Y= screen. We defined the list L₁ to be {0.2, 0.4, 0.6, 0.8, 1} and defined the function $Y = L₁ 1.2^\theta$, which gave us five equations and five graphs with one entry in the Y= screen. The results are shown in Figure 11.87.

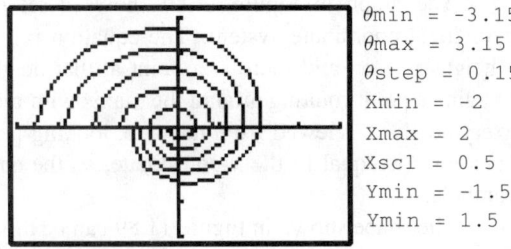

θmin = -3.15
θmax = 3.15
θstep = 0.15
Xmin = -2
Xmax = 2
Xscl = 0.5
Ymin = -1.5
Ymin = 1.5

Figure 11.87

Cylindrical Coordinates

In spherical coordinates we extended the idea of two-dimensional polar coordinates by introducing a second angle, ϕ. In the **cylindrical coordinate system** we also begin with polar coordinates and introduce a second linear dimension, the vertical height z. This system is used as a coordinate system in the programs that control some robots.

As in two-dimensional polar coordinates, we begin with a point called the **pole** and a ray called the **polar axis**, located in the xy-plane. To that we add the z-coordinate, the height above the xy-plane. Any point in space can be located by giving its angle of rotation θ, its distance from the pole in the xy-plane r, and its height above the xy-plane. Each point in the cylindrical coordinate system is described by the ordered triple (r, θ, z). The cylindrical coordinate system is demonstrated in Figure 11.88.

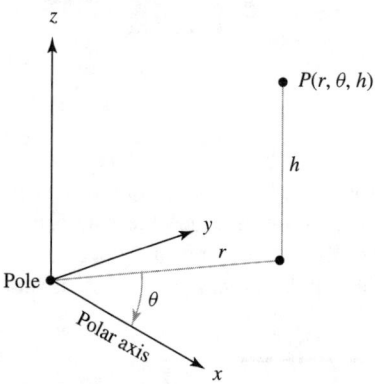

Figure 11.88

Locating a Point in Cylindrical Coordinates

To locate the point (r, θ, z) in cylindrical coordinates,

1. Move a distance $|r|$ along the polar axis. Move in the positive or negative direction, depending on the sign of r.
2. Rotate an angle θ from the positive x-axis in the xy-plane.
3. Move a distance $|z|$ parallel to the z-axis. If z is positive, move up; if z is negative, move down.

The graph in Figure 11.89 shows a set of points that fit an equation in a cylindrical coordinate system. The equation is $z = r^2$. Given a value of r, there is a height $z = r^2$, and there is a point at that height and radius for every value of θ. Imagine a point rotating around the z-axis with radius r. Therefore there is a circle at every height z. Viewed from the side, looking perpendicularly to the xz-plane, each value of r is equal to the x-coordinate, so the equation in the xz-plane is $z = x^2$, a parabola.

The shape shown in Figure 11.89 can be thought of as a parabola rotated around the z-axis. Such an object is called a **paraboloid**, a three-dimensional parabola similar to the ellipsoid we encountered in Chapter 6 in the lithotripter kidney stone

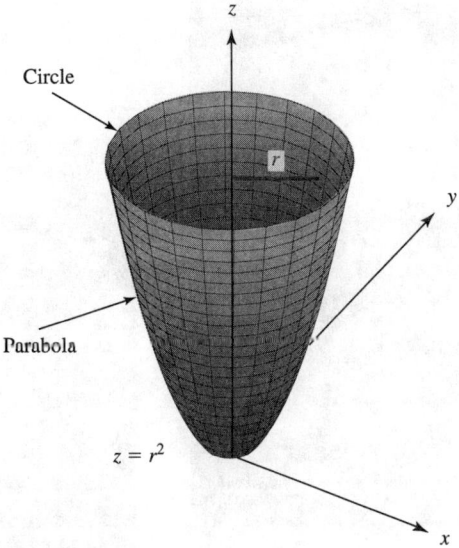

Figure 11.89

machine. Like the ellipsoid, the paraboloid has unique reflective properties. All waves coming in parallel to the z-axis are reflected to the paraboloid's focus. Figure 11.90 demonstrates the reflective properties of a parabola or a paraboloid.

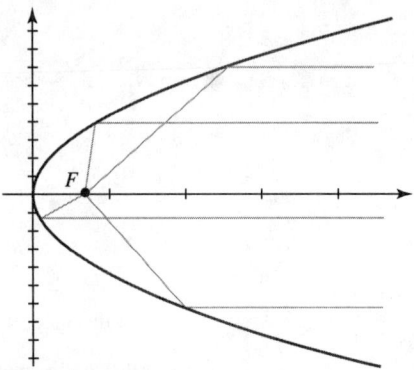

Figure 11.90
Incoming sound waves are reflected to the focus of the parabola. Outgoing light waves from the focus are reflected into a beam.

Since incoming light or sound is focused into a single point (the focus), the paraboloid is the shape used for microphones such as those used along the sidelines at football games. Paraboloids are also used for television dish-antennas (Figure 11.91) and giant astronomical telescopes (Figure 11.92).

This reflective property of paraboloids also works in the reverse direction. If waves are generated at the focus, they will be sent out from the paraboloid in straight lines. This property makes the paraboloid the shape used in loudspeakers, flashlights, and large spotlights. In World War II paraboloid-shaped spotlights were used by ground forces to see low flying aircraft at night. Now such spotlights are employed to attract attention at commercial events.

Figure 11.91

Figure 11.92
Courtesy of William B. Stine, Ph.D., California State Polytechnic University, and Richard B. Diver, Ph.D., Sandia National Laboratories; for the United States Department of Energy by Sandia Corporation.

Section 11.6 Exercises

In Exercises 1–4 use the parametric equations

$$\begin{cases} x = \cos t \\ y = t \\ z = \sin t \end{cases}$$

1. Describe the graph of these parametric equations.

2. Sketch a side and a 3-D view of the curve formed by these parametric equations.

3. If P is $t = -2$ and Q is $t = 21$, what is the straight line distance between P and Q?

4. What is the average speed along the straight line from P to Q?

In Exercises 5–8 use the parametric equations

$$\begin{cases} x = \cos t \\ y = 1 + 2\sin t \\ z = t \end{cases}$$

5. Describe the graph of these parametric equations.

6. Sketch a side and a 3-D view of the curve formed by these parametric equations.

7. If R is $t = 3$ and S is $t = 18$, what is the straight line distance between R and S?

8. What is the average speed along the straight line from R to S?

In Exercises 9–12 describe the graphs of the given equations in spherical coordinates.

9. $r = 5$

10. $\theta = \frac{\pi}{3}$

11. $\phi = \frac{\pi}{6}$

12. $3 \leq r \leq 5$

13. Consider the equation $z = 2\sin\theta - \cos 3\phi$. Its graph in spherical coordinates is given in Figure 11.93. Use your calculator to plot the graph when $\theta = \frac{\pi}{2}$ and then describe the curve.

 14. Use your calculator to plot a family of curves for the spherical equation of Exercise 15, $z = 2\sin\theta - \cos 3\phi$ where $\theta = 0, 0.5, 1.0, 1.5, 2.0, 2.5$, and 3.0.

 15. Consider the equation $z = 2\sin\theta - \cos 3\phi$. Use your calculator to plot the graph when $\phi = \frac{\pi}{4}$ and then describe the curve.

 16. Use your calculator to plot a family of curves for the spherical equation of Exercise 15, $z = 2\sin\theta - \cos 3\phi$, where $\phi = 0, 0.5, 1.0, 1.5, 2.0, 2.5$, and 3.0.

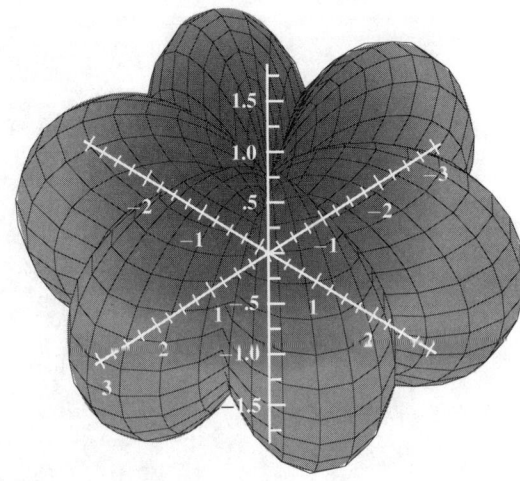

Figure 11.93

In Exercises 17–20 describe the graphs of the given equations in cylindrical coordinates.

17. $r = 5$

18. $\theta = \frac{\pi}{3}$

19. $z = 4$

20. $3 \le r \le 5$

21. Figure 11.94 shows a star map produced with the program *Skymap*. This map is for the sky in San Jose, California on July 25, 1997, and shows an alternate form of spherical coordinates. The center of the map is directly above the observer, and the angle that we would call ϕ is zero at the horizon instead of being zero straight up as in spherical coordinates in mathematics. The angle that we have called θ is zero at the southern horizon (bottom of the map) and opens up to the east. The distance from the observer, what we have called r, is ignored on elementary sky maps like this. The arc sketched across the map is the path of the Sun, and the Sun itself is a circle in the third quadrant of the map.

What are the coordinates (ϕ, θ) of each of these objects:

(a) Sun

(b) Mercury

(c) Venus

(d) Mars

(e) Pluto

(f) Southern horizon

(g) Northern horizon

[*Note: Skymap* is distributed by JASCO software and more information is available at http://www.skymap.com. (Copyright ©C. A. Marriott 1992–1996)

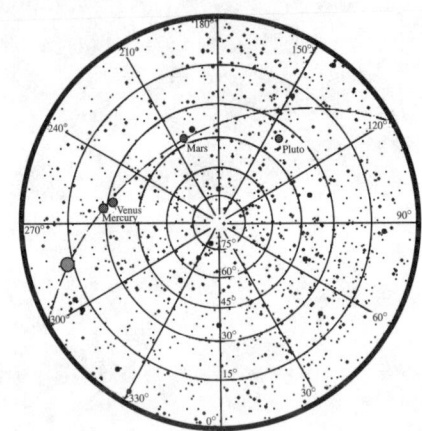

Figure 11.94

11.7 Rotations and Translations in Three Dimensions

In this chapter matrices have proven useful for solving systems of equations and for working with some geometric transformations. We moved from two to three dimensions in the linear programming problem; now we'll move to three dimensions in the geometric transformations.

Matrix representations of rotation and scaling in two dimensions are summarized as follows.

Matrices and 2-D Transformations

Point $P(x, y)$ is transformed to the point $P_1(a, b)$.

▸ **Scaling** by the factor m in the x-direction and n in the y-direction. The equations are

$$a = mx$$

$$b = ny$$

with matrix representation

$$\begin{pmatrix} m & 0 \\ 0 & n \end{pmatrix} \begin{pmatrix} x \\ y \end{pmatrix} = \begin{pmatrix} a \\ b \end{pmatrix}$$

▸ **Rotation** through the angle α. The equations are

$$a = x \cos \alpha - y \sin \alpha$$

$$b = x \sin \alpha + y \cos \alpha$$

with matrix representation

$$\begin{pmatrix} \cos \alpha & -\sin \alpha \\ \sin \alpha & \cos \alpha \end{pmatrix} \begin{pmatrix} x \\ y \end{pmatrix} = \begin{pmatrix} a \\ b \end{pmatrix}$$

▸ **Combination of scaling and rotation.**
  • Scaling, then rotation:

$$\begin{pmatrix} \cos \alpha & -\sin \alpha \\ \sin \alpha & \cos \alpha \end{pmatrix} \begin{pmatrix} m & 0 \\ 0 & n \end{pmatrix} \begin{pmatrix} x \\ y \end{pmatrix} = \begin{pmatrix} a \\ b \end{pmatrix}$$

  • Rotation, then scaling:

$$\begin{pmatrix} m & 0 \\ 0 & n \end{pmatrix} \begin{pmatrix} \cos \alpha & -\sin \alpha \\ \sin \alpha & \cos \alpha \end{pmatrix} \begin{pmatrix} x \\ y \end{pmatrix} = \begin{pmatrix} a \\ b \end{pmatrix}$$

Scaling in Three Dimensions

Suppose the coordinates of $P(x, y, z)$ are each multiplied by different factors m, n, and p and moved to the point $P_1(a, b, c)$. The equations are quite similar to the two-dimensional case:

$$a = mx$$

$$b = ny$$

$$c = pz$$

Therefore it should be no surprise that the matrix representation is also very similar to the two-dimensional result:

$$\begin{pmatrix} m & 0 & 0 \\ 0 & n & 0 \\ 0 & 0 & p \end{pmatrix} \begin{pmatrix} x \\ y \\ z \end{pmatrix} = \begin{pmatrix} a \\ b \\ c \end{pmatrix}$$

Example 11.42 **Application**

The point $P(2, -3, 5)$ is scaled with the factors $m = 0.5$, $n = -1.5$, and $p = 2.5$. Write the matrix form of the transformation and give the coordinates of the new location.

Solution   The matrix representation is

$$\begin{pmatrix} 0.5 & 0 & 0 \\ 0 & -1.5 & 0 \\ 0 & 0 & 2.5 \end{pmatrix} \begin{pmatrix} 2 \\ -3 \\ 5 \end{pmatrix} = \begin{pmatrix} 1 \\ 4.5 \\ 12.5 \end{pmatrix}$$

The coordinates of the new location are $P_1(1, 4.5, 12.5)$.   ■

Rotation in Three Dimensions

There are two kinds of rotation that might happen to a point in three dimensions. Just as there are two angles in spherical coordinates, so there are two directions in which a point can be rotated. We will refer to these angles as α and β. The rotation through α swings the point through space parallel to the xy-plane. The rotation through β swings the point through space toward or away from the z-axis, in a plane that is perpendicular to the xy-plane.

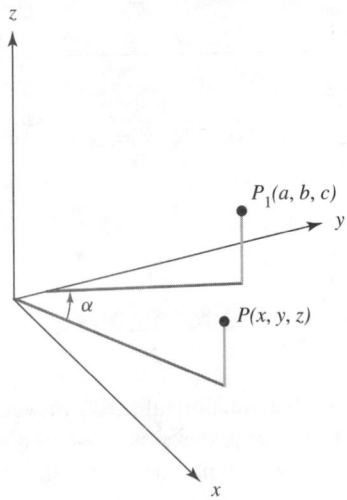

Figure 11.95

Figure 11.95 shows the result of a rotation of point $P(x, y, z)$ through an angle α. The first thing to notice is that points P and P_1 have the same coordinate in the z-direction. Therefore the only changes are in the x- and y-coordinates. Since only x and y change, we can use the two-dimensional equations for them. The three equations, then, are

$$a = x \cos \alpha - y \sin \alpha$$

$$b = x \sin \alpha + y \cos \alpha$$

$$c = z$$

These equations can be represented with matrices by

$$\begin{pmatrix} \cos\alpha & -\sin\alpha & 0 \\ \sin\alpha & \cos\alpha & 0 \\ 0 & 0 & 1 \end{pmatrix} \begin{pmatrix} x \\ y \\ z \end{pmatrix} = \begin{pmatrix} a \\ b \\ c \end{pmatrix}$$

Multiply the two matrices on the left-hand side of the equation to see if you get the three equations above.

Example 11.43   Application

If point $P_1(2, -6, 4)$ is the result of a rotation of $270°$ in the xy-plane, what are the coordinates of the original point?

Solution   The problem here is to find $P(x, y, z)$. The matrix equation is

$$\begin{pmatrix} \cos 270° & -\sin 270° & 0 \\ \sin 270° & \cos 270° & 0 \\ 0 & 0 & 1 \end{pmatrix} \begin{pmatrix} x \\ y \\ z \end{pmatrix} = \begin{pmatrix} 2 \\ -6 \\ 4 \end{pmatrix}$$

The solution is

$$\begin{pmatrix} x \\ y \\ z \end{pmatrix} = \begin{pmatrix} \cos 270° & -\sin 270° & 0 \\ \sin 270° & \cos 270° & 0 \\ 0 & 0 & 1 \end{pmatrix}^{-1} \begin{pmatrix} 2 \\ -6 \\ 4 \end{pmatrix}$$

$$= \begin{pmatrix} 6 \\ 2 \\ 4 \end{pmatrix}$$

Thus rotating the point $P(6, 2, 4)$ through an angle of $270°$ in the xy-plane results in the point $P'(2, -6, 4)$. ∎

Activity 11.12
Rotation in Three Dimensions

1. In Example 11.43 a rotation of $270°$ moved the point $P(6, 2, 4)$ to the point $P_1(2, -6, 4)$. In this case the x- and y-coordinates switched and the y-coordinate changed sign. Draw a graph that shows this rotation and write an explanation of why you would expect this to occur.
2. Show on a graph where the point $P(x, y, z)$ would have been if it moved to $P_1(2, -6, 4)$ after a rotation of $125°$.
3. Follow the pattern of calculations in Example 11.43 to compute the location of $P(x, y, z)$ and compare your answer with your graphical estimate.

Although the equations and matrix representation for rotation through the angle α are very similar to the two-dimensional case, it turns out that rotation through the angle β toward or away from the z-axis is not so easy to understand. The answer is not so difficult to write, but the result does not follow easily from the two-dimensional case, and the equations and matrices are quite complicated. Therefore we will omit discussion of this rotation in this book.

Section 11.7 Exercises

1. *Computer Aided Design (CAD)* Suppose that the point $P(1, -4, 5)$ is scaled with factors $m = 1.2$, $n = 0.75$, and $p = -2.5$. Determine the coordinates of P', the point that results after the scaling.

2. *Computer Aided Design (CAD)* Suppose that the point $Q(-2, 3, -6)$ is scaled with factors $m = -3.1$, $n = 1.5$, and $p = -0.25$. Determine the coordinates of Q', the point that results after the scaling.

3. *Computer Aided Design (CAD)* If point $R'(5, -2, -3)$ is the result of a point R being scaled with factors $m = -2$, $n = 1.5$, and $p = 3.2$, what are the coordinates of the original point R?

4. *Computer Aided Design (CAD)* If point $S'(-2.8, 3.1, 4.2)$ is the result of a point S being scaled with factors $m = -2.4$, $n = 1.6$, and $p = -0.24$, what are the coordinates of the original point S?

5. *Computer Aided Design (CAD)* Consider the segment $\overline{AB}$ where $A = (2, 1, -1)$ and $B = (6, 1, 2)$, and suppose $\overline{AB}$ is scaled with factors $m = 2.4$, $n = 1.75$, and $p = 1.05$. The result is segment $A'B'$.
   (a) What is the length AB?
   (b) What are the coordinates of A' and B'?
   (c) What is the length $A'B'$?

6. *Computer Aided Design (CAD)* Consider the segment $\overline{CD}$ where $C = (4, -2, 6)$ and $D = (-2, 5, -3)$, and suppose $\overline{CD}$ is scaled with factors $m = 4.2$, $n = 5.1$, and $p = 2$. The result is segment $C'D'$.
   (a) What is the length CD?
   (b) What are the coordinates of C' and D'?
   (c) What is the length $C'D'$?

7. *Computer Aided Design (CAD)* Consider the square $ABCD$ where $A = (2, 4, 3)$, $B = (12, 4, 3)$, $C = (12, -6, 3)$, and $D = (2, 5-6, 3)$, and suppose $ABCD$ is scaled with factors $m = n = p = 2$. The result is another square $A'B'C'D'$.
   (a) What is the area of $ABCD$?
   (b) What are the coordinates of A', B', C', and D'?
   (c) What is the area of $A'B'C'D'$?
   (d) Discuss the relationship between m^2 and your answers to parts (a) and (c).

8. *Computer Aided Design (CAD)* Consider the square $EFGH$ where $E = (-3, 5, 2)$, $F = (8, 5, 2)$, $G = (8, -6, 2)$, and $H = (-3, -6, 2)$, and suppose $EFGH$ is scaled with factors $m = n = p = 2.5$. The result is another square $E'F'G'H'$.
   (a) What is the area of $EFGH$?
   (b) What are the coordinates of E', F', G', and H'?
   (c) What is the area of $E'F'G'H'$?

(d) Discuss the relationship between m^2 and your answers to parts (a) and (c).

9. *Computer Aided Design (CAD)* The point $P(-3, 5, 1)$ is rotated through an angle of $125°$ in the xy-plane to the point P'. Determine the coordinates of P'.

10. *Computer Aided Design (CAD)* The point $Q(2.3, -6.5, -3.1)$ is rotated through an angle of $-73°$ in the xy-plane to the point Q'. Determine the coordinates of Q'.

11. *Computer Aided Design (CAD)* If point $R'(5, -2, -3)$ is the result of a rotation of $135°$ in the xy-plane, what are the coordinates of the original point?

12. *Computer Aided Design (CAD)* If point $S'(-3.6, 4.2, -2.8)$ is the result of a rotation of $135°$ in the xy-plane, what are the coordinates of the original point?

13. *Computer Aided Design (CAD)* Suppose that the point $P(-3, 5, 1)$ is rotated through an angle of $142°$ in the xy-plane and scaled with factors $m = 2$, $n = 2.5$, and $p = 3$.
   (a) Determine the coordinates of the resulting point P_1 if P is first rotated and then scaled.
   (b) Determine the coordinates of the resulting point P_2 if P is first scaled and then rotated.

14. *Computer Aided Design (CAD)* Suppose that the point $Q(4, 1.5, -2)$ is rotated through an angle of $-57°$ in the xy-plane and scaled with factors $m = 1.5$, $n = 2.5$, and $p = 1$.
   (a) Determine the coordinates of the resulting point Q_1 if Q is first rotated and then scaled.
   (b) Determine the coordinates of the resulting point Q_2 if Q is first scaled and then rotated.

15. *Computer Aided Design (CAD)* Suppose that a tetrahedron has vertices $A(4, 2, 1)$, $B(6, 5, 1)$, $C(2, 7, 1)$, and $D(4, 4.5, 6)$, is rotated through an angle of $30°$ in the xy-plane, and scaled with factors $m = 2$, $n = -1$, and $p = 3$.
   (a) Determine the coordinates of the resulting points A_1, B_1, C_1, and D_1 if the tetrahedron is first rotated and then scaled.
   (b) Determine the coordinates of the resulting points A_2, B_2, C_2, and D_2 if the tetrahedron is first scaled and then rotated.

16. *Computer Aided Design (CAD)* Suppose that a tetrahedron has vertices $A(3, -2, 1)$, $B(4, 3, 2)$, $C(-1, 4, 5)$, and $D(3, 1, 4)$, is rotated through an angle of $-60°$ in the xy-plane, and scaled with factors $m = 2$, $n = 3$, and $p = 1$.
   (a) Determine the coordinates of the resulting points A_1, B_1, C_1, and D_1 if the tetrahedron is first rotated and then scaled.

(b) Determine the coordinates of the resulting points A_2, B_2, C_2, and D_2 if the tetrahedron is first scaled and then rotated.

17. In the Chapter Project memo from management you were asked to prepare a report that answers the following three questions.

▶ What is the mathematics behind rotation and scaling objects in two dimensions?

▶ How does a matrix contribute to the solution?

▶ How does this work in three dimensions?

Write a memo to the management describing the progress you have made toward answering these questions.

●●●●●●●●● Chapter 11 Summary and Review

Topics You Learned or Reviewed

▶ Figures can be graphed in three dimensions using Cartesian coordinates, parametric equations, cylindrical coordinates, or spherical coordinates.

- Plotting points and visualizing graphs
- Cartesian coordinates locate a point in space using the x-, y-, and z-axes, where each pair of axes is perpendicular to the other.
- The distance d between the points $P_1(x_1, y_1, z_1)$ and $P_2(x_2, y_2, z_2)$ is

$$d = P_1 P_2 = \sqrt{(x_2 - x_1)^2 + (y_2 - y_1)^2 + (z_2 - z_1)^2}$$

- **Parametric equations** can be used to locate a three-dimensional point where the coordinates of x, y, and z are given in terms of the parameter t.
- Any point in space can be located by giving its angle of rotation θ, its distance from the pole in the xy-plane r, and its height above the xy-plane. Each point in the **cylindrical coordinate system** is described by the ordered triple (r, θ, ϕ).
- In the **spherical coordinate system** we locate points in space with two angles and a distance r from a fixed point called the pole, as in two-dimensional polar coordinates. The angle θ is in the xy-plane and is measured from the x-axis (the polar axis). The angle ϕ is measured down from the z-axis.

▶ The trigonometric expressions for sine and cosine of angle sums are

$$\sin(\theta + \alpha) = \sin \alpha \cos \theta + \sin \theta \cos \alpha$$

$$\cos(\theta + \alpha) = \cos \theta \cos \alpha - \sin \theta \sin \alpha$$

▶ Matrix algebra

- A **matrix** is a two-dimensional list of data that contains numbers or variables in rows and columns. A matrix with m rows and n columns has dimension $m \times n$.
- The product of two matrices A and B is AB or BA. In many cases $AB \neq BA$.
- An **identity matrix** I is a square matrix of the form $\begin{pmatrix} 1 & 0 \\ 0 & 1 \end{pmatrix}$ or $\begin{pmatrix} 1 & 0 & 0 \\ 0 & 1 & 0 \\ 0 & 0 & 1 \end{pmatrix}$. An identity matrix has a 1 for each entry along the diagonal from the upper left to the lower right of the matrix and a 0 for every other entry.
- If A is a square matrix, its inverse, denoted A^{-1}, is the matrix where $AA^{-1} = A^{-1}A = I$.
- A system of two, three, and more equations can be solved using matrices. If A represents an $n \times n$ matrix of the nonconstant coefficients of the n equations, X is an $n \times 1$ matrix of the variables, and C is an $n \times 1$ matrix of the constants, then the solution of the matrix equation $AX = C$ is $X = A^{-1}C$.
- The matrix $\begin{pmatrix} m & 0 \\ 0 & n \end{pmatrix}$ is used to scale the point (x, y) with an x-scaling factor of m and a y-scaling factor of n.

The matrix $\begin{pmatrix} m & 0 & 0 \\ 0 & n & 0 \\ 0 & 0 & p \end{pmatrix}$ is used to scale the point (x, y, z) with an x-scaling factor of m, a y-scaling factor of n, and a z-scaling factor of p.

▶ The matrix $R = \begin{pmatrix} \cos\alpha & -\sin\alpha \\ \sin\alpha & \cos\alpha \end{pmatrix}$ is used to rotate the point (x, y) through an angle α with respect to the origin of a two-dimensional coordinate system. The matrix $\begin{pmatrix} \cos\alpha & -\sin\alpha & 0 \\ \sin\alpha & \cos\alpha & 0 \\ 0 & 0 & 1 \end{pmatrix}$ will rotate the point (x, y, z) through an angle α in the xy-planes of a three-dimensional coordinate system.

▶ The reflective properties of a paraboloid mean that incoming waves parallel to the axis are reflected to the focus of the paraboloid. Outgoing waves from the focus are reflected into a beam.

▶ Matrices on a *TI-83*.
  • They can be created and edited by first pressing MATRX ◀.
  • Two matrices can be multiplied once the matrix is entered in the calculator by using MATRX to select the matrices.
  • The *TI-83* can be used to compute the inverse of a square matrix.

▶ You used a list within a function definition to create a family of functions and their graphs.

Review Exercises

1. Using an x-scaling factor of 3 and a y-scaling factor of -0.5, calculate the missing coordinates in the table at right.

Original Point	Transformed Point
$(-2.5, 6)$	$(\_\_, \_\_)$
$(1.5, 4)$	$(\_\_, \_\_)$
$(0, 3)$	$(\_\_, \_\_)$
$(1.25, -2)$	$(\_\_, \_\_)$
$(\_\_, \_\_)$	$(4, 3)$
$(\_\_, \_\_)$	$(6, -2)$

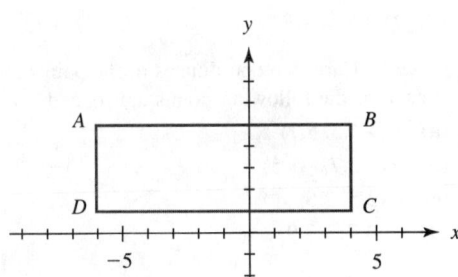

Figure 11.96

In Exercises 2 and 3 use Figure 11.96.

2. (a) What is the area of $ABCD$?
   (b) Use an x-scaling factor of 4 and a y-scaling factor of 0.25 to transform $ABCD$ into rectangle $A'B'C'D'$.
   (c) What is the area of $A'B'C'D'$?

3. (a) Use an x-scaling factor of 2 and a y-scaling factor of 1.5 to transform $ABCD$ into rectangle $EFGH$.
   (b) What is the area of $EFGH$?

In Exercises 4 and 5 use Figure 11.97.

4. (a) What is the area of $ABCD$?
   (b) Use an x-scaling factor of 2 and a y-scaling factor of 0.2 to transform $ABCD$ into rectangle $A'B'C'D'$.
   (c) What is the area of $A'B'C'D'$?

5. (a) Use an x-scaling factor of -2 and a y-scaling factor of -0.5 to transform $ABCD$ into rectangle $EFGH$.
   (b) What is the area of $EFGH$?

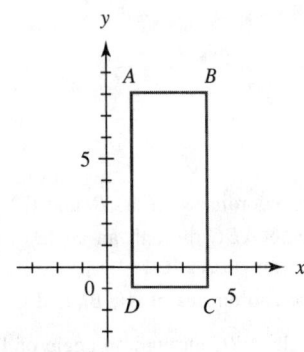

Figure 11.97

In Exercises 6 and 7 use Figure 11.98.

6. **(a)** What is the area of triangle $\triangle ABC$?

   (b) Use an x-scaling factor of -2 and a y-scaling factor of 2 to transform triangle $\triangle ABC$ into triangle $\triangle A'B'C'$.

   (c) Describe the effect of these scaling factors on triangle $\triangle ABC$.

   (d) What is the area of $A'B'C'$?

7. **(a)** Use an x-scaling factor of 2 and a y-scaling factor of -2 to transform triangle $\triangle ABC$ into triangle $\triangle DEF$.

   (b) Describe the effect of these scaling factors on triangle $\triangle ABC$.

   (c) What is the area of $\triangle DEF$?

8. Give the polar coordinates of the points that result if each of the following points are rotated $117°$.

   (a) $A(9, 37°)$

   (b) $B(2, 135°)$

   (c) $C(5, 216°)$

   (d) $D(7, 345°)$

9. Give the polar coordinates of the points that result if each of the following points are rotated $-123°$.

   (a) $E(6.5, 57°)$

   (b) $F(3.7, 153°)$

   (c) $G(2.5, 261°)$

   (d) $H(4.7, 297°)$

In Exercises 12–15 use Figure 11.99.

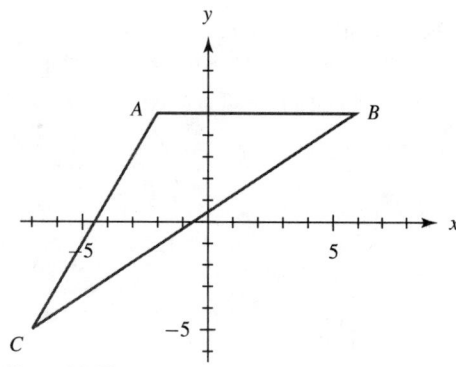

Figure 11.99

12. **(a)** What are the coordinates of A, B, and C?

   (b) Rotate triangle ABC through an angle of $90°$ to form a new triangle $A'B'C'$.

   (c) What are the coordinates of A', B', and C'?

13. **(a)** Rotate triangle ABC through an angle of $180°$ to form a new triangle DEF.

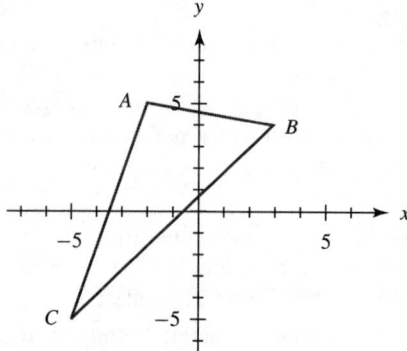

Figure 11.98

10. Give the Cartesian coordinates of the points that result if each of the following points are rotated $117°$.

   (a) $A(9, 5)$

   (b) $B(2, -6)$

   (c) $C(-5, 7)$

   (d) $D(-7, -3)$

11. Give the Cartesian coordinates of the points that result if each of the following points are rotated $-123°$.

   (a) $E(-6.5, 5.7)$

   (b) $F(-3.7, -1.5)$

   (c) $G(2.5, 2.6)$

   (d) $H(4.7, -7.9)$

   (b) What are the coordinates of D, E, and F?

14. **(a)** Rotate triangle ABC through an angle of $45°$ to form a new triangle $D'E'F'$.

   (b) What are the coordinates of D', E', and F'?

15. **(a)** Rotate triangle ABC through an angle of $-45°$ to form a new triangle GHI.

   (b) What are the coordinates of G, H, and I?

16. For this exercise use Figure 11.100.

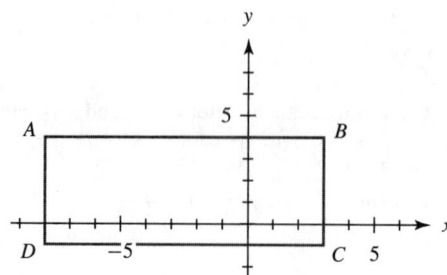

Figure 11.100

(a) Rotate rectangle $ABCD$ through an angle of $45°$ to form a new rectangle $A'B'C'D'$.

(b) Use an x-scaling factor of -1 and a y-scaling factor of 2.5 to transform $A'B'C'D'$ into rectangle $A''B''C''D''$.

(c) Use an x-scaling factor of -1 and a y-scaling factor of 2.5 to transform $ABCD$ into rectangle $EFGH$.

(d) Rotate rectangle $EFGH$ through an angle of $45°$ to form a new rectangle $E'F'G'H'$.

(e) Compare rectangles $A''B''C''D''$ and $E'F'G'H'$. How are they alike? How are they different? How do they compare to rectangle $ABCD$?

In Exercises 17–20 use matrix $B = \begin{pmatrix} 4 & -2.4 & 1.8 \\ 0.5 & 0 & -3.2 \\ 0 & -0.1 & 1.4 \\ 3.5 & 7.1 & 5.4 \end{pmatrix}$.

17. What are the dimensions of B?

18. $B(4, 2) =$

19. $b_{23} =$

20. $b_{32} =$

21. What is the dimension of AB if A is a 5×7 matrix and B is a 7×3 matrix?

22. What is the dimension of Y if X is a 4×6 matrix and XY is a 4×2 matrix?

In Exercises 23–28 use matrix $C = \begin{pmatrix} 2.4 & 0.1 & 1.2 \\ 4.2 & 5.7 & -3.2 \\ -1.3 & 0 & 12 \end{pmatrix}$ and $D = \begin{pmatrix} -0.2 & 4.3 & 2.8 \\ -5.2 & 3.6 & -6.7 \\ 1 & 0 & 2.5 \end{pmatrix}$.

23. What are the dimensions of $E = CD$?

24. $e_{12} =$

25. $e_{23} =$

26. $e_{32} =$

27. $CD =$

28. $DC =$

In Exercises 29 and 30 use Figure 11.101.

29. *Computer Aided Design (CAD)*

(a) What are the coordinates of A, B, and C?

(b) Rotate triangle ABC through an angle of $75°$ to form a new triangle $A'B'C'$.

(c) What are the coordinates of A', B', and C'?

(d) Graph $\triangle ABC$ and $\triangle A'B'C'$ on the same set of axes.

30. *Computer Aided Design (CAD)*

(a) Rotate $\triangle ABC$ through an angle of $-15°$ to form a new triangle DEF.

(b) What are the coordinates of D, E, and F?

(c) Graph $\triangle ABC$ and $\triangle DEF$ on the same set of axes.

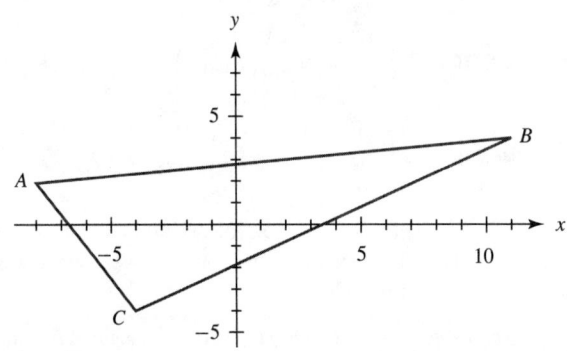

Figure 11.101

In Exercises 31 and 32 use Figure 11.102.

31. *Computer Aided Design (CAD)*

(a) What are the coordinates of D, E, and F?

(b) Rotate triangle DEF through an angle of $140°$ and then use an x-scaling factor of 1.25 and a y-scaling factor of -0.75 to form a new triangle $\triangle D_1 E_1 F_1$.

(c) What are the coordinates of D_1, E_1, and F_1?

(d) Graph $\triangle DEF$ and $\triangle D_1 E_1 F_1$ on the same set of axes.

32. *Computer Aided Design (CAD)*

(a) First apply an x-scaling factor of 1.25 and a y-

scaling factor of -0.75 to $\triangle DEF$ and then rotate the result through an angle of $140°$ to form a new triangle $\triangle D_2 E_2 F_2$.

(b) What are the coordinates of D_2, E_2, and F_2?

(c) Graph $\triangle DEF$ and $\triangle D_2 E_2 F_2$ on the same set of axes.

(d) Compare $\triangle D_2 E_2 F_2$ with $\triangle D_1 E_1 F_1$ from Exercise 31.

33. What is the multiplicative inverse of 2.5?

34. What is the multiplicative inverse of $\frac{3}{17}$?

35. If $A = \begin{pmatrix} 5 & 2 \\ 4.75 & 2.5 \end{pmatrix}$, calculate the determinant of A.

 36. If $B = \begin{pmatrix} 3.2 & -5.4 & 1.4 \\ 0 & 9.5 & 2 \\ 0 & 0 & -1 \end{pmatrix}$, calculate the determinant of B.

37. If $A = \begin{pmatrix} 2.3 & -25 & 0.1 & 7.1 \\ 0 & -1.6 & 3.8 & 19.4 \end{pmatrix}$, determine A^T.

38. If C is a 5×2 matrix, what is the dimension of C^T?

39. Explain the function of the selection `3:dim(` in the MATRIX Math menu.

40. Explain the function of the selection `1:det(` in the MATRIX Math menu.

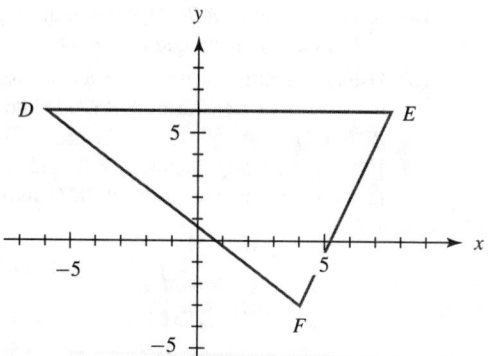

Figure 11.102

In Exercises 41–44 use matrices to solve each of the following systems of linear equations.

41. $\begin{cases} 9x - 5y = -57 \\ -3x + 5y = 27 \end{cases}$

42. $\begin{cases} 2 - 5x + 2.4y = 25.08 \\ 2.4x - y = -11.4 \end{cases}$

43. $\begin{cases} 2.1x + 3.2y - 1.4z = -5.44 \\ 3.5x - y + 0.2z = -11.98 \\ x - 5z = -28.5 \end{cases}$

44. $\begin{cases} 2.5x + 3.5y - 4.5z = -19.5 \\ 3.5x + 4.5y - 1.5z = -2.5 \\ 4.2x + 3.8y + 8z = 49.2 \end{cases}$

45. If $A = \begin{pmatrix} 3 & 2 & -1 \\ 4 & 3 & -2 \\ 0 & 2 & 1 \end{pmatrix}$, calculate A^2.

46. If $B = \begin{pmatrix} 5 & 0 & -1 \\ -2 & 0 & 8 \\ 0 & -2.5 & 1 \end{pmatrix}$, calculate B^4.

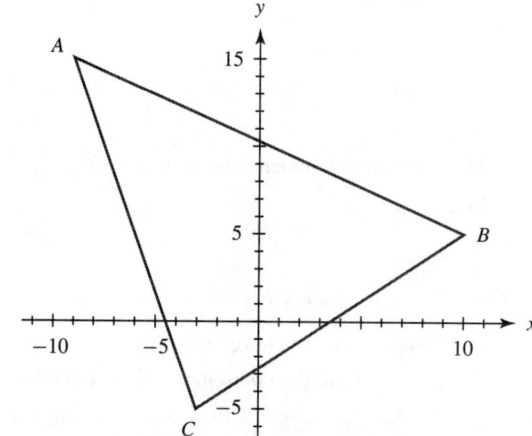

Figure 11.103

47. *Computer Aided Design (CAD)* Suppose that a rotation of $48°$ produced the new point $(2, -3)$. What was the original point?

48. *Computer Aided Design (CAD)* Triangle ABC in Figure 11.103 is the result of transforming $\triangle A_1 B_1 C_1$ through a rotation of $75°$ and then applying an x-scaling factor of 1.5 and a y-scaling factor of -2.5.

(a) What are the coordinates of A, B, and C?

(b) Write the matrices that were used to transform $\triangle A_1 B_1 C_1$ into $\triangle ABC$.

(c) Write the matrices that you would use to transform $\triangle ABC$ back into $\triangle A_1 B_1 C_1$.

(d) Graph ABC and $A_1 B_1 C_1$ on the same set of axes.

In Exercises 49–52 let the point $(0, 0, 0)$ represent the location of your eyes in a three-dimensional coordinate system. If you are looking straight ahead, then you are looking along the positive x-axis. The positive y-axis is to your left and the positive z-axis is above. Describe the location of each of the following points if all units are in meters.

49. $P(-2, 0, 0)$

50. $Q(-2, 2, 0)$

51. $R(-2, 2, -2)$

52. $T(1, -2, 3)$

In Exercises 53 and 54 compute the distance between the given point and the origin.

53. $P(-3, 4, -5)$

54. $Q(1.5, -2.5, 6.2)$

In Exercises 55 and 56 determine the distance between the two given points.

55. $P(2, 3.2, -3)$ and $Q(-8, -6.8, 7)$

56. $C(2.1, -4.6, 1.5)$ and $D(2.6, 1.25, 3.5)$

57. *Transportation* An airplane taking off from Chicago, Illinois, flying with a constant velocity at an altitude of 25,500 feet, is 45 miles west and 50 miles north of Chicago O'Hare Airport. The components of its velocity are 375 mph west, 52 mph north, and 2 mph up.

(a) How far is the plane from Chicago?

(b) After flying for ten minutes at this rate, what is the plane's location and distance from Chicago?

(c) How far did the plane travel during this ten-minute period?

(d) What was the airplane's actual speed in miles per hour over this ten-minute period?

In Exercises 58–61 analyze the equation $x = 2.5$ in a three-dimensional coordinate system.

58. Give four points that satisfy the equation.

59. At what point(s) will the graph cross the x-axis.

60. At what point(s) will the graph cross the y-axis.

61. Describe the graph of $x = 2.5$.

In Exercises 62–65 analyze the graph of $z = 10 - 2x + y$.

62. List four points that are on this graph.

63. At what point(s) will the graph cross the y-axis?

64. At what point(s) will the graph cross the z-axis?

65. Describe the intersection between the graph of this equation and the xz-plane.

66. (a) Describe the intersection of $z = 10 - 2x + y$ and $x = 2.5$.

(b) Give the coordinates of four points that lie on the intersection of $z = 10 - 2x + y$ and $x = 2.5$.

67. (a) Describe the intersection of $z = 10 - 2x + y$ and $y = 5$.

(b) Give the coordinates of four points that lie on the intersection of $z = 10 - 2x + y$ and $y = 5$.

In Exercises 68–71 use the parametric equations $\begin{cases} x = t \\ y = \cos t \\ z = \sin t \end{cases}$

68. Describe the graph of these parametric equations.

69. Sketch a side- and a 3-D view of the curve formed by these parametric equations.

70. If P is $t = -2.5$ and Q is $t = 16.25$, what is the straight line distance between P and Q?

71. What is the average speed along the straight line from P to Q?

In Exercises 72 and 73 use the parametric equations $\begin{cases} x = \cos t \\ y = 2 + \sin 3t \\ z = t \end{cases}$

72. If R is $t = 0$ and S is $t = 5$, what is the straight line distance between R and S?

73. What is the average speed along the straight line from R to S?

In Exercises 74–77 describe the graphs of the given equations in spherical coordinates.

74. $r = 4$

75. $\theta = \frac{\pi}{4}$

76. $\phi = \frac{\pi}{3}$

77. $2 \le r \le 6$

 78. Use your calculator to plot a family of curves for the spherical equation $z = \sin 1.5\theta - 3\cos 2\phi$ where $\theta = 0$, 0.5, 1.0, 1.5, 2.0, 2.5, and 3.0.

79. Use your calculator to plot a family of curves for the spherical equation of Exercise 78: $\sin 1.5\theta - 3\cos 2\phi$ where $\phi = 0$, 0.5, 1.0, 1.5, 2.0, 2.5, and 3.0.

In Exercises 80–83 describe the graphs of the given equations in cylindrical coordinates.

80. $r = 2.5$

81. $\theta = \frac{\pi}{6}$

82. $z = -3$

83. $2.5 \le r \le 6$

84. *Computer Aided Design (CAD)* Suppose that the point $P(2, 3, 5)$ is scaled with factors $m = 1.75$, $n = -0.75$, and $p = 1.25$. Determine the coordinates of P', the point that results after the scaling.

85. *Computer Aided Design (CAD)* Suppose that the point $Q(3, 1, 4)$ is scaled with factors $m = 2.4$, $n = -3.2$, and $p = 0.1$. Determine the coordinates of Q', the point that results after the scaling.

86. *Computer Aided Design (CAD)* If point $R'(4, 3, -2)$ is the result of a point R being scaled with factors $m = -3$, $n = 1.5$, and $p = 4$, what are the coordinates of the original point R?

87. *Computer Aided Design (CAD)* If point $S'(1.6, -5.2, 4.8)$ is the result of a point S being scaled with factors $m = -4$, $n = 2$, and $p = -0.25$, what are the coordinates of the original point S?

88. *Computer Aided Design (CAD)* Consider the segment $\overline{AB}$ where $A = (1, -5, 4)$ and $B = (6, 1, 9)$ and suppose the $\overline{AB}$ is scaled with factors $m = 2.5$, $n = 0.75$, and $p = 1.5$. The result is segment $A'B'$.

   (a) What is the length AB?

   (b) What are the coordinates of A' and B'?

   (c) What is the length $A'B'$?

89. *Computer Aided Design (CAD)* The point $P(2.5, 3.1, 7.2)$ is rotated through an angle of $65°$ in the xy-plane to the point P'. Determine the coordinates of P'.

90. *Computer Aided Design (CAD)* The point $Q(-5.2, 4.6, 8.0)$ is rotated through an angle of $-125°$ in the xy-plane to the point Q'. Determine the coordinates of Q'.

91. *Computer Aided Design (CAD)* If point $R'(4, 8, 12)$ is the result of a rotation of $-55°$ in the xy-plane, what are the coordinates of the original point?

92. *Computer Aided Design (CAD)* If point $S'(-6, 4, 2)$ is the result of a rotation of $125°$ in the xy-plane, what are the coordinates of the original point?

93. *Computer Aided Design (CAD)* Suppose that the point $P(8, 4, -6)$ is rotated through an angle of $24°$ in the xy-plane and scaled with factors $m = 2$, $n = 2.5$, and $p = -1$.

   (a) Determine the coordinates of the resulting point P_1 if P is first rotated and then scaled.

   (b) Determine the coordinates of the resulting point P_2 if P is first scaled and then rotated.

94. *Physics* Masses of $9\,\text{kg}$ and $11\,\text{kg}$ are attached to a cord that passes over a frictionless pulley. When the masses are released, the acceleration a of each mass [in meters per second squared (m/s^2)] and the tension T in the cord [in newtons (N)] are related by the system

$$\{T - 75.4 = 9a \quad 100.0 - T = 11a$$

Find a and T.

95. *Business* A computer company makes three types of computers: a personal computer (PC), a business computer (BC), and a technical computer (TC). There are three parts in each computer that they have difficulty getting: RAM chips, EPROMS, and transistors. The number of each part needed by each computer is shown in the table.

	RAM	EPROM	Transistor
PC	4	2	7
BC	8	3	6
TC	12	5	11

If the company is guaranteed 1,872 RAM chips, 771 EPROMS, and 1,770 transistors each week, how many of each computer can be made?

96. Given the equations

$$\left\{ x' = \frac{1}{2}\left(x + y\sqrt{3}\right) \quad y' = \frac{1}{2}\left(-x\sqrt{3} + y\right) \right.$$

and

$$x'' = \frac{1}{2}\left(-x' + y'\sqrt{3}\right)$$
$$y'' = -\frac{1}{2}\left(x'\sqrt{3} + y'\right)$$

write each set as a matrix equation and then solve for x'' and y'' in terms of x and y by multiplying matrices.

97. The equations in Exercise 96 represent rotations of axes in two directions. If the angle of rotation is θ we know that

$$x'' = x\cos\theta + y\sin\theta$$
$$y'' = -x\sin\theta + y\cos\theta$$

What was the rotation angle for the equations in Exercise 96?

98. *Optics* The following matrix product is used in discussing two thin lenses in air:

$$M = \begin{pmatrix} 1 & -\dfrac{1}{f_2} \\ 0 & 1 \end{pmatrix} \begin{pmatrix} 1 & 0 \\ d & 1 \end{pmatrix} \begin{pmatrix} 1 & -\dfrac{1}{f_1} \\ 0 & 1 \end{pmatrix}$$

where f_1 and f_2 are the focal lengths of the lenses and d is the distance between them. Evaluate M.

99. *Optics* In Exercise 98, element M_{12} of M is $-\dfrac{1}{f}$ where f is the focal length of the combination. Determine $\dfrac{1}{f}$.

●●●●●●●● Chapter 11 Test

1. Using an x-scaling factor of 2 and a y-scaling factor of -1.5, calculate the transformed point if the original point is $A(-2, 5)$.

2. Using an x-scaling factor of 2.5, and a y-scaling factor of 0.25, calculate the original point if the transformed point is $A(-5, 6)$.

3. Give the polar coordinates of the point that results if the point $P(5, 153°)$ is rotated $24°$.

4. Give the Cartesian coordinates of the point that results if the point $B(1, 4)$ is rotated $66°$.

5. Refer to Figure 11.104.

   (a) What are the coordinates of A, B, and C?

   (b) Rotate triangle $\triangle ABC$ through an angle of $60°$ to form a new triangle $A'B'C'$.

   (c) What are the coordinates of A', B', and C'?

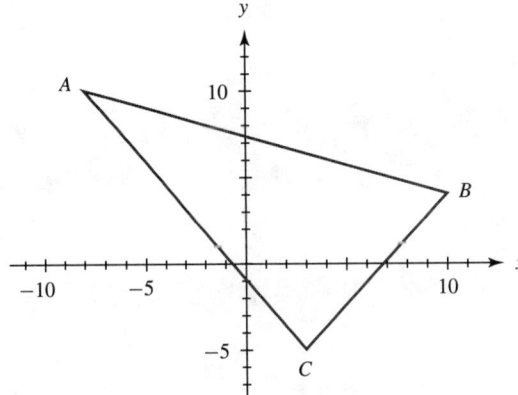

Figure 11.104

In Exercises 6–11 use matrix $A = \begin{pmatrix} 4.5 & 5.7 & 9.2 \\ 0.5 & 0 & -1.2 \\ 0 & -0.5 & 0 \end{pmatrix}$ *and* $B = \begin{pmatrix} 1.5 & 0 & 1.2 \\ 0.25 & 0 & -0.75 \\ 0 & 5 & 1 \end{pmatrix}$.

6. What are the dimensions of A?

7. $a_{23} =$

8. What is the determinant of A?

9. What is the dimension of AB if A is a 3×7 matrix and B is a 7×4 matrix?

10. $AB =$

11. $BA =$

12. *Computer Aided Design (CAD)* Suppose that $\triangle ABC$ has coordinates $A(1, 2)$, $B(-5, 4)$, and $C(6, 3)$.

    (a) First apply an x-scaling factor of 0.25 and a y-scaling factor of 2 to $\triangle ABC$; then rotate the result through an angle of $75°$ to form a new triangle $\triangle A_1 B_1 C_1$.

    (b) What are the coordinates of A_1, B_1, and C_1?

13. What is the multiplicative inverse of 0.4?

14. Solve the following system of linear equations.

$$\begin{cases} -x + 6y - 2z = -18 \\ 3x - y + 4x = 27 \\ 4x + 3y + 6z = 35 \end{cases}$$

15. Consider the parametric equations

$$\begin{cases} x = 1 - \cos t \\ y = t \\ z = \sin 2t \end{cases}$$

If P is $t = 0$ and Q is $t = 10$, what is the straight line distance between P and Q?

16. Use your calculator to plot a family of curves for the spherical equation $z = 1 - 2 \sin 3\theta - \cos \phi$ where $\theta = 0$, 0.5, and 3.

17. Suppose that the point $P(2, 4, 6)$ is scaled with factors $m = 1$, $n = 0.75$, and $p = 2.5$. Determine the coordinates of P', the point that results after the scaling.

18. *Electronics* Find the current of the system in Figure 11.105.

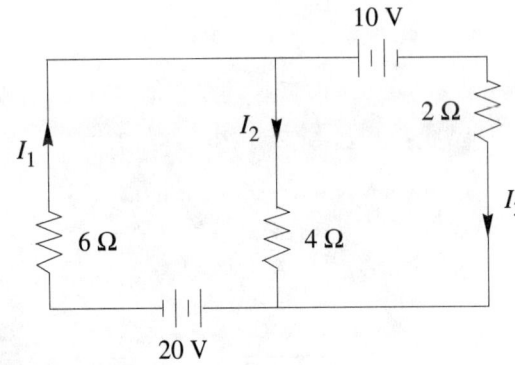

Figure 11.105

CHAPTER

12

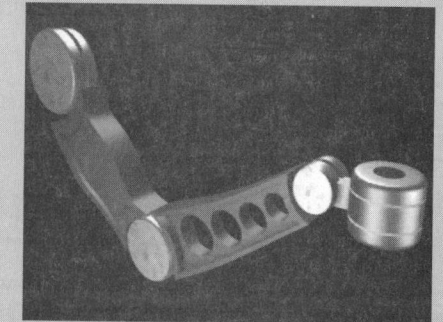

Modeling with Probability and Statistics

Topics You'll Learn or Review

- Data analysis
  - mean, median, quartiles
  - frequency table and histogram
  - box plot
- Probability
  - definition
  - odds
  - expected value
  - application to games of chance
- Probability of more than one event
- Counting combinations
- Relative error
- Law of large numbers

Mathematics Skills You'll Use

- Converting fractions to decimals
- Adding and multiplying fractions

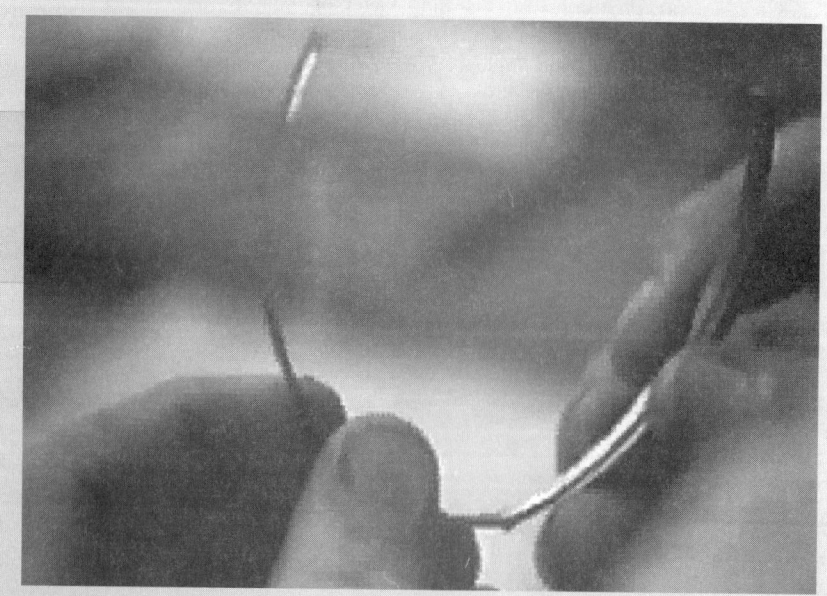

```
(50 nCr 4)*0.1^4
*0.9^46
        .1809045009
(50 nCr 3)*0.1^3
*0.9^47
        .138565 1496
(50 nCr 2)*0.1^2
*0.9^48
```

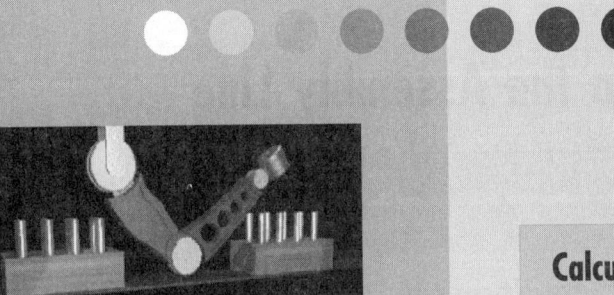

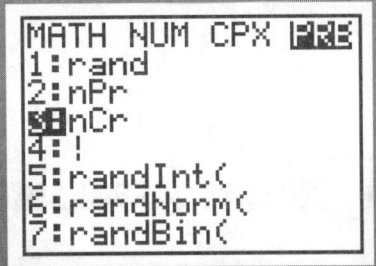

```
MATH NUM CPX PRB
1:rand
2:nPr
3:nCr
4:!
5:randInt(
6:randNorm(
7:randBin(
```

Calculator Skills You'll Learn

▶ One variable statistical analysis: (1-Var Stat)
▶ Plotting histograms and box plots: (Stat Plot)
▶ Computing numbers of combinations: ($_nC_r$)
▶ Producing random integers with randInt(

```
Plot1  Plot2  Plot3
On Off
Type:
Xlist:FAIL1
Freq:1
Mark: ▪  +  .
```

Calculator Skills You'll Need

▶ Knowledge of how to run a program
▶ Entering data into a list
▶ Graphing with STAT PLOT

What You'll Do in This Chapter

In this chapter you will see how mathematics can help to clarify the understanding of chance occurrences. With the mathematics of probability you will learn how to predict the likelihood of uncertain events and the number of times an uncertain event is likely to occur in the long run. We will use probability to analyze your chance of winning some games of chance and to monitor the quality of products coming off an assembly line.

```
P1:L2,L1
min=9
max<10              n=16
```

How many combinations
have two failures out of four?

First one is F

Three combinations have
one F out of three

```
F   F P P
F   P F P
F   P P F
```

Chapter Project—Assessing Quality on the Assembly Line

Custom Circuits Corporation
"Your design is our command"

To: Research department
From: Management
Subject: Designing board tests

Our new contract with Ace PCs contains a penalty clause. If more than
10% of our circuit boards fail Ace's rigorous testing, we risk losing this
profitable business.

Manufacturing has promised that they have developed new methods that
will bring the failure rate under 10%. Now we need a testing method that
can help us monitor the failure rate of the boards as they come off the line.
Of course we already do our own testing of all boards, but it would be
unprofitable to test all boards with the expensive equipment Ace requires.

Please design a daily testing process that will tell me with 95% certainty
that we are under the 10% failure rate. How many should we test (that is,
how big should the sample be), and how sure can we be from the result of
each sample that the new manufacturing process is meeting its target?

For example, if we decided to test 50 boards and 4 of them were found to be
defective, how sure could we be that the failure rate is really less than 10%?

Preliminary Analysis

"Quality Is Job One," says an advertisement for a car company. All companies try to
minimize the number of products that do not meet the quality requirements of their
customers. The problem of estimating the percent of a company's products that do
not meet the required quality standards can be difficult. Although a company could
test every single item that comes off the assembly line, this is usually impractical.
For example, how would you test a light bulb to determine the number of hours until
it burns out? Whatever the test, it leaves you with a useless bulb. Such destructive

testing must be done sparingly. For products that require destructive testing, the approach is to test a **representative sample** of the items. A representative sample is a group of items that we assume has the same level of quality as the entire **population** of items.

We are led, then, to questions like these:

1. How big should this sample be? That is, what **sample size** should be used?
2. How can we estimate the failure rate in the population from the measured failure rate in the sample?
3. With what confidence can we estimate the failure rate in the population?

In this chapter we will study some methods of probability and statistics that can be used to answer such questions.

The Chapter Project states that the failure rate on the manufacturing line was 10% and that management ordered that the process be improved. You are in charge of testing to see if the process has improved or if it is still 10%. You look at 50 items and find 4 failures. Can you be sure that the failure rate is now below 10%?

In the following activities we will model quality testing in two ways. In Activity 12.1 you will take samples directly, and in Activity 12.2 you will use a calculator program to generate the samples.

In Activity 12.1 you will use a bag of M&Ms® to begin to study the problem of estimating the failure rate of products from an assembly line. We'll use a 1.69 oz. bag of plain M&Ms, which contains about 60 candies in six different colors. Because there is a variation in the number of each color in a bag, we are uncertain about the percent of each color in an unopened bag. Therefore the bag can be used to represent a group of products being manufactured. In this experiment the bag represents a population of products and the unknown percent of one color can represent the unknown defective rate on the assembly line.

There are two benefits to using M&Ms in this experiment. First we can look inside the bag at the end of the experiment and find out the true percents. In real life we usually cannot know the truth about a population, so we must use samples and the mathematics of probability and statistics to make guesses about the population. The second benefit of experimenting with M&Ms is that you can eat them when you have finished.

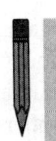

Activity 12.1
Modeling Quality Control With M&Ms

We'll choose one color to represent a defective product. By looking at different groups (samples) of candies from the bag we can begin to see what it is like to investigate the failure rate in a large population.

Follow these steps to complete this activity:

1. Open the bag, but don't spill out the contents or look inside.
2. Begin by taking a random sample of five candies from the bag. To make sure that the sample is random (that is, any group of five candies is just as likely to be picked as any other group of five), do the following:

   ▶ Don't look as you select the five candies in the sample.
   ▶ Mix up the bag each time before choosing your sample.

3. When you have removed your sample of five candies, count the number of red candies and record your answer in Table 12.1 under sample number 1.

Table 12.1

Sample #*	1	2	3	4	5	6	7	8	9	10	Average
Number of Reds											
Percent of Reds											

*Sample Size = 5

4. Return your first sample of five candies to the bag, mix up the contents, and repeat the sampling process nine more times, each time using Table 12.1 to record the number of red candies in each sample.

5. Now let's see what kind of prediction you can make from this information.

   (a) What percent of the candies in your bag do you think are red? What is your reason for selecting this percent?

   (b) Write a sentence that describes your chances of picking a red candy from the bag. If you are accustomed to gambling, what odds would you give in a bet that one candy picked from the bag would be red?

   (c) What could you do to increase the certainty of your answers to the first two questions, without opening the bag and counting all the pieces?

   (d) Compare your table with two other people or groups. Why do you think other people's answers are different?

   (e) If there were 100 candies in the bag, how many do you think would be red?

   (f) If there were 57 candies in the bag, how many do you think would be red?

6. Repeat the experiment by taking ten samples of ten candies each and counting the number of yellow M&Ms. Record your results in Table 12.2.

Table 12.2

Sample #*	1	2	3	4	5	6	7	8	9	10	Average
Number of Yellows											
Percent of Yellows											

*Sample Size = 10

7. For your study of the yellow candies, answer again the questions from part 5. Are you more sure or less sure of your guesses about the yellow candies in your bag, compared with your guesses about the red candies? Why?

8. Now empty the bag and determine the actual numbers.

   N, number of candies in the bag, = _____
   R, number of red candies in the bag, = _____
   Y, number of yellow candies in the bag, = _____
   Actual percent of red candies in the bag = _____
   Actual percent of yellow candies in the bag = _____

9. Which of your estimates was more accurate, the estimate of the percent of yellows or the estimate of the percent of reds? Why do you think this happened?

   We would expect different groups to get different results from the experiments in Activity 12.1 for two important reasons that always occur in sampling experiments like this. Each of these reasons will be important for us to consider when we think about making estimates of failure rate by looking at a sample.

▶ Each bag of M&Ms is different, therefore each group was sampling from a different population.

- In estimating the failure rate of products from an assembly line, we will need to be aware of whether or not the products in one sample come from the same population as another sample, for example one taken on a different day or at a different plant site.

▶ Even if the bags were all identical, there still would be different results from each random sample, as you saw from your own ten samples.

- In estimating the failure rate on an assembly line, we must be aware of the varying results that are possible in different samples from the same population.

Finally, you noticed a variation among your samples. We will discuss this variation in Activity 12.2 and throughout this chapter, looking at the effect of the size of the sample on the accuracy of the estimates you can make from it.

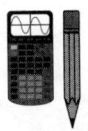

Activity 12.2
Modeling Quality Control With a Calculator Program

The *TI-83* program SAMPLE5 models the process of looking at five representative samples from the assembly line to see how many items fail the test.

For this activity, students should work in pairs with each pair having its "Student A" and "Student B."

▶ Student A's job is to select a failure rate (to the nearest 10%) for the population and decide the maximum number of items that can be tested.

▶ Student B's job is to guess the failure rate in the population by looking at the results of the five samples.

When the program is executed, Student A sets the failure rate as a percentage, in multiples of 10, without letting Student B know the number. The pair then decide how many total items can be tested in this run, and Student A enters that number. (Begin with a total of 50 items, and increase this number if Student B is having trouble guessing the failure rate entered by Student A.)

Student B then takes the calculator and decides on the sample size that will be used for each of the five samples (the size of each sample will be the same). There is a limit to the number of items Student B can enter. If N is the total number of items, then Student B must enter a number not larger than $\frac{N}{5}$.

The program then looks at each of the items and randomly decides whether each meets the quality requirement, rejecting items according to the failure rate entered by Student A*, and then reporting the number of items that failed in each of the five samples. From the resulting samples, Student B tries to guess the value of the failure rate.

After one run of the program, Student B can decide either to guess at the failure rate or to request another sample. If Student B is having a hard time guessing the rate, Student A should raise the maximum number of items that can be tested.

---

*The program uses a random number generator to decide whether to pass or fail an item. A random number between 0 and 100 is produced and if the number is less than or equal to the established failure rate, that item is a failure.

After Student B has an opportunity to guess the failure rate several times, the players should switch roles.

Answer the following questions based on your experience with this activity.

1. How does your ability to guess the failure rate change as the sample size changes?
2. How does your confidence about the failure rate change as the sample size changes?
3. Suppose that your job depends on your being 99% sure of your estimate of the failure rate. How big should the samples be? You can only say "yes" or "no", and you must be 99% confident that you are right. Based on your experience with the calculator:

▶ Do you think 5 samples of 10 would be good enough?
▶ Do you think 5 samples of 50 would be good enough?
▶ Do you think 5 samples of 100 would be good enough?
▶ Do you think 5 samples of 1000 would be good enough?

12.1

Looking at Data

In the Preliminary Analysis of the Chapter Project you looked at data and made some guesses based on what you saw. In this section you will learn some ways to organize and summarize data that can help you make better-informed decisions about the information contained in the numbers.

Numerical Analysis—Failure Rate

We will begin with the results of testing 20 samples, each containing 10 items. A total of 200 items were tested. The test determined whether each item passed the quality standards. Table 12.3 shows how many items in each sample of 10 failed the test.

Table 12.3

Sample Number	1	2	3	4	5	6	7	8	9	10
Number of Failures	7	4	3	2	2	3	7	3	5	4

Sample Number	11	12	13	14	15	16	17	18	19	20
Number of Failures	4	4	3	2	3	6	3	4	6	3

There are several numerical ways to analyze the data in Table 12.3. If you had to guess the percentage of failures in the population, what would it be? Make your guess now before you read further.

What is the average number of failures across the 20 samples? This is the **mean** of the data. The mean, or **average**, is the sum of the number of failures divided by the number of samples. The sum of the numbers of failures in Table 12.3 is 78 and the number of samples is 20, so the mean is $\frac{78}{20} = 3.9$. If you put the numbers of failures in order from lowest to highest, which number would be in the middle? This is the **median** of the data. Table 12.4 shows the data arranged in order of increasing numbers of failures.

Table 12.4

Sample Number	4	5	14	3	6	8	13	15	17	20
Number of Failures	2	2	2	3	3	3	3	3	3	3

Sample Number	2	10	11	12	18	9	16	19	1	7
Number of Failures	4	4	4	4	4	5	6	6	7	7

Since there are an even number of samples, 20, there is no middle number. In this case we take the average of the two middle numbers. For the case of 20 numbers we take the average of the 10th and the 11th numbers. The 10th number is 3 and the 11th number is 4. Therefore the median is 3.5.

What percent of the number of failures are higher than the mean? The mean is 3.9, and there are 10 entries higher than the mean. This is half the entries, or 50%. What percent of the number of failures are less than 5? There are 15 numbers less than 5, a percentage of $\frac{15}{20} = 75\%$.

The median divides the ranked numbers into two equal-sized parts, with half the numbers above the median and half below the median. The three **quartiles**, denoted Q_1, Q_2, and Q_3, divide the ranked numbers into four equal-sized parts. In general Q_1 separates the bottom quarter (or 25%) of the scores from the top three-fourths (or 75%), Q_2 is the median, and Q_3 separates the top 25% from the bottom 75% of the numbers.

Which numbers are below Q_1? That is, if you divide the ranked data into groups that each comprise 25% of the entries, which numbers are in the lowest quarter? Dividing 20 entries into quarters gives 5 entries in each quarter. Q_1 will be the average of the 5th and 6th scores. Since these are both 3, $Q_1 = 3$. From the ordered list you can see that the numbers 2, 2, 2, 3, and 3 are in the lowest quarter.

The second quartile, Q_2, is the median, so $Q_2 = 3.5$. The third quartile is the average of the 5th and 6th numbers from the top of the ranked data. Since these numbers are 4 and 5, we see that $Q_3 = \frac{4+5}{2} = 4.5$.

Make a **frequency table** for the numbers of failures. A frequency distribution shows how many times each number appears. There are three 2's, seven 3's, etc. A frequency table is shown in Table 12.5. Notice that the frequency table contains all the possible numbers of failures in a sample of 10, even though in this experiment those numbers did not occur.

Table 12.5

Number of Failures	0	1	2	3	4	5	6	7	8	9	10
Frequency	0	0	3	7	5	1	2	2	0	0	0

Now that you have done some more analysis of this data, what would be your guess of the percentage of failures in the population? Has it changed from your earlier quess?

The program that produced the data in Table 12.3 was given the value 40% for the failure rate. You might be surprised that in two cases there were 7 failures out of 10. Unexpected occurrences will always be with you whenever you take samples from a population—That's what makes people gamble. Our job in this chapter is to learn how to analyze random variation like this and learn how to minimize the chances of making incorrect conclusions from the data.

Example 12.1

The data in Table 12.6 contains the results of testing 15 samples, each containing 20 items. The test determined whether each item passed the test.

(a) What is the average number of failures in these 15 samples?
(b) What is the median of the data?
(c) What are the quartiles?

Table 12.6

Sample Number	1	2	3	4	5	6	7	8	9	10
Number of Failures	5	4	5	5	11	7	6	8	5	3

Sample Number	11	12	13	14	15
Number of Failures	6	8	7	6	7

Solutions (a) The total number of items that failed is 93. Since there were 15 samples, the average number of failures in these samples is $\frac{93}{15} = 6.2$.

(b) In order to determine the quartiles, and hence the median, we need to rank the numbers of failures. In increasing order they are

$$3, 4, 5, 5, 5, 5, 6, 6, 6, 7, 7, 7, 8, 8, 11$$

The median is the midpoint of these 15 samples. To determine the midpoint of an odd number of samples, in this case 15, divide the number of samples in half and round up: $\frac{15}{2} = 7.5$, which rounds up to 8. The median is the 8th of the ranked scores, in this case 6.

(c) In (b) we found that Q_2, the median, is 6. To find the other two quartiles, divide the number of samples by 4 and round up. There were 15 samples, and $\frac{15}{4} = 3.75$, which rounds up to 4. Thus Q_1 is the 4th ranked number from the bottom, so $Q_1 = 5$; Q_3 is the 4th ranked number from the top, so $Q_3 = 7$. ∎

Example 12.2

Recent real estate transactions listed in a local newspaper included the following prices (in dollars):

57,000	47,000	102,000	76,500	69,000	81,500
117,000	45,000	63,000	399,000	89,509	107,425
44,500	90,000	740,000	164,900	116,900	103,150

Determine the mean and median for these real estate transactions.

Solution The sum of these 18 real estate transactions is $2,513,384, and so the mean is $\frac{\$2,513,384}{18} \approx \$139,632$. The median is $89,754.50. ∎

Notice that in Example 12.2 there is almost $50,000 difference between the mean and the median. The two high-priced houses costing $740,000 and $399,500 cause the mean to be raised quite a bit. The other 16 real estate transactions have a mean of $85,899 and a median of $85,504.50. You will often find that census and real estate data report the median of the data rather than the mean.

Numerical Analysis—Throwing Two Dice

The next experience in data analysis deals with a common tool used in many games of chance: a pair of dice. When one *die* (the singular of *dice*) is thrown, one of the numbers 1 through 6 lands face up. If you roll two dice, the sum of the two face-up numbers is recorded.

Activity 12.3
First Thoughts about Dice

Think about the possible outcomes when two dice are thrown.

1. List the possible values for the sums of the two dice.
2. Which sums are *least* likely to appear? Why?
3. Which sums are *most* likely to appear? Why?
4. The five ways that the sums of the two dice can be 6 are listed in the following table:

First Die	1	2	3	4	5
Second Die	5	4	3	2	1

Complete the table below to show how many ways there are to get each of the other possibilities.

Sum	2	3	4	5	6	7	8	9	10	11	12
How Many Ways					5						

5. (a) If you rolled two dice 100 times, about how many times would you expect that the sums would turn out to be 12?
   (b) If you rolled two dice 100 times, about how many times would you expect that the sums would turn out to be 7?

In Example 12.3 we'll analyze 20 rolls of two dice. The results of this experiment are shown in Table 12.7.

Example 12.3

Answer the following questions for the data in Table 12.7.

(a) What is the average (mean) of the sums across the 20 tosses?
(b) What is the median of the data?
(c) What percent of the sums are higher than the mean?

Table 12.7

Toss Number	1	2	3	4	5	6	7	8	9	10
Sum	2	7	9	11	8	8	3	10	7	9

Toss Number	11	12	13	14	15	16	17	18	19	20
Sum	7	4	10	7	9	6	9	6	5	10

(d) What percent of the sums are less than 5?

(e) What are the quartiles?

(f) Make a frequency table for the sums.

Solutions   (a) The average, or mean, is the sum of the numbers of failures divided by the number of samples. The sum is 147 and the number of samples is 20, so the mean is $\frac{147}{20} = 7.35$.

(b) Table 12.8 shows the data arranged in order of increasing numbers of failures. The number of tosses is 20, an even number. To find the median of a set of 20 numbers, we take the average of the 10th and the 11th numbers. The 10th entry in the ordered list is 7 and the 11th number is 8. Therefore the median is 7.5.

Table 12.8

Toss Number	1	7	12	19	16	18	2	9	11	14
Sum	2	3	4	5	6	6	7	7	7	7

Toss Number	5	6	3	10	15	17	13	20	8	4
Sum	8	8	9	9	9	9	10	10	10	11

(c) The mean is 7.35 and there are 10 entries higher than the mean—50% of the entries.

(d) There are 3 sums less than 5, a percentage of $\frac{3}{20} = 15\%$.

(e) Dividing 20 entries into quarters gives 5 entries in each quarter. Q_1 is the average of the 5th and 6th numbers from the bottom, so $Q_1 = \dfrac{6+6}{2} = 6$; Q_3 is the average of the 5th and 6th numbers from the largest of the ranked numbers, so $Q_3 = \dfrac{9+9}{2} = 9$. In (b) we found that the median is 7.5, so $Q_2 = 7.5$.

(f) The frequency table is shown in Table 12.9.

Table 12.9

Sum	2	3	4	5	6	7	8	9	10	11	12
Frequency	1	1	1	1	2	4	2	4	3	1	0

Again, the frequency table contains all the possible sums, even though in this experiment some of the sums did not occur.   ∎

Analyzing Data with the *TI-83*

The calculator will produce many of the results from the previous examples. The instruction to use is called 1-Var Stats, found on the ▇STAT▇ CALC menu (Fig-

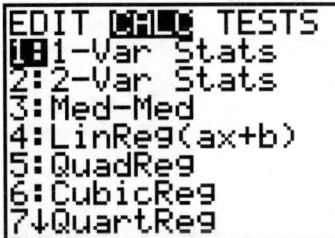

Figure 12.1

ure 12.1), the same menu that you have used whenever you needed to fit a regression line to data.

To use `1-Var Stats`, select it to bring that instruction into the home screen, then enter the name of the list. (For the data in Table 12.3 this list is called `FAIL1`). When you press (ENTER) you will see Figure 12.2, which shows the first part of the results, and Figure 12.3, which shows the second part. Press the ▼ or ▲ key to move between the two screens that contain the results of the `1-Var Stats` instruction.

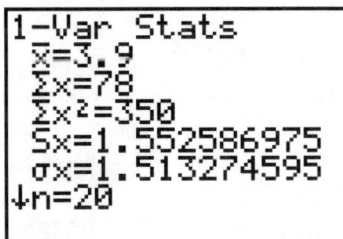

Figure 12.2

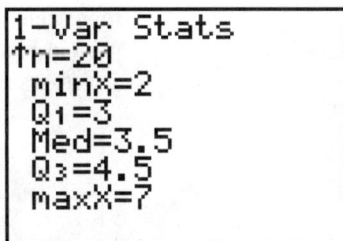

Figure 12.3

There are several new symbols on these two screens. In the first screen you will see the following:

▶ $\bar{x}$ represents the mean of the data.
▶ Σx and Σx^2 represent the sum of the numbers and the sum of the squares of the numbers. These are useful in some calculations, but we will not use them.
▶ Sx and σx are two forms of what is called the **standard deviation** of the set of numbers. The more the data is spread out, the larger the standard deviation. We will make use of the standard deviation in a later section.
▶ n is the number of entries in the set of data.

The symbols on the second screen relate to the quartiles.

▶ `minX` and `maxX` are the minimum and maximum values in the data set.
▶ Q_1, `Med`, and Q_3 refer to the quartiles, with `Med` being the median or Q_2.

Representing Data Graphically—The Histogram

In previous chapters you worked with calculator tables that gave the values of a function, and you have seen that the graph from such a table can give valuable information about the function. Likewise, although you can read a lot of information from a table, a glance at a graph can tell you a lot about the data. We will look at two kinds of graphs of data: the **histogram** and the **box plot**.

Histograms

One of the most common statistical graphs is the histogram. It is built from a frequency table. For example, the frequency table of failure rates in Table 12.5 is reproduced as Table 12.10. The histogram for this frequency table is shown in Figure 12.4. You can see at a glance that there were more failure rates of 3 out of 10 than any other, and that 4 was next most common.

Table 12.10

Number of Failures	0	1	2	3	4	5	6	7	8	9	10
Frequency	0	0	3	7	5	1	2	2	0	0	0

The Scale on a Histogram

Notice in Figure 12.4 how the bars are drawn with respect to the horizontal scale. Putting the number in the middle of the bar makes it clear which number the bar represents. However there is another way to write the horizontal scale. The scale used in Figure 12.5 is the one you will usually see if a calculator or computer draws a histogram.

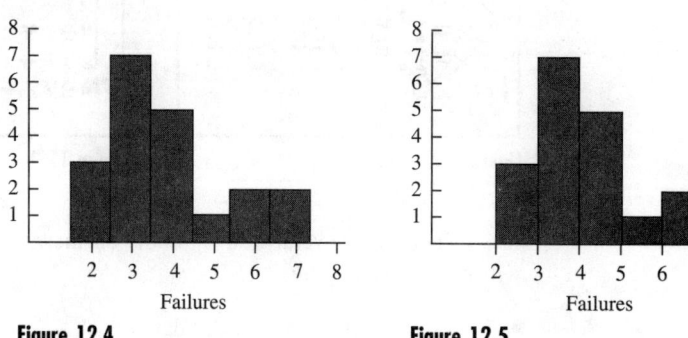

Figure 12.4   Figure 12.5

In experiments where the data might not be whole numbers, it is important to understand how to interpret the scale of the histogram you are using. In Figure 12.4 the bar above 2 represents all the failures F where $1.5 \leq F < 2.5$, and in Figure 12.5 the bar between 2 and 3 represents all the numbers $2 \leq F < 3$. Notice that in both cases the bar does not include the right-hand boundary number.

Plotting Histograms Using the TI-83

The *TI-83* can be used to plot histograms. Example 12.4 will use the *TI-83* to draw a histogram for the data from Table 12.10.

Example 12.4

Plot a histogram for the data from Table 12.10 with your calculator.

Solution The data for Table 12.10 is stored in a list named `FAIL1`. Access the STAT PLOTS menu by pressing `2nd` [STAT PLOT] and press `1` to activate the `Plot1` screen. In the `Plot1` menu, select the symbol for the histogram, as shown in the third row of Figure 12.6. The value for `Freq` is taken from the value of `Xscl` from the WINDOW menu.

In setting the dimensions of the graph window we could have used the window settings $0 \leq x \leq 11$ and $0 \leq y \leq 10$, because the numbers in the frequency table fit within those ranges. However we used `Xmin=-3` so that the Trace information would not cover part of the histogram. The graph is shown in Figure 12.7. Notice that the bar marked by the Trace cursor in the figure includes any values between 3 and 4, but not including 4, and that there are seven of these.

Figure 12.6

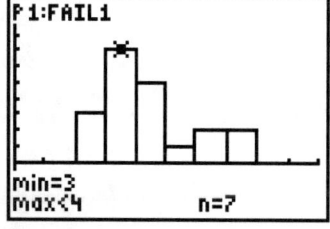

Figure 12.7

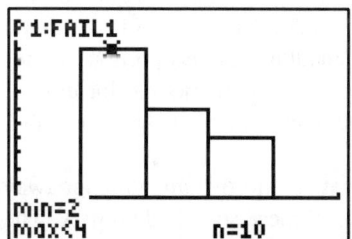

Figure 12.8

The bars on a histogram do not all have to have a width of 1. For example, in Figure 12.8 each of the bars has a width of 2. This histogram was generated using the same data used to generate Figure 12.7. Notice how the bar marked by the Trace cursor in the figure includes any values between 2 and 4, but not including 4, and that there are ten of these. If you look at Figure 12.7 you will see that there were three values between 2 and 3, but not including 3, and seven values between 3 and 4, but not including 4.

Activity 12.4
Thinking about Histograms

The data used in Tables 12.3 and 12.10 were collected from a calculator program named SAMPLEH. In this run we assumed that the average failure rate was 40%. Figure 12.9 shows the results of running the program SAMPLEH with 400 samples of 10 products, assuming a failure rate of 40%. This total of 4000 items to test took the calculator about four minutes to complete.

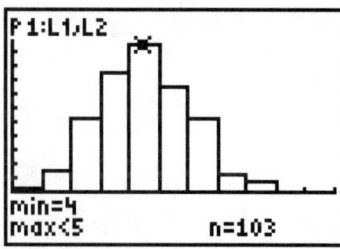

Figure 12.9

1. Look at the histograms in Figures 12.7 and 12.9.

   (a) How are they different?

   (b) How are they the same?

   (c) What more do you learn from doing 4000 tests as compared with doing 200 tests?

2. Sketch the histogram that you would predict to result from 4000 tests if the calculator program assumed a failure rate in the population of 60%.

   (a) Run the program SAMPLEH for 60% failure rate and for 400 samples of 10 products each; compare it with your answer. The program will run for about four minutes before the task is completed.

   (b) Explain any differences between your results and your sketch.

3. Sketch the histogram that you would predict to result from 4000 tests if the calculator program assumed a failure rate in the population of 10%.

(a) Run the program SAMPLEH for 10% failure rate and 400 samples of 10 products; compare it with your answer.

(b) Explain any differences between the results of the program and your sketch.

Representing Data Graphically—The Box Plot

Although a box plot may be less familiar to you than the histogram, it is quite useful, particularly when you want to compare more than one set of data. Figure 12.10 shows a box plot for the data in Table 12.10, plus some explanation. The *box* portion extends from Q_1 to Q_3, and the vertical dividing line within the box marks the location of the median, Q_2. The horizontal line segments (sometimes called *whiskers*) on either side of the box show the range of the numbers.

Some computer and calculator box plots treat data points that are quite far away from the median as **outliers**. Figure 12.10 shows one of these so-called outliers. The normal definition of *outlier* is a data point that is farther than 1.5 times the length of the box from either end of the box. The length of the box in Figure 12.10 is $Q_3 - Q_1 = 4.5 - 3 = 1.5$ and $(1.5)(1.5) = 2.25$. The upper end of the box is at $Q_3 = 4.5$ and $2.25 + 4.5 = 6.75$. Since $7 > 6.75$, the scores of 7 are treated as outliers—indicated by the mark to the right of the right-hand whisker.

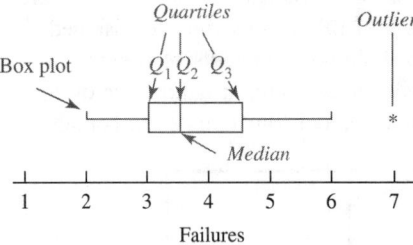

Figure 12.10

Plotting Box Plots Using the TI-83

Just as a *TI-83* can be used to plot a histogram, it can also be used to draw a box plot. The *TI-83* allows you to draw box plots that either do or do not show outliers.

Example 12.5

Plot a box plot for the data from Table 12.10 with your calculator and indicate the outliers.

Solution   The data for Table 12.10 are stored in a list named FAIL1. The box plot type we have selected (Figure 12.11) shows the outliers as in Figure 12.10. (The other box plot selection does not show outliers, it just extends the whiskers out to the lowest and highest values.) The window settings are the same as for the histogram, although when it plots a box plot, the calculator ignores the values of YMin and YMax. The graph is shown in Figure 12.12.

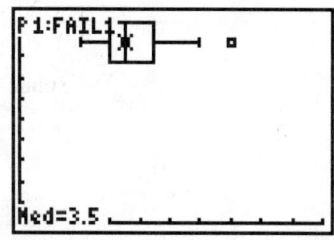

Figure 12.11

Figure 12.12

Figure 12.13 shows how the box plot can reveal the differences between sets of data. The first plot is the one for FAIL1; the second plot is from an analysis of the failure rate for 20 sets of 10 samples. You can see that these two plots have differences and similarities, and are likely to be from a similar experiment. The third box plot shows the data for the 20 tosses of 2 dice we studied in Example 12.3. If you didn't know the source of the data, the box plot would give a good hint that it came from a different kind of experiment.

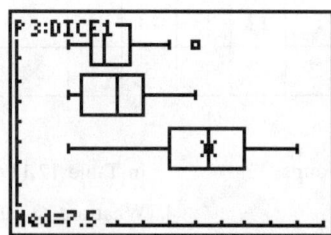

Figure 12.13

Activity 12.5
Thinking about Box Plots

Figure 12.14 shows box plots from three analyses of failure of products.

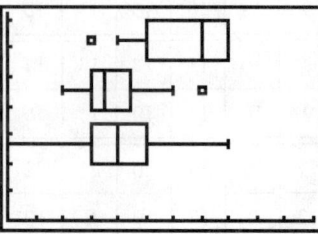

Figure 12.14

1. Which of the following tests does each box plot describe?
   (a) 400 samples of 10 products, with 40% failure rate in the population
       (*Hint:* In general if you increase the number of samples or the sample size, the distribution of the data should be more symmetric. That is, the values in the bottom quarter should more closely match the values in the top quarter, and likewise for the middle two quarters.)
   (b) 400 samples of 10 products, with 60% failure rate in the population

(c) 20 samples of 10 products, with 40% failure rate in the population

(d) 20 samples of 10 products, with 60% failure rate in the population

2. Explain your selections.

Section 12.1 Exercises

In Exercises 1–6 use the data in Table 12.11. The data in Table 12.11 show the results of testing 20 groups with 10 items in each group. Each item was tested, and Table 12.11 shows the number of items in each group that failed the test.

Table 12.11

Sample Number	1	2	3	4	5	6	7	8	9	10
Number of Failures	5	3	7	6	3	4	5	4	2	2
Sample Number	11	12	13	14	15	16	17	18	19	20
Number of Failures	3	5	3	5	4	2	6	4	2	2

1. What is the mean number of failures for the 20 groups in Table 12.11?

2. What percent of the number of failures are higher than the mean?

3. What is the median number of failures for the 20 groups

in Table 12.11?

4. What are Q_1 and Q_3?

5. Draw a histogram of the data in Table 12.11.

6. Make a box plot for the data.

In Exercises 7–12 use the data in Table 12.25. The data in Table 12.25 show the results of testing 24 groups with 40 items in each group. Each item was tested and Table 12.25 shows the number of items in each group that failed the test.

Table 12.12

Sample Number	1	2	3	4	5	6	7	8
Number of failures	9	12	10	15	13	10	9	9
Sample Number	9	10	11	12	13	14	15	16
Number of failures	9	16	15	10	11	9	16	9
Sample Number	17	18	19	20	21	22	23	24
Number of failures	12	9	17	19	9	10	8	9

7. What is the mean number of failures for the 20 groups in Table 12.25?

8. What percent of the number of failures are higher than the mean?

9. What is the median number of failures for the 20 groups

in Table 12.25?

10. What are Q_1 and Q_3?

11. Draw a histogram of the data in Table 12.25.

12. Make a box plot for the data.

In Exercises 13–18 use the data in Table 12.13. The data in Table 12.13 show the results of testing 20 groups with 10 items in each group. Each item was tested, and Table 12.13 shows the number of items in each group that failed the test.

Table 12.13

Sample Number	1	2	3	4	5	6	7	8	9	10
Number of Failures	16	5	9	9	13	13	15	8	12	9
Sample Number	11	12	13	14	15	16	17	18	19	20
Number of Failures	13	8	6	6	15	13	11	14	10	9

13. What is the mean number of failures for the 20 groups in Table 12.13?

14. What percent of the number of failures are higher than the mean?

15. What is the median number of failures for the 20 groups in Table 12.13?

16. What are Q_1 and Q_3?

17. Draw a histogram of the data in Table 12.13.

18. Make a box plot for the data.

In Exercises 19–22 use Figure 12.15. Figure 12.15 shows the results of testing 20 groups with 10 items in each group. Each item was tested, and the figure shows the number of items in each group that failed the test.

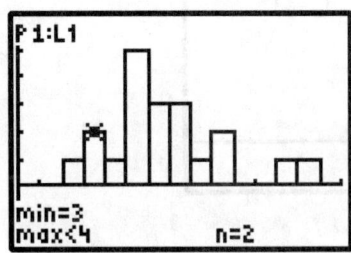

Figure 12.15

19. What is the mean for the 20 groups in Figure 12.15?

20. What is the median number of failures for the 20 groups in Figure 12.15?

21. What are Q_1 and Q_3?

22. Make a box plot for the data.

In Exercises 23–26 use Figure 12.16. Figure 12.16 shows the results of testing 20 groups with 10 items in each group. Each item was tested, and the figure shows the number of items in each group that failed the test.

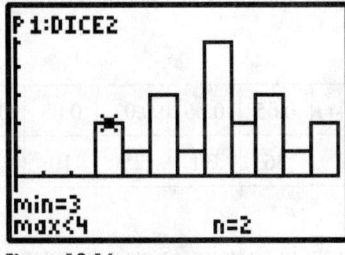

Figure 12.16

23. What is the mean for the 20 groups in Figure 12.16?

24. What is the median number of failures for the 20 groups in Figure 12.16?

25. What are Q_1 and Q_3?

26. Make a box plot for the data.

In Exercises 27 and 28 use Figure 12.17. Figure 12.17 shows the results of tossing a pair of dice 30 times and recording the 30 sums.

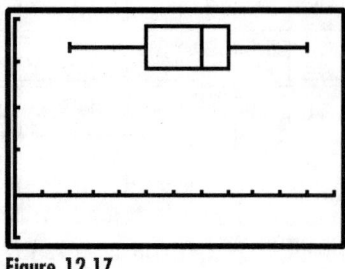

Figure 12.17

27. What is the median?

28. What are Q_1 and Q_3?

In Exercises 29 and 30 use Figure 12.18. Figure 12.18 shows the results of tossing a pair of dice 50 times and recording the 50 sums.

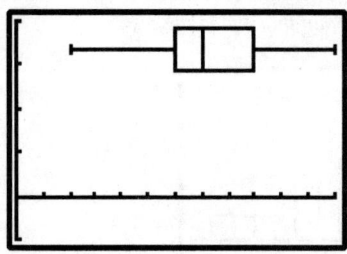

Figure 12.18

29. What is the median?

30. What are Q_1 and Q_3?

31. Determine Q_1 and Q_3 for the real estate data in Example 12.2.

32. Make a box plot for the real estate data in Example 12.2.

33. Is it possible to determine the mean for the dice tosses in Figure 12.18? If so, determine the mean. If it is not possible, explain why and tell what other information you would need to determine the mean.

34. Explain why you cannot make a frequency diagram from a box plot. Be sure to discuss any additional information that is needed and explain why it is needed.

35. *Industrial technology* A technician was measuring the thickness of a plastic coating on some pipe and obtained the following data.

Thickness (mm)	0.01	0.02	0.03	0.04	0.05	0.06	0.07	0.08	0.09
Frequency	1	5	40	50	36	30	25	10	3

Form a histogram of this data.

36. *Industrial technology* Form a histogram for the data in Exercise 35 over the intervals 0.01–0.02, 0.03–0.04, and so on.

37. *Industrial technology* Determine the mean, median, and quartiles for the data in Exercise 35.

38. *Industrial technology* Draw a box plot for the data in Exercise 35.

39. *Electrical technology* A technician tested an electric circuit and found the following values in milliamperes on successive trials:

5.24,   5.31,   5.42,   5.26,   5.31,   5.47,   5.27,   5.29,   5.35,   5.44
5.35,   5.31,   5.45,   5.46,   5.39,   5.34,   5.35,   5.46,   5.26,   5.27
5.47,   5.34,   5.28,   5.39,   5.34,   5.42,   5.43,   5.46,   5.34,   5.29

Form a frequency distribution for this data.

40. *Electrical technology* For the data in Exercise 39, draw a histogram for the intervals 5.21–5.25, 5.26–5.30, and so on.

41. *Electrical technology* Determine the mean and median for the data in Exercise 39.

42. *Electrical technology* Determine the quartiles and draw a box plot for the data in Exercise 39.

43. *Police science* A patrol officer using a laser gun recorded the following speeds for motorists driving through a 55 mph speed zone.

52	57	62	59	67	54
55	64	65	59	63	72

(a) Determine the mean, median, and quartiles for the given data.

(b) Draw a box plot for the given data.

44. *Environmental science* An environmental officer measured the carbon monoxide emissions (in g/m) for several vehicles. The results are shown in the following table.

5.02	12.36	13.46	6.92	7.44	8.52	12.82
11.92	14.32	12.06	8.02	11.34	6.66	9.28

(a) Determine the mean, median, and quartiles for the given data.

(b) Draw a box plot for the given data.

45. *Energy science* The carbon monoxide emissions (in g/m) were measured for several vehicles. The results are shown in the following table.

892	673	534	437	449	524
627	735	892	923	1024	905
865	704	624	535	432	495
572	625	655	684	532	484

Determine the mean, median, and quartiles for the given data.

46. *Insurance* The measurements of the blood alcohol content of 15 drivers convicted of and imprisoned for causing fatal accidents are given below.

0.14	0.16	0.21	0.10	0.13
0.19	0.26	0.22	0.13	0.09
0.11	0.18	0.12	0.24	0.27

Determine the mean, median, and quartiles for the given data.

47. *Business* A store's daily sales in dollars for one 31-day month are shown below.

24,562	38,646	43,988	15,122	14,321	17,479	19,478
25,625	39,476	45,353	15,972	13,793	17,457	18,681
20,562	38,606	53,788	15,122	10,321	13,037	17,038
25,625	37,036	45,353	15,732	13,373	13,053	18,681
21,903	41,775	52,117				

Determine the mean, median, and quartiles for the given data.

48. Explain *mean* and *median*. Include in your answer how they are alike, how they are different, and when it is better to use one rather than the other.

12.2 Modeling Uncertainty with Probability

Probability is a numerical measure of the chance that an **event** will occur. We often talk in general about uncertainty, using phrases like, "I'm pretty sure ...," "It looks like rain," "He wins more than half the time," "It's as certain as death and taxes," or "That's never going to happen!" With probability, mathematicians try to convert statements like these to numbers. We'll begin with a definition that will become clearer as we work through examples in this section.

The first step toward measuring the chance of an event is to consider all the other events that could occur in the situation. For example, a coin toss is usually modeled as resulting in one of two events: heads (H) or tails (T).

In the more complicated situations, we can often model the events of interest to us as being composed of **elementary events**. For example, we usually model the possible outcomes of two tosses of a coin to the following elementary events: HH, HT, TH, and TT. Furthermore we often model these elementary events as being equally likely. Since there are just four possibilities in this case, the chance of any one of them would be $\frac{1}{4}$. If the event we are interested in is that of getting one head and one tail, we see that exactly two of the elementary events yield this outcome: HT and TH.

Definition: Probability

The probability, P, of an event is equal to the fraction $\frac{M}{N}$, where M is the number of ways that the event can happen and N is the number of ways for the event or any other event to happen. This definition requires that all N events are equally likely.

Example 12.6

What is the probability of drawing an ace out of a 52-card deck of well-shuffled playing cards?

Solution   There are 4 aces, so $M = 4$, and 52 different cards could be drawn, so $N = 52$. The probability of drawing an ace out of a 52-card deck, or

$$P(\text{drawing an ace}) = \frac{M}{N} = \frac{4}{52} \qquad \blacksquare$$

From the definition you can see that P is a number between 0 and 1, inclusive. That is

$$0 \leq P \leq 1$$

If $P = 0$, the event is impossible, and if $P = 1$, the event is certain.

Example 12.7

What is the probability of drawing a fourteen of hearts from a 52-card deck of well-shuffled playing cards?

Solution   This probability is zero because there is no fourteen of hearts in a deck of playing cards. Since there is no fourteen of hearts $M = 0$ and

$$P(\text{fourteen of hearts}) = P(14\heartsuit) = \frac{M}{N} = \frac{0}{52} = 0 \qquad \blacksquare$$

Example 12.8

What is the probability of drawing a heart, spade, diamond, or club from a 52-card deck of playing cards?

Solution   This probability is 1 because each card is either a heart, spade, diamond, or club. Thus, $M = 52$ and $\dfrac{M}{N} = \dfrac{52}{52} = 1$.   ■

Translating English to Mathematics

The English statements in the left column in Table 12.14 are converted to mathematical statements on the right. Notice that the parentheses following P describe the event.

Table 12.14

English Statement	Mathematics Statement
The failure rate in this product line is 10%.	$P(\text{defective product}) = 0.1$
The chance of rain today is less than 50%.	$P(\text{rain today}) < 0.5$
I'm 100% sure of that.	$P(\text{I am right}) = 1$
I never win lotteries.	$P(\text{I will win a lottery}) = 0$
Your chances of winning are one in a million.	$P(\text{You will win}) = 0.000001$

Activity 12.6
Thinking about Probabilities

Make up some English statements about your job, school, or social life and rewrite them in the mathematical language of probabilities.

Example 12.9

Use the definition of probability to calculate the probability of each of the following events. In each case, give the values of M (the number of ways the event can happen) and N (the total number of possibilities). Write the probability as a fraction and as a decimal to two places.

From an ordinary 52-card deck of well-shuffled playing cards, what is the probability of drawing

(a) an ace of spades?
(b) a two of clubs?
(c) an ace of any kind?
(d) a heart?

(e) a red card?

(f) anything but a three?

Solutions   In all cases we are drawing one card out of a deck, so there are 52 possibilities and $N = 52$.

(a) There is 1 ace of spades, so $M = 1$ and $P(\text{ace}\spadesuit) = \frac{1}{52} \approx 0.02$.

(b) There is 1 two of clubs, so $M = 1$ and $P(\text{deuce}\clubsuit) = \frac{1}{52} \approx 0.02$.

(c) There are 4 different aces, so $M = 4$ and $P(\text{an ace}) = \frac{4}{52} \approx 0.08$.

(d) There are 13 hearts, so $M = 13$ and $P(\text{one }\heartsuit) = \frac{13}{52} = 0.25$.

(e) There are 26 red cards, so $M = 26$ and $P(\text{one red card}) = \frac{26}{52} = 0.50$.

(f) There are 48 cards that are not threes, so $P(\text{not a three}) = \frac{48}{52} \approx 0.92$. ∎

Example 12.10

Use the definition of probability to calculate the probability of each of the following events. In each case, give the value of M (the number of ways the event can happen) and N (the total number of possibilities). Write the probability as a fraction and as a decimal to two places.

In a bag of M&Ms you discover that there are a total of 57 candies and that 14 are yellow.

(a) If I reach in and pull out 1 M&M, what is the probability that it is yellow?

(b) In the same bag, the first one was not yellow. I ate it. What is the probability that the next one will be yellow?

(c) Later, if I have eaten 5 M&Ms and given away 30, and a total of 10 of them were yellow, what is the probability that the next one I pull out will be yellow?

(d) If I have eaten 50 M&Ms from the same bag and 14 of them were yellow, what is the probability that the next one I pull out will be yellow?

Solutions   In each case the value of N is equal to the number of M&Ms remaining in the bag.

(a) There are 57 M&Ms, so $N = 57$. There are 14 yellow M&Ms, so $M = 14$. Thus, $P(\text{yellow M\&M}) = \frac{14}{57} \approx 0.25$.

(b) Since I ate 1 M&M there are only 56 M&Ms left, so $N = 56$. There are still 14 yellow M&Ms, so $M = 14$. Thus, $P(\text{yellow M\&M}) = \frac{14}{56} = 0.25$.

(c) There are 22 left in the bag, so $N = 22$, and 4 of them are yellow ($M = 4$). $P = \frac{4}{22} \approx 0.18$.

(d) Only 7 M&Ms remain, so $N = 7$. All the yellow M&Ms have been removed, which means that $M = 0$. $P = \frac{0}{7} = 0.00$. ∎

How Many Ways Can It Happen—Adding Probabilities

Often you can solve a probability problem by counting the possibilities. Next you will use this approach to calculate the probability of getting a particular sum in a toss of two dice.

Activity 12.7
Probabilities with Dice

In Activity 12.3 you made a table of the number of different ways that two dice can produce each sum. For example, there are five ways that the sum can equal 6.

1. Complete that table if you have not already done so.
2. Add the numbers in the second row. The result is the number N, the total number of possibilities when two dice are thrown. Make sure your value of N is correct before going on to the next part.
3. Use the value of N and the entries in the second row of the table (M) to compute the probability of each sum. Do not reduce any of your fractions. Use a table like the following.

Sum	2	3	4	5	6	7	8	9	10	11	12
How many ways (M)					5						
Probability $\left(\frac{M}{N}\right)$					$\frac{5}{-}$						
Probability (nearest hundredth)											

4. Add up the probabilities across the third and fourth rows. Think about what probabilities mean and explain why your answers make sense.

Your answers to Activity 12.7 will help you understand the following properties of probabilities.

Properties of Probabilities—Addition

▶ If you add up the probabilities of all the possibilities in an experiment, the sum is equal to 1.
▶ If the probability of an event is p and the probability of the event *not* happening is q, then

$$p + q = 1$$

▶ If there is more than one way for an event to happen, the probability of the event is the sum of the individual probabilities.

Example 12.11

The probability of tossing a coin and getting heads is $\frac{1}{2}$ and the probability of tails is $\frac{1}{2}$. The sum of these probabilities is equal to 1, assuming there are no other possibilities. ∎

Example 12.12

If the probability of finding a defective product is 0.15, what is the probability of finding a good product?

Solution   The probability of finding a good product is $1 - 0.15 = 0.85$.   ■

Example 12.13

What is the probability of rolling 2 dice and getting a sum of 4?

Solution   The probability of getting a sum of 4 with 2 dice is

$$P(\text{sum is } 4) = P(3 \text{ and } 1) + P(1 \text{ and } 3) + P(2 \text{ and } 2)$$
$$= \frac{1}{36} + \frac{1}{36} + \frac{1}{36} = \frac{3}{36}$$

■

Example 12.14

If the probability of a coin landing on its edge is one in a million (probability $= 0.000001$), what is the probability of getting heads?

Solution   Now there are three possibilities—heads, tails, and edge. If the probability of getting heads is h and the probability of getting tails is t, then, assuming that heads and tails are equally likely, $h = t$, and we have the equation

$$h + t + 0.000001 = 1$$
$$h + h + 0.000001 = 1$$
$$2h = 0.999999$$
$$h = 0.4999995$$

■

Next we will look at how we can use probabilities to predict experimental results.

Expected Number: How Well Can We Predict?

We have studied probability as a fraction. Next you will see how to use that fraction to estimate the outcomes of an experiment, and how to judge the accuracy of the predictions. We will deal mostly with dice, but the ideas will be generally applicable to other experiments, whether with M&Ms, sampling from products on an assembly line, or thinking about whether the odds in a game are fair.

Example 12.15

If you roll two dice 100 times and record the sum each time, how many times to the nearest whole number would you expect the sum to be 6?

Solution   In Activity 12.7 you saw that the probability of getting a sum of 6 in one roll of two dice is $\frac{5}{36}$. If x is the number of 6's that would result in 100 rolls, we need to solve the proportion

$$\frac{x}{100} = \frac{5}{36}$$

$$x = 100 \times \frac{5}{36} \approx 14$$

Therefore in 100 rolls of two dice we would expect that 14 times the sum would be 6.   ■

Expected Number

If the probability of an event is p, and we have N identical experiments in which the event might occur, then the expected number of times the event would occur is

$$E(p, N) = pN$$

In Activity 12.8 you will see how the expected value varies for different values of the sum of two dice.

Activity 12.8
Graphing the Expected Number

In Activity 12.7 you made a table of the probabilities for each possible sum of two dice, 2 through 12. In this activity you will look at the graph of these probabilities.

1. Use the table of probabilities in Activity 12.7 to complete Table 12.15, expected numbers of each sum in 100 rolls of two dice.

Table 12.15   Expected Numbers in 100 Rolls of Two Dice

Sum:	2	3	4	5	6	7	8	9	10	11	12
Probability											
Expected Number											

2. Plot the Expected Number versus the Sum on a graph like the one shown in Figure 12.19.
3. Which sum is the most expected? Which sums are next highest in expected number?
4. What are the slopes of the two lines in the graph?

The probabilities and graph in Activities 12.7 and 12.8 are only true in theory. What happens when you actually throw 100 dice? In Calculator Lab 12.1 you will use the calculator program TWODICE to simulate many rolls of two dice. The program allows you to compare the actual number of occurrences of each sum with the expected number for that sum.

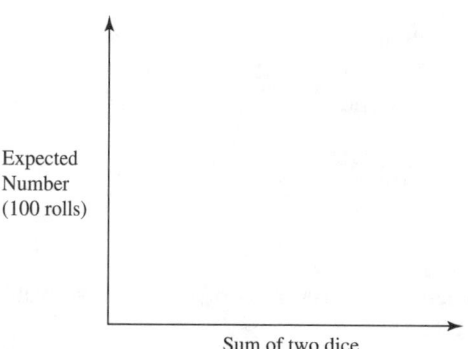

Expected
Number
(100 rolls)

Sum of two dice

Figure 12.19

Calculator Lab 12.1

Expected vs. Actual Results

In Activities 12.7 and 12.8 you explored the probabilities with two dice. The program TWODICE simulates any number of rolls of two dice. The program then allows you to compare the actual and the predicted values in a table and a graph.

Procedures

1. Start the program called TWODICE.
2. Enter 100 for the number of rolls.
   The calculator will simulate 100 rolls of two dice and make a frequency table similar to the one in Activity 12.8. The three lists displayed in Figure 12.20 are the sums (top row), actual results (middle row), and expected, or predicted, numbers (third row).

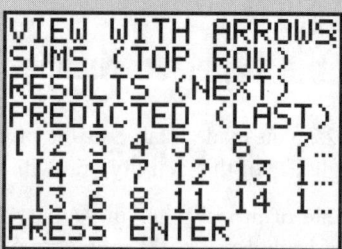

Figure 12.20

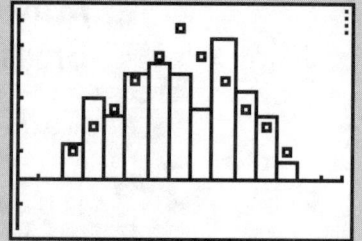

Figure 12.21

You can look at each of the entries in the lists by pressing the right or left arrow keys. When you have reviewed the results, press **ENTER** to see a graph of the same information.

3. Our graph screen is shown in Figure 12.21. Your histogram of actual results will be different from ours, but the expected values plotted with squares will be the same as ours and the same as your plot in Activity 12.8.
4. Press **ENTER** to exit the program, then press **TRACE** to look at the numbers. Figure 12.22 shows that our results are really quite strange, since the most common sum over the 100 rolls was 9. The numbers at the bottom of the screen show that there were 16 rolls whose sum was 9. In this experiment, 7, which is predicted to be the most common sum, occurred only 12 times.
5. Run the program again for 100 rolls and compare the results with your previous results. Is the same number the most common? Do the sums 2 and 12 appear about an equal number of times? What odds would you take that 9 would not be the most common sum in 100 rolls?
6. Run the program with 1000 rolls (it takes about two minutes to complete this experiment). You will see that the actual and expected values are much closer,

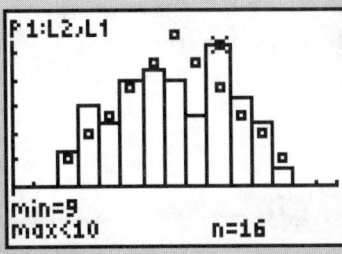

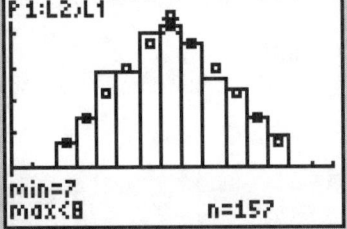

Figure 12.22

Figure 12.23
1000 rolls of two dice

as shown in Figure 12.23. However, there always will be some error. For example the sum 4 occurred quite a bit more often than expected. Notice that the sum of 7 occurred 157 times.

How Accurate Were the Predictions?

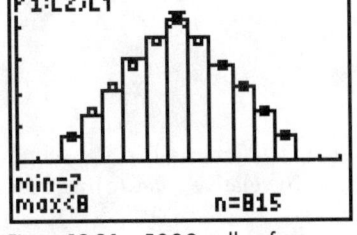

Figure 12.24   5000 rolls of two dice

We'll make one more experiment before looking at the errors. Figure 12.24 shows that for 5000 rolls (10 minutes on the calculator) the actual and expected results are even closer. Note that the sum of 7 occurred 815 times.

Let's look at how well our probability model predicted the number of times that the sum 7 occurred in each of the experiments in Calculator Lab 12.1. Table 12.16 shows the expected numbers and the results.

Table 12.16

Number of Rolls	Expected Number of 7's	Actual Results	Absolute Error
100	$\frac{6}{36} \times 100 \approx 17$	12	$\|12 - 17\| = 5$
1000	$\frac{6}{36} \times 1000 \approx 167$	157	$\|157 - 167\| = 10$
5000	$\frac{6}{36} \times 5000 \approx 833$	815	$\|815 - 833\| = 18$

You can see that the absolute error between the expected numbers and the actual results increases as the number of rolls increases. However you can see from the graphs that the experimental results are actually getting better as the number of rolls increases. For this reason, the absolute error is not the best measure of accuracy. The **relative error** or **percent error** are actually the preferred methods of evaluating the accuracy of an experiment.

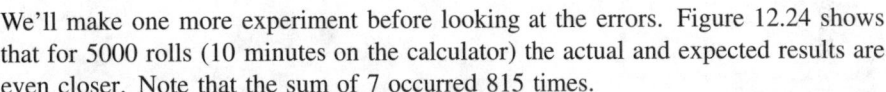

Relative Error

Relative error in an experiment is defined as the absolute difference between the experimentally observed value and the expected value, divided by the expected value.

$$R = \frac{|\text{observed value} - \text{expected value}|}{\text{expected value}}$$

Relative error can be expressed as a fraction, a decimal, or a percent. When it is expressed as a percent it is often called the **percent error**. The absolute difference between the experimentally observed value and the expected value is called the **absolute error**.

Example 12.16

Compare an absolute error of $300 on one family's monthly budget of $3000 in an absolute error of $300 to a small company's budget of $30,000.

Solution   An absolute error of $300 on one family's monthly budget of $3000 represents a relative error of $\dfrac{\$300}{\$3000} = 0.1$ or a percent error of 10%. The same absolute error in a small company's budget of $30,000 is $\dfrac{\$300}{\$30,000} = 0.01$ or a percent error of 1%.

Example 12.17

The outside diameter of a valve stem was measured as 7.127 mm when the actual diameter was 7.1346 mm. What are the (a) absolute, (b) relative, and (c) percent error?

Solutions   (a) Here the observed value is 7.127 mm and the expected value is 7.1346 mm. Thus, the absolute error is

$$\text{Absolute error} = |\text{observed value} - \text{expected value}|$$
$$= |7.127 \text{ mm} - 7.1346 \text{ mm}|$$
$$= 0.0076 \text{ mm}$$

(b) The relative error is absolute error divided by the expected value. In part (a) we found the absolute error to be 0.0076 mm. Thus

$$\text{Relative error} = \frac{|\text{observed value} - \text{expected value}|}{\text{expected value}}$$
$$= \frac{\text{absolute error}}{\text{expected value}}$$
$$= \frac{0.0076 \text{ mm}}{7.1346 \text{ mm}}$$
$$= 0.00107$$

(c) Expressing the relative error as a percent we get the percent error as $0.00107 \times 100\% = 0.107\%$.

The absolute error is 0.0076 mm, the relative error is 0.00107, and the absolute error is 0.107%.

Example 12.18

(a) Compute the relative and percent errors for the sum 7 for the experiments in Calculator Lab 12.1.

(b) On your calculator or on paper, graph the relative error vs. the number of rolls.

Solutions The information below comes from Table 12.16.

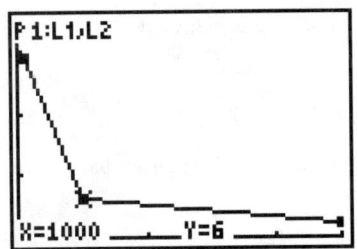

Figure 12.25

(a)

Number of rolls	Relative error	Percent error
100	$\frac{5}{17} \approx 0.29$	29%
1000	$\frac{10}{167} \approx 0.06$	6%
5000	$\frac{18}{833} \approx 0.02$	2%

(b) To make a calculator graph we entered the number of rolls into calculator list L_1 and the relative errors into list L_2. The graph is shown in Figure 12.25. We used graph window dimensions of 0 to 5000 for the *x*-axis and 0 to 35 for the *y*-axis. ■

The three graphs shown in Calculator Lab 12.1, and the decreasing relative error with larger numbers of rolls, demonstrate what mathematicians call the **law of large numbers**, which says that the more trials we include in an experiment (that is, the larger the sample size) the more our results will look like the expected numbers. Stated another way, the law of large numbers predicts that as $n \to \infty$, the relative or percent error approaches zero.

How Does the Calculator Throw Dice?

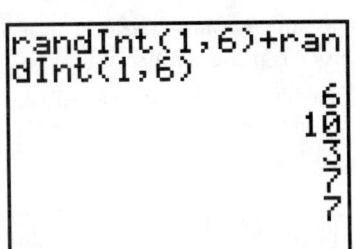

Figure 12.26

Here are the two lines in the program TWODICE that simulate the rolling of two dice:

$$\texttt{randInt(1,6)} \to \texttt{A}$$
$$\texttt{randInt(1,6)} \to \texttt{B}$$

The sum A+B is used to represent the result from one roll of the dice.

The instruction `randInt(1,6)` produces a random integer from the list {1, 2, 3, 4, 5, 6}. You can throw two dice with your calculator by entering the single command shown in Figure 12.26 and then pressing **ENTER** to execute the command as many times as you want. The `randInt(` function is found on the PRB submenu of the MATH menu.

A variation of `randInt(` allows you to simulate many rolls of the dice with one instruction. The instruction `randInt(1,6,10)+randInt(1,6,10)` produces ten pairs of random numbers and adds them up to produce a list of ten sums. This method is good for small experiments, but because lists on the *TI-83* cannot have more than 999 elements, the approach used in the program TWODICE is to look at the rolls one at a time and build the frequency table as we go. Therefore the list where the data is kept in the program needs to contain only 11 elements, for the sums from 2 to 12.

Now that you have seen how to add individual probabilities to compute the probability of an event that can happen in several different ways, we will move to the next section, in which we'll study more complex events. In Section 12.3 you'll learn how to compute the probability that a family of four children will have exactly two boys (it's not 0.5), and the probability that an assembly line with a failure rate equal to 10% can produce a sample that has one-third failures (it's surprisingly likely).

Section 12.2 Exercises

In Exercises 1 and 2 indicate which of the following numbers cannot be the probability of some event.

1. (a) $-\frac{3}{4}$

 (b) 0

 (c) $\frac{11}{15}$

 (d) 2.5

2. (a) -0.5

 (b) 0.98

 (c) $\frac{11}{7}$

 (d) 1

In Exercises 3–8 determine the probability of drawing the indicated event from a well-shuffled deck of 52 playing cards.

3. a seven

4. a six of clubs

5. a five or a queen

6. a black card

7. a red card or a jack

8. a black card smaller than four (where an ace has a value of 1)

In Exercises 9–12 assume that you are rolling a single die. What is the probability of rolling the following?

9. 5

10. 2 or 5

11. even number

12. anything except 3

13. *Aeronautics* On a two-engine jet plane, the probability of one engine failing is 0.001. What is the probability that one engine will not fail?

14. *Insurance* Insurance company tables show that for a married couple of a certain age group the probability the husband will be alive 25 years from now is 0.7. What is the probability that the husband will die within 25 years?

15. *Industrial technology* A machine produces 25 defective parts out of every 1000.

 (a) What is the probability that a defective part will be produced?

 (b) What is the probability that a defective part will not be produced?

16. *Medical technology* A new medicine cures a specific illness in 740 out of 960 people who have the illness.

 (a) What is the probability that this medicine will cure a person who has this particular illness?

 (b) What is the probability that this medicine will not cure a person who has the illness?

In Exercises 17–22 convert the given English statement into a mathematical statement.

17. The failure rate of this product is 4%.

18. She hits 82.5% of her free throws.

19. Your chance of winning the grand prize is less than one in 25 million.

20. The chance of rain tomorrow is 75%.

21. This storm is expected to leave at least 8 in. of snow.

22. I win something half the time I play the lottery.

In Exercises 23–28 convert the given mathematical statement into an English statement.

23. $P(2 \text{ yellow M\&Ms}) = 0.6$

24. $P(\text{we are lost}) = 0.25$

25. $P(\text{home run}) = 0.018$

26. $P(\text{high today of } 85°) < 0.45$

27. $P(\text{good product}) = 0.95$

28. $P(\text{seed will germinate}) > 0.875$

29. *Finance* Table 12.17 shows the number of credit cards sales at a large department store on a particular day.

Table 12.17

Size of Purchase	Type of Credit Card Used			
	MasterCard	VISA	American Express	Store Card
< $25	78	82	45	47
≥ $25 but ≤ $100	67	48	21	38
> $100	26	32	45	17

(a) What is the probability that a credit card sale selected at random will be an American Express card sale?

(b) What is the probability that a credit card sale selected at random will **not** be an American Express card sale?

(c) What is the probability that a credit card sale selected at random will be between $25 and $100 inclusive?

(d) What is the probability that a credit card sale selected at random will be an American Express card sale **and** between $25 and $100 inclusive?

30. Table 12.18 shows the number of students in three different degree programs and whether they are graduate or undergraduate students.

Table 12.18

	Undergraduate	Graduate	Total
Business	150	50	200
Engineering	150	25	175
Arts & Sciences	100	25	125
Total	400	100	500

(a) What is the probability that a randomly selected student is undergraduate?

(b) What percentage of students are engineering majors?

(c) If we know that a selected student is an undergraduate, what is the probability that he or she is a business major?

(d) If we know that a selected student is in Arts & Sciences, what is the probability that he or she is an undergraduate?

(e) What is the probability that a randomly selected student is a graduate Business major?

31. *Medical technology* Six vitamin and three sugar tablets identical in appearance are in a box. One tablet is taken at random and given to Person *A*. A tablet is then selected and given to Person *B*. What is the probability that

(a) Person *A* was given a vitamin tablet?

(b) Person *B* was given a sugar tablet, if you know that Person *A* was given a vitamin tablet?

(c) Neither was given a vitamin tablet?

(d) Both were given a vitamin tablet?

(e) Person *A* was given a sugar tablet and Person *B* was given a vitamin tablet?

(f) Person *A* was given a vitamin tablet and Person *B* was given a sugar tablet?

32. A survey of the gender and marital status of students enrolled in a certain mathematics course at a certain school provided the information in Table 12.19.

Table 12.19

	Male	Female
Single	65	50
Married	27	78

(a) What is the probability of finding a single female student?

(b) What is the probability of finding a married male student?

(c) If a student is male, what is the probability that he is married?

(d) What percent of the students are female?

(e) If a student is female, what is the probability that she is single?

33. *Construction* The actual length of an I-beam is 14.458 m. An engineer measures the beam as 14.45 m. What are the absolute, relative, and percent errors in this measurement?

34. *Computer science* A microcomputer chip is supposed to measure 24 mm long, 7.8 mm wide, and 3.2 mm thick. One chip, when measured with a micrometer, is 23.78 mm long, 7.78 mm wide, and 3.18 mm thick. What are the absolute, relative, and percent errors of each of these measurements?

 35. Run the program TWODICE for 100, 200, 300, ..., 1000 rolls. Compute and plot the relative error in the prediction of the number of times the sum is 12 vs. the number of rolls.

 36. Modify the program TWODICE to make a program HEADTAIL that will simulate the tossing of a coin. Run your program to simulate flipping a coin 100 times. How close did you get to the expected values of 50 heads and 50 tails?

 37. Modify the program TWODICE to make a program MMCANDY that will simulate the distribution of the colors of plain M&Ms. The actual distribution of M&M candies produced is 10% blue, 10% orange, 10% green, 20% red, 20% yellow, and 30% brown.

(a) What is the expected value for each color?

(b) Run your program to simulate selecting 500 candies. How close did you get to the expected values?

38. In the Chapter Project memo from management, you were asked to design a daily testing process that will tell, with 95% certainty, that we are under the 10%

failure rate. You were asked to determine how many items should be tested and to indicate how sure we can be from the result of each sample that the new manufacturing process is meeting its target. Write a memo to management describing the progress you have made toward meeting these goals.

Probability of Binary Events

In this section we will look at the probability that a specific combination of events will occur. We will restrict our discussion to the type of event that can have only two outcomes, for example heads or tails, boy or girl, defective or good product. These kind of events are called **binary events**. We will begin with the different possibilities of families with more than one child, and the discussion will move into a detailed study of defective products off an assembly line.

Independent Events—Multiplying Probabilities

The gender order of births in a family of three children can occur in several ways. For example, a boy born first, then a girl, and then another girl is one of the combinations of three binary events that would produce one boy. You can probably think of other combinations. Example 12.19 shows how to compute the probability of a family of three children having exactly one boy. Later we'll use a similar analysis of another set of binary events, the testing of products that either pass or fail a test.

Example 12.19

What is the probability of a family with three children having only one boy?

Solution For our first attempt at solving this problem we will list all the possibilities. In the list below, the symbol BGG means boy first, then girl, then girl. The eight possibilities are listed below and the three that have one boy are marked with an X:

BBB
BBG
BGB
BGG   X
GBB
GBG   X
GGB   X
GGG

There are three families of three children that have exactly one boy, so the probability of only one boy in a family of three is $\frac{3}{8}$. ∎

Let's look at one of the entries in the table, for example BGG. The probability of this particular birth order is $\frac{1}{8}$ because it is one of the eight possibilities. Another way to arrive at this probability is to look at the BGG event as three separate events whose probabilities are not affected by previous results. That is, the probability of a boy being born does not depend on how many boys or girls are already in the family. If that assumption is true, then the births are **independent events**. Then, as with coin flipping, we can assume that the probability of a boy is $\frac{1}{2}$ with every birth. Another way to calculate the probability of the BGG event is to multiply the probabilities of each individual event. That is,

$$P(BGG) = P(B) \times P(G) \times P(G)$$
$$= \frac{1}{2} \times \frac{1}{2} \times \frac{1}{2} = \frac{1}{8}$$

Finally, to get the probability of only one boy in a three-child family, we add the probabilities of each of the ways that can happen:

$$P(\text{one boy}) = P(BGG) + P(GBG) + P(GGB)$$
$$= \frac{1}{8} + \frac{1}{8} + \frac{1}{8} = \frac{3}{8}$$

In summary, we have the multiplication property of probabilities:

Properties of Probabilities—Multiplication

- ▶ **Independent events** are events whose probability is not affected by previous results of the same experiment. Events that are not independent are said to be **dependent events**.
- ▶ The probability of getting a specific sequence of independent events is the product of the individual probabilities.

A good example of dependent events is drawing more than one card from a deck of cards if you do not replace the first card before the second one is drawn. Before the first card is drawn, the probability of selecting the king of hearts is $\frac{1}{52}$ and that is also the probability of drawing the queen of clubs or any other specific card. Suppose that the king of hearts was drawn on the first try and then set aside. Now the probability of drawing the queen of clubs is $\frac{1}{51}$ because only 51 cards remain in the deck. Here, drawing two cards represent dependent events.

Example 12.20

Suppose that you have flipped a coin five times and gotten a head each time. What is the probability of getting a head on the next toss?

Solution   The result of flipping a coin is an independent event, and so what happened on the first five coin flips will not affect the next coin toss. If you flip a coin five times and get all heads, the probability of getting a head on the sixth flip is still $\frac{1}{2}$.   ■

Example 12.21

What is the probability of flipping a coin six times and getting the sequence $HHTTHT$?

Solution   Since the result of tossing a coin is an independent event, the probability of getting this specific order of tosses is the product of the individual probabilities. Since $P(H) = \frac{1}{2}$ and $P(T) = \frac{1}{2}$, the probability of flipping a coin six times and getting the sequence $HHTTHT$ is

$$\frac{1}{2} \times \frac{1}{2} \times \frac{1}{2} \times \frac{1}{2} \times \frac{1}{2} \times \frac{1}{2} = \frac{1}{64}$$ ∎

The result in Example 12.21 should not surprise you. In fact, $\frac{1}{64}$ is the probability of getting any of the 64 possible sequences of 6 coin flips. That is, the sequence $HHHHHH$ has the same probability as the sequence $HTHTHT$.

Example 12.22

If the probability of a defective item is 0.1, what is the probability of looking at three items and getting exactly one defective item?

Solution   We need to make the assumption that these are independent events. You might want to think of these parts as if they were all gathered together and you were going to randomly select one of them. If $\frac{1}{10}$ of them are known to be defective then the probability of selecting a defective item is 0.1.

This problem starts out just like finding the probability of only one boy in a three-child family. The ways that you can get one defective item are listed below, where F indicates a defective item because it fails a quality test and P stands for a product that passes the test.

$$FPP$$
$$PFP$$
$$PPF$$

However, unlike coin flipping, the probabilities of the individual binary events are not equal. We assume that there are only two possibilities whenever a item is tested; it fails the test or it passes the test. If the probability of failure is $P(F) = 0.1$, then the probability of passing is $P(P) = 1 - 0.1 = 0.9$.

Therefore the three probabilities are

$$P(FPP) = 0.1 \times 0.9 \times 0.9 = 0.081$$

$$P(PFP) = 0.9 \times 0.1 \times 0.9 = 0.081$$

$$P(PPF) = 0.9 \times 0.9 \times 0.1 = 0.081$$

And the probability of exactly one failure is the sum of these three probabilities:

$$P(\text{one failure out of three}) = 3 \times 0.081 = 0.243$$

That is, almost 25% of the time you will get one failure out of three tests ($33\frac{1}{3}\%$ failure rate), even if the probability of failure in the population is 10%.   ∎

In this section so far we have discussed the following topics:

▶ Binary events, which can have one of only two outcomes. Binary events can have equal probability, like heads or tails in the flip of a coin, or different probabilities, like defective or good products off an assembly line.

▶ Independent events, which are events whose probability is not affected by previous results.

- Successive flips of a coin are independent events.
- Choosing a card from a deck and then choosing another card, without replacing the first one, are not independent events, because if the first is the ace of spades, the probability that the second card is a black card is no longer $\frac{1}{2}$, but is now $\frac{25}{51}$ because there is one less black card in the deck (25) and one less card in the whole deck (51).
- Binary events can be independent (coin flipping) or dependent (red or black cards taken from the deck and not replaced).

Now we will use these ideas to continue our study of the mathematics that will lead to a solution of the chapter problem.

In Section 12.2 we studied an experiment in which two dice were tossed many times. In the next experiment, a number of coins will be tossed many times and the number of heads will be counted each time. We'll begin with the experiment of tossing three coins.

Example 12.23

If three coins are tossed,

(a) What is the probability of getting 0 heads, 1 head, 2 heads, and 3 heads?
(b) What is the expected number of times each number of heads will appear in 50 tosses of these three coins?

Solution   This problem is similar to the one presented in Example 12.19, which analyzed the possibilities in a family with three children. If we label the coins a, b, and c, then Table 12.20 shows a list of the possibilities of tossing three coins

Table 12.20

a	H	T	H	T	H	T	H	T
b	H	H	T	T	H	H	T	T
c	H	H	H	H	T	T	T	T

Each of these eight possibilities consists of three independent events, since heads on one coin does not change the probability of heads on another. Therefore the probability of each possible of the eight possibilities is $\frac{1}{2} \times \frac{1}{2} \times \frac{1}{2} = \frac{1}{8}$.

(a) From the table you can see that there is one way to get no heads, so the probability of no heads is $P(0) = \frac{1}{8}$. Likewise there are three ways to get one head, so $P(1) = \frac{3}{8}$. In a similar way we find that $P(2) = \frac{3}{8}$ and $P(3) = \frac{1}{8}$.
(b) The expected numbers of each number of heads is the probability times the number of × the three coins are tossed:

$$E(0) = \frac{1}{8} \times 50 \approx 6$$

$$E(1) = \frac{3}{8} \times 50 \approx 19$$

$$E(2) = \frac{3}{8} \times 50 \approx 19$$

$$E(3) = \frac{1}{8} \times 50 \approx 6$$

■

Next you'll see how to simulate the experiment of tossing three coins on your calculator.

Calculator Lab 12.2	**Tossing Three Coins**

We saw in Section 12.2 that the calculator function `randInt(` produces random integers. The instruction `randInt(0,1)` produces 0 or 1 with equal probability, so it is useful for simulating the toss of one coin. The instruction `randInt(0,1,3)` produces three random integers that are each 0 or 1 with equal probability, so it is useful for simulating the toss of three coins. We need to execute this instruction 50 times to simulate tossing three coins 50 times.

Procedures

1. At the Home Screen, enter the instruction `randInt(0,1,3)` and press `ENTER`.

2. Press `ENTER` five more times so thatyou see something like Figure 12.27, which shows the simulated result of tossing three coins six times.

3. For each toss of the three coins, count the number of 1's and make a mark in the appropriate place in a table like Table 12.21. We'll think of the 1's as heads. Your results will differ, but for Figure 12.27 the table will look like this:

4. Continue pressing `ENTER` and making marks in your frequency table until you have made the equivalent of 50 tosses of three coins. Count the marks and enter the number for each. Compare your numbers with the expected values calculated in Example 12.23.

5. Draw a graph that includes the expected number and the observed number of each number of heads. Make the observed number a histogram and use dots for the expected number, as you saw in the program `TWODICE`. Use a graph like the one in Figure 12.28.

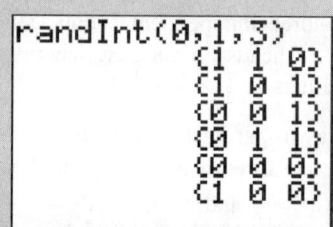

Figure 12.27

Table 12.21

No. of Heads	How Many Times
0	/
1	//
2	///
3	

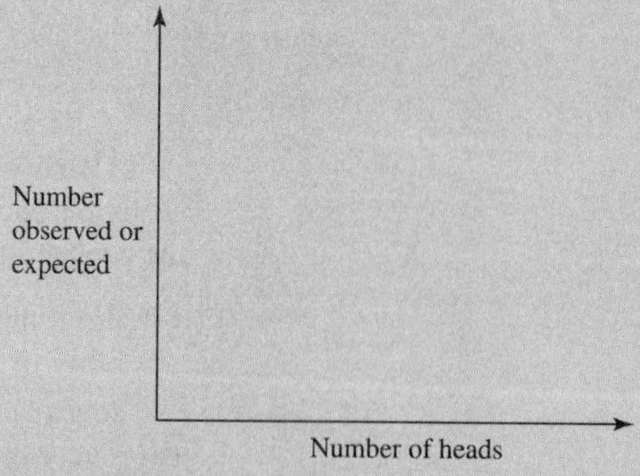

Number observed or expected

Number of heads

Figure 12.28

6. Answer these questions:

   (a) What is the relative error for the prediction of the number of times resulting in two heads?

(b) What kind of experiment could you run that would probably show a lower relative error?

(c) Does the number of times you got zero heads equal the number of times you got three heads? If not, why do you think this did not occur?

The calculator program COINTOSS allows you to simulate tossing some coins a large number of times. In Calculator Lab 12.3 you will use this program to run an experiment that would be quite time-consuming to run with real coins.

Calculator Lab 12.3

Tossing Ten Coins

First we will run the program with 50 tosses of three coins, and compare the results with what you got in Calculator Lab 12.2. Then we will run the program for 100 tosses of ten coins.

Procedures

1. Run the program COINTOSS.
2. Enter 3 for number of coins and 50 for number of tosses.
3. You will see a screen like Figure 12.29, which is similar to the output of the program TWODICE. In the third row of numbers you can see the expected numbers computed in Example 12.23.

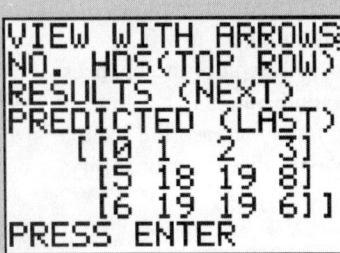

Figure 12.29

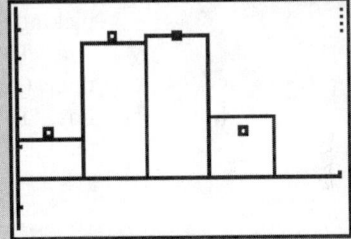

Figure 12.30

4. Press [ENTER] to see a graph of the observed numbers in a histogram and the expected numbers plotted with square points. The graph we got is shown in Figure 12.30; your graph will differ.
5. Press [ENTER] to exit the program
6. Press [TRACE] and move around the graph until you have all the information you need to compute the relative error for each of the four events. Complete the following table:

Number of Heads	Relative Error of the Prediction
0	
1	
2	
3	

7. Run the program again with 100 tosses of ten coins.
8. From the graph on the calculator screen, sketch the curve of the expected numbers for each of the possible numbers of heads and draw the histogram of observed values. Use a graph like Figure 12.28.

   (a) What number of heads is expected to appear the most number of times?
   (b) What number of heads is expected to appear the least number of times?

9. Press **ENTER** to exit the program.
10. Press **TRACE** and move around the graph until you have all the information you need to compute the relative error for the prediction of the number of times that six heads appeared.
11. How could you improve the relative error of that prediction?

It turns out that the expected number of times that six heads will occur out of 100 tosses of ten coins is 21. To compute this number, you would need the probability of getting six heads when ten coins are tossed. One way to get that probability is to list all of the outcomes, from $HHHHHHHHHH$ to $TTTTTTTTTT$, and count the number of outcomes that contain exactly six H's. The problem with this method is that there are over 1000 different possible outcomes.

Next we'll explore the mathematics needed to compute the number of ways to get six heads out of ten coins. The answer will be useful for solving the Chapter Project and also for analyzing games of chance, particularly state lotteries that involve picking the right numbers to win.

Then we will explore the chances of getting an average of one failure out of three ($33\frac{1}{3}\%$ failure rate), if we look at more than three items from the assembly line. That is, if we increase the sample size, how does that affect the chances of thinking mistakenly that the probability of defective parts is $\frac{1}{3}$?

Counting Events

In Example 12.22 we saw that if there are 10% defective parts, you could still get one defective part out of a sample of three parts almost 25% of the time. That is, the chances are pretty good (almost one out of four) that you would make a serious mistake and think that the failure rate was one-third when it was really only one-tenth. How can we reduce the chance of this error?

One way you can reduce your chances of making inaccurate predictions like this is to increase the number of parts that you inspect—by increasing the sample size.

In Example 12.27 we'll look at the probability of getting two failures out of six (that is, the chance that one-third of the parts fail the test). However, first we need a method for counting the number of ways to get exactly two failures out of six. When we were looking only at a sample of three parts, it was easy to list the ways of getting one failure. With six parts it's more difficult, and with 20 parts it's almost impossible to make an accurate list. Activity 12.9 gives you practice listing different possibilities, and Example 12.26 shows you how to use the calculator to make these counts.

Activity 12.9
Counting the Combinations

Count each of the following combinations by making lists of the possibilities. You'll need to develop a strategy for counting to make sure you get them all. The first one is done for you to show one strategy.

1. How many ways could you get two failed items when you test a sample of four? Figure 12.31 illustrates one counting strategy to use.

How many combinations
have two failures out of four?

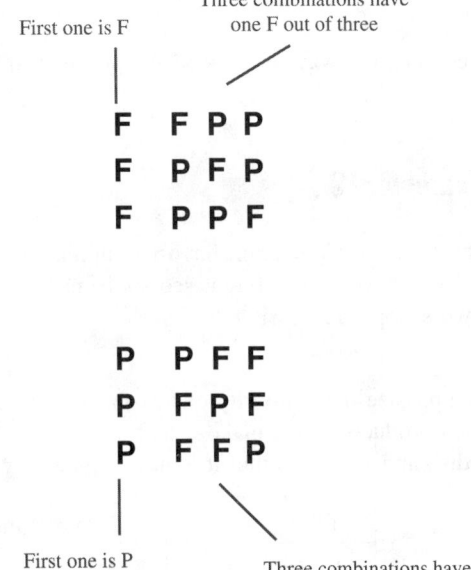

Three combinations have
First one is F   one F out of three

F   F P P
F   P F P
F   P P F

P   P F F
P   F P F
P   F F P

First one is P
                  Three combinations have
                  two F's out of three

Figure 12.31

We begin the list by assuming that the first item we look at is defective, and then looking at the combinations of the remaining three that have one failure. But that's a simpler problem, which we have solved in Example 12.22; the answer is three. Therefore there are three possibilities that begin with F. Then we count those combinations that begin with P, and list the number of ways that you can get two failures out of three. But that's the same number as one pass out of three, so there are also three combinations that begin with P.

As the numbers get larger you will need to adopt a counting strategy in order to get all of the possible combinations.

2. How many ways can you get two failures out of three?
3. How many ways can you get one failure out of four?
4. How many ways can you get one failure out of five?
5. How many ways can you get two failures out of five?
6. How many ways can you get two failures out of six?

We now turn to the general form of the question posed in Activity 12.9:

▶ How many combinations are there of n products, in which there are r failures among them?

This is identical to the question that was raised in our work with coins: "How many combinations of n coins are there that have r heads?"

For example, in the first part of Activity 12.9, we saw that if there are four products ($n = 4$), then there are six different combinations of these products that have two F's ($r = 2$). These six combinations are:

> FFPP
> FPFP
> FPPF
> PFFP
> PFPF
> PPFF

Here is another way to think of the problem of counting combinations.

Example 12.24

If there are four ice cream flavors (vanilla, strawberry, chocolate, and butter pecan) and I can have two different scoops in my dish, how many different combinations of two scoops are possible?

Solution   To emphasize the similarity of this problem to counting combinations of failed and passed products, we'll make a table in which Y means that a flavor is included in the dish and N means that it is not included.

Vanilla	Strawberry	Chocolate	Butter Pecan	Combination
Y	Y	N	N	VS
Y	N	Y	N	VC
Y	N	N	Y	VB
N	Y	Y	N	SC
N	Y	N	Y	SB
N	N	Y	Y	CB

Notice that the order of the Y's and N's is the same as the F's and P's shown above. The two problems that seem different have the same mathematical solution. ■

It may seem surprising that the problem of counting the number of different two-scoop combinations is the same as counting how many combinations of products can be failures. The similarity is clearer when you think of the idea of choosing two products out of four to classify as failures. Or you can think of classifying the ice cream flavors as chosen or not chosen.

The following table summarizes the answers to the questions in Activity 12.9 using this new terminology.

Problem	n	r	Number of Combinations
1	4	2	6
2	3	2	3
3	4	1	4
4	5	1	5
5	5	2	10
6	6	2	15

You probably had difficulty finding all 15 of the combinations in part six of Activity 12.9. Fortunately there is a mathematical way to compute the number of combinations. Most advanced calculators contain that computation as a built-in feature. The following summary gives the symbolism, and Example 12.26 shows you how to do the computation on a *TI-83*.

Counting Combinations

Many different counting questions can be solved by thinking of them as combinations. The general question—"If you have a choice between two outcomes, A and B, and you make that choice n times, how many combinations of choices will have A appearing exactly r times?"—is usually restated in mathematics books as

What is the number of combinations of n things taken r at a time?

Different books use different symbolism for this idea. We'll use the symbolism found on the *TI-83*. The number of combinations of n things taken r at a time is equal to

$$_nC_r$$

Example 12.25

What is the number of combinations of six choices between A and B that contain two A's?

Solution $_6C_2$ is the number of combinations of six choices between A and B that contain two A's. $_6C_2$ is also equal to the number of combinations of six choices between A and B that contain two B's.   ■

Example 12.26

Give three interpretations of the symbol $_6C_2$.

Solutions   Three questions that have the answer $_6C_2$ are:

> ▶ How many different ways can a family of six children have exactly two boys?
> ▶ If you flip a coin six times, how many different ways can exactly two heads come up?
> ▶ If you test six parts off an assembly line, how many different combinations of test results could contain exactly two defective parts?   ■

$_nC_r$ is only a symbol, like $\sqrt{2}$ or $\sin 25°$. As with those symbols, you might understand the meaning behind the symbol without knowing how to do the computation that it stands for. For example, while most people know the meaning of $\sqrt{2}$, very few know how to compute its value to four decimal places. We'll do the same for $_nC_r$—we'll emphasize the meaning of the symbol and show you how to get the calculator to do the computation.

> ### Example 12.27

Use the calculator to compute the number of combinations of six objects taken two at a time.

Solution $_6C_2$ is computed as follows with the *TI-83*. First notice that $_nC_r$ is a function of two variables, n and r. The calculator requires that you enter n first and then r.

> ▶ Go to the home screen
> ▶ Press 6, the value of n
> ▶ Press **MATH** and then select the PRB (Probability) menu, which is highlighted on the top row in Figure 12.32.

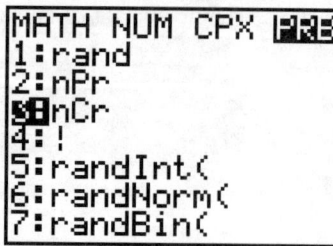

Figure 12.32

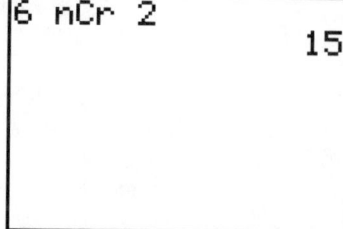

Figure 12.33

> ▶ Select item 3 to bring the nCr instruction into home screen
> ▶ Press 2, the value of r, and then press **ENTER** to complete the computation. Figure 12.33 shows the result, which is the same as your answer to part 6 of Activity 12.9.   ■

The idea of counting combinations and the technique of getting the answer with your calculator will be used in the next two sections. In Section 12.4 we will analyze games of chance, and in Section 12.5 we will use combinations to produce one solution to the chapter problem—assessing the true failure rate on an assembly line.

Section 12.3 Exercises

In Exercises 1–6 use your calculator to evaluate the given symbol.

1. $_8C_5$

2. $_8C_3$

3. $_{10}C_5$

4. $_6C_3$

5. $_6C_1$

6. $_{10}C_1$

7. What is the probability of a family having four children born in the gender order boy, girl, boy, girl?

8. What is the probability of a family having four children born in the gender order boy, boy, girl, girl?

9. Use the process from Activity 12.9 and the birth-order list in Example 12.19 to determine the probability of a family with four children having exactly two girls.

10. If a die is rolled once and a coin is tossed once what is the probability that the die will show a four and the coin will come up tails?

11. *Industrial technology* Sandy is having trouble with the disk drive of his computer. He is also using cheap discs which are often defective. The probability that the disk drive is not operating properly is $\frac{4}{145}$. The probability that the disk drive is not operating properly *and* a cheap disk is defective is $\frac{3}{325}$. What is the probability that a cheap disk is defective?

12. *Industrial technology* Nifty Duplo Company operates two photocopying machines. The probability that on any given day the first machine will not work properly is 0.17, and the probability that the second machine will not work properly is 0.12. What is the probability that on any given day both machines will not work properly?

13. Twenty people are asked to report for jury duty. In how many different ways can 12 of them be selected to serve on the jury?

14. *Medical technology* A medical researcher needs six mice for an experiment. A particular laboratory has 18 mice available. How many different ways can six mice be selected?

15. *Industrial technology* Out of every 75 microwave ovens produced seven are to be selected for radiation leakage.
    (a) In how many different ways can seven microwave ovens be selected to be checked?
    (b) In how many different ways can 68 microwave ovens be selected to *not* be checked?
    (c) How do your answers in (a) and (b) compare? Explain your answer.

16. *Industrial technology* Out of every 100 light bulbs manufactured, a company removes five and tests them to see how long they will run. In how many different ways can these five bulbs we selected?

 17. Run the program COINTOSS for 100 tosses of five coins.
    (a) From the graph on the calculator screen, sketch the curve of the expected numbers for each of the possible numbers of heads, and draw the histogram of observed values.
    (b) What number of heads is expected to appear the most number of times?
    (c) Compute the relative and percent error for each of the six events.

 18. Run the program COINTOSS for 100 tosses of six coins.
    (a) From the graph on the calculator screen, sketch the curve of the expected numbers for each of the possible numbers of heads, and draw the histogram of observed values.
    (b) What number of heads is expected to appear the most number of times?
    (c) Compute the relative and percent error for each of the seven events.

12.4   Analyzing Games of Chance

In this section you will learn how to figure out the probabilities in some games of chance and to estimate the expected return you can get on a bet. You will also learn why betting in casinos always favors the casinos. We will conclude by using what you have learned about counting combinations to study a type of state-sponsored lottery that is held in 30 states.

Card Games

We will look at some examples for the game of draw poker. At the start of a game, each player receives five cards and then can discard up to four of the cards (sometimes the number is three) in hopes of improving the hand. We will study the probability of getting a hand called a *straight*, which consists of five cards in a row such as 2, 3, 4, 5, 6 or 8, 9, 10, Jack, Queen, or a *pair*, which is two cards of the same value, such as a 2 and 2 or Queen and Queen.

Example 12.28

You have been dealt the cards 4, 5, 6, 7, and Jack. You discard the Jack in hopes of getting either a 3 or an 8 to make the straight. What is the probability that you will be successful?

   You have seen five cards. The number of cards that you don't know about is $52 - 5 = 47$. Eight different cards can complete your straight. They are the four 3's and the four 8's. Therefore the probability of completing the straight is $\frac{8}{47} \approx 0.17$ or 17%. You will be successful about 1 time out of 6.   ■

Example 12.29

In Example 12.28, what is the probability of improving your hand by either completing a straight or matching one of the cards you have, to get a pair?

Solution   There are six ways to improve this hand: drawing a 3 or an 8 to make a straight or by drawing a 4, 5, 6, or 7 to make a pair. Again there are 47 cards you haven't seen. The probability of matching any one of the four cards you have is $\frac{3}{47}$ because there are only three of the cards left in the deck. Therefore the probability of improving the hand is

$$p = \frac{8}{47} \text{ (getting a straight)} + 4 \cdot \frac{3}{47} \text{ (4 ways to make a pair)}$$

$$= \frac{20}{47} \approx 0.43$$

That means that you will be successful about 43%, or somewhat less than half, of the time.   ■

Dice Games, Odds, and Expected Return

In a casino there often are dice tables where a game called *craps* is played. Craps is played with two dice and the sum of the two dice tells which bettors have won or lost. This is a complicated game with many types of bets, so we will concentrate on the simplest bets offered—the bets on a single sum coming up. We studied the probability of each sum from 2 to 12 in Section 12.2. If you want to remember those probabilities, count up all the different possibilities for getting a sum and divide by 36, the total number of possibilities.

Figure 12.34
Courtesy T.R. King and Company, Inc.
http://www.alcasoft.com/trking/dice.html

Payoffs on bets are described as ratios like 4 : 1 and 7 : 2. A payoff of 4 : 1 means that you get your bet back *plus* $4 for every $1 you bet. A payoff of 7 : 2 means that you get your bet back plus $3.50 for every $1 you bet. These are sometimes called odds, but we will see later that there are three different kinds of odds.

Example 12.30

At the beginning of the 1996–7 professional football season, you could have placed a bet for a 12 : 1 payoff that the Green Bay Packers would win the Super Bowl. In January 1997 they did win. How much would be returned to you if you had bet $35 on them at the beginning of the season?

Solution   The payoff is 12 : 1 so you would get $12 \cdot \$35 = \$420$ plus your original bet of $35 for a total of $455.   ■

Example 12.31

At the dice or craps table, if you bet on the sum being 7, and you win, the payoff at most casinos is 4 : 1.

(a) What is the probability of winning this bet?
(b) If you make this bet 100 times, what is the expected number of times that you will win, to the nearest whole number?
(c) How much money would you expect to have left after 100 bets if you bet $1 each time?

Solutions   You studied these ideas in Section 12.2.

(a) The probability of winning is $\frac{6}{36} \approx 0.17$ because there are six ways of getting a sum of 7.
(b) The expected number of times that the 7 will occur in 100 rolls is $100 \cdot \frac{6}{36} \approx 17$.
(c) Since you get $5 returned each time you win, you would expect to have about $17 \cdot \$5 = \$85$ at the end of 100 bets. That is, you would have lost about $15.■

The actual loss in Example 12.31(c) is a little more than $15 since we rounded off $100 \cdot \frac{6}{36}$ to get the 17. The expected number of wins is actually 16.67, so the money returned is $5(16.67) = $83.33. The actual expected loss on 100 bets is therefore about $16.67.

If you made 100 bets as described in Example 12.31, how likely would it be that your loss would be $16.67? Actually your loss could *never* be that amount because the bets and payoffs are whole dollar amounts. This just represents the average over a large number of bets. You have seen throughout this chapter that actual results vary considerably from the expected values, especially for a small number of trials. Sometimes you could be winning for many times in a row, and then there might be a sequence of losses.

However, the law of large numbers guarantees that in the long run you will lose this bet about 16.67% of the time. Remember, the casino is a business, and in all the possible bets at the dice table you will lose if you keep playing. People who win in casinos generally are lucky enough to get ahead in the beginning and then smart enough to quit when they are ahead.

Expected Return

One way to compute the expected return was shown in Example 12.31. In that example we saw that $100 became about $83 over the course of 100 bets. A more usual way to think of the return is to define it as the amount of gain that is expected over the long run. In this way of looking at the return, if your expected return is positive you are winning, and if the gain is negative you are losing. Here is a definition of expected return that gives your expected gain or loss over time.

Expected Return

The **expected return** on a bet of $1 is the probability of winning times the amount won, plus the probability of losing times the amount lost. That is, if the probability of winning is p and the payoff is $D : 1$, then the expected return is:

$$E_R = p \cdot D + (1 - p)(-1)$$

That is, with probability p you win D and with probability $1 - p$ you lose $1. The sum is the expected return.

The expected return after N identical bets is $N \cdot E_R$

Example 12.32

Compute the expected return on the 100 bets on the sum being 7 at a dice or craps table described in Example 12.31.

Solution   We will assume, as in Example 12.31, that the payoff is $4 : 1$. In Example 12.31 we saw that the probability of winning is $\frac{6}{36}$. Thus, the values of the variables are

$$p = \frac{6}{36}$$

$$1 - p = \frac{30}{36}$$

$$D = 4$$

Then

$$E_R = \frac{6}{36} \cdot 4 + \frac{30}{36} \cdot (-1)$$

$$= \frac{-6}{36} \approx -0.17$$

That is, for every bet there is an expected loss of about $0.17, or 17 cents, and after 100 similar bets the expected return would be about

$$N \cdot E_R = 100 \cdot \frac{-6}{36} \approx -\$16.67$$

■

Odds

There are three different ways that people look at odds:

▶ Odds in favor (mathematical odds)

   The mathematical definition of the odds that something will happen (the odds in favor of it happening) is the ratio of the probability that it will happen (p) to the probability that it will not happen ($1 - p$). Written as a fraction, this ratio is $\dfrac{p}{1 - p}$.

▶ Odds against (normally used in stating probabilities in gambling)

   The odds against something happening is the reciprocal of the odds in favor, or $\dfrac{p - 1}{p}$. For example, the odds against getting a royal flush in a five-card poker hand are listed as 649,739 to 1.*

▶ Odds as payoffs

   The payoffs that we have been discussing are sometimes incorrectly called odds. They are not actually odds, although the word *odds* is often used to describe payoffs. A **payoff** is just a statement of how much someone is willing to pay for a winning bet. The person considers true probabilities in setting these payoffs, but generally adjusts the payoffs so that he or she will make money over the long run, just as the casino in the dice examples makes $16.67 on the average for every $100 bet on a 7 at the craps table.

Example 12.33

One of the possible bets on the craps table is of getting a 4, 6, 8, or 10 the *hard way*. You bet that the dice will have any one of those sums, but only with the same numbers showing on each of the dice. What are the (a) odds in favor of and (b) odds against winning this bet?

Solutions There are four ways of winning this bet, 2 and 2, 3 and 3, 4 and 4, and 5 and 5. So the probability of winning is $p = \frac{4}{36} = \frac{1}{9}$ and the probability of losing is $1 - p = \frac{8}{9}$.

(a) The odds in favor of winning are $\frac{1}{9} : \frac{8}{9} = 1 : 8$.
(b) The odds against winning this bet are $8 : 1$.

■

World Almanac and Book of Facts, Funk & Wagnalls, 1994, p. 285.

Example 12.34

The payoff on the hard way bet is 7 : 1. What is the expected return on a $10 bet?

Solution   For each dollar bet, the probability that you win $7 is $\frac{1}{9}$ and the probability that you lose $1 is $\frac{8}{9}$. Therefore the expected return on a $1 bet is

$$\frac{1}{9} \cdot 7 + \frac{8}{9}(-1) = \frac{-10}{9} \approx -\$0.111$$

The expected return on a $10 bet is 10 times this answer of $10(-0.111) = -\$1.11$. You can see that this is a better bet than betting on a 7, which loses $1.67 for every bet of $10. ■

Roulette

In the game of Roulette there is a spinning horizontal wheel that looks something like Figure 12.35. The numbers 1 through 36 appear on the wheel in either red or black, and also the numbers 0 and 00 in green. A stainless steel ball is tossed into the spinning wheel and when the wheel stops the ball comes to rest on one of the 38 locations.

European roulette wheels do not have a 00, so there are only 37 numbers in all.

Figure 12.35
Courtesy Magic Math Roulette Strategies.
http://www.4-1-1.com/magic.htm

Many bets are taken in roulette. A roulette table is laid out something like Figure 12.36, and bets can be placed on any of the rectangles. One can bet on any individual number, on red or black, on odd or even, the first or second half, and so on. In some casinos one can even place a chip on one of the intersecting lines and select 4 numbers. For example you can bet on the numbers 7, 8, 10, 11 by placing a chip on the corner shared by the four rectangles containing these numbers. However, 0 and 00 are not red or black, not even or odd, not in the first or second half. In fact, they are not included in any of the bets except a bet on one of them specifically.

These two possibilities are the way the casino makes money on roulette, as we shall see.

00	3	6	9	12	15	18	21	24	27	30	33	36	C3
	2	5	8	11	14	17	20	23	26	29	32	35	C2
0	1	4	7	10	13	16	19	22	25	28	31	34	C1

1st Dozen		2nd Dozen		3rd Dozen	
1-18	Even	Red	Black	Odd	19-36

Figure 12.36

Example 12.35

What are the odds in favor of getting each of these outcomes in the roulette wheel?

(a) The number 14 or any single number
(b) A red number
(c) An odd number

Solutions   There are 38 possibilities, so each probability has 38 in the denominator.

(a) The probability of getting a 14 is $\frac{1}{38}$, so odds in favor are 1 : 37.
(b) There are 18 red numbers, so the probability of getting a red number is $\frac{18}{38}$ and the odds in favor of getting a red number are 18 : 20 or 9 : 10.
(c) There are 18 odd numbers, so the probability of getting an odd number is $p = \frac{18}{38}$ and the odds in favor of getting an odd number are 18 : 20 or 9 : 10. ■

The payoffs offered at most casinos are listed below in Table 12.22.*

---

*Silberstang, Edwin. *The Winner's Guide to Casino Gambling*. Penguin Books, New York, 1993.

Table 12.22

Bet	Payoff
Any single number	35 : 1
A red number	1 : 1
An odd number	1 : 1
Either half	1 : 1
Any of the dozens	2 : 1
Two numbers	16 : 1
Four numbers	8 : 1

Activity 12.10
Which Bet Is Best?

1. For each of the bets and payoffs listed in Example 12.35, compute the expected return on a bet of $100 on that outcome.
2. Use these results to determine which bet is the best for the bettor and which is best for the casino.
3. Explain why the numbers 0 and 00 make roulette profitable for the casino.
4. Is the European roulette wheel, with only one 0, a better deal for the bettor or a better deal for the casino?

State Lotteries

Many states sponsor lotteries of various kinds. In the most common kind of lottery, contestants pick three to six numbers and win if some of their numbers are selected. The biggest prizes are awarded when a bettor chooses all six numbers correctly. Table 12.23 shows the games with the biggest payoffs in some of the states that have lotteries.* For example, in West Virginia (WV) a bettor can pick six different numbers from the numbers 1 through 25 and will win the top prize if all six of the numbers are correct.

Example 12.36

In the West Virginia lottery, what is the probability of winning with one ticket?

Solution   You have one ticket with one set of six numbers, so there is one set of numbers that can win for you. The question is, how many different combinations of six numbers are possible out of 25 numbers? This is a straight computation of $_{25}C_6$, which is 177100. Therefore the probability of your set of six numbers being selected is $\frac{1}{177100}$.

---

*Ben E. Johnson, *Getting Lucky: Answers to Nearly Every Lottery Question You Can Ask*, Bonus Books, Chicago, 1994, pp. 161–62.

Table 12.23

State	Needed to Win
WV	Pick 6 out of 25
VT	Pick 6 out of 30
KS	Pick 6 out of 33
DE, NH, WI	Pick 6 out of 36
DC, IA	Pick 6 out of 39
Tri-State (VT, ME, NH)	Pick 6 out of 40
Tri-West (ID, MT, ND)	Pick 6 out of 41
AZ, CO, MA	Pick 6 out of 42
CT, IN, LA, OR, VA	Pick 6 out of 44
GA, NJ	Pick 6 out of 46
OH	Pick 6 out of 47
PA, MO	Pick 6 out of 48
FL, KY, MA, MD, WI, MI, WA	Pick 6 out of 49
TX	Pick 6 out of 50
CA	Pick 6 out of 51
IL, NY	Pick 6 out of 54

Next we will look at the prizes that are awarded in this kind of game.

Example 12.37

In the Ohio lottery the minimum prize for picking six numbers correctly is $4 million. A single ticket costs $1. What is the expected return on your $1 investment?

Solution To win the grand prize in Ohio you must match all six numbers chosen from the numbers 1 through 47. The probability of winning $4 million is $\frac{1}{47C_6} = \frac{1}{10737573}$. Therefore the expected return to two decimal places is

$$E_R = \frac{1}{10737573} \cdot 4000000 + \frac{10737572}{10737573} \cdot (-1) \approx -\$0.63$$

If you compare the expected return in Ohio's lottery (−$0.63) with the expected returns at the dice table or roulette wheel (−$0.05 to −$0.17), you can see that betting in casinos offers a better return on your money (even though it's still negative!) than this state lottery. The argument in favor of state lotteries is that the lottery business sends money to schools.

Since you could win $4 million with a one-dollar ticket, some people get the idea of buying lots of tickets. Some even form a betting group to pool their money and buy a large number of tickets. Let's look at that strategy.

According to figures reported in 1997 on http://www.calottery.com, the California Lottery has given approximately 34% of its revenue to the public schools of California, or about $9 billion since the lottery began in 1985.

Example 12.38

What is the probability of winning and the expected return per person if 100 people each invest $1000 in a betting group that buys 100,000 tickets for the Ohio lottery, for $1 each? Assume the prize is $4 million.

Solution   The probability of winning with one ticket was computed in Example 12.37 to be $\frac{1}{10737573}$. The probability that one of the 100,000 tickets is the winner is $\frac{100000}{10737573} \approx 0.009$. The group has multiplied each person's chances of winning by 100,000 but the probability is still less than one in a hundred.

To compute the expected return for each person we need to remember that each person invested $1000 and will split the winnings, so each person would receive $4000000/100 = \$40,000$. The individual expected return is

$$\frac{100000}{10737573} \cdot 40000 + \frac{10637572}{10737573} \cdot (-1000) \approx -\$618.16$$

On a per-dollar basis, the expected return is about the same as for the individual bettor in Example 12.37. Remember that you *only* lose $618.16 per $1000 bet in the long run. Since the probability of winning is only 0.009 you will most likely lose every time you bet this way.   ■

Finally, let's consider the chances of winning $4,000,000 if your betting club bets 10 different times on the Ohio lottery. By betting 10 different times we mean that the group buys 100,000 tickets the next 10 times the lottery is offered. By then each member will have spent $10,000.

Example 12.39

What is the probability of winning the Ohio lottery one or more times in 10 trials as described in Examples 12.37 and 12.38?

Solution   Many people might think that the answer is 10 times the probability of winning it once. *This is not true.* If this reasoning were correct then the probability of getting one head in three flips of a coin would be $3 \cdot \frac{1}{2} = 1.5$, which is an impossible value for a probability. In fact, as you can see from our studies of three coins in Sections 12.2 and 12.3, there is a $\frac{7}{8}$ probability of getting at least one head. So what is the answer? We are essentially flipping 10 coins that have a probability of heads equal to about 0.009 and asking for the probability that heads will come up at least once.

There are two ways to do this:

(a) Look at the probabilities of one win, two wins, three wins, and so on, and add them all up. Let's take three wins and compute its probability and then you'll see the pattern in the solution.

Three wins can come about in many ways. For example, you can win the first three times and then lose the next seven times. We will write this result as $WWWLLLLLLL$. The probability of this particular combination is $0.009^3 0.991^7 \approx 0.0000007$. Of course this is a small probability, because it's very unlikely to win three times out of ten in this situation. But there are many ways to win three times in ten tries, for example, $LLWLLWLLWL$ is another.

The probabilities multiply because each run of the lottery is assumed to be independent of previous runs.

The total number of ways to win three times in ten tries is $_{10}C_3 = 120$. So the probability of winning three times is

$$_{10}C_3(0.009^3)(0.991^7) \approx 0.00008$$

which is better, but still small. All the possibilities follow this pattern, as shown in Table 12.24.

Table 12.24

Number of Wins in Ten Tries	Probability
1	$_{10}C_1(0.009^1) \cdot (0.991^9)$
2	$_{10}C_2(0.009^2) \cdot (0.991^8)$
3	$_{10}C_3(0.009^3) \cdot (0.991^7)$
4	$_{10}C_4(0.009^4) \cdot (0.991^6)$
5	$_{10}C_5(0.009^5) \cdot (0.991^5)$
6	$_{10}C_6(0.009^6) \cdot (0.991^4)$
7	$_{10}C_7(0.009^7) \cdot (0.991^3)$
8	$_{10}C_8(0.009^8) \cdot (0.991^2)$
9	$_{10}C_9(0.009^9) \cdot (0.991^1)$
10	$_{10}C_{10}(0.009^{10}) \cdot (0.991^0)$

The chances of winning at least once is the sum of all the numbers in the Probability column. We could do that computation if necessary, but let's not be greedy; suppose we just wanted to win once. That probability is $_{10}C_1(0.009^1) \cdot (0.991^9) \approx 0.083$.

(b) There is another method that is quicker, and it is exact. We'll approach the problem in a different way and ask, "What is the probability of *not* winning any of the ten bets?" That answer is $0.991^{10} \approx 0.913$, so the probability of winning once in the ten tries is about $1 - 0.913 = 0.087$.   ∎

Probability of at Least One Success

If the probability of success on one trial is p, the probability of at least one success in n independent trials is

$$p_n = 1 - (\text{probability of no successes})$$
$$= 1 - (1 - p)^n$$

Each bettor would win \$40,000 because the prize of \$4 million would be split among the 100 members of the club.

Now you can do this kind of computation quite quickly. What if the club decided to bet in 50 different runnings of the lottery? The probability of winning at least once is $1 - (1 - 0.009)^{50} = 1 - 0.991^{50} \approx 0.36$. That is, each person would have spent $50 \cdot 1000 = \$50,000$ and would have only a 36% chance of winning \$40,000 at least once.

Spending $50,000 for about a one-in-three chance to win $40,000 is typical of these lotteries. And even if you are betting alone and buying only single $1 tickets, the numbers would work out in a similar way. That's why you might hear the saying, "A few people get rich in the lottery, but most people get poor."

The mathematics of figuring the probability of winning the lottery one or more times in ten tries is going to be essential in the final section when we return to the opening problem of the chapter, which asks how we can be sure from testing a sample of products that the failure rate on the assembly line is less than 10%.

Section 12.4 Exercises

1. In the game of draw poker you have been dealt the cards 4, 4, 5, 10, and Queen. You discard the 5. What is the probability of getting a card and having two pairs?

2. In the game of draw poker you have been dealt the cards 2, 5, 5, 10, and Queen. You discard the 2 and the 10. What is the probability of getting a three of a kind or two pairs?

3. In the game of draw poker you have been dealt the cards 2♡, 5♡, 7♡, Jack♡, and Queen♣. You discard the Queen♣. What is the probability of getting a flush, that is, all five cards are hearts?

4. In the game of draw poker you have been dealt the cards 4, 6, 6, Jack, and King. You discard the 4 and Jack. What is the probability of getting a full house where a full house is two cards of one denomination and three of another?

5. At a dice or craps table a roll of 2, 3, or 12 is called "craps." The payoff on rolling craps is 7 : 1.
   (a) What is the probability of winning this bet?
   (b) If you make this bet 100 times, what is the expected number of times that you will win, to the nearest whole number?
   (c) How much money would you expect to have left after 100 bets if you bet $1 each time?

6. What are the odds against rolling craps? (See Exercise 5.)

7. What are the odds against rolling a 7 in dice or craps?

8. A "horn bet" in craps is a bet that the next roll will be a 2, 3, 11, or 12.
   (a) What is the probability of winning this bet?
   (b) If you make this bet 100 times, what is the expected number of times that you will win, to the nearest whole number?

9. (a) What are the odds in favor of winning if you bet on the first half of a roulette wheel? The first half is 1–18.
   (b) Use Table 12.22 to compute the expected return on a bet of $100 on that outcome.

10. (a) What are the odds in favor of winning if you bet on the second dozen of a roulette wheel? The second dozen are 13–24.

11. (a) What are the odds in favor of winning on a roulette wheel if you bet on two numbers sharing a line segment, for example 7 and 8?

12. (a) What are the odds in favor of winning on a roulette wheel if you bet on four numbers sharing a vertex, for example 7, 8, 10, 11?

13. What is the probability of winning with one ticket in the Kansas (KS) lottery?

14. What is the probability of winning with one ticket in the Florida (FL) lottery?

15. What is the probability of winning with the Texas (TX) lottery one or more times in ten trials?

16. What is the probability of winning with the New York (NY) lottery one or more times in 52 trials?

12.5 Solving the Chapter Project

In this section we will complete our study of the mathematics of probability by presenting a solution to the folowing question: How sure can we be that the assembly line failure rate is less than 10% given that, in a sample of 50 items, there were 4 failures (8%)?

As with our analysis of the chances of winning in a lottery, once again the probability of independent events and the method of counting combinations ($_nC_r$) will be very useful. We'll begin with the computation of how likely we are to get a failure rate of one-third from samples of three different sizes. We'll begin with the sample of three that we analyzed thoroughly in Example 12.22 on page 860 and move on to a sample of six and a sample of 30.

Example 12.40

If the population has an actual failure rate of 10%, what is the probability of getting a failure rate of one-third ($33\frac{1}{3}\%$) if the sample size is:

(a) $n = 3$ (that is, what is the probability of getting one failure out of three)?
(b) $n = 6$ (that is, what is the probability of getting two failures out of six)?
(c) $n = 30$ (that is, what is the probability of getting ten failures out of thirty)?

Solutions   Figure 12.37 shows the computation of the three needed values of $_nC_r$.

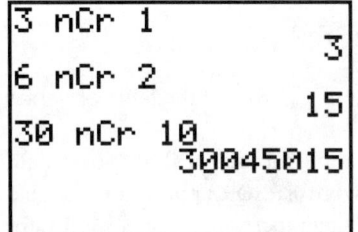

Figure 12.37

(a) In Example 12.40 this solution was given as $3 \times 0.1 \times 0.9^2$. The number 3 is the number of ways that one failure could occur in a sample size of three. The symbol for it is $_3C_1$. Again, the probability of getting one failure out of three tests is about 0.24.

(b) There are 15 ways to get two failures in a sample size of six, because $_6C_2 = 15$. The probability of getting any one of those combinations is $0.1^2 \times 0.9^4 = 0.006561$. There are 15 different combinations, so the probability of getting two failures out of a sample of six is

$$_6C_2 \times 0.1^2 \times 0.9^4 = 15 \times 0.0006561$$

$$\approx 0.098$$

Notice that now you have less than one chance in 10 of getting a failure rate of one-third. Increasing the sample size gives you a much better chance of avoiding the terrible error of thinking that the failure rate was about 33% when, in fact, it was 10%.

(c) The last line of Figure 12.37 shows that there are over 30 million ways to get 10 failures in a sample of 30. The probability of any one of these combinations is $0.1^{10} \times 0.9^{20}$ (that is, 10 items fail and 20 items pass), so the probability of getting 10 failures out of 30 is

$$_{30}C_{10} \times 0.1^{10} \times 0.9^{20} \approx 0.00037 \approx \frac{1}{2700} \qquad \blacksquare$$

In Example 12.40 you can see that by raising the sample size to 30 we have given ourselves an extremely small chance of getting results that make us think that the failure rate is 33%. Before this chapter is finished we'll give an answer to the question, "How sure are you that the failure rate is still at 10%?" and the analysis in Example 12.40 is a step toward that goal.

How big should your sample size be? Remember that sampling costs money and that, in some products, it is not possible to test every single item. The sample size must be a balance between the cost of testing a large number of items and the cost of making a bad decision. We'll end with one way of stating the sample size question.

Example 12.41

If the true failure rate is in fact 10%, and you wanted to have less than 1 chance in 100 of getting a sample that showed a one-third failure rate, how big should the sample size be?

Solutions

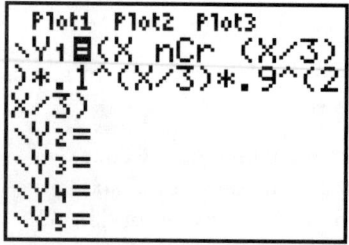

Figure 12.38

We would have to do calculations like those in Example 12.40 until we got a probability that was lower than 0.01. For simplicity we need only test values of n that are divisible by 3. The next ones to test are $n = 9$ and $n = 12$. The probability of getting three failures in a sample of nine are:

$$P(3 \text{ failures out of } 9) = {}_9C_3 \times 0.1^3 \times 0.9^6 \approx 0.04$$

and the probability of getting four failures in a sample of twelve are:

$$P(4 \text{ failures out of } 12) = {}_{12}C_4 \times 0.1^4 \times 0.9^8 \approx 0.02$$

We could continue until the probability went below 0.01. Another way to approach the problem is to look at the pattern in the expressions above and set up a general formula:

$$P\left(\frac{n}{3} \text{ failures out of } n\right) = {}_nC_{n/3} \times 0.1^{n/3} \times 0.9^{2n/3}$$

X	**Y1**	
12	.02131	
13	ERROR	
14	ERROR	
15	.01047	
16	ERROR	
17	ERROR	
18	.00524	

Y1◼(X nCr (X/3)...

Figure 12.39

We entered that formula into the calculator (see Figure 12.38) and the table of values of Y1 is shown in Figure 12.39.

You can see from the table that to get below a probability of 0.01, you would need to have a sample size of more than 15. The word ERROR appears in the table because ${}_nC_r$ is only defined if n and r are whole numbers, and for $n = 13$ for example, our formula is trying to use $r = \frac{13}{3}$. If we let TblStart $= 0$ and ΔTbl$= 3$ we obtain the result in Figure 12.40. ■

X	Y1	
0	1	
3	.243	
6	.09842	
9	.04464	
12	.02131	
15	.01047	
18	.00524	

X=0

Figure 12.40

We'll conclude this section with a calculation that is related to the Chapter Project. In that problem, we wanted to know whether the failure rate had decreased from 10%. Our sample of 50 items produced four failures, for a failure rate of $\frac{4}{50} = 8\%$. What are the chances that the failure rate is *still* 10%? That is, what is the probability of getting a failure rate this small if the population rate is still 10%? Normally, the problem is stated a little differently: "What is the probability of getting a failure rate this small *or smaller* if the population failure rate is still 10%?" Example 12.42 takes up this question.

Example 12.42

Our job is to determine whether the failure rate in the assembly line is still at 10%. We sample 50 items and find four failures. What are the chances of getting this rate *or less* if the population failure rate is actually still 10%?

Solution

We want to know the probability of getting four failures *or less* in a sample of 50 (a measured failure rate of 8%), if the population failure rate is in fact 10%. Therefore, we need to calculate the probability of getting each of the following

▶ 4 failures out of a sample of 50
▶ 3 failures out of a sample of 50
▶ 2 failures out of a sample of 50

▸ 1 failure out of a sample of 50
▸ 0 failures out of a sample of 50

These are five different events, and we want the probability of getting any *one* of these five results. The task, then, is to compute the five probabilities and add them.

$$P(4 \text{ out of } 50) = {}_{50}C_4 \times 0.1^4 \times 0.9^{46}$$

$$P(3 \text{ out of } 50) = {}_{50}C_3 \times 0.1^3 \times 0.9^{47}$$

$$P(2 \text{ out of } 50) = {}_{50}C_2 \times 0.1^2 \times 0.9^{48}$$

$$P(1 \text{ out of } 50) = {}_{50}C_1 \times 0.1^1 \times 0.9^{49}$$

$$P(0 \text{ out of } 50) = {}_{50}C_0 \times 0.1^0 \times 0.9^{50}$$

Figure 12.41 shows how to compute the first three probabilities. The results are $P(4 \text{ or } 3 \text{ or } 2 \text{ or } 1 \text{ or } 0 \text{ failures out of } 50) \approx 0.181 + 0.139 + 0.078 + 0.029 + 0.005 \approx 0.432$. ■

```
(50 nCr 4)*0.1^4
*0.9^46
          .1809045009
(50 nCr 3)*0.1^3
*0.9^47
          .1385651496
(50 nCr 2)*0.1^2
*0.9^48
```

Figure 12.41

Given that the probability of getting four or fewer failures in a sample of 50 items is 0.432, what can we conclude about the failure rate in the population? The answer must be carefully stated. It is this:

Suppose the failure rate has not changed and it still is at 10%. If that were the case, the probability that we would get our number of failures *or fewer* in a sample of 50 is about 0.43. That is, about 43% of the time we would get a sample at least this good if the population failure rate has not changed from 10%. Therefore, four out of 50 is not surprising for a 10% population failure rate. This is not strong evidence that the rate has declined.

Would you want to take this chance? The answer depends on how important the decision is and how costly it would be to make an error. Often you hear of people wanting to reduce the chance of making an incorrect conclusion such as this to 5% or even 1%, depending on the consequences of being wrong.

Throughout this chapter you have done experiments and solved problems showing that strange things can and will happen when the outcomes are uncertain:

▸ The number 7 can come up fewer times than the number 9 in a roll of two dice 100 times.
▸ Looking at 10 M&Ms gives little information about how many yellow candies are in the bag.
▸ A few people win millions of dollars in state lotteries, but
▸ The vast majority of bettors never win the big prize no matter how many tickets they buy.
▸ If the failure rate of products from an assembly line is actually 10%, you could still get one failure out of a sample of three about 25% of the time.

With so much uncertainty, what can we be sure of? You have seen that your estimates of the actual failure rate on an assembly line can be improved by looking at larger samples. You have also seen that the tools of probability can tell you how confident you are in a statement such as "the failure rate on this line is now less than 10%." As in so many of the applications in this book, there is no one right answer. However, the mathematical tools and the knowledge of how to apply them can help you to make an informed and trustworthy response to a complicated question.

Section 12.5 Exercises

1. *Industrial technology* A certain manufacturing line is known to have a failure rate of 5%. What is the probability of getting a failure rate of 25% if the sample size is 4?

2. *Industrial technology* A certain manufacturing line is known to have a failure rate of 5%. What is the probability of getting a failure rate of 25% if the sample size is 20?

3. *Industrial technology* A certain manufacturing line is known to have a failure rate of 1%. What is the probability of getting a failure rate of 25% if the sample size is 4?

4. *Industrial technology* A certain manufacturing line is known to have a failure rate of 8%. What is the probability of getting a failure rate of 20% if the sample size is 10?

5. *Industrial technology* Suppose a true defective rate is 5% and you want to have less than one chance in 100 of getting a sample that showed a 25% rate for defective items. What size sample should you pick?

6. *Industrial technology* Suppose a true defective rate is 1% and you want to have less than one chance in 100 of getting a sample that showed a 25% rate for defective items. What size sample should you pick?

7. *Industrial technology* Suppose a true defective rate is 8% and you want to have less than one chance in 100 of getting a sample that showed a 20% rate for defective items. What size sample should you pick?

8. *Industrial technology* Suppose a true defective rate is 5% and you want to have less than one chance in 100 of getting a sample that showed a 15% rate for defective items. What size sample should you pick?

9. *Industrial technology* Suppose you select a sample of 50 items and find that four of them are defective. What are the chances of getting this rate or less if the population's defective rate is actually 5%?

10. *Industrial technology* Suppose you select a sample of 50 items and find that three of them are defective. What are the chances of getting this rate or less if the population's defective rate is actually 8%?

11. *Industrial technology* Suppose you select a sample of 100 items and find that eight of them are defective.
    (a) What is the probability of getting this number of defective items if the actual defective rate is 10%?
    (b) Compare this result with Example 12.42 and comment on how much difference the change in the sample size made.

12. In the Chapter Project memo from management, you were asked to design a daily testing process that will tell, with 95% certainty, that we are under the 10% failure rate. You were asked to determine how many items should be tested and indicate how sure we can be from the result of each sample that the new manufacturing process is meeting its target. Write a memo to the management describing what you have done to meet these goals.

●●●●●●●● Chapter 12 Summary and Review

Topics You Learned or Reviewed

▶ Data analysis
  • The **mean**, or **average**, is the sum of the scores divided by the number of scores.
  • If a group of numbers are arranged in order from lowest to highest, the number in the middle, or the average of the two middle numbers, is called the **median**.
  • The three **quartiles**, denoted Q_1, Q_2, and Q_3, divide the ranked numbers into four equal-sized parts. In general, Q_1 separates the bottom quarter (or 25%) of the scores from the top three-fourths (or 75%), Q_2 is the median, and Q_3 separates the top 25% from the bottom 75% of the numbers.
  • A **frequency table** or **frequency distribution** shows how many times each number appears.
  • A **histogram** is built from a frequency table and is a bar-like graph that shows how often each item or group of items occurred.
  • A **box plot** draws a box from Q_1 to Q_3, with a vertical dividing line within the box at the median, Q_2. Horizontal line segments (sometimes called *whiskers*) are drawn on either side of the box to show the range of the numbers.

▶ **Probability** is a numerical measure of the chance that an **event** will occur.

▶ A **payoff** is a statement of how much someone is willing to pay for a winning bet.

▶ There are three different ways that people look at **odds**.

- Odds in favor (mathematical odds): The mathematical definition of odds that something will happen (the odds in favor of it happening) is the ratio of the probability that it will happen (p) to the probability that it will not happen ($1 - p$). Written as a fraction, this ratio is $\dfrac{p}{1 - p}$.

- Odds against (normally used in stating probabilities in gambling): The odds against something happening is the reciprocal of the odds in favor, or $\dfrac{p - 1}{p}$.

- Odds as payoffs: The payoffs are sometimes incorrectly called odds. They are not actually odds, although the word *odds* is often used to describe payoffs.

▶ The **expected return** on a bet of $1 is the probability of winning times the amount won, plus the probability of losing times the amount lost.

▶ You saw how probability was applied to games of chance such as dice or craps, draw poker, roulette, and lotteries.

▶ You computed the probability of more than one event happening.

▶ Combinations can be counted using the symbol $_nC_r$.

▶ **Relative error** in an experiment is defined as the absolute difference between the experimentally observed value and the expected value, divided by the expected value. Relative error can be expressed as a fraction, a decimal, or a percent. When it is expressed as a percent it is often called the **percent error**. The absolute difference between the experimentally observed value and the expected value is called the **absolute error**.

▶ The **law of large numbers** says that the more trials that are included in an experiment, the more that the results will look like the expected numbers.

▶ A *TI-83* can be used to compute a one-variable statistical analysis by using `1-Var Stat`.

▶ A *TI-83* can plot histograms and box plots using `Stat Plot`.

▶ A *TI-83* can compute numbers of combinations using the command `nCr`.

▶ A *TI-83* can produce random integers with `randInt(`.

Review Exercises

In Exercises 1–4 use your calculator to evaluate the given symbol.

1. $_9C_5$

2. $_9C_3$

3. $_{10}C_7$

4. $_6C_5$

In Exercises 5–10 use the data in Table 12.25, which shows the results of testing 24 groups with 30 items in each group. Each item was tested, and Table 12.25 shows the number of items in each group that failed the test.

Table 12.25

Sample No.	1	2	3	4	5	6	7	8
No. of Failures	9	6	8	6	7	5	7	9

Sample No.	9	10	11	12	13	14	15	16
No. of Failures	4	1	7	3	7	5	6	4

Sample No.	17	18	19	20	21	22	23	24
No. of Failures	4	6	5	6	7	4	1	4

5. What is the mean number of failures for the 24 groups in Table 12.25?

6. What percent of the number of failures are higher than the mean?

7. What is the median number of failures for the 24 groups in Table 12.25?

8. What are Q_1 and Q_3?

9. Draw a histogram of the data in Table 12.25.

10. Make a box plot for this data.

In Exercises 11–14 use Figure 12.42, which shows the results of testing 30 groups with 15 items in each group. Each item was tested, and the figure shows the number of items in each group that failed the test.

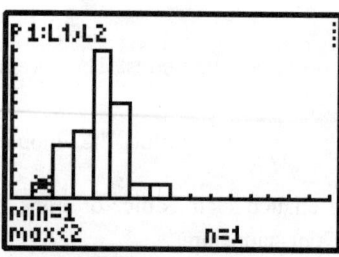

Figure 12.42

11. What is the mean for the 30 groups in Figure 12.42?

12. What is the median number of failures for the 30 groups in Figure 12.42?

13. What are Q_1 and Q_3?

14. Make a box plot for this data.

In Exercises 15 and 16 use Figure 12.43, which shows the results of tossing a pair of dice 50 times and recording the sum of the numbers on the top of the two dice.

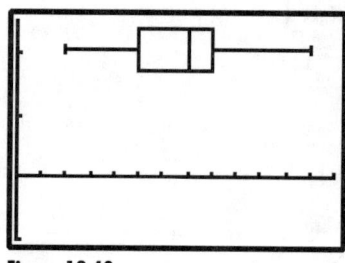

Figure 12.43

15. What is the median?

16. What are Q_1 and Q_3?

17. *Electrical technology* A technician tested an electrical circuit and found the following values in milliamperes on successive trials:

5.35, 5.31, 5.53, 5.36, 5.31, 5.52, 5.32, 5.37, 5.35, 5.55
5.35, 5.31, 5.55, 5.56, 5.37, 5.35, 5.35, 5.56, 5.36, 5.32
5.52, 5.35, 5.39, 5.37, 5.35, 5.53, 5.53, 5.56, 5.35, 5.37

Form a frequency distribution for this data.

18. *Electrical technology* For the data in Exercise 17, draw a histogram for the intervals 5.31–5.35, 5.36–5.40, and so on.

19. *Electrical technology* Determine the mean and median

for the data in Exercise 17.

20. *Electrical technology* Determine the quartiles and draw a box plot for the data in Exercise 17.

21. *Police science* A patrol officer using a laser gun recorded the following speeds for motorists driving through a 55 mph speed zone:

64	57	64	69	67	54	72
56	64	65	59	63	74	49

(a) Determine the mean, median, and quartiles for the given data.

(b) Draw a box plot for the given data.

22. *Environmental science* The annual composite average of carbon monoxide pollution in parts per million (ppm) is shown in the following table:

Year	1985	1986	1987	1988	1989	1990	1991	1992	1993	1994
Pollution	6.97	6.85	6.69	6.38	6.34	5.87	5.55	5.18	4.86	5.01

   (a) Determine the mean, median, and quartiles for the annual composite average during this 10-year period.

   (b) Draw a box plot for the given data.

23. *Environmental science* The national air pollution emissions of carbon monoxide (in thousands of tons) for the 10-year period 1985–1994 is given in the table below.

Year	1985	1986	1987	1988	1989	1990	1991	1992	1993	1994
Pollution	114,690	109,199	108,012	115,849	103,144	100,650	97,376	94,043	94,133	98,017

Determine the mean, median, and quartiles for the annual totals.

In Exercises 24–27 determine the probability of drawing the indicated event from a well-shuffled deck of 52 playing cards.

24. a six

25. an eight of hearts

26. a face card (Jack, Queen, or King)

27. a black card or a seven

In Exercises 28–29 assume that you are rolling a single die. What is the probability of rolling the following?

28. 3

29. 1 or 2

30. *Insurance* Insurance company tables show that for a married couple of a certain age group the probability that the husband will be alive 25 years from now is 0.65. What is the probability that the husband will die within 25 years?

31. *Industrial technology* A certain machine produces 125 defective parts out of every 10,000.

   (a) What is the probability that a defective part will be produced?

   (b) What is the probability that a defective part will not be produced?

In Exercises 32–35 convert the given English statement into a mathematical statement.

32. The defective rate of manufacturing this product is 7%.

33. He hits 73.8% of his free throws.

34. His batting average is 0.287.

35. The chance of snow tomorrow is 30%.

In Exercises 36–39 convert the given mathematical statement into an English statement.

36. $P(2 \text{ blue M\&Ms}) = 0.3$

37. $P(\text{makes her three-point shot}) = 0.35$

38. $P(\text{low tonight of } 27°) = 0.75$

39. $P(\text{he will get well}) \geq 0.50$

40. *Medical technology* Table 12.26 shows the results of a research study into the relationship between smoking and heart disease of 1500 men over 50 years of age.

   (a) What is the probability that a person in the study and selected at random has heart disease?

   (b) What is the probability that a person in the study and selected at random is a smoker?

   (c) What is the probability that a person in the study and selected at random is a smoker with heart disease?

   (d) What is the probability that a nonsmoker in the study has heart disease?

Table 12.26

	Smoker	Nonsmoker	Total
Heart disease	350	180	530
No heart disease	250	720	970
Total	600	900	1500

41. A survey of the gender and type of school in which students were enrolled in 1993 is provided in Table 12.27.

Table 12.27

	Type of School		
	2-year	4-year	Graduate
Male	1,522	3,356	1,152
Female	2,127	3,299	1,264

(a) What is the probability of finding a male student at a two-year college?

(b) What is the probability of finding a male student at one of these three types of schools?

(c) If a student is female, what is the probability that she is in graduate school?

(d) What percent of the students are female?

42. *Construction* A room in a house under construction is supposed to measure 24 ft long and 13.5 ft wide. The contractor measured the framed room and found that is is 24.18 ft long and 13.35 ft long. What are the absolute, relative, and percent errors of each of these measurements?

43. Run the program TWODICE for 100 rolls. Compute and plot the relative error in the prediction of the number of times the sum is 9 vs. the number of rolls.

44. What is the probability of a family having four children born in the gender order boy, girl, girl, girl?

45. What is the probability of a family with four children having three girls and one boy?

46. If a die is rolled once and a coin is tossed once, what is the probability that the die will show a five or a six and the coin will come up heads?

47. *Industrial technology* An Emergency 911 office has two telephone systems. The probability on any given day that the primary system is not operating properly is 0.015. The probability that the secondary system is not operating properly is 0.008. If these two systems are independent what is the probability that both of them are not operating properly?

48. Thirty people are asked to report for jury duty. In how many different ways can 14 of them be selected to serve as regular or alternate members of the jury?

49. *Manufacturing* A company needs to select two out of every five bottles from a filling line to make sure they have been filled properly. How many different ways can these two bottles be selected?

50. *Industrial technology* Out of every 125 computer monitors produced, 35 are to be selected and tested for radiation leakage.

(a) In how many different ways can 35 monitors be selected to be checked?

(b) In how many different ways can 90 monitors be selected to *not* be checked?

(c) How do your answers in (a) and (b) compare? Explain your answer.

51. *Industrial technology* Out of every 200 tires manufactured, a company removes nine and tests them to see how many miles they will last before they blow out. In how many different ways can these nine tires be selected?

52. Run the program COINTOSS for 100 tosses of seven coins.

(a) From the graph on the calculator screen, sketch the curve of the expected numbers for each of the possible numbers of heads, and draw the histogram of observed values.

(b) What number of heads is expected to appear the most times?

(c) Compute the relative and percent error for each of the eight events.

53. Run the program COINTOSS for 100 tosses of eight coins.

(a) From the graph on the calculator screen, sketch the curve of the expected numbers for each of the possible numbers of heads, and draw the histogram of observed values.

(b) What number of heads is expected to appear the most times?

(c) Compute the relative and percent error for each of the nine events.

54. In the game of draw poker you have been dealt the cards 3, 4, 6, 9, and Jack. You discard the 3. What is the probability of getting a card and having a pair?

55. In the game of draw poker you have been dealt the cards 3, 7, 7, 9, and King. You discard the 3 and the 9. What is the probability of getting three of a kind?

56. In the game of draw poker you have been dealt the cards 3♡, 6♡, 9♡, Jack♠, and Queen♣. You discard the Jack♠ and Queen♣. What is the probability of getting a flush, that is, all five cards are hearts?

57. What are the odds in favor of rolling craps? (See Section 12.4, Exercise 5 on page 880.)

58. What are the odds against rolling a 6 in dice or craps?

59. (a) What are the odds in favor of winning if you bet on the reds of a roulette wheel?

(b) Use Table 12.22 to compute the expected return on a bet of $100 on that outcome.

60. (a) What are the odds in favor of winning on a roulette wheel if you bet on the numbers in the row labeled C2 in Figure 12.35?

61. What is the probability of winning with one ticket in the Pennsylvania (PA) lottery?

62. What is the probability of winning with the Georgia (GA) lottery one or more times in 10 trials?

63. *Industrial technology* A certain manufacturing line is known to have a failure rate of 5%. What is the probability of getting a failure rate of $33\frac{1}{3}\%$ if the sample size is 6?

64. *Industrial technology* Suppose a true defective rate is 4% and you want to have less than one chance in 100 of getting a sample that showed a 25% rate for defective items. What size sample should you pick?

65. *Industrial technology* Suppose you select a sample of 75 items and find that four of them are defective. What are the chances of getting this rate or less if the population's defective rate is actually 5%?

66. *Industrial technology* Suppose you select a sample of 100 items and find that 12 of them are defective. What is the probability of getting this number of defective items if the actual defective rate is 8%?

Chapter 12 Test

1. Use your calculator to evaluate the symbol $_8C_3$.

In Exercises 2–6 use the data in Table 12.28, which shows the results of testing 10 groups with 30 items in each group. Each item was tested, and Table 12.28 shows the number of items in each group that failed the test.

Table 12.28

Sample No.	1	2	3	4	5	6	7	8	9	10
No. of Failures	3	1	8	3	7	5	6	4	1	6

2. What is the mean number of failures for the 10 groups in Table 12.28?

3. What is the median number of failures for the 24 groups in Table 12.28?

4. What are Q_1 and Q_3?

5. Draw a histogram of the data in Table 12.28.

6. Make a box plot for this data.

7. Determine the probability of drawing a Jack from a well-shuffled deck of 52 playing cards.

8. Determine the probability of drawing an eight or a face card (Jack, Queen, or King) from a well-shuffled deck of 52 playing cards.

9. A certain machine produces 65 defective parts out of every 1,500. What is the probability that a defective part will be produced?

10. Convert the English statement "Your chance of getting to class on time is 50-50" into a mathematical probability statement.

A survey of the gender and age of students were enrolled in college in 1994 is provided in Table 12.29.

Table 12.29

	Age 18–24	Age 25–34	Age ≥ 35
Male	4,152	1,589	958
Female	4,576	1,830	1,766

11. What is the probability of finding a male student between 25 and 34 years old?

12. If a student is 18–24, what is the probability that the student is female?

13. Describe any additional information that you think should be added to Table 12.29 to make it easier to use.

14. A driveway is supposed to measure 42.5 ft long. The contractor measured the driveway after the cement was poured and found that it was 41 ft long. What is the relative error of this measurement?

15. Two out of every 250 ovens that a company manufactures are removed and tested for durability. In how many different ways can these two ovens be selected?

16. Run the program COINTOSS for 50 tosses of five coins.

  (a) What number of heads is expected to appear the most number of times?

  (b) Compute the relative and percent error for each of the six events.

17. In the game of draw poker you have been dealt the cards 3, 6, 6, 9, and Jack. You discard the 3. What is the probability of getting a card and having two pairs?

18. Suppose a true defective rate is 3%, and you want to have less than one chance in 100 of getting a sample that showed 25% rate for defective items. What size sample should you pick?

19. Suppose you select a sample of 100 items and find that 3 of them are defective. What is the probability of getting this number of defective items if the actual defective rate is 5%?

INDEX OF APPLICATIONS

This index lists the categories of the applications problems and examples in this book alphabetically. After each topic, the index lists the text page number followed by the problem and example numbers in parentheses. Italic numbers indicate example and unlabeled problem numbers. For example, if you are looking for an application to architecture, you would turn to page 393 and look for example 6.24 or to page 415 and look for problem number 27.

Aeronautical engineering, 438(44)

Aeronautics, 856(13)

Aerospace technology, 262(30)

Agriculture, 224(23, 24), 273(41), 617(6)

Air traffic control, 58(20), 384(5, 6)

Archaeology, 483(16, 17), 511(62)

Archeology, 476(*7.37*), 477(*7.38*)

Architecture, 393(*6.24*), 415(27), 670(*12*), 685(18), 715(36)

Astronomy, 245(*4.24*), 253(27), 303(13, 14), 351(38), 388(*6.20*), 389(*6.21*), 390(*6.22*), 396(1, 2, 4), 415(24, 25), 417(*13*), 629(32), 668(25)

Automobile technology, 507(8)

Automotive technology, 18(*1.8*), 203(12, 13), 207(54), 262(22), 275(*15*), 303(15, 16), 449(27, 28), 686(22), 702(24), 786(31), 804(22, 25)

Biology, 635(*9.20, 9.21*), 645(3, 4), 665(6), 668(27), 670(44), 702(22)

Business, 45(28), 251(*4.28*), 415(26), 429(*7.6*), 437(37, 38), 512(67), 646(17), 669(32), 703(30), 715(31), 726(19, 20, 21, 22), 733(11, 12, 13, 14, 15), 736(47, 48, 49, 52), 769(9, 10), 781(*11.21*), 786(33), 803(20), 824(95), 845(47)

Chemistry, 686(21)

Civil engineering, 239(*4.19*), 240(15), 241(17, 19, 20), 253(24), 270(11), 273(43), 274(45, 47)

Commercial design, 224(16, 17, 18)

Communications, 252(*4.29*)

Computer Aided Design (CAD), 763(*11.12*), 767(*11.14*), 770(19, 20, 21, 22, 23, 24), 771(25, 26, 27, 28, 29, 30), 785(25, 26, 27, 28, 29), 815(*11.42*), 816(*11.43*), 817(1, 2, 3, 4, 5, 6, 7, 8, 9, 10, 11, 12, 13, 14, 15, 16), 821(29, 30, 31, 32), 822(47, 48), 823(84), 824(85, 86, 87, 88, 89, 90, 91, 92, 93), 825(5, 12, 17)

Computer science, 261(21), 413(6, 7), 417(*14*), 665(3, 4), 715(34), 857(34)

Computer technology, 510(50)

Construction, 36(*1.27*), 37(*1.28*), 45(30), 57(18), 58(19), 59(*11*), 253(19, 20, 25), 273(37, 38), 275(*14*), 438(40), 497(26), 507(7), 512(65, 69), 665(1, 2), 669(43), 783(*11.22*), 791(*11.29*), 857(33), 888(42), 890(*14*)

Construction technology, 804(23)

Ecology, 507(6), 612(*9.1, 9.2*), 616(1, 2, 3, 4), 645(5, 6), 646(19), 655(*9.34*), 667(1, 2, 3), 668(28), 669(*33*), 670(*1, 2, 9*)

Electrical engineering, 703(29), 786(32)

Electrical technology, 655(*9.35*), 844(39), 845(40, 41, 42), 886(17, 18, 19, 20)

Electricity, 45(27), 131(22, 23), 253(22), 260(*4.38*), 645(10), 646(11), 658(15), 668(30), 685(17), 736(50)

Electronics, 30(24), 149(41, 42, 43, 44, 45, 46), 163(21), 175(7, 8, 9, 10, 11), 176(17, 18), 188(22, 23), 203(15, 16, 18), 207(40, 53), 208(58, 59, 61), 241(21, 22), 262(24, 25, 26, 27, 28, 29), 274(48), 318(*5.26*), 319(*5.27*), 320(*5.28*), 322(21, 22, 23, 24), 333(*5.43*), 340(*5.51*), 341(25, 26, 27, 28, 29), 342(30, 31, 32, 33), 351(43, 44, 45, 46, 47, 48, 49, 50), 353(21, 19, 20), 369(23, 24), 412(4), 413(5), 416(28, 29, 33), 470(*7.31*), 471(*7.32*), 472(*7.33*), 473(*7.34*), 482(10), 483(11, 12, 13), 497(24), 507(10), 511(59, 60), 512(72), 513(*15*), 638(*9.24*), 646(15), 658(19, 20), 669(31, 40), 696(*10.14*), 702(25, 26, 27, 28), 710(*10.24*), 802(*11.39*), 803(17, 18, 19), 825(18)

Energy science, 845(45)

Energy technology, 271(14), 715(32)

Engineering technology, 45(23, 24, 25, 26)

Environmental Science, 437(39)

Environmental science, 130(18, 19), 438(45, 46), 458(14), 459(16), 507(9), 511(53), 512(71), 628(23), 668(22), 670(*6*), 804(24), 845(44), 887(22, 23)

Finance, 428(*7.5*), 437(35, 36), 459(15), 461(*7.25*), 462(*7.26, 7.27*), 463(*7.28*), 464(*7.29*), 482(1, 2, 3, 4, 5, 6), 493(*7.52*), 494(*7.53*), 497(23, 27), 511(56, 57), 512(64),

513(*14*), 631(*9.17*), 634(*9.19*), 637(*9.23*), 645(1, 2), 668(26), 670(*8*), 856(29)

Food technology, 628(24, 25)

Forestry, 241(16), 274(44), 686(23), 715(38, 39)

Hydrology, 30(23)

Industrial design, 224(19, 20, 21, 22), 225(27, 28, 29, 30), 270(9, 10), 273(39, 40, 42), 275(*16*)

Industrial engineering, 253(23), 270(12), 274(50)

Industrial management, 31(26), 726(24, 26), 733(17), 734(18)

Industrial technology, 844(35, 36, 37, 38), 856(15), 869(11, 12, 15, 16), 884(10, 11, 1, 2, 3, 4, 5, 6, 7, 8, 9), 887(31), 888(47, 50, 51), 889(63, 64, 65, 66, *9*), 890(*15, 18, 19*)

Insurance, 845(46), 856(14), 887(30)

Land management, 242(23, 24)

Landscape architecture, 228(*4.8*), 253(21)

Lighting technology, 591(*8.38*), 592(*8.39*), 593(*8.40, 8.41*), 617(8), 646(16), 715(35)

Machine design, 253(26)

Machine technology, 482(9)

Manufacturing, 888(49)

Material science, 715(40)

Mechanical engineering, 391(*6.23*), 686(19)

Mechanics, 270(13)

Medical technology, 163(22, 23), 274(49), 396(8, 9, 10, 11, 12, 13, 14), 438(41), 458(13), 459(17, 18), 483(18), 497(25), 511(54, 55), 595(31), 603(56), 617(7), 624(*9.10*), 628(26), 645(9), 646(13, 14), 665(5), 666(11, 12), 668(29), 671(*13*), 686(20), 715(33), 856(16), 857(31), 869(14), 887(40)

Metallurgy, 804(21, 26)

Meteorology, 130(21), 163(17, 18, 24), 176(12, 13, 14, 15, 16), 208(56), 646(18), 658(13, 14)

Music, 187(17, 18), 203(17), 478(*7.39*), 480(*7.41*), 483(19, 20), 512(63)

Navigation, 38(*1.31*), 45(29), 274(46), 289(18), 300(*5.14*), 302(*5.15*), 304(19), 311(*5.23*), 312(*5.24*), 321(13), 322(14, 15, 16, 17), 351(37, 40, 41, 42)

Nuclear technology, 475(*7.35, 7.36*), 476(*7.37*), 477(*7.38*), 483(14, 15), 639(*9.26*), 645(7, 8), 646(12)

Nutrition, 726(23), 733(16)

Optics, 148(39, 40), 824(98, 99)

Petroleum technology, 694(*10.12*), 702(23), 786(30)

Physics, 31(25), 104(7, 8, 9, 10, 12, 13), 113(1), 115(*11, 12, 13*), 438(42, 43), 468(*7.30*), 482(7, 8), 497(21, 22), 511(58), 539(*8.9*), 658(16), 715(37), 824(94)

Police science, 13(16), 207(52), 595(32), 845(43), 886(21)

Radio technology, 396(6, 7)

Recreation, 241(18), 288(11), 289(12, 13, 14, 15, 16, 17), 350(33, 34, 35, 36), 352(*17*), 353(*18*), 512(66), 547(5, 6), 601(17)

Seismology, 629(29, 30, 31), 668(24)

Sound, 187(13, 14, 15, 16, 19, 20, 21), 203(14, 19, 20), 204(21, 22), 207(55), 208(60, 63)

Sound technology, 628(27, 28), 668(23), 670(7)

Space technology, 396(5), 538(*8.8*), 544(*8.12, 8.13*), 547(7, 8, 13, 14), 601(19), 700(*10.16*)

Sports management, 617(5)

Sports technology, 12(3, 9, 10), 20(*1.10*), 56(13), 57(14), 58(*7, 8*), 105(18), 115(15, 16, 17), 262(23), 303(17, 18), 351(39)

Temperature, 24(*1.18*), 46(*1.38*)

Thermodynamics, 665(7, 8, 9), 666(10), 670(45), 671(*15*)

Transportation, 7(*1.2*), 12(1, 2, 4, 5, 6, 7, 8), 13(11, 12, 13, 14), 19(*1.9*), 57(16), 68(19), 69(20), 114(5), 115(*3*), 130(17), 207(51), 224(15), 736(51), 794(17, 18), 823(57)

Waste technology, 511(61)

Index

Absolute error, 854, 885
Algebraic function, 599
Alternating current, 141, 188–191
Amplitude, 152–157
Angle
    associated with a complex number, 335, 349
    complement of, 124
    of rotation, 742
    phase, 166–167
    terminal side of, 372
Angles
    complementary, 124
    negative, definition, 139
    positive, definition, 139
Angular speed, 259
Annuity, 493
Arc, 251
Arc length, 251, 252
Area
    parallelogram, 226
    rectangle, 226
    surface, *See* Surface area
    triangle, 226–229
astronomical unit (A.U.), 297
Asymptote, 532, 582
    horizontal, 532, 582, 599
    vertical, 533, 583, 599
Average, 832, 884
    of two numbers, 89
Axis
    polar, 810

Beat frequency, 201
Binary, 711
    events, 858, 860, 861
    logic system, 711
    number system, 711
Boole, George, 711
Boolean algebra, 711
Box plot, 837, 840–842, 884

Boyle's law, 658

Calculator
    when to use to solve equations, 678–679
Calculator command
    and, 712
    ClrDraw, 398
    Δ Tbl, 374
    DrawInv, 626
    ENTRY, 399
    FOR, 497
    intersect, 689, 691, 692, 709, 729, 734
    LOGIC, 712
    MATH, 711
    or, 712
    Prompt, 406
    Pt-On, 399, 408
    TblStart, 374
    TEST, 711
Calculator function
    ≥Frac, 432
    Format, 375
    logic, 712
    randInt(, 855
    seq, 431, 508, 565
    test, 712
    TEST, 711
    Trace, 46
Calculator graph style
    path, 293
    above, 720
    below, 720
    dot, 384
    path, 366
Calculator mode
    a+bi, 330
    dot, 713
    Func, 359
    Par, 290
    Pol, 374, 375

real, 330
Seq, 435, 440, 441, 508
    problems with, 451, 457
Simul, 298
Calculator programs
    AREAMID, 503
        how it works, 504–506
    BALLDROP, 75, 102, 103
    BOUNCE, 69
    CIRCLE, 406
    COINTOSS, 863, 869, 888, 890
    CONICS, 417
    GRAPH, 405–406, 416
    GUESSLIN, 31
    HEADTAIL, 857
    HIKER, 5, 6, 13, 14, 51, 63, 67, 103, 114
    HYPERBLA, 416
    MMCANDY, 857
    OTHERFUN, 644, 647, 659
    POINTS, 407, 409, 412
    POWERFUN, 644
    QUADFORM, 406
    ROBOTDRW, 411–412
    SAMPLEH, 839, 840
    SAMPLE5, 831
    SLINKY, 186, 483, 647
    SOUND, 187
    TWODICE, 851, 857, 862, 863, 888
    VOLUME, 401–404
    XYGRABBR, 397, 399–401, 406
Calculator windows
    ZoomFig, 82
    ZSquare, 25, 375, 416
    ZTrig, 406
Calculator-Based Laboratory™, 4
Capacitor
    charging, 469–474
    discharging, 469
    time constant, 469
Card games, 870
Cartesian coordinate system, 648, 649, 667
    quadrants, 132
CBL, *See* Calculator-Based Laboratory™
Central angle, 251
Chaos theory, 442
Charging a capacitor, 469–474
Chord, 191
    modeling, 194–202
Circle, 242
    arc, 251
    center, 242
    circumference, 248
    diameter, 242
    parametric equations, 294
    radius, 242
    sector, 251

arc length, 251
    central angle, 251
    three equations for, 373
Circuit
    RC, 191
Circuits
    RC, 316–320
    RL, 342
Coefficient of absorption, 646
Comet, 385, 387–388
    Chiron, 396
    Giacobini-Zinner, 396
    Hale-Bopp, 387
    Halley, 387, 389, 390
    Hyakutake, 387
    Kahoutek, 417
    Temple-Tuttle, 415
Complementary angles, 124
Complex plane, 334
Complex number, 330
    angle associated with, 335, 349
    imaginary part, 330
    nonreal, 331
    real part, 330
Complex numbers
    absolute value, 334, 349
    addition of, 332
    division of, 339
    geometry of, 334–337
    multiplication of, 332, 339, 349
    polar form for, 338
Components of vectors, 281
Cone, 266
    frustum, 266
    lateral surface area, 266
    slant height, 266
    total surface area, 266
    volume, 266
Consecutive differences method, 562, 600
Constraints, 727
Continuous growth, 633
Continuous variation, 474
Controlling the variables, 91
Converges, 490
Converting
    between polar and rectangular equations, 380–383
    parametric equations to x-y equations, 360
Coordinate system
    cylindrical, 810, 818
    spherical, 807, 818
Cosine law, 237–240
Cruciform, 384
Cube, 217
Cubic
    function, 549
    regression, 562

Current
    alternating, 141, 188–191
    direct, 141
    peak values, 153
Curves
    Lissajous, 367, 414
    parabolic, 83
    tangents to, 281
Cycle, 149–151
    definition, 149
Cycloid, 303
Cylinder
    definition, 221
    lateral surface area, 266
    right, 221
    right circular, 221, 265
    slant circular, 265
    total surface area, 266
    volume, 265
Cylindrical coordinate system, 810, 818

Damped simple harmonic motion, 640
Damping factor, 641
Dantzig, George, 728
Data, 6
    fitting curves to, 23
    modeling, 23
    transforming, 29
Declination, 18
Decrease
    without bound, 440
Degrees, 177–179
Dependent events, 859
Dependent variable, 219, 267
Determinant, 782–783
Dice games, 870–874
Dimension of a matrix, 756
Direct current, 141
Discharging, 469
Discrete growth, 633
Discriminant, 327, 348
Distance
    between two points, 235
    directed, 370
Diverges, 490
Domain, 76
    of the model, 219, 267
    of the parameter, 290
    practical, 219, 267
Domain of a function, 583
Draw poker, 870
DTMF, See Touch tone system

Earthquakes, 516
    epicenter, 517
    faults, 517
    plates, 516

Eccentricity, 385–387
    of Comet Hale-Bopp, 387
    of Comet Halley, 387
    of Comet Hyakutake, 387
    of conics, 417
    of planets, 387
Element of a matrix, 756
Elementary events, 846
Ellipse, 294
    foci, 294
    major axis, 294, 361, 385, 395
    minor axis, 294, 361, 385
    parametric equations, 295, 363, 385, 414
    polar equation, 389
    polar equations, 388–390, 414
    rectangular equation, 363, 385, 414
Ellipsoid, 395
Ellipsoidal reflector, 395
Entry of a matrix, 756
Equal functions method for solving an equation, 679
Equations
    circle, 373
    exponential, 464
    parametric, 289–302
    polar, 369
        plotting, 370–377
    rectangular, 359
    slope-intercept of a line, 23
    system of, 686
Error, 661, 667
    absolute, 662, 854, 885
    absolute percent, 662
    percent, 661, 853, 885
    relative, 853
Even function, 528, 599
Event, 845, 884
    elementary, 846
Events
    binary, 858, 860, 861
    dependent, 859
    independent, 858, 859, 861
Expected
    number, 851, 871
    return, 872–873, 885
Exponential
    damping, 182, 640
    equations
        solving, 635–637
    function, 612, 666
Exponential equations, 464
Extraneous roots, 685

Faces, 217
Feasible point, 727
Feasible region, 727

Filter
  band-elimination, 413
  high-pass, series, M-derived, 416
  low-pass, constant-$k\pi$, 412
Flow chart, 570
Formula
  quadratic, 86
Frequency, 151
  beat, 201
  distribution, 884
  table, 833, 884
Frustum of a
  cone, 266
  pyramid, 263
Function
  algebraic, 523–524, 599
  common logarithmic, 621, 666
  cubic, 549
  definition, 75
  domain, 76, 583
  even, 148, 528, 599
  exponential, 612, 666
  inverse, 625
  linear, 549
  natural logarithmic, 621, 666
  objective, 728
  odd, 148, 528, 599
  polynomial, 548, 599
  power, 522–535, 599
  quadratic, 83, 549
  quartic polynomial, 552
  range, 76, 583
  rational, 579, 599
  root, 84
  secant, 248
  slope, 573
  step, 712
  zero, 84
Functions
  periodic, 141, 149
    definition, 149
Future value, 493

General form of
  quadratic function, 85
Grade of a hill, 18
Grapevine Pass, 128–129
Graphs
  of inequalities, 717
  quadratic equations, 90

Half-life, 474
Harmonics, 204
Helix, 805
Hertz, 151
Histogram, 837–840, 884
Horizontal

asymptote, 532, 582, 599
  shift, 167–171
  translation, 167–171
Hyperbola
  polar equation, 417
  rectangular equation, 416
Hyperbolic spiral, 384
Hypotenuse, 121

i, 328
Identity
  matrix, 774–778
  multiplicative, 774
Imaginary
  axis, 334, 349
  numbers, 328–331, 348
  unit, 328, 348
Inclination, 18
Increase
  without bound, 440
Independent events, 858, 859, 861
Independent variable, 219, 267
Inequalities
  in one variable, 703–715
Inequality
  graph of, 717
Infinity
  negative, 440
  positive, 439
Integers, 323
  positive, 323
Intercept
  x, 21
  y, 15
Intercepts
  of a graph, 21–23
Interest
  compounded annually, 459
Inverse
  function, 625
  matrix, 775, 778–780
  multiplicative, 774
  relation, 625, 667
Irrational numbers, 323
Isotopes, 474

j, 328

Kirchhoff's
  Laws of circuit analysis, 695–697
  Laws of current law, 695
  Laws of electric circuits, 695
  Laws of voltage law, 695
Kirchhoff, Gustav, 695
Knots, 38

Large numbers, law of, 855, 885

Lateral surface area
    cone, 266
    cylinder, 266
    definition, 214
    prism, 262
    pyramid, 263
Law of
    cosines, 237–240
    sines, 229–233
        activity on, 229
Law of large numbers, 855, 885
Least common multiple, 193
Leonardo of Pisa (Fibonacci), 433
Limaçon, 384
Limit, 438
    oscillate without, 440
Linear
    function, 549
    programming, 727–734
        history, 728
        Simplex method, 728
    regression, 46
    system, 686
Lissajous curves, 367, 414
Lithotripsy, 394
Lithotripter, 395
Log-log axes, 653, 654
Log-log graph, 653, 667
Logarithmic
    common function, 621, 666
    coordinate axes, 624, 648
    natural function, 621, 666
    scale, 648
Logical connections, 711

Maclaurin series, 646
Mathematical model, 219, 267
Matrix, 755, 818
    determinant of, 782–783
    dimension, 756
    element, 756
    entry, 756
    identity, 774–778
    inverse, 775, 778–780
    multiplication, 756–761
        on TI-83, 759–760
        which can be, 761
    singular, 782
    square, 777, 780
Max/min problems, 219
    solving, 219, 267
Maximum, 219
Mean, 662, 884
Median, 832, 884
Method
    consecutive differences, 562, 600

Minimum, 219
Model, 2
    mathematical, 2
Modeling
    chords, 194–202
    data, 23
    impedance in RC circuits, 316–320
    motion affected by gravity, 100
    musical notes, 194–202
    sounds, 182–186
    vector sums
        with parallelograms, 308–312
        with triangles, 306–307
    vibrations, 180–182
    wave combinations, 188–204
Motion
    sub-orbital, 299
Motion Detector, 4
Multiplicative
    identity, 774
    inverse, 774

Navigation, 304–306
Negative
    angles, 139
    reciprocal, 237
Newton's Law of Cooling, 663, 667, 671
Newton, Isaac, 74
Nonsinusoidal, 141
Numbers
    complex, See Complex numbers
    imaginary, 348

Objective function, 728
Odd function, 528, 599
Odds, 873–874, 885
Ohm's law, 316, 320
Oscillate without limit, 440
Oscilloscope, 365
Outlier, 840

Parabola, 83, 358
    polar equation, 417
    vertex, 89
Parabolic curve, 83
Paraboloid, 810
Parallelogram, 226
Parameter, 289
    domain, 290
Parametric equations, 289–302
    and the oscilloscope, 365–367
    circle, 294
    converting to x-y equations, 360
    ellipse, 295
    graphing, 290, 358–365
Payoff, 871, 873, 885
Peak values, 153

Pendulum, 518
   bob, 519
   natural frequency, 518
   natural period, 518
Pentagon, 262
Percent error, 853, 885
Period, 149–151, 641
   definition, 149
   of a pendulum, 518
Periodic function, 141, 149
   definition, 149
Perpendicular lines
   slopes of, 237
Phase
   angle, 166–167
   shift, 165–166
Phase relationships
   in phase, 189
   lags, 166
   leads, 166
   out of phase, 166
   trails, 166
Phasor, 340
Planes, 787
Plotting
   polar coordinates, 369–370
   polar equations, 370–377
Point of tangency, 569
Polar
   axis, 369
   coordinates, 369–385
   curves
      cardioid, 377, 379, 409, 412
      circle, 373–376, 382, 384, 409
      limaçon, 384, 412, 413
      rose, *See* Rose
      trisectrix, 377, 384, 409, 413
   equation, 369
      circle, 417
      conic, 417
      ellipse, 417
      hyperbola, 417
      rectangular form of, 381
      straight line, 382, 383
   $\frac{\pi}{2}$-axis, 369
Polar axis, 810
Polar equations
   converting to rectangular equations, 380–383
Polar form for a complex number, 338
PolarGC, 375
Pole, 369, 810
Polynomial
   complex, 548
   degree, 549
   first degree, 549
   fourth degree, 552

quartic, 552
   real, 548
   second degree, 549
   third degree, 549
Polynomial function, 548, 599
Population, 829
Positive
   angles, 139
Power function, 599
Prism, 262
   lateral surface area, 262
   pentagonal, 262
   total surface area, 262
   volume, 262
Probability, 845, 884
Program Editor, 401
Programming
   editing, 401
      an existing program, 402–406
      entering a program, 401–402
      restarting the program, 398
      running a program, 397–399
      stopping the program, 398
Projectile
   motion, 93–105
Projectiles, 93
Pseudocode, 357
Pyramid, 263
   frustum, 263
   lateral surface area, 263
   right, 263
   total surface area, 263
   vertex, 263
   volume, 263
Pythagorean
   theorem, 233–235
   trigonometry identities, 246–247

Quadrants, 132
Quadratic
   formula, 86
   function, 549
   regression, 73–74
Quadratic equation
   discriminant, 327, 348
   graphs, 90
Quadratic function
   definition, 83
   factored form, 85
   factors, 85–90
   general form, 85
   maximum, 83
   minimum, 83
   roots, 84–90
Quartic
   function, 552

Quartile, 833, 884
  lowest, 833
Quasiperiod, 641

Radian, 254
Radians, 177–179
Range, 76
Range of a function, 583
Ratio
  of cubic dimensions, 538
  of linear dimensions, 538
  tangent, 34
Rational
  function, 579, 599
Rational numbers, 323
RC circuits, 316–320
RCcircuit, 191
Real numbers, 324
Reciprocal, 237, 774
Rectangular coordinate system, 648, 667
Rectangular equations
  converting to polar equations, 380–383
Rectangular form of polar equations, 381
Recursive, 432
Refraction
  index of, 146
  second law of, 146
Regression
  cubic, 562
  linear, 46
  quadratic, 73–74
Relative error, 853, 885
Representative sample, 829
Riemann sums, 497–500
Right
  angle, 121
  triangle, 121
RL circuit, 342
Roof pitch, 18
Root of a function, 84
Roots
  extraneous, 685
  of trigonometric functions, 147, 159–161
  quadratic functions, 84–90
Roots method for solving an equation, 679
Rose
  8-leafed, 377
  5-leaved, 384, 415
  4-leaved, 417
  hybrid, 384
  n-leafed, 378–379
  n-leaved, 409
  patterns in, 378–379
  3-leaved, 384
Rotation, 742, 745–752
  angle of, 742

Roulette, 874–876

Sample
  representative, 829
  size, 829
Scale
  logarithmic, 648
Scaling, 742–745, 747–749
Scaling factor
  x, 742
  y, 742
Scatterplot, 6
Secant, 109, 248
Secant function, 248
Sector, 251
  arc length, 251, 252
  area, 252
  central angle, 251
Seismograph, 516
Semi-log axes, 654
Semi-log graph, 648, 667
  on a graphing calculator, 657
Semilogarithmic graph, 648, 667
Sequence, 424
  arithmetic, 424
    common difference, 424
    direct definition, 430
    recursive definition, 434
  discrete, 474
  Fibonacci, 433
  fixed point, 442
  geometric, 425
    common ratio, 425
    direct definition, 431
    recursive definition, 434
  recursive, 432
  terms of, 424
Sequences
  and infinity, 439
  limits of, 438–447
Series, 484
  arithmetic, 484, 508
  geometric, 484, 508
  Maclaurin, 558, 646
  Taylor, 558
Shift
  horizontal, 167–171
  phase, 165–166
  vertical, 164–165
Simplex method, 728–729
Sine law, 229–233
Sine wave, 141
Singular matrix, 782
Sinusoidal, 141, 204
  data, 171

Sinusoidal curve
  fitting data to, 171–175
Skid pad, 262
Slide rule, 623
Slinky™, 180, 483
Slope, 15–21
  and velocity, 18
  applications, 17–21
  definition, 15
  function, 573, 643–645
  of perpendicular lines, 237
  perpendicular lines, 237
Slope-intercept equation of a line, 23
Snell's law, 146
Solving
  equations
    Equal functions method, 679
    Roots method, 679
    using graphs, 679
    using tables, 680
  equations with a
    calculator, 679–685
  equations without a
    calculator, 677–678
  exponential equations, 635–637
  inequalities with
    algebra, 705–709
    binary logic, 711–713
    calculator, 709–711
  max/min problems, 219, 267
  trigonometric equations, 142–147, 157–159
  two equations
    addition method, 687–688
    graphical method, 688–690
    substitution method, 690–692
Sounds
  modeling, 182–186
Speed
  angular, 259
  common units, 9
  constant, 7
Spherical coordinate system, 807, 818
Standard deviation, 837
State lotteries, 876–880
Surface area, 214
  lateral
    cone, 266
    cylinder, 266
    definition, 214
    prism, 262
    pyramid, 263
  total
    cone, 266
    cylinder, 266
    definition, 214
    prism, 262

  pyramid, 263
Symmetry
  about the y-axis, 373, 413
  about the polar axis, 384
  with respect to the origin, 528
System
  linear, 686
  nonlinear, 691
  of equations, 686

Tangency
  point of, 569
Tangent
  ratio, 34
Tangents
  to curves, 281
θstep, 375, 377
Time constant, 469
Total surface area
  cone, 266
  cylinder, 266
  definition, 214
  prism, 262
  pyramid, 263
Touch tone
  keypad, 118–119, 185
  phone, 118–119
  system, 118
Touch tones
  picturing, 201–202
Trajectory, 281
  planetary, 295–297
Transformation, 742
Transformed graphs
  comparing, 99
Transforming data, 29, 95–96
  horizontally, 99
  vertically, 99
Translation, 742
  horizontal, 167–171
  vertical, 164–165
Triangle
  area, 226
  area using trigonometry, 227–229
  right, 121
    adjacent side, 121
    hypotenuse, 121
    legs, 121
    opposite side, 121
    side adjacent, 121
    side opposite, 121
Triangular pulse, 203
Trigonometric equations
  solving, 142–147, 157–159
Trigonometric functions
  definitions, 120

right triangle definitions, 121, 204
roots, 147, 159–161

Universal Gravitational Constant, 544

Variable
   dependent, 219, 267
   independent, 219, 267
Variation
   direct, 535–539
   inverse, 544, 599
Varies
   directly, 539, 599
   inversely, 544, 599
Vector, 281–283
   components, 281
   definition, 281
Vectors, 304–320
   addition of
      parallelogram method, 308
      triangle method, 306
   and gravity, 313–316
   equal, 308
   in Navigation, 304–306
Velocity, 9
   and slope, 18
   at an instant, 105–113
      definition, 111
   average, 10–11
   components, 281, 283
   escape, 299
   horizontal, 281
   initial, 112
   planetary, 297–298
   resultant, 281
   vertical, 281

Vertex
   of a parabola, 89
Vertical
   asymptote, 533, 583, 599
   shift, 164–165
   translation, 164–165
Vibrations
   modeling, 180–182
Volume, 214
   cone, 266
   cylinder, 265
   frustum of a cone, 266
   frustum of a pyramid, 263
   prism, 262
   pyramid, 263

Wave
   nonsinusoidal, 141
   rectified, 141
   sawtooth, 141, 208
   sine, 141
   sinusoidal, 141
   square, 141, 203
   triangular, 141
Whispering galleries, 393
Whole numbers, 323

x-intercept, 21
x-scaling factor, 742

y-intercept, 15
y-scaling factor, 742

Zero of a function, 84